COMPLETE ORDINARY LEVEL

MATHEMATICS

PASSPORT

Nji Emmanuel Ndi

First Edition

Printed by CreateSpace, an Amazon.com Company

eStore address: www.CreateSpace.com/5720147

Available from Amazon.com, CreateSpace.com, and other retail outlets

Available on Kindle and other retail outlets

DEDICATION

Dedicated to all those who appreciate the Mathematical nature of the universe; love, do and use Mathematics to reveal the unknowns of creation; bringing the invisible to light.

TABLE OF CONTENTS

DEDICATION i

ACKNOWLEDGEMENT iv

FORWARD... v

INTRODUCTION vi

NOTATIONS USED IN THIS BOOK vii

TOPIC 1: NUMBERS AND NUMERALS 1

TOPIC 2: SETS OF NUMBERS 11

TOPIC 3: FRACTIONS, DECIMALS, PERCENTAGES, RATIOS AND PROPORTIONS 25

TOPIC 4: ESTIMATIONS, APPROXIMATIONS AND STANDARD FORM 47

TOPIC 5: WEIGHTS AND MEASURES 55

TOPIC 6: FINANCIAL ARITHMETIC 67

TOPIC 7: DIRECTED NUMBERS.. 77

TOPIC 8: BASIC ALGEBRAIC EXPRESSIONS 85

TOPIC 9: SIMPLE LINEAR EQUATIONS... 93

TOPIC 10: EXPANSIONS AND FACTORISATION 103

TOPIC 11: SIMULTANEOUS LINEAR EQUATIONS.. 111

TOPIC 12: QUADRATIC EQUATIONS... 119

TOPIC 13: ALGEBRAIC FRACTIONS 129

TOPIC 14: TRANSPOSITION OF FORMULAE 137

TOPIC 15: THE REMAINDER AND FACTOR THEOREMS 143

TOPIC 16: INEQUALITIES.. 149

TOPIC 17: INDICES, LOGARITHMS AND SURDS... 159

TOPIC 18: NUMBER BASES 171

TOPIC 19: COORDINATE GEOMETRY 181

TOPIC 20: SET THEORY AND LANGUAGE 197

TOPIC 21: BINARY OPERATIONS... 209

TOPIC 22: LOGIC 217

TOPIC 23: BINARY RELATIONS AND FUNCTIONS.. 229

TOPIC 24: SEQUENCES AND SERIES 247

TOPIC 25: POINTS, LINES AND ANGLES.. 257

TOPIC 26: VOCABULARY OF PLANE FIGURES 267

TOPIC 27: POLYGON THEOREMS 275

TOPIC 28: TRIGONOMETRY 281

TOPIC 29: APPLICATIONS OF TRIGONOMETRY 297

TOPIC 30: VOCABULARY OF SOLID FIGURES 309

TOPIC 31: MENSURATION OF PLANE FIGURES 315

TOPIC 32: MENSURATION OF SOLID FIGURES 333

TOPIC 33: SIMILARITY AND CONGRUENCY 347

TOPIC 34: GRAPHS OF FUNCTIONS 359

TOPIC 35: VARIATIONS (PROPORTIONS) 377

TOPIC 36: GRAPHS OF INEQUALITIES 385

TOPIC 37: STATISTICS 391

TOPIC 38: PROBABILITY 417

TOPIC 39: CONSTRUCTIONS AND LOCI 433

TOPIC 40: CIRCLE GEOMETRY 443

TOPIC 41: SOLID GEOMETRY AND TRIGONOMETRY 453

TOPIC 42: NETWORKS 461

TOPIC 43: MATRICES 475

TOPIC 44: VECTORS 489

TOPIC 45: TRANSFORMATIONS AND SYMMETRY 507

TOPIC 46: TRANSFORMATIONS WITH MATRICES 517

TOPIC 47: FLOW DIAGRAMS 529

APPENDIX -THE SCIENTIFIC CALCULATOR 543

ANSWERS TO STRUCTURAL EXERCISES 547

ANSWERS TO APPENDIX EXERCISES 600

ANSWERS TO MULTIPLE CHOICE EXERCISES 601

GLOSSARY 607

INDEX 615

ACKNOWLEDGEMENT

Many thanks go to Mr. Akoko Godfred Amana of GBHS Kedjom-Keku, the North West Regional Pedagogic Inspector for Mathematics Mr. Nfor Samuel Ndi who edited the manuscript and gave ample advice, which went a long way to reshape the document. I heartily thank the Former North West Regional Pedagogic Inspector for Mathematics Mr. Nji Samuel Tatah who made a very commendable effort to edit the Mathematics content of the book. I cannot forget the encouragements and advice, which the National inspector of Mathematics Mme Babila Emilia inspired and gave me.

I equally pay much tribute to my students on which this material was tested.

I cannot end here without thanking my wife Nji Irene Nfih and my children who encouraged and supported me in one way or the other during the course of the work. Many thanks go to the North West Mathematics Pedagogic Office, the Mathematics Teacher's Association (MTA) and the Teacher's Resource Centre (TRC), the WAEC and the CGCE Board for allowing their past questions to be used directly or indirectly.

NJI EMMANUEL NDI

G.B.H.S. MANKON, BAMENDA

NORTH WEST REGION

CAMEROON

TEL: (+237) 76 68 40 50

E-mail: manuelndike@gmail.com

FORWARD

Syllabuses are most often reviewed to meet up with new challenges in Education. The Cameroon General Certificate of Education Board recently reviewed her Examination syllabuses and their Examinations will be based on these new syllabuses. There was therefore need for a corresponding review of textbooks used in teaching subjects examined at these Examinations. The "Complete Ordinary Level Mathematics Passport" is one of those books that have been written for this purpose. This book is equally useful for examinations examined by other Examination bodies. The book has included new topics such as Logic, Networks, Flow diagrams etc that have been introduced into the new syllabus for Ordinary level Mathematics.

The book is highly recommended as a companion to Mathematics teachers and students offering Ordinary Level Mathematics.

NFOR SAMUEL NDI

RPI MATHEMATICS

NORTH WEST REGION

CAMEROON

INTRODUCTION

Mathematics is a way of observing, studying and classifying patterns found in things around us. "Patterns" here covers almost any kind of regularity that can be recognized by the mind. Pattern is that property of an object, which distinguishes the object from all others. Without patterns, it will not be possible for anyone to distinguish between any two things, say a horse and a cow or a man and a woman.

Mathematics is a very useful subject and is applied in almost every aspect of life-games, poetry, science, the social sciences, geography, history, economics, music etc. In fact, there is no discipline in life, which does not involve some form of Mathematics.

Apart from the role played by mathematics in other disciplines, mathematics generally teaches people how to reason. This explains why most mathematicians are almost all round, having a broad base knowledge in disciplines that many people may feel are outside their domain.

STUDY MATHEMATICS AND YOU CAN FIT ALMOST ANYWHERE!

HOW TO STUDY MATHEMATICS SUCCESSFULLY

Work less but work frequently

Mathematics is very hierarchical and every new concept relies on some previously learned elementary ones. If any of the previous knowledge required was not well established and consolidated by the learner, the new one may never be understood. This explains why many students find Mathematics difficult. When work is accumulated, new knowledge becomes a misery. It is better to study one-hour everyday than studying seven hours in one day.

Concentrate

Always study in a very quiet environment. Avoid serving two masters at a time, such as eating and reading, studying and watching television or listening to the radio.

Read and Write

Simply reading, Mathematics yields very poor results. Read the notes, Study the worked examples and solve all the exercises. Never just, read a problem and assume that you can solve it. Solve the problem, and then refer to the answers at the end of the book to confirm that you are right. If not read over the notes again and go back to the worked examples until you are able to solve the problem. However, the answers at the back of the book should be used only to check that you are right rather than for finding out how to solve the problem. For the multiple choice exercises do not try to memorize the answers. Do them with utmost care especially because the detractors are often so close to the correct answer that it is liable to find your wrong answer among the given alternatives.

Assignments and Class Exercises

Many students are weak simply because they do not do their assignments and exercises. Even in class, some students have a habit that when the teacher puts a problem on the board for the students to attempt, they are reluctant to solve the problem and only wait to copy. This bad spirit should be highly discouraged.

NOTATIONS USED IN THIS BOOK

$\{...\}$	The set of or the unordered list of
$n(A)$	The number of element in set A
$\{\ x:\quad\ \}$	The set of all x such that
$\in$	Is an element of
$\notin$	Is not an element of
$\{\ \ \}$ or	The empty set
$\mathscr{E}$	The universal set
$\cup$	The union set
$\cap$	The intersection set
$\subseteq$	Is a subset of
$\subset$	Is a proper subset of
$A\backslash B$	The difference between the sets A and B. i.e. $A\backslash B = A\cap B'$
$(a, b, c,...)$	An ordered list of elements $a, b, c,...$
$\{a,b,c,...\}$	The set of elements a,b,c,... or an unordered list of elements $a, b, c, ...$
$\mathbb{N}$	The set of integers, $\{0, \pm1, \pm2, \pm3, \pm4...\}$
$\mathbb{Z}$	The set of all positive integers and zero, $\{0, 1, 2, 3, 4 \cdots$
$\mathbb{Z}^+$	The set of positive integers $\{+1, +2, +3, +4 ...\}$
$\mathbb{Q}$	The set of rational numbers
$\mathbb{Q}^+$	The set of positive rational numbers
$\mathbb{R}$	The set of all real numbers $x \in \mathbb{R}$
$\mathbb{R}^+$	The set of positive real numbers $\{x \in \mathbb{R} : x > 0\}$
$f(x)$	f of x. The image of x under the function f
f^{-1}	The inverse function of the function f
fg or $f \circ$	The function f of the function f
$=$	Is equal to
$\neq$	Is not equal to
$\approx$	Is approximately equal to
$<$	Is less than
$>$	Is greater than
$\nless$	Is not less than
$\ngtr$	Is not greater than
$\leq$	Is less than or equal to
$\geq$	Is greater than or equal to
$a < x < b$ or $]a, b[$ or (a, b)	An open interval on the number line
$a \leq x \leq b$ or $[a, b]$	A closed interval on the number line
$\{x : a < x < b\}$	The set of elements x such that a is less than x and x is less than b
$\mathbf{a}$	The vector $\mathbf{a}$
$\mathbf{AB}$	The vector represented in magnitude and direction by AB
$\|x\|$	The modulus or absolute value of x. i.e.$\{x$ for $x > 0, -x$ for $x < 0, x \in \mathbb{R}\}$
$\boldsymbol{a \cdot b}$	The dot or scalar product of the vectors $\mathbf{a}$ and $\mathbf{b}$

A^{-1}	The inverse of the non-singular matrix A
A^T	The transpose of the matrix A
lg x or log x	The common logarithm of x
x^n	The number x, raised to the power n
$\propto$	Is proportional
∞	Infinity
$\sqrt{\ }$	The positive square root
$-\sqrt{\ }$	Negative square root
$\sqrt[n]{a}$	The nth root of a
p:	The statement or preposition p
T or 1 in truth tables	True
F or 0 in truth tables	False
$\sim p$ or p' or $\neg p$	The negation of a statement p
$p \wedge q$	The conjunction of the statements p and q
$p \wedge q$	The disjunction of the statements p and q
$A \cap B$ or $\{x : x \in A \wedge x \in B\}$	The intersection of sets A and B.
$A \cup B$ or $\{x : x \in A \wedge x \in B\}$	The union of sets A and B.
$p \Rightarrow q$ or $p \rightarrow q$	p implies q or p is sufficient for q or p only if q or q is necessary for p
$p \Leftrightarrow q$ or $p \leftrightarrow q$ or p iff q	p is a necessary and sufficient condition for q or p implies and is implied by q or p if and only if q
$\forall x$	For all or for every element x
$\exists x$	There exists or for at least one or for some element x
$\exists! x$	There exists one and only one element x
$\equiv$	Is equivalent or is congruent to
///	Is similar to
$\perp$	Is perpendicular to
$\parallel$	Is parallel to
$G = (V, E)$	The graph of the set V of vertices together with the set E of edges
$D = (V, A)$	The directed graph (digraph) of the set of vertices V and the set of ordered edges A.
$G = (V, E, A)$	The mixed graph of the set of vertices V, unordered edges E and ordered edges A.
$n: =$	The store n takes the value…
$\dfrac{dy}{dx}$ or $f'(x)$	The derivative of $y = f(x)$ with respect to x

TOPIC 1

NUMBERS AND NUMERALS

OBJECTIVES

At the end of this topic, the learner should be able to:

1. Represent numbers using tallies and vice versa.
2. Recognise numerals as symbols used in representing numbers and that numerals are not numbers.
3. Represent numbers using Egyptian, Roman and Hindu-Arabic numerals.
4. Determine the place value of a digit in a given Hindu-Arabic numeral.
5. Read/write in words a number given in symbols and vice versa.
6. Identify the commutative, associative, distributive, identity element and inverse element properties of numbers in various operations.
7. Perform mixed operations involving BEODMAS.

Satyajit Ray, one of India's leading filmmakers, wrote a fantasy called The Alien. In this script a character called Mohan exclaims, "One-two-three-four-five-six-seven-eight-nine—zero! You put a zero after one and it becomes more than nine. Every zero added multiplies it ten times. Isn't that wonderful? Well we discovered that zero—an unknown Indian."

Huynh Cong/AP/Wide World Photos

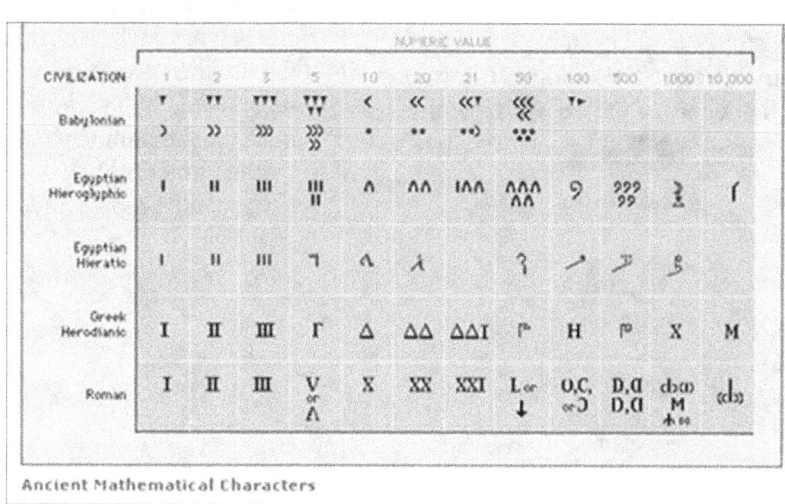

Ancient Mathematical Characters

1

Historical Development of Counting Systems

The system of counting of the ancient civilizations was very limited. For instance in China about 5000 BC, Chang who had three sheep said he had 'many' sheep. At that time, the Chinese system of counting and that of many other civilizations consisted only of one, two, and many. Mesopotamia, which was more civilized, had a counting system, which consisted of one, two, three and many. In Egypt where civilization started their counting system was even more developed and consisted of one, two, three, four, five and many.

As years went by the Chinese, found out that they could use their one and two to count even up to five. Thus, three pumpkins were counted as two and one

Four pumpkins were counted as two twos

Five pumpkins were counted as two twos and one

About 3000 BC, many civilizations began to write. They carved notches on stone or clay to represent their ideas. Numbers were represented by marking strokes on stones or clay or tying stones or sticks in bags. Figure 1:1 shows how three cattle could have been represented by people of the early civilization.

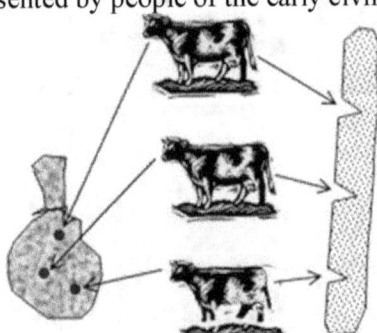

Figure 1:1

As civilizations grew, the method of carving notches on stones and clay to represent numbers was developed and is still used today in many disciplines such as statistics. Instead of carving notches on stones or clay, strokes are made on a piece of paper to represent how many has been counted. These strokes are called **tallies** and the method is called **tallying**. In tallying, a stroke is made to represent one item. The fifth stroke is drawn horizontal across the first four to stand for five things. For instance, seven items will be represented as ⲎⲎ‖ .

Example 1:1
Use tallies to represent the number 'fourteen'.

Solution
ⲎⲎ ⲎⲎ ‖‖

Example 1:2
What number does ⲎⲎ ⲎⲎ ⲎⲎ ⲎⲎ ‖ represent?

Solution
The number is 'twenty two'.

Notion of Numbers and Numerals

A **number** is an idea that expresses a quantity, or that expresses how many things have been counted. Being an idea a number, can neither be seen nor touched. Thus, one can see or touch two pencils, but cannot touch the number two. Numbers are usually represented by symbols called **numerals**. For instance the number 'one', can be represented by the numerals 1, I, or i while the number, 'two' can be represented by the numerals 2, II or ii.

EXERCISE 1:1

1. By 3000 BC, the Mesopotamian counting system consisted of one, two, three and many while the Egyptian system consisted of one, two, three, four, five and many. Explain using an illustration in each case,
   (a) How Laban a Mesopotamian might have expressed the idea that he has 8 boats.
   (b) How Amenes an Egyptian might have expressed the idea that he has thirteen palms.

2. Represent the following using tallies.
   (a) 33   (b) 17   (c) 28   (d) 41 (e) 15

3. What numbers do the following tallies represent?
   (a) 卌 卌 卌 卌 卌 ||
   (b) 卌 卌 卌 |||
   (c) 卌 卌 卌 卌 卌 卌 卌 ||||
   (d) 卌 卌 卌 卌 |
   (e) 卌 卌 卌 卌 卌 卌 ||

4. State whether each of the following is a number or a numeral.
   (a) Five oranges   (b) 卌   (c) 5

The Egyptian Numerals

As time went by the Egyptian system of numeration was developed and they counted in groups of tens. Table 1:1 shows the symbols, which, the Egyptians used.

Number	Numeral
One		
Ten	∧
One hundred	9
One thousand	⚡
Ten thousand	ſ

Table 1:1

A system such as that of the Egyptians in which counting is done in groups of tens is called a **base ten system**, a **denary system** or a **decimal system**. The order in which the symbols were written in the Egyptian system was of no significance. Also in this system, some large numbers were written using only very few symbols while some small numbers were written using so many symbols.

Example 1:3
Write the following using Egyptian numerals.
(a) two hundred and forty nine
(b) Ten thousand one hundred.
(c) Twenty-three thousands four hundred and twenty one.

Solution
(a) 9∧∧||||
     9∧∧|||||
(b) ſ9
(c) ſ⚡⚡99∧
    ſ⚡99∧|

EXERCISE 1:2

1. Write the following using Egyptian numerals.
   (a) 179   (b) 3478   (c) 6351   (d) 421
   (e) 2456   (f) 1334   (g) 18
   (h) 1888   (i) 4478   (j) 286

2. What do the following Egyptian numerals stand for?
   (a) ∧∧|
      ∧∧|
   (b) |||| ⚡9
      ||⚡⚡9
   (c) ⚡⚡∧9
      ∧∧∧||
   (d) ||||99∧⚡
      ||||99∧⚡
   (e) ∧∧∧∧
      ∧∧⚡91
   (f) ∧∧9⚡|
      ∧99|||
   (g) 9999∧||
      999∧∧||
   (h) ⚡⚡⚡9999∧∧
      ⚡⚡⚡999||∧∧
   (i) ⚡⚡⚡999||
      ⚡⚡∧∧991
   (j) ⚡⚡⚡ſ99∧∧||
      ⚡⚡⚡9 9∧∧∧|||

3. Add the following like an Egyptian.
   (a) ⚡⚡⚡99∧ and ⚡⚡99∧|
      ⚡⚡⚡9∧∧| ⚡⚡99∧|
   (b) ⚡⚡⚡9999∧∧ and ⚡⚡⚡ 999∧||
      ⚡⚡⚡999||∧∧   ⚡⚡∧∧99∧|

4. An Egyptian newspaper reported that during a flood of the R. Nile, out of the

⌐⫶⫶⫶∧‖
9999∧‖ people along the R. Nile,

many people died leaving only

⌐⫶⫶⫶99∧∧‖
9⫶⫶⫶99∧∧‖ . How many people

died during the flood?

The Roman Numerals

Despite the fact that the Roman system of numeration is very difficult to use in calculations, it is still in great use today, and can be found on some clock faces, documents and numbering of some lists, exercises and chapters of some textbooks. Table 1:2, shows these symbols developed by the Romans to represent how much they counted.

Number	Numeral
One	I
Five	V
Ten	X
Fifty	L
One hundred	C
Five hundred	D
One thousand	M

Table 1:2

From the table it can be noticed that they use mostly capital letters of the English alphabet. The small letters i, v and x are used to represent one, five and ten.

The Romans use these symbols in different combinations to represent numbers even more than a million. For instance,

1998 is written as MCMXCVIII

754 is written as DCCLIV

Remarks

1. For the Romans unlike the Egyptians, the order in which the symbols are written is very important. For instance, while the Egyptians write ∧9 or 9∧ to represent one hundred and ten, for the Romans, CX, represented one hundred and ten and XC, represented ninety.
2. The Roman system of numeration is built on "five", probably because they used the five fingers of one hand to count. Thus, they write V instead of IIIII, L instead of XXXXX, D instead of CCCCC and so on. Such a system whereby counting is done in groups of fives is called a **base five system** of numeration.
3. In the Roman system a letter placed after another of greater value, adds to its value e.g. VI = 5+1 = 6. On the other hand, a letter placed before another letter of greater value, subtracts from its value e.g. IV=5−1=4.

Some Uses of Roman Numerals

Though Egyptian numerals are not readily used in our day-to-day life, Roman numerals are used very often to:

(i) Number pages, lists of items or exercises in books as in this list has been numbered.
(ii) Number some clock faces.
(iii) Number result sheets and articles of constitutions.
(iv) Matriculate cars.

EXERCISE 1:3

1. Write the following using Roman numerals
   - (a) 29
   - (b) 48
   - (c) 4874
   - (d) 3993
   - (e) 1338
   - (f) 1990
   - (g) 3000
   - (h) 2432
   - (i) 549
   - (j) 1428
   - (k) 4488
   - (l) 286

2. Write the following Roman numerals in words.

   (a) CLIV   (b) DCIX

   (c) MCXIV   (d) DCCLXI

   (e) DLXXXIX   (f) MCMXCIX

   (g) LIX   (h) MDIX

   (i) MMMCDXLIX   (j) DCCCI

   (k) MMCDL   (l) MCDXLVIII

3. List the Roman numbers in increasing order of magnitude from one to fifty.

4. What Roman numerals do the following Egyptian numerals represent?

   (a) (b)

   (c) (d)

   (e) (f)

5. What Egyptian Numerals stand for the following Roman numerals?

   (a) MDCCLVIII   (b) LXXXVII

   (c) XLV   (d) DCLXV

   (e) MMCCLXXI   (f) MCMXCVIII

The Hindu-Arabic Numerals

Our present number system is called the Hindu-Arabic system. These numerals, which were developed about 200 AD, were brought to Africa and Europe by Arabic traders but were not widely accepted until the sixteen century.

Originally, these numerals were so different from what exist today. Even today, the Arabic numerals used in the Islamic world are very different from the ones used internationally. Compare the two different types in Table 1.3.

Hindu-Arabic	1	2	3	4	5	6	7	8	9	0
Islamic world	١	٢	٣	٤	٥	٦	٧	٨	٩	٠

Table 1.3

Though the order in which the symbols were written was of great importance, the Hindus were unable to distinguish clearly between certain numbers such as six thousand and five, six hundred and five and sixty-five. This led to the invention of the symbol called **zero,** which means '**nothing**', initially represented by a dot (·). Zero was called **sifr**, later **cipher** and today it is called naught and is represented by 0.

With zero, six thousand and five was written 6005 and sixty-five was written 65. The 0 here is called a placeholder.

The Place Value System

Each of the symbols 0, 1, 2, 3, 4, 5, 6, 7, 8 and 9 is called a **digit**. The value of each digit is equal to the product of the digit and its place value in the numeral. For instance, the value of 3 in 846302 is three hundred and in 8463020 is three thousand.

For this reason, the Hindu-Arabic system is a place value system.

The Hindu-Arabic system is also built on 'ten'. Thus, this system is also a base ten, denary or decimal system like the Egyptian system, but the Egyptian system is not a place value system.

It should be noted that, a place-value system can use any base, one of the symbols must be zero and the number of digits must be equal to the base used.

The place value system makes it possible for numbers to be written without any confusion. Thus, bills such as water and electric bills are easily written and read by the public. Without the place value system, it will be very difficult to distinguish say a bill of four thousands and fifteen (4015) francs from a bill of four hundred and fifteen (415) francs. Table 1.4 gives the place values of whole numbers.

Billions			Millions			Thousands			Ones		
Hundred Billions	Ten Billions	Billions	Hundred Millions	Ten Millions	Millions	Hundred Thousands	Ten Thousands	Thousands	Hundreds	Tens	Units
8	6	3	4	0	1	7	4	9	3	0	2

Table 1.4

5

Table 1.4 can be used to read and write number names up to hundreds of billions. To facilitate the reading of numbers, the digits are often grouped in threes, counting from the right and separated with commas.

Example 1:4

Write out the following in words.

(a) 56,454   (b) 897,543

Solution

(a) Fifty-six thousand four hundred and fifty four.

(b) Eight hundred and ninety seven thousand five hundred and forty three.

Example 1:5

Write in figures

(a) Three million seven hundred and eighty two thousand four hundred and sixteen

(b) Ninety million three hundred and one thousand and five

Solution

$$
\begin{array}{rr}
\text{(a)} \quad 3,000,000 & \text{(b)} \quad 90,000,000 \\
700,000 & 300,000 \\
82,000 & 1,000 \\
+ \quad 416 & + \quad 5 \\
\hline
3,782,416 & 90,301,005 \\
\end{array}
$$

Example 1:6

What is the value of 6 in the number 356,789,743?

Solution

$6 \times 1,000,000 = 6,000,000$ i.e. Six million

EXERCISE 1:4

(1) Give the value of ٥ in the following Islamic numerals.

(a) ٨٤٠٥   (b) ٩٥٧١

(c) ٣١٨٥٤   (d) ٣٥٨٠٢٩

(2) What modern Hindu-Arabic numerals are represented by the following Islamic numerals?

(a) ٣٤٠٥   (b) ٩٨٧٠

(c) ٧١٨٩٤   (d) ٣٥٨٠٢٧

(3) Write the Islamic numerals, which stand for the following.

(a) 34956 (b) 30478 (c) 3260 (d) 97203

(4) State the value of 3 in each of the following.

(a) 37570   (b) 613004   (c) 9320161

(5) What makes 68 and 86 different?

(6) Write in words

(a) 5578   (b) 50448

(c) 893261   (d) 17204

(7) Write each of the following using Modern Hindu-Arabic numerals

(a) Five hundred and thirty eight thousand and one.

(b) Seventeen thousand and four

(c) Nine thousand nine hundred and nine

(d) Two hundred and thirty two

(e) One hundred and eleven thousand one hundred and one

(f) Eight thousand and Eighty

(g) Ten thousand and ten.

8. Write out the following in words.

(a) 7,564   (b) 644,325

(c) 29,576,532   (d) 6,435,553

(e) 56,442,443

9. Write the following in figures

(a) Five million two hundred and forty six thousand eight hundred and thirty one

(b) Forty three million two hundred and seven thousand and nineteen

10. What is the value of 9 in the 59,432,762?

11. What is the value of 4 in 1645,632?

The Four Basic Operation Rules

The four basic operations of number manipulation are:

Addition (+), is also referred to as the **sum or plus**. The numbers to be added are called **addends**. The result of the addition is called the **sum**.

$$
\underset{\text{term}}{5} + \underset{\text{term}}{7} = \underset{}{12}
$$

addend   addend   sum

Subtraction (−), is also called the **difference**. The number from which another is to be subtracted is called the **minuend** and the number to be subtracted is called the **subtrahend**. The result of the subtraction is called the **difference**.

$$\underset{\text{term}}{\underset{\text{minuend}}{18}} \quad - \quad \underset{\text{term}}{\underset{\text{subtrahend}}{13}} \quad = \quad \underset{\text{difference}}{5}$$

In addition and subtraction the addends, the minuend and the subtrahend are also called the **terms**.

Multiplication (×), is also called **times** or the **product**. In multiplication, the number being multiplied is called the **multiplicand** and the number by which the multiplication is to be performed is called the **multiplier**. The result of the multiplication is also called the **product**.

$$\underset{\text{multiplicand}}{4} \quad \times \quad \underset{\text{multiplier}}{6} \quad = \quad \underset{\text{product}}{24}$$

Division (÷) is also called the **quotient**. The number to be divided is called the **dividend** and the one to be divided by is called the **divisor**. The result of the division is called the **quotient**.

$$\underset{\text{dividend}}{14} \quad \times \quad \underset{\text{divisor}}{2} \quad = \quad \underset{\text{quotient}}{7}$$

Properties of Operations

Inverse Property

Consider the following

$$4 + (-4) = (-4) + 4 = 0$$

$$\frac{3}{5} \times \frac{5}{3} = \frac{5}{3} \times \frac{3}{5} = 1$$

In the first case, which deals with addition, the result is 0; the identity element for addition. In the second case, which deals with multiplication, the result is 1; the identity element for multiplication.

For any operation, any two elements which when operated together give the identity element are said to be **inverses** of each other. Thus for addition, 4 is the inverse of − 4 and vice versa while for multiplication, $\frac{3}{5}$ is the inverse of $\frac{5}{3}$ and vice versa.

Properties of Addition

Investigative Exercise 1:1

Evaluate the following

(1)	(a)	2+5	(b)	5+2
(2)	(a)	0+8	(b)	8+0
(3)	(a)	(3+5) + 2	(b)	3+ (5+2)

In each case compare your answers in (a) and (b). Hence, draw a conclusion for each case.

(i) From (1) it can be seen that, *interchanging the addends does not change the sum*. This property is known as the **commutative property of addition**.

(ii) From (2) it can be seen that, *the sum of any number and 0 is the number*. This property is known as the **identity property of zero.** 0 is said to be the identity element for addition.

(iii) From (3) it can be seen that, *changing the grouping of the addends does not change the sum*. This property is known as the **associative property of addition.**

Properties of Multiplication

Investigative Exercise 1:2

Evaluate the following.

(1)	(a)	2×5	(b)	5×2
(2)	(a)	1×8	(b)	8×1
(3)	(a)	(3×5) ×2	(b)	3× (5×2)
(4)	(a)	0×9	(b)	9×0

In each case, compare your answers in (a) and (b). Hence, draw a conclusion for each case.

(i) From (1) it can be seen that, *interchanging the multiplicand and the multiplier does not change the product.* This property is known as the

commutative property of multiplication.

(ii) From (2) it can be seen that, *the product of any number and 1 is the number.* This property is known as the **identity property of one.** 1 is said to be the identity element for multiplication.

(iii) From (3) it can be seen that, *changing the grouping of the numbers does not change the product.* This property is known as the **associative property of multiplication.**

(iv) From (4) it can be seen that, multiplying any number by 0, the result is 0. This is called the **multiplicative property of zero.**

The Distributive Property

Consider the simplification of $3(4+2)$,
This can be done in any of the two ways below

$$3(4+2) = 3(6) \quad \text{or} \quad 3(4+2) = 3(4) + 3(2)$$
$$=18 \qquad\qquad\qquad = 12+6 = 18$$

The second way uses the concept of the distributive property of multiplication over addition.

Applications of the Properties of Numbers

The properties of numbers are very useful in simplifying calculations as illustrated in the below

(a) Numbers can be grouped to make multiples of 10 since they are much easier to manipulate.
$$36 - 42 + 14 = 36 + 14 - 42 \quad \text{(commutative property)}$$
$$= 50 - 42 = 8$$

(b) Inverse elements can be grouped to simplify numerical expressions. Thus

$$9 + 17 - 9 = 9 - 9 + 17 \qquad \text{(commutative property)}$$
$$= 0 + 17 \qquad \text{(additive inverse property)}$$
$$= 17 \qquad \text{(additive property of zero)}$$

(c) Numbers can be broken into units that can easily be simplified. Thus
$$\begin{aligned}(i) \quad 8 \times 23 &= 8(20 + 3) \\ &= 8 \times 20 + 8 \times 3 \\ &\quad \text{(Distributive property)} \\ &= 160 + 24 \\ &= 184\end{aligned}$$

$$\begin{aligned}(ii) \quad 97 \times 16 &= (100 - 3)16 \\ &= 100 \times 16 - 3 \times 16 \quad \text{(distributive property)} \\ &= 1600 - 48 \\ &= 1552\end{aligned}$$

(d) Common factors can be factored out to simplify expressions. Thus
$$\begin{aligned}54 + 36 &= (9 \times 6) + (9 \times 4) \\ &= 9(6 + 4) \quad \text{(Distributive property)} \\ &= 9 \times 10 \\ &= 90\end{aligned}$$

(e) The associative property can be used to regroup numerical expressions to ease simplification. Thus

$$\begin{aligned}72 + (28 + 89) &= (72 + 28) + 89 \quad \text{(associative property)} \\ &= 100 + 89 \\ &= 189\end{aligned}$$

EXERCISE 1:5

Without solving or evaluating, each of the following, state the property used in each case.

(a) $5 + (3 + 4) = (5 + 3) + 4$

(b) $3 \times 0 = 0$

(c) $9 \times (2 + 8) = 9 \times 2 + 9 \times 8$

(d) $17 + 5 = 5 + 17$

(e) $20 \times 5 = 5 \times 20$

(f) $(8 \times 3) \times 2 = 8 \times (3 \times 2)$

(g) $0 + 12 = 12$

(h) $1 \times 18 = 18$

(i) $5 + (-5) = 0$

(j) $\frac{1}{7} \times 7 = 1$

Sequence of Operations (BEODMAS)

When two or more of the operations +,−, ×, ÷, exponents and brackets occur in one expression, perform them in the following order.

Brackets
Exponents
Of
Division
Multiplication
Addition
Subtraction

The contraction BEODMAS may help to recall this to memory.

Note!

1. Division and multiplication have the same preferences. However, it is advisable to do division before multiplication. By doing multiplication first the numbers, get larger and make, a calculation complicated. By doing division, first the numbers get smaller and make a calculation less complicated.

2. Addition and subtraction have the same preferences. However, it is advisable to add all the numbers with a plus sign before them, and add all the numbers with a minus sign before them before subtracting.

Example 1:7

Evaluate; $16 + 8 \div 4 - 3 \times 2 + (9 - 3)$

Solution

$16 + 8 \div 4 - 3 \times 2 + (9 - 3) = 16 + 2 - 6 + 6 = 18$

EXERCISE 1:6

Evaluate the following.

(1) $15 - 8 \div 2 + 5$
(2) $9 - 6 + 4 \times 6 \div 3$
(3) $7 + 6 \times 15 - 12 \div 4$
(4) $7 \times 14 - 12 \div 4 + 5$
(5) $13 - 9 \div 3 + 3 \times 8 - 3$
(6) $3 \times 24 - 9 \div 3 + 4$

MULTIPLE CHOICE EXERCISE 1

1. The statement that refers to numeral is:
   [A] When counting 7 comes before 8
   [B] The sum of 2 and 6 is 8
   [C] In 78, 7 comes before 8
   [D] 78 is the sum of 50 and 28
2. The statement that refers to numbers is:
   [A] In 23, 2 comes before 3
   [B] When counting, 2 comes before 3
   [C] 2 combined with 3 is either 23 or 32
   [D] 23 consist of 2 and 3
3. In 5+0 = 5, the property that applies is:
   [A] The commutative property of zero
   [B] The associative property of zero
   [C] The distributive property of zero
   [D] The additive identity property of zero
4. In 2×0+7 = 7, the property of 0 that applies is:
   [A] The multiplicative property
   [B] The additive property
   [C] The distributive property
   [D] The identity property
5. In 3×1+2=5, the property of 1 that applies is:
   [A] The distributive property
   [B] The additive property
   [C] The multiplicative property
   [D] The identity property
6. $(6 \times 2) \times 50 = 6 \times (2 \times 50)$. The property used is:
   [A] The commutative law of multiplication
   [B] The associative law of multiplication
   [C] The distributive law of multiplication
   [D] The multiplicative property of numbers
7. $(73 + 25) + 75 = 73 + (25 + 75)$. The property applied is:
   [A] The associative law of addition
   [B] The commutative law of addition
   [C] The distributive law of addition
   [D] The addition property of numbers
8. The Hindu Arabic numeral representing "Two hundred and four thousand and four" is:
   [A] 20404   [B] 240004
   [C] 24400   [D] 204004
9. One million three hundred and fifty four is written as:
   [A] 1000354   [B] 1030054
   [C] 1300054   [D] 1354000

10. Ninety nine thousand and ninety nine written in figures is:
   [A] 990099   [B] 9999
   [C] 99099   [D] 90999

11. The number 605, 080 is read:
   [A] Sixty thousand and five thousand and eighty
   [B] Six hundred and five hundred and eighty
   [C] Six thousand and five hundred and eighty
   [D] Six hundred and five thousand and eighty

12. The amount 2,300,240 francs is read:
   [A] Twenty three million four hundred and twenty francs
   [B] Two million, three hundred thousand four hundred and twenty francs
   [C] Two million three hundred thousand two hundred and forty francs.
   [D] Twenty three million two hundred and forty francs

13. The value of the digit 6 in the number 726251 is:
   [A] six hundred   [B] six hundredth
   [C] six thousandth   [D] six thousand

14. The value of 5 in 2753 is:
   [A] Tenth   [B] tens
   [C] Hundredth   [D] hundredth

15. When the value of 6 in 5624 is divided by the value of 3 in 2639, the result is:
   [A] 2   [B] 5   [C] 20   [D] 16

16. The product of the value of 7 in 2721 and the value of 3 in 5837 is:
   [A] 21000   [B] 26677
   [C] 21   [D] 2100

17. The sum of eleven thousand and one thousand hundred is:
   [A] 11100   [B] 12100
   [C] 11110   [D] 111000

18. In 6,367,804, the value of the underlined digit is:
   [A] 7   [B] 700   [C] 7,000   [D] 70,000

19. Four million and six is represented by:
   [A] 4,000,600   [B] 4,000,006
   [C] 4,600   [D] 4,006

20. In $3 \times 7 = 21$, 3 is called:
   [A] the multiplier   [B] the multiplicand
   [C] the minuend   [D] the dividend

21. In $8 - 5 = 3$, 5 is called:
   [A] the difference   [B] the minuend
   [C] the dividend   [D] the subtrahend

22. $8 + 5(4 - 2)$ is equal to:
   [A] 12   [B] 50   [C] 18   [D] 22

23. The number represented by the Roman numeral CDXXXVII is:
   [A] 437   [B] 637   [C] 187   [D] 487

24. The number represented by the Roman numeral CMXXII is:
   [A] 5220   [B] 1922   [C] 922   [D] 918

25. The number 40 in Roman numerals is:
   [A] XV   [B] XL   [C] XLX   [D] LX

26. 1167 in Roman numerals is:
   [A] MCLXII   [B] DCLXVII
   [C] MCLXVII   [D] MCXLVII

27. As a Roman numeral 2598 is the same as:
   [A] IIMVCIXVIII   [B] MMDXCVIII
   [C] MMIIDC   [D] XXVIIC

28. In Roman numerals, 548 can be written as:
   [A] VIIILD   [B] IIXLD
   [C] DLIIX   [D] DXLVIII

29. The number 1200 in Egyptian numerals is:
   [A] 𓏺𓏏𓏏 [B] 𓆼𓎆𓎆 [C] 𓆼𓎆𓐍 [D] 𓏺𓐍𓐍

30. The Egyptian numeral 𓆼𓎆𓏺𓏺𓏺 stands for the Hindu-Arabic numeral:
   [A] 91200   [B] 991
   [C] 11111   [D] 1112

31. The Hindu Arabic numeral representing "Two hundred and four thousand and four" is:
   [A] 20404   [B] 240004
   [C] 24400   [D] 204004

TOPIC 2

SETS OF NUMBERS

OBJECTIVES

At the end of this topic, the learner should be able to:

1. Recognise $\mathbb{N}$ as the set of natural numbers and distinguish between counting numbers and natural numbers.
2. Identify odd and even numbers.
3. Express normal form numbers in index form and vice versa.
4. Express a product of same factors in index form.
5. Determine the value of a number expressed in index form.
6. List the factors and multiples of any given number.
7. Identify prime and composite numbers.
8. Find the common factors and common multiples of numbers in $\mathbb{N}^*$.
9. Express any member of $\mathbb{N}$ as a product of prime factors.
10. Find the HCF and LCM of a set of at most four numbers.
11. Use prime factorization to find the square root and/or cube root of a number.
12. Express simple perfect squares and perfect cubes in index form.
13. Recognise by inspection and/or verification whether a number is divisible by 2, 3, 4, 5, 6, 7, 8, 9, 10, 11, 12, 25, and 50,100.
14. Make triangular, square or rectangular dot patterns of numbers in $\mathbb{N}^*$.
15. List the sets of triangular, square and rectangular numbers.
16. Identify elements of the sets $\mathbb{N}, \mathbb{Z}, \mathbb{Q}, \mathbb{Q}'$ and $\mathbb{R}$ and determine to which of the sets a given number belongs.
17. Use set notation, tree diagrams and Venn diagrams to show the relationship between the various sets of numbers.

Augustin Louis Cauchy

French mathematician Augustin Louis Cauchy provided logical foundations to calculus by developing the theory of limits. Cauchy became a leader in 19th-century mathematics and was one of the first mathematicians to publish his work extensively.

Roger Viollet/Getty Images

NATURAL NUMBERS

The set of Natural numbers usually denoted by $\mathbb{N}$ is the set of all positive whole numbers including zero.

$$\mathbb{N} = \{0,1,2,3 \dots\}.$$

The set of natural numbers can be partitioned into three sets namely; the set containing the element zero, the set of odd numbers,

$O = \{1,3,5,7,9,..\}$ and the set of even

numbers $E = \{2, 4, 6, 8, 10,...\}$.

Thus, $\mathbb{N} = \{0\} \cup O \cup E$.

Operations with Natural Numbers
Index Notation (Index Form)

3^4 means 'multiply 3 by itself four times'.
  i. e. $3^4 = 3 \times 3 \times 3 \times 3$

5^7 means 'multiply 5 by itself seven times'.
  i. e. $5^7 = 5 \times 5 \times 5 \times 5 \times 5 \times 5 \times 5$

8^3 means 'multiply 8 by itself three times'.
  i. e. $8^3 = 8 \times 8 \times 8$

Study the following:

$$5^{6 \leftarrow \text{exponent or index or power}}_{\quad \leftarrow \text{base}}$$

5^6 is called the power. In 5^6, 6 is called the **index** or **exponent** or may also be loosely referred to as **power** while 5 is called the **base**.

5^6 is read 'five to the sixth power' or 'five to the power six' or 'five, raised to the sixth power' or simply 'five to the sixth' or five exponent six.

A number such as 5^6 is said to be in **index form** or **index notation** or **exponential notation**.

Any real number other than 0 raised to the power 1 is the number. Any real number other than 0 raised to the power 0 is 1. 0^0 is meaningless.

1. State the meaning of the following.
   (a) 7^9 (b) 2^3 (c) 8^2 (d) 4^5
2. Write the following using index notation.
   (a) Multiply 6 by 6 five times.
   (b) Multiply 12 by 12 three times.
3. Write the following using index notation.
   (a) $7 \times 7 \times 7 \times 7 \times 7 \times 7 \times 7$
   (b) $13 \times 13 \times 13 \times 13 \times 13$

Powers of Ten

Study Table 2:1

Number	Number as product of tens	Index form	Number of zeros
1	No factor of 10	10^0	Non
10	10	10^1	1
100	10×10	10^2	2
1000	$10 \times 10 \times 10$	10^3	3
10000	$10 \times 10 \times 10 \times 10$	10^4	4
100000	$10 \times 10 \times 10 \times 10 \times 10$	10^5	5
1000000	$10 \times 10 \times 10 \times 10 \times 10 \times 10$	10^6	6

Table 2:1

From Table2:1, it should be clear that:

To write any power of ten in index form, simply raise 10 to a power equivalent to the number of zeros in the power of ten.

Example 2:1

Write in exponential (index) form.
(a) 100,000   (b) 10,000,000
(c) 1,000,000,000
Hint: count the number of zeros in each case.

Solution

(a) $100,000 = 10^5$
(b) $10,000,000 = 10^7$
(c) $1,000,000,000 = 10^9$

Multiplication Law of Indices

Consider the following:
(i) $10^3 \times 10^4 = (10 \times 10 \times 10) \times (10 \times 10 \times 10 \times 10)$
$\qquad\qquad = 10^7$
(ii) $2^3 \times 2^5 = (2 \times 2 \times 2) \times (2 \times 2 \times 2 \times 2 \times 2)$
$\qquad\qquad = 10^8$

Notice that if in (i) the indices 3 and 4 are added, the result will be 7. Similarly, if in (ii) 3 and 5 are added, the result will be 8. Thus,

(i) $10^3 \times 10^4 = 10^{3+4} = 10^7$

(ii) $2^3 \times 2^5 = 2^{3+5} = 10^8$

Therefore,

To multiply together quantities written to the same base in index form, raise the base to the sum of the exponent or indices.

EXERCISE 2:2

Compute the following products writing your answer in index form.

1. $2^3 \times 2^6$
2. $7^3 \times 7^5$
3. $3^{11} \times 3^2 \times 3^4$
4. $10^3 \times 10^2 \times 10^5$
5. $10 \times 10^{11} \times 10^4$

Division Law of Indices

Consider the following:

(a) $\dfrac{10^7}{10^5} = \dfrac{\cancel{10} \times \cancel{10} \times \cancel{10} \times \cancel{10} \times \cancel{10} \times 10 \times 10}{\cancel{10} \times \cancel{10} \times \cancel{10} \times \cancel{10} \times \cancel{10}} = 10^2$

(b) $\dfrac{3^9}{3^5} = \dfrac{\cancel{3} \times \cancel{3} \times \cancel{3} \times \cancel{3} \times \cancel{3} \times 3 \times 3 \times 3 \times 3}{\cancel{3} \times \cancel{3} \times \cancel{3} \times \cancel{3} \times \cancel{3}} = 3^4$

Notice that the results in (a) and (b) above could have been obtained by subtracting the indices in each case. Thus,

(a) $\dfrac{10^7}{10^5} = 10^{7-5} = 10^2$

(b) $\dfrac{3^9}{3^5} = 3^{7-3} = 3^4$

Therefore,

To divide quantities written to the same base in index form, raise the base to the difference of the exponents or indices.

EXERCISE 2:3

Evaluate leaving your answer in index form.

(a) $10^8 \div 10^3$ (b) $10^{14} \div 10^9$ (c) $\dfrac{10^{17}}{10^{13}}$ (d) $\dfrac{10^6}{10^4}$

(e) $3^{22} \div 3^{19}$ (f) $5^9 \div 5^5$ (g) $\dfrac{6^{13}}{6^{11}}$ (h) $\dfrac{13^4}{13^2}$

FACTORS AND MULTIPLES

Consider the following:

dividend	divisor	quotient		multiplicand	multiplier	product
21	÷ 7	= 3	or	3	× 7	= 21
multiple	factor	factor		factor	factor	multiple

Provided there is no remainder, the divisor and the quotient are also referred to as **factors** of the dividend while the dividend is also referred to as a **multiple** of the divisor and the quotient. In other words, the multiplicand and the multiplier are also referred to as **factors** of the product while the product is also referred to as a **multiple** of the multiplicand and the multiplier.

Therefore, every number is a factor of itself and 1 is a factor of every number, since for instance $1 \times 21 = 21$.

To summarise,

A multiple of a number is the product of the number and any non-zero whole number.
A factor of a number is the quotient of the number and any non-zero whole number less than or equal to the number.

Example 2:2

Write down the set of all the factors of 36.

Solution

Factors of 36 = {1, 2, 3, 4, 6, 9, 12, 18, 36}
Every number is **divisible** by each of its factors. Thus, 36, is divisible by 1, 2, 3, 4, 6, 9, 12, 18, 36.

EXERCISE 2:4

1. List the set of factors of
   (a) 24   (b) 60   (c) 120
   (d) 72   (e) 105   (f) 75
2. List the first 5 multiples of each of the following numbers
   (a) 2   (b) 4   (c) 5   (d) 7   (e) 8   (f) 12
3. Which of the following are factors of 48?
   1, 2, 3, 4, 5, 6, 7, 8, 9
4. Which of the following are multiples of 6?
   18, 20, 21, 22, 24, 26, 27, 28, 30, 32, 34, 35, 36, 38.

PRIME AND COMPOSITE NUMBERS

Number	Set of factors	Number of factors
1	1	1
2	1,2	2
3	1,3	2
4	1,2,4	3
5	1,5	2
6	1,2,3,6	4
7	1,7	2
8	1,2,4,8	4
9	1,3,9	3
10	1,2,5,10	4
11	1,11	2
12	1,2,3,4,6,12	6
13	1,13	2
14	1,2,7,14	4

Table 2.2

Table 2.2, shows some counting numbers and their factors. It can be seen that some natural numbers have only two factors while others have more than two factors. A **prime number** is a natural number with exactly two factors, the number itself and one. A **composite number** on the other hand is a number that has more than two factors.

Note that the number 1 is neither a prime number nor a composite number because it has only one factor, only 1 itself.

EXERCISE 2:5

1. Which of the following are prime numbers?
   5, 27, 29, 39, 24, 47, 49
2. Which of the following are composite numbers?
   11, 17, 49, 35, 24, 19
3. How many prime numbers are there between 1 and 50? List all these prime numbers.
4. List the set of all the factors of 72. State the number of factors of 72 that are composite numbers.

PRIME FACTORIZATION

Every composite number can be expressed as a product of prime factors. This process of expressing a composite number as a product of prime factors is known as **prime factorization**.

Example 2:3

Express each of the following as a product of prime factors.
(a) 30   (b) 24

Solution

(a) $30 = 2 \times 15 = 2 \times 3 \times 5$
(b) $24 = 2 \times 12 = 2 \times 2 \times 6 = 2 \times 2 \times 2 \times 3 = 2^3 \times 3$

It is not often as easy as in the above example to express a number as a product of prime factors. Therefore, it is necessary to study some techniques of prime factorization. These techniques include the peeling method and the factor tree method.

The Peeling Method

In this method, divide the number repeatedly by prime factors until the result is 1.

Example 2:4

Write each of the following as a product of prime factors, using the peeling method.
(a) 110   (b) 7290

Solution

(a)

$$\begin{array}{r|r} 2 & 110 \\ \hline 5 & 55 \\ \hline 11 & 11 \\ \hline & 1 \end{array}$$

$$\therefore 110 = 2 \times 5 \times 11$$

(b)

$$\begin{array}{r|r} 2 & 7290 \\ \hline 3 & 3645 \\ \hline 3 & 1215 \\ \hline 3 & 405 \\ \hline 3 & 135 \\ \hline 3 & 45 \\ \hline 3 & 15 \\ \hline 5 & 5 \\ \hline & 1 \end{array}$$

$$\therefore 7290 = 2 \times 3^6 \times 5$$

The Factor Tree

Examples 2:5

Do example 3:4 using the factor tree method.

Solution

(a)

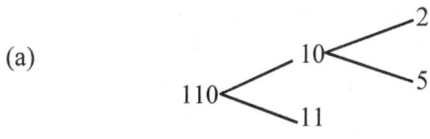

$$\therefore 110 = 2 \times 5 \times 11$$

(b)

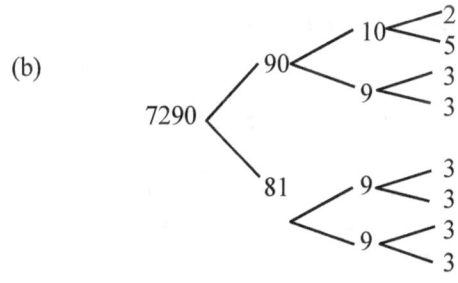

$$\therefore 7290 = 2 \times 3^6 \times 5$$

Note that the branches of the factor trees end with prime factors.

> ## EXERCISE 2:6

1. Write the following as a product of prime factors in index form.

   (a) 32 (b) 81 (c) 60
   (d) 72 (e) 45 (f) 63
   (g) 51 (h) 48 (i) 243

2. Decompose the following into the product of their prime factors.

   (a) 630 (b) 2200 (c) 2,002
   (d) 1728 (e) 5280

COMMON FACTORS

As the name suggest, the common factors of two or more numbers are the factors that are common (or belong) to the numbers.

Examples 2:6

List the common factors of 18 and 24.

 Solution

Factors of 18 = {1, 2, 3, 6, 9, 18}
Factors of 24 = {1, 2, 3, 4, 6, 8, 12, 24}
∴ Common factors of 18 and 24 = {1, 2, 3, 6}

THE HIGHEST COMMON FACTOR

The highest common factor (HCF) also called the greatest common divisor (GCD) is the greatest of the common factors of given numbers. Therefore, the common factors of whole numbers are the factors of their HCF.

Examples 2:7

Find the HCF of 42 and 48.

Solution

Factors of 42 = {1, 2, 3, 6, 7, 14,21, 42}
Factors of 48 = {1, 2, 3, 4, 6, 8, 12, 16,24, 48}
Common factors of 42 and 48 = {1, 2, 3, 6}
∴ The HCF of 42 and 48 = 6

COMMON MULTIPLES

Again, the name suggests that the common multiples of two or more whole numbers are the multiples of these numbers that are common (or belong) to the numbers.

Example 2:8

List the first 4 common multiples of 2 and 3.

Solution

Let M_2 = multiples of 2
and M_3 = multiples of 3
Then,
$M_2 = \{2,4,6,8,10,12,14,16,18,20,22,24,26,...\}$
$M_3 = \{3, 6, 9,12,15,18,21,24,27,..\}$
∴ First 4 common multiples = {6, 12, 18, 24}

THE LEAST COMMON MULTIPLE

The Least common multiple (LCM) of two or more members is the smallest of their common multiples. The LCM of the denominators of fractions is sometimes referred to as the least common denominator (LCD).

Example 2:9
Find the LCM of 3 and 4

Solution
Multiples of 3 = {3, 6, 9, 12, 15, 18, 21, 24, 27,...}
Multiples of 4 = {4, 8, 12, 16, 20, 24, 28,...}

Common multiples of 3 and 4 = {12, 24,...}
∴ LCM of 3 and 4 = 12

The method of listing factors or multiples as in Example 2:6, Example 2:7, Example 2:8 and Example 2:9 called the **roster method**, is good for finding the HCF and LCM of simple cases. Cases that are more difficult may be done using the prime factorization methods.

HCF and LCM by Prime Factorization

Generally;
1. *The HCF is obtained by multiplying together the common prime factors.*
2. *The LCM on the other hand is obtained by finding the product of all the uncommon prime factors and the HCF.*

HCF = Product of highest powers of common prime factors
LCM = HCF × product of uncommon prime factors

Example 2:10
Find the HCF and LCM of 28 and 42.

Solution by the Peeling Method
Simultaneously peel all the numbers as shown below. The common prime factors appear on the leftmost column, while the uncommon prime factors appear at the base of the table.

2	28	42
7	14	21
	2	3

HCF = 2×7 = 14
LCM = 2×7×2×3 = 84

Solution by product of prime Factors Method

Write each of the numbers as a product of its prime factors. In the solution that follows, the common factors have been mapped with two-way arrows. The uncommon prime factors are unmapped.

$$28 = 2 \times 2 \times 7$$
$$\updownarrow \qquad \updownarrow$$
$$42 = 2 \times 3 \times 7$$

HCF = 2 × 7 = 14
LCM = HCF× product of uncommon prime factors
⇒ LCM = 14 × 2 × 3 = 84

Example 2:11
Find the HCF and LCM of 15 and 20.

Solution
This problem is solved below in the various methods in order to acquaint the reader on the application of all the methods.

Solution by the peeling method

5	15	20
	3	4

HCF = 5   LCM = 5×3×4 = 60

Solution by product of prime Factors Method

$$15 = 3 \times 5$$
$$20 = 2 \times 2 \times 5 = 2^2 \times 5$$
HCF = 5
LCM = 2 × 2 × 3 × 5 = 60

Example 2:12
Find the LCM and HCF of 8,12 and 20.

Solution
Using the peeling method

2	8	12	20
2	4	6	10
	2	3	5

HCF = 2 × 2 = 4
LCM = 2 × 2 × 2 × 3 × 5 = 120

Using the method of product of primes
$$8 = 2 \times 2 \times 2$$
$$12 = 2 \times 2 \times 3$$

$$20 = 2 \times 2 \times 5$$

HCF = product of common prime factors
$$= 2 \times 2 = 4$$

LCM = HCF × product of uncommon prime factors
$$= 4 \times 2 \times 3 \times 5$$
$$= 120$$

Example 2:13
Find the LCM and HCF of 16, 20 and 24

Solution

Using the peeling method

2	16	20	24
2	8	10	12
2	4	5	6
	2	5	3

To find the LCM, by the peeling method, peel the numbers until no two of them have common factors.

$$HCF = 2 \times 2 = 4$$
$$LCM = 2 \times 2 \times 2 \times 2 \times 3 \times 5 = 240$$

Using the method of product of primes
$$8 = 2 \times 2 \times 2$$
$$12 = 2 \times 2 \times 3$$
$$20 = 2 \times 2 \times 5$$

HCF = product of common prime factors
$$= 2 \times 2 = 4$$

LCM = HCF × product of uncommon prime factors
$$= 4 \times 2 \times 2 \times 3 \times 5$$
$$= 240$$

The LCM can also be defined as the product of the highest powers of the prime factors of given numbers.

Example 2:14
Find the LCM of 27, 63, and 75.

Solution
$27 = 3^3$, $63 = 3^2 \times 7$ and $75 = 3 \times 5^2$
The highest powers of each of the prime factors are 3^3, 5^2, and 7.
$\therefore LCM = 3^3 \times 7 \times 5^2 = 4,725$

1. Find the HCF and LCM of
   (a) 12 and 18   (b) 18 and 16
   (c) 96 and 72   (d) 24 and 21
   (e) 12 and 8   (f) 4,6 and 8
   (g) 18, 24 and 36   (h) 15,21 and 105
   (i) 15,25 and 75   (j) 216, 288 and 360
2. The HCF of three numbers is 18. List the common factors of the numbers.
3. State the four smallest numbers whose HCF is 9.

Squares and Square Roots

The square of a number is the product of the number and itself.
Thus in the set $\mathbb{R}$, $5 = 5 \times 5 = 25$

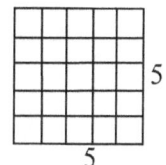

A whole number, which is the square of another whole number, is called a **perfect square**. e.g. (i.e. 25 is a perfect square because $25 = 5^2$, 49 is a perfect square because $49 = 7^2$.

Square Roots of Whole Numbers
A number is the square root of a second number provided it can be multiplied by itself to give the second number. Thus;
4 is a square root of 16 since $4 \times 4 = 16$

$\sqrt{16}$ denotes the positive square root of 16

$-\sqrt{16}$ denotes the negative square root of 16

Square Roots by Prime Factorization
To find the square root of a number
(i) Peel the given number completely using prime factors.
(ii) Pair up the repeated prime factors
(iii) Select one factor from each pair
(iv) Multiply the selected factors together.

Example 2:15

Find (a) $\sqrt{64}$ (b) $\sqrt{900}$ (c) $\sqrt{196}$

Solution
(a) 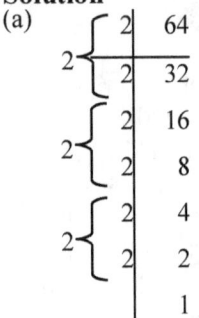 (b)

$\therefore \sqrt{64} = 2 \times 2 \times 2 = 8$ $\therefore \sqrt{900} = 2 \times 3 \times 5 = 30$

(c)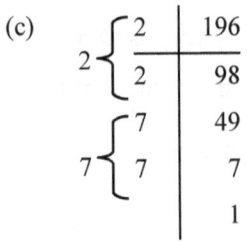

$\therefore \sqrt{196} = 2 \times 7 = 14$

CUBES OR CUBIC NUMBERS

To cube a number means to multiply the number by itself three times. This product is called the **cube** of the number.

7^3 is read as '7 cube' or '7 to the 3rd power'

Example 2:16

Evaluate (i) 7^3 (ii) 13^3

Solution
(i) $7^3 = 7 \times 7 \times 7 = 49 \times 7 = 343$
(ii) $13^3 = 13 \times 13 \times 13 = 169 \times 13 = 2197$

CUBE ROOTS

If a number is a cube of a second number, the second number is its cube root.

$2 \times 2 \times 2 = 8$ means 2 is the cube root of 8. The cube root of 8 is denoted by $\sqrt[3]{8}$.

Cube Roots by Prime Factorization

To find the cube root of a number,
(a) Peel the given number completely using prime factors,

(b) Group the repeated prime factors into three's,
(c) Select one factor from each group of three,
(d) Multiply together the selected factors.

Alternatively,
(a) Express the number as a product of prime factors using the index notation.
(b) Divide each index by 3 and find the product of the result.

Example 2:17
Find the cube root of
(a) 729   (b)   1000

Solution
(a) 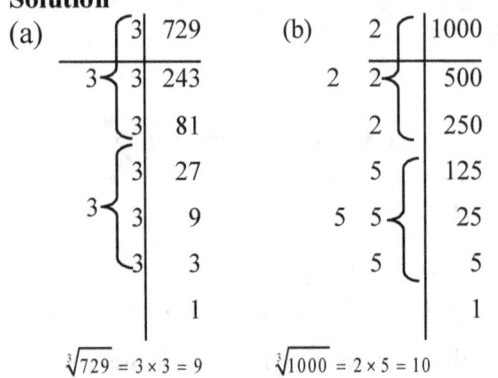   (b)

$\sqrt[3]{729} = 3 \times 3 = 9$ $\sqrt[3]{1000} = 2 \times 5 = 10$

or $\sqrt[3]{729} = 3^{6 \div 3} = 3^2 = 9$ $\sqrt[3]{1000} = 2^{3 \div 3} \times 5^{3 \div 3} = 2 \times 5 = 10$

EXERCISE 2:8

1. Square each of the following
    (a) 13   (b) 31   (c) 17   (d) 20
2. Evaluate each of the following
    (a) 18^2 (b) 28^2 (c) 19^2 (d) 32^2
3. Find the square root of each of the following numbers.
    (a) 784   (b) 289   (c) 6,400   (d) 4624
4. Evaluate each of the following
    (a) $\sqrt{625}$ (b) $\sqrt{529}$
    (c) $\sqrt{1,024}$ (d) $\sqrt{2,809}$
5. Cube the following
    (a) 4   (b) 7   (c) 13   (d) 20
6. Evaluate each of the following
    (a) 5^3 (b) 11^3 (c) 8^3 (d) 10^3
7. What is the smallest number, which can be multiplied by $(2^3) \times 3$ to give a perfect square?

Divisibility

Divisibility tests are very useful tools for finding the factors of numbers.

Investigative Exercise

The following table is a multiplication table for the numbers 2, 3, 4, 5, 6, 7, 8, 9, 10, 11, 12, 25 and 50.

×	51	52	53	54	55	56	57	58	59	60
2	102	104	106	108	110	112	114	116	118	120
3	153									
4	204									
5	255									
6	306									
7	357									
8	408									
9	459									
10	510									
11	561									
12	612									
25	1275									
50	2550									3000

Table 3:3

1. Complete the table. You may use a calculator.
2. What is common about all the numbers in row 2 (Multiplication table for 2)? What is the general name given to these types of numbers?
3. Add the digits of each of the numbers in row 3 (Multiplication table for 3). Is the sum of the digits of each number exactly divisible by 3?
4. Add the digits of each of the numbers in row 9 (Multiplication table for 9). Is the sum of the digits of each number exactly divisible by 9?
5. Are the numbers in row 6 (Multiplication table for 6) all divisible by both 2 and 3?
6. Is the number formed by the last two digits in row 4 (Multiplication table for 4) divisible by 4?
7. Is the number formed by the last three digits in row 8 (Multiplication table for 8) divisible by 4?
8. What is the last digit of each of the numbers in row 10 (Multiplication table for 10)?
9. Do the numbers in row 13 (Multiplication table for 25), end either in 25, 50, 75 or 00?
10. Do the numbers in row 14 (Multiplication table for 50), end either in 50 or 00?
11. In row 7 (Multiplication table for 7), subtract twice the last digit from the number formed by the remaining digits. Is your answer 0 or is it divisible by 7?
12. In row 11 (Multiplication table for 11), find the difference between the sum of the odd digits and the sum of the even digits. Is your answer 0 or is it divisible by 11?
13. Is the number formed by the last two digits in row 12 (Multiplication table for 12) divisible by 4? Add the digits of each of the numbers in this row. Is the sum of the digits of each number exactly divisible by 3?

Divisibility Rules

- A number is divisible by **2** if its last digit is 0, 2, 4, 6 or 8. For instance 64744635$\underline{8}$, is divisible by 2.
- A number is divisible by **4** if the number determined by the last two digits is divisible by 4. For instance, 2356$\underline{56}$ is divisible by 4 since 56 is divisible by 4.
- A number is divisible by **8** if the number determined by the last three digits is divisible by 8. For instance, 276$\underline{248}$ is divisible by 8 since 248 is divisible by 8.
- A number is divisible by **3** if the sum of its digits is divisible by 3. For instance 87146328 is divisible by 3 since $8+7+1+4+6+3+2+8 = 39$ and 39 is divisible by 3.
- A number is divisible by **9** if the sum of its digits is divisible by 9. For instance 6465357 is divisible by 9 since $6 + 4 + 6 + 5 + 3 + 5 + 7 = 36$ and 36 is divisible by 9.
- A number is divisible by **6** if its last digit is 0, 2, 4, 6, 8 and the sum of its digits is divisible by 3. For instance 8714632$\underline{8}$, is divisible by 6 since it last digit is 8 and $8 + 7 + 1 + 4 + 6 + 3 + 2 + 8 = 39$ and 39 is divisible by 3.
- A number is divisible by **5** if its last digit is 5 or 0.

- A number is divisible by **50** if its last two digits are 50 or 00.
- A number is divisible by **25** if its last two digits are 25, 50, 75 or 00.
- A number is divisible by **10** or a power of ten if the number of zeros in the number is at least equal to the number of zeros in the power of 10. for instance 70, 700, 7000 are divisible by 10, 100, and 1000 respectively since their last, last two and last three digits are, 0, 00, and 000 respectively.
- A number is divisible by **7** if the difference between the twice the last digit and the rest of the number is 0 or is divisible by 7. If the difference is not 0, this rule can be applied repeatedly until it is clearly determined that the difference between the twice the last digit and the rest of the number is 0 or is divisible by 7. For instance 37821 is divisible by 7 since $3782 - 2(1) = 3780$, $378 - 2(0) = 378$, $37 - 16 = 21$ and 21 is divisible by 7.
- A number is divisible by **11** if the difference between the sum of the odd digits and the sum of the even digits is 0 or is divisible by 11. For instance 17578 is divisible by 11 since $(1 + 5 + 8) - (7 + 7) = 14 - 14 = 0$.
- A number is divisible by **12** if the number is divisible by 3 and 4. For instance 684 is divisible by 12 since 84 is divisible by 4 and $6 + 8 + 4 = 18$. 18 is divisible by 3.

EXERCISE 2:9

1. Copy and complete Table 3:4 by marking X where applicable.

Number	Divisible by										
	2	3	4	5	6	8	9	10	25	50	100
2644750											
74319275											
861425											
6671456300											
925675435											

Table 2:4

2. Say giving reasons which of the following decimal numbers is divisible by
   (a) 2   (b) 3   (c) 4   (d) 5
   (e) 6   (f) 7   (g) 8   (h) 9
   (i) 10   (j) 11   (k) 12

64665, 3689, 85564, 21342, 97965, 76445, 4378, 23490, 6936.

NUMBERS AS DOTS PATTERNS

When dots or pebbles are arranged to represent whole numbers, many geometrical shapes can be produced. The numbers 3, 6, 10, and so on form triangles and are therefore called **triangular numbers**.

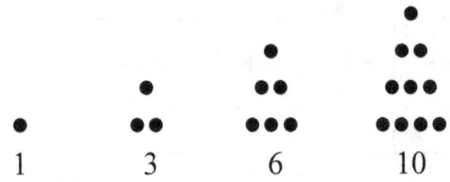

The numbers 4, 9, 16, and so on form squares and are therefore called **square numbers**.

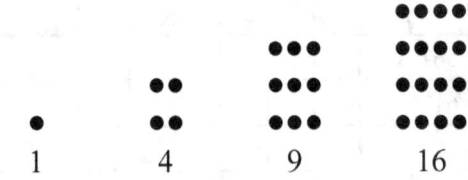

All composite natural numbers can be arranged as rectangular patterns.
The numbers 2, 4, 6, 8, 10, 12… and so on form rectangles and are therefore called **rectangular numbers**.

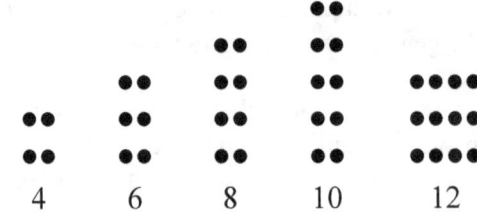

Apart from 2, no other prime number can be arranged as a rectangular pattern.
Though it may not be easy to use dots or pebbles to appreciate that 0 and 1 belong to all the three sets through logical intuition, this can be ascertained as follows.
Since a square has equal sides, 0 and 1 can be considered squares of side 0 and 1 respectively. In addition, since a square is a special rectangle, all squares are rectangles. Therefore, both 0 and 1 are rectangular numbers. It follows that 1 and 0 are

equilateral triangles of side 1 and 0 respectively.

From the above, let

T = All triangular numbers
S = All square numbers
R = All rectangular numbers

Then,

$T = \{0, 1, 3, 6, 10, 15, 21, 28,...\}$
$S = \{0, 1, 4, 9, 16, 25, 36, 49,...\}$
$R = \{0, 1, 2, 4, 6, 8, 9, 10, 12, 14, 15,...\}$

The relationship between these three sets is illustrated in Figure 2:1.

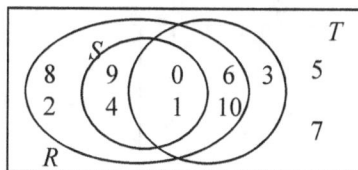

Figure 2:1

From simple geometrical arrangements, many properties of whole numbers immerge. For instance

-The sum of two consecutive triangular numbers is always a square number.
E.g. $1 + 3 = 4$, $3 + 6 = 9$ etc.

-A **perfect number** is a positive integer that equals the sum of its positive proper divisors. For example

$6 = 1 + 2 + 3$,
$28 = 1 + 2 + 4 + 7 + 14$
$496 = 1 + 2 + 4 + 8 + 16 + 31 + 62 + 124 + 248$

Even and Odd Numbers

Even numbers can be represented as rectangular patterns with two rows or two columns. Odd numbers cannot be represented as rectangular patterns with two rows or two columns.

Odd numbers

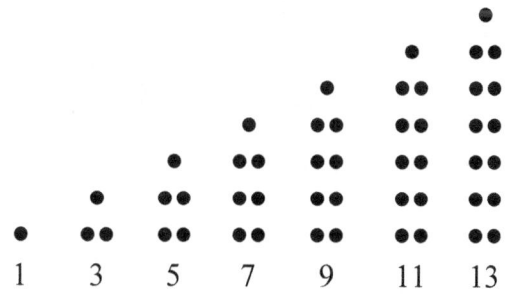

Even numbers

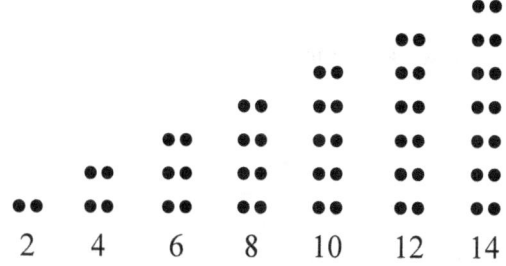

EXERCISE 2:10

1. The sum of any number of consecutive triangular numbers, beginning with the smallest is a pyramidal number. Given that 0 is conventionally considered an element of this set. List the first 7 pyramidal numbers.

2. Given that,

T = All triangular numbers
S = All square numbers
R = All rectangular numbers

(i) List the first three elements of the following sets
  (a) $R \cap S$ (b) $R \cap T$ (c) $T \cap S$

(ii) The following illustrates the meaning of the difference between consecutive members of S.

$$1 - 0 = 1$$
$$4 - 1 =$$
$$9 - 4 =$$
$$16 - 9 =$$
$$25 - 16 =$$
$$36 - 25 =$$

(a) Copy and complete the pattern.
(b) State the name of the set of numbers obtained as your results.
(c) Compose similar tables for the following and use it to state the name of the set of numbers obtained as in (b).
  (i) the sum of two consecutive members of T.
  (ii) the difference between two consecutive members of T.
(d) Given that the sets in (c) (i) and (c) (ii) are A and B respectively. State the

relation between any element x of A, and a corresponding element y of B?

3. The dots below represent the first three triangular numbers.

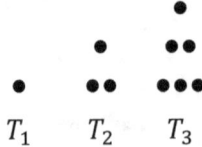

T_1 T_2 T_3

Draw dot patterns for T_4 and T_5 and obtain their values.

INTEGERS, $\mathbb{Z}$

This set denoted by $\mathbb{Z}$ is the set of all positive and negative whole numbers including zero. The set of integers is partitioned into three sets namely; the set containing the element zero, the set of positive whole numbers denoted by $\mathbb{Z}^+$ and the set of negative whole numbers denoted by $\mathbb{Z}$. Thus,

$$\mathbb{Z} = \{0, \pm 1, \pm 2, \pm 3 ...\}.$$
$$\mathbb{Z}^+ \text{ or } \mathbb{Z}^* = \{+1, +2, +3 ...\} = \{x \in \mathbb{Z} : x > 0\}.$$
$$\mathbb{Z}^- = \{-1, -2, -3 ...\} = \{x \in \mathbb{Z} : x < 0\}.$$
$$\mathbb{Z} = \mathbb{Z}^- \cup \{0\} \cup \mathbb{Z}^+.$$

Numbers are often represented on a horizontal or vertical line called the **number line** as shown in Figure 2:2. The relative position of numbers on a horizontal number line is such that the larger number is always to the right.

From Figure 2:2, it should be appreciated that -2, say, is greater than -3, 0 is greater than -2, -3 is less than $+1$ etc.

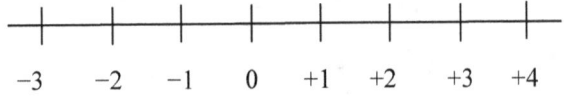

-3 -2 -1 0 $+1$ $+2$ $+3$ $+4$

Figure 2:2

RATIONAL NUMBERS, $\mathbb{Q}$

A **rational number** is any number, which can be expressed as the quotient of two integers. The set of rational number is denoted by $\mathbb{Q}$. Thus,

$$\mathbb{Q} = \left\{ x : x = \frac{p}{q} ; p, q \in \mathbb{Z}, q \neq 0 \right\}$$

For instance

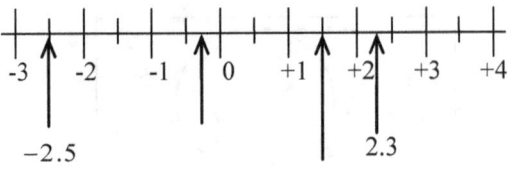

$$4 = \frac{4}{1}, -5 = \frac{-5}{1}, 2.3 = \frac{23}{10}, 1\frac{1}{2} = \frac{3}{2}, -3\frac{3}{4} = \frac{-15}{4}$$

Therefore, $4, -5, 2.3, 1\frac{1}{2}, -3\frac{3}{4}$ are all rational numbers. On the number line in Figure 2:3, some rational numbers are plotted.

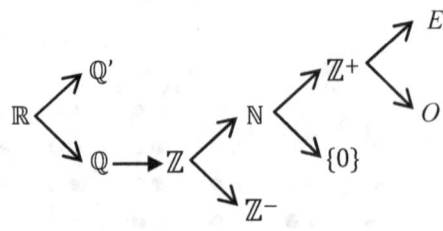

Figure 2:3

IRRATIONAL NUMBERS, $\mathbb{Q}'$

An irrational number is a number, which cannot be expressed as a ratio of two integers. Some examples of irrational numbers are $0.117111711117111117...,\pi, \sqrt{2}$, 0.41586237 etc.

REAL NUMBERS, $\mathbb{R}$

All the rational numbers and all the irrational numbers constitute the set of real numbers denoted by $\mathbb{R}$.

The set of all positive real numbers and zero is denoted by $\mathbb{R}^+ = \{x \in \mathbb{R} : x \geq 0\}$. Any real number can be represented by a point on a number line and vice versa.

Figure 2:4

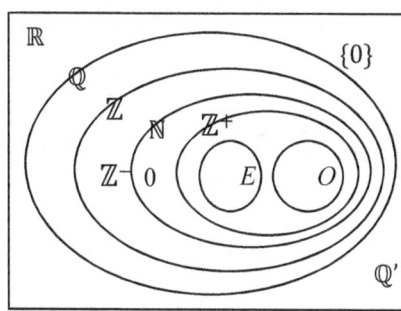

Figure 2:5

◆ **MULTIPLE CHOICE EXERCISE 2** ◆

1. The number of prime numbers between 1 and 20 is:
   [A] 9   [B] 8   [C] 7   [D] 6
2. The first four prime numbers are:
   [A] 1, 2, 3, 4   [B] 1, 3, 5, 7
   [C] 2, 3, 5, 7   [D] 2, 4, 6, 8
3. 84 as a product of prime factors is:
   [A] $2^2 \times 3^2 \times 7^2$ [B] $2^2 \times 3^3 \times 7$
   [C] $2^2 \times 3 \times 7$ [D] $2^2 \times 3^2 \times 7$
4. As a product of prime factors 200 is equal to:
   [A] 2×5 [B] $2^3 \times 5^2$
   [C] $2^2 \times 5^3$ [D] $2^2 \times 5^2$
5. Leaving your answer as a product of prime factors in index form, 72 equals:
   [A] $2^2 \times 3^2$ [B] $2^2 \times 3^3$
   [C] $2^3 \times 3^3$ [D] $2^3 \times 3^2$
6. The number that is not a prime number is:
   [A] 9   [B] 7   [C] 3   [D] 2
7. The number of factors a prime number has is:
   [A] 0   [B] 1   [C] 2   [D] 3
8. The prime number among the following is:
   [A] 15   [B] 13   [C] 9   [D] 1
9. 17 is a prime number because:
   [A] it is not divisible by 2
   [B] it is a sieve of Erasthodenes
   [C] it has no factor other than itself
   [D] it has only two factors
10. The prime number among the following is:
    [A] 57   [B] 61   [C] 63   [D] 69
11. Two of the numbers 11, 21, 31, 77, 112 are prime numbers. The number lying

exactly half way between these two prime numbers are:
[A] 54   [B] 26   [C] 21   [D] 16

12. Three of the numbers 11, 21, 31, 77, 112 have a common factor. The common factor is:
    [A] 14   [B] 11   [C] 7   [D] 2
13. The number that is the product of two consecutive prime numbers is:
    [A] 8   [B] 15   [C] 18   [D] 21
14. The LCM of 6 and 14 is:
    [A] 42   [B] 24   [C] 14   [D] 84
15. The LCM of 8,9, and 12 is:
    [A] 29   [B] 72   [C] 96   [D] 108
16. The result of dividing the LCM of 8 and 12 by 3 is:
    [A] 12   [B] $10\frac{2}{3}$ [C] $9\frac{1}{3}$ [D] 8
17. The HCF of the numbers 30,120 and 125 is:
    [A] 3   [B] 5   [C] 10   [D] 15
18. The HCF of 18,24, and 36 expressed as a product of prime factors is:
    [A] $2^2 \times 3$ [B] 2×3^2
    [C] 2×3 [D] $2^2 \times 3^2$
19. Dividing the LCM of 24 and 30 by their HCF gives:
    [A] 2   [B] 20   [C] 24   [D] 25
20. The result of dividing the LCM of 12,16 and 24 by their HCF is:
    [A] 12   [B] 11   [C] 10   [D] 9
21. The result of squaring the number 6 is:
    [A] 12   [B] 26   [C] 36   [D] 62
22. $\sqrt{7744}$ in index form is:
    [A] 26   [B] $2^3 \times 13$
    [C] $2^3 \times 9$ [D] $2^3 \times 11$
23. $\sqrt{3136}$ in index form is:
    [A] 2×7^2 [B] $2^3 \times 7$
    [C] $2^2 \times 7$ [D] 2×7
24. The smallest number by which $3^2 \times 5$ can be multiplied to give a perfect square is:
    [A] 5   [B] 6   [C] 15   [D] 25
25. The least number, which multiplies 54 to make a perfect square, is:
    [A] 3   [B] 4   [C] 6   [D] 8
26. It is true to say that:
    [A] $\sqrt{5} \in \mathbb{R}$ [B] $\sqrt{2} \in \mathbb{Q}$
    [C] $-5 \in \mathbb{N}$ [D] $\sqrt{-3} \in \mathbb{Q}'$
27. It is not true to say that:
    [A] $16 \in \mathbb{N}$ [B] $\sqrt{9} \in \mathbb{R}$

[C] $\pi \in \mathbb{R}$ [D] $\frac{1}{3} \in \mathbb{Z}$

28. It is not true to say that:
    [A] $-2 \in \mathbb{Z}$ [B] $7 \in \mathbb{Z}$
    [C] $\pi \in \mathbb{Q}$ [D] $3 \in \mathbb{Q}$

29. The number, which does not belong to the set $\mathbb{R}$ of real numbers, is:

    [A] $\sqrt{-9}$ [B] $\frac{7}{11}$ [C] π [D] -3

30. The number, which belongs to all the four sets, $\mathbb{R}, \mathbb{Q}, \mathbb{N}$ and $\mathbb{Z}$ is:

    [A] $\sqrt{-9}$ [B] $\frac{7}{11}$ [C] π [D] 3

31. The number, which does not belong to the set $\mathbb{Q}$ of rational numbers, is:

    [A] 5   [B] $\frac{7}{11}$ [C] π [D] -3

32. The set inclusion statement, which is correct, is:
    [A] $\mathbb{Q} \subset \mathbb{N} \subset \mathbb{Z} \subset \mathbb{R}$ [B] $\mathbb{N} \subset \mathbb{Z} \subset \mathbb{Q} \subset \mathbb{R}$
    [C] $\mathbb{N} \subset \mathbb{Q} \subset \mathbb{Z} \subset \mathbb{R}$ [D] $\mathbb{Z} \subset \mathbb{N} \subset \mathbb{Q} \subset \mathbb{R}$

33. The correct Venn diagram in Figure 2:6 is:

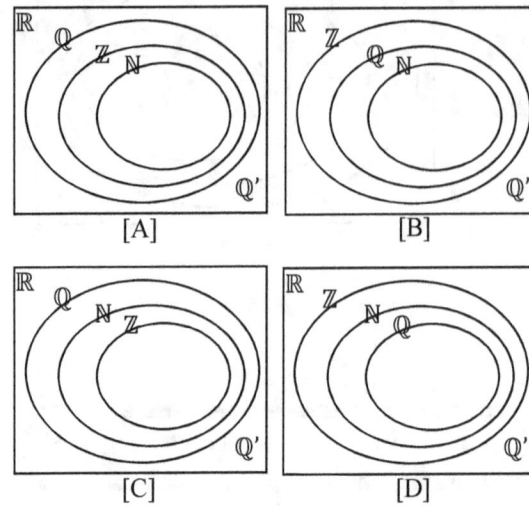

Figure 2:4

24

TOPIC 3

FRACTIONS, DECIMALS, PERCENTAGES, RATIOS AND PROPORTIONS

OBJECTIVES

At the end of this topic, the learner should be able to:

1. Identify and distinguish between the various types of fractions.
2. Read/write fractions in the form $\dfrac{a}{b}$ not a/b.
3. Appreciate the value of a fraction when both numerator and denominator zero 0.
4. Identify equivalent fractions and compare fractions by expressing them as equivalent fractions with a common denominator.
5. Simplify fractions to their lowest terms.
6. Change improper fractions to mixed numbers and vice versa.
7. Add, subtract, multiply and divide fractions, decimals and percentages.
8. Express one quantity as a fraction of another.
9. Appreciate that a decimal is simply a fraction with a power of 10 as denominator.
10. State the value of a digit in a given decimal and compare decimals by the use of place value.
11. Multiply and divide decimals with powers of 10 by moving the decimal marker.
12. Identify which fraction is likely to be recurring and classify given decimals as terminating, recurring or non-recurring.
13. Appreciate that a percentage is simply a fraction with denominator 100.
14. Interconvert percentages, vulgar fractions and decimals.
15. Find a percentage of a given quantity.
16. Express one quantity as a percentage of another.
17. Express two quantities in a given ratio.
18. Determine whether or not two ratios are equal.
19. Solve practical problems involving ratios and proportions. e.g. Use ratios to determine best buy.

VULGAR FRACTIONS

Meaning of a Fraction

A **fraction** is a number that expresses part of a whole quantity. Dividing one quantity by another, may lead to a fraction. Examples are $\frac{2}{3}, \frac{4}{3}, 1\frac{1}{3}$.

The fraction $\frac{2}{3}$ means, 'divide 2 things by 3' or 'divide a quantity into 3 parts and take 2 parts'. In fractions such as $\frac{2}{3}$ and $\frac{4}{3}$, the top number is called the **numerator** or **dividend**, while the bottom number is called the **denominator** or **divisor**.

$$\frac{4}{3} \begin{array}{l} \leftarrow \text{numerator} \\ \leftarrow \text{denominator} \end{array}$$

Practical examples of Fractions

1. A man sliced a loaf of bread into three equal slices and ate two slices. The fraction eaten is $\frac{2}{3}$ (shaded) and the fraction left is $\frac{1}{3}$ (not shaded).

2. If three children share four fruits equally then each of the children will take the fraction $\frac{4}{3}$.
    This means, they will take a fruit each and still share a full fruit.

    Therefore, $\frac{4}{3} = 1\frac{1}{3}$.

Types of Fractions

Proper fractions are fractions such as $\frac{1}{5}, \frac{3}{4}, \frac{2}{3}$ etc whose numerators are less than their denominators.

Improper fractions are fractions such as $\frac{3}{2}, \frac{7}{3}, \frac{8}{5}$ etc whose numerators are greater than their denominators.

A **mixed number** or **mixed fraction** is a number with a whole number part and a proper fractional part. For instance, $1\frac{1}{3}$ read 'one and one third' is a mixed number.

Fractions with zero denominators

Any fraction with numerator zero and a denominator other than zero is equal to zero. For instance, $\frac{0}{7} = 0$. On the other hand, any fraction with denominator zero has no meaning. For instance, $\frac{0}{0}, \frac{7}{0}$ are meaningless.

Equivalent Fractions

Examine the following rectangles.

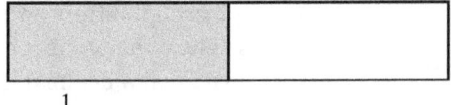

$\frac{1}{2}$ of the rectangle is shaded.

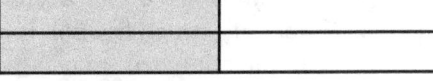

$\frac{2}{4}$ of the rectangle is shaded.

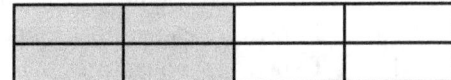

$\frac{4}{8}$ of the rectangle is shaded.

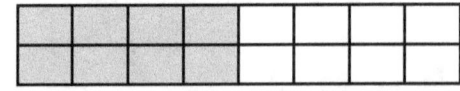

$\frac{8}{16}$ of the rectangle is shaded.

Since exactly the same quantity has been shaded in each case, it means $\frac{1}{2} = \frac{2}{4} = \frac{4}{8} = \frac{8}{16}$. The fractions $\frac{1}{2}, \frac{2}{4}, \frac{4}{8}$ and $\frac{8}{16}$ are examples of

equivalent fractions because they represent exactly the same quantity. Thus

$$\frac{2}{4} = \frac{1}{2} \times \frac{2}{2}, \frac{4}{8} = \frac{1}{2} \times \frac{4}{4}, \frac{8}{16} = \frac{1}{2} \times \frac{8}{8}$$

Therefore,

The value of a fraction remains the same on multiplying both numerator and denominator by the same quantity.

When two fractions are equivalent, the symbol ≡, in the strictest sense is used though = is often used.

Thus, $\frac{1}{2} \equiv \frac{8}{16}$.

Example 3:1

Find the missing number in $\frac{2}{3} \equiv \frac{?}{15}$.

Solution
What can be multiplied by 3 to give 15? Obviously 5. Therefore, multiply the numerator of the LHS by 5 since denominator has been multiplied by 5. Thus,

$$\frac{2}{3} = \frac{2}{3} \times \frac{5}{5} = \frac{10}{15}$$

Simplifying fractions to their lowest terms

A fraction is in its **lowest terms**, if the numerator and denominator have no common factors.

Examples are $\frac{2}{3}, \frac{4}{9}, \frac{8}{15}$ etc.

To simplify a fraction to its lowest terms divide both numerator and denominator by their HCF.

Example 3:2
Simplify the following:

(a) $\frac{28}{36}$ (b) $\frac{15}{30}$ (c) $\frac{48}{120}$

Solutions
(a) Since the HCF of 28 and 36 is 4,
$$\frac{28}{36} = \frac{7 \times 4}{9 \times 4} = \frac{7}{9}$$

(b) Since the HCF of 15 and 30 is 15,
$$\frac{15}{30} = \frac{1 \times 15}{2 \times 15} = \frac{1}{2}$$

(c) Since the HCF of 48 and 120 is 24
$$\frac{48}{120} = \frac{2 \times 24}{5 \times 24} = \frac{2}{5}$$

EXERCISE 3:1

1. Find the missing number, which makes the RHS an equivalent fraction to the LHS.

(a) $\frac{3}{5} = \frac{?}{100}$ (b) $\frac{3}{2} = \frac{?}{24}$ (c) $\frac{5}{8} = \frac{?}{40}$

(d) $\frac{2}{3} = \frac{24}{?}$ (e) $\frac{7}{3} = \frac{28}{?}$ (f) $\frac{4}{9} = \frac{60}{?}$

2. Simplify:

(a) $\frac{33}{120}$ (b) $\frac{72}{48}$ (c) $\frac{64}{80}$

(d) $\frac{24}{18}$ (e) $\frac{28}{21}$ (f) $\frac{60}{81}$

Some Order Symbols

< means 'is less than'
Example: 2 < 5 is read '2 is less than 5'.
> means 'is greater than'
Example: 9 > 5 is read '9 is greater than 5'.

Comparing fractions

To compare fractions having common denominators, compare the numerators.

Example 3:3
Compare the following using >, < or =.

(i) $\frac{5}{8}$ and $\frac{3}{8}$ (ii) $\frac{7}{10}$ and $\frac{9}{10}$

Solution

(i) $\dfrac{5}{8} > \dfrac{3}{8}$ since $5 > 3$ (ii) $\dfrac{7}{10} < \dfrac{9}{10}$ since $7 < 9$

To compare fractions with different denominators, first convert the fractions to equivalent fractions with the LCM as their denominators then compare the numerators.

Example 3:4

Compare the following using $>$, $<$ or $=$.

(i) $\dfrac{3}{4}$ and $\dfrac{7}{10}$ (ii) $\dfrac{1}{6}$ and $\dfrac{3}{8}$

(iii) $\dfrac{2}{3}$ and $\dfrac{4}{5}$

Solution

(i) Since the LCM of 4 and 10 is 20,

$$\dfrac{3}{4} = \dfrac{3}{4} \times \dfrac{5}{5} = \dfrac{15}{20} \text{ and } \dfrac{7}{10} = \dfrac{7}{10} \times \dfrac{2}{2} = \dfrac{14}{20}$$

Since $15 > 14$, $\dfrac{3}{4} > \dfrac{7}{10}$

(ii) Since the LCM of 6 and 8 is 24,

$$\dfrac{1}{6} = \dfrac{1}{6} \times \dfrac{4}{4} = \dfrac{4}{24} \text{ and } \dfrac{3}{8} = \dfrac{3}{8} \times \dfrac{3}{3} = \dfrac{9}{24}$$

Since $4 < 9$, $\dfrac{1}{6} < \dfrac{3}{8}$.

(iii) Since the LCM of 3 and 5 is 15,

$$\dfrac{2}{3} \times \dfrac{5}{5} = \dfrac{10}{15} \text{ and } \dfrac{4}{5} = \dfrac{4}{5} \times \dfrac{3}{3} = \dfrac{12}{15}$$

Since $10 < 12$, $\dfrac{2}{3} < \dfrac{4}{5}$

An alternative method is to convert all the fractions to decimals before comparing.

> **EXERCISE 3:2**

1. Compare the following using the symbols $>$, $<$ or $=$.

(a) $\dfrac{7}{9} \quad \dfrac{6}{9}$ (b) $\dfrac{2}{3} \quad \dfrac{4}{5}$

(c) $\dfrac{3}{6} \quad \dfrac{2}{4}$ (d) $\dfrac{3}{4} \quad \dfrac{7}{9}$

2. Arrange each of the following groups of fractions in order of increasing magnitude.

(a) $\dfrac{3}{6}, \dfrac{5}{6}, \dfrac{7}{6}, \dfrac{4}{6}, \dfrac{2}{6}$ (b) $\dfrac{6}{8}, \dfrac{7}{8}, \dfrac{2}{8}, \dfrac{4}{8}$

(c) $\dfrac{9}{7}, \dfrac{9}{2}, \dfrac{9}{5}, \dfrac{9}{11}, \dfrac{9}{8}$ (d) $\dfrac{11}{3}, \dfrac{12}{4}, \dfrac{10}{5}, \dfrac{16}{4}$

3. Arrange the following in ascending order of magnitude.

$$\dfrac{2}{3}, \dfrac{1}{2}, \dfrac{7}{8}, \dfrac{1}{4}, \dfrac{3}{4}, \dfrac{1}{3}, \dfrac{2}{5}, \dfrac{4}{5}, \dfrac{8}{9}, \dfrac{6}{7}$$

Converting Mixed Numbers to Improper Fractions

Example 3:5

Convert the following mixed numbers to improper fractions.

(i) $1\dfrac{4}{5}$ (ii) $5\dfrac{7}{8}$

Solutions

(i) Appreciating that $1 = \dfrac{1}{1}$ and $1\dfrac{4}{5} = 1 + \dfrac{4}{5}$.

Then $1\dfrac{4}{5} = \dfrac{1}{1} + \dfrac{4}{5} = \dfrac{5}{5} + \dfrac{4}{5} = \dfrac{9}{5}$

(ii) Similarly, $5\dfrac{7}{8} = \dfrac{5}{1} + \dfrac{7}{8} = \dfrac{40}{8} + \dfrac{7}{8} = \dfrac{47}{8}$

Alternatively, this procedure can be summarized as below. The working is ordinarily done mentally.

(i) $1\dfrac{4}{5} = \dfrac{5 \times 1 + 4}{5} = \dfrac{9}{5}$

(ii) $5\dfrac{7}{8} = \dfrac{8 \times 5 + 7}{8} = \dfrac{47}{8}$

> **EXERCISE 3:3**

1. Compare the following using the symbols $>$, $<$ or $=$.

(a) $\dfrac{17}{9} \quad 1\dfrac{6}{9}$ (b) $\dfrac{32}{30} \quad \dfrac{4}{5}$

(c) $\dfrac{33}{66}\quad\dfrac{22}{44}$ (d) $1\dfrac{3}{4}\quad 1\dfrac{7}{9}$

2. Convert the following improper fractions into mixed numbers
   (a) $\dfrac{11}{3}$ (b) $\dfrac{15}{2}$ (c) $\dfrac{9}{4}$ (d) $\dfrac{13}{5}$

3. Convert the following mixed numbers into improper fractions
   (a) $2\dfrac{2}{3}$ (b) $5\dfrac{1}{2}$ (c) $1\dfrac{3}{4}$ (d) $2\dfrac{3}{5}$

4. Arrange the following in descending order of magnitude.
   $1\dfrac{2}{3}, 3\dfrac{1}{2}, \dfrac{17}{8}, 3\dfrac{1}{4}, \dfrac{13}{3}, 5\dfrac{1}{3}, 4\dfrac{2}{5}, \dfrac{15}{5}, \dfrac{88}{6}, \dfrac{20}{3}$

Addition and subtraction of fractions

Fractions with equal denominators

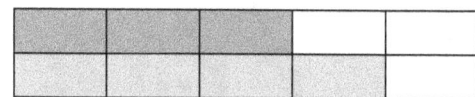

State the fraction of the rectangle that is
(a) coloured dark
(b) coloured gray
What fraction of the rectangle is coloured? Explain how you would use the diagram to evaluate

(i) $\dfrac{3}{10}+\dfrac{4}{10}$ (ii) $\dfrac{7}{10}-\dfrac{3}{10}$

From this, it can be deduced that,

To add or subtract fractions with equal denominators, keep the common denominator and add or subtract the numerators.

Example 3:6
Evaluate the following,

(i) $\dfrac{3}{5}+\dfrac{1}{5}$ (ii) $\dfrac{9}{10}+\dfrac{2}{10}$

(iii) $\dfrac{3}{5}-\dfrac{1}{5}$ (iv) $\dfrac{9}{10}-\dfrac{2}{10}$

Solutions

(i) $\dfrac{3}{5}+\dfrac{1}{5}=\dfrac{3+1}{5}=\dfrac{4}{5}$

(ii) $\dfrac{9}{10}+\dfrac{2}{10}=\dfrac{9+2}{10}=\dfrac{11}{10}$

(iii) $\dfrac{3}{5}-\dfrac{1}{5}=\dfrac{3-1}{5}=\dfrac{2}{5}$

(iv) $\dfrac{9}{10}-\dfrac{2}{10}=\dfrac{9-2}{10}=\dfrac{7}{10}$

Fractions with unequal denominators

To add or subtract fractions with unequal denominators, first convert each of the fractions to equivalent fractions with the LCM of the denominators as their denominators.

Example 3:7
Evaluate:

(i) $\dfrac{5}{6}+\dfrac{3}{4}$ (ii) $\dfrac{2}{3}+\dfrac{3}{5}+\dfrac{1}{6}$

(iii) $\dfrac{7}{8}-\dfrac{3}{12}$ (iv) $\dfrac{5}{6}-\dfrac{2}{3}-\dfrac{1}{10}$

Solutions

(i) $\dfrac{5}{6}+\dfrac{3}{4}=\dfrac{10}{12}+\dfrac{9}{12}=\dfrac{19}{12}=1\dfrac{7}{12}$

(ii) $\dfrac{2}{3}+\dfrac{3}{5}+\dfrac{1}{6}=\dfrac{20+18+5}{30}=\dfrac{43}{30}=1\dfrac{13}{30}$

(iii) $\dfrac{7}{8}-\dfrac{3}{12}=\dfrac{21-6}{24}=\dfrac{15}{24}=\dfrac{5}{8}$

(iv) $\dfrac{5}{6}-\dfrac{2}{3}-\dfrac{1}{10}=\dfrac{25-20-3}{30}=\dfrac{2}{30}=\dfrac{1}{15}$

EXERCISE 3:4

1. Evaluate the following sums.
   (a) $\dfrac{4}{7}+\dfrac{2}{7}$ (b) $\dfrac{6}{11}+\dfrac{4}{11}$ (c) $\dfrac{3}{8}+\dfrac{5}{8}$

   (d) $\dfrac{1}{9}+\dfrac{4}{9}$ (e) $\dfrac{3}{10}+\dfrac{1}{10}$

2. Evaluate the following differences.
   (a) $\dfrac{4}{7}-\dfrac{2}{7}$ (b) $\dfrac{6}{11}-\dfrac{4}{11}$ (c) $\dfrac{7}{8}-\dfrac{5}{8}$

   (d) $\dfrac{9}{13}-\dfrac{4}{13}$ (e) $\dfrac{3}{3}-\dfrac{1}{5}$

3. Add the following

(a) $\dfrac{2}{3}+\dfrac{3}{4}$ (b) $\dfrac{6}{7}+\dfrac{4}{9}$ (c) $\dfrac{3}{5}+\dfrac{5}{8}$

(d) $\dfrac{1}{6}+\dfrac{4}{5}$ (e) $\dfrac{3}{5}+\dfrac{1}{10}$

4. Compute the following

(a) $\dfrac{2}{3}-\dfrac{4}{9}$ (b) $\dfrac{8}{9}-\dfrac{4}{7}$ (c) $\dfrac{7}{8}-\dfrac{5}{11}$

(d) $\dfrac{9}{13}-\dfrac{3}{7}$ (e) $\dfrac{3}{5}-\dfrac{1}{4}$

5. Evaluate and simplify the following where appropriate.

(a) $\dfrac{3}{7}+\dfrac{1}{7}+\dfrac{2}{7}$ (b) $\dfrac{2}{5}+\dfrac{2}{3}+\dfrac{3}{4}$

(c) $\dfrac{6}{7}+\dfrac{2}{3}-\dfrac{4}{9}$ (d) $\dfrac{4}{5}-\dfrac{3}{8}+\dfrac{3}{4}$

(e) $\dfrac{11}{13}-\dfrac{4}{13}-\dfrac{3}{13}$ (f) $\dfrac{8}{9}-\dfrac{2}{7}-\dfrac{1}{5}$

Adding and subtracting mixed numbers

To add or subtract mixed numbers, convert the mixed numbers to improper fractions before adding or subtracting.

Example 3:8

Evaluate the following.

(a) $1\dfrac{3}{10}+2\dfrac{1}{10}$ (b) $4\dfrac{5}{6}-1\dfrac{1}{6}$

(c) $2\dfrac{2}{5}+3\dfrac{1}{3}$ (d) $3\dfrac{2}{5}-2\dfrac{3}{4}$

Solution

(a) $1\dfrac{3}{10}+2\dfrac{1}{10}=\dfrac{13}{10}+\dfrac{21}{10}=\dfrac{\overset{17}{\cancel{34}}}{\underset{5}{\cancel{10}}}=3\dfrac{2}{5}$

(b) $4\dfrac{5}{6}-1\dfrac{1}{6}=\dfrac{29}{6}-\dfrac{7}{6}=\dfrac{\overset{11}{\cancel{22}}}{\underset{3}{\cancel{6}}}=3\dfrac{2}{3}$

(c) $2\dfrac{2}{5}+3\dfrac{1}{3}=\dfrac{12}{5}+\dfrac{10}{3}=\dfrac{36}{15}+\dfrac{50}{15}=\dfrac{86}{15}=5\dfrac{11}{15}$

(d) $3\dfrac{2}{5}-2\dfrac{3}{4}=\dfrac{17}{5}-\dfrac{11}{4}=\dfrac{68}{20}-\dfrac{55}{20}=\dfrac{13}{20}$

Alternatively, compute the whole number part and the fractional part, and then combine them.

(a) $1\dfrac{3}{10}+2\dfrac{1}{10}=3\dfrac{4}{10}=3\dfrac{2}{5}$

(b) $4\dfrac{5}{6}-1\dfrac{1}{6}=3\dfrac{4}{6}=3\dfrac{2}{3}$

(c) $2\dfrac{2}{5}+3\dfrac{1}{3}=5\dfrac{6+5}{15}=5\dfrac{11}{15}$

(d) $3\dfrac{2}{5}-2\dfrac{3}{4}=1+\dfrac{8-15}{20}=\dfrac{20+8-15}{20}=\dfrac{13}{20}$

N.B! $1=\dfrac{1}{1}=\dfrac{20}{20}$

Example 3:9

Evaluate the following

(a) $6\dfrac{1}{3}+2\dfrac{1}{6}+2\dfrac{2}{7}$ (b) $4\dfrac{3}{5}-2\dfrac{1}{2}+1\dfrac{1}{10}$

Solution

(a) $6\dfrac{1}{3}+2\dfrac{1}{6}+2\dfrac{2}{7}=10\dfrac{14+7+12}{42}=10\dfrac{33}{42}=10\dfrac{11}{14}$

(b) $4\dfrac{3}{5}-2\dfrac{1}{2}+1\dfrac{1}{10}=3\dfrac{6-5+1}{10}=3\dfrac{2}{10}=3\dfrac{1}{5}$

> ## EXERCISE 3:5

Evaluate and simplify the following.

(1) $7\dfrac{2}{5}+2\dfrac{4}{5}$ (2) $4\dfrac{2}{7}-3\dfrac{6}{7}$ (3) $5\dfrac{2}{5}-3\dfrac{4}{5}$

(4) $4\dfrac{3}{5}-2\dfrac{5}{7}$ (5) $3\dfrac{6}{7}+4\dfrac{2}{7}+2\dfrac{2}{7}$

(6) $7\dfrac{2}{5}-2\dfrac{4}{5}-3\dfrac{3}{5}$ (7) $6\dfrac{1}{3}-2\dfrac{1}{6}-3\dfrac{2}{5}$

(8) $3\dfrac{3}{5}+3\dfrac{3}{8}+5\dfrac{3}{4}$ (9) $7\dfrac{1}{2}-3\dfrac{1}{2}+1\dfrac{1}{4}$

(10) $6\dfrac{3}{8}+2\dfrac{1}{2}-3\dfrac{3}{4}$ (11) $8\dfrac{1}{6}-3\dfrac{3}{4}+2\dfrac{1}{2}$

(12) $9\dfrac{1}{12}-5\dfrac{3}{8}-1\dfrac{3}{4}$ (13) $5\dfrac{6}{7}-3\dfrac{3}{5}+2\dfrac{5}{14}$

Multiplying fractions

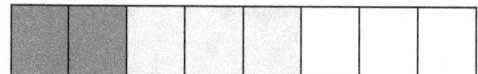

By counting squares, what fraction of the whole rectangle is coloured? What fraction of the coloured portion is dark? What fraction of the whole rectangle is gray?

By counting squares, it can be seen that:

$\dfrac{5}{8}$ of the whole rectangle is coloured.

$\dfrac{2}{5}$ of the shaded portion is dark. i.e. $\dfrac{2}{5}$ of $\dfrac{5}{8}$.

Also, the gray portion is $\dfrac{2}{8}$ or $\dfrac{1}{4}$ of the whole rectangle.

Multiplying the numerators together and the denominators together in $\dfrac{2}{5} \times \dfrac{5}{8}$ gives

$$\frac{2}{5} \times \frac{5}{8} = \frac{10}{40} = \frac{1}{4} \quad\ldots\ldots\ldots\ldots\ldots\ldots\ldots\textcircled{1}$$

Hence, it can be deduced that,

$$\frac{2}{5} \times \frac{5}{8} = \frac{2}{5} \text{ of } \frac{5}{8} = \frac{1}{4}$$

Instead of multiplying the numerators together and the denominators together as in ①, it is better to simplify by dividing the common factors as in ② before multiplying.

$$\frac{\cancel{2}^{1}}{\cancel{5}_{1}} \times \frac{\cancel{5}^{1}}{\cancel{8}_{4}} = \frac{1}{4} \quad\ldots\ldots\ldots\ldots\ldots\ldots\ldots\textcircled{2}$$

Therefore,

To multiply two fractions, multiply the numerators together and the denominators together.

Example 3:10

Evaluate $\dfrac{18}{35} \times \dfrac{14}{27}$

Solution

$$\frac{18}{35} \times \frac{14}{27} = \frac{\cancel{18}^{2}}{\cancel{35}_{5}} \times \frac{\cancel{14}^{2}}{\cancel{27}_{3}} = \frac{4}{15}$$

Example 3:11

Manka took $\dfrac{2}{3}$ of her money to school. At school, she used $\dfrac{3}{4}$ of what she took. What fraction of her money did she use?

Solution

$$\frac{3}{4} \text{ of } \frac{2}{3} = \frac{\cancel{3}^{1}}{\cancel{4}_{2}} \times \frac{\cancel{2}^{1}}{\cancel{3}_{1}} = \frac{1}{2}$$

Multiplying fractions by whole numbers

A whole number can be written as a fraction with denominator 1. In this way, the method in ② above can be employed. Thus,

$$\frac{3}{5} \times 15 = \frac{3}{\cancel{5}} \times \frac{\cancel{15}^{3}}{1} = 9$$

Multiplying the numerator by the whole number and maintaining the denominator the result will be

$$\frac{3}{5} \times 15 = \frac{45}{5} = 9$$

Therefore,

To multiply a fraction by a whole number, multiply the numerator of the fraction by the whole number and maintain the denominator of the fraction.

Example 3:12

Evaluate the following.

(i) $\dfrac{3}{11} \times 2$ (ii) $\dfrac{5}{9} \times 4$

Solution

(i) $\dfrac{3}{11} \times 2 = \dfrac{3 \times 2}{11} = \dfrac{6}{11}$ (ii) $\dfrac{5}{9} \times 4 = \dfrac{5 \times 4}{9} = \dfrac{20}{9}$

Multiplying Fractions by Mixed Numbers

To multiply with mixed numbers, first convert the mixed numbers to improper fractions.

Example 3:13
Compute the following products

(a) $5\dfrac{3}{8} \times 4$ (b) $3\dfrac{2}{15} \times 5$

Solution

(a) $5\dfrac{3}{8} \times 4 = \dfrac{43}{8} \times 4 = 21\dfrac{1}{2}$

(b) $3\dfrac{2}{15} \times 5 = \dfrac{47}{15} \times 5 = 15\dfrac{2}{3}$

Example 3:14

Calculate $\dfrac{3}{4}$ of 300

Solution

$\dfrac{3}{4}$ of $300 = \dfrac{3}{4} \times 300 = 225$

Example 3:15
Mrs. Mundi had 357,000 Frs. in the credit union. She signed out $\dfrac{2}{3}$ of it. How much did she sign out?

Solution

Amount signed out $= \dfrac{2}{3}$ of $357,000$

$\Rightarrow$ Amount signed out $= \dfrac{2}{3} \times 357,000 = 238,000$ Frs.

Example 3:16

(i) $2\dfrac{1}{2} \times 1\dfrac{1}{4}$ (ii) $4\dfrac{2}{3} \times 2\dfrac{3}{5}$

Solution

(i) $2\dfrac{1}{2} \times 1\dfrac{1}{4} = \dfrac{5}{2} \times \dfrac{5}{4} = \dfrac{25}{8} = 3\dfrac{1}{8}$

(ii) $4\dfrac{2}{3} \times 2\dfrac{3}{5} = \dfrac{14}{3} \times \dfrac{13}{5} = \dfrac{182}{15} = 12\dfrac{2}{15}$

EXERCISE 3:6

Evaluate the following.

1. $\dfrac{2}{3}$ of 60 2. $\dfrac{3}{4}$ of 180 3. $\dfrac{5}{8}$ of $\dfrac{3}{4}$

4. $\dfrac{2}{5}$ of $\dfrac{75}{80}$ 5. $8 \times \dfrac{3}{4}$ 6. $12 \times \dfrac{1}{4}$

7. $\dfrac{2}{3} \times 9$ 8. $\dfrac{5}{7} \times 14$ 9. $\dfrac{3}{4} \times \dfrac{8}{11}$

10. $\dfrac{1}{5} \times \dfrac{2}{3} \times \dfrac{1}{4}$ 11. $\dfrac{5}{6} \times \dfrac{7}{10}$ 12. $\dfrac{4}{5} \times \dfrac{15}{16}$

13. $\dfrac{3}{5} \times \dfrac{2}{9}$ 14. $\dfrac{3}{4} \times \dfrac{1}{6} \times \dfrac{2}{5}$

15. $\dfrac{1}{2} \times \dfrac{9}{10} \times \dfrac{2}{3}$ 16. $2\dfrac{1}{2} \times 4$

17. $1\dfrac{2}{5} \times 10$ 18. $12 \times 6\dfrac{1}{2}$

19. $8 \times 3\dfrac{3}{4}$ 20. $2\dfrac{1}{2} \times 1\dfrac{1}{4}$

21. $3\dfrac{1}{3} \times 1\dfrac{3}{4}$ 22. $4\dfrac{2}{3} \times 2\dfrac{3}{5}$

23. $6 \times 5\dfrac{3}{5} \times 1\dfrac{2}{3}$ 24. $6\dfrac{2}{3} \times 7 \times 1\dfrac{1}{5}$

25. $2\dfrac{1}{6} \times 1\dfrac{7}{8} \times 5\dfrac{1}{3}$ 26. $2\dfrac{3}{5} \times 1\dfrac{5}{6} \times 1\dfrac{1}{2}$

Reciprocals
Study the following.

(i) $\dfrac{3}{4} \times \dfrac{4}{3} = \dfrac{3}{4} \times \dfrac{4}{3} = 1$

(ii) $\dfrac{5}{7} \times \dfrac{7}{5} = \dfrac{5}{7} \times \dfrac{7}{5} = 1$

(iii) $3 \times \dfrac{1}{3} = 3 \times \dfrac{1}{3} = 1$

(iv) $\dfrac{3}{5} \times 5 = \dfrac{3}{5} \times 5 = 1$

If two numbers are such that their product is 1, one is said to be the **reciprocal** or the **multiplicative inverse** of the other.

Division by Fractions

Consider $\dfrac{4}{5} \div \dfrac{2}{3}$. Another way of writing this is $\dfrac{\frac{4}{5}}{\frac{2}{3}}$.

$\dfrac{\frac{4}{5}}{\frac{2}{3}}$, is an example of a **complex fraction.**

Multiplying numerator and denominator of any fraction by the same quantity does not change the value of the fraction. Thus, numerator and denominator can be multiplied by the multiplicative inverse of $\dfrac{2}{3}$ i.e. $\dfrac{3}{2}$

$$\frac{\frac{4}{5}}{\frac{2}{3}} = \frac{\frac{4}{5} \times \frac{3}{2}}{\frac{2}{3} \times \frac{3}{2}} = \frac{\cancel{4}}{5} \times \frac{3}{\cancel{2}} = \frac{6}{5} = 1\frac{1}{5}$$

Appreciate that $\dfrac{\frac{4}{5}}{7}$ is also a complex fraction which should be interpreted as

$$\frac{\frac{4}{5}}{\frac{7}{1}} = \frac{\frac{4}{5} \times \frac{1}{7}}{\frac{7}{1} \times \frac{1}{7}} = \frac{4}{5} \times \frac{1}{7} = \frac{4}{35}$$

It can thus be deduced that,

To divide by a fraction, multiply by the reciprocal of the fraction.

Example 3:17

Evaluate (i) $\dfrac{2}{3} \div \dfrac{3}{4}$ (ii) $\dfrac{9}{10} \div \dfrac{3}{5}$

Solution

Evaluate (i) $\dfrac{2}{3} \div \dfrac{3}{4} = \dfrac{2}{3} \times \dfrac{4}{3} = \dfrac{8}{9}$

(ii) $\dfrac{9}{10} \div \dfrac{3}{5} = \dfrac{\cancel{9}^{3}}{\cancel{10}_{2}} \times \dfrac{\cancel{5}^{1}}{\cancel{3}_{1}} = \dfrac{3}{2} = 1\dfrac{1}{2}$

In dividing fractions involving mixed numbers, first convert the mixed numbers to improper fractions.

Example 3:18

Evaluate (i) $1\dfrac{4}{5} \div 3$ (ii) $5 \div \dfrac{16}{5}$

Solution

(i) $1\dfrac{4}{5} \div 3 = \dfrac{\cancel{9}^{3}}{5} \times \dfrac{1}{\cancel{3}_{1}} = \dfrac{3}{5}$

(ii) $5 \div \dfrac{16}{5} = 5 \times \dfrac{5}{16} = \dfrac{25}{16}$

EXERCISE 3:7

1. State the reciprocal of each of the following

   (a) $\dfrac{7}{3}$ (b) $\dfrac{1}{9}$ (c) $\dfrac{3}{4}$ (d) $\dfrac{5}{8}$

   (e) $\dfrac{1}{6}$ (f) 8   (g) 3

2. Evaluate

   (a) $9 \div \dfrac{3}{7}$ (b) $\dfrac{9}{10} \div 6$

3. Simplify

   (a) $\dfrac{1}{10} \div \dfrac{3}{5}$ (b) $\dfrac{3}{4} \div \dfrac{7}{8}$

   (c) $\dfrac{\frac{2}{3}}{\frac{6}{9}}$ (d) $\dfrac{\frac{5}{6}}{\frac{1}{2}}$ (e) $\dfrac{7}{\frac{10}{11}}$

4. Compute

   (a) $3\dfrac{1}{2} \div 4\dfrac{1}{2}$ (b) $\dfrac{5}{6} \div 1\dfrac{1}{9}$

   (c) $\left(\dfrac{9}{10} \times \dfrac{5}{8} \right) \div \dfrac{3}{8}$

5. Evaluate

   (a) $\left(3\dfrac{2}{3} \div 5\dfrac{1}{2} \right) \div \left(4\dfrac{1}{2} \div \dfrac{3}{4} \right)$

   (b) $\left(\dfrac{3}{5} \div \dfrac{1}{3} \right) \div \left(\dfrac{3}{4} - \dfrac{7}{10} \right)$

Expressing One Quantity as a Fraction of Another

To express one quantity as a fraction of the other, put the quantity over the other and simplify the resulting fraction.

Example 3:19
Express 25 as a fraction of 80

Solution

25 as a fraction of $80 = \dfrac{25}{80} = \dfrac{5}{16}$

Example 3:20
Out of the 13464 candidates who sat for the G.C.E 1986, 7854 passed. What fraction of the candidates passed?

Solution

Fraction that passed $= \dfrac{7854}{13464} = \dfrac{7}{12}$

EXERCISE 3:8

1. In Table 3:1, express the number in column A as a fraction of that in column B.

	A	B
(a)	20	25
(b)	35	90
(c)	280	150
(d)	525	400

2. During the 2009 GCE examination a school registered 400 candidates, 16 candidates abstained and 280 passed.
   (a) How many students failed?
   (b) Calculate the fraction of students who
       (i) Passed   (ii) failed   (iii) abstained
   (c) Express the number of students who failed as a fraction of the number of students who passed.

DECIMALS

A **decimal fraction** or **decimal** is a fraction whose denominator is a power of 10.
The following are examples of decimal fractions.

$$\frac{1}{10}, \frac{348}{100}, \frac{257}{10000}, \frac{75}{10}, \frac{32}{1000} \text{ etc}.$$

It saves time and space to write decimals or decimal fractions as shown in the following examples.

$$\frac{1}{10} = 0.1, \frac{348}{100} = 3.48, \frac{257}{10000} = 0.0257, \frac{75}{10} = 7.5 \text{ etc}.$$

The dot, which separates the whole number part from the fractional part, is called a **decimal point** or **decimal marker**. In a whole number, the decimal point is to the right of the last digit. In writing decimals in this way, first count the number of zeros in the power of 10, then move the decimal point the corresponding number of places to the left. Sometimes a zero is written to the left of the decimal point when a number is less than one.

For example, $\dfrac{1}{4}$ can be written as either .25 or 0.25.

Example 3:21
Rewrite the following decimals using a decimal marker.

(i) $\dfrac{7}{10}$ (ii) $\dfrac{931}{100}$ (iii) $\dfrac{573}{10000}$ (iv) $\dfrac{765}{10}$ (v) $\dfrac{347}{100}$

Solution

(i) $\dfrac{7}{10} = 0.7$ (power of ten has one zero)

(ii) $\dfrac{931}{100} = 9.31$ (power of ten has two zero)

(iii) $\dfrac{573}{10000} = 0.0573$ (power of ten has four zero)

(iv) $\dfrac{765}{10} = 76.5$ (power of ten has one zero)

(v) $\dfrac{347}{100} = 3.47$ (power of ten has two zero)

Therefore,

To divide by a power of 10, count the zeros in the power of 10 and move the decimal point the corresponding number of decimal places to the left.

Place Value System and Decimals

Each digit to the right of the decimal point represents a number of tenths, hundredths, thousandths etc as shown in Table 3:2 below.

Place Value of Digits	Millions (1,000,000)	Hundred thousand (100,000)	Ten thousands (10,000)	Thousands (1,000)	Hundreds (100)	Tens (10)	Units (1)	DECIMAL POINT	Tenths (0.1)	Hundredths (0.01)	Thousands (0.001)	Ten thousandths (0.0001)	Hundred thousandths (0.00001)	Millionths (0.000001)
Number	3	5	2	8	4	1	3	•	6	4	7	1	5	9
Value of Digits	3,000,000	500,000	20,000	8,000	400	10	3		0.6	0.04	0.007	0.0001	0.00005	0.0000009
	Whole Number Part								Decimal Part					

Table 3:2

Example 3:22
State the value of 3 in each of the following
(a) 0.03   (b) 0.483   (c) 51.25431

Solution
(a) three hundredth   (b) three thousandth
(c) thirty thousandth

Example 3:23
What is the value of 4 in the number 6.75487?

Solution
Four thousandth

EXERCISE 3:9

1. Rewrite the following decimals using a
   decimal marker.
   (a) $\dfrac{675}{1,000}$ (b) $\dfrac{586}{100}$ (c) $\dfrac{398896}{1,000,000}$
   (d) $\dfrac{76587}{10,000}$ (e) $\dfrac{45350}{100,000}$

2. What is the value of the underlined digit?
   (a) 0.689097̲6̲   (b) 0.084̲2̲
   (c) 0.3̲2̲547   (d) 4.81̲3̲

Changing Fractions to Decimals

To change a fraction to a decimal, first write the fraction as an equivalent fraction whose denominator is a power of 10, and then write the decimal using a decimal marker.

Example 3:24

Convert $\dfrac{4}{25}$ to a decimal.

Solution
$$\frac{4}{25} = \frac{4}{25} \times \frac{4}{4} = \frac{16}{100} = 0.16$$

Alternatively use long division to divide.

$$\frac{4}{25} = 25\overline{)\begin{array}{l} 0.16 \\ 40 \\ \underline{25} \\ 150 \\ \underline{150} \end{array}}$$

If the fraction is a mixed number, first change the fractional part to a decimal and combine the two.

Example 3:25

Convert $2\dfrac{1}{25}$ to decimal.

Solution

$$2\frac{1}{25} = \frac{51}{25} = 25\overline{)\begin{array}{l} 2.04 \\ 51 \\ 50 \\ 100 \\ \underline{100} \\ 0 \end{array}}$$

$$\therefore 2\frac{1}{25} = 2.04 \quad \text{or}$$

$$\therefore 2\frac{1}{25} = 2 + \frac{1 \times 4}{25 \times 4} = 2 + \frac{4}{100} = 2.04$$

Changing Decimals to Fractions
To change a decimal to a fraction, count the number of decimal places in the number then, write the number as a decimal fraction with denominator a power of 10 having the same number of zeros as the number of decimal places counted. Simplify the result.

Example 3:26
Change the following fractions to decimals
(i) 0.125   (ii) 0.6

Solution

(i) $0.125 = \dfrac{125}{1000} = \dfrac{1}{8}$

(ii) $0.6 = \dfrac{6}{10} = \dfrac{3}{5}$

EXERCISE 3:10

1. Convert the following fractions to decimals.

(a) $\dfrac{3}{4}$ (b) $\dfrac{1}{2}$ (c) $\dfrac{2}{5}$ (d) $\dfrac{5}{8}$

(e) $\dfrac{3}{2}$ (f) $\dfrac{5}{4}$ (g) $2\dfrac{1}{4}$ (h) $5\dfrac{12}{25}$

2. Convert the following decimals to fractions.

(a) 0.8   (b) 0.65   (c) 1.23   (d) 3.75

Operations with decimals

Addition and Subtraction of Decimals

To add or subtract decimals, arrange the numbers vertically so that the decimal point is aligned. Add or subtract as with whole numbers.

Example 3:27
Compute the following without using a calculator.
(i) $6.04 + 3.23$ (ii) $6.4163 + 7.3187 + 5.4128$

Solution

(i)
```
    6.04
  + 3.23
    9.27
```

(ii)
```
    6.4163
    7.3187
  + 5.4128
   19.1478
```

Example 3:28
Evaluate
(i) $0.957 - 0.831$ (ii) $8.90 - 2.47$

Solution

(i)
```
    0.9 5 7
  - 0.8 3 1
    0.1 2 6
```

(ii)
```
    8.9 0
  - 2.4 7
    6.4 3
```

EXERCISE 3:11

Without using a calculator, evaluate the following.

(1)
```
    98.95
  + 45.35
```

(2)
```
     0.89
  + 2.083
```

(3)
```
    6.75
    8.78
  + 8.68
```

(4)
```
    8.316
    2.492
  + 3.542
```

(5) 3.238   (6) $89.7675 + 86.9847$
(7) $5.45 + 6.76 + 4.65$
(8) $76.743 + 6.467 + 67.58$

(9)
```
    7.3
  - 0.78
```

(10)
```
    6.3419
  - 2.4834
```

(11) $8.7234 - 2.6009$ (12) $24.88 - 9.48$

Multiplying Decimals by Whole Numbers

Using the idea of multiplication as repeated addition.
$2.6 \times 3 = 2.6 + 2.6 + 2.6 = 7.8$
But $26 \times 3 = 78$

Therefore,

To multiply a decimal by a whole number, multiply as with whole numbers, and affix the corresponding number of decimal places as is in the decimal.

Example 3:29
Evaluate   (i) 0.14×6 (ii) 136.8×47
Solution

(i)
```
    0.14
  ×    6
    0.84
```

(ii)
```
      136.8
  ×      47
      9576
     5472
    6429.6
```

Multiplying Decimals by Powers of 10

To multiply by a power of 10, count the number of zeros in the power of 10, and move the decimal point the corresponding number of places to the right.

Example 3:30

Evaluate:

(i) 10×2.581 (ii) 100×2.581
(iii) 1000×2.581

Solution

(i) $10 \times 2.581 = 25.81$
(ii) $100 \times 2.581 = 258.1$
(iii) $1000 \times 2.581 = 2581$

EXERCISE 3:12

Compute the following
1. 6×0.008 2. 3×0.012
3. 100×63 4. 100×9.34
5. 1000×2.75 6. $10,000 \times 0.0065$

Multiplying Decimals by Decimals

To multiply decimals by decimals, multiply as with whole numbers. The number of decimal places in the product is the sum of the number of decimal places in the numbers.

Example 3:31
(i) 0.236×0.3 (ii) 0.6×0.8

Solution

(i)
$$\begin{array}{rl} 0.236 & \text{3 decimal places} \\ \times \quad 0.3 & +\ \underline{\text{1 decimal place}} \\ \hline 0.0708 & \text{4 decimal places} \end{array}$$

(ii)
$$\begin{array}{rl} 0.6 & \text{1 decimal place} \\ \times 0.8 & +\ \underline{\text{1 decimal place}} \\ \hline 0.48 & \text{2 decimal places} \end{array}$$

EXERCISE 3:13

Without using a calculator, perform the following products.

1.
$$\begin{array}{r} 4.5 \\ \times\ \underline{0.3} \\ \hline = \end{array}$$

2.
$$\begin{array}{r} 7.86 \\ +\ \underline{8.97} \\ \hline = \end{array}$$

3.
$$\begin{array}{r} 5.643 \\ \times\ \underline{4.37} \\ \hline = \end{array}$$

4.
$$\begin{array}{r} 45.65 \\ +\ \underline{4.37} \\ \hline = \end{array}$$

5. 6.05×45.36 6. 0.453×0.436
7. 0.065×4.3 8. 6.74×34.23
9. 7.656×4.32 10. 0.035×34.57

Dividing Decimals

To divide decimals, determine which of the dividend or divisor has the greater number of decimal places. Multiply both the dividend and the divisor by an equivalent power of 10, to get rid of the decimals. Simplify or do short or long division as with whole numbers.

Example 3:32

Compute the following without using a calculator.

(a) $3\overline{)0.9}$ (b) $6\overline{)17.022}$

Solution

(a) $3\overline{)0.9} = \dfrac{0.9}{3} \times \dfrac{10}{10} = \dfrac{\overset{3}{\cancel{9}}}{\underset{1}{\cancel{3}}} \times \dfrac{1}{10} = 0.3$

(b) $6\overline{)17.022} = \dfrac{17.022}{6} \times \dfrac{1000}{1000}$

$$= \dfrac{\overset{2837}{\cancel{17022}}}{\underset{1}{\cancel{6}}} \times \dfrac{1}{1000} = 2.837$$

Alternatively, using long or short division.

(a) $3\overline{)0.9} = 3\overline{)\overset{0.3}{0.9}}$

$\underline{0.9}$

$\Rightarrow 3\overline{)0.9} = 0.3$

(b) $6\overline{)17.022} = \overline{)\overset{2.837}{17.022}}$

$\underline{12}$
$\ 50$
$\ \underline{48}$
$\ \ 22$
$\ \ \underline{18}$
$\ \ \ 42$
$\ \ \ \underline{42}$
$\ \ \ \ 0$

$\Rightarrow 6\overline{)17.022} = 2.837$

Example 3:33

Evaluate

(a) $0.5\overline{)3.5}$ (b) $0.00025\overline{)0.4}$

Solution

(a) $0.5\overline{)3.5} = 5\overline{)35} = 7$

(b) $0.00025\overline{)0.4} = 25\overline{)40000} = 1600$

EXERCISE 3:14

Evaluate the following.

1. $5\overline{)3.58}$ 2. $25\overline{)0.475}$

3. $3\overline{)3.9}$ 4. $2\overline{)4.674}$

5. $7\overline{)3.514}$ 6. $0.025\overline{)0.004}$

7. $0.5\overline{)6.5}$ 8. $0.05\overline{)0.564}$ 9. $0.06\overline{)23.4}$

10. $4.3\overline{)0.44548}$ 11. $6.08 \div 4$

12. $0.55085 \div 2.3$ 13. $14.688 \div 4.32$

14. $1.575 \div 4.5$

Terminating and Non-terminating Decimals

Sometimes on division, the process continues indefinitely. Such decimals are called **non-terminating, recurring, or repeating decimals**. For instance

$$1 \div 3 = 3\overline{)1.0} \; \overset{0.3333\overline{3}}{} = 0.3333\overline{3}$$

0.33333 is a repeating decimal and the bar over the 3 indicates that 3 repeats endlessly.

These types of non-terminating repeating decimals are called **recurring decimal**. Non-terminating decimals that do not repeat are called **non-terminating non-recurring decimals**. All other decimals that have been met in this section have been **terminating decimals**.

Any fraction that cannot be expressed as an equivalent fraction with denominator 10, results in a non-terminating decimal when expressed as a decimal.

Note that every rational number can be expressed as either a terminating or a non-terminating decimal.

Most fractions whose denominators are multiples of prime numbers other than 2 and 5 will result in non-terminating decimals.

e.g. $\frac{1}{3}, \frac{2}{7}, \frac{5}{11}$ etc.

EXERCISE 3:15

1. Which of the following fractions are likely to result in terminating decimals?

(a) $\frac{1}{4}$ (b) $\frac{7}{9}$ (c) $\frac{3}{8}$

(d) $\frac{3}{2}$ (e) $\frac{3}{11}$ (f) $\frac{11}{25}$

(g) $\frac{1}{6}$ (h) $\frac{13}{40}$ (i) $\frac{4}{7}$

(j) $\frac{3}{20}$ (k) $\frac{25}{6}$ (l) $\frac{7}{10}$

2. Express each of the fractions in (1) as a decimal.

PERCENTAGES

A percentage is a fraction whose denominator is 100. Examples of percentages are $\frac{25}{100}, \frac{130}{100}$.

It saves time and space to write percentages as in the following examples

$\frac{25}{100} = 25\%$, read '25 percent'

$\frac{130}{100} = 130\%$, read '130 percent'

Changing Percentages to Fractions

To change a percentage to a fraction, write the number as a fraction with denominator 100 and simplify.

Example 3:34

Express the following percentages as fractions.

(i) 75%   (ii) 66%

(iii) 250%   (iv) 100%

Solution

(i) $75\% = \dfrac{75}{100} = \dfrac{3}{4}$

(ii) $66\% = \dfrac{66}{100} = \dfrac{33}{50}$

(iii) $250\% = \dfrac{250}{100} = \dfrac{5}{2}$

(iv) $100\% = \dfrac{100}{100} = 1$

Changing Fractions to Percentages

To change a fraction to a percentage, multiply the given fraction by 100 or change the fraction to an equivalent fraction with denominator 100 and then write down the numerator followed by the symbol %.

Example 3:35

Change $\dfrac{3}{4}$ to a percentage.

Solution

$\dfrac{3}{4} = \dfrac{3 \times 25}{4 \times 25} = \dfrac{75}{100} = 75\%$ or

$\dfrac{3}{4} = \dfrac{3}{4} \times 100 = 75\%$

EXERCISE 3:16

1. Convert the following percentages to fractions.
   (a) 25%   (b) 72%   (c) 82%

   (d) 95%   (e) $13\dfrac{1}{2}\%$ (f) $\dfrac{1}{20}\%$

   (g) $7\dfrac{1}{4}\%$ (h) $34\dfrac{3}{4}\%$ (i) 400%

   (j) 250%   (k) 600%   (l) 750%

2. Convert the following fractions to percentages.

   (a) $\dfrac{3}{4}$ (b) $\dfrac{4}{5}$ (c) $\dfrac{300}{100}$

   (d) $\dfrac{17}{50}$ (e) $\dfrac{450}{1000}$ (f) $\dfrac{13}{25}$

Changing Percentages to Decimals

Recall the rule for division by powers of 10. Using this rule, percentages can be changed to decimals by simply moving the decimal point two places to the left.

Example 3:36

Convert the following percentages to decimals.
   (a) 75%   (b) 66%   (c) 250%

Solution

(a) $75\% = \dfrac{75}{100} = 0.75$

(b) $66\% = \dfrac{66}{100} = 0.66$

(c) $250\% = \dfrac{250}{100} = 2.5$

Changing Decimals to Percentages

Since this is the reverse of changing percentages to decimals, simply move the decimal point two places to the right and inserting the symbol %.

Example 3:37

Convert the following decimals to percentages.
   (a) 0.41   (b) 3.42

Solution

(a) $0.41 = \dfrac{41}{100} = 41\%$ (b) $3.42 = \dfrac{342}{100} = 342\%$

EXERCISE 3:17

1. Convert the following decimals to percentages
   (a) 0.4   (b) 0.75
   (c) 2.5   (d) 40.35

2. Convert the following percentages to decimals
   (a) 20%   (b) 35%
   (c) 115%   (d) 250%

Finding a Percentage of a Given Quantity

To find a percentage of a given quantity, write the percent as a fraction with denominator 100 and multiply by the given quantity.

Example 3:38

Evaluate 32% of 200

Solution

$32\% \text{ of } 200 = \dfrac{32}{100} \times 200 = 64$

Example 3:39

Out of the 312 students who took part in the elections for the school senior prefect, Ngeh had 75%. How many votes did he have?

Solution

$75\% \text{ of } 312 = \dfrac{75}{100} \times 312 = 234$

Expressing One Quantity as a Percentage of Another

To express one quantity as a percentage of another express the quantity as a fraction of the other and multiply by 100. Affix the percentage symbol %.

$$\text{Quantity } A \text{ as \% of } B = \frac{\text{Quantity } A}{\text{Quantity } B} \times 100$$

Example 3:40
What percent, of 500 Frs. is 175 Frs.?

Solution

$$\frac{175}{500} \times 100 = 35\%$$

Example 3:41
A shoe company produced 270 shoes on Monday and 210 on Tuesday. What percentage of the shoes was produced on Monday?

Solution

Total number of shoes produced = 270 + 210
$$= 480$$

$$\text{Percentage on Monday} = \frac{270}{480} \times 100 = \frac{225}{4} = 56.25\%$$

EXERCISE 3:18

1. Evaluate:

   (a) 25% of 60   (b) $33\frac{1}{2}$% of 16500

   (c) 0.07% of 8000   (d) 120% of 67500

2. In a conference, 60% of the participants are women. Given that there are 90 participants, how many men attended the conference?
3. What percent of 25 is 4?
4. Express 0.5 as a percentage of 20.
5. How many percent of 5 is 0.4?
6. Out of the 60 students in a class, 40 passed in a mathematics test. How many percent of the students failed?

RATIOS AND PROPORTIONS

Ratios

A ratio $p : q$ or $\frac{p}{q}$ is a means of comparing two quantities p and q which have the same units or which are similar.

Example 3:42
In a certain class, there are 30 boys and 40 girls. Find and simplify the ratio of the number of boys to the number of girls in the class.

Solution

Ratio of boys to girls $= 30 : 40 = \dfrac{30}{40} = \dfrac{3}{4} = 3 : 4$

Other ways of expressing the ratio in example 3:42 are as follows:

The ratio of boys to girls $= \dfrac{3}{4}$ or 75% or 0.75.

Note that, to find the ratio of one quantity to another, the two quantities must have the same units.
Ratios have no units because the dividend units cancel the divisor units.

Equality of Ratios
Two ratios are equal if they can be written as equivalent fractions. For instance

$$6 : 8 = 3 : 4, \text{ since } \frac{6}{8} = \frac{3}{4}$$

Example 3:43
The heights of two poles are in the ratio 8: 5. If the second pole is 120 cm what is the height of the first?

Solution

$$\frac{\text{height of first pole}}{120} = \frac{8}{5}$$

$$\Rightarrow \text{ Height of first pole} = \frac{120 \times 8}{5} = 192 \text{ cm}$$

Proportions
A proportion is a statement involving equal ratios. For instance, the ratio 6:8 is the same as 3:4. This is expressed symbolically as 6:8 :: 3:4 and read "6 is to 8 as 3 is to 4." Another way of writing the proportion

6:8 :: 3:4 is $\dfrac{6}{8} = \dfrac{3}{4}$ or 6:8 = 3:4.

The names of the terms of a proportion are shown in the following sketch.

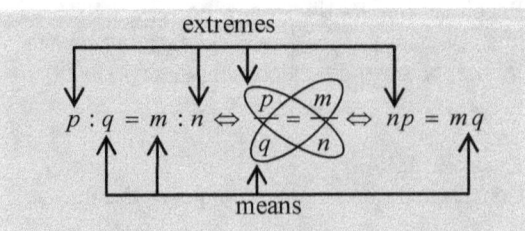

The middle terms of a proportion are called the **means** while the outside terms are called the **extremes**. Thus in the ratio 6:8 =3:4,

    6 and 4 are the extremes
    8 and 3 are the means

The following formula called the proportion rule is often very useful in problem solving involving ratios and proportion.

> Product of means = Product of extremes

Thus, $6 : 8 = 3 : 4 \Leftrightarrow 6(4) = 24$ and $8(3) = 24$

Example 3:44
Solve the following proportions:

(a) $2 : 3 = x : 6$ (b) $\dfrac{x}{8} = \dfrac{3}{2}$

Solution
    Product of means = product of extremes
(a) Product of extremes = $2(6) = 12$
    Product of means = $3(x) = 12$.
    Clearly, the only number that can be multiplied by 3 to have 12 is 4. $\therefore x = 4$.
(b) Product of means = $8(3) = 24$.
    Product of extremes = $2(x) = 24$
    Clearly, the only number that can be multiplied by 2 to have 24 is 12. $\therefore x = 12$.

Proportional parts

A   B   C

Figure 3:1
Figure 3:1 is a line segment AC of length 60 mm, which is divided in the ratio 1:3. This means that if the portion AB is 1 unit, the portion BC is 3 units. If so, obviously the line AC must be 1 +3 = 4 units. Therefore, the total number of parts is 4. Thus, 60 mm represent 4 parts. The total number of parts is called the **sum of ratio**.

Example 3:45
A woman's salary is 80,000 FRS. The ratio of the money she saves per month to that which she spends is 1:3. Calculate the amount which she,
(a) saves per month   (b) spends per month.

Solution
    Sum of ratio =1 + 3 = 4
(a) Amount she saves = $80000 \times \dfrac{1}{4}$
                     = 20000 FRS.
(b) Amount she saves = $80000 \times \dfrac{3}{4}$
                     = 60000 FRS.

Example 3:46
Mrs. Aba sells 4 buckets of rice for 13,000 FCFA while Mrs. Nsah sells 3 buckets of rice for 10,500 FCFA. From whom will you prefer to buy?

Solution
    Mrs. Aba's unit price : Mrs. Aba's unit price
    $= \dfrac{13000}{4} : \dfrac{10500}{3} = 65 : 21$.
$\therefore$ Therefore the preference is Mrs. Aba.

Rates and Scales
Rates
A rate is the comparing of two quantities usually using the word "per" which means "for every". The second unit is the **standard of comparison.**

Scales
A scale is a rate that compares the size of an actual object to the size of its picture or drawing.

Speed as a Rate
The speed at which an object moves is the rate of change of distance.

$$s = \frac{d}{t} \qquad\qquad t = \frac{d}{s} \qquad\qquad d = st$$

Where d = distance, s = speed and t = time
The d-s-t triangle) can be used as a mental aid.

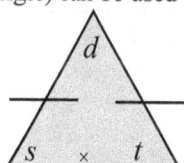

Figure 3:1: The d-s-t triangle

Example 3:47
The distance from town A to town B is 210 km. A car leaves town A to town B travelling at an average speed of 60 km/h. Calculate the time it will take.

Solution
$$t = \frac{d}{s} = \frac{210}{60} = 3.5 \text{ hours.}$$

Example 3:48
A man running at 6.25 m/s completed a race in 8 minutes. Determine the distance of the race.

Solution
    $d = s \times t = 6.25 \times 8 \times 60 \text{ m} = 3000 \text{ m.}$

EXERCISE 3:19

1. Find the missing number in the following proportions

   (a) $5 : 4 = \square : 8$ (b) $8 : 2 = 12 : \square$ (c) $4 : \square = 3 : 6$

   (d) $\square : 9 = 4 : 6$ (e) $\dfrac{\square}{6} = \dfrac{12}{9}$ (f) $\dfrac{3}{\square} = \dfrac{4}{2}$

   (g) $\dfrac{9}{6} = \dfrac{12}{\square}$ (h) $\dfrac{3}{4} = \dfrac{\square}{2}$ (i) $\dfrac{13}{\square} = \dfrac{78}{42}$

2. During a European Champions League, Eto'o played 6 matches and did not play 4. Express as a ratio.
   (a) The number of matches he played to the total number of matches.
   (b) The number of matches he did not play to the number he played.

3. A class is made up of 18 boys and 24 girls
   Find the ratio of
   (a) Boys to girls
   (b) Boys to the total number of students
   (c) Girls to the total number of students.

4. Simplify the following ratios
   (a) 6:9   (b) 5:10   (c) 20:30

5. Divide 2000 FRS to two children in the ratio 2:3.

6. Bih and Manka share 70 kg of rice in the ratio of 2:5. What quantity does each get?

7. Mr. Ngala shares 24000 FCFA to his three sons in the ratio 5: 4: 3. How much does each get?

8. Ndi, Shey and Nfor share 54 oranges in the ratio 2: 3: 4. How many does each get?

9. Given that a man divided his wealth to his three children in the ratio of their ages 10 years, 15 years and 20 years and that the youngest received 750000 CFA. Find
   (a) The total amount
   (b) The amount received by each of the elderly children.

10. Eight men can dig a trench in 4 hours. How long will 5 men dig a similar trench, working at the same rate?

11. A team's ratio of the games won to the games played was 3 to 8. If the team played 88 games, how many games did the team win?

12. 18 bags of corn costs 36,000 Frs. 36 bags of the same corn costs 54,000 Frs.
    (a) Find the unit price for each buy.
    (b) Explain which is the better buy.

MULTIPLE CHOICE EXERCISE 3

1. The fraction which is equivalent to $\dfrac{5}{6}$ is:

   [A] $\dfrac{35}{36}$ [B] $\dfrac{10}{15}$ [C] $\dfrac{30}{36}$ [D] $\dfrac{110}{120}$

2. The fraction, which is not equivalent to, $\dfrac{3}{4}$ is:

   [A] $\dfrac{35}{36}$ [B] $\dfrac{10}{15}$ [C] $\dfrac{30}{36}$ [D] $\dfrac{110}{120}$

3. The fraction which is equivalent to $\dfrac{5}{4}$ is:

   [A] $\dfrac{14}{12}$ [B] $\dfrac{14}{8}$ [C] $\dfrac{20}{18}$ [D] $\dfrac{30}{24}$

4. The fraction $\dfrac{12}{16}$ is the same as:

   [A] $\dfrac{2}{6}$ [B] $\dfrac{1}{4}$ [C] $\dfrac{3}{4}$ [D] $\dfrac{1}{3}$

5. In its lowest terms $\dfrac{60}{108}$ is:

   [A] $\dfrac{7}{9}$ [B] $\dfrac{5}{9}$ [C] $\dfrac{15}{27}$ [D] $\dfrac{20}{36}$

6. The largest among the following fractions is:

   [A] $\dfrac{2}{3}$ [B] $\dfrac{11}{15}$ [C] $\dfrac{7}{10}$ [D] $\dfrac{5}{6}$

7. The fraction, which is an improper fraction, is:

   [A] $\dfrac{3}{4}$ [B] $\dfrac{7}{8}$ [C] $\dfrac{4}{3}$ [D] $\dfrac{1}{5}$

8. $7\dfrac{4}{5}$ expressed as an improper fraction is:

   [A] $\dfrac{35}{5}$ [B] $\dfrac{27}{5}$ [C] $\dfrac{20}{5}$ [D] $\dfrac{39}{5}$

9. As a mixed number $\dfrac{40}{3}$ is:

   [A] $13\dfrac{2}{3}$ [B] $13\dfrac{1}{3}$ [C] $13\dfrac{1}{4}$ [D] $13\dfrac{3}{4}$

10. The sum of $\dfrac{1}{5}$ and $\dfrac{3}{10}$ is:

    [A] $\dfrac{4}{5}$ [B] $\dfrac{1}{2}$ [C] $\dfrac{4}{10}$ [D] $\dfrac{4}{15}$

11 In its lowest terms $\dfrac{5}{6} - \dfrac{5}{8}$ is:

[A] $\dfrac{5}{12}$ [B] $\dfrac{10}{48}$ [C] $\dfrac{35}{24}$ [D] $\dfrac{5}{24}$

12. On simplification $1 - \left(\dfrac{1}{4} + \dfrac{2}{3} \right)$ gives:

[A] 0 [B] $\dfrac{11}{12}$ [C] $\dfrac{1}{12}$ [D] $\dfrac{4}{7}$

13. The value of $6\dfrac{1}{5} - 2\dfrac{2}{3} + 1\dfrac{1}{6}$ is:

[A] $4\dfrac{3}{10}$ [B] $4\dfrac{7}{10}$ [C] $5\dfrac{7}{10}$ [D] $5\dfrac{9}{10}$

14. A jar is $\dfrac{4}{5}$ full of water. If Nfor drinks $\dfrac{5}{9}$ of the water, the fraction of the water left will be:

[A] $\dfrac{11}{45}$ [B] $\dfrac{4}{9}$ [C] $\dfrac{5}{9}$ [D] $\dfrac{16}{45}$

15. A man gave $\dfrac{5}{8}$ of his money to his wife and $\dfrac{1}{4}$ to his son. His fraction of the money left is:

[A] $\dfrac{7}{8}$ [B] $\dfrac{1}{4}$ [C] $\dfrac{5}{8}$ [D] $\dfrac{1}{8}$

16. $\dfrac{2}{3}$ and $\dfrac{1}{4}$ of a floor are covered by tiles and carpet respectively. The fraction of the floor not covered is:

[A] $\dfrac{1}{12}$ [B] $\dfrac{5}{12}$ [C] $\dfrac{11}{12}$ [D] $\dfrac{3}{4}$

17. The value of $\dfrac{1}{2} \times \dfrac{1}{3}$ is:

[A] $\dfrac{1}{5}$ [B] $\dfrac{5}{6}$ [C] $\dfrac{1}{6}$ [D] $\dfrac{2}{3}$

18. $\dfrac{7}{9}$ of $6\dfrac{3}{7}$ is equal to:

[A] 5 [B] 10 [C] 12 [D] 15

19. The result of multiplying $1\dfrac{2}{11}$ by $1\dfrac{7}{26}$ is:

[A] $\dfrac{11}{2}$ [B] $\dfrac{11}{3}$ [C] $1\dfrac{1}{2}$ [D] $\dfrac{2}{3}$

20. Three quarters of 12 is:
   [A] 16 [B] 8 [C] 7 [D] 9

21. $2\dfrac{1}{4} \times 3\dfrac{1}{2}$ is:

[A] $5\dfrac{3}{4}$ [B] $6\dfrac{1}{8}$ [C] $6\dfrac{7}{8}$ [D] $7\dfrac{7}{8}$

22. A student ate $\dfrac{1}{2}$ of $\dfrac{2}{3}$ of food he preserved for super. The fraction of the food he preserved left is:

[A] $\dfrac{1}{2}$ [B] $\dfrac{2}{3}$ [C] $\dfrac{1}{3}$ [D] $\dfrac{3}{5}$

23. The value of the quotient $2\dfrac{1}{5} \div \dfrac{1}{5}$ is:
   [A] 10 [B] 11 [C] 12 [D] 13

24. $\dfrac{1}{2} + \dfrac{1}{2} \times \dfrac{1}{2}$ is equal to:

[A] $\dfrac{1}{8}$ [B] $\dfrac{1}{2}$ [C] $\dfrac{3}{4}$ [D] $1\dfrac{1}{2}$

25. The value of $3 \times \left(\dfrac{1}{2} + \dfrac{1}{4} \right)$ is:

[A] 1 [B] $\dfrac{3}{8}$ [C] $3\dfrac{3}{4}$ [D] $2\dfrac{1}{4}$

26. When evaluated $\dfrac{2\dfrac{2}{3} \times 1\dfrac{1}{2}}{4\dfrac{4}{5}}$ gives:

[A] $\dfrac{5}{12}$ [B] $9\dfrac{3}{5}$ [C] $13\dfrac{1}{5}$ [D] $\dfrac{5}{6}$

27. $\left(\dfrac{2}{3} \times \dfrac{1}{4} \right) - \dfrac{1}{12}$ is the same as:

[A] $\dfrac{1}{12}$ [B] $\dfrac{1}{6}$ [C] $\dfrac{1}{4}$ [D] $\dfrac{1}{3}$

28. $\dfrac{4}{5}$ of $\left(\dfrac{1}{2} + \dfrac{3}{4} \right)$ has the value:

[A] 1 [B] 2 [C] 3 [D] 4

29. When simplified, the result of $\dfrac{\dfrac{1}{2} - \dfrac{1}{3}}{\dfrac{1}{2} + \dfrac{1}{3}}$ is:

[A] $\dfrac{1}{10}$ [B] $\dfrac{1}{15}$ [C] $\dfrac{1}{5}$ [D] $\dfrac{1}{20}$

30. On simplification $2\dfrac{3}{5} \div 1\dfrac{1}{5} + \dfrac{1}{2}$ gives:

[A] $1\dfrac{1}{3}$ [B] $2\dfrac{2}{3}$ [C] $2\dfrac{1}{2}$ [D] $3\dfrac{1}{2}$

31. The product of $\dfrac{1}{6}$ and the sum of $\dfrac{2}{5}$ and $1\dfrac{1}{3}$ is:

[A] $\dfrac{13}{45}$ [B] $\dfrac{14}{45}$ [C] $\dfrac{15}{45}$ [D] $\dfrac{17}{45}$

32. When simplified the value of $5 + \dfrac{5}{4} - 5 \times \dfrac{5}{4}$ is:

    [A] $\dfrac{25}{16}$ [B] $\dfrac{5}{16}$ [C] 0   [D] 1

33. The value of $\dfrac{\frac{3}{4}}{1\frac{1}{4}} \times \left(1\dfrac{1}{2} - \dfrac{2}{3}\right)$ is:

    [A] $\dfrac{25}{32}$ [B] $\dfrac{7}{24}$ [C] $\dfrac{9}{25}$ [D] $\dfrac{1}{2}$

34. When simplified, the result of

$5\dfrac{1}{4} \div \left(1\dfrac{2}{3} - \dfrac{1}{2}\right)$ is:

    [A] $1\dfrac{3}{4}$ [B] $3\dfrac{1}{4}$ [C] $4\dfrac{1}{2}$ [D] $8\dfrac{1}{2}$

35. A labourer's monthly salary is 48000 FRS. If he saves $\dfrac{1}{5}$ of this amount, in one year he must have saved:

[A] 9,600 FRS   [B] 115,20 FRS
[C] 180,000 FRS   [D] 115,200 FRS

36. A man spends $\dfrac{3}{4}$ of his monthly salary on food and $\dfrac{1}{2}$ of the remainder on rent. If he has 15000 FRS left, he surely earns:

[A]   60,000 FRS   [B] 90,000 FRS
[C] 105,000 FRS   [D] 120,000 FRS

37. Given that a pole has $\dfrac{1}{3}$ of its length in mud, $\dfrac{2}{5}$ of the remainder in water and the rest 6 m long above the surface of the water, the length of the pole is:

[A] 12 m   [B] 10 m   [C] 15 m   [D] 16 m

38. The fraction, which can be subtracted from the sum

of $2\dfrac{1}{6}$ and $2\dfrac{2}{12}$ to give $3\dfrac{1}{4}$ is:

    [A] $\dfrac{1}{3}$ [B] $\dfrac{1}{2}$ [C] $1\dfrac{1}{2}$ [D] $1\dfrac{1}{6}$

39. A student spends $\dfrac{1}{4}$ of her pocket money on books and $\dfrac{1}{3}$ on a dress. The fraction of her pocket money remaining will be:

    [A] $\dfrac{5}{12}$ [B] $\dfrac{7}{12}$ [C] $\dfrac{5}{6}$ [D] $\dfrac{1}{6}$

40. A man spent $\dfrac{3}{8}$ of his salary on rent and $\dfrac{1}{3}$ of the remainder on cloths. The fraction of his salary left is:

    [A] $\dfrac{23}{24}$ [B] $\dfrac{5}{6}$ [C] $\dfrac{19}{24}$ [D] $\dfrac{5}{12}$

41. The reciprocal of 21 is:

    [A] 21^2 [B] 0.21   [C] $\sqrt{21}$ [D] $\dfrac{1}{21}$

42. The reciprocal of 0.02 is:
[A] 500   [B] 50   [C] 0.5   [D] 0.05

43. The reciprocal of 0.0002 is:
[A] 50   [B] 500   [C] 5000   [D] 50,000

44. To one decimal place, the reciprocal of 0.625 is:
[A]   1.6   [B]   0.6   [C]   6.3   [D]   62.5

45. The reciprocal of $\dfrac{x}{y}$ is:

    [A] $-\dfrac{y}{x}$ [B] $-\dfrac{x}{y}$ [C] $\dfrac{y}{x}$ [D] $\dfrac{1}{xy}$

46. $\dfrac{5}{16}$ as a decimal is:
[A]   0.3125   [B] 0.5125
[C]   0.4125   [D] 0.2725

47. 25 out of 200 pineapples, are bad. As a decimal, the fraction of pineapples that are bad is:
[A] 0.875   [B] 0.185
[C] 0.225   [D] 0.125

48. 0.375 as a fraction is:

    [A] $\dfrac{1}{8}$ [B] $\dfrac{3}{8}$ [C] $\dfrac{5}{16}$ [D] $\dfrac{11}{16}$

49. 0.72 is equivalent to:

    [A] $\dfrac{18}{25}$ [B] $\dfrac{7}{10}$ [C] $7\dfrac{1}{5}$ [D] $\dfrac{71}{100}$

50. 5 − 0.003 equals:
[A] 0.002 [B] 4.003 [C] 4.007 [D] 4.997

51. The value of 4.7−1.9 + 2.1 is:
[A] 5.9   [B]   8.7   [C]   1.7   [D]   4.9

52. 0.93 + 0.08 is equals to:
[A] 1.01 [B] 1.1 [C] 1.11 [D] 0.101

53. 0.1 ×0.2 ×0.3 is equal to:
[A] 0.06 [B] 0.006 [C] 0.05 [D] 0.005

54. The value of $\dfrac{2.4}{4}$ is:
[A]   0.6   [B]   6   [C]   60   [D]   2.1

55. The value of 136×47 is 6392. The value of 1.36×4.7 is:
    [A] 0.6392   [B] 6.392
    [C] 63.92   [D] 639.2

56. The value of 136×47 is 6392. The value of $\dfrac{63.92}{13.6}$ is:
    [A] 47 [B] 0.047 [C] 0.47 [D] 4.7

57. $(0.12)^2$ is equal to:
    [A] 1.44 [B] 0.144 [C] 0.0144 [D] 0.24

58. The value of $\dfrac{1}{0.2} + \dfrac{1}{0.25}$ is:
    [A] 45   [B] 4.5   [C] 2.5   [D] 9

59. $78.75 \div 0.35$ is:
    [A] 0.225 [B] 0.25 [C] 22.5 [D] 225

60. $\dfrac{3}{40}$ as a percentage is:
    [A] 3% [B] 7.5% [C] 40% [D] 75%

61. $37\dfrac{1}{2}\%$ expressed as a fraction is:
    [A] $\dfrac{7}{8}$ [B] $\dfrac{5}{8}$ [C] $\dfrac{3}{8}$ [D] $\dfrac{1}{8}$

62. 0.125 as a percentage is:
    [A] 125%   [B] 12.5%
    [C] 1.25%   [D] 0.125%

63. As a decimal 172% to 2 decimal places is:
    [A] 1.68 [B] 1.70 [C] 1.72 [D] 1.66

64. Ambe's monthly salary is 300, 000 FRS. Given that he spends 15 % on rents, the amount spent on rent is:
    [A] 60,000 FRS   [B] 45,000 FRS
    [C] 40,000 FRS   [D] 30,000 FRS

65. A boy scored 70% in a test. If the maximum mark is 40, then the boy's mark is:
    [A] 4 [B] 10 [C] 28 [D] 30

66. If 20 % of a sum of money is 4000 FCFA, then the whole sum of money is:
    [A] 200,000 FCFA [B] 80,000 FCFA
    [C] 20,000 FCFA   [D] 40,000 FCFA

67. In a certain class, 21 students are boys and 19 are girls. The percentage of the class who are boys is:
    [A] 21%   [B] 40%
    [C] 47%   [D] 52.5%

68. The percentage of 4, which is 5, is:
    [A] 20% [B] 80% [C] 120% [D] 125%

69. Given that out of the 500 students who sat for an examination 150 students failed. The percentage that failed is:
    [A] 70% [B] 30% [C] 60% [D] 40%

70. A girl obtains 60 marks out of 75 on an examination. This is equivalent to a percentage of:
    [A] 60% [B] 30% [C] 90% [D] 80%

71. Given that $\dfrac{5}{8}$ of the pupils in a school are boys, then the ratio of boys to girls is:
    [A] 5:3 [B] 5:8 [C] 3:5 [D] 8:5

72. 63000 FCFA is divided into two parts so that the first part is $\dfrac{3}{4}$ of the second. The value of the larger part is:
    [A] 24,000 FCFA   [B] 27,000 FCFA
    [C] 28,000 FCFA   [D] 36,000 FCFA

73. The ratio of A's share to B's share in the profit of a business is 5:4. If the total profit is 450, 000 FCFA, then A's share is:
    [A] 90,000 FCFA   [B] 200,000 FCFA
    [C] 225,000 FCFA   [D] 250,000 FCFA

74. If 120,000 FCFA is divided in the ratio 2 to 3 then the smaller share is:
    [A] 48,000 FCFA   [B] 80,000 FCFA
    [C] 72,000 FCFA   [D] 60,000 FCFA

75. The ratio of the shares of two partners A and B in the profit of a business is 5 : 3 respectively. The amount A will receive when B receives 48,000 FRS is:
    [A] 80,000 FRS   [B] 30,000 FRS
    [C] 28,800 FRS   [D] 30,000 FRS

76. A sum of money was divided in the ratio of 2:3 between two children. The fraction of the money received by the child who received the smaller amount is:
    [A] $\dfrac{2}{5}$ [B] $\dfrac{3}{5}$ [C] $\dfrac{1}{2}$ [D] $\dfrac{2}{3}$

77. The ratio of the number of men to the number of women in a 20-member committee is 3:1. The number of women who must be added to the 20-member committee to make the ratio of men to women 3:2 is:
    [A] 2   [B] 5   [C] 7   [D] 9

78. Three children share 10500 FRS among themselves in the ratio 6:7:8. The largest share is:
    [A] 3,000 FRS   [B] 3,500 FRS
    [C] 4,000 FRS   [D] 4,500 FRS

79. An amount of money is shared among P, Q and R such that for every 3 P gets, Q gets 2 and for every 1 Q gets R gets 4. The ratio $P : Q : R$ is:
    [A] 3:1:4 [B] 3:2:1 [C] 3:2:4 [D] 3:2:8

80. If $a : b = 4 : 9$ and $b : c = 3 : 5$, $a : b : c$ is:
    [A] 4:9:15   [B] 3:9:14
    [C] 4:9:20   [D] 4:12:45

81. The sum of 2400 FCFA is shared between Neba, Ndeh and Ambe. If Neba receives 600 FCFA and the remaining amount is shared between Ndeh and Ambe in the ratio 4:5 respectively. In FCFA Ambe receives:
    [A] 800 [B] 900 [C] 1,000 [D] 1,200

82. The value of the underlined digit in 5.41$\underline{7}$ is:
    [A] 7 [B] 0.7 [C] 0.07 [D] 0.007
83. The value of 5 in the number 47.581 is:
    [A] 5 tens   [B] 5 units
    [C] 5 tenth   [D] 5 hundredth
84. Ngalim buys 25 oranges, 32 pineapples, and 9 pears. In three ways, the ratio pineapples to oranges is:

    [A] 32:9, $\dfrac{32}{9}$, 32 to 9

    [B] 25:32, $\dfrac{25}{32}$, 9 to 25

    [C] 9:25, $\dfrac{9}{32}$, 9 to 25

    [D] 32:25, $\dfrac{32}{25}$, 32 to 25

85. A sports club has 28 balls and 8 bats. The ratio of balls to bats in simplest form is:
    [A] 7 : 2   [B] 7 : 14
    [C] 4 : 14   [D] 4 : 2
86. 294 km in 6 h is the same as:
    [A] 98 km/h   [B] 49 km/h
    [C] 51 km/h   [D] 50 km/h
87. 520 FCFA for 8 cans is equal to:
    [A] 55 FCFA per can   [B] 60 FCFA per can
    [C] 65 FCFA per can   [D] 70 FCFA per can
88. The ratio, which is the same as 10: 19, is:
    [A] 107 to 193   [B] 100 to 190
    [C] 107 to 190   [D] 100 to 193
89. At an average rate of 33 km an hour, the distance, which can be travelled in 10 hours, is:
    [A] 340 km   [B] 374 km
    [C] 363 km   [D] 330 km
90.   The pair of ratios that can form a proportion is:

    [A] $\dfrac{3}{5}, \dfrac{18}{45}$ [B] $\dfrac{3}{5}, \dfrac{27}{35}$

    [C] $\dfrac{3}{5}, \dfrac{21}{35}$ [D] $\dfrac{3}{5}, \dfrac{24}{30}$

91.   The pair of ratios, which CANNOT form a proportion, is:

    [A] $\dfrac{20}{70}, \dfrac{2}{7}$ [B] $\dfrac{2}{7}, \dfrac{4}{14}$

    [C] $\dfrac{2}{7}, \dfrac{4}{21}$ [D] $\dfrac{2}{7}, \dfrac{6}{21}$

92. The ratio that can form a proportion with $\dfrac{2}{3}$ is:

    [A] $\dfrac{8}{9}$ [B] $\dfrac{6}{12}$ [C] $\dfrac{8}{15}$ [D] $\dfrac{18}{27}$

93. A scale drawing of a town park has a scale of 2 cm : 300 m. The actual length of each cm in the drawing is:
    [A] 300 m   [B] 75 m
    [C] 150 m   [D] 37.5 m
94. The scale for a model of a cow is 1 mm : 4 cm. The cow is 184 cm tall. The height of the cow in the model is:
    [A] 46 cm   [B] 4.6 mm
    [C] 181 mm   [D] 46 mm
95. The scale of a 3 cm tall model of a 1.5 m tall man is:
    [A] 1 : 80   [B] 1 : 70
    [C] 1 : 50   [D] 1 : 60
96. In the number 8.6792 the value of the digit 7 is:

    [A] 70   [B] $\dfrac{7}{10}$ [C] $\dfrac{7}{100}$ [D] 700

97. The digit 8 occupies the second place to the right of the decimal point. Its value is:
    [A] 8 hundred   [B] 8 tens
    [C] 8 tenth   [D] 8 hundredth

98. The square root of $2\dfrac{7}{9}$ is:

    [A] $1\dfrac{2}{9}$ [B] $1\dfrac{2}{3}$ [C] $1\dfrac{2}{5}$ [D] $1\dfrac{2}{7}$

99. The value of $\sqrt{5\dfrac{4}{9}}$ is:

    [A] $2\dfrac{1}{3}$ [B] $\dfrac{13}{18}$ [C] $5\dfrac{2}{3}$ [D] $16\dfrac{1}{3}$

100. The square root of 0.0036 is:
    [A] 0.6   [B] 0.06
    [C] 0.006   [D] 0.0006

TOPIC 4

ESTIMATIONS, APPROXIMATIONS AND STANDARD FORM

OBJECTIVES

At the end of this topic, the learner should be able to:

1. Round down and round up numbers.
2. Estimate sums and differences in whole numbers and decimals.
3. Estimate products and quotients in whole numbers and decimals.
4. Approximate a given number to a given number of decimal places or significant figures.
5. Write a normal form number in standard form and vice versa.
6. Perform calculations in standard form.

Thomas Malthus
Thomas Malthus' studies on the growth of population led to the development of the field of demography. Malthus (1766-1834) believed that the population would naturally increase faster than the amount of food that could be produced to feed them. He advocated sexual abstinence or restraint to control population increases and acknowledged the role of plagues, wars, and epidemics in containing overpopulation. Malthus specifically suggested that people marry later and have small families. Due to these ideas, economics earned its name as "the dismal science."

According to the U.S. Census Bureau, the world's population was 6.5 billion in February 2006. The Bureau **estimated** that 249 people are born and 108 people die every minute, meaning that the world's population grows by 141 each minute of 2006. The total was expected to reach 7 billion in 2012.

Rounding Down and Rounding Up

For ease of calculations, numbers are usually given less exact and smaller values (**rounded down**) or larger values (**rounded up**). This process is known as **approximation**. To round up a number, the last digit required for the given degree of accuracy is increased by 1 if the digit after it is 5, 6, 7, 8 or 9. To round down a number, the last digit required for the given degree of accuracy is maintained if the digit after it is 0, 1, 2, 3 or 4 and the digits after this last digit are ignored in the case of decimals or written as zeros in the case of whole numbers. To round to the nearest ten, hundred, thousand…, replace the number by the closest multiple of 10, 100, 1000…, respectively. To round to the nearest tenth, hundredth, thousandth, etc write the number to 1, 2, 3, etc decimal places.

ESTIMATIONS

An **estimate** is an intelligent guess on the dimensions of an object or the result of a calculation. When an estimate is made, only an approximate (not exact) value is obtained. To estimate the result of a calculation, round up or round down each number in the calculation to a reasonable or given degree of accuracy.

Estimating sums and differences in whole numbers

Example 4:1
Estimate the following
(i) $245 + 350 + 570$
(ii) $431 + 53$
(iii) $748 - 394$

Solution
(i) $245 + 350 + 570 \approx 200 + 400 + 600$
≈ 1200
(ii) $431 + 53 \approx 430 + 50 \approx 480$
(iii) $748 - 394 \approx 700 - 400 \approx 300$

Estimating Sums and Differences in Decimals

Example 4:2
Estimate to the nearest whole number.
(i) $4.68 + 0.71$
(ii) $6.7234 - 3.5138$

Solution

(i)	$4.68 \approx$	5	(ii)	$6.7234 \approx$	7
	$+\ 0.71 \approx$	$+\underline{1}$		$-3.5138 \approx$	$-\underline{4}$
		$\underline{\underline{6}}$			$\underline{\underline{3}}$

Example 4:3
Estimate to the nearest tenth.
(a) $3.623 + 0.29 + 5.386$
(b) $4.86 - 3.456$

Solution
(a) $3.623 + 0.29 + 5.386$
$\approx 3.6 + 0.3 + 5.4 = 9.3$
(b) $4.86 - 3.456 \approx 4.9 - 3.5 = 1.4$

EXERCISE 4:1

1. Estimate the following sums and differences
   (a) $844 + 239$ (b) $464 + 37$
   (c) $7982 + 1486$ (d) $5391 - 247$
   (e) $26212 - 14084$ (f) $748 - 394$
   (g) $509 - 42$
   (h) $13486 + 4842 + 29072$
2. Estimate to the nearest whole number.
   (a) $0.89 + 2.083$
   (b) $6.4139 + 2.8238 + 3.2500$
   (c) $4.9 - 0.87$
   (d) $8.7234 - 2.6006$
3. Estimate to the nearest tenth.

(a)	0.835		(b)	7.428
	$+\ \underline{0.27}$			6.37
				$+\ \underline{0.843}$
(c)	0.82		(d)	12.34
	$-\ \underline{0.243}$			$-\ \underline{8.45}$

Estimating products and quotients in whole numbers

Example 4:4
Estimate the following products

(i)	58×24	(ii)	653×56
(iii)	48830×750	(iv)	275×24

Solution

(i) $58 \times 24 \approx 60 \times 20 \approx 1200$

(ii) $653 \times 56 \approx 700 \times 60 \approx 42,000$

(iii) $48,830 \times 750 \approx 50,000 \times 800 \approx 4,000,000$

(iv) $275 \times 24 \approx 300 \times 20 \approx 6,000$

Example 4:5

Estimate the following quotients

(i) $\dfrac{562}{18}$ (ii) $\dfrac{68}{27}$ (iii) $\dfrac{62}{24}$

Solution

(i) $\dfrac{562}{18} \approx \dfrac{600}{20} \approx 30$ (ii) $\dfrac{68}{27} \approx \dfrac{70}{30} \approx 2.3$

(iii) $\dfrac{62}{24} \approx \dfrac{60}{20} \approx 3$

Estimation of Products and Quotients in Decimals

Example 4:6

Give an estimate for the following.

(i) 4.755×0.5 (ii) $\dfrac{89.93}{4.1}$

Solution

(i) $4.755 \times 0.5 \approx 5 \times 0.5 \approx 2.5$

(ii) $\dfrac{89.93}{4.1} \approx \dfrac{899.3}{41} \approx \dfrac{900}{40} \approx 22.5$

EXERCISE 4:2

1. Estimate the following products.

(a) $\begin{array}{r} 21 \\ \times \underline{45} \end{array}$ (b) $\begin{array}{r} 763 \\ \times \underline{53} \end{array}$ (c) 4.8×6

(d) 0.21×3.81 (e) 8.6×5.9 (f) 3.1×8.2

2. Estimate the following quotients.

(a) $32\overline{)85}$ (b) $91\overline{)873}$ (c) $3\overline{)6.31}$

(d) $4.2\overline{)0.439}$ (e) $0.31\overline{)6.125}$ (f) $5.7\overline{)0.483}$

APPROXIMATIONS

Decimal Places

The terms tenth, hundredth, thousandth etc respectively mean one, two, three etc decimal places. Thus rounding off to the nearest tenth, hundredth, thousandth etc is the same as rounding off to 1, 2, 3 etc decimal places.

Significant Figures

The significant figures (s.f) of a given number are the figures (or digits) required to express the number to a given degree of accuracy. The number of significant figures is always counted from the first non-zero digit to the last non-zero digit.

Example 4:7

Round 6526 to the nearest:

(a) ten   (b) hundred   (c) thousand.

In each case, state the number of significant figures.

Solution

(a) 6526 to the nearest ten = 6530
                                    [3 s.f.]

(b) 6526 to the nearest hundred = 6500
                                    [2 s.f.]

(c) 6526 to the nearest thousand = 7000
                                    [1 s.f.]

Example 4:8

Round 1.045 to:

(a) 2 decimal places   (b) 1 decimal place

In each case, state the number of significant figures.

Solution

(a)   1.045 to 2 d.p.s. = 1.05   [3 s.f.]

(b)   1.045 to 1 d.p. = 1.0   [1 s.f.]

Example 4:9

Round 0.01027 to:

(a) 4 decimal places.   (b) 3 decimal places.

(c) 2 decimal places.   (d) 1 decimal place.

In each case, state the number of significant figures.

Solution

(a)   0.01027 to 4 d.p.s. = 0.0103 [3 s.f.]

(b)   0.01027 to 3 d.p.s. = 0.010 [1 s.f.]

(c)   0.01027 to 2 d.p.s. = 0.01   [1 s.f.]
(d)   0.01027 to 1 d.p.s. = 0.0   [no s.f.]

EXERCISE 4:3

1. Complete Table 4:1.

Number		Number of significant figures			
		1	2	3	4
a	0.0068398				
b	2.0068398				
c	4.69768				
d	1.006127				

Table 4:1

2. Complete Table 4:2.

Number		Number of Decimal places			
		1	2	3	4
a	0.0068398				
b	2.0068398				
c	4.69768				
d	1.006127				

Table 4:2

3. Express to 3 decimal places.
   (a) 0.003646   (b) 0.4567
   (c) 0.5046
4. Express to 2 decimal places.
   (a) 14.9028   (b) 23.1058
   (c) 6.0381
5. Express to 3 significant figures.
   (a) 0.02485   (b) 4.027956
6. Express to 2 significant figures.
   (a) 547.53   (b) 59.81798
   (c)   5382
7. Express to 1 significant figure.
   (a) 0.009238   (b) 5.097
   (c)   0.2309

STANDARD FORM

In science, very large or very small numbers occur. For instance, the mass of an election is 0.000,000,000,000,000,000,000,000,000,911 g; the velocity of electromagnetic waves in air is 300,000,000 m/s. Numbers written in this way are said to be in **normal or decimal form**. Writing numbers in this form, especially when there are so many zeros often lead to errors.
The **standard form** or **scientific notation** provides an easier way of writing such large or small numbers.

To express a number in standard form:
(a) Move the decimal point to the right of the first non-zero digit, counting the number of steps.
(b) Multiply the resulting number by 10 raised to the power corresponding to the number of steps moved.
(c) If the decimal point was originally to the left of its current position the power is negative and if it was originally to the right, the power is positive.

Summarily, the number is written in the form

$$N = A \times 10^n, \text{ where } n \in \mathbb{N} \text{ and } 1 \le |A| < 10$$

If $|N| < 1, n < 0$
If $|N| = 1, n = 0$
If $|N| > 1, n > 0$

Example 4:10
Express the following in standard form
(a) 0.000,000,000,000,000,000,000,000,000,911
(b) 300,000,000
(c) 0,00048

Solution
(a) 0.000,000,000,000,000,000,000,000,000,911
$$= 9.11 \times 10^{-28}$$

(b) $300,000,000 = 3 \times 10^8$

(c) $0,00048 = 4.8 \times 10^{-4}$

EXERCISE 4:4

Express the following numbers in standard form.

(a) 5000
(b) 480
(c) 10200
(d) 700000
(e) 0.0032
(f) 0.000073
(g) 0.925
(h) 0.001
(i) 0.5600
(j) 3000×10^{-8}
(k) 19.6×10^{-4}
(l) 0.034×10^{-8}

Calculations in Standard Form

In performing calculations in standard form, the powers of 10 are manipulated using the multiplication, division and power laws of indices. Thus,

$$10^{n} \times 10^{m} = 10^{n+m}$$

$$10^{n} \div 10^{m} = 10^{n-m}$$

$$\left(10^{m}\right)^{n} = 10^{mn}$$

EXERCISE 4:5

1. Evaluate giving your result in standard form.

(a) $9.5 \times 10^{7} - 3.08 \times 10^{6}$
(b) $\dfrac{0.45 \times 0.91}{0.0117}$

(c) 0.06×0.09
(d) $\dfrac{0.24}{0.012}$

(e) $\dfrac{8.75}{0.025}$
(f) $\sqrt{\dfrac{0.81 \times 10^{5}}{2.25 \times 10^{7}}}$

(g) $\dfrac{9.6 \times 10^{8}}{0.24 \times 10^{5}}$
(h) $\dfrac{0.9687}{0.001}$

(i) $\dfrac{0.203 \times 0.55}{3.05}$
(j) $\sqrt{\dfrac{1.44 \times 10^{5}}{8.1 \times 10^{4}}}$

(k) $\sqrt{\dfrac{0.0016 \times 0.0081}{0.36}}$

(l) $\sqrt{\dfrac{76.42 \times 10^{-1}}{0.004 \times 10^{2}}}$

2. Given that, $a = 6 \times 10^{3},$ $b = 2 \times 10^{-4}$, $c = 4 \times 10^{-5}$. Find the value of $\dfrac{a \times b}{c}$ expressing your answer in the standard form.

3. The weight of a single atom of hydrogen is about 0.00000000000000000000000017 grams. Express this weight in standard form.

4. Evaluate $\dfrac{1946 \times 10^{-1}}{2 \times 10^{2}}$, giving the answer
   (a) in standard form.
   (b) correct to 1 decimal place.
   (c) correct to 2 significant figures.

5. Evaluate $\dfrac{12.78 \times 10^{-3}}{9 \times 10^{-1}}$, expressing your answer
   (a) In standard form
   (b) Correct to two significant figures
   (c) Correct to three decimal places

MULTIPLE CHOICE EXERCISE 4

1. 0.015849 expressed correct to three significant figures is:
   [A] 0.0158 [B] 0.0159
   [C] 0.0160 [D] 0.020

2. Given that $x = 0.0102$, correct to three significant figures. The value, which cannot be the actual value of x, is:
   [A] 0.01021 [B] 0.01014
   [C] 0.01015 [D] 0.01016

3. 6474 correct to three significant figures is:
   [A] 647 [B] 648
   [C] 6470 [D] 6480

4. The number 25.973 correct to three significant figures is:
   [A] 25.973 [B] 25.97
   [C] 25.9 [D] 26.0

5. The number of people attending a football match is quoted as 27000, correct to 2 significant figures. The greatest possible attendance shown by this figure is:
   [A] 27,000 [B] 27499

[C] 27599   [D] 26999
6. 0.0063 correct to 2 decimal places is:
    [A] 0.006   [B] 0.01
    [C] 0.06   [D] 0.10
7. After evaluating 2.35 × 0.48, the answer to 2 decimal places is:
    [A] 11.28   [B] 1.13
    [C] 1.128   [D] 1.10
8. The value of 3.769 ÷ 0.7 to the nearest tenth is:
    [A] 5.41   [B] 5.0   [C] 10   [D] 5.4
9. To the nearest whole number, the result of $\frac{6.6 \times 1.8}{5.4}$ is:
    [A] 2.2   [B] 3   [C] 2   [D] 22
10. 0.000252 ÷ 0.007 to two decimal places is:
    [A] 0.04   [B] 0.03   [C] 0.36   [D] 0.40
11. By evaluating $\frac{7 + 3.32}{9.91 - 5.11}$, the answer to one decimal place will be:
    [A] 21.5   [B] 2.1   [C] 22.0   [D] 2.2
12. 0.44734 ÷ 0.01, evaluated to the nearest hundredth is:
    [A] 44.70   [B] 45   [C] 44.73   [D] 44.00
13. $\frac{6.3 \times 60 \times 0.2}{3.6 \times 1.4}$, when simplified, the answer to the nearest ten is:
    [A] 15   [B] 20   [C] 10   [D] 1.5
14. 930,000,000 in standard form is:
    [A] 93.0×10^9 [B] 9.3×10^8
    [C] 9.3×10^7 [D] 9.3×10^{-7}
15. 5238, expressed in standard form is:
    [A] 5.238×10^3 [B] 5.238×10^2
    [C] 5.238×10^1 [D] 5.238×10^0
16. Expressed in standard form 435600 is:
    [A] 4.536×10^7 [B] 4.536×10^6
    [C] 4.536×10^5 [D] 4.536×10^4
17. When expressed in standard form 2789 equals:
    [A] 2.789×10^{-3} [B] 2.789×10^2
    [C] 2.789×10 [D] 2.789×10^3
18. In standard form, 52006 can be written as:
    [A] 5.2006×10^3 [B] 5.2006×10^{-4}
    [C] 5.2006×10^4 [D] 5.2006×10^{-3}
19. 120,000 written in standard form is:
    [A] 1.2×10^2 [B] 1.2×10^3

[C] 1.2×10^4 [D] 1.2×10^5
20. The number 36700 written in standard form is:
    [A] 3.67×10^3 [B] 3.67×10^5
    [C] 3.67×10^4 [D] 3.67×10^2
21. 325,000 in standard form is:
    [A] 3.25×10^6 [B] 3.25×10^{-6}
    [C] 3.25×10^5 [D] 3.25×10^{-5}
22. 0.00562 in standard form is:
    [A] 5.62×10^{-3} [B] 0.562×10^{-2}
    [C] 5.62×10^{-2} [D] 5.62×10^2
23. Expressed in standard form 0.0462 is:
    [A] 0.462×10^{-1} [B] 0.462×10^{-2}
    [C] 4.62×10^{-1} [D] 4.62×10^{-2}
24. 0.000834 in standard form is:
    [A] 8.34×10^{-4} [B] 8.34×10^{-5}
    [C] 8.34×10^3 [D] 8.34×10^4
25. 0.0000027 in standard form is:
    [A] 2.7×10^6 [B] 2.7×10^{-6}
    [C] 2.7×10^5 [D] 2.7×10^{-5}
26. 0.000,000,070,2 in standard form is:
    [A] 7.02×10^{-5} [B] 7.5×10^{-6}
    [C] 7.02×10^{-7} [D] 7.5×10^{-8}
27. Expressed in standard form 0.000,082,3 becomes:
    [A] 0.823×10^5 [B] 0.823×10^{-5}
    [C] 8.23×10^5 [D] 8.23×10^{-5}
28. Written in standard form 0.000370 is:
    [A] 3.7×10^{-1} [B] 7.5×10^{-2}
    [C] 3.7×10^{-3} [D] 7.5×10^{-4}
29. 46 × 900 expressed in standard form is:
    [A] 4.14×10^3 [B] 4.14×10^5
    [C] 4.14×10^4 [D] 4.14×10^6
30. When 4 hours is converted to seconds and expressed in standard form, the result is:
    [A] 1.44×10^3 [B] 1.44×10^{-3}
    [C] 1.44×10^4 [D] 1.44×10^{-4}
31. 258 km when expressed to mm and expressed in standard form becomes:
    [A] 2.58×10^8 [B] 2.58×10^7
    [C] 2.58×10^6 [D] 2.58×10^5
32. $\frac{8.75}{0.025}$, expressed in standard form is:
    [A] 3.5×10^{-3} [B] 3.5×10^{-2}

[C] 3.5×10^1 [D] 3.5×10^2

33. Given that $0.000208 = 2.08 \times 10^x$, the value of x is:

[A] 4   [B] -4 [C] 5   [D] -5

34. The product of 0.06 and 0.09 in standard form is:

[A] 5.4×10^2 [B] 5.4×10^{-3}

[C] 5.4×10^1 [D] 5.4×10^{-2}

35. When $0.009 \div 0.012$ is evaluated, the answer in standard form is:

[A] 7.5×10^2 [B] 7.5×10^{-3}

[C] 7.5×10^{-1} [D] 7.5×10^{-2}

36. 5.7×10^4 in ordinary form is:

[A] 5700   [B] 57000

[C] 7500   [D] 75000

37. As a decimal fraction 8.2×10^{-5} is:

[A] 0.0082   [B] 0.00082

[C] 0.000082   [D] 0.0000082

38. In normal form 3.746×10^{-3} is:

[A] 0.003746   [B] 0.0003746

[C] 0.03746   [D] 3746

39. 9.258×10^{-3} in normal form correct to 3 significant figures is:

[A] 926   [B] 0.093

[C] 0.009   [D] 0.00926

40. 7.15×10^5 in normal form is:

[A] 71500   [B] 715000

[C] 7150   [D] 7150000

41. 4.5×10^3 is a number in standard form. The number is:

[A] 0.0045   [B] 0.045

[C] 4500   [D] 45000

42. The sum of 728.93 and 0.46 expressed in standard form to three significant figures is:

[A] 72.9×10 [B] 7.29×10^2

[C] 728×10^0 [D] 7.28×10^0

43. The sum of 2.48×10^3 and 5.9×10^4 is:

[A] 6.148×10^1 [B] 6.148×10^4

[C] 6.148×10^3 [D] 6.148×10^2

44. $4 \times 10^2 \times 2 \times 10^{-4}$ is equal to:

[A] 8×10^6 [B] 8×10^{-2}

[C] 8×10^{-8} [D] 6×10^{-2}

45. $7.580 \times 10^9 + 7.677 \times 10^9$ is equal to:

[A] 1.5257×10^{10} [B] 1.5257×10^8

[C] 1.5257×10^9 [D] 1.5257×10^7

46. Given that

$x = 5.7 \times 10^6, y = 1.8 \times 10^6, x - y = :$

[A] 3.9×10^{-6} [B] 3900000×10^6

[C] 3.9×10^6 [D] 39×10^5

47. 2.52×10^5 is equal to:

[A] 2.52×10^5 [B] 2.52×10^6

[C] 2.52×10^4 [D] 2.52×10^{-5}

48. The population of two towns A and B is given as 5.77×10^6 and 3.66×10^6 respectively. The difference in the population of A and B is:

[A] 2.11×10^0 [B] 2.11×10^{-2}

[C] 2.11×10^4 [D] 2.11×10^6

49. $7.42 \times 10^{-6} - 4.33 \times 10^{-7}$ is the same as:

[A] 6.987×10^{-6} [B] 5.987×10^{-6}

[C] 4.987×10^{-6} [D] 3.987×10^{-6}

50. $5.72 \times 10^3 - 2.37 \times 10^2$ in the normal form, is equal to:

[A] 4483   [B] 4473   [C] 5483   [D] 3450

51. The product of 0.012 and 0.0008 in standard form is:

[A] 9.6×10^6 [B] 9.6×10^{-6}

[C] 9.6×10^5 [D] 9.6×10^{-5}

52. When simplified, $\left(2 \times 10^{-5}\right) \times \left(4 \times 10^{-2}\right)$ becomes:

[A] 8.0×10^{-7} [B] 8.0×10^7

[C] 8.0×10^3 [D] 8.0×10^{-3}

53. On simplification $8 \times 10^5 + 2 \times 10^3$ gives:

[A] 4×10^8 [B] 4×10^2

[C] $4 \times 10^{\frac{5}{3}}$ [D] $4 \times 10^{-\frac{5}{3}}$

54. 450×70 is:

[A] 3.15×10^5 [B] 3.15×10^4

[C] 3.15×10^3 [D] 3.15×10^2

55. The result of $\dfrac{0.126}{36}$ is:

[A] 3.5×10^4 [B] 3.5×10^{-4}

[C] 3.5×10^3 [D] 3.5×10^{-3}

56. The simplified value of

$\dfrac{\left(12 \times 10^8\right)\left(16 \times 10^{-6}\right)}{1 \times 10^{-2}}$ in standard form is:

[A] 1.92×10^{6} [B] 192×10^{4}

[C] 1.92×10^{4} [D] 1.92×10^{2}

57. On evaluation $\sqrt{\dfrac{0.81 \times 10^{-5}}{2.25 \times 10^{7}}}$ equals:

[A] 3.6×10^{-13} [B] 6.0×10^{-7}

[C] 3.6×10^{13} [D] 6.0×10^{7}

58. Given that

$p = 3.6 \times 10^{-3}$ and $q = 2.25 \times 10^{6}$ the

value of $\sqrt{\dfrac{p}{q}}$ is:

[A] 1.6×10^{-9} [B] 4.0×10^{-5}

[C] 1.6×10^{9} [D] 4.0×10^{5}

TOPIC 5

WEIGHTS AND MEASURES

OBJECTIVES

At the end of this topic, the learner should be able to:

1. Appreciate the need for standard units of measurement.
2. Convert one SI unit of length to another.
3. Convert one SI unit of mass to another.
4. Convert one SI unit of area to another.
5. Convert one SI unit of volume to another.
6. Read, write and understand time using BC and AD.
7. Convert time from the 12 hour clock to the 24 hour clock and vice versa.
8. Convert from one time unit to another.
9. Appreciate the need for larger units for a larger number of years.
10. Write temperatures in degrees and Celsius.
11. Convert from degrees to Celsius and vice versa.
12. Name and identify currencies of different countries or zones.
13. Convert from one currency to another for a given exchange rate.

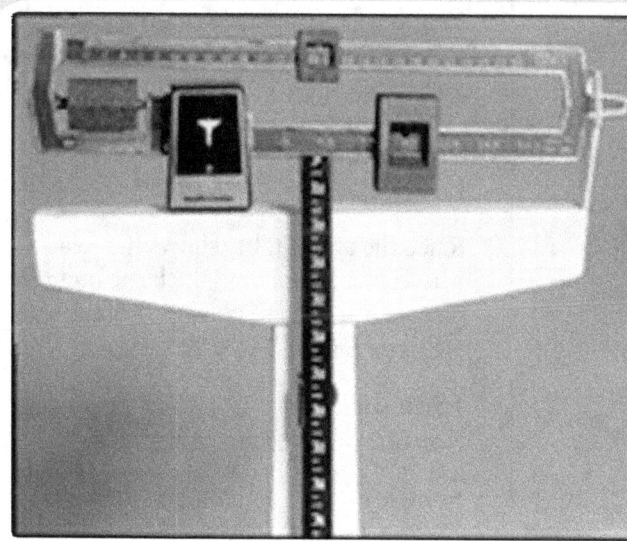

Beam Scale
Often found in doctor's offices, the beam scale uses small adjustable weights called poises to balance the load. The weight is measured from markings on the beam.

The Need for Standard Units

Long ago, people used their arms, feet and reaches to measure length. This is still true of some traditional societies today.

An arm is the length between one's shoulder and his hand while a reach is the length between a person's hands, when the arms are fully stretched. These types of units of measurement are very inconsistent since people's arms are not of the same length. In our society today, tomato tins are still cut to any sizes and used in selling groundnuts, kernels, salt etc. These types of units hinder trade immensely.

Even in scientific works, until about 1800, different countries still used different systems of units. For instance, while the imperial units (inches, feet, yards, fathoms, miles etc) were used in Britain, the M.K.S systems of units were used in countries such as France. This made trade between different parts of the world very difficult. It also made scientific works difficult to be interpreted in different countries of the world. Designing, building and taxation were greatly hindered. For these reasons, it was very necessary to standardize the units.

The SI Units

In 1960 the General Conference on weights and measures, met and resolved on the unanimous use of the Système International d'Unités (abbreviated S.I.) which was derived from the M.K.S system, so called because it is based on the metre (m), kilogram (kg) and seconds (s).

The advantages of the SI units are that they are accepted internationally and are easy to use. Their easy usage depends upon the fact that the SI units except the time unit are metric units. In other words, the SI units follow the decimal system of numeration.

Prefixes in the SI Units

Greek prefixes are used for multiples of the basic units, while Latin prefixes are used for sub-multiples of the standard units as shown in Table 5:1 below.

Linear Measure

Length is the measure of distance. The basic unit of length is the metre (m). Table 5:2, shows the relationship between other units and the basic unit.

Unit	Symbol	Meaning
Mega metre	Mm	1, 000,000 m
Kilometre	Km	1, 000 m
Hectometre	Hm	100 m
Decametre	Dam	10 m
Decimetre	dm	0.1 m
Centimetre	cm	0.01 m
Millimetre	mm	0.001 m
Micrometre	μm	0.000001 m

Table 5:2

Of the units above, the most commonly used are the Kilometre, the metre, the centimeter, Millimetre. the Kilometre is used for long distances such as distances on the road. The metre is used for distances such as distances on a field or plot. The Centimetre is used for short distances such as length of a table or height of a person and the Millimetre is used for very short distances such as length of figures drawn on a page.

Some types of lengths are width (breadth), height (altitude) and depth.

Conversion of Units of Length

Since the above table shows the relationship between each unit and the basic unit (the metre), it is easy to convert from one unit to the basic unit and vice versa.

Example 5:1
Convert 3000 cm to metres.

Solution
$$1 \text{ cm} = 0.01 \text{ m}$$
$$\Rightarrow 3000 \text{ cm} = 3000 \times 0.01 \text{ m} = 30 \text{ m}$$

*Note that in the layout of the solution, the metres are on the right and the centimetres are on the left since the conversion is from cm to m.

Example 5:2
Convert
(a) 6 km to metres (b) 4.315 dam to m
Solution
(a)   1 km = 1000 m
    $\Rightarrow$ 6 km = 6 × 1000 m = 6000 m
(b)   1 dam = 10 m
    $\Rightarrow$ 4.315 dam = 4.315 × 10 m
                = 43.15 m

To convert a unit other than the basic unit to another unit other than the basic unit, first convert to the basic unit and then convert the result to the required unit.

Example 5:3
Change
(a) 17238 cm to kilometres
(b) 2.38 Hm to mm

Solution
(a)   1 cm = 0.01 m
    17238 cm = 17238 × 0.01 m
            = 172.38 m
    But 1000 m = 1 km
    $\Rightarrow$ 1 m = 0.001 km
    $\therefore$ 172.38 m = 0.001 × 172.38 km
            = 0.17238 km
(b)   1 Hm = 100 m
    2.38 Hm = 2.38 × 100 m = 238 m
    But 1 m = 1000 mm
    $\Rightarrow$ 238 m = 1000 × 238 mm
            = 238000 mm

EXERCISE 5:1

1. Convert each of following to metres:
    (a) 6.38 km   (b) 823 dm
    (c) 14.352 km   (d) 24.35 mm
    (e) 28 km   (f) 5.297 Dam
    (g) 7 mm   (h) 249 mm
    (i) 3128 mm   (j) 4351 cm
    (k) 72 cm   (l) 1379 cm
2. Convert to millimetres:
    (a) 0.382 m   (b) 7.3423 m
    (c) 22.1 cm   (d) 0.82 km
    (e) 437 cm
3. Convert to kilometres:
    (a) 1794300 mm   (b) 78 m
    (c) 14873 cm   (d) 23 m
    (e) 371 cm
4. Convert to centimetres:
    (a) 8.312 m   (b) 5128 mm
    (c) 45 mm   (d) 3.24 km
    (e) 0.991 m
5. Estimate the length of the following:
    (a) Your foot   (b) your height
    (c) your desk
    (d) the blackboard of your classroom
    (e) the door of your classroom.
6. Measure the length of the following:
    (a) Your foot   (b) your height
    (c) your desk
    (d) the blackboard of your classroom
    (e) the door of your classroom
7. Which unit is most suitable to measure
    (a) the distance from Bamenda to Yaounde.
    (b) the length of the sides of a triangle.
    (c) the length of a hall.
    (d) the radius of the cross–section of a piece of chalk.
8. Estimate first and measure the length of the given item. Copy Table 5:3 and use it to record your data.

S/N	Item	Estimate	Measurement
(a)	Length of your Maths textbook		
(b)	Width of your Maths textbook		
(c)	Your pen		
(d)	Length of your Maths exercise book		
(e)	Width of your Maths exercise book		

Table 5:3

8. In your exercise book, copy Table 5:4. Then estimate each of the given lengths.

Cut a staff one metre long and use it to measure the lengths.

S/N	Item	Estimate	Measurement
(a)	Football field		
(b)	Handball field		
(c)	Football pole		
(d)	Handball pole		

Table 5:4

Mass Measure

The **mass** of a body often referred to erroneously in daily life as weight is the amount of matter contained in the body. The basic unit of measurement of mass is the kilogram. Table 5:5, shows the subunits of the kilogram and their relationship with one of the units for measuring mass called-the gram.

	Relationship to the gram
1 Kilogram (kg)	1000 g
1 Hectogram (Hg)	100 g
1 Decagram (dag)	10 g
1 gram (g)1 Decigram	1 g
(dg)	0.1 g
1 Centigram (cg)	0.01 g
1 Milligram (mg)	0.001 g

Table 5:5

For very large masses, the ton is used.
$$1 \text{ ton} = 1000 \text{ kg}$$
The most commonly used units of mass are the ton, the kilogram, the gram and the milligram. The ton is used for measuring very large quantities such as the mass of heap of stones. The kilogram is used for measuring quantities such as the mass of meat or fish. The gram is used for measuring quantities such as the mass of paper, pins or buttons and the milligram is used for measuring quantities

such as the mass of a chemical in a tablet of a drugs.

Conversion of Units of Mass

The methods of converting units of mass are the same as those for converting units of length.

Example 5:4
Convert:
(a) 3000 cg to grams
(b) 6 kg to grams
(c) 4.315 dag to g
(d) 17238 cg to kg
(e) 2.38 Hg to mg

Solution

(a)
$$1 \text{ cg} = 0.01 \text{ g}$$
$$\Rightarrow 3000 \text{ cg} = 3000 \times 0.01 \text{ g}$$
$$= 30 \text{ g}$$

(b)
$$1 \text{ kg} = 1000 \text{ g}$$
$$\Rightarrow 6 \text{ kg} = 6 \times 1000 \text{ g}$$
$$= 6000 \text{ g}$$

(c)
$$1 \text{ dag} = 10 \text{ g}$$
$$\Rightarrow 4.315 \text{ dag} = 4.315 \times 10 \text{ g}$$
$$= 43.15 \text{ g}$$

(d)
$$1 \text{ cg} = 0.01 \text{ g}$$
$$17238 \text{ cg} = 17238 \times 0.01 \text{ g}$$
$$= 172.38 \text{ g}$$
But 1000 g = 1 kg
$$172.38 \text{ g} = 1 \text{ kg} \div 1000 \text{ g} \times 172.38 \text{ g}$$
$$= 0.17238 \text{ kg}$$

(e)
$$1 \text{ Hg} = 100 \text{ g}$$
$$2.38 \text{ Hm} = 2.38 \times 100 \text{ g} = 238 \text{ g}$$
But 1 g = 1000 mg
$$\Rightarrow 238 \text{ g} = 1000 \times 238 \text{ mg} = 238000 \text{ mg}$$

Example 5:5
Convert to kilograms.
(a) 16 tons   (b) 0.25 tons

Solution

(a)
$$1 \text{ ton} = 1000 \text{ kg}$$
$$\Rightarrow 16 \text{ tons} = 16 \times 1000 \text{ kg} = 16000 \text{ kg}$$

(b)
$$1 \text{ ton} = 1000 \text{ kg}$$
$$\Rightarrow 0.25 \text{ tons} = 0.25 \times 1000 \text{ kg} = 250 \text{ kg}$$

Example 5:6

Convert to tons.

(a) 3500 kg (b) 320 kg

Solution

$$1000 \text{ kg} = 1 \text{ ton}$$

$$1 \text{ kg} = \frac{1}{1000} \text{ tons}$$

(a) $\Rightarrow 3500 \text{ kg} = \frac{1}{1000} \times 3500 \text{ tons} = 3.5 \text{ tons}$

(b) $\Rightarrow 320 \text{ kg} = \frac{1}{1000} \times 320 \text{ tons} = 0.32 \text{ tons}$

EXERCISE 5:2

1. Convert 5.264 grams to milligrams.
2. How many kilograms are there in 18 tons?
3. Convert 7148 mg to grams
4. Convert 0.56 tons to kilograms.
5. Change 7342 kg to tons.
6. Estimate your mass in kilograms.
7. Estimate the mass of
   (a) A magi cube   (b) A 25 francs coin
   (c) Ten litres of water.

Quick Conversion of Metric Units

Conversion of metric units can be done faster by simply moving the decimal point a number of places to the right or left as the case may be. The sketch below can quickly be jotted and used.

Metres
Kilo—Hecto—Deka or deci—centi—milli
grams

Using the sketch,
To convert to a unit, that is to the left, move the decimal point the corresponding number of decimal places to the left.

Example 5:7

Convert the following
(a) 25 mm to decimetres.
(b) 3000 dg to kilograms.
(c) 1200 cm to metres.
(d) 42100 mg to hectograms.

Solution

(a) $25 \text{ mm} = 2\overset{\curvearrowleft\curvearrowleft}{5} = 0.25 \text{ dm}$

(b) $3000 \text{ dg} = \overset{\curvearrowleft\curvearrowleft\curvearrowleft\curvearrowleft}{3000} = 0.3 \text{ kg}$

(c) $1200 \text{ cm} = 1\,2\overset{\curvearrowleft\curvearrowleft}{00} = 12 \text{ m}$

(d) $42100 \text{ mg} = \overset{\curvearrowleft\curvearrowleft\curvearrowleft\curvearrowleft\curvearrowleft}{42100} = 0.421 \text{ hg}$

To convert to a unit that is to the right, move the decimal point the corresponding number of decimal places to the right.

Example 5:8

Convert the following.
(a) 23 km to centimetres.
(b) 0.31 dag to decigrams.
(c) 7 Hm to millimetres.
(d) 0.04 g to milligrams

Solution

(a) 23 km = 2,300,000 cm
        (5 d.ps. to the right)

(b) 0.31 dag = 31 dg
        (2 d.ps. to the right)

(c) 7 Hm = 700,000 mm
        (5 d.ps. to the right)

(d) 0.04 g = 40 mg
        (3 d.ps. to the right)

Area measure

The **area of a plane figure** is the number of square units that it covers. Area is measured in square units and its basic unit is the square metre (abbreviated m^2). One square metre (Figure 5:1) is defined as the area covered by a square whose length (side) is 1 m.

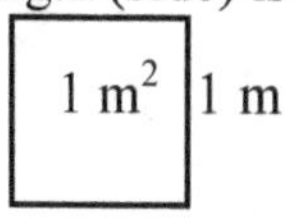

$$\boxed{1 \text{ m}^2}\, 1 \text{ m}$$

Figure 5:1

Table 5:6, shows other common units of area in relation to the square metre.

Area	In square metres
1 Square kilometre ($1km^2$)	1,000,000 m^2
1 Hectare (1 Ha)	10,000 m^2
1 Are (a)	100 m^2
1 Square decimetre (dm^2)	0.01 m^2
1 Square centimetre (cm^2)	0.0001 m^2
1 Square millimetre (mm^2)	0.000001 m^2

Table 5:6

Conversion of Units of Area

Units of area can equally be converted easily by using the format below.

$$Km^2 - Ha - a - m^2 - dm^2 - cm^2 - mm^2$$

To convert to a unit of area that is to the left, count the number of steps and multiply by 2. Next, move the decimal point the resulting number of decimal places to the left.

Example 5:9

Convert 2,000 mm^2 to m^2

Solution

The unit m^2 is 3 steps to the left of mm^2.
Since 3 × 2 = 6, count 6 decimal places to the left. Thus

2,000 mm^2 = .0 0 2 0 0 0 = 0.002 m^2

Example 5:10

Convert 12,700 dm^2 to a.

Solution

The unit a, is 2 steps to the left of dm^2.
Since 2 × 2 = 4, count 4 decimal places to the left.

12,700 mm^2 = 1.2 7 0 0 = 1.27 a.

To convert to a unit of area that is to the right, count the number of steps and multiply by 2. Next, move the decimal point the corresponding number of decimal places to the right.

Example 5:11

Convert 3.5 km^2 to m^2.

Solution

m^2 is 3 steps to the right of km^2.
Since 3 × 2 = 6, count 6 decimal places to the right.

3.5 km^2 = 3.5 0 0 0 0 0 = 3,500,000 m^2

Example 5:12

Convert 0.42 Ha to cm^2.

Solution

The unit cm^2, is 4 steps to the right of ha.
Since 4 × 2 = 8, count 8 decimal places to the right.

0.42 Ha = 0.4 2 0 0 0 0 0 0 = 42,000,000 cm^2

Example 5:13

A football field has a length of 100 m and a width of 60 m. calculate its area in hectares.

Solution

Given l = 100 m and w = 60 m

$\Rightarrow$ Area=100× 60 =6000 m^2 = 0.6 Ha

(counting 4 d.p.s. to the left)

EXERCISE 5:3

1. Convert to cm^2
   (a) 2 m^2 (b) 3 mm^2
   (c) 1 m^2 (d) 400 mm^2
2. Convert to km^2
   (a) 900 m^2 (b) 4 m^2 (c) 500 m^2
3. Convert to mm^2
   (a) 6 cm^2 (b) 12 m^2
   (c) 4 cm^2 (d) 1 m^2
4. Convert to m^2
   (a) 30,000 cm^2 (b) 2 cm^2
   (c) 3 ha   (d) 5a
   (e) 4km^2 (f) 400 cm^2
   (g) 53 mm^2 (h) 7,000,000 mm^2
5. Convert to cm^2
   (a) 70,000 m^2 (b) 40,000 m^2
6. Convert to a
   (a) 700 m^2 (b) 4 km^2
7. The length of a tabletop is 1.2 m. If the width is 80 cm what is it area in Ares?
8. A square plot has sides 15 m. What is the area of the plot in cm^2?
9. Find the area of a piece of plywood measuring 172 cm by 110 cm. Express your answer in square metres.
10. Measure the length and width of your classroom, and hence find the area of the floor of your classroom.
11. Measure the length and width of your mathematics exercise book. What is the area of your mathematics exercise book?
12. Estimate the area of the school
    (a) Tennis court   (b) handball pitch

Volume Measure (Cubic Measure)

The **volume of a solid figure** is the amount of space the solid occupies. Volume is measured in cubic units and the basic unit of volume is the cubic metre (m^3). One cubic metre (Figure 5:2) is the volume occupied by a cube whose edge is 1m. All the edges of a cube are equal.

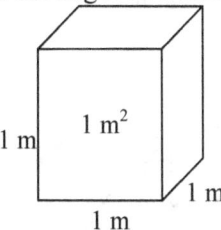

Figure 5:2

The units most commonly used in measuring volume are shown in Table 5:7 below.

Unit of Volume	Relationship to mm^3
Cubic metre m^3	$1000,000,000 \ mm^3$
Cubic decimetres dm^3	$1000,000 \ mm^3$
Cubic centimetres cm^3	$1000 \ mm^3$
Cubic millimetres mm^3	$1 \ mm^3$

Table 5:7

Volume of Cubes and Cuboids

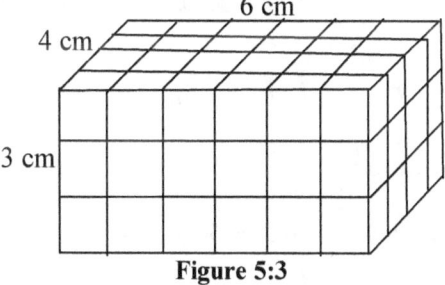

Figure 5:3

Figure 5:3, shows a cuboids whose length, width and height are 6 cm, 4 cm and 3 cm respectively. This cuboids is made up of a number of cubes of volume 1 cm^3 each. How many such cubes are there altogether? What is the result of multiplying the length , width and height of the cuboids? Is it equal to the number of cubes?

From above, it can be seen that the volume of a cuboids is given by

Volume of Cuboid = length × width × height

Therefore for the cuboids above,

Volume = length × width × height

$\Rightarrow$ Volume = 6 cm × 4 cm × 3 cm = 72 cm^3

A cube is a special cuboids with all edges equal.

Example 5:14

The length, width and height of a classroom are 7 m,   5 m and 4 m respectively. What is the volume of the classroom?

Solution

Volume of classroom = *lwh*

$$= (7)(5)(4)$$
$$= 140 \ m^2$$

Capacity

The volume of a liquid is known as **capacity**. Thus, all the units of volume are used for capacity. In addition, the following units based on the litre are also used.

$$1 \text{ litre } (l) = 1000 \ cm^3$$
$$1 \text{ millilitre } (m \, l) = 1 \ cm^3$$
$$= 0.001 \text{ litre}$$
$$10 \text{ millilitres } (ml) = 1 \text{ centilitre } (cl)$$
$$10 \ cl = 1 \text{ decilitre } (dl)$$
$$10 \ dl = 1 \ l$$
$$1000 \ cl = 1 \ l$$

EXERCISE 5:4

1. Convert to cubic metres
   (a) 700,000 cm^3 (b) 50,300 cm^3
   (c) 8,000,000 mm^3 (d) 504,000 dm^3
2. Convert to cubic centimetres
   (a) 400 m^3 (b) 3,140,000 mm^3
   (c) 3,000 dm^3 (d) 64 m^3
3. Convert to litres
   (a) 200 cm^3 (b) 3,500 cm^3
   (c) 7,000 cm^3 (d) 4800 cm^3
4. Convert to millilitres
   (a) 17.6 cm^3 (b) 850 cm^3
   (c) 350 cm^3 (d) 0.174 cm^3
5. A container can take 3 litres of water. How much water can four such containers take? Give your answer in cm^3.
6. 3000 cm^3 of water is poured in to a container, which already contained 7 litres

of water. What is the total amount of water in the container? Give your answer in litres.

7. 9000 cm³ of oil are removed from a tin, which originally contained 20 litres of oil. How much oil in cm³ is left in the tin?

8. A quantity of oil is shared to 24 women. If each has 3 litres what is the total quantity of oil in cm³?

9. Convert to cm³
   (a) 7 *l*   (b) 40 *l*   (c) 30 m*l*   (d) 17.4 m*l*

10. Without measuring, pour into a bucket an amount of water, which is about
    (a)   3 litres   (b)   Half a litre

11. (a) Estimate the length, width and height of your classroom.
    (b) Use your estimates to calculate
       (i) The area of your classroom.
       (ii) The volume of your classroom.

Time Measure
Historical Time

Julius Caesar

In 45 BC, Julius Caesar established the first Calendar based on a solar year of 365 and a quarter days. This Calendar Has stood the test of time, as it is still the universal one used today.

Historical time is often written with the suffixes B.C. which means Before Christ (was born) or A.D., which means anno Domini in Latin, translated in English as in the year of our Lord. The number line in Figure 5:4, illustrates that 100 B.C. when Julius Caesar was born is earlier than the year 00 estimated as the birth year of Jesus Christ.
Cameroon had its independence 1961 years after Jesus Christ was born.

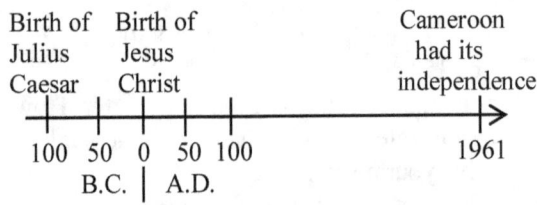

Figure 5:4

Time Units

Below is the relationship between time units.

60 seconds = 1 minute (min)
60 minutes = 1 hour (h)
24 hours   = 1 day (d)
7 days   = 1 week (wk)
4 weeks   = 1 month (mth)
12 months = 1 year
   30 days = 1 month (mth)
  365 days = 1 year
  52 weeks = 1 year
   10 years = 1 decade
  100 years = 1 century
 1000 years = 1 millennium

Conversion of Time Units

To convert from a larger unit to a smaller one, multiply.

Example 5:15
Convert 40 minutes to seconds

Solution
40 minutes = 40 × 60 minutes = 2400 seconds

Example 5:16
Convert 3 hours to seconds.

Solution
  3 hours = 3 × 60 minutes
⇒3 hours = 3 × 60 × 60 seconds
       = 10800 seconds

Example 5:17
Convert 2 days to seconds.

Solution
2 days = 2 × 24 hours = 2 × 24 × 60 minutes
⇒2 days = 2 × 24 × 60 × 60 seconds
       = 172,800 seconds

Example 5:18
Convert 4 years to seconds.

Solution
4 years = 4 × 365 days
       = 4 × 365 × 24 hours
       = 4 × 365 × 24 × 60 minutes
       = 4 × 365 × 24 × 60 × 60 seconds
       = 126,144,000 seconds
       ≈ 1.26×10^8 seconds

To convert from a smaller unit to a larger one, divide.

Example 5:19

Convert 18921600 seconds to years.

Solution

$$18921600 \text{ seconds} = \frac{18921600}{60} \text{ minutes}$$

$$= \frac{18921600}{60 \times 60} \text{ hours}$$

$$= \frac{18921600}{60 \times 60 \times 24} \text{ days}$$

$$= \frac{18921600}{60 \times 60 \times 24 \times 365} \text{ years}$$

$= 0.6$ years

Time as a Non-Metric S.I. Unit

Time is the only S.I. unit, which does not use the metric system. This means that, the units of time do not follow the decimal system. Can it be that one day time will be metricated? Can it be possible? Suppose that, 1 year equals 10 months. How is it going to look like? Can it be possible? Think over this and give reasons why it can be possible or why it cannot be possible.

EXERCISE 5:5

1. Convert $\frac{3}{8}$ of a day to
   (a) Hours (b) minutes (c) seconds
2. How many days are there in 20 weeks 4 days?
3. How many seconds are there in 12 minutes 10 seconds?
4. Find the value of $\frac{3}{4}$ hours 20 minutes in minutes.
5. Auk took 35 minutes to walk to school. He reached school at 7.25 a.m., at what time did he leave his house?
6. What is the value of $\frac{1}{6}$ a week in:
   (a) Hours (b) minutes (c) seconds
7. A man did a piece of job for 1 year 3 months 3 weeks 2 days. How long in days did it take him?
8. Convert 7 days to minutes.

Clocks and Watches

Analog clocks and watches have faces labeled in hours from 1 to 12. Unfortunately, a day is 24 hours. This often poses some problems, especially from noon to midnight.
To take care of this confusion, the suffix a.m. (ante meridiem) before 12 noon and p. m. (post meridiem) after 12 noon are added to the analog time. For instance, if an analog watch shows 6 O'clock in the evening, the time will be read or written as 6 p.m. but if it shows 6 O'clock in the morning, the time will be read or written as 6 a.m. On the other hand, 24-hour clocks and watches automatically take care of this confusion. Most 24 hour clocks and watches are digital. The day begins at 12 midnight and ends at 12 midnight. Thus, 12 midnight is considered 0 O'clock or 24 O'clock while 12 noon is just 12 O'clock.
For example if a 24 hour clock or watch reads 19:00, it is interpreted that the time is 7 p.m. Table 5:8, shows how time based on the 12 hour and 24 hour clocks is written.

12 hour clock	24 hour clock
1:40 p.m.	13:40
1:30 a.m.	01:30

Table 5:8

Figure 5:5: A 24 hour digital clock

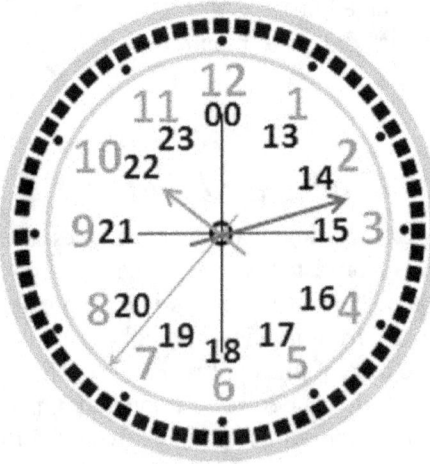

Figure 5:6 : A 24 hour analog clock

Figure 5:7: A 12 hour analog clock

Example 5:20

Change the following to 24 hour time.

(a)   10 p.m.   (b)   2.45 p.m.

Solution

(a) 10 p.m. = 12 + 10 = 22:00

(b) 2.45 p.m. = 12 + 2.45 = 14:45

Example 5:21

Change the following to 12 hour time.

(a)   12:00   (b)   16:22

Solution

(a) 12:00 = 12 a.m.

(b) 16:22 =16:22–12:00 = 4:22 p.m.

1. State the following as 12 hour time
   (a) 17:25   (b) 23:36
   (c) 14:20   (d) 11:10
2. State the following as 24 hour time.
   (a)   8: 34 p.m.   (b)   5:56 a.m.
   (c)   12:45 p.m.   (d) 12:45 a.m.
3. Calculate the time between
   (a) 10:42 a.m. and 8:12 p.m.
   (b) 8:12 a.m. and 10:42 p.m.
   (c) 8:16 a.m. and 7:20 p.m.
   (d) 22:33 and 9:15
4. What is the meaning of
   (a) a.m.   (b) p.m.

Temperature

Temperature is the degree of hotness or coldness of an object. It is measured using a thermometer. The units of temperature are the degree Fahrenheit (°F) and the degree Celsius or Centigrade (°C).

Some significant temperatures worth noting are:

Boiling point of water = 100 °C

Normal body temperature = 37 °C

Normal room temperature = 20 °C

Freezing point of water = 0 °C

Inter-conversion of Temperature

The following formulae are used to interconvert degrees Celsius and degrees Fahrenheit.

$$F = \frac{9}{5}C + 32 \Leftrightarrow C = \frac{(F - 32)5}{9}$$

Example 5:22

Convert (a) 85 °C from °F (b) 149 °F to °C

Solution

(a) $F = \frac{9}{5}C + 32 = \frac{9}{5}(85) + 32 = 185°C$

(b) $C = \frac{(F - 32)5}{9} = \frac{(149 - 32)5}{9} = 65°F$

EXERCISE 5:7

1. Convert to degrees Fahrenheit.
   - (a) 60 °C
   - (b) 35 °C
   - (c) 20 °C
   - (d) 40 °C
   - (e) 75 °C
2. Convert to degrees Celsius.
   - (a) 122 F°
   - (b) 93.2 F°
   - (c) 62.6 F°
   - (d) 199.4 F°
   - (e) 113 F°
3. State which of the following is cold, warm or hot.
   - (a) 94 °C
   - (b) 35 °C
   - (c) 5 °C
   - (d) 40 °C
   - (e) –8 °C
   - (f) 15 °C
   - (g) 212°F
   - (h) 32°F
   - (i) 100°F

The Celsius and Fahrenheit Thermometers

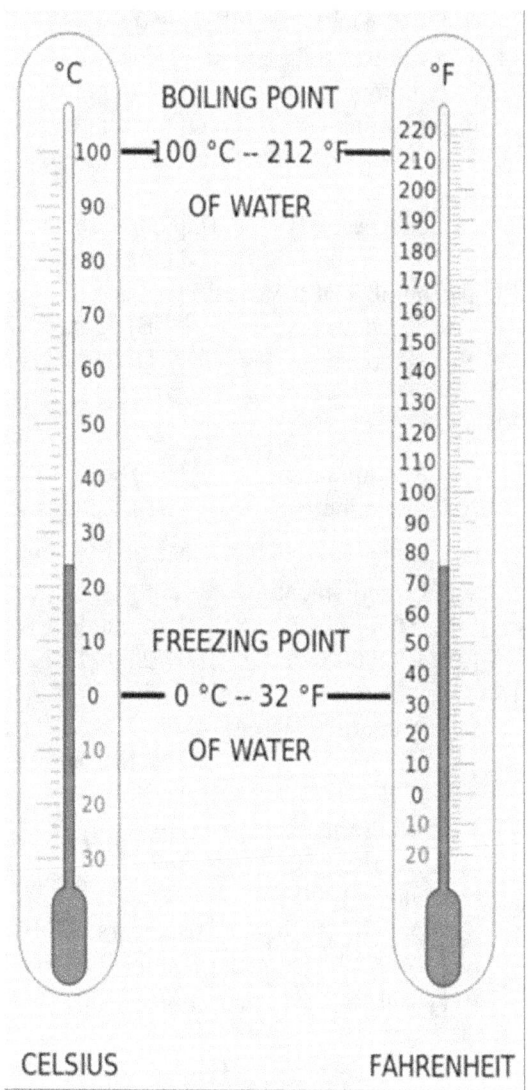

Figure 5:8

Money Measure ## *(The Decimal Currency)*

The currency of many countries is subdivided into units of 100 called cents. Such a currency is known as the decimal currency.
Some decimal currencies are shown in Table 5:9.

Country	Basic unit	Sub unit
Europe	Euro (€)	Centime
Nigeria	Naira (₦)	Kobo (k)
USA	Dollars ($)	Cents (c)
Cameroon	Francs (FCFA)	-

Table 5:9

In each case 100 subunits = 1 basic unit
Amounts less than the basic unit are expressed either as decimals to 2 decimal places or using the sub units. Thus,

$$28 p = £ 0.28$$

$$200 \text{ naira}, 40 \text{ Kobo} = ₦ 200.4$$

$$52 \text{ dollars}, 3 \text{ cents} = \$52.03$$

Conversion of Currency

When travelling from one country to another, it is necessary for one to change the currency, which he has to that of the destination country. International travelers must be able to interconvert currency because exchange rates though they exist, keep on fluctuating from time to time.

Example 5:23
The rate of exchange is such that 1 FF = 120 FCFA and 45 Belgian Franc (BF) = 1 FF. Find in FCFA the cost of a motorcycle, which cost 180,000 BF.

Solution

$$45 \text{ BF} = 1 \text{ FF}$$

$$\therefore 180,000 \text{ BF} = \frac{1}{45} \times 180,000 \text{ FF} = 4,000 \text{ FF}$$

$$1 \text{ FF} = 120 \text{ FCFA}$$

$$\therefore 4,000 \text{ FF} = 4,000 \times 120 \text{ FCFA} = 480,000 \text{ FCFA}$$

EXERCISE 5:8

1. In September 1986, the exchange rates were as follows: $1 = 340 FCFA,

£1 = $1.146. A student was required to pay £2000 as tuition for the 1986/87 academic year in a British university. Determine how much money was required in FCFA.

2. The rate of exchange is such that 1FB = 7.83 FCFA and $1 = 313.52 FB. Find in F**CFA**, to the nearest thousand francs, the cost of a car marked at $1000.

3. An Englishman bought a car from France for 90,000 FF. The exchange rate was then, £1= 11.77 FF. He paid for the car when the exchange rate was £1= 16.68 FF. How much did he lose to the nearest £?

4. State the currency used in the following countries:
   (a) Great Britain   (b) Ghana
   (c) South Africa

MULTIPLE CHOICE EXERCISE 5

1. Four packets have weights marked: 2 kg, 250 g, 500 g, and 3.5 kg .Their total weight in kilograms is:
   [A] 5.25   [B] 6.25   [C] 6   [D] 13

2. 400 grams are taken out of a bucket containing 25 kg of rice. The number of kilograms left in the bucket is:
   [A] 2.46   [B] 24.6   [C] 246   [D] 2460

3. A metre rule was broken into 4 equal pieces. Each piece is:
   [A] 25 cm   [B] 15 cm
   [C] 10 cm   [D] 50 cm

4. 42,600 m expressed in kilometres is:
   [A] 0.426 km   [B] 4.26 km
   [C] 42.6 km   [D] 426 km

5. 650 mm written in metres is:
   [A] 6500 m   [B] 65 m
   [C] 0.065 m   [D] 0.65 m

6. When 48 km is added to 20 dm the result in metres is:
   [A] 48.002 m   [B] 480.02 m
   [C] 4800.02 m   [D] 48002 m

7. The number of square centimetres in a square metre is:
   [A] 1,000 cm^2 [B] 10,000 cm^2
   [C] 100,000 cm^2 [D] 1,000,000 cm^2

8. $\frac{2}{3}$ of 3 hours is:
   [A] 1200 minutes   [B] 120 minutes
   [C] 1200 seconds   [D] 120 seconds

9. The difference between $3x$ minutes and $40x$ seconds is:
   [A] 140 seconds   [B] 140x seconds
   [C] 37 seconds   [D] 37x seconds

10. Bih took 35 minutes to go to school. If she arrived at the school at 7:25 a.m., the time she left her house was:
    [A] 6:50 a.m.   [B] 6:55 a.m.
    [C] 6:45 a.m.   [D] 6:50 a.m.

11. The value of 0.75 hours: 20 minutes is:
    [A] 85 minutes   [B] 75 minutes
    [C] 65 minutes   [D] 55 minutes

12. The number of seconds in 12 minutes 10 seconds is:
    [A] 720 seconds   [B] 730 seconds
    [C] 490 seconds   [D] 630 seconds

13. The number of days in 30 weeks 5 days is
    [A] 305 days   [B] 35 days
    [C] 180 days   [D] 215 days

14. $\frac{3}{8}$ of a day is:
    [A] 324 minutes   [B] 540 seconds
    [C] 540 minutes   [D] 324 seconds

15. Given that 1000 FCFA = ₦ 240. ₦ 30,000 will be equivalent to:
    [A] 250,000 FCFA   [B] 80,000 FCFA
    [C] 125,000 FCFA   [D] 72,000 FCFA

16. Given that $3 = 720 FCFA. 600,000 FCFA in dollars ($) is:
    [A] $2,500   [B] $14,400,000
    [C] $277   [D] $200 00

17. At the current exchange rate, 1 dollar ($) = 500 FCFA and 1 pound = 1000 FCFA.
    200 pounds, exchanged to dollars will be:
    [A] $400   [B] $10.00
    [C] $100   [D] $2.00

TOPIC 6

FINANCIAL ARITHMETIC

OBJECTIVES

At the end of this topic, the learner should be able to:

1. Define the terms, principal, interest, time, rate and amount.
2. Identify the principal, interest, time, rate and amount in problems.
3. Solve problems on simple interest.
4. Distinguish between compound interest and simple interest.
5. Calculate compound interest for at most three successive years.
6. Calculate the depreciation and present value of an article after a period of usage.
7. Solve problems involving percentage change (e.g. error, difference, discount, profit and loss)
8. Determine profit or loss on an article.
9. Calculate profit/loss as a percentage of cost price
10. Manipulate with cost price, selling price, profit and loss.
11. Solve problems involving taxable and in-taxable income.

Most nations have their own system of money and print their own currency. Made of paper, these pieces of currency have very little intrinsic value. As fiat money, however, the paper notes represent a specific monetary value decreed by the government and accepted by the people. The notes pictured here are examples of fiat money from all over the world.

William Taufic/Corbis

Definition of Basic Terms

Principal (P)
Principal is the amount of money, which is invested (lent or borrowed).

Interest (I)
Interest is the charge by an investor or lender on his investment or loans. The one who borrows the money incurs the interest.

Time (T)
Time is the duration or period for which an investment is made.

Rate (R)
Rate is the ratio of the interest to the principal, expressed as a percentage for a given period. Though the period is usually one year (per annum), it could be any other period say per month or per week. For instance, 12% per annum, 5% per month, etc

Amount (A)
Amount is the sum of the principal and the interest after a given period.

$$A = I + P$$

SIMPLE INTEREST

Simple interest is the interest charged on the principal for a given period. With simple interest, unless the investor increases the principal, the principal will remain the same no matter the number of years the investment (or loan) lasts. Interest increases with the principal, the time and the rate.

Thus,

$$I = \frac{PRT}{100}$$

$$\Rightarrow P = \frac{100I}{RT}, T = \frac{100I}{PR}, R = \frac{100I}{PT}.$$

Example 6:1
Calculate the simple interest, which Mrs. Fube will get on 640,000 FRS for 2 years 6 months at the rate of $4\frac{1}{2}$ % per annum.

Solution
$$P = 640000 \text{ FRS}, T = 2\frac{1}{2} \text{ years}, R = 4\frac{1}{2}\%$$

$$I = \frac{PRT}{100} = \frac{640000 \times 4.5 \times 2.5}{100} = 72000 \text{ FRS}.$$

Example 6:2
Mrs. Danjuma paid a simple interest of 132, 000 FRS. for money borrowed at 8% per annum after 3 years. Calculate the sum of money she borrowed.

Solution
$I = 132,000$ Frs., $T = 3$ years, $R = 8\%$ per annum.

$$P = \frac{100I}{RT} = \frac{100 \times 132000}{3 \times 8} = 550,000 \text{ Frs}.$$

Example 6:3
Mr. Jaiy invested 700,000 Frs. at 4% per annum simple interest. How long will the amount reach 784000 Frs.?

Solution
$A = 784,000$ Frs., $P = 784,000$ Frs., $R = 4\%$

$$I = A - P = 784,000 - 700,000 = 84,000 \text{ Frs}.$$

$$T = \frac{100I}{PR} = \frac{100 \times 84,000}{700,000 \times 4} = 3 \text{ years}$$

Example 6:4
At what rate must Pa Fru invest 2,500,000 Frs. for 4 years to yield a simple interest of 500,000 Frs.?

Solution
$I = 500,000$ Frs., $P = 2,500,000$ Frs., $T = 4$ years

$$R = \frac{100I}{PT} = \frac{100 \times 500,000}{2,500,000 \times 4} = 5 \text{ % per annum}.$$

Sometimes the rate is given per annum while the time is given in months or days. In such a case, it is necessary to convert the time to years by dividing by 12 or 365 as the case may be.

Example 6:5
Mr. Nfor invested 700,000 Frs. in a Credit Union for 36 months at 6% per annum. Calculate his simple interest for this period.

Solution

$T = 36 \text{ months} = \dfrac{36}{12} \text{ years} = 3 \text{ years}$

$P = 700,000 \text{ Frs}, R = 6\%$

$I = \dfrac{PRT}{100} = \dfrac{700,000 \times 6 \times 3}{100} = 126,000 \text{ Frs}.$

EXERCISE 6:1

1. Find the simple interest on:
   (a) 300,000 FCFA at 12 % per annum for 5 years.
   (b) 160,000 FCFA at 8 % per annum for 9 months.
   (c) 400,000 FCFA at 10 % per annum for 18 months.
2. What principal will generate an interest of 15,000 FCFA for two months at 9 % per annum simple interest?
3. At what rate per annum can 73,000 FCFA yield an interest of 4800 FCFA for 240 days?
4. How long will 1,000,000 FCFA yield an interest of 40,000 FCFA at 8 % per annum?
5. In Table 6:1, the interest is per annum. Solve for the missing item.

	P (FCFA)	R	T	I (FCFA)
a	500,000	?	6 months	30,000
b	?	7 %	1.5 years	21,000
c	5,000,000	10 %	?	1,000,000

Table 6:1

COMPOUND INTEREST

Compound interest is the money paid on the principal and the interest accumulated from the past years or months as the case may be. While the simple interest remains the same over the whole period, the compound interest increases from year to year as the amount increases.

Progressive Method for Compound Interest

To calculate compound interest using the progressive method, the principal at the beginning of each year is calculated progressively and used to compute the interest. The compound interest is then the sum of all the interest for all the years.

Example 6:6
Mr. Mofor invested the sum of 3,000,000 Frs. for 3 years at 10% per annum compound interest. Calculate his interest at the end of the 3 years.

Solution
Let the principals for first, second and third years be $P_1, P_2, P_3,$ and interest for first, second and third years be I_1, I_2 and I_3 respectively.
Then,

$R = 10\%., \quad P = 3,000,000 \text{ Frs}, \quad I = PRT = \dfrac{10}{100}P$

$P_1 = 3,000,000 \text{ Frs} \Rightarrow I_1 = \dfrac{10 \times 3,000,000}{100} = 300,000 \text{ Frs}.$

$P_2 = P_1 + I_1 = 3,000,000 + 300,000 = 3,300,000 \text{ Frs}.$

$\Rightarrow I_2 = \dfrac{10 \times 3,300,000}{100} = 330,000 \text{ Frs}.$

$P_3 = P_2 + I_2$
$= 3,300,000 + 330,000$
$= 3,630,000 \text{ Frs}$

$\Rightarrow I_3 = \dfrac{10 \times 3,630,000}{100}.$
$= 363,000 \text{ Frs}$

Compound interest $= I_1 + I_2 + I_3$
$= 300,000 + 330,000 + 363,000$
$= 993,000 \text{ Frs}.$

Alternatively,
The amount after three years is
$A_3 = I_3 + P_3 = 3,630,000 + 363,000$
$\Rightarrow A_3 = 3,993,000 \text{ Frs}.$
Compound interest $I = A - P$
$= 3,993,000 - 3,000,000$
$= 993,000 \text{ Frs}.$

The Compound Interest Formula

Suppose that an amount P is invested at the rate of r % per annum. Then the amount after t years will be given by:

$$A = P\left(1 + \dfrac{r}{100}\right)^t \text{ and } I = A - P$$

Example 6:7
Solve Example 6:6 using the compound interest formula.

Solution

$$A = P\left(1 + \frac{r}{100}\right)^t$$

$$= 3,000,000\left(1 + \frac{10}{100}\right)^3$$

$$= 3,000,000(1.1)^3$$

$$= 3,993,000 \text{ Frs.}$$

Compound interest $I = A - P$
$$= 3,993,000 - 3,000,000$$
$$= 993,000 \text{ Frs.}$$

Compound Interest with Varying Principal

Sometimes, the principal increases due to loan addition or decreases due to partial repayment. In this case, the progressive method can be used and at each instance that a loan addition or a partial repayment is made, the amount is added to or subtracted from the total amount before calculating the interest.

Example 6:8
Mr. Nformi borrows 6,000,000 FCFA from his Credit Union at 3% per annum compound interest. He repays 2,000,000 FCFA at the end of each year. How much does he still owe at the end of the third repayment?

Solution

Amount borrowed = 6,000,000 FCFA

$$I_1 = \frac{3}{100} \times 6,000,000 = 180,000 \text{ FCFA}$$

Amount owed at end of year 1 = 6,180,000 FCFA
First repayment = 2,000,000 FCFA
Amount owed after first repayment
$$= 4,180,000 \text{ FCFA}$$

$$I_2 = \frac{3}{100} \times 4,180,000 = 125,400 \text{ FCFA}$$

Amount owed at end of year 2 = 4,305,400 FCFA
Second repayment = 2,000,000 FCFA
Amount owed after 2nd repayment
$$= 2,305,400 \text{ FCFA}$$

$$I_3 = \frac{3}{100} \times 2,305,400 = 69,162 \text{ FCFA}$$

Amount owed at end of year 3
$$= 2,374,562 \text{ FCFA}$$
Third repayment = 2,000,000 FCFA
Amount owed after 3rd repayment
$$= 374,562 \text{ FCFA}$$

Therefore, the amount owed at the end of the third year is 374,562 FCFA.

EXERCISE 6:2

1. Mr. Ngala borrowed 2,510,000 FCFA from his credit union at 10 % per annum compound interest. How much will he pay back at the end of 3 years?
2. Calculate the compound interest on 500,000 FCFA for 2 years at 6 % per annum.
3. Mrs. Teboh took a building loan of 7,000,000 FCFA. She is charged a compound interest of 8% per annum. She repairs 960,000 FCFA at the end of each year. How much does she still owe at the end of 3 years?
4. Miss Bih invests 100,000 FCFA on first January of each year at 5% per annum compound interest. Find to the nearest franc the total amount of her investment at the end of 3 years.
5. Mr. Ndumbe borrows 7,000,000 FCFA at 4% per annum compound interest. He repays 2,000,000 FCFA each year. In how many years will the loan be cleared?
6. A man invest 500,000 FCFA each year at 4% per annum compound interest. Calculate his investment just before he invest the fourth time.
7. A woman borrows 6,000,000 FCFA from a credit union at 3% per annum. She repays 2,000,000FCFA at the end of each year. Calculate her debt after the third repayment.

Depreciation

Assets such as equipment or buildings lose some of their market value as time passes. This loss in value is called **depreciation**. Suppose an asset whose original value is P,

depreciates at the rate of r % per annum then its value V, after t years will be given by
The depreciation after this period will then be given by

$$V = P\left(1 - \frac{r}{100}\right)^t \text{ and } D = P - V$$

Example 6:9
A car, which was bought at 6,000,000, FCFA depreciates at the rate of 10% per annum. What will be the value of this car after 3 years?

Solution

$$\begin{aligned} V &= P\left(1 - \frac{r}{100}\right)^t \\ &= 6,000,000\left(1 - \frac{10}{100}\right)^3 \\ &= 3,000,000(0.9)^3 \\ &= 4,374,000 \text{ Frs.} \end{aligned}$$

This problem can also be solved using the progressive method as follows.
Let V_1, V_2, and V_3, be the values after the first, second and third years. Then

$$V_1 = 6,000,000\left(\frac{90}{100}\right) = 5,400,000 \text{ FCFA}$$

$$V_2 = 5,400,000\left(\frac{90}{100}\right) = 4,860,000 \text{ FCFA}$$

$$V_3 = 4,860,000\left(\frac{90}{100}\right) = 4,374,000 \text{ FCFA}$$

Therefore, the value after 3 years is 4,374,000 FCFA.

EXERCISE 6:3

1. A car bought at 5,000,000 FCFA depreciates at the rate of 20 % per annum. Find the cost of the car at the end of the 4th year.
2. A machine cost 160,000 FRS. If this machine depreciates by $\frac{1}{5}$ of its value each year, what will be its value at the end of the third year?

Percentage Change (Difference)

Percentage change is the ratio of a change in value to the original value expressed as a percentage. There are so many different types of percentage changes and all these are calculated using the following formula.

$$\text{Percentage change} = \frac{\text{Change in value}}{\text{Original value}} \times 100\%$$

Some common types of percentage changes are discount, percentage profit (gain) or loss, percentage error, percentage passed or failed, percentage devaluation, percentage increase or decrease etc.

Example 6:10
A student measured the length of a pole as 34.5 m instead of 34.8 m. Calculate the percentage error in the measurement of the pole.

Solution

$$\text{Percentage error} = \frac{\text{Difference}}{\text{Real value}} \times 100\%$$

$$\Rightarrow \text{Percentage error} = \frac{34.8 - 34.5}{34.8} = 0.86\%$$

Example 6:11
In 1985 the number television owners in a certain town was 3000. In 1995 the number increased by 80%. How many people had televisions in 1995?

Solution

$$\text{Percentage increase} = \frac{\text{increase}}{\text{original}} \times 100\%$$

$$80\% = \frac{\text{No. in 1995} - 3000}{3000} \times 100\%$$

$$\Rightarrow \text{No. in 1995} = 80 \times 30 + 3000 = 5400 \text{ people.}$$

Alternatively,

$$\text{Increase in TV owners} = \frac{80}{100} \times 3000 = 2400$$

$$\therefore \text{No. in 1995} = 3000 + 2400 = 5400 \text{ people.}$$

Or % of TV owners in 1995 = 100 + 80 = 180%
No. of TV owners in 1995= $\frac{180}{100} \times 3000$
= 5400 people.

BUYING AND SELLING

Profit And Loss

When a businessperson buys goods for resale, the price at which the goods are bought is called the **cost price** while the price at which the goods are sold is called the **selling price**. Though his objective is to sell the goods at a selling price, which is higher than the cost price, he might end up selling them at a selling price, which is less than the cost price.

If his selling price is more than the cost price, he is said to have made a **profit** or **gain**. Thus,

> Profit = Selling Price – Cost Price

If his selling price is less than the cost price, he is said to have made a **loss**.

> Loss = Cost Price – Selling Price

Example 6:12
A dealer bought a car at 3,200,000 Frs. and sold it at 3,900,000 Frs. What was his profit?

Solution
Profit = selling price – cost price
$\Rightarrow$ Profit = 3,900,000 – 3,200,000
$\qquad$ = 700,000 Frs.

Example 6:13
A trader bought 15 bags of rice at 12,500 Frs. each. If she sold all these bags for 180,000 Frs., what profit or loss, did she make?

Solution
Total cost price for 15 bags = 12,500 $\times$ 15
$\qquad\qquad\qquad\qquad$ = 187,500 Frs.
Total selling price for 15 bags = 180,000 Frs.

Cost price is greater than the selling price so she made a loss.
$\qquad$ Loss = cost price – selling price
$\Rightarrow$ Loss = 187,500 – 180,000 = 7,500 Frs.

Percentage Profit And Loss
The percentage profit or loss is the ratio of the profit or loss to the cost price expressed as a percentage.

Let the cost price be CP and the selling price be SP.

$$\text{Percentage profit} = \frac{\text{profit}}{CP} \times 100\% = \frac{SP - CP}{CP} \times 100\%$$

$$\text{Percentage loss} = \frac{\text{loss}}{CP} \times 100\% = \frac{CP - SP}{CP} \times 100\%$$

It follows that:

(1) A profit of p % means the SP is $(100 + p)$ % of the CP. In this case,

$$SP = \frac{(100 + p)}{100} \times CP \text{ and } CP = \frac{100}{(100 + p)} \times SP$$

(2) A loss of l % means the SP is $(100 - l)$ % of the CP. In this case,

$$SP = \frac{(100 - l)}{100} \times CP \text{ and } CP = \frac{100}{(100 - l)} \times SP$$

Example 6:14
A Woman bought a car at 4,500,000 Frs. and sold it at a profit of 15%. What was its selling price?

Solution
$$SP = \frac{100 + p}{100} \times CP$$

$$\Rightarrow SP = \frac{100 + 15}{100} \times 4,5500,000 = 5,175,000 \text{ Frs.}$$

Example 6:15
A man sold a house for 3,600,000 Frs., incurring a loss of 40 %. Calculate the amount he paid for the house.

Solution
$$CP = \frac{100}{(100 - l)} \times SP$$

$$\Rightarrow CP = \frac{100}{(100 - 40)} \times 3,600,000 = 6,000,000$$

$\therefore$ Amount he paid = 6,000,000 Frs.

Example 6:16

A trader bought a television at 250,000 Frs. and sold it for 280,000 Frs. Calculate his percentage profit or loss.

Solution

He made profit because his selling price is higher than the cost price.

$$\text{Percentage profit} = \frac{SP - CP}{CP} \times 100\%$$

$$= \frac{280,000 - 250,000}{250,000} \times 100\%$$

$$= 12\%$$

EXERCISE 6:4

1. In 2006, the population of a certain village was 12,500 people. In 2009, the population was 15,500 people. Find the percentage increase in the population of the village.

2. A businessperson had 720 customers last year. Due to poor management, the number of customers reduced to 600 this year. Calculate the percentage decrease in the number of customers.

3. An article costing 24,000 FCFA depreciated to 17,000 FCFA at the end of the year. By what percentage did it depreciate?

4. To convert 4 km/h to m/s, a student obtained an answer 1 m/s. Calculate his percentage error.

5. After a slim course the weight of a woman who originally weighed 89 kg 600 g, reduced to 78 kg 400 g. Calculate the percentage decrease in her weight.

6. The enrolment of a school in 1995 was 18,525 students and in 2005, it was 22,750 students. Calculate the percentage increase in the population of the school for this period correct to 1 decimal place.

7. During a Physics practical, a student measured the length of a wire as 18 mm instead of 20 mm.
What is his percentage error in the measurement of the wire?

8. Yaje bought a bag of huckleberry at 6,000 FCFA and retailed it for 5,800 FCFA. Calculate her percentage profit or loss.

9. The enrolment of a school increased by 10 %. If the enrolment was originally 3000 students, how many new students were admitted?

10. A man bought a goat for 20,000 FCFA and sold it for 22,000 FCFA. What was his percentage profit?

11. A shoe dealer bought a pair of shoes for 25,000 FCFA and sold it for 21,000 FCFA. What is his percentage loss?

12. Mr. Tamfu bought a car for 6,000,000 FCFA and later sold it at a loss of 6 %. Calculate the selling price of the car.

13. A woman buys a house for 5,000,000 FCFA and intends to make a profit of 15 %. Given that, she made repairs up to the cost of 200,000 FCFA, how much must she sell the house?

14. By selling a raffia bag for 3,600 FCFA a man makes a profit of 80 %. How much did he pay for the bag?

15. An article bought for 1,000 FCFA was sold at a profit of 30 %. What was the selling price?

DISCOUNT

Sometimes traders decide to encourage their customers by reducing a certain amount of money from the selling price, because the customer has paid cash or has paid at a stipulated time. This reduction is called **discount**. Discount is usually expressed as a percentage of the normal price or marked price. A discount of 7 % means that the customer pays only 93 % of the selling price.

$$\text{Discount} = \text{Marked price} - SP$$

$$\text{Percentage discount} = \frac{\text{Marked price} - SP}{\text{Marked price}} \times 100\%$$

Example 6:17

A shirt marked 2800 Frs. was sold at a discount of 5 %. What was its selling price?

Solution

$$SP = \% \text{ discount} \times \text{marked price}$$

$$\Rightarrow SP = \frac{95}{100} \times 2800 = 2660 \text{ Frs.}$$

Example 6:18
A pair of shoes has a marked price of 16000 Frs. A customer buys it at 15000 FRS, what is his percentage discount?

Solution

Discount = 16000 −15000 Frs. =1000 Frs.

$$\text{Percentage discount} = \frac{\text{Discount}}{\text{Marked price}} \times 100\%$$

$$\Rightarrow \text{Percentage discount} = \frac{1000}{16000} \times 100 = 6.25\%$$

Example 6:19
A book is sold at 7200 Frs. Given that, the discount on the book is 25 %. Calculate the marked price.

Solution

$$\text{Marked price} = \frac{\text{Percetage discount}}{100} \times SP$$

$$\Rightarrow \text{Marked price} = \frac{75}{100} \times 7,200 = 9,600 \ FRS$$

> **EXERCISE 6:5**

1. Awa and Sons Enterprise decide to offer a discount of 30 % on all goods bought from them. What will be the selling price of a camera marked 120,000 FCFA?

2. A pair of shoes marked 10,000 FCFA was sold at 8,000 FCFA during an auction sale. Find the rate of discount as a percent.

3. A suit marked 32,000 FCFA was sold at 24,000 FCFA. Find the rate of discount as a percent.

TAXES

Every government has an obligation to render certain services to its citizens. Such services include security, health, education, road maintenance etc. In order to undertake this task, government raises money through many sources one of which is taxes. A tax is an amount of money levied by a government or association on its citizens or members and used to run the government, the country, a state, a county, a municipality or association. Some types of taxes common in most countries are the income tax, land tax, cattle tax, business tax, custom duty, purchase tax, value added tax, etc.

Calculation of Taxes

Taxes are usually calculated as a percentage of the earnings of each citizen for a certain period usually a month or a year.

$$\text{Tax} = \frac{\text{Tax rate}}{100} \times \text{earnings}$$

Example 6:20
A man earns 136,000 FRS. per month. If he pays a tax of $3\frac{1}{2}\%$ of his salary, calculate his tax in FRS.

Solution

$$\text{Tax} = \frac{\text{Tax rate}}{100} \times \text{earnings}$$

$$\Rightarrow \text{Tax} = \frac{3.5}{100} \times 136,000 = 4760 \ FRS.$$

Taxable and Nontaxable Income

Civil servants and state agents often receive certain allowances. Such allowances include family allowance, rent allowance, research allowance etc. Most governments do not impose taxes on these allowances. Therefore, these allowances are usually subtracted before the remainder is taxed.

Example 6:21
A civil servant has a monthly salary of 189,000 Frs. If his allowances total 42,000 Frs. Calculate his income tax given that, the rate is 21 % and his allowances are not taxed.

Solution

Taxable income = 189,000 −42,000
= 147,000 Frs.

$$\text{Tax} = \frac{\text{Tax rate}}{100} \times \text{earnings}$$

$$\Rightarrow \text{Tax} = \frac{21}{100} \times 147,000 = 30870 \ FRS.$$

EXERCISE 6:6

1. A woman's monthly salary is 250,000 FCFA. Given that, a tax of 8 % is imposed on her. Calculate,
   (a) The amount of tax she pays per year.
   (b) Given that she saves 15 % of the remainder, what amount will be recorded in her savings passbook at the end of the year?
2. A civil servant earns a monthly salary of 280,000 FCFA allowances inclusive. Given that, 20 % of his pay package is nontaxable allowances and that a tax of 12% is imposed on the rest. Calculate his net monthly salary.

MULTIPLE CHOICE EXERCISE 6

1. The cost of 4 articles at 720 FCFA each is:
   [A] 2920 FCFA [B] 2880 FCFA
   [C] 2820 FCFA [D] 2900 FCFA
2. The cost of 2.5 kg of tomatoes at 360 FCFA per kg is:
   [A] 900 FCFA [B] 860 FCFA
   [C] 1100 FCFA [D] 720 FCFA
3. The cost of 2 metres of material at 1200 FCFA per metre and 3 metres at 1500 FCFA per metre is:
   [A] 6900 FCFA [B] 13500 FCFA
   [C] 6600 FCFA [D] 7100 FCFA
4. The cost of 2000 articles at 25 FCFA each is:
   [A] 5,000 FCFA [B] 25,000 FCFA
   [C] 50,000 FCFA [D] 500,000 FCFA
5. The change from 1000 FCFA after buying 18 buttons at 5 FCFA each is:
   [A] 990 FCFA [B] 890 FCFA
   [C] 810 FCFA [D] 910 FCFA
6. The cost of 200 articles at 30 FCFA each is:
   [A] 7000 FCFA [B] 6000 FCFA
   [C] 5000 FCFA [D] 5700 FCFA
7. While doing his Physics practical, Che recorded a reading as 1.12 cm, instead of 1.21 cm. The percentage error he made is:
   [A] 1.17% [B] 6.38%
   [C] 7.44% [D] 8.5%

8. Miss Yaje sold an article for 7500 FRS instead of 12750 FRS. Her percentage error correct to one decimal place is:
   [A] 41.2% [B] 5.3%
   [C] 1.7% [D] 1.4%
9. A student measured the length and breadth of a rectangular lawn as 59.6 cm and 40.3 cm respectively instead of 60 cm and 40 cm. The percentage error in his calculation of the perimeter of the lawn is:
   [A] 1.4% [B] 0.1%
   [C] 0.2% [D] 0.7%
10. A boy estimated his transport fare for a journey as 1900 FRS instead of 2000 FRS. The percentage error in his estimate is:
    [A] 95% [B] 47.5%
    [C] 5.26% [D] 5%
11. To two significant figures, the percentage error in approximating 0.375 to 0.4 is:
    [A] 6.7% [B] 6.6%
    [C] 2.5% [D] 2.0%
12. A furniture maker estimated that the cost of making a cupboard would be about 250000 FRS. He bought the materials and found that the cost came to 275000 FRS. The percentage increase in the estimate is:
    [A] 5% [B] 10%
    [C] 15% [D] 20%
13. The word discount means:
    [A] Money reduced from the price of a hired article.
    [B] Money taken out of your wage.
    [C] Money taken off the price of an article.
    [D] Money owed to a businessperson.
14. During a sale, a shop reduced the price of every article by 10%. The selling price of an article originally priced at 4300 FCFA is:
    [A] 4300 FCFA [B] 3400 FCFA
    [C] 3870 FCFA [D] 3970 FCFA
15. For his holidays, a man put aside 10% of his 15000 FCFA weekly wage for 40 weeks in the year. The amount saved for his holidays is:
    [A] 60000 FCFA [B] 30000 FCFA
    [C] 15000 FCFA [D] 150000 FCFA
16. A businessperson decided to give 10 % discount on all the purchases from his store. The price a customer pays for a shirt that was marked at 5,400FRS is:

[A] 5,940 FCFA   [B] 5,500 FCFA
[C] 5,300 FCFA   [D] 4,860 FCFA

17. A girl bought a record for 1500 FCFA and sold it for 1200 FCFA. Her loss as a percentage of the cost price is:
[A] 15%   [B] 20%
[C] 60%   [D] 75%

18. A woman bought a dress for 15000 FCFA. To make a profit of 20% the dress should be sold at:
[A] 19,000 FRS   [B] 18,000 FRS
[C] 16,000 FRS   [D] 14,400 FRS

19. A trader made a profit of 50% on his cost price by selling a radio at 15,000 FCFA. The record cost him:
[A] 5,000 FRS   [B] 7,500 FRS
[C] 10,000 FRS   [D] 25,000 FRS

20. This year the budget of a development association decreased by 4 % given that its annual budget for last year was 32,000,000 FRS. Its budget for this year is:
[A] 30,720,000 FRS   [B] 25,000 FRS
[C] 33,280,000 FRS   [D] 2,240,000 FRS

21. A house bought for 1,000,000 FRS was later auctioned for 800,000 FRS. The percentage loss was:
[A] 20%   [B] 30%
[C] 40%   [D] 25%

22. By selling some crates of soft drinks for 6000 FRS, a dealer makes a profit of 50 %. The amount, which the dealer pays for the drinks, is:
[A] 12,000 FRS   [B] 25,000 FRS
[C] 4,500 FRS   [D] 4,000 FRS

23. A trader makes a loss of 15 % when selling an article. The ratio selling price: cost price is:
[A] 3:20   [B] 3:17
[C] 17:20   [D] 20:23

24. A man made a loss of 15 % by selling an article for 59500 FRS. The cost price of the article was:
[A] 60,000 FRS   [B] 70,000 FRS
[C] 68,425 FRS   [D] 89,250 FRS

25. Mr. Anyang bought a piece of land for 2.5 million FCFA and sold it for 3 million FCFA. His percentage profit is:
[A] $18\frac{2}{3}$ %   [B] 20 %   [C] 16.7 %   [D] 25 %

26. If the simple interest on 200,000 FRS after 9 months is 6000 FRS, the interest rate per annum is:

[A] $2\frac{1}{4}$ %   [B] 6 %   [C] 5 %   [D] 4 %

27. A cooperative society charges an interest of $5\frac{1}{2}$ % per annum on any amount borrowed by its members. If a member borrows 125000 FRS, the amount he pays back after 1 year will be:
[A] 136,875 FRS   [B] 131,875 FRS
[C] 128,750 FRS   [D] 126,250 FRS

28. The compound interest on 400,000 FRS for 2 years at 8 % per annum is:
[A] 32,000 FRS   [B] 34,560 FRS
[C] 66,560 FRS   [D] 43,200 FRS

29. A car brand under usage depreciates at the rate of 10 % per annum. Given that, a new brand of this car cost 4,000,000 FRS. The cost of such a car that has been used for 3 years will be:
[A] 3,440,000 FRS   [B] 2,916,000 FRS
[C] 3,600,000 FRS   [D] 3,240,000 FRS

30. The exchange rate for FCFA is 1000 FCFA for 238 Naira. 2,856,000 FCFA changed into Naira would be:
[A] 12   [B] 1200
[C] 679728   [D] 6797280

31. A computer can be bought for 189500 FCFA cash or on hire purchase for 9 monthly payments of 23700 FCFA each. The hire purchase method cost is greater by:
[A] 22,100 FCFA   [B] 23,800 FCFA
[C] 29,500 FCFA   [D] 33,200 FCFA

32. A car worth 10,000,000 FCFA depreciates by 10% yearly. The value after 2 years will be:
[A] 1,200,000 FCFA   [B] 8,000,000 FCFA
[C] 9,000,000 FCFA   [D] 8,100,000 FCFA

TOPIC 7

DIRECTED NUMBERS

OBJECTIVES

At the end of this topic, the learner should be able to:

1. Read and write negative numbers and "not minus numbers".
2. Represent real numbers on the number line.
3. Identify the relative position of real numbers on the number line.
4. Determine the absolute value of any real number.
5. Determine if a given number is greater than, less than or equal to another given number.
6. Compare real numbers using <, >, =, ≥ or ≤.
7. Evaluate numerical expressions involving directed numbers.
8. Perform in correct order mixed directed number operations involving multiplication, division, addition and subtraction with real results.

Mount Everest, the highest mountain in the world, with a height of 8,850 m above sea level, rises in the Himalayas on the frontier of Nepal and Tibet. Numerous groups tried to reach the summit before the successful attempt by two members of a British expedition on May 29, 1953.

Keren Su/Tony Stone Images

The lowest point in the western hemisphere, 86 m below sea level, is located in Death Valley, California. The valley is part of the Great Basin in the western United States.

ML Sinibaldi/Corbis

Meaning of Directed Numbers

Directed numbers are numbers with a positive (+) or negative (−) sign attached to them.
Examples are   −12, +2, +4.5, −5.2, −π.
   +2 is read 'positive two' <u>not</u> 'plus two',
   −12 is read 'negative twelve' <u>not</u> 'minus twelve'.

A directed number with a (+) sign is called a positive number, while a directed number with a (−) sign is called a negative number. Below are examples of situations where directed numbers are used.

Some Practical Uses of Directed Numbers

1. The Celsius or Centigrade Thermometer

Temperatures above freezing point are *positive*. Hence (+) indicates a temperature which is above the freezing point while (−) indicates a temperature which is below the freezing point.

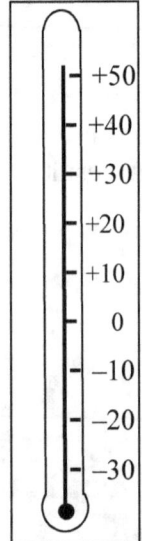

Figure 7:1

2. Distances

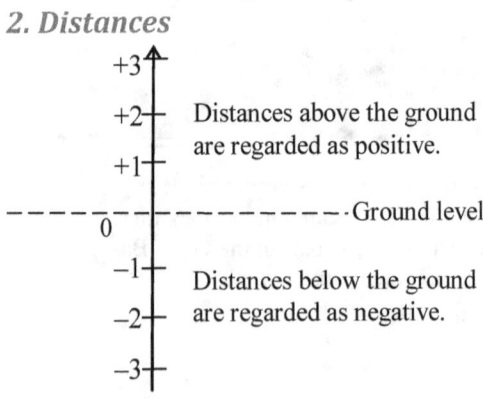

Figure 7:2

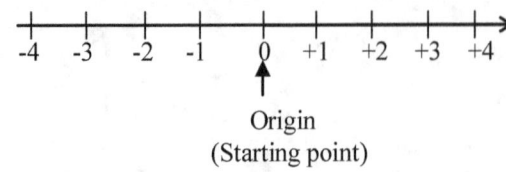

Origin
(Starting point)

Figure 7:3

Distances to the left are considered negative (−) while distances to the right are considered positive (+).

3. Banking-Saving and Borrowing

In banking, savings (deposits) are regarded as positive while loans (withdrawals) are regarded as negative. It is often very useful when adding and subtracting directed numbers to use this analogy thereby thinking of positive number as savings and negative numbers as a loans.

The Absolute Value of a Real Number

The absolute value of a real number usually written by placing the number between vertical bars is the value of the number when the sign is ignored. Thus, the absolute value of +3 is 3 and that of −3 is 3.

   i.e. $|-3| = 3$, $|+3| = 3$

Example 7:1

Find the absolute value of each of the following.
(a) +8   (b) −17   (c) +100
(d) − 120   (e) 0   (f) −π

Solution

(a) $|+8| = 8$ (b) $|-17| = 17$ (c) $|+100| = 100$

(d) $|-120| = 120$ (e) $|0| = 0$ (f) $|-\pi| = \pi$

Algebra Tiles

*Algebra tiles are t*iles labeled ⊕ and ⊖ (where ⊕ = +1 and ⊖ = −1). Combining ⊕ and ⊖ gives ⊕⊖ whose sum is zero. ⊕⊖ is called a zero pair.

OPERATIONS WITH DIRECTED NUMBERS

Many students at the early stage find many difficulties to manipulate directed numbers. However, these difficulties can be eliminated by the use of a number line, the savings and loans analogy or algebra tiles. After enough skills and facility have been developed, the way the numbers are manipulated will not be necessary and can be aborted at that stage.

Addition and Subtraction of Directed Numbers

Adding Two Positive Numbers

Example 7:2
Evaluate +5 + (+3) using
(a) A number line.
(b) The savings and loans analogy.
(c) Algebra tiles.

Solution
(a) **Using the number line.**
  Count +5 and from +5 count 3 units to the right.

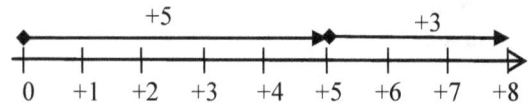

Figure 7:4

∴ +5 + (+3) = +8

(b) **Using the savings and loans analogy**

  If one saves 5 and saves 3, the net amount in his account will be 8.
  ∴ +5 + (+3) = +8

(c) **Using algebra tiles**

  ⊕⊕⊕⊕⊕ + ⊕⊕⊕ = ⊕⊕⊕⊕⊕⊕⊕⊕
  ∴ +5 + (+3) = +8

In general, *the sum of two or more positive numbers is a positive number.*

Adding two negative numbers

Example 7:3
Find the sum −2 + (−3) using the three methods introduced in Example 7:1.

Solution
(a) **Using the number line,**
  Count −2 and from −2 count 3 units to the left.

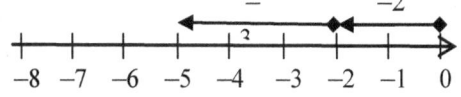

Figure 7:5

∴ −2 + (−3) = −5

(b) **Using the savings and loans analogy**

  If one loans 2 and loans 3 the net loan will be 5.
  ∴ −2 + (−3) = −5

(c) **Using algebra tiles**

  ⊖⊖ + ⊖⊖⊖ = ⊖⊖⊖⊖⊖
  ∴ −2 + (−3) = −5

In general, *the sum of two or more negative numbers is a negative number.*

The two conclusions above can be combined as follows,

To add several numbers with the same sign, ignore the signs and add the numbers together. The sign of the sum is the same as the sign of each of the numbers.

Note that a + sign which stands for positive is usually omitted. Thus, except for emphasis, +5, is written as 5 and +5 + (+3) simply as 5 + 3.

Adding a Positive number and a Negative number

Example 7:4
Find the value of +5 + (−2) using the three methods.

Solution
(a) **Using the number line,**
  From +5 count 2 units to the left.

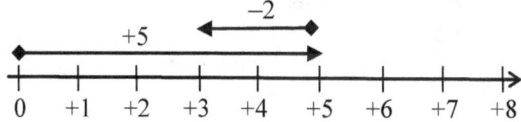

Figure 7:6

∴ +5 + (−2) = +3

(b) **Using the savings and loans analogy**
  If one saves 5 and loans 2, the net amount in his account will be 3.
  ∴ +5 + (−2) = +3

(c) **Using algebra tiles**

  ⊕⊕⊕⊕⊕ + ⊖⊖ =   ⊕⊖⊕⊖   + ⊕⊕⊕
               (Zero pairs = 0)
  ∴ +5 + (−2) = +3

Example 7:5
Find − 4 + (+3) using the three methods.

Solution

(a) Using the number line,
   From -4 count 3 units to the right.

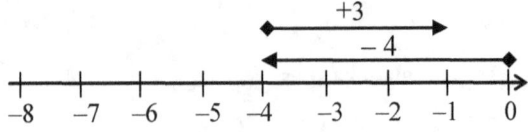

Figure 7:7

$\therefore -4 + (+3) = -1$

(b) Using the savings and loans analogy
   If one loans 4 and saves 3, he owes 1.
   $\therefore -4 + (+3) = -1$

(c) Using algebra tiles

$\ominus\ominus\ominus\ominus + \oplus\oplus\oplus = \ominus\oplus\ \ominus\oplus\ \ominus\oplus\ +\ \ominus$
   (Zero pairs = 0)
   $\therefore -4 + (+3) = -1$

Example 7:6
Evaluate $+4 + (-7)$ using the three methods.

Solution

(a) Using the number line,

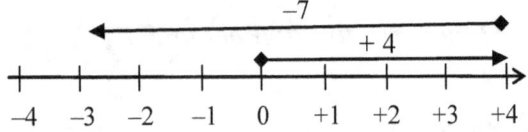

Figure 7:8

$+4 + (-7) = -3$

(b) Using the savings and loans analogy
   A saving of 4 and a loan of 7 is a debt of 3.
   $\therefore +4 + (-7) = -3$

(c) Using algebra tiles
$\oplus\oplus\oplus\oplus + \ominus\ominus\ominus\ominus\ominus\ominus\ominus = \oplus\ominus\ \oplus\ominus\ \oplus\ominus\ \oplus\ominus\ + \ominus\ominus\ominus$
   (Zero pairs = 0)
   $\therefore +4 + (-7) = -3$

Example 7:7
Determine the value of $-1 + (+4)$.

Solution

(a) Using the number line,

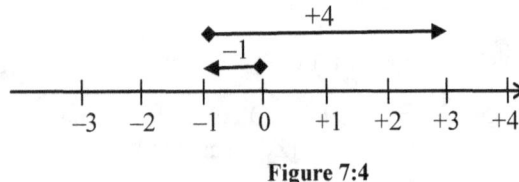

Figure 7:4

$\therefore -1 + (+4) = +3$

(b) Using the savings and loans analogy
   A loan of 1 and a saving of 4 is a saving of 3.
   $\therefore -1 + (+4) = +3$

(c) Using algebra tiles

$\ominus + \oplus\oplus\oplus\oplus = \ominus\oplus + \oplus\oplus\oplus$
   $\therefore -1 + (+4) = +3$

Notice that in adding a positive and negative number the sign of the result is always the sign of the number with the larger absolute value.

In general, *to add a positive and a negative number, subtract the number with the smaller absolute value from the one with the larger absolute value and retain the sign of the number with the larger absolute value.*

If there are more than two directed numbers to be added, first combine numbers with like sign.

Example 7:8
Evaluate $+5 + (-6) + (-8)$

Solution
$+5 + (-6) + (-8) = +5 + (-14) = -9$

Example 7:9
Compute $-7 + (-6) + (-4)$

Solution
$-7 + (-6) + (-4) = -13 + (-4) = -17$

When a (+) sign and a (−) sign occur together, the (−) sign takes preference.

Example 7:10
What is the value of $3 + (-4) - (+5)$?

Solution
$3 + (-4) - (+5) = 3 - 4 - 5 = 3 - 9 = -6$

Though the examples above involved only integers, the principles apply to all real numbers.

Example 7:11

Evaluate $+2\frac{3}{4} + \left(-5\frac{1}{2}\right)$.

Solution

$+2\frac{3}{4} + \left(-5\frac{1}{2}\right) = -\left(\left|-5\frac{1}{2}\right| - \left|+2\frac{1}{2}\right|\right) = -\left(5\frac{1}{2} - 2\frac{3}{4}\right) = -2\frac{3}{4}$

Some Order Symbols

Recall that,

< means 'is less than' e.g. $-4 < 3$

> means ' is greater than'. e.g. $7 > 2$

≤ means 'is less than or equal to' e.g. $-2 \leq 5$

≥ means 'is greater than or equal to' e.g. $9 \geq -5$

EXERCISE 7:1

1. State the absolute value of each of the following.
   (a) $+20$ (b) -13 (c) $+5.8$ (d) π
2. Fill the blank spaces with the symbol >, < or =.
   (a) $4\_\_\_12$ (b) $-7\_\_\_5$

   (c) $|-16|\_\_14$ (d) $14\_\_-6$

   (e) $7.2\_\_|+7.2|$ (f) $-8\_\_-11$

   (g) $|-1|\_\_|+1|$ (h) $|+9|\_\_|-9|$

   (i) $|-5|\_\_|+3|$ (j) $\frac{3}{7}\_\_\left|-\frac{3}{7}\right|$

   (k) $\left|\frac{1}{2}\right|\_\_\left|-\frac{2}{3}\right|$ (l) $\left|-\frac{2}{9}\right|\_\_\left|-\frac{3}{7}\right|$

3. Answer true or false.
   (a) $2 < 8$ (b) $-5 > 2$ (c) $-5 \leq 5$
   (d) $7 > -4$ (e) $3.6 \geq 3.6$ (f) $-7 < -9$
   (g) $|-4| < |+4|$ (h) $|-2| = |+2|$ (i) $|-2| \leq |+2|$
   (j) $\frac{5}{2} > \left|-\frac{5}{2}\right|$ (k) $\left|\frac{12}{5}\right| \geq \left|-\frac{12}{5}\right|$ (l) $\left|-\frac{3}{4}\right| \leq \left|\frac{3}{4}\right|$

4. Evaluate the following.
   (a) $(+20)+(-7)$ (b) $(-17)+(+3)$
   (c) $(-5)+(-15)$ (d) $(+5)+(-15)$
   (e) $(+8)+(+20)$ (f) $(-8)+(+8)$
   (g) $\left(+10\frac{1}{2}\right)+\left(7\frac{1}{2}\right)$ (h) $(-1.4)+(-2.5)$
   (i) $\left(+6\frac{1}{4}\right)+\left(-3\frac{1}{2}\right)$ (j) $(-0.23)+(+0.18)$
   (k) $\left(-7\frac{3}{4}\right)+\left(+4\frac{1}{2}\right)$ (l) $(-5.5)+(+8.2)$

5. Evaluate the following.
   (a) $(+5)+(+3)+(+12)$
   (b) $(-1)+(-3)+(-2)$
   (c) $(+23)+(-13)+(+12)$
   (d) $(+5)+(-8)+(+17)$
   (e) $(+12)+(-12)+(+19)$
   (f) $(-42)+(+42)+(-85)$
   (g) $\left(+3\frac{1}{4}\right)+\left(+8\frac{1}{2}\right)+\left(-40\frac{3}{4}\right)$
   (h) $(-1.48)+(-2.54)+(+12)$
   (i) $(+11)+(+3)+(-9)+(-8)$
   (j) $(+23)+(-16)+(-22)+(+16)$

The Additive Inverse of a number

The counterpart of a real number such that their sum is zero is called its **additive inverse**. For instance, since the sum of -4 and $+4$ is 0. i.e. $-4+4=0$, -4 is said to be the additive inverse of $+4$ and vice versa. This is a property of all real numbers.

Example 7:12
State the additive inverse of each of the following.
(a) $+17$ (b) -15 (c) 30 (d) $-\pi$
(e) $+\frac{4}{7}$ (f) $-\frac{9}{13}$ (g) $\frac{12}{5}$ (h) $-\frac{\sqrt{2}}{3}$

Solution
(a) -17 (b) 15 (c) -30 (d) π
(e) $-\frac{4}{7}$ (f) $\frac{9}{13}$ (g) $-\frac{12}{5}$ (h) $\frac{\sqrt{2}}{3}$

Subtraction of Directed numbers

Subtracting a directed number is the same as adding its additive inverse. Equally it was stated earlier that when a (+) sign and a (−) sign occur together, the (−) sign takes preference. These two rules can be applied to subtract directed numbers easily.

Example 7:13
Evaluate
(a) $13-(+5)$ (b) $+15-(-18)$ (c) $1.97-2.35$
(d) $\frac{3}{14}-\frac{4}{7}$ (e) $\frac{3}{4}-\frac{2}{3}+\left(-\frac{5}{6}\right)$

Solution
(a) $13-(+5)=13-5=8$

(b) $+15-(-18)=15+18=33$

(c) $1.97-2.35=-(2.35-1.97)=-0.38$

(d) $\frac{3}{14}-\frac{4}{7}=\frac{3}{14}-\frac{8}{14}=-\left(\frac{8}{14}-\frac{3}{14}\right)=-\frac{5}{14}$

(e) $\frac{3}{4}-\frac{2}{3}+\left(-\frac{5}{6}\right)=\frac{3}{4}-\frac{2}{3}-\frac{5}{6}$

$$=\frac{9-8-10}{12}=\frac{-9}{12}=-\frac{3}{4}$$

EXERCISE 7:2

1. State the additive inverse of each of the following.
   (a) $+25$ (b) -53 (c) 42 (d) 2π
   (e) $+\frac{5}{13}$ (f) $-\frac{4}{11}$ (g) $\frac{26}{19}$ (h) $-\frac{\sqrt{3}}{2}$
2. Evaluate the following

(a) $(+20) - (-7)$

(b) $(-17) - (+3)$

(c) $(-5) - (-15)$

(d) $(+5) - (-15)$

(e) $(+8) - (+20)$

(f) $(-8) - (+8)$

(g) $\left(+10\frac{1}{2}\right) - \left(7\frac{1}{2}\right)$

(h) $(-1.4) - (-2.5)$

(i) $\left(+6\frac{1}{4}\right) - \left(-3\frac{1}{2}\right)$

(j) $(-0.23) - (+0.18)$

(k) $\left(-7\frac{3}{4}\right) - \left(+4\frac{1}{2}\right)$

(l) $(-5.5) - (+8.2)$

3. Evaluate the following.
   (a) $(+5) - (+3) - (+12)$
   (b) $(-1) - (-3) - (-2)$
   (c) $(+23) - (-13) - (+12)$
   (d) $(+5) - (-8) - (+17)$
   (e) $(+12) - (-12) - (+19)$
   (f) $(-42) - (+42) - (-85)$
   (g) $\left(+3\frac{1}{4}\right) - \left(+8\frac{1}{2}\right) - \left(-40\frac{3}{4}\right)$
   (h) $(-1.48) - (-2.54) - (+12)$
   (i) $(+11) - (+3) - (-9) - (-8)$
   (j) $(+23) - (-16) - (-22) - (+16)$

Multiplication of Directed Numbers

Multiplication is repeated addition i.e.
$(+4) \times 5 = (+4) + (+4) + (+4) + (+4) + (+4)$

$$= +20 = 20$$

$(-4) \times 5 = (-4) + (-4) + (-4) + (-4) + (-4) = -20$

Algebra tiles can also be used to model the product -4×5 as follows.

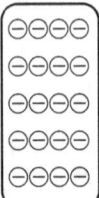 There are 5 groups of -4. By counting all these 5 groups of -4, it can clearly be seen that $-4 \times 5 = -20$.

Since multiplication is commutative
$$-4 \times 5 = 5 \times (-4) = -20$$

Generally, *the product of a negative number and a positive number is negative.*

Example 7:14
Evaluate the following.

(a) -7×3

(b) $6 \times (-8)$

(c) $8 \times -\frac{5}{16}$

(d) $-5 \times \frac{3}{17}$

(e) $\frac{5}{21} \times (-7)$

(f) $-\frac{4}{11} \times 3$

Solution

(a) $-7 \times 3 = -21$

(b) $6 \times (-8) = -48$

(c) $8 \times -\frac{5}{16} = -\frac{5}{2}$

(d) $-5 \times \frac{3}{17} = -\frac{15}{17}$

(e) $\frac{5}{21} \times (-7) = -\frac{5}{3}$

(f) $-\frac{4}{11} \times 3 = -\frac{12}{11}$

What is the sign of the product of two negative real numbers?

Consider $-4 \times (-5)$.
Since $-4 = -(+4)$,
$-4 \times (-5) = -(+4) \times (-5) = -(+4 \times (-5)) = -(-20)$

$-(-20)$ is the opposite of count 20 steps to the left i.e. count 20 steps to the right. Thus $-(-20) = +20$

Therefore, $-4 \times (-5) = +20$

In general, *the product of two negative real numbers is a positive real number.*

Example 7:15
Evaluate (a) $-13(-5)$ (b) $(-15)(-8)$
Solution
(a) $-13(-5) = 65$ (b) $(-15)(-8) = 120$

Division of Directed Numbers

Consider the following.

(a) Since $7 \times 3 = 21$, $\frac{21}{3} = 7$

(b) Since $-7 \times 3 = -21$, $\frac{-21}{3} = -7$

(c) Since $-7 \times (-3) = 21$, $\frac{21}{-3} = -7$

(d) Since $7 \times (-3) = -21$, $\frac{-21}{-3} = 7$

Therefore, *the quotient of two negative real numbers is a positive real number.*

Summary on Multiplication and Division

Table 7.1, is a summary of the sign of the product or quotient of two real numbers.

Like signs	Unlike signs
$(+)(+) = (+)$ $(-)(-) = (+)$	$(+)(-) = (-)$ $(-)(+) = (-)$
$\frac{(+)}{(+)} = (+)$ and $\frac{(-)}{(-)} = (+)$	$\frac{(+)}{(-)} = (-)$ and $\frac{(-)}{(+)} = (-)$

Table 7:1

In general,

The sign of the product or quotient of directed numbers is negative if the number of negative signed numbers is odd and positive if the number of negative signed numbers is even.

Sign of Products and Quotients

Number of negative signs	Sign of result
Odd	Negative (−)
Even	Positive (+)

Table 7:2

EXERCISE 7:3

1. Multiply the following.

   (a) $(+2)(-3)$ (b) $(-4)(+3)$ (c) $(-5)(-3)$
   (d) $(+5)(-6)$ (e) $(+8)(+2)$ (f) $(-8)(+8)$
   (g) $\left(+2\frac{1}{4}\right)\left(+3\frac{1}{2}\right)$ (h) $(-1.4)(-2.5)$
   (i) $\left(+2\frac{1}{4}\right)\left(-3\frac{1}{2}\right)$ (j) $(-0.23)(+0.18)$
   (k) $\left(-1\frac{3}{4}\right)\left(3\frac{1}{2}\right)$ (l) $(-1.5)(+3.2)$

2. Multiply the following.

   (a) $(+5)(+3)(+1)$ (b) $(-1)(-3)(-2)$
   (c) $(+3)(-1)(+2)$ (d) $(+5)(-8)(+1)$
   (e) $(+2)(-1)(+4)$ (f) $(-2)(+4)(-5)$
   (g) $\left(+\frac{1}{4}\right)\left(+\frac{1}{2}\right)\left(-\frac{3}{4}\right)$
   (h) $(-1.4)(-2.4)(+2)$ (i) $(+1)(+3)(-4)(-2)$
   (j) $(+3)(-1)(-2)(+6)$

3. Evaluate the following.

   (a) $\dfrac{+36}{+6}$ (b) $\dfrac{-36}{+6}$ (c) $\dfrac{-36}{-6}$
   (d) $\dfrac{+36}{-6}$ (e) $\dfrac{+0.36}{+6}$ (f) $\dfrac{-0.36}{+6}$
   (g) $\dfrac{-0.36}{-0.6}$ (h) $\dfrac{+36}{-0.6}$

4. Evaluate.

   (a) $\dfrac{(+20)(+12)}{(+3)(+5)}$ (b) $\dfrac{(+20)(-12)}{(-3)(-5)}$
   (c) $\dfrac{(+3)(+6)(+18)}{(+12)(-3)}$ (d) $\dfrac{(-3)(-6)(+18)}{(+12)(+3)}$
   (e) $\dfrac{0}{(-17)(-24)}$ (f) $\dfrac{(-120)(+31)(0)}{(-5)(-8)(-9)}$

(g) $\dfrac{(+3)(+6)(+1.8)}{(+1.2)(-0.3)}$ (h) $\dfrac{(+0.1)(++0.2)(-30)}{(+0.4)(-0.1)}$

MULTIPLE CHOICE EXERCISE 7

1. The greatest of the following is:
   [A] −2 [B] −4 [C] −6 [D] −8
2. As an integer, 32°F below zero degrees is:
   [A] −31° [B] 32° [C] −32° [D] 31°
3. The opposite of the integer −3 is:
   [A] 1 [B] 3 [C] −3 [D] $\dfrac{1}{3}$
4. The absolute value of 5 is:
   [A] 5 [B] −5 [C] $\dfrac{1}{5}$ [D] $-\dfrac{1}{5}$
5. The correct order of the integers 8, 15, −1, 6, −6 from least to greatest is:
   [A] −6, −1, 6, 8, 15 [B] −1, −6, 15, 6, 8
   [C] 8, 6, −6, −1, 15 [D] 15, 6, 8, −1, −6
6. The sum 1 and 5 is:
   [A] 6 [B] 4 [C] −4 [D] −6
7. The difference −5 − (−2) is:
   [A] 3 [B] −3 [C] −7 [D] 7
8. When evaluated, −7 + 5 − 2 + 11 − 8 equals:
   [A] +1 [B] −33 [C] +33 [D] −1
9. The value of −6 − (−6) is:
   [A] −12 [B] 0 [C] 12 [D] 36
10. The value of $(-4)(-3)$ is:
    [A] 7 [B] −7 [C] 12 [D] −12
11. $(-2) \times (-3)$ is equal to:
    [A] −6 [B] 6 [C] −5 [D] 5
12. The value of $(+4)(-2) - (-2)$ is:
    [A] −6 [B] −4 [C] +4 [D] +6
13. −12 + 8.3 equals:
    [A] −4.7 [B] −4.36 [C] −3.7 [D] 3.7
14. The result of $\dfrac{(-1)(-2)(+3)(-4)}{(+6)(-8)}$ is:
    [A] 2 [B] −2 [C] $-\dfrac{1}{2}$ [D] $\dfrac{1}{2}$
15. The value of $\left(-\dfrac{6}{7}\right) \times \left(-\dfrac{7}{18}\right) \times \left(-\dfrac{3}{4}\right)$ is:
    [A] $-\dfrac{1}{4}$ [B] $\dfrac{1}{4}$ [C] −1 [D] 1
16. $1\dfrac{3}{4} \div \left(-2\dfrac{1}{4}\right)$ equals:
    [A] $\dfrac{7}{9}$ [B] $\dfrac{8}{63}$ [C] $-\dfrac{7}{9}$ [D] $-\dfrac{16}{63}$
17. A fish was swimming at 180 m below sea

level. Then it descended to 315 m below sea level. As an integer in meters, the change in the fish's depth is:

[A] 135   [B] −135   [C] −495   [D] 495

18. As arctic air moved into a region where the temperature was originally 25°C, the temperature began falling at a steady rate of 4°C per hour. The temperature after 9 hours would be:

[A] −7°C   [B] −11°C
[C] −36°C   [D] −15°C

19. 8 ÷ (−4) is:

[A] −2   [B] $\dfrac{1}{2}$ [C] 2   [D] $-\dfrac{1}{2}$

20. Anye owns a small business. There was a loss of 14 FCFA on Thursday and a loss of 12 FCFA on Friday. On Saturday, there was a loss of 11 FCFA, and on Sunday, there was a profit of 16 FCFA. The total profit or loss for the four days is:

[A] 31 FCFA loss   [B] 25 FCFA profit
[C] 21 FCFA loss   [D] 53 FCFA profit

21. Using table 7:1, Pa Tangwe's profit or loss for the month of January is:

Pa Tangwe Income and Expenditure

Month	Income	Expenses
January	1,817 FCFA	−2,338 FCFA
February	2,271 FCFA	−2,315 FCFA
March	3,243 FCFA	− 1,530 FCFA
April	3,929 FCFA	−1,167 FCFA
May	3,477 FCFA	−1,101 FCFA
June	3,077 FCFA	−834 FCFA

Table 7:1

[A] −4,155   [B] 4,155   [C] 521   [D] −521

22. Simplifying $\dfrac{5}{3}$ of $\left(1\dfrac{1}{5} - 2\dfrac{1}{2}\right) \div 3\dfrac{1}{3}$ leads to:

[A] $-\dfrac{2}{20}$ [B] $-\dfrac{7}{20}$ [C] $-\dfrac{13}{20}$ [D] $-\dfrac{39}{20}$

23. The value of 8 + 30 ÷ 10 − 2 × 5 is:

[A] 38   [B] 1   [C] $\dfrac{95}{4}$ [D] $\dfrac{19}{20}$

24. The result after evaluating 5 × 4 − 18 ÷ 6 is:

[A] $\dfrac{17}{3}$ [B] $-\dfrac{35}{3}$ [C] $-\dfrac{28}{3}$ [D] 17

25. 8 × 6 − 10 ÷ 5 + 12 after simplification gives:

[A] 58   [B] 34   [C] $\dfrac{28}{5}$ [D] $-\dfrac{32}{7}$

26. |−1| is equal to:

[A] 0   [B] −1   [C] 1   [D] $\dfrac{1}{2}$

27. The symbol, which makes |−14| ▓ |10| true is:
[A] =   [B] >   [C] <   [D] =

28. It is true to say that:
[A] |−7| > |5| [B] |−7| < |5|
[C] |−7| < |7| [D] |0| > |−5|

TOPIC 8

BASIC ALGEBRAIC EXPRESSIONS

OBJECTIVES

At the end of this topic, the learner should be able to:

1. Use letters to represent real objects and numbers.
2. Distinguish between a variable and a pro-numeral.
3. Substitute variables in an expression and evaluate the expression.
4. Translate verbal expression into algebraic expressions.
5. Distinguish between an unknown, a variable and a constant.
6. Identify the numerical and literal coefficients of a term.
7. Identify like and unlike terms.
8. Combine like terms.
9. Multiply and Divide simple algebraic terms.
10. Simplify simple algebraic expressions in not more than three variables.
11. Identify common factors.
12. Find the HCF and LCM of Literal Expressions.

The German mathematician Carl Friedrich Gauss contributed to many areas of mathematics, including algebra. In his doctoral thesis he proved a theory which became known as the fundamental theory of algebra.

Science Source/Photo Researchers, Inc.

Symbolic Expressions

In addition to the numerals and operations +, −, ×, ÷ used in arithmetic, algebra also employs letters as numbers to generalize situations or as shorthand symbols to stand for objects. For instance, the statement

3 books +2 books = 5 books, can simply be written as **3b + 2b = 5b**.

6 apples − 4 apples = 2 apples, can simply be written as **6p − 4p = 2p**.

If one packet of pens contains 54 pens
2 packets of pens will contain 2 × 54 pens
3 packets of pens will contain 3 × 54 pens
Consequently, any number of packets will contain that number multiplied by 54 pens.
A letter can be used to stand for any number.
Hence, *n packets of pens will contain n × 54 pens.*

In algebra, the use of the multiplication sign '×' is generally avoided because it is easily confused with the letter *x*, which is very often employed. Hence *n × 54 is written as 54n* and is read 'fifty four *n*'.

In an algebraic product consisting of a numeral and a letter symbol, the numeral is always written before the letter symbol. Thus if *a* = 7, 2*a* = 14 not 27.

Therefore, the correct way to write *n* × 54 in algebra is 54*n* not *n*54.

1*b* is written simply as *b*, while − 1*b* is written simply as − *b*.

Variables

A symbol or letter such as *n*, which stands for any number, is called a **variable**. A variable can be replaced by any number and the statement will still be true. The numbers used to replace or substitute a variable are called the **values** of the variable. Any symbol or letter, which stands for a numeral, is called a **pro-numeral**.

Algebraic Sentences (*Expressions*)

Algebraic sentences or expressions are formed by combining pro-numerals and numerals using arithmetical operations such as +, −, ×, ÷, square root, and so on.

From English to Algebra

Many problems that will rather be very difficult to solve can easily solved when they are translated to algebra. The ability to translate word phrases to algebraic expressions cannot therefore be over emphasized.

In translating English sentences to algebraic sentences, certain key words always indicate the operation, which is required.
Table 8:1 is an outline of the most common of the key words.

Operation	Key Words
Addition	Add (to) Sum Plus More than Increased by
Subtraction	Subtract (from) Difference Minus Less than Decreased by Reduced by Take away Less
Multiplication	Multiply (by) Product (of) Times Twice, thrice Of (fractions/percentages)
Division	Divide Quotient Share

Table 8:1

Table 8:2 shows how word expressions can be translated into algebraic expressions.

	WORD EXPRESSION	ALGEBRAIC EXPRESSION
1	Multiply a number by 2	$2y$
2	Twice a number	$2x$
3	3 more than a number	$x + 3$
4	A number increased by 3	$n + 3$
5	The sum of 5 and a number	$m + 5$
6	4 less than a number	$p - 4$

	WORD EXPRESSION	ALGEBRAIC EXPRESSION
7	Multiply a number by 5 and add 2	$5u + 2$
8	The square root of a number	$\sqrt{x}$
9	3 more than twice a number	$2p + 3$
10	A number divided by 4	$\dfrac{w}{4}$
11	The quotient of a number and 7	$\dfrac{x}{7}$
12	The product of 2 numbers	xy
13	A number decreased by 8	$n - 8$
14	The quotient of two numbers	$\dfrac{a}{b}$
15	The square of a number	x^2
16	Twice the square of a number	$2x^2$
17	3 less than half a number	$\dfrac{x}{2} - 3$
18	The difference between two numbers	$b - n$
19	The sum of two numbers	$x + y$

Table 8:2

Expressions that are even more complicated can be written such as $\dfrac{x-2}{x^2-1}, 3xy - 8xy$ etc.

EXERCISE 8:1

Translate the following into algebraic expressions.

1. Two times a number.
2. Three more than a number.
3. Five less than a number.
4. A number increased by seven.
5. The sum of a number and eight.
6. The difference between a number and 9.
7. Twice a number.
8. Half of a number.
9. A number decreased by ten.
10. Seven more than twice a number.
11. A number divided by two.
12. The product of a number and four.
13. The quotient of a number with 4 as the dividend.
14. The square of a number.
15. The cube of a number.
16. Four less than three times a number.
17. The square root of twice a number.
18. The square of a number increased by two.
19. Fifteen decreased by a number.
20. Two times a number decreased by eight.
21. Seven less than half a number.
22. Half a number increased by seven.

Variable Substitution

If a and b are variables, then there exist different values of $a + b$ for different values of a and b. The process of finding different values of an expression for different values of the variables is known as **variable substitution**.

Example 8:1

Find $a + b$ when $a = 2$ and $b = 8$

Solution

When $a = 2$ and $b = 8$, $a + b = 2 + 8 = 10$

Example 8:2

If $t = s + u$, find t when $s = 6$ and $u = -3$

Solution

$t = s + u = 6 + (-3) = 3$

EXERCISE 8:2

1. Find the value of $2x + 1$, when $x = 3$.
2. If $y = -1$ and $x = 4$, find the value of $x + 3y$.
3. The area of a rectangle is given by $A = lw$. Find the area when $l = 5$ cm and $w = 2$ cm.
4. What is the difference between $2x$ and 25 when $x = 5$?
5. The perimeter of a field is given by $p = 2(l + w)$. Find p when $l = 8$ and $w = 10$.
6. The volume, V of a cylinder is given by the formula $V = \pi r^2 h$.
   Find the volume of a cylinder whose radius r is 2 cm and whose height h is 7 cm.
7. The radius r of a circle is 7 cm and the area is given by πr^2. Calculate the area A of the circle. Take $\pi = \dfrac{22}{7}$.

Unknowns and Constants

Consider the statement $3x + 2 = 5$. The only value, which can be substituted for x, for the statement to remain true is 1. Any other value will make the statement false. Such a pro-numeral, which can take only particular values, is called an **unknown**. x in this case is not a variable since its value does not change. A **constant** is a letter or symbol, which has a fixed value. An unknown can be a constant but a constant is not necessarily an unknown. Examples of constants are 2, π and -1. In the formula $A = \pi r^2$, π is a constant but not an unknown.

For a circle with radius $r = 2$ cm, the area A is an unknown since $r = 2$ cm can be substituted in the formula to find A. If on the other hand, there are many different circles with different radii, then both r and A are variables since substituting different values of r will give different values of A.

Terms and Coefficients

The **terms** of an expression are the different parts of the expression that are linked together by '+' or '−' signs.

Thus the expression $3xy^2 - 8xy + \dfrac{7}{y} + x + 2$

has terms $3xy^2$, $-8xy$, $\dfrac{7}{y}$, x and 2.

An expression with only one term is called a **monomial**, one with two terms a **binomial**, one with three terms a **trinomial** and one with many terms a **polynomial**. A term may simply be a constant called the **constant term** or a variable or a product or quotient of constants and variables.
Each of the numbers or variables that are multiplied together to make up the term is a factor of the term and is called the **coefficient** of the other factors. For instance, the factors of the term $3xy^2$ are 3, x, y, $3x$, $3y$, $3y^2$, xy, xy^2 and $3xy^2$.
3 is the coefficient of xy^2 and because 3 is a numeral, 3 is called the **numerical coefficient** of xy^2. On the other hand x, y, y^2, xy, xy^2 are called **literal coefficients** or **non-numerical coefficients** because they are composed of letters.

EXERCISE 8:3

(1) Write down the terms of the following expressions:

(a) $6x - 2xy + 3y$ (b) $px + \dfrac{p}{y}$

(c) $w + 5pz + r$ (d) $3pt - \dfrac{8x}{y} + \dfrac{1}{w}$

(2) State the numerical coefficients of the following terms:

(a) $7px$ (b) $-\dfrac{2}{y}$ (c) $5pw$

(3) State the literal coefficients of the following terms.

(a) $3pt$ (b) $-\dfrac{8x}{y}$ (c) $5pw$

Like and Unlike Terms

If two or more terms have the same literal coefficient, they are said to be **like terms**. On the other hand, if the literal coefficients of two terms are not the same, the terms are said to be **unlike terms**.

Examples of like terms are:

(a) $3x$, $5x$ and $\dfrac{2x}{7}$ (b) $2pq$, $\dfrac{4pq}{3}$, $-19pq$

(c) $13x$, $-x$, $-\dfrac{2x}{3}$ (d) $4ab$, $\dfrac{3ab}{5}$

(e) $\dfrac{ab}{c}$, $\dfrac{17ab}{c}$, $\dfrac{8ab}{5c}$

Algebraic Rules

Algebra obeys all the laws of arithmetic and any operation or manipulation, which works in arithmetic, works in algebra. Any operation or manipulation, which does not work in arithmetic, does not work in algebra. Therefore, the properties of numbers discussed in Topic 1, which are summarised in Table 8.2 are all true for both arithmetic and algebra. These properties are used to simplify algebraic expressions.

Properties of addition	For all values of x, y, z
Commutative	$x + y = y + x$
Associative	$(x + y) + z = x + (y + z)$
Additive identity property of zero	$x + 0 = x$ and $0 + x = x$

Distributive property	For all values of x, y, z
Over addition	$x(y + z) = xy + xz$
Over subtraction	$x(y - z) = xy - xz$

Properties of multiplication	For all values of x, y, z
Commutative	$xy = yx$
Associative	$(xy)z = x(yz)$
Multiplicative property of zero	$0(x) = 0$ and $x(0) = 0$

Table 8:2

Combining Like Terms

Like terms, of an algebraic expression can be added or subtracted and written as single terms following the algebraic rules. Unlike terms on the other hand cannot be added or subtracted.

Example 8:3

Compute   (a) $a + 2a + 7a$

        (b) $5p + \dfrac{3p}{8} - \dfrac{5p}{8}$

Solution

(a) $a + 2a + 7a = 10a$

(b) $5p + \dfrac{3p}{8} - \dfrac{5p}{8} = \dfrac{38p}{8} = \dfrac{19p}{4}$

Example 8:4

Evaluate the following.

(a) $19x - 15x - 7x + 4x$

(b) $3x + 2y - 5x + 7y$

Solution

(a) $19x - 15x - 7x + 4x = x$

(b) $3x + 2y - 5x + 7y = (3x - 5x) + (2y + 7y)$

                      $= -2x + 9y$

EXERCISE 8:4

Evaluate the following.

1. $8x - 4y + 13x - 13y$
2. $6s + 7t - 8t - 5s$
3. $12p - 9q - 13q - 7p$
4. $32a + 6b - 12a + 4b$
5. $2u - 5v - u + 2v$
6. $32a + 6b - 12a + 4b$
7. $-7x + 8y - 2 + 9x - 10y + 4$
8. $a + (-a) + b$
9. $15x - 12y + 8z - 11x + 12y - 8z$
10. $2pq - 3qr - 7pq + 5rq + 4pq$

Multiplication and Division of Terms

1. **Multiplication**

To multiply algebraic terms,
(a) Multiply the numerical coefficients
(b) Multiply the literal coefficients
(c) Write the result as the product of the numerical and literal coefficients placing the numerical coefficient first.

Example 8:5

Evaluate each of the following.

(i) $(4x)(2x)$ (ii) $(3x)(5y)$ (iii) $(-2ab)(3a)$

Solution

(i) $(4x)(2x) = 8x^2$ (ii) $(3x)(5y) = 15xy$

(iii) $(-2ab)(3a) = -6a^2b$

2. **Division**

To divide algebraic terms,
(a) Divide the numerical coefficients
(b) Divide the literal coefficients
(c) Write the result as the product of the numerical and literal coefficients.

Example 8:6

Simplify the following.

(i) $\dfrac{x}{x}$ (ii) $\dfrac{2x^2y}{xy} + \dfrac{5xy}{xy}$ (iii) $\dfrac{20pq}{4p^2} + \dfrac{16q}{4p}$

Solution

(i) $\dfrac{x}{x} = 1$

(ii) $\dfrac{2x^2 y}{xy} + \dfrac{5xy}{xy} = 2x + 5$

(iii) $\dfrac{20pq}{4p^2} + \dfrac{16q}{4p} = \dfrac{5q}{p} + \dfrac{4q}{p} = \dfrac{9q}{p}$

EXERCISE 8:5

1. Simplify the following.

   (a) $4(2b)$ (b) $3x(x)$

   (c) $4p(5q)$ (d) $2ab(9b)$

   (e) $3x(8xy)$ (f) $4xy(7x)$

   (g) $10a(5b)$ (h) $3a(10a)$

   (i) $5x(4xy)$ (j) $9uv(3uv)$

2. Simplify the following.

   (a) $28xy \div 4$ (b) $6pq \div q$

   (c) $\dfrac{7x^2}{x}$ (d) $\dfrac{54xy}{9y}$

   (e) $\dfrac{33ab}{3a}$ (f) $\dfrac{42pq}{7q}$

   (g) $\dfrac{7x^2 y}{8x}$ (h) $\dfrac{48u^2 v}{12uv}$

   (i) $\dfrac{40xy^2}{8xy}$ (j) $\dfrac{60m^2 n}{20mn}$

   (k) $\dfrac{12p^2 q}{3q^2 p} + \dfrac{28pq^2}{7p^2 q}$ (l) $\dfrac{15xy}{3y} - \dfrac{24x}{6}$

The HCF and LCM of Literal Expressions

The HCF and LCM of literal expressions can be found using repeated division, just in the same way as that of numerical expressions. Repeated division is done until there are no common factors left. The HCF is the product of all the common factors, while the LCM is the product of the common factors and the uncommon factors.

Example 8:7

Find the HCF and LCM of $24x^2 y$ and $27xy^2$.

Solution

3	$24x^2 y$	$27xy^2$
x	$8x^2 y$	$9xy^2$
y	$8xy$	$9y^2$
	$8x$	$9y$

$\therefore \mathrm{HCF} = 3 \times x \times y = 3xy$

and $\mathrm{LCM} = 3xy \times 8x \times 9y = 216x^2 y^2$

Example 8:8

Find the HCF and LCM of

$8pq^2 r^3, 6p^2 q^2, 10pq^2 r^2$

Solution

2	$8pq^2 r^3$	$6p^2 q^2$	$10pq^2 r^2$
p	$4pq^2 r^3$	$3p^2 q^2$	$5pq^2 r^2$
q^2	$4q^2 r^3$	$3pq^2$	$5q^2 r^2$
r^2	$4r^3$	$3p$	$5r^2$
	$4r$	$3p$	5

$\therefore \mathrm{HCF} = 2 \times p \times q^2 = 2pq^2$

and $\mathrm{LCM} = 2 \times p \times q^2 \times r^2 \times 4r \times 3p \times 5 = 120p^2 q^2 r^3$

Note that the factors of the HCF are those factors that divide all the numbers or expressions.

EXERCISE 8:6

Find the HCF and LCM of the following.

(1) $3a^2$ and $12a$

(2) $10a^2$ and $8ab$

(3) ax, ay and ab

(4) $15a, 12a$ and $14a^2$

(5) $8a^2$ and $6a^2 b^2$

(6) $5ab^2$ and $3a^2 b$

(7) $2a$ and $7a^2$

(8) $9p^2 q$ and $15pq^2$

(9) $12x^2 y$ and $8x^3 y^2$

(10) $6a^2 b^2, 24ab^2$ and $40a^2 b^2 c^2$

《 MULTIPLE CHOICE EXERCISE 8 》

1. The pair of expressions which are like terms are:

    [A] $2xy$ and $3x$

    [B] $5xy^2$ and yx^2

    [C] $3x^2y$ and $7yx^2$

    [D] $4y$ and $2yx$

2. The pair of expressions which are unlike terms are:

    [A] $2ab^5$ and $3b^5a$

    [B] $5xy^2$ and yx^2

    [C] $3x^2y$ and $7yx^2$

    [D] $4xy$ and $2yx$

3. The expressions, which are like terms, are:

    [A] $\frac{1}{5}xy$ and $\frac{1}{5}x$

    [B] $6xy^2$ and $6yx^2$

    [C] $\frac{1}{5}x^2y$ and $\frac{1}{5}x$

    [D] $6y$ and $6yx$

4. The expressions, which are unlike terms, are:

    [A] $\frac{1}{7}ab^3$ and $7b^3a$

    [B] $7xy^2$ and $7yx^2$

    [C] $7x^2y$ and $7yx^2$

    [D] $4xy$ and $4yx$

5. In the expansion $a(b+c) = ab + ac$, the law used is:

    [A] The associative law

    [B] The distributive law

    [C] The commutative law

    [D] The multiplicative law

6. In $31 + (52 + 23) = (31 + 52) + 23$, the law used is:

    [A]   Associative law of addition

    [B]   Commutative law of addition

    [C]   Distributive law of addition

    [D]   Identity law of addition

7. In $\left(50 \times \frac{1}{3}\right) \times 24 = 50 \times \left(\frac{1}{3} \times 24\right)$, the law used is:

    [A]   Associative law of multiplication

    [B]   Commutative law of multiplication

    [C]   Distributive law of multiplication

    [D]   Multiplicative identity law

8. In $3\frac{2}{5} \times x = x \times 3\frac{2}{5}$, the law used is:

    [A]   Associative law of multiplication

    [B]   Commutative law of multiplication

    [C]   Distributive law of multiplication

    [D]   Multiplicative identity law

9. In $x(7) = 7x$, the law used is:

    [A] Associative law

    [B] Commutative law

    [C] Distributive law

    [D] Multiplicative identity law

10. In $\left(p \times \frac{3}{4}\right) \times 68 = p \times (p \times 68)$, the law used is:

    [A]   Associative law of multiplication

    [B]   Commutative law of multiplication

    [C]   Distributive law of multiplication

    [D]   Identity law of multiplication

11. Given that $p = 1, q = -1$ and $r = 0$. The value of $p + q + r$ is:

    [A] -2 [B] -1 [C] 0 [D] 1

12. Given that $p = 1, q = -1$ and $r = 0$. The value of pq is:

    [A] -2 [B] -1 [C] 0 [D] 1

13. Given that $p = 1, q = -1$ and $r = 0$. The value of $p(q + r)$ is:

    [A] -2 [B] -1 [C] 0 [D] 1

14. If $x = 2$, the value of $2x^2 + 3$ is:

    [A] 11 [B] 19 [C] 14 [D] 24

15. When $b = -1$, the value of $5 - b - b^2$:

    [A] 5 [B] 3 [C] 2 [D] 0

16. If $\frac{1}{v} = \frac{1}{f} - \frac{1}{u}$ then the value of v when $f = 2$ and $u = 3$ is:

    [A] -1 [B] $+5$ [C] $+6$ [D] -6

17. When $x = -1$, the value of $\frac{x^2 + x - 2}{x^2 + x - 3}$ is:

    [A] -2 [B] -9 [C] $-\frac{1}{2}$ [D] 1

18. Given that $p = 2, q = 5$ and $r = -4$. When evaluated $3p^2 - q^2 - r^3$ gives:

    [A] -77 [B] 77 [C] 51 [D] 101

19. Given that $x = -3$ and $y = -7$, $\frac{x^2 - y}{y^2 - x}$ has value:

[A] $-\dfrac{1}{11}$ [B] $\dfrac{1}{23}$ [C] $\dfrac{4}{13}$ [D] $\dfrac{12}{17}$

20. The expression, which is not equal to,

$\dfrac{1}{2}pq$ is:

[A] $\dfrac{pq}{2}$ [B] $\dfrac{p \times q}{2}$

[C] $\dfrac{1}{2}qp$ [D] $\dfrac{1}{2}p \times q$

21. If $\dfrac{1}{2}p = x + y$, then p equals:

[A] $2x + y$ [B] $\dfrac{1}{2}(x + y)$

[C] $x + 2y$ [D] $2(x + y)$

22. The LCM of $9p^2q$ and $15pq^2$ is:

[A] $45p^2q^2$ [B] $45p^2q$

[C] $45pq^2$ [D] $45pq$

23. The LCM of $2a$ and $7a^2$ is:

[A] $14a$ [B] $14a^2$ [C] $28a$ [D] $28a^2$

24. The LCM of $6a^2b^2, 24ab^2, 40a^2bc^2$ is:

[A] $120a^2b^2c^2$ [B] $124a^2b^2c^2$

[C] $144a^2b^2c^2$ [D] $96a^2b^2c^2$

25. Given that $\dfrac{7}{4x} - \dfrac{5}{x} + \dfrac{4}{3x}$. The L C M of the denominators is:

[A] $12x^2$ [B] $12x$

[C] 12 [D] $12x^{-2}$

26. The HCF of $12a^2b$ and $10ab$ is:

[A] $5ab$ [B] $4ab$

[C] $2ab$ [D] $2a^2b$

27. The HCF of $8ab^2c^3, 6a^2b^2, 10ab^2c^2$ is:

[A] $8ab^2c^3$ [B] $10ab^2c^2$

[C] $2ab$ [D] $2ab^2$

28. If $b \in \mathbb{R}$, the positive square root of $9b^2$ is:

[A] $9b$ [B] 9 [C] $3b$ [D] $3b^2$

29. The factor (s) of $4x^2y$ is/are:

I. $2x$ II. xy III. $4x$

[A] I only   [B] II only

[C] III only   [D] I, II and III

30. The coefficient of x in $(x - 2)(x + 9)$ is:

[A] -2 [B] 7 [C] 9 [D] -18

TOPIC 9

SIMPLE LINEAR EQUATIONS

OBJECTIVES

At the end of this topic, the learner should be able to:

1. Define an equation and distinguish between equations and expressions.
2. Construct simple linear equations from given situations.
3. Solve one-step simple linear equations.
4. Solve two-step simple linear equations.
5. Solve multi-step simple linear equations.
6. Solve simple linear equations involving fractions and decimals.

Albert Einstein
Mathematical relationships are expressed in symbols. Equations express the most fundamental mathematical relationship: equality. Physicist Albert Einstein produced one of the most famous equations, $E = mc^2$, which expressed the equivalence of matter and energy. Although he later became politically active, supporting pacifism and zionism, he claimed, "Equations are more important to me, because politics is for the present, but an equation is something for eternity."

What is an equation?

An equation is a statement of equality between two expressions. Examples of equations are:

$$2 + 3 = 5 \dots\dots\dots\dots\dots\dots① $$

$$2(4.5) = 9 \dots\dots\dots\dots\dots\dots② $$

$$1 + x = 4 \dots\dots\dots\dots\dots\dots③ $$

$$\frac{n}{6} = 3 \dots\dots\dots\dots\dots\dots④ $$

Equations① and ② are examples of numerical equations. Equations ③and ④ which involve a letter or symbol (or sometimes more than one letter or symbol) are called **algebraic equations**.

Equations are so important because they are used in almost all branches of pure and applied mathematics and in the physical, biological, and social sciences.

Simple Linear Equations

An equation in which the power(s) of the unknown(s) is (are) unity is called a **simple linear equation**. Equations ③and ④above are examples of linear or simple equations. The solution of the following problem is very easy if it is translated into a simple (linear) equation.

The cost of a pen is 50 FRS. Nde needed to buy 7 books and 3 pens, but the amount he had was short by 200 FRS, so he decided to buy 5 books and 8 pens for all the money he had. Determine
(a) the price of a book
(b) the amount he had.

The above problem can very easily be solved if it is translated into a simple algebraic equation. Revise the section "*From English to Algebra*" treated on pages 80 to 81. That section together with Table 9:1 provide the basic skills needed to be able to build up simple linear equations from real life situations. Table 9:1 is a list of some examples of simple linear equations and their equivalent algebraic translations. Study them carefully and use the skill acquired to translate the simple equation above.

	Word Equation	Algebraic Equation
1	Three more than a certain number is 20	$x + 3 = 20$
2	Twice a number is 8	$2n = 8$
3	When a number is multiplied by 5 and 2 is added, the result is 22	$5y + 2 = 22$
4	A man shares some mangoes equally to his 4 children and each child has 2 mangoes	$\frac{x}{4} = 2$
6	I think of a number, double it and add 1. The result is 7	$2a + 1 = 7$
7	Njuh buys 2 pens at 50 FRS each and 3 books at y FRS each for 400 FRS.	$3y + 100 = 400$

Table 9:1

Using our knowledge, let us translate Nde's problem to a simple linear equation.

Let the cost of a book be b.
Then cost of 7 books and 3 pens $= 7b + 3(50)$
$$= 7b + 150$$
Cost of 5 books and 8 pens $= 5b + 8(50)$
$$= 5b + 400$$
But cost of 5 books and 8 pens is 200 FRS less than cost of 7 books and 3 pens.
Therefore $7b + 150 - 200 = 5b + 400$
This is the required simple linear equation.
Watch out for the solution to this equation!

EXERCISE 9:1

Write algebraic equations to represent each of the following:
1. Kanjo is travelling at 3 km/h and Ndi is travelling at x km/h. Their average speed is 4 km/h.
2. A tailor whose daily wage is d FCFA obtains 42,000 FCFA every week.
3. A boy will be 14 years in three years time.
4. During a football competition, a team, which played 16 matches, altogether, lost three times as many matches as it won.

5. Three times the sum of a number and six is 33.
6. A number increased by 20 equals three times the same number.
7. A number increased by 5 equals twice the same number decreased by 4.
8. Twelve times h equals one hundred and twenty minus thirty-six.
9. MTN charges 250 Frs. per minute for calls. After making a call that last *m* minutes, Konyuy is charged a bill of 450 Frs.
10. Mbianda bought a radio that costs 16960 FCFA. With tax, *t* she pays 17810 FCFA.

Difference between Equations and Expressions

Recall that an **expression** is several terms joined by positive or negative signs. It may consist of only one term. Examples of expressions are $2x + 3y - 4$, $3x + 2$. Expressions do not contain an equality sign. An **Equation** on the other hand contains an equal to sign. Examples of equations are $3x + 2 = 5$ and $2x + 3y = 4$. In an expression such as $3x + 2$, x is a **variable** but in the equation such as $3x + 2 = 5$, x is an **unknown**. An equation may contain one or more unknowns. For instance, $2x + 3y = 7$ contains two unknowns x and y.

Additive Inverses

If the sum of two numbers is zero, then one is the **additive inverse** of the other and vice versa. For instance;

The additive inverse of 12 is -12 because $12 + (-12) = 0$.

The additive inverse of -4 is 4 because $(-4) + 4 = 0$.

The additive inverse of $-\frac{1}{2}$ is $\frac{1}{2}$ because $\frac{1}{2} + \left(-\frac{1}{2}\right) = 0$.

The additive inverse of $\frac{5}{8}$ is $-\frac{5}{8}$ because $\left(-\frac{5}{8}\right) + \frac{5}{8} = 0$.

Inverse Operations

Inverse operations are operations, which can undo each other. Consider the following:

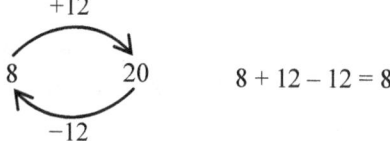 $8 + 12 - 12 = 8$

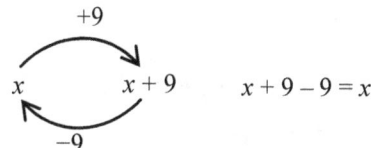 $x + 9 - 9 = x$

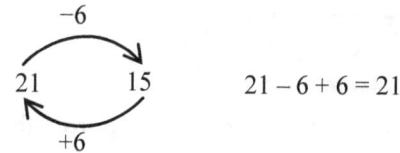 $21 - 6 + 6 = 21$

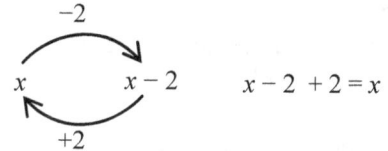 $x - 2 + 2 = x$

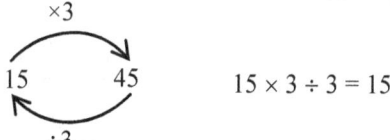 $15 \times 3 \div 3 = 15$

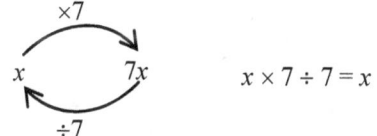 $x \times 7 \div 7 = x$

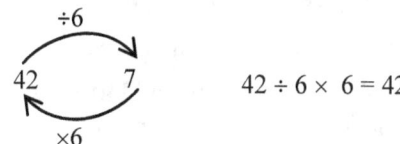 $42 \div 6 \times 6 = 42$

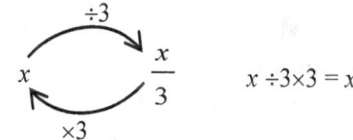 $x \div 3 \times 3 = x$

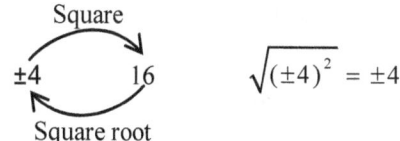 $\sqrt{(\pm 4)^2} = \pm 4$

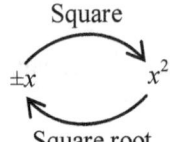

Square

$\sqrt{(\pm x)^2} = \pm x$

Square root

From the above examples, it can be seen that:
1. Addition of a quantity undoes subtraction of the same quantity.
2. Multiplication by a quantity undoes division by the same quantity.
3. Squaring a quantity undoes finding the square root of the resulting quantity.

EXERCISE 9:2

1. What number should be subtracted from the expression to have x?
   (a) $x + 5$ (b) $x + 8$
2. What number should be added to the expression to have x?
   (a) $x - 10$ (b) $x - 17$
3. Say what to do so that the result is x in the following cases.

   (a) $3x$ (b) $7x$ (c) $\dfrac{x}{5}$

   (d) $\dfrac{x}{14}$ (e) $0.5x$ (f) $\dfrac{x}{0.02}$

   (g) $x + 5$ (h) $\dfrac{x-2}{13}$

Solving Simple Linear Equations

The process of finding the value of the unknown in an equation or the unknowns in a series of equations is called **solving the equation or equations**. The set of values of the unknown is called the **solution set**. Simple equations can be solved using the **guess method** or the **flow chart method** but the most authentic method for solving equations is the **balancing method**.

The Guess Method

Example 9:1
If $2x = 14$, what number gives 14 on multiplying by 2?

Solution
Obviously, the number is 7. Therefore $x = 7$.
Checking, $2(7) = 14$ (true)

Example 9:2
Given that $x - 2 = 9$, what number gives 9 on subtracting 2?

Solution
Obviously, the number is 11.
Therefore, $x = 11$.
Checking, $11 - 2 = 9$ (true)

EXERCISE 9:3

Solve the following equations using the guess method.

1. $3x = 18$ 2. $6t = 42$

3. $\dfrac{x}{2} = 5$ 4. $\dfrac{3a}{2} = 12$

5. $x + 3 = 10$ 6. $y - 4 = 6$

7. $2 + x = 11$ 8. $16 = 2x + 8$

9. $2y + 4 = 12$ 10. $7x - 5 = 2x + 11$

11. $2(n + 5) = 18$ 12. $\dfrac{2x-5}{3} = 25$

The Balancing Method

When weighing something such as fish or meat on the scale balance, a mass is placed on one side of the balance and an equal mass of fish or meat is then placed on the other side. If a mass of 1kg is placed on one side, 1kg mass worth of fish or meat must be placed on the other side. When the two sides are equal, the scale balance will be horizontal.

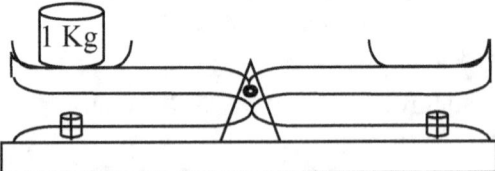

1 Kg

Figure 9:1 Equations are Balanced System.

Suppose a 0.5 kg mass is added to the left side, of a scale balance, a 0.5 kg mass worth of fish must be added to the right hand side to maintain equilibrium. Suppose fish of mass 1kg is removed from the right side, the 1kg mass must be removed from the left side to ensure balance. If the mass is tripled, the mass of fish must be tripled to maintain balance.

Thus, an equation is a balanced system and can be compared to a scale balance.
1. Adding the same quantity to both sides does not destroy the balance.
2. Subtracting the same quantity from both sides does not destroy the balance.
3. Multiplying or dividing both sides by the same non-zero number does not destroy the balance.

One Step Simple Linear Equations

In one step simple linear equations, simply perform one of the following operations
(a) Add the same quantity to both sides to undo subtraction or subtract the same quantity from both sides to undo addition.
(a) Multiply both sides by the same quantity to undo division or divide both sides by the same quantity to undo multiplication.

Example 9:3
Solve the following equations:

1. $3x = 18$ 2. $26 = 2p$

3. $-2y = 6$ 4. $\dfrac{x}{5} = 2$

5. $\dfrac{t}{6} = -6$ 6. $6 = \dfrac{t}{6}$

7. $\dfrac{2}{3} = \dfrac{x}{15}$ 8. $x + 3 = 9$

9. $13 = y + 2$ 10. $u - 3 = 11$

Solutions

1. $3x = 18$

Divide both sides by 3 $\left(\text{or multiply both sides by } \dfrac{1}{3} \right)$

$x = 6$

2. $26 = 2p$

Divide both sides by 2 $\left(\text{or multiply both sides by } \dfrac{1}{2} \right)$

$p = 13$

3. $-2y = 6$

Divide both sides by -2 $\left(\text{or multiply both sides by } -\dfrac{1}{2} \right)$

$y = -3$

4. $\dfrac{x}{5} = 2$

Multiply both sides by 5
$x = 10$

5. $\dfrac{t}{6} = -6$

Multiply both sides by 6
$t = -36$

6. $6 = \dfrac{t}{6}$

Multiply both sides by 6
$t = 36$

7. $\dfrac{2}{3} = \dfrac{x}{15}$

Multiply both sides by the GCD (i.e. 15)
$x = 10$

8. $x + 3 = 9$

Subtract 3 from both sides
$x = 6$

9. $13 = y + 2$

Subtract 2 from both sides
$11 = y$

10. $u - 3 = 11$

Add 3 to both sides
$u = 14$

> ### EXERCISE 9:4

Solve the following equations.

1. $4x = 32$ 2. $15 = 5t$

3. $-3p = 18$ 4. $\dfrac{u}{7} = 6$

5. $4 = \dfrac{z}{9}$ 6. $\dfrac{w}{-3} = 11$

7. $-9 = \dfrac{r}{3}$ 8. $\dfrac{-5q}{2} = 15$

The solution is sometimes written using set builder notation as follows.

Solution = $\{x : x = 6\}$

Solution = $\{y : y = -3\}$

Solution = $\{q : q = 18\}$

Two-Step Simple Linear Equations

Most of these types of equations have like terms on both sides and the unknown is usually multiplied or divided by some constant. To solve them, perform the inverse of the operations identified aiming at isolating the unknown.

Examples 9:4

Find the solution set of each of the following equations.

(a) $1 - 2x = -7$ (b) $2r - 5 = 8$

(c) $9 = 4m - 7$ (d) $7p = 8 + 3p$

(e) $5x + 9 = 2x$ (f) $\dfrac{v}{4} - 6 = 2$

Solutions

(a) $1 - 2x = -7$

    Subtract 1 from both sides

    $-2x = -8$

    Divide both sides by -2

    $x = 4$

    Solution set = $\{x : x = 4\}$

Explaining the steps taken is not always necessary. Wherever the explanations are given, this is to make the learner understand the procedure. Thus;

(b) $2r - 5 = 8$

    $2r = 13 \Rightarrow r = \dfrac{13}{2}$

    Solution set = $\left\{r : r = \dfrac{13}{2}\right\}$

(c) $9 = 4m - 7$

    $16 = 4m \Rightarrow 4 = m$

    Solution set = $\{m : m = 4\}$

(d) $7p = 8 + 3p$

    $4p = 8 \Rightarrow p = 2$

    Solution set = $\{p : p = 2\}$

(e) $5x + 9 = 2x$

    $9 = -3x \Rightarrow -3 = x$

    Solution set = $\{x : x = -3\}$

(f) $\dfrac{v}{4} - 6 = 2$

    $\dfrac{v}{4} = 8 \Rightarrow v = 32$

Solution set = $\{v : v = 32\}$

EXERCISE 9:5

Solve the following equations.

1. $2x + 5 = 9$ 2. $4t + 11 = 21$

3. $20 = 3q + 8$ 4. $13 = 6 + 7p$

5. $2a - 5 = 9$ 6. $4x - 11 = 21$

7. $60 = 10d - 20$ 8. $11 = 6y - 16$

9. $\dfrac{u}{4} + 3 = 7$ 10. $\dfrac{m}{5} + 2 = 10$

11. $17 = \dfrac{n}{2} + 15$ 12. $25 = \dfrac{x}{10} + 2$

13. $\dfrac{k}{4} - 3 = 7$ 14. $\dfrac{z}{5} - 2 = 10$

15. $2(n + 5) = 18$ 16. $5u = 2u + 27$

17. $5x = 40 - 3x$ 18. $2x = 90 - 7x$

19. $10t - 11 = 8t$ 20. $18 - 5a = a$

21. $9u = 16u - 105$ 22. $13y = 15 + 3y$

23. $4y + 5 = 5y - 30$ 24. $3t + 10 = 2t + 20$

Multi-Step Simple Linear Equations

Most of these types of equations have the unknown on both sides. To solve them, perform the inverse of the operations identified aiming at combining like terms and bringing the unknown on one side.

Examples 9:6

Solve the equations

(1) $5x - 2 = 3x + 4$

(2) $-7 - 4a = 48 + 7a$

Solutions

(1) $5x - 2 = 3x + 4$

    Add 2 to both sides

    $5x = 3x + 6$

    $2x = 6 \Rightarrow x = 3$

(2) $-7 - 4a = 48 + 7a$

    Add 7 to both sides

$-4a = 55 + 7a$
Subtract $7a$ from both sides
$-11a = 55 \Rightarrow a = -5$

Now enough resources have been built to solve Nde's problem.
We were required to determine
(a) the price of a book and
(b) the amount Nde had.

$7b + 150 - 200 = 5b + 400$
$\Rightarrow 7b - 50 = 5b + 400$
Adding 50 to both sides,
$7b = 5b + 450$
Subtracting $5b$ from both sides,
$2b = 450$
Dividing both sides by 2,
$b = 225$
(a) the price of a book 225 FRS
(b) the amount Nde had $5b + 400$
$$= 5(225) + 400.$$
$$= 1525) \text{ FRS}$$

EXERCISE 9:6

1. $5p + 3 - 2p = p + 8$
2. $7m + 10 - 2m = 16m - 12$
3. $9y - 19 = 5y + 21$
4. $3x - 7 = 5 - x$
5. $6x - 3 = 7 + x$
6. $5 - 6x = x - 9$
7. $36 = -2(m + 3) + m$
8. $6 = 2(x + 8) - 5x$
9. $3(t + 7) = t - 19$
10. $5(4x - 2) = x + 9$
11. $-6p - 21 = 3p - 12$
12. $3x + 5 = 21 - x$

Simple Linear Equations Involving Fractions and Decimals

When an equation involves fractions, multiply both sides by the LCM of the denominators to get rid of the fractions before solving. If it involves decimals, multiply both sides by the power of ten, which has the same number of zeros as the decimal with the highest number of decimal places.

Examples 9:7
Solve the following equations.

(a) $1.1x + \dfrac{1}{5}x = 12.2 - 3\dfrac{1}{10}$

(b) $0.17x - 10.966 = 1\dfrac{7}{20}x + 36.234$

Solution

(a) $1.1x + \dfrac{1}{5}x = 12.2 - 3\dfrac{1}{10}$

$1.1x + \dfrac{1}{5}x = 12.2 - \dfrac{31}{10}$

Multiply both sides by 10.

$1.1x + \dfrac{1}{5}x = 12.2 - \dfrac{31}{10}$

$11x + 2x = 122 - 31$

$13x = 91$

$\Rightarrow x = 7$

(b) $0.17x - 10.966 = 1\dfrac{7}{20}x + 36.234$

$0.17x - 10.966 = 1.35x + 36.234$

Multiply both sides by 1,000.

$170x - 10966 = 1350x + 36234$

$-1180x = 47200$

$x = -40$

EXERCISE 9:7

Solve the following equations.

1. $0.6x - 0.4 = 0.4x + 0.6$

2. $p - 0.1p + 0.9 = 0.2(p + 1)$

3. $\dfrac{3}{4}x + \dfrac{1}{5}x = \dfrac{1}{2}x - \dfrac{3}{10}$

4. $\dfrac{14}{3} - 7x = 8$ 5. $3 = \dfrac{w}{12} - 7\dfrac{1}{4}$

6. $5\dfrac{1}{2} = \dfrac{x}{8} - 4$ 7. $\dfrac{2x-5}{3} = 25$

8. $3y + \dfrac{1}{2}y - \dfrac{2}{5}y = \dfrac{y}{10} + \dfrac{7}{10}$

9. $100 + 3\dfrac{1}{2}x = 23\dfrac{1}{2}x$

10. $\dfrac{4x}{3} + 2 = \dfrac{5x}{2} - \dfrac{3}{2}$

Word Problems on Simple Linear Equations

Nde's problem is an example of a real life application of simple linear equations. The following are more examples.

Examples 9:8
Think of a number, divide it by 15 then subtract 3, the result is $\frac{1}{3}$. What is the number?

Solution
$$\frac{x}{15} - 3 = \frac{1}{3}$$
Multiply both sides by 15.
$$x - 45 = 5$$
Add 45 to both sides
$$x = 50$$

Examples 9:9
A girl has three times an amount of money as her friend. If their total sum is 700 FRS, find how much each of them has.

Solution
Let the girl's friend have x FRS.
Then, the girl has $3x$ FRS.
$$3x + x = 700$$
$$4x = 700$$
Divide both sides by 4
$$x = 175$$
Therefore, the girl has 525 FRS and her friend has 175 FRS.

EXERCISE 9:7

1. The length of a rectangle is 2 cm longer than its width. Find the length of the rectangle if the perimeter is 48 cm.
2. A woman bought some mangoes and shared to her four children, each child having 2. How many mangoes did she buy?
3. A student buys 2 pens at 50 FRS each and 3 exercise books at y FRS each. If her total expenditure on these is 400 FRS, find the value of y.
4. Kanjo and Ndi are travelling at x km/h and $3x$ km/h respectively. Find the value of x if the arithmetic mean of their speeds is 4 km/h.
5. When the difference between 3 and a number is divided by 4, the result is the number divided by two. What is the number?

6. Auk is travelling at 3 km/h and Tabi is travelling at x km/h. Their average speed is 4 km/h. Find the value of x.
7. A tailor whose daily wage is d FCFA obtains 42,000 FCFA every week. What is his daily wage?
8. A boy will be 14 years three years from now. How many years was he 4 years ago?
9. During a football competition a team, which played 16 matches, altogether lost three times as many matches as it won. How many marches did the team win?
10. Three times the sum of a number and six is 33. Find the number.
11. A number increased by 20 equals three times the same number. What is the number?
12. Twelve times h equals one hundred and twenty minus twelve. Find h.
13. MTN charges 250 Frs. per minute for calls. After making an MTN call that last m minutes, Konyuy is charged a bill of 1750 Frs. For how many minutes did Konyuy make the call?
14. Mbianda bought a radio that costs 16960 FCFA. With tax, t she pays 17810 FCFA. Find the value of t.
15. Tanto intends to buy a bicycle, which costs 27000 FCFA. He has 6000 FCFA and saves 3000 FCFA every week for n weeks. Write down an equation and use it to calculate the number of weeks he saves 3000 FCFA.
16. A number increased by 5 equals twice the same number decreased by 4. Find the number.
17. Use the Pythagoras theorem (stated below), to find the value of x in the given right-angled triangle.

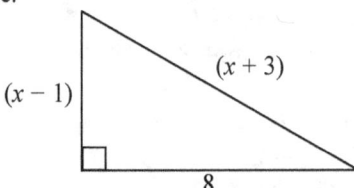

Figure 9:2
The Pythagoras theorem states that '*the square on the hypotenuse is equal to the sum of the squares on the two arms of any right angle triangle*'.

MULTIPLE CHOICE EXERCISE 9

1. x is an unknown in:

    [A] $3x + 5 = 0$ [B] $x^2 + x + 5$

    [C] $y = x^2 + x + 5$ [D] $y = 3x + 5$

2. x is a variable in:

    [A] $3x + 5 = 0$ [B] $2x^2 + x + 5$

    [C] $(2x + 1)(x + 1)$ [D] $y = 3x + 5$

3. The additive inverse of $\dfrac{7}{2}$ is:

    [A] $\dfrac{7}{2}$ [B] $\dfrac{2}{7}$

    [C] $-\dfrac{7}{2}$ [D] $-\dfrac{2}{7}$

4. The additive inverse of -14 is:

    [A] -14 [B] 14

    [C] $\dfrac{1}{14}$ [D] $-\dfrac{1}{14}$

5. The multiplicative inverse of 12 is:

    [A] -12 [B] 12

    [C] $-\dfrac{1}{12}$ [D] $\dfrac{1}{12}$

6. The multiplicative inverse of $\dfrac{2}{15}$ is:

    [A] $-\dfrac{2}{15}$ [B] $\dfrac{15}{2}$

    [C] $\dfrac{2}{15}$ [D] $-\dfrac{15}{2}$

7. Given that $8(x + 8) = 40$. The statement, which best interprets this equation is:

    [A] Eight times the sum of a number and 8 is 40

    [B] Eight times a number less than eight is 40

    [C] Eight less than a number is 40

    [D] The product of eight and a number is 40

8. The root of the equation $3x + 4x = 42$ is:

    [A] $x = 4$ [B] $x = 6$

    [C] $x = 8$ [D] $x = 7$

9. The root of the equation $3n + 14 = 47$ is:

    [A] $n = 8$ [B] $n = 9$

    [C] $n = 10$ [D] $n = 11$

10. The root of the equation $6y - 48 = 2y$ is:

    [A] $y = 8$ [B] $n = 9$

    [C] $n = 10$ [D] $n = 12$

11. The value of x for which $2x + 8 = 0$ is:

    [A] -4 [B] 4 [C] -6 [D] 6

12. Given that $33 = 6y + 3$. The value of y must be:

    [A] 3 [B] 5 [C] 6 [D] 8

13. If $3x - 7 = 10$, then the value of x is:

    [A] $\dfrac{3}{17}$ [B] $-\dfrac{17}{3}$

    [C] $\dfrac{17}{3}$ [D] $-\dfrac{3}{17}$

14. If $6x + 4 = -20$, then the value of x is:

    [A] $-\dfrac{8}{3}$ [B] $\dfrac{8}{3}$ [C] -4 [D] 4

15. The value of x in the equation $5x + 1 = 31$ is:

    [A] 1 [B] 5 [C] 25 [D] 6

16. The value of y, which satisfies the equation, $4(y - 4) = 20$ is:

    [A] 1 [B] 24 [C] 6 [D] 9

17. The solution of $5(x - 4) - 4(x + 1) = 0$ is:

    [A] 16 [B] -24

    [C] 24 [D] -16

18. The root of the equation $6(x - 4) + 3(x + 7) = 3$ is:

    [A] $\dfrac{3}{2}$ [B] $\dfrac{1}{3}$ [C] $\dfrac{1}{2}$ [D] $\dfrac{2}{3}$

19. $\dfrac{x - 2}{3} = 8$ only if:

    [A] 26 [B] 22 [C] 24 [D] 19

20. When $\dfrac{x}{4} = \dfrac{5}{2}$ the value of x is:

    [A] 2 [B] 4 [C] 5 [D] 10

21. Given that $\dfrac{1}{3}(x + 1) = 6$ the value of x is:

    [A] 19 [B] 17 [C] 5 [D] 3

22. Given that $\dfrac{2}{x} = \dfrac{3}{6}$, the value of x is:

    [A] 6 [B] 4 [C] 3 [D] 2

23. The only condition for which $\dfrac{5}{x + 1}$ is equal to 4 is that:

    [A] $x = 4$ [B] $x = 8$

    [C] $x = \dfrac{1}{8}$ [D] $x = \dfrac{1}{4}$

24. Given that, $\dfrac{3 - 2y}{4} - \dfrac{2y}{6}$ the value of y is:

    [A] $\dfrac{10}{9}$ [B] $\dfrac{9}{10}$ [C] 3 [D] -3

25. The root of the equation $\dfrac{2x + 7}{6} + \dfrac{x - 5}{3} = 0$ is:

[A] $x = -\dfrac{3}{5}$ [B] $x = \dfrac{3}{4}$ [C] $x = \dfrac{1}{4}$ [D] $x = \dfrac{2}{5}$

26. Given that $\dfrac{3x - 2}{6} - \dfrac{2x + 7}{9} = 2$. x is equal to:

   [A] $x = -\dfrac{36}{5}$ [B] $x = \dfrac{36}{5}$ [C] $x = -\dfrac{16}{5}$ [D] $x = \dfrac{16}{5}$

27. The value of x which satisfies the expression

   $\dfrac{1}{x} + \dfrac{4}{3x} - \dfrac{5}{6x} + 1 = 0$ is:

   [A] $\dfrac{1}{6}$ [B] $\dfrac{1}{4}$ [C] $-\dfrac{3}{2}$ [D] $-\dfrac{7}{8}$

28. Using the relation $C = \dfrac{5}{9}(F - 32)$, the value

   of F when $C = 40$ is:

   [A] 67 [B] 77 [C] 81 [D] 104

29. The value of t which satisfies

   $\dfrac{3t}{4} + \dfrac{1}{3}(21 - t) = 11$ is:

   [A] $9\dfrac{3}{5}$ [B] $3\dfrac{9}{13}$ [C] 5 [D] $\dfrac{9}{13}$

30. If $8x - 4 = 6x - 10$, the value of $5x$ is:

   [A] 7 [B] −15 [C] −3 [D] 3

31. The value of x which satisfies the equation

   $5(x - 7) = 7x - 5$ is:

   [A] $x = 6$ [B] $x = -30$
   [C] $x = -15$ [D] $x = -6$

32. If $2(x + 1) = 4x + 3$, x equals:

   [A] 2 [B] −2 [C] $\dfrac{1}{2}$ [D] $-\dfrac{1}{2}$

33. The value of v which satisfies the equation

   $\dfrac{12}{v} = \dfrac{15}{v + 4}$ is:

   [A] 10 [B] 12 [C] 14 [D] 16

34. The value of $3(p + 7)$ for which

   $6p + 5 = 4p + 11$ is:

   [A] 15 [B] 20 [C] 25 [D] 30

35. The equation $\dfrac{2}{3}(x + 5) = \dfrac{1}{4}(5x - 3)$ has

   root:

   [A] $1\dfrac{1}{7}$ [B] 7 [C] 3 [D] $4\dfrac{3}{7}$

36. Given that $\dfrac{m}{3} + \dfrac{1}{2} = \dfrac{3}{4} + \dfrac{m}{4}$, the value of m

   is:

   [A] −3 [B] −2 [C] 2 [D] 3

37. The only condition for $0.6x - 0.4$ to be equal

to $1.2x + 0.8$ is that:

   [A] $x = -2$ [B] $x = 2$
   [C] $x = -0.5$ [D] $x = -0.5$

38. Given that $0.9n - 0.7 = 0.3n - 0.1$, then n must be:

   [A] 4 [B] 3 [C] 2 [D] 1

39. A man is 23 years older than his son this year. Given that his son will be 12 years old in ten years time, the man will be:

   [A] 25 years [B] 35 years
   [C] 33 years [D] 37 years

40. Abe bought two packets of sugar at x FRS each and four tins of milk at 400 FRS each. If the total cost is 3400 FRS, the price of a packet of sugar is:

   [A] 400 FRS [B] 1800 FRS
   [C] 900 FRS [D] 1600 FRS

41. 8 more than thrice a number is 35. The number is:

   [A] 27 [B] 43 [C] 9 [D] 14.3

42. A trader bought 400 liters of palm oil and sold $8x$ liters. If 160 liters are left, the number of liters he sold is:

   [A] 30 [B] 240 [C] 5 [D] 20

43. Given that $x = 3$. The number which can be added to $12x$ to make 57 is:

   [A] 21 [B] $\dfrac{19}{4}$ [C] 45 [D] 33

44. The sum of three consecutive numbers is 42. The largest of the numbers is:

   [A] 13 [B] 14 [C] 15 [D] 16

45. Three quarters of a certain number is 15. The number is:

   [A] 20 [B] 16 [C] 15 [D] 12

46. The ratio of $2x$ to $x + 1$ is $5 : 3$ only if x is:

   [A] $x = 5$ [B] $x = 4$ [C] $x = 3$ [D] $x = 2$

47. The ages of two people are in the ratio 5: 9. If the elder is 8 years older, the age of the younger is:

   [A] 45 years [B] 10 years
   [C] 8 years [D] 2 years

TOPIC 10

EXPANSIONS AND FACTORISATION

OBJECTIVES

At the end of this topic, the learner should be able to:

1. Expand the product of a monomial and a binomial.
2. Expand the product of two binomials.
3. Recognise and expand a perfect square.
4. Recognise and expand the square of a binomial and use it to evaluate the squares of numerical expressions.
5. Apply the expansion of a perfect square to square numbers.
6. State the condition for the expression $x^2 + bx + c$ to be a perfect square and use it to determine whether an expression of the form $x^2 + bx + c$ is a perfect square.
7. Complete the square of the trinomial $x^2 + bx + c$.
8. Factorise linear expressions by grouping.
9. Recognise and factorise or expand the difference of two squares expression and use it to evaluate numerical expressions involving the difference of two squares.
10. Factorise quadratic expressions in one or two variables.
11. Identify simple identities and distinguish between identities and equations.

Sir Isaac Newton
Isaac Newton's work represents one of the greatest contributions to science ever made by an individual. Most notably, Newton derived the law of universal gravitation, invented the branch of mathematics called calculus, and performed experiments investigating the nature of light and color. The binomial theorem was formulated by Sir Isaac Newton and the Swiss mathematician Leonhard Euler, but Niels Henrik Abel gave it a more comprehensive generalization, including the cases of irrational and imaginary exponents.

Rex Features, Ltd.

EXPANSIONS

To expand an algebraic expression, an algebraic product is changed into an algebraic sum.

(1) *The product of a Monomial and a Binomial*

A **binomial expression** is an expression containing exactly two terms (monomials) separated by + or −.
$a + b$, $ax + y$, $5x + 7y$, $5y - 2xy$ are examples of binomials.

Investigatory Exercise 13.1

(a) Bame bought 3 pencils at 20 francs each and 3 pens at 50 francs each. Calculate in two different ways, the total cost of the pencils and the pens.

Method 1	Method 2
$= 3 \times 20 + 3 \times 50$	$= 3 \times (20 + 50)$
$= 60 + 150$	$= 3 \times 70$
$= 210$ francs	$= 210$ francs

(b) A man built a store 3 m by 4 m. Later he decided to extend the length of the store by 2 m. Calculate in two different ways the total area of the floor of the store.

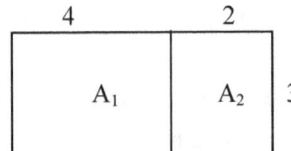

Figure 10:1

Method 1	Method 2
$= 4 \times 3 + 2 \times 3$	$= (4 + 2) \times 3$
$= 12 + 6$	$= 6 \times 3$
$= 18$ square m	$= 18$ square m

From (a) and (b), it can be deduced that,

(1) $a(b + c) = ab + ac$ or $(b + c)a = ab + ac$

In like manner,

(2) $a(b - c) = ab - ac$ or $(b - c)a = ab - ac$

These expansions use the **distributive law** and the process is known as **removal of brackets**.

Example 10:1
Expand the following
(a) $2(4x + 3)$ (b) $x(2 + y)$
(c) $u(3x-1)$ (d) $2x(x + 1)$

Solution
(a) $2(4x + 3) = 2(4x) + 2(3) = 8x + 6$
(b) $x(2 + y) = x(2) + xy = 2x + xy$
(c) $u(3x-1) = u(3x) - u(1) = 3ux - u$
(d) $2x(x + 1) = 2x(x) + 2x(1) = 2x^2 + 2x$

Example 10:2
Expand (a) $(3b - 2c)4$ (b) $(24a - 1)a$

Solution
(a) $(3b - 2c)4 = 3b(4) - 2c(4) = 12b - 8c$
(b) $(24a - 1)a = 24a^2 - a$

(2) *The Product of Two Binomials*

Using the idea of the expansion of the product of a monomial and a binomial above, the expansion of the product of two binomials can be done as follows.
$$(a + b)(c + d) = a(c + d) + b(c + d)$$
$$= ac + ad + bc + bd$$

$$\therefore (a + b)(c + d) = ac + bc + ad + bd$$

Example 10:3
Expand
(a) $(x + 1)(x + 2)$ (b) $(y - 2)(y + 3)$

Solution
(a) $(x + 1)(x + 2) = x(x + 2) + 1(x + 2)$
$$= x^2 + 2x + x + 2$$
$$= x^2 + 3x + 2$$
(b) $(y - 2)(y + 3) = y(y + 3) - 2(y + 3)$
$$= y^2 + 3y - 2y - 6$$
$$= y^2 + y - 6$$

(3) *The Square of a Binomial*
$$(a + b)^2 = (a + b)(a + b)$$
$$= a(a + b) + b(a + b)$$
$$= a^2 + ab + ab + b^2$$
$$= a^2 + 2ab + b^2$$

$$\therefore (a + b)^2 = a^2 + 2ab + b^2 \ldots\ldots\ldots\ldots ①$$

(b) $(a - b)^2 = (a - b)(a - b)$
$$= a(a - b) - b(a - b)$$
$$= a^2 - ab - ab + b^2$$

$$= a^2 - 2ab + b^2$$

$$\therefore (a+b)^2 = a^2 + 2ab + b^2 \ldots\ldots\ldots ②$$

The identities ① and ② are examples of **perfect squares**. A perfect square is a number or expression, which has an exact square root. Numerical examples of perfect squares are 1, 4, 9, 16, 25, 36, 49, etc. The above expansions are very useful in computing numerical perfect squares.

Example 10:4
Evaluate   (a) 17^2 (b) 99^2

Solution
(a) $17^2 = (10+7)(10+7)$
   $\quad = 100 + 140 + 49 = 289$
(b) $99^2 = (100-1)(100-1)$
   $\quad = 10000 - 200 + 1 = 9801$

> **EXERCISE 10:1**

(a) Expand the following.
   1. $3(x+5)$ 2. $(y+2)4$
   3. $x(4+y)$ 4. $(3+u)v$
   5. $2(x-1)$ 6. $(y-3)5$
   7. $3p(2-q)$ 8. $(4-s)t$
   9. $2x(5-x)$ 10. $(x+1)(x+3)$
   11. $(x-1)(x+2)$ 12. $(p+3)(p-1)$
   13. $(x+3)^2$ 14. $(x-2)^2$
   15. $(4+a)^2$ 16. $(3-y)^2$

(b) Evaluate the following.
   (1) 49^2 (2) 101^2
   (3) 198^2 (4) 442^2

The Trinomial Perfect Square Test

From the expansions of the perfect squares in ① and ② it can be seen that,
$$(x \pm k)^2 = x^2 \pm 2kx + k^2.$$

Comparing the RHS with $x^2 + bx + c$ it can be seen that,
$$2k = b \Rightarrow k = \frac{b}{2} \ldots\ldots\ldots ③$$
$$\text{and } c = k^2 \ldots\ldots\ldots ④$$

Substituting ③ in ④
$$c = \left(\frac{b}{2}\right)^2$$

Therefore, the necessary and sufficient condition for the quadratic expression $x^2 + bx + c$ to be a perfect square is that, $c = \left(\frac{b}{2}\right)^2$. Hence, to make $x^2 + bx$ a perfect square it is necessary to add $\left(\frac{b}{2}\right)^2$ to $x^2 + bx$.

This process is known as **completing the square**.

Example 10:5
Complete the square of $x(x-5)$

Solution
$$x(x-5) = x^2 - 5x$$
$$c = \left(\frac{b}{2}\right)^2, b = -5$$
$$\Rightarrow c = \left(\frac{-5}{2}\right)^2 = \frac{25}{4}$$
$$\Rightarrow \text{the required perfect square is } x^2 - 5x + \frac{25}{4}$$

> **EXERCISE 10:2**

1. Complete the square in each of the following cases.
   (a) $p^2 - 20p$ (b) $y^2 - 14y$
   (c) $x^2 + 5x$ (d) $u^2 + \frac{4}{3}u$
   (e) $x(x+18)$ (f) $x^2 - 4x$

2. Write the following perfect squares as a square of a binomial.
   (a) $t^2 + 12t + 36$ (b) $x^2 - 18x + 81$
   (c) $y^2 + 7y + \frac{49}{4}$ (d) $u^2 - 11u + \frac{121}{4}$
   (e) $x^2 + 20x + 100$ (f) $m^2 - 30m + 225$

FACTORISATION

Factorisation is the reverse process of expansion. To factorise, extract the HCF of all the terms. For instance, since $a(b+c) = ab+ac$, a is the HCF of ab and ac. So to factorise $ab + ac$, remove a from the bracket. Thus,

$$ab + ac = a(b + c)$$

Example 10:6

Factorise the following:

(a) $2x + 2y$ $\qquad$ (b) $4xy + 12xz$

(c) $\dfrac{1}{4}pqr - \dfrac{5}{12}pqs$

Solution

(a) Since HCF of $2x$ and $2y$ is 2,
$$2x + 2y = 2(x + y)$$

(b) Since the HCF of $4xy$ and $12xz$ is $4x$,
$$4xy + 12xz = 4x(y + 3z)$$

(c) Since HCF of $\dfrac{1}{4}pqr$ and $-\dfrac{5}{12}pqs$ is $\dfrac{1}{4}pq$,

$$\frac{1}{4}pqr - \frac{5}{12}pqs = \frac{1}{4}pq\left(r + \frac{5}{3}s\right)$$

EXERCISE 10:3

Factorise:

1. $3x + xy$ $\qquad$ 2. $2p + 6q$

3. $5x - 10y$ $\qquad$ 4. $4xy + 8y$

5. $6uv - 3uf$ $\qquad$ 6. $\dfrac{1}{4}xy + \dfrac{1}{4}px$

7. $\dfrac{1}{2}ax - \dfrac{1}{2}bx$ $\qquad$ 8. $\dfrac{2}{3}y - \dfrac{1}{3}x$

Factorising by Grouping

In factorising more than three terms, similar terms are usually grouped.

Example 10:7

Factorise the following:

(a) $px - py + qx - qy$

(b) $x - y + xy - 1$

(c) $6x - 6y + 3ax - 3ay$

Solution

(a) $px - py + qx - qy = p(x - y) + q(x - y)$
$$= (x - y)(p + q)$$

(b) Rearrange
$$x - y + xy - 1 = xy - y + x - 1$$
$$= (xy - y) + (x - 1)$$
$$= y(x - 1) + 1(x - 1)$$
$$= (x - 1)(y + 1)$$

(c) $6x - 6y + 3ax - 3ay = (6x - 6y) + (3ax - 3ay)$
$$= 6(x - y) + 3a(x - y)$$
$$= (x - y)(6 + 3a)$$

EXERCISE 10:4

Factorise

(a) $xy + x + y + 1$

(b) $ax + a + 3x + 3$

(c) $6px + 4p + 3x + 2$

(d) $3x - 6xy + 2 - 4y$

(e) $6y - 9x - 4y + 6$

(f) $5u - ut - 5v + tv$

(g) $an + am - 3m - 3n$

(h) $pr + 3ps - 2qr - 6qs$

(i) $x^2 - 4x + 6xy - 24y$

(j) $2q^3 - 14q^2 + 3q - 21$

(k) $24x + 2ab - 3bx - 16a$

(l) $p^2 - 5 - p^2q + +5q$

The Difference of Two Squares

Consider the expansion of $(a + b)(a - b)$

Therefore, $(a + b)(a - b) = a^2 - b^2$

An expression such as $a^2 - b^2$ is called the **difference of two squares**. The difference of two squares is very useful in evaluating numerical expressions such as $19^2 - 13^2$, $925^2 - 725^2$.

Example 10:8

Using the idea of the difference of two squares, evaluate the following.

(a) $4^2 - 3^2$ $\qquad$ (b) $25^2 - 16^2$

(c) $19^2 - 13^2$ (d) $925^2 - 725^2$

Solution

(a) $4^2 - 3^2 = (4-3)(4+3) = 1(7) = 7$

(b) $25^2 - 16^2 = (25-16)(25+16) = 9(41) = 369$

(c) $19^2 - 13^2 = (19-13)(19+13) = 6(32) = 192$

(d) $925^2 - 725^2 = (925-725)(925+725)$

$$= 200(1650)$$

$$= 330,000$$

Example 10:9

Factorise the following:

(a) $x^2 - y^2$ (b) $(ax)^2 - (by)^2$

(c) $25a^2 - 9b^2$ (d) $a^2 - b^2c^2$

Solution

(a) $x^2 - y^2 = (x-y)(x+y)$

(b) $(ax)^2 - (by)^2 = (ax - by)(ax + by)$

(c) $25a^2 - 9b^2 = (5a)^2 - (3b)^2$
$$= (5a - 3b)(5a + 3b)$$

(d) $a^2 - b^2c^2 = a^2 - (bc)^2 = (a - bc)(a + bc)$

EXERCISE 10:5

1. factorise:

(a) $x^2 - y^2$ (b) $1 - 9y^2$

(c) $36 - 9x^2$ (d) $4a^2 - 1$

(e) $4y^2 - 9x^2$ (f) $4u^2 - 25y^2$

(g) $36a^2 - 49b^2$ (h) $x^2 - y^2z^2$

2. Evaluate the following:

(a) $41^2 - 40^2$ (b) $124^2 - 120^2$

(c) $625 - 525^2$ (d) $3.8^2 - 3.7^2$

(e) $1375^2 - 1325^2$ (f) $0.003^2 - 0.002^2$

Factorising Quadratic Expressions

Consider the following expansion

$(2x + 1)(x + 2) = 2x^2 + 5x + 2$

The RHS is of the form $ax^2 + bx + c$, where a, b, c are constants and $a \neq 0$. This is the general form of a quadratic expression. The power of the unknown (x) is 2.

To factorise such an expression, the following steps may be taken:

(i) Multiply a by c to have ac

(ii) Find the pair of integral factors p and q of ac whose sum or difference is b.

(iii) Substitute the middle term bx with the sum or difference of px and qx in $ax^2 + bx + c$.

(iv) Factorise the expression by grouping.

Example 10:10

Factorise:

1. $6x^2 + 7x - 3$ 2. $2x^2 - x - 1$

3. $12x^2 + 8x - 15$

Solutions

1) $6x^2 + 7x - 3$

$ac = (6)(-3) = -18$ and $b = 7$

Pairs of factors of 18 are {1, 18}, {2, 9}, {3, 6}.

Since 9–2 = 7, it means –2 and 9 are the required factors.

$\Leftrightarrow 6x^2 + 7x - 3 = 6x^2 + 9x - 2x - 3$

Factorising by grouping

$6x^2 + 7x - 3 = (6x^2 + 9x) + (-2x - 3)$

$$= 3x(2x + 3) - 1(2x + 3)$$

$$= (2x + 3)(3x - 1)$$

2) $2x^2 - x - 1$

$ac = -2$

Pair of factors of $2 = \{1, 2\}$ and

$-2 + 1 = -1$

$2x^2 - x - 1 = 2x^2 - 2x + x - 1$

Factorising by grouping

$2x^2 - x - 1 = (2x^2 - 2x) + (x - 1)$

$$= 2x(x - 1) + 1(x - 1)$$

$$= (x - 1)(2x + 1)$$

3) $12x^2 + 8x - 15 = 12x^2 + 18x - 10x - 15$

$$= 6x(2x + 3) - 5(2x + 3)$$

$$= (2x + 3)(6x - 5)$$

When a quadratic is factorable, a binomial factor of the first bracket during grouping is always certainly a factor of the second bracket. For instance in Example 10:10 (3) above $2x + 3$ is a binomial factor of the first and second bracket of the expression $6x(2x + 3) - 5(2x + 3)$.
Advantage should always be taken of this.

When the coefficient of x^2 is unity

When the coefficient of x^2 is unity, i.e. $a = 1$, the factorisation becomes simpler.

Example 10:11

Factorise:

(1) $x^2 + x - 2$ (2) $x^2 + 9x + 18$)

(3) $3x^2 - 12x - 63$

Solution

(1) Factors of –2 whose sum is +1 are, +2 and –1

$$x^2 + x - 2 = (x + 2)(x - 1)$$

(2) Factors of +18 whose sum is +9 are, +6 and +3

$$x^2 + 9x + 18 = (x + 6)(x + 3)$$

(3) $3x^2 - 12x - 63 = 3\{x^2 - 4x - 21\}$

Factors of –21 whose sum is – 4 are, –7 and +3.

$$\Rightarrow 3x^2 - 12x - 63 = 3\{(x - 7)(x + 3)\}$$

Example 10:12

Factorise $3 - 8y + 4y^2$

Solution

$$3 - 8y + 4y^2 = 3 - 6y - 2y + 4y^2$$
$$= 3(1 - 2y) - 2y(1 - 2y)$$
$$= (1 - 2y)(3 - 2y)$$

Quadratic Expressions in Two Variables

For a quadratic expression of the form $ax^2 + bxy + cy^2$, follow steps (i) and (ii) as in the preceding section, but in (iii) substitute bxy with the sum or difference of pxy and qxy and then proceed to (iv)

Example 10:13

Factorise $3x^2 + xy - 10y^2$

Solution

$ac = -30$ and factors of –30 whose sum is +1 are –5 and + 6

$$3x^2 + xy - 10y^2 = 3x^2 - 5xy + 6xy - 10y^2$$
$$= (3x^2 - 5xy) + (6xy - 10y^2)$$
$$= x(3x - 5y) + 2y(3x - 5y)$$
$$= (x + 2y)(3x - 5y)$$

EXERCISE 10:6

Factorise the following.

(1) $x^2 + x - 6$ (2) $p^2 + 12p + 11$

(3) $y^2 - 7y + 12$ (4) $x^2 + 6x - 16$

(5) $x^2 - 2x - 15$ (6) $10x^2 - x - 3$

(7) $4a^2 - 3a - 10$ (8) $5 - 16y + 12y^2$

(9) $3 - x - 2x^2$ (10) $10 - 11x + 3x^2$

(11) $x^2 + xy - 12y^2$ (12) $2x^2 + 7xy - 15y^2$

(13) $6k^2 + 17kx - 3x^2$ (14) $12p^2 - 16pq + 5q^2$

Identities

Consider $3x = x + x + x$.

The right hand side (RHS) is simply another way of writing the left hand side (LHS). In addition, any number can be used to substitute x on both sides and the statement will still be true. For instance:

If $x = 0$, then LHS $= 3(0) = 0$ and
$$\text{RHS} = 0 + 0 + 0 = 0$$
If $x = -7$, then LHS $= 3(-7) = -21$ and
$$\text{RHS} = -7 - 7 - 7 = -21$$
If $x = 7$, then LHS $= 3(7) = 21$ and
$$\text{RHS} = 7 + 7 + 7 = 21$$

Thus, the statement is true for all values of x. Such a statement, which is true for all values of the variable is called an **identity** and the RHS substitutes the LHS and vice versa. When it is clear that a statement is an identity, the symbol '≡', read 'identical to' is used instead of '='.

The following are more examples of identities:

1. $(1 + x)^2 = 1 + 2x + x^2$
2. $3x^2 + 2x - 1 = 3(x^2 + 2x + 5) - 4(x + 4)$
3. $4x^2 - 1 = (2x + 1)(2x - 1)$
4. $(2x + 1)(x - 1) = 2x^2 - x - 1$

Identities differ from equations in that equations are true only for particular values of the unknown but identities are true for all values of the variable.

MULTIPLE CHOICE EXERCISE 10

1. The expression, which is a perfect square, is:

   [A] $(x +1)(x - 1)$ [B] $x^2 + 2x + 1$

   [C] $x^2 - y^2$ [D] $x^2 - 2x - 1$

2. The expression, which is not a difference of two squares, is:

   [A] $(x +1)(x - 1)$ [B] $x^2 - 1$

   [C] $x^2 - y^2$ [D] $(x - y)^2 = 6$

3. An identity among the following is:

   [A] $(x +1)(2x + 3) = 2x^2 + 5x + 3$

   [B] $(3p - 1)(2p + 1) = 3p^2 - 5p - 1$

   [C] $2y + 7 = 3y - 5$

   [D] $x^2 + 2x +1 = 6$

4. The statement, which is not an identity, is:

   [A] $(x + 2)(x - 1) = x^2 + x - 2$

   [B] $(5x - 1)(x + 1) = 5x^2 + 4x - 1$

   [C] $(3x + 1)(2x - 1) = 6x^2 - x - 1$

   [D] $(3x - 1)(2x + 1) = 3x^2 - 5x - 1$

5. When simplified, $5yx - 7xy + 4yx$ equals:

   [A] $9yx$ [B] $- 9xy$

   [C] $8xy$ [D] $2xy$

6. The simplified form of $6p + 7q - 8q - 5p$ is:

   [A] $p - q$ [B] $q - p$

   [C] $p + q$ [D] $11p - 5q$

7. $-7x + 8y - 2 + 9x - 10y + 4$ can be simplified to have:

   [A] $2(x - y + 1)$ [B] $2(x - y - 1)$

   [C] $2(x + y - 1)$ [D] $2(x + y + 1)$

8. Simplifying $15x - 12y + 8z - 14x + 12y - 8z$ leads to:

   [A] x [B] y [C] z [D] $x - y + z$

9. By simplification $x + (- x) + y$ is exactly:

   [A] $-y$ [B] y [C] $2x - y$ [D] $2x + y$

10. $32e + 6f - 12e + 4f$ can also be:

   [A] $38e + 6f$ [B] $30e + 8f$

   [C] $28e + 9f$ [D] $20e + 10f$

11. When expanded $- 2a(3a^2b + 4b^2)$ gives:

   [A] $- 6ab^2 - 8a^2b$ [B] $- 6ab^2 - 4ab^2$

   [C] $- 6ab^2 + 8a^2b$ [D] $- 6a^3b - 8ab^2$

12. On Simplification $13x - (2x - 4x - 3x)$ becomes:

   [A] $8x$ [B] $18x$ [C] $-8x$ [D] $-18x$

13. $9x - (5x - 3y) - y$ is equal to:

   [A] $4x - 2y$ [B] $4x + 2y$

   [C] $5x - 2y$ [D] $5x + 2y$

14. $-2a - 5b - (8b - 5a) =$

   [A] $- 8a + 13b$ [B] $3a + 3b$

   [C] $3a - 13b$ [D] $7a - 13b$

15. $(2x + y) + (x - 2y)$ is the same as:

   [A] $3x + y$ [B] $x - 3y$

   [C] $x + 3y$ [D] $3x - y$

16. $(2x + y) - (x - 2y)$ is equal to:

   [A] $3x - y$ [B] $x + 3y$

   [C] $x - 3y$ [D] $3x + y$

17. $(2x - 3) - (2 - 3x)$ is equal to:

   [A] $5x - 5$ [B] $5x - 1$

   [C] $x - 5$ [D] $x - 1$

18. Adding $(2x + y)$ and $(x - 2y)$ gives:

   [A] $3x + y$ [B] $x - 3y$

   [C] $x + 3y$ [D] $3x - y$

19. Given the statement

   $x - 13y + 5z - 4m = x - ($ $)$

   The expression required in the bracket is:

   [A] $- 13y + 5z - 4m$ [B] $- 13y + 5z$

   [C] $- 13y + 5z - 4x$ [D] $13y - 5z + 4m$

20. On expansion $(2x - 5)(x - 3)$ gives:

   [A] $x^2 - 11x - 15$ [B] $2x^2 - 11x + 15$

   [C] $2x^2 - 5x - 8$ [D] $x^2 - 5x + 15$

21. Given that $p = 3 - 2y$ and $q = 4 + 3y$. The value of pq is:

   [A] $- 6y^2 - y - 12$ [B] $6y^2 - y - 12$

   [C] $- 12 + y + 6y^2$ [D] $12 + y - 6y^2$

22. $(2x + y)(x - 2y)$ is equal to:

   [A] $2x^2 - 2y^2$ [B] $2x^2 + 3xy + 2y^2$

   [C] $2x^2 - 3xy - 2y^2$ [D] $2x^2 + 3xy - 2y^2$

23. $(2x - 1)(x + 2)$ is equal to:

   [A] $2x^2 - 2$ [B] $2x^2 + x - 2$

   [C] $2x^2 - x - 2$ [D] $2x^2 - 3x - 2$

24. On expansion $(4x - y)(x - 3y)$ becomes:

   [A] $4x^2 + 13xy - 3y^2$ [B] $6x^2 - 13xy + 3y^2$

   [C] $4x^2 - 13xy + 3y^2$ [D] $6x^2 + 13xy - 3y^2$

25. The product of $x - 1$ and $x + 1$ is:

   [A] 2 [B] $2x$

   [C] $x^2 + 2x - 1$ [D] $x^2 - 1$

26. If $(a + b)^2 = a^2 + 2ab + b^2$ the value of $(2a + 1)^2$ is:

   [A] $4a^2 + 4a - 1$ [B] $4a^2 + 4a + 1$

[C] $4a^2 - 4a - 1$ [D] $4a^2 - 4a + 1$

27. The square of $x - 8$ is equal to:

[A] $x^2 - 16x - 64$ [B] $x^2 + 16x - 64$

[C] $x^2 - 16x + 64$ [D] $x^2 - 32x + 64$

28. The coefficient of x in the expansion of

$(x + 9)(x + 3)$ is:

[A] –12 [B] 12 [C] 3 [D] –3

29. The coefficients of x and x^2 in the expansion

of $(x–3)^2$ are respectively:

[A] – 6, 1 [B] 6, 1

[C] –1, 6 [D] 1, –6

30. The coefficient of xy in the expansion of

$(3x + 2y)(4x - 2y)$ is:

[A] –2 [B] –14 [C] 2 [D] 14

31. When factorised $3x(4 - y) - m(y - 4)$

becomes:

[A] $(3x + m)(4 - y)$ [B] $(3x - m)(4 - y)$

[C] $(3x + m)(y - 4)$ [D] $(3x - m)(y + 4)$

32. The expression $x(a - c) + y(c - a)$ can be

factorised to obtain:

[A] $(a - c)(y + x)$ [B] $(a - c)(x - y)$

[C] $(a + c)(x - y)$ [D] $(a - c)(y - x)$

33. By factorising $m(2a - b) - 2n(b - 2a)$, the

result is:

[A] $(2a - b)(2n - m)$ [B] $(2a - b)(m - 2n)$

[C] $(2a - b)(m + 2n)$ [D] $(2a - b)(m - 2n)$

34. The difference between the squares of the

numbers 21 and 11 is:

[A] 20 [B] 100

[C] 220 [D] 320

35. The value of $13^2 - 12^2$ is:

[A] 25 [B] 5 [C] 1^2 [D] 125

36. $32x^3 - 8xy^2$ when factorised gives:

[A] $4(4x + y)(2x - y)$ [B] $(16x - y)(2x + y)$

[C] $8x(2x - y)$ [D] $8x(2x + y)(2x - y)$

37. By factorising $27p^2x^2 - 48y^2$ the result is:

[A] $3(3px - 4y)(3px + 4y)$ [B] $9(3px - 4y)^2$

[C] $9(px - 4y)(3px + 4y)$ [D] $3(3px - 4y)^2$

38. $(x - 2)(x + 3)$ are the factors of:

[A] $x^2 - 9$ [B] $x^2 - 6$

[C] $x^2 - x - 6$ [D] $x^2 + x - 6$

39. The result of factorising $x^2 + 4x - 192$ is:

[A] $(x - 4)(x + 48)$ [B] $(x + 48)(x + 4)$

[C] $(x - 12)(x + 16)$ [D] $(x - 12)(x - 16)$

40. When factorised the expression $2x^2 + x - 15$

equals:

[A] $(2x + 5)(x - 3)$ [B] $(2x - 5)(x + 3)$

[C] $(2x - 5)(x - 3)$ [D] $(2x - 3)(x + 5)$

41. The quadratic $2e^2 - 3e + 1$ when factorised,

becomes:

[A] $(2e - 1)(e - 1)$ [B] $(e^2 - 3)(2e - 1)$

[C] $(2e + 3)(e - 2)$ [D] $(2e - 3)(e - 1)$

42. Factorising $3a^2 - 11a + 6$ leads to:

[A] $(3a - 2)(a - 3)$ [B] $(2a - 2)(a - 3)$

[C] $(3a - 2)(a + 3)$ [D] $(3a + 2)(a - 3)$

43. $2x^2 - 9x - 45$, written as the product of two

factors is:

[A] $(2x - 9)(x - 5)$ [B] $(2x - 15)(x + 3)$

[C] $(2x + 15)(x - 3)$ [D] $(2x - 15)(x - 3)$

44. On factorisation $6x^2 + 7x - 20$ becomes:

[A] $(6x - 5)(x + 4)$ [B] $2(3x - 5)(x + 2)$

[C] $(3x + 4)(2x - 5)$ [D] $(3x - 4)(2x + 5)$

45. The quadratic $3p^2 + 2p - 1$ can be factorised

to have:

[A] $(3p + 2)(p - 1)$ [B] $(3p - 1)(p + 1)$

[C] $(3p + 1)(p - 1$ [D] $(3p - 2)(p + 1)$

46. Factorising the expression $2y^2 + xy - 3x^2$

gives rise to:

[A] $(x - y)(2y + 3x)$ [B] $(2y - x)(2y + x)$

[C] $(3x - 2y)(x - y)$ [D] $(2y + 3x)(y - x)$

47. $6x^2 + 7xy - 5y^2$ when factorised leads to:

[A] $(6x + 5y)(x - y)$ [B] $(2x + 5y)(3x - y)$

[C] $(3x + 5y)(2x - y)$ [D] $(2x + y)(3x - 5y)$

48. The result of factorising $mn - xy - nx + my$

is:

[A] $(n + y)(m - x)$ [B] $(n - y)(m + x)$

[C] $(x - m)(n - y)$ [D] $(n - y)(m - x)$

TOPIC 11

SIMULTANEOUS LINEAR EQUATIONS

OBJECTIVES

At the end of this topic, the learner should be able to:

1. Construct simultaneous equations in two unknowns from a given situation.
2. Use the method of elimination or substitution to solve:
   (a) Uniform coefficient simultaneous equations in two unknowns.
   (b) Non-uniform coefficient simultaneous equations in two unknowns.
   (c) simultaneous equations involving fractions and decimals
3. Translate and solve worded simultaneous equations in two unknowns.

 George Boole (1815-1864), British mathematician and logician, who developed Boolean algebra. Largely self-educated, in 1849 Boole was appointed professor of mathematics at Queen's College (now University College) in Cork, Ireland. In 1854, in *An Investigation of the Laws of Thought,* Boole described an algebraic system that later became known as Boolean algebra. In Boolean algebra, logical propositions are denoted by symbols and can be acted on by abstract mathematical operators that correspond to the laws of logic. Boolean algebra is of prime importance to the study of pure mathematics and to the design of modern computers.

The Concept of Simultaneous Equations

Consider the following problem.

A seamstress bought 2 razors and 3 needles at 70 FRS and her apprentice bought 3 razors and 5 needles at 110 FRS from the same store. How much does a tailor need to buy 5 razors and 7 needles?

A good method of solving this problem is to translate the statements into two equations as follows.
Let r represent the cost of a razor and n represent the cost of a needle. Then,

$$2r + 3n = 70 ....① $$
$$3r + 5n = 110 ....② $$

It can be shown that the cost of a razor and needle are 20 FCFA and 10 FCFA respectively. Substitute these values into the equations to confirm.

These types of equations, which are usually solved together, are called **simultaneous equations**. Simultaneous equations are solved in a variety of ways but for now only two methods, the method of substitution and elimination will be examined.

Simultaneous Equations with uniform Coefficients

Method of Elimination

In this method, the equations are added or subtracted to eliminate one of the unknowns. The value of the remaining unknown is then found and substituted into any of the equations to find the other unknown.

Example 11:1
Solve the following simultaneous equations using the method of elimination.

(a)　$7x + y = 22$　　(b)　$3a - b = 21$
　　　$5x + y = 14$　　　　$2a + b = 4$

(c)　$5m + 3n = 17$　(d)　$4p - 2q = -14$
　　　$m + 3n = 1$　　　　$4p - 5q = -32$

Solutions
(a)　$7x + y = 22 ....① $
　　　$5x + y = 14 ....② $

Subtract equation ② from equation ①
$$2x = 8 \Rightarrow x = 4$$
Substitute in equation ②
$$5(4) + y = 14$$
$$20 + y = 14$$
Subtract 20 from both sides
$$\Rightarrow y = -6$$

(b)　$3a - b = 21 ....① $
　　　$2a + b = 4 ....② $
Add equation ① to equation ②.
$5a = 25$ and $a = 5$
Substitute in equation ②
$$2(5) + b = 4$$
$$10 + b = 4 \Rightarrow b = -6$$

Alternatively $a = 5$ could have been substituted in equation ① instead of equation ② as follows.
$$3(5) - b = 21$$
$$-b = 6 \Rightarrow b = -6$$

(c)　$5m + 3n = 17 ....① $
　　　$m + 3n = 1 ....② $
①$-$②: $4m = 16 \Rightarrow m = 4$
Substitute in ②
$$4 + 3n = 1$$
$$3n = -3 \Rightarrow n = -1$$

(d)　$4p - 5q = -32 ....① $
　　　$4p - 2q = -14 ....② $
②　① 　,　$8 \Rightarrow q = 6$
Substitute in equation ②
$$4p - 2(6) = -14$$
$$4p = -2 \Rightarrow p = -\frac{1}{2}$$

EXERCISE 11:1

Solve the following simultaneous equations using the method of elimination.

1.　$x + y = 3$　　　　2.　$y + x = 4$
　　$3x - y = 1$　　　　　　$2y - x = 5$

3.　$a = 2b - 1$　　　4.　$x + y = 4$
　　$2a = 2b + 1$　　　　　$x - y = 2$

5.　$2x - 3y = 5$　　　6.　$6x + 5y = 11$
　　$3x + 2y = 14$　　　　　$5x - 2y = 5$

7. $7x + 3y = 6$ 8. $y = 4x$

$5y - 9x = 10$ $3x + y = 21$

Method of Substitution

In this method, one of the unknowns is made the subject of one of the equations and substituted into the equation to find the value of the other. This value is then substituted in the subject equation of the other unknown to find it.

Example 11:2

Solve the simultaneous equations in example 11.1 using the method of substitution.

(a) $7x + y = 22$ .... ①

$5x + y = 4$ .... ②

From equation ②

$y = 14 - 5x$ .... ③

Substitute in equation ①

$7x + 14 - 5x = 22$

$2x + 14 = 22$

$2x = 8 \Rightarrow x = 4$

Substitute in equation ③

$y = 14 - 5(4) \Rightarrow y = -6$

(b) $3a - b = 21$ .... ①

$2a + b = 4$ .... ②

From equation ②

$b = 4 - 2a$ .... ③

Substitute in equation ①

$3a - (4 - 2a) = 21$

$5a - 4 = 21$

$5a = 25 \Rightarrow a = 5$

Substitute in equation ③

$b = 4 - 2(5) \Rightarrow b = -6$

(c) $5m + 3n = 17$ .... ①

$m + 3n = 1$ .... ②

From equation ② ,

$m = 1 - 3n$ .... ③

Substitute in equation ①

$5(1 - 3n) + 3n = 17$

$5 - 15n + 3n = 17$

$-15n + 3n = 12$

$-12n = 12 \Rightarrow n = -1$

Substitute in equation ③

$m = 1 - 3(-1) \Rightarrow m = 4$

(d) $4p - 5q = -32$ .... ①

$4p - 2q = -14$ .... ②

From equation ②

$2p - q = -7$

$-q = -7 - 2p$

$q = 7 + 2p$ .... ③

Substitute in equation ①

$4p - 5(7 + 2p) = -32$

$4p - 35 - 10p = -32$

$-6p = 3 \Rightarrow p = -\dfrac{1}{2}$

Substitute in equation ③

$q = 7 + 2\left(-\dfrac{1}{2}\right) \Rightarrow q = 6$

Notice that though any simultaneous equation in two unknowns can be solved using either method, *simultaneous equations with uniform coefficients are easier to solve using the method of elimination.*

EXERCISE 11:2

Solve the simultaneous equations in Exercise 11:1 above using the method of substitution.

Simultaneous Equations with Non-uniform Coefficients

Method of Elimination

To solve these types of simultaneous equations by elimination, the objective is to eliminate one of the unknowns. To do this, make the coefficients of one of the unknowns' uniform by multiplying both equations by factors of the LCM of the coefficients of one of the unknowns. In this way, the LCM will be the uniform coefficient. In order that the problem does not become very complicated with large numbers, it is advisable that the LCM should be that of the smaller coefficients.

Example 11:3

Solve the following simultaneous equations using the method of elimination.

(a) $x + 2y = 5$ (b) $2b + 3n = 70$

$3x - y = 1$ $3b + 5n = 110$

(c) $2x + 3y = 7$

$\quad\;\; 5x - 2y = 8$

Solution

(a) $x + 2y = 5$.....................①

$\quad\;\; 3x - y = 1$.....................②

To eliminate y, multiply ② by 2

$\quad\;\; 6x - 2y = 2$.................③

Add equations ① and ③

$\qquad\; 7x = 7 \Rightarrow x = 1$

Substitute in equation ②

$\qquad\; 3 - y = 1 \Rightarrow y = 2$

(b) $2b + 3n = 70$.....................①

$\quad\;\; 3b + 5n = 110$.....................②

To eliminate b, multiply ① by 3 and ②
by 2, since the LCM of 2 and 3 is 6.

$\quad\;\; 6b + 9n = 210$.....................③

$\quad\;\; 6b + 10n = 220$.....................④

Subtract equation ③ from equation ④

$\qquad\; \Rightarrow n = 10$

Substitute in equation ①

$2b + 3(10) = 70$

$\quad\; 2b + 30 = 70$

$\qquad\; 2b = 40 \Rightarrow b = 20$

(c) $2x + 3y = 7$.....................①

$\quad\;\; 5x - 2y = 8$.....................②

Since the LCM of 2 and 3 is 6, multiply
equation ① by 2 and equation ② by 3

$\quad\;\; 4x + 6y = 14$.....................③

$\quad\;\; 15x - 6y = 24$.....................④

③ + ④ : $19x = 38 \Rightarrow x = 2$

Substitute in equation ①

$\quad\; 2(2) + 3y = 7$

$\qquad\; 4 + 3y = 7$

$\qquad\qquad\; 3y = 3 \Rightarrow y = 1$

Method of Substitution

Example 11:4

Solve the simultaneous equations in example
11:3 using the method of substitution.

Solutions

(a) $x + 2y = 5$.....................①

$\quad\;\; 3x - y = 1$.....................②

Make x the subject in equations ①

$\qquad\; x = 5 - 2y$.....................③

Substitute in equation ②

$\quad\; 3(5 - 2y) - y = 1$

$\quad\;\; 15 - 6y - y = 1$

$\qquad\; -7y = -14 \Rightarrow y = 2$

Substitute in equation ③

$\quad\; x = 5 - 2(2) \Rightarrow x = 1$

(b) $2b + 3n = 70$.....................①

$\quad\;\; 3b + 5n = 110$.....................②

From equation ①, $2b = 70 - 3n$

$\qquad\; b = \dfrac{70 - 3n}{2}$ ....③

Substitute in equation ②

$\quad\; 3\left(\dfrac{70 - 3n}{2}\right) + 5n = 110$

$\quad\; 210 - 9n + 10n = 220$

$\qquad\qquad\; \Rightarrow n = 10$

Substitute in equation ③

$\qquad\; b = \dfrac{70 - 3(10)}{2}$

$\qquad\quad\; = \dfrac{40}{2} \Rightarrow b = 20$

(c) $2x + 3y = 7$.....................①

$\quad\;\; 5x - 2y = 8$.....................②

From equation ①, $2x = 7 - 3y$

$\quad\; \Rightarrow x = \dfrac{7 - 3y}{2}$.....................③

Substitute in equation ②

$\quad\; 5\left(\dfrac{7 - 3y}{2}\right) - 2y = 8$

$\quad\; 5(7 - 3y) - 4y = 16$

$\quad\; 35 - 15y - 4y = 16$

$\qquad\; -19y = -19 \Rightarrow y = 1$

Substitute in equation ③

$\qquad\; x = \dfrac{7 - 3(1)}{2} \Rightarrow x = 2$

EXERCISE 11:3

1. Solve the following simultaneous equations
using the method of elimination.

(a) $\quad p - 3q = 10$ (b) $\quad s = 2t - 1$

$\qquad 3p - 2q = 16$ $\qquad\quad 2s = 3t + 2$

(c) $5x - 2y = 14$
$2x + 2y = 14$

(d) $4m + 4n = 3$
$m + 2n = 1$

(e) $2x + y = 7$
$3x - 2y = 7$

(f) $3u - 7v = 1$
$2u + v = 12$

2. Solve the simultaneous equations in problem 1, using the method of substitution.

3. Determine which method is better and use it to solve each of the following simultaneous equations.

(a) $3a + 5b = 4$
$2a + 3b = 4$

(b) $x + y = 1$
$x - y = 3$

(c) $3x + 4y = 0$
$x = 2y - 5$

(d) $y = x + 1$
$x + y = 3$

(e) $2x + y = 7$
$3x - 2y = 7$

(f) $3u - 7v = 1$
$2u + v = 12$

(g) $x + 2y = 8$
$x - 3y = 3$

(h) $x - 2y = 0$
$x + 3y = -10$

(i) $4x = y + 7$
$3x + 4y + 9 = 0$

(j) $x + y = 7$
$x - y = 1$

(k) $x + y = 4$
$-x + y = 2$

(l) $5a + 3b = 12$
$a - 3b = 6$

(m) $x + y = 7$
$x - y = 3$

(n) $x - y = 3$
$2x + y = 12$

(o) $5a + 3b = 1$
$2a + 3b = -5$

(p) $x + y = 11$
$x - y = 5$

(q) $x + y = -2$
$x - y = 0$

(r) $3r + 5s = 21$
$7r - 2s = 8$

Simultaneous Equations Involving Fractions and Decimals

Method of Elimination

In solving simultaneous equations involving fractions and decimals, first get rid of the fractions by multiplying each term by the LCM of the denominators (or a required power of 10 in the case of decimals). After this, the method of elimination or substitution can be employed.

Example 11:5

Solve the following simultaneous equations:

(a) $\dfrac{x}{3} + \dfrac{y}{4} = \dfrac{1}{12}$

$\dfrac{3x}{2} - \dfrac{y}{3} = -4$

(b) $0.03x + 0.05y = 51$
$0.8x - 0.7y = 140$

Solutions

To solve these equations by either method, it is advisable to first eliminate the fractions or decimals.

(a) $\dfrac{x}{3} + \dfrac{y}{4} = \dfrac{1}{12}$ ....①

$\dfrac{3x}{2} - \dfrac{y}{3} = -4$ ....②

Since the LCM of 3, 4 and 12 is 12 and the LCM of 2 and 3 is 6, multiply equation ① by 12 and equation ② by 6:

$4x + 3y = 1$ ....③

$9x - 2y = -24$ ....④

The simultaneous equations may now be solved using any of the methods as desired.

By Method of Elimination

Multiply equation ③ by 2 and equation ④ by 3.

$8x + 6y = 2$ ....⑤

$27x - 6y = -72$ ....⑥

⑥ ⑤ $x = -70 \Rightarrow x = -2$

Substitute in ③

$4(-2) + 3y = 1$

$3y = 9 \Rightarrow y = 3$

By Method of Substitution

From ③, $y = \dfrac{1 - 4x}{3}$ ....⑦

Substitute in ④

$9x - 2\left(\dfrac{1 - 4x}{3}\right) = -24$

Multiply both sides by 3

$27x - 2(1 - 4x) = -72$

$27x - 2 + 8x = -72$

$35x = -70 \Rightarrow x = -2$

Substitute in equation ⑦

$y = \dfrac{1 - 4(-2)}{3} \Rightarrow y = 3$

(b) $0.03x + 0.05y = 51$.................①

$0.8x - 0.7y = 140$.................②

Multiply equation ① by 100 and equation ② by 10.

$3x + 5y = 5100$.................③

$8x - 7y = 1400$.................④

The simultaneous equations are then solved using any of the methods as follows.

By Method of Elimination

Multiply equation ③ by 7 and equation ④ by 5

$21x + 35y = 35700$.....................⑤

$40x - 35y = 7000$.....................⑥

Add equation ⑥ and equation ⑤

$61x = 42700 \Rightarrow x = 700$

Substitute in ③,

$3(700) + 5y = 5100$

$2100 + 5y = 5100$

$5y = 3000 \Rightarrow y = 600$

By Method of Substitution

From ③, $5y = 5100 - 3x$

$y = \dfrac{5100 - 3x}{5}$.............................⑦

Substitute in equation ④

$8x - 7\left(\dfrac{5100 - 3x}{5}\right) = 1400$

$40x - 7(5100 - 3x) = 7000$

$40x - 35700 + 21x = 7000$

$\qquad\qquad 61x = 42700 \Rightarrow x = 700$

Substitute in equation ⑤

$y = \dfrac{5100 - 3(700)}{5} \Rightarrow y = 600$

Solve the following simultaneous equations.

1. $\dfrac{x}{3} - 2y = 1$

$x + 3y = 12$

2. $\dfrac{x}{3} + \dfrac{y}{4} = \dfrac{1}{12}$

$2x - 3y = -4$

3. $3a - b = 500$

$0.7a + 0.2b = 550$

4. $x - 2y = 500$

$0.03x + 0.02y = 51$

5. $m = 4n - 100$

$0.06m = 0.05n + 32$

6. $0.03x + 0.04y = 44$

$0.04x + 0.02y = 42$

7. $0.8p - 0.7q = 140$

$0.03p + 0.05q = 51$

8. $0.05(u + 2000) = 0.3(v + 3000)$

$u = \dfrac{v}{2} + 500$

Simultaneous Linear Equations in Real Life

In Topic 9, techniques of translating verbal equations to algebraic equations were studied. Simultaneous linear equations occur in many instances in real life. Some of these include problems on marketing, movement, ages, mixtures etc. The following examples illustrate some real life applications of simultaneous linear equations.

Example 11:6

A credit union gave Mr. Ngong 16000 francs consisting of 500 francs coins and 100 francs coins. The number of 100 francs coins is three times the number of 500 francs coins. How many of each type of coin was Mr. Ngong given?

Solution

Let h = number of 100 francs coins and

f = number of 500 francs coins

Then $100h + 500f = 16000$

$\Rightarrow h + 5f = 160$.....................①

Also $h = 3f$.........................②

Substitute ② in ①:

$\Rightarrow 3f + 5f = 160$

$8f = 160 \Rightarrow f = 20$

Substitute in ②: $h = 3(20) = 60$

Number of 100 francs coins = 60

Number of 500 francs coins = 20

Example 11:7

A father gave his twin children 350 francs each. One bought 3 packets of biscuits and 2 sweets and the other bought 2 packets of biscuits and 6 sweets. Find the cost of:
(a) a packet of biscuits   (b) a single sweet.

Solution

$$3b + 2s = 350 \dots \dots \dots \text{①}$$

$$2b + 6s = 350 \dots \dots \dots \text{②}$$

From ②, $b + 3s = 175$

$$\Rightarrow b = 175 - 3s \dots \dots \dots \text{③}$$

Substitute in ①: $3(175 - 3s) + 2s = 350$

$$\Rightarrow 525 - 9s + 2s = 350$$

$$525 - 7s = 350$$

$$-7s = -175 \Rightarrow s = 25 \text{ francs}$$

Substitute in ③:

$$b = 175 - 3(25) = 100 \text{ francs}$$

(a) a packet of biscuits = 100 francs

(b) a single sweet = 25 francs .

Example 11:8

Two cars leave Makenene at the same time travelling in opposite directions. One is travelling at 80 km/h and the other is travelling at 70 km/h. How long will it take the two cars to be 300 km apart?

Solution

Makenene

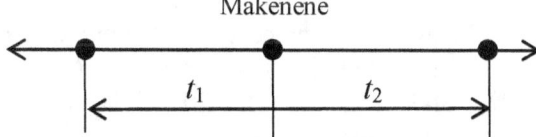

Let the times taken by the two cars be t_1 and t_2.

Then $t_1 = t_2 \dots \dots \dots \dots \text{①}$

Since distance = speed × time

$$80t_1 + 70t_2 = 300$$

$$\Rightarrow 8t_1 + 7t_2 = 30 \dots \dots \dots \dots \text{②}$$

Substitute ① in ②: $8t_1 + 7t_1 = 30$

$$15t_1 = 30 \Rightarrow t_1 = 2$$

Therefore, it will take 2 hours for the two cars to be 300 km apart.

1. A student bought 3 books and 2 pencils for 340 FRS. Another student bought 2 books and 3 pencils of same kind for 260 FRS. Find the cost of a book and a pencil.

2. The value of the expression $mx^2 + nx$ is 8 when $x = 2$, and 27 when $x = 3$. Determine the value of n.

3. The sum of two numbers is 26 and their difference is 28. Find them.

4. The average of two numbers is 13. The difference of the two numbers is 6. Find the two numbers.

5. A bottle and a cork together weigh 18g. Given that, the bottle weighs 16g more than the cork, finds the weight of each.

6. 3 nuts and 4 bolts have a mass of 72g. 4 nuts and 5 bolts have a mass of 94g. Find the mass of
    (a) a nut   (b) a bolt

7. Three coconuts and two oranges cost 430 FRS. One coconut and four oranges cost 210 FRS. Find the cost of each coconut and each orange.

8. A man is 9 times as old as his son. In four years time, he would be 5 times as old as his son. What is the present age of both of them in years?

9.   A hotel has first-class rooms, which cost 4500 francs, and second-class rooms, which cost 2500 francs. On a certain day, 32 rooms were given out at a total cost of 130,000 francs.
    (a) How many first-class rooms and second-class rooms were given out on that day?
    (b) Calculate the total amount collected
        (i) for first-class rooms.
        (ii) for second-class rooms.

10. A tailor takes 8 hours to stitch a trouser, which cost 4800 francs. The tailor takes 6 hours to stitch a shirt, which cost 4000 francs. The tailor used 120 hours to stitch shirts and trousers, which he sold for 78400 francs. How many shirts and how many trousers did he stitch?

11. A man left his house jogging at 8 km/h. On his return, he was tired and could only run at 5 km/h. The total time used for this exercise was 1 and a half hours. How far did he go from his house?

MULTIPLE CHOICE EXERCISE 11

1. The pair of values of x and y which satisfy the simultaneous equations $x + y = 3$ and $3x - y = 1$ are:
   [A] $(1, -2)$ [B] $(2, -1)$
   [C] $(1, 2)$ [D] $(-2, 1)$

2. Given that $x = 2y - 1$ and $2x = 2y - 1$. The value of x is:
   [A] -9 [B] -4 [C] 9 [D] 0

3. If $x = 2y - 1$ and $2x = 3y + 2$, then y equals:
   [A] 4 [B] 9 [C] -4 [D] -9

4. The roots of the simultaneous equations $x + y = 4$ and $2y - x = 5$ are:
   [A] $(-1, 3)$ [B] $(-1, -3)$
   [C] $(1, -3)$ [D] $(1, 3)$

5. The values of x and y that satisfy the simultaneous equations $2x + y = 7$ and $3x - 2y = 7$ are:
   [A] $(-1, 3)$ [B] $(3, 1)$
   [C] $(-1, -3)$ [D] $(3, -1)$

6. Given that $2p - m = 6$ and $2p + m = 1$. The value of $4p + 3m$ is:
   [A] 1 [B] 3 [C] 5 [D] 7

7. If $x + y = \dfrac{3}{2}$ and $x - y = \dfrac{5}{2}$, then $2y + x$ equals:
   [A] -2 [B] 1 [C] $\dfrac{1}{2}$ [D] -1

8. If $x + 2y = 1$ and $x - y = 2$, the value of $x + y$ is:
   [A] $1\dfrac{1}{3}$ [B] 1 [C] -1 [D] $-1\dfrac{1}{3}$

9. Given that $x + y = 7$ and $3x - y = 5$. When evaluated $\dfrac{y}{2} - 3$ gives:
   [A] 3 [B] 1 [C] -1 [D] 4

10. If $2x + y = 7$ and $3x - y = 3$, then $7x$ is greater than 10 by:
    [A] 1 [B] 7 [C] 3 [D] 10

11. Given the equations $4y - 5x = 14$ and $y = 3x$. The values of x and y are respectively:
    [A] $(2, 6)$ [B] $(-2, -6)$
    [C] $(2, -6)$ [D] $(-2, 6)$

12. The values of x and y which satisfy the simultaneous equations $4x - y = 11$ and $5x + 2y = 4$ are:
    [A] $x = 2, y = 3$ [B] $x = -2, y = -3$
    [C] $x = -2, y = 3$ [D] $x = 2, y = -3$

13. If $3p - q = 6$ and $2p + q = 4$, then q is equal to:
    [A] 0 [B] $\dfrac{1}{2}$ [C] $\dfrac{2}{3}$ [D] 1

14. The values of $x - y$, which satisfy the simultaneous equations $4x - 3y = 7$ and $3x - 2y = 5$ are:
    [A] -3 [B] 3 [C] 2 [D] -2

15. $\dfrac{1}{3}x + y = 3$ and $x + \dfrac{1}{2}y = 4$ provided:
    [A] $x = 3, y = 3$ [B] $x = 3, y = -3$
    [C] $x = -3, y = 3$ [D] $x = -3, y = -3$

16. If the solutions of the pair of equations $2x + 3y = p$ and $3x - y = q$ are $x = -1$ and $y = 2$, the values of p and q must be:
    [A] $p = -4, q = 5$ [B] $p = -5, q = 4$
    [C] $p = -4, q = -5$ [D] $p = 4, q = -5$

17. An exercise book and a pencil cost 180 F. If the exercise book costs 140 F more than the pencil, then the pencil costs:
    [A] 30 F [B] 25 F [C] 20 F [D] 15 F

18. Ambe is four times as old as Ndeh. If the sum of their ages is 20 years, the difference in their ages in years is:
    [A] 16 [B] 12 [C] 8 [D] 4

19. 2 nuts and 3 bolts have a mass 28 g. 3 nuts and a bolt have a mass of 21 g. The mass of a bolt is:
    [A] 4 g [B] 5 g [C] 6 g [D] 7 g

TOPIC 12
QUADRATIC EQUATIONS

OBJECTIVES

At the end of this topic, the learner should be able to:

1. Identify the standard form quadratic equation.
2. Solve quadratic equations using the method of factorisation, completing the square or the quadratic formula.
3. Identify and solve non-standard forms and incomplete quadratic equations.
4. Determine the nature of roots of a quadratic equation.
5. Construct quadratic equations from a given situations.
6. Solve simultaneous equations one linear, one quadratic.

Leonhard Euler
Although hindered by loss of sight, Leonhard Euler was an important contributor to both pure and applied mathematics. Euler is best known for his analytical treatment of mathematics and his discussion of calculus concepts, but he is also credited for work in acoustics, mechanics, astronomy, and optics.
The binomial theorem was formulated by Sir Isaac Newton and the Swiss mathematician Leonhard Euler, but Niels Henrik Abel gave it a more comprehensive generalization, including the cases of irrational and imaginary exponents.

Culver Pictures

Concept of Quadratic Equations

Consider the following problem.
The cost of a carton of macaroni is normally 4000 FRS. Due to a 20 FRS discount per sachet, a housewife buys 10 more sachets at the same amount. Determine
(a) the number of sachets in a carton.
(b) the cost of each discounted sachet.

Solution

Let x be the number of sachets per carton.

Then cost of each sachet from the cartoon $= \dfrac{4000}{x}$ FRS.

If 10 more sachets are bought at the same amount (4000 FRS),

Then cost of each discounted sachet $= \dfrac{4000}{x+10}$ FRS.

But cost per discounted sachet is 20 FRS less

$\therefore \quad \dfrac{4000}{x} - \dfrac{4000}{x+10} = 20$

Multiply both sides by the LCM (i.e. by $x(x+10)$)

$4000(x+10) - 4000x = 20x(x+10)$
$4000x + 40000 - 4000x = 20x^2 + 200x$
$40000 = 20x^2 + 200x$

Dividing by 20 and rearranging leads to

$x^2 + 10x - 2000 = 0$

Eh! What a complicated equation? Don't panic! The solutions are easy and many too. Get the bullets!

Standard Form Quadratic Equations

A quadratic equation is any second-degree polynomial equation (i.e. any polynomial equation where the highest power of the unknown is 2). The standard form of a quadratic equation is $ax^2 + bx + c = 0$, where $a \neq 0$, b and c are constants. Standard form quadratic equations can be solved in four basic methods, which include the factorisation method, the method of completing the square, the formula method and the graphical method. The first three methods shall be examined in this topic and the graphical method shall be examined in Topic 34.

(a) Factorisation method

If the expression $ax^2 + bx + c$ can be factorised, the result will be of the form $(px + q)(rx + s) = 0$.

For the product of two numbers to be zero, at least one of them must be zero.

The fundamental theorem for solving the equation $(px + q)(rx + s) = 0$ is therefore,
Either $(px + q) = 0$ or $(rx + s) = 0$
$$\Rightarrow x = -\frac{q}{p} \quad \text{or} \quad x = -\frac{s}{r}$$

Example 12:1
Find the value of x if $(x - 1)(x + 2) = 0$.

Solution
$$(x - 1)(x + 2) = 0$$
Either $x - 1 = 0 \Rightarrow x = 1$
or $x + 2 = 0 \Rightarrow x = -2$.

Example 12:2
Solve the following quadratic equations using the factorisation method.
(1) $6x^2 + 7x - 3 = 0$ (2) $x^2 + 9x + 18 = 0$

Solution
(1) $6x^2 + 7x - 3 = 0$
$6x^2 + 9x - 2x - 3 = 0$
$3x(2x + 3) - 1(2x + 3) = 0$
$(3x - 1)(2x + 3) = 0$
$\therefore 2x + 3 = 0$ or $3x - 1 = 0$
$$x = -\frac{3}{2} \quad \text{or} \quad x = \frac{1}{3}$$

(2) $\qquad x^2 + 9x + 18 = 0$
$(x + 3)(x + 6) = 0$
$\therefore (x + 3) = 0$ or $(x + 6) = 0$
$\qquad x = -6$ or $x = -3$

EXERCISE 12:1

Solve the following quadratic equations using the factorisation method.
(1) $x^2 - 2x - 8 = 0$
(2) $x^2 + x - 2 = 0$

(3) $x^2 - 5x + 6 = 0$

(4) $2x^2 - x - 1 = 0$

(5) $3x^2 - 12x - 63 = 0$

(6) $6x^2 - 5x - 6 = 0$

(7) $12x^2 + 8x - 15 = 0$

(8) $2x^2 - 9x - 5 = 0$

(b) *Method of Completing the Square*

Sometimes the quadratic expression is not factorable. In such a case, the method of completing the square discussed in Topic 10 is used.

Example 12:3

Solve the equation $2x^2 - 5x - 5 = 0$, using the method of completing the square.

Solution

$$2x^2 - 5x - 5 = 0$$

Add 5 to both sides.

$$2x^2 - 5x = 5$$

Divide each term by the coefficient of x^2

$$x^2 - \frac{5}{2}x = \frac{5}{2}$$

Add the square of half the coefficient of x to both sides

$$x^2 + \left(-\frac{5}{2}\right)x + \left(-\frac{5}{4}\right)^2 = \frac{5}{2} + \left(-\frac{5}{4}\right)^2$$

The LHS is now a perfect square and can now be factorised.

$$\therefore \quad \left(x - \frac{5}{4}\right)^2 = \frac{5}{2} + \frac{25}{16}$$

Summing the RHS and finding the square roots of both sides.

$$x - \frac{5}{4} = \pm\frac{\sqrt{65}}{4}$$

(Note that every positive number has two square roots, with the same absolute value but one positive and the other negative.)

Adding $\frac{5}{4}$ to both sides and rearranging gives:

$$x = \frac{5 \pm \sqrt{65}}{4}$$

$$\Rightarrow x = \frac{5 + \sqrt{65}}{4} \quad \text{or} \quad x = \frac{5 - \sqrt{65}}{4}$$

Example 12:4

Solve the equation $x^2 + 2x - 7 = 0$, using the method of completing the square.

Solution

$$x^2 + 2x - 7 = 0$$
$$x^2 + 2x = 7$$

Completing the square

$$x^2 + 2x + 1 = 8$$
$$(x + 1)^2 = 8$$
$$x + 1 = \pm 2\sqrt{2}$$
$$x = -1 \pm 2\sqrt{2}$$

Example 12:5

Solve the equation $3x^2 + 2x - 7 = 0$, using the method of completing the square.

Solution

$$3x^2 + 2x - 7 = 0$$

Add 7 to both sides

$$3x^2 + 2x = 7$$

Divide both sides by 3

$$x^2 + \frac{2}{3}x = \frac{7}{3}$$

Add (half the coefficient of x)^2 to both sides.

$$x^2 + \frac{2}{3}x + \left(\frac{1}{3}\right)^2 = \frac{7}{3} + \left(\frac{1}{3}\right)^2$$

$$\left(x + \frac{1}{3}\right)^2 = \frac{22}{9}$$

$$x = \frac{1 \pm \sqrt{22}}{3} \Leftrightarrow x = \frac{1 + \sqrt{22}}{3} \quad \text{or} \quad x = \frac{1 - \sqrt{22}}{3}$$

EXERCISE 12:2

Solve the following quadratic equations using the method of completing the square.

1. $x^2 + 12x - 45 = 0$ 2. $u^2 + 20u + 19 = 0$

3. $p^2 + 7p - 8 = 0$ 4. $y^2 - 18y + 65 = 0$

5. $a^2 - 11a + 28 = 0$ 6. $x^2 - 30x - 99 = 0$

7. $x^2 + 14x - 32 = 0$ 8. $x^2 + 18x - 19 = 0$

The Quadratic Formula

When the general form of the quadratic equation is solved using the method of completing the square, this leads to the quadratic formula derived below.

Derivation of the quadratic formula

Consider the general form of the quadratic equation

$$ax^2 + bx + c = 0$$

Subtract c from both sides.

$$ax^2 + bx = -c$$

Divide both sides by a.

$$x^2 + \frac{b}{a}x = -\frac{c}{a}$$

To make the LHS a perfect square add

$\left(\dfrac{b}{2a}\right)^2$ to both sides

$$x^2 + \frac{b}{a}x + \left(\frac{b}{2a}\right)^2 = \left(\frac{b}{2a}\right)^2 - \frac{c}{a}$$

Factorising the LHS and rearranging the RHS,

$$\left(x + \frac{b}{2a}\right)^2 = \frac{b^2 - 4ac}{4a^2}$$

Finding the square roots of both sides and rearranging,

$$x + \frac{b}{2a} = \pm\frac{\sqrt{b^2 - 4ac}}{2a}$$

Subtracting $\dfrac{b}{2a}$ from both sides and rearranging,

$$x = \frac{-b \pm \sqrt{b^2 - 4ac}}{2a}$$

i.e. $x = \dfrac{-b + \sqrt{b^2 - 4ac}}{2a}$ or $x = \dfrac{-b - \sqrt{b^2 - 4ac}}{2a}$

Therefore, if $ax^2 + bx + c = 0$, where a, b and c are constants and $a \neq 0$, then

$$x = \frac{-b \pm \sqrt{b^2 - 4ac}}{2a}.$$

This formula called the **quadratic formula** can be used to find the roots of any quadratic equation.

Take note that the division bar goes across!

i.e. $x \neq -b \pm \dfrac{\sqrt{b^2 - 4ac}}{2a}$

Remarks!

In the quadratic formula $x = \dfrac{-b \pm \sqrt{b^2 - 4ac}}{2a}$;

1. The sum of the roots $= -\dfrac{b}{a}$. This is a very useful check to solutions of quadratic equations.

2. If $b^2 - 4ac$ called the discriminant is equal to zero, the expression $ax^2 + bx + c$ is a perfect square.

Example 12:6

Use the quadratic formula to solve the equation $2x^2 - 5x - 5 = 0$.

Solution

$$x = \frac{-b \pm \sqrt{b^2 - 4ac}}{2a}$$

$$a = 2, b = -5, c = -5$$

$$x = \frac{-(-5) \pm \sqrt{(-5)^2 - 4(2)(-5)}}{2(2)} = \frac{5 \pm \sqrt{65}}{4}$$

$$x = \frac{5 - \sqrt{65}}{4} \quad \text{or} \quad x = \frac{5 + \sqrt{65}}{4}$$

EXERCISE 12:3

Solve the following quadratic equations using the formula method, leaving your answer in surd form.

1. $2x^2 - 5x - 4 = 0$
2. $x^2 + 7x + 5 = 0$
3. $2x^2 + 5x + 1 = 0$
4. $x^2 + 6x - 10 = 0$
5. $3x^2 - 4x - 2 = 0$
6. $6x^2 - 10x + 3 = 0$
7. $5x^2 - 10x + 4 = 0$
8. $9x^2 + x - 2 = 0$

Non-Standard Quadratic Equations

Sometimes a quadratic equation is not in the standard form. In this case, first rearrange it to be in standard form before solving.

Example 12:7

Solve the equation $x - 6 = \sqrt{x}$.

Solution

Squaring both sides

$$x^2 - 12x + 36 = x$$
$$\Rightarrow x^2 - 13x + 36 = 0$$
$$(x - 9)(x - 4) = 0$$
$$\Rightarrow x = 9 \text{ or } x = 4$$

Testing in the equation, it is clear that $x = 9$ and $x \neq 4$. $x = 4$ Could have been a solution if the original equation were $x - 6 = \pm\sqrt{x}$.

EXERCISE 12:4

Rearrange the following and solve by factorisation.

1. $u^2 + 6u = -8$ 2. $a^2 = 5a + 24$

3. $x(3x + 2) = 7$ 4. $5 = m(2m + 3)$

5. $p + 8 = \dfrac{20}{p}$ 6. $\dfrac{7}{x} = 9 - 2x$

7. $2 + \dfrac{5}{x} = \dfrac{12}{x^2}$ 8. $(a - 8)(2a - 3) = 34$

Incomplete Quadratic Equations

Quadratic equations of the form

$ax^2 + bx = 0$ or $ax^2 + c = 0$ which lack the constant term or the term in x are called incomplete quadratic equations and are solved differently from the standard quadratic equation form as shown below.

1. When $c = 0$, the quadratic equation
   $ax^2 + bx + c = 0$ becomes $ax^2 + bx = 0$.
   $$x(ax + b) = 0$$
   $$\therefore x = 0 \text{ or } ax + b = 0$$

   Hence, $ax^2 + bx = 0 \Leftrightarrow x = 0 \text{ or } x = -\dfrac{b}{a}$.

It is unwise to divide both sides by any multiple of the unknown for this leads to the loss of the solution $x = 0$.

Example 12:8

Solve the quadratic equation $3x^2 + 4x = 0$.

Solution

$$3x^2 + 4x = 0 \Rightarrow x(3x + 4) = 0$$
$$\Leftrightarrow x = 0 \text{ or } x = -\frac{4}{3}$$

2. When $b = 0$, the quadratic equation
   $ax^2 + bx + c = 0$ becomes

   $$ax^2 + c = 0 \Leftrightarrow x^2 = -\frac{c}{a}.$$

   This equation will have real solutions only if $\dfrac{c}{a} \leq 0$.

   Hence, if $ax^2 + c = 0$ and $\dfrac{c}{a} \leq 0$ then $x = \pm\sqrt{-\dfrac{c}{a}}$

Example 12:9

Solve the equation $4x^2 - 1 = 0$.

Solution

$$4x^2 - 1 = 0 \Leftrightarrow x^2 = \frac{1}{4} \Leftrightarrow x = \pm\frac{1}{2}$$

Alternatively, using the difference of two squares identity,

$4x^2 - 1 = (2x)^2 - 1^2 = (2x+1)(2x-1)$

$4x^2 - 1 = 0 \Leftrightarrow (2x+1)(2x-1) = 0 \Leftrightarrow x = \pm\dfrac{1}{2}$

Note that any positive number has two square roots and care should be taken not to forget one of the square roots. For instance, $(-2)^2 = 4$ and $2^2 = 4$ so the square roots of 4 are 2 or –2.

EXERCISE 12:5

Solve the following equations.

1. $x^2 - 64 = 0$ 2. $4y^2 = 28y$ 3. $4a^2 = 1$

4. $9p^2 = p$ 5. $x^2 - 4x = 0$ 6. $y^2 - 9 = 0$

7. $3b^2 = 18$ 8. $\dfrac{8p}{25} = \dfrac{16}{p}$

Nature of roots of a Quadratic Equation

In the quadratic formula above, $b^2 - 4ac$ is called the **discriminant** of the expression

$ax^2 + bx + c$ denoted by Δ. Thus,

$\Delta = b^2 - 4ac$.

The nature of Δ can give rise to three different types of roots as follows:

1. If $\Delta > 0$, the equation has two real and distinct roots.
2. If $\Delta = 0$, the equation has repeated or equal real roots.
3. If $\Delta < 0$, the equation has no real roots i.e. the roots are complex or imaginary.

It is necessary to always check the discriminant of a quadratic equation to see if the equation has real roots or if it can be solved.

EXERCISE 12:6

Determine the nature of the roots of each of the following quadratic equations.

1. $x^2 + 2x + 3 = 0$ 2. $u^2 - 4u - 4 = 0$

3. $4x^2 - 4x = -9$ 4. $2p^2 + 5p = -8$

---

5. $3y^2 - 12y = -6$ 6. $9x + 12 + \dfrac{4}{x} = 0$

7. $2x^2 - 7x = -13$ 8. $2x^2 + 8x = 17$

9. $1 - 4x + 4x^2 = 0$ 10. $4x^2 - 9x + 9 = 0$

11. $2x^2 - 7x = -13$ 12. $x^2 + 1 = -2x$

Worded Problems that lead to Quadratic Equations

Many real life problems such as the macaroni problem at the beginning of this topic involve quadratic equations and until the solving begins, it will not be easy to predict. Before continuing, go back now and solve the macaroni problem haven had the bullets. What remark or observation do you have concerning the solutions of the equation?

Example 12:10
A man wants to buy a piece of land. He is told that the area of the piece of land is 80 square metres and that the length is 2 metres longer than the width. The man hires you to work out the length and width for him. Go ahead and do your job.

Solution
Let the width be x m, then the length will be $(x + 2)$ m.

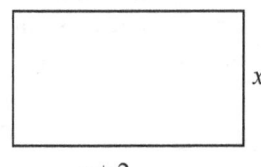

$x(x+2) = 80 \Leftrightarrow x^2 + 2x = 80$

$\Leftrightarrow x^2 + 2x - 80 = 0$ or $(x-8)(x+10) = 0$

$\Leftrightarrow x = 8$ or $x = -10$

Since $x > 0$, so the solution $x = -10$ should be discarded. Clearly if the width is 8 m, the length will be 10 m, since $8 \times 10 = 80$.

EXERCISE 12:7

Translate and solve the following equations.
1. The sum of the squares of two consecutive integers is 13. Find the numbers.
2. The square of a number is 12 more than the

number. Find the number and its square.

3. At the beginning of a school year, a man bought for his two children $(x - 5)$ exercise books at $(3x + 75)$ FCFA each. Given that, he spent 2025 FCFA altogether, find the number of exercise books he bought and the exact price of each.

4. A woman bought a piece of land in Bamenda. While in Yaounde, she decided to sell it. The buyer was interested in knowing the dimensions of the piece of land. What the woman could remember was that its area is 247.5 square metres and the length is 1.5 m more than the width. Help the woman out of this problem.

5. The formula $T_n = \dfrac{n(n + 1)}{2}$ is a formula for finding any triangular number. Find the value of n for which $T_n = 78$.

6. The product of two positive integers is 27. One is three more than twice the other is. Find the numbers.

7. The capacity of a hall is 144 persons. The number of rows is 7 more the number of seats in each row. Determine the number of rows of seats in the hall.

8. The area and perimeter of a rectangular farm are 24 square metres and 20 metres respectively. Find the length and width of the farm.

9. The height of a triangular flowerbed is 4 m less than the base. The area of the flowerbed is 48 m^2. Calculate the height and the length of the base of the flowerbed.

10. The product of a positive integer and a number three less than the integer is equal to the integer increased by 32. What is the integer?

11. Two positive integers differ by 8. The sum of the square of the larger and the smaller is 124. Find the numbers.

12. During a procession, the signboard bearer and the choir leader are alone on their own rows. The flag bearer and shield bearer occupy the next row. The arrangement of the rest of the choir members is such that the number of choristers on each row exceeds the number of rows by three. There are altogether 112

choristers. Find the total number of rows on this procession.

Simultaneous Equations (One Linear-One Quadratic)

Simultaneous equations, one linear one quadratic are usually solved using the method of substitution. To do this, one of the unknowns is made a subject, using the linear equation and substituted into the quadratic equation.

Example 12:11
Solve the simultaneous equations $y = 2x - 3$ and $y = x^2 + 5x - 7$.

Solution
$y = 2x - 3$ ....①

$y = x^2 + 5x - 7$ ....②
Substitute ① in ②.
$x^2 + 5x - 7 = 2x - 3$
$x^2 + 3x - 4 = 0$
$(x + 4)(x - 1) = 0$
∴ $x = -4$ or $x = 1$
Substitute in ①,
When $x = -4$, $y = -11$ or when $x = 1$, $y = -1$.
∴ $x = -4$ and $y = -11$ or $x = 1$ and $y = -1$.

Example 12:12
Solve the simultaneous equations
$2x^2 + 7x - y^2 = 21$ and $3x + 2y = 8$.

Solution
$3x + 2y = 8$ ....①

$2x^2 + 7x - y^2 = 21$ ....②
From ①: $y = \dfrac{8 - 3x}{2}$ ....③
Substitute ③ in ②
$2x^2 + 7x - \left(\dfrac{8-3x}{2}\right)^2 = 21$

$2x^2 + 7x - \left(\dfrac{64-48x+9x^2}{4}\right) = 21$
Multiply all through by 4,
$8x^2 + 28x - (64 - 48x + 9x^2) = 84$
$x^2 - 76x + 148 = 0$
$(x - 74)(x - 2) = 0$
∴ $x = 2$ or $x = 74$

Substitude in ③:

When $x = 2$, $y = \dfrac{8 - 3(2)}{2} \Rightarrow y = 1$

When $x = 74$, $y = \dfrac{8 - 3(74)}{2} \Rightarrow y = -107$

∴ $x = 2$ and $y = 1$ or $x = 74$ and $y = -107$.

EXERCISE 12:8

Solve the following simultaneous equations.

1. $x - 2y = 7$, $x^2 + 4y^2 = 37$

2. $y + 7x = 3x^2$, $y - 11x + 16 = 0$

3. $x + y = 7$, $xy = 12$

4. $2x - y = 2$, $4x^2 + y^2 = 2$

5. $3x - 4y + 9 = 0$, $2x^2 + y^2 = 11$

6. $2x - 3y + 11 = 0$, $2x^2 - xy = 36$

7. $5x + 3y + 7 = 0$, $3y^2 = x^2 - 4y + 3$

8. $x(x + y) = 3$, $y - x = 1$

9. $\dfrac{x}{2} + \dfrac{y}{3} = 2$, $\dfrac{x^2}{4} + \dfrac{y^2}{9} = 2$

10. $x + 2y = 5$, $2x^2 + y^2 = 6$

MULTIPLE CHOICE EXERCISE 12

1. The function, which is not a quadratic in x, is:
   [A] $y = 2x^2 - 5x$ [B] $y = x(x - 5)$
   [C] $y = x^2 - 5$ [D] $y = 5(x - 1)$

2. The quadratic equation whose roots are $x = -2$ and $x = 7$ is:
   [A] $x^2 + 2x - 7 = 0$ [B] $x^2 - 2x + 7 = 0$
   [C] $x^2 - 5x - 14 = 0$ [D] $x^2 + 5x - 14 = 0$

3. A quadratic equation has roots $-\dfrac{2}{3}$ and $-\dfrac{1}{4}$. The required equation is:
   [A] $x^2 - \dfrac{11}{12}x + 2 = 0$ [B] $12x^2 + 11x + 2 = 0$
   [C] $12x^2 - 11x + 2 = 0$ [D] $12x^2 - 11x - 2 = 0$

4. The only quadratic equation, which can be constructed with roots $-\dfrac{1}{2}$ and 2 is:
   [A] $3x^2 - 3x + 2 = 0$ [B] $3x^2 + 3x + 2 = 0$
   [C] $2x^2 - 3x + 2 = 0$ [D] $2x^2 - 3x + 2 = 0$

5. The quadratic equation whose roots are 3 and $\dfrac{2}{3}$ is:
   [A] $3x^2 - 11x + 6 = 0$ [B] $x^2 - 11x + 6 = 0$
   [C] $3x^2 - 11x + 2 = 0$ [D] $3x^2 - 11x - 2 = 0$

6. If the roots of a quadratic equation are $\dfrac{1}{4}$ and 3 then the quadratic equation is:
   [A] $4x^2 - 13x + 3 = 0$ [B] $4x^2 - 13x - 3 = 0$
   [C] $4x^2 + 13x - 3 = 0$ [D] $4x^2 + 13x + 3 = 0$

7. The equation whose roots are 4 and -5 is:
   [A] $x^2 - x - 20 = 0$ [B] $x^2 + x + 20 = 0$
   [C] $x^2 - x + 20 = 0$ [D] $x^2 + x - 20 = 0$

8. The equation whose roots are $\dfrac{2}{3}$ and $-\dfrac{1}{4}$ is:
   [A] $12x^2 + 11x + 2 = 0$ [B] $12x^2 - 11x + 2 = 0$
   [C] $12x^2 - 5x - 2 = 0$ [D] $12x^2 - 11x - 2 = 0$

9. The roots of a quadratic equation in x are $-m$ and $2n$. The equation is:
   [A] $x^2 + x(m - 2n) - 2mn = 0$
   [B] $x^2 - x(m - 2n) - 2mn = 0$
   [C] $x^2 - x(m - 2n) + 2mn = 0$
   [D] $x^2 + x(m - 2n) + 2mn = 0$

10. The equation whose roots are -8 and 5 is:
    [A] $x^2 + 3x + 40 = 0$ [B] $x^2 - 3x - 40 = 0$
    [C] $x^2 + 3x - 40 = 0$ [D] $x^2 - 3x + 40 = 0$

11. If $x^2 + 15x + 50 \equiv ax^2 + bx + c = 0$, it is not true to say that:
    [A] $x = 5$ [B] $x = 10$
    [C] $x + 10 = 0$ [D] $bc = 750$

12. If $2x^2 + kx - 14 = (x + 2)(2x - 7)$ then, the value of k is:
    [A] -3 [B] 5 [C] 9 [D] 11

13. Given that
    $5x^2 + 4x + 3 = a + b(x + 1) + cx(x + 1)$.
    The value of the constants a, b, and c are respectively:

[A] 7, 4, 5   [B] 4, −1, 5

[C] −1, 5, 4   [D] 7, 5, 4

14. The equation $(x + 2)(x − 7) = 0$ has roots:

   [A] − 2 and 7   [B] 2 and − 7

   [C] − 2 and − 7   [D] 2 and 7

15. If the roots of the equation $(3x − 1)(x + 2) = 0$ are p and q, then, the value of $p + q$ is:

   [A] $2\frac{1}{2}$ [B] $1\frac{2}{3}$ [C] $−1\frac{2}{3}$ [D] $−2\frac{1}{2}$

16. The values of x, which satisfy, $x^2 + 2x + 1 = 25$ are:

   [A] −6, −4   [B] 6, −4

   [C] 6, 4   [D] −6, 4

17. The values of x, which satisfy the equation, $x^2 − 2x − 3 = 0$ are:

   [A] (−3, 1)   [B] (−1, 3)

   [C] (−3, 1)   [D] (−1, −3)

18. The smaller value of x for which $x^2 − 3x + 2 = 0$ is:

   [A] 1   [B] 2   [C] − 1   [D] − 2

19. $6x^2 − 7x − 5 = 0$, only if:

   [A] $x = \frac{1}{2}$ or $− 2\frac{1}{2}$ [B] $x = \frac{1}{3}$ or $2\frac{1}{2}$

   [C] $x = 1\frac{2}{3}$ or $−\frac{1}{2}$ [D] $x = −1\frac{2}{3}$ or $\frac{1}{2}$

20. $2x^2 − 3x − 2 = 0$ is true if and only if:

   [A] $x = −2$ or $\frac{1}{2}$ [B] $x = 1$ or 8

   [C] $x = −\frac{1}{2}$ or 2 [D] $x = −1$ or 2

21. $2a^2 − 3a − 27 = 0 \Leftrightarrow a = $:

   [A] $− 3, −\frac{9}{2}$ [B] $−\frac{2}{3}, 9$

   [C] $3, \frac{9}{2}$ [D] $− 3, \frac{9}{2}$

22. The equation $3x^2 + 25x − 18 = 0$ has roots:

   [A] $−3, 2$ [B] $−9, \frac{2}{3}$

   [C] $−\frac{3}{2}, 9$ [D] $−2, 3$

23. The sum of the roots of the equation $2x^2 + 3x − 9 = 0$ is:

   [A] −18   [B] −6   [C] $\frac{9}{2}$ [D] $−\frac{3}{2}$

24. The two values of x, which satisfy the equation $5x^2 − 4x − 1 = 0$ are:

   [A] $1, −\frac{1}{5}$ [B] $−1, −\frac{1}{5}$

   [C] $−1, \frac{1}{5}$ [D] $1, \frac{1}{5}$

25. The solutions of the equation $3 + 5x − 2x^2 = 0$ are:

   [A] $−\frac{1}{2}, −3$ [B] $2, 3$

   [C] $−2, 3$ [D] $−\frac{1}{2}, 3$

26. Given that $10 − 3x − x^2 = 0$. The values of x are:

   [A] $x = 2$ or $x = −5$

   [B] $x = −2$ or $x = 5$

   [C] $x = −1$ or $x = 10$

   [D] $x = 2$ or $x = 5$

27. The equation $3a + 10 = a^2$ gives rise to the roots:

   [A] $a = 5$ or $a = 2$ [B] $a = −5$ or $a = 2$

   [C] $a = −5$ or $a = −2$ [D] $a = 5$ or $a = −2$

28. One of the roots of the equation $6x^2 = 5 − 7x$ is:

   [A] $−\frac{1}{2}$ [B] $−\frac{1}{3}$ [C] $\frac{1}{2}$ [D] $2\frac{1}{2}$

29. The values of y, which satisfy the equation $3y^2 = 3y$ are:

   [A] $y = −3$ or $y = 9$ [B] $y = 0$ or $y = 9$

   [C] $y = −3$ or $y = 3$ [D] $y = 3$ or $y = 9$

30. The equation $7y^2 = 27y$ has roots:

   [A] $y = 3$ and $y = 7$ [B] $y = 0$ and $y = 7$

   [C] $y = 0$ and $y = \frac{3}{7}$ [D] $y = 0$ and $y = 9$

31. A root of the equation $x^2 + 6x = 0$ is:

   [A] 0   [B] 6   [C] 2   [D] 3

32. If $x^2 − k^2 = 0$ where k is an integer, the truth about the roots of the equation is:

   [A] The two roots are equal.

   [B] The sum of the two roots is zero.

[C] The difference of the two roots is $2x$.
[D] The sum of the two roots is zero and
    the difference of the two roots is $2x$.

33. The value of k, which makes the expression $m^2 - 8m + k$ a perfect square is:
[A] 2   [B] 4   [C] 8   [D] 16

34. For $x^2 - 6x$ to be a perfect square:
[A] add 9   [B] add 36
[C] 1   [D] add 3

35. The number that must be added to the expression $x^2 - 8x$ to make a perfect square is:
[A] 36   [B] 9   [C] 25   [D] 16

36. If five times a certain integer is subtracted from twice the square of the integer, the result is 63. The integer is:
[A] 21   [B] 7   [C] 9   [D] 4

37. The area of a rectangle is the product of its length and breadth. The length and breadth of a rectangle are given as $(x-3)$ cm and $(x-5)$ cm. The area of the rectangle is 24 cm^2 only if:
[A] $x = -9$ or -1 [B] $x = 9$ or 1
[C] $x = 9$ or -1 [D] $x = -9$ or 1

38. The nature of the roots of the quadratic equation $x^2 - 3x + 2 = 0$ is that the roots are:
[A] Real and equal
[B] Real and distinct
[C] Imaginary
[D] Imaginary and equal

39. The sum of the squares of two natural numbers is 29. One of the numbers is three more than the other is. The larger of the numbers is:
[A] −5   [B] −2   [C] 2   [D] 5

40. 120 soldiers are standing in rows. There are 2 more soldiers in each row than there are rows. The number of rows is:
[A] 8   [B] 10   [C] 12   [D] 55

41. The difference between two numbers is 6 and the difference between their squares is 132. The numbers are:
[A] 8,14   [B] −8,−14
[C] −8,14   [D] 8,−14

42. A man uses 50 m of fencing to fence a garden at his back yard with his house as one side of the fence. If the area of the garden is 300 square metres, the length of the garden is:
[A] 6   [B] 10   [C] 15   [D] 25

TOPIC 13

ALGEBRAIC FRACTIONS

OBJECTIVES

At the end of this topic, the learner should be able to:

1. Simplify algebraic fractions to their lowest terms.
2. Multiply and Divide algebraic fractions.
3. Add and subtract algebraic fractions.
4. Write an algebraic fraction containing many terms as a single fraction with a common denominator.
5. Determine the domain of any given algebraic fraction.
6. Find the set of values for which an expression is undefined.

Norwegian mathematician Niels Henrik Abel proved that it was impossible to solve an algebraic equation containing a polynomial higher than the fourth degree. In 1825 the Norwegian government granted Abel a scholarship that enabled him to travel throughout Europe. During his travels he published papers and met with other mathematicians to compare theories. Abel's contacts with other leading mathematicians of his day inspired his work on elliptical functions and the solving of equations with radical numbers. Abel gave the binomial theorem a more comprehensive generalization, including the cases of irrational and imaginary exponents.

Roger Viollet/Getty Images

Simplifying Algebraic Fractions

The same rules used in simplifying and evaluating numerical fractions apply to algebraic fractions.

First, factorise the numerators and denominators completely. Common factors can then be cancelled.

Example 13:1

Evaluate and simplify the following.

1. $\dfrac{12x^3y^2}{24x^2y}$ 2. $\dfrac{9p^2}{6p^2-3pq}$

3. $\dfrac{5m+5n}{5n^2-5m^2}$ 4. $\dfrac{(x+y)^2}{-wyz-wxz}$

5. $\dfrac{3-3a}{6ab-6b}$ 6. $\dfrac{(x-y)^2}{y^2-x^2}$

7. $\dfrac{y^2-9y+20}{4y-y^2}$ 8. $\dfrac{x^2+2x-15}{2x^2-12x+18}$

Solution

1. $\dfrac{12x^3y^2}{24x^2y} = \dfrac{\overset{1}{\cancel{12}}\,x^{\overset{x}{\cancel{3}}}\,y^{\overset{y}{\cancel{2}}}}{\underset{2}{\cancel{24}}\,\underset{1}{\cancel{x^2}}\,\underset{1}{\cancel{y}}} = \dfrac{xy}{2}$

2. $\dfrac{9p^2}{6p^2-3pq} = \dfrac{\overset{3}{\cancel{9}}\,p^{\overset{p}{\cancel{2}}}}{\underset{1}{\cancel{3}}\,\underset{1}{\cancel{p}}\,(2p-q)} = \dfrac{3p}{2p-q}$

3. $\dfrac{5m+5n}{5n^2-5m^2} = \dfrac{\overset{1}{\cancel{5}}\,\overset{1}{\cancel{(m+n)}}}{\underset{1}{\cancel{5}}\,\underset{1}{\cancel{(n+m)}}(n-m)} = \dfrac{1}{n-m}$

4. $\dfrac{(x+y)^2}{-wyz-wxz} = \dfrac{\overset{x+y}{\cancel{(x+y)^2}}}{-wz\,\underset{1}{\cancel{(y+x)}}} = -\dfrac{x+y}{wz}$

5. $\dfrac{3-3a}{6ab-6b} = \dfrac{\overset{1}{\cancel{3}}\,\overset{-1}{\cancel{(1-a)}}}{\underset{2}{\cancel{6}}\,b\,\underset{1}{\cancel{(a-1)}}} = -\dfrac{1}{2b}$

Note that $a-1=-(a-1)$.

6. $\dfrac{(x-y)^2}{y^2-x^2} = \dfrac{\overset{-1}{\cancel{(x-y)}}(x-y)}{\underset{1}{\cancel{(y-x)}}(y+x)} = \dfrac{y-x}{y+x}$

7. $\dfrac{y^2-9y+20}{4y-y^2} = \dfrac{(y-5)\,\overset{-1}{\cancel{(y-4)}}}{y\,\underset{1}{\cancel{(4-y)}}} = \dfrac{5-y}{y}$

8. $\dfrac{x^2+2x-15}{2x^2-12x+18} = \dfrac{(x+5)\,\overset{1}{\cancel{(x-3)}}}{2\,\underset{1}{\cancel{(x-3)}}(x-3)} = \dfrac{x+5}{2(x-3)}$

EXERCISE 13:1

Evaluate and simplify the following.

1. $\dfrac{3xy}{12xz}$ 2. $\dfrac{10x^2}{5x^3}$

3. $\dfrac{8a-4b}{12g}$ 4. $\dfrac{10pq}{5p^2+20p}$

5. $\dfrac{3x-12}{9x-36}$ 6. $\dfrac{64xyz-48xy}{32xyz-24xy}$

7. $\dfrac{3x-3y}{6x-6y}$ 8. $\dfrac{x^2-9}{7x+21}$

9. $\dfrac{(p-5)^2}{4p-20}$ 10. $\dfrac{x^2-y^2}{(x-y)^2}$

11. $\dfrac{px+qx}{q^2+pq}$ 12. $\dfrac{ab+5b}{5x+ax}$

13. $\dfrac{6-x}{5x-30}$ 14. $\dfrac{x^2-3x}{21-7x}$

15. $\dfrac{p^2-4}{2x-px}$ 16. $\dfrac{2y-3xy}{9x^2-4}$

17. $\dfrac{1-2a}{2a^2-a}$ 18. $\dfrac{3u-3v}{2v^2-2u^2}$

19. $\dfrac{mx^2-my^2}{y^2-x^2}$ 20. $\dfrac{(a-b)^2}{b^2-a^2}$

21. $\dfrac{x^2+5x}{x^2+12x+35}$ 22. $\dfrac{y^2+6y-7}{y^2-49}$

23. $\dfrac{4a^2-b^2}{4a^2-4ab+b^2}$ 24. $\dfrac{9-3x}{x^2-5x+6}$

25. $\dfrac{2u^2 - 2}{u^2 - 4u - 5}$ 26. $\dfrac{2y^2 - 14y + 24}{12y^2 - 32y - 12}$

27. $\dfrac{2x^2 - 9x - 5}{6x^2 + 7x + 2}$ 28. $\dfrac{2r^2 - r - 3}{6r^2 - 11r + 3}$

29. $\dfrac{2x^2 - 5x - 12}{6x^2 - 12x - 48}$ 30. $\dfrac{2y^2 - y - 3}{4y^2 - 12y + 9}$

31. $\dfrac{12y^2 + 6y - 18}{(72y + 108)^2}$

Multiplying Algebraic Fractions

To multiply algebraic fractions, first factorise the numerators and denominators completely. Common factors can then be cancelled.

Example 13:2
Simplify the following.

1. $\left(\dfrac{z}{x + y}\right)\left(\dfrac{3x + 3y}{2z^2 + 2z}\right)$

2. $\left(\dfrac{9a}{3a - 15}\right)\dfrac{(a - 5)}{2(5 - a)}$

3. $\left(\dfrac{n^2 + 6n + 5}{7n^2 - 63}\right)\dfrac{7n + 21}{(5 + n)^2}$

4. $\left(\dfrac{x + 4}{4x}\right)\left(\dfrac{2x - 8}{4 + x}\right)\left(\dfrac{x^2 - 4}{24 - 12x}\right)\left(\dfrac{4x^2}{4 - x}\right)$

Solution

1. $\left(\dfrac{z}{x + y}\right)\left(\dfrac{3x + 3y}{2z^2 + 2z}\right) = \dfrac{\cancel{z}}{(\cancel{x + y})} \cdot \dfrac{3\,(\cancel{x + y})}{2\,\cancel{z}\,(z + 1)}$

$= \dfrac{3}{2(z + 1)}$

2. $\left(\dfrac{9a}{3a - 15}\right)\dfrac{(a - 5)}{2(5 - a)} = \dfrac{\cancel{9}a}{\cancel{3}\,(\cancel{a - 5})} \cdot \dfrac{(\cancel{a - 5})}{2(5 - a)}$

$= \dfrac{3a}{2(5 - a)}$

3. $\left(\dfrac{n^2 + 6n + 5}{7n^2 - 63}\right)\dfrac{7n + 21}{(5 + n)^2}$

$= \dfrac{(\cancel{n + 5})(n + 1)}{\cancel{7}(n - 3)(\cancel{n + 3})} \cdot \dfrac{\cancel{7}\,(\cancel{n + 3})}{(\cancel{5 + n})(5 + n)}$

$= \dfrac{n + 1}{(n - 3)(n + 5)}$

4. $\left(\dfrac{x + 4}{4x}\right)\left(\dfrac{2x - 8}{4 + x}\right)\left(\dfrac{x^2 - 4}{24 - 12x}\right)\left(\dfrac{4x^2}{4 - x}\right)$

$= \dfrac{(\cancel{x + 4})}{\cancel{4}x} \cdot \dfrac{\cancel{2}\,(\cancel{x - 4})}{(\cancel{4 + x})} \cdot \dfrac{(x + 2)\,(\cancel{x - 2})}{\cancel{12}\,(\cancel{2 - x})} \cdot \dfrac{\cancel{4}\cancel{x^2}}{(\cancel{4 - x})}$

$= \dfrac{x(x + 2)}{6}$

EXERCISE 13:2

Simplify the following.

1. $\left(\dfrac{x + y}{5}\right)\left(\dfrac{15}{7x + 7y}\right)$

2. $\left(\dfrac{x^2}{y}\right)\left(\dfrac{y^2}{3}\right)\left(\dfrac{1}{x^3}\right)$

3. $\left(\dfrac{6p + 12q}{2pq - 4p}\right)\left(\dfrac{7q - 14}{3p + 6q}\right)$

4. $\left(\dfrac{15a + 15b}{16y^2 - 9}\right)\left(\dfrac{8y + 6}{5a + 5b}\right)$

5. $\dfrac{(x + y)^2}{(ax + ay)}\left(\dfrac{a^2}{3x + 3y}\right)$

6. $\left(\dfrac{a^4 - b^4}{5a + 5b}\right)\left(\dfrac{25}{2a^2 + 2b^2}\right)$

7. $\left(\dfrac{5ax + 10a}{ap - 10p}\right)\left(\dfrac{200 - 2a^2}{ax + 2a}\right)$

8. $\dfrac{x^2 - 4}{(x - 2)^2}\left(\dfrac{x^2 - 9x + 14}{x^3 + 2x^2}\right)$

9. $\left(\dfrac{2v+u}{3-y}\right)\left(\dfrac{3u^2-3uv}{uv+2v^2}\right)\left(\dfrac{5uv-10v^2}{u^2-3uv+2v^2}\right)$

10. $\left(\dfrac{x^2-y^2}{3-y}\right)\left(\dfrac{x^2+3xy}{4y-4x}\right)\left(\dfrac{xy-3y}{x^2+4xy+3y^2}\right)$

Dividing by Algebraic Fractions

To divide by an algebraic fraction multiply by the reciprocal of the fraction as with numerical fractions.

Example 13:3

Simplify the following.

1. $\left(\dfrac{5a}{7}\right)\left(\dfrac{2b}{a}\right)\div\dfrac{b^3}{21}$

2. $\dfrac{4y^2-1}{9y-3y^2}\div\dfrac{2y^2-7y-4}{y^2-7y+12}$

Solution

1. $\left(\dfrac{5a}{7}\right)\left(\dfrac{2b}{a}\right)\div\dfrac{b^3}{21}=\dfrac{5\overset{1}{\cancel{a}}}{\cancel{7}}\cdot\dfrac{2\overset{1}{\cancel{b}}}{\cancel{a}}\cdot\dfrac{\overset{3}{\cancel{21}}}{\underset{b^2}{\cancel{b^3}}}=\dfrac{30}{b^2}$

2. $\dfrac{4y^2-1}{9y-3y^2}\div\dfrac{2y^2-7y-4}{y^2-7y+12}$

$=\dfrac{(2y-1)\overset{1}{\cancel{(2y+1)}}}{3y\underset{1}{\cancel{(3-y)}}}\cdot\dfrac{\overset{-1}{\cancel{(y-3)}}\overset{1}{\cancel{(y-4)}}}{\underset{1}{\cancel{(2y+1)}}\underset{1}{\cancel{(y-4)}}}=\dfrac{1-2y}{3y}$

EXERCISE 13:3

Evaluate and simplify the following.

1. $3p\div\dfrac{12p}{7}$ 2. $\dfrac{5}{x}\div\dfrac{7}{x}$ 3. $\dfrac{uv}{4}\div\dfrac{v^2}{6}$

4. $\dfrac{y}{2}\div\dfrac{ay}{6}$ 5. $\dfrac{a^2b}{16x}\div\dfrac{ab^2}{8x^2}$ 6. $\dfrac{2a}{b}\div\dfrac{10a}{3b}$

7. $\dfrac{3a}{a+2}\div\dfrac{5a}{a+2}$ 8. $\dfrac{2(x+3)}{7}\div\dfrac{8(x+3)}{63}$

9. $\dfrac{2p-10}{3q-21}\div\dfrac{p-5}{4q-28}$

10. $\dfrac{3(x^2+5)}{y(x^2-3)}\div\dfrac{18(x^2+5)}{7(x^2-3)}$

11. $\dfrac{a^2-1}{3a+9}\div\dfrac{a^2b-b}{ab+3b}$

12. $\dfrac{p^2-pq}{3p^2+3pq}\div\dfrac{7p-7q}{p^2-q^2}$

13. $\dfrac{2x^2+2y^2}{x^2-4y^2}\div\dfrac{x^2+y^2}{x+2y}$

14. $\dfrac{b^2+bc}{ab-bc}\div\dfrac{b^2-c^2}{a^2-c^2}$

15. $\dfrac{x^3-64x}{2x^2+16x}\div\dfrac{x^2-9x+8}{x^2+4x-5}$

16. $\dfrac{u^3-uv^2}{6u+6v}\div\dfrac{u^2-4uv+3v^2}{12u-36v}$

17. $\dfrac{3y+y^2}{12+3y}\div\dfrac{9y-y^3}{12-y-y^2}$

18. $\dfrac{x^2+16x+64}{2x^2-128}\div\dfrac{3x^2+30x+48}{x^2-6x-16}$

Sum and Difference of Algebraic Fractions

To express the fractions as a single fraction find the LCM of the denominators and use it to evaluate the sums and differences. After evaluating an algebraic fraction, simplify the single fraction to its lowest terms.

Example 13:4

Simplify the following algebraic fractions, writing your answer as a single fraction.

(a) $\dfrac{1}{x+3}-\dfrac{1}{x-1}$ (b) $\dfrac{1}{x+3}-\dfrac{1}{x+4}$

(c) $\dfrac{x+5}{x-2}-\dfrac{x-5}{2}$ (d) $\dfrac{3}{3x+1}-\dfrac{2}{2x+3}$

(e) $\dfrac{x+1}{2}-\dfrac{x+2}{3}+\dfrac{x-4}{4}$ (f) $\dfrac{x+2}{3}-\dfrac{2x-1}{5}$

Solutions

(a) $\dfrac{1}{x+3} - \dfrac{1}{x-1} = \dfrac{(x-1)-(x+3)}{(x+3)(x-1)}$

$= \dfrac{x-1-x-3}{(x+3)(x-1)}$

$= \dfrac{-4}{(x+3)(x-1)}$

(b) $\dfrac{1}{x+3} - \dfrac{1}{x+4} = \dfrac{(x+4)-(x+3)}{(x+3)(x+4)}$

$= \dfrac{x+4-x-3)}{(x+3)(x+4)}$

$= \dfrac{1}{(x+3)(x+4)}$

(c) $\dfrac{x+5}{x-2} - \dfrac{x-5}{2} = \dfrac{2(x+5)-(x-2)(x-5)}{2(x-2)}$

$= \dfrac{2x+10-(x^2-7x+10)}{2(x-2)}$

$= \dfrac{2x+10-x^2+7x-10}{2(x-2)}$

$= \dfrac{9x-x^2}{2(x-2)}$

$= \dfrac{x(9-x)}{2(x-2)}$

(d) $\dfrac{3}{3x+1} - \dfrac{2}{2x+3} = \dfrac{3(2x+3)-2(3x+1)}{(3x+1)(2x+3)}$

$= \dfrac{6x+9-6x-2}{(3x+1)(2x+3)}$

$= \dfrac{7}{(3x+1)(2x+3)}$

(e) $\dfrac{x+1}{2} - \dfrac{x+2}{3} + \dfrac{x-4}{4} = \dfrac{12(x+1)-8(x+2)+6(x-4)}{24}$

$= \dfrac{12x+12-8x-16+6x-24}{24}$

$= \dfrac{10x-28}{24}$

$= \dfrac{5x-14}{12}$

(f) $\dfrac{x+2}{3} - \dfrac{2x-1}{5} = \dfrac{5(x+2)-3(2x-1)}{3(5)}$

$= \dfrac{5x+10-6x+3}{15}$

$= \dfrac{13-x}{15}$

EXERCISE 13:4

Simplify the following algebraic fractions, writing your answer as a single fraction with a common denominator.

1. $\dfrac{x}{4} + \dfrac{5x}{4}$
2. $\dfrac{7a}{5} - \dfrac{3a}{5}$
3. $\dfrac{10}{x} + \dfrac{3}{x}$

4. $\dfrac{17}{3p} - \dfrac{14}{3p}$
5. $\dfrac{5x}{y^2} - \dfrac{3x}{y^2}$
6. $\dfrac{5u}{v} + \dfrac{u}{v} - \dfrac{3}{x}$

7. $\dfrac{x+2}{4} + \dfrac{7x-2}{4}$
8. $\dfrac{2a-7}{5} + \dfrac{2a-10}{5}$

9. $\dfrac{8}{p} + \dfrac{2}{p} - \dfrac{3p+10}{p}$
10. $\dfrac{5x}{x-y} - \dfrac{5y}{x-y}$

11. $\dfrac{6}{2y-3} - \dfrac{4y}{2y-3}$
12. $\dfrac{3a}{a^2-4} - \dfrac{a-4}{a^2-4}$

13. $\dfrac{2}{x+1} + \dfrac{3}{x-1}$
14. $\dfrac{y}{2-y} - \dfrac{y}{2+y}$

15. $\dfrac{a}{8} + \dfrac{b}{3} - \dfrac{c}{6}$
16. $\dfrac{4}{3t} + \dfrac{3}{2t} + \dfrac{1}{6t}$

17. $\dfrac{10}{x^2y^2} - \dfrac{4}{xy^2} + \dfrac{5}{x^2y}$

18. $\dfrac{2}{3x+3y} - \dfrac{3}{5x+5y}$

19. $\dfrac{4x}{x^2-36} - \dfrac{4}{x+6}$

20. $\dfrac{3b^2}{9b^2-16a^2} - \dfrac{b}{3b+4a}$

21. $\dfrac{\dfrac{x}{4}}{\dfrac{7}{8} - \dfrac{x}{2}}$
22. $\dfrac{\dfrac{x}{y} + \dfrac{y}{x}}{\dfrac{x}{y} - \dfrac{y}{x}}$

Undefined Expressions

A fraction with denominator 0 is undefined. This idea can be used to find the values of x for which an algebraic fraction is undefined or does not have real values. To do this, simply equate the denominator to zero and solve this equation. The set of values for which an expression is undefined is also called the restricted domain or asymptotic values.

Example 13:5

Find the value(s) of x for which each of the following expressions is undefined.

(a) $\dfrac{x^2 + 15x + 50}{x - 5}$ (b) $\dfrac{2x^2 - 3x + 1}{6x^2 + 5x + 1}$

Solutions

(a) For the expression $\dfrac{x^2 + 15x + 50}{x - 5}$ to be undefined $x - 5 = 0$, so $x = 5$.

(b) $\dfrac{2x^2 - 3x + 1}{6x^2 + 5x + 1} = \dfrac{(2x - 1)(x - 1)}{(2x + 1)(3x + 1)}$

For the expression $\dfrac{2x^2 - 3x + 1}{6x^2 + 5x + 1}$ to be undefined, $(2x + 1)(3x + 1) = 0$,

Therefore, $x = -\dfrac{1}{2}$ or $-\dfrac{1}{3}$.

Always factorise the numerator to detect any common factors, which are cancellable.

EXERCISE 13:5

(a) Find the value(s) of x for which each of the following expressions undefined.

1. $\dfrac{4}{x}$ 2. $\dfrac{x + 3}{x - 3}$ 3. $\dfrac{x - 3}{x + 3}$

4. $\dfrac{7}{xy}$ 5. $\dfrac{x(x + 5)}{y(x - 5)}$ 6. $\dfrac{y(x - 5)}{x(x + 5)}$

7. $\dfrac{(x - 1)(x + 3)}{(x + 1)(x - 3)}$ 8. $\dfrac{(x - 4)(x + 5)}{(x + 4)(x - 5)}$

9. $\dfrac{x^2 + 3x}{x^2 + 12x + 35}$ 10. $\dfrac{a^2 + b^2}{4a^2 - 4ax + x^2}$

(b) When is each of the following quotients impossible?

1. $\dfrac{2x - 16}{x^2 - 11x + 24}$ 2. $\dfrac{2x + 1}{x^2 + 6x + 5}$

3. $\dfrac{x + 2}{x^2 - 3x - 10}$ 4. $\dfrac{3x^2 + 1}{x^2 + 2x - 8}$

5. $\dfrac{2x^2 + x - 1}{(x - 1)(6x^2 - x - 1)}$ 6. $\dfrac{2y^2 - 9y - 5}{6y^2 + 7y + 2}$

(c) When is each of the following quotients impossible?

1. $\dfrac{6}{xy}$ 2. $\dfrac{5a}{4mn}$ 3. $\dfrac{15}{x - y}$

4. $\dfrac{z}{x - 2y}$ 5. $\dfrac{4}{2x - 4}$ 6. $\dfrac{3}{9 - x}$

MULTIPLE CHOICE EXERCISE 13

1. As a single fraction, $\dfrac{x - 3}{6} - \dfrac{x + 3}{5}$ equals:

[A] $\dfrac{-x + 33}{30}$ [B] $-\dfrac{x + 33}{30}$

[C] $\dfrac{x - 33}{30}$ [D] $\dfrac{x + 33}{30}$

2. When $\dfrac{t + 2}{2} + \dfrac{t + 3}{3}$ is simplified, the result is:

[A] $\dfrac{5t + 5}{6}$ [B] $\dfrac{5t + 12}{6}$

[C] $\dfrac{2t + 5}{5}$ [D] $\dfrac{t^2 + 6}{6}$

3. $\dfrac{x - 2}{6} - \dfrac{x - 7}{4}$ is equal to:

[A] $\dfrac{8 - 5x}{6}$ [B] $\dfrac{17 - 5x}{6}$

[C] $\dfrac{x - 17}{12}$ [D] $\dfrac{17 - x}{12}$

4. When simplified, $\dfrac{1}{u} + \dfrac{1}{v} = \dfrac{1}{f}$ equals:

[A] $\dfrac{vf^2 + uf^2 - uv}{fuv}$ [B] $\dfrac{vf + uf^2 - uv}{fuv}$

[C] $\dfrac{vf + uf - uv}{fuv}$ [D] $\dfrac{vf + uf - uv}{u}$

5. The result of simplifying $\dfrac{x^2 - y^2}{x + y}$ is:

[A] $x + y$ [B] $x^2 + y^2$

[C] $x - y$ [D] $(x - y)^2$

6. When simplified $\dfrac{2 - 18m^2}{1 + 3m}$ becomes:

[A] $(1 + 3m)$ [B] $2(1 + 3m^2)$

[C] $2(1 - 3m)$ [D] $2(1 - 3m^2)$

7. In its simplest form $\dfrac{x^2 - 3x + 2}{x^2 - 4}$ is:

[A] $\dfrac{x + 1}{x + 2}$ [B] $\dfrac{x + 2}{x - 2}$

[C] $\dfrac{x + 1}{x - 2}$ [D] $\dfrac{x - 1}{x + 2}$

8. As a single fraction $\dfrac{5}{1 - x} + \dfrac{2}{1 + x}$ is:

[A] $\dfrac{7 + 3x}{1 - x^2}$ [B] $\dfrac{7 + 3x}{(1 - x)^2}$

[C] $\dfrac{7 + 3x}{(1 + x)^2}$ [D] $\dfrac{7 - 3x}{1 - x^2}$

9. $\dfrac{3}{6r} + \dfrac{3}{4r}$ as a single fraction is:

[A] $\dfrac{1}{12r}$ [B] $\dfrac{12}{r}$ [C] $\dfrac{1}{6r}$ [D] $\dfrac{5}{4r}$

10. The simplified form of $\dfrac{5}{x - y} - \dfrac{4}{y - x}$ is:

[A] $\dfrac{9}{x - y}$ [B] $\dfrac{9}{y - x}$ [C] $\dfrac{1}{x - y}$ [D] $\dfrac{9}{y - x}$

11. After expressing as a single fraction,

$\dfrac{x}{x - 2} - \dfrac{x + 2}{x + 3}$ becomes:

[A] $\dfrac{3x - 4}{(x - 2)(x + 3)}$ [B] $\dfrac{2x^2 + 3x - 4}{(x - 2)(x + 3)}$

[C] $\dfrac{2}{(x - 2)(x + 3)}$ [D] $\dfrac{3x + 4}{(x - 2)(x + 3)}$

12. Simplifying $\dfrac{4}{x + 1} - \dfrac{3}{x - 1}$ leads to:

[A] $\dfrac{x + 7}{x^2 - 1}$ [B] $\dfrac{x - 1}{x^2 + 1}$

[C] $\dfrac{x - 7}{x^2 - 1}$ [D] $\dfrac{x - 11}{x^2 - 1}$

13. $\dfrac{1}{x - 3} - \dfrac{3(x - 1)}{x^2 - 9}$ as a single fraction is:

[A] $\dfrac{x - 1}{x - 3}$ [B] $\dfrac{-2}{x + 3}$

[C] $\dfrac{x - 1}{x + 3}$ [D] $\dfrac{4x}{x^2 - 9}$

14. In its simplest form, $\dfrac{2x - 1}{3} - \dfrac{x + 3}{2}$ is:

[A] $\dfrac{x + 7}{6}$ [B] $\dfrac{x + 8}{6}$

[C] $\dfrac{x - 11}{6}$ [D] $\dfrac{x - 4}{6}$

15. By simplifying $\dfrac{1}{1 - x} + \dfrac{2}{1 + x}$, the result is:

[A] $\dfrac{x - 3}{1 - x^2}$ [B] $\dfrac{x - 3}{1 + x^2}$

[C] $\dfrac{3 - x}{1 - x^2}$ [D] $\dfrac{3 + x}{1 + x}$

16. Simplifying $\dfrac{1}{x - 3} - \dfrac{2}{x + 4}$, gives rise to:

[A] $\dfrac{1 - x}{(x - 3)(x + 4)}$ [B] $\dfrac{7 - x}{(x - 3)(x + 4)}$

[C] $\dfrac{10 - x}{(x - 3)(x + 4)}$ [D] $\dfrac{-(x + 2)}{(x - 3)(x + 4)}$

17. The result of simplifying $1 - \dfrac{12 - 3x^2}{2x^2 - 8}$ is:

[A] $\dfrac{2 - x}{x - 2}$ [B] $\dfrac{3x}{4}$ [C] $\dfrac{3}{2}$ [D] $\dfrac{5}{2}$

18. $4 - \dfrac{y - x}{x}$, expressed as a single fraction

is:

[A] $\dfrac{5x-y}{x}$ [B] $\dfrac{3x-y}{x}$

[C] $\dfrac{4-y+x}{x}$ [D] $\dfrac{4-y-x}{x}$

19. When simplified $\dfrac{3x-2}{x-5} - \dfrac{2x+3}{2(x-5)}$ equals:

[A] $\dfrac{1}{x-5}$ [B] $\dfrac{1}{2(x-5)}$

[C] $\dfrac{4x-7}{2(x-5)}$ [D] $\dfrac{4x-7}{x-5}$

20. $\dfrac{2}{a+b} - \dfrac{b}{a^2-b^2}$ can be simplified to obtain:

[A] $\dfrac{3}{a+b}$ [B] $\dfrac{a-3b}{a^2-b^2}$

[C] $\dfrac{3a+b}{a^2-b^2}$ [D] $\dfrac{2a-3b}{a^2-b^2}$

21. The fraction that is equal to $\dfrac{4}{a-3} - \dfrac{1}{a^2-b^2}$

is:

[A] $\dfrac{3a+11}{(a+3)(a-2)}$ [B] $\dfrac{3a+11}{(a+3)(a-2)}$

[C] $\dfrac{3a-11}{(a-3)(a+2)}$ [D] $\dfrac{3a+11}{(a-3)(a-2)}$

22. The expression $\dfrac{a-b}{x-y}$ is undefined when:

[A] $a=b$ [B] $x=y$

[C] $a-b=0$ [D] $\dfrac{a-b}{x-y}=0$

23. The quotient $\dfrac{2x+1}{x^2 6x+5}$ is impossible when:

[A] $x=1$ [B] $x=5$
[C] $x=1\text{ or }x=5$ [D] $x=-1\text{ or }x=-5$

24. The restricted domain of $\dfrac{2x^2 3x+1}{6x^2-x-1}$ is:

[A] $x=-\dfrac{1}{3}\text{ or }x=\dfrac{1}{2}$

[B] $x=\dfrac{1}{3}\text{ or }x=-\dfrac{1}{2}$

[C] $x=-\dfrac{1}{3}\text{ or }x=-\dfrac{1}{2}$

[D] $x=\dfrac{1}{3}\text{ or }x=\dfrac{1}{2}$

25. A value of y, which makes the expression

$\dfrac{y+2}{y^2-3y-10}$ undefined, is:

[A] $y=10$ [B] $y=2$ [C] $y=3$ [D] $y=5$

26. The values of x, which renders the expression

$\dfrac{x-5}{x^2-2x-3}$ undefined, is:

[A] $3,-2$ [B] $-1,-3$ [C] $-1,3$ [D] $1,-3$

27. The value of x for which the expression

$\dfrac{x^2+15x+50}{x-5}$ is undefined is:

[A] -10 [B] -5 [C] 0 [D] 5

28. The values of x, which make the expression

$\dfrac{6x-1}{x^2+4x-5}$ undefined are:

[A] $+4\text{ and }+1$ [B] $-5\text{ and }+1$

[C] $+4\text{ and }-1$ [D] $+5\text{ and }-1$

29. The pair of values of x for which

$\dfrac{1}{2x^2-13x+15}$ is undefined are:

[A] $5\text{ or }\dfrac{3}{2}$ [B] $1\text{ or }\dfrac{15}{13}$

[C] $2\text{ or }15$ [D] $13\text{ or }15$

30. The pair of values of x for which the

expression $\dfrac{3x-2}{4x^2+9x-9}$ is undefined is:

[A] $-\dfrac{3}{4}\text{ or }3$ [B] $-\dfrac{2}{3}\text{ or }-3$

[C] $\dfrac{2}{3}\text{ or }3$ [D] $\dfrac{3}{4}\text{ or }-3$

31. The restricted domain of $\dfrac{2y-16}{y^2-11y+24}$ is:

[A] 8 [B] -3 [C] 3 [D] $3\text{ or }8$

32. $\dfrac{2x^2-9x-5}{6x^2+7x+2}$ is impossible when:

[A] $\dfrac{2}{3}$ [B] $-\dfrac{2}{3}$ [C] 5 [D] -5

TOPIC 14

TRANSPOSITION OF FORMULAE

OBJECTIVES

At the end of this topic, the learner should be able to:

1. Make a subject of a formula which does not contain square roots.
2. Make a subject of a formula which contains square roots.
3. Make a subject of a formula which involves quadratics.

Pierre de Fermat

French mathematician Pierre de Fermat formulated a theorem that mathematicians spent centuries trying to prove. The theorem, called Fermat's Last Theorem, stipulates that the Pythagorean equation $(a^2 + b^2 = c^2)$ is true only for squares and that no positive integers can be found to satisfy the equation when the exponent is greater than two. While the unique theorem has a limited practical use, attempts to prove it have generated important discoveries in algebra and analysis.

Corbis

Making a Subject of a Formula

The methods used in making a subject of a formula are the same as those used in solving equations. The only difference is that the subject will be in terms of the other unknowns in the formula.

Formulae without Square Roots

Solve for the subject in the same way as equations are solved. In some cases, factorisation is required.

Example 14:1

Make x the subject of the formula $Cx - R = T$.

Solution

$$Cx - R = T$$

Adding R to both sides

$$Cx = T + R$$

Dividing both sides by C

$$C = \frac{T+R}{x}$$

Example 14:2

Make y the subject of the formula

$a(y + B) = C$.

Solution

$$a(y + B) = C$$

Expanding the LHS

$$ay + aB = C$$

Subtracting aB from both sides

$$ay = C - aB$$

Dividing both sides by a

$$y = \frac{C - aB}{a} \quad \text{or} \quad y = \frac{c}{a} - B$$

Example 14:3

Make n the subject of the formula

$f = ea - nh$.

Solution

$$f + nh = ea$$

$$nh = ea - f \iff n = \frac{ea - f}{h}$$

Example 14:4

Make a the subject of the formula $\frac{t}{a} = \frac{b}{e}$.

Solution

$$\frac{t}{a} = \frac{b}{e}$$

$$et = ab \iff a = \frac{et}{b}$$

Example 14:5

Make x the subject of the formula $r - \frac{m}{x} = p^2$.

Solution

$$r - \frac{m}{x} = p^2$$

$$rx - m = p^2 x$$

$$rx - p^2 x - m = 0$$

$$rx - p^2 x = m$$

$$x(r - p^2) = m \iff x = \frac{m}{r - p^2}$$

Example 14:6

Make y the subject of each of the following formulae.

(a) $Ny + B = C - Ny$ (b) $\frac{p - y}{p + y} = q$

Solution

(a) $\quad Ny + B = C - Ny$

$$2Ny + B = C$$

$$2Ny = C - B \iff y = \frac{C - B}{2N}$$

(b) $\quad \frac{p - y}{p + y} = q$

$$p - y = q(p + y)$$

$$p - y - pq = qy$$

$$p - pq = qy + y$$

$$p - pq = y(q + 1) \iff y = \frac{p - pq}{q + 1}$$

EXERCISE 14:1

1. Express w in terms of a, b, u and T given

    that $T = \dfrac{wa}{(u + w)b}$.

2. The formula for converting temperatures from Celsius ($C°$) to Fahrenheit ($F°$) is

    given by $F = \dfrac{9}{5}C + 32$. Find the temperature at which the Celsius and Fahrenheit thermometers record the same values. Make C the subject.

3. Given that $S = \dfrac{T}{5} - \dfrac{2R}{W}$. Make W the subject of the formula and find W when $T = 10.5,$ $S = 1$ and $R = 5.5$.

4. Given that $V = \dfrac{\pi r^2 h}{c}$.

    (a) Find the numerical value of V, giving your answer in standard form, given that $r = 0.01, c = 528, h = 25.2$ and

    $\pi = \dfrac{22}{7}$.

    (b) Express h in terms of $V, c, r,$ and π.

5. Given that $\dfrac{2x + y}{5y + 3x} = \dfrac{1}{4}$. Find the value $\dfrac{x}{y}$

6. Express p in terms of m and n given that

    $\dfrac{n}{4} = \dfrac{m}{2} - \dfrac{p}{8}$.

7. (i) In the formula $L = \dfrac{gh}{(g - h)k}$, make g the subject.

    (ii) Find the value of k if $g = 12, h = 4$ and $L = 3$.

Formulae Containing Square Roots

If there is a square root sign, first square both sides to eliminate the square root sign.

Example 14:7

Make x the subject of each of the following formulae.

(a) $\sqrt{Ax + B} = \sqrt{C}$ (b) $t = \sqrt{m + x^2}$

(c) $a\sqrt{k^2 - x} = b$ (d) $\sqrt{\left(\dfrac{a}{x} - b\right)} = c$

Solution

(a) $\sqrt{Ax + B} = \sqrt{C}$

Squaring both sides

$Ax + B = C$

$Ax = C - B \Leftrightarrow x = \dfrac{C - B}{A}$

(b) $t = \sqrt{m + x^2}$

Squaring both sides

$t^2 = m + x^2$

$t^2 - m = x^2 \Leftrightarrow x = \pm\sqrt{t^2 - m}$

Remember the $\pm$ sign or else a solution is lost.

(c) $a\sqrt{k^2 - x} = b$

Squaring both sides and expanding.

$a^2(k^2 - x) = b^2$

$a^2 k^2 - a^2 x = b^2$

$-a^2 x = b^2 - a^2 k^2 \Leftrightarrow x = \dfrac{a^2 k^2 - b^2}{a^2}$

(d) $\sqrt{\left(\dfrac{a}{x} - b\right)} = c$

Squaring both sides

$\dfrac{a}{x} - b = c^2$

$a - bx = c^2 x$

$a = (c^2 + b)x$

$\Leftrightarrow x = \dfrac{a}{(c^2 + b)}$

> **EXERCISE 14:2**

1. Given that $V = \dfrac{\pi n}{\sqrt{(A-3)}}$, express A in terms of V, π and n.

2. Given that $a = p\left(1 + \sqrt{\dfrac{r}{t}}\right)$, express t in terms of a, p and r.

3. Make p the subject of the formula

   $$m = 2n\pi\sqrt{\left(\dfrac{k+1}{p}\right)}.$$

   Hence, or otherwise, and taking π as $\dfrac{22}{7}$, find the value of p when $m = 22$, $n = 7$ and $k = 15$.

4. Given that $p = \sqrt{\dfrac{nq}{na+b}}$, make n the subject of the formula.

5. (a) Make v the subject of the formula

   $$y = \dfrac{k}{\sqrt[n]{v}}.$$

   (b) Given that $k = 16$ and $y = 4$, find v when $n = \dfrac{1}{2}$.

6. Make l the subject of the formula

   $$T = \dfrac{1}{2\pi}\sqrt{\dfrac{l}{g}}.$$

7. Make a the subject of the formula

   $$\sqrt{\dfrac{ra=t}{s}} = v.$$

8. Given that, $p = \sqrt{\dfrac{m+n}{m}}$ express m in terms of p and n.

9. The formula for finding the volume of a certain composite figure is

   $V = \dfrac{1}{3}\pi r^2\left(h + r\right)$. Make h the subject of the formula.

Formulae Containing Quadratics

When a formula contains a quadratic, the subject often has two different values. The factorisation method, the quadratic formula or the method of completing the square can be employed to find the two values.

Example 14:8

Make x the subject of the formula
$2x^2 + 7xyz - 15y^2z^2 = 0$. Find the values of x for which $y = 2$ and $z = -7$

Solution

$$2x^2 + 7xyz - 15y^2z^2 = 0$$

$$2x^2 + 7x(yz) - 15(yz)^2 = 0$$

$$(2x - 3yz)(x + 5yz) = 0$$

$$x = \dfrac{3}{2}yz \text{ or } x = -5yz$$

Alternatively, the quadratic formula

$x = \dfrac{-b \pm \sqrt{b^2 - 4ac}}{2a}$ may be used.

Thus, $a = 2, b = 7yz, c = -15y^2z^2$

$$x = \dfrac{-7yz \pm \sqrt{(7yz)^2 - 4(2)(-15y^2z^2)}}{2(2)}$$

$$x = \dfrac{-7yz \pm \sqrt{169y^2z^2}}{2(2)}$$

$$x = \dfrac{-7yz \pm 13yz}{4}$$

$$x = \dfrac{3yz}{2} \text{ or } x = -5yz$$

When $y = 2$ and $z = -7$

$$x = \dfrac{3}{2}(2)(-7) \text{ or } x = -5(2)(-7)$$

$$\Leftrightarrow x = -21 \text{ or } x = 70$$

EXERCISE 14:3

1. Make the letter in bracket a subject.

(a) $A = \pi rl + \pi r^2$ (r)

(b) $S = ut + \dfrac{1}{2}at^2$ (t)

(c) $ax^2 + c = bx$ (x)

(d) $A = \pi \{r + t\}^2 - r^2$ (t)

2. Given that $r^2 + 3s^2 = 4rs$, find the two values of $\dfrac{r}{s}$.

In the problem 3 to 5, make the letter in bracket the subject of the given formula.

3. $T^2 + TG = 12G^2$ (G)

4. $6p^2 + 17pqr = 3q^2r^2$ (r)

5. $12m^2 - 16mn + 5n^2 = 0$ (n)

MULTIPLE CHOICE EXERCISE 14

1. Given that $v = u + at$. In terms of u, v and t, a will be:

[A] $\dfrac{v + u}{t}$ [B] $\dfrac{v + t}{u}$ [C] $\dfrac{v - u}{t}$ [D] $\dfrac{t + u}{v}$

2. The sum of the interior angles of a polygon of n sides is given by $S = 180(n - 2)$. The number of sides in that polygon is given by:

[A] $n = \dfrac{s}{180} - 2$ [B] $n = \dfrac{s}{180}$

[C] $n = \dfrac{2s}{180}$ [D] $n = \dfrac{s}{180} + 2$

3. The volume V of a cylinder is given by $V = \pi r^2 h$ where r and h are the radius and height of the cylinder respectively. The height h is given by:

[A] $h = V\pi r^2$ [B] $\dfrac{V}{\pi r^2}$ [C] $\dfrac{\pi r^2}{V}$ [D] $\dfrac{\pi}{Vr^2}$

4. As a subject of the formula $A = 2\pi rh$, r is:

[A] $\dfrac{A}{2\pi h}$ [B] $\dfrac{\pi}{2Ah}$ [C] $\dfrac{\pi h}{2A}$ [D] $2Ah\pi$

5. The volume of a cone is given by the formula $V = \dfrac{1}{3}\pi r^2 h$. In terms of v, r and π, the height h is:

[A] $\dfrac{\pi r^2}{3}$ [B] $\dfrac{3Vr^2}{\pi}$ [C] $\dfrac{3V}{\pi r^2}$ [D] $\dfrac{\pi^2 r}{3}$

6. In the temperature conversion formula $F = \dfrac{9}{5}C + 32$, C is given by:

[A] $\dfrac{5F - 160}{9}$ [B] $\dfrac{5F + 160}{9}$

[C] $\dfrac{9F + 160}{5}$ [D] $\dfrac{9F - 160}{9}$

7. If $I = \dfrac{PRT}{100}$, the value of T when $P = 450$, $R = 12$ and $I = 90$ is:

[A] $\dfrac{3}{5}$ [B] $\dfrac{5}{6}$ [C] $\dfrac{5}{3}$ [D] 15

8. If $y = \dfrac{a + p}{a - p}$, then :

[A] $\dfrac{2a - y}{a + y} = p$ [B] $\dfrac{ay - 1}{y + 1} = p$

[C] $\dfrac{a(y - 1)}{y + 1} = p$ [D] $\dfrac{2y - 1}{y - 1} = p$

9. If $h(m + n) = m(h + r)$, h in terms of m, n and r is:

[A] $h = \dfrac{mr}{2m + n}$ [B] $h = \dfrac{mr}{2m - n}$

[C] $h = \dfrac{m + r}{n}$ [D] $h = \dfrac{mr}{n}$

10. If q is made the subject of the relation $t = \sqrt{\dfrac{pq}{r} - r^3 q}$, the relation will now be:

[A] $q = \dfrac{t^2}{p - r^4}$ [B] $q = \dfrac{rt^2}{p - r^4}$

[C] $q = \dfrac{rt}{p - r^4}$ [D] $q = \dfrac{p - r^4}{rt^2}$

11. Making S the subject of the formula

$$V = \frac{K}{\sqrt{T - S}} \text{ gives:}$$

[A] $\quad S = T - \dfrac{K^2}{V^2}$ $\qquad$ [B] $\quad S = \dfrac{K^2}{V^2} - T$

[C] $\quad S = T - \dfrac{V^2}{K^2}$ $\qquad$ [D] $\quad S = T\left(\dfrac{V^2 - K^2}{V^2}\right)$

12. If $y = \sqrt{ax - b}$, x in terms of y, a and b is:

[A] $x = \dfrac{y^2 - b}{a}$ $\qquad$ [B] $x = \dfrac{y + b}{a}$

[C] $x = \dfrac{y - b}{a}$ $\qquad$ [D] $x = \dfrac{y^2 + b}{a}$

13. $k = m\sqrt{\dfrac{t - p}{r}}$, as subject of the formula, t is equal to:

[A] $\quad \dfrac{rk^2 + p}{m^2}$ $\qquad$ [B] $\quad \dfrac{rk^2 + pm^2}{m^2}$

[C] $\quad \dfrac{rk^2 - p}{m^2}$ $\qquad$ [D] $\quad \dfrac{rk^2 - p^2}{m^2}$

14. Given that $U = \dfrac{T}{5} - \dfrac{2R}{Q}$, then T expressed in terms of U, R and Q is:

[A] $\quad T = 5U + \dfrac{2R}{Q}$ $\qquad$ [B] $\quad T = 5U + \dfrac{10R}{Q}$

[C] $\quad T = 5\left(U - \dfrac{2R}{Q}\right)$ $\qquad$ [D] $\quad T = 5U - \dfrac{2R}{Q}$

TOPIC 15

THE REMAINDER AND FACTOR THEOREMS

OBJECTIVES

At the end of this topic, the learner should be able to:

1. Obtain the roots or zero values of a polynomial.
2. Perform long division in algebra.
3. State and use the remainder theorem.
4. State and use the factor theorem.

German mathematician Carl Friedrich Gauss contributed to many areas of mathematics, including probability theory, algebra, and geometry. In his doctoral thesis he proved that every polynomial has at least one root, or solution; this theory became known as the fundamental theory of algebra. Gauss also applied his mathematical work to theories of electricity and magnetism. The magnetic unit of intensity is named in his honor. He considered that his greatest discovery was the method of constructing a regular seventeen sided polygon. This was not of the slightest use outside the world of mathematics, but was a great achievement of the human mind.

Science Source/Photo Researchers, Inc.

Introduction

In Topic 2, it was mentioned that, provided there is no remainder, the divisor and the quotient are factors of the dividend. Thus,

dividend   divisor   quotient

$\underbrace{\qquad}_{\text{multiple}}$ $\underbrace{\quad\quad\quad}_{\text{factors}}$

dividend   divisor   quotient

$$25 \div 7 = 3 \quad \text{Remainder } 4$$

$$\Rightarrow 25 = 7 \times 3 + 4$$

Zero Values and Factors of a Polynomial

Consider the polynomial,
$$f(x) = (x-1)(x+3)(x-2)$$
If $f(x) = 0$ then either $x - 1 = 0 \Leftrightarrow x = 1$

or $x + 3 = 0 \Leftrightarrow x = -3$

or $x - 2 = 0 \Leftrightarrow x = 2$

The values $x = 1, x = -3$ and $x = 2$, which make the polynomial $f(x)$ equal to zero, are called **zero values** or the **roots** of the equation $f(x) = 0$, while $(x - 1)$, $(x + 3)$ and $(x - 2)$ are called the factors of the expression.

Example 15:1

Given that $f(x) = x^3 - 2x^2 - x - 2$, find $f(-1)$.

Solution

$$f(-1) = (-1)^3 - 2(-1)^2 - (-1) - 2$$
$$= -1 - 2 + 1 - 2$$
$$\Rightarrow f(-1) = -4$$

Example 15:2

Given that $f(x) = x^3 + 2x^2 - x - 2$, find $f(-1)$.

Solutions

$$f(-1) = (-1)^3 + 2(-1)^2 - (-1) - 2$$
$$= -1 + 2 + 1 - 2$$
$$\Rightarrow f(-1) = 0$$

Division of Polynomials

Long division can be done in algebra using the same rules as in arithmetic.

Example 15:3

Divide $x^3 - 2x^2 - x - 2$ by $x + 1$.

Compare the remainder with the result in Example 15:1.

Solutions

$$\begin{array}{r} x^2 - 3x + 2 \\ (x+1)\overline{\smash{\big)}\, x^3 - 2x^2 - x - 2} \\ \underline{-(x^3 + x^2)} \\ -3x^2 - x - 2 \\ \underline{-(3x^2 - 3x)} \\ 2x - 2 \\ \underline{-(2x + 2)} \\ -4 \end{array}$$

$$\therefore \frac{x^3 - 2x^2 - x - 2}{x + 1} = x^2 - 3x + 2 \text{ remainder } -4.$$

Therefore, the remainder is −4 which is equal to $f(-1)$.

Example 15:3

Divide $x^3 - 2x^2 - x - 2$ by $x + 1$.

Compare the remainder with the result in Example 15:2.

Solutions

$$(x+1)\overline{)\,x^3+2x^2-x-2\,}$$

$$\begin{array}{r} x^2+x-2 \\ \hline -(x^3+x^2) \\ \hline x^2-x-2 \\ -(x^2+x) \\ \hline -2x-2 \\ -(2x-2) \\ \hline 0 \end{array}$$

$$\therefore \frac{x^3+2x^2-x-2}{(x+1)}=x^2+x-2 \text{ remainder } 0.$$

Hence $x+1$ is a factor of x^3+2x^2-x-2. moreover, the remainder is 0, which is equal to $f(-1)$.

Example 15:5

If $f(x)=x^3+3x-4$, find $f(4)$. Also, find the remainder when $f(x)$ is divided by $x-4$. What relationship exist between the remainder and $f(4)$?

Solutions

$$f(4)=4^3+3(4)-4 = 64+12-4 = 72$$

$$x-4\overline{)\,x^3+0x^2+3x-4\,}$$

$$\begin{array}{r} x^2+4x+19 \\ \hline -(x^3-4x^2) \\ \hline 4x^2+3x-4 \\ -(4x^2-16x) \\ \hline 19x-4 \\ -(19x-76) \\ \hline 72 \end{array}$$

$$\therefore \frac{x^3+3x-4}{x-4}=x^2+4x+19 \text{ remainder } 72.$$

Notice that the remainder is equal to $f(4)$. From above it can be seen that any polynomial $f(x)$ can be expressed in the form

$$f(x)=(ax+b)Q(x)+R$$

$Q(x)$ is called the **quotient**, $ax+b$ is

called the **divisor** and R is called the **remainder**.

The conclusions in the above examples can be summarized in two theorems called the **remainder theorem and** the **factor theorem**.

Remainder theorem

The **remainder theorem** states that,

When a polynomial $f(x)$ is divided by $ax+b$, the remainder is $f\left(-\dfrac{b}{a}\right)$

Note that the remainder is always of a lower degree than the divisor.

Factor theorem

The remainder theorem can be extended to the **factor theorem**, which states that,

If the remainder when $f(x)$ is divided by $ax+b$ is zero, then $ax+b$ is a factor of $f(x)$.

Conversely, if $f\left(-\dfrac{b}{a}\right)=0$ then $ax+b$ is a factor of $f(x)$.

Example 15:6

Given that $f(x)=6x^3-5x^2-17x+7$.

(a) Find $f\left(-\dfrac{3}{2}\right)$

(b) Divide $f(x)$ by $2x+3$

Solution

(a) $f\left(-\dfrac{3}{2}\right)=6\left(-\dfrac{3}{2}\right)^3-5\left(-\dfrac{3}{2}\right)^2-17\left(-\dfrac{3}{2}\right)+7.$

$$=-\dfrac{81}{4}-\dfrac{45}{4}+\dfrac{51}{2}+7$$

$$=1$$

(a) $(2x+3)\overline{\smash{\big)}\,6x^3-5x^2-17x+7}$ with quotient $3x^2-7x+2$

$$\underline{-(6x^3+9x^2)}$$
$$-14x^2-17x+7$$
$$\underline{-(-14x^2-21x)}$$
$$4x+7$$
$$\underline{-(4x+6)}$$
$$1$$

$$\therefore \frac{6x^3-5x^2-17x+7}{2x+3}=3x^2-7x+2$$

remainder 1.

Again the remainder is $f\left(-\dfrac{3}{2}\right)$.

Trial and Error Method

Bearing in mind the factor theorem, the factors of $f(x)$ are often obtained by trial and error method.

Example 15:7
Determine the factors of
$f(x) = 2x^3 + x^2 -7x -6$.

Solution
By trial and error method,

$f(1) = 2(1)^3 + (1)^2 - 7(1) - 6 = 2+1-7-6 \neq 0$

$f(-1) = 2(-1)^3 + (-1)^2 - 7(-1) - 6 = -2+1+7-6 = 0$

Therefore, $(x+1)$ is a factor of $f(x)$.

$f(2) = 2(2)^3 + (2)^2 - 7(2) - 6 = 16+4-14-6 = 0$

Therefore, $(x-2)$ is a factor of $f(x)$.

$\Rightarrow f(x) = (x+1)(x-2)(ax+3)$

Since the coefficient of x^3 and the independent terms are 2 and -6 respectively, then $a = 2$.

$\Rightarrow f(x) = (x+1)(x-2)(2x+3)$

EXERCISE 15

1. Determine the factors of the following polynomials.
   (a) $x^3 - 2x^2 - x + 2$
   (b) $2x^3 + x^2 - 2x - 1$

2. Solve the equations
   (a) $2x^3 - x^2 - 3x + 2 = 0$
   (b) $6x^3 - x^2 - 6x + 1 = 0$

3. Find the remainder when $x^3 + 5x^2 + 7x - 6$ is divided by $x + 2$.

4. Given that $x + 3$ is a factor of $f(x) = x^3 + 6x^2 + kx + 6$, find the value of k. Hence, factorise the expression completely.

5. Given that $x - 2$ is a factor the polynomial $f(x) = ax^2 + bx + c$ and that the remainders when $f(x)$ is divided by $x - 2$ and $x + 1$ respectively are -4 and 6. Find the values of a, b and c.

6. Find the equation whose roots are -1, -2 and 3.

7. The expression $ax^2 + 7x + b$ leave remainders 8 and 10 when divided by $x + 1$ and $x + 2$ respectively. Find the values of a and b.

8. If $x-2$ is a factor of $x^2 + kx + 6$, find the value of k. Using this value of k find the roots of the equation $x^2 + kx + 6 = 0$.

9. Find the value of k for which $x^3 + kx + 6 = f(x)$ has a root -3. Hence, factorise $f(x)$ completely.

10. Given that $f(x) = x^3 + kx^2 - x - 2$ and that $x + 1$ is a factor of $f(x)$. Find the value of k and hence factorise $f(x)$ completely.

11. Find the remainder when $x^3 + 5x^2 - 7x - 6$ is divided by $x + 1$. Hence, determine the number, to be added to $x^3 + 5x^2 - 7x - 6$, to make the result divisible by $x + 1$.

MULTIPLE CHOICE EXERCISE 15

1. The factor theorem states that if the remainder when $f(x)$ is divided by $ax + b$ is 0 then:

   [A] $f\left(\dfrac{a}{b}\right) = 0$ [B] $f\left(-\dfrac{a}{b}\right) = 0$

   [C] $f\left(-\dfrac{b}{a}\right) = 0$ [D] $f\left(\dfrac{b}{a}\right) = 0$

2. The remainder theorem states that when a polynomial $f(x)$ is divided by $ax + b$ the remainder is:

   [A] $f\left(\dfrac{a}{b}\right)$ [B] $f\left(-\dfrac{a}{b}\right)$

   [C] $f\left(-\dfrac{b}{a}\right)$ [D] $f\left(\dfrac{b}{a}\right)$

3. If $f(x)$ is divided by $ax - 1$ the remainder is:

   [A] $f(a)$ [B] $f(-a)$

   [C] $f\left(-\dfrac{1}{a}\right)$ [D] $f\left(\dfrac{1}{a}\right)$

4. The remainder when $x^3 + 2x^2 + 1$ is divided by x is:

   [A] x [B] 1
   [C] 2   [D] -1

5. $f(x) = -x^2 + kx - 6$ has a factor $(x + 2)$. The value of k is:

   [A] -3 [B] -7
   [C] -5 [D] 7

6. Given that $f(x) = 2x^2 + 3x - 2$. It is true to say that:

   [A] $(x - 2)$ is a factor of $f(x)$.
   [B] $(x + 2)$ is a factor of $f(x)$.
   [C] $f(x)$ leaves a remainder -2 when divided by $(x + 2)$.
   [D] $(2x + 1)$ is a factor of $f(x)$.

7. The remainder when $x^3 - 2x^2 - x - 2$ is divided by $x + 1$ is:

   [A] -2 [B] 4
   [C] -4 [D] 2

8. If $f(x) = x^3 - 3x - 4$. The remainder when $f(x)$ is divided by $x - 4$ is:

   [A] -56 [B] 48
   [C] -80 [D] 72

9. The remainder when $x^3 + 3x - 4$ is divided by $x + 1$ is:

   [A] -2 [B] -8
   [C] 0   [D] -7

10. Given that $f(x) = 2x^3 - 3x^2 - 3x + 11$. The remainder when $f(x)$ is divided by $2x + 1$ is:

    [A] 1.5   [B] 2.5
    [C] -1.5 [D] -2.5

11. The remainder when $3x^3 - x^2 - 2x + 13$ is divided by $x + 2$ is:

    [A] 4   [B] -3
    [C] -4 [D] 3

12. Given that when divided by $x + 1$ and $x + 2$ the expression $ax^2 + bx + 3$ leaves remainders 6 and 9 respectively. The values of a and b are:

    [A] $a = -2, b = 1$ [B] $a = -3, b = 0$
    [C] $a = 0, b = -3$ [D] $a = 2, b = -1$

13. The remainder when $x^3 + 3x^2 - 5x - 6$ is divided by $x + 2$ is:

    [A] -12 [B] 8
    [C] 4   [D] 12

14. $x - 1$ and $x - 2$ are both factors of $x^3 + ax^2 + bx - 6$ when:

    [A] $a = -6, b = 11$ [B] $a = 6, b = 11$
    [C] $a = 6, b = -11$ [D] $a = -6, b = -11$

15. If $(x + 2)$ is a factor of $x^3 + kx^2 - 2x + 4$. The value of k is:

    [A] 2   [B] -2
    [C] 1   [D] 0

16. $x^3 - 3x^2 + 6x - 2$ has remainder 2 when divided by:

    [A] $x - 1$ [B] $x + 1$
    [C] $x + 2$ [D] $2x - 1$

17. $x^3 - 3x^2 + 2x - 6$ has a factor:

    [A] $x - 4$ [B] $x - 2$
    [C] $x - 3$ [D] $x + 3$

18. Given that $f(x) = 3x^3 + 4x^2 - 3x - 4$. One of the factors of $f(x)$ is:

    [A] $3x - 4$ [B] $4x + 7$
    [C] $x - 1$ [D] $x - 4$

19. A factor of $x^3 + 2x^2 - 5x - 6$ is:

    [A] $x + 2$ [B] $x - 1$
    [C] $x + 1$ [D] $x - 2$

20. A factor of $x^3 + 3x^2 - 4x - 12$ is:

   [A] $x - 4$ [B] $x + 4$

   [C] $x - 3$ [D] $x + 3$

21. $x - 2$ is a factor of:

   [A] $x^3 - 3x^2 - 4x + 12$

   [B] $x^3 + 3x^2 + 4x + 12$

   [C] $x^3 - 3x^2 + 4x + 12$

   [D] $x^3 + 3x^2 - 4x + 12$

TOPIC 16

INEQUALITIES

OBJECTIVES

At the end of this topic, the learner should be able to:

1. Order and/or compare real numbers using the symbols <, > and =.
2. Read mathematical inequalities written using the symbols <, >, ≤ and ≥.
3. Identify, represent and notate closed intervals, open intervals and half closed intervals.
4. Solve simple inequalities and quadratic inequalities accurately.
5. Represent solutions of inequalities on number lines.
6. Construct inequalities from given practical situations.
7. Distinguish between absolute inequalities and identities.

In the late 1960s women began to work for equal rights. They wanted to end discrimination against women at home and work. To accomplish this, women began taking part in marches, working for the passage of the Equal Rights Amendment (ERA), and generally speaking out against **inequality**. The women's rights movement had a significant effect on the family, as more women entered the workforce and began to expect their husbands or partners to help with domestic duties.

Meaning of an Inequality

A statement that two real quantities or expressions are not equal is called an **inequality**.

< means 'is less than'
> means 'is greater than'
≤ means 'is less than or equal to'
≥ means 'is greater than or equal to'

Ordering

$\forall a, b, c \in \mathbb{R}$,
1. Either $a < b$ or $a > b$ or $a = b$.
2. $a < b \Leftrightarrow a > b$.
3. If $a < b$ and $b < c$ then $a < c$. This is known as the **transitive property** of inequalities.
4. If $a < b$ and $c > 0$ then $ac < bc$ and if $a < b$ and $c < 0$ then $ac > bc$.

Inequations

An inequality, which involves an unknown, is called an **inequation**.

Representation of inequalities

On a real number line, if $a < b$ then the point a is to the left of the point b. When representing inequalities on a number line an open circle o is used when the boundary point is not included and a fill-in circle ● is used when the boundary point is included.

Open Intervals

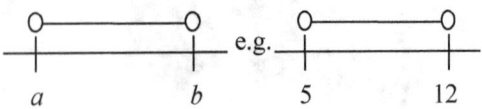

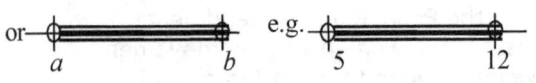

or 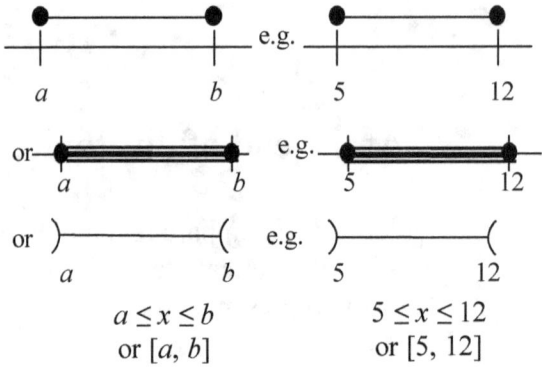   e.g.

$a < x < b$ or $5 < x < 12$ or
]a, b[or (a, b)   ]5, 12[or (5, 12)

The boundary points a and b are not included.

Closed Intervals

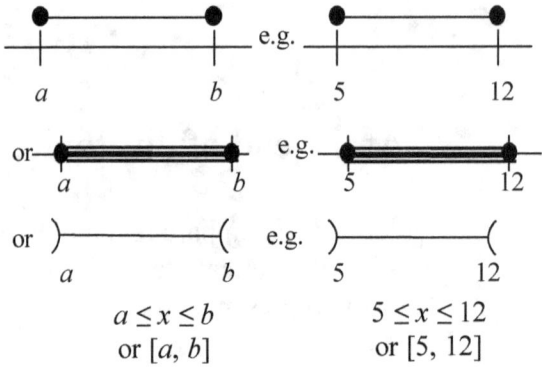

$a \leq x \leq b$ $5 \leq x \leq 12$
or [a, b]   or [5, 12]

The boundary points a and b are included.

Half-Open or Half Closed Intervals

Open-Closed Interval

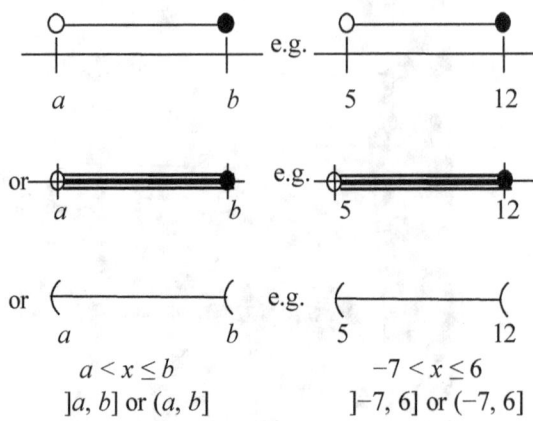

$a < x \leq b$ $-7 < x \leq 6$
]a, b] or (a, b]   ]-7, 6] or (-7, 6]

The boundary point a is not included but b is included.

Closed-open Interval

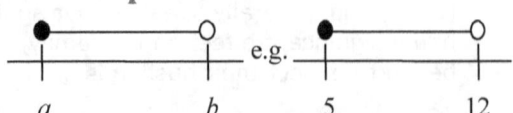

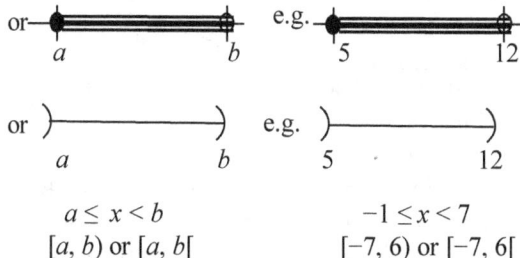

$$a \le x < b$$
$$[a, b) \text{ or } [a, b[$$

$$-1 \le x < 7$$
$$[-7, 6) \text{ or } [-7, 6[$$

The boundary point a is included but b is not included.

Unbounded Intervals
An interval that extends indefinitely in one or both directions is called an unbounded interval.

Right-unbounded Open Intervals

or   or
$$x > a \text{ or } (a, \infty)$$

E.g.

or   or
$$x > 2 \text{ or } (2, \infty)$$

The boundary point a is not included and all points to the right of a are included.

Left-unbounded Open Intervals

or   or
$$x < a \text{ or } (-\infty, a)$$

E.g.

or   or
$$x < 2 \text{ or } (-\infty, 2)$$

Right-unbounded Closed Interval

or   or
$$x > a \text{ or } (a, \infty)$$

E.g.

or   or
$$x > 2 \text{ or } (2, \infty)$$

The boundary point a is included and all points to the left of a are included.

Intervals may also be unbounded on both ends.

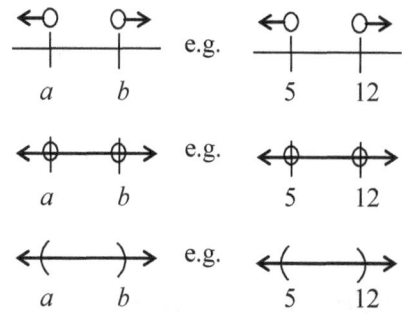

The boundary points a and b are not included. Points between a and b are not included.

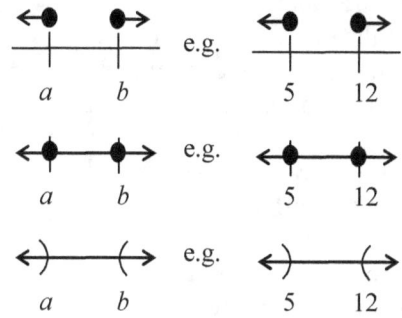

The boundary points a and b are included. Points between a and b are not included.

EXERCISE 16:1

1. Represent each of the following intervals on a real number line.
   - (a) $(4, 6)$
   - (b) $[7, 3]$
   - (c) $[2, 4[$
   - (d) $]-3, 9[$
   - (e) $\{x: -11 < x \le 5\}$

2. State which of the following intervals are closed, open or half open (half-closed) giving reasons for your answer.

   $P = \{17 \le x \le 19, x \in \mathbb{R} \qquad [1, 3],$

   $R = (-1, 2] \qquad S = \{x : -2 \le x < 2, x \in \mathbb{R}$

   $T = [2, 5], \qquad U =]-2, 0[,$

   $V =]5, 8[, \qquad W = \{x : -7 < x < -2, x \in \mathbb{R}$

3. Represent each of the intervals in question 2, above on a number line.

4. Give the other notations for each of the following intervals.
   - (a) $\{x: -2 < x \le 2\}$
   - (b) $[98, 101]$
   - (c) $(2, 5)$
   - (d) $\{x: 1 \le x < 3\}$

151

(e) $]-6, -1[$ (f) $(14, 19]$ (g) $[0, 7)$
(h) $[-3, 1[$ (i) $] 11, 13]$

5. Classify the following intervals as closed, open or half open (half closed).

(a)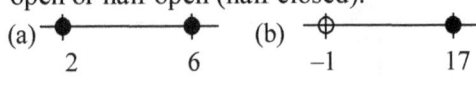

     2   6   (b)   −1   17

(c) $(-2, 5]$ (d) $(-10, -6)$
(e) $[0, 8]$ (f) $-3 < x < 8$
(g) $\{x : -2 \leq x \leq 0, x \in \mathbb{R}\}$

Conditional Inequalities or Inequations

Conditional inequalities are inequalities that are sometimes true and sometimes false depending on the range of values of the variable. Another name for conditional inequalities is inequations. Inequations are analogous to equations because they can be solved to obtain the range of values of the variable. The mean difference between inequalities and inequations is that inequalities are always true (e.g. $-5 < 8$, $13 > 4$) while inequations are sometimes true and sometime false (e.g. $x^2 - 2x - 3 \geq 0$), $x + 2 < 5$

Building up Inequations

Inequations are built in the same way in which equations are built but for the fact that the inequality sign is used instead of the equal to sign.

Example 16:1

1. The product of 2 and a number is at least 8.
2. When 5 is added to a number, the result is greater than 9.
3. When an amount of money is increased by 70, FCFA the total amount is at most 200 FCFA.
4. Four girls shared a number of oranges and each had at least 8 oranges.

Solutions

1. Let the number be n. Then, $2n \geq 8$.
2. Let the number be x. Then $x + 5 > 9$.
3. Let the amount of money be m.
   Then, $m + 70 \leq 200$.
4. Let the basket contain b oranges.
   Then, $\frac{b}{4} \geq 8$.

EXERCISE 16:2

Build inequations from the following situations

1. When a number is increased by 8 the resulting sum is equal to or less than 14.
2. When 40, is divided by a certain number, the resulting quotient is greater than 4.
3. Thrice the sum of a number and 7 is not more than 27.
4. The product of a number and 4 is at least 20.
5. 60 % of a number decreased by 10 is less than 12.
6. When half of a number is subtracted from three times the number, the difference is less than the number increased by 6.
7. Twice the sum of a number and 7 is at most 12 more than the number.
8. Three times the sum of a number and 5 is less than four times the number increased by 2.
9. Thirty minus five times a certain number is at least 4.
10. Three boys earn at most 600,000 FCFA. Ambe earns 20,000 FCFA less than Ambe earns Anye and Ndoh earns twice as much as Ambe.

Laws of Inequalities

Addition and Subtraction of Inequalities

Consider the inequality 8>6.

Adding a positive number e.g. +2 to both sides,
LHS = $8 + (+2) = 10$ and RHS = $6 + (+2) = 8$
Since 10>8, the inequality is still true.

Adding a negative number e.g. –4 to both sides,
LHS = $8 + (-4) = 4$ and RHS = $6 + (-4) = 2$

Since 4>2, the inequality is still true.

Subtracting a positive number e.g. +7 from both sides,
LHS = $8 - (+7) = 1$ and RHS = $6 - (+7) = -1$
Since, 1> –1 the inequality is still true

Subtracting a negative number e.g. –9 from

both sides,
LHS = $8 - (-9) = 17$ and RHS = $6 - (-9) = 15$

Since $17 > 15$, the inequality is still true.

Therefore,

If the same real quantity is added to or subtracted from both sides of an inequality, the inequality remains valid.

Multiplication and Division of Inequalities

Again consider the inequality $8 > 6$

Multiplying both sides by a positive number e.g. $+2$,
LHS = $8 \times (+2) = 16$ and RHS = $6 \times (+2) = 12$
Since $16 > 12$, the inequality is still true.
Multiplying both sides by a negative number e.g. -3

LHS = $8 \times (-3) = 42$ and RHS = $6 \times (-3) = -18$

But $-24 > -18$ is false,
So $-24 < -18$ is instead the true inequality.

Dividing both sides by a positive number e.g. $+2$,

LHS = $\dfrac{8}{+2} = 4$ and RHS = $\dfrac{6}{+2} = 3$
Since $4 > 3$, the inequality is still true.

Dividing both sides by a negative number, e.g. -2;

LHS = $\dfrac{8}{-2} = -4$ and RHS = $\dfrac{6}{-2} = -3$
But $-4 > -3$ is false,
So $-4 < -3$ is instead the true inequality.
Therefore,

If both sides of an inequality are multiplied or divided by any positive real quantity, the inequality remains valid.

But it should be re-iterated that,
If both sides of an inequality are multiplied or divided by any negative real quantity, the inequality changes sense from < to > or ≤ to ≥ and vice versa.

Solving Inequations
Inequations are solved in the same way as equations but for the fact that when multiplying or dividing both sides by a negative number, the inequality sign changes from < to > or from ≤ to ≥ and vice versa.

Linear Inequalities

Example 16:2
Solve the following inequalities and represent the solution on a number line.

1. $n + 5 > 9$ 2. $y + 7 < 12$
3. $x - 3 \geq -1$ 4. $x - 8 \leq 3$
5. $2x \geq 8$ 6. $-32x < 64$
7. $\dfrac{x}{5} \leq 3$ 8. $-\dfrac{x}{3} > 4$
9. $2x + 5 \leq 13$ 10. $\dfrac{2}{3}n + 5 > 11$

Solution

1.
$n + 5 > 9$
$\Rightarrow n > 4$

4

2.
$y + 7 < 12$
$\Rightarrow y < 5$

5

3.
$x - 3 \geq -1$
$\Rightarrow x \geq 2$

2

4. $x - 8 \leq 3$
$\Rightarrow x \leq 11$

11

5. $2x \geq 8$
$\Rightarrow x \geq 4$

4

6.
$-32x < 64$
$\Rightarrow n > 4$

-2

7. $\dfrac{x}{5} \leq 3$
$\Rightarrow x \leq 15$

15

8. $-\dfrac{x}{3} > 4$
$\Rightarrow x < -12$

-12

9.
$2x + 5 \leq 13$
$\Rightarrow x \leq 4$

4

10. $\dfrac{2}{3}n + 5 > 11$

$\Rightarrow \dfrac{2}{3}n > 6$

$2n > 18$

$n > 9$

9

Compound Inequalities

An inequality which consist of two inequalities joined by '*and*' or '*or*' is called a **compound inequality**. The following are some examples of compound inequalities.

- $3 \le 6 + 3x \le 9$ read '$6x + 3$ is greater than or equal to 3 <u>and</u> less than or equal to 9.
- $3x - 4 < x \le 5x + 12$ read 'x is greater than $3x - 4$ <u>and</u> less than or equal to $5x + 12$.

Such inequalities are solved by either performing the same operation to each expression or separating them into two entities before solving them.

Example 16:3

Solve the following inequalities and represent your result on a number line.

(a) $3 \le 6 + 3x \le 9$ (b) $3x - 4 < x \le 5x + 12$

Solution

(a) $3 \le 6 + 3x \le 9$

Subtract 6 from each expression.

$-3 \le 3x \le 3$

Divide through by 3.

$-1 \le x \le 1$

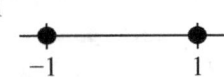

(b) $3x - 4 < x \le 5x + 12$.

This can better be solved by separating into the two entities $3x - 4 < x$ and $x \le 5x + 12$.

$3x - 4 < x$

Subtract x from both sides

$2x - 4 < 0$

Add 4 to both sides

$2x < 4$

Divide both sides by 2.

$x < 2$

$x \le 5x + 12$.

Subtract $5x$ from both sides

$-4x \le 12$

Divide both sides by -4 and reverse inequality sign.

$x \ge -3$

The results can now be combined to have $-3 \le x < 2$.

Example 16:4

Solve the inequality $2x + 3 < -1$ or $3x - 5 > -2$ and represent your solution on a real number line.

Solution

$2x + 3 < -1$ or $3x - 5 > -2$

$\qquad 2x < -4$ or $3x > 3$

$\qquad\quad x < -2$ or $x > 1$

Absolute Value Inequalities

An inequality, which involves absolute values, is called an **absolute value inequality**. For example, $|3x + 2| \le 5$.

To solve an absolute value inequality, it is necessary to appreciate that two numbers with opposite signs have the same absolute value. Thus, if $|n| < k$ then, $-k < n < k$.

Example 16:5

Solve the inequality $|3x + 2| \le 5$ and represent your solution on a real number line.

Solution

$$|3x + 2| \le 5$$

$$-5 \le 3x + 2 \le 5$$

$$-7 \le 3x \le 3$$

$$-\frac{7}{3} \le x \le 1$$

EXERCISE 16:2

Solve the following inequalities and represent your solution on a real number line.

1. $2x < 4$
2. $2 + x > 11$
3. $16 \le 2x + 8$
4. $2y + 4 \ge 12$

5. $7x - 5 \leq 2x + 11$

6. $2(n + 5) > 18$

7. $\dfrac{2x - 5}{3} < 25$

8. $5x > 40 - 3x$

9. $4y + 5 \leq 5y - 30$

10. $3t + 10 < 2t + 20$

11. $5u < 2u + 27$

12. $2x \geq 90 - 7x$

13. $10t - 11 > 8t$

14. $18 - 5a \leq a$

15. $13y \leq 15 + 3y$

16. $100 + 3\frac{1}{2}x > 23\frac{1}{2}x$

17. $9u \geq 16v - 105$

18. $5p + 3 - 2p < p + 8$

19. $7m + 10 - 2m > 16m - 12$

20. $20 \geq 3x - 1$

21. $0.6x - 0.4 \geq 0.4x + 0.6$

22. $3x - 2 \leq 2 - x$

23. $15 \geq 2 - 4x \geq 2$

24. $|2w - 7| \leq 9$

25. $x - 1 < 3x + 1 < x + 5$

26. $|2x - 1| < 3$

27. $|4x - 2| \leq 6$

Quadratic Inequations

A quadratic inequation is an inequation, which involves a quadratic expression.

Consider the inequation $(x-2)(x+3) < 0$.

$(x-2)(x+3) = 0$ is the related quadratic equation and its roots are 2 and –3. 2 and –3 are called the **critical values** of the inequality $(x-2)(x+3)<0$. These values can be plotted on a number line as follows.

These boundary points define three intervals

$x < -3, -3 < x < 2$ and $x > 2$ of which at least one must satisfy the inequality. By testing a point within each interval, the range of values, which satisfies the inequation, can easily be determined.

Testing the point –4 in the interval $x < -3$, LHS $= (-4 - 2)(-4 + 3) = (-6)(-1) = +6$ and $+6 \not< 0$

Testing the point 0 in the interval $-3 < x < 2$, LHS $= (0 - 2)(0 + 3) = (-2)(+3) = -6$ and $-6 < 0$

Testing the point 4 in the interval $x > 2$, LHS $= (4 - 2)(4 + 3) = (+2)(+7) = +14$ and $+14 \not< 0$

So clearly the interval which is satisfied by the inequality $-3 < x < 2$ is, shaded in the number line below.

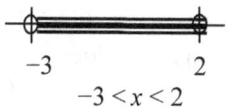

$$-3 < x < 2$$

The range of values which satisfy the inequality always lie either within the limits of the interval or outside the limits, but not both. So testing a point within or without suffices. If the interval within satisfies the interval outside does not and vice versa. Advantage is often taken of the origin since computing with zero can be very easy.

Example 16:6
Find the solution set of the inequation $x^2 - 5x + 6 < 0$ and represent it on a number line.

Solution
$$x^2 - 5x + 6 < 0$$
$$(x - 2)(x - 3) < 0$$

$$2 < x < 3$$

Critical values are 2 and 3
Solution set $= \{x: 2 < x < 3, x \in \mathbb{R}\}$

Example 16:7
Find the set of real values of x, which satisfy the inequality $x^2 + 2x - 3 \geq 0$. Represent your solution on a number line.

Solution

$x^2 + 2x - 3 \geq 0$

$(x + 3)(x - 1) \geq 0$

Critical values are –3 and 1. Testing a value the range outside is satisfied.

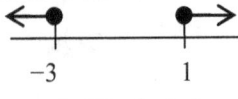

$$-3 \qquad 1$$

$x \leq -3$ and $x \geq 1$

Solution set = $\{x: x \leq -3 \text{ or } x \geq 1, x \in \mathbb{R}\}$

Absolute Inequalities

An **absolute inequality** is an inequality, which is always true for all real values of the variables involved. In fact, an absolute inequality is analogous to an identity. Examples of absolute inequalities are:

$x^2 \geq 0, x \in \mathbb{R}; \ (x - 1)2 \geq 0, x \in \mathbb{R};$

$(x \pm y)2 \geq 0, x, y \in \mathbb{R}.$

Notice that in these examples provided the variables are real, the value of the quantity on the LHS is always positive no matter the value of the variables-positive or negative. The concept of an absolute inequality should not be confused with the absolute value inequality, which is true only for values of the variable within a certain range.

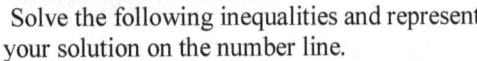

EXERCISE 16:4

Solve the following inequalities and represent your solution on the number line.

1. $u^2 + 6u < -8$

2. $a^2 > 5a + 24$

3. $x(3x + 20) \geq 7$

4. $5 \leq m(2m + 3)$

5. $p + 8 > \dfrac{20}{p}$

6. $\dfrac{7}{x} \geq 9 - 2x$

7. $2 + \dfrac{5}{x} < \dfrac{12}{x^2}$

8. $(a - 8)(2a - 3) \leq 34$

MULTIPLE CHOICE EXERCISE 16

1. The solution of the inequality $3m + 3 > 9$ is:

   [A] $m > 2$ [B] $m > 3$
   [C] $m > 4$ [D] $m > 6$

2. If x is positive, the range of values of x for which $4 + 3x < 10$ is:

   [A] $0 < x < 2$ [B] $x < 2$
   [C] $1 < x < 2$ [D] $0 > x > 2$

3. The inequality $\dfrac{1}{3}(2x - 1) < 5$ is true only if:

   [A] $x < -5$ [B] $x > -5$
   [C] $x < 7$ [D] $x > 8$

4. p and q are two positive real numbers such that $p > 2q$. The inequality which is not true is:

   [A] $-p < -2q$ [B] $-p > 2q$
   [C] $-p < 2q$ [D] $-p < \dfrac{1}{2}q$

5. The solution of the inequality $(y - 3) < \dfrac{y}{3}$ is:

   [A] $y > -\dfrac{9}{2}$ [B] $y < 3$
   [C] $y > 4$ [D] $y < \dfrac{9}{2}$

6. The solution of the inequality $3x - 8 \geq 5x$ is:

   [A] $x \geq 4$ [B] $x \geq 1$
   [C] $x \geq -4$ [D] $x \geq -1$

7. The range of values of x which satisfy the inequality $2x + 3 < 5x$ are:

   [A] $x > 1$ [B] $x < \dfrac{3}{7}$
   [C] $x < \dfrac{3}{7}$ [D] $x > -1$

8. The smallest whole number which satisfies the inequality $9 - 2x < 5x - 12$ is:

   [A] 1   [B] 2   [C] 3   [D] 4

9. A distance d metres which is more than 18 m, but not more than 23 m can be represented by:

   [A] $18 \leq d \leq 23$
   [B] $18 < d \leq 23$
   [C] $18 \leq d < 23$

[D] $d < 18$ or $d > 23$

10. Nfor had x oranges. He ate 2 and shared the remainder equally with Ngala. In terms of x, the inequality which represents the information that Ngala's share is at least 5 oranges is:

[A] $\frac{x}{2} - 2 \le 5$ [B] $\frac{x}{2} - 2 \ge 5$

[C] $\frac{(x-2)}{2} \ge 5$ [D] $\frac{(x-2)}{2} \le 5$

11. The range of values of x for which

$\frac{1}{2}(4x+2) - (x-5) \le \frac{1}{4}(3x-1)$ is:

[A] $x \ge 25$ [B] $x \le 25$
[C] $x \ge -25$ [D] $x \le -25$

12. In Figure 16:1, the solution to the

inequality $\frac{x}{3} - \frac{(x-3)}{2} < 1$ is represented

by the line:

[A] 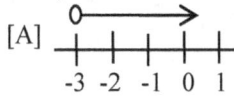   [B]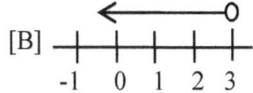

[C] 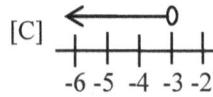   [D]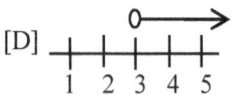

Figure 16:1

13. If x varies over the set of real numbers, the inequality illustrated in Figure 16:2 is:

[A] $-3 < x \le 2$ [B] $-3 \le x < 2$
[C] $-3 < x < 2$ [D] $-3 \le x \le 2$

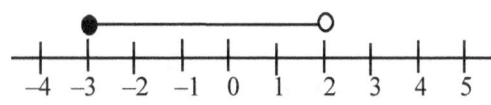

Figure 16:2

14. If x is a real number the inequality represented in Figure 16:3 is:

[A] $\{x: -5 < x \le 3\}$ [B] $\{x: -5 \le x \le 3\}$
[C] $\{x: -5 \le x < 3\}$ [D] $\{x: -5 < x < 3\}$

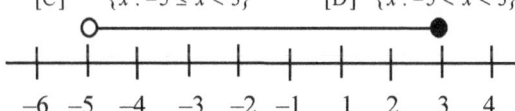

Figure 16:3

15. If x is a real number, the inequality more illustrated in the number line in Figure 16:4 is:

[A] $x < 4$ [B] $x > -2$
[C] $-2 < x \le 4$ [D] $-2 \le x < 4$

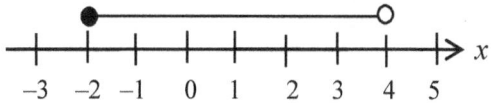

Figure 16:4

16. If x is a real number, the inequality more illustrated in the number line in Figure 16:5 is:

[A] $x < 4$ [B] $x > -2$
[C] $-2 < x \le 4$ [D] $-2 \le x < 4$

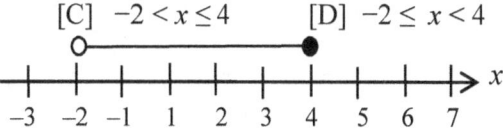

Figure 16:5

17. If x varies over the set of real numbers, the inequality illustrated in Figure 16:6 is:

[A] $-2 \le x < 3$ [B] $-2 < x \le 3$
[C] $-2 < x < 3$ [D] $-2 \le x \le 3$

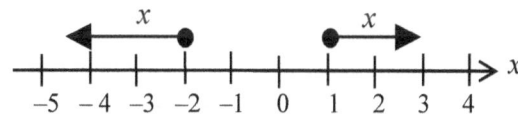

Figure 16:6

18. The pairs of inequalities is represented on the number line in Figure 16:7 are:

[A] $x < -2$ and $x \ge 1$

[B] $x \le -2$ and $x > 1$

[C] $x \le -2$ and $x < 1$

[D] $x < -2$ and $x > 1$

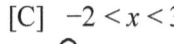
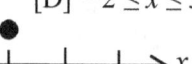

Figure 16:7

19. The number lines in Figure 16:8 which represents the inequality $2 \le x < 9$ is:

[A]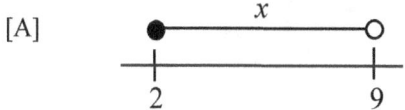

[B]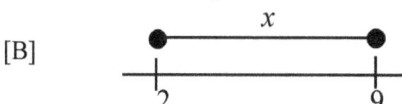

[C]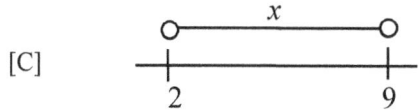

[D]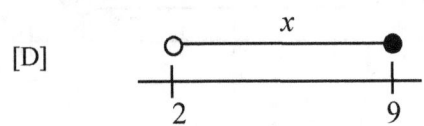

x

2   9

Figure 16:8

20. The number line in Figure 16:9
    represents:
    [A] $0 < x < -7$ [B] $-7 < x < 1$
    [C] $-7 < x \leq -1$ [D] $-7 \leq x < -1$

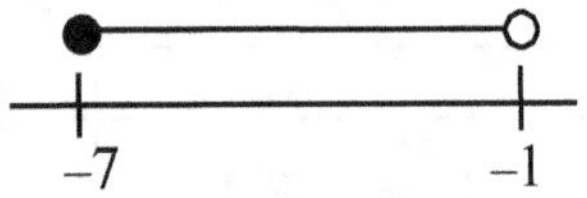

    −7   −1

Figure 16:9

21. Given that, $x \in \mathbb{R}$. An absolute
    inequality among the following is:
    [A] $|3x + 2| \leq 5$
    [B] $(x-1)^2 \geq (x+1)^2$
    [C] $(x-1)^2 \geq 0$
    [D] $-2 < x < 0$

22. The inequalities represented on the
    number line in figure 16:10 above are:
    [A] $-2 < x \leq 0$ or $x > 3$

[B] $-2 \leq x < 0$ or $x \geq 3$
[C] $-2 \leq x < 0$ or $x > 3$
[D] $-2 \leq x \leq 0$ or $x > 3$

−2   0   3

Figure 16:10

23. Given that, $x \in \mathbb{R}$. An absolute value
    inequality among the following is:
    [A] $|7x - 16| \geq 0$
    [B] $(x+3)^2 \geq (x-3)^2$
    [C] $(x-3)^2 \geq 0$
    [D] $-4 < x + 2 < 10$

24. A conditional inequality is an inequality
    which is:
    [A] always true
    [B] sometimes true
    [C] always false
    [D] sometimes false

TOPIC 17

INDICES, LOGARITHMS AND SURDS

OBJECTIVES

At the end of this topic, the learner should be able to:

1. Interpret and evaluate positive, negative, zero and fractional indices.
2. State and apply the laws of indices to solve simple problems involving indices.
3. Define logarithm and use the definition to solve simple problems involving logarithms.
4. State and apply the laws of logarithms in problem solving.
5. State and use the change of base formula to change the base of a logarithmic expression.
6. Solve simple exponential and logarithmic equations.
7. State and apply the laws of Surds in evaluating simple surd expressions.
8. State the conjugate of a surd expression and use it to rationalise the denominators of rational expressions containing surds.

A child prodigy, American mathematician Norbert Wiener earned his Ph.D. from Harvard University at the age of 18. Wiener contributed to many areas of mathematics and science, but he became famous for developing the field of science called cybernetics. Interdisciplinary in nature, cybernetics is concerned with communication and control systems in living organisms, machines, and organizations.

UPI/THE BETTMANN ARCHIVE

THEORY OF INDICES
Meaning of b^p

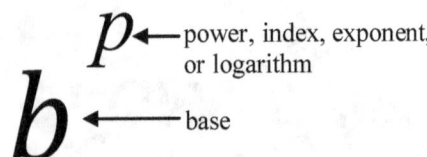

p — power, index, exponent, or logarithm

b — base

In Topic 2 the meaning of the index notation was explained. Just to recall, another name for **power** is **index, exponent** or **logarithm**.
Note! Any base with exponent 1 is itself.

Law of Indices
Multiplication Law of Indices
Consider the following

(a) $3^5 \times 3^3 = (3 \times 3 \times 3 \times 3 \times 3) \times (3 \times 3 \times 3)$

$\qquad = 3 \times 3 \times 3 \times 3 \times 3 \times 3 \times 3 \times 3$

$\qquad = 3^8$

(b) $4^3 \times 4^2 = (4 \times 4 \times 4) \times (4 \times 4)$

$\qquad = 4 \times 4 \times 4 \times 4 \times 4$

$\qquad = 4^5$

Notice that in (a) and (b) the results can be obtained by adding the indices. Thus,

$$b^m \times b^n = b^{m+n} \quad\ldots\ldots\ldots\ldots\ldots\ldots\ldots\ldots①$$

Example 17:1
Evaluate the following, allowing your answer in index form
(i) $5^2 \times 5^4 \times 5^1 \times 5^3$
(ii) $b^4 \times b^2 \times b^3 \times b^0$

Solution
(i) $5^2 \times 5^4 \times 5^1 \times 5^3 = 5^{2+4+1+3} = 5^{10}$
(ii) $b^4 \times b^2 \times b^3 \times b^0 = b^{4+2+3+0} = b^9$

Division Law of Indices
Consider the following

(c) $3^7 \div 3^5 = \dfrac{\cancel{3} \times \cancel{3} \times \cancel{3} \times \cancel{3} \times \cancel{3} \times 3 \times 3}{\cancel{3} \times \cancel{3} \times \cancel{3} \times \cancel{3} \times \cancel{3}} = 3^2$

(d) $4^5 \times 4^2 = \dfrac{\cancel{4} \times \cancel{4} \times 4 \times 4 \times 4}{\cancel{4} \times \cancel{4}} = 4^3$

Notice that in (c) and (d), the results could be obtained by subtracting the indices. Thus,

$$b^m \div b^n = b^{m-n} \quad\ldots\ldots\ldots\ldots\ldots\ldots\ldots②$$

Example 17:2
Evaluate the following, allowing your answer in index form where necessary.

(a) $81(x)^9 \div (3x)^3$ (b) $\dfrac{(2p) \times (3p)^2}{36p^4}$

Solution
(a) $\quad 81(x)^9 \div (3x)^3 = 3^4(x^9) \div 3^3(x^3) = 3x^6$

(b) $\dfrac{(2p) \times (3p)^2}{36p^4} = \dfrac{2^3 \times p^3 \times 3^2 \times p^2}{2^2 \times 3^2 \times p^4} = \dfrac{2^3 \times 3^2 \times p}{2^2 \times 3^2} = 2p$

EXERCISE 17:1

Evaluate the following.

1. 6^3
2. 8^3
3. $2^3 \times 3^2$
4. $3^3 \times 5^2$
5. $y^3 \times y^2 \times y^5$
6. $x^2 \times 4x^5 \times 2x^3$
7. $3x \times 4x^3 \times 5x^5$
8. $\dfrac{a^2 b^4 c^7}{c^7 a^2 b^4}$
9. $\dfrac{10a^6}{5a^3}$
10. $\dfrac{2x^2 y}{3xy^2}$
11. $\dfrac{12a^2}{18a^2 b}$
12. $a^4 \div a^6$
13. $25x^5 \div 5x^4$
14. $20a^3 b^5 \div 5a^2 b^3$
15. $(a^2 b^5) \div (a^3 b^4)$
16. $\dfrac{-5x^2}{15x}$
17. $16a^5 \div 8a^3$
18. $\dfrac{5x^2}{x}$

Other laws can be derived from equation ② as follows.

The Zero index law

Suppose in equation ②, $m = n$, then

$$\frac{b^m}{b^n} = b^{m-n} = b^{m-m} = b^0 \quad \text{or} \quad \frac{b^m}{b^n} = \frac{b^m}{b^m} = 1$$

($b \neq 0$, since $\dfrac{0}{0}$ is meaningless)

$$\Rightarrow b^0 = 1, \quad b \neq 0 \quad \cdots\cdots\cdots\cdots \quad ③$$

Therefore, any base other than 0 with exponent zero is 1.

Example 17:3

Simplify $16x^5 \times 3x^3 \div 96x^8$

Solution

$$16x^5 \times 3x^3 \div 96x^8 = \frac{48x^{8-8}}{96} = \frac{1}{2}x^0 = \frac{1}{2}$$

The negative index law

Suppose in equation ②, $m = 0$

$$\frac{b^m}{b^n} = b^{0-n} = b^{-n}$$

$$\text{Also} \quad \frac{b^m}{b^n} = \frac{b^0}{b^n} = \frac{1}{b^n}$$

$$\therefore \quad b^{-n} = \frac{1}{b^n} \quad \cdots\cdots\cdots\cdots\cdots\cdots \quad ④$$

Example 17:4

Simplify $4y^6 \times 9y^{-10} \div 72y^{-3}$

Solution

$$4y^6 \times 9y^{-10} \div 72y^{-3} = \frac{4 \times 9 \times y^{6-10}}{72y^{-3}} = \frac{y^{6-10+3}}{2} = \frac{1}{2y}$$

The product index law

$$(bm)^n = b^m \times b^m \times \ldots \times b^m \text{ up to } n \text{ times}$$

$$= b^{m+m+\ldots+m} \text{ up to } n \text{ times}$$

But $m + m + m + \ldots + m$ up to n times $= mn$

$$\Rightarrow (b^m)^n = b^{mn}$$

Since multiplication is commutative $mn = nm$.

$$\Rightarrow (b^m)^n = b^{mn} = b^{nm} = (b^n)^m \quad \cdots\cdots\cdots\cdots ⑤$$

Example 17:5

Simplify $\left(\dfrac{2a^2 b^4}{c^3} \right)^2$

Solution

$$\left(\frac{2a^2 b^4}{c^3} \right)^2 = \frac{4a^4 b^8}{c^6}$$

The fractional index laws

Consider $\left(\sqrt[n]{b} \right)^n = b$

Since $\left(b^{\frac{1}{n}} \right)^n = b^{\frac{n}{n}} = b^1 = b \quad \ldots\ldots\ldots ⑥$

Then,

$$b^{\frac{1}{n}} = \sqrt[n]{b} \quad \cdots\cdots\cdots\cdots\cdots ⑦$$

Also from ⑥, if m is substituted for the n outside the bracket,

Then, $b^{\frac{m}{n}} = \left(\sqrt[n]{b} \right)^m = \sqrt[n]{b^m} \quad \cdots\cdots\cdots ⑧$

Example 17:6

Simplify $\left(\dfrac{625}{144} \right)^{-\frac{3}{2}}$

Solution

$$\left(\frac{625}{144} \right)^{-\frac{3}{2}} = \left(\frac{144}{625} \right)^{\frac{3}{2}} = \left(\frac{12}{25} \right)^3 = \frac{1728}{15625}$$

Product of Numbers with the same Power

Consider the following

$$3^2 \times 4^2 = 9 \times 16 = 144$$

$$(3 \times 4)^2 = 12^2 = 144$$

$$\therefore \quad a^m b^m = (ab)^m \quad \cdots\cdots\cdots\cdots ⑨$$

To multiply numbers with the same power, multiply the numbers and retain the power.

Exponential or Index Equations

Equations involving indices are called **exponential or index equations** and are solved by applying the laws of indices.

Example 17:7

Find the value of x for which $64^x = 16^{2x+1}$.

Solution

$$64^x = 16^{2x+1}$$

Since $64 = 2^6$ and $16 = 2^4$ the equation can be expressed in terms of the common base 2. Thus,

$$2^{6x} = 2^{4(2x+1)}$$

Equating exponents

$$6x = 4(2x+1)$$
$$\Rightarrow 6x = 8x + 4$$
$$-2x = 4$$
$$\Rightarrow x = -2$$

EXERCISE 17:2

(i) Simplify the following.

1. $64^{\frac{1}{3}}$

2. $\left(\dfrac{16}{25}\right)^{-\frac{1}{2}}$

3. $\dfrac{5^3 \times 5^0}{25}$

4. $\left(216^{\frac{1}{3}}\right)^{-2}$

5. $\dfrac{81^{0.25}}{32^{0.2}}$

6. $6^{-3} \times 2^5 \times 3^3$

7. $\left(\dfrac{4}{25}\right)^{-\frac{1}{2}} \times 2^4 \div \left(\dfrac{15}{2}\right)^{-2}$

8. $5\dfrac{2}{5} \times \left(\dfrac{2}{3}\right)^2 \div \left(1\dfrac{1}{2}\right)^{-1}$

9. $\left(\dfrac{1}{343}\right)^{\frac{1}{3}} + 64^{-\frac{1}{3}} - \left(\dfrac{4}{9}\right)^{-\frac{1}{2}}$

10. $27^{-\frac{1}{3}} \times (64)^{-\frac{1}{3}} - \left(\dfrac{4}{9}\right)^{-\frac{1}{2}}$

11. $27^{-\frac{1}{3}} \times 64^{-\frac{1}{2}} \times 4^{\frac{1}{2}}$

12. $125^{-\frac{1}{3}} \times 64^{-\frac{1}{3}} \times 81^0$

13. $3^{y+x} = 9^{y-x}$ and $2^{x-y} = 8^{x-3}$

14. $2^{x+y} = 8; \ 3^{2x-y} = 27$

15. $4^{x+1} - 9(2^x) = -2$

(ii) Solve the following equations.

1. $4^x = 2^{\frac{1}{2}} \times 8$ 　　 2. $9 \times 3^{3+x} = 27^{-x}$

3. $3 \times 9^{1+x} = 27^{-x}$ 　　 4. $x^4 = (0.25)^2$

5. $9^{2x+1} = \dfrac{81^{x-1}}{3^x}$ 　　 6. $\dfrac{9^{2x-1}}{3^{x+3}} = 1$

7. $9^x = 729$ 　　 8. $9^{2x-1} \times 3^{x+1} = 27^{x+3}$

9. $27^{2x+1} = 3^{x-1}$ 　　 10. $2^{x-1} = 32$

11. $3^{2y} = \dfrac{1}{27}$ 　　 12. $8^{x-1} = 16$

13. $3^{y+x} = 9^{y-x}$ and $2^{x-y} = 8^{x-3}$

14. $2^{x+y} = 8; \ 3^{2x-y} = 27$

15. $4^{x+1} - 9(2^x) = -2$

THEORY OF LOGARITHMS

Definition of Logarithms

The logarithm of a positive number n to a base b, is the power p to which b must be raised to give the number n.

Using algebraic notation;

$$\log_b n = p \Leftrightarrow n = b^p, \ n > 0, \ n \in \mathbb{R}$$

The above definition can be used to transform logarithmic equations to exponential equations and vice versa.

Example 17:8

Write down the following in logarithmic form

(a) $3^4 = 81$ 　　 (b) $x^y = z$

(c) $(ab)^{pq} = xz$

Solution

(a) $\log_3 81 = 4$ 　　 (b) $\log_x z = y$

(c) $\log_{(ab)} xz = pq$

Example 17:9

Write down the following in exponential form

(a) $a = \log_b c$ 　　 (b) $\log_{10} 100 = 2$

(c) $\log_{x^2} 3y = n$

Solution

(a) $b^a = c$ 　　 (b) $10^2 = 100$

(c) $(x^2)^n = 3y$ or $x^{2n} = 3y$

N. B: The logarithm of negative numbers and zero do not exist.

Example 17:10

Use the definition of logarithms to simplify the following.

(i) $\log_{10} 100$ (ii) $\log_{10} 10000$

(iii) $\log_{10}\left(\dfrac{1}{100}\right)$ (iv) $\log_b b$

(v) $\log_b b^4$

Solution

(i) Let $\log_{10} 100 = x \Leftrightarrow 10^x = 100$

$$10^x = 10^2$$
$$\Rightarrow x = 2$$

(ii) Let $\log_{10} 10000 = x \Leftrightarrow 10^x = 10000$

$$10^x = 10^4$$
$$\Rightarrow x = 4$$

(iii) Let $\log_{10}\left(\dfrac{1}{100}\right) = p \Leftrightarrow 10^p = \dfrac{1}{100}$

$$10^p = 10^{-2}$$
$$\Rightarrow p = -2$$

Let $\log_b b = n \Leftrightarrow b^n = b$

$$b^n = b^1$$
$$n = 1$$

Hence, $\boxed{\log_b b = 1}$

(iv) Let $\log_b b^4 = u \Leftrightarrow b^4 = b^u$

$$\Rightarrow u = 4$$
$$\log_b b^4 = 4$$

Hence, $\boxed{\log_b b^n = n}$

Logarithmic Equations

Equations, which involve logarithms, are called logarithmic equations. Many logarithmic equations can simply be solved using the definition of logarithms.

Example 17:11

Use the definition of logarithms to solve the equation $\log_8 x = 3$.

Solution

$\log_8 x = 3 \Rightarrow x = 8^3 = 512$

EXERCISE 17:3

(i) Evaluate the following.

1. $\log_x x^n$ 2. $\log_8 64$

3. $\log_5\left(\dfrac{1}{5}\right)$ 4. $\log_5 125$

5. $\log_{125} 5$ 6. $\log_3 81$

7. $\log_3 243$ 8. $\log_8 8\sqrt{8}$

9. $\log_4 8$ 10. $\log_x \sqrt[3]{x}$

11. $\log_{14}\left(\sqrt{2}\right)\left(\sqrt{7}\right)$ 12. $\log_{20} 20\sqrt{20}$

(ii) Express the following as exponential equations.

13. $\log_n y = 3$ 14. $\log_{10} 3 = x$

(iii) Find the value of the unknown in each of the following.

15. $\log_4 n = 0$ 16. $\log_{10} n = 1$

17. $\log_a y = 0$ 18. $\log_4 x = 2$

19. $\log_{10} x = -1$ 20. $\log_3 81 = x + 1$

Laws of Logarithms

The Addition Law of logarithms

$$\log_b xy = \log_b x + \log_b y$$

Proof

Let $\log_b x = t$, $\log_b y = u$ and $\log_b xy = v$

Then by definition of logarithms

$$x = b^t \dots\dots\dots\dots①$$
$$y = b^u \dots\dots\dots\dots②$$
$$\text{and } xy = b^v \dots\dots\dots\dots③$$

Multiply equation ① by equation ②

$$xy = \left(b^t\right)\left(b^u\right)$$

Using the multiplication law of indices

$$xy = b^{t+u} \dots\dots\dots④$$

Substituting equation ④ in equation ③

$$b^v = b^{t+u}$$

Comparing powers

$$v = t + u$$
$$\Rightarrow \log_b xy = \log_b x + \log_b y$$

The Subtraction Law of logarithms

$$\log_b\left(\frac{x}{y}\right) = \log_b x - \log_b y$$

Proof

Let $\log_b x = t$, $\log_b y = u$ and $\log_b\left(\frac{x}{y}\right) = w$

Then by definition of logarithms

$$x = b^t \dots\dots\dots\dots\dots\dots\dots① $$

$$y = b^u \dots\dots\dots\dots\dots\dots\dots②$$

and $\dfrac{x}{y} = b^w \dots\dots\dots\dots\dots\dots③$

Dividing equation ① by equation ②

$$\frac{x}{y} = \frac{b^t}{b^u}$$

Using the division law of indices

$$\frac{x}{y} = b^{t-u} \dots\dots\dots\dots\dots④$$

Substituting equation ④ in equation ③

$$b^w = b^{t-u}$$

Comparing powers,

$$w = t - u$$

$$\Rightarrow \log_b\left(\frac{x}{y}\right) = \log_b x - \log_b y$$

The Exponential Law of Logarithms

$$\log_b x^n = n\log_b x$$

Proof

Let $\log_b x = t$, $\log_b y = u$ and $\log_b x^n = r$
Then by definition of logarithms

$$x = b^t \dots\dots\dots\dots\dots\dots.①$$

and $x^n = b^r \dots\dots\dots\dots\dots\dots②$

Substituting equation ① in equation ②

$$b^{nt} = b^r$$

Comparing powers, $r = nt$

Since $r = \log_b x^n$ and $t = \log_b x$

$\log_b x^n = n\log_b x$ as required.

Examples 17:12
Without using tables or calculators, simplify the following.

1. $\log_{10} 8 + \log_{10} 125$
2. $\log_3 21 - \log_3 7$
3. $\log_5 0.25 + \log_5 100$
4. $\dfrac{\log_{10} 81}{\log_{10} 27}$
5. $\dfrac{\log_{10}\sqrt[3]{13}}{\log_{10} 13}$
6. $\log_{10} 75 + 2\log_{10} 2 - \log_{10} 3$
7. $\log_{10} 64 + 2\log_{10} 5 - 2\log_{10} 40$
8. $\log_{10}\left(\frac{15}{8}\right) - 2\log_{10}\left(\frac{5}{9}\right) + \log_{10}\left(\frac{400}{243}\right)$
9. $\dfrac{\log_{10} 6 - \log_{10} 3}{\log_{10} 8 - \log_{10} 4} \div \dfrac{\log_{10} 5}{\log_{10} 0.2}$
10. $\log_{10}\sqrt{35} + \log_{10}\sqrt{2} - \log_{10}\sqrt{7}$

Solutions

1. $\log_{10} 8 + \log_{10} 125 = \log_{10}(8)(125) = \log_{10} 1000 = 3$

2. $\log_3 21 - \log_3 7 = \log_3 \dfrac{21}{7} = \log_3 3 = 1$

3. $\log_5 0.25 + \log_5 100 = \log_5 (0.25)(100) = \log_5 25$
   $\Rightarrow \log_5 0.25 + \log_5 100 = \log_5 5^2 = 2\log_5 5 = 2$

4. $\dfrac{\log_{10} 81}{\log_{10} 27} = \dfrac{\log_{10} 3^4}{\log_{10} 3^3} = \dfrac{4\log_{10} 3}{3\log_{10} 3} = \dfrac{4}{3}$

5. $\dfrac{\log_{10}\sqrt[3]{13}}{\log_{10} 13} = \dfrac{\log_{10}(13)^{\frac{1}{3}}}{\log_{10} 13} = \dfrac{\frac{1}{3}\log_{10} 13}{\log_{10} 13} = \dfrac{1}{3}$

6. $\log_{10} 75 + 2\log_{10} 2 - \log_{10} 3 = \log_{10}\left(\frac{75}{3}\right) + \log_{10} 2^2$
   $= \log_{10}(25)(4) = \log_{10} 100 = 2$

7. $\log_{10} 64 + 2\log_{10} 5 - 2\log_{10} 40$
   $= \log_{10} 64 + \log_{10} 25 - \log_{10} 1600$
   $= \log_{10} \frac{64(25)}{1600} = \log_{10} 1 = 0$

8. $\log_{10}\left(\frac{15}{8}\right) - 2\log_{10}\left(\frac{5}{9}\right) + \log_{10}\left(\frac{400}{243}\right)$
   $= \log_{10}\left(\frac{15}{8}\right)\left(\frac{400}{243}\right) - \log_{10}\left(\frac{5}{9}\right)^2$
   $= \log_{10}\left(\frac{15}{8}\right)\left(\frac{400}{243}\right)\left(\frac{81}{25}\right) = \log_{10} 10 = 1$

9. $\dfrac{\log_{10} 6 - \log_{10} 3}{\log_{10} 8 - \log_{10} 4} \div \dfrac{\log_{10} 5}{\log_{10} 0.2} = \dfrac{\log_{10}\left(\frac{6}{3}\right)}{\log_{10}\left(\frac{8}{4}\right)} \times \dfrac{\log_{10} 0.2}{\log_{10} 5}$

   $= \dfrac{\log_{10} 2}{\log_{10} 2} \times \dfrac{\log_{10}\left(\frac{1}{5}\right)}{\log_{10} 5} = \dfrac{\log_{10} 5^{-1}}{\log_{10} 5} = \dfrac{-1\log_{10} 5}{\log_{10} 5} = -1$

10. $\log_{10} \sqrt{35} + \log_{10} \sqrt{2} - \log_{10} \sqrt{7}$

$= \log_{10}\left(\dfrac{(35)(2)}{(7)}\right)^{\frac{1}{2}} = \dfrac{1}{2}\log_{10}10 = \dfrac{1}{2}$

The Change of Base Formula

$$\log_b x = \frac{\log_a x}{\log_a b}$$

Proof

Let $\log_b x = t$

Then by definition of logarithms

$x = b^t$

Taking the log to the base a of both sides

$\log_a b^t = \log_a x$

Applying the exponential law

$t \log_a b = \log_a x$

$t = \dfrac{\log_a x}{\log_a b}$

$\Rightarrow \log_b x = \dfrac{\log_a x}{\log_a b}$

The Change of base formula can be used to change the logarithm of a number to one base b to the logarithm of the number to another base a

Example 17:13

Given that $\log_{10} 3 = 0.4771$ and $\log_{10} 4 = 0.6021$. Find the value of $\log_4 3$ to three significant figures.

Solution

$\log_4 3 = \dfrac{\log_{10} 3}{\log_{10} 4} = \dfrac{0.4771}{0.6021} = 0.792$

The Logarithm of 1

$$\log_b 1 = 0$$

The logarithm of 1 to any base is zero.

Proof

Let $\log_b 1 = p \Rightarrow$ by definition of logarithms $b^p = 1$.

But $b^0 = 1 \Rightarrow b^p = b^0$

Equating exponents, $p = 0 \Rightarrow \log_b 1 = 0$.

EXERCISE 17:4

(1) Simplify the following without using tables or calculators.

(a) $\dfrac{\log_{10} 16 - \log_{10} 2}{\log_{10} 2 - \log_{10} 1}$ (b) $\dfrac{\log_{10} 81 + \log_{10} 27}{\log_{10} 9 - \log_{10} 3}$

(c) $\log_{10} 125 + \log_{10} 8$ (d) $\log_{10} 30 - \log_{10} 3$

(e) $2\log_{10} 8 + 2\log_{10} 5 - 4\log_{10} 2$

(f) $\dfrac{\log_{10} a + \log_{10} a^2}{\log_{10} a}$

(g) $\log_{10} 15 + 3\log_{10} 2 - 4\log_{10} 1.2$

(h) $3\log_{10} 2 + \log_{10} 200 - \log_{10} 16$

(i) $\dfrac{\log_{10} 81}{\log_{10}\left(\frac{1}{3}\right)}$ (j) $\dfrac{\log_{10} 8}{\log_{10} 12 - \log_{10} 3}$

(k) $\log_{10} 10 - 2\log_{10}\left(\dfrac{1}{5}\right) - \log_{10} 2.5$

(l) $\log_2 8 - \log_2 0.25 + 2\log_2\left(\dfrac{1}{8}\right)$

(m) $\log_{10} 64 + \log_{10} 25 - 2\log_{10} 4$

(n) $\log_2 8 - \log_3\left(\dfrac{1}{27}\right)$

(2) Given that $\log_{10} 8 = 0.9030$, write down without the use of tables, the values of

(a) $\log_{10} 64$ (b) $\log_{10} 2$ (c) $\log_{10} \sqrt{2}$

(3) Given that $\log 3 = 0.4771$ and $\log 2 = 0.3010$, find the values of the following without using tables or calculators.

(i) $\log_{10} \sqrt{6}$ (ii) $\log_{10} \sqrt[3]{0.3}$

(4) Given that $\log_{10} 5 = 0.699$ and $\log_{10} 3 = 0.477$, find $\log_{10} 45$ without using tables or calculators.

SURDS

A surd is an expression, which involve roots.

Laws of Surds

1. $\sqrt[n]{ab} = \sqrt[n]{a} \times \sqrt[n]{b}$

2. $\sqrt[n]{\dfrac{a}{b}} = \dfrac{\sqrt[n]{a}}{\sqrt[n]{b}}$

3. $\sqrt[n]{a^m} = \left(\sqrt[n]{a}\right)^m$

4. $\sqrt[nm]{a^{lm}} = \sqrt[n]{a^l}$

5. $\sqrt[mn]{a} = \sqrt[n]{\sqrt[m]{a}}$

Note!!

$\sqrt[n]{a} + \sqrt[n]{b} \neq \sqrt[n]{a+b}$ and $\sqrt[n]{a} - \sqrt[n]{b} \neq \sqrt[n]{a-b}$

Example 17:14
Evaluate the following.

(a) $\sqrt{1296}$

(b) $\sqrt{\dfrac{342225}{38025}}$

Solution

(a) $\quad \sqrt{1296} = \sqrt{(16)(81)}$

$\quad \Rightarrow \sqrt{1296} = \left(\sqrt{16}\right)\left(\sqrt{81}\right) = 4 \times 9 = 36$

(b) $\quad \sqrt{\dfrac{342225}{38025}} = \sqrt{\dfrac{25 \times 81 \times 169}{25 \times 9 \times 169}} = \sqrt{9} = 3$

Example 17:15
(a) $\sqrt[6]{729}$

(b) $\sqrt[6]{4096}$

Solution

(a) $\quad \sqrt[6]{729} = \sqrt[6]{9 \times 9 \times 9} = \sqrt[6]{3^6} = 3$

(b) $\quad \sqrt[6]{4096} = \sqrt[6]{64 \times 64} = \sqrt[6]{4^6} = 4$

The Conjugate of a Surd

The **conjugate** of $a + \sqrt{b}$ is $a - \sqrt{b}$
To multiply a surd by its conjugate, advantage can be taken of the expansion of the factors of the difference of two squares. Thus,

$$\left(a + \sqrt{b}\right)\left(a - \sqrt{b}\right) = a^2 - \left(\sqrt{b}\right)^2 = a^2 - b$$

Rationalizing the Denominator

The process by which the radical sign is removed from the denominator of a surd expression in order to simplify or evaluate it is known as **rationalizing the denominator**. Usually when the denominator is of the form $a + \sqrt{b}$ or $a - \sqrt{b}$ both numerator and denominator are multiplied by the conjugate.

Example 17:16
Simplify (a) $\dfrac{2}{\sqrt{3}}$ (b) $\dfrac{1}{\sqrt{5}}$

Solution

(a) $\dfrac{2}{\sqrt{3}} = \dfrac{2}{\sqrt{3}} \times \dfrac{-\sqrt{3}}{-\sqrt{3}} = \dfrac{2\sqrt{3}}{3}$

(b) $\dfrac{1}{\sqrt{5}} = \dfrac{1}{\sqrt{5}} \times \dfrac{-\sqrt{5}}{-\sqrt{5}} = \dfrac{\sqrt{5}}{5}$

Example 17:17
Rationalize the denominator of $\dfrac{\sqrt{7}}{2\sqrt{7} - 5}$

Solution

$\dfrac{\sqrt{7}}{2\sqrt{7} - 5} = \dfrac{\sqrt{7}}{2\sqrt{7} - 5}\left(\dfrac{-2\sqrt{7} - 5}{-2\sqrt{7} - 5}\right)$

$\quad = \dfrac{2(7) + 5\sqrt{7}}{4(7) - 25}$

$\quad = \dfrac{14 + 5\sqrt{7}}{3}$

Surd Equations

Equations involving surds are called **surd equations** or **radical equations** and are solved by applying the laws of surds.

Example 17:18
Solve the following equations

(a) $\sqrt{2x - 1} = 3$

(b) $\sqrt{4x + 1} = x - 5$

Solution

(a) $\quad \sqrt{2x - 1} = 3$

$\qquad$ Squaring both sides

$\qquad 2x - 1 = 9$

$\qquad\quad 2x = 10$

$\qquad\quad\ x = 5$

(b) $\sqrt{4x+1} = x - 5$

Squaring both sides

$$4x + 1 = (x-5)^2$$

$$4x + 1 = x^2 - 10x + 25$$

$$x^2 - 14x + 24 = 0$$

$$(x - 12)(x - 2) = 0$$

$$x = 12 \text{ or } x = 2$$

EXERCISE 17:4

1. Evaluate the following.

   (a) $3\sqrt{4} + \sqrt{25} - 2\sqrt{9}$ (b) $\sqrt{\dfrac{25}{9}} + \sqrt{\dfrac{64}{9}}$

   (c) $\sqrt[4]{\dfrac{81}{4}}$ (d) $\dfrac{1}{3}\sqrt{36} + \dfrac{2}{5}\sqrt{100}$

2. Rationalize the denominator in each of the following.

   (a) $\dfrac{12}{\sqrt{6}}$ (b) $\dfrac{\sqrt{3}}{\sqrt{10}}$ (c) $\dfrac{24}{\sqrt{8}}$

   (d) $\dfrac{3 + \sqrt{2}}{\sqrt{2}}$ (e) $\dfrac{3}{\sqrt{3}}$ (f) $\dfrac{c^3}{\sqrt{c^3}}$

   (g) $\dfrac{7}{\sqrt{7}}$ (h) $\dfrac{20}{\sqrt{50}}$

3. Solve the following equations

   (a) $\sqrt{7x+5} = 3$ (b) $2\sqrt{x-8} = 3$

   (c) $\sqrt{8 - 4x} = x + 1$ (d) $\sqrt{-x-5} = x + 5$

4. Find the value of each of the following

   (a) $\sqrt{81} - \sqrt{25}$ (b) $2\sqrt{9} + 3\sqrt{4}$

   (c) $\sqrt{\dfrac{4}{9}} + \sqrt{\dfrac{1}{9}}$ (d) $\sqrt[8]{\dfrac{1}{16}} + \sqrt[4]{\dfrac{9}{16}}$

   (e) $\sqrt{16} + \sqrt{1600} + \sqrt{160,000}$

   (f) $\left(-\sqrt{\dfrac{4}{9}}\right)\left(-\sqrt{\dfrac{81}{100}}\right)$

5. Simplify the following.

   (a) $\dfrac{1}{5 - \sqrt{3}} + \dfrac{1}{5 + \sqrt{3}}$ (b) $\sqrt[3]{216}$

   (c) $\sqrt[3]{\dfrac{9}{27}}$ (d) $\sqrt[3]{64^2}$

6. Rationalize the denominator in each of the following.

   (a) $\dfrac{1}{2\sqrt{3} - 1}$ (b) $\dfrac{1}{\sqrt{3} - \sqrt{2}}$

MULTIPLE CHOICE EXERCISE 17

1. The value $(2^3)^2$ is:

   [A] 16 [B] 32 [C] 36 [D] 64

2. The value $2^2 + 3^3$ is:

   [A] 13 [B] 25 [C] 31 [D] 36

3. The value $64^{\frac{1}{3}}$ is:

   [A] 16 [B] 8 [C] 4 [D] 2

4. 5^4 has the value of is:

   [A] 9 [B] 20 [C] 125 [D] 625

5. The value $2^0 - 2^{-2}$ is:

   [A] 1 [B] $-\dfrac{1}{4}$ [C] $\dfrac{3}{4}$ [D] $\dfrac{1}{4}$

6. $\dfrac{2^3 \times 2^4}{2^2}$ is equal to:

   [A] 2^4 [B] 2^3 [C] 2^2 [D] 2^5

7. The value $\left(\dfrac{196}{225}\right)^{-\frac{1}{2}}$ is:

   [A] $\dfrac{17}{14}$ [B] $\dfrac{15}{14}$ [C] $\dfrac{14}{15}$ [D] $\dfrac{14}{17}$

8. On simplification $16^{\frac{1}{2}}\left(4^{-1} + 5^0\right)$ gives:

   [A] 5 [B] $5\dfrac{1}{2}$ [C] $4\dfrac{1}{2}$ [D] $4\dfrac{1}{4}$

9. $(0.7)^3$ equals:

   [A] 2.1 [B] 0.49 [C] 3.43 [D] 0.343

10. When simplified $\left(27^2\right)^{\frac{1}{3}}$ equals:

    [A] 81 [B] 6 [C] 9 [D] 8

11. After evaluating $36^{\frac{1}{2}} \times 64^{-\frac{1}{2}} \times 5^0$ equals:

    [A] $\dfrac{3}{4}$ [B] $\dfrac{1}{24}$ [C] $\dfrac{2}{3}$ [D] $1\dfrac{1}{2}$

12. When $\dfrac{9^{-\frac{1}{2}}}{27^{\frac{2}{3}}}$ is simplified the result is:

    [A] $\dfrac{1}{2}$ [B] $\dfrac{1}{9}$ [C] $\dfrac{1}{18}$ [D] $\dfrac{1}{27}$

13. Simplifying $125^{-\frac{1}{3}} \times 49^{-\frac{1}{2}} \times 10^0$ gives:

[A] 350 [B] 35 [C] $\dfrac{1}{35}$ [D] $\dfrac{1}{350}$

14. $\left(\dfrac{1}{4}\right)^{-1\frac{1}{2}}$ when simplified is:

   [A] $\dfrac{1}{8}$ [B] $\dfrac{1}{4}$ [C] 8 [D] 4

15. On evaluation $\left(\dfrac{16}{81}\right)^{\frac{1}{4}}$ becomes:

   [A] $\dfrac{8}{27}$ [B] $\dfrac{1}{3}$ [C] $\dfrac{4}{9}$ [D] $\dfrac{2}{3}$

16. The result of simplifying $\dfrac{4^{-\frac{1}{2}} \times 16^{\frac{3}{4}}}{4^{\frac{1}{2}}}$ is:

   [A] $\dfrac{1}{4}$ [B] 0 [C] 1 [D] 2

17. On evaluation, the result of $\dfrac{27^{\frac{1}{3}}}{16^{-\frac{1}{4}}}$ is:

   [A] 6 [B] 2 [C] 4 [D] 3

18. $16^{\frac{5}{4}} \times 2^{-3} \times 3^{0}$ is equal to:
   [A] 20 [B] 2 [C] 4 [D] 10

19. The result of evaluating $0.027^{-\frac{1}{2}}$ is:

   [A] $3\dfrac{1}{3}$ [B] 3 [C] $\dfrac{3}{10}$ [D] $\dfrac{1}{3}$

20. $\dfrac{8^{\frac{2}{3}} \times 27^{-\frac{1}{3}}}{64^{\frac{1}{3}}}$ simplifies to:

   [A] $\dfrac{1}{3}$ [B] $\dfrac{1}{9}$ [C] $\dfrac{16}{3}$ [D] $\dfrac{27}{8}$

21. After evaluating $5\dfrac{2}{3} \times \left(\dfrac{2}{3}\right)^{2} \div \left(1\dfrac{1}{2}\right)^{-1}$, the

   result is:

   [A] $\dfrac{12}{5}$ [B] $\dfrac{8}{5}$ [C] $3\dfrac{3}{5}$ [D] $4\dfrac{1}{8}$

22. The result of simplifying $\dfrac{2^{\frac{1}{2}} \times 8^{\frac{1}{2}}}{4}$:

   [A] 1 [B] 2 [C] 4 [D] 16

23. $\left(\dfrac{1}{4}\right)^{-1\frac{1}{2}}$ is equal to:

   [A] 8 [B] 4 [C] $\dfrac{1}{4}$ [D] $\dfrac{1}{16}$

24. The result of evaluating

25. $\left(\dfrac{16}{81}\right)^{-\frac{3}{4}} \times \left(\dfrac{100}{81}\right)$ is:

   [A] $\dfrac{80}{243}$ [B] $\dfrac{20}{27}$

   [C] $\dfrac{25}{6}$ [D] $\dfrac{15}{4}$

25. $\left(3a^{2}\right)^{3}$ is equal to:

   [A] $3a^{6}$ [B] $9a^{6}$
   [C] $27a^{2}$ [D] $27a^{6}$

26. $a^{2} \times b \times a^{4} \times b^{2}$ simplifies to:
   [A] $a^{6}b^{2}$ [B] $a^{8}b^{2}$
   [C] $a^{3}b^{3}$ [D] $a^{6}b^{3}$

27. When simplified $\dfrac{x^{3}y^{4}z^{7}}{x^{2}y^{6}z^{7}}$ to:

   [A] $\dfrac{x}{y^{2}}$ [B] $\dfrac{y^{2}}{x}$ [C] $\dfrac{x^{2}}{y}$ [D] $\dfrac{1}{y}$

28. $\left(m^{2}n^{5}\right) \div \left(m^{3}n^{4}\right)$ equals:

   [A] mn^{-1} [B] $m^{-1}n$
   [C] $m^{5}n^{9}$ [D] mn

29. When $10a^{6}$ is divided by $5a^{3}$ the result
   is:
   [A] $3a^{3}$ [B] $2a$ [C] a^{3} [D] $2a^{3}$

30. If $2^{x} \times 3^{2} = 144$, the value of x is:
   [A] 7 [B] 5 [C] 4 [D] 8

31. If $x^{2} \times 3^{2} \times 1^{2} = 144$, the value of x is:
   [A] -4 [B] 2 [C] -2 [D] 16

32. When $3^{x} = 81$, the value of x is:
   [A] 2 [B] 3 [C] 4 [D] 27

33. The value of x for which $3^{x} = 243$ is:
   [A] 6 [B] 5 [C] 4 [D] 3

34. If $3^{x} + 6 = 87$, the value of x is:
   [A] 1 [B] 2 [C] 3 [D] 4

35. If $3^{2x} = 27$ the value of x is:
   [A] 1 [B] 1.5 [C] 4.5 [D] 18

36. The value of t for which $\dfrac{64}{27} = \left(\dfrac{3}{4}\right)^{t-1}$ is:

   [A] -4 [B] 2 [C] 4 [D] -2

37. Given that $27^{(1+x)} = 9$. The value of x is:

   [A] -3 [B] $-\dfrac{1}{3}$ [C] $\dfrac{1}{3}$ [D] 2

38. If $16(4)^{2x} = \left(\dfrac{1}{2}\right)^{x}$, the value of x is:

[A] -3 [B] $-\dfrac{4}{5}$ [C] $-\dfrac{4}{3}$ [D] $\dfrac{4}{3}$

39. $\left(\dfrac{1}{4}\right)^{2-y} = 1$, the value of y is:

[A] -2 [B] $-\dfrac{1}{2}$ [C] $\dfrac{1}{2}$ [D] 2

40. The solution of the equation $2\sqrt{x} = 4$ is:
[A] -2 [B] 2 [C] 4 [D] 6

41. The value of x for which $2^{-6x} = 8^{(1-x)}$ is true is:

[A] $-\dfrac{7}{3}$ [B] $\dfrac{1}{3}$ [C] -1 [D] $\dfrac{7}{9}$

42. The value of n which satisfies the

equation $\dfrac{3^{(1-n)}}{9^{-2n}} = \dfrac{1}{9}$ is:

[A] $-\dfrac{3}{2}$ [B] $\dfrac{1}{3}$ [C] -3 [D] 1

43. When simplified $\dfrac{1}{4}\left(2^n - 2^{n+2}\right)$ becomes:

[A] 2^{n-1}-2^n [B] $2^{n-2}\left(1-2^n\right)$ [C] $2^{n+2}+2$ [D] 2^n

44. When $56x^{-4} \div 14x^{-8}$ is simplified the result is:

[A] $2x^{-12}$ [B] $4x^{-4}$

[C] $4x^{+4}$ [D] $4x^{-3}$

45. Given that $81 \times 2^{2n-2} = k$, the value of $\sqrt{k}$ which satisfies this equation is:

[A] 4.5×2^n [B] 4.5×2^{2n} [C] $9 \times 2^{n-1}$ [D] 9×2^n

46. The word that is not another name for logarithm is:
[A] Power [B] exponent
[C] Base [D] index

47. If $\log_{10} 5.444 = 0.7359$, then

$\log_{10} 54440$ is:
[A] 3.7359 [B] 4.7359
[C] 5.7359 [D] 6.7359

48. If $\log_{10} x = 2.8765$, the value of x is:

[A] Greater than 100
[B] between 1 and 10
[C] less than 10
[D] between 10 and 100

49. The value of

$\log_{10} 6 + \log_{10} 45 - \log_{10} 27$ is:
[A] 0 [B] 1 [C] 1.1738 [D] 10

50. On simplification $\dfrac{\log \sqrt{8}}{\log 8}$ is equal to:

[A] $\dfrac{1}{3}$ [B] $\dfrac{1}{3}\log \sqrt{8}$

[C] $\dfrac{1}{3}\log \sqrt{2}$ [D] $\dfrac{1}{2}$

51. Simplifying $\dfrac{\log 27^{\frac{1}{3}}}{\log 81}$ gives:

[A] 14 [B] $\dfrac{3}{8}$ [C] $\dfrac{1}{2}$ [D] $\dfrac{3}{4}$

52. The value of $\log_{10} 25 + \log_{10} 32 - \log_{10} 8$ is:
[A] 0.2 [B] 2 [C] 100 [D] 409

53. The value of $2\log_3 6 + \log_3 16$ is:
[A] $4 - \log_3 2$ [B] $3 + \log_3 2$
[C] $2 + 6\log_3 2$ [D] $3 - \log_3 2$

54. On evaluation, $\log_{10} 4 + \log_{10} 25$ becomes:
[A] 1 [B] 2 [C] 3 [D] 4

55. $\log_3 9 + \log_3 15 - \log_3 5$ is equal to:
[A] $\log_3 19$ [B] $\log 3$ [C] 3 [D] 1

56. Simplifying $\log_7 8 - \log_7 2 + \log_7 4$, gives:
[A] 0 [B] 2
[C] $2\log_2 7$ [D] $4\log_2 7$

57. The value of $\log_{10} 5 + \log_{10} 20$ is:
[A] 2 [B] 3 [C] 4 [D] 5

58. If $\log_{10} 2 = 0.3010$ and $\log_{10} 2^y = 1.8062$. The value of y to the nearest whole number is:
[A] 4 [B] 6 [C] 5 [D] 2

59. If $3\log_{10} a = \log_{10} 64$, the value of a is:
[A] 4 [B] 6 [C] 8 [D] 16

60. Given that $\log p = 2\log x + 3\log q$ .The correct expression of p in terms of x and q is:

[A] $p = 6xq$ [B] $p = x^2 q^3$

[C] $p = x^2 + q^3$ [D] $p = 2x + 3q$

61. The solution of the equation

$\log_8 x - 4\log_8 x = 2$ is:

[A] $\dfrac{1}{4}$ [B] $\dfrac{1}{2}$ [C] 4 [D] 2

62. $7^{x-1} = \log_5 5$, then x is equal to:
[A] 1 [B] 7 [C] -1 [D] -7

63. Given that $\dfrac{1}{3}\log_{10} p = 1$. The value of p is:

[A] 3 [B] 10 [C] 100 [D] 1000

64. $\log_2 a = \log_8 4$ only if a is equal to:

[A] $2^{\frac{1}{2}}$ [B] $2^{\frac{2}{3}}$ [C] $4^{\frac{2}{3}}$ [D] $4^{\frac{1}{3}}$

65. If $\log_a x = p$, then in terms of a and p, x is equal to:

[A] a^p [B] $\frac{a}{p}$ [C] p^a [D] ap

66. The value of p for which $\frac{1}{2}\log_{10} p = 1$ is true is:

[A] 10^{-1} [B] 10^3

[C] 10^2 [D] 10^1

67. Given that $\log_4 x = -3$. The value of x is:

[A] $\frac{1}{81}$ [B] $\frac{1}{64}$ [C] 64 [D] 81

68. On simplification $\frac{4\sqrt{18}}{\sqrt{8}}$ becomes:

[A] 2 [B] 3 [C] 6 [D] 12

69. The value of $\sqrt{96} + \sqrt{54} - \sqrt{24}$ is:

[A] $\sqrt{6}$ [B] $2\sqrt{6}$

[C] $3\sqrt{6}$ [D] $5\sqrt{6}$

70. $\sqrt{32} - \sqrt{98} + 5\sqrt{2}$ is equal to

[A] $\frac{1}{2}\sqrt{2}$ [B] $2\sqrt{2}$

[C] $3\sqrt{2}$ [D] $4\sqrt{2}$

71. The value of $3\sqrt{12} + 10\sqrt{3} - \frac{6}{\sqrt{3}}$ is:

[A] $7\sqrt{3}$ [B] $10\sqrt{3}$

[C] $14\sqrt{3}$ [D] $18\sqrt{3}$

72. Given that $\frac{1}{\sqrt{2}} = 0.7071$. Then, $\frac{3\sqrt{2}}{2}$ is greater than $\frac{1}{\sqrt{2}}$ by:

[A] -3 [B] -1.4142

[C] 1.4142 [D] 3

73. Evaluating $\sqrt{20} \times \left(\sqrt{5}\right)^3$ gives:

[A] 10 [B] 20 [C] 25 [D] 50

74. If $K\sqrt{28} + \sqrt{63} - \sqrt{7} = 0$ the value of K is:

[A] -2 [B] -1 [C] 1 [D] 2

75. $\sqrt{128} + \sqrt{18} - \sqrt{k} = 7\sqrt{2}$, then k must be:

[A] 8 [B] 16 [C] 32 [D] 48

76. When the denominator is rationalized, $\frac{10}{\sqrt{32}}$ becomes:

[A] $\frac{5}{4}\sqrt{2}$ [B] $\frac{4}{5}\sqrt{2}$

[C] $\frac{5}{16}\sqrt{2}$ [D] $\frac{16}{5}\sqrt{2}$

77. The number $\frac{6}{\sqrt{2}}$ is equal to:

[A] $4\sqrt{2}$ [B] $3\sqrt{2}$ [C] $2\sqrt{2}$ [D] 2

TOPIC 18

NUMBER BASES

OBJECTIVES

At the end of this topic, the learner should be able to:

1. List the digits used in different number bases from base two to ten.
2. List at least the first twenty consecutive numbers in different number bases.
3. Write expanded and condensed forms in different number bases.
4. Convert from a non-denary base to base ten and vice versa.
5. Convert from one non-denary base to another non-denary base.
6. Appreciate the role of the binary system in computer systems.
7. Use scientific calculators to perform calculations in different number bases.

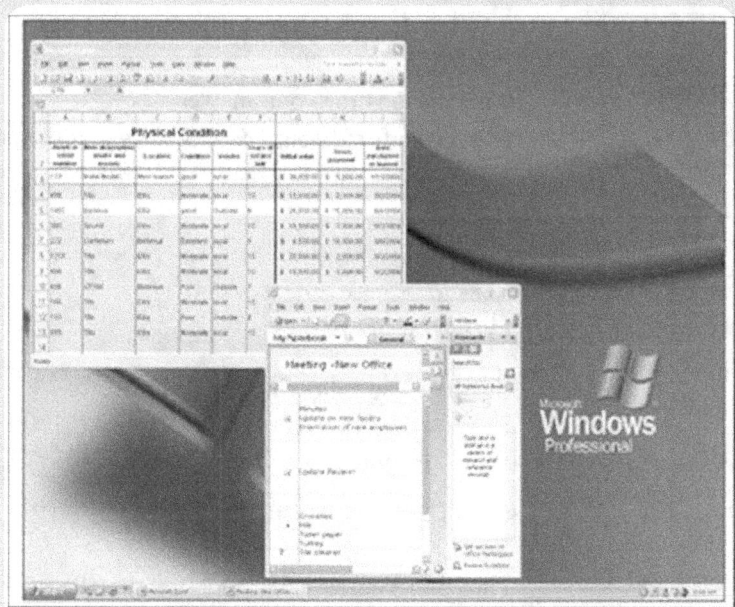

Arithmetic and logic form the basis of all computer software—the instructions that tell computers what to do. Shown on this computer screen are programs running on the Windows XP operating system, the software that allows a computer's other software to run.

Grace Hopper
A pioneer in data processing, Grace Hopper received credit for creating the first compiler in 1952. Hopper helped to develop two computer languages and to make computers attractive to businesses. One of the most prominent women in the computer industry, Hopper died in 1992.

Number System Vocabulary

A number system in which counting is done in groups of ten is called a **base ten system**, a **denary system** or a **decimal system**. The digits used in base ten are 0, 1, 2, 3, 4, 5, 6, 7, 8, and 9. The Table 18:1(a) shows the names given to number bases less than ten and the digits used and Table 18:1(b) shows some more number bases and their names.

Number base	Name	Digits Used
Base two	Binary system	0,1
Base three	Ternary system or tertiary system	0,1,2
Base four	Quadrinal system	0,1,2,3
Base five	Quinary system	0,1,2,3,4
Base six	Hexal system or senary system	0,1,2,3,4,5
Base seven	Heptademal system	0,1,2,3,4,5,6
Base eight	Octal system	0,1,2,3,4,5,6,7
Base nine	Nonal system	0,1,2,3,4,5,6,7,8

(a)

Number base	Name
Base eleven	Unidecimal system
Base twelve	Duodecimal system or duodenary system
Base sixty	Sexagenary system or sexagesimal system
Base one hundred	Centenary system or centesimal system

(b)
Table 18:1

In base eleven, the following digits 0, 1, 2, 3, 4, 5, 6, 7, 8, 9, t, where t means ten may be used.
In base twelve the digits used are 0, 1, 2, 3, 4, 5, 6, 7, 8, 9, t, e, where t and e represent ten and eleven respectively. Notice that the highest digit in any number base is less than the number base by one.
To show the base to which a numeral is written, a subscript spelling the base is used after the numeral. For instance, if 1652 is written to base seven this is written as 1652_{seven}, and read ''one-six-five-two base seven' and **not** 'one thousand six hundred and fifty two base seven'. Note that it is wrong to write 1652_7 because the numeral 7 does not

exist in base seven.
Often the base subscript is omitted for base ten, numerals. Thus in base ten, two hundred and seventy eight, is written, 278_{ten} or simply 278. This notation must never be omitted for any numeral written in a base other than ten.

EXERCISE 18:1

1. Write 'Yes' if it is possible to write the given number in the base indicated, or 'No', if it is not possible. In each case give reasons for your answer.
   (a) 324_{six} (b) 471_{seven}
   (c) 1011_{two} (d) 897_{nine}
   (e) 952_{ten} (f) 782_{nine}
   (i) 3420_{five} (j) 2102_{three}
   (k) 5604_{six} (l) 2389_{ten}
2. Write the numbers from one to thirty in your native language.
3. What is the base used in counting in your native language?
4. Count in French from one to one hundred. What do you suggest is the number base used in French?
Give reasons for your answer.
5. List the first twenty consecutive numbers of each of the number bases two, three, four, five... twelve.

Powers of Ten

Consider the following
$10 \times 10 \times 10 \times 10 = 10000$, $10 \times 10 \times 10 = 1000$, $10 \times 10 = 100$.

A short way of writing the product 10×10 is 10^2

Thus $10 \times 10 \times 10 = 10^3$ and $10 \times 10 \times 10 \times 10 = 10^4$.

10^4 is read 'ten to the fourth power' or simply 'ten to the fourth'. 10^2 is read 'ten to the second power' or 'ten to the second' or 'ten squared'.
Using this notation, $10^1 = 10$ and $10^0 = 1$.

Expanded Forms in Base Ten

Example 18:1

Write 2579_{ten} in expanded form.

Solution

$$2579_{ten} = 2000 + 500 + 70 + 9$$
$$= 2 \times 1000 + 5 \times 100 + 7 \times 10 + 9$$
$$\therefore 2579_{ten} = 2 \times 10^3 + 5 \times 10^2 + 7 \times 10^1 + 9 \times 10^0 \ldots \circledast$$

The LHS of $\circledast$ is called the **condensed form** while the RHS is called the **expanded form**.

N.B.: $2000 = 2 \times 1000$ i.e. '2 thousands' or 20×100 i.e. twenty hundreds' or 200×10 i.e. two hundred tens' or '2000×1 i.e. 'two thousand ones'.

Example 18:2

Write the following in expanded form.
(a) 268   (b) 4783

Solution

(a) $268 = 200 + 60 + 8 = 2 \times 100 + 6 \times 10 + 8$
$$= 2 \times 100 + 6 \times 10 + 8$$
$$2 \times 10^2 + 6 \times 10^1 + 8$$

(b) $4783 = 4000 + 700 + 80 + 3$
$$= 4 \times 1000 + 7 \times 100 + 8 \times 10 + 3$$
$$\therefore 4783 = 4 \times 10^3 + 7 \times 10^2 + 8 \times 10 + 3$$

The expanded forms can be written directly without passing through the first two steps as in Example 18:2 (i) and (ii).

Example 18:3

Write the expanded forms of the following.
(i) 42384   (ii) 735

Solution

(i) $42384 = 4 \times 10^4 + 2 \times 10^3 + 3 \times 10^2 + 8 \times 10 + 4$
(ii) $735 = 7 \times 10^2 + 3 \times 10 + 5$

> ### EXERCISE 18:2

Write down the expanded forms of each of the following base ten numbers.
(1) 87   (2) 124   (3) 845   (4) 1374
(5) 14891 (6) 735   (7) 430   (8) 6007
(9) 3204 (10) 96800

Expanded Forms in other Bases

Expanded forms can also be written in bases other than ten.

Example 18:4

Write the expanded forms of the following.
(a) 324_{five} (b) 2517_{eight} (c) 4702_{nine}

Solution

(a) $324_{five} = 3 \times 5^2 + 2 \times 5 + 4$
(b) $2517_{eight} = 2 \times 8^3 + 5 \times 8^2 + 1 \times 8 + 7$
(c) $4702_{nine} = 4 \times 9^3 + 7 \times 9^2 + 0 \times 9 + 2$
$$= 4 \times 9^3 + 7 \times 9^2 + 2$$

> ### EXERCISE 18:3

Write the expanded forms of the following.
(1) 3214_{five} (2) 1011101_{two}
(3) 52301_{seven} (4) 43021_{six}
(5) 51632_{eight} (6) 24162_{seven}
(7) 20121_{three} (8) 34721_{eight}
(9) 7481_{nine} (10) 1101_{two}
(11) 4836_{nine} (12) 2311_{four}
(13) 421_{six} (14) 7104_{eight}
(15) 13221_{four} (16) 2102_{three}
(17) 1102_{three} (18) 43251_{six}

Conversion of Number Bases

Converting None Denary to Denary

To convert a number written in some base other than ten to base ten, the number is first written in expanded form. Secondly, the expanded form is evaluated completely.

Example 18:5

Convert to base ten
(i) 123_{six} (ii) 5724_{eight}

Solution

(i) $123_{six} = 1 \times 6^2 + 2 \times 6 + 3$
$$= 36 + 12 + 3$$
$$= 51_{ten}$$

(ii) $5724_{eight} = 5 \times 8^3 + 7 \times 8^2 + 2 \times 8 + 4$
$$= 5 \times 512 + 7 \times 64 + 16 + 4$$
$$= 2560 + 448 + 16 + 4$$
$$= 3028_{ten}$$

173

EXERCISE 18:4

Convert the following to base ten numbers.

1. 142_{five}
2. 214_{five}
3. 435_{six}
4. 2542_{seven}
5. 1126_{seven}
6. 1011101_{two}
7. 3247_{eight}
8. 45631_{seven}
9. 11011_{two}
10. 3487_{nine}
11. 11201_{three}
12. 2103_{four}
13. 35201_{six}
14. 1824_{nine}
15. 2112101_{three}
16. 587234_{nine}

Converting Denary to Non-Denary

Consider $20 \div 6 = 3$ remainder 2.
20, is called the **dividend**; 6, is called the **divisor**; 3, is called the **quotient** and 2, is called the **remainder**. Two methods shall be examined.

Repeated Division Method

To change a base ten number to a number in a different base, repeated division by the base to which the number is to be converted is done. In each step the remainder is written until the dividend is zero. The result is made up of the digits of the remainders taken in order from the last to the first.

Example 18:6

Convert the following to the specified bases.
(a) 327 to base six   (b) 7985 to base eight.

Solution

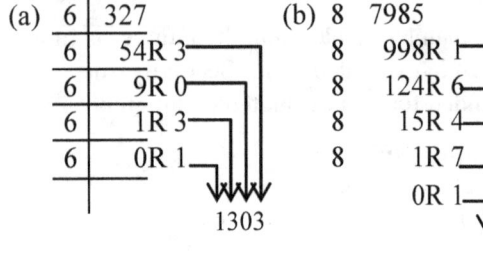

$\therefore 327 = 1303_{\text{six}}$ $\therefore 7985 = 17461_{\text{eight}}$.

Expanded Form Method

In this method, the power of the base to which the number has to be converted, is used. The greatest power of the base is that which is closest to the given number. This method is the reverse process of the method used in expressing numbers in the expanded form.

Example 18:7

Convert the following to the specified bases using the expanded form method.
(a) 327 to base six   (b) 7985 to base eight
(c) 29 to a base two.

Solution

(a) The powers of 6 are

$$6^1 = 6, \ 6^2 = 36, \ 6^3 = 216$$

$$327_{\text{ten}} = 216 + 111$$

$$= 216 + 108 + 3$$

$$= 6^3 + 3 \times 3^6 + 3$$

$$= 1 \times 6^3 + 3 \times 6^2 + 0 \times 6^1 + 3$$

$$\downarrow \qquad \downarrow \qquad \downarrow \qquad \downarrow$$

$$1 \qquad 3 \qquad 0 \qquad 3$$

$$\therefore 327_{\text{ten}} = 1303_{\text{six}}$$

Similarly,

(b) The powers of 8 are

$$8^1 = 8, \ 8^2 = 64, \ 8^3 = 512, \ 8^4 = 4096$$

$$\therefore 7985 = 4096 + 3889$$

$$= 4096 + 3584 + 305$$

$$= 4096 + 3584 + 256 + 49$$

$$= 4096 + 3584 + 256 + 48 + 1$$

$$= 1 \times 8^4 + 7 \times 512 + 4 \times 64 + 6 \times 8 + 1$$

$$= 1 \times 8^4 + 7 \times 8^3 + 4 \times 8^2 + 6 \times 8 + 1$$

$$\therefore 7985_{\text{ten}} = 17461_{\text{eight}}$$

(c) The powers of 2 are

$$2^1 = 2, \ 2^2 = 4, \ 2^3 = 8, \ 2^4 = 16$$

$$29 = 16 + 8 + 4 + 1$$

$$= 1 \times 2^4 + 1 \times 2^3 + 1 \times 2^2 + 0 \times 2^1 + 1$$

$$= 11101_{\text{two}}$$

EXERCISE 18:5

Convert the following base ten numbers to the bases specified.

1. 789 to base two.
2. 4899 to base eight.
3. 316 to base six.
4. 3057 to base four.
5. 1349 to base three.
6. 537 to base nine.
7. 92 to base two.
8. 8403 to base six.
9. 792 to base eight.
10. 2740 to base seven.

Converting Non-Denary to Non-Denary

To convert a number from a non-denary base to another non-denary base, first convert the numeral to base ten, then change the base ten number to the destination base.

Example 18:8

Convert (a) 276_{eight} to base four.

(b) 1346_{seven} to a base five.

Solution

(a) 276_{eight} to base ten   190 to base four.

Changing 276_8 to base ten

$276_{eight} = 2 \times 8^2 + 7 \times 8 + 6$

$= 128 + 56 + 6$

$= 190_{ten}$

4	190
4	47R 2
4	11R 3
4	2R 3
	0R 2

$\therefore 276_{eight} = 2332_{four}$

(b) Change 1346_{seven} to base ten

$1346_{seven} = 1 \times 7^3 + 3 \times 7^2 + 4 \times 7 + 6$

$= 1 \times 343 + 3 \times 49 + 4 \times 7 + 6$

$= 343 + 147 + 28 + 6$

$= 524_{ten}$

524_{ten} is then changed to base five

5	524
5	104R 4
5	20R 4
5	4R 0
	0R 4

$\therefore 1346_{seven} = 4044_{five}$

> **EXERCISE 18:6**

(i) In Table 18:2, convert to the base given in column B.

	Column A	Column B
1	142_{five}	seven
2	11011_{two}	eight
3	45631_{seven}	five
4	421_{six}	four
5	7481_{nine}	three
6	1322_{four}	nine
7	3472_{eight}	seven
8	2416_{seven}	six
9	43021_{six}	four
10	2102_{three}	seven
11	3214_{five}	eight
12	2311_{four}	six
13	24162_{seven}	four
14	1011101_{two}	nine
15	7104_{eight}	three

Table 18:2

(ii) Change 1346_{eight} km to:

(a) meters in base ten. (b) meters in base five.

Arithmetic in Non-Denary Bases

In order to do successfully arithmetic in any number base, the place value of each digit in the number base must be well understood. For instance, in base seven the only digits used are 0, 1, 2, 3, 4, 5, 6. Numbers greater than 6 are interpreted in terms of the number of sevens and these digits only. Table 18:3 interprets some two digit base seven numbers in terms of seven and the base seven digits.

Base seven number	Meaning
10	1 seven + 0 units
11	1 seven + 1 unit
12	1 seven + 2 units
13	1 seven + 3 units
14	1 seven + 4 units
15	1 seven + 5 units
16	1 seven + 6 units
20	2 sevens + 0 units
43	4 sevens + 3 units
65	6 sevens + 5 units

Table 18:3

Addition and Subtraction

Addition and subtraction are done in a way similar to base ten. The numbers are arranged one below the other, ensuring that the unit digits fall in the same column. The other digits follow this alignment from right to left.

Example 18:9

Evaluate the following

(a) $235_{seven} + 421_{seven}$ (b) $637_{eight} - 325_{eight}$

Solution

(a)
$$
\begin{array}{r}
2\ \ 3\ \ 5_{\text{seven}} \\
+\ \ 4\ \ 2\ \ 1_{\text{seven}} \\
\hline
6\ \ 5\ \ 6_{\text{seven}} \\
\hline
\end{array}
$$

(b)
$$
\begin{array}{r}
6\ \ 3\ \ 7_{\text{eight}} \\
-\ \ 3\ \ 2\ \ 5_{\text{eight}} \\
\hline
3\ \ 1\ \ 2_{\text{eight}} \\
\hline
\end{array}
$$

In adding numbers if the sum of the digits in a column is greater than the largest digit in the number base divide the sum by the number base and write down the remainder in the column. The quotient is then added with the numbers in the next left column.

Example 18:10

Compute $103_{\text{four}} + 213_{\text{four}}$

Solution

(a)
$$
\begin{array}{r}
1\ \ 0\ \ 3_{\text{four}} \\
+2\ \ 1\ \ 3_{\text{four}} \\
\hline
3\ \ 2\ \ 2_{\text{four}} \\
\hline
\end{array}
$$

Explanation

In column 3: $3_{\text{four}} + 3_{\text{four}} = 12_{\text{four}}$
Write the 2 under column 3 and add the 1 to the sum in column 2.

To understand subtraction well, it should be appreciated that the place value of 1 in any column has a value equal to the base in the next right column. Thus, in 13_{six}, the 1 in column 1 is equal to six in column 2. With this in mind, subtraction can then be done in a way similar to base ten.

Example 18:11

Evaluate $853_{\text{nine}} - 237_{\text{nine}}$.

Solution

(b)
$$
\begin{array}{r}
8\ \ 5\ \ 3_{\text{nine}} \\
-\ 2\ \ 3\ \ 7_{\text{nine}} \\
\hline
6\ \ 1\ \ 5_{\text{nine}} \\
\hline
\end{array}
$$

Explanation

Since $7 > 3$, subtract 1 nine from the 5 in column 2 and add to the 3 column 3.
1 nine $+ 3 = 5 + 7$ and $5 + 7 - 7 = 5$.
4 is left in the minuend in column 2.
$4 - 3 = 1$ and $8 - 2 = 6$

Multiplication

Multiplication in non-denary bases is very similar to that in base ten. The steps are as follows:

(i) After multiplying any two digits, divide this product by the number base and write down the remainder.

(ii) Add the quotient to the product in the next left column in the same way as in base ten.

(iii) Continue as in (i) and (ii) until the multiplication is over.

(iv) If the multiplier has more than one digit, there will be more than one row of the product of each digit and the multiplicand. Add these rows.

Example 18:13

Evaluate

(a) $53_{\text{seven}} \times 6_{\text{seven}}$ (b) $42_{\text{five}} \times 34_{\text{five}}$

Solution

(a)
$$
\begin{array}{r}
5\ \ 2_{\text{seven}} \\
\times\ \ \ \ \ 6_{\text{seven}} \\
\hline
4\ \ 3\ \ 5_{\text{seven}} \\
\hline
\end{array}
$$

Explanation

$6 \times 2 \div 7 = 1\ R\ 5$. Write 5 and take over 1.
$6 \times 5 \div 7 = 4\ R\ 2$. Add the 1 taken over to the remainder and take over 4 to the next left column.

(b)
$$
\begin{array}{r}
4\ \ 2_{\text{five}} \\
\times\ \ \ \ \ 3\ \ 4_{\text{five}} \\
\hline
3\ \ 2\ \ 3_{\text{five}} \\
+\ 23\ \ 1\ \ 0_{\text{five}} \\
\hline
31\ \ 3\ \ 3_{\text{five}} \\
\hline
\end{array}
$$

Explanation

$4 \times 2 \div 5 = 1\ R\ 3$. Write 3 and take over 1.
$4 \times 4 \div 5 = 3\ R\ 1$. Add the 1 taken over to the remainder and take over 3.
$3 \times 2 \div 5 = 1\ R\ 1$. Write 1 and take over 1.
$3 \times 4 \div 5 = 2\ R\ 2$. Add the 1 taken over to the remainder and take over 2.
Add the two rows of products.

Division

To do division effectively in a non-denary base it is necessary to determine first the products of the base digits and the divisor. After this has been done, long division can then be carried out with great ease.

Example 18:14

Evaluate $4346_{eight} \div 42_{eight}$.

Solution

Product of base digits and divisor

$1 \times 42_{eight} = 42_{eight}$
$2 \times 42_{eight} = 104_{eight}$
$3 \times 42_{eight} = 1446_{eight}$
$4 \times 42_{eight} = 210_{eight}$
$5 \times 42_{eight} = 252_{eight}$
$6 \times 42_{eight} = 314_{eight}$
$7 \times 42_{eight} = 356_{eight}$

Long division can now be done referring factors from the above product of base digits.

$$
\begin{array}{r}
103 \\
42_{eight}\overline{)4346_{eight}} \\
42_{eight} \\
\hline
146_{eight} \\
146_{eight} \\
\hline
0
\end{array}
$$

Explanation
$43 \div 42 = 1$, $1 \times 42_{eight} = 42_{eight}$, $43-42 = 1$
Bring down 4. Now 42 cannot divide 14 so write 0 in the answer and bring down 6.
$146 \div 42 = 3$, $3 \times 42_{eight} = 146_{eight}$
$146 - 146 = 0$

EXERCISE 18:7

Evaluate the following.
1. $322_{four} + 31_{four}$
2. $637_{eight} + 56_{eight}$
3. $112_{three} + 112_{three} + 212_{three}$
4. $315_{six} + 24_{six} + 52_{six}$
5. $413_{five} + 324_{five}$
6. $321_{five} - 232_{five}$
7. $4374_{eight} - 2537_{eight}$
8. $2122_{three} - 1022_{three}$
9. $311_{four} - 232_{four}$
10. $212_{three} \times 21_{three}$
11. $423_{five} \times 34_{five}$
12. $177_{eight} \times 27_{eight}$
13. $564_{seven} \times 65_{seven}$
14. $10211_{eight} \div 123_{eight}$
15. $15311_{six} \div 125_{six}$
16. $44341_{seven} \div 145_{seven}$

The Binary System

As it has been seen the binary system otherwise referred to as base two is a number system consisting only of the two digits, 0 and 1. These digits are called **bits**. In this system, the place values are arranged based on powers of 2 in a format called the 16-8-4-2-1. Thus
$$11011_{two} = 1 \times 2^4 + 1 \times 2^3 + 0 \times 2^2 + 1 \times 2^1 + 1 \times 2^0$$
$$= 16 + 8 + 0 + 2 + 1$$
$$= 27_{ten}$$

Table 18:4, shows the first twenty base ten numerals and their corresponding base two counterparts.

Decimal	Binary	Decimal	Binary
0	0	10	1010
1	1	11	1011
2	10	12	1100
3	11	13	1101
4	100	14	1110
5	101	15	1111
6	110	16	10000
7	111	17	10001
8	1000	18	10010
9	1001	19	10011
10	1010	20	10100

Table 18:3

In the binary system, the **binary point** or **binary marker** separates the whole number part from the fractional part and the powers of 2 to the right of the binary point are negative. Thus,
$$11.01_{two} = 1 \times 2^1 + 1 \times 2^0 + 0 \times 2^{-1} + 1 \times 2^{-2}$$
$$= 2 + 1 + 0 + \frac{1}{4}$$
$$= 3.25_{ten}$$

Binary numbers containing the binary point are called **bicimals**.

Importance of the Binary System

Since switches are two state devices (ON and OFF), they operate in the binary system.

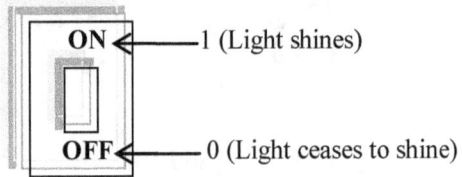

ON ←——— 1 (Light shines)

OFF ←——— 0 (Light ceases to shine)

The ON position can be represented by 1 and the OFF position by 0.

Table 18:4, shows some other situations, which can be represented by the binary bits 0 and 1.

Up	Down
Odd	Even
Left	Right
Good	Bad

Table 18:4

In fact, the binary system can be employed in all two state devices. Give six more situations, which can be represented by bits.

The importance of the binary system in electronic calculators, computers and other electronic devices, which equally use the 16-8-4-2-1 format, cannot be over emphasized. Because electronic devices are made up of a combination of so many switches, they work in a binary system. When information is fitted into them, they convert the information to base two, process it and reconvert the result to base ten or ordinary language. Due to the usefulness of the binary system, it shall be examined in detail.

Inter Binary-Decimal Conversion

This is done using the very methods discussed above.

Example 18:15

Convert 87_{ten} to a binary number.

Solution

By repeated division

2	87	
2	43	R 1
2	21	R 1
2	10	R 1
2	5	R 0
2	2	R 1
2	1	R 0
	0	R 1

$$\Rightarrow 87_{ten} = 1010111_{two}$$

Alternatively, the expanded form method can be used as follows

	2^6	2^5	2^4	2^3	2^2	2^1	2^0
	64	32	16	8	4	2	1
87	1×2^6	0×2^5	1×2^4	0×2^3	1×2^2	1×2^1	1×2^0

Table 18:5

$$\Rightarrow 87_{ten} = 1010111_{two}$$

Example 18:16

Convert 111011_{two} to a denary number.

Solution

$$111011_{two} = 1 \times 2^5 + 1 \times 2^4 + 1 \times 2^3 + 1 \times 2 + 1$$
$$= 32 + 16 + 8 + 2 + 1$$
$$111011_{two} = 59_{ten}$$

Example 18:17

Evaluate the following

(a) $100101_{two} + 110101_{two}$

(b) $10110101_{two} - 110101_{two}$

(c) $1001_{two} \times 101_{two}$

(d) $1000001_{two} \div 1101_{two}$

Solution

(a)
$$\begin{array}{r} 100101_{two} \\ + \ 110101_{two} \\ \hline 1011010_{two} \\ \hline \end{array}$$

(b)
$$\begin{array}{r} 1011010_{two} \\ - \ 110101_{two} \\ \hline 100101_{two} \\ \hline \end{array}$$

(c)
$$
\begin{array}{r}
1001_{two} \\
\times \quad 101_{two} \\
\hline
100100_{two} \\
1001_{two} \\
\hline
101101_{two} \\
\hline
\end{array}
$$

(d)
$$
1101_{two} \overline{)1000001_{two}} \\
\underline{1101_{two}} \\
1101_{two} \\
\underline{1101_{two}} \\
1101_{two}
$$

Example 18:18

Evaluate the following

(a) $101.1_{two} \times 11_{two}$ (b) $1011.1_{two} \times 11.1_{two}$

Solution

Ignore the binary point and compute the product. Now count the total number of binary points and insert.

(a) $101.1_{two} \times 11_{two} = 10000.1_{two}$

(b) $1011.1_{two} \times 11.1_{two} = 101000.01_{two}$

EXERCISE 18:8

1. Convert the following decimal numbers to a binary number

   (a) 34 (b) 57 (c) 46 (d) 63 (e) 72

2. Convert the following binary numbers to denary numbers.

   (a) 110101_{two} (b) 11011_{two}

   (c) 1101101_{two} (d) 1010101_{two}

   (e) 11101_{two}

3. Evaluate the following in the binary scale

   (a) $101_{two} + 11_{two}$ (b) $101011_{two} + 10110_{two}$

   (c) $11000_{two} + 10100_{two}$ (d) $1011100_{two} - 110011_{two}$

   (e) $110110_{two} - 10111_{two}$ (f) $1010111_{two} - 111010_{two}$

   (g) $101_{two} \times 110_{two}$ (h) $10101_{two} \times 1110_{two}$

   (i) $1100_{two} \times 1101_{two}$ (j) $101110_{two} \div 10011_{two}$

   (k) $11011_{two} \div 1011_{two}$ (l) $101011_{two} \div 1010_{two}$

MULTIPLE CHOICE EXERCISE 18

1. 75_{ten} is the same as:

   [A] 300_{five} [B] 400_{five}

   [C] 500_{five} [D] 600_{five}

2. When expressed as a binary number 27_{ten} is equal to:

   [A] 1110_{two} [B] 1111_{two}

   [C] 11011_{two} [D] 1001_{two}

3. The denary (base ten) number 37, written in binary (base two) is:

   [A] 100011_{two} [B] 100111_{two}

   [C] 100001_{two} [D] 100101_{two}

4. The denary number 39 written in binary is:

   [A] 100111_{two} [B] 100101_{two}

   [C] 1001_{two} [D] 10001_{two}

5. The value of the decimal number 89 as a binary number is:

   [A] 101101_{two} [B] 1011001_{two}

   [C] 1001001_{two} [D] 1001101_{two}

6. 35 in base two is:

   [A] 1000_{two} [B] 10011_{two}

   [C] 100011_{two} [D] 110010_{two}

7. The equivalence of 11111_{two} in base ten is:

   [A] 9   [B] 17   [C] 19   [D] 31

8. 101101_{two} in expanded form is:

   [A] $1 \times 2^{-6} + 1 \times 2^{-4} + 1 \times 2^{-3} + 1 \times 2^{-1}$

   [B] $1 \times 2^{-5} + 1 \times 2^{-3} + 1 \times 2^{-2} + 1 \times 2^{0}$

   [C] $1 \times 2^{5} + 1 \times 2^{3} + 1 \times 2^{2} + 1 \times 2^{0}$

   [D] $1 \times 2^{6} + 1 \times 2^{4} + 1 \times 2^{3} + 1 \times 2^{1}$

9. As a number in base ten 321_{five} is equal to:

   [A] 85_{ten} [B] 86_{ten}

   [C] 32_{ten} [D] 43_{ten}

10. The value of $3310_{five} - 1442_{five}$ is:

    [A] 2131_{five} [B] 1313_{five}

    [C] 1103_{five} [D] 4302_{five}

11. If $540_{seven} - 253_{seven} = x_{seven}$, x must be:

    [A] 457 [B] 475 [C] 254 [D] 284

12. The base of the addition

    $324 + 135 = 503$ is:

    [A] 3   [B] 4   [C] 5   [D] 6

13. Evaluating $2002_{three} - 202_{three}$ gives:

    [A] 100_{three} [B] 101_{three}

    [C] 1100_{three} [D] 1010_{three}

14. 42_{five} is equivalent to:

    [A] 21_{ten} [B] 10101_{two}

    [C] 112_{four} [D] 212_{three}

15. 24_{five} is equivalent to:

    [A] 40_{three} [B] 112_{three}

    [C] 11000_{two} [D] 120_{ten}

16. The possible binary number in the following is:

    [A] 112 [B] 101 [C] 102 [D] 211

17. The sum of 1111_{two} and 111_{two} is:

    [A] 101100_{two} [B] 101101_{two}

    [C] 10110_{two} [D] 10011_{two}

18. Given that 1101, 11011 and 11 are binary numbers, their sum will be:

    [A] 11001_{two} [B] 111001_{two}

    [C] 10110_{two} [D] 101011_{two}

19. The difference between 1110_{two} and 101_{two} is:
   [A] 1010_{two} [B] 1001_{two}
   [C] 101_{two} [D] 1011_{two}
20. When 1101_{two} is subtracted from 110011_{two}, the result is:
   [A] 10110_{two} [B] 111000_{two}
   [C] 100110_{two} [D] 110011_{two}
21. The square of 101_{two} is:
   [A] 1010_{two} [B] 1111_{two}
   [C] 1011_{two} [D] 11001_{two}
22. $111_{four}, 22_{eight}, 11011_{two}$ in ascending order of size is:
   [A] $11_{four}, 22_{eight}, 11011_{two}$
   [B] $22_{eight}, 111_{four}, 11011_{two}$
   [C] $11011_{two}, 22_{eight}, 111_{four}$
   [D] $11011_{two}, 111_{four}, 22_{eight}$
23. When evaluated, $101_{two} \times 11_{two}$ equals:
   [A] 1111_{two} [B] 1110_{two}
   [C] 1101_{two} [D] 1011_{two}
24. The average of 1011_{two} and 111_{two} is:
   [A] 1000_{two} [B] 10010_{two}
   [C] 1001_{two} [D] 10001_{two}
25. When 101_{two} and 10_{two} are multiplied the answer is:
   [A] 1001_{two} [B] 1010_{two}
   [C] 1011_{two} [D] 1110_{two}
26. $1001_{two} \times 101_{two}$ is equal to:
   [A] 101011_{two} [B] 101101_{two}
   [C] 110110_{two} [D] 110101_{two}
27. On dividing 1010_{two} by 101_{two}, the result is:
   [A] 101_{two} [B] 11_{two}
   [C] 10_{two} [D] 100_{two}
28. In base ten 11.1011_{two} is the same as:
   [A] $\frac{59}{4}$ [B] $\frac{59}{8}$
   [C] $\frac{59}{16}$ [D] $\frac{59}{32}$
29. When converted to base ten, 10.1001_{two} becomes:
   [A] $2\frac{9}{16}$ [B] $2\frac{5}{16}$ [C] $2\frac{1}{16}$ [D] $2\frac{3}{16}$
30. Two numbers 24_x and 36_y are equal in value when converted to base ten. The true equation under this condition is:
   [A] $2x - 3y = 2$ [B] $3y - 2x = 2$
   [C] $3y = x + 2$ [D] $3y = 2 - x$
31. If $104_x = 68$. The value of x is:
   [A] 5   [B] 7   [C] 8   [D] 9
32. Given that $4P4_{five} = 119_{ten}$, the value of P, which makes this equation true, is:
   [A] 0   [B] 1   [C] 2   [D] 3
33. If $M5_{ten} = 1001011_{two}$, the value of M is:
   [A] 5   [B] 6   [C] 7   [D] 8
34. Given that x is a denary number and $x = 111101_{two}$, the value of x in base ten is:
   [A] 29   [B] 61   [C] 62   [D] 63

TOPIC 19

COORDINATE GEOMETRY

OBJECTIVES

At the end of this topic, the learner should be able to:

1. Draw and label the Cartesian plane.
2. Use a scale to graduate and calibrate the axes.
3. Plot and read points on the Cartesian plane.
4. Calculate the distance between two given points.
5. Find the coordinate of a point that divides a line segment in a given ratio internally or externally.
6. Find the coordinates of the midpoint of a line segment
7. Find the gradient of a straight line.
8. Find the equation of a straight line.
9. Recognise the equation of a straight line in the gradient and one point form, the gradient form, the double intercept form and the two point form.
10. State and use the condition for two straight lines to be parallel.
11. State and use the condition for two straight lines to be perpendicular.

René Descartes

The first modern philosopher, René Descartes believed science and mathematics could explain and predict events in the physical world. Descartes developed the Cartesian coordinate system for graphing equations and geometric shapes. Modern maps use a grid system that can be traced back to Cartesian graphing techniques.

Hulton Deutch

The Concept of Coordinates

Figure 19:1 is a plan of a classroom. Each student has a separate table. The tables are arranged into columns and rows. The columns and rows are numbered, so that it is possible to identify a student by his column number and row number.

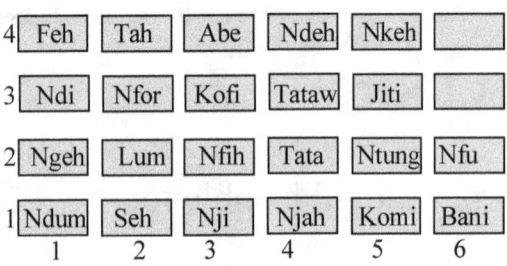

Figure 19:1

Identify the student who sits on
(a) Column 2, row 3   (b) Column 6, row 2
(c) Column 4, row 1   (d) Column 3, row 4.

Instead of writing column 3, row 2; (3,2) in that order can be written, to stand for column 3, row 2, making it a rule to write the column number first and row number second in the pair of numbers. Following this rule, (3, 4) and (4,3) do not mean the same. The order in which the numbers are written is therefore very important. A pair of numbers written in this way is called an **ordered pair**, because they are in pairs and their order is very important. The notation (a, b) for ordered pairs should not be confused with the open interval notation, which is written in the same way.

EXERCISE 19:1

1. The ordered pairs stand for column number and row number in that order. With reference to Figure 19:1, write down the meaning of the following.
    (a) (4, 3)   (b) (4, 1)
    (c) (2, 2)   (d) (3, 4)
    (e) (1, 4)   (f) (4, 2)
2. Say whether or not (3, 4) and (4, 3) mean the same giving reasons for your answers.

Plotting of Points

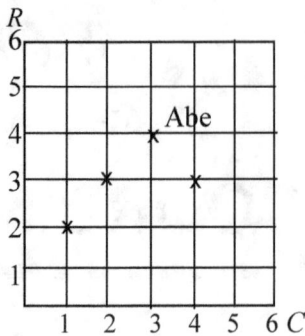

Figure 19:2

Figure 19:1 can be simplified as shown in Figure 19:2, C standing for column and R for row. To represent the position of a student, a cross ×, is marked at the point where his column number intersects his row number. The position of Abe has been done as an example.

EXERCISE 19:2

1. In Figure 19:2, write down the ordered pairs representing the three unlabelled points and label against each the names of the students who sit there.
2. Draw another diagram and mark the points where Ndum, Tata, Feh, Bani sit, using the first letters of their names.
3. Figure 19:3 is a map of West Cameroon. Using the grid lines on the Map, write down the ordered pairs representing the following towns: Bamenda, Limbe, Mbengwi, Mutengene, Kumba, Fundong, Nkambe, Buea, Nguti, Eyumojock. The first number is that of the vertical grid line while the second is that of the horizontal grid line. For instance the ordered pair representing Akwaya is (8, 22).
4. Using Figure 19:3, state the town represented by the following ordered pairs. (12, 26), (15, 22), (10, 16), (6, 8), (17, 22).

Real Life Applications of Coordinates

The idea of coordinates is very useful in real life. In plantations crops are usually planted in rows and columns. Surveyors use it to plan towns and architects use it to plan houses. Games such as draught and chess are designed based on this principle. The idea is used in locating places on the earth's surface.

Map of West Cameroon

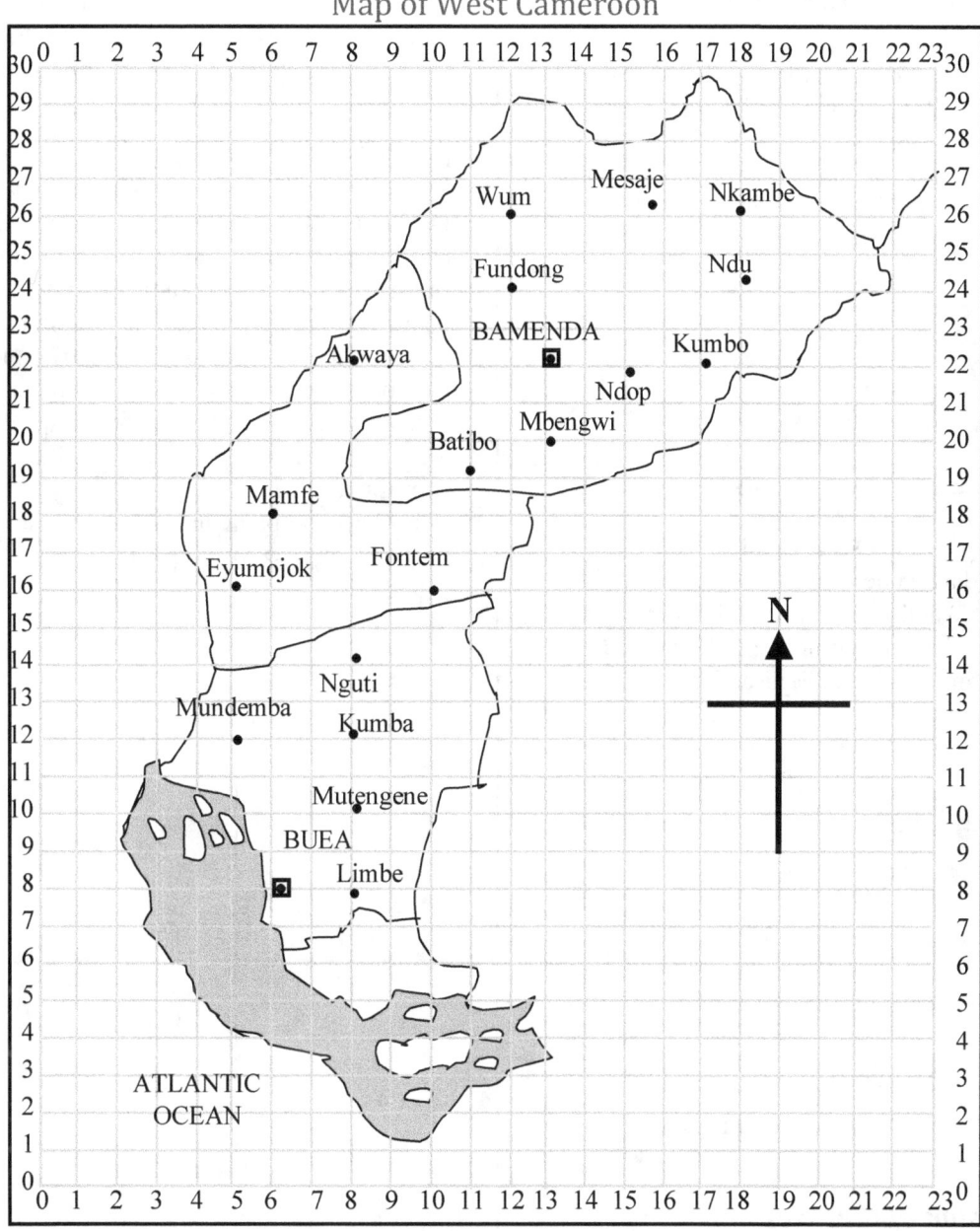

Figure 19:3

The Cartesian (or Coordinate) Plane

A diagram such as Figure 19:2 is called a **Cartesian plane** or **coordinate plane,** or the **x-y plane,** invented by the French Mathematician René Descartes (1596-1650). Generally the vertical bold line labelled R is labelled y, and called the **y-axis** while the horizontal bold line labeled C is labeled x, and called the **x-axis**. The x- and the y- axes meet at a point O, called the **origin**.

In this particular case, the first number in each ordered pair called the **x-coordinate** or **abscissa** is always taken from the x-axis, while the second called the **y-coordinate** or **ordinate** is taken from the y-axis. The ordered pairs are called **coordinates**. Notice that these axes are **perpendicular** to each other (i.e. they meet at 90^0 to each other). Another name for perpendicular is **orthogonal**.

Note that the numbers written against the x-axis refer to the vertical lines, while those written against the y-axis refer to the

183

horizontal lines. Do not be confused with these lines!

Figure 19:4 shows the three points (2, 1), (4, 2) and (2, 6) plotted and labeled.

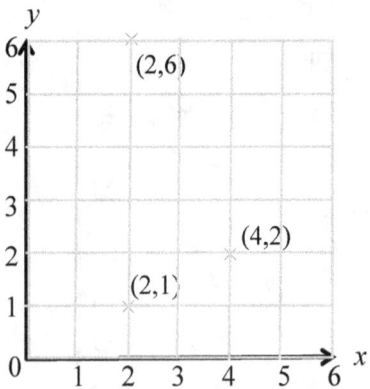

Figure 19:4

It should again be emphasized that the points (2,1),(4,2) and (2,6) in Figure 19:4 are different from the points (1,2), (2,4) and (6,2), plotted and labeled in Figure 19:5.

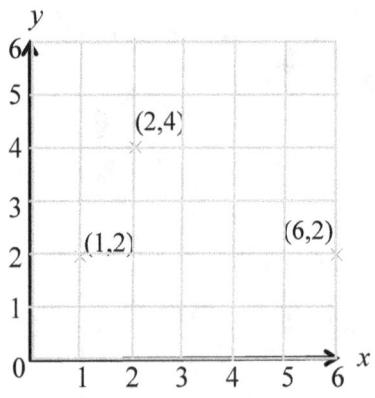

Figure 19:5

Extension of the Axes

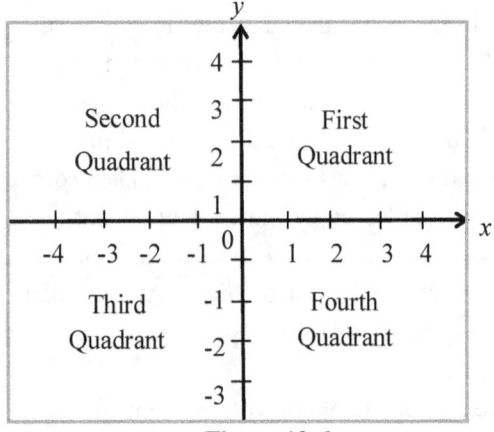

Figure 19:6

The *x*- and *y*- axes can be thought of as vertical and horizontal number lines drawn to intersect at their origins. Each of these axes can be extended in both directions as long as possible as shown in Figure 19:6.

The *x*-axis and the *y*-axis divide the *x*-*y* plane into four sections called the first, second, third and fourth quadrants, as shown in Figure 19:6. In the first quadrant, both the *x*- and *y*-coordinates are positive, in the second quadrant the *x*-coordinate is negative and the *y*- coordinate is positive in the third quadrant both are negative and in the fourth quadrant the *x*- coordinate is positive while the *y*-coordinate is negative.

The Cartesian plane in Figure 19:7, shows the points (2, 5), (−3, 0), (−4, −1), and (3, −2).

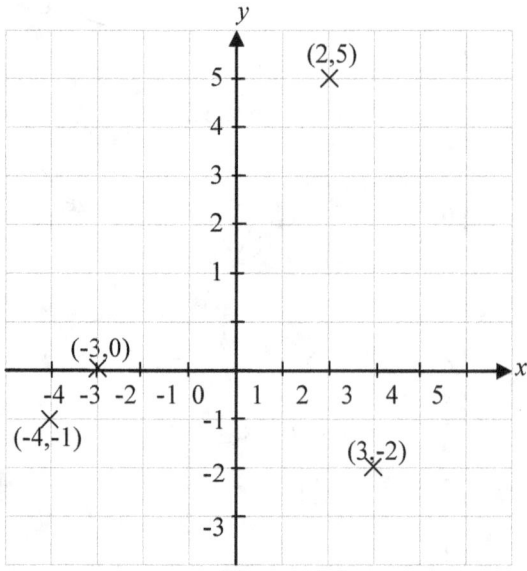

Figure 19:7

EXERCISE 19:3

1. Plot the points $A(1,4)$, $B(4,1)$, $C(3,0)$ and $D(0,3)$ on the same Cartesian plane.
2. Plot on the Cartesian plane the points $W(0,-3)$, $X(-3,0)$, $Y(1,-2)$, $Z(-2,1)$.
3. Plot the points $P(-1,-1)$, $Q(-2,-3)$, $R(-4,-1)$ and connect them with straight lines.
4. Write down the coordinates of the points lettered in Figure 19:8.

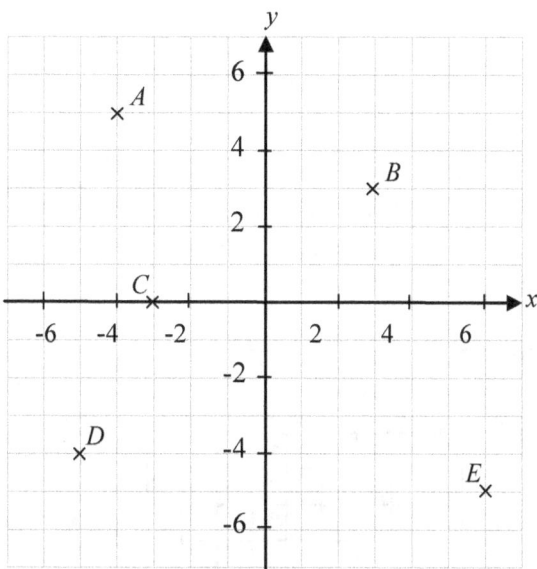

Figure 19:8

The Line Segment

A **line segment** denoted by AB is the set of points between A and B inclusively. The length of this line segment is denoted by AB or $d(AB)$.

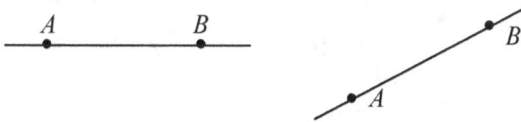

Figure 19:9

The Pythagoras Theorem

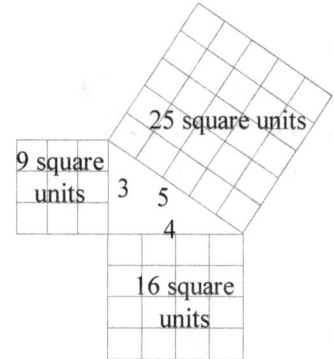

Figure 19:10

The Pythagoras theorem states that the sum of the squares on the two arms of a right-angled triangle is equal to the square on the hypotenuse.

Example 19:1

Find the length AB of the longest side of the right-angled triangle ABC with the other sides AC and BC given as 3 cm and 4 cm

respectively.

Solution

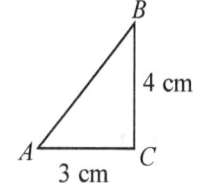

Figure 19:11

$$AB^2 = BC^2 + AC^2$$
$$= 4^2 + 3^2$$
$$AB^2 = 25$$
$$\Rightarrow AB = 5 \text{ cm}$$

EXERCISE 19:4

Find the unknown in each of the following triangles.

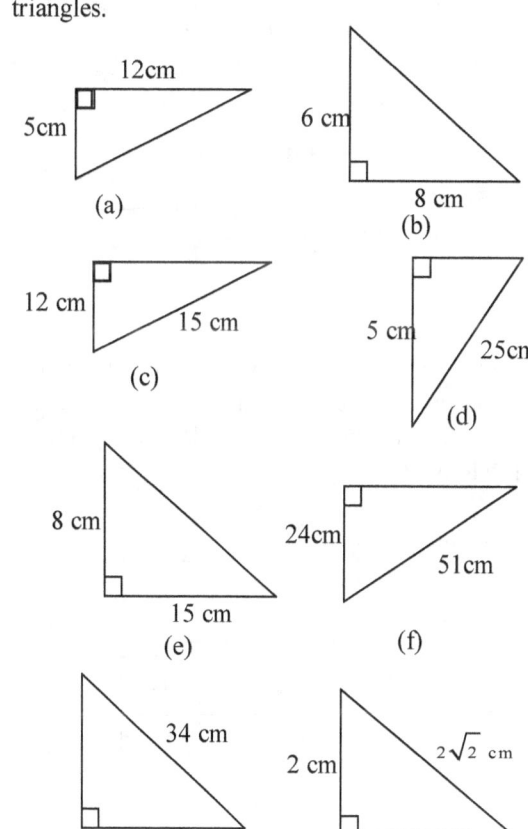

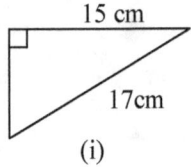

(i)

Figure 19:11

The Distance between Two Points

If $A(x_1, y_1)$ and $B(x_2, y_2)$ are any two points, the distance AB between A and B, can be found using the Pythagoras theorem.

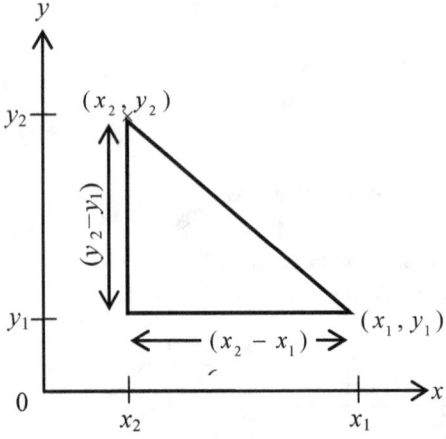

Figure 19:12

Thus,

$$AB = \sqrt{(x_2 - x_1)^2 + (y_2 - y_1)^2}$$

Example 19:2

Plot on a graph the points $A(1,0)$ and $B(5,3)$. By completing the right-angled triangle with AB as one of the sides, calculate the length of the line AB.

Solution

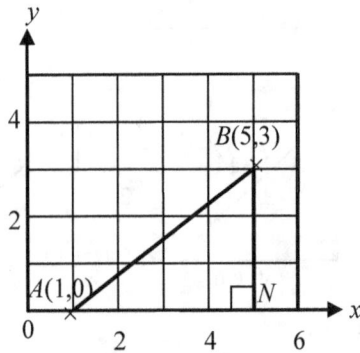

Figure 19:13

$$AB^2 = AN^2 + BN^2 = (5-1)^2 + (3-0)^2 = 25$$

$$\Rightarrow AB = \sqrt{25} = 5 \text{ units}$$

Example 19:3

On a Cartesian plane, plot the points $Q(8,6)$ and $R(16,12)$ and calculate the distance between them.

Solution

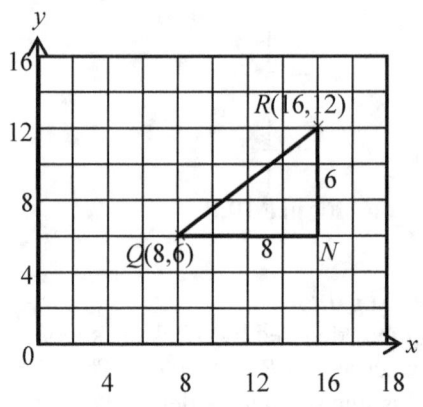

Figure 19:14

$$QR^2 = QN^2 + RN^2$$
$$= (16-8)^2 + (12-6)^2$$
$$QR^2 = 8^2 + 6^2$$
$$= 100$$
$$QR = \sqrt{100}$$
$$QR = 10 \text{ units}$$

EXERCISE 19:5

1. Find The distance between the following pairs of points;
   (a) $(-2, 3)$ and $(-1, 4)$
   (b) $(1, 2)$ and $(6, 14)$
   (c) $(3, -4)$ and $(0, 0)$
   (d) $(4, -2)$ and $(-10, 1)$
   (e) $(-12, -15)$ and $(-14, -15)$
2. Triangle ABC has vertices $A(-5,5)$, $B(7,0)$ and $C(12,5)$. Calculate the length of each side.
3. Given that the triangle with vertices $(0,6)$, $(k,-k)$ and $(-6,0)$ is equilateral. Find the value of k.
4. The point $A(x, y)$ is equidistant from the point $B(3,3)$ and $C(7,5)$. Find an equation, which connects x and y.

5. Calculate the distance between the points $A(at, a)$ and $(-at, at^2)$.

6. The point (x, y) is 4 units from the point $(3,4)$. Find an equation, which connects x and y.

7. The points $A(1,2)$, $B(5,-6)$ and $C(k, k)$ are such that angle ABC is a right angle. Find the value of k.

8. Given that the triangle with vertices $(5,7)$, $(14, -5)$ and $(x, -5)$ is equilateral. Find the possible values of x.

The Midpoint of a Line

Consider the points $A(x_1,y_1)$ and $B(x_2,y_2)$ in Figure 19:15. If $M(x, y)$ is the midpoint of the line segment AB then M divides AB in the ratio 1: 1.

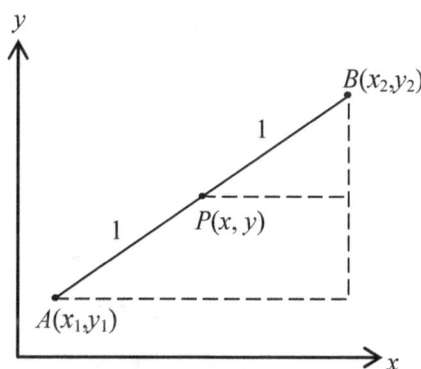

Figure 19:15

Then using similar triangles

$$\frac{y_2 - y}{y_2 - y_1} = \frac{1}{2} \Leftrightarrow y = \frac{y_2 + y_1}{2}$$

Similarly;

$$\frac{x_2 - x}{x_2 - x_1} = \frac{1}{2} \Leftrightarrow x = \frac{x_2 + x_1}{2}$$

Therefore, if $A(x_1,y_1)$ and $B(x_2,y_2)$ are any two points, the midpoint $M(x, y)$ of the line segment is given by

$$M(x, y) = \left(\frac{x_1 + x_2}{2}, \frac{y_1 + y_2}{2} \right)$$

Example 19: 6
Find the mid-point between $A(8,2)$ and $B(-2,-6)$.

Solution

$$M(x, y) = \left(\frac{x_1 + x_2}{2}, \frac{y_1 + y_2}{2} \right)$$

$$= \left(\frac{8 + (-2)}{2}, \frac{2 + (-6)}{2} \right)$$

$$= (3, -2)$$

EXERCISE 19:6

1. In the following, find the coordinates of the midpoint of the line segment AB
   (a) $A(-2, 7)$ and $B(-2, -11)$
   (b) $A(-5, -4)$ and $B(-10, -4)$

2. Write down formulae, which can be used to find the midpoint of
   (a) A vertical line segment
   (b) A horizontal line segment

3. Find the coordinates of the midpoints of the line segments with the following end points
   (a) $(6, 0)$ and $(10, 2)$
   (b) $(-5, 6)$ and $(6, -5)$
   (c) $(0, -7)$ and $(-7, 0)$
   (d) $\left(\frac{3}{4}, -\frac{5}{3} \right)$ and $\left(\frac{3}{4}, \frac{2}{3} \right)$
   (e) $\left(\sqrt{2}, -\sqrt{3} \right)$ and $\left(\sqrt{8}, \sqrt{75} \right)$

4. Given the points, $P(7,13)$, $Q(10,5)$ and $R(-6,-3)$ and that T is the midpoint of QR, calculate the distance PT.

Internal and External Division of a Line Segment

Consider the line segment AB in Figure 19:16. P divides AB internally while Q divides AB. Externally

Figure 19:16

Internal division

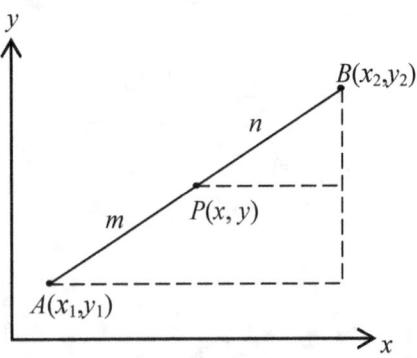

Figure 19:17

Suppose $P(x, y)$ divides AB internally in the ratio $m{:}n$, where A is the point (x_1, y_1) and B is the point (x_2, y_2) as shown in Figure 19:17.

Then using similar triangles

$$\frac{y_2 - y}{y_2 - y_1} = \frac{n}{m + n} \Leftrightarrow y = \frac{my_2 + ny_1}{m + n}$$

Similarly;

$$\frac{x_2 - x}{x_2 - x_1} = \frac{n}{m + n} \Leftrightarrow x = \frac{mx_2 + nx_1}{m + n}$$

Therefore,

$$P(x, y) = \left(\frac{mx_2 + nx_1}{m + n}, \frac{my_2 + ny_1}{m + n} \right)$$

External division

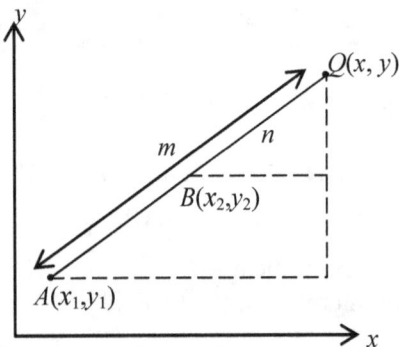

Figure 19:18

Suppose $Q(x, y)$ divides $A(x_1, y_1)$ and $B(x_2, y_2)$ externally in the ratio $m{:}n$, as shown in Figure 19:18.

Again using similar triangles,

$$\frac{y - y_2}{y - y_1} = \frac{n}{m} \Leftrightarrow y = \frac{my_2 - ny_1}{m - n}$$

Similarly,

$$\frac{x - x_2}{x - x_1} = \frac{n}{m} \Leftrightarrow y = \frac{mx_2 - nx_1}{m - n}$$

Therefore,

$$P(x, y) = \left(\frac{mx_2 - nx_1}{m - n}, \frac{my_2 - ny_1}{m - n} \right)$$

Example 19: 4

Find the coordinates of the point which divides the line joining the points $A(-2, 5)$ and $B(4, 2)$ in the ratio 2:1.
(a) internally   (b) externally.

Solution

Let the coordinate of the point that divides AB be (x, y). Then,

(a) Internally $(x, y) = \left(\dfrac{mx_2 + nx_1}{m + n}, \dfrac{my_2 + ny_1}{m + n} \right)$

$$= \left(\frac{2(4) + 1(-2)}{2 + 1}, \frac{2(2) + 1(5)}{2 + 1} \right)$$

$$\Rightarrow (x, y) = (2, 3)$$

(b) Externally $(x, y) = \left(\dfrac{mx_2 - nx_1}{m - n}, \dfrac{my_2 - ny_1}{m - n} \right)$

$$= \left(\frac{2(4) - 1(-2)}{2 - 1}, \frac{2(2) - 1(5)}{2 - 1} \right)$$

$$\Rightarrow (x, y) = (10, -1)$$

A closer look at the formulae reveals that,

(a) For internal division, formula ① can be used directly.
(b) For external division, the ratio can be written as $m{:} -n$ and used in formula ① directly.

Example 19: 5

The point $P(x, y)$ divides the line AB in the ratio 3:2 where A is the point $(-1, 6)$ and B is the point $(3, -2)$.
(a) internally   (b) externally.
Find the coordinates of the point P in each case.

Solution

$$\text{Internally } (x, y) = \left(\frac{mx_2 + nx_1}{m + n}, \frac{my_2 + ny_1}{m + n} \right)$$

(a) For internal division, the ratio is

$$m : n = 3 : 2$$

$$\Rightarrow P(x, y) = \left(\frac{3(3) + 2(-1)}{3 + 2}, \frac{3(-2) + 2(6)}{3 + 2} \right)$$

$$= \left(\frac{7}{5}, \frac{6}{5} \right)$$

(b) For internal division, the ratio is

$$m : n = 3 : -2$$

$$\Rightarrow P(x, y) = \left(\frac{3(3) + (-2)(-1)}{3 + (-2)}, \frac{3(-2) + (-2)(6)}{3 + (-2)} \right)$$

$$= (11, -18)$$

EXERCISE 19:7

Find the coordinates of the point, which divides AB in the given ratio in each of the following cases.
1. $A(2, 4)$, $B(-3, 9)$; 1:4 internally
2. $A(-3, -4)$, $B(3, 5)$; 3:1 externally
3. $A(1, 5)$, $B(8, -2)$; 4:3
4. $A(5, 2)$, $B(-2, 8)$; 3: -2

Gradient

Some slopes are steeper than others. The measure of the steepness of a slope is called the **gradient** of the slope. The gradient of a slope compares the vertical distance (rise) and the horizontal distance (run) when one ascends a slope. Gradient is denoted by m.

$$m = \frac{\text{rise}}{\text{run}}$$

$$\text{or } m = \frac{\text{vertical distance}}{\text{horizontal distance}}$$

$$m = \frac{v}{h}$$

Example 19:7

A wall is 6 m tall and a ladder leans on the wall, with the top of the ladder touching the top of the wall. If the distance from the wall to the base of the ladder is 2 m, calculate the gradient of the ladder.

Solution

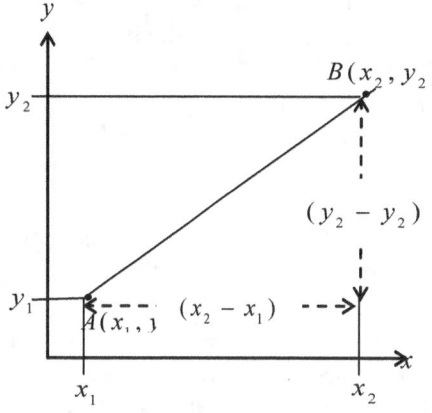

$$m = \frac{v}{h} = \frac{6}{2} = 3$$

Figure 19:19

The Gradient of a Straight line

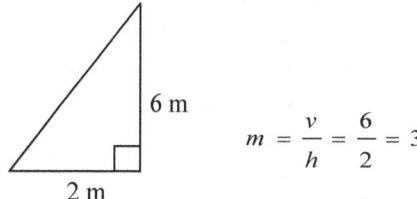

Figure 19:20

Consider the straight line passing through the points and as shown in Figure 19:20.
Then from the definition of gradient,

$$m = \frac{v}{h} \Rightarrow m = \frac{y_2 - y_1}{x_2 - x_1}$$

Example 19:8
Find the gradient of the line joining the points $P(4,11)$ and $Q(7,2)$.

Solution

$$m = \frac{y_2 - y_1}{x_2 - x_1} = \frac{2 - 11}{7 - 4} = \frac{-9}{3} = -3$$

Example 19:9
Calculate the steepness of the line joining the points $A(2,1)$ and $B(-5,3)$.

Solution

$$m = \frac{y_2 - y_1}{x_2 - x_1} = \frac{1 - 3}{2 - (-5)} = -\frac{2}{7}$$

1. Find the gradient of the line which passes through each of the following pair of points
   (a) $(-4, 2)$ and $(3, 5)$
   (b) $(-4, 7)$ and $(-4, 6)$
   (c) $(3, 3)$ and $(-3, -3)$
   (d) $(-3, -3)$ and $(5, 3)$
   (e) $\left(\dfrac{1}{2}, \dfrac{1}{3}\right)$ and $\left(\dfrac{1}{4}, \dfrac{1}{5}\right)$
   (f) $(0, 0)$ and $(-4, -3)$

2. Given that the gradient of the line which passes through the points $(7, -6)$ and $(k, 2)$ is 2. Find the value of k?

3. Given that the gradient of the line which passes through the points $(k, 5)$ and $(0, -11)$ is k. Find the value of k.

4. Plot the points $P(1, 3)$, $Q(3, 7)$ and $R(-2, -3)$ and connect these points with a straight line.
   (a) Make a remark concerning these points.
   (b) Where does the line cut the y-axis?
   (c) Where does the line cut the x-axis?
   (d) Calculate the gradient of the line using the points.
      (i) P and Q (ii) P and R (iii) Q and R
   (e) What conclusion do you draw and what does this suggest?

Intercepts

The y-coordinate of the point where a line cuts the y-axis is called the **intercept with the y-axis** and at that point $x = 0$. The x-coordinate of the point where a line cuts the x-axis is called the intercept with the x-axis and at that point $y = 0$.
From problem 4, it is clear that the gradient of a straight line can be calculated using the coordinates of any, two points that lie on the line.

The Equation of a Straight Line

The various forms of the equation of a straight line are; the **gradient and one point form**, the **gradient–intercept form**, the **double intercept form**, the **two-point form** and the **general form**.

Gradient and one point form

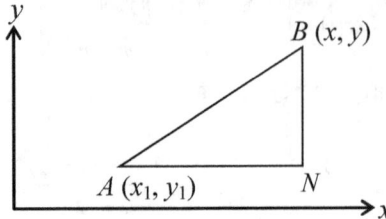

Figure 19:21

Consider the straight line whose gradient is m, and which passes through the point $A(x_1, y_1)$ and any other point $B(x, y)$ as shown in Figure 19:21.

Therefore, from Figure 19:21 and by the definition of gradient, the gradient of the line is given by

$$m = \frac{y - y_1}{x - x_1} \quad \text{.....................①}$$

Rearranging equation ①,

$$y - y_1 = m(x - x_1) \quad \text{.....................②}$$

This form is known as **Gradient and one point form** of the equation of a Straight Line. This form of the equation of a straight line is useful when, the gradient of the line and one point on the line are known.

Example 19:10
Find the equation of the line which passes through the point $(2, -5)$ and whose gradient is -2.

Solution
$$y - y_1 = m(x - x_1)$$
$$x_1 = 2, \ y_1 = -5 \text{ and } m = -2$$
$$y - (-5) = m(x - 2)$$
$$y = -2x - 1$$

Gradient –Intercept form

Consider the straight line whose gradient is m, and which cuts the y-axis at the point $A(0, c)$ and passes through any other point $B(x, y)$ as shown in Figure 19:22.

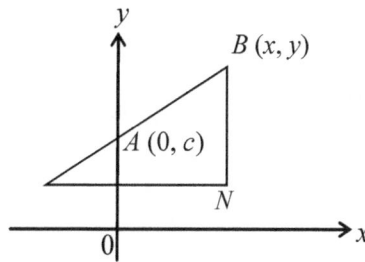

Figure 19:22

Then, by the definition of gradient, the gradient of the line is given by

$$\frac{y - c}{x - 0} = m$$

This can be rearranged, as $y = mx + c$

$$\Rightarrow \quad y = mx + c \quad \dots\dots\dots\dots\dots ③$$

Equation ③ is known as the **Gradient-intercept form** of the equation of a Straight Line and is useful when the gradient of the line and the intercept with the y-axis are known.

For this form, the gradient is m and the intercepts can be obtained by substituting

$x = 0$ or $y = 0$ in the equation.

Thus, when $x = 0$, $y = c$ and when $y = 0$,

$$x = -\frac{c}{m}.$$

Example 19:11

A line whose gradient is $-\dfrac{3}{4}$, is known to cut the y-axis at the point $(0, 5)$. Find the equation of the line.

Solution

$y = mx + c$, $\quad m = -\dfrac{3}{4}$ and $c = 5$

$$\Rightarrow \quad y = -\frac{3}{4}x + 5$$

The Double Intercept Form

Consider the straight line which cuts the x- and y-axes at the points $(a, 0)$ and $(0, b)$ respectively and passes through any other point $B\ (x, y)$ as shown in Figure 19:23.

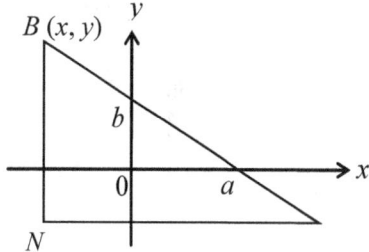

Figure 19:23

Using the points $(a, 0)$, $(0, b)$ and $B\ (x, y)$ and the definition of the gradient of a straight line,

$$m = \frac{b - 0}{0 - a}$$

$$\Rightarrow \quad m = -\frac{b}{a} \dots\dots\dots\dots ④$$

Similarly using the points $(a, 0)$ and $B\ (x, y)$ and the definition of the gradient of a straight line,

$$m = \frac{y - 0}{x - a}$$

$$\Rightarrow \quad m = \frac{y}{x - a} \dots\dots\dots\dots ⑤$$

Equating ④ and ⑤

$$\frac{y}{x - a} = -\frac{b}{a}$$

Cross multiplying,

$$-ay = bx - ab$$

$$\Rightarrow \quad -bx - ay = -ab$$

Dividing both sides by $-ab$

$$\frac{x}{a} + \frac{y}{b} = 1 \quad \dots\dots\dots ⑥$$

This form of the equation of a straight line is known as the double intercept form and is useful when the intercepts with the x- and y-axes are known.

Equation ⑥ is known as the double intercept form because it makes use of the x and y intercepts.

For this form, the gradient is $-\dfrac{b}{a}$ and the intercepts can be obtained by substituting $x = 0$ and $y = 0$ in the equation. Hence, When $x = 0$, $y = b$ and when $y = 0$, $x = a$.

Two point form

Consider the straight line that passes through the points $A\ (x_1, y_1)$ and $B\ (x_2, y_2)$ respectively

191

and passes through any other point $B(x, y)$

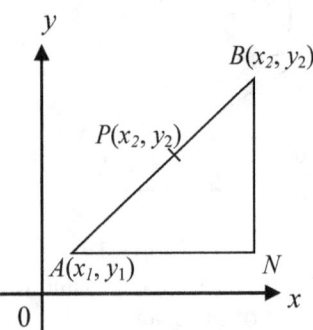

Figure 19:24

Then using the points (x_1, y_1) and (x_2, y_2) and the definition of the gradient of a straight line,

$$m = \frac{y_2 - y_1}{x_1 - x_1} \dots\dots\dots\dots\dots\dots\dots⑦$$

Similarly using the points (x_1, y_1) and $B(x, y)$ and the definition of the gradient of a straight line,

$$m = \frac{y - y_1}{x - x_1} \dots\dots\dots\dots\dots\dots\dots⑧$$

Equating ⑦ and ⑧

$$\frac{y - y_1}{x - x_1} = \frac{y_2 - y_1}{x_2 - x_1} \dots\dots\dots\dots\dots\dots⑨$$

This form of the equation of a straight line is known as the two-point form and is useful when two points that lie on the straight line are known.

General form

Cross-multiplying equation ⑨ becomes,

$$(y - y_1)(x_2 - x_1) = (y_2 - y_1)(x - x_1)$$

$$yx_2 - yx_1 - y_1x_2 + y_1x_1 = y_2x - y_2x_1 - y_1x + y_1x_1$$

$$(y_1 - y_2)x + (x_2 - x_1)y + (x_1y_2 - x_2y_1) = 0$$

Substituting a, b and k for the constants $x_2 - x_1$, $y_1 - y_2$ and $x_1y_2 - x_2y_1$ leads to

$$ax + by + c = 0 \dots\dots\dots⑩$$

Equation ⑩ is known as the general form of the equation of a straight line. For this form, the gradient is $-\dfrac{a}{b}$ and the intercepts can be

obtained by substituting $x = 0$ and $y = 0$ in the equation.

Thus, when $x = 0$, $y = -\dfrac{c}{b}$, and when $y = 0$,

$$x = -\frac{c}{a}.$$

Example 19:12

Given the line $\dfrac{x}{2} + \dfrac{y}{5} = 1$, find

(i) the intercept with the x-axis
(ii) the intercept with the y-axis
(iii) The gradient of the line.

Solution
(i) Intercept with the x-axis is 2, when $y = 0$.
(ii) Intercept with the y-axis is 5, when $x = 0$.
(iii) $m = -\dfrac{5}{2}$

The Table 19:1 is a summary of the four forms of the equation of a straight line.

Form	Gradient	Intercept	
		x-axis	y-axis
Gradient/ Intercept Form $y = mx + c$	m	c	c
General Form $ax + bx + k = 0$	$-\dfrac{a}{b}$	$-\dfrac{k}{a}$	$-\dfrac{k}{b}$
Double/Intercept Form $\dfrac{x}{a} + \dfrac{y}{b} = 1$	$-\dfrac{b}{a}$	a	b
Gradient/ one point form $y - y_1 = m(x - x_1)$	m	$\dfrac{mx_1 - y_1}{m}$	$y_1 - mx$

Table 19:1

Parallel and Perpendicular Lines

Consider the lines l_1 and l_2 whose gradients are m_1 and m_2 respectively.
(a) The lines are parallel if and only if their gradients are equal.

$$l_1 \parallel l_2 \Leftrightarrow m_1 = m_2$$

(b) The lines are perpendicular if and only if the product of their gradients is -1.

$$l_1 \perp l_2 \Leftrightarrow m_1 m_2 = -1$$

Example 19:13

Find the equation of the straight line which passes through the point $(-2, 5)$ and is perpendicular to the line $y = 2x+3$.

Solution

Let the gradient of the line $y = 2x + 3$ be $m_1 = 2$ and that of the perpendicular line be m_2.

Then $m_1 m_2 = -1 \Rightarrow 2 m_2 = -1$

$$m_2 = -\frac{1}{2}$$

Therefore, the equation of the perpendicular line is

$$y - y_1 = m_2 (x - x_1)$$

Substitute $(-2, 5)$ and $m_2 = -\dfrac{1}{2}$ into the equation.

$$y - 5 = -\frac{1}{2}\left(x - (-2)\right)$$

$$\therefore \ y - 5 = -\frac{1}{2} x - 1 \quad \text{or} \quad y = -\frac{1}{2} x + 4$$

Example 19:14

A line is known to pass through the point $(1, 3)$ and is parallel to the line $y = 3x+2$. Find the equation of the line.

Solution

Lines are parallel so $m_1 = m_2 = 3$.

Therefore, the equation of the parallel line is

$$y - y_1 = m_2 (x - x_1)$$

Substitute $(1, 3)$ and $m_2 = 3$ into the equation.

$$y - 3 = 3 (x - 1)$$

$$y - 3 = 3x - 3$$

$$\therefore \ y = 3x$$

> **EXERCISE 19:9**

1. Find the equation of the line, which passes through the following pair of points.
   (a) $(4, 1)$ and $(-2, 5)$
   (b) $(-6, -7)$ and $(7, 8)$
   (c) $\left(-1, \dfrac{3}{4}\right)$ and $\left(-\dfrac{1}{2}, 4\right)$

   (d) $(2, 7)$ and $(-8, 5)$
   (e) $\left(\sqrt{2}, \sqrt{8}\right)$ and $\left(-\sqrt{8}, -\sqrt{2}\right)$

2. Find the equation of the line, which passes through the point $(6, -1)$ and is perpendicular to the line $y - 3x - 1 = 0$.

3. Find the equation of the line, which passes through the point $(3, 1)$ and is parallel to the line $2x + 5y - 4 - 0$.

4. Find the equation of the straight line, which passes through the point $(2, -1)$ and
   (a) is perpendicular to the line $3x + y = 7$.
   (b) is parallel to the line $3x + y = 7$.

5. A line l_1 passes through the points $(1, 8)$ and $(-2, 1)$. Find the equation of the line that passes through the point $(2, 1)$ and is
   (a) perpendicular to l_1.
   (b) parallel to l_1.

6. A straight line has gradient $-\dfrac{12}{5}$ and passes through the point $(2, 1)$. Find its equation.

7. Given the points $X(-2, 7)$, $Y(6, -1)$ and $Z(9, 4)$ and that P is the midpoint of the line segment $[XY]$. Calculate:
   (a) the coordinates of P
   (b) the numerical value of $XY : PZ$.
   (c) the equation of the line YZ.

8. A straight line whose gradient is -4 passes through the point $(2, 3)$. Find the equation of the line.

9. Given that the line which passes through the points $(b, 3)$ and $(-2, 1)$ is parallel to the line, which passes through the points $(5, b)$ and $(1, 0)$. Find the value(s) of b.

10. Given that the line which passes through the points $(3m, 0)$ and $(2n, 0)$ is parallel to the line which passes through the points $(0, n)$ and $(0, 6m)$. Find a relation between m and n.

11. If the line that passes through the points $(0, 0)$ and (n, m) is perpendicular to the line that passes through the points $(0, 0)$ and $(-a, b)$. Find the relation between a, b, m and n.

MULTIPLE CHOICE EXERCISE 19

1. A triangle with vertices at the points with coordinates (–4,4), (4,4), (1,–1) is:
   [A] Right–angled   [B] Equilateral
   [C] Isosceles   [D] Scalene

2. The statement which is true about the points $P(–1, –4)$, $Q(6, –5)$, $R(–1,5)$ and $S(3,2)$ is:
   [A] R and Q are in the second and third quadrants respectively.
   [B] P and S are in the fourth and third quadrants respectively.
   [C] S and R are in the first and second quadrants respectively.
   [D] P and Q are in the second and fourth quadrants respectively.

3. The coordinates of the midpoint of $P(–4,5)$ and $Q(2,1)$ are:
   [A] (–1,3)   [B](–1,–3)
   [C] (1,–3)   [D] (1,3)

4. A straight line passes through the points (5,3) and (8,4). The line:
   [A] is parallel to the x-axis
   [B] slopes from left to right
   [C] is parallel to the x-axis
   [D] slopes from right to left

5. Among the following, the straight line, which slopes from left to right, is the line, which passes through the points:
   [A] (–1,5) and (2,0) [B] (3,–5) and (3,3)
   [C] (–2,1) and (1,1)   [D] (0,0) and (3,6)

6. Among the following, the straight line, which is parallel to the x-axis, is the line that passes through the points:
   [A] (–1,5) and (2,0) [B] (3,–5) and(3,3)
   [C] (–2,1) and (1,1)   [D] (0,0) and (3,6)

7. Among the following, the straight line, which slopes from right to left, is the line, which passes through the points:
   [A] (–2,5) and (–2,3)
   [B] (3,–5) and (8,–5)
   [C] (–2,1) and (1,1)
   [D] (0,0) and (3,6)

8. The point, which lies on the line $y = 2x–5$ is:
   [A] (1,3) [B] (2,5) [C] (3,–1) [D] (3,1)

9. The gradient of the line $2x+3y = 12$ is:
   [A] $-\dfrac{2}{3}$ [B] $-\dfrac{3}{2}$
   [C] $\dfrac{2}{3}$ [D] $\dfrac{3}{2}$

10. The intercept of the line $2x+3y = 12$ with the x-axis is:
    [A] 2   [B] 3   [C] –6   [D] 6

11. The intercept of the line $2x+3y = 12$ with the y–axis is:
    [A] 2   [B] 3   [C] 4   [D] –4

12. The area of the triangle OAB, where O is the origin and A and B are the points where the line cuts the x- and y-axes respectively is:
    [A] 12 un^2 [B] 24 un^2
    [C] 6 un^2 [D] 18 un^2

13. The gradient of the line $2x+y = 8$ is:
    [A] 4   [B] 2   [C] –2   [D] –8

14. Given that the lines $2x+y = 8$ and $6y–mx = 3$ are parallel. The value of m is:
    [A] 2   [B] –2   [C] $\dfrac{1}{2}$ [D] $-\dfrac{1}{2}$

15. Given that the lines $2x+y = 8$ and $6y–mx = 3$ are perpendicular. The value of m is:
    [A] 2   [B] –2   [C] $\dfrac{1}{2}$ [D] $-\dfrac{1}{2}$

16. In the form $y = mx + c$, the equation of the line $2x–3y+5 = 0$ is:
    [A] $y = -\dfrac{2}{3}x + \dfrac{5}{3}$ [B] $y = \dfrac{2}{3}x - \dfrac{5}{3}$
    [C] $y = \dfrac{2}{3}x + \dfrac{5}{3}$ [D] $y = -\dfrac{2}{3}x - \dfrac{5}{3}$

17. The gradient of the line $2x–3y+5 = 0$ is:
    [A] $-\dfrac{5}{3}$ [B] $\dfrac{5}{3}$ [C] $-\dfrac{2}{3}$ [D] $\dfrac{2}{3}$

18. The intercept of the line $2x–3y + 5 = 0$ with the x–axis is:
    [A] $-\dfrac{5}{2}$ [B] $\dfrac{5}{2}$ [C] $-\dfrac{5}{3}$ [D] $\dfrac{5}{3}$

19. The coordinates of the midpoint of (3,2) and (–1,0) are:
    [A] (–1,–1)   [B] (1,1)
    [C] (–1,1)   [D](1,–1)

20. The gradient of the straight line which passes through the points (–1,0) and (0,–2) is:
    [A] –2   [B] 2   [C] $-\dfrac{1}{2}$ [D] $\dfrac{1}{2}$

21. The distance between the points (3,2) and (0,–2) is:
    [A] 3   [B] $\sqrt{3}$ [C] 5   [D] $\sqrt{5}$

22. The coordinate of the point of intersection of the lines $x–y = 3$ and $x+2y = 6$ is:

[A] (4,1)   [B] (6,1)

[C] $\left(\dfrac{9}{4},\dfrac{3}{4}\right)$ [D] (1,4)

23. Given the straight lines

$L_1 : 2x = 3y + 5$ $L_2 : 2y = 4x + 3$

$L_3 : 2x + 4y = 5$ $L_4 : 3x - 2y = 3$

The lines, which are parallel, are:
[A] L_1 and L_4 [B] L_2 and L_4
[C] L_2 and L_3 [D] L_1 and L_4

24. Given the straight lines

$L_1 : 2x = 3y + 5$ $L_2 : 2y = 4x + 3$

$L_3 : 2x + 4y = 5$ $L_4 : 3x - 2y = 3$

The lines are perpendicular are:
[A] L_1 and L_4 [B] L_2 and L_4
[C] L_2 and L_3 [D] L_1 and L_4

25. Given the straight lines

$L_1 : 2x = 3y + 5$ $L_2 : 2y = 4x + 3$

$L_3 : 2x + 4y = 5$ $L_4 : 3x - 2y = 3$

The line, which passes through the point (−1,−3) is:
[A] L_1 [B] L_2 [C] L_3 [D] L_4

26. P is the midpoint of the line segment joining the points $A(2,-3)$ and $B(4,5)$. C is the point $(5,9)$. The equation of the line PC is:
[A] $y = -4x + 11$ [B] $y = 4x - 11$
[C] $y = 4x + 11$ [D]
$y = -4x - 11$

27. Given the straight lines

$L_1 : y - 2x = 5$ $L_2 : y = 2x + 3$

$L_3 : 4y = -2x - 6$ $L_4 : 2y = x - 5$

The lines, which are parallel, are:
[A] L_1 and L_2 [B] L_2 and L_3
[B] [C] L_1 and L_4 [D] L_3 and L_4

28. Given the straight lines

$L_1 : y - 2x = 5$ $L_2 : y = 2x + 3$

$L_3 : 4y = -2x - 6$ $L_4 : 2y = x - 5$

One of the pair of perpendicular lines are:
[A] L_1 and L_2 [B] L_2 and L_3
[C] L_1 and L_4 [D] L_3 and L_4

29. Given that,

$L_1 : y = 2x + 3$ $L_2 : 2x + y = 5$

$L_3 : 4y = 2x + 3$ $L_4 : x + \dfrac{y}{-2} = 1$

The lines, which are parallel, are:
[A] L_1 and L_4 [B] L_2 and L_4
[C] L_2 and L_3 [D] L_1 and L_4

30. Given that,

$L_1 : y = 2x + 3$ $L_2 : 2x + y = 5$

$L_3 : 4y = 2x + 3$ $L_4 : x + \dfrac{y}{-2} = 1$

The lines are perpendicular are:
[A] L_1 and L_4 [B] L_2 and L_4
[C] L_2 and L_3 [D] L_1 and L_2

31. The point of intersection of the lines $y - 2x = 5$ and $3y = 4x + 9$ is:
[A] (3,−1)   [B] (3,1)
[C] (−3,1)   [D] (−3,−1)

32. The line $3x - 5y = 15$ cuts the y-axis at:
[A] (0,5)   [B] (0,−3)
[C] (5,0)   [D] (−3,0)

33. The line $3x - 5y = 15$ cuts the x-axis at:
[A] (0,5)   [B] (0,−3)
[C] (5,0)   [D] (−3,0)

34. The gradient of the line $3x - 5y = 15$ is
[A] $-\dfrac{5}{3}$ [B] $\dfrac{5}{3}$ [C] $-\dfrac{3}{5}$ [D] $\dfrac{3}{5}$

35. The value of m for which the lines $y = 2x - 13$ and $y - mx = -3$ are parallel is:
[A] −2   [B] 2   [C] $\dfrac{1}{2}$ [D] $-\dfrac{1}{2}$

36. The value of m for which the lines $y = 2x - 13$ and $y - mx = -3$ are perpendicular is:
[A] −2   [B] 2   [C] $\dfrac{1}{2}$ [D] $-\dfrac{1}{2}$

37. Given that the lines $y = 2x - 13$ and $y - mx = -3$ are perpendicular. Their point of intersection is:
[A] (−4,−5)   [B] (4,5)
[C] (−4,5)   [D] (4,−5)

38. The gradient of the lines of the line $x + 2y = 2$ is:
[A] $-\dfrac{1}{2}$ [B] $\dfrac{1}{2}$ [C] 2   [D] −2

39. The intercept of the line $x + 2y = 2$ with the x-axis is:
[A] (0,−1)   [B] (0,1)   [C] (2,0)   [D] (−2,0)

40. The intercept of the line $x + 2y = 2$ with the y-axis is:
[A] (0,−1)   [B] (0,1)
[C] (2,0)   [D] (−2,0)

41. The area of the triangle between the line $x + 2y = 2$ and the coordinate axes is:
[A] 1 un^2 [B] 2 un^2
[C] 5 un^2 [D] $\sqrt{5}$ un^2

42. $P(3,0)$ and $Q(5,2)$ are two points on a straight line. The equation of the line PQ is:

[A] $x + y = 3$ [B] $y = x–3$
[C] $y = x+3$ [D] $y = –x –3$

43. Given the lines
    L_1: $y = 2x–4$, L_2: $2y+x–6 = 0$,

    L_3: $y = \dfrac{1}{3}x + 7$ L_4: $3y = x–5$.

    The two perpendicular lines are:
    [A] L_1 and L_3 [B] L_2 and L_4
    [C] L_3 and L_4 [D] L_1 and L_2

44. Given the lines
    L_1: $y = 2x–4$, L_2: $2y+x–6 = 0$,

    L_3: $y = \dfrac{1}{3}x + 7$ L_4: $3y = x–5$.

    The two parallel lines are:
    [A] L_1 and L_3 [B] L_2 and L_4
    [C] L_3 and L_4 [D] L_1 and L_2

45. Given that the points $P(2,2)$, $Q(3,5)$ and $R(4,a)$ are collinear, the coordinates of R are:
    [A] $(4,–8)$ [B] $(3,–7)$
    [C] $(4,7)$ [D] $(4,8)$

46. The y intercept on the line $6x+3y–7 = 0$ is:
    [A] $\dfrac{7}{6}$ [B] $–7$ [C] $\dfrac{7}{3}$ [D] $–2$

47. The x intercept on the line $6x+3y–7 = 0$ is:
    [A] $\dfrac{7}{6}$ [B] $–7$ [C] $\dfrac{7}{3}$ [D] $–2$

48. The gradient of the line $6x+3y –7 = 0$ is:
    [A] $\dfrac{7}{6}$ [B] $–7$ [C] $\dfrac{7}{3}$ [D] $–2$

49. The straight line parallel to the y-axis passes through the points:
    [A] $(–1,5)$ and $(2,0)$ [B] $(3,–5)$ and $(3,3)$
    [C] $(–2,1)$ and $(1,1)$ [D] $(0,0)$ and $(3,6)$

50. P is the midpoint of the line segment joining the point $A(2,–3)$ and $B(4,5)$. C is the point $(5,9)$. The gradient of the line PC is:
    [A] $–4$ [B] 4 [C] 11 [D] $–11$

51. P is the midpoint of the line segment joining the point $A(2,–3)$ and $B(4,5)$. C is the point $(5,9)$. The intercept of the line PC with the y-axis is:
    [A] 11 [B] $–11$ [C] $\dfrac{11}{4}$ [D] $–\dfrac{11}{4}$

52. P is the midpoint of the line segment joining the point $A(2,–3)$ and $B(4,5)$. C is the point $(5,9)$. The intercept of the line PC with the x-axis is:

[A] 11 [B] $–11$ [C] $\dfrac{11}{4}$ [D] $–\dfrac{11}{4}$

53. In Figure 19:25, the intercept of the line l_1 with the y-axis is:
    [A] $–\dfrac{5}{4}$ [B] $–\dfrac{4}{5}$ [C] 4 [D] $–5$

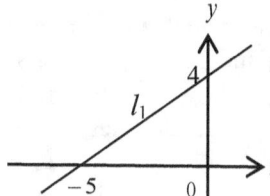

Figure 19:25

54. In Figure 19:24, the intercept of the line l_1 with the x-axis is:
    [A] $–\dfrac{5}{4}$ [B] $–\dfrac{4}{5}$ [C] 4 [D] $–5$

55. In Figure 19:24, the gradient of the line l_1 is:
    [A] $–\dfrac{5}{4}$ [B] $–\dfrac{4}{5}$ [C] $–4$ [D] $–5$

56. In Figure 19:24, the equation of the line l_1 is:
    [A] $5y = 4x – 20$ [B] $4y = 5x – 20$
    [C] $5y = –4x + 20$ [D] $4y = –5x + 20$

57. In Figure 19:26, the intercept of the line l_2 with the y-axis is:
    [A] $–\dfrac{3}{5}$ [B] $\dfrac{3}{5}$ [C] 6 [D] 10

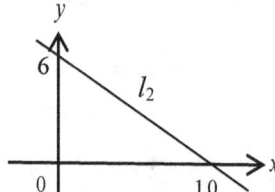

Figure 19:26

58. In Figure 19:25, the intercept of the line l_2 with the x-axis is:
    [A] $–\dfrac{3}{5}$ [B] $\dfrac{3}{5}$ [C] 6 [D] 10

59. In Figure 19:25, the gradient of the line l_2 is:
    [A] $–\dfrac{3}{5}$ [B] $\dfrac{3}{5}$ [C] 6 [D] 10

60. In Figure 19:25, the equation of the line l_2 is:
    [A] $5y = 3x + 30$ [B] $3y = 5x – 30$
    [C] $5y = 3x – 30$ [D] $3y = –5x + 30$

TOPIC 20

SET THEORY AND LANGUAGE

OBJECTIVES

At the end of this topic, the learner should be able to:

1. Appreciate the concept of sets.
2. Recognise and use special symbols related to sets properly.
3. Use membership notation to denote the idea that an element belongs to or does not belong to a given set.
4. State the cardinality of a set.
5. Define sets using the roster method, the rule definition method and the set builder notation.
6. Name the different types of sets.
7. Appreciate that the universal set is the set that contains all the elements under consideration.
8. Identify the proper and improper subsets of a given set.
9. Identify equal sets.
10. Identify equivalent sets.
11. Use set inclusion and Venn diagrams to show the relationship between sets.
12. Find the intersection of given sets.
13. Find the union of given sets.
14. Identify disjoint sets.
15. State the complement of a given set.
16. List the power set of a given set with at most three elements.
17. Determine the cardinality of a power set without listing the elements.
18. Solve logical problems on cardinality of sets.

Georg Cantor

In the 19th century German mathematician Georg Cantor developed set theory. Set theory facilitates clear definition of mathematical concepts, and it became the basis of subsequent mathematical analysis. Cantor also developed a theory of irrational numbers.

Library of Congress/Corbis

General Concept of a Set

Consider the following lists:

A = Monday, Tuesday, Wednesday, Thursday, Friday, Saturday, Sunday.

B = blue, green, yellow, red, white, grey, black, violet, orange.

Clearly, the first list A, is made up of days of the week and the second list B, is made up of colours. Also January is not a day of the week, so it cannot be included in the list A. Equally a blackboard is not a colour so it cannot be included in the list B. The above lists are examples of sets. Such a list, which represents a set, is usually enclosed in braces, as below

A = {Monday, Tuesday, Wednesday, Thursday, Friday, Saturday, Sunday}.

B = {blue, green, yellow, red, white, grey, black, violet, orange}.

Therefore, a set is a collection of objects, ideas, people, numbers, animals, days, colours, etc, that have similarities or some common property or function or behavior or that are related in one way or the other. The objects, ideas, people numbers etc, that make up the set are called the members or elements of the set.

The elements of a set must be such that one can say exactly whether any given element belongs to the set. In other words, a set must be well defined.

Set Notation and Definition of Sets

Capital letters A, B, C... are used to denote sets.

There are three basic methods for describing or defining sets.

The Roster Method

This is done by listing the elements of a set, separating them by commas and enclosing them with braces.

Example 20:1

Let V = English vowels. List the elements of V.

Solution

V = {a, e, i, o, u}.

If the list is so long, the first few elements are listed and three dots are used to show that the list continues.

Example 20:2

Let E = positive even numbers (i.e. positive numbers that can be divided exactly by 2). List the elements of E.

Solution

E = {2, 4, 6, 8, 10...}

The Rule Definition Method

In this case, a rule, which qualifies any element as a member of the set, is stated or defined.

Example 20:3

Let C = {North, West, South, East}
R = {Red, Orange, Yellow, Green, Blue, Indigo, Violet}
Use the rule definition method to define C and R.

Solution

C = cardinal points

R = Colours of the rainbow.

In using the rule definition method, it is unwise to use braces, for,

Colours of the rainbow $\neq$ {Colours of the rainbow}

The explanation of this follows. The left hand side (L.H.S) means {Red, Orange, Yellow, Green, Blue, Indigo, Violet} but the right hand side (R.H.S) is a set containing only one element. Its sole element is the phrase "Colours of the rainbow".

To make this point clearer, consider the set.

P = {He ate the mangoes, I drank the wine}.

Clearly, P is a set of two elements. Its two elements are the phrases "He ate the mangoes" and "I drank the wine". A comma separates these two elements.

Set Builder Notation Method

Example 20:4

Given that W is a set of days of the week. Use set builder notation to write down this statement.

Solution

W = {x:x is a day of the week}
or W = {x/x is a day of the week}

This is read, "W is the set of all elements, x, such that x is a day of the week".

x:x or x/x is read, 'x such that x'

Example 20:5

Write in full the meaning of
P = {(x, y): y = $2x + 5$}.

Solution

P is the set of all points (x, y) which satisfy the line $y = 2x + 5$.

Example 20:6

State the meaning of $A = \{x: 0 \leq x \leq 3\}$.

Solution

A is the set of all points which lie between the points 0 and 3 inclusively.

Membership Notation

Let $V = \{x \mid x$ is an English vowel$\}$
Clearly, a, belongs to the set V, or in other words, a, is a member of V or a, is an element of the set V. This fact is denoted by $a \in V$ and is read as "a belongs to V" or "a is an element of V" or "a is a member of V"
In mathematics, cancelling a sign changes its meaning to the opposite one. For instance, $\neq$ means, "is not equal to". Thus, b is not an element of V is written mathematically as $b \notin V$, read "b is not an element of V" or "b does not belong to V".

EXERCISE 20:1

1. Identify the odd element in the following list of objects:
   (a) boy, girl, man, pen, woman
   (b) rice, ink, beans, maize, plantain
   (c) Cameroon, Nigeria, London, Egypt, France
   (d) h, j, i, p, q, w.
   (e) Black, orange, yellow, white paper, green
2. Write the following statements using set notation:
   (a) x belongs to A (b) y is not a member of B
   (c) G has 3 members
3. Given that $F = \{a, b\}$. State which of the following statements is correct or incorrect giving reasons for your answer.
   (a) $b \in F$ (b) $\{b\} \in F$
4. Which of the following collection of objects form a set? Say why or why not.
   A = Months of the year with 30 days.
   B = Beautiful girls in form one.
   X = Form three book list.
   G = Good people in our school.
   P = Cripples in our school.
   I = Intelligent students in form two.
   C = Black, yellow, white paper, green.
5. Given that $X = \{x : 3x = 12\}$ and $y = 4$. Is y equal to X? Give a reason for your answer.
6. Write the following sets using the roster method
   M = multiples of 2 less than 20

F = suits of playing cards
$V = \{x : x$ is a vowel of the English alphabet$\}$
$A = \{y : y$ is a factor of 36$\}$
$N = \{n \mid n$ is an integer between 1 and 10$\}$

7. Define a rule to qualify the members of each of the following sets.
   $A = \{1,3,5,7,9,11,13\}$
   $B = \{2,4,6,8,10,12,14,16,18,20\}$
   $C = \{$Mangoes, oranges, pineapples, bananas,...$\}$
   $D = \{$tall, great, intelligent, small,...$\}$
   $E = \{$football, table tennis, volleyball, athletics...$\}$
8. Write the following using set builder notation.
   $X = \{2,3,5,7, 11,13,17,19\}$
   $M = \{1,2,3,6,7,14,21,42\}$
   $C = \{$Momo, Mezam, Menchum, Boyo, Bui, Donga/Mantung, Ngohkentungia$\}$
   $Y = \{$Manyu, Sanaga, Ndian, Wouri, Mungo, Shari, Munaya, Mbam, Katsina$\}$
   $T = \{3,6,9, 12,15,18,21,24,27,30\}$

The Cardinality of a Set

The number of elements in a set is known as its cardinality or the cardinal number of a set or the order of a set. For example, the cardinality of $V = \{a, e, i, o, u\}$ is 5.
This is denoted mathematic by $n(V) = 5$

TYPES OF SETS

Singleton or unit set

Consider the sets $X = \{4\}$ and $N = \{$Paul$\}$, then $n(X) = 1$ and $n(N) = 1$. A set whose cardinality is 1 or in other words a set that has only one element such as X and N above is called a **singleton**.

Doubleton or Pair set

Consider the set $Y = \{3,5\}$. Y contains two elements or $n(Y) = 2$. Such a set whose cardinality is two is called a doubleton or a pair set.

A Trebleton

Let $C = \{2,3,5\}$. C contains 3 elements or $n(C) = 3$. A set whose cardinality is 3 is called a trebleton.

Empty or null set

Let A = people who shall never die
and B = Students of form five who are four years old

Clearly A and B have no elements. Such sets, which have no elements, are called empty sets or

null sets. An empty set or null set is denoted as Ø
or { }.

Thus, $A = Ø = \{ \}$, $B = Ø = \{ \}$
$\Rightarrow n(A) = 0, n(B) = 0$.

Note that it is wrong to write:
$Ø = \{0\}$ or $Ø = \{Ø\}$ or $Ø$ or $\{...\}$
$\{Ø\}$ and $\{0\}$ are singletons, containing the
elements Ø and 0 respectively. While, $\{...\}$, is a
set whose elements have been omitted, or
voluntarily left out.

Finite sets

It is possible to list all the elements of some sets.
For such a set, one can say exactly how many
elements it contains. Such a set is called a Finite
set.

Example 20:7
Let F = factors of 30. Explain whether F is a finite
set.

Solution

$F = \{1, 2, 3, 5, 6, 10, 15, 30\}$ and $n(F) = 8$
Therefore, F is a finite set because it has an exact
countable number of elements.

Infinite sets

The elements of some sets cannot all be listed
because are inexhaustible. Such a set is called an
infinite set. The cardinality of an infinite set is
infinity, meaning "extremely large". Infinity is
denoted by the symbol 8.

Example 20:8
Let P = all the people since creation
And $E = \{x:x$ is an even number$\}$
What property has P and E in common?

Solution
Both P and E are infinite sets because one cannot
list all the elements of P and E.

EXERCISE 20:2

1. Classify the following sets as singleton,
   doubleton, trebleton, empty set or none of the
   above.
   (a) $\{1,3,5\}$ (b) $\{1\}$ (c) $\{0\}$
   (d) $\{3,8\}$ (e) $\{0, Ø\}$ (f) $\{0, Ø, 1\}$
   (g) $\{2,4\}$ (h) $\{a,e,i,o,u\}$ (i) $\{...\}$
   (j) $\{ \}$ (k) Ø
2. State the cardinality of each of the following
   sets.

(a) $\{1,3,5,7,9,11,13\}$ (b) $\{3,6\}$ (c) $\{ \}$
(d) {Students of your class}   (e) $\{0\}$
(f) S = students of your class
(g) Factors of 42
(h) Multiples of 5 less than 30
(i) Even numbers less than 30
(j) Odd numbers less than 40
3. State whether the set is finite or infinite.
   (a) The set of whole numbers. (b) Ø (c) $\{0\}$
   (d) The set of even numbers
   (e) The letters of the English alphabet
   (f) The set of odd numbers
   (g) $\{x/x$ is a prime number less than 1000$\}$
   (h) $\{x:x$ is a planet$\}$
   (i) The set of teachers in your school
   (j) $\{1,2,4,8,16,32,64,...\}$
4. Fill in the blank spaces with ∈ or ∉
   (a) 2___$\{1,3,5,7,9,11,...\}$
   (b) 14___$\{2,4,6,8,10,...\}$
   (c) 30___$\{x:x$ is a multiple of 6$\}$
   (d) 17___$\{1,2,4,7,11\}$
   (e) Chalk___{paper, pencil, pen, ruler}
   (f) Beautiful___{go, take, come, eat, write}

The Universal Set
When dealing with sets it is often necessary to
define the sphere of discuss. This sphere of discuss
is called the universal set and contains all the
elements under consideration. The universal set is
usually denoted by $\mathscr{E}$ or U.

Subsets
If all the elements of a set A are found in another
set B, then A is said to be a subset of B, denoted by
$A \subset B$ or $B \supset A$. A is not a subset of B is written
$A \not\subset B$.

Example 20:9
Let $A = \{1,2,3,4,5,6,7,8\}$ and $B = \{2,5,7\}$.
Write three statements relating A and B using the
symbols (a) ⊂ (b) ⊃   (c) ⊄.
Which of these statements are the same?

Solution
(a) $B \subset A$ (b) $A \supset B$ (c) $A \not\subset B$
The statement in (a) is the same as statement (b).

Example 20:10
$\mathscr{E} = \{1,2,3,4,5,6,7,8\}$
$A = \{2,4,6,8\}$
$B = \{1,3,5,7,9\}$
and $C = \{1,2,3,4\}$
Which of the sets A, B and C is or is not a subset
of $\mathscr{E}$? Explain.

Solution
The sets A and C are both subsets of the universal set $\mathscr{E}$ because all their elements are found in $\mathscr{E}$, while the set B is not because $9 \notin \mathscr{E}$.

Example 20:11
Given that, A = Multiples of 3. List the elements of A when $\mathscr{E}$ = Whole numbers less than 30

Solution
$$\mathscr{E} = \{1,2,3,4,5,6,7,89,10...,29\}$$
$$A = \{3,6,9,12,15,18,21,24,27\}$$

Example 20:12
Given that,
$\mathscr{E} = \{x{:}x$ is an odd number less than 30$\}$
$C = \{ x{:}x$ is a multiple of 3$\}$
List the elements of set C.

Solution
$$\mathscr{E} = \{1,3,5,7,9,11,13,15,17,19,21,23,25,27,29\}$$
$$C = \{3,9,15,21,27\}$$
This is so since 6, 12, 18 and 24 are not odd numbers, hence not elements of C.

Equality of Sets

Two sets X and Y are said to be equal if they have the same elements. This means that all the elements in set X are found in set Y and vice versa. It follows that $X \subset Y$ and $Y \subset X$. By implication $n(X) = n(Y)$. If two sets X and Y are equal, this is written mathematically as $X = Y$.

Note here that *repeating elements of a set does not change the set*.

Thus, $\{2, 4, 6, 8\} = \{8, 2, 6, 4, 6, 8\}$

Equivalent Sets

If two non empty sets X and Y are such that $n(X) = n(Y)$ but $X \not\subset Y$, then X and Y are said to be equivalent denoted by $X \sim Y$. In simple terms, equivalent sets have the same number of elements.

Example 20:13
If $X = \{3,6,9,12\}$ and $Y = \{2,4,6,8\}$. Say, giving reasons whether X and Y are equivalent sets.

Solution
$n(X) = n(Y) = 4$ and $X \not\subset Y$. Therefore X and Y are equivalent sets.

Proper and Improper Subsets

If two sets X and Y are not equal and all the elements of X are found in Y, then X is a proper subset of Y and Y is said to be the superset of X. On the other hand if two sets X and Y are equal then X is an improper subset of Y. Equally Y is an improper subset of X. This means that every set is an improper subset of itself.
The empty set $\varnothing$ is conventionally an improper subset of every set. Therefore, every set has two improper subsets, the empty set $\varnothing$, and the set itself.

Example 20:14
Let $X = \{2, 4, 6, 8\}$
$Y = \{6, 8, 2, 4, 10\}$
$Z = \{1, 2, 3, 4, 5, 6, 7, 8, 9, 10\}$
Write symbolic statements linking $\varnothing$ and each of the sets X, Y and Z and explain the meaning of these statements in ordinary English.
Which of the sets are improper subsets of each other? State the proper subsets of each of the sets.

Solution
$\varnothing \subset X$, $\varnothing \subset Y$ and $\varnothing \subset Z$

$\varnothing$ is an improper subset of each of the sets X, Y and Z. $X \subset Y$ and $Y \subset X$. Therefore, X and Y are improper subsets of each other.

Equally, X is an improper subset of X, and Y is an improper subset of Y. X and Y are proper subsets of Z.
By implication if A is an improper subset of B and $B \neq \varnothing$, then B is an improper subset of A.
Therefore, A must be equal to B.

EXERCISE 20:3

1. Complete the following statements using the symbols $\subset, \not\subset, =, \neq, \sim$.

   (a) $\{a, e, i, o, u\}$ ___$\{x{:} x$ is a letter of the alphabet$\}$

   (b) e ___ $\{a, e, i, o, u\}$

   (c) 3___$\{x{:} x$ is an even number$\}$

   (d) $\{3,7,2,5,1,6,4\}$___$\{1,2,3,4,5,6,7\}$

   (e) $\{2, 4, 6\}$ ___ $\{1, 3, 5, 7, 9\}$

   (f) $\{a, b, c\}$___$\{1,2,3\}$

2. Let $\mathscr{E}$ = universal set and A, B, C, be any other sets such that
   $\mathscr{E} = \{1,2,3,...,49,50\}$
   A = multiples of 5
   $B = \{1,2,3,4,5\}$
   $C = \{1,2,3\}$
   $D = \{3,5,1,4,2\}$
   $E = \{3,4,5\}$

State whether each of the following is true or false.

(a) $A \subset E$ (b) $B \subset E$ (c) $B \in E$

(d) $C \sim$ (e) $B \not\subset D$ (f) $E \subset B$

(g) $C \not\subset D$ (h) $55 \in A$ (i) $17 \notin E$

(j) $B \supset C$

(k) $B = D$ (m) $A \not\subset E$

3. List all the subsets of the set $P = \{1, 2, 3\}$. How many subsets, has P?

4. Which of the following sets are equal?
$\{a, b, c\}, \{c, a, b, c\}, \{b, a, b, c\}, \{a, c, b, c\}$

5. Using only $=$ or $\sim$, write statements between the sets where applicable.

(a) $\{1, 2, 3, 4, 5\}$ (b) $\{a, b, c, d, e\}$

(c) $\{2, 4, 6\}$ (d) $\{4, 2, 5, 1, 3\}$

(e) {Biology, Chemistry, Mathematics}

6. How many proper subsets, has an empty set?

7. Given that $A = \{4\}$, $B = \{3,4\}$, $C = \{1,2,3\}$, $D = \{1,2\}$ and $E = \{1,2,4\}$. State whether each of the following statements is true or false.

(a) $D \subset B$ (b) $B \neq E$ (c) $A \subset C$

(d) $B = D$ (e) $B = C$ (f) $B \subset C$

(g) $B \sim D$ (h) $C \sim E$

8. Given that (i) $A \subset B, B \subset C$

(ii) $a \in A, b \in B$ and $c \in C$

(iii) $d \notin A, e \notin B$ and $f \notin C$

Which of the following statements must be true?

(a) $a \in C$ (b) $b \in A$ (c) $c \notin A$

(d) $d \in B$ (e) $e \notin A$ (f) $f \notin A$

9. State which of the following sets are equivalent or equal?

(a) $A = \{a, b, c, d, e\}$

(b) $B = \{1, 2, 3, 4, 5\}$

(c) $C = \{4, 2, 3, 5, 1\}$

(d) $D = \{b, a, d, c, e\}$

(e) $E = \{a, 5, 3, b, 4\}$

(f) $F = \{b, 3, a, 5, 4\}$

Intersection of Sets

The intersection of sets is denoted by $\cap$. Thus, $A \cap B$ is read as 'A intersection B' or 'the intersection of A and B' or 'A cap B'. The intersection of two sets A and B is the set, which consist of all the elements that are common to both set A and set B. i.e.

$A \cap B = \{x: x \in A \text{ and } x \in B\}$.

Example 20:15

$\mathscr{E} = \{0,1,2,3,4,5,6,7,8,9,10,11,12\}$

$A = \{2, 4, 6, 8, 10, 12\}$

$B = \{1, 2, 3, 6, 9\}$

Find $A \cap B$

Solution

$A \cap B = \{2,6\}$

If there are many sets, the intersection of these sets will be the set that consist of all the elements common to all these sets.

Union of Sets

The union of sets is denoted by $\cup$. Thus, $A \cup B$ is read 'A union B' or the union of 'A and B' or A cup B'. The union of two sets A and B is a set containing all the elements found in set A and all the elements found in set B, put together.

Example 20:16

$\mathscr{E} = \{0,1,2,3,4,5,6,7,8,9,10,11,12\}$

$A = \{2,4,6,8,10,12\}$

$B = \{1,2,3,6,9\}$

Find $A \cup B$

Solution

$A \cup B = \{1, 2, 3, 4, 6, 8, 9, 10, 12\}$

If there are, many sets the union of all these sets is a set containing all the elements of all these sets put together.

Disjoint Sets

If two non empty sets A and B have no elements in common, they are said to be disjoint. This can be expressed symbolically as

$A \cap B = \varnothing$ or $n(A \cap B) = 0$

Representation of Sets-Venn Diagrams

Venn diagrams were invented by an English Mathematician John Venn Euler (1834-1923). They are used to represent relationships between sets. The usual way is to use a large rectangle to represent the universal set and circles to represent its subsets, (though any other, plane figure can be used)

The intersection of sets can be illustrated using Venn diagrams as in Figure 20:1

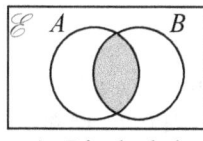

$A \cap B$ is shaded

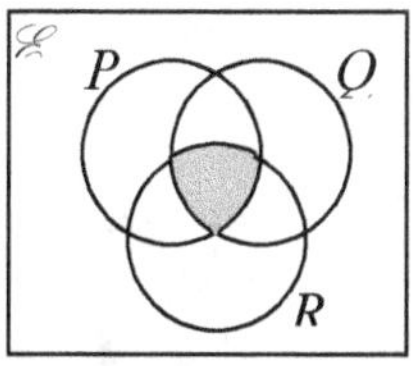

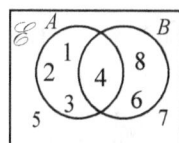

$P \cap Q \cap R$ is shaded

Figure 20:1

The Venn diagram in Figure 20:2, above shows the universal set $\mathscr{E}$ and the set B, which is a subset of another set A.

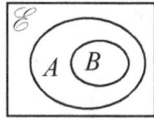

Figure 20:2

The Venn diagrams in Figure 20:3 illustrate the union of sets. The shaded portion is the required region.

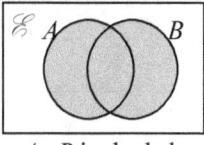

$A \cup B$ is shaded

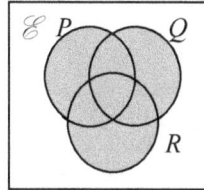

$P \cup Q \cup R$ is shaded

Figure 20:3

The Venn diagram in Figure 20:4 shows two disjoint sets A and B.

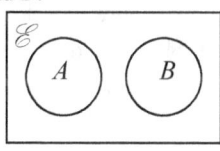

$A \cap B = \varnothing$

Figure 20:4

Example 20:17

$\mathscr{E} = \{1,2,3,4,5,6,7,8,9,10\}$

$A = \{1, 3, 4, 5, 7\}$

$B = \{3, 4, 8, 9\}$

Draw a Venn diagram to illustrate the relationship between these sets.

Using this Venn diagram, list the elements of $A \cap B$.

Solution

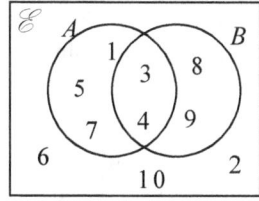

Figure 20:5

Example 20:18

$\mathscr{E} = \{1,2,3,4,5,6,7,8\}$

$A = \{1,2,3,4\}$

$B = \{4,6,8\}$

Draw a Venn diagram to show the relationship between these sets, listing the elements in each region. Using this Venn diagram, find $A \cup B$.

Solution

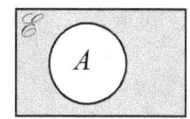

Figure 20:6

From the Venn diagram in Figure 20:6,

$A \cup B = \{1,2,3,4,6,8\}$

Example 20:19

$\mathscr{E}$= Students of form one

A = All form one boys

B = All form one girls

State whether the sets A and B are disjoint.

Solution

A and B are disjoint.

The Complement of a Set $C_{\mathscr{E}}^{A}$ or A'

The complement of a set A is denoted by A' or $C_{\mathscr{E}}^{A}$ and is the set containing all the elements of the universal set $\mathscr{E}$ that do not belong to the set A.

Example 20:20

Let $\mathscr{E}$ = Whole numbers from 1 to 10 and $B = \{2,4,6,8,10\}$. Find B'.

Solution

$B' = \{1, 3, 5, 7, 9\}$

The following Venn diagrams in Figure 20:7 represent the complement of sets. The shaded portion is that described under each Venn diagram.

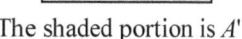

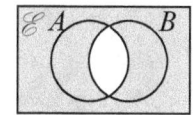

The shaded portion is A' The shaded portion is $(A \cap B)' = A' \cup B'$

203

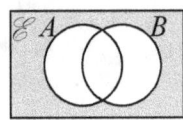

The shaded portion is
$(A \cup B)' = A' \cap B'$

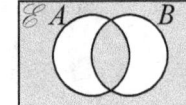

The shaded portion is
$(A \cap B) \cup (A \cup B)'$

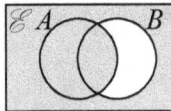

The shaded portion is
$A \cup B'$

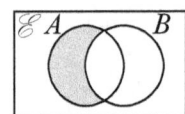

The shaded portion is
$A \cap B'$

Figure 20:7

The Power or Derived Set of a Set

The power or derived set $P(A)$ of a set A is a set, which consists of all the subsets of A.

Example 20:21
If $A = \{a, b\}$, find the power set of A.

Solution
$P(A) = \{\emptyset, \{a\}, \{b\}, \{a, b\}\} = \{\emptyset, \{a\}, \{b\}, A\}$

Example 20:22
Find the derived set of $B = \{1, 3, 5\}$

Solution
$P(B) = \{\emptyset, \{1\}, \{3\}, \{5\}, \{1, 3\}, \{1, 5\}, \{3, 5\}, B\}$

Cardinality of a Power or Derived Set

Consider the following sets.
$\emptyset = \{\}, A = \{a\}, B = \{a, b\}$ and $C = \{a, b, c\}$.

$P(\emptyset) = \{\}$ or $\emptyset$.

$P(A) = \{\emptyset, A\}$.

$P(B) = \{\emptyset, \{a\}, \{b\}, B\}$.

$P(C) = \{\emptyset, \{a\}, \{b\}, \{c\}, \{a, b\}, \{a, c\}, \{b, c\}, C\}$.
By counting elements in each of the above sets, it can be seen that,

$n(\emptyset) = 0$ and $n(P()) = 1 = 2^0$

$n(A) = 1$ and $n(P()) = 2 = 2^1$

$n(B) = 2$ and $n(P()) = 4 = 2^2$

$n(C) = 3$ and $n(P()) = 8 = 2^3$

Hence, $n(Y) = x \Leftrightarrow n(P(Y)) = 2^x$

Thus,

The cardinality of the power or derived set of a set with n elements is 2^n.

Careful observations of the above power or

derived sets reveal that each power set contains the improper subsets $\emptyset$ and the set itself.

Therefore, if only the proper subsets are considered, the cardinality of the power set of a set with n elements will be $2^n - 2$.

EXERCISE 20:4

1. Describe the shaded region in each Venn diagrams in Figure 20:8 using set notation.

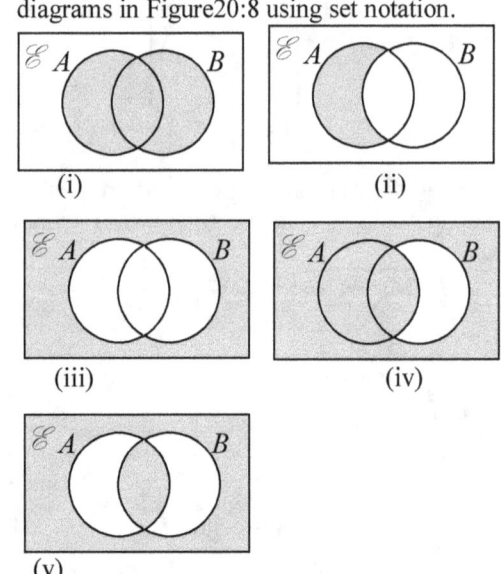

Figure 20:8

2. Given that A and B are subsets of $\mathscr{E}$ the universal set. Shade on different Venn diagrams
   (i) A (ii) A' (iii) $A \cap B$
   (iv) $(A \cap B)'$ (v) $A \cap B'$ (vi) $A \cup B$
   (vii) $(A \cup B)'$ (viii) $A \cup B'$

3. Represent the following on a Venn diagram
   (i) A and B have no elements in common.
   (ii) All the elements in A are found in B.

4. Use only set notation to write down the following statements:
   (i) A and B have no elements in common.
   (ii) All the elements in A are found in B.

5. Represent the following on a Venn diagram
   $\mathscr{E} = \{0, 1, 2, 3, 4, 5, 6, 7, 8, 9\}$
   $A = \{2, 4, 6, 8\}$
   Where, $\mathscr{E}$ is the universal set.

6. Let $\mathscr{E} = $ letters of the English alphabet
   $X = \{a, b, c, d, e, f, g, h, i, j\}$
   $Y = \{a, e, i\}$
   Show by use of a Venn diagram, the relationship between $\mathscr{E}, X$ and Y

7. Given that,

$\mathscr{E}$ = {0,1,2,3,4,5,6,7,8,9,10}

P = {2, 4, 6, 8, 10}

Q = {1, 2, 3, 6, 9}

Represent $\mathscr{E}$, P and Q on a Venn diagram.

8. If $\mathscr{E}$ = Natural numbers from 1 to 20

A = factors of 18

B = Multiples of 3

C = Prime numbers

Draw a Venn diagram to represent the relationship between these sets. Hence, find

(a) $A \cap B \cap C$ (b) $A \cap B$ (c) $B \cap C$ (d) $A \cap C$

9. $\mathscr{E}$ = Whole numbers greater than zero but less than or equal to 124.

A = Multiples of 10

B = Multiples of 15

C = multiples of 25

(a) Draw a Venn diagram to represent the relationship between these sets.

(b) Hence, find $n\,(A \cup B)$ and $n\,(A \cap B \cap C)$.

Algebraic Laws of Sets

Under the operations of union, intersection and complement, sets obey the following laws, which are referred to as the **algebraic laws of sets**.

1	Idempotent Laws	$A \cup A = A$; $A \cap A = A$
2	Associative Laws	$A \cup (B \cup C) = (A \cup B) \cup C$ $A \cap (B \cap C) = (A \cap B) \cap C$
3	Commutative Laws	$A \cup B = B \cup A$; $A \cap B = B \cap A$
4	Distributive Laws	$A \cup (B \cap C) = (A \cup B) \cap (A \cup C)$ $A \cap (B \cup C) = (A \cap B) \cup (A \cap C)$
5	De-Morgan's Laws	$(A \cup B)' = A' \cap B'$ $(A \cap B)' = A' \cup B'$
6	Complement Laws	$A \cup A' = \mathscr{E}$; $A \cap A' = \varnothing$ $\mathscr{E}' = \varnothing$; $\varnothing' = \mathscr{E}$; $(A')' = A$
7	Identity Laws	$A \cup \varnothing = A$; $A \cap \varnothing = \varnothing$ $A \cup \mathscr{E} = \mathscr{E}$; $A \cap \mathscr{E} = A$

Table 20:1

Cardinality Logic

Example 20:23

In a certain class, 10 students do both arts and science, 24 do science and 22 do arts. Given that every student does at least one of these options, how many students are there in the class?

Solution

Let $\mathscr{E}$ = All the students

S = Science students

And A = Arts students

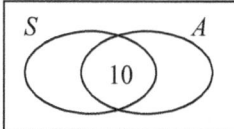 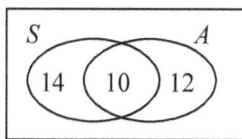

Figure 20:9

$n(\mathscr{E}) = 14 + 10 + 12 = 36$

Therefore, there are 36 students in the class.

Example 20:24

During a conference attended by 30 participants, 18 of them ate rice, 13 ate beans and 5 of them ate both rice and beans. How many ate neither rice nor beans?

Solution

If $\mathscr{E}$ = All participants at a conference

R = Participants who ate rice

B = Participants who ate beans

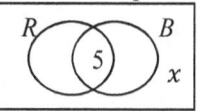 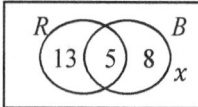

Figure 20:10

$x = n(R' \cap B')$

$n(\mathscr{E}) = n(R) + n(B) - n(R \cap B) + n(R' \cap B')$

$\Leftrightarrow 30 = 18 + 13 - 5 + n(R \cup B)'$

$\therefore n(R \cup B)' = 4$

Therefore, 4 participants ate neither beans nor ice.

Example 20:25

A newsagent sells three papers: the Post, the Messenger and the Standard. Customers buy one of the three papers. It is found that:

80 customers buy the Post,

70 customers buy the Messenger,

60 customers buy the Standard,

21 customers buy the Post and the Messenger,

14 customers buy the Messenger and the Standard,

16 customers buy the Standard and the Post,

6 customers buy all three newspapers.

Draw a Venn diagram to illustrate this information. Hence, find the number of customers who buy only

(a) The Post.

(b) The Messenger.

(c) The standard.

(d) Find also the total number of customers for the three papers.

Solution

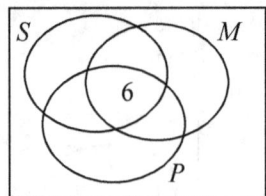

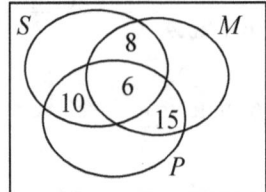

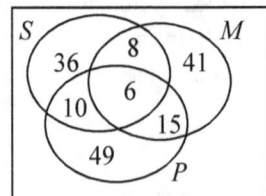

Figure 20:11

(a) Post customers only = $n((M \cup S)') = 49$.

(b) Messenger customers only = $n((P \cup S)') = 41$.

(c) Standard customers only = $n((M \cup P)') = 36$.

(d) Total number of customers = $n(M \cup P \cup S)$

$$= 36 + 41 + 49 + 8 + 10 + 15 + 6$$
$$= 144$$

Example 20:26

Out of 200 students in a school, 130 go for holidays in Douala and 120 go for holidays in Yaounde. Calculated the number of students who go to:

(i) Both Douala and Yaounde, (ii) Douala only,

(iii) Yaounde only,

Solution

$\mathscr{E}$ = All the 120 students

D = those who go to Douala,

Y = those who go to Yaounde

Suppose x students go to both Douala and Yaounde.

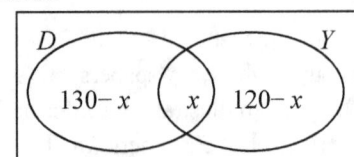

Figure 20:12

$$n(\mathscr{E}) = 200 = n(D \cap Y) + n(D') + n(Y')$$

$$\Rightarrow 200 = x + (130 - x) + (120 - x)$$

$$200 = 250 - x \Rightarrow x = 50$$

1. In a survey of the number of people who visit a restaurant, which serves beans (B), rice (R), and achu (A), the following information was recorded.

   5 people did not eat any of the items served.
   Everyone who ate beans ate rice.
   3 people ate all the three items.
   10 people ate beans.
   4 people ate only rice
   20 people ate only achu
   10 people ate rice and not beans

   By using a Venn diagram or otherwise, find the number of people who

   (a) Ate achu

   (b) ate rice and beans but not achu

   (c) Visited the restaurant.

2. Out of a group of 20 persons, 7 like njang music, 12 like makossa music and 10 like bikutsi music; furthermore, 3 like both njang and bikutsi, 2 like both njang and makossa and 2 like all three kinds of music. Draw a Venn diagram and find how many of the 20 persons like makossa and bikutsi but not njang. Assume all the 20 persons like at least one of the three kinds of music.

3. A universal set $\mathscr{E}$ includes subsets A, B and C. There are 5 members of $\mathscr{E}$ who are not in any of these subsets. Every member of B is a member of A but C contains members who do not belong either to A or B. Draw a Venn diagram to incorporate these features.
   Given further that,

   (a) 20 members belong to C but not to A or B,

   (b) 3 members belong to A, B and C,

   (c) 13 members belong to B,

   (d) 27 members belong to A but not to C,

   (e) 10 members belong to both A and C.

   Insert appropriate numbers in the regions of your diagram. Hence, calculate,

   (i) $n(A)$, (ii) $n(A \cap C \cap B')$, (iii) $n(\mathscr{E})$.

4. Out of 93 students in a lower sixth class, 56 offer History and 59 offer geography. Only 8 students offer History but not geography. Find how many of the students offer

   (a) Geography but not History

   (b) Neither Geography nor History.

MULTIPLE CHOICE EXERCISE 20

1. Let $P = \{2, 4, 6, 8, 10\}$, $Q = \{1, 3, 5, 7\}$,

   $R = \left\{\dfrac{1}{2}, \dfrac{1}{4}, \dfrac{1}{6}, \dfrac{1}{8}\right\}$ and $S = \{10, 6, 2, 8, 4\}$

   It is true to say that:
   [A] P and Q are equivalent sets.
   [B] Q and R are equivalent sets
   [C] P and S are equivalent sets
   [D] Q and R are equal sets

2. Let $P = \{2, 4, 6, 8, 10\}$, $Q = \{1, 3, 5, 7\}$,

   $R = \left\{\dfrac{1}{2}, \dfrac{1}{4}, \dfrac{1}{6}, \dfrac{1}{8}\right\}$ and $S = \{10, 6, 2, 8, 4\}$

   It is not true to say that:
   [A] P and Q are equivalent sets
   [B] Q and R are equivalent sets
   [C] P and S are equal sets
   [D] P and $P \cap S$ are equal sets.

3. Given that $H = \varnothing$. It is true to say that:

   [A] $0 \in H$ [B] $n(H) = \varnothing$
   [C] $H = \{0\}$ [D] $n(H) = 0$

4. Given the universal set $\mathscr{E} = \{x: 0 < x < 10, x \in \mathbb{Z}\}$.
   The complement of the set
   $P = \{x: x \in \mathscr{E}, x \text{ is not divisible by 4}\}$ is:
   [A] $\{4\}$ [B] $\{4, 8\}$
   [C] $\{1, 2, 3\}$ [D] $\{1, 2, 3, 5, 6, 7, 9\}$

5. If $A = \{a, b, c\}$, $B = \{a, b, c, d, e\}$ and
   $C = \{a, b, c, d, e, f\}$, $(A \cup B) \cap (A \cup C)$ is equal to:
   [A] $\{a, b, c, d\}$ [B] $\{a, b, c, d, e\}$
   [C] $\{a, b, c, d, e, f\}$ [D] $\{a, b, c\}$

6. Let J be the set of positive integers. If
   $H = \{x: x^2 < 3, x \neq 0\}$, then
   [A] $H = \{1\}$ [B] H is an infinite set
   [C] $H = \{0, 1\}$ [D] $H = \{\ \}$

7. Given that

   $P = \left\{2, 1, 3, 9, \dfrac{1}{2}\right\}$, $Q = \left\{1, 2\dfrac{1}{2}, 3, 7\right\}$,

   $R = \left\{5, 4, 2\dfrac{1}{2}\right\}$ and $P \cup Q \cup R$ is equal to:

   [A] $\left\{5, 4, 2\dfrac{1}{2}\right\}$ [B] $\{1, 2, 3, 4, 5, 6, 7\}$

   [C] $\{1, 9\}$ [D] $\left\{\dfrac{1}{2}, 1, 2, 2\dfrac{1}{2}, 3, 4, 5, 7, 9\right\}$

8. Given that

   $P = \left\{2, 1, 3, 9, \dfrac{1}{2}\right\}$, $Q = \left\{1, 2\dfrac{1}{2}, 3, 7\right\}$,

   $R = \left\{5, 4, 2\dfrac{1}{2}\right\}$ and $P \cap Q \cap R$ is equal to:

   [A] $\{5, 7, 9\}$ [B] $\varnothing$ [C] $\{1, 3, 7\}$ [D] $\{4\}$

9. Let $\mathscr{E} = \{1, 2, 3, 4\}$, $P = \{2, 3\}$ and $Q = \{2, 3, 4\}$.
   $P \cap Q$ is equal to:
   [A] $\{1, 2, 3\}$ [B] $\{1, 3, 4\}$ [C] $\{2, 3\}$ [D] $\{1, 3\}$

10. $S = \{1, 2, 3, 4, 5, 6\}$, $T = \{2, 4, 5, 7\}$ and $R = \{1, 4, 5\}$. Then $(S \cap T) \cup R$ is:
    [A] $\{1, 4, 5\}$ [B] $\{2, 4, 5\}$
    [C] $\{2, 3, 4, 5\}$ [D] $\{1, 2, 4, 5\}$

11. If $R = \{2, 4, 6, 7\}$ and $S = \{1, 2, 4, 8\}$ then $R \cup S$ equals:
    [A] $\{1, 2, 4, 6, 7, 8\}$ [B] $\{1, 2, 4, 7, 8\}$
    [C] $\{1, 4, 7, 8\}$ [D] $\{2, 6, 7\}$

12. If $P = \{3, 5, 6\}$ and $Q = \{4, 5, 6\}$ then $P \cap Q$ equals:
    [A] $\{3, 6\}$ [B] $\{4, 5\}$ [C] $\{4, 6\}$ [D] $\{5, 6\}$

13. If $S \subset R$, then
    [A] $S \cap R = R$ [B] $S \cap R = S$
    [C] $S \cup R = S'$ [D] $S \cup R = S$

14. If $P = \{3, 7, 11, 13\}$, $Q = \{2, 4, 8, 16\}$. It follows that:
    [A] $(P \cap Q)' = \{2, 3, 4, 13\}$ [B] $n(P \cup Q) = 4$
    [C] $P \cup Q = \varnothing$ [D] $P \cap Q = \varnothing$

15. Given that $P = \{b, d, e, f\}$ and $Q = \{a, c, f, g\}$
    are subsets of the universal set
    $U = \{a, b, c, d, e, f, g\}$. $P' \cap Q$ is equal to:
    [A] $\{a, c\}$ [B] $\{a, c, d, g\}$
    [C] $\{c, d, g\}$ [D] $\{a, c, g\}$

16.
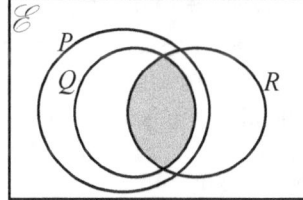

Figure 20:13

In the Venn diagram in figure 20:13, the shaded region is described by
[A] $Q \cap R$ [B] $Q \subset R$ [C] $P \cap Q \cup R$ [D] $P \cup Q \cap R$

17. Given the universal set $\mathscr{E} = \{1, 2, 3, 4, 5\}$,
    $P = \{1, 2\}$, $Q = \{2, 3, 4\}$, then $(P \cap Q)'$ is
    [A] $\{2\}$ [B] $\{5\}$ [C] $\{1, 3, 4, 5\}$ [D] $\{1, 2, 3, 5\}$

18.
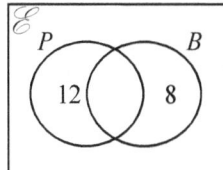

Figure 20:14

Figure 20:14, given that $\dfrac{4}{9}$ of the total number of students offering Physics (P) or Biology (B)

207

offer Physics only. The number of students offering both subjects is:

[A] 27   [B] 20   [C] 7   [D] 18

19. In figure 20:15, $n(P \cap Q)$ is:

[A] 1   [B] 2   [C] 4   [D] 6

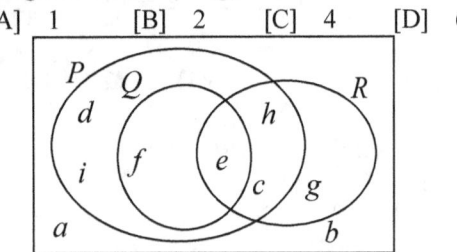

Figure 20:15

20. In Figure 20:15, $Q' \cap R$ is equal to:

[A] $\{e\}$ [B] $\{c, h\}$
[C] $\{c, g, h\}$ [D] $\{c, e, g, h\}$

21. In Figure 20:16, the shaded portion is:

[A] $P' \cap Q$ [B] $(R \cap Q) \cup (P' \cap R)$
[C] $P' \cap Q \cap R$ [D] $(P \cup Q)' \cap R$

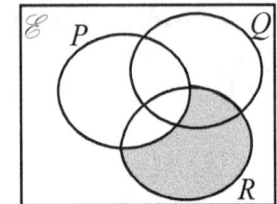

Figure 20:16

22. In a class of 80 students, every student study Economics or Geography. If 65 students study Economics and 50 study Geography, the number of students who study both subjects is:

[A] 15   [B] 30   [C] 35   [D] 45

23. A and B are two sets. The number of elements in $A \cup B$ is 49. The number in A is 22 and the number in B is 34. The number of elements in $A \cap B$ is:

[A] 7   [B] 27   [C] 15   [D] 12

24. If $n(P) = 21$, $n(R) = 33$ and $n(P \cup R) = 46$. $n(P \cap R)$ is equal to:

[A] 8   [B] 34   [C] 58   [D] 100

25. The Venn diagram in Figure 20:17, shows the number of students who studied physics P, chemistry C, and mathematics M in a certain school. How many students studied at least

two of the three subjects?

[A] 165   [B] 160   [C] 155   [D] 135

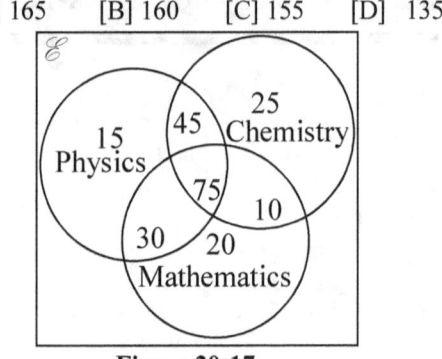

Figure 20:17

26. Let $\mathscr{E}$ = All men, H = honest persons, B = Businesspersons, S = Successful persons The set of dishonest, unsuccessful businesspersons is represented by:

[A] $H \cap S \cap B$ [B] $H \cap S \cap B$
[C] $H' \cup S' \cup B$ [D] $H' \cap S' \cap B$

27. Let M = mammals, V = vertebrates, I = insects, H = horned animals. The statement, 'All mammals are vertebrates' is denoted by:

[A] $M \cap V \neq \varnothing$ [B] $M \subset V$
[C] $M \cup V = \varnothing$ [D] $M \cup H \neq \varnothing$

28. Let M = mammals, V = vertebrates, I = insects, H = horned animals. $V \cap I = \varnothing$ in ordinary English means:

[A] Vertebrates and insects are empty sets.
[B] Some vertebrates are insects
[C] Vertebrates cannot join with insects.
[D] No insect is a vertebrate.

29. Let M = mammals, V = vertebrates, I = insects, H = horned animals. Using set notation the statement 'Some mammals are horned animals', can be written as:

[A] $H \cap M \neq \varnothing$ [B] $M \cap H = \varnothing$
[C] $M \cup H \neq \varnothing$ [D] $M \cup H = \varnothing$

30. The statement $A \subset B$ is the same as:

[A] $A \cap B = B$ [B] $A \cap B = A$
[C] $A \cup B = B$ [D] $A \cup B = A$

TOPIC 21

BINARY OPERATIONS

OBJECTIVES

At the end of this topic, the learner should be able to:

1. Combine elements on an operation table using binary operations.
2. Identify the closure, associative, commutative, identity and inverse element properties in a binary operations.
3. State and use the properties of a group.
4. Determine whether or not a group is Abelian (or commutative) group.

Évariste Galois (1811-1832), French mathematician is best known for his development of group theory. Several of his constructs, now termed Galois group, Galois field, and Galois theory, remain fundamental concepts in modern algebra. Évariste Galois died in 1832 at the age of 21 after being mortally wounded in a duel. While he was alive, his mathematical theories were criticized as obscure and undeveloped, and he twice failed the entrance examination for the École Polytechnique. When his work was published in 1846 and 1870, Galois gained recognition as an important mathematician.

Science Source/Photo Researchers, Inc.

What is an Operation?

An operation is a definite rule of combination of elements of a set or sets of a universal set. Thus $+, -, \times, \div$, and combination of functions and matrices are examples of operations.

Example 21:1

Let $A = \{0, 1, 2, 3\}$. Then if the operation addition $(+)$ is defined in A, we have:

$0 + 0 = 0 \quad 0 + 1 = 1 \quad 0 + 2 = 2 \quad 0 + 3 = 3$
$1 + 0 = 1 \quad 1 + 1 = 2 \quad 1 + 2 = 3 \quad 1 + 3 = 4$
$2 + 0 = 2 \quad 2 + 1 = 3 \quad 2 + 2 = 4 \quad 2 + 3 = 5$
$3 + 0 = 3 \quad 3 + 1 = 4 \quad 3 + 2 = 5 \quad 3 + 3 = 6$

Operation or Combination Tables

Operation tables often simplify the combination of the elements like those in Examples 21:1 especially when one has to combine all the elements of the set. The operation addition above is shown on the operation table (Table 21:1).

+	0	1	2	3
0	0	1	2	3
1	1	2	3	4
2	2	3	4	5
3	3	4	5	6

Table 21:1

Defined Operations

An infinite number of rules for combining elements of a set can be made. These types of operations are symbolized using special symbols such as $*, \circ, \sim, \Delta$, etc, and are called defined operations

Example 21:2

An operation is defined on the set $A = \{1, 3, 5, 7, 9\}$ by $x * y =$ the last digit in the product xy.
Draw a combination table for this operation.

Solution

*	1	3	5	7	9
1	1	3	5	7	9
3	3	9	5	1	7
5	5	5	5	5	5
7	7	1	5	9	3
9	9	7	5	3	1

Table 21:2

Example 21:3

If the operation $a \sim b$ gives the numerical difference between a and b e.g. $8 \sim 5 = 3$, $7 \sim 4 = 3$. Draw an operation table for the set S where, $S = \{0, 4, 8, 12\}$

Solution

~	0	4	8	12
0	0	4	8	12
4	4	0	4	8
8	8	4	0	4
12	12	8	4	0

Table 21:3

Example 21:4

An operation is defined on the set $\mathbb{Z}$ of integers by $x * y = x + 2y$ find:

(i) $2 * 11$ (ii) $4 * -2$

Solution

(i) $2 * 11 = 2 + 2(11) = 24$
(ii) $4 * -2 = 4 + 2(-2) = 0$

Closure

The set S is said to be closed under an operation $*$, if the result of performing the operation on any two elements in the set gives another element also belonging to the set. i.e. if $a, b \in S \Rightarrow a * b \in S$. In other words no new element is introduced in the cause of operation.
The set S in Example 21:3 above is closed because no new element is introduced as seen from the operation table (Table 21:3).
However not all sets are closed under an operation. For instance, the set A in Example 21:1 above is not closed under the operation since new elements are introduced.
Consider the operation $\ominus$, which means ordinary subtraction defined over the set of natural numbers $\mathbb{N}$. It can clearly be seen that the set $\mathbb{N}$ is not close under the operation e.g. $2 \ominus 3 = -1, -1 \notin \mathbb{N}$.
Also, consider the operation $\ominus$ defined in set $\mathbb{Z}$ of integers. Clearly, $\mathbb{Z}$ is closed under this operation.
Consider the operation $\div$, which means ordinary division defined on the set $\mathbb{Z}$.
Clearly, $\mathbb{Z}$ is not closed under the operation since $3 \div 4 = \frac{3}{4}$ and $\frac{3}{4} \notin \mathbb{Z}$.

Under addition, subtraction, multiplication and division, the set $\mathbb{R}$ is closed.

Binary Operations

A binary operation is an operation in which any two elements of a set or any two sets of a power set combine to give an element of the given set or power set. This means that a binary operation is always closed in the set in which it is defined.

Associativity

If an operation is independent of the way the elements are grouped, the operation is said to be associative.

Example 21:5

1) The operations addition and multiplication are associative on the set $\mathbb{R}$ of real numbers, since
   $(4+3)+2 = 4 + (3+2) = 9$ and
   $(4\times3) \times 2 = 4 \times (3\times2) = 24$
2) On the other hand, the operations of subtraction and division are not associative on the set $\mathbb{R}$ of real numbers since,
   $(4 - 3) - 2 \neq 4 - (3 - 2)$ and
   $(8 \div 2) \div 3 \neq 8 \div (2 \div 3)$

Commutativity

An operation $*$ defined on a set S, is said to be commutative if interchanging the elements does not change the result. i.e. $a * b = b * a$, $\forall a, b \in S$

Addition and multiplication are commutative on the set $\mathbb{R}$ but subtraction and division are not.

Thus,

$5 + 2 = 2 + 5 = 7$ and $2 \times 5 = 5 \times 2 = 10$

But, $6 - 2 = 4$ and $2 - 6 = -4$;

$4 \div 2 = 2$ and $2 \div 4 = \dfrac{1}{2}$

On operation tables, *symmetry of elements about the leading diagonal indicates that the operation is commutative.*

Example 21:6

State with reasons whether any of the operations $\circ$ in Table 21:4 (i) and (ii) defined on the set $X = \{a, b, c, d\}$ is commutative.

$\circ$	a	b	c	d
a	b	a	d	c
b	d	c	a	b
c	a	b	c	d
d	c	d	b	a

(i)

$\circ$	a	b	c	d
a	b	d	a	c
b	d	c	b	a
c	a	b	c	d
d	c	a	d	b

(ii)

Table 21:4

Solution

The operation in (i) is not commutative because there is no symmetry along the leading diagonal. On the other hand, the operation in (ii) is commutative because there is symmetry along the leading diagonal as in Table 21:5. (The leading diagonal is the diagonal, which runs from upper left to lower right).

$\circ$	a	b	c	d
a	b	d	a	c
b	d	c	b	a
c	a	b	c	d
d	c	a	d	b

Table 21:5

Identity or Neutral Element

Consider the following operations

(i) $0 + 5 = 5 + 0 = 5$

(ii) $30 + 0 = 0 + 30 = 30.$

(iii) $1 \times 7 = 7 \times 1 = 7$

(iv) $1 \times (-2) = (-2) \times 1 = -2$

Notice that for addition in $\mathbb{R}$, when 0 is operated with any other element, the result is the element. 0 is said to be the identity or neutral element for addition in $\mathbb{R}$.
 Similarly, for multiplication in $\mathbb{R}$, when 1 is operated with any other element, the result is the element. 1 is said to be the identity or neutral element for multiplication in $\mathbb{R}$.

Generally, for any operation $*$ defined on a given set S, the identity or neutral element usually denoted by e, is that element which when operated with any other element a of S leaves the element a unchanged. i.e.
   $a * e = e * a = a$, $\forall a \in S.$

On an operation table, *the identity element is the element on the intersection of the row and column whose elements are in the same order as the header row and header column.*

Example 21:7
State the identity element for the operation in Table 21:4(ii).

Solution
The identity element is *c*. This is because the row and column, which are marked with arrows and intersect at *c*, are arranged in the same order as the header row and column respectively.

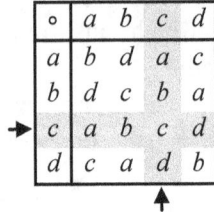

Table 21:6

The Inverse Element

Consider the following operations in $\mathbb{R}$, the set of real numbers.
(i) $2 + (-2) = (-2) + 2 = 0$
(ii) $5 + (-5) = (-5) + 5 = 0$
Notice that in each case two elements in the set $\mathbb{R}$ of real numbers are operated to give 0, the identity element for addition in $\mathbb{R}$. In (i), 2 is said to be the additive inverse of -2 and vice versa, in (ii), 5 is said to be the additive inverse of -5 and vice versa. State the additive inverse of (a) -13 (b) 57.

Generally, for an operation $*$ defined in a set *S*, the inverse of an element *a* is denoted by a^{-1}, read '*a* inverse' or 'the inverse of *a*'. a^{-1} is the element which satisfies the condition that $a * a^{-1} = a^{-1} * a = e$, where *e* is the identity element. In other words, an element *a* is the inverse of another a^{-1} and vice versa if the result after operating the two elements is the identity element *e*.

Example 21:8

For addition in $\mathbb{R}$ the inverse of the element $a \in \mathbb{R}$ is $-a$ since $a + (-a) = (-a) + a = 0$ and 0 is the identity for addition in $\mathbb{R}$.
For multiplication in $\mathbb{R}$ the inverse of any non-zero element *a*, is it's reciprocal

$\dfrac{1}{a}$, since $\dfrac{1}{a} * a = a * \dfrac{1}{a} = 1$ and 1 is the identity element for multiplication in $\mathbb{R}$.

Therefore, on an operation table, *any two elements, which head the row and column, which intersect on the identity element are inverses of each other.*

Example 21:9
In Table 21:4(ii), state the inverse of each element.

Solution
Row *a* intersects with column *d* at *c* and vice versa.
Row *b* intersects with column *b* at *c* and vice versa.
Row *c* intersects with column *c* at *c* and vice versa.
Therefore, $a^{-1} = d$, $d^{-1} = a$, $b^{-1} = b$ and $c^{-1} = c$. An element, which is an inverse of itself such as b and c in this example, is said to be **self-inverse**.

Residue Classes

Consider the following:
$$6 \div 3 = 2, \text{ remainder } 0$$
$$7 \div 3 = 2, \text{ remainder } 1$$
$$8 \div 3 = 2, \text{ remainder } 2$$
$$9 \div 3 = 2, \text{ remainder } 0$$
$$10 \div 3 = 2, \text{ remainder } 1$$
$$11 \div 3 = 2, \text{ remainder } 2$$

Generally in arithmetic modulo *n*, the remainder belongs to the set $\{0, 1, 2 ... (n-1)\}$.

The above can be rewritten as follows:

$$6 = 2 \times 3 + 0 \quad \Rightarrow \quad 6 = 0 \bmod 3$$
$$7 = 2 \times 3 + 1 \quad \Rightarrow \quad 7 = 1 \bmod 3$$
$$8 = 2 \times 3 + 2 \quad \Rightarrow \quad 8 = 2 \bmod 3$$
$$9 = 3 \times 3 + 0 \quad \Rightarrow \quad 9 = 0 \bmod 3$$
$$10 = 3 \times 3 + 1 \quad \Rightarrow \quad 10 = 1 \bmod 3$$
$$11 = 3 \times 3 + 2 \quad \Rightarrow \quad 11 = 2 \bmod 3$$
$$m = kn + r \quad \Rightarrow \quad \therefore m = r \bmod n$$

Numbers *a*, *b*, *c*,... which have the same residue, are called **equivalent numbers** and are said to be **congruent modulo m**.

Example 21:7

1) The binary operation ∗ is defined on the set S where $S = \{1, 2, 3, 4\}$ as, $a * b =$ the remainder when ab is divided by 5.

(a) Form the operation table for the set S.

(b) Name the identity element and the inverse of each element.

(c) Is the operation commutative or associative?

Solution

(a)

∗	1	2	3	4
1	1	2	3	4
2	2	4	1	3
3	3	1	4	2
4	4	3	2	1

Table 21:7

(b) The identity element $e = 1$.

$1 * 1 = 1$ and $4 * 4 = 1$, i.e 1 and 4 are self-inverse.

$2 * 3 = 1$ and $3 * 2 = 1$.

$\Rightarrow 2^{-1} = 3$ and $3^{-1} = 2$.

(c) From the table the operation is commutative since there is symmetry about the leading diagonal.

$(2 * 3) * 4 = 1 * 4 = 4$ and

$2 * (3 * 4) = 2 * 2 = 4$

Therefore, the operation is also associative.

Example 21:8

The operation ∗ is defined over the set $S = \{0, 1, 2, 3, 4\}$ as multiplication modulo 5. Find

(i) $3 * 4$ (ii) $2 * 4$ (iii) $2 * 3$.

Is S closed under this operation?

Solution

$$3 * 4 = \frac{3(4)}{5} = 2R2 = 2$$

$$2 * 4 = \frac{2(4)}{5} = 1R3 = 3$$

$$2 * 3 = \frac{2(3)}{5} = 1R1 = 1$$

Since 2, 3, 1 are in S, the set S is closed under ∗.

EXERCISE 21:1

1. The operation ∗ is defined on the set ℝ of real numbers by $a * b = a + b - ab$, where $a, b \in ℝ$.
   (a) Evaluate $(-2 * 3) * 5$
   (b) Find the solution set of $a * a = -3$.

2. Determine which of the following operations is
   (i) Commutative
   (ii) Associative over the set
   $ℕ* = \{1, 2, 3, 4, 5...\}$.
   (a) Addition (b) multiplication
   (c) Subtraction (d) division

3. The operation ∗ is defined over the set $A = \{2, 3, 4\}$ as multiplication mod 5. Find
   (a) $3 * 4$ (b) $2 * 4$ (c) $2 * 3$
   Is A closed under this operation?

4. The operation ∗ is defined over the set ℤ of integers. If $a, b \in ℤ$ and $a * b$ is the quotient when a is divided by b. For example $13 * 4 = 3$ and $23 * 4 = 5$.
   Evaluate
   (a) $35 * 5$ (b) $100 * 9$ (c) $100 * 8$
   (d) $33 * 4$ (e) $80 * 6$
   (f) Solve the equation $100 * a = 9$

5. The operation ∗ is defined over ℝ the set of real numbers by $a * b = a + b^2$. Find
   (a) $2*3$ (b) $3*2$ (c) $3*5$
   (d) $2*(3*5)$ (e) $(2*3)*5$.
   Is R closed under this operation? Is the operation associative or commutative?

6. The operation × is defined on the universal set 𝓔 by $A × B = \{(x, y): x \in A \text{ and } y \in B\}$.
   Where A and B are subsets of 𝓔.
   If $A = \{-1, 2, 5\}$ and $B = \{4, 1\}$. List the elements of $A × B$.

7. The operation ∗ is defined on a set S. State the property of the operation ∗ given by each of the following statements:
   (a) $x * y \in S$, whenever $x, y \in S$.
   (b) $x * y = y * x$ for all $x, y \in S$.
   (c) $(x * y) * z = x * (y * z)$ for all $x, y \in S$.
   (d) $x * e = x$ where $e, x \in S$.

8. Table 20:5, is an operation table for the set $S = \{a, b, c, d\}$ under the operation ∗.

∗	a	b	c	d
a	b	c	a	d
b	c	d	b	a
c	a	b	c	d
d	d	a	d	c

Table 21:5

(a) State with reasons whether or not S is closed under $*$.

(b) Write down
   (i) The identity element of S under $*$.
   (ii) The inverse of b with respect to $*$.

9. The operation $*$ is defined on the set $\mathbb{R}$ of real numbers, by $a * b = \dfrac{ab}{2}$, where $a, b \in \mathbb{R}$. Given that the identity element of $\mathbb{R}$ under $*$ is 2,

(a) Determine the inverse of a general element $x \in \mathbb{R}$.

(b) Hence state which real number has no inverse with respect to the operation $*$.

10. Let $S = \{A, B, C\}$, where $A \neq \varnothing, B \neq \varnothing$ and $C \neq \varnothing$ are sets in the set S. The operation $*$ is defined on S as intersection i.e. $A*B = A \cap B$. Draw a pair of well shaded Venn diagrams to illustrate that $(A*B)*C = A*(B*C)$. If the operation $*$ is defined as union, draw another pair of well-shaded Venn diagrams to show that $(A*B)*C = A*(B*C)$.

11. Show that the set $\mathbb{N}^* = \{1, 2, 3, 4, 5 \dots\}$ is closed under the operation of :
   (a) Addition   (b) multiplication,
   but not closed under the operation of
   (a) Subtraction   (b) division.

12. The operation $*$ is defined on the set $\mathbb{R}$ of real numbers, by $x * y = 2xy + x + y$.
   Find (a) the identity element.
       (b) The inverse of 4 under this operation.

13. The operation $*$ defined over the set $S = \{2,3,8\}$, is such that $a * b$ denotes the smallest integer greater than $\dfrac{a}{b}$.
   (a) State with reasons whether S is closed.
   (b) Determine whether S is commutative.

14. $x * y$ denotes the greatest integer less than $\dfrac{x}{y}$.
   (a) Find (i) $3 * 2$ (ii) $8 * 3$ (iii) $3 * 8$.
   (b) Is the operation commutative?
   (c) Find $(3 *2) * 3$

15. The operation $*$ is defined on $\mathbb{Q}$, the set of rational numbers by $a * b = \dfrac{2a + 7b}{4}$.

Given that, $-4 * x = -23$, find the value of x.

16. The operation $*$ is defined on the set $\mathbb{R}$ of real numbers, by $a * b = 2a - b$.
   (a) Find the value of (i) $3 * 4$ (ii) $4 *{-}4$
   (b) Show that the operation $*$ is not commutative over $\mathbb{R}$.
   (c) If $2 * x = 6$, find the value of x.

ELEMENTARY GROUP THEORY

A non empty set, S, together with a binary operation $*$ is said to be a group $(S, *)$ if:
 (i) The set is Closed with respect to $*$.
   i.e. $\forall a, b \in S, a* b \in S$.

(ii) There is an identity element
   i.e. $\exists! e \in S: a * e = e * a = a$

(iii) Every element $a \in S$ has an inverse $a^{-1} \in S$
   i.e. $\forall a \in S, \exists a^{-1} \in S: a^{-1} * a = a * a^{-1} = e$

(iv) The operation $*$ is associative over S
   $\forall a, b, c \in S, (a * b) * c = a * (b * c)$

The mnemonic *CL-ID-IN-AS* may help to bring these properties to memory.

Therefore if one is required to proof that a set S together with an operation $*$ defined on the set S forms a group, one must show that:
(a) the set is closed under the operation $*$
(b) there is an identity element
(c) every element has an inverse
(d) the operation $*$ is associative over S

Abelian (or Commutative) Group

An Abelian or commutative group is a group, for which the commutative law also holds. i.e. together with the above points $a * b = b * a$, $\forall a, b \in S$.

The mnemonic *CL-ID-IN-AS-CO* may help to bring these properties to memory.

EXERCISE 21:2

1. The operation $*$ is defined over the set $S = \{1, 2, 3, 5\}$ as multiplication mod 6. Draw up an operation table for this operation. Is $(S, *)$ a group?

2. The operation $*$ is defined on the set S, where $S = \{0, 1, 2, 3, 4, 5\}$ as follows; $\forall a, b, c \in S$, $a*b =$ the remainder when the sum of a and b is divided by 6.
   (a) Form an operation table for the set S under $*$.
   (b) Say whether S forms an Abelian group under $*$.

3. The operation $\otimes$ is defined as multiplication modulo 8 and the operation $\oplus$ is defined as addition modulo 8. e.g. $3 \otimes 7 = 5$ and $3 \oplus 7 = 2$
   (a) Draw up a combination table for $\otimes$ and $\oplus$ on the set S, where $S = \{1, 3, 5, 7\}$.
   (b) Explain why the set S is closed under the operation $\otimes$ but not under the operation $\oplus$.
   (c) Name the identity element for the operation $\otimes$.
   (d) State the inverse of each element under the operation $\otimes$.
   (e) Determine whether the operation $\otimes$ is associative or commutative.
   (f) Does $(S, \otimes)$ form a group? Why?

MULTIPLE CHOICE EXERCISE 21

1. An operation $*$ is defined on $\mathbb{Z}$ the set of integers, by $a*b = a^2 b - b^2 a$. $3*5$ is equal to:
   [A] -45 [B] -30
   [C] 45 [D] 30

2. An operation $*$ is defined on $\mathbb{Z}$ the set of integers, by $a*b = a^2 b - b^2 a$. The operation $*$ is:
   [A] Commutative because,
   $$a*b \neq b*a, \forall a, b \in \mathbb{Z}.$$
   [B] Not commutative because,
   $$a*b \neq b*a, \forall a, b \in \mathbb{Z}.$$
   [C] Commutative because,
   $$a*b = b*a, \forall a, b \in \mathbb{Z}.$$
   [D] Not commutative because,
   $$a*b = b*a, \forall a, b \in \mathbb{Z}.$$

3. An operation $*$ is defined on the set $\mathbb{R}$, of real numbers, as $x*y = xy - (x + y)$. The value of $2*3$ is:
   [A] 6 [B] -5 [C] 1 [D] 5

4. An operation $*$ is defined on the set $\mathbb{R}$, of real numbers, as $x*y = xy - (x + y)$. Given that $x*2 = 8$, the value of x is:
   [A] 6 [B] -6 [C] -10 [D] 10

5. An operation $*$ is defined on $\mathbb{Z}$ as follows: $\forall a, b \in \mathbb{Z}, a * b = a + b - 4$. The identity element for the operation is:
   [A] 1 [B] -1 [C] -4 [D] 4

6. An operation $*$ is defined on $\mathbb{Z}$ as follows; $\forall a, b \in \mathbb{Z}, a * b = a + b - 4$. The inverse of 5 is:
   [A] $\dfrac{1}{5}$ [B] -5 [C] 3 [D] -3

7. The operation $*$ is defined on $\mathbb{Q}$, the set of rational numbers, by $a * b = \dfrac{a + 3b}{2}$. Given that $5*u = 20$, the value of u is:
   [A] $u = \dfrac{35}{3}$ [B] $u = \dfrac{8}{3}$
   [C] $u = \dfrac{5}{3}$ [D] $u = \dfrac{1}{3}$

8. If the operation $a \sim b$ gives the numerical difference between a and b, e.g. $8 \sim 5 = 3$ and $7 \sim 11 = 4$. The operation table, which best represents the operation $\sim$ for the set $S = \{0, 4, 8, 12\}$ is:

[A]
$\sim$	0	4	8	12
0	0	8	4	12
4	4	0	4	8
8	8	4	0	4
12	12	4	8	0

[B]
$\sim$	0	4	8	12
0	0	4	8	12
4	4	0	4	8
8	8	4	0	4
12	12	8	4	0

[C]
$\sim$	0	4	8	12
0	12	8	4	0
4	8	4	0	4
8	4	0	12	8
12	0	8	4	12

[D]
$\sim$	0	4	8	12
0	12	8	4	0
4	8	4	0	4
8	4	0	12	8
12	0	4	8	12

Table 21:6

9. If the operation $a \sim b$ gives the numerical difference between a and b, e.g. $8 \sim 5 = 3$ and $7 \sim 11 = 4$. The identity element is:
   [A] 12 [B] 8 [C] 4 [D] 0

10. The operation $*$ is defined on $\mathbb{R}$, the set of real numbers, by $a*b = a + b + 2ab$. Given that $x* x = 12$, the possible value(s) of x are:
    [A] -3 and 2 [B] -3 and -2
    [C] 3 and -2 [D] 3 and 2

215

11. The operation $*$ is defined on the set $\mathbb{N}$ of natural numbers by $x*y$ is the absolute value of x–y. e.g. $2*2=0$, $5*8=3$ etc. The identity element is:

   [A]   0   [B]   1   [C]   –1

   [D]   there is no identity element

12. The operation $*$ is defined on the set $\mathbb{N}$ of natural numbers by $x*y$ is the absolute value of x–y. e.g. $2*2=0$, $5*8=3$ etc. The inverse of each element is:

   [A]   Its reciprocal.

   [B]   The element itself.

   [C]   The negation of the element.

   [D]   The square of the element.

13. The operation $*$ is:

   [A]   Not associative because,

       $a*b \neq b*a, \forall a, b\in\mathbb{N}$.

   [B]   Associative because,

       $(a*b)*c = a*(b*c), \forall a, b, c\in\mathbb{N}.$

   [C] Not associative because,

       $(a*b)*c \neq a*(b*c), \forall a, b, c\in\mathbb{N}.$

   [D] Not associative because,

       $(a*b)*c = a*(b*c), \forall a, b, c\in\mathbb{N}.$

14. Table 20:7 is a combination table for the set $S =\{a, b, c, d\}$ under a given operation$*$. The identity element is:

   [A] a [B] b [C] c [D] d

$*$	a	b	c	d
a	b	c	a	d
b	c	d	b	a
c	a	b	c	d
d	d	a	d	c

Table 20:7

15. Table 20:7 is a combination table for the set $S = \{a, b, c, d\}$ under a given operation$*$.

   For this operation, it is true to say that:

   [A] $a^{-1}=b,\ b^{-1}=c,\ c^{-1}=d,\ d^{-1}=c$

   [B] $a^{-1}=d,\ b^{-1}=c,\ c^{-1}=a,\ d^{-1}=b$

   [C] $a^{-1}=c,\ b^{-1}=d,\ c^{-1}=a,\ d^{-1}=b$

   [D] $a^{-1}=b,\ b^{-1}=a,\ c^{-1}=c,\ d^{-1}=d$

16. Table 20:7 below is a combination table for the set $S = \{a, b, c, d\}$ under a given operation$*$. The operation $*$ is:

   [A]   Not associative because,

       $a*b \neq b*a, \forall a, b\in S$.

   [B]   Not associative because,

       $(a*b)*c = a*(b*c), \forall a, b, c\in S.$

   [C]   Not associative because,

       $(a*b)*c \neq a*(b*c), \forall a, b, c\in S.$

   [D]   Associative because,

       $(a*b)*c = a*(b*c), \forall a, b, c\in S.$

17. A binary operation $*$ is defined on $\mathbb{R}$ by $x*y = xy + 2x$. $-3*4$ is equal to:

   [A] -4 [B] -18 [C] -12 [D] 8

18. A binary operation $*$ is defined on $\mathbb{R}$ by $x*y = xy + 2x$. The operation $*$ is:

   [A] Not commutative because,

       $a*b \neq b*a, \forall a, b\in\mathbb{R}.$

   [B] Commutative because,

       $a*b \neq b*a, \forall a, b\in\mathbb{R}.$

   [C] Not commutative because,

       $a*b = b*a, \forall a, b\in\mathbb{R}.$

   [D] Commutative because,

       $a*b = b*a, \forall a, b\in\mathbb{R}.$

19. The operation $*$ on the set of real numbers is defined by $a * b = \dfrac{8a + 5b}{3}$. The value of b for which $2 * b = -6$ is:

   [A] $\dfrac{34}{5}$ [B] $-\dfrac{34}{5}$ [C] $-\dfrac{2}{5}$ [D] $\dfrac{2}{5}$

20. The operation ∇ is defined on the set $\{0, 2, 4, 6, 8\}$ by $x \nabla y = (x-y)^2$. The property or properties valid for the operation is/are:

   [A] closure

   [B] commutativity

   [C] associativity

   [D] closure, commutativity and associativity.

21. A binary operation is defined on $\mathbb{R}$ the set of real numbers as follows $a*b = a^2 + ab + b^2$. The value of $-3*2$ is:

   [A] -6 [B] -12 [C] 4 [D] 7

22. The operation $*$ on the set $\mathbb{R}$ of real numbers is defined by $a * b = \dfrac{a^2 + b^2}{2ab}$. The value of $3*-4$ is:

   [A] -25 [B] $-\dfrac{24}{25}$ [C] -24 [D] $-\dfrac{25}{24}$

23. The operation $*$ on the set $\mathbb{R}$ of real numbers is defined by $a * b = \dfrac{a^2 + b^2}{2ab}$. Given that $2*x = 1$, the value of x is:

   [A] 4 [B] -4 [C] -2 [D] 2

TOPIC 22

LOGIC

OBJECTIVES

At the end of this topic, the learner should be able to:

1. Appreciate that a statement or proposition is a sentence that is either true or false but not both.
2. State the truth value of a statement.
3. Identify closed and open statements.
4. Identify the variable and domain of a statement.
5. State the negation of a given statement.
6. Construct and use truth tables to determine the truth value of a statement.
7. Identify composite statements and construct composite statements from given statements.
8. Distinguish between conjunctions and disjunctions and build up truth table for each.
9. Appreciate that a conditional statement is always true if both statements are true.
10. Appreciate that a biconditional statement is always true if both statements are true or if both statements are false.
11. Distinguish between conditional statements and biconditional statements.
12. Identify and use the unitary connector ~ and the binary connectors.
13. Identify tautologies and contradictions and distinguish between them.
14. Make mathematical statements and read statements with quantifiers.
15. Draw conclusions from given hypotheses.
16. Identify syllogisms and make statements using the four forms of syllogisms.

In the early 20th century British mathematician and philosopher **Bertrand Russell** (left), along with British mathematician and philosopher **Alfred North Whitehead** (right), attempted to demonstrate that mathematics and numbers can be understood as groups of concepts, or classes. Russell and Whitehead tried to show that mathematics is closely related to logic and, in turn, that ordinary sentences can be logically analyzed using mathematical symbols for words and phrases. This idea resulted in a new symbolic language, used by Russell in a field he termed philosophical logic, in which philosophical propositions were reformulated and examined according to his symbolic logic.

THE BETTMANN ARCHIVE/Corbis

The Concept of Logic

Logic is a science, which deals with the principles of valid reasoning and argument. Logic is very important in disciplines such as mathematics, law, computer science etc.

Statements or Propositions

A **statement** or **proposition** is a sentence that is either true or false but not both.

This definition lays emphasis on the fact that a statement must contain enough information for anyone to decide whether it is true or false. This information may be buried in the statement itself or in the context, time and place parameters. Questions, commands and remarks are therefore not statements because one cannot decide whether they are true or false. Statements are usually denoted by lower case letters a, b, c,....

Example 22:1

State giving reasons whether or not, the following are statements.

p: Bamenda is not in Cameroon.
q: In which country is Bamenda found?
r: Let us eat the rice!
s: That is exceptional of you!
t: Time for breakfast! Bring tea and bread!

Solution

Among the above, the only statement, which though is false, is "p".

"q" is not a statement. It is a question.
"r" is not a statement. It is a command.
"s" is not a statement. It is a remark. Its truth depends on the judge and the addressee.
"t" is not a statement, It is a reminder and a command.

Questions, commands, remarks and reminders are not statements because they are neither true nor false.

The Truth Value of a Statement

The truthfulness or falsity of a statement is called its **truth-value**. The truth-value of a true statement is denoted by T or 1, while that of a false statement is denoted by F or 0. This is often summarized on a table called the truth table. Thus for a statement p, the truth table could be as follows.

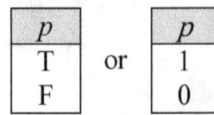

Established statements, which have been established by means of a proof, are called **theorems** and their truth-value is T or 1. The following are some examples of theorems.

t: The sum of the interior angles of a triangle is $180°$.
s: The exterior angle of a triangle is equal to the sum of the two interior opposite angles.

Closed and Open Statements

A closed statement is a statement concerning a definite object. An open statement on the other hand is a statement that does not concern a definite object.

Consider the following statements.
a: Mr. Tantoh has four Children.
b: He is a teacher.
c: $4 + 5 = 9$
d: $2 \times 3 = 8$
e: $3x - 1 = 11$
f: $4p \geq 1 = 11$

Statements c, d, e and f are examples of mathematical statements. Statements a, c and d are closed while statements b, e and f are open. The "he" in statement b may stand for any member in a group of male teachers. This statement can be closed by replacing the "he" by the name of any male teacher in the reference group. The statement will then be either true or false. The x and p in statements e and f can be substituted with numbers to close them and render them true or false.

Domain and Variable

The open part, which must be closed in an open statement to make it true or false, is called the **variable** while the set of all the possible values, which the variable can take, is called the **domain** or the **replacement set** of the variable. For instance;

If $x = 4$, statement e is true, otherwise it is false.

If $p \geq 3$, $p \in \mathbb{R}$, statement f will be true, otherwise it is false. The set of values of a variable, which makes a statement true, is called the **truth** or **solution set** of the statement. Thus, the truth set of statements e and f are $\{x = 4\}$ and $\{p \geq 3, p \in \mathbb{R}\}$ respectively.

Examples 22:2

State the domain and the truth set for each of the following statements.
(a) $5x + 2 = 13$ (b) $7y < 42$
(c) $(x + 1)(x - 2) = 0$

Solution
(a) The domain is $\mathbb{Q}$, the set of rational numbers and the truth set is
$$x: x = \frac{11}{5}, x \in \mathbb{Q}.$$
(b) The domain is $\mathbb{R}$, the set of natural numbers and the truth set is
$$\{y : y < 6, y \in \mathbb{R}\}.$$
(c) The domain is $\mathbb{Z}$, the set of integers and the truth set is $\{-1, 2\}$.

EXERCISE 22:1

1. Which of the following is a true, false or open statement?
   (a) $3 + 3 = 3 \times 2$ (b) $8 \times 1 \neq 8 + 1$
   (c) $21 + 0 = 21 \times 1$ (d) $3 + 7 = 7 + 3$
   (e) The sum of 25 and 0 is 25.
   (f) The sum of a whole number and 0 is the number.
   (g) The sum of a whole number and 7 is 7.
   (h) The product of 67 and 1 is 68.
   (i) If 3 is subtracted from 20 , the difference is the same as when 20 is subtracted from 3.
   (j) The product of 8 and 4 is the same as the product of 4 and 8.
   (k) If 8, is divided by 4, the quotient is the same as when 4 is divided by 8.
2. Given that the domain of the variable x is $A = \{1, 2, 3, 4, 5, 6, 7, 8, 9\}$. State the truth or solution set of each of the following statements.

   (a) $x + 3 = 10$ (b) $x > 7$
   (c) x is an even number   (d) $x^2 = 25$
   (e) $x + 10 = 10 + x$ (f) $x - 8 = 8 - x$
   (g) $0 \times x = x$ (h) $0 \times x = 0$
   (i) $2x = x + 5$ (j) $2x \neq x + 5$
   (k) $\{x : x \geq 7\}$ (l) $\{x : x > 4\}$
3. Sort out the sentences, which are statements and state whether they are true or false.
   (a) Give me that pencil.
   (b) There is an Ocean in Bamenda.
   (c) All teachers are lazy.
   (d) All Bamenda people eat Achu.
   (e) Make sure your uniforms are very neat.
   (f) Is Nigeria an African Country?
   (g) History is a science subject.
   (h) Ahidjo was a president of Cameroon.
4. Classify the following statements as closed or open statements.
   (a) Mr. Fonche died two years ago.
   (b) $3 + 5 = 9$
   (c) They are lazy.
   (d) All Bamenda people eat Achu.
   (e) $4 \times 3 = 12$
   (f) Nigeria is an African Country
   (g) History is a science subject.
   (h) He was the president of Cameroon.
   (i) Everyone loves Mr. Paul Biya.
   (j) She comes from Nkambe.
   (k) $2x - 1$
5. Given that $0 \leq x \leq 10$, $x \in \mathbb{N}$. State the truth set for the following statements.
   (a) $x > 5$.
   (b) x is a factor of 10.
   (c) x is an odd number.
   (d) x is an even number.
   (e) x is a prime number
   (f) x is a multiple of 3.
6. State the truth-value of each of the following statements.
   (a) 37 is a prime number.
   (b) 105 is a multiple of 3.
   (c) All Bamenda people eat Achu.
   (d) History is a science subject.
   (e) $4 \times 3 = 12$
   (f) In the set of real numbers $a - b = b - a$
   (g) Nigeria is an African Country.

Negation, $\sim p$

The negation of a statement p, is the statement formed from p by inserting the word "not" into p or placing the phrases "it is false that"…or "it is not true that"…before the statement p. The negation of a statement p is denoted by $\sim$ $\neg$ or p' and read 'not p'.

Example 22:3
Let p be the statement "Bamenda is in Cameroon". State the negation $\sim p$ of p.

Solution
$\sim p$: Bamenda is not in Cameroon or
$\sim p$: It is not true that Bamenda is in Cameroon or
$\sim p$: It is not true to say that Bamenda is in Cameroon or
$\sim p$: It is false that Bamenda is in Cameroon.

The fundamental property of a statement p and $\sim p$ is that,

If p is, true $\sim p$ is false; if p is false, $\sim p$ is true.

This means that p and $\sim p$ cannot be both true and both false. This fundamental property can be summarized on a truth table or Venn diagram as follows,

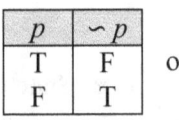 or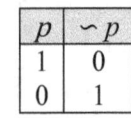

p	$\sim p$
T	F
F	T

p	$\sim p$
1	0
0	1

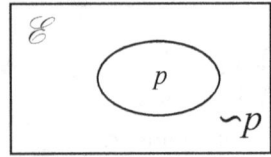

Example 22:4
State the negation of the following statements.
q: It is raining.
r: All Cameroonians speak both English and French.
s: 5 is a multiple of 3.
t: $2x + 1 = 0$, for all values of x.

Solution
$\sim q$: It is not raining.
$\sim r$: It is not true that all Cameroonians speak both English and French.
$\sim s$: 5 is not a multiple of 3.
$\sim t$: It is not true that $2x + 1 = 0$, for all values of x.

EXERCISE 22:2
1. State the negation of the following statements.
    (a) Mr. Fonche died two years ago.
    (c) They are lazy.
    (d) All Bamenda people eat Achu.
    (e) Loh cannot drive.
    (f) Nigeria is an African Country
    (g) History is a science subject.
    (h) He was the president of Cameroon.
    (i) Everyone loves Mr. Paul Biya.
    (j) She comes from Nkambe.
    (k) Science has done more harm than good.
2. Write the negation of the following statements in symbolic language
    (a) $4y > 12$ (b) $3 + 5 = 9$ (c) $3p - 1 \geq 17$
    (d) $4 \times 3 = 12$ (e) $6x + 1 < 19$ (f) $2x - 1 = 0$
    (g) $A \not\subset B$ (h) $A \cap B \neq \varnothing$

Compound or Composite Statements

A compound or composite statement is a statement, which is made by combining two or more statements as in the following cases.

Conjunction

When a composite statement is made by combining two statements with the use of the preposition "and" the resulting composite statement is called a **conjunction**. Symbolical a conjunction composed of the two statements p and q is denoted by $p \wedge q$.
This fundamental property can be summarized on a truth table or Venn diagram as follows,

The fundamental property of the conjunction $p \wedge q$ is that $p \wedge q$ can only be true if both p and q are true otherwise the conjunction is false.

p	q	$p \wedge q$
T	T	T
T	F	F
F	T	F
F	F	F

or

p	q	$p \wedge q$
1	1	1
1	0	0
0	1	0
0	0	0

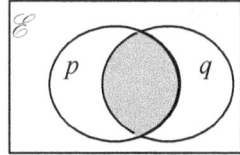

p ∧ q is shaded

Example 22:5

Given the statements

  x: Ngwa ate rice.

  y: Ngwa ate beans.

Make a conjunction involving *x* and *y*.

Solution

  $x \wedge y$: Ngwa ate rice and beans.

The symbol ∧ is often used to define the intersection of sets. Thus,

$A \cap B = \{x : x \in A \wedge x \in B\}$.

Disjunction

When a composite statement is made by combining two statements with the use of the preposition "or" the resulting composite statement is called a **disjunction**. Sometimes it is necessary to use the words 'either...or' for disjunctions. Symbolical a disjunction composed of the two statements *p* and *q* is denoted by $p \vee q$.

Consider the composite statement *a* below;

a: At 8 a.m. I shall be teaching in school or resting in my house.

Is it possible for someone to be teaching in school and resting in his house simultaneously? It is impossible!

Therefore, only one of the following simple statements *b* and *c* can be true of *a*.

b: At 8 a.m., I shall be teaching in school.
c: At 8 a.m., I shall be resting in my house.

A disjunction composed of two statements *p* and *q*, which cannot both be true, is called an **exclusive disjunction**.

The truth table and Venn diagram below represent exclusive disjunctions.

p	*q*	*p* ∨ *q*
T	T	F
T	F	T
F	T	T
F	F	F

or

p	*q*	*p* ∨ *q*
1	1	0
1	0	1
0	1	1
0	0	0

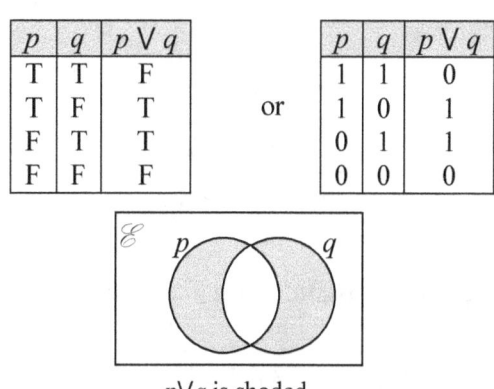

p∨*q* is shaded

Now consider the statement *r* below.

r: All the students who passed in English or French shall be given prizes.

Is it possible to pass in both English and French? Obviously! Therefore, any student who passes in both English and French should stand even a better chance to receive prizes. *r* equally carries the meaning in the statements *s* and *t* below.

s: All the students who passed in English shall be given prizes.
t: All the students who passed in French shall be given prizes.

Therefore, *s* is true, *t* is true and *s* and *t* are true.

A disjunction composed of two statements *p* and *q*, which can both be true, is called an **inclusive disjunction**. In logic a disjunction generally refers to inclusive disjunction where "or" is used in the sense of "and/or".

The fundamental property of the disjunction $p \vee q$ is that *p* is true or *q* is true or both *p* and *q* are true otherwise, the conjunction is false.

This fundamental property can be summarized on a truth table or Venn diagram as follows.

p	*q*	*p* ∨ *q*
T	T	T
T	F	T
F	T	T
F	F	F

or

p	*q*	*p* ∨ *q*
1	1	1
1	0	1
0	1	1
0	0	0

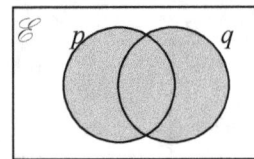

p V *q* is shaded

The symbol V is often used to define the union of sets. Thus,

$$A \cup B = \{x : x \in A \vee x \in B\}.$$

Conditional Statements, $p \rightarrow q$

A **Conditional Statement** is a composite statement obtained by joining two statements *p* and *q* in such a way that if *p* is true *q* must be true. The conditional statement carrying the meaning "if *p* is true then *q* is true" is written symbolically as $p \rightarrow q$. The conditional statement $p \rightarrow q$ can be read in the following ways

(i) *p* implies *q*   (ii) *p* is sufficient for *q*

(iii) *p* only if *q*   (iv) *q* is necessary for *p*

If $p \rightarrow q$ is a theorem, the conditional statement is called an **implication** and the symbol $p \Rightarrow q$ is used.

The fundamental property of a conditional statement
$p \rightarrow q$ is that $p \rightarrow q$ is always true except *p* is true and *q* is false.

This fundamental property can be summarized on a truth table or Venn diagram as follows.

p	*q*	$p \rightarrow q$
T	T	T
T	F	F
F	T	T
F	F	F

or

p	*q*	$p \rightarrow q$
1	1	1
1	0	0
0	1	1
0	0	0

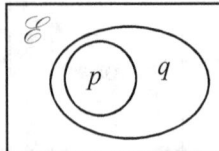

Example 22:6

Which of the following statements is false?

p: Cameroon is in Africa $\longrightarrow$ $3 \times 2 = 7$.

q: Cameroon is in Europe $\longrightarrow$ $3 \times 2 = 6$.

r: Cameroon is in Africa $\longrightarrow$ $3 \times 2 = 6$.

s: Cameroon is in Europe $\longrightarrow$ $3 \times 2 = 7$.

Solution

By the fundamental property of conditional statements, only *p* is false and *q*, *r* and *s* are all true.

Biconditional Statement, $p \Leftrightarrow q$ or $p \leftrightarrow q$

A **Biconditional Statement** is a proposition in logic involving two statements one of which is true if, and only if, the other is true. The biconditional statement involving *p* and *q* is written symbolically as $p \Leftrightarrow q$ or $p \leftrightarrow q$. The biconditional statement $p \Leftrightarrow q$ or $p \leftrightarrow q$ can be read in the following ways

(i) *p* is a necessary and sufficient condition for *q*.

(ii) *p* implies and is implied by *q*

(iii) *p* if and only if *q*

The biconditional statement $p \Leftrightarrow q$ or $p \leftrightarrow q$ is sometimes written as *p* iff *q*.

The fundamental property of a biconditional statement is that, either both statements are true or both are false, otherwise, the statement is false.

This fundamental property can be summarized on a truth table as follows.

p	*q*	$p \leftrightarrow q$
T	T	T
T	F	F
F	T	F
F	F	T

or

p	*q*	$p \leftrightarrow q$
1	1	1
1	0	0
0	1	0
0	0	1

Example 22:7

Which of the following statements is true?

p: Cameroon is in Africa $\leftrightarrow$ $3 \times 2 = 7$.

q: Cameroon is in Europe $\leftrightarrow$ $3 \times 2 = 6$.

r: Cameroon is in Africa $\leftrightarrow$ $3 \times 2 = 6$.

s: Cameroon is in Europe $\leftrightarrow$ $3 \times 2 = 7$.

Solution
By the fundamental property of biconditional statements, r and s are true and p and q are false.

Logical Equivalence

Two statements p and q are **logically equivalent** if and only if they have the same truth tables. The statement p and q are logically equivalent is written symbolically as $p \equiv q$. The importance of truth tables cannot therefore be over emphasized. Apart from the use of truth tables to summarize composite statements, they are also used to determine if two statements are logically equivalent.

Example 22:8
Draw truth tables showing the following:
(a) $p \wedge q$ (b) $p \vee q$ (c) $q \wedge p$ (d) $q \vee p$.
Hence, deduce that:
(i) $p \wedge q \equiv q \wedge p$ (ii) $p \vee q \equiv q \vee p$

Solution
(a)

p	q	$p \wedge q$
T	T	T
T	F	F
F	T	F
F	F	F

(b)

p	q	$p \vee q$
T	T	T
T	F	T
F	T	T
F	F	F

(c)

p	q	$q \wedge p$
T	T	T
T	F	F
F	T	F
F	F	F

(d)

p	q	$q \wedge p$
T	T	T
T	F	T
F	T	T
F	F	F

It can be seen that the truth tables for $p \wedge q$ and $q \wedge p$ are the same.
Therefore, (i) $p \wedge q \equiv q \wedge p$
Similarly, the truth tables for) $p \vee q$ and $q \vee p$ are the same.
Therefore, (ii) $p \vee q \equiv q \vee p$

Example 22:8
Draw the truth tables for the following.
(a) $\sim p \vee \sim q$ (b) $\sim (p \wedge q)$
(c) $\sim p \wedge \sim q$ (d) $\sim (p \vee q)$.
 Deduce that:
 (i) $\sim p \vee \sim q \equiv \sim (p \wedge q)$
 (ii) $\sim p \wedge \sim q \equiv \sim (p \vee q)$

Solution
(a)

p	q	$\sim p$	$\sim q$	$\sim p \vee \sim q$
T	T	F	F	T
T	F	F	T	T
F	T	T	F	T
F	F	T	T	F

(b)

p	q	$\sim (p \wedge q)$
T	T	T
T	F	T
F	T	T
F	F	F

(c)

p	q	$\sim p$	$\sim q$	$\sim p \wedge \sim q$
T	T	F	F	F
T	F	F	T	F
F	T	T	F	F
F	F	T	T	T

(d)

p	q	$p \vee q$	$\sim (p \vee q)$
T	T	T	F
T	F	F	F
F	T	F	F
F	F	F	T

From the tables it can be seen that the truth tables for $\sim p \wedge \sim q$ and $\sim (p \vee q)$ are the same.
Therefore, (i) $\sim p \wedge \sim q \equiv \sim (p \vee q)$
Similarly, the truth tables for $\sim p \vee \sim q$ and $\sim (p \wedge q)$ are the same.
Therefore, (ii) $\sim p \vee \sim q \equiv \sim (p \wedge q)$

De Morgan's Laws

The logically equivalent statements in example 22:8 are called the **De Morgan's laws**. These laws are restated below.

The De Morgan's laws of logic state that:
(i) $\sim p \wedge \sim q = \sim (p \vee q)$
(ii) $\sim p \vee \sim q = \sim (p \wedge q)$

Connectors

The logical symbols $\sim$, $\wedge$, $\vee$, $\rightarrow$ and $\leftrightarrow$ are called connectors and have ', $\cap$, $\cup$, $\rightarrow$ and $=$ as their respective set algebra equivalents. $\sim$ is called a unitary connector because it affects only one statement while $\wedge$, $\vee$, $\rightarrow$ and $\leftrightarrow$ are called binary connectors because they combine two statements.

> ### EXERCISE 22:3

1. Let p be the statement "She is lazy" and q be the statement "She is beautiful". Write down the following using symbolic language.
   (i) She is not lazy.
   (ii) She is beautiful and lazy.
   (iii) She is neither lazy nor beautiful.
   (iv) She is either lazy or beautiful.
   (v) It is not true that she is beautiful and lazy.
   (vi) It is false that she is lazy and beautiful.
2. Decompose the following composite statements into its components.
   (a) Mrs. Ngwa and Mrs. Tayong visited me.
   (b) Nfor likes rice and beans.
   (c) Mr. Nkwain is a Cameroonian ambassador.
   (d) Bamenda and Bafoussam are big cities.
3. Let p and q be two statements defined as follows.

   p: Nfor is hungry.
   q: Nfor is thirsty.

   Write out the following in English.
   (i) $\sim p$ (ii) $p \wedge q$
   (iii) $p \vee q$ (iv) $p \Rightarrow q$
   (v) $p \Leftrightarrow q$ (vi) $p \Rightarrow \sim q$
   (vii) $\sim p \wedge \sim q$ (viii) $p \Leftrightarrow \sim q$

4. Make a conditional statement by combining the following pair of statements
   (a) Fombe is rich. Fombe is happy.
   (b) He was drunk. He drank alcohol.
   (c) It is night in Cameroon. Places are dark.
   (d) She performed well. She had a prize.
5. Make a biconditional statement by combining the pair of statements in question 4.
6. Draw the truth tables for the propositions
   $(p \rightarrow q) \wedge (q \rightarrow p)$ and $p \leftrightarrow q$.
   Use your tables to determine whether the statement
   $(p \rightarrow q) \wedge (q \rightarrow p) \equiv p \leftrightarrow q$ is true.
7. Draw the truth tables for
   $p \rightarrow q$ and $\sim p \vee q$ and say whether or not $p \rightarrow q \equiv \sim p \vee q$.
8. Proof the following:
   (a) the associative law of logic
       i.e. $(p \wedge q) \wedge r \equiv p \wedge (q \wedge r)$
   (b) the distributive law of logic
       i.e. $p \vee (q \wedge r) \equiv (p \vee q) \wedge (p \vee r)$
9. Draw up the truth tables for the following compound statements.
   (a) $(p \vee q) \rightarrow (p \wedge q)$ (b) $(p \vee q) \wedge \sim p$
   (c) $p \rightarrow (p \wedge \sim q)$ (d) $(p \rightarrow q) \wedge q$
   (e) $(p \rightarrow q) \rightarrow \sim p$ (f) $p \wedge (p \rightarrow q)$
10. Determine which of the following statements are equivalent.
    (a) $(p \rightarrow q) \wedge r$; $p \rightarrow (q \wedge r)$
    (b) $p \vee (q \wedge r)$; $(p \vee q) \wedge (p \vee r)$
    (c) $(p \wedge q) \rightarrow r$; $p \rightarrow (q \rightarrow r)$
11. Given that water can flow through two taps A and B as shown below. Draw the truth table for A or B using T for open and F for closed.

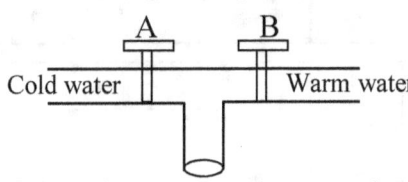

12. The following shows a TV connected to a socket. S is the power button of the swicth and P is the power button of the TV. Draw the truth table for S and T using 1 for on and 0 for off.

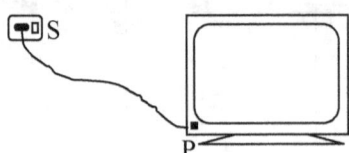

Tautologies

A tautology is a proposition, which is always true. Therefore, the last column of the truth table of a tautology is made up of all Ts or 1s. For instance, the proposition "p or not p" is a tautology.

p	$\sim p$	$p \vee \sim p$
T	F	T
F	T	T

Contradictions

A contradiction is a proposition, which is always false. Therefore, the last column of the truth table of a contradiction is made up of all Fs or 0s. For instance, the proposition "p and not p" is a contradiction.

p	$\sim p$	$p \wedge \sim p$
T	F	F
F	T	F

Quantifiers

Logical operators which are used to stand for words such as for all, for every, for some, for each, for any, there exist, for at least etc. are called **Quantifiers** because they suggest the idea of quantity. The common quantifiers are universal, existential and unitary quantifiers.

Universal Quantifier

The symbol for the universal quantifier is $\forall$, read "for all" or "for every". Thus if $p(x)$ is a true statement concerning all the elements in a set A, this idea can be expressed symbolically as follows.

$$\forall x \in A, p(x) \text{ or } (\forall x \in A) \, p(x).$$

This is read as "for all x belonging to A, $p(x)$ is true" or "for every element x of A, $p(x)$ is true".

The fundamental property of the universal quantifier is as follows.

(1) True: if for all x in A, $p(x)$ is true.
(2) False: If there is at least one element in A, for which $p(x)$ is false.

Existential Quantifier

The symbol for the existential quantifier is $\exists$, read "there exists" or "for at least one" or "for some". Thus if $p(x)$ is a true statement concerning at least one but not all the elements in a set A, this idea can be expressed symbolically as follows.

$$\exists x \in A, p(x) \text{ or } (\exists x \in A) \, p(x).$$

Unitary Existential Quantifier

The symbol for the unitary existential quantifier is $\exists!$ read, "there exists one and only one". Thus if $p(x)$ is a true statement concerning only one element in a set A, this idea can be expressed symbolically as follows.

$$\exists! x \in A, p(x) \text{ or } (\exists! x \in A) \, p(x).$$

This is read as "there exist one and only one element x belonging to A, for which $p(x)$ is true".

The fundamental property of the unitary existential quantifier is as follows.

(1) True: If there is at least one element x in A, for which $p(x)$ is true.
(2) False: If for all, x in A, $p(x)$ is false.

Example 22:9

In each of the following cases write statements using a quantifier and explain each in English.

(1) Let $A = \{1,2,3,4,5\}$ and $p(x): 0 \leq x - 1 < 5$.
(2) Let $E = \{2,4,6,8,10\}$ and $p(x): x$ is even.
(3) Let $E = \{2,4,6,8,10\}$ and $p(x): x$ is a multiple of 2.
(4) $\mathbb{R}$ is the set of real numbers and $p(a): a + 1 = 1 + a = a$.
(5) Let $A = \{1,2,3,4,5\}$ and $p(x): x - 1$ is odd.
(6) Let $E = \{2,4,6,8,10\}$ and $p(x): x$ is a multiple of 4.
(7) $\mathbb{R}$ is the set of real numbers and $p(a): a \times 0 = 0 \times a = 0$.
(8) $\mathbb{R}$ is the set of real numbers and $p(a): a \times 1 = 1 \times a = a$.

Solution

(1) $\forall x \in A, p(x)$ or $(\forall x \in A)\, p(x)$.

This statement means "for all values of x belonging to the set A, $0 \le x - 1 < 5$.

(2) $\forall x \in E, p(x)$ or $(\forall x \in E)\, p(x)$.

This statement means, "for all values of x belonging to the set E, x is even".

(3) $\forall x \in E, p(x)$ or $(\forall x \in E)\, p(x)$.

This statement means, "for all values of x belonging to the set E, x is a multiple of 2".

(4) $\forall a \in \mathbb{R}, \sim p(a)$ or $(\forall a \in \mathbb{R}) \sim p(a)$.

This statement means "for all real values of a, $a + 1 = 1 + a = a$.

(5) $\exists x \in A, p(x)$ or $(\exists x \in A)\, p(x)$.

This statement means "There exist at least one element x belonging to the set A, for which $p(x)$ is true".

(6) $\exists x \in E, p(x)$ or $(\exists x \in E)\, p(x)$.

This statement means, "There are some elements in the set E, which are multiples of 4".

(7) $\exists!\, x \in \mathbb{R}, p(x)$ or $(\exists!\, x \in \mathbb{R})\, p(x)$.

There exists one and only one element a belonging to $\mathbb{R}$, which satisfies the condition $a \times 0 = 0 \times a = 0$.

(8) $\exists!\, x \in A, p(x)$ or $(\exists!\, x \in A)\, p(x)$.

There exists one and only one element belonging to $\mathbb{R}$, which satisfies the condition $a \times 1 = 1 \times a = a$.

EXERCISE 22:4

Using variables of your choice to represent the sets involved, write out the following using the quantifiers $\forall$, $\exists$ or $\exists!$

1. 2 is the only even prime number.
2. All doctors are scientists.
3. Some footballers play volleyball.
4. No student passed the examination.
5. In $\mathbb{R}$, 0 is the additive inverse for addition.
6. All fishes swim.
7. Some parallelograms are rectangles
8. There is no real number such that $x^2 + 1 = 0$

Syllogisms

A **syllogism** is an argument made up of statements in one of four forms:

(a) Universal affirmative
   Form: All A's are B's.
   e.g. p: all cows eat grass.
(b) Universal negative
   Form: No A's are B's.
   E.g., No historians are chemists.
(c) A particular affirmative
   Form: Some A's are B's.
   e.g., some polygons are rectangles.
(d) A particular negative
   Form: Some A's are not B's.
   E.g., some students are not studious.

The common nouns, such as cows, grass, historians, chemists etc are called the terms of the syllogism.

Hypotheses and Conclusions

Consider the following statements.
p: All even numbers are divisible by 2.
q: 14 is an even number
   Therefore,
r: 14 is divisible by 2.

In logic, statements such as p and q are called the **premises** or **hypotheses** while r is called the **conclusion**.
A well-formed syllogism consists of two premises and a conclusion, each premise having one term in common with the conclusion and one in common with the other premise.
It follows that if the premises or hypotheses are true then the conclusion is bound to be true.

EXERCISE 22:5

Write a conclusion r which follows from the given premises p and q.

1. p: All politicians are liars.
   q: Ngoh is a politician
2. p: Some polygons are parallelograms.
   q: Some parallelograms are rectangles.
3. p: $x > 7$
   q: $x < 4$
4. p: The rich are never happy.
   q: Nfor is rich.
5. p: Bamenda is in Cameroon.

q: Cameroon is in Europe.

6. p: x is a multiple of 4 $\Rightarrow$ x is a multiple of 2.
   q: 18 is a multiple of 4.

7. p: If each angle in a quadrilateral is a right angle, then the quadrilateral is a rectangle.
   q: The quadrilateral $ABCD$ is a rectangle.

8. p: All rhombuses have equal sides.
   q: $ABCD$ is not a rhombus.

9. p: The beautiful ones are not yet born.
   q: Bih was born in 1998.

10. p: $A > B$
    q: $B < A$

MULTIPLE CHOICE EXERCISE 22

1. Given the truth table

p	q
T	F
F	T

The correct relationship between the statements p and q is:

[A] $p \Rightarrow q$ [B] $p \Leftrightarrow q$

[C] $\sim p = q$ [D] $p = q$

2. The closed statement among the following is:

[A] $4 + 4 = 8$ [B] He ate rice

[C] $2x - 1 \leq 0$ [D] $2x - 1 = 0$

3. The open statement among the following is:

[A] $7 + 7 = 14$

[B] Her makeup is good

[C] $3 + 4 \leq 0$

[D] Ngwa is a footballer

4. Which of the following is not a statement?

[A] $3p - 7 = 5$. [B] Roses are red.

[C] Tse is sick. [D] Read your Bible everyday.

5. A statement among the following is:

[A] $5 + 8$

[B] Hit the iron while hot

[C] $ax^2 + bx + c$

[D] $x = -b \pm \dfrac{\sqrt{b^2 - 4ac}}{2}$

6. The truth set for the statement
   $X = \{x: 2x - 1 \leq 7, x \in \mathbb{N}\}$ is:

[A] $X = \{0, 1, 2, 3, 4\}$

[B] $X = \{x: x < 4, x \in \mathbb{N}\}$

[C] $X = \{x: 0 \leq x < 4, x \in \mathbb{N}\}$

[D] $X = \{x: 0 \leq x \leq 4, x \in \mathbb{N}\}$

7. A statement among the following is:

[A] To God be the glory.

[B] $10 - 3$

[C] Stand up when the visitor arrives

[D] $5x > -20$

8. If p: Bih is in Bamenda.
   q: Bamenda is in Cameroon.
   Then,

[A] $q \Rightarrow p$ [B] $\sim p \Rightarrow q$

[C] $p \Rightarrow q$ [D] $\sim q \Rightarrow p$

9. Given that,
   p: $\triangle ABC$ is equilateral.
   q: $\triangle ABC$ is equiangular.
   Then,

[A] $\sim p \Rightarrow q$ [B] $\sim q \Leftrightarrow p$

[C] $p \Leftrightarrow q$ [D] $\sim p \Rightarrow q$

10. The truth table for $p \wedge q$ is:

p	q	$p \wedge q$
T	T	T
T	F	F
F	T	F
F	F	F

[A]

p	q	$p \wedge q$
T	T	T
T	F	T
F	T	T
F	F	F

[B]

p	q	$p \wedge q$
T	T	T
T	F	F
F	T	T
F	F	T

[C]

p	q	$p \wedge q$
T	T	T
T	F	T
F	T	F
F	F	F

[D]

11. The truth table for $p \vee q$ is:

p	q	$p \vee q$
T	T	T
T	F	F
F	T	F
F	F	F

[A]

p	q	$p \vee q$
T	T	T
T	F	T
F	T	T
F	F	F

[B]

p	q	$p \vee q$
T	T	T
T	F	F
F	T	T
F	F	T

[C]

p	q	$p \vee q$
T	T	T
T	F	T
F	T	F
F	F	F

[D]

12.

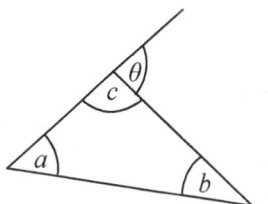

The following statements refer to Figure 22:1

$p: a + b = \theta$

$q: c + \theta = 180°$

$r: a + b + c = 180°$

The arguments which proves that the exterior angle of a triangle is equal to the sum of the two interior opposite angles is:

[A] $q \Rightarrow r \Rightarrow p$ [B] $r \Rightarrow p \Rightarrow q$

[C] $r \Rightarrow q \Rightarrow p$ [D] $q \Rightarrow p \Rightarrow r$

13. Which of the following is a composite statement?

[A] He studied Mathematics at E.N.S.

[B] The kola nut has been eaten.

[C] He can drive.

[D] There is no chalk in the school.

14. Which of the following statements is not a composite statement?

[A] Bamenda is in the NW region of Cameroon.

[B] Ngala was dead and buried.

[C] Fonjong eats coco yams.

[D] The sun is shining when it is raining.

15. The truth table that represents a tautology is:

p	q	$p \wedge q$
T	T	T
T	F	T
F	T	T
F	F	T

[A]

p	q	$p \wedge q$
T	T	T
T	F	F
F	T	F
F	F	F

[B]

p	q	$p \wedge q$
T	T	T
T	F	F
F	T	F
F	F	T

[C]

p	q	$p \wedge q$
T	T	F
T	F	F
F	T	F
F	F	F

[D]

16. The truth table that represents a contradiction is:

p	q	$p \wedge q$
T	T	T
T	F	T
F	T	T
F	F	T

[A]

p	q	$p \wedge q$
T	T	T
T	F	F
F	T	F
F	F	F

[B]

p	q	$p \wedge q$
T	T	T
T	F	F
F	T	F
F	F	T

[C]

p	q	$p \wedge q$
T	T	F
T	F	F
F	T	F
F	F	F

[D]

17. An inclusive disjunction among the following is:

[A] Nursing requires a mastery of physics or chemistry.

[B] At 6 a.m. I shall go to Bamenda or Douala.

[C] I shall be sleeping or eating.

[D] In high school, I shall offer LS1 or LA1.

18. An exclusive disjunction among the following is:

[A] $-5 < 0$ or $5 > 2$.

[B] $4 > 1$ or $4 < 6$.

[C] $\triangle ABC$ is equilateral or a right-angled triangle.

[D] $0 < x \leq 20$ or $x > 3$.

19. The symbol, which is not a logical connector, is:

[A] $\wedge$ [B] $>$ [C] $\vee$ [D] $\rightarrow$

20. The unitary connector among the following logical symbols is:

[A] $\sim$ [B] $>$ [C] $\vee$ [D] $\rightarrow$

21. The statement, which is not a syllogism, is:

[A] All cows eat grass.

[B] No historians are chemists.

[C] I love yams.

[D] Some polygons are rectangles.

22. The negation of the statement "All Anglophones speak English" is:

[A] Some Anglophones speak English.

[B] Not all Anglophones do speak English.

[C] No Anglophone speaks English.

[D] All Anglophones do not speak English.

23. The negation of the statement "All x are not y" is:

[A] Some x are y.

[B] Some x are not y.

[C] No x are y.

[D] All x are not y.

24. The negation of the statement "Not all that glitters is gold" is:

[A] All that glitters is not gold.

[B] Some of what glitters is gold.

[C] Nothing that glitters is gold.

[D] All that glitters is gold.

25. Given the following statements:

p: It is raining.

q: I will go to school by car.

The statement $p \wedge q$ is:

[A] It is raining or I will go to school by car.

[B] It is raining and I will go to school by car.

[C] It is raining and if I will go to school by car.

[C] It is raining and if and only if I will go to school by car.

TOPIC 23

BINARY RELATIONS AND FUNCTIONS

OBJECTIVES

At the end of this topic, the learner should be able to:

1. Appreciate the concept of a relation.
2. Find the Cartesian product of two given sets.
3. State the domain and range of a mathematical relation.
4. Identify the different notations used to denote relations.
5. Define or redefine relations using set builder notation, table of values, sets of ordered pairs, formulas, rule definition, graphs and arrow diagrams.
6. State the inverse of a given relation and represent it on graphs and arrow diagrams.
7. State and use the properties of relations in a set.
8. Determine whether or not a relation is an equivalence relation.
9. Determine whether or not a given relation is a function.
10. Identify functions written using various functional notations.
11. Representation functions on graphs and arrow diagrams.
12. Use flow charts to analyze functions.
13. Identify the different types of mappings and represent them graphically or using arrow diagrams.
14. Analyze a given function using flow charts.
15. Find the inverse of a given rational function.
16. Find the composite function of two or more functions.

Antarctic Giant Petrel

Sometimes known as the giant fulmar, the Antarctic giant petrel is a migratory offshore seabird **related** to the albatross. This photo shows the horny tubular nostrils on top of the bill that characterize members of the Shearwater family, giving rise to their nickname "tubenose." One of the **functions** of these nostrils is to detect changes in air currents.

The Idea of a Relation

Suppose that Mr. Ngala from family A is married to Mafor who is a sister to Mr. Ndoh of family B. Then the following statements can be made concerning these three people.
(i) Mr. Ngala is the husband of Mafor.
(ii) Mr. Ngala is a brother-in-law to Mr. Ndoh.
(iii) Mr. Ndoh is the brother of Mafor.

Statement (i) and (ii) relate Mr. Ngala who is from family A to Mafor and Mr. Ndoh who are from family B. statement (iii) relates Mr. Ndoh to Mafor who are both from family B. Statements (i) and (ii) illustrate relations from one family or set A called the **domain** to another family or set B called the **codomain**. On the other hand, statement (iii) illustrates a relation within only one set called a **relation in a set**.

The Cartesian product of Two Sets

The Cartesian product of two sets A and B is denoted by $A \times B$, read "A cross B" and is the set of all ordered pairs (a, b), formed by choosing the first element from the set A and the second element from the set B.

$$A \times B = \{(a,b) : a \in A \text{ and } b \in B\}$$

In any ordered pair (a, b); the order of the elements is very important.
Thus, $(a, b) \neq (b, a)$
Again, the ordered pair (a, b) should not be confused with the same notation used to represent an open interval.

Example 23:1
Given that $A = \{1,2,3\}$ and $B = \{x, y\}$. Compute the Cartesian products
  (a) $A \times B$ (b) $B \times A$

Solution
(a) $A \times B = \{(1,x),(1,y),(2,x),(2,y),(3,x),(3,y)\}$
(b) $B \times A = \{(x,1),(x,2),(x,3),(y,1),(y,2),(y,3)\}$
Therefore, $A \times B \neq B \times A$

Mathematical Relation

Relations also exist outside human families. A relation that exists between sets of numbers is called a **mathematical relation** or **number relation**.

Example 23:2

A Mathematical relation "is a factor of" is defined from a set $X = \{2, 3, 4\}$ to another set $Y = \{1, 2, 3, 4, 5, 6, 7, 8\}$. Illustrate this relation diagrammatically.

Solution

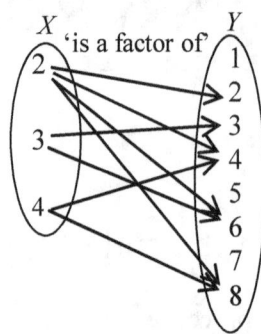

Figure 23:1

Notice that some elements in the codomain Y have not been used up. The set of elements of the codomain, which are used up, is called the **range**. The illustrations above lead us to the definition of a relation.

A **relation** R is a rule that assigns an element $x \in A$ to another element $y \in A$ (relations in a set) or an element $x \in A$ to another element $y \in B$ (relations from one set to another). Therefore, a relation is a set of ordered pairs. Hence, the relation "is a factor of" above defined from set $X = \{2,3,4\}$ to set $Y = \{1,2,3,4,5,6,7,8\}$ may simply be written as
$$\Re = \{(2,2),(2,4),(2,6),(2,6),(3,3),(3,6),(4,4),(4,8)\}$$
Comparing $\Re$ with the Cartesian product $X \times Y$, it can be seen that a relation is a subset of a Cartesian product.

Notation

The statement 'a relates b' is denoted by $a \Re b$ or (a, b).

Ways of Defining Relations
Relations may be defined in the following ways.
(i) Rule definition Method
(ii) Formula Method
(iii) Ordered pair Method
(iv) Table of values Method
(v) Graphical Method
(vi) Arrow or pappy diagrams Method
(vii) Set builder notation Method

Example 23:3

Let $\Re$ be the relation "is half of" defined from the set $A = \{1, 2, 3, 4\}$ to $B = \{2, 4, 6, 8\}$. Redefine the relation using,
(i) Formula Method
(ii) Arrow or pappy diagrams Method
(iii) Set builder notation Method
(iv) Ordered pair Method
(v) Table of values Method
(vi) Graphical Method

Solution

(i) Using the formula Method R can be redefined

as $x = \dfrac{1}{2}y, x \in A, y \in B$

(ii) Using an arrow or pappy diagram the relation can be represented by,

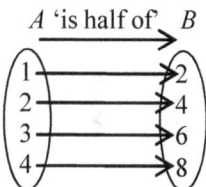

Figure 23:2

(iii) By set builder notation method the relation can be defined as

$$\Re = \left\{ (x, y) : x = \frac{1}{2}y, x \in A, y \in B \right\}$$

(iv) As ordered pairs, the relation is

$$\Re = \{(1,2),(2,4),(3,6),(4,8)\}$$

(v) Using a table of values the relation is

x	1	2	3	4
y	2	4	6	8

(vi) On a graph the relation can be represented as

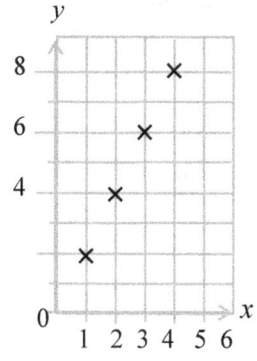

(i)

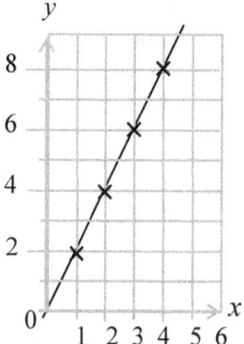

Figure 23:3

Notice that the graph in Figure 23:3 (i) is made up of discrete points. If the relation was defined on the set $\mathbb{R}$ of real numbers, a line as in Figure 23:3 (ii) will then connect these points.

This type of relation is known as a **linear relation**. A linear relation is usually expressed as a linear equation of the form $y = mx + c$, where m and c are constants.

Other forms of expressing a linear relation as an equation are treated in Topic 19.

Example 23:4
A relation is defined on the set $X = \{a, b, c, d, e\}$ of children of the same family, as 'is a brother of'. Given that b and d are boys and a, c and e are girls. Draw an arrow diagram to illustrate this relation.

Solution
Since this relation is defined in a set, the representation will be as in figure 23.4

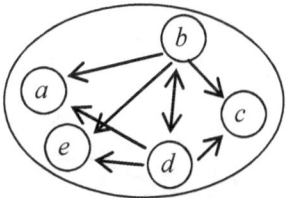

Figure 23:4

Example 23:5
A relation is defined on that set $X = \{1,2,3,4,5,6\}$ as 'is a factor of'. Represent this relation on an arrow diagram.

Solution

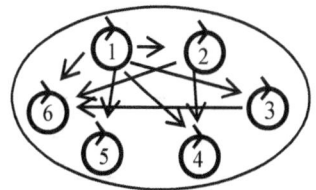

Figure 23:5

231

The loops Ö round each of the numbers is to indicate that each number is a factor of itself.

Example 23:4 and Example 23:5 are examples of relations in a set.

Inverse Relation

Example 23:6

Consider the relation 'is a wife of' from the set
$W = \{a, b, c, d\}$ of women to a set
$M = \{w, x, y, z\}$ of men represented by the arrow diagram in Figure 23.6. Define a relation from the set M to the set W and represent this relation on an arrow diagram.

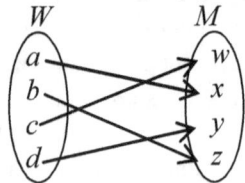

Figure 23:6

Solution

Clearly the relation "is the husband of" can be defined from the set M to the set W. This can be represented on arrow diagrams as in Figure 23:7.
(a) or (b)

(a)

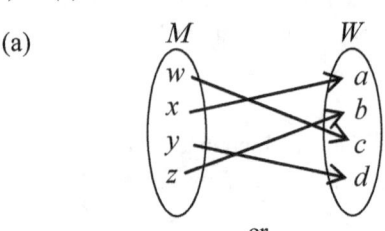

or

(b)

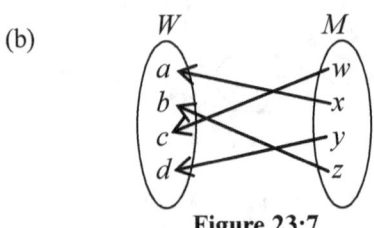

Figure 23:7

Example 23:7

Find the inverse of each of the following relations

(a) $\Re = \{(-4,4),(-2,2),(0,0),(2,-2),(4,-4)\}$

(b)

x	0	1	2	3	4
y	0	2	4	6	8

Solution

(a) $\Re = \{(4,-4),(2,-2),(0,0),(-2,2),(-4,4)\}$

(b)

x	0	1	2	3	4
y	0	1	2	3	4

Example 23:8

Draw the graph of the inverse relations in example 23:7.

Solution

(a)

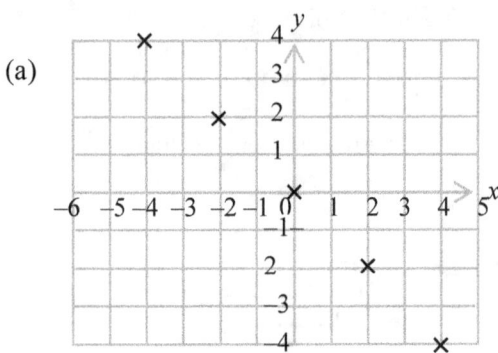

(b)

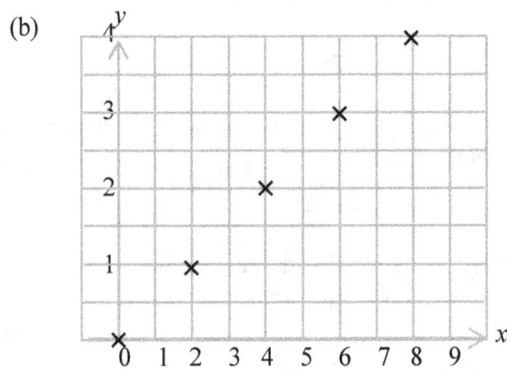

Figure 23:8

Example 23:9

Bih who is the sister of Ambe is married to Anye and they are blessed with two children Manka and Neba. Ambe has a son called Fube.

(i) Define a relation from the set of children of Bih to the son of Ambe

(ii) Illustrate this relation on an arrow diagram.

(iii) State the inverse of the relation in (i) and represent it on an arrow diagram.

Solution

The relation from the set of children of Bih to the son of Ambe is 'is a cousin of'.

A 'is a cousin of' B

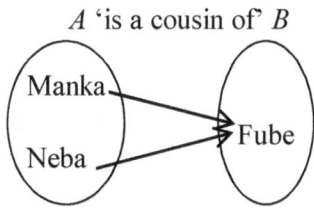

The inverse of the relation is 'is a cousin of'.

A 'is a cousin of' B

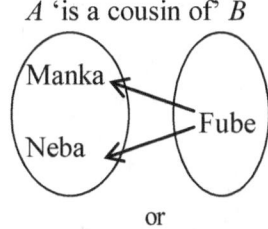

or

B 'is a cousin of' A

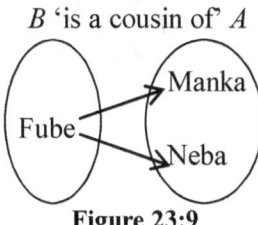

Figure 23:9

A relation such as 'is a cousin of' Example 23:9, whose inverse is still 'is a cousin of' is said to be **self inverse**.
Examples of mathematical relations, which are self-inverse, are,

(a) $\Re = \{(x,y): x + y = 8, x, y \in \mathbb{R}\}$

(b) $\Re = \{(-2,-2),(-1,-1),(0,0),(1,1)(2,2)\}$

EXERCISE 23:1

1. State the domain D and the range R of each of the following relations
   (a) $\{(0, 0), (1, 3), (2, 6), (3, 9)\}$
   (b) $\{(0, 0), (1, 1), (1,-1), (4, 2), (4,-2)\}$
   (c) $\{(1, 15), (1, 20), (2, 20), (3, 25)\}$

2. Given the domain $\{1, 2, 3, 4\}$, determine the set of ordered pairs (x, y) of the relation defined by each of the following.
   (a) $y = x + 1$ (b) $y = 4 - x$
   (c) $y = 2(x - 3)$ (d) $y = x^2 + 1$

3. Given the domain $\{-4, -2, 0, 2, 4\}$, list the set of ordered pairs (x, y) of the relation defined by each of the following.
   (a) $\{(x, y) : x = -y\}$ (b) $\{(x, y) : y = 2x\}$
   (c) $\{(x, y) : y = |x| - 2\}$ (d) $\{(x, y) : y = x + 1\}$

4. If W is the set of all real numbers, represent each of the following relations graphically.

   (a) $\{(x, y) : y = 9 - 2x, x, y \in W\}$
   (b) $\{(x, y) : y = 2x, x, y \in W\}$
   (c) $\{(x, y) : x + y = 4, x, y \in W\}$
   (d) $\{(x, y) : y = x + 1, x, y \in W\}$
   (e) $\{(x, y) : y = x, x, y \in W\}$
   (f) $\{(x, y) : x = y + 2, x, y \in W\}$
   (g) $\{(x, y) : y = x^2, x, y \in W\}$

5. Draw a graph to represent the inverse of each of the relations in question 4 above.

6. Find the Cartesian product $A \times B$ of the sets $A = \{1, 3, 5\}$ and $B = \{2, 4, 6\}$.

7. Given that $A = \{-2, 1, 4\}$ and $B = \{10, 20\}$. List the ordered pairs of each of the following.
   (a) $A \times B$ (b) $B \times A$
   (c) $A \times A$ (d) $B \times B$

8. Given the domain $\{-1, 1, 3, 5\}$, draw arrow or pappy diagrams for the relations defined by each of the following formulae.
   (a) $y = x + 1$ (b) $y = 4 - x$
   (c) $y = 2(x - 3)$ (d) $y = x^2 + 1$

9. State the inverse of each of the following relations:
   (a) 'is a wife of'
   (b) 'is a multiple of'
   (c) 'is the teacher of'
   (d) 'is five times'
   (e) 'is two more than one-third of'

10. Figure 23:10 illustrates the relation, 'is a factor of' in the set of numbers represented by $\{u, v, w, x, z\}$ which are integers.
    (a) Explain why each number has a loop.

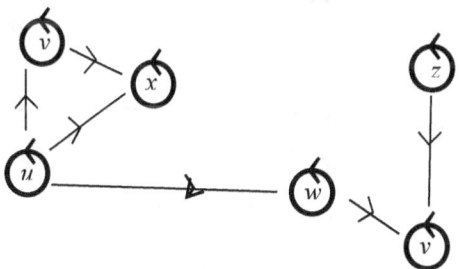

Figure 23:10

    (b) Given that $y = 8$ and $z = 9$, determine the values of u, v, w and x given that u, v, w and x are as small as possible.

Properties of Relations in a Set

Reflexive

A relation $\Re$ in a set A is reflexive if every element in A is related to itself.

i.e. $\forall x \in A, x\Re x$

Example 23:10

Let $E = \{1, 2, 3, 4\}$ and $\Re$ a relation in E defined by $(x\Re y) \Leftrightarrow (x + x)$ is even. Show that R is reflexive.

Solution

$1 + 1 = 2$ and 2 is even, $2 + 2 = 4$ and 4 is even, $3 + 3 = 6$ and 6 is even, $4 + 4 = 8$ and 8 is even.

$\therefore x + x = 2x$, and $2x$ is even. Hence, $x\Re x$

$\forall x \in E \Rightarrow$ relation is reflexive.

Symmetric

A relation $\Re$ in a set A, is symmetric if, x relates y implies y relates x for all x, y belonging to A.

i.e. $\forall x, y \in A, x\Re y \Rightarrow y\Re x$

Examples of symmetric relations are the relations 'is a friend of' and 'is a cousin of' defined over the set of people. For if, A is a friend to B then B is a friend to A and if X is a cousin to Y then Y is a cousin to X.

Example 23:11

Let $E = \{1,2,3,4\}$ and $\Re$ a relation in E defined by $(x\Re y) \Leftrightarrow (x + y)$ is even. Show that R is symmetric.

Solution

$\Re = \{(1,1),(1,3),(2,2),(2,4),(3,1),(3,3),(4,2),(4,4)\}$.

Observation of $\Re$ shows that $\forall x, y \in E, x\Re y \Rightarrow y\Re x$. Therefore, $\Re$ is symmetric.

Transitive

A relation $\Re$ in a set A is transitive if, x relates y and y relates z implies x relates z for all x, y, z, belonging to A.

i.e. $\forall x, y, z \in A, x\Re y$ and $y\Re z \Rightarrow x\Re z$

Example 23:12

Show that the order relation 'is less than' in $\mathbb{R}$ is transitive.

Solution

Any 3 real numbers x, y and z are such that, if $x < y$ and $y < z$ then, $x < z$. For instance, $2 < 5$ and $5 < 8 \Rightarrow 2 < 8$. Hence, the order relation 'is less than' in $\mathbb{R}$ is transitive.

Anti-symmetric

A relation $\Re$ in a set A is anti-symmetric if, x is related to y and y is related to x implies x is equal to y.

i.e. $\forall x, y \in A, x\Re y$ and $y\Re x \Rightarrow x = y$

Example 23:13

Let P be a set containing many sets and $\Re$ be the inclusion relation $\subseteq$. Show that the relation $\Re$ is anti-symmetric.

Solution

$A \subseteq B$ and $B \subseteq A \Rightarrow A = B$. Therefore, $\Re$ is anti-symmetric.

Equivalence relation

A relation $\Re$ in a set, which is reflexive, symmetric and transitive, is called an **equivalence relation**.

Example 23:14

Let T be the set of all triangles and $\Re$ the relation 'is similar to'. Show that $\Re$ is an equivalence relation.

Solution

(i) Every triangle is similar to itself. $a\Re a \Rightarrow \Re$ is reflexive.
(ii) If a $\triangle ABC$ is similar to $\triangle XYZ$, then $\triangle XYZ$ must be similar to $\triangle ABC$. Therefore, $a\Re b \Leftrightarrow b\Re a$. Hence $\Re$ is symmetric.
(iii) If a $\triangle ABC$ is similar to $\triangle PQR$ and $\triangle PQR$ is similar to $\triangle XYZ$ then $\triangle ABC$ must be similar to $\triangle XYZ$. Therefore, $a\Re b$ and $bc \Rightarrow a\Re c$ and $\Re$ is transitive.

Since the relation $\Re$ on T is reflexive, symmetric and transitive, it means $\Re$ is an equivalence relation.

EXERCISE 23:2

1. The relation 'is less than or equal to' is defined on the set $A = \{1,2,3,4\}$.
   (i) Draw an arrow diagram to show the above relation on the set A.
   (ii) Explain why the relation is not an equivalence relation on the set A.
2. (i) Determine which of the properties: reflexive, symmetric, transitive, anti-symmetric apply to each of the following relations.
   (a) $'a$ divides $b'; a, b \in \mathbb{Z}$
   (b) $a \leq b; a, b \in \mathbb{Z}$
   (c) $|x| \leq |y|; x, y \in \mathbb{R}$
   (d) $x - y \leq 0; x, y \in \mathbb{R}$
   (e) $x + y \leq 1; x, y \in \mathbb{R}$
   (f) $x^2 = y^2; x, y \in \mathbb{R}$
   (g) $x - y$ is a multiple of $2\pi; x, y \in \mathbb{R}$
   (h) $x^2 + y^2 = 1; x, y \in \mathbb{R}$
   (ii) State which of the relations in (i) are equivalence relations.

Order Relation

A relation $\Re$ in a set A, which is **reflexive**, **transitive** and **anti-symmetric**, is called an **order relation**.

Example 23:14

Define a relation $\leq$ in the set $\mathbb{R}$ of real numbers as 'is less than or equal to'. Show that $\Re$ is an order relation.

Solution

(i) $\forall a \in \mathbb{R}, a \leq a \Rightarrow \Re$ is reflexive.

(ii) $\forall x, y, z \in \mathbb{R}$ if $x \leq y$ and $y \leq z$ then, $x \leq z$. For instance, $2 \leq 5, 5 \leq 8$ and $2 \leq 8$ hence, the relation '$\leq$' defined in $\mathbb{R}$ is transitive.

(iii) $\forall x, y \in A, x\Re y$ and $y\Re x \Rightarrow x = y$. Hence, the relation '$\leq$' defined in $\mathbb{R}$ is anti-symmetric. Since the relation '$\leq$' defined in $\mathbb{R}$ is reflexive, transitive and anti-symmetric then the relation is an order relation.

MAPPINGS AND FUNCTIONS

The Idea of a Function or a Mapping

Though the words function and mapping are often used interchangeably, there is a slight difference in their meanings. A function is necessarily a mapping but some mappings may not be functions. A function (sometimes for emphases, referred to as a "numerical function") usually refers to a mapping which involves the set of numbers. A mapping on the other hand may relate elements, which are neither numbers nor pro-numerals. Therefore, the set of all functions is a subset of the set of mappings.

Functions or mappings are special kinds of relations in which every element, x of the domain A is related to one and only one element y of the codomain B. To simplify, a function or mapping is a relation in which one and only one arrow leaves each element in the domain. A mapping is analogous to a relation 'is married to' defined from a set W of women to a set M of men in a society in which all women are married, polygamy is allowed but polyandry is forbidden. It should be appreciated that this definition implies that for a mapping,

(i) Each element of the domain must be mapped to one element of the codomain.

(ii) Not all elements of the codomain should necessarily be mapped to.

(iii) An element of the codomain may be mapped to by more than one or even all the elements of the domain.

In the arrow diagrams in Figure 23:11, (i) is a mapping (function) from set A to set B but (ii) is not. If a, b, c, and d are real numbers then, (i) is a mapping but (ii) is not.

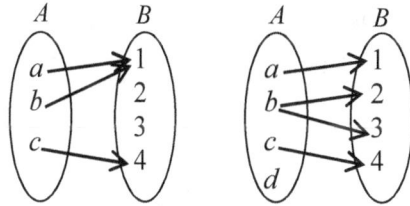

Figure 23:11

Function Notation

The two common notations used are:
(a) $f(x) = 5x + 4$ or $y = 5x + 4$
(b) $f: x \mapsto 5x + 4$

In the notation in (b) distinguish between the arrow ↦, which is used to map elements of two sets, and →, which is used to map a set to another. For example $f: A \rightarrow B$.

Representation of Functions

Since a function is a relation it is usually represented using horizontal or vertical arrow diagrams or graphically by plotting the Cartesian coordinates.

Example 23:14

Let $A = \{1, 2, 3, 4\}$ and $B = \{2, 4, 6, 8, 10\}$. Given the function $f: x \longmapsto 2x$ defined from A to B. Represent this function using:
(a) A horizontal arrow diagram,
(b) A Vertical arrow diagram,
(c) Cartesian coordinates.

Solution

(a)
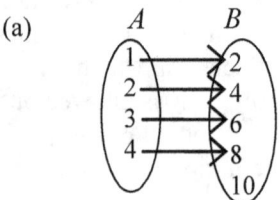

(b)
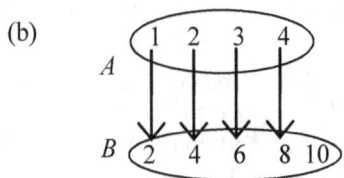

Figure 23:12

Note!

In arrow diagrams, the arrows of a mapping or function are parallel or convergent and never divergent.

(c)
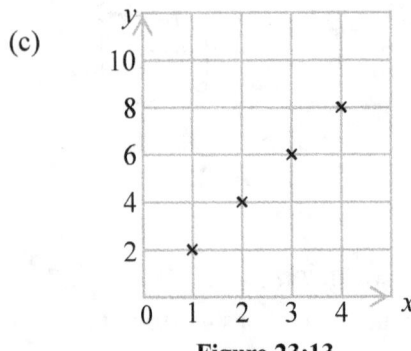

Figure 23:13

Example 23:15

Which of the following relations are functions?

(a) $\{(0,1),(0,2)\}$ (b) $\{(1,0),(2,0)\}$

(c) $\{(-3,1),(-3,2),(-3,3)\}$

(d) $\{(1,-3),(2,-3),(3,-3)\}$

(e) $\left\{(1,1),\left(\frac{1}{2},1\right),\left(\frac{1}{4},1\right),\left(-\frac{1}{4},1\right)\right\}$

(f) $\{(-1,1),(-2,2),(-3,3),(2,-1),(3,-3),(1,-1)\}$

Solution

With the aid of arrow diagrams, the solution is clearer.

(a)
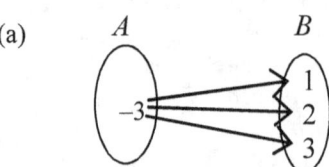

(b)
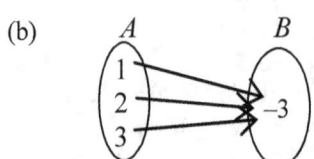

(c)
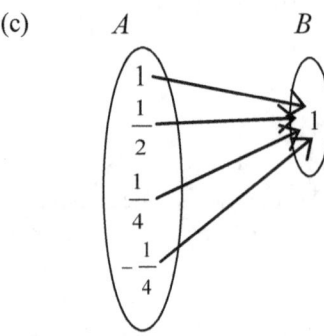

(d)
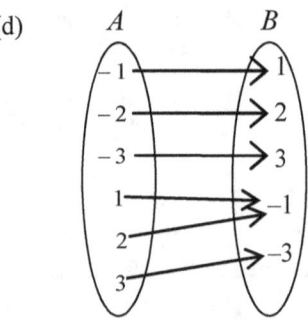

(e)

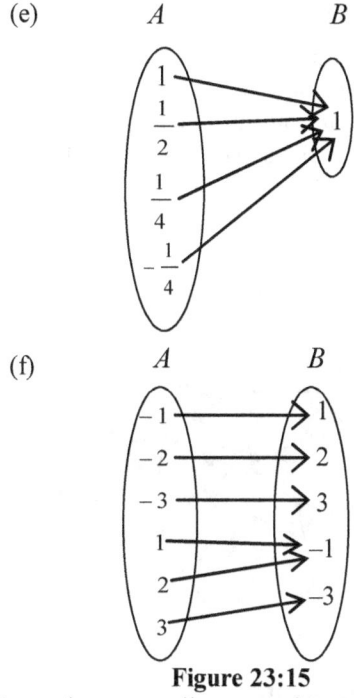

(f)

Figure 23:15

From the arrow diagrams, it can be seen that in (b), (d), (e) and (f), the arrows are convergent. Hence, (b), (d), (e) and (f) are functions. (a) and (c) are not functions.

Alternatively, draw the graphs of the relations as shown in Figure 23:16. In this method if it is possible to draw a vertical straight line through any two points on the graph, then the relation is **not** a function.

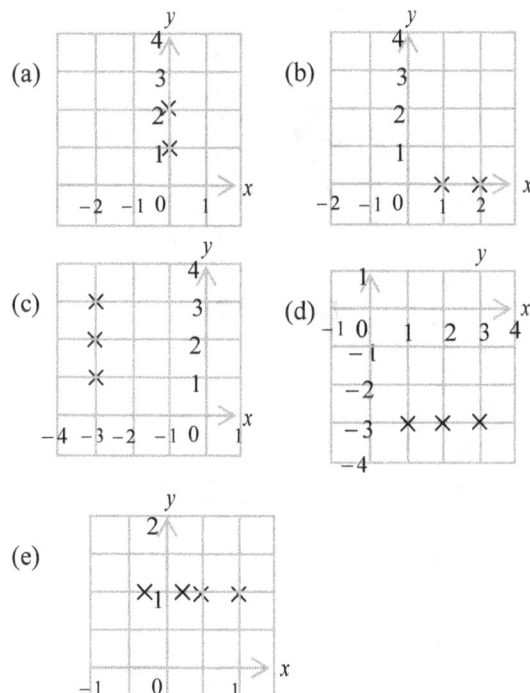

(f)

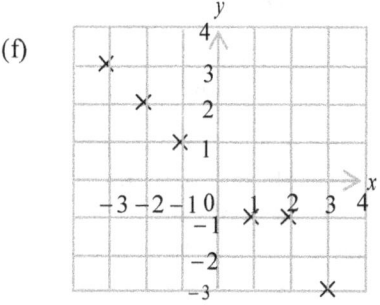

Figure 23:16

From the graphs, in Figure 23:16 it can be seen that in (b), (d), (e) and (f), a vertical line cannot be drawn to pass through any two of the points. On the other hand in (a) and (c) a vertical line can be drawn to pass through two or more points. Hence, (b), (d), (e) and (f) are functions while (a) and (c) are not.

EXERCISE 23:3

1. State which of the relations in Figure 23:17 (a) to (i) are mappings.

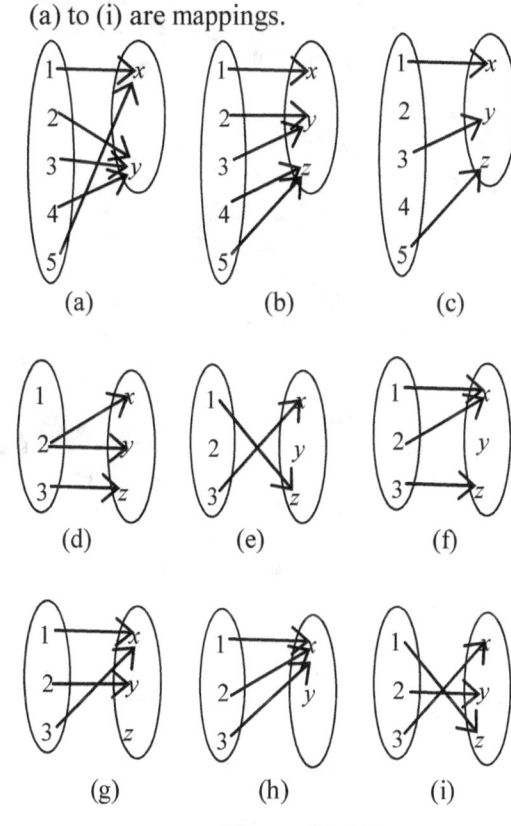

Figure 23:17

2. Let $A = \{x, y, z\}$ and $B = \{0, 1\}$.

   (a) Draw arrow diagrams of all the possible mappings from A into B.

   (b) How many mappings are there altogether?

3. Let $X = \{-2, -1, 0, 1, 2\}$ and given the function $f: X \longrightarrow \mathbb{Z}$ defined as
   $f: x \longmapsto 2x - 3$.
   (i) Represent this function using
       (a) A horizontal arrow diagram,
       (b) A Vertical arrow diagram,
       (c) Cartesian coordinates.
   (ii) State the domain and range of the function.

Types of Mapping

Mappings can be classified as one-one, onto, into or many-one mappings.

One-one mapping

A mapping in which one and only one element in the domain A, maps to one and only one element in the codomain B, is called a "one-one" mapping. For instance, the function f, which maps every country to its capital city, is a one-one mapping, since each country has only one capital and there is no city, which is the capital of two different countries. The mapping in Figure 23:18 is a simple example of a one-one mapping.

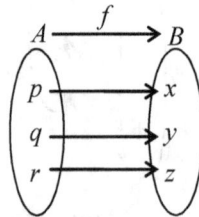

Figure 23:18

Onto mappings

A mapping which is such that every element in the codomain is an image of at least one element in the domain is called an "onto" mapping. In other words an "onto" mapping is a mapping in which all the elements of the codomain are "used up". For instance;
If $A = \{a, b, c, d\}$, $B = \{x, y, z\}$ and $f : A \rightarrow B$ is defined as in Figure 23:18 then, f is an "onto" mapping.

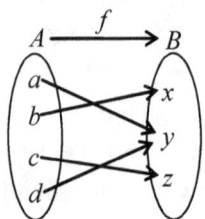

Figure 23:19

Into Mappings

A mapping which is such that every element in the codomain is not an image of one element in the domain is called an "into" mapping. In other words an "into" mapping is a mapping in which all the elements of the codomain are not "used up". For instance;

If $A = \{a, b, c, d\}$, $B = \{x, y, z\}$ and $f : A \rightarrow B$ is defined as in Figure 23:20, then, f is an "into" mapping.

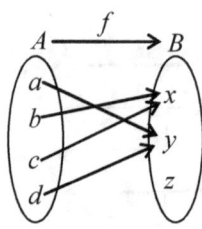

Figure 23:20

Many-one mapping

If a mapping is such that two or more elements in the domain give rise to the same image in the codomain, such a mapping is said to be a many-one mapping. For instance, the function $f: x \longmapsto x^2$ defined in the set $\mathbb{Z}$ of integers is a many-one mapping because each element in the codomain is the image of two elements in the domain.

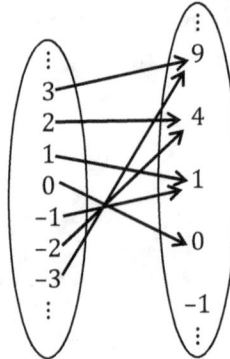

Figure 23:21

Mappings can also be classified as injections, surjections or bijections.

Injections

An injection or injective mapping or function is a mapping, which is such that every element $x \in A$ has a unique image $y \in B$. In other words, an injection is a one-to-one mapping of two sets such that each element of each set corresponds to only one element of the other set. For instance;
Let $A = \{1, 2, 3, 4\}$ and $B = \{1, 2, 3, 4, 5, 6\}$. The function $f: A \longrightarrow B$, defined by $f: x \longmapsto x + 1$,

shown in Figure 23:20, below is injective.

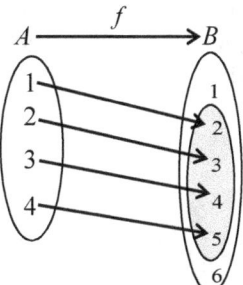

Figure 23:22

In figure 23:22, the range is shaded.
Notice that some elements in B may not be images of elements in A. Therefore, the range R is a proper subset of the codomain B. Hence for an injection, the cardinality of the domain A is always less than or equal to that of the codomain B.

$$n(A) \leq n(B)$$

Surjections

A surjection or surjective mapping or function is a function, which is such that every element $y \in B$ is the image of at least some element $x \in A$. In other words, all elements of the codomain B are "used up". For a surjection, there need not be a one-one correspondence between the elements of the domain and those of the range.
Hence for a surjection, the cardinality of the codomain B is always less than or equal to that of the domain A.

$$n(B) \leq n(A)$$

For instance;

Let $A = \{-3, -2, -1, 0, 1, 2, 3\}$ and B = $\{0, 1, 4, 9\}$. The relation $f:A \to B$, defined by $f : x \mapsto x$ shown in Figure 23:23 is surjective.

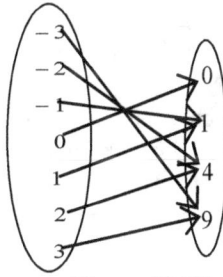

Figure 23:23

Let $f : \mathbb{N} \to E$, where $E = \{e, o\}$. Define a relation from $\mathbb{N}$ to E as follows,

$$f(n) = \begin{cases} o \text{ if } n \text{ is odd} \\ e \text{ if } n \text{ is even} \end{cases}$$

Since every element in E is an image of an element in $\mathbb{N}$, $f(n)$ is a surjection of $\mathbb{N}$ onto E.

Bijections

A bijection or bijective mapping or function is a function that is both injective (into) and surjective (onto). Therefore, a bijection is a one-one mapping which uses up all the elements in the codomain. A mapping between two sets in which every element in each sets corresponds to only one element of the other sets for mapping in either direction.
For instance;

Let $A = \{1,2,3,4\}$ and $B = \{2,3,4,5\}$. The function $f : A \to B$ defined by $f : x \mapsto x + 1$, shown in Figure 23:24, below is injective and surjective.

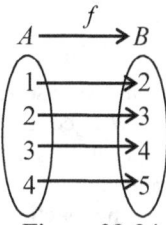

Figure 23:24

▷ EXERCISE 23:4

1. Classify the following functions in Figure 23:25
   as (a) one-one   (b) many-one
       (c) onto   (d) into.

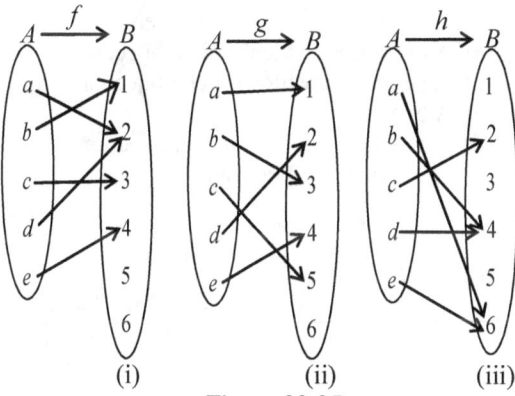

Figure 23:25

2. State whether each of the following mappings is one-one, many-one.
   (i) Assign to each person in your village a number, which corresponds to his age.
   (ii) Assign to each city in your country, the corresponding number of people in the city.

(iii) Assign to each book written by only one author its author.

(iv) Assign to each country with a prime minister its prime minister.

3. In exercise 23:3 (3), state the number of mappings, which are:
(i) one-one.   (ii) one-many.

4. Let $= A = \{1, 2, 3, 4, 5\}$. Figure 23:26, shows the function $f : A \rightarrow A$.

(a) State the range of the function f.

(b) Is f an "onto" or "into" function?

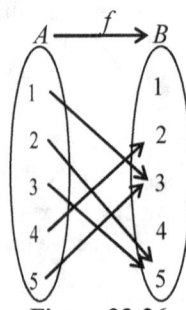

Figure 23:26

5. Classify the functions in Exercise 23:3 (2) as injections, surjections or bijections.

6. A correspondence $f: \mathbb{N} \longrightarrow \mathbb{Z}$, is defined as follows. In each case determine whether f is a function, then classify the functions as injective or surjective.

(a) $f(x) = \frac{1}{5}x + 3$ (b) $f(x) = x^2 - x$

(c) $f(x) = x^2$ (d) $f(x) = \frac{1}{2}x^2 + 4$

(e) $f(x) = x + 2$ (f) $f(x) = |x|$

7. Let $A = \{1, 2, 3, 4\}$, $B = \{w, x, y, z\}$ and $C = \{a, b, c\}$.

Draw an arrow diagram to represent:

(a) An injection, which is not a bijection.

(b) A surjection, which is not a bijection.

(c) A bijection.

(d) A function, which is not an injection, a surjection or a bijection.

Flow Charts

Functions are often analyzed using flow charts. More on flow charts have been treated in Topic 47. The reader may delay this section if necessary.

Example 23:16

Draw a flow chart to represent the function

$$f(x) \longmapsto \frac{9-7x}{6}$$

Solution

$$x \rightarrow \boxed{\times(-7)} \xrightarrow{-7x} \boxed{+9} \xrightarrow{9-7x} \boxed{\div 6} \rightarrow \frac{9-7x}{6}$$

Notice that in flow charts the order of operation is of utmost importance. For instance in the function in Example 23:16, it is imperative to first multiply x by -7 before adding 9 to the result.

Inverse Function

The inverse f^{-1} of a bijective function f is a function that performs the reverse process of what the function f does. For instance, if $g: x \longmapsto x - 2$ then the inverse of g denoted by g^{-1} is $g^{-1}: x \longmapsto x + 2$.

Note that only bijections have inverses

Finding the Inverse of a function

Suppose that $f: x \mapsto f(x)$. To find f^{-1}, equate $y = f(x)$ and solve for x, then substitute x for y to have $f^{-1}(x)$.

Example 23:17

Given that $f: x \longmapsto \frac{2x-7}{5}$, find the inverse of f.

Solution

$$\text{Let } \frac{2y - 7}{5} = x$$

$$\Rightarrow 2y - 7 = 5x$$

$$2y = 5x + 7$$

$$y = \frac{5x + 7}{2}$$

$$\therefore f^{-1}: x \longmapsto \frac{5x+7}{2}$$

Alternatively, a flow chart of the function is drawn.

$$x \rightarrow \boxed{\times 2} \xrightarrow{2x} \boxed{-7} \xrightarrow{2x-7} \boxed{\div 5} \rightarrow \frac{2x-7}{5}$$

An inverse flow chart is drawn from the first. This is usually drawn in the opposite direction replacing each operation by its inverse.

$$\frac{5x + 7}{2} \leftarrow \boxed{\div 2} \xleftarrow{5x+7} \boxed{+7} \xleftarrow{5x} \boxed{\times 5} \leftarrow x$$

Hence $f^{-1}: x \longmapsto \frac{5x+7}{2}$

Composite Functions

A **composite function** $f \circ g$ or fg, is a function, which is made up of two simpler functions f and g. A composite function such as $f \circ g \circ h$ by convention is operated from right to left. Figure 24:27 shows $h = f \circ g(x) = 2x + 5$, the result of composing set A using the function $g: x \mapsto 2x$ followed by $f: x \mapsto x + 5$.

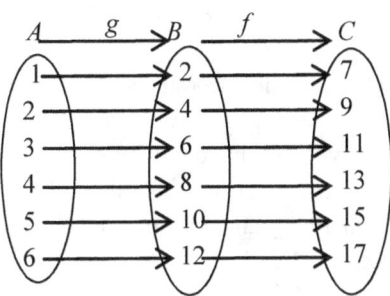

Figure 23:27

The composition of a function and its inverse is the **identity function**, usually denoted by $I : x \mapsto x$. The identity function always maps every element to itself.

Example 23:18
Given that $f: x \mapsto 2x + 1$ and $g: x \mapsto x + 3x^2$, express $f \circ g$ in the form $f \circ g: x \mapsto \ldots$

Solution
$$f \circ g(x) = f(3x^2) = 2(3x^2) + 1 = 6x^2 + 1$$
$$\Rightarrow f \circ g: x \mapsto 6x^2 + 1$$

Example 23:19
The functions f and g are defined on $\mathbb{R}$, the set of real numbers, by
$$f: x \mapsto x + 4, g: x \mapsto \frac{1}{x+2}, x \neq -2.$$

Evaluate $f \circ g(2)$

Solution
$$f \circ g(2) = f\left(\frac{1}{2+2}\right) = f\left(\frac{1}{4}\right) = \frac{1}{4} + 4 = \frac{17}{4}$$

Alternatively,
$$f \circ g(x) = f\left(\frac{1}{x+2}\right) = \frac{1}{x+2} + 4 = \frac{1 + 4x - 8}{x+2}$$
$$f \circ g(2) = \frac{1 + 4(2) - 8}{2 + 2} = \frac{17}{4}$$

Restricted Domain and Restricted Function

Consider the function $f: x \mapsto x^2$. The domain of definition of this function is $\mathbb{R}$ to $\mathbb{R}$. Clearly, but for the element 0 there will be two values of x for each image $f(x)$ as shown in the arrow diagram in Figure 23:28(i).

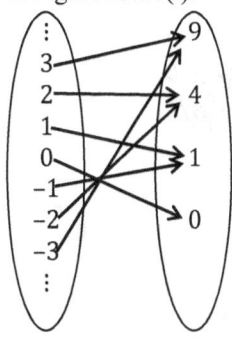 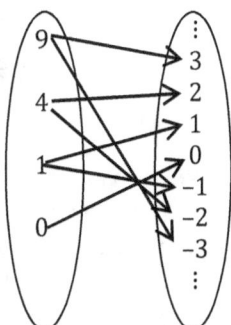

Figure 23:28

The arrow diagram representing the inverse relation will be as shown in Figure 23:28(ii). Notice that though the relation is a function, the inverse relation is not a function. Hence, the function $f: \mathbb{R} \longrightarrow \mathbb{R}$; $f: x \mapsto x^2$ has no inverse. Suppose the function is redefined as $f: \mathbb{R}^+ \longrightarrow \mathbb{R}$; $f: x \mapsto x^2$ the arrow diagram for the relation and the inverse relation will be as shown in Figure 23:29 (i) and (ii).

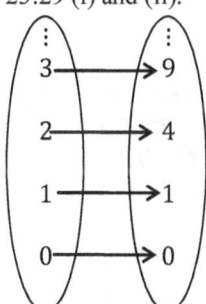 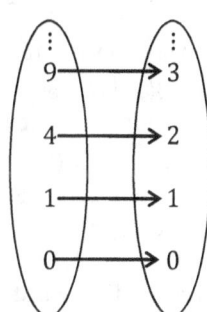

Figure 23:29

Clearly both the relation and the inverse relation are functions. The new domain $\mathbb{R}^+ = [0, +\infty)$ of the function f is called the **restricted domain** of the function. A restricted domain is therefore the use of a domain for a function that is smaller than the function's domain of definition. Restricted domains are commonly used to specify a one-to-one section of a function. A function with a restricted domain is called a **restricted function**.

Another example is the function $f: x \mapsto \frac{1}{x}$. The set $\mathbb{R}$ of real numbers cannot be the domain of this function, since there is no value for $f(0)$. Therefore the domain of this function is the restricted domain $\mathbb{R} - \{0\}$

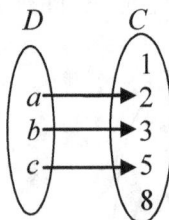

EXERCISE 23:5

1. The following diagram represents a mapping from set D to set C.

Figure 23:30

State (a) the element whose image is 2.
(b) the range set.

2. The functions f and g, are defined in $\mathbb{R}$ such that

$$f(x) = 2x + 1 \text{ and } g(x) = \frac{1}{x+2}, x \neq -2.$$

Find,
(i) The function $f^{-1}(x)$ that is the inverse of f.
(ii) The value of $f \circ g(4)$.

3. Given the functions
$f: x \mapsto 3x-1, x \in \mathbb{R}$ and $g: x \mapsto x^2 -1, x \in \mathbb{R}$.
Solve the equation $f \circ g(x) = 0$.

4. f and g are functions defined on the set $\mathbb{R}$ of real numbers by

$$f: x \mapsto 3x - 1 \text{ and } g: x \mapsto \frac{2x+1}{3}.$$

Express in a similar manner
(a) $g \circ f(x)$ (b) g^{-1}

5. Given the functions $f: s \mapsto 4s-7$ and $g: s \mapsto s^2$ for $s \in \mathbb{R}^+$. Find (i) $f(2)$ (ii) $g^{-1} \circ f(4)$

6. The function f is defined on the set $\mathbb{R}$ of real numbers, by $f: x \mapsto (x+1)(x+2)$.
(i) $f(-3)$ (ii) Solve the equation $f(x) = 0$.

7. The function g is defined from a subset of $\mathbb{R}$ to a subset of $\mathbb{R}$ by $g(x) = x^2-3x$.
(a) Find the range of g when the domain is $A = \{-2,0,2\}$.
(b) Find the domain of g when the range is $B = \{0, -2\}$.

8. The function f is defined in the set of real numbers $\mathbb{R}$ by $f: x \mapsto 4-x^2$.
(a) $f(3)$ (b) Solve the equation $f(x) = 0$.

9. The functions f, g, and h are defined on $\mathbb{R}$ the set of real numbers as $f: x \mapsto x^2-3$, $g: x \mapsto x + 2$, $h: x \mapsto px + q$, where p and q are constants.
(a) Evaluate $f(-2)$.
(b) Find the composite function $g \circ f(x) = ...$
(c) Express g^{-1} in the form $g^{-1}: x \mapsto ....$
(d) Given that $g \circ f(x) = 3x +1$, determine the values of p and q.

10. The functions f, g, and h are defined on $\mathbb{R}$ the set of real numbers as follows:
$f: x \mapsto x +2, g: x \mapsto x^2 -1, h: x \mapsto 7x^2 -x +2$.
(a) Find a similar expression for f^{-1}.
(b) Evaluate $g\left(\sqrt{2}\right)$ and $h \circ f(0)$.
(c) Find the value of x for which $h(x) = f(x)$.

11. The functions f, g and h are defined over the set $\mathbb{R}$ of real numbers by:
$f: x \mapsto x^2 - 4, g: x \mapsto 3x- 6, h: x \mapsto (3x-8)(3x-4)$.
(i) Prove that $f \circ g(x) = h$.
(ii) Calculate of $g \circ f(x)$.
(iii) Express the quadratic equation $f \circ g(x) = g \circ f(x)$ in the form $ax^2 + bx + c = 0$.
(iv) Solve the equation
$f \circ g(x) - g \circ f(x) - 50 = 0$.

12. The functions f and g are defined on the set $\mathbb{R}$ of real numbers by
$$f: x \mapsto \frac{x+3}{x-2} \text{ and } g: x \mapsto 2x - 1$$
(i) Determine the domain and range of the function f.
(ii) Find $g^{-1}(x)$. Hence or otherwise
(iii) Find the function $h(x)$ such that
$$hg(x) = \frac{2x+1}{2(x-1)}.$$

13. The functions p, q and r are defined in the set $\mathbb{R}$ of real numbers as follows:
$$p: x \mapsto x^2 + 1, \ q: x \mapsto \frac{x}{3} + 1, r: x \mapsto 2x$$
(a) Find expressions for $pr(x)$ and $rp(x)$.
(b) Solve correct to two decimal places, the equation $qp(x) = x +1$.
(c) State the conditions necessary for p^{-1} to exist.
(d) Find expressions for the inverses p^{-1} and q^{-1} and express each one in a similar form to p, q.
(e) Explain why p^{-1} is not a function.

14. The functions f and g are defined in the set $\mathbb{R}$ of real numbers by
$$f: x \mapsto 1 - x \text{ and } g: x \mapsto \frac{1}{2+x}, x \neq -2$$
(a) Evaluate $g(-3)$ (b) Find $f^{-1}(x)$.
(c) Find to two significant figures the values of x for which $f \circ g(x) = g \circ f(x)$.

15. In each of the following functions, the codomain is the set of real numbers. State the domain of each function.
(a) $f: x \mapsto \frac{\sqrt{x}}{x^2-9}$ (b) $f: x \mapsto \sqrt{x^2 - 16}$
(c) $f: x \mapsto 5x + 2$ (d) $f: x \mapsto \frac{1}{x}$
(e) $f: x \mapsto \frac{x-1}{(x+3)(x-2)}$ (f) $f: x \mapsto |2x|$

MULTIPLE CHOICE EXERCISE 23

1. Given that $A = \{1,2,3,4,5\}$ and
   $B = \{-1,0,1,\ldots,12\}$. A relation $\Re$ is
   defined from A to B as $a \, \Re \, b$ means
   $b = 3a - 4$. e.g. $_1 \Re _{-1}$ since, $-1 = 3(1) - 4$.
   The set of ordered pairs (a, b) of the
   relation R is:
   [A] $\{(1, -1),(2,2),(3,5),(4,8),(5,11)\}$
   [B] $\{(-1,1),(2,2),(5,3),(8,4),(11,5)\}$
   [C] $\{(2,2),(3,5)\}$
   [D] $\{(2,2),(5,3)\}$

2. Given that $A = \{1,2,3,4,5\}$ and
   $B = \{-1,0,1,\ldots,12\}$. A relation $\Re$ is
   defined from A to B as $a \, \Re \, b$ means
   $b = 3a - 4$. e.g. $_1 \Re _{-1}$ since, $-1 = 3(1) - 4$. It
   is true that:
   [A] $\Re$ is an onto relation since, $\forall a \in A$,
        $b \in B, (a,b) \in A \times B$.
   [B] $\Re$ is an onto relation since, $\forall a \in A, b \in B$,
        $(a,b) \notin A \times B$.
   [C] $\Re$ is an onto relation since, $\forall a \in A, b \in B$,
        $(a,b) \supset A \times B$.
   [D] $\Re$ is an onto relation since, $\forall a \in A, b \in B$,
        $(a,b) \subset A \times B$.

3. The properties, which both satisfy the
   relation 'is less than' are:
   [A] reflexive and transitive.
   [B] reflexive and symmetric.
   [C] anti-symmetric and transitive
   [D] symmetric and transitive.

4. Figure 23:31, shows a mapping involving
   two sets A and B described by the
   relation:
   [A] $x \mapsto x$ [B] $x \mapsto x - 1$
   [C] $x \mapsto x + 1$ [D] $x \mapsto 2x + 1$

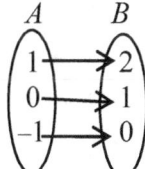

Figure 23:31

5. The ages of five children Cha, Abe, Eme,
   Bep and Dah, are such that Cha is older
   than Abe but younger than Eme. Eme is
   younger than Bep. Dah is older than Bep.
   The correct arrangement of the Children in
   the ascending order of their ages is:
   [A] Dah, Bep, Eme, Cha, Abe
   [B] Abe, Cha, Eme, Bep, Dah
   [C] Eme, Cha, Abe, Dah, Bep
   [D] Eme, Bep, Dah, Abe, Cha

6. A relation is defined on the set
   $X = \{1, 2, 3, 4, 5, 6\}$ as 'is a factor of'. The
   diagrams in Figure 23:32, which represents
   this relation, is:

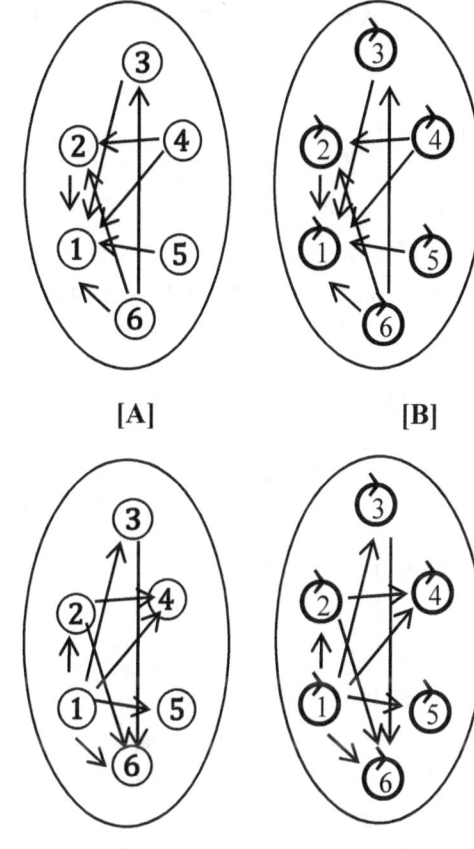

Figure 23:32

7. A relation $\Re$ is defined in a set
   $A = \{x, y, z\}$. $\Re$ is such that $x \Re y$ and
   $y \Re z \Rightarrow x \Re z$. Therefore:
   [A] $\Re$ is reflexive. [B] $\Re$ is symmetric.
   [C] $\Re$ is transitive. [D] $\Re$ is anti-Symmetric.

8. A relation $\Re$ is defined in a set
   $A = \{x, y, z\}$. If $\Re$ is said to be anti-
   symmetric. This means that:
   [A] $\forall x, y \in A, x \Re y \Rightarrow y \Re x$.
   [B] $\forall x, y, z \in A, x \Re y$ and $y \Re z \Rightarrow x \Re z$.
   [C] $\forall x, y \in A, x \Re y$ and $y \Re x \Rightarrow x = y$.
   [D] $\forall x \in A, x \Re x$.

9. A relation $\Re$ is defined in a set
   $A = \{x, y, z\}$. $\Re$ is such that, $x \Re x, \forall x \in A$.
   Therefore:

[A] $\Re$ is reflexive. [B] $\Re$ is symmetric.

[C] $\Re$ is transitive. [D] $\Re$ is anti-Symmetric.

10. A relation $\Re$ is defined in a set
$A = \{x, y, z\}$. $\Re$ is symmetric means:

[A] $\forall x, y \in A, x \Re y \Rightarrow y \Re x$.

[B] $\forall x, y, z \in A, x \Re y$ and $y \Re z \Rightarrow x \Re z$.

[C] $\forall x, y \in A, x \Re y$ and $y \Re x \Rightarrow x = y$.

[D] $\forall x \in A, x \Re x$.

11. A relation $\Re$ is defined in a set
$A = \{x, y, z\}$. $\Re$ is such that, $x \Re y \Rightarrow y \Re x$,
$\forall x, y \in A$. Therefore, $\Re$ is:

[A] $\Re$ is reflexive.

[B] $\Re$ is symmetric.

[C] $\Re$ is transitive.

[D] $\Re$ is anti-Symmetric.

12. A relation $\Re$ is defined in a set
$A = \{x, y, z\}$. $\Re$ is transitive means:

[A] $\forall x, y \in A, x \Re y \Rightarrow y \Re x$.

[B] $\forall x, y, z \in A, x \Re y$ and $y \Re z \Rightarrow x \Re z$.

[C] $\forall x, y \in A, x \Re y$ and $y \Re x \Rightarrow x = y$.

[D] $\forall x \in A, x \Re x$.

13. A relation $\Re$ is defined in a set
$A = \{x, y, z\}$. $\Re$ is such that, $x \Re y$ and
$y \Re x \Rightarrow x = y$, $\forall x, y \in A$. Therefore, $\Re$ is:

[A] $\Re$ is reflexive.

[B] $\Re$ is symmetric.

[C] $\Re$ is transitive.

[D] $\Re$ is anti-Symmetric.

14. A relation $\Re$ is defined in a set
$A = \{x, y, z\}$. $\Re$ is reflexive means:

[A] $\forall x, y \in A, x \Re y \Rightarrow y \Re x$.

[B] $\forall x, y, z \in A, x \Re y$ and $y \Re z \Rightarrow x \Re z$.

[C] $\forall x, y \in A, x \Re y$ and $y \Re x \Rightarrow x = y$.

[D] $\forall x \in A, x \Re x$.

15. The diagram in Figure 23:33, which does
not represent a function from set P to Q is:

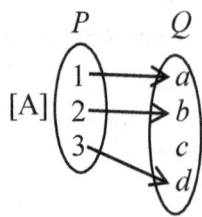
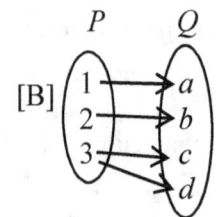

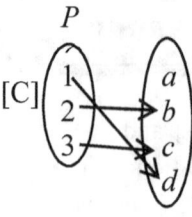
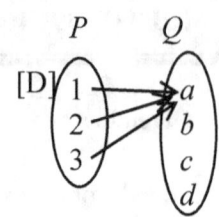

Figure 23:33

16. A function $f: \mathbb{R} \to \mathbb{R}$ is defined by
$f: x \longmapsto 2(x^2 + 1)$. The elements of the
domain whose image is 10 are:

[A] -1 or 3 [B] 1 or 3

[C] -1 or -3 [D] -2 or 2

17. The function $f: \mathbb{R} \to \mathbb{R}$ is defined by
$$f: x \longmapsto \begin{cases} -x \text{ when } x < -1 \\ 1 \text{ when } -1 \le x \le 1 \\ 2x - 1 \text{ when } x > 1 \end{cases}$$

The graph of f in Figure 23:34 is:

[A] 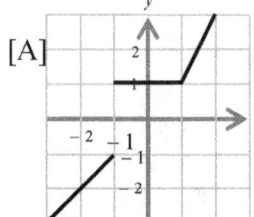   [B]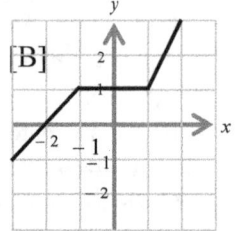

[C] 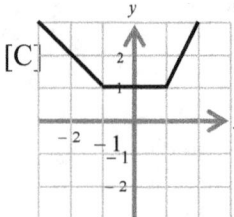   [D]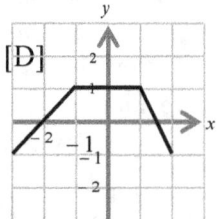

Figure 23:34

18. A function $f: \mathbb{R} \longrightarrow \mathbb{R}$ is defined by
$f: x \mapsto x + 2$. $f\left(\frac{1}{2}\right)$ is equal to:

[A] $\frac{5}{4}$ [B] 1 [C] $\frac{3}{2}$ [D] $\frac{5}{2}$

19. A function $f: \mathbb{R} \longrightarrow \mathbb{R}$ is defined by
$f: x \mapsto x + 2$. $f \circ f(8)$ is:
[A] 6 [B] 8 [C] 12 [D] 4

20. A function $f: \mathbb{R} \longrightarrow \mathbb{R}$ is defined by
$f: x \mapsto x + 2$. If $f\left(\frac{3}{a}\right) = f \circ f(a)$, the
value of a must be:
[A] $-\frac{3}{2}$ [B] $\frac{3}{2}$ [C] 3 [D] -3

21. The function g is defined in $\mathbb{R}$, the set of
real numbers, by $g: x \longmapsto x + 4$. $g^{-1}: x$ is:
[A] $g^{-1}: x \longmapsto \frac{4}{x}$ [B] $g^{-1}: x \longmapsto x - 4$

[C] $g^{-1}: x \mapsto \frac{x}{4}$ [D] $g^{-1}: x \mapsto \frac{1}{x-4}$

22. The function f is defined in $\mathbb{R}$, the set of real numbers, by $f: x \mapsto \frac{1}{x+2}, x \neq -2$. $f(-4)$ is equal to:

[A] $-\frac{1}{2}$ [B] $\frac{1}{2}$ [C] $-\frac{1}{6}$ [D] $\frac{1}{6}$

23. The functions f and g are defined in $\mathbb{R}$, the set of real numbers, by

$f: x \mapsto \frac{1}{x+2}, x \neq -2$ and $g: x \mapsto x + 4$.

$g \circ f(2)$ is equal to:

[A] $\frac{3}{2}$ [B] $\frac{13}{2}$ [C] $\frac{17}{4}$ [D] $\frac{25}{6}$

24. Given the function f defined in the set $\mathbb{R}$ of real numbers by $f: x \mapsto x^2 - 3x + 2$, $f(-3)$ is equal to:

[A] 2 [B] 20 [C] 7 [D] 5

25. Given the function f defined in the set $\mathbb{R}$ of real numbers by $f: x \mapsto x^2 - 3x + 2$. The values of x for which $f(x) = 0$ are:

[A] 2 and 1   [B] -2 and -1
[C] 2 and -1 [D] -2 and 1

26. The functions f and g are defined in $\mathbb{R}$ by $f: x \mapsto \sin x^{\circ}$ and $g: x \mapsto 2x$. $f \circ g(45)$ is equal to:

[A] 0 [B] $\frac{\sqrt{2}}{2}$ [C] 1 [D] $\frac{1}{2}$

27. The functions f and g are defined in $\mathbb{R}$ by $f: x \mapsto \sin x^{\circ}$ and $g: x \mapsto 2x$. $g \circ f(30)$ is equal to:

[A] 0 [B] $\frac{\sqrt{2}}{2}$ [C] 1 [D] $\frac{1}{2}$

28. The function f is defined on $\mathbb{Z}$, the set of integers, by $f: x \mapsto 1-2x$. $f(-4)$ is equal to:

[A] -7 [B] 7 [C] -9 [D] 9

29. The functions f and g are defined on $\mathbb{Z}$, the set of integers, by $f: x \mapsto 1-2x$ and $g: x \mapsto 5x - k$. Given that $f \circ g(x) = g \circ f(x)$, the value of k is:

[A] -4 [B] 4 [C] $-\frac{4}{3}$ [D] $\frac{4}{3}$

30. The number of simple functions that make up the composite function $f: x \mapsto (2x + 5)^2$ is:

[A] 2 [B] 3 [C] 4 [D] 5

31. A function is defined on $\mathbb{Z}$ the set of integers as $f: x \mapsto 3 + x$. The image of -4 is:

[A] -1 [B] 1 [C] 4 [D] -4

32. The relation from set P to set Q in the arrow diagram in Figure 23:35 is:

[A] 'is the square root of' [B] 'is double'
[C] 'is the square of' [D] 'is 4 times'

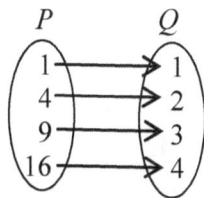

Figure 23:35

33. Given that $f: x \mapsto \frac{1}{x-2}$, then $f^{-1}: x \mapsto$:

[A] $\frac{1}{x} - 2$ [B] $\frac{1}{x} + 2$ [C] $\frac{1}{x+2}$ [D] $\frac{1}{x-2}$

33. Given that $g(x) = x^2 - 6, x \in \mathbb{Z}$, then the value of m when $g(2m) = 10$ is:

[A] $\sqrt{2}$ [B] 4 [C] -2 or 2 [D] 0 or 4

34.

x	-2	-1	0	1
y	-1	2	1	0

Table 23:1

The correct graph and assertion about the relation in Table 23:1 is:

[A]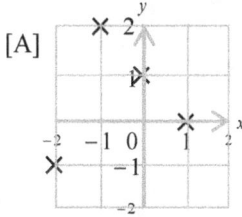

Not a function

[B]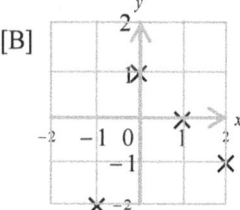

A function

[C]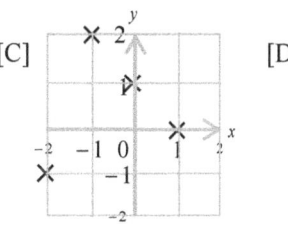

A function

[D]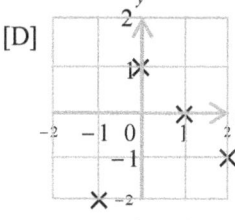

Not a function

Figure 23:34

35.

x	-1	0	1	1
y	1	-2	-1	2

The correct graph and assertion about the relation in Table 23:2 is:

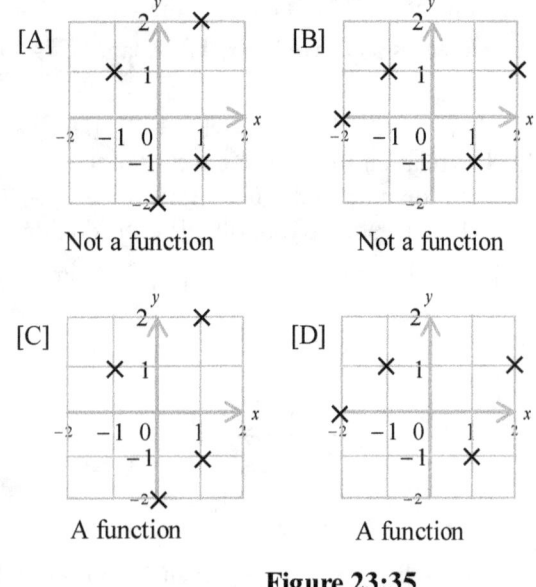

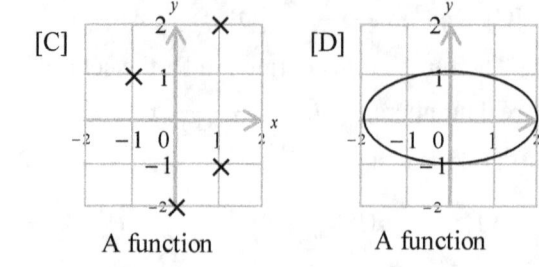

Figure 23:36

37. The table, which represents a relation, which is a function, is:

[A]

x	–2	–1	0	1
y	–1	2	1	–1

[B]

x	–2	–1	0	1
y	–2	0	–1	–2

[C]

x	–2	–1	0	1
y	–2	–1	0	–1

[D]

x	–2	–1	0	–2
y	–1	2	1	–2

Figure 23:35

36. The graph, which represents a function, is:

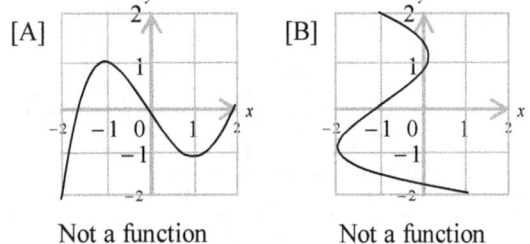

TOPIC 24

SEQUENCES AND SERIES

OBJECTIVES

At the end of this topic, the learner should be able to:

1. Recognise patterns and sequences of numbers such as triangular numbers, square numbers, cubes, simple APs and GPs etc and state the missing terms.
2. Find the nth term of a simple sequence.
3. Find the sum of the first n terms of a simple sequence.
4. Explain the meaning of an arithmetic progression (*A.P.*)
5. Find the nth term of a given *A.P.*
6. Find the sum of the first n terms of a given *A.P.*
7. Find the arithmetic mean of a given set of numbers.
8. Insert a given number of arithmetic means between two given numbers.
9. Explain the meaning of geometric progression (*G.P.*)
10. Find the nth term of a given *G.P.*
11. Find the sum of the first n terms of a given *G.P.*
12. Find the sum to infinity given converging *G.P.*
13. Find the geometric mean of two given numbers.
14. Insert a given number of geometric means between two given numbers.

Leonardo Fibonacci

Italian mathematician Leonardo Fibonacci made advances in number theory and algebra. He is especially known for his work on series of numbers, including the Fibonacci series. Each number in a Fibonacci series is equal to the sum of the two numbers that came before it.

Corbis

The Concept of a number Sequence

A **number sequence** is a group of numbers given in a specified order. A sequence may or may not have any definite rule. Examples of sequences are,

(a) 1, 5, 9, 13...   (b) 1, 5, 25, 125...
(c) 1,2,4,7...   (d) 62, 59, 57, 54, 52...
(e) 5, 7, 13, 16...

Each of the numbers in a sequence is called a **term**.
A sequence, which has a last term, is called a **finite sequence**. On the other hand, a sequence, which continues indefinitely, is called an **infinite sequence**.

Sequence recognition
For sequences that have a definite rule, it is often required to find the rule and write the next few terms

Example 24:1
Determine the rule and use it to write down the next three terms in each of the following sequences.

(a) 1, 5, 9, 13 ...   (b) 1, 5, 25, 125 ...
(c) 1,2,4,7 ...   (d) 62, 59, 57, 54, 52 ...

Solution
(a) By observation, the rule for this sequence is 'add 4 to a term to obtain the next term'.

$$1 \xrightarrow{+4} 5 \xrightarrow{+4} 9 \xrightarrow{+4} 13 \xrightarrow{+4} 17$$

Therefore the next three terms are,
$13 + 4 = 17,\ 17 + 4 = 21$ and $21 + 4 = 25$.
Therefore, the sequence is
1,5,9,13,17,21,25 ...

(b) By observation, the rule is 'multiply a term by 5 to obtain the next term'

$$1 \xrightarrow{\times 5} 5 \xrightarrow{\times 5} 25 \xrightarrow{\times 5} 125 \xrightarrow{\times 5} 625$$

Therefore the next three terms are,
$125 \times 5 = 625, 625 \times 5 = 3125$ and $3125 \times 5 = 15625$. Therefore, the sequence is 1, 5, 25, 125, 625, 3125, 15625 ...

(c) By observation, the rule is add 1,2,3 etc to the first, second, third terms, respectively.

$$1 \xrightarrow{+1} 2 \xrightarrow{+2} 4 \xrightarrow{+3} 7 \xrightarrow{+4} 11$$

Therefore the next three terms are
$7 + 4 = 11, 11 + 5 = 16$ and $16 + 6 = 22$.
Therefore, the sequence is 1,2,4,7,11,16,22...

(d) The rule is alternating thus; subtract 3 from the first term, subtract 2 from the second term, subtract 3 from the third term, subtract 2 from the fourth term, and so on.

$$62 \xrightarrow{-3} 59 \xrightarrow{-2} 57 \xrightarrow{-3} 54 \xrightarrow{-2} 52$$

Therefore, the sequence is 62, 59, 57, 54, 52, 49, 47, 44...

EXERCISE 24:1

Determine the rule and use it to write down the next three terms in each of the following sequences

(1) 2,4,6,8 ...   (2) 3,9,15,21, ...
(3) 1,4,9,16 ...   (4) 2,5,8,11,14,17, ...
(5) 160,80,40,20 ...   (6) 1,2,4,7, ...
(7) 81,27,9,3 ...   (8) 1,4,7,10, ...
(9) 2,12, 22, 32, ...   (10) 41,37,33,29 ...
(11) 3,13,10,20 ...   (12) 0, – 4, – 8, – 12 ...
(13) 3,5,9,15,23 ...   (14) – 1,3,7,11,15,..

The nth Term of a Sequence
Sometimes instead of listing terms of sequences as above, a formula is usually given which defines the general term U_n of the sequence. If instead of separating terms with commas, the terms are separated with (+) signs, such an ordered arrangement of numbers is called a **series**. For instance, $1+2+3+4+5+6+...$ is a series. Using the given general term of a sequence or series, every other term can be obtained. The first term is sometimes denoted by a instead of U_1.

Example 24:2
The nth term U_n of a series is given by $U_n = 2n^2 + 1$, where $n \in \mathbb{N}$. List the first 5 terms of the series.

Solution

Therefore the first five terms of the series are,
$3 + 9 + 19 + 33 + 51 + \cdots$

Example 24:3
Write down the sixth term U_6 of a sequence whose n^{th} term is given by $U_n = n(n + 1)$, where $n \in \mathbb{N}$.

Solution
$$U_n = n(n + 1) \Rightarrow U_6 = 6(6 + 1) = 42$$

Example 24:4
Find the nth term of each of the following sequences.
(a) 1, 4,7,10, 13, 16 ...   (b) 2, 5, 10, 17 ...

Solution

(a) $1, 4, 7, 10, 13, 16 \ldots$

$1 + 3(0) = 1$
$1 + 3(1) = 4$
$1 + 3(2) = 7$
$1 + 3(3) = 10$
$1 + 3(4) = 13$
$1 + 3(5) = 16$
$\therefore U_n = 1 + 3(n - 1) = 3n - 2$

(b) $2, 5, 10, 17 \ldots$

$1 + 1^2 = 2$

$1 + 2^2 = 5$

$1 + 3^2 = 10$

$1 + 4^2 = 17$

$\therefore U_n = 1 + n^2$

Sum of the First n Terms of a Sequence

Just as a formula can be given for the nth term U_n of a sequence, a formula can equally be given for the sum S_n, of the first n terms of a sequence, where $n \in \mathbb{N}$.

Example 24:5

The sum S_n of the first n terms, of a sequence is given by $S_n = 3n^2 - 2n$, where $n \in \mathbb{N}$. Find

(a) S_1, S_2 and S_{10}.   (b) U_1, U_2 and U_{10}.

Solution

$$S_n = 3n^2 - 2n$$

(a) $S_1 = 3(1)^2 - 2(1) = 1$

$S_2 = 3(2)^2 - 2(2) = 8$

$S_{10} = 3(10)^2 - 2(10) = 280$

(b) $U_1 = S_1 = 1$

$U_2 = S_2 - S_1 = 8 - 1 = 7$

$U_{10} = S_{10} - S_9$

But $S_9 = 3(9)^2 - 2(9) = 225$

$U_{10} = 280 - 225 = 55$

EXERCISE 24:2

1. Write down the nth term of the following sequences:
   (a) $3, 6, 12, 24, \ldots$ (b) $5, 8, 11, 14, \ldots$

   (c) $1, 8, 27, 64, \ldots\ldots$ (d) $10, 40, 160, 640 \ldots$
   (e) $1, \dfrac{1}{3}, \dfrac{1}{4}, \dfrac{1}{8}, \ldots$ (f) $\dfrac{1}{2}, \dfrac{2}{3}, \dfrac{3}{4}, \dfrac{4}{5}, \ldots$

2. Write down the first three terms in the sequences whose nth terms are given below:
   (a) $u_n = n(n + 4)$ (b) $u_n = n(n + 1)$

   (c) $u_n = \dfrac{n}{10} + \dfrac{1}{n}$ (d) $u_n = n^2 - 5n$

   (e) $u_n = 3(2)^n$

3. The sum of the first n terms of a sequence is $\dfrac{1}{3} n(n + 1)(n + 2)$. Find the first four terms and the nth term, where $n \in \mathbb{N}$.

4. The sum of the first n terms of a sequence is $n(n + 3)$, where $n \in \mathbb{N}$. Find the first three terms and the nth term.

5. The sum of the first n terms of a sequence is $\dfrac{1}{6} n(n + 1)(2n + 1)$, where $n \in \mathbb{N}$. Find $S_1, S_2,$ S_3, S_4 and S_5. Hence, find the first five terms. Write down the nth term of the sequence.

6. The sum of the first n terms of a series is $\dfrac{1}{2} n(n + 1)$, where $n \in \mathbb{N}$, Find the first three terms and the nth term.

7. Give, with reasons, the next two terms in each of the following sequences:
   (a) $1, 3, 7, 13, 21, \ldots$

   (b) $1, \dfrac{1}{2}, \dfrac{1}{6}, \dfrac{1}{24}, \dfrac{1}{120}, \ldots$

8. The nth term U_n of a certain sequence is given as $U_n = (4 - n)^2 - 1$, where $n \in \mathbb{N}$.
   Write down the first five terms of the sequence.

9. The sum of the first n terms of a sequence is given by $n.2^n$, where n is a positive integer. Find the third term of the sequence.

10. The sum of the first n terms of a sequence of number is given by $3n^2 + 2n$, where $n \in \mathbb{N}$.
    (a) Find the first 3 terms of the sequence.
    (b) Find the sum of first 12 terms of the sequence.

11. The first term of a sequence of numbers is -10 and each succeeding term of the sequence is obtained by adding 4 to the previous term.
    (a) Write down the first 5 terms of the sequence.
    (b) Given that the sum of the first n terms of the sequence is $2n^2 + 12n$, where $n \in \mathbb{N}$, calculate the sum of the first 11 terms.

ARITHMETIC AND GEOMETRIC PROGRESSIONS

The arithmetic and geometric progressions are two very important types of sequences in daily life especially in fields like banking and Economics. These progressions are so rampant that they require some special focus. This section will give many Economic students, an opportunity to appreciate better, the so-called Malthusian theory published sometimes in 1798. What did Malthus, Thomas Roberts really mean by population was growing at a geometric progression while food production was growing at an arithmetic progression?

ARITHMETIC PROGRESSION (A.P.)

In an arithmetic progression (*A.P.*), the difference between a term and the preceding one is called the **common difference** and denoted by *d* is constant. In other words, each term is obtained by adding the common difference *d* to the preceding one. Some examples of arithmetic progression are, $14, 17, 20, 23 \dots$ and $4, 2, 0, -2, -4 \dots$

Example 24:6

In the following *A.P*'s, write down the first term *a* and the common difference *d*.

   (a) $14, 17, 20, 23, \dots$ (b) $4, 2, 0, -2, -4, \dots$
   (c) $1, 2, 3, \dots$ (d) $4, 7, 10, \dots$

   (e) $1\frac{1}{2}, \ -1\frac{1}{2}, \ -4\frac{1}{4}, \dots$

Solution

(a) $a = 14, d = 3$ (b) $a = 4, d = -2$ (c) $a = 1, d = 1$

(d) $a = 4, d = 3$ (e) $a = 1\frac{1}{2}, \ d = -3$

Example 24:7

Insert four numbers between 7 and 67 so that the sequence is an A.P.

Solution

Let the common difference of the *A.P.* be *d* then consecutive terms of the *A.P.* are

$$7, 7+d, 7+2d, 7+3d, 7+4d, 67.$$

$$\Rightarrow d = 67 - (7 + 4d)$$

$$d = 67 - 7 - 4d$$

$$5d = 60 \Rightarrow d = 12$$

Therefore the four numbers are
$7 + 12, 7 + 2(12), 7 + 3(12), 7 + 4(12)$ i.e.
$19, 31, 43$ and 55 in that order.

The nth term of an A.P.

Consider an *A.P.* whose first term is *a* and whose common difference is *d*. Then this *A.P.* can be written as $a, a + d, a + 2d, a + 3d, \dots$

$$\Rightarrow U_1 = a + (1 - 1)d = a$$

$$U_2 = a + (2 - 1)d = a + d$$

$$U_3 = a + (3 - 1)d = a + 2d$$

$$U_4 = a + (4 - 1)d = a + 3d$$

And so on

$$\therefore U_n = a + (n - 1)d \text{ where } n \in \mathbb{N}.$$

Hence the general term of an A.P. is

$U_n = a + (n - 1)d$.

Example 24:8

Find the 200th term of the A.P. $4, 2, 0, -2, -4 \dots$

Solution

$$U_n = a + (n - 1)d$$

$$a = 4, \ d = -2 \text{ and } n = 200$$

$$U_{200} = 4 + (200 - 1)(-2) = 4 + 199(-2) = -394$$

Example 24:9

The nth term of the *A.P.* $\frac{1}{2}, \frac{3}{2}, \frac{5}{2}, \frac{7}{2}, \dots$ is $\frac{649}{2}$.

Find the value of *n*.

Solution

$$U_n = a + (n - 1)d$$

$$a = \frac{1}{2}, \ d = 1 \text{ and } U_n = \frac{649}{2}$$

$$\frac{1}{2} + (n - 1)(1) = \frac{649}{2}$$

$$n - 1 = 324$$

$$n = 325$$

Sum of the first n terms of an A.P.

Example 24:10

Find the sum of the first 5 terms of the *A.P.* 3, 7, 11, 15, 19....

Solution

$$S_5 = 3 + 7 + 11 + 15 + 19 = 55$$

Alternatively;

$$S_5 = 3 + 7 + 11 + 15 + 19 = 55$$

$$S_5 = 3 + 7 + 11 + 15 + 19 \dots\dots\dots\dots\dots① $$

Or $S_5 = 19 + 15 + 11 + 7 + 3$ ....②

①+②:

$2S_5 = 22 + 22 + 22 + 22 + 22$

$2S_5 = 110$

Divide both sides by 2,

$S_5 = 55$

This could have been done more easily by adding the first and last terms, multiplying by the number of terms, then dividing by 2.

When the numbers of terms is large, the alternative method is a better method for finding the sum of the first n terms of an A.P.

Example 24:11
Find the sum of the first 75 terms of the A.P.
3, 7, 11, 15, 19 ....

Solution
To list all these terms before adding them is cumbersome. So let us use the alternative method.

$$a = 3 = U_1, \ d = 4$$

$$U_n = a + (n-1)d$$

$$U_{75} = 3 + (75-1)4 = 299$$

$$2S_{75} = 75(U_1 + U_{75})$$

$$\Rightarrow 2S_{75} = 75(3 + 299)$$

$$\Rightarrow 2S_{75} = 22650$$

$$\therefore S_{75} = 11325$$

To generalize, suppose the first term of an A.P. is a and the common ratio is d, then the nth term of the A.P. is

$$U_n = a + (n-1)d$$

$$2S_n = n(U_1 + U_n)$$

$$\Rightarrow 2S_n = n\{2a + (n-1)d\}$$

$$\Rightarrow S_n = \frac{n}{2}\{2a + (n-1)d\}$$

Therefore if $n \in \mathbb{N}$, then the sum of the first n terms of an A.P. is given by;

$$S_n = \frac{n}{2}\{2a + (n-1)d\} \ \text{or} \ S_n = \frac{n}{2}(a + l)$$

Example 24:12
Find the sum of the first 20 terms of the A.P. 3, 7, 11, 15.

Solution
$$a = 3, \ d = 4, \ n = 20$$

$$S_n = \frac{n}{2}\{2a + (n-1)d\}$$

$$S_{20} = \frac{n}{2}\{2(3) + (20-1)4\} = 10\{6 + 19(4)\} = 820$$

The Arithmetic Mean

If p, q, r are consecutive terms of an A.P. then, the common difference of the A.P. will be

$$d = q - p = r - q$$
$$\Rightarrow 2q = p + r$$
$$q = \frac{p + r}{2}$$

q is called the arithmetic mean of p and r and is often referred to as the average and denoted by . Hence, the arithmetic mean of two numbers x_1 and x_2 is given by

$$\bar{x} = \frac{x_1 + x_2}{2}$$

Generally, the arithmetic mean of a set of n numbers $x_1, x_2, x_3, .., x_n$ is defined as the quotient of the sum of the numbers and the number of numbers. i.e.

$$\bar{x} = \frac{x_1 + x_2 + x_3 + ... + x_n}{n}$$

If one is requested to insert n arithmetic means between two numbers a and b, then n numbers x_1, x_2, $x_3, .., x_n$ must be inserted between a and b so that $a, x_1, x_2, x_3, .., x_n, b$ are consecutive terms of an A.P.

Example 24:13
Find the arithmetic mean of 13 and 17.

Solution
$$\bar{x} = \frac{a + b}{2} = \frac{13 + 17}{2} = 15$$

EXERCISE 24:3

1. Find the 24th term of the sequence 7, 13, 19, ...
2. The common difference of an arithmetic progression is 2 and the sum of the first 28 terms is 0. Find the first term and the 28th term.
3. Given that, the nth term of the sequence 6, 11, 16, 21 ... is 121. Find the value of n.

4. Find the first four terms of the arithmetic progression with first term 5 and the seventh term is -11.
5. The first and last terms of an arithmetic progression are -3 and 25. The sum of all the terms is 1837. Find (a) The number of terms; (b) The common difference, and (c) The middle term.
6. Insert four arithmetic means between the numbers 9 and 29. [Hint: The four arithmetic means are numbers such that the sequence is an A.P.]
7. Find the sum of the first ten terms of the sequence 3,7,11…. What is the sum of the next ten terms?
8. The sum of the first n terms of the sequence 7,11,15,… is 250. Find the value of n.
9. The sum of five consecutive numbers of an arithmetic progression is 50, and the sum of their squares is 590. Find the numbers.
10. The first term of an A.P. is 4 and the common difference is 7. How many terms are required in order that the sum may exceed 500?
11. Insert three arithmetic means between a and b. In an A.P., the first term is a and the common difference is $2a$. Show that the sum to $2n$ terms is always equal to the sum of the n terms.

GEOMETRIC PROGRESSION (G.P.)

In a Geometric progression, the ratio of a term and the preceding one called the **common ratio** r is constant. In other words, each term is obtained by multiplying the preceding one by the common ratio r. Some examples of geometric progressions are,

(a) 1, 4, 16,…   (b) $4, 2, \dfrac{1}{2}…$

Example 24:15
In the following $G.P$'s, write down the first term a and the common ratio r.

(a) 1, 4, 16,…   (b) $4, 2, \dfrac{1}{2}…$

(c) 6, 18, 54,…   (d) 7, 14, 28,…

(e) $\dfrac{1}{9}, \dfrac{1}{3}, 1, 3, \cdots$

Solution

(a) $a = 1, r = 4$ (b) $a = 4, r = \dfrac{1}{2}$

(c) $a = 6, r = 3$ (d) $a = 7, r = 2$

(e) $a = \dfrac{1}{9}, r = \dfrac{1}{3}$

The nth term of a G.P.
Consider a $G.P.$ whose first term is a and whose common ratio is r. Then this $G.P.$ can be written as

$$a, ar, ar^2, …$$

$$\Rightarrow U_1 = ar^{1-1} = a$$

$$U_2 = ar^{2-1} = ar$$

$$U_3 = ar^{3-1} = ar^2$$

$$U_4 = ar^{4-1} = ar^3$$

And so on

$$\therefore U_n = ar^{n-1}$$

Hence, the general term of a $G.P.$ is $U_n = ar^{n-1}$, where $n \in \mathbb{N}$.

Example 24:16

Find the 30th term of the $G.P.$ $4, 2, 1, \dfrac{1}{2}, … ,$ leaving your answer in index form.

Solution

$$U_n = ar^{n-1}$$

$$a = 4, r = \dfrac{1}{2} \text{ and } n = 30$$

$$\Rightarrow U_{30} = 4\left(\dfrac{1}{2}\right)^{30-1} = 4\left(\dfrac{1}{2}\right)^{29} = 4(2)^{-29} = 2^{-27}$$

Example 24:17
The nth term of the $G.P.$ 2, 4, 8, 16… is 2048. Find the value of n.
Solution

$$U_n = ar^{n-1}$$

$$a = 2, r = 2 \text{ and } U_n = 2048$$

$$\Rightarrow 2(2)^{n-1} = 2048$$

$$2^n = 2^{11}$$

$$\Rightarrow n = 11$$

The Sum of n terms of a G.P.

Example 24:18
Find the sum of the first 5 terms of the $G.P.$ 2, 6, 18, 54, 162 ....

Solution

$S_5 = 2 + 6 + 18 + 54 + 162 = 242$

Alternatively;

$S_5 = 2 + 6 + 18 + 54 + 162 \dots\dots\dots① $

The common ratio $r = 3$
Multiply both sides by the common ratio 3

$3S_5 = 6 + 18 + 54 + 162 + 486 \dots\dots\dots②$

②– ①:

$2S_5 = 486 - 2$

$\Rightarrow 2S_5 = 484$

Divide both sides by 2,

$S_5 = 242$

Notice the way the terms cancel out on subtracting ① from ②, leaving only the last term of ② and the first term of ①.
When the number of terms is large, the alternative method is a better method for finding the sum of the first n terms of a G.P.

Example 24:19
Find the sum of the first 18 terms of the G.P.

$\dfrac{1}{4}, \dfrac{1}{2}, 1, 2, 4, 8, \dots$

Solution
To list all these terms before adding them is cumbersome. So let us use the alternative method.

$a = \dfrac{1}{4}, \; r = 2, \; n = 18, \; U_n = ar^{n-1}$

$U_{18} = \frac{1}{4}(2^{18-1}) = 2^{15}$

$S_{18} = \dfrac{1}{4} + \dfrac{1}{2} + 1 + 2 + 4 + \dots + 2^{15} \dots\dots\dots①$

Multiply both sides by $r = 2$

$2S_{18} = \dfrac{1}{2} + 1 + 2 + 4 + \dots + 2^{16}$

$2S_{18} - S_{18} = 2^{16} - \dfrac{1}{4}$

$S_{18} = \dfrac{4\left(2^{16}\right) - 1}{4} = 65535\dfrac{3}{4}$

To generalize, suppose the first term of a G.P. is a and the common ratio is r, then the nth term of the G.P. is

$U_n = ar^{n-1}$

$S_n = a + ar + ar^2 + \dots + ar^{n-1} + \dots\dots\dots①$

Multiply both sides by r.

$rS_n = ar + ar^2 + \dots + ar^n + \dots\dots\dots②$

②– ①:

$rS_n - S_n = ar^n - a$

$rS_n - S_n = ar^n - a$

$S_n(r - 1) = a(r^n - 1)$

$\Rightarrow S_n = \dfrac{a(r^n - 1)}{(r - 1)}, |r| > 1$ or $S_n = \dfrac{a(1 - r^n)}{(1 - r)}, |r| < 1$

Depending on the individual or problem, either of the forms of this formula may be used. It is not necessary to memorize the two since one can easily be obtained from the other and whichever is used the final result is the same.

Therefore, the sum of the first n terms of a G.P. is given by

$$S_n = \dfrac{a(r^n - 1)}{(r - 1)} \quad \text{or} \quad S_n = \dfrac{a(1 - r^n)}{(1 - r)}$$

Example 24:20
Find the sum of the first 20 terms of the G.P.
$3, 6, 12, 24, \dots$

Solution

$S_n = \dfrac{a(r^n - 1)}{(r - 1)}$

$a = 3, r = 2$ and $n = 20$

$\Rightarrow S_{20} = \dfrac{3\left(2^{20} - 1\right)}{2 - 1} = 3(1048576 - 1) = 3145725$

Sum to infinity, S_∞

Consider a G.P. with $|r| < 1$. As the number of terms, n increases (i.e. $n \to \infty$) the value of r^n decreases (i.e. $r^n \to 0$) and the sum S_n of the G.P. approaches a limiting value called **sum to infinity** denoted by S_∞.

$$S_\infty = \dfrac{a}{(1 - r)}, \quad |r| < 1$$

Example 24:21
Find the sum to infinity of the G.P. whose first term is 81 and whose common ratio is $\dfrac{1}{3}$.

Solution

$$S_\infty = \frac{a}{1-r}, \quad a = 81, r = \frac{1}{3}$$

$$S_\infty = \frac{81}{1-\frac{1}{3}} = \frac{81}{\frac{2}{3}} = \frac{243}{2}$$

Geometric Mean

Let a, b, c be three consecutive terms of a G.P.
Then by the definition of a G.P., the common ratio
r is given by

$$r = \frac{b}{a} \text{ or } r = \frac{c}{b} \Leftrightarrow \frac{b}{a} = \frac{c}{b}$$

$$b^2 = ac$$

$$b = \pm\sqrt{ac}$$

b is said to be the **geometric mean** of a and c,
usually denoted by GM.

$$GM = \pm\sqrt{ac}$$

Generally, the geometric mean of a set of n
numbers $x_1, x_2, x_3, ..., x_n$ is defined as the nth root
of the product of the n numbers. i.e.

$$GM = \pm\sqrt[n]{x_1 x_2 ... x_n}$$

n numbers can be inserted between two numbers a
and b, so that the sequence is a G.P.

Example 24:22

Find the geometric mean of 6 and $13\frac{1}{2}$.

Solution

$$GM = \pm\sqrt{ac} = \pm\sqrt{6\left(\frac{27}{2}\right)} \pm\sqrt{81} = \pm 9$$

Example 24:23

Insert 2 numbers between -6 and $20\frac{1}{4}$, so that

the sequence is a G.P.

Solution
Let the common ratio of the G.P. be r then
consecutive terms of the G.P. are

$$-6, -6r, -6r^2, 20\frac{1}{4}.$$

$$\Rightarrow \frac{-6r}{-6} = \frac{20\frac{1}{4}}{-6r^2}$$

$$\Rightarrow r^3 = -\frac{81}{24} = -\frac{27}{8} \Rightarrow r = -\frac{3}{2}$$

Therefore, the 2 geometric means are $-6\left(-\frac{3}{2}\right)$

and $-6\left(-\frac{3}{2}\right)^2$ i.e. 9 and $-\frac{3}{2}$ in that order.

EXERCISE 24:4

1. The first three terms of a geometric progression
   are $2x, x+6, 4x+1\frac{1}{2}$. Find
   (a) The two possible values of x.
   (b) The common ratio and the fourth term of
       that progression for which x is positive.

2. The second term of a G.P. is -6 and the fifth
   term is $20\frac{1}{4}$. Find the common ratio and the
   nth term.

3. The third term of a geometric sequence exceeds
   the second by $\frac{9}{14}$ and the second exceeds the
   first by $\frac{3}{7}$. Find the common ratio.

4. The first term of a geometric progression is $\frac{3}{5}$
   and the fourth term is $9\frac{3}{8}$. Find the 12th term
   and the sum of the first 12 terms.

5. Find the geometric means of
   (a) 6 and 24   (b) 36 and $20\frac{1}{4}$ (c) x^2 and x^4

6. Insert three numbers between the following so
   that the sequence is a G.P.
   (a) 2 and 32   (b) 16 and $\frac{1}{16}$

7. In Table 24:1, find the sum of the series in
   column A to the number of terms indicated in
   column B.

	A	B
(a)	$1 + 2 + 4 + 8 + \ldots$	10
(b)	$4 - 8 + 16 - 32 + \ldots$	8
(c)	$6 + 2 + \dfrac{2}{3} + \dfrac{2}{9} + \ldots$	7
(d)	$1 + 3a + 9a^2 + 27a^3 + \ldots$	10

Table 24:1

8. In both an *A.P.* and a *G.P.* the first and fourth terms are 4 and $\dfrac{1}{2}$ respectively. Calculate

   (a) The common difference of the *A.P.*
   (b) The common ratio of the *G.P.*
   (c) The second and fifth terms of each progression.
   (d) The number of terms of the *A.P.*, which have a sum of $3\frac{1}{2}$.

═══════════════════════════════
◁ **MULTIPLE CHOICE EXERCISE 24** ▷
═══════════════════════════════

1. The next 2 terms in the sequence 1, 2, 4, 7, 11, 16… are:
   [A] 17, 29   [B] 29, 24
   [C] 22, 29   [D] 29, 40

2. The next term in the sequence 1, 4, 9, 16,…is:
   [A] 20   [B] 25   [C] 23   [D] 27

3. The next term in the sequence 2, 5, 11, 23, 47…is:
   [A] 95   [B] 93   [C] 71   [D] 27

4. In the sequence 1,3,7,15,31 the number that must be added to 31 to give the next term is:
   [A] 4   [B] 8   [C] 16   [D] 32

5. The number represented by * in the arithmetic progression $14, -3, *, -37$ is:
   [A] 11   [B] -14 [C] 17   [D] -20

6. The 4th term of an *A.P.* whose first term is 2 and whose common difference is 0.5 is:
   [A] 3   [B] $\dfrac{7}{2}$ [C] $\dfrac{11}{2}$ [D] 5

7. The first term of an *A.P.* is equal to twice the common difference *d*. In terms of *d* the 5th term of the *A.P.* is:
   [A] $4d$ [B] $5d$ [C] $6d$ [D] $a + 5d$

8. The 9th term of the Arithmetic progression 18, 12, 6, 0,-6,... is:
   [A] -54 [B] -30 [C] 30   [D] 42

9. The eleventh term of an *A.P.* is 25 and its first term is 3. Its common difference is:
   [A] $1\dfrac{9}{10}$ [B] $2\dfrac{1}{5}$ [C] $2\dfrac{4}{5}$ [D] $2\dfrac{1}{2}$

10. If the first term of an *A.P.* is 4 and the 5th term is 12, then the mean of the first five terms is:
    [A] 4   [B] 6   [C] 8   [D] 10

11. The nth term U_n of the *A.P.* 11, 4, -3… is:
    [A] $U_n = 19 + 7n$ [B] $U_n = 19 - 7n$
    [C] $U_n = 18 - 7n$ [D] $U_n = 18 + 7n$

12. The nth term U_n of the sequence 4, 10, 16,… is:
    [A] $2(3n - 1)$ [B] $2(2 + 3^{n-1})$
    [C] $2^2 + 2$ [D] $2(3n + 1)$

13. In an *A.P.*, the first term is 2, and the sum of the first and the 6th terms is $16\dfrac{1}{2}$. The 4th term is:
    [A] $5\dfrac{1}{2}$ [B] $9\dfrac{1}{2}$ [C] 8   [D] 7

14. The sum of the 1st and 2nd terms of an *A.P.* is 4 and the 10th term is 19. The sum of the 5th and 6th terms is:
    [A] 11   [B] 22   [C] 21   [D] 20

15. It is observed that:
    If $1 + 3 + 5 + 7 + 9 + 11 + 13 + 15 = p^2$, then the value of *p* is:
    [A] 6   [B] 7   [C] 8   [D] 9
    $$1 + 3 = 2^2$$
    $$1 + 3 + 5 = 3^2$$
    $$1 + 3 + 5 + 7 = 4^2$$

16. The common ratio of a *G.P.* is 2. If the 5th term is greater than the first by 45, the 5th term will be:
    [A] 3   [B] 6   [C] 45   [D] 48

17. The fifth and seventh terms of the geometric progression $-2, -3, -\dfrac{9}{2}, -\dfrac{27}{4}\cdots$ are respectively:
    [A] $-\dfrac{81}{16}, -\dfrac{729}{32}$ [B] $-\dfrac{8}{81}, \dfrac{72}{18}$
    [C] $-\dfrac{21}{8}, \dfrac{32}{618}$ [D] $-\dfrac{27}{16}, -\dfrac{79}{18}$

18. The 6th term of a geometric progression is $-\dfrac{2}{27}$ and the first term is 18. The common ratio is:
    [A] $-\dfrac{1}{2}$ [B] $-\dfrac{1}{3}$ [C] $\dfrac{1}{4}$ [D] 3

19. The common ratio of the *G.P.* log 3, log 9, log 81,… is:
    [A] 1   [B] 2   [C] 3   [D] 6

20. The 16th term of the *G.P.* 2,6,18,... is given by:

[A] 2×3^{12} [B] 2×3^{13}

[C] 2×3^{15} [D] 2×3^{16}

21. The sum of the first 5 terms of the *G.P.* 2,6,18,...is:
    [A] 121 [B] 243 [C] 242 [D] 130

22. If the second and fourth terms of a *G.P.* are 8 and 32 respectively, the sum of the first four terms will be:
    [A] 28   [B] 40   [C] 48   [D] 60

23. If the second and 5th terms of a *G.P.* are –6 and 48 respectively, the sum of the first four terms is:
    [A] –45   [B] –15   [C] 15 [D] 45

24. The nth term of a sequence is represented by $3 \times 2^{(2-n)}$. The first three terms of the sequence are respectively:

    [A] $\frac{3}{2},3,6$ [B] $6,3,\frac{3}{2}$

    [C] $\frac{3}{2},3,\frac{1}{3}$ [D] $\frac{2}{3},3,\frac{8}{3}$

25. The nth term of a sequence is $2^{(2n-1)}$. If $U_n = 2^9$, n is:
    [A] 3   [B] 4   [C] 5   [D] 6

26. The next two terms of the sequence 1,5,14,30,55,... are respectively:
    [A] 61,110   [B] 67,116
    [C] 81,140   [D] 91,140

27. The nth term of a sequence is given by $(-1)^{(n-2)}$. The sum of the second and third terms is:
    [A] 0 [B] 1 [C] 2 [D] 6

28. The sum of the first n terms of a sequence is given by $S_n = 17n - 3n^2$. The fourth term of the sequence is:
    [A] 20   [B] –10   [C] –4   [D] 10

29. The nth term of a sequence is given as $\dfrac{n^2 + n}{2}$. The 7th term of the sequence is:
    [A] 36   [B] 28   [C] 21   [D] 14

30. The common difference in the sequence 10, 2, –6, –14, . . .is:
    [A] 10   [B] 8   [C] –16   [D] –8

31. The next three terms and the rule that describe the sequence 10, 20, 40, 80 . . . . are:
    [A] 82, 84, 86; start with 10 and add 2 repeatedly.
    [B] 90, 100, 110; start with 10 and add 10 repeatedly.
    [C] 320, 1280, 5120. Start with 10 and multiply by 4 repeatedly.
    [D] 160, 320, 640; start with 10 and multiply by 2 repeatedly.

32. The next three terms in the sequence 3,12,21,30,... are:
    [A] 40, 50, 60   [B] 38, 46, 54

[C] 39, 48, 57   [D] 36, 32, 39

33. The next four terms in the sequence –4, –1, 2, 5... are:
    [A] 5, 8, 11, 14   [B] 8, 11, 14, 17
    [C] 3, 6, 9, 12   [D] 0, 8, 11, 14

34. If n points are marked on a circle, where n is a whole number greater than 1, $\frac{1}{2}n^2 - \frac{1}{2}n$ segments can be drawn to connect these points. The number of segments that can be drawn if 8 points are marked on the circle is:
    [A] 28 segments   [B] 32 segments
    [C] 60 segments   [D] 56 segments

35. If the pattern continues, the number of squares in the 8th diagram is:
    [A] 16 squares   [B] 18 squares
    [C] 14 squares   [D] 15 squares

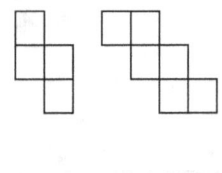 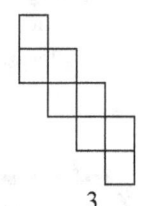

     1   2   3

36. The number of dots in the ninth figure is:
    [A]   25 dots   [B]   27 dots
    [C]   19 dots   [D]   26 dots

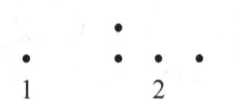 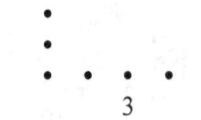

   1   2   3

37. The diagram below represents a five by five square of black and white floor tiles. The number of black tiles, which can be added to the existing pattern to make it six by six, are:
    [A]   8 black tiles   [B]   6 black tiles
    [C]   5 black tiles   [D]   4 black tiles

TOPIC 25

POINTS, LINES AND ANGLES

OBJECTIVES

At the end of this topic, the learner should be able to:

1. Appreciate that a point has no size and is represented with a dot and named with a capital letter.
2. Define and identify a straight line
3. Distinguish and draw straight zigzag and curved lines.
3. Appreciate that a line containing points *A* and *B* extends infinitely.
4. Distinguish, identify, draw, represent and denote lines, line segments and rays.
5. Understand and explain the meaning of an angle.
6. Name and write angles using letter notation.
7. Identify, name and distinguish between the various types of angles (full turn, half turn, quarter turn etc).
8. Identify vertically opposite angles, adjacent angles and angles on a straight line.
9. Identify parallel and perpendicular lines.
10. Identify angles between intersecting lines or pairs of parallel lines cut by a transversal.

Euclid was a Greek mathematician who lived around 300 B.C. He wrote Elements, a 13-volume work on the principles of geometry and properties of numbers.

THE BETTMANN ARCHIVE

Definition of terms

A Point: A point usually designated by a capital letter, *P*, is a size less representation of position. Though a dot is often used to represent a point, it is not a point.

. A Point

Figure 25:1

A Line: A line is the path of a series of points. It has a length but no thickness, and may be straight (as in Figure 25:2(i) and (ii)), zigzag (as in Figure 25:2(iii) or curved (as in Figure 25:2(i) and (iv)). A line is often designated by two capital letters representing its extreme points (as in Figure 25:2(i) and (iv)) or by a small letter (as in (ii)).

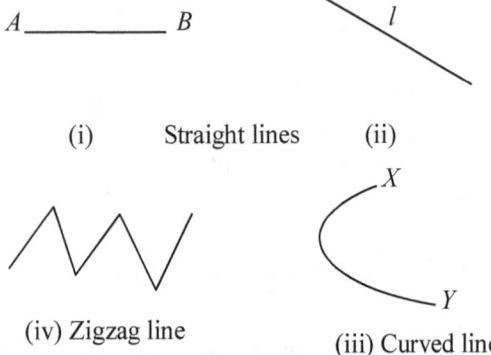

(i)   Straight lines   (ii)

(iv) Zigzag line   (iii) Curved line

Figure 25:2

A line containing the points *A* and *B* is denoted by (*AB*). A line containing two points *A* and *B* extends infinitely to both ends.

A **straight line** is the shortest distance between two points. A straight line should be looked at as the path of a point moving in the same direction.

Line Segment

A **line segment** is part of a line consisting of two points, called endpoints, and all the points that lie between these endpoints.

A   *B*

Figure 25:3

Line segments are often named according to their endpoints. A line segment with endpoints *A* and *B* is denoted by [*AB*] or $\overline{AB}$. The length of a line segment [*AB*] is denoted as *AB*. Thus, if the length of the line segment [*AB*] is 4 cm, then *AB* = 4 cm.

Ray

A **ray** is a line, which is endless in one direction. It has a definite beginning but no definite end. Rays are often designated according to their endpoints and a point through which they pass; a ray with endpoint *A* that passes through point *B* (Figure 25:4(i)) is denoted [*AB*) or $\overrightarrow{AB}$ while a ray with endpoint *B* that passes through point *A* (Figure 25:4(ii)) is be denoted (*AB*] or $\overleftarrow{AB}$.

A   *B*   *A*   *B*

(i)   (ii)

Figure 25:4

A Plane

A plane is a surface such that any straight line, connecting any two of its points lies entirely in it. A plane has length and width but no depth. A plane is similar to an infinitely large tabletop that has no edges. There exists one and only one plane that contains any three points that do not all lie on a single straight line.

EXERCISE 25:1

1. Write down four objects bounded by straight lines.
2. Write down four objects bounded by curved lines.
3. Which of the following objects are bounded by
    (a) straight lines?
    (b) curved lines?
        a ruler, a textbook, a ball, a protractor, the walls of your classroom, an unused piece of chalk, the moon, a star, the sun, a wire, a wheel.
4. State the meaning of the each of the following notations.
    (a) *AB*   (b) (*AB*]   (c) [*AB*)
    (d) (*AB*)   (e) [*AB*]   (f) $\overline{AB}$
5. Using the number line in Figure 25:5, find
    (a) *AB*   (b) *BD*   (c) *AC*
    (d) *AD*   (e) *AB* + *CD*

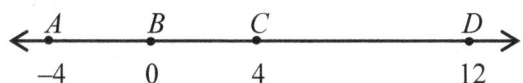

Figure 25:5

6. Identify and name each of the following using the appropriate notation.

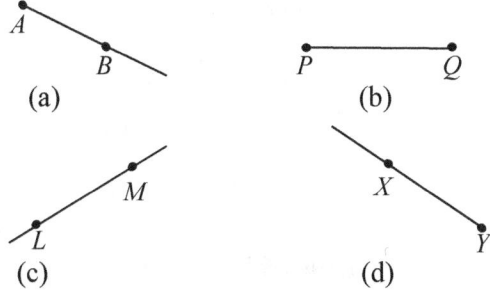

Figure 25:6

7. How many line segments bound each of the following figures?

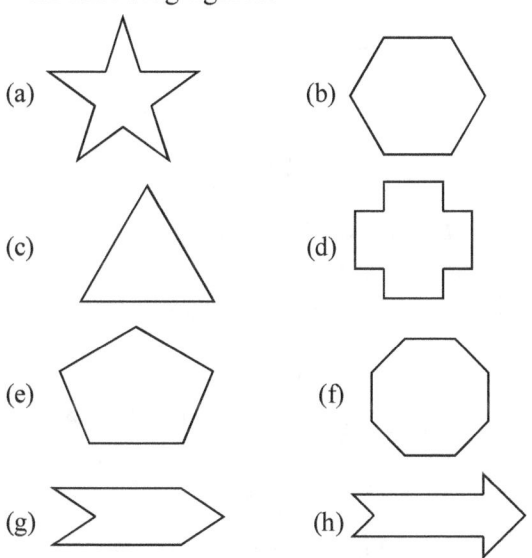

Figure 25:7

ANGLES

An **angle** (Figure 25:8) is the space between two lines or surfaces that meet. An angle may be generated by the rotation of two lines with a fixed endpoint. Thus, an angle can also be defined as the measure of the amount of turn between two lines that meet. It is measured in degrees, minutes and seconds, or in radians.

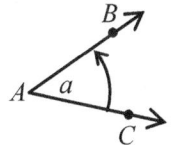

Figure 25:8

The various ways of naming the angle in Figure 25:8 are Angle A, angle a, $\hat{a}$, $\angle BAC$, $\angle CAB$, $B\hat{A}C$, $C\hat{A}B$, angle BAC, or angle CAB.

Types of Angles

1. A full turn or a revolution

A **full turn** or a **revolution** is an angle, which is equal to $360°$. Another name for a full turn is **angles at a point**.

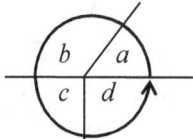

Figure 25:9

Angles at a point sum up to $360°$. Thus in Figure 25:9, $a + b + c + d = 360°$

Examples 25:1
Find y in Figure 25:10.

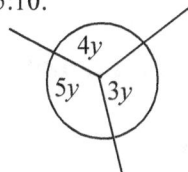

Figure 25:10

Solution

$$5y + 4y + 3y = 360°$$
$$\Rightarrow 12y = 360°$$
$$y = 30°$$

2. A Half Turn

Another name for a half turn is **adjacent angles on a straight line.** A half turn is $180°$ or adjacent angles on a straight line, sum up to $180°$.

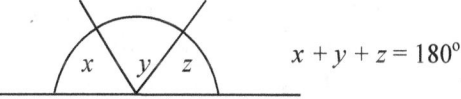

$x + y + z = 180°$

Figure 25:11

Examples 25:2

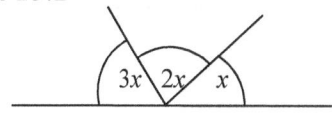

Figure 25:12

259

In Figure 25:12, above, find the value of x.

Solution

$$3x + 2x + x = 180°$$
$$\Rightarrow 6x = 180°$$
$$x = 30°$$

Two angles, whose sum is $180°$, are said to be **supplementary**.

Examples 25:3

Given that $3x = 2y$ and that x and y are supplementary. Find the values of x and y.

Solution

x and y are supplementary $\Rightarrow x + y = 180°$

Also $y = \dfrac{3x}{2} \Rightarrow x + \dfrac{3x}{2} = 180°$

$$5x = 360°$$
$$x = 72°$$
$$\Rightarrow y = \tfrac{3}{2}(72) = 108°$$

3. A Quarter Turn or a Right Angle

An angle whose measure is $90°$ is called a **quarter turn** or a **right angle**.

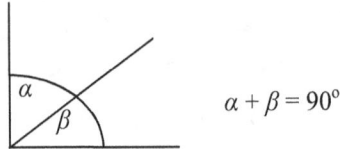

$$\alpha + \beta = 90°$$

Figure 25:13

Two angles whose sum is $90°$ are said to be **complementary**.

Examples 25:4

Using Figure 25:14, write down an equation that satisfies p and q. Given also that $p - q = 40°$, find the values of p and q.

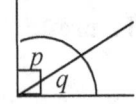

Figure 25:14

Solution

The required equation is $p + q = 90°$......①

Since $p - q = 40°$......②

①+②: $2p = 130° \Rightarrow p = 65°$

Substitute in ①, $65 + q = 90° \Rightarrow q = 25°$

4. An acute angle

An **acute angle** (Figure 25:15) is an angle, whose value is less than $90°$.

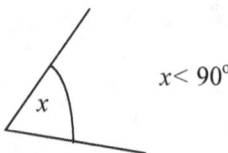

$x < 90°$

Figure 25:15

5. An obtuse angle

An **obtuse angle** (Figure 25:16) is an angle, whose value lies between $90°$ and $180°$.

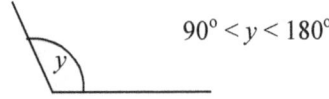

$90° < y < 180°$

Figure 25:16

6. A reflex angle

A **reflex angle** (Figure 25:17) is an angle, which lies between $180°$ and $360°$

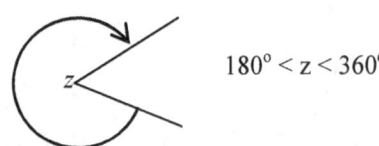

$180° < z < 360°$

Figure 25:17

EXERCISE 25:2

1. Find the value of the unknowns designated by the letter in each of the following.

(a)

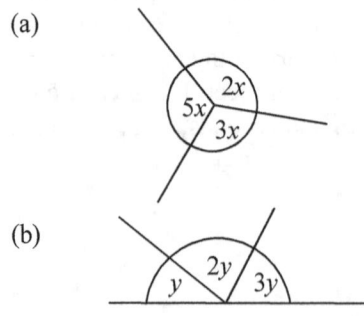

(b)

260

(c)

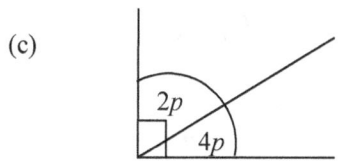

(d)

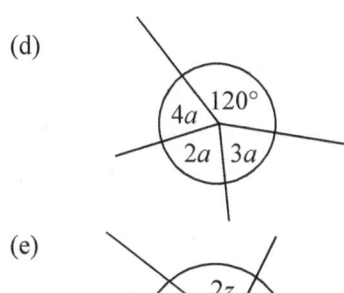

(e)

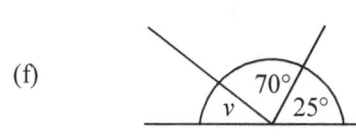

(f)

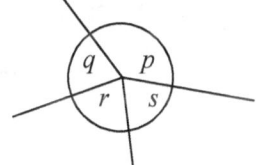

Figure 25:18

2. Given that y and $2y$ are complementary angles, find the value of y.

3. Find the value of p if the angles p, $3p$ and $4p$ are supplementary.

4. In Figure 25:19, write down an equation that satisfy p, q, r and s. Find the values of p, q, s and r, given that $p = 2q = 3r = 4s$

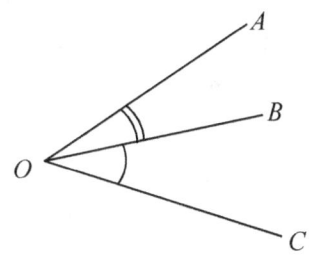

Figure 25:19

5. State the range of values of x, if angle x is:
   (a) An obtuse angle   (b) a reflex angle,

Adjacent and Non-Adjacent angles

(a) Adjacent angles

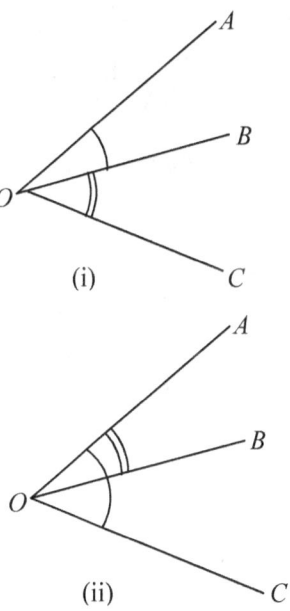

Figure 25:20

In Figure 25:20, the rays $[OA]$, $[OB]$ and $[OC]$ have a common origin. $\angle AOB$ and $\angle BOC$ share the common ray $[OB]$ which, lies between them and forms one side of each of the angles. In such a case, $\angle AOB$ and $\angle BOC$ are said to be **adjacent angles**.

For $\angle AOB$ and $\angle BOC$ to be adjacent,
$$\angle AOB + \angle BOC = \angle AOC.$$

Example 25:5

Given that in Figure 25:21, POQ and QOR are adjacent angles and that $\angle POR$ is equal to $59°$, find the size of

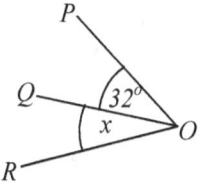

Figure 25:21

Solution

$\angle x - \angle POR - \angle POQ == 59° - 32° = 27°$

(b) **Non-Adjacent Angles**
In Figure 25:22 (i), the angles marked are not adjacent because the rays do not emanate from one point.   In Figure 25:22(ii), the common side (ray) is not between the other two rays.

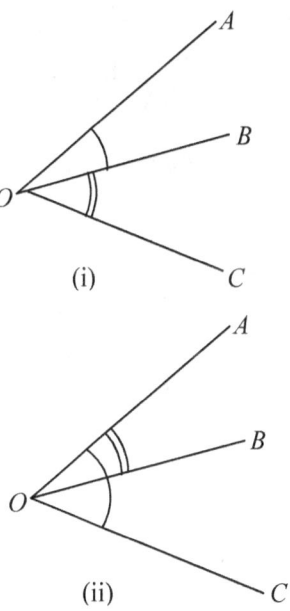

Figure 25:22

Angles between Intersecting Lines

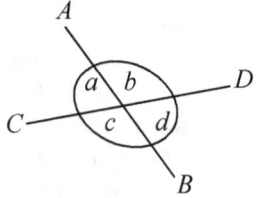

Figure 25:23

Vertically opposite angles are formed when two lines intersect. In the figure above a and d, b and c are pairs of vertically opposite angles.

Vertically opposite angles are equal

i.e. $a = d$ and $b = c$.

EXERCISE 25:3

1. In Figure 25:24(a) and (b), determine the value of the lettered symbols.

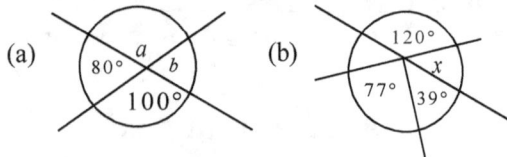

Figure 25:24

2. Using Figure 25:25, write down an equation connecting a, b and c.

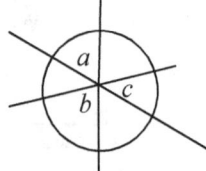

Figure 25:25

Perpendicular Lines

If the angle between two intersecting lines is $90°$, the lines are said to be perpendicular. To represent two perpendicular lines on a plane, a square is usually used as shown Figure 25:26.

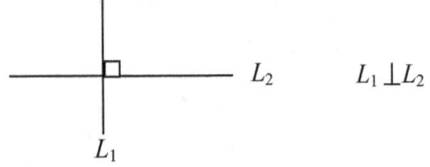

Figure 25:26

Parallel Lines and Transversals

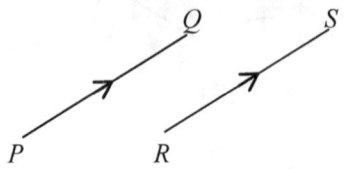

Figure 25:27

Two lines are said to be **parallel** if they have the same direction. Parallel lines never meet. Figure 25:27 above shows, that the lines PQ and RS are parallel.

A **transversal** is a line, which intersects two or more parallel lines. (Figure 25:28)

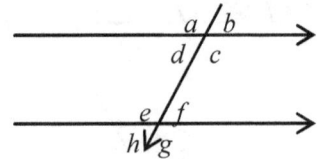

Figure 25:28

When a transversal intersects parallel lines as shown in Figure 25:28 above, pairs of **corresponding angles, alternate angles** (sometimes referred to as Z-angles) and **co-interior angles** are formed as tabulated below.

Pairs of alternate interior angles	d and f, c and e
Pairs of alternate exterior angles	a and g, b and h
Pairs of corresponding angles	d and f, c and e, d and h, c and g
Pairs of co-interior angles	d and e, c and f

Table 25:1

Theorem

When a transversal intersects parallel lines:
Alternate interior angles are equal

i.e. $d = f$, $c = e$.

Alternate exterior angles are equal

i.e. $a = g$, $b = h$.

Corresponding angles are equal

i.e. $a = e$, $b = f$, $d = h$, $c = g$

Adjacent angles are supplementary

i.e. $d + e = 180°$, $c + f = 180°$

Conversely, if a transversal intersects two straight lines, the line will be parallel if anyone of the following statements is fulfilled.

(i)   Two corresponding angles are equal,
(ii) Two adjacent angles are supplementary,
(iii) Two alternate interior (Z-) angles are equal.
(iv) Two alternate exterior angles are equal.

Note that in the diagram above, all the acute angles are equal and all the obtuse angles are equal.

Examples 25:6

In Figure 25:29, below, *AB*, *l* and *DE* are parallel. Find *x*.

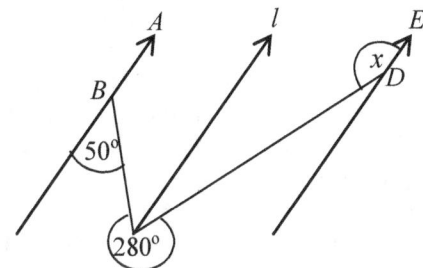

Figure 25:29

Solution

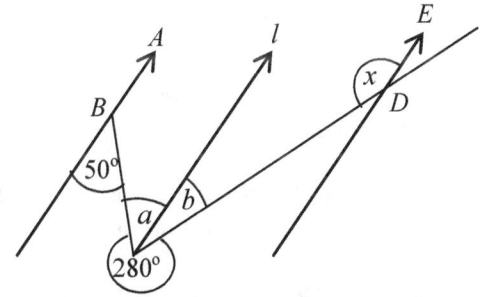

Figure 25:30

$$a = 50° \quad \text{[alternate angles]}$$
$$a + b + 280° = 360° \quad \text{[angles in a circle]}$$
$$\Rightarrow 50° + b + 280° = 360°$$
$$\Rightarrow b = 30°$$
$$b + x = 180°$$
$$\Rightarrow 30° + x = 180°$$
$$x = 150°$$

EXERCISE 25:4

Find the value of the unknowns designated by the letter in Figure 25:31 (1) to (5) below.

(1)
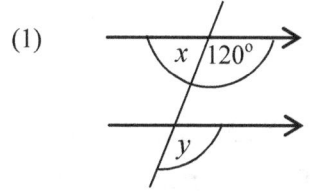

(2)
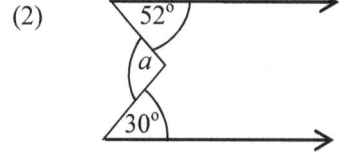

(3)
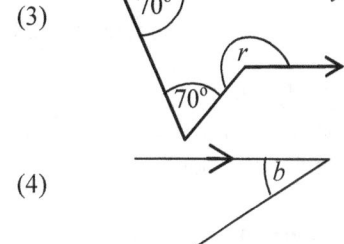

(4)
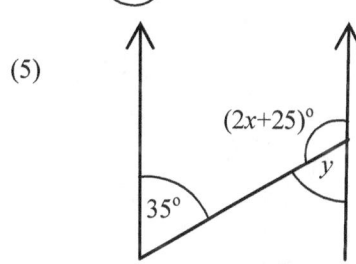

(5)

Figure 25:31

MULTIPLE CHOICE EXERCISE 25

1. 141° is an example of:
   [A] an acute angle
   [B] an obtuse angle
   [C] a reflex angle
   [D] an alternate angle

2. An angle which is between 180° and 360° is called:
   [A] a complementary angle
   [B] an acute angle
   [C] an obtuse angle

[D] a reflex angle

3. The angle which is not a reflex angle is:
   [A] 317°   [B] 258°
   [C] 193°   [D] 116°

4. The angle θ shown in Figure 25:32 is:
   [A] acute   [B] right
   [C] obtuse   [D] reflex

Figure 25:32

5. A reflex angle is:
   [A] < 90°
   [B] > 90° but < 180°
   [C] > 180° but < 360°
   [D] equal to 90°

6. The complement of 47° is
   [A] 133° [B] 43° [C] 313° [D] −47°

7. The angle which is supplementary to 128° is:
   [A] 52° [B] 42° [C] 32° [D] 22°

8. Two angles whose sum is 180° are called:
   [A] complementary angles
   [B] alternate angles
   [C] supplementary angles
   [D] corresponding angles

9. In Figure 25:33, *AO* is perpendicular to *OB*. The value of *x* is:
   [A] 75°   [B] 15°
   [C] 22.5°   [D] 30°

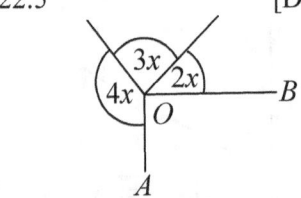

Figure 25:33

10. In Figure 25:34, *TRQ* is a straight line. If
    $P = \dfrac{1}{3}(a + b + c)$, then *p* equals:
    [A] 45°   [B] 60°
    [C] 90°   [D] 120°

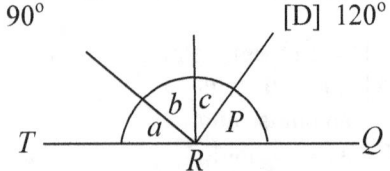

Figure 25:34

11. In Figure 25:35, *AB*, *CD* and *XY* are straight lines intersecting at *W*. The value of ∠*CWX* is:
    [A] 80°   [B] 100°

[C] 120°   [D] 140°

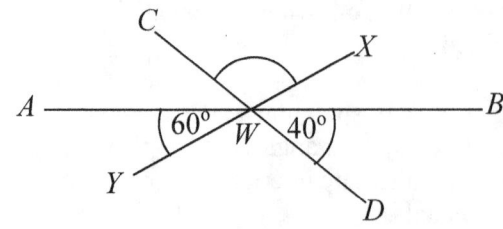

Figure 25:35

12. In Figure 25:36, *PR* and *QS* are straight lines. The value of the angle *y* is:
    [A] 30° [B] 56° [C] 87° [D] 93°

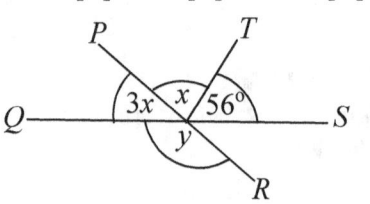

Figure 25:36

13. In Figure 25:37, *XY* and *PQ* intersect at *O*. Angle *TOP* = 64°, implies that *m* is equal to:
    [A] 93°   [B] 96°
    [C] 116°   [D] 151°

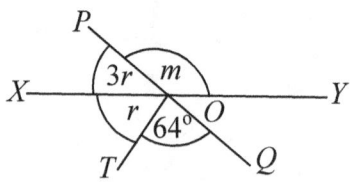

Figure 25:37

14. The angle *a*, shown in Figure 25:38 is equal to:
    [A] 60° [B] 120° [C] 30° [D] 90°

15. The angle *b* shown in Figure 25:38 is equal to:
    [A] 60° [B] 120° [C] 30° [D] 90°

16. The angle *c* shown in Figure 25:38 is equal to:
    [A] 60° [B] 120° [C] 30° [D] 90°

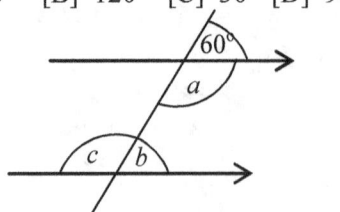

Figure 25:38

17. In Figure 25:39, it is true that:
    [A] *a* = *d*   [B] *a* = *b*
    [C] *e* = *b*   [D] *a* = *c*

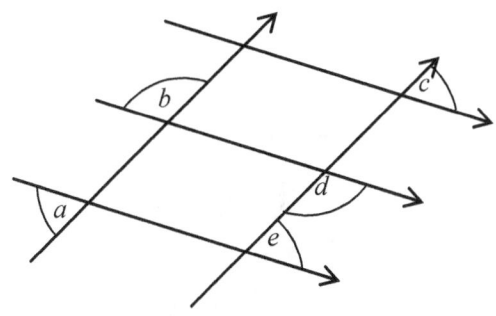

Figure 25:39

18. In Figure 25:40, it is true that:
    [A] $q = p+r$ [B] $p + q + r = 180°$
    [C] $q = r - p$ [D] $q = 360° - p - r$.

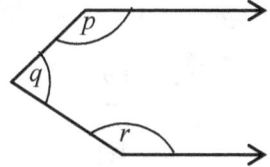

Figure 25:40

19. In Figure 25:41, the relation between x and y is:
    [A] $x = y$ [B] $x = y + 180°$
    [C] $x = y - 180°$ [D] $x + y = 180°$.

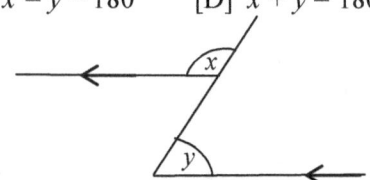

Figure 25:41

20. The value of x in Figure 25:42 is:
    [A] $76°$ [B] $104°$
    [C] $14°$ [D] $36°$

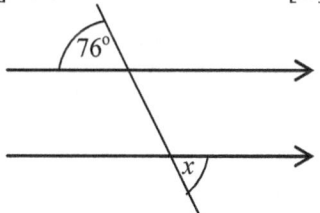

Figure 25:42

21. In Figure 25:43, $LK \parallel PQ$. $\angle KLM = 241°$ and $\angle QPM = 89°$. The value of $\angle LMP$ is:
    [A] $61°$ [B] $30°$
    [C] $119°$ [D] $150°$

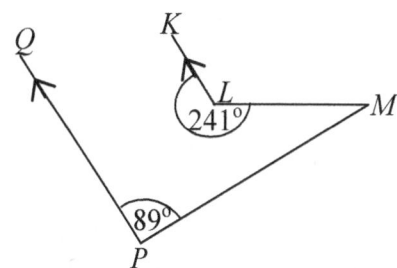

Figure 25:43

22. In Figure 26.44, $ML \parallel PQ$ and $NP \parallel QR$. Given that $\angle LMN = 40°$ and $\angle MNP = 55°$ then $\angle RQP$ equals:
    [A] $15°$ [B] $25°$ [C] $35°$ [D] $40°$

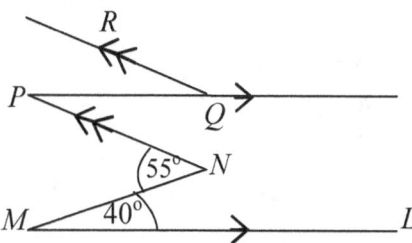

Figure 25:44

23. In Figure 25:45, $PQ \parallel RS$ and the angles are shown. The size of x is:
    [A] $145°$ [B] $150°$
    [C] $155°$ [D] $165°$

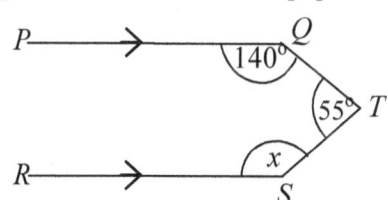

Figure 25:45

24. In Figure 25:47, $\angle PQU = 36°$, $\angle QRT = 29°$, $PQ \parallel RS$ and $UQ \parallel RT$. $\angle PQR$ should be:
    [A] $94°$ [B] $65°$ [C] $61°$ [D] $54°$

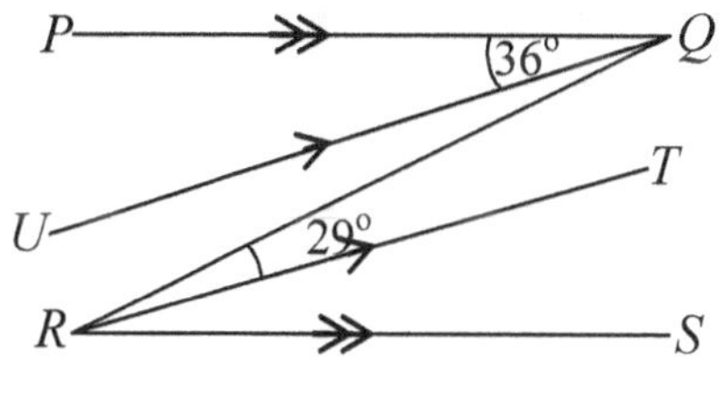

Figure 25:47

25. Given that in Figure 25:51, the lines L_L and L_2 are parallel, then the value of x is:
    [A] $45°$ [B] $30°$
    [C] $36°$ [D] $60°$

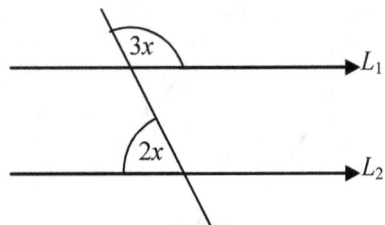

Figure 25:51

26. The diagram in Figure 25:52 which shows an impossible situation is:

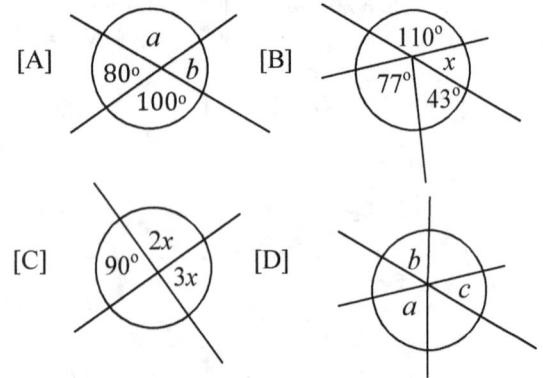

Figure 25:52

27. The fraction of the circle which has been shaded in Figure 25:53 is:

[A] $\frac{7}{9}$ [B] $\frac{2}{9}$ [C] $\frac{3}{8}$ [D] $\frac{5}{8}$

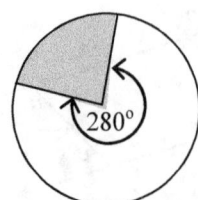

Figure 25:53

28. The pie chart (Figure 25:54) illustrates how groups of children travel to school. The percentage of children who walk to school is:
[A] 90% [B] 20% [C] 25% [D] 10%

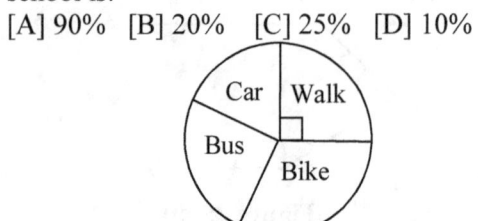

Figure 25:5

TOPIC 26

VOCABULARY OF PLANE FIGURES

OBJECTIVES

At the end of this topic, the learner should be able to:

1. Identify, draw and name different types of polygons according to the number of sides they contain.

2. Distinguish between regular and irregular polygons.

3. Distinguish between concave and convex polygons.

4. Classify triangles by measure of their angles and by measure of their sides.

5. Name, draw and state the properties of the different types of quadrilaterals.

6. Use tree diagrams and /or Venn diagrams to show the relationship between the different types of quadrilaterals.

7. Name, draw and describe the different parts of a circle.

French mathematician Gaspard Monge began teaching physics at Collège de la Trinité at the age of 16. He invented descriptive geometry, which became the basis of mechanical drawing and architectural plans. His ideas were formally published in 1795, in his *Géométrie descriptive.* This led to a new branch of mathematics known as descriptive geometry.

Roger Viollet/Getty Images

What is a Polygon?

A **polygon** is a plane figure with three or more sides.

Naming Polygons

Polygons are named according to the number of sides they contain. (see Table 26:1)

Number of sides	Name of polygon	Diagram
3	Triangle	
4	Quadrilateral	
5	Pentagon	
6	Hexagon	
7	Heptagon	
8	Octagon	

Table 26:1

Below are some more Polygons with different sides

Nonagon............... 9 sides
Decagon................10 sides
Hendecagon............11 sides
Dodecagon.............12 sides
n-gon.....................*n* sides

Regular and Irregular Polygons

An **equilateral polygon** is a polygon with all the sides equal. An **equiangular polygon** is a polygon with all angles equal. A **regular polygon** (Figure 26:1(i)) is an equilateral and equiangular polygon. An **irregular polygon** on the other hand (Figure 26:1(ii)) is one with at least one of its sides not equal to the others.

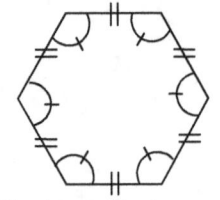

(i) A regular hexagon

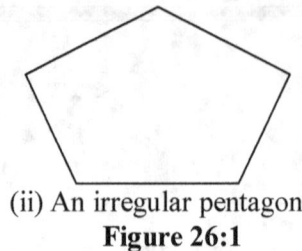

(ii) An irregular pentagon
Figure 26:1

Convex and Re-entrant Polygons

A **convex polygon** (Figure 26:2(i)) is a polygon with none of its interior angles greater than 180°. A **concave** or **re-entrant polygon** (Figure 26:2(ii)) is a polygon with at least one of its interior angles greater than 180°.

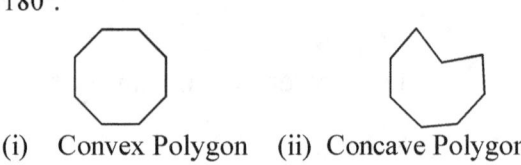

(i)   Convex Polygon   (ii) Concave Polygon
Figure 26:2

> **EXERCISE 26:1**

1. In Table 26:2, mark 'X' if the property holds for the given polygon.

Property	\multicolumn{6}{c}{Name of Polygon}

Property	Equilateral	Equiangular	Regular	Irregular	Convex	Re-entrant
All the sides are equal						
All the angles are equal						
All the angles are less than 180°						
At least one angle is greater than 180°						

Table 26:2

2. Name the polygons in Figure 26:3 by the number of sides.

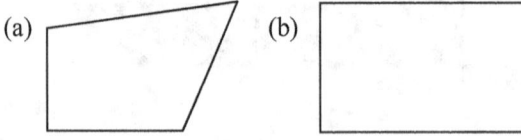

(a)   (b)

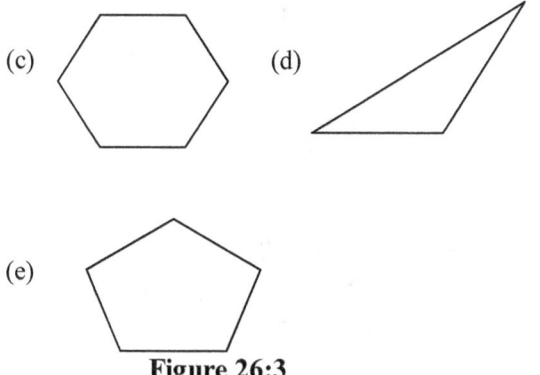

(c)   (d)

(e)

Figure 26:3

Quadrilaterals and Their Properties

A quadrilateral is a four sided plane figure. Below are the six major types of quadrilaterals:

Trapezium

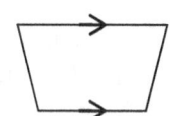

Figure 26:4

A trapezium is a quadrilateral with two parallel sides.

Kite

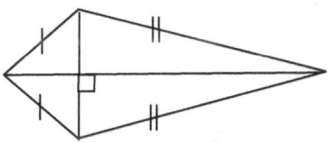

Figure 26:5

Two pairs of adjacent sides of a kite are equal. Diagonals of a kite intersect each other at right angles.

Parallelogram

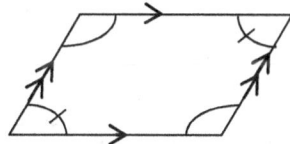

Figure 26:6

The properties of a parallelogram are:
Opposite sides are equal.
Opposite sides are parallel.
Opposite angles are equal.
Adjacent angles are supplementary (sum up to 180°)
Diagonals bisect each other.
Diagonals bisect the parallelogram.

The rhombus, the rectangle and the square that follow are all special types of parallelograms.

Rhombus

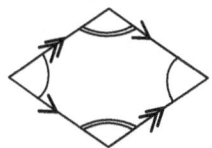

Figure 26:7

The properties of a rhombus are:
All the four sides are equal.
Opposite sides are parallel.
Opposite angles are equal.
Diagonals bisect each other at right angles.
Diagonals bisect the rhombus.

Rectangle

Figure 26:8

The properties of a rectangle are:
Opposite sides are equal.
Opposite sides are parallel.
Each angle is equal to 90°.
Diagonals bisect each other.
Diagonals are equal in length.
Diagonals bisect the rectangle.

Square

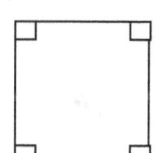

Figure 26:9

The properties of a square are:
All sides are equal.
Opposite sides are parallel.
All angles are equal, each equal to 90°.

Diagonals bisect each other at right angles.
Diagonals bisect the square.
Diagonals are equal in length.

In thinking about the properties of quadrilaterals, it may be useful to think of the nature of the sides, angles, and diagonals. The mnemonic **SAD** may be a useful mental aid.

Relationship between Quadrilaterals

The relationship between quadrilaterals can be illustrated using the quadrilateral family tree (Figure 26:10) or a Venn diagram (Figure 26:11).

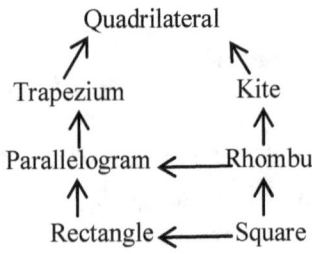

Figure 26:10

Let $\mathscr{E}$ = All quadrilaterals
P = All parallelograms
K = All kites
R_h = All rhombuses
R_e = All rectangles
S = All squares
T = All trapeziums

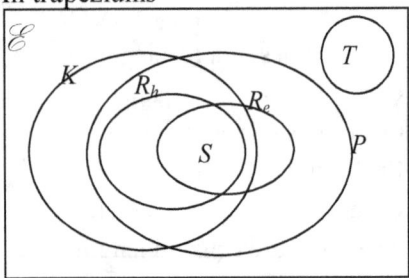

Figure 26:11

EXERCISE 26:2

1. If P = parallelograms, R_e = rectangles, S = squares, R_h = rhombuses, K = kites, then $R_h \subset P$.
    (i) Write down all other inclusion relations between these sets.
    (ii) Illustrate by a single Venn diagram the relationship between all these sets.
    (iii) Hence or otherwise, list the elements of
        (a) $K \cap P$ (b) $(K \cap P)'$

2. Figure 26:12 is a quadrilateral. Calculate the values of x and y.

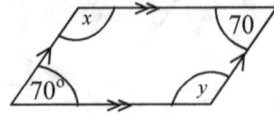

Figure 26:12

3. Let $\mathscr{E}$ = {x:x is a polygon}
    A = {x:x is a regular polygon}
    B = {x:x is a quadrilateral}
    (a) List the elements of $A \cap B$.
    (b) List the elements of $A' \cap B$.

4. Name the quadrilaterals in Figure 26:13.

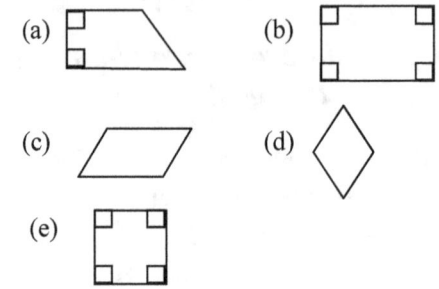

Figure 26:13

5. In Table 26:3, put 'Y' in the appropriate space if the property holds for the given quadrilateral and 'N' if it does not hold.

	Property	Square	Rectangle	Rhombus	Parallelogram	Trapezium
a	All the sides are equal					
b	All the diagonals are equal					
c	Diagonals bisect each other					
d	Diagonals are perpendicular					
e	Diagonals bisect opposite angles					
f	Adjacent sides are equal					
g	Opposite sides are equal and parallel					
h	Only two sides are parallel					
i	Adjacent sides are perpendicular					

Table 26:3

6. Let $\mathscr{E}$ = {x:x is a quadrilateral}
    Q_s = {x:x is a quadrilateral with at least one equal pair of adjacent sides}
    Q_r = {x:x is a quadrilateral with at least one right angle}
    P = {x:x is a parallelogram}
    Write down each of the following as a

single set and draw a Venn diagram to represent the relationship in each case.

(i) $P \cap Q_s$ (ii) $P \cap Q_r$

(iii) $P \cap Q_s \cap Q_r$

Types of Triangles

Triangles are classified in two ways.

(a) *By the Measures of their Angles*

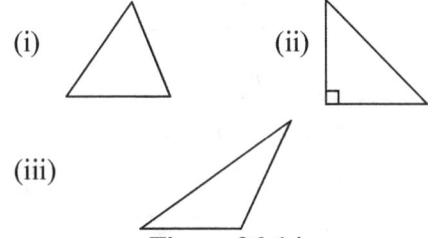

Figure 26:14

(1) **An Acute angle triangle** (Figure 26:14(i)) has all the angles each less than $90°$.

(2) A **right-angled triangle** (Figure 26:14(ii)) has one of its angles equal to $90°$.

(3) **Obtuse angle triangle** (Figure 26:14(iii)) has one of its angles between $90°$ and $180°$.

(b) *By the Measures of their Sides*

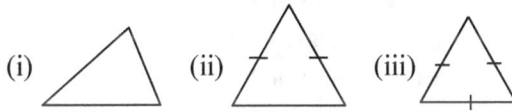

Figure 26:15

(1) **A Scalene triangle** (Figure 26:15(i)) has no sides equal.

(2) **An Isosceles triangle** (Figure 26:15(ii)) **has** two sides equal.

(3) **An Equilateral triangle** (Figure 26:15(iii)) has all the sides equal.

EXERCISE 26:3

(1) Let $\mathscr{E}$ = All triangles

A = acute-angled triangles

R = right-angled triangles

O = obtuse-angled triangles

S = scalene triangles

I = isosceles triangles

E = equilateral triangles

(i) Draw a diagram of one element of:

(a) $A \cap S$ (b) $A \cap I$

(c) $R \cap S$ (d) $R \cap I$

(e) $O \cap S$ (f) $O \cap I$

(ii) State the super set(s) of E.

(iii) Which set(s) is/are disjoint with E?

2. Classify the triangles in Figure 26:16 by the measures of their angles.

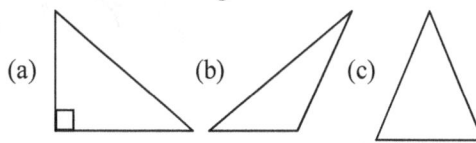

Figure 26:16

3. In the following, two angles of a triangle are given. Name the type of triangle.

(a) $50°, 50°$ (b) $30°, 60°$ (c) $24°, 85°$

4. Classify the triangles in Figure 26:17 by the measures of their sides.

(a)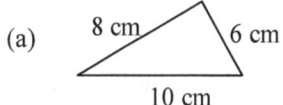

(b)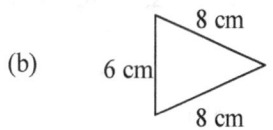

(c)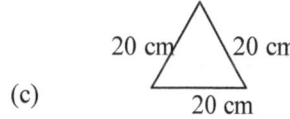

Figure 26:17

(d) 10 m, 10 m, 10 m   (e) 4 m, 6 m, 8 m

(f) 3 m, 4 m, 3 m

Vocabulary Associated with Circles

A circle is a plane figure bounded by points, which are equidistant from a fixed point called the **centre** of the circle.

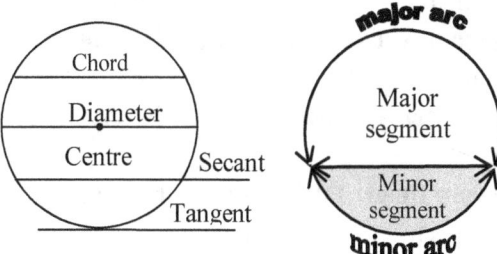

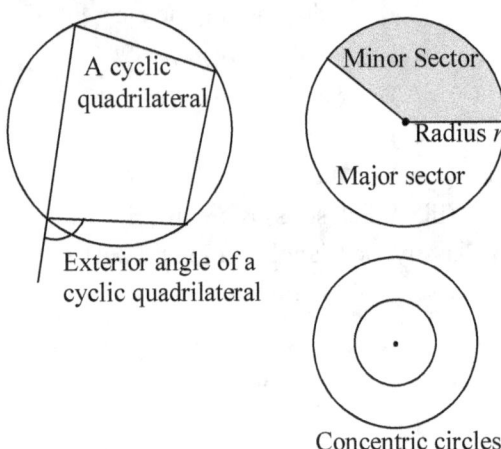

A cyclic quadrilateral

Exterior angle of a cyclic quadrilateral

Minor Sector

Radius *r*

Major sector

Concentric circles

Figure 26:19

The **circumference** of a circle denoted by *C* is the distance round the circle.

The **radius** denoted by *r* is the distance from the centre of a circle to a point on the circle.

A chord is a line segment joining any two points on the circle.

The **diameter** denoted by *d* is the longest chord of a circle and its length is twice that of the radius. A diameter always passes through the centre.

A **tangent** is a line that intersects or touches the circle at exactly one point and is perpendicular to the radius at the point of contact.

A **secant** is a line that cuts through a circle, intersecting it at two points.

An **arc** is part of the circumference. The longer of the arcs is called the **major arc** while the shorter is called the **minor arc**.

A **sector** is a portion of a circle bounded by two radii and an arc of the circle. The larger of the sectors is called the **major sector** while the smaller is called the **minor sector**.

A **segment** is a portion of a circle bounded by a chord of the circle and an arc of the circle. The smaller segment is called the **minor segment** while the larger segment is called the **major segment**.

Concentric circles are circles with the same centre.

A **cyclic quadrilateral** is a quadrilateral, which is inscribed in a circle.

EXERCISE 26:4

1. In Figure 26:19,
    (i) Name the set of points represented by the line segment
    (a) *AC*   (b) *OC*   (c) *FD*   (d) *ABC*
    (e) *ACX*   (d) *EY*   (g) *AEC*
    (ii) Name the set of points in the region.
    (a) *DEFD*   (b) *CODC*   (c) *ABCA*

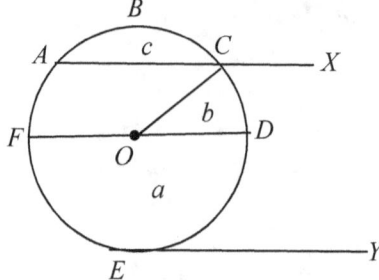

Figure 26:19

2. Figure 26:20 shows a circle with centre *O* and four points *A*, *B*, *C* and *D* on the circle. The line segment *ON* is perpendicular to *AB*.

   Give a name for each of the following:
    (a) The set of points on *ACB*.
    (b) The set of points on *OA*.
    (c) The triangle *OAB*.
    (d) The set of points on *AB*.
    (e) The set of points in the region *R*.

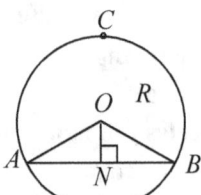

Figure 26:20

3. With respect to Figure 26:21, where *O* is the centre of the circle, name:
    (a) The shaded region *G*.
    (b) The set of points on the curve *CD*.
    (c) The set of points on the line segment *OD*.
    (d) The set of points on the line segment *PC*.
    (e) The set of points on the line segment *AB*.
    (f) The shaded region *F*.

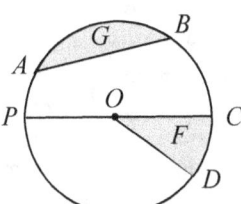

Figure 26:21

⟨ **MULTIPLE CHOICE EXERCISE 26** ⟩

1. The statement, which is always true of a rhombus, is:
   [A] All the angles are complementary
   [B] All the sides are equal
   [C] The adjacent angles are equal
   [D] All the angles are equal

2. The plane shape, which is not a quadrilateral, is:
   [A] A kite   [B] a rhombus
   [C] A pentagon   [D] a parallelogram

3. The plane shape, which is not an example of a quadrilateral, is:
   [A] Square   [B] Trapezium
   [C] Rhombus   [D] Triangle

4. The statement, which is not true of a parallelogram, is:
   [A] It has more than 4 sides
   [B] Opposite sides are parallel.
   [C] It has exactly 4 sides.
   [D] The sum of its angles is 360°.

5. A seven sided plane figure is called:
   [A] An octagon   [B] a pentagon
   [C] A hexagon   [D] a heptagon

6. A polygon with all its interior angles less than 180° is definitely:
   [A] a convex polygon
   [B] a regular polygon
   [C] a re-entrant polygon
   [D] a quadrilateral

7. A triangle with vertices (–4, 4), (4,4) ,(0,–1) is:
   [A] Right-angled
   [B] equilateral
   [C] Isosceles
   [D] scalene

8. The largest angle of any triangle:
   [A] Must always be an acute angle.
   [B] Can sometimes be an acute angle.
   [C] Can never be a right-angle.
   [D] Must always be an obtuse angle.

9. A quadrilateral with one pair of sides equal is:
   [A] A rhombus   [B] a parallelogram
   [C] A rectangle   [D] a trapezium

10. The assertion about a rhombus, which may not be true, is:
    [A] The opposite angles are equal.
    [B] The diagonals bisect the angles through which they pass.
    [C] The diagonals are equal.
    [D] Opposite sides are equal.

11. A quadrilateral whose diagonals bisect at right angles is:
    [A] A rectangle   [B] a parallelogram
    [C] A trapezium   [D] a rhombus

12. The property/properties, which do not characterise a rectangle, is/are:
    I. The diagonals bisect at right angles
    II. Opposite sides are equal and parallel
    III. Each of its angles is a right angle
    [A] I only   [B] II only
    [C] III only   [D] II and III only

13. In Figure 26:22, the value of the angle marked y is:
    [A] 28° [B] 62° [C] 118° [D] 152°

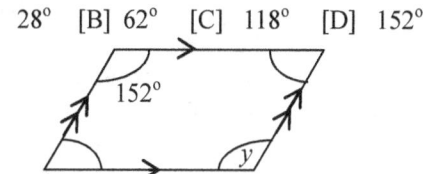

Figure 26:22

14. The value of the angle marked x in Figure 26:23 is:
    [A] 140° [B] 130° [C] 110° [D] 70°

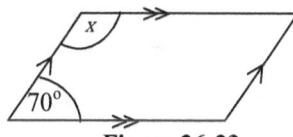

Figure 26:23

15. The value of the angle marked y in Figure 26:24 is:
    [A] 270° [B] 210° [C] 190° [D] 95°

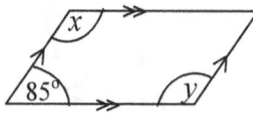

Figure 26:24

16. In Figure 26:24, the sum of x and y is:
    [A] 270° [B] 210° [C] 190° [D] 95°

17. The values of x, y, and z in Figure 26:25 are respectively:
    [A] 130°, 50°, 130°
    [B] 140°, 40°, 140°
    [C] 150°, 30°, 150°
    [D] 120°, 60°, 120°

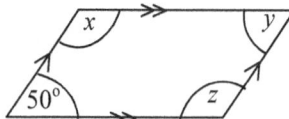

Figure 26:25

18. The name given to Figure 26:26 is:

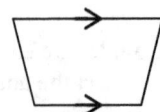

Figure 26:26

[A] A parallelogram
[B] a trapezium
[C] A rhombus
[D] a rectangle

19. Let $\mathscr{E}$ = {x:x is a polygon}
    A = {x:x is a regular polygon}
    B = {x:x is a quadrilateral}
    Then an element of $A \cap B$ is:
    [A] A square   [B] a rhombus
    [C] A trapezium   [D] a rectangle

20. Let $\mathscr{E}$ = {x:x is a polygon}
    A = {x:x is a regular polygon}
    B = {x:x is a quadrilateral}
    Then an element of $A' \cap B$ can never be:
    [A] A square   [B] a trapezium
    [C] A rhombus   [D] a rectangle

21. The sum of the angles of a square is:
    [A] 90°   [B] 120°   [C] 180°   [D] 360°

22. The shaded portion in Figure 26:27 is called:
    [A] A minor segment [B] a major segment
    [C] A minor sector   [D] a major sector

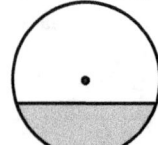

Figure 26:27

23. By the number of sides the polygon in Figure 26:28 is:
    [A] An octagon   [B] a pentagon
    [C] A quadrilateral   [D] a hexagon

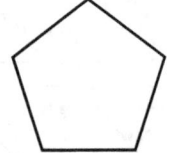

Figure 26:28

24. The polygon, which has the shape of the Cameroon flag, is:
    [A] An octagon   [B] a pentagon
    [C] A hexagon   [D] a quadrilateral

25. A shape that has two more sides than a football field is:
    [A] Pentagonal   [B] octagonal
    [C] Hexagonal   [D] decagonal

26. A Polygon has 4 more sides than a rectangle. The polygon is:
    [A] Pentagon   [B] decagon
    [C] Hexagon   [D] octagon

27. The real name of the plane Figure 26:29 and some of its possible names are:
    [A] parallelogram; quadrilateral, rhombus.
    [B] Rhombus; quadrilateral, trapezium.
    [C] Trapezium; quadrilateral, polygon.
    [D] Quadrilateral, parallelogram, polygon.

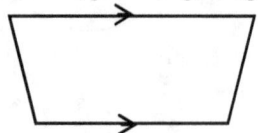

Figure 26:29

28. The best name of the quadrilateral, which has 4 congruent sides, is:
    [A] Parallelogram   [B] rhombus
    [C] Trapezoid   [D] rectangle

29. A triangle has sides 3, 5, 8 and angles 25°, 85°, 70°. By the measure of its angles and its sides the triangle is:
    [A] Isosceles, obtuse   [C] isosceles, acute
    [B] Scalene, obtuse   [D] scalene, acute

30. The diagram in Figure 26:30 which is a polygon is:

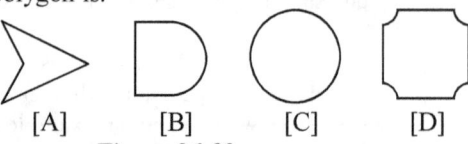

[A]   [B]   [C]   [D]

Figure 26:30

31. The property, which makes a rhombus different from every other parallelogram, is:
    [A] All sides are equal.
    [B] Opposite sides are parallel.
    [C] Opposite angles are equal.
    [D] Diagonals bisect each other at right angles.

32. The property, which makes a square a unique rectangle, is:
    [A] All sides are equal.
    [B] Opposite sides are parallel.
    [C] Opposite angles are equal.
    [D] Diagonals bisect each other at right angles.

33. The property, which makes a square a unique rhombus, is:
    [A] All sides are equal.
    [B] Opposite sides are parallel.
    [C] Opposite angles are equal.
    [D] Diagonals bisect each other at right angles.

34. Among the properties, the property, which makes a rectangle a special parallelogram, is:
    [A] Opposite sides are equal.
    [B] Opposite sides are parallel.
    [C] Diagonals bisect each other.
    [D] Diagonals are equal in length.

TOPIC 27

POLYGON THEOREMS

OBJECTIVES

At the end of this topic, the learner should be able to:

1. Identify interior and exterior angles of polygons and distinguish between them.
2. Use the angle sum property of a triangle correctly to calculate unknown angles.
3. Identify the exterior angle and corresponding interior angles of a polygon and use the exterior angle property.
4. Recall and use the angle sum formula to determine the angle sum of any polygon or number of sides of a polygon.

Euclid was a Greek mathematician who lived around 300 B.C. He wrote Elements, a 13-volume work on the principles of geometry and properties of numbers.

THE BETTMANN ARCHIVE

Interior and exterior angles of polygons

(i)

(ii)

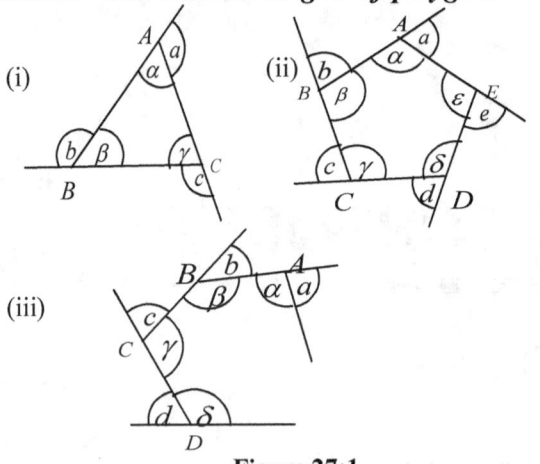

(iii)

Figure 27:1

In Figure 27:1, the angles α, β, γ, δ, ε, shown are called the **interior angles** of the polygons, because they are inside the polygons. On the other hand, the angles a, b, c, d, e formed outside the polygons when AC, BA, CB, DC, ED respectively are produced are called the **exterior angles** of the polygons. Notice that each interior angle and the corresponding exterior angle are supplementary since they are angles on a straight line.

Chasles' Theorem

1. The exterior angle of a triangle is equal to the sum of the two interior opposite angles.
   i.e. $\theta = \alpha + \beta$
2. The sum of the interior angles of a triangle is two right angles (or 180°).
   i.e. $\alpha + \beta + \gamma = 180^\circ$

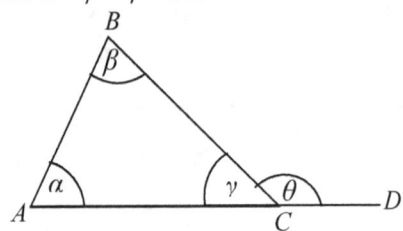

Figure 27:2

Practical Proof of Chasles's Theorem

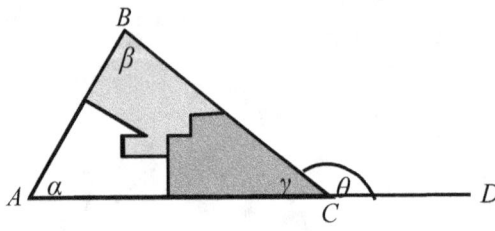

Figure 27:3

If the vertices A and B of a triangle ABC whose angles are α and β respectively as in Figure 27:3 are cut and arranged at C as in Figure 27:4 they will fit exactly onto the angle θ the exterior angle. This shows that the exterior angle of a triangle is equal to the sum of the two interior opposite angles. In addition the three vertices α, β and γ fit exactly on a straight line proving that the sum of the interior angles of a triangle is two right angles (or 180°).

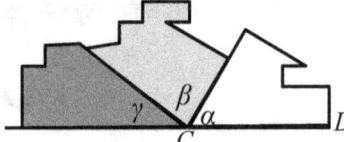

Figure 27:4

Mathematical Proof of Chasles's Theorem

In Figure 27:2, construct CE parallel to AB as in Figure 27:5. Then,

$\alpha = y$......① [corresponding $\angle$s, $AB\|CE$]
$\beta = x$...... .② [alternate $\angle$s, $AB\|CE$]

① + ②: $\alpha + \beta = x + y$....................③

Therefore, the exterior angle of a triangle is equal to the sum of the interior opposite angles.

$\gamma + x + y = 180^\circ$.......................................④

Substitute ③ in ④: $\alpha + \beta + \gamma = 180^\circ$

∴ The sum of the interior angles of a triangle is two right angles (i.e. 180°).

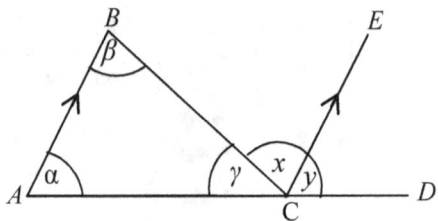

Figure 27:5

> **EXERCISE 27:1**

1. Find the value of the lettered angles in Figure 27:6 (a) to (g).

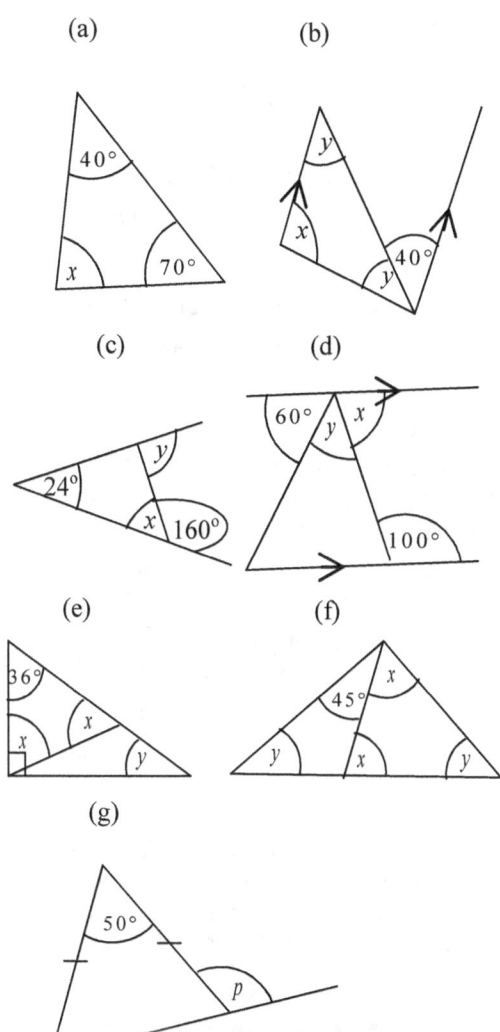

(a)   (b)

(c)   (d)

(e)   (f)

(g)

Figure 27:6

2. (a) Draw the following polygons; a triangle, a quadrilateral, a pentagon, a hexagon and a heptagon and label the vertices using the letters *A, B, C...*
   (b) Draw lines from one particular vertex say *A* to all the other vertices.
   (c) Count the number of triangles into which the polygon has been divided and record your result in Table 27:2.
        Use the following example, which has been done for an octagon to help you.

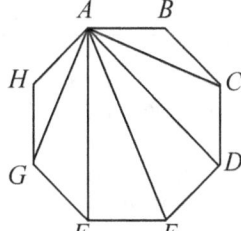

Figure 27:7

No of sides of polygon	No of triangles insides polygon	Sum of interior angles
3		
4		
5		
6		
7		
8	6	1080°

Table 27:2

(d) Study results in Table 27:2 carefully and;
   (i) Predict the number of triangles inside a polygon with 9 sides.
   (ii) Predict the sum of all the interior angles in a convex polygon with 9 sides.
        Write down an expression for
   (iii) The number of triangles insides a polygon with *n* sides.
   (iv) the sum of interior angles of a polygon with *n* sides.
   (v) Deduce the sum of the exterior angles of a polygon.

3. (a) Repeat exercise 2(a)
   (b) Draw lines from the vertex to the centre of each polygon as in Figure 27:8.

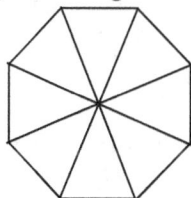

Figure 27:8

   (c) Count the number of triangles into which the polygon has been divided and record your result in Table 27:3.
   (d) Study your results in Table 27:3 carefully and;
   (i) Predict the number of triangles inside a polygon with 9 sides.
   (ii) What is the sum of the angles of all the triangles in a convex polygon with 9 sides
   (iii) What is the sum of the angles at the centre of the polygon?
   (iv) Use (ii) and (iii) to find the sum of interior angles of a polygon with 9 sides.
   (v) What is the number of triangles inside a polygon with *n* sides?
   (vi) What is the sum of the angles of all the triangles in a convex polygon with *n* sides?
   (vii) Write down an expression for the sum of interior angles of a polygon with *n* sides.

(viii) Deduce the sum of exterior angles of a polygon.

No. of sides of polygon	No. of triangles insides polygon	Sum of interior angles
3		
4		
5		
6		
7		
8	8	$1080°$

Table 27:3

Problems 2 and 3 of Exercise 27:1 lead us into the polygon theorems that follow. The polygon theorems are actually extensions of Chasles' Theorem.

Polygon Theorems

In a convex polygon with n sides,
(1) The sum of the interior angles is
$(2n - 4)$ right angles or $(n - 2)180°$.

(2) The sum of the exterior angles is 4 right angles (or $360°$) no matter the value of n.

Examples 27:6
Each angle of a regular polygon is $170°$. Find the number of sides of the polygon.

Solution
Let the number of sides of the polygon $= n$

Each interior angle $= 170°$

$\therefore$ Each exterior angle $= 180° - 170° = 10°$

$\Rightarrow$ Sum of exterior angles of the polygon $= 10n°$

But sum of exterior angles of any polygon $= 360°$

$\Rightarrow 10n = 360$

$\Rightarrow n = 36$

Alternatively;

Sum of interior angles of the polygon $= 170n°$

But sum of interior angles of polygon $= (n - 2)180°$

$\therefore 170n° = (n - 2)180°$

$170n = 180n - 360$

$-10n = -360$

$n = 36$

Therefore, the polygon has 36 sides.

Examples 27:7
One angle of a hexagon is $140°$ and five angles are equal. Find the value of each of the five angles.

Solution
Let the value of each of the five angles be p.

$\therefore$ Sum of the angles of the hexagon $= 5p + 140°$

But sum of interior angles of polygon $= (n - 2)180°$

$\therefore$ Sum of the angles of the hexagon $= (6 - 2)180°$

$\Rightarrow 5p + 140° = (6 - 2)180°$

$5p = (6 - 2)180° - 140°$

$5p = 720° - 140°$

$5p = 580°$

$\Rightarrow p = 116°$

Alternatively;
Let the value of each of the 5 angles be p and each of the 5 exterior angles be y.

The exterior angle corresponding to $140°$

$= 180° - 140° = 40°$

But sum of exterior angles of any polygon $= 360°$

$\therefore$ Sum of the 5 exterior angles $= 360° - 40° = 320°$

$\Rightarrow 5y = 320°$

$\Rightarrow y = 64°$

$\therefore p = 180° - 64° = 116°$

Note!
In solving problems on polygons, it is often easier to use the theorem on sum of exterior angles of a polygon rather than that on the sum of interior angles, though both lead to the same answer.

EXERCISE 27:2

1. Four angles of a hexagon are $130°$, $160°$, $112°$, and $140°$. If the remaining angles are equal, find the size of each.
2. Each of the exterior angles of a regular polygon is $100°$ less than the interior angle. Calculate the size of the exterior angle.
3. The sum of the interior angles of an n-sided convex polygon is double the sum of the exterior angles. Find the value of n.
4. The angles of a pentagon are $x°$, $2x°$, $(x+30)°$, $(x-10)°$, and $(x+40)°$. Find the value of x.
5. The sum of the angles of a polygon is $1800°$. Calculate the number of sides of the polygon
6. How many sides, has a convex polygon with each interior angle equal to $150°$?
7. How many sides, has a convex polygon with each exterior angle equal to $20°$?
8. Find the size of an interior angle of a regular ten-sided polygon.
9. Determine whether it is possible to have a regular convex polygon with exterior angles,
   (a) $20°$ (b) $16°$ (c) $15°$.

If so state the number of sides of the polygon.

10. One exterior angle of a polygon is 54°, and other exterior angles are each 34°. Calculate the number of sides of the polygon.

11. Four of the sides of a pentagon are 72°, 100°, 120°, and 140° in that order. Find the remaining angle; hence prove that two of the sides are parallel.

12. *ABCDEFGH* is a regular octagon. Calculate ∠*AFC*.

13. *AB*, *BC*, *CD* are three consecutive sides of a regular polygon. If ∠*ACB* =15°, calculate the number of sides of the polygon and ∠*ACD*.

14. *ABCDEFGHI* is a regular nonagon. Calculate
    (a) The angles *ABC*, *CAE*, *ACG*
    (b) The angles between *AE* and *CG*

15. *P*, *Q*, *R*, *S* are four adjacent vertices of a regular octagon. Calculate the angles *PQR* and *RSP*.

16. *AB* and *BC* are adjacent sides of a regular dodecagon. The perpendicular from *C* meets AB produced at *D*. Calculate ∠*BCD*.

17. By dividing an *n*-sided convex polygon into isosceles triangles each with one vertex at the centre and a side of the polygon as its base, prove that:
    (a) The sum of the interior angles of any *n*-sided convex polygon is (2*n*-4) right angles or (*n*-2)180°.
    (b) The sum of the exterior angles of any *n*-sided convex polygon is 360°.

◀ MULTIPLE CHOICE EXERCISE 27 ▶

1. In Figure 27:10, *AB*∥*CD*. The size of the angle marked *x* is:
   [A] 103° [B] 93° [C] 62° [D] 52°

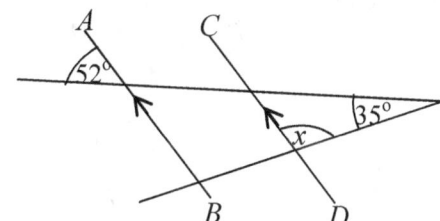

Figure 27:10

2. In Figure 27:11, the value of *x* is:

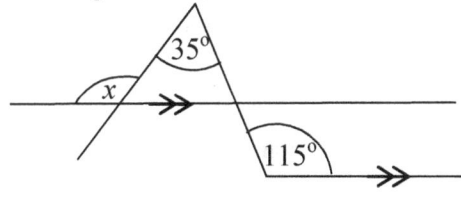

Figure 27:11

[A] 35° [B] 80° [C] 100° [D] 115°

3. In Figure 27:12, *PQ*∥*ST*. The value of *x* is:
   [A] 82° [B] 108° [C] 124° [D] 164°

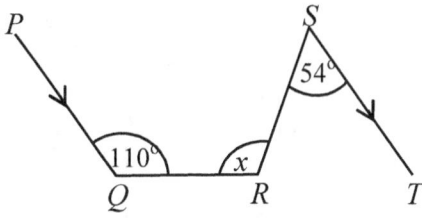

Figure 27:12

4. The number of sides in a regular polygon whose interior angle is 135° is:
   [A] 7   [B] 8   [C] 10   [D] 12

5. The angles of a pentagon are $x°$, $2x°$, $(x+60)°$, $(x+10)°$, $x°$, $(x-10)°$. The value of *x* is:
   [A] 80° [B] 75° [C] 60° [D] 40°

6. The number of sides in a regular polygon with each of its interior angles equal to 108° is:
   [A] 4   [B]   5   [C] 6   [D]   7

7. The number of sides in a regular polygon with each of its interior angles equal to 140° is:
   [A] 7   [B] 8   [C] 9   [D] 10

8. The number of sides in a regular polygon with each of its interior angles 120° is:
   [A] 4   [B] 6   [C] 7   [D]   8

9. In Figure 27:13, the true relation is:
   [A] $a+b+x=180°$ [B] $a=b+x$
   [C] $a-b=180°-x$ [D] $a+b=x+180°$

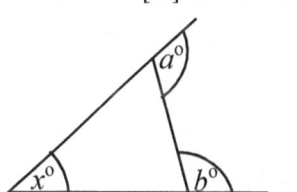

Figure 27:13

10. In Figure 27:14, *x* is equal to:
    [A] $a+b+c$ [B] $360°-(a+b+c)$
    [C] $a+b+c+180°$ [D] $360°-a+b+c$

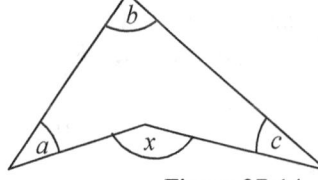

Figure 27:14

11. In Figure 27:15, *y* is equal to:
    [A] 80° [B] 70° [C] 40° [D] 100°

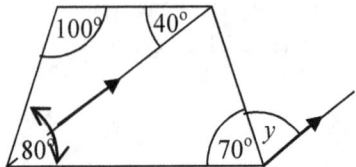

Figure 27:15

12. In Figure 27:16, the size of angle *ACB* is:
[A] 40° [B] 50° [C] 60° [D] 80°

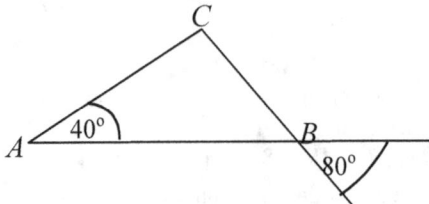

Figure 27:16

13. In Figure 27:17, $|PQ| = |PR| = RS$ and $\angle RPS = 32°$. The value of $\angle QPR$ is:
[A] 64° [B] 52° [C] 32° [D] 26°

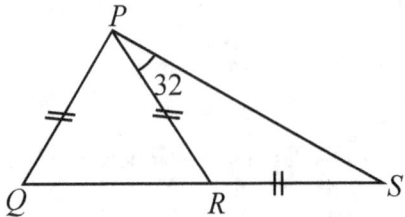

Figure 27:17

14. In Figure 27:18, *QRS* is a straight line, *QP∥RT*, $\angle PQR = 56°$, $\angle QPR = 84°$, $\angle TRS = x°$. The value of *x* is:
[A] 28° [B] 40° [C] 44° [D] 84°

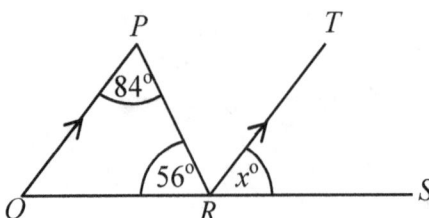

Figure 27:18

15. In Figure 27:20, *ABC* is a triangle, *BC* is produced to *D*, $|AB| = |AC|$, $\angle BAC = 50°$. The value of $\angle ACD$ is:
[A] 115° [B] 65° [C] 60° [D] 50°

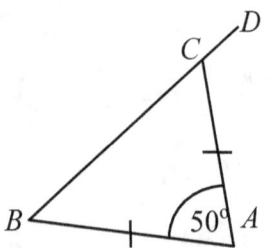

Figure 27:20

16. The number of sides in a regular polygon with one interior angle 160° is:
[A] 10 [B] 36 [C] 18 [D] 20

17. In Figure 27:21, *PQ* is parallel to *RST*.
$\angle PQR = 35°$ and $|RS| = |SQ| = |TQ|$.
The size of $\angle STQ$ is:
[A] 35° [B] 40° [C] 70° [D] 110°

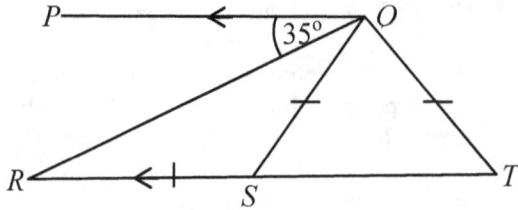

Figure 27:21

18. In Figure 27:22, *WXYZ* is a rhombus and $\angle WYZ = 20°$
The value of angle *XZY* is:
[A] 20° [B] 30° [C] 60° [D] 70°

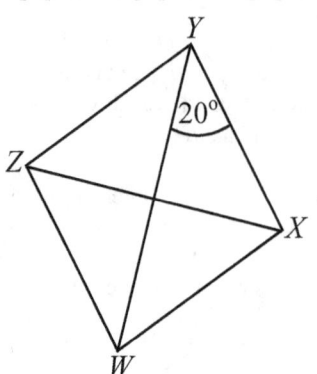

Figure 27:22

19. A regular polygon centre *O* can be sub-divided into isosceles triangles identical to triangle *POQ* (Figure 27:23). The number of such triangles in the polygon is:
[A] 12 [B] 10 [C] 9 [D] 8

20. The value of angle *t* in figure 27:24 is:
[A] 115° [B] 120° [C] 125° [D] 145°

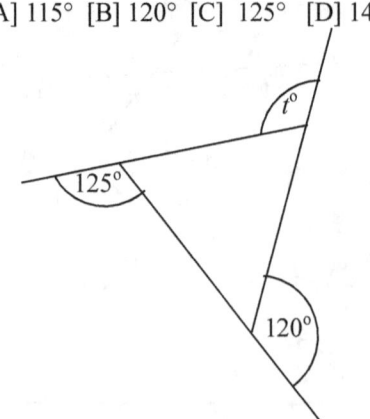

Figure 27:24

TOPIC 28

TRIGONOMETRY

OBJECTIVES

At the end of this topic, the learner should be able to:

1. Appreciate the meaning of trigonometry.
2. Label the angles and sides of a triangle.
3. State the properties of a right-angled triangle.
4. State and use the Pythagoras theorem to solve right-angled triangles.
5. Identify Pythagorean triples.
6. Find and use the trigonometric ratios of a right-angled triangle.
7. Refer acute angle trigonometric ratios and inverse trigonometric ratios from tables.
8. Find trigonometric ratios and inverse trigonometric ratios using calculators.
9. Find other trigonometric ratios given another.
10. State the complement of an angle and find the trigonometric ratio of complementary angles.
11. State and find the reciprocal of any trigonometric ratio.
12. Derive and memorise the trigonometric ratios of the special angles $0°$, $30°$, $45°$, $60°$ and $90°$.
13. Find the trigonometric ratios and inverse trigonometric ratios for angles in the range $0° \leq x \leq 360°$.
14. Appreciate the meaning of negative angles and find their trigonometric ratios.

Considered the first true mathematician, **Pythagoras** in the 6th century BC emphasized the study of mathematics as a means to understanding all relationships in the natural world. His followers, known as Pythagoreans, were the first to teach that the Earth is a sphere revolving around the Sun. This detail showing Pythagoras surrounded by his disciples comes from a fresco known as the School of Athens (1510-1511), by Italian Renaissance painter Raphael.

Archivo Fotografico Oronoz

Meaning of Trigonometry

The word trigonometry comes from three Latin words:

'Tri' means 'three'

'gon' means 'angle'

'metry' means 'measure'

Therefore trigonometry is the study of the relationship between the angles and sides of triangles or simply the study of the measure of triangles.

Standard Notation for Triangles

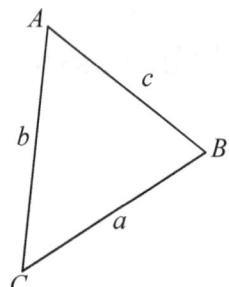

Figure 28:1

In labelling a triangle, uppercase or capital letters are used to label the vertices (corners) as shown in Figure 28:1. The side opposite a vertex is labelled using its corresponding lowercase or small letter. For instance the side opposite the vertex *A* is labelled using the letter *a*. The triangle above is referred to as triangle *ABC*. The order of the letters does not matter.

The Right-Angled Triangle

Recall from page 227 that a **right-angled triangle** is a triangle with one of its angles equal to 90°. The other two angles are each less than 90°. An angle, which is equal to 90°, is called a **right angle**. A right angle is usually marked using a small square as shown in Figure 28:2. An angle, which is greater than or equal to 0° but less than 90° is called an **acute angle**. In a right-angled triangle Figure 28:2, the longest side *AC* is called the **hypotenuse** and is the side opposite to the right angle. The two shorter sides *AB* and *BC* are next to the right angle and are called the **arms** or the **legs** of the right-angled triangle.

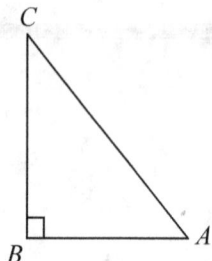

Figure 28:2

In Figure 28:2, the sides *a* and *c* can be reckoned in the following ways;

Side *a* is opposite to angle *BAC* or *a* is adjacent or next to angle *ACB*.

Side *c* is opposite to angle *ACB* or *c* is adjacent or next to angle *BAC*.

The Pythagoras Theorem

Investigative Exercise

Figure 28:3, shows a right-angled triangle *ABC* with   *AB* = 3 squares and *BC* = 4 squares.

1. Count the number of squares on each arm of the triangle and record your result in Table 28:1.
2. Also count the number of squares on the hypotenuse and record your result in Table 28:1.

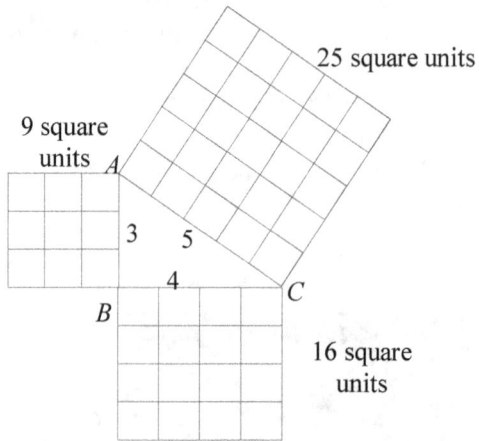

Figure 28:3

3. On a square paper draw a right-angled triangle *ABC* right angle at *B*, with *AB* = 5 squares and *BC* = 12 squares.
4. Cut out part of the square paper and place on the hypotenuse *AC* of the triangle as in Figure 28:3.
5. What is the length of the hypotenuse in terms of the number of squares?
6. Draw the large squares on each side of the

282

triangle.

7. Count the number of squares on each arm of the triangle and record your result in Table 28:1.
8. Also count the number of squares on the hypotenuse and record your result in Table 28:1.
9. Study the table and make a conclusion on your investigation.

Number of squares on side			Sum of squares on *AB* and *BC*
AB = c	*BC = a*	*AC = b*	

Table 28:1

The above investigation leads us to the following theorem called the Pythagoras theorem.

The Pythagoras theorem states that in a right-angled triangle, the number of squares on the hypotenuse is equal to the sum of the squares on the other two sides.

Thus for the triangle in Figure 28:2

$$AC^2 = AB^2 + BC^2$$
$$\Rightarrow AC = \sqrt{AB^2 + BC^2}$$
$$\Rightarrow AB = \sqrt{AC^2 - BC^2}$$
$$\Rightarrow BC = \sqrt{AC^2 + AB^2}$$

To confirm further, this relationship draw right-angled triangles with sides of different lengths, measure their sides and test in the formula.

Test Rule for a Right-Angled Triangle

If in a triangle whose sides are known, the sum of the squares on two of the sides is equal to the square on the other side, the triangle is a right-angle triangle. Otherwise the triangle cannot be a right-angled triangle.

Pythagorean Triples

Any three whole numbers, which can form sides of a right-angled triangle, are called Pythagorean triples. Examples of Pythagorean triples are 3,4,5; 6,8,10; 5,12,13; 8,15,17; 7,24,25 etc

Multiples of Pythagorean Triples

Consider the Pythagorean triple 3, 4, 5. Table 28:2, shows the result of multiplying each of the numbers in the triplet by the same factor *n*.

n	*AB*	*BC*	*AC*	AB^2	BC^2	AC^2	$AB^2 + BC^2$
1	3	4	5	9	16	25	25
2	6	8	10	36	64	100	100
3	9	12	15	81	144	225	225
4	12	16	20	144	256	400	400
5	15	20	25	225	400	625	625

Table 28:2

From Table 28:2, it can be seen that multiples of the Pythagorean triple 3, 4, 5 are also Pythagorean triples.
Generally,

Multiples of Pythagorean triples are also Pythagorean triples.

EXERCISE 28:1

1. Find the hypotenuse of a right-angled triangle whose arms are 6 cm and 8 cm.
2. One side of a right-angled triangle is 3 cm. If the hypotenuse is 5 cm, find the length of the other side.
3. Find the diagonal of a rectangle whose sides are 40 cm by 9 cm long.
4. The diagonal of a rectangle is 30 cm and one of its sides is 24 cm; find the other side of the rectangle.
5. In a right-angled triangle whose hypotenuse is 20 cm, the ratio of the two arms is 3 : 4. Find each arm.
6. Determine which of the following triplets are Pythagorean triples.
   (a) 17, 15, 8   (b) 6, 9, 11   (c) 40, 41, 9
   (d) 3, 2, 5   (e) 5, 12, 13   (f) 7, 9, 12
   (g) 14, 17, 20   (h) 6, 8, 10   (i) 3, 6, 8

Trigonometric Ratios

The right-angled triangles in figure 28:5 have been drawn using the Pythagorean triples 3, 4, 5 and 6, 8, 10 respectively. Use a protractor to measure and confirm that

$\angle A = \angle A' = 36.9°$ and $\angle B = \angle B' = 53.1°$

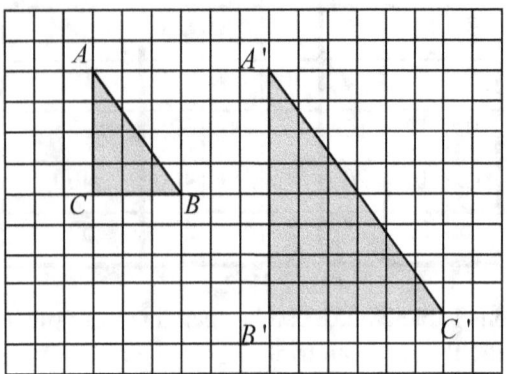

Figure 28:5

By calculation show that the ratios of the sides of these triangles are as recorded in Table 28:3

a	b	c	$\dfrac{a}{c}$	$\dfrac{b}{c}$	$\dfrac{a}{b}$
3	4	5	0.6	0.8	0.75
6	8	10	0.6	0.8	0.75

Table 28:3

The ratios $\dfrac{a}{c}$, $\dfrac{b}{c}$ and $\dfrac{a}{b}$ are called

trigonometric ratios or simply **trig ratios** and depend only on the values of $\angle A$ and $\angle B$, of any right-angled triangle. These ratios are given the special names sine (sin), cosine (cos) and tangent (tan) defined as follows;

$$\sin \hat{A} = \frac{\text{side opposite to angle } A}{\text{hypotenuse}} = \frac{\text{opp}}{\text{hyp}} = \frac{O}{H}$$

$$\cos \hat{A} = \frac{\text{side adjacent to angle } A}{\text{hypotenuse}} = \frac{\text{adj}}{\text{hyp}} = \frac{A}{H}$$

$$\tan \hat{A} = \frac{\text{side opposite to angle } A}{\text{side adjacent to angle } A} = \frac{\text{opp}}{\text{adj}} = \frac{O}{A}$$

Similarly;

$$\sin \hat{B} = \frac{\text{side opposite to angle } B}{\text{hypotenuse}} = \frac{\text{opp}}{\text{hyp}} = \frac{O}{H}$$

$$\cos \hat{B} = \frac{\text{side adjacent to angle } B}{\text{hypotenuse}} = \frac{\text{adj}}{\text{hyp}} = \frac{A}{H}$$

$$\tan \hat{B} = \frac{\text{side opposite to angle } B}{\text{side adjacent to angle } B} = \frac{\text{opp}}{\text{adj}} = \frac{O}{A}$$

Remember that!!

Adjacent to $\angle A$' means 'next to $\angle A$' and 'opposite to $\angle A$' means 'on the other side of $\angle A$'.

The mnemonic **RAT-SOH-CAH-TOA** may help to commit the definitions to mind. Thus

RAT – means in any **R**ight – **A**ngled **T**riangle

SOH – stands for $S = \dfrac{O}{H}$ i.e. $\sin = \dfrac{Opposite}{Hypotenus}$

CAH – stands for $C = \dfrac{A}{H}$ i.e. $\cos = \dfrac{Adjacent}{Hypotenus}$

TOA – stands for $T = \dfrac{O}{A}$ i.e. $\tan = \dfrac{Opposite}{Adjacent}$

Example 28:1

Use Figure 28:6 to write down as fractions the value of the given trigonometric ratios.

(a) $\sin A$ (b) $\cos C$ (c) $\sin C$
(d) $\cos A$ (e) $\tan A$ (f) $\tan C$

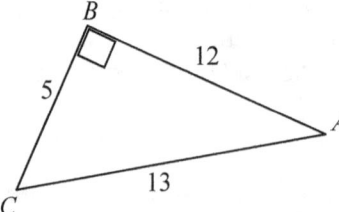

Figure 28:6

Solution

(a) $\sin A = \dfrac{\text{opp}}{\text{hyp}} = \dfrac{5}{13}$ (b) $\cos C = \dfrac{\text{adj}}{\text{hyp}} = \dfrac{5}{13}$

(c) $\sin C = \dfrac{\text{opp}}{\text{hyp}} = \dfrac{12}{13}$ (d) $\cos A = \dfrac{\text{adj}}{\text{hyp}} = \dfrac{12}{13}$

(e) $\tan A = \dfrac{\text{opp}}{\text{adj}} = \dfrac{5}{12}$ (f) $\tan C = \dfrac{\text{opp}}{\text{adj}} = \dfrac{12}{5}$

EXERCISE 28:2

1.

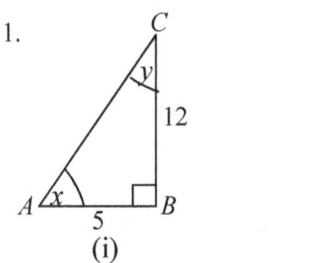

Figure 28:7

In Figure 28:7(i) and (ii), find
(a) $\tan x$ (b) $\sin x$ (c) $\cos x$
(d) $\tan y$ (e) $\sin y$ (f) $\cos y$

2. In Figure 28:8, find the value of
(a) $\tan x$ (b) $\sin x$ (c) $\cos x$
(d) $\tan y$ (e) $\sin y$ (f) $\cos y$

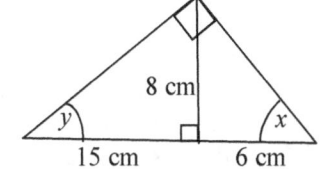

Figure 28:8

3. If α is acute and $\tan \alpha = \dfrac{7}{24}$, find the value of:
   (a) $\sin \alpha$ (b) $\cos \alpha$

4. Using Figure 28:9(i), write down the values of:
   (i) $\sin \theta$ (ii) $\cos \theta$ (iii) $\tan \theta$

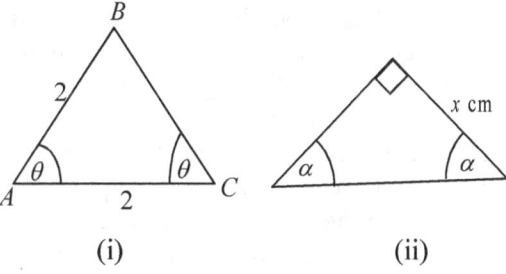

Figure 28:9

5. Using Figure 28:9(ii), write down the value of:
   (i) $\sin \alpha$ (ii) $\cos \alpha$ (iii) $\tan \alpha$

6. Given that x is an acute angle and that $\sin x = \dfrac{m}{n}$, find $\cos x$ and $\tan x$ in terms of m and n.

7. If α is acute and $\cos A = \dfrac{4}{5}$, find the value of

   (a) $\sin A$ (b) $\tan A$

8. In Figure 28:10, find
   (a) $\cos w$ (b) $\sin w$ (c) $\tan w$
   (d) $\cos x$ (e) $\sin x$ (f) $\tan x$
   (g) $\cos y$ (h) $\sin y$ (i) $\tan y$
   (j) $\cos z$ (k) $\sin z$ (l) $\tan z$

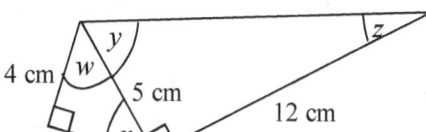

Figure 28:10

Acute Angle Trigonometric Ratios

It has already been seen that the trigonometric ratios are the same for any given angle, no matter the size of the triangle. Table 28:4 shows the trigonometric ratios for integral values of θ for some angles from $0°$ to $90°$.

θ	$\sin \theta$	$\cos \theta$	$\tan \theta$
0	0.0000	1.0000	0.0000
5	0.0872	0.9962	0.0875
10	0.1736	0.9848	0.1763
15	0.2588	0.9659	0.2679
20	0.3420	0.9397	0.3640
25	0.4226	0.9063	0.4663
30	0.5000	0.8660	0.5774
35	0.5736	0.8192	0.7002
40	0.6428	0.7660	0.8391
45	0.7071	0.7071	1.0000
50	0.7660	0.6428	1.1918
55	0.8192	0.5736	1.4281
60	0.8660	0.5000	1.7321
65	0.9063	0.4226	2.1445
70	0.9397	0.3420	2.7475
75	0.9659	0.2588	3.7321
80	0.9848	0.1736	5.6713
85	0.9962	0.0872	11.4301
90	1.0000	0.0000	8

Table 28:4

Notice that both $\sin \theta$ and $\tan \theta$ increase as θ increases, while $\cos \theta$ decreases as θ increases. Thus;

as $\theta \rightarrow 90°$, $\sin \theta \rightarrow 1$, $\tan \theta \rightarrow \infty$ while $\cos \theta \rightarrow 0$

as $\theta \rightarrow 0°$, $\sin \theta \rightarrow 0$, $\tan \theta \rightarrow 0$ while $\cos \theta \rightarrow 1$

Therefore, $\sin \theta$ and $\tan \theta$ are increasing functions, while $\cos \theta$ is a decreasing function.

Trigonometric Ratios from Tables

Trigonometric ratios of angles from $0°$ to $90°$ can be obtained from four figure tables. To do this, always make sure that the heading is 'Natural sines' when referring sines, 'Natural cosines' when referring cosines or 'Natural tangents' when referring tangents.

(i) For an angle with an exact number of degrees, look under the column headed 0'.

(ii) For an angle with multiple of 6 as the minutes, look under the column headed by that multiple of 6. If the minutes are not a multiple of 6 subtract the lower but closest multiple of 6 to obtain the difference heading. The difference is added for sine and tangent, because these are increasing functions but subtracted for cosine because it is a decreasing function.

Example 28:2
Use tables to find
(i) $\sin 16°$ (ii) $\sin 16°24'$ (iii) $\sin 16°29'$

Solution
Make sure the heading of your table is 'Natural sines'.
(i) Look for $\sin 16°$ under 0'. This gives 0.2756.
(ii) Look for $\sin 16°$ under 24'. This gives 0.2823.
(iii) Add the difference 14 found under the difference column 5' to the result in (ii) above. Thus
   $0.2823 + 14 = 0.2837$.

Notice that the difference 14 is actually an abbreviation for 0.0014.

Example 28:3
Use tables to find
 (i) $\tan 16°$ (ii) $\tan 16°24'$ (iii) $\tan 16°29'$

Solution
Make sure the heading of your table is 'Natural tangents'. The procedure is as in example 28.2.

(i) $\tan 16° = 0.2867$ (ii) $\tan 16°24' = 0.2943$
(iii) $\tan 16°29' = 0.2959$

Example 28:4
Use tables to find
 (i) $\cos 16°$ (ii) $\cos 16°24'$ (iii) $\cos 16°29'$

Solution
Make sure the heading of your table is 'Natural cosines'.
The procedure is as in example 28.2 and 28.3 but for the fact that the difference is subtracted. Thus,
(i) $\cos 16° = 0.9613$ (ii) $\cos 16°24' = 0.9593$
(iii) $\cos 16°29' = 0.9589$.

Note!
Some tables have the natural sines and cosines on one page. On such tables, the angles for cosines are usually on the right most columns, written in descending order and the minute row for cosines at the base of the table also written in descending order.

Inverse Trigonometric Ratios

Given a trigonometric ratio, the corresponding angle can equally be found. To do this from four figure tables, refer this trigonometric ratio from the body of the table and copy the corresponding angle. If the trigonometric ratio is not found, copy out the angle corresponding to the trigonometric ratio just lower than the one given. Subtract the trigonometric ratio copied from the one given and look for the difference under the remnant. The difference is added in the case of sine and tangent and subtracted in the case of cosine. The corresponding angle to a given trigonometric ratio is called the **arc trigonometric ratio** of the angle. For instance, arc $\sin^{-1} 0.5$.

Arc $\sin 0.5$ is written simply as $\sin^{-1} 0.5 = 30°$.

Example 28:5
Use tables to find
(i) $\sin^{-1} 0.2756$ (ii) $\sin^{-1} 0.2823$ (iii) $\sin^{-1} 0.2837$

Solution
(i) Look for 0.2756 in the body of the table. This would be found under 16° 0'. Therefore, arcsine 0.2756 is 16°.

(ii) Look for 0.2823. This would be found under $16°24'$. Therefore arcsine 0.2756 is $16°24'$.

(iii) 0.2837 is not seen but the next smaller value 0.2823 is seen under $16°24'$. The difference between them is 14 found under 5'. Add this 5' to 24'. Therefore, $\sin^{-1} 0.2837$ is $16°29'$.

Example 28:6
Use tables to find

(i) $\tan^{-1} 0.2867$ (ii) $\tan^{-1} 0.2943$ (iii) $\tan^{-1} 0.2959$

Solution
The procedure is the same as in Example 28:5.

(i) $\tan^{-1} 0.2867 = 16°$ (ii) $\tan^{-1} 0.2943 = 16°24'$

(iii) $\tan^{-1} 0.2959 = 16°29'$

Example 28:7
Use tables to find

(i) $\cos^{-1} 0.9613$ (ii) $\cos^{-1} 0.9593$ (iii) $\cos^{-1} 0.9593$

Solution
The procedure is the same as in example 28.5 and 28.6 but for the fact that the difference is subtracted. Thus

(i) $\cos^{-1} 0.9613 = 16°$ (ii) $\cos^{-1} 0.9593 = 16°24'$

(iii) $\cos^{-1}(0.9589) = 16°29'$

EXERCISE 28:3

1. Use tables to find the following trigonometric ratios

   (a) $\sin 34°$ (b) $\sin 72°18'$ (c) $\sin 81°20'$

   (d) $\cos 63°$ (e) $\cos 54°36'$ (f) $\cos 63°43'$

   (g) $\tan 46°$ (h) $\tan 27°6'$ (i) $\tan 82°16'$

2. Use tables to find the angles whose sines are given below.

   (a) 0.1564   (b) 0.9135   (c) 0.9880

   (d) 0.802   (e) 0.9814   (f) 0.7395

   (g) 0.0500   (h) 0.2700

3. Use tables to find the angles whose cosines are given below.

   (a) 0.9135   (b) 0.3420   (c) 0.9673

   (d) 0.4289   (e) 0.9586   (f) 0.0084

   (g) 0.2611   (h) 0.4700

4. Use tables to find the angles whose tangents are given below.

(a) 0.4452   (b) 3.2709   (c) 0.0769

(d) 0.3977   (e) 0.3568   (f) 1.9251

(g) 0.0163   (h) 0.8263

Finding other Trig Ratios Given Another

Example 28:8
Given that $\cos \theta = \dfrac{15}{17}$, find the values of $\sin\theta$ and $\tan\theta$ without using tables or calculators.

Solution

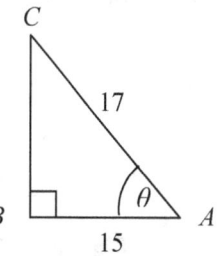

Figure 28:11

Using the Pythagoras theorem in Figure 28.11,

$$BC = \sqrt{AC^2 - AB^2}$$

$$\Rightarrow BC = \sqrt{17^2 - 15^2} = \sqrt{64} = 8 \text{ units}$$

$$\Rightarrow \sin \theta = \frac{BC}{AC} = \frac{8}{17} \text{ and } \tan \theta = \frac{BC}{AB} = \frac{8}{15}$$

Example 28:9
Given that $\sin \theta = 0.6$, find the values of $\cos \theta$ and $\tan \theta$ without using tables or calculators.

Solution

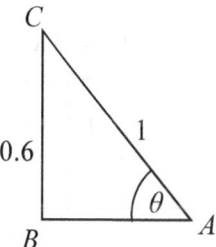

Figure 28:12

Using the Pythagoras theorem in Figure 28.12,

$$AB = \sqrt{AC^2 - BC^2}$$

$$\Rightarrow AB = \sqrt{1^2 - 0.6^2} = \sqrt{0.64} = 0.8 \text{ units}$$

$$\cos \theta = \frac{AB}{AC} = \frac{0.8}{1} = 0.8 \text{ and } \tan \theta = \frac{BC}{AB} = \frac{0.6}{0.8} = 0.75$$

EXERCISE 28:4

1. If $\cos \alpha = 0.64$, find the values of $\sin \alpha$ and $\tan \alpha$ without using tables or calculators.

2. Given that $\sin x = \dfrac{5}{13}$. Find $\cos x$ and $\tan x$ without using tables or calculators.

3. If $\tan A = \dfrac{4}{3}$, find the values of $\sin A$ and $\cos A$ without using tables or calculators.

Complementary Angles

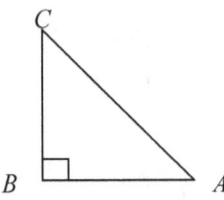

Figure 28:13

If the angles of any right-angled triangle ABC as in Figure 28:13 are measured, it will be noticed that in every case,

$$\angle A + \angle C = 90°$$

$\Rightarrow$ If $\angle A = x°$, then $\angle C = 90 - x°$

Two angles (such as $\angle A$ and $\angle C$) which sum up to $90°$ are called **complementary angles** and are said to be complementary.
There is a relationship between the trig ratios of complementary angles. Use tables or calculators to confirm this relationship in the following examples;

$\sin 60° = 0.8660$ and $\cos 30° = 0.8660$
$\sin 30° = 0.5000$ and $\cos 60° = 0.5000$
$\sin 13° = 0.2250$ and $\cos 77° = 0.2250$

$\sin 77° = 0.9744$ and $\cos 13° = 0.9744$
$\sin 45° = 0.7071$ and $\cos 45° = 0.7071$

Notice that all the pairs of angles are complementary (sum up to $90°$) and that in each case the cosine of one angle is equal to the sine of its complement. Thus
$\sin 60° = 0.8660$ and $\cos 30° = 0.8660$

Generally,

> *The sine of any acute angle is equal to the cosine of its complement and vice versa.*

Symbolically, this is expressed as,

$$\left. \begin{array}{l} \sin \theta = \cos(90° - \theta) \\ \cos \theta = \sin(90° - \theta) \end{array} \right\}, \quad 0 \le \theta \le 90°$$

Example 28:10
Two angles x and y are complementary. Find the value of x if $y = 60°$.

Solution
$$x = 90° - y \text{ and } y = 60°$$
$$\Rightarrow x = 90° - 60° = 30°$$

EXERCISE 28:5

1. Given that x, y and z are angles of a right-angled triangle. Find the missing value.

	$x°$	$y°$	$z°$
(i)	72	90	——
(ii)	25	——	90
(iii)	43.5	——	90
(iv)	——	57	33
(v)	90	22.5	——
(vi)	——	13	77

2. From Figure 28:14, find $\cos \varnothing$.

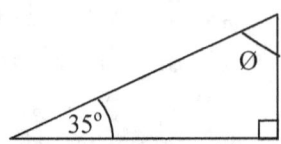

Figure 28:14

3. Write down the value of the missing angle in Figure 28:15. (a) to (d)

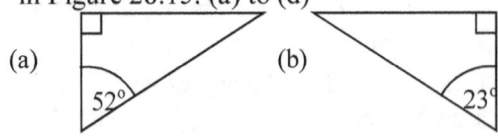

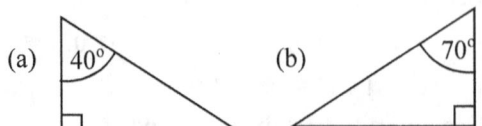

Figure 28:15
State the complement of each of the

following angles.

(a) 60° (b) 34° (c) 45° (d) 90° (e) 0°

5. Given that sin 48° = 0.7431. Find the value of cos 42° without using tables or calculator.

6. Given that cos 63 = 0.4540, without using tables or calculator find the value of sin 27°.

7. Write down 20 possible pairs of values of *x* and *y* in Figure 28:16.

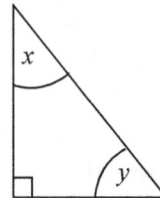

Figure 28:16

8. State two possible angles of a right-angled triangle and draw the triangle marking the value of each of its angles.

Reciprocal Trigonometric Ratios

Sometimes it is easier to solve problems using the reciprocals of the familiar trigonometric ratios sine, cosine and tangent studied earlier. These reciprocals are as follows;

1. The cosecant (cosec) is the reciprocal of sine,

$$\operatorname{cosec}\theta = \frac{1}{\sin\theta} \Rightarrow \operatorname{cosec}\theta = \frac{\text{hyp}}{\text{opp}}$$

2. The secant (sec) is the reciprocal of cosine,

$$\sec\theta = \frac{1}{\cos\theta} \Rightarrow \sec\theta = \frac{\text{hyp}}{\text{adj}}$$

3. The cotangent (cot) is the reciprocal of tangent,

$$\cot\theta = \frac{1}{\tan\theta} \Rightarrow \cot\theta = \frac{\text{adj}}{\text{opp}}$$

EXERCISE 28:6

1. In Figure 28:17(i) and (ii), find
    (a) cot *x*   (b) cosec *x*   (c) sec *x*
    (d) cot *y*   (e) cosec *y*   (f) sec *y*

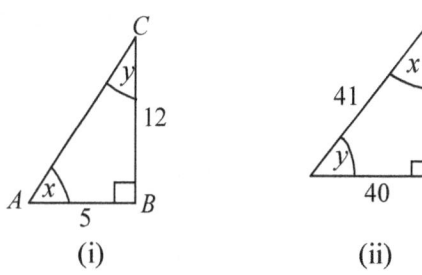

Figure 28:17

2. In Figure 28:18, find the value of
    (a) cot *x*   (b) cosec *x*   (c) sec *x*
    (d) cot *y*   (e) cosec *y*   (f) sec *y*

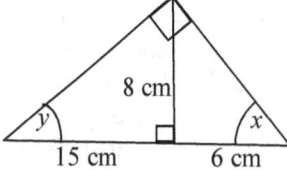

Figure 28:18

3. If *α* is acute and $\cot\alpha = \frac{24}{7}$, find the value of (a) cosec *α*   (b) sec *α*

4. Using Figure 28:19(i), write down the values of
    (i) cosec *θ*   (ii) sec *θ*   (iii) cot *θ*

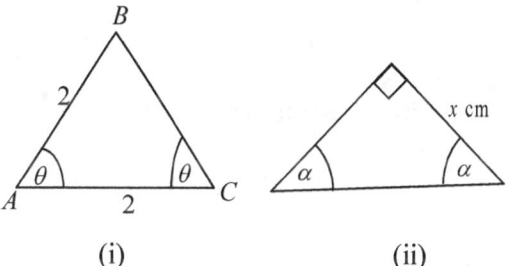

Figure 28:19

5. Using Figure 28:19(ii), write down the value of
    (i) cosec *α*   (ii) sec *α*   (iii) cot *α*

6. Given that *x* is an acute angle and that $\operatorname{cosec} x = \frac{m}{n}$, find sec *x* and cot *x* in terms of *m* and *n*.

7. If *α* is acute and $\sec A = \frac{4}{5}$, find the value of (a) cosec *A*   (b) cot *A*

8. In Figure 28:20(i) and (ii), find
    (a) sec *w*   (b) cosec *w*   (c) cot *w*
    (d) sec *x*   (e) cosec *x*   (f) cot *x*
    (g) sec *y*   (h) cosec *y*   (i) cot *y*
    (j) sec *z*   (k) cosec *z*   (l) cot *z*

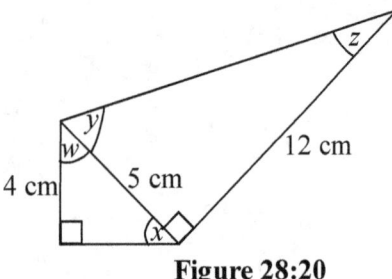

Figure 28:20

Further Trig Ratios from Tables

It was earlier mentioned that when using tables to refer the trigonometric ratios sine, cosine and tangent, the difference is added for both sine and tangent because these are increasing functions, while the difference is subtracted for cosine because it is a decreasing function.

Using tables to refer the trigonometric ratios cosecant, secant and cotangent is similar noting that, the cosecant and the cotangent are decreasing functions while the secant is an increasing function. Therefore, the difference is added for secant and subtracted for cosecant and cotangent.

Generalising,

> To refer trigonometric ratios from tables, the difference is;
> 1. Subtracted for all trigonometric ratios starting with the prefix **co-**
> 2. Added for all other trigonometric ratios.

EXERCISE 28:7

1. Use tables to find the following trigonometric ratios
   (a) cosec 34° (b) cosec 72°18' (c) cosec 81°
   (d) sec 52°   (e) sec 54°36'   (f) sec 63°43'
   (g) cot 48°   (h) cot 27°6'   (i) cot 82°16'
2. Use tables to find the angles whose cosecants are given below.
   (a) 1.6243   (b) 3.5843   (c) 1.0803
   (d) 5.9554   (e) 1.0105   (f) 2.6260
   (g) 1.4746   (h) 4.1824
3. Use tables to find the angles whose secants are given below.
   (a) 1.2062   (b) 1.0589   (c) 4.3684
   (d) 1.0033   (e) 2.1116   (f) 1.1262
   (g) 1.1753   (h) 11.1045
4. Use tables to find the angles whose

cotangents are given below.
(a) 1.3270   (b) 2.4506   (c) 0.5147
(d) 6.5606   (e) 0.4935   (f) 2.2062
(g) 1.4994   (h) 0.1257

Trig Ratios of Special Angles

The trigonometric ratios of certain angles are used so frequently that they need to be given special attention. These are the trigonometric ratios of 0°, 30°, 45°, 60° and 90°.

Trigonometric Ratios of 30° and 60°

Consider an equilateral triangle ABC (Figure 28:21) with each side equal to 2 units.

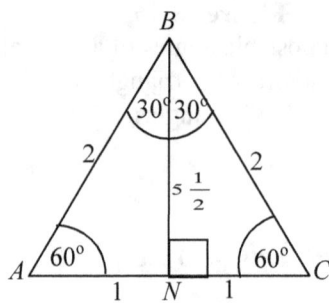

Figure 28:21

If a perpendicular bisector is drawn from vertex B to the opposite side AC, the 60° angle ABC will be bisected into two 30° angles ABN and CBN, and the opposite side AC will be bisected into AN and NC each equal to 1 unit. Using the Pythagoras theorem, the height BN of triangle ABC can then be calculated.

From either of the two right-angled triangles formed, the trigonometric ratios of 60° and 30° can then be found as follows;

$$\sin 60° = \frac{\sqrt{3}}{2} \qquad \cos 30° = \frac{\sqrt{3}}{2}$$

$$\sin 30° = \frac{1}{2} \qquad \cos 60° = \frac{1}{2}$$

$$\tan 60° = \frac{\sqrt{3}}{1} = \sqrt{3} \qquad \cot 30° = \frac{\sqrt{3}}{1} = \sqrt{3}$$

$$\tan 30° = \frac{1}{\sqrt{3}} = \frac{\sqrt{3}}{3} \qquad \cot 60° = \frac{1}{\sqrt{3}} = \frac{\sqrt{3}}{3}$$

Trigonometric Ratios of 45°

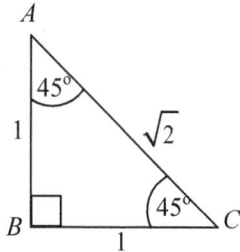

Figure 28:22

Consider a right-angled isosceles triangle ABC (Figure 28:22) with each of the legs AB and BC equal to 1 unit. The base angles BAC and BCA will therefore be 45° each.

Using the Pythagoras theorem, the hypotenuse AC of triangle ABC can then be calculated. Thus;

$$AC = \sqrt{1^2 + 1^2} = \sqrt{2}$$

The trigonometric ratios of 45° can then be found as follows;

$$\sin 45° = \frac{1}{\sqrt{2}} = \frac{\sqrt{2}}{2} \qquad \cos 45° = \frac{1}{\sqrt{2}} = \frac{\sqrt{2}}{2}$$

$$\tan 45° = \frac{1}{1} = 1 \qquad \cot 45° = \frac{1}{1} = 1$$

$$\sec 45° = \frac{\sqrt{2}}{1} = \sqrt{2} \qquad \csc 45° = \frac{\sqrt{2}}{1} = \sqrt{2}$$

Trigonometric Ratios of 0° and 90°

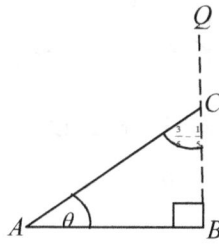

Figure 28:23

Consider the right-angled triangle ABC (Figure 28:23) with angle $BAC = \theta$ and angle $BCA = \alpha$. Suppose that C is a moving point along BQ, then as C moves along BQ, the line AC rotates about A as centre. Thus, the angle can be made as small as possible (or nearly 0°). At the same time as θ is getting smaller and smaller, α is getting larger and larger and approaching the value of 90°. Under these

conditions, the length AC approximates the length AB and the length BC approximates 0.

Symbolically;

as $\theta \to 0°$, $\alpha \to 90°$, $AC \to AB$ and $BC \to 0$

$$\text{Since } \sin\theta = \frac{BC}{AC}, \qquad \csc\theta = \frac{AC}{BC},$$

$$\cos\theta = \frac{AB}{AC}, \qquad \sec\theta = \frac{AC}{AB},$$

$$\tan\theta = \frac{BC}{AB} \qquad \tan\theta = \frac{AC}{BC}$$

$$\text{also } \sin\alpha = \frac{AB}{AC}, \qquad \csc\alpha = \frac{AC}{AB},$$

$$\cos\alpha = \frac{BC}{AC}, \qquad \sec\alpha = \frac{AC}{BC},$$

$$\tan\alpha = \frac{AB}{BC} \qquad \tan\alpha = \frac{BC}{AB}$$

The limiting values are

$\theta = 0°$, $\alpha = 90°$, $AC = AB$ and $BC = 0$

Substituting these limiting values,

$$\sin 0° = \frac{0}{AB} = 0, \qquad \csc 0° = \frac{AB}{0} = \infty,$$

$$\cos 0° = \frac{AB}{AB} = 1, \qquad \sec 0° = \frac{AB}{AB} = 1,$$

$$\tan 0° = \frac{0}{AB} = 0, \qquad \tan 0° = \frac{AB}{0} = \infty$$

$$\sin 90° = \frac{AB}{AB} = 1, \qquad \csc 90° = \frac{AB}{AB} = 1,$$

$$\cos 90° = \frac{0}{AB} = 0, \qquad \sec 90° = \frac{AB}{0} = \infty,$$

$$\tan 90° = \frac{AB}{0} = \infty, \qquad \tan 90° = \frac{0}{AB} = 0$$

These results are summarized in Table 28:5

θ	$\sin\theta$	$\cos\theta$	$\tan\theta$	$\csc\theta$	$\sec\theta$	$\cot\theta$
0°	0	1	0	∞	1	∞
30°	$\dfrac{1}{2}$	$\dfrac{\sqrt{3}}{2}$	$\dfrac{\sqrt{3}}{3}$	2	$\dfrac{2\sqrt{3}}{3}$	$\sqrt{3}$
45°	$\dfrac{\sqrt{2}}{2}$	$\dfrac{\sqrt{2}}{2}$	1	$\sqrt{2}$	$\sqrt{2}$	1
60°	$\dfrac{\sqrt{3}}{2}$	$\dfrac{1}{2}$	$\sqrt{3}$	$\dfrac{2\sqrt{3}}{3}$	2	$\dfrac{\sqrt{3}}{3}$
90°	1	0	∞	0	∞	0

Table 28:5

EXERCISE 28:8

1. Compute the following without using tables or calculators. Where appropriate leave your answer in surd form.

   (a) $\cos^2 60° + \tan^2 45°$ (b) $\sin^2 30° - \cos^2 60°$

   (c) $\dfrac{\sin 65°}{\cos 25°}$ (d) $2 - \cos 60°$

   (e) $\sec^2 45° + \tan^2 45°$ (f) $\dfrac{\cos ec\ 40°}{\sec 50°}$

   (g) $1 + \tan^2 30°$ (h) $\sin^2 30° + \cos^2 30°$

   (i) $\dfrac{\cot 20°}{\tan 70°}$ (j) $\cos ec^2 30° + \cot^2 30°$

2. Without tables or calculators, evaluate the following leaving your answers in surd form.

   (a) $\cos 60° + \sin 30°$ (b) $\sin 60° + \cos 30°$

The General Angle

Recall from Topic 19 that, the x-axis and the y-axis divide the x-y plane into four sections called the first, second, third and fourth quadrants.

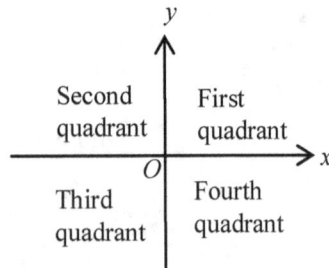

Figure 28:24

To find the trigonometric ratios of angles greater than 90°, consider a point P (Figure 28:25), which starts from the line Ox, where O is the origin and moves on a circle of radius r in the anticlockwise sense. The coordinates of P when P is in each of the four quadrants will be either positive or negative though the length r of OP remains positive as shown.

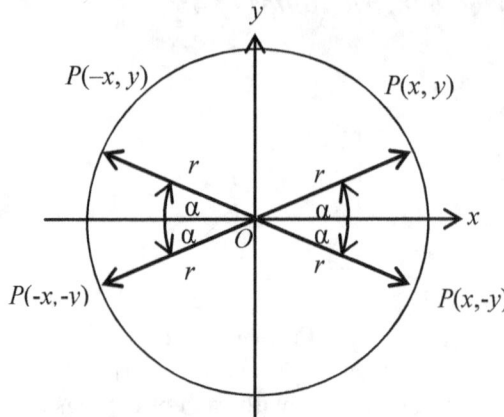

Figure 28:25

The angle α which OP makes with the x-axis remains acute (i.e. $0° \le α \le 90°$), no matter the angle of rotation $θ$ of P from Ox. α is called the **associated angle**.

The values of α in the first, second, third and fourth quadrants are $θ$, $180° - θ$, $θ - 180°$ and $360° - θ$ respectively. It can be proven (though beyond the scope of this book) that in each position, the trigonometric ratios of the angle $θ$ are numerically equal in magnitude to the trigonometric ratios of the associated angle α.

In the first quadrant $0° \le θ \le 90°$ and $α = θ$ (figure 28:26).

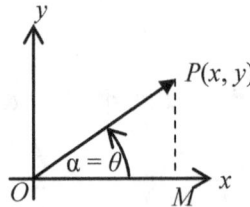

Figure 28:26

$$\sin θ = \frac{y}{r} = \sin α$$

$$\cos θ = \frac{x}{r} = \cos α$$

$$\tan θ = \frac{y}{x} = \tan α$$

In the second quadrant $90° \le θ \le 180°$ and $α = 180° - θ$ (figure 28:27).

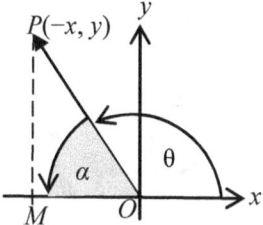

Figure 28:27

$$\sin \theta = \frac{y}{r} = \sin \alpha$$

$$\cos \theta = \frac{-x}{r} = -\cos \alpha$$

$$\tan \theta = \frac{y}{-x} = -\tan \alpha$$

In the third quadrant $180° = \theta \leq 270°$ and $\alpha = \theta - 180°$ (Figure 28:28).

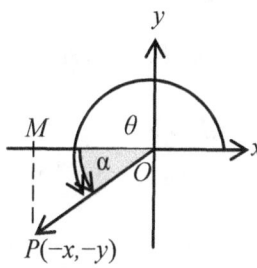

Figure 28:28

$$\sin \theta = \frac{-y}{r} = -\sin \alpha$$

$$\cos \theta = \frac{-x}{r} = -\cos \alpha$$

$$\tan \theta = \frac{-y}{-x} = \tan \alpha$$

In the fourth quadrant $270° = \theta \leq 360°$ and $\alpha = 360° - \theta$ (figure 28:29).

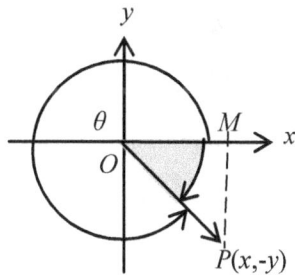

Figure 28:29

$$\sin \theta = \frac{-y}{r} = -\sin \alpha$$

$$\cos \theta = \frac{x}{r} = \cos \alpha$$

$$\tan \theta = \frac{-y}{x} = -\tan \alpha$$

Figure 28:30 summarizes the positive trigonometric ratios, in each quadrant.

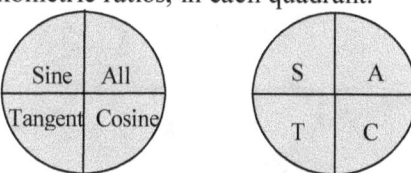

Figure 28:30

Thus,
In the first quadrant, ALL the three are positive.
In the second quadrant, only the SINE is positive.
In the third quadrant, only the TANGENT is positive.
In the fourth quadrant, only the COSINE is positive.

The mnemonic **ACTS** may help to recall these facts to memory. Note that the mnemonic is read from the fourth quadrant.

EXERCISE 28:9

1. Evaluate the following without using tables or calculators.
   (a) $\cos 300°$ (b) $\tan 210°$ (c) $\tan 120°$
   (d) $\sin 270°$ (e) $\sin 135°$ (f) $\cos 150°$
2. Evaluate the following using tables
   (a) $\cos 100°$ (b) $\tan 235°$ (c) $\tan 125°$
   (d) $\sin 260°$ (e) $\sin 160°$ (f) $\cos 170°$
3. Evaluate the following without using tables or calculators.
   (a) $\operatorname{cosec} 300°$ (b) $\cot 210°$ (c) $\cot 120°$
   (d) $\sec 270°$ (e) $\sec 135°$ (f) $\operatorname{cosec} 150°$
4. Evaluate the following using tables
   (a) $\operatorname{cosec} 100°$ (b) $\cot 235°$ (c) $\cot 125°$
   (d) $\sec 260°$ (e) $\sec 160°$ (f) $\operatorname{cosec} 170°$

The Meaning of Negative Angles

Conventionally angles measured in the anticlockwise sense are considered positive while angles measured in the clockwise sense are considered negative. Consider again the point P, which is moving on the circle.

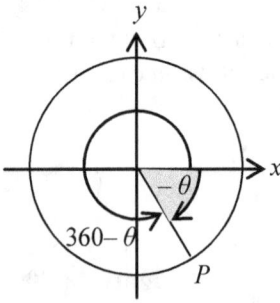

Figure 28:31

Suppose P is on any position, which is not along Ox. P could attain this position by either moving clockwise or anticlockwise. It follows that the angle $-\theta$ is equivalent to the angle $(360 - \theta)$ as shown in Figure 28:31. Hence the trigonometric ratios of $-\theta$ are equal to the trigonometric ratios of $(360 - \theta)$.

$$-\theta \equiv (360 - \theta)$$

From the above explanations,

$$-30° \equiv (360 - 30)° \equiv 330°$$
$$-120° \equiv (360 - 120)° \equiv 240°$$
$$-45° \equiv (360 - 45)° \equiv 315°$$
$$-270° \equiv (360 - 270)° \equiv 90°$$

Generalizing,

$$(-\theta) \equiv (360 - \theta)$$
$$\Leftrightarrow \sin(-\theta) \equiv \sin(360 - \theta) = -\sin\theta$$
$$\Leftrightarrow \cos(-\theta) \equiv \cos(360 - \theta) = \cos\theta$$
$$\Leftrightarrow \tan(-\theta) \equiv \tan(360 - \theta) = -\tan\theta$$

Example 28:11
Evaluate the following
(a) $\sin 330°$ (b) $\tan 315°$ (c) $\cos 240°$

Solution

(a) $\sin 330° = \sin(360 - 330)° = -\sin 30° = -\dfrac{1}{2}$

(b) $\tan 315° = \tan(360 - 315)° = -\tan 45° = -1$

(c) $\cos 240° = \cos(360 - 240)° = \cos 120°$

Since $120°$ is in the second quadrant, the associated angle is $180° - 120° = 60°$ and the cosine is negative.

$$\Rightarrow \cos 240° = -\cos 60° = -\frac{1}{2}$$

1. Evaluate the following without using tables or calculators.

    (a) $\cos (-300)°$ (b) $\tan (-210)°$

    (c) $\tan (-120)°$ (d) $\sin (-270)°$

    (e) $\sin (-135)°$ (f) $\cos (-150)°$

2. Evaluate the following using tables.

    (a) $\cos (-100)°$ (b) $\tan (-235)°$

    (c) $\tan (-125)°$ (d) $\sin (-260)°$

    (e) $\sin (-160)°$ (f) $\cos (-170)°$

3. Evaluate the following using calculators.

    (a) $\cos (-100)°$ (b) $\tan (-235)°$

    (c) $\tan (-125)°$ (d) $\sin (-260)°$

    (e) $\sin (-160)°$ (f) $\cos (-170)°$

4. Evaluate the following without using tables or calculators.

    (a) $\csc (-100)°$ (b) $\cot (-235)°$

    (c) $\cot (-125)°$ (d) $\sec (-260)°$

    (e) $\sec (-160)°$ (f) $\csc (-170)°$

5. Evaluate the following using tables.

    (a) $\csc (-100)°$ (b) $\cot (-235)°$

    (c) $\cot (-125)°$ (d) $\sec (-260)°$

    (e) $\sec (-160)°$ (f) $\csc (-170)°$

6. Evaluate the following using calculators.

    (a) $\csc (-100)°$ (b) $\cot (-235)°$

    (c) $\cot (-125)°$ (d) $\sec (-260)°$

    (e) $\sec (-160)°$ (f) $\csc (-170)°$

MULTIPLE CHOICE EXERCISE 28

1. A triangle has sides 8 cm, 15 cm and 17 cm. Therefore, the best name for it is:
   [A] a equilateral triangle.
   [B] an obtuse triangle.
   [C] a right-angled triangle.
   [D] an isosceles triangle.

2. The pair of trigonometric ratios, which are equal, is:
   [A] sin 50° and cos 50°
   [B] sin 50° and tan 50°
   [C] sin 50° and tan 40°
   [D] sin 50° and cos 40°

3. The pair of trigonometric ratios such that one is the inverse of the other is:
   [A] sin θ and cosec θ
   [B] cos θ and cot θ
   [C] sin θ and sec θ
   [D] cos θ and cosec θ

4. Given the triangle, in Figure 28:32, the incorrect relation is:
   [A] $r^2 = p^2 + q^2$ [B] $p^2 = r^2 - q^2$
   [C] $q^2 = r^2 - p^2$ [D] $q^2 = r^2 + p^2$

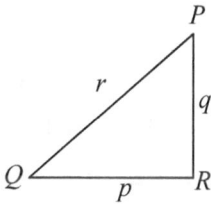

Figure 28:32

5. Given the triangle, in Figure 28:31, the value of p is:
   [A] $\sqrt{q^2 - r^2}$ [B] $\sqrt{r^2 - q^2}$
   [C] $\sqrt{q^2 + r^2}$ [D] $\sqrt{r^2 + q^2}$

6. A Pythagorean triple among the following is:
   [A] 6, 9, 11   [B] 7, 9, 12
   [C] 8, 15, 17   [D] 14, 17, 20

7. The triplet, which does not represent the lengths of the sides of a right-angled triangle is:
   [A] 6, 8, 10   [B] 5, 12, 10
   [C] 8, 15, 17   [D] 7, 23, 24

8. 36.4° is equal to:
   [A] 36°34'   [B] 36°60'
   [C] 36°24'   [D] 36°40'

9. 41°27' in degrees as a decimal is:
   [A] 41.45°   [B] 41.33°
   [C] 41.25°   [D] 41.60°

10. The ratio, which is equal to sin 60o, is:
    [A] cos 30°   [B] sin 30°
    [C] cos 60°   [D] tan 30°

11. The ratio, which is equal to sin 30o, is:
    [A] 0.0500   [B] 0.5050
    [C] 0.866   [D] 0.5000

12. The value of tan 45° is:
    [A] 2.0000   [B] 0.5000
    [C] 1.0000   [D] 1.5000

13. The cosine of $60°\frac{1}{2}$, the value of $28 - 20\cos 60°$ is:
    [A] 18   [B] 10   [C] 4   [D] 24

14. If $\sin \theta = \frac{3}{5}$, the value of tan θ for 0<θ<90° is:
    [A] $\frac{4}{5}$ [B] $\frac{3}{4}$ [C] $\frac{5}{8}$ [D] $\frac{1}{2}$

15. If $\cos \theta = \frac{5}{13}$ the value of tan θ for 0<θ<90° is:
    [A] $\frac{5}{12}$ [B] 5   [C] $\frac{13}{5}$ [D] $\frac{12}{5}$

16. Given that $\tan x = \frac{5}{12}$. The value of sin x + cos x is:
    [A] $\frac{5}{13}$ [B] $\frac{7}{13}$ [C] $\frac{17}{13}$ [D] $\frac{5}{12}$

17. If $\tan x = 2\frac{1}{2}$ the value of sin x for 0°<x<90° is:
    [A] $\frac{\sqrt{29}}{5}$ [B] $\frac{5\sqrt{29}}{29}$ [C] $\frac{2\sqrt{29}}{29}$ [D] $\frac{\sqrt{29}}{2}$

18. Given that $\sin P = \frac{5}{13}$ where P is acute. The value of cot P − tan P is:
    [A] $\frac{79}{156}$ [B] $\frac{95}{156}$ [C] $\frac{5}{13}$ [D] $\frac{13}{12}$

19. Given that sin θ = − 0.5000, where 0° ≤ θ ≤ 90°, θ is:
    [A] 30°   [B] 120°   [C] 150°   [D] 210°

20. If sin x = cos 50° then, x equals:
    [A] 40°   [B] 45°   [C] 50°   [D] 90°

21. If $\sin (x + 30)° = \cos 40°$, the value of $x°$ is:
    [A] 15°   [B] 20°   [C] 60°   [D] 90°

22. If 10 tan 60° = 20 tan x. Correct to the nearest degree x is equal to:
    [A] 30°   [B] 40°   [C] 41°   [D] 60°

23. If sin x = cos 70°, x equals:

[A] 110°   [B] 70° [C] 30°   [D] 20°

24. If cos x=sin 27.3°, the value of x where $0° \leq x \leq 90°$ is:

   [A] 27.3°[B] 35.4° [C] 54.6° [D] 62.7°

25. Without using tables or calculators, the value of $\dfrac{\sin 20°}{\cos 70°} + \dfrac{\cos 25°}{\sin 65°}$ is:

   [A] 2   [B] 1   [C] –2   [D] –1

26. If $\sin x = \dfrac{12}{13}$, where 0°<x<90°. The value of $1 - \cos^2 x$ is:

   [A] $\dfrac{25}{169}$ [B] $\dfrac{64}{169}$ [C] $\dfrac{105}{169}$ [D]$\dfrac{144}{169}$

27. Using four figure tables, the sine of 70° is:
   [A] 0.9390   [B] 0.9394
   [C] 0.9397   [D] 0.9399

28. Using four figure tables cos 80° is:
   [A] 0.173   [B] 0.1736
   [C] 0.1740   [D] 0.1744

29. Using four figure tables, the angle whose sine is 0.841 to one decimal place is:
   [A] 57.2°   [B] 56.8°
   [C] 32.8°   [D] 32.2°

30. Given that cos x = 0.5321, then x is equal to:
   [A] 56.2°   [B] 57.9°
   [C] 33.2°   [D] 32.1°

31. If sin θ = 0.6088, two possible values of θ are:
   [A] 37°30', 142°30'   [B] 52°30', 127°30'
   [C] 37°30', 127°30'   [D] 52°30', 142°30'

32. Using Mathematical tables cos 40° – sin 30° equals:
   [A] – 0.2660   [B] 0.2660
   [C] – 0.0266   [D] 0.0266

33. If $\cos 60° = \dfrac{1}{2}$, The angle whose cosine is equal to $\dfrac{1}{2}$ is:
   [A] 120°   [B] 150°
   [C] 210°   [D] 300°

34. If cos$x$ is negative and sin$x$ is negative, it is true that:
   [A] 0°<x<90°   [B] 90°<x<180°
   [C] 180°<x<270°   [D] 270°<x<360°

35. If$\sin \theta = \dfrac{\sqrt{3}}{2}$ and $\cos \theta = -\dfrac{1}{2}$. The value of θ is:

[A] 30° [B] 60° [C] 90°   [D] 120°

36. The value of tan 315° is:

   [A] –1   [B] $-\dfrac{\sqrt{2}}{2}$ [C] 0   [D] 1

37. The value of sin210° is:

   [A] $\dfrac{1}{2}$ [B] $-\dfrac{\sqrt{3}}{2}$ [C] $-\dfrac{1}{2}$ [D] $\dfrac{\sqrt{3}}{2}$

38. cos75° has the same value as:
   [A]   cos 115°   [B]   cos 255°
   [C]   cos 285°   [D] –cos 255°

39. If sinθ =cosθ for $0° \leq \theta \leq 360°$. The possible values of θ are:
   [A] 45°, 225°   [B] 135°, 315°
   [C] 45°, 315°   [D] 135°, 225°

40. cos 57° has the same value as:
   [A] sin 213°   [B] cos 303°
   [C] cos 127°   [D] cos 137°

41. If $\sin \theta = -\dfrac{1}{2}$. The values of θ between 0° and 360° are:
   [A] 120°,240°   [B] 120°,180°
   [C] 210°,330°   [D] 210°,300°

42. cos 65° has the same value as:
   [A] sin 65°   [B] cos 25°
   [C] cos 205°   [D] cos 295°

43. In $\triangle PQR, \angle PQR$ is a right angle $|QR| = 2$ cm and $\angle PRQ = 60°$. $|PR|$ is equal to:

   [A] $4\sqrt{3}$ cm   [B] 4 cm

   [C] $2\sqrt{3}$ cm   [D] 1 cm

44. If $\cos x = \dfrac{5}{8}$ for $0° \leq x \leq 180°$, the value of x is:
   [A] 141.3°   [B] 128.7°
   [C] 51.3°   [D] 48.7°

45. 25° 45' as a decimal is:
   [A] 25.75°   [B] 25.55°
   [C] 25.45°   [D] 25.15°

46. In surd form $\sin 45° \cos 30° + \cos 45° \sin 30°$ is equal to:

   [A] $\dfrac{\sqrt{2}}{2}$ [B] $\dfrac{\sqrt{3}}{2}$

   [C] $\sqrt{2}$ [D] $\dfrac{\sqrt{6} + \sqrt{2}}{4}$

TOPIC 29

APPLICATIONS OF TRIGONOMETRY

OBJECTIVES

At the end of this topic, the learner should be able to:

1. Apply trigonometry to solve problems involving right-angled triangles.
2. Distinguish between angle of elevation and angle of depression and solve problems involving angle of elevation and angle of depression.
3. Understand the convention for denoting bearings and solve problems on bearings in two dimensions.
4. State and use the Sine and Cosine Formulae to find the unknown sides or angles of any triangle.

Surveying Team

Surveyors and civil engineers apply the principles of geometry and trigonometry in determining the shape, measurements, and position of features on or beneath the surface of the Earth. Such topographic surveys are useful in the design of roads, tunnels, dams, and other structures. In this picture, a civil engineer peers through the eyepiece of a surveying theodolite to a marked rod held by another engineer down the road. Surveying measurements include changes in ground elevation from the rod to the theodolite, horizontal distances, and both vertical and horizontal angles. A third member of the surveying team, (right) records the survey data.

Blair Seitz/Photo Researchers, Inc.

Egyptian Groma

Surveying is believed to have originated in Egypt, primarily because constructing buildings as massive as the pyramids requires an ability to measure angles and calculate long distances. Early surveying equipment like this Egyptian groma were of somewhat limited use, but were apparently sufficient for flat terrain and a small range of angles. The groma consists of stones hanging from sticks set at right angles to one another. Distant objects could be marked out against the position of the stones in a horizontal plane.

Dorling Kindersley

The trigonometric ratios studied in TOPIC 28 are used to solve trigonometric problems involving right-angled triangles. There are three cases to consider.

Given One Side and One Acute Angle

Example 29:1

In figure 29.1, find to 2 decimal points
(a) AB (b) AC

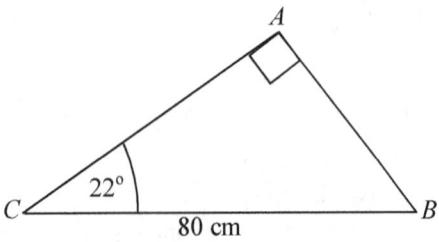

Figure 29:1

Solution

(a) $AB = 80 \sin 22 = 80(0.3746) = 29.97$ cm

(b) $AC = 80 \cos 22 = 80(0.9272) = 74.18$ cm

Example 29:2

In the triangle ABC, angle BAC is equal to $23°35'$. Given that $BC = 60$ cm, find
(a) AB (b) AC.

Solution

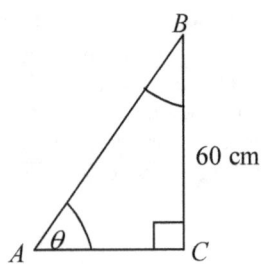

Figure 29:2

(a) $AB = \dfrac{BC}{\sin \theta}$, where $\theta = 23°35'$

$\Rightarrow AB = \dfrac{60}{\sin 23°35'} = \dfrac{60}{0.4}$ cm $= 150$ cm

(b) $AC = \dfrac{BC}{\tan \theta}$, where $\theta = 23°35'$

$\Rightarrow AC = \dfrac{60}{\tan 23°35'} = \dfrac{60}{0.4369}$ cm $= 136$ cm

In solving problems involving triangles, there are often a number of alternative methods as demonstrated below.

$AB = BC$ cosec $23°35' = 60(2.5) = 150$ cm
$AC = BC$ cot $23°35' = 60(2.2937) = 137.3$ cm
AB and AC can be found using $\angle B$ as follows.
$\angle B = 90° - \theta = 90° - 23°35' = 66°25'$.
$AB = BC$ sec $66°25' = 60(2.5) = 150$ cm
$AC = BC$ tan $66°25' = 60(2.29) = 137.4$ cm

EXERCISE 29:1

1. Find the length of the two unknown sides of the triangles in Figure 29:3 (a) and (b) to two decimal places.

(a)

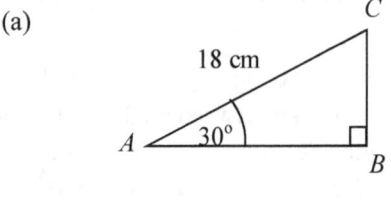

(b)

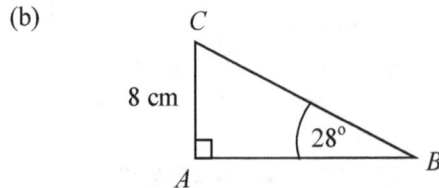

Figure 29:3

2. Find to the nearest degree each acute angle of a triangle whose sides are in the ratio;
    (a) 5:12:13   (b) 8:15:17
    (c) 7:24:25   (d) 11:60:61

3. A ladder leans against a vertical wall and makes an angle of $70°$ with the horizontal ground. The foot of the ladder is 3 m from the wall. How far up the wall to the nearest m does the ladder reach?

4. Figure 29:4 shows the cross-section of roof. How long must the carpenter cut the plank support BN? What is the length of the base AC?

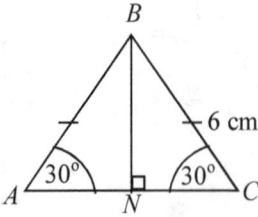

Figure 29:4

5. A road rises 105 m for every x m of the horizontal. If the angle between the slope and the horizontal is $6°$, find the value of x.

6. The roof of a bamboo hurt is a triangular prism. If the altitude of the roof from the ceiling is, 3 m and the equal slanting sides make each an angle of 30° with the horizontal. Calculate:
   (a) The length of the slanting sides
   (b) The angle made by the slanting sides at the upper edge.
   (c) The length of the base between the two slanting sides.

7. A plane takes off at A and ascends at a fixed angle of 22° with the level ground. After it flies 3000 m, find to the nearest metre;
   (a) The altitude of the plane
   (b) The distance covered horizontally.

Given Two Sides of a Right-Angled Triangle

When two sides of a right-angled triangle are given, the third side can be found using the Pythagoras theorem. The angles can be found using the trig ratios of the sides.

Example 29:3
Given the triangle in Figure 29:5, find
(i) a, leaving your answer in surd form.
(ii) Angle A (iii) Angle C

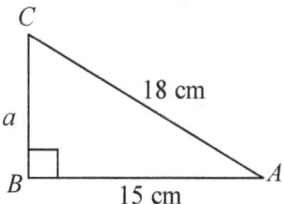

Figure 29:5

Solution
(i) Using the Pythagoras theorem,
$$a^2 = 18^2 - 15^2 \Rightarrow a = 3\sqrt{11} \text{ cm}$$
(ii) $\cos A = \dfrac{15}{18} \Rightarrow A = \cos^{-1} 0.55278 = 33.6°$
(iii) Angle $C = 90 - 33.6 = 56.4°$

Example 29:4
Given that $\hat{B} = 90°, a = 8.1 \text{ cm}$ and $c = 7.5 \text{ cm}$.
Find (a) angle A (b) angle C (c) AC

Solution

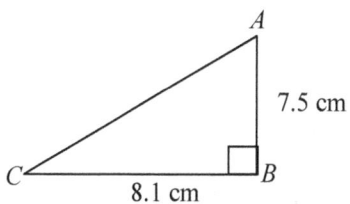

Figure 29:6

(a) $\tan A = \dfrac{8.1}{7.5} \Rightarrow \angle A = \tan^{-1} 1.08 = 47.2°$
(b) $\angle C = 90 - 47.2 = 42.8°$
(c) By the Pythagoras theorem,
$$AC = \sqrt{8.1^2 + 7.5^2} = 11.04 \text{ cm}$$

EXERCISE 29:2

1. In Figure 29:7 (a) to (f) find,
   (i) The unknown side leaving your answer in surd form.
   (ii) The angles X and Y to 1 decimal place.

(a)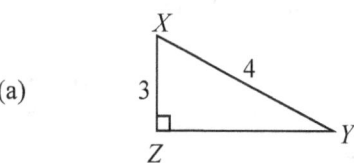

(b)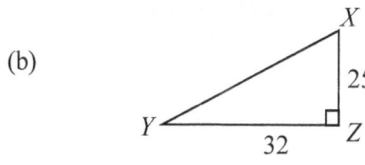

(c)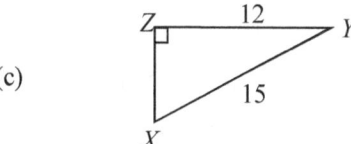

(d)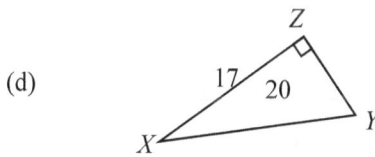

(e)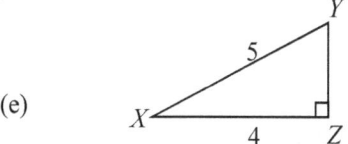

(f)

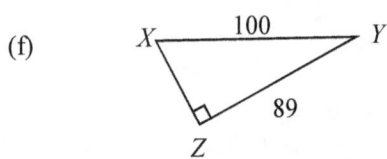

Figure 29:7

2. A helicopter flew 70 km east from A to C. It flew 100 km north to B. Find the angle of turn, to the nearest degree that it must make at B to return to A.

3. In Figure 29:8, find the side marked y and determine the values of $\angle X$ and $\angle Z$

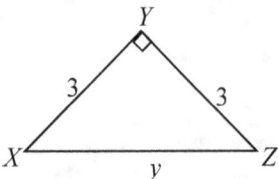

Figure 29:8

Given an Isosceles Triangle with Given Sides

To solve an isosceles triangle with given sides, draw a perpendicular from the vertex between the equal sides to the opposite side.

Example 29:5``

Find the angles of the triangle ABC with $AB = 9.6$ cm and $AC = BC = 8.2$ cm .

Solution

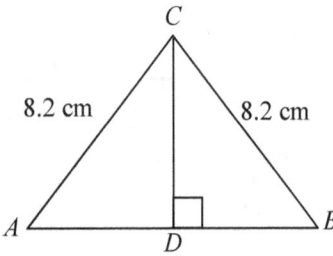

Figure 29:9

Since CD bisects AB,

$$AD = \frac{9.6}{2} = 4.8 \text{ cm}$$

From $\triangle ADC$ $\cos A = \dfrac{4.8}{8.2}$

$$\Rightarrow \angle A = \cos^{-1} \frac{4.8}{8.2} = 54°10', \angle B = 54°10'$$

and $\angle C = 180° - 2(54°10') = 71°40'$

---

1. In Figure 29:10, find AC to 3 significant figures.

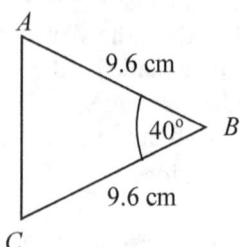

Figure 29:10

2. An isosceles triangle ABC has $AB = BC = 6$ cm and a base 4 cm long. Find the altitude of the triangle, leaving your answer in surd form. Also find the angles of the triangle to 3 significant figures.

3. Given that, all the triangles in Figure 29:11(a) to (c) are isosceles, find the perpendicular distance from B to AC, leaving your answer in surd form.   Also find the angles A, B and C of the triangles.

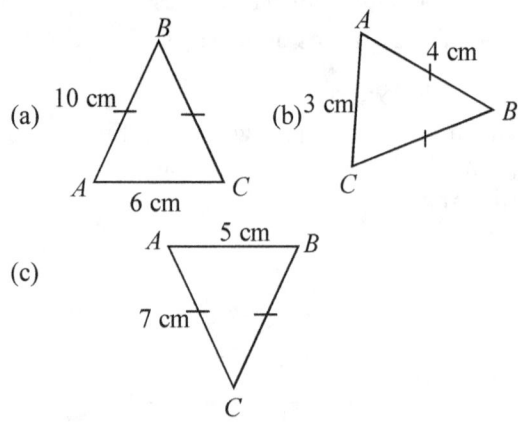

Figure 29:11

4. The base angle of an isosceles triangle is $28°$ and each arm is 45 cm. Find to the nearest cm
   (a) The altitude of the line drawn to the base
   (b) The length of the base

5. An isosceles triangle ABC has $AB = AC = 8$ cm, and $BC = 5$ cm. Calculate the size of its angles.

ANGLES OF ELEVATION AND DEPRESSION

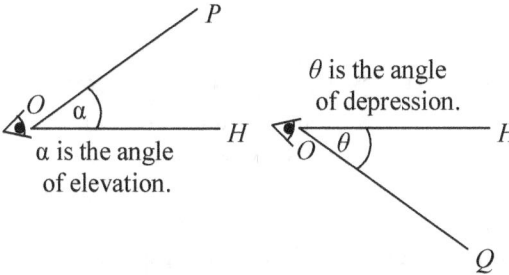

Figure 29:12

The angles of elevation and depression give the direction of one point with respect to another in a vertical plane. Thus α is the angle of elevation of P with respect to the horizontal OH, while θ is the angle of depression of Q with respect to the horizontal OH.

Example 29:6

The angle of elevation of the top of a house is $30°$ to a man lying 11 m away from the house. Calculate the height of the house.

Solution

Let OT be the tree of height h and OM the distance of the man from the tree. Then from

Figure 29:13, $h = 11 \tan 30° = \dfrac{11\sqrt{3}}{3}$ m

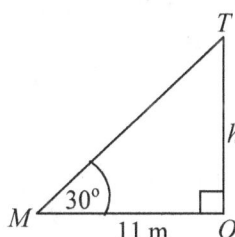

Figure 29:13

Example 29:7

A road surveyor measures the difference in altitude of two points to be 2 m. If the distance between these two points is 12 m, calculate the angle of elevation of the higher point from the lower one. Hence, find the angle of depression of the lower point from the higher one.

Solution

Let the lower point be P, the higher point be Q and the difference in altitude between the points be OQ as shown in Figure 29:14. Then the angle θ of elevation of Q from P is equal

to the angle α of depression of P from Q [alternate angles between parallel lines].

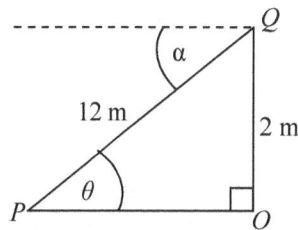

Figure 29:14

$\sin \theta = \dfrac{2}{12} = \dfrac{1}{6} \Rightarrow \theta = \sin^{-1}\left(\dfrac{1}{6}\right) \approx 9.6°.$ Hence $\alpha = 9.6°$

Example 29:8

Two points A and B 15 m apart are in line and on the same level with a tree. If the angles of elevation of the top of the tree from the points A and B are $16°$ and $25°$ respectively, calculate the height of the tree and the distance of each of the points from the tree.

Solution

Let the height OT of the tree be h.
The point B with the greater angle of elevation is closer to the tree than the point A.

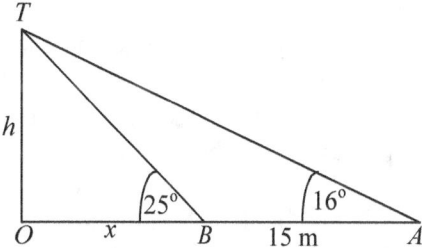

Figure 29:15

$$h = x\tan 25° \ldots\ldots\ldots\ldots\ldots\ldots\ldots ①$$
$$h = (x+15)\tan 16° \ldots\ldots\ldots\ldots ②$$
$$② - ①: \ (x+15)\tan 16° - x\tan 25° = 0$$

$$x = \frac{15 \tan 16°}{\tan 25° - \tan 16°} = \frac{15(0.29)}{0.47 - 0.29} = 23.95$$

Therefore, $h = 23.95 \tan 25° = 11.17$ m.
Hence, distance of B from the tree $= 23.95$ m and that of $A = 23.95 + 15 = 38.95$ m.

EXERCISE 29:4

1. The angle of elevation of the top of a pole from a point 50 m (away from the top) is 21°. Find to the nearest metre.
   (a) The height of the pole
   (b) The distance between the pole and the point

2. An electrician climbed on a straight pole 13 m tall and sighted his bag at an angle of depression of 30°. Calculate,
   (a) The distance of the bag from the bottom of the pole.
   (b) The distance of the bag from the face of the electrician.

3. From the top of a cliff 18 m high, the angle of depression of a fish in the sea is 24°. Find to the nearest metre the distance of the fish from the bottom of the cliff.

4. An observer at the top of a hill 75 m above the level of a lake sighted two fishermen directly in line. Find to the nearest metre the distance between the fishermen, if the angles of depressions noted by the observer were:
   (a) 20° and 15°   (b) 35° and 24°
   (c) 9° and 6°

5. The angle of elevation of the sun at a certain time is 42°. Calculate to the nearest metre;
   (a) The height *h* of a tree whose shadow is 25 m long.
   (b) The shadow of the tree along the level ground if its height is 35 m.

6. Standing on a platform 200 m away from a building, Ndoh found that the angle of elevation of the top of a building is 21°. Given that, the heights of the platform and that of Ndoh are 3.3 m and 1.7 m respectively. Find the height of the building to the nearest metre.

7. Sitting at the top of a tree 20 m tall, a boy sighted a bird and a goat directly beneath the bird. Given that, the angle of elevation of the bird is 25° and the angle of depression of the goat is 32°. Find to the nearest metre;
   (a) The distance of the goat from the tree
   (b) The height of the bird above the ground

8. A man standing at the top of a telephone transmitter pole 50 m tall, sights two cars on a horizontal road. Find the distance between the two cars if the angles of depression noted by the man were 21° and 36°.

BEARINGS IN TWO DIMENSIONS

Bearings deal with the angular direction of one point from another. There are main two ways of expressing the bearing of one point from another. These are the cardinal point or compass bearing and the three digit bearings.

Cardinal Points or Compass Bearing

This method uses the four cardinal points North, South, West and East. In this method bearings are measured from North or South to East or West. Notice that the angle from north to east or north to west is only 90°. Similarly from south to east or south to west is only 90°. Thus N 35° E means an angle of 35° from the North towards the East, S 42° W means an angle of 42° from the South towards the West.

Three Digit Bearings

Conventionally and for analytical purposes bearings are measured in the clockwise direction from the north which is considered 0°. In this method N 35° E will be written as 035°, S 30° W will be written as 210°. By this convention, all bearings are given to three digits. This minimizes errors. Thus, east will be represented by 090°, South West by 225°, North West by 315°, west by 270° etc.

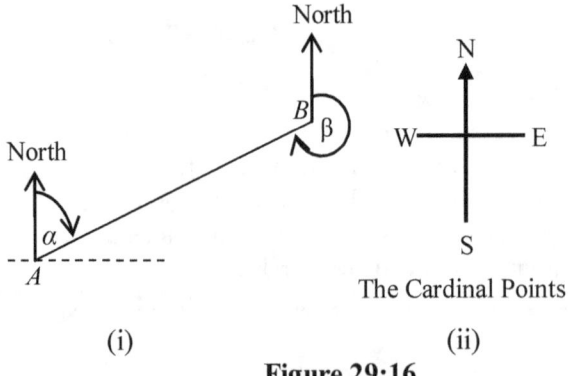

(i)   (ii)

Figure 29:16

Three digit bearings can be converted to compass bearings and vice versa.

Example 29:9
Convert the following 3 digit bearings to

compass bearings.
(a) 075°   (b) 324° (c) 138°   (d) 249°

Solution

(a)
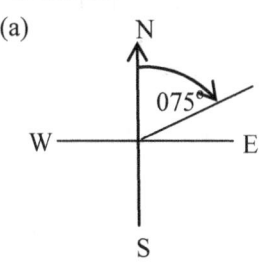

(b)
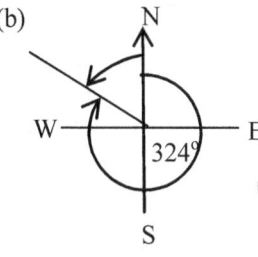

$075° = N 75° E$

$324° = 360° - 324°$
$\quad\quad = N 36° W$

(c)
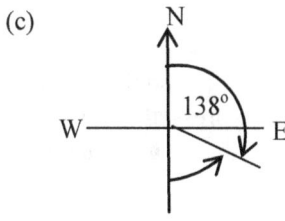

(d)
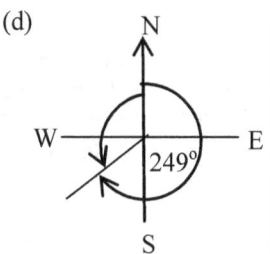

Figure 29:17

$138° = 180° - 138°$
$\quad\quad = S 42° E$

$249° = 270° - 249°$
$\quad\quad = S 21° W$

Example 29:10
Convert the following compass bearings 3 digit bearings to (a) N 85° E   (b)   S 28° W

Solution

(a)
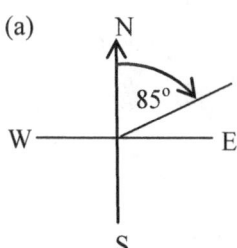

(b)
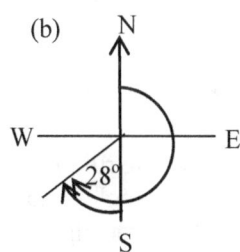

N 85° E = 085°

S 28° W =180° + 28°
$\quad\quad\quad = 208°$

Figure 29:18
In Figure 29:16(i) above, α is the bearing of the point B from the point A, and β is the bearing of the point A from the point B. Thus if the bearing of a point B from another point A is θ, then the bearing of the point A from B will be $(180° + \theta)$.

Example 29:11
A boy starts from a point A and moves on a

bearing of 020° to a point B, which is 5 km away from A. He then changes his course to a bearing of 110° and moves to a point C, which is 12 km from B. Find the distance and bearing of the point C from the point A, giving your answer to one decimal place.

Solution

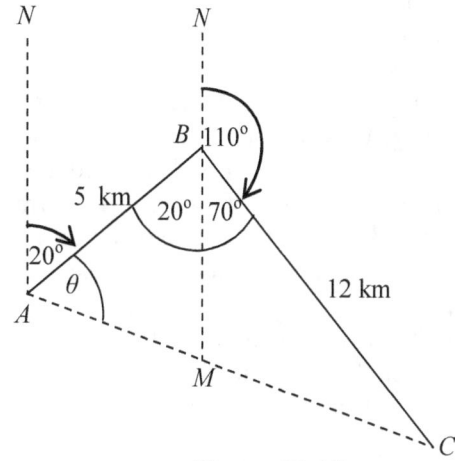

Figure 29:19
From Figure 29:19,
$\angle ABM = 20°$ (alternate angles)
$\angle CBM = 180° - 110°$ (angles on straight line)
$\quad\quad\quad = 70°$
$\angle ABC = 70° + 20° = 90°$
Using the Pythagoras theorem,
$$AC = \sqrt{AB^2 + BC^2} = \sqrt{5^2 + 12^2} = \sqrt{169} = 13 \text{ m}$$

$$\sin\theta = \frac{12}{13} \Rightarrow \theta = \sin^{-1}\left(\frac{12}{13}\right) \approx 67.38°$$

Bearing of C from A = $20° + 67.38° \approx 87.4°$, and the distance of C from A is 13m.

EXERCISE 29:5

1. Anye walks 3 km from a point P on a bearing of 023°. He then walks 4 km on a bearing of 113° to a point Q. Calculate
   (a) The distance of Q from P
   (b) The bearing of Q from P
2. From a point A, a boy walks 2 km on a bearing of 017°. He then walks 3 km on a bearing of 107° to C. What is the distance and bearing of C from A?
3. A mail van travels from its head office in

Yaounde to a town B 210 km away on a bearing of 055° to deliver a mail. It then changes course and moves to its branch office in Yaounde on a bearing of 220°. If the branch office is directly east of the head office, calculate correct to 3 significant figures
(i) The distance between the branch office and the head office;
(ii) How far is town B from the branch office?

4. A King sent two messengers to deliver a message to two of his notables Shey and Nformi. Shey lives 20 km away from the palace on a bearing of 205° while Nformi lives 15 km from the palace on a bearing of 060°. Calculate to the nearest whole number
   (a) The distance apart of Nformi's home from Shey's home.
   (b) The bearing of Nformi's home from Shey's home.

5. In Figure 29:20, $|XY|$ = 8 cm, $|YZ|$ = 13 cm, the distance of Y from X is 050° and the bearing of Z from Y is 130°.
   (a) Calculate correct to 3 significant figures,
   (i) $|XZ|$ (ii) the bearing of Z from X.
   (b) (i) Calculate the shortest distance between point Y and XZ.
   Hence, find the area of triangle XYZ.

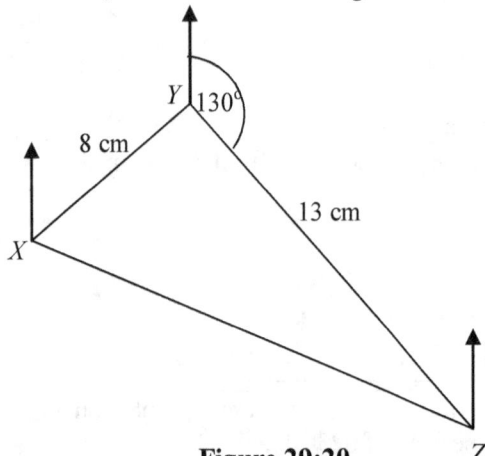

Figure 29:20

6. Three hills P, Q and R are such that Q is 24 km away from P on bearing of 150° and R is 32 km away from P, and P is on a bearing of 015° from Q. Calculate to one decimal place,
   (i) The distance between the hills Q and R.
   (ii) The bearing of hill R from hill Q.

SINE AND COSINE FORMULAE

Sometimes problems in trigonometry involve triangles that are not right-angled. In such a case, the sine or the cosine formulae (sometimes referred to as the sine and cosine rules) are used.
Consider the triangles in Figure 29:21.

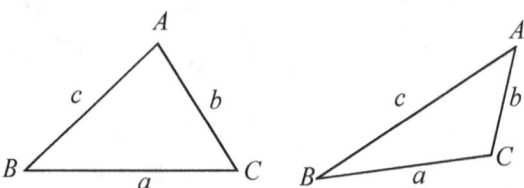

Figure 29:21

Each side of a triangle is proportional to the opposite side. This fact is expressed by the **sine formula**, which states that in any triangle

$$\frac{a}{\sin A} = \frac{b}{\sin B} = \frac{c}{\sin C}$$

This form is used when a side is required. When an angle is required, the following equivalent form is used

$$\frac{\sin A}{a} = \frac{\sin B}{b} = \frac{\sin C}{c}$$

The **cosine formula** states that in any triangle

$$a^2 = b^2 + c^2 - 2bc \cos A$$
$$b^2 = a^2 + c^2 - 2ac \cos B$$
$$c^2 = a^2 + b^2 - 2ab \cos C$$

The sine rule is used when:
(i) Two angles and one side opposite one of the given angles are given.
(ii) Two sides and an angle opposite one of the given sides are given.

The cosine rule is used when:
(i) Two sides and the included angle are given.
(ii) All the three sides are given.

Example 29:12
In Figure 29:22, find AC and AB.

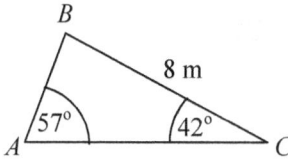

Figure 29:22

Solution

$$\angle B = 180 - (57 + 42) = 81°$$

Using the sine rule,

$$\frac{a}{\sin A} = \frac{b}{\sin B} = \frac{c}{\sin C}$$

$$\Rightarrow \frac{c}{\sin 42} = \frac{b}{\sin 81} = \frac{8}{\sin 57}$$

$$\Rightarrow b = \frac{8\sin 81}{\sin 57} = 9.4 \text{ m and } c = \frac{8\sin 42}{\sin 57} = 6.4 \text{ m}$$

Example 29:13
In Figure 29:23, find *BC*.

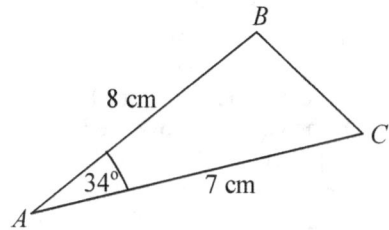

Figure 29:23

Solution

$$a^2 = b^2 + c^2 - 2bc\cos A$$

$$a^2 = 7^2 + 8^2 - 2(7)(8)\cos 34°$$

$$\Rightarrow a = \sqrt{20.15} = 4.5 \text{ cm}$$

Sometimes both the sine and the cosine rules are required to solve some problems.

Examples 29:14
In Figure 29:24, calculate
(a) *AC*   (b) $\angle BAC$ (c) $\angle ACB$

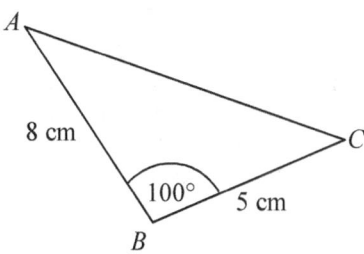

Figure 29:24

Solution
(a) By the cosine rule,

$$b^2 = a^2 + c^2 - 2ac\cos B$$

$$b^2 = 5^2 + 8^2 - 2(5)(8)\cos 100°$$

$$\Rightarrow b = \sqrt{102.89} = 10.14 \text{ cm}$$

(b) By the sine rule,

$$\frac{\sin A}{5} = \frac{\sin 100°}{10.14}$$

$$\Rightarrow \sin A = \frac{5\sin 100°}{10.14} = 0.4856$$

$$\Rightarrow \angle BAC = \sin^{-1} 0.4856 = 29.1°$$

(c) $\angle ACB = 180 - (100 + 29.1°) = 50.9°$

EXERCISE 29:6

1. In Figure 29:25, find to 1 decimal place *BC* and *AC*.

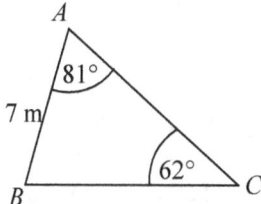

Figure 29:25

2. A ship steams 3 km from a port *P* on a bearing of 080° and then 4 km on a bearing of 047°. Find to 1 decimal place its distance and bearing from *P*.

3. A ship sails 2 km due north and then 3 km on a bearing of 060°. Find its distance and bearing from the original position.

4. A ship sails 110 km from a port *X* on a bearing of 035° and then 116 km on a bearing of 105° to another point *Y*. Calculate to 1 decimal place:
    (a) the distance *XY*   (b) the bearing of *Y* from *X*.

5. Figure 29:26 shows the position of two schools *G* and *H* the inspectorate *I*. G is 8 km on a bearing of 290° from *I* and G is from *H* 10 km on a bearing of 050°.

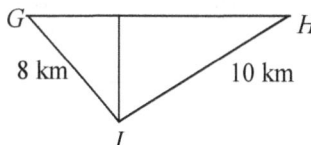

Figure 29:26

(a) Determine ∠*GIH*
(b) Calculate correct to 1 decimal place,
 (i) The distance of *GH*.
 (ii) The bearing of *H* from *G*.

MULTIPLE CHOICE EXERCISE 29

1. In the triangle shown in Figure 29:26 the length of the side marked *x* is:
 [A] 7sin 56° [B] 7tan 56°
 [C] 7tan 34° [D] 7cos 34°

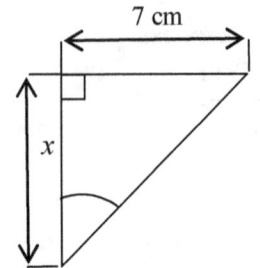

Figure 29:26

2. In Figure 29:27, *AC* = 42 cm, *AD* = 12 cm and *BD* = 16 cm. As a vulgar fraction tan *A* is:
 [A] $\dfrac{4}{5}$ [B] $\dfrac{3}{5}$ [C] $\dfrac{5}{6}$ [D] $\dfrac{4}{3}$

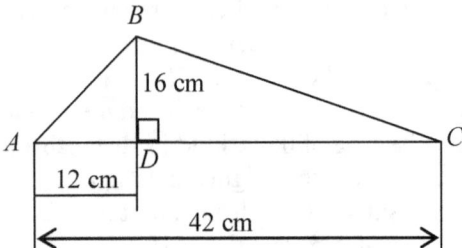

Figure 29:27

3. In Figure 29:27, *AC* = 42 cm, *AD* = 12 cm and *BD* = 16 cm. Sin *A* as a decimal fraction is:
 [A] 0.8 [B] 1.3 [C] 1.6 [D] 0.6

4. In Figure 29:27, *AC* = 42 cm, *AD* = 12 cm and *BD* = 16 cm. The size of *C* to the nearest degree is:
 [A] 27° [B] 32° [C] 28° [D] 58°

5. In Figure 29:27, *AC* = 42 cm, *AD* = 12 cm and *BD* = 16 cm. The perimeter of triangle *BDC* in cm is:
 [A] 70 [B] 80 [C] 90 [D] 92

6. *ABCD* is a trapezium (Figure 29:28).

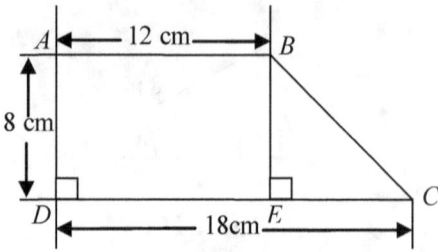

Figure 29:28

Angle *D* is a right angle and *BE* is perpendicular to *DC*. *AB* = 12 cm and *AD* = 8 cm. The value of *DC* =18 cm. The value of tan *C* is:
 [A] $\dfrac{4}{9}$ [B] $\dfrac{2}{3}$ [C] $\dfrac{4}{3}$ [D] $\dfrac{4}{5}$

7. A ladder 9 m long leans against a vertical wall, making an angle of 64° with the horizontal ground. To one decimal place, the distance of the foot of the ladder from the wall is:
 [A] 3.9 m [B] 5.8 m [C] 7.9 m [D] 8.1 m

8. When helicopter is 800 m above the ground, its angle of elevation from a point *P* on the ground is 30°. The distance of the helicopter from *P* by the line of sight is:
 [A] 400 m [B] 800 m
 [C] 1600 m [D] 1700 m

9. The angle of elevation of *X* from *Y* is 30°. If *XY* = 40 m the height of *X* above the level of *Y* is:
 [A] 10 m [B] 20 m [C] 40 m [D] 50 m

10. If the shadow of a pole 7 m high is equal to half its length, the angle of elevation of the sun; correct to the nearest degree is:
 [A] 63° [B] 0° [C] 60° [D] 26°

11. The angle of elevation of the top of a tree 39 m away from the point on the ground is 30°. The height of the tree is:
 [A] $39\sqrt{3}$ m [B] $13\sqrt{3}$ m
 [C] $\dfrac{13}{\sqrt{3}}$ m [D] $\dfrac{\sqrt{3}}{13}$ m

12. A ladder leans against the wall at an angle 60° to the wall. If the foot of the ladder is 5 m away from the wall, the length of the ladder is:
 [A] $\dfrac{5\sqrt{3}}{3}$ m [B] 5 m
 [C] $5\sqrt{3}$ m [D] $\dfrac{10\sqrt{3}}{3}$ m

13. The angle of elevation of the top of a tower from a point on the horizontal ground, 40 m away from the foot of the tower is 30°. The height of the tower is:
 [A] 20 m [B] $\dfrac{40\sqrt{3}}{3}$ m
 [C] $20\sqrt{3}$ m [D] $40\sqrt{3}$ m

14. At a point 500 m from the base of a water-tank, the angle of elevation of the top of the tank is 45°. The height of the tank is:
    [A] 500m [B] 353 m [C] 354 m [D] 250 m

15. A ladder 6 m long leans against a vertical wall, so that it makes an angle of 60° with the wall. The distance of the foot of the ladder from the wall is:
    [A] 3 m [B] 6 m
    [C] $2\sqrt{3}$ m [D] $3\sqrt{3}$ m

16. The angle of elevation of the top X of a vertical pole from a point P on a level ground is 60°. The distance from P to the foot of the pole is 55 m. Without using tables, the height of the pole is:
    [A] $\dfrac{50}{3}$ m [B] 50 m
    [C] $55\sqrt{3}$ m [D] 60 m

17. The angle of elevation of a point T on a tower from a point U on the horizontal ground is 30°. If $TU = 54$ m, the height of T above the ground is:
    [A] 108 m [B] 72 m [C] 31.2 m [D] 27 m

18. In Figure 29:29, $\angle PQR$ is a right angle $|QR| = 2$ cm and $\angle PRQ = 60°$. $|PR|$ is equal to:
    [A] 1 m [B] 4 m [C] $2\sqrt{3}$ m [D] $\dfrac{4}{3}$ m

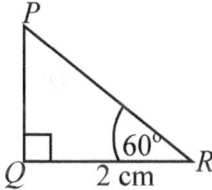

Figure 29:29

19. A ladder 5 m long rest against a wall so that it foot makes an angle of 30° with the horizontal. The distance of the foot of the ladder from the wall is:
    [A] $\dfrac{5\sqrt{3}}{3}$ m [B] $2\dfrac{1}{2}$ m
    [C] $\dfrac{5\sqrt{3}}{2}$ m [D] $\dfrac{10\sqrt{3}}{3}$ m

20. A pole of length l leans against a vertical wall so that it makes an angle of 60° with the horizontal ground. If the top of the pole is 8 m above the ground, l must be:
    [A] $16\sqrt{3}$ m [B] $\dfrac{\sqrt{3}}{16}$ m
    [C] 16 m [D] $\dfrac{16\sqrt{3}}{3}$ m

21. The angle of depression of a point on the ground from the top of a building is 20.3°. If the distance of the point from the foot of the building is 40 m, the height of the building, correct to one decimal place is:
    [A] 13.9 m [B] 28.1 m
    [C] 27.8 m [D] 14.8 m

22. From the top of a cliff 20 m high, a boat can be sighted at sea 75 m from the foot of the cliff. The angle of depression of the boat from the top of the cliff is:
    [A] 14.9° [B] 15.5° [C] 74.5° [D] 75.1°

23. From the top of a building 10 m high, the angle of depression of a stone lying on the ground is 69°. Correct to one decimal place, the distance of the stone from the foot of the building is:
    [A] 3.6 m [B] 3.8 m [C] 6.0 m [D] 9.3 m

24. A cliff on the bank of a river is 300 metres high. If the angle of depression of a point on the opposite side of the river is 60°, the width of the river is:
    [A] 100 m [B] $75\sqrt{3}$ m
    [C] $100\sqrt{3}$ m [D] $200\sqrt{3}$ m

25. From the top of a cliff, the angle of depression of a boat on the sea is 60°; if the top of the cliff is 25 m above the sea level, the horizontal distance from the bottom of the cliff to the boat is:
    [A] $\dfrac{\sqrt{3}}{25}$ m [B] $25\sqrt{3}$ m
    [C] $\dfrac{25\sqrt{3}}{3}$ m [D] $\dfrac{25}{3}$ m

26. The angle of depression of a point Q from a vertical tower PR, 30 m high, is 40°. If the foot P of the tower is on the same level ground as Q, correct to two decimal places, $|PQ|$ should be:
    [A] 35.75 m [B] 25.00 m
    [C] 22.98 m [D] 19228 m

27. In Figure 29:30, AB is a vertical pole and BC is horizontal. If $|AC| = 10$ m and $|BC| = 5$ m. The angle of depression of C from A is:

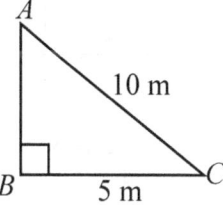

Figure 29:30

    [A] 63° [B] 60° [C] 45° [D] 30°

28. In Figure 29:31, QRT is a straight line. If $\angle PTR = 90°$, $\angle PRT = 60°$, $\angle PQR = 30°$ and

$|PQ| = 6\sqrt{3}$ m . $|RT|$ is equal to:

[A] 0.3 m [B] $\dfrac{\sqrt{3}}{2}$ m

[C] 3 m [D] $3\sqrt{3}$ m

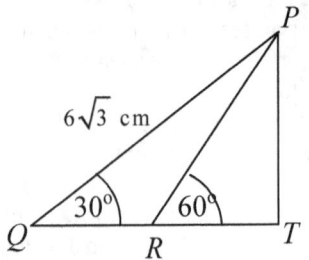

Figure 29:31

29. The true bearing of $250°$ as a compass bearing is:
   [A] S 70° E [B] N 70° E
   [C] S 70° W [D] N 70° W

30. The bearing, which is equivalent to S 50o, W is:
   [A] 230° [B] 220° [C] 130° [D] 040°

31. The bearing S 40° E is the same as:
   [A] 040° [B] 050° [C] 130° [D] 140°

32. The bearing S 50° W is the same as:
   [A] 050° [B] 130° [C] 140° [D] 230°

33. Town P is 150 km from a town Q in the direction 050°. The bearing of Q from P is:
   [A] 050° [B] 150° [C] 230° [D] 310°

34. The bearing of P from Q is x, where $270° < x < 360°$. The bearing of Q from P is:
   [A] $(x - 90)°$ [B] $(x + 180)°$
   [C] $(x - 135)°$ [D] $(x - 180)°$

35. Figure 29:32 shows the position of three ships A, B and C at sea. B is due north of C such that $|AB| = |BC|$ and the bearing of B from A is 040°. The bearing of A from C is:
   [A] 040° [B] 070° [C] 110° [D] 290°

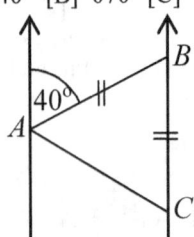

Figure 29:32

36. A tree is 8 km due south of a building. Ambe is standing 8 km west of the tree. The distance of Ambe from the building is:

[A] $4\sqrt{2}$ km [B] 8 km
[C] $8\sqrt{2}$ km [D] 16 km

37. A tree is 8 km due south of a building. Ambe is standing 8 km west of the tree. The bearing of Ambe from the building is:
   [A] 315° [B] 270° [C] 225° [D] 135°

38. Three observation posts P, Q and R are such that Q is due east of P and R is due north of Q, if $|PQ| = 5$ km and $|PR| = 10$ km, $|QR|$ equals:
   [A] 50 km [B] 9.5 km
   [C] 7.6 km [D] 8.7 km

39. From Figure 29:33, the bearing of Q from P is:
   [A] 236° [B] 214°
   [C] 146° [D] 124°

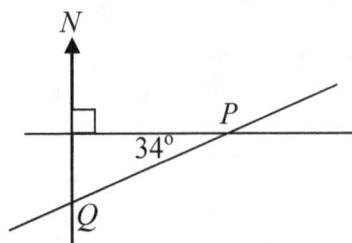

Figure 29:33

40. The bearing of a town Q from a town P is 215°. P is 80 km north of R while R is due east of Q. The distance between Q from R correct to the nearest km is:
   [A] 46 km [B] 56 km
   [C] 38 km [D] 98 km

41. From the diagram in Figure 29:34, the bearing of C from B is:
   [A] 060° [B] 090° [C] 120° [D] 240°

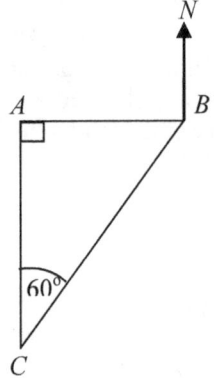

Figure 29:34

TOPIC 30
VOCABULARY OF SOLID FIGURES

OBJECTIVES

At the end of this topic, the learner should be able to:

1. Identify, draw and name different prisms by the nature of their cross-section.
2. Identify, draw and appreciate that a cylinder may be right circular cylinder, solid or open at one or both ends.
3. Identify, draw and classify pyramids according to the nature of their bases.
4. Identify and draw cones and frustums.
5. Identify and draw a tetrahedron noting that it is a solid figure with triangular faces.
6. Draw and distinguish between a sphere and a hemisphere.
7. Identify and recognise nets of various solids including cubes, cuboids, prisms, cylinders, pyramids and cones.

French mathematician Gaspard Monge began teaching physics at Collège de la Trinité at the age of 16. He invented descriptive geometry, which became the basis of mechanical drawing and architectural plans. His ideas were formally published in 1795, in his *Géométrie descriptive*. This led to a new branch of mathematics known as descriptive geometry.

Roger Viollet/Getty Images

Cylinders

A cylinder is a three dimensional figure with a uniform circular cross-section. When the curved surface of a cylinder is smooth and the two circular faces are parallel, it is described as a **right circular** cylinder and when it is closed at both ends; it is described as a **solid** cylinder. The curved surface of a cylinder is called the **lateral surface** of the cylinder.

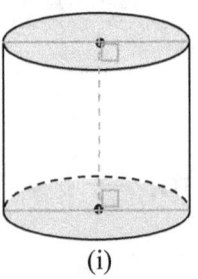 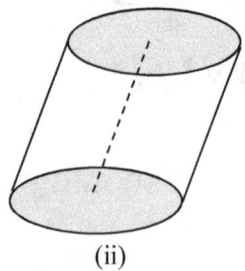

| (i) | (ii) |

Figure 30:1

A cylinder is called a right cylinder if the line connecting the centers of the two bases is perpendicular to the bases (Figure 30:1(i)); otherwise, it is called an oblique cylinder (Figure 30:1(ii)).

Prisms

A prism is a solid with uniform polygonal cross-sectional area. This definition means that the lateral surfaces prisms are made up of rectangles.

Prisms are named from the nature of their cross-sections as shown in Table 30:1 and (Figure 30:2(i) and (ii)).

Nature of cross-section	Name of Prism	Diagram of Prism
Triangle	Triangular Prism	
Square	Square Prism or cube	
Rectangle	Rectangular Prism or cuboids	
Pentagon	Pentagonal Prism	
Hexagon	Hexagonal Prism	

Table 30:1

A **right-angled triangular prism** (Figure 30:2(i))

is a prism whose cross-section is a right-angled triangle.

A **trapezoidal prism** (Figure 30:2(ii)) is a prism whose cross-section is a trapezium.

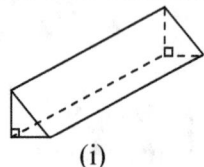 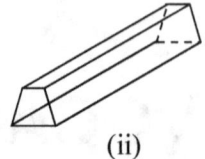

| (i) | (ii) |

Right-angled triangular prism   Trapezoidal prism

Figure 30:2

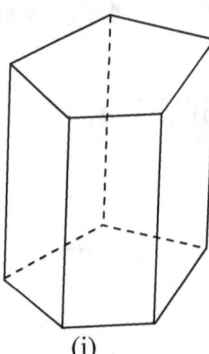 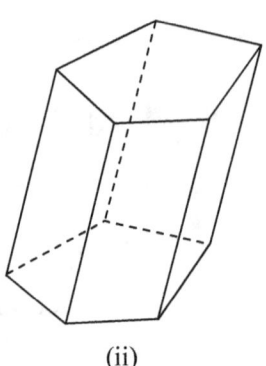

| (i) | (ii) |

A prism is called a **right prism** if the lateral surfaces are made up of rectangles (Figure 30:3(i)); otherwise, it is called an oblique prism (Figure 30:3(ii)).

Pyramids

A **pyramid** (Figure 30:4) is a solid with triangular faces and a polygonal base. Pyramids derive their names from the nature of their bases as shown in Table 30:2.

Nature of Base	Name of Pyramid
Triangle	Triangular Pyramid
Square	Square pyramid
Rectangle	Rectangular pyramid
Pentagon	Pentagonal pyramid
Hexagon	Hexagonal pyramid

Table 30:2

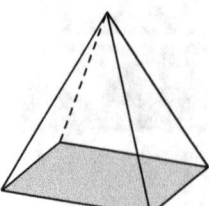 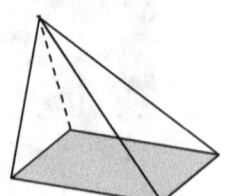

Figure 30:4

A pyramid is called a **right pyramid** if the

line connecting the apex and the center of the base is perpendicular to the base (Figure 30:4(i)); otherwise, it is called an oblique prism (Figure 30:4(ii)).

Cones

A **cone** (Figure 30:5) is a solid whose base is circular and whose lateral surface reduces gradually to the vertex.

 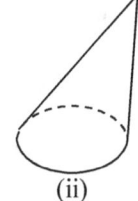
(i)   (ii)

Figure 30:5

A cone is called a **right cone** if the line connecting the apex and the center of the base is perpendicular to the base (Figure 30:5(i)); otherwise, it is called an oblique cone (Figure 30:5(ii)).

Pyramids and cones have a common property in that they both have a sharp apex.

Frustums

When the smaller end of a cone or pyramid is cut off through its cross-section by a plane parallel to the base as shown in Figure 30:6 the portion left is called a **frustum**.

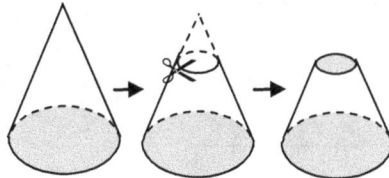

Figure 30:6

Frustums are named from the nature of their bases. Thus, the frustum in Figure 30:6 is called a conical frustum while that in Figure 30:7 is called a pentagonal pyramidal frustum.

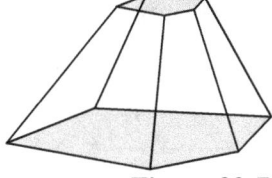

Figure 30:7

Polyhedrons

Polyhedrons are solid figures whose faces are made up of polygons. The faces of a regular polyhedron are equal in size and shape. There are only five regular polyhedrons. These are shown in Figure 30:8, with their names and number of sides.

The commonest of the five regular polyhedrons are the cube (which is also a prism) and the tetrahedron.

A tetrahedron is a solid figure with four triangular congruent faces. A tetrahedron can be looked upon as a special pyramid because it has triangular faces which are congruent (i.e. the faces have the same shape and size).

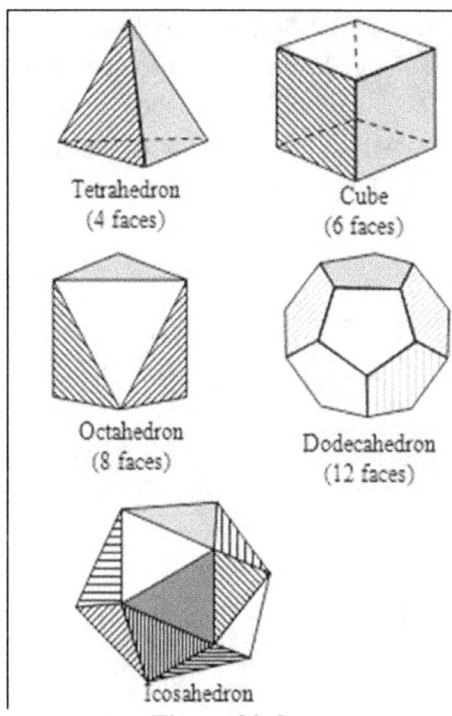

Tetrahedron
(4 faces)

Cube
(6 faces)

Octahedron
(8 faces)

Dodecahedron
(12 faces)

Icosahedron

Figure 30:8

The Sphere and the Hemisphere

A sphere (Figure 30:9(a)) is a solid in which all the points on the surface are a fixed distance from a particular fixed point called the centre of the sphere.

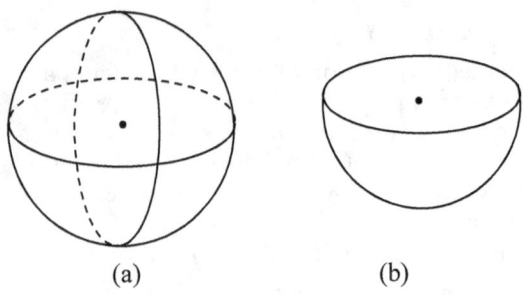

Figure 30:9

A hemisphere (Figure 30:9(b)) is half of a sphere.

Nets of Solid Figures

Consider the shapes in Figure 30:10 called **nets of the solid figures**. If each is cut and folded along, the dotted lines so that the ends meet, (i) and (ii) will form cubes, while (iii) and (iv) will form a triangular prism and a square pyramid respectively.

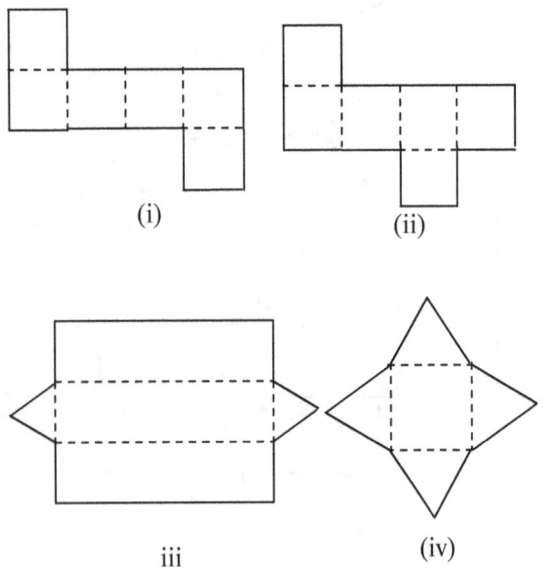

Figure 30:10

```
EXERCISE 30
```

1. How many faces bound the following?
   (i) A sphere
   (ii) A cube
   (iii) A cone
   (iv) A cylinder

(v) A square pyramid
(v) A trapezoidal prism
(vii) A tetrahedron

2. Figure 30:11, shows the net of a solid in which all the faces are equilateral triangles. If a solid is made from this net,
   (i) To which line will *AB* be joined?
   (ii) Which other point(s) will meet at the same vertex *A*?
   (iii) How many edges and vertices will the solid have?
   (iv) Suggest a name for the solid.

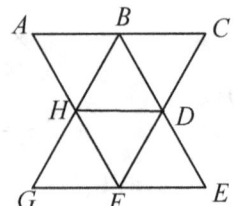

Figure 30:11

3. A square piece of paper *ABCD* is folded so that *B* falls on *D* and then it is folded so that *A* falls on *C* (Figure 30:12). The corner *BD* is now cut off along the line *MN*, at right angles to *DN*, *M* being the midpoint of the sloping side. The paper is then opened out. Outline the shape of the remaining paper.

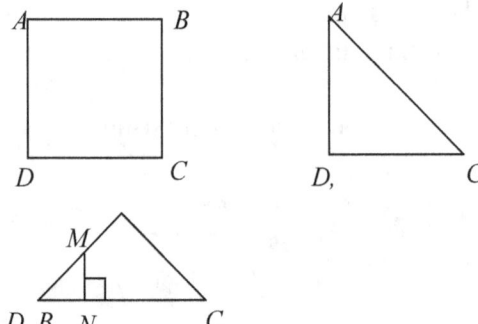

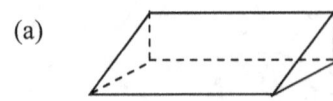

Figure 30:12

4. In Figure 30:13 (a) to (c), list the shapes that make up the net for each solid, write the number of times each shape is used and draw a net for each solid.

(a)

(b)

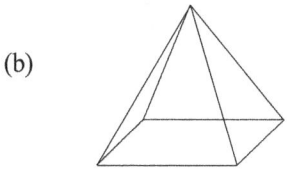

(c)

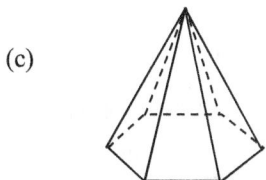

Figure 30:13

5. Explain the differences between a triangular prism and a triangular pyramid.

6. Name the solid that can be formed from each of the nets in Figure 30:14 (a) and (b).

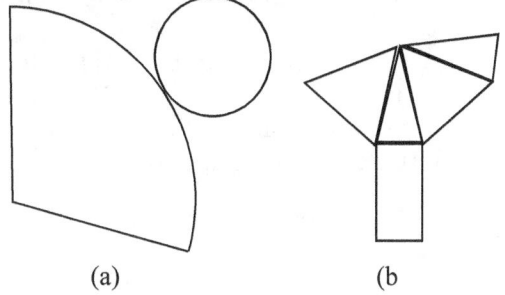

(a) (b)

Figure 30:14

7. Name the solid figure, which can be obtained from each of the nets in Figure 30:14 (a) to (d).

(a)

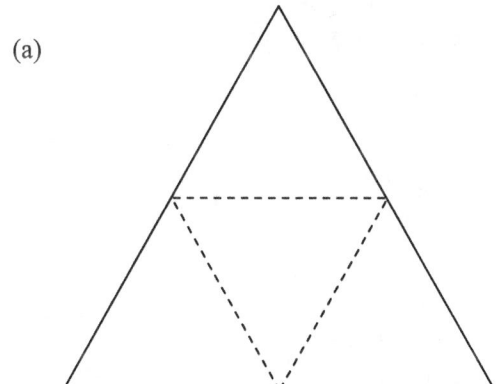

(b)

(c)

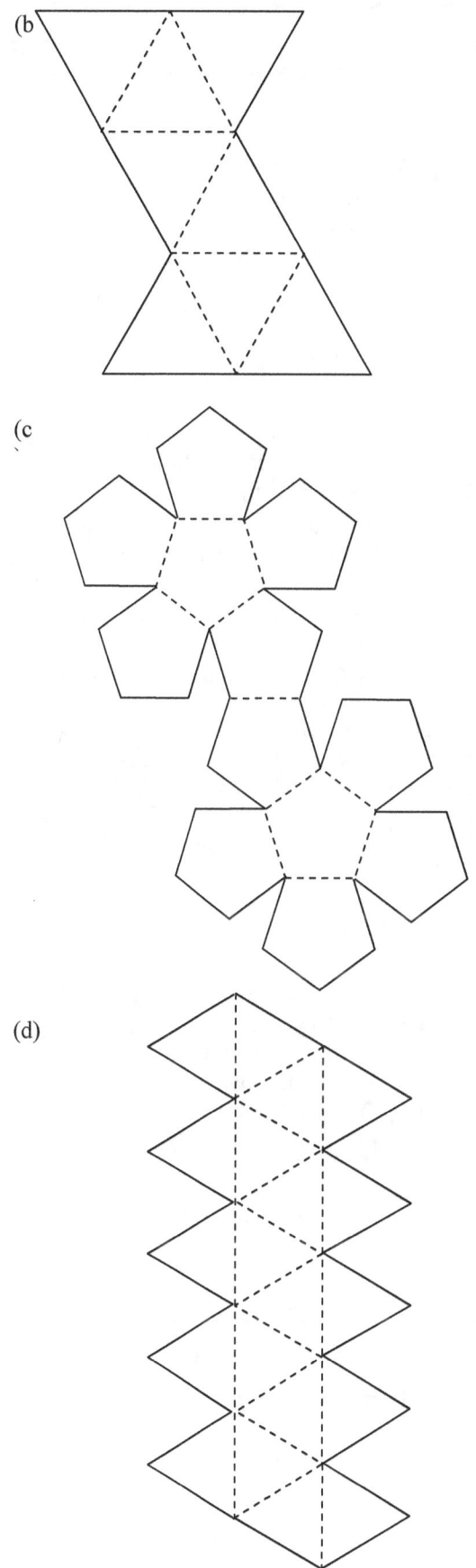

(d)

Figure 30:15

MULTIPLE CHOICE EXERCISE 30

1. The net in Figure 30:16 is a net of
   [A] A tetrahedron
   [B] a pyramid
   [C] A cone
   [D] a triangular prism

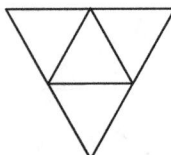

Figure 30:16

2. A prism is a solid figure with:
   [A] Regular faces
   [B] Uniform cross-sectional area.
   [C] Triangular faces
   [D] A square base and regular triangular faces.

3. In Figure 30:17 the figure that is certainly not a prism is:

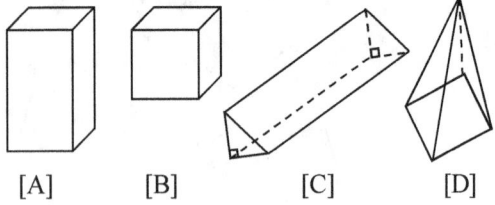

[A]   [B]   [C]   [D]

Figure 30:17

4. Figure 30:18 is called:
   [A] A rhombus
   [B] a triangular prism
   [C] A triangular pyramid
   [D] a cone

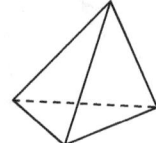

Figure 30:18

5. The solid formed by the net in Figure 30:19 is:
   [A] Hexagonal prism
   [B] rectangular pyramid
   [C] Hexagonal pyramid
   [D] rectangular prism

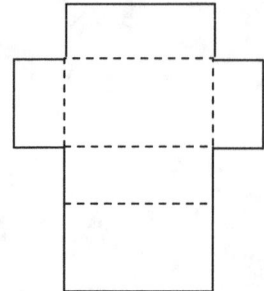

Figure 30:19

6. The figure, which has one rectangular base and four lateral triangular surfaces, is:
   [A] Square pyramid
   [B] rectangular pyramid
   [C] Cone
   [D] rectangular prism

7. A solid with two parallel and congruent bases *cannot* be:
   [A] Cone   [B] prism
   [C] cylinder   [D] cube

8. None of the diagrams in Figure 30:20 and 30:21 are drawn to scale.

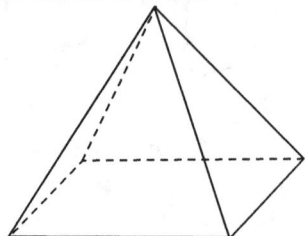

Figure 30:20

The net in Figure 30:21 which corresponds to Figure 30:20 is:

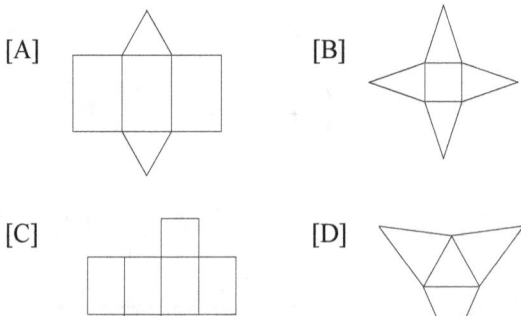

[A]   [B]

[C]   [D]

Figure 30:21

TOPIC 31

MENSURATION OF PLANE FIGURES

OBJECTIVES

At the end of this topic, the learner should be able to:

1. Define perimeter and area of a plane figure.
2. Find the perimeter and area of squares, rectangles, parallelograms, triangles, trapeziums and rhombuses.
3. Find the areas of parallelograms and triangles which have a common base and are between two parallels lines.
4. Define and find circumference and area of a given circle.
5. Calculate the arc length, area of a sector and area of a segment of a circle.
6. Find the perimeters/areas of composite plane figures made up of identifiable plane figures.
7. Find the area of any plane figure by counting squares.

Archimedes of Samos (287-212 BC), famous Greek mathematician and inventor, who wrote important works on plane and solid geometry, arithmetic, and mechanics and anticipated many of the discoveries of modern science, such as the integral calculus, through his studies of the areas and volumes of curved solid figures and the areas of plane figures. He also proved that the volume of a sphere is two-thirds the volume of a cylinder that circumscribes the sphere. Among his works on mathematics that survive, is *Measurement of the Circle, Spirals,* and *Sphere and Cylinder.* They all exhibit the rigor and imaginativeness of his mathematical thinking.

Culver Pictures

The vocabulary associated with all plane figures in this topic was earlier treated in Topic 26.

Perimeter

Perimeter is the length of the distance all round a plane figure. Being a length, perimeter is measured in units such as metres (m), centimetres (cm). Perimeter is often denoted by P.

Area

Area is the amount of surface (or the number of square units) covered by a plane figure. Area is measured in square units such as square metres (m^2), square centimetres (cm^2). Area is usually denoted by A.

The Rectangle

The shorter side is called the width or the breadth denoted by w and the longer side is called the length denoted by l.

Length (l)

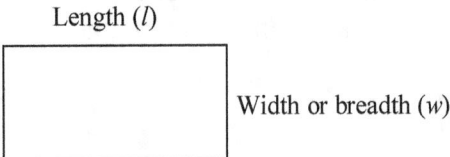

Width or breadth (w)

Figure 31:1

Perimeter and Area of Rectangle

11units

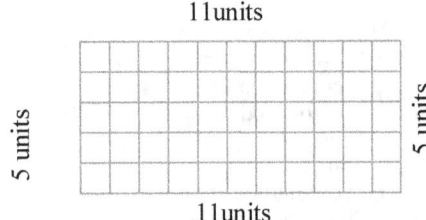

5 units   5 units

11units

Figure 31:2

Perimeter = Distance all round the rectangle
$$= (5 + 11 + 5 + 11) \text{ units}$$
$$= 32 \text{ units}$$
From the above it can be seen that for a rectangle of length l and width w,

Perimeter, $P = l + w + l + w = 2l + 2w$

$$\Rightarrow P = 2(l + w)$$

$$\Rightarrow l = \frac{P}{2} - w \text{ and } w = \frac{P}{2} - l$$

From Figure 31:2 and by counting squares
Area of rectangle = Number of square units
= 55 square units

Notice that this area could have easily been obtained by multiplying 5 units by 11 units. Therefore, for a rectangle of length l and width w, the area is given by

$$A = lw \Rightarrow l = \frac{A}{w} \text{ and } w = \frac{A}{l}$$

Example 31:1

A rectangular floor has sides 4 m by 6 m. Calculate
(a) the perimeter   (b) area of the floor

Solution

(a) $P = 2(l + w)$ (b) $A = lw$
$\quad\quad = 2(4 + 6)$ $= 4(6)$
$\quad\quad = 20 \text{ cm}$ $= 24 \text{ cm}^2$

The Square

A square is a special type of rectangle in which all the sides are equal. i.e. the length and width are equal.

l

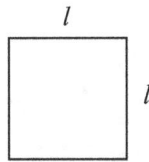

l

Figure 31:3

Perimeter of square, $P = 2(l + l)$
$\Rightarrow$ Perimeter of square, $P = 4l$.

Area of square, $A = l(l)$
$\Rightarrow$ Area of square, $A = l^2$

Example 31:2

A square has side 11 cm. Find
(a) the perimeter   (b) the area

Solution
(a) $P = 4l = 4(11) = 44 \text{ cm}$
(b) $A = l^2 = (11)^2 = 121 \text{ cm}^2$

EXERCISE 31:1

1. Find (a) the area   (b) the perimeter, of a rectangular plot with the dimensions,
   (i) 50 m by 35 m   (ii) 90 m by 60 m
   (iii) 32 m by 16 m   (iv) 20 m by 15 m

2. A rectangular carpet measures x cm by 5 cm. Find the value of x if the area is 20 cm^2.

3. The perimeter of a hall is 33.9 m. Find the length l if the width is 7.6 m.

4. A plank has an area of 1.8 m^2 and a width of 40 cm. Find its length in cm^2.

5. A square room of area 64 m^2 is extended on one side by x m. Find the value of x if the new area is 104 m^2.

6. The length of a rectangle is 3 m greater than its breadth. If the area of the rectangle is 180 square metres, calculate its length and its perimeter.

7. Find the length of a rectangle, which has an area of 99 cm^2 and a breadth 9 cm.

8. The area of a rectangle is 40 m^2 and its breadth is 5 m, find its perimeter.

9. The perimeter of a rectangle is 42 cm and its length is x cm, find its breadth.

10. Figure 31:4, shows a rectangle with its dimensions shown. Find   (a) the value of x (b) the area   (c) perimeter, given that the dimensions are in cm.

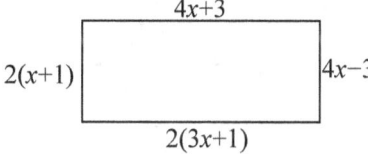

$4x+3$

$2(x+1)$ ⎸ ⎹ $4x-3$

$2(3x+1)$

Figure 31:4

11. Find (a) the area (b) the perimeter, of a square of side
    (i) 51 cm (ii) 12 cm (iii) 15 m (iv) 30 m

12. Calculate the perimeter of a square whose area is 196 cm^2.

13. Given that the dimensions of the square in Figure 31:5 are in centimetres, calculate
    (a) the area
    (b) the perimeter, of the square.

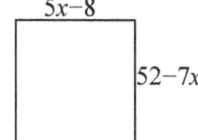

$5x-8$

$52-7x$

Figure 31:5

14. The diagonal of a square is 15 cm and the area is 108 cm^2. Calculate
    (a) the length
    (b) perimeter of the square.

The Parallelogram

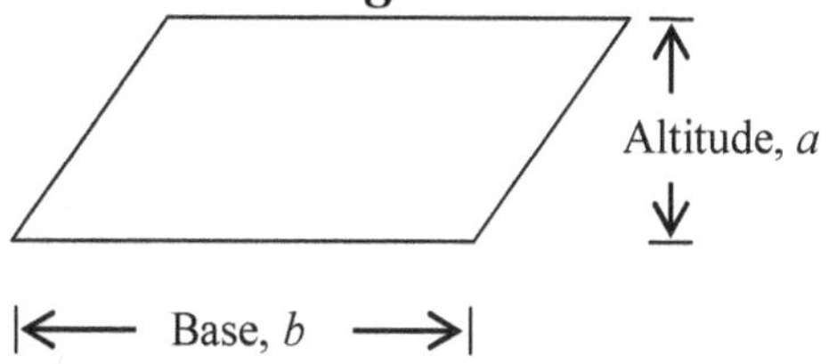

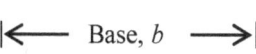

Altitude, a

|← Base, b →|

Figure 31:6

A parallelogram is a quadrilateral with opposite sides equal and parallel. Adjacent sides may or may not be perpendicular. Therefore, a rectangle is a special parallelogram with adjacent sides' perpendicular.

Area of a Parallelogram

If the shaded portion in Figure 31:7(i), is cut off and fitted to the right of the same figure, a rectangle whose length and width are b and h respectively, will be formed as in Figure 31:7 (ii). Therefore, the area of the parallelogram in Figure 31:7 (i) can conveniently be found by calculating the area of the rectangle in Figure 31:7(ii).

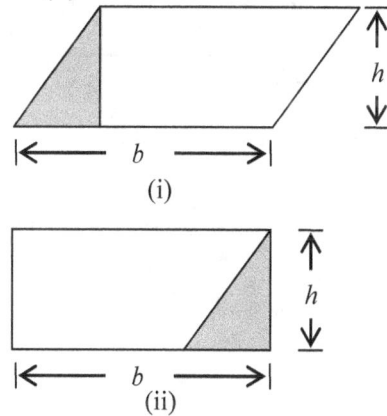

h

|← b →|

(i)

h

|← b →|

(ii)

Figure 31:7

Area of parallelogram= Area of rectangle

Area of parallelogram= base × altitude

$$A = bh \implies b = \frac{A}{h} \text{ and } h = \frac{A}{b}$$

Example 31:3

A parallelogram has an altitude of 13 cm and a base 15 cm, calculate its area.

Solution

$A = bh = (13)(15) = 195 \text{ cm}^2$

Example 31:4

Find the altitude of a parallelogram with base 9 cm and area 45 cm².

Solution

$$h = \frac{A}{b} = \frac{45}{9} = 5 \text{ cm}$$

The Triangle

A triangle is a three-sided polygon.

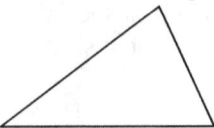

Figure 31:8

Consider the parallelogram $OABC$ (Figure 31:9), which has been cut along AC into two congruent (equal) triangles OAC, shaded and ACB, un-shaded. The area of each of the triangles is half the area of the parallelogram.

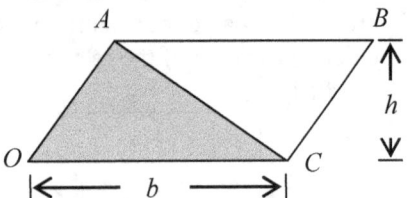

Figure 31:9

Hence,

$$\text{Area of triangle} = \frac{1}{2} \times \text{Area of parallelogram}$$

$$\Rightarrow \text{Area of triangle} = \frac{1}{2}\text{base} \times \text{height} = \frac{1}{2}bh$$

Example 31:5

A triangular lawn has a base of 8 m and a height of 4 m. What is its area?

Solution

$$A = \frac{1}{2}bh = \frac{1}{2}(8 \text{ m})(4 \text{ m}) = 16 \text{ m}^2$$

It can also be shown that the area of a triangle with sides a, b and c can be found from the following formula called Hero's formula. [The derivation of the formula is beyond the scope of this book].

$$A = \sqrt{p(p - a)(p - b)(p - c)}$$

Where, $p = \frac{1}{2}$ of the perimeter of the triangle

$$p = \frac{1}{2}(a + b + c)$$

Example 31:6

Calculate the area of a triangle whose sides are 5 cm, 7 cm and 10 cm.

Solution

$$p = \frac{1}{2}(a + b + c) = \frac{1}{2}(5 + 7 + 10) = 11 \text{ cm}$$

$$A = \sqrt{p(p - a)(p - b)(p - c)}$$
$$= \sqrt{11(11 - 5)(11 - 7)(11 - 10)}$$
$$= \sqrt{(11)(6)(4)(1)}$$
$$= \sqrt{264} = 16.25 \text{ cm}^2$$

The Rhombus

A rhombus is a parallelogram with all sides equal. It can be viewed as four congruent right-angled triangles carefully arranged as in Figure 31:10.

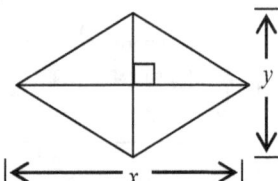

Figure 31:10

Therefore, if the diagonals of a rhombus are x and y cm, the height and base of each of the triangles will be $\frac{1}{2}x$ cm and $\frac{1}{2}y$ cm respectively.

∴ Area of rhombus = 4 × Area of triangle

$$A = 4 \times \frac{1}{2}\left(\frac{1}{2}x\right)\left(\frac{1}{2}y\right)$$

$$A = \frac{1}{2}xy$$

Area of rhombus = Half the product of diagonals.

Example 31:7

Find the area and perimeter of a rhombus

whose diagonals are 12 cm by 16 cm.

Solution

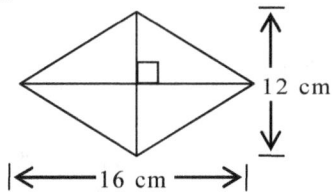

Figure 31:11

$$A = \frac{1}{2}xy = \frac{1}{2}(12)(16) = 96 \text{ cm}^2$$

From Figure 31:11, the rhombus is made up of four congruent triangles with arms 6 cm and 8 cm.

$$\therefore \text{Hypotenus} = \sqrt{6^2 + 8^2} = 10 \text{ cm}$$

$$\Rightarrow \text{Perimeter} = 4 \times 10 \text{ cm} = 40 \text{ cm}$$

Common Base Parallelograms and Triangles between Two Parallels Lines

(i) The areas of triangles, which are on the same base and between same parallels, are equal.

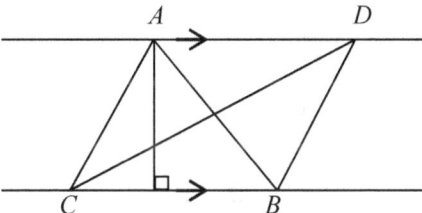

Figure 31:12

(ii) The areas of parallelograms, which are on the same base and between same parallels, are equal.

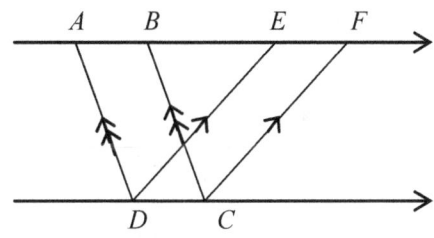

Figure 31:13

Area of *ABCD* = Area of *CDEF*

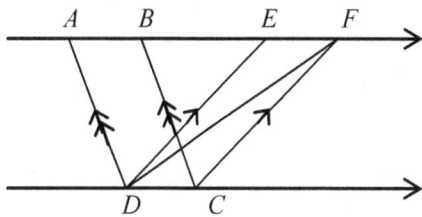

Figure 31:14

(iii) The area of any parallelogram, which is on the same base and between same parallels as a triangle, is twice the area of the triangle.

Area of *ABCD* = 2 × Area of Δ*CDF*

Figure 31:14

Example 31:8

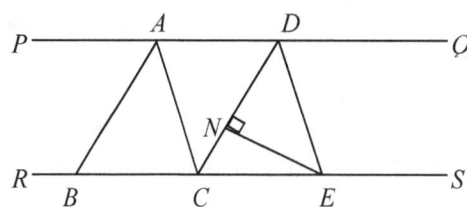

Figure 31:15

In Figure 31:15, *PQ*∥*RS*, *AB*∥*DC* and *AC*∥*DE*. Given that *CD* = 8 cm and *EN* = 6 cm, calculate the area of the parallelogram *ABCD*.

Solution

Area, *A* of *ABDC* = Area of *ACED*

[Equal base between ∥ lines]

$$= 2 \times \text{Area of } \Delta CDE$$

$$\Rightarrow A = 2 \times \frac{1}{2}(DC)(EN) = 2 \times \frac{1}{2}(8)(6) = 48 \text{ cm}^2$$

Trapezium

A trapezium is a quadrilateral with one pair of parallel sides.

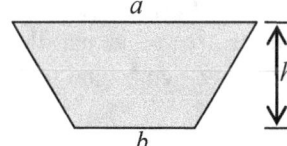

Figure 31:16

If two congruent trapezoids with parallel sides of length *a* and *b* and altitude *h* are joined end to end as in Figure 31:17, a parallelogram with base *a* + *b* and altitude *h* is formed.

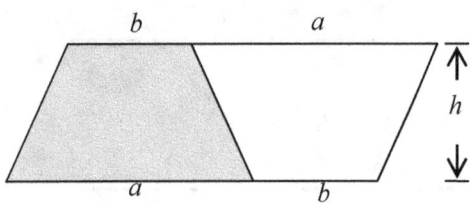

Figure 31:17

Area of trapezium $= \dfrac{1}{2} \times$ Area of parallelogram

$$= \dfrac{1}{2} \text{ base} \times \text{ altitude}$$

$$A = \dfrac{1}{2}(a + b)h$$

Area of trapezium $= \dfrac{1}{2} \times$ sum of parallel sides $\times$ height

Example 31:9
Find the area of a trapezium whose bases are 6 cm and 4 cm with the distance between parallel sides 5 cm.

Solution
$$A = \dfrac{1}{2}(a + b)h = \dfrac{1}{2}(6 + 4)5 = 25\,\text{cm}^2$$

EXERCISE 31:2

In this exercise, where appropriate, allow your answer in surd form.
1. Find the area of the parallelogram in Figure 31:18

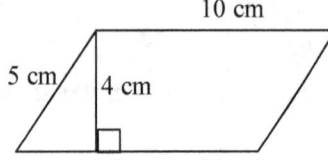

Figure 31:18
2. Given that the area of the parallelogram in Figure 31:20 is 54 cm². Find the value of x.

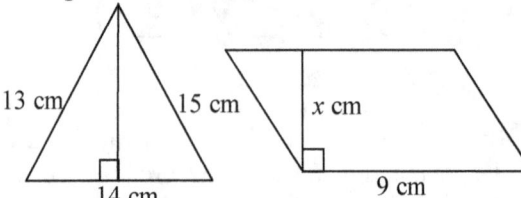

Figure 31:19   **Figure 31:20**
3. Find the area of the triangle in Figure

31:19.
4. The perimeter of an isosceles triangle is 11 cm and the length of each of the two equal sides is 4 cm. Find
   (a) the length of the third side
   (b) The altitude of the triangle
   (c) its area.
5. Calculate the area of a triangle whose sides are 12 cm, 13 cm and 5 cm long.
6. The area of a triangle is 36 cm². Given that the base is 12 cm, find the height of the triangle.
7. Calculate the area of an equilateral triangle whose sides are 8 cm.
8. The diagonals of a rhombus are 20 cm and 16 cm. Calculate
   (a) its area   (b) its side.
9. The side of a rhombus is 26 cm and one diagonal is 36 cm. Find
   (a) its area
   (b) The length of the other diagonal.
10. Find the length of one of the parallel sides of a trapezium given that the height is 8 cm, the length of the other parallel is 10 cm and the area is 124 cm².
11. In Figure 31:21, calculate

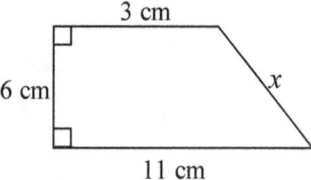

Figure 31:21
   (a) The value of x.
   (b) The area of the trapezium.
12. In Figure 31:22, calculate
   (a) The value of y.
   (b) The area of the trapezium.

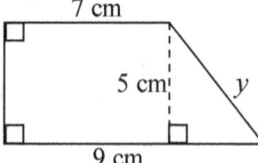

Figure 31:22
13. The parallel sides of a trapezium are 13 cm and 9 cm each. If the distance between them is 12 cm, find the area of the trapezium.
14. Calculate the area of the trapezium in Figure 31:23 and the length BC.

320

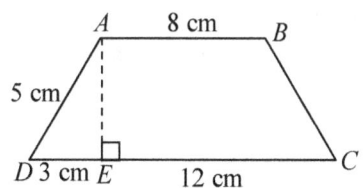

Figure 31:23

15. In Figure 31:24, the area of trapezium *ABCE* is 54 cm^2 (a) Find the value of *x*.

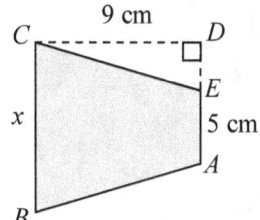

Figure 31:24

16. In Figure 31:25, *WX* ‖ *YZ*, *AB*‖*CD* and

BC ‖ *DE*. Given that *BC* =16 cm and

AN = 12 cm.
  (i) Calculate the area of
     (a) triangle *ABC*.
     (b) Parallelogram *ABDC*.
  (ii) Deduce the area of
     (c) Triangle *BDC*.
     (d) Triangle *CDE*.
     (e) Parallelogram *ABDC*.
     (f) Trapezium *ABDE*.

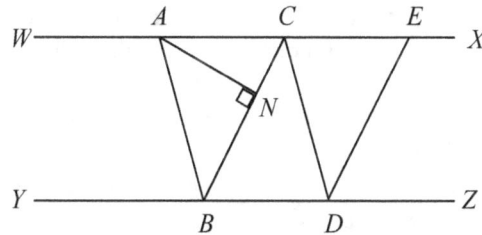

Figure 31:25

Circle Mensuration

Pi (π)

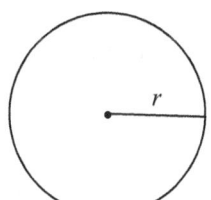

Figure 31:26

The relationship between the circumference and the diameter of a circle can be investigated using tins of different sizes, a

sticky tape and a ruler.

The circumferences of the tins may be measured by rapping a sticky tape around each tin in such a way that it surrounds the tin and overlaps. It can then be cut with a razor peeled off, pasted on a flat board and measured with a ruler. The diameter of the tin is measured by placing it in-between two blocks and measuring the distance between them as shown in Figure 31:27.

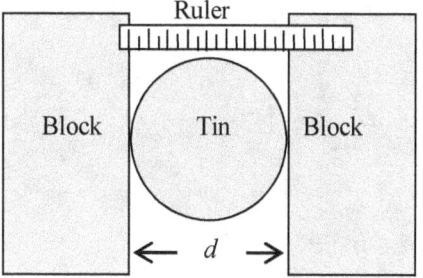

Figure 31:27

Table 31:1, shows the result of one such experiment.

Tin	Circumference, C (cm)	Diameter, d (cm)	$\dfrac{C}{d}$
Tin 1	22.5	7	3.2
Tin 2	28.6	9.2	3.1
Tin 3	12.7	4	3.2
Tin 4	20	6.4	3.1
Tin 5	40	12.6	3.2
Tin 6	9.1	3	3.0
Tin 7	15.4	5	3.1
Tin 8	27.8	8.8	3.2

Table 31:1

The results clearly show that the ratio $\dfrac{C}{d}$ is almost the same for all the tins.

More measurements that are accurate show that this ratio is approximately 3.142, sometimes approximated to $\dfrac{22}{7}$.

This constant is called **Pi** denoted by π.

Thus, $\pi \approx 3.142 \approx \dfrac{22}{7}$

Since $\pi = \dfrac{C}{d} \Leftrightarrow C = \pi d$

But $d = 2r \Rightarrow C = 2\pi r \Rightarrow r = \dfrac{C}{2\pi}$

321

Example 31:10

Calculate the circumference of a circle whose radius is 21 cm. Take $\pi = \dfrac{22}{7}$.

Solution

$$C = 2\pi r = 2\left(\frac{22}{7}\right)(21) = 132 \text{ cm}$$

Examples 31:11

Find the radius of a circle whose circumference is 11 cm, taking π as 3.142.

Solution

$$r = \frac{C}{2\pi} = \frac{11}{2(3.142)} = 1.75 \text{ cm}$$

Area of a Circle

A circle cut into many equal sectors and arranged as shown in Figure 31:28, will form a figure similar to a parallelogram.

If the sectors, are many and smaller the figure can be approximated to a parallelogram whose base is πr (half of the circumference) and whose altitude is r.

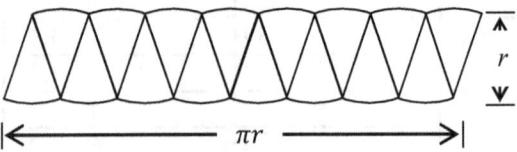

Figure 31:28

$\therefore$ area of circle $\approx$ area of the parallelogram
$$= \text{base} \times \text{height}$$
$$= \pi r(r)$$

$$A = \pi r^2 \Rightarrow r = \sqrt{\frac{A}{\pi}}$$

Examples 31:12

Calculate the area of a circle whose radius is 14 cm. $\left(\text{Take } \pi = \frac{22}{7}\right)$.

Solution

$$A = \pi r^2 = \frac{22}{7}(14)^2 = 616 \text{ cm}^2$$

Examples 31:13

Calculate to the nearest centimeter the radius of a circle whose area is 2464 cm². (Take = 3.142)

Solution

$$r = \sqrt{\frac{A}{\pi}} = \sqrt{\frac{2464}{3.142}} = 28 \text{ cm}$$

Example 31:14

Find the area of a circle with circumference 44 cm. $\left(\text{Take } \pi = \frac{22}{7}\right)$.

Solution

$$A = \pi r^2 \text{ and } C = 2\pi r \Rightarrow r = \frac{C}{2\pi}$$

$$\Rightarrow A = \frac{C}{4\pi} = \frac{44^2}{4\left(\frac{22}{7}\right)} = \frac{44^2}{4}\left(\frac{7}{22}\right) = 154 \text{ cm}^2$$

EXERCISE 31:3

Where necessary in this exercise, take $\pi = \dfrac{22}{7}$.

1. Find the area of a circle whose radius is 3.5 cm.
2. Calculate the area a goat will graze if it is tied to a pole by a rope 7 m from its neck.
3. The area of a circle is 38.5 cm². What is its radius?
4. Find the diameter of a circle whose area is 616 m².
5. Calculate the circumference of a circle whose radius is 3.5 cm.
6. Find the circumference of a circle whose radius is 14 cm.
7. Calculate the radius of a circle whose circumference is 44 cm.
8. Calculate the area of a circle whose circumference is 154 cm.
9. Find the distance covered by a bike, which runs once round a circular track of radius 21 m.
10. Find the distance between two concentric circles with areas 804 cm² and 1661 cm².
11. Calculate to the nearest whole number the area of the shaded portion in Figure 31:29.

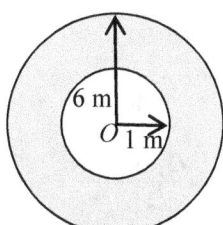

Figure 31:29

12. Two circles have radii 3.5 cm and 7 cm respectively. Find the area between them given that they have the same centre.

Arc Length

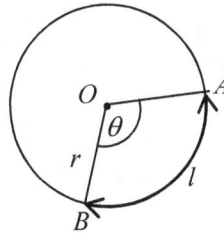

Figure 31:30

Since a revolution is 360°, the length l of an arc AB of a circle subtending an angle θ at the centre of the circle is the fraction $\dfrac{\theta}{360}$ of the circumference of the circle.

Hence, arc length l is given by $l = \dfrac{\theta}{360} \times C$, where C is the circumference of the circle.

Since $C = 2\pi r$, length l of arc AB is

$$l = \frac{\theta}{360} \times 2\pi r$$

$$\Rightarrow r = \frac{360l}{2\pi\theta} \text{ and } \theta = \frac{360l}{2\pi r}.$$

Example 31:15
An arc of a circle of radius 21 cm subtends an angle of 120° at the centre of a circle.
Calculate the length of the arc. $\left(\text{Take } \pi = \frac{22}{7}\right)$.

Solution

$$l = \frac{\theta}{360} \times 2\pi r = \frac{120}{360} \times 2(\tfrac{22}{7})(21) = 44 \text{ cm}$$

Example 31:16
Given that the length of the major arc of a circle of radius 14 cm is 70 cm, calculate the angle subtended at the centre by the arc. $\left(\text{Take } \pi = \frac{22}{7}\right)$.

Solution

$$\Rightarrow \theta = \frac{360l}{2\pi r} = \frac{360(70)(7)}{2(22)(14)} = 286.4°$$

Example 31:17
An arc of length 48 cm subtends an angle of 55° at the centre of a circle. Find the radius of the circle. $\left(\text{Take } \pi = \frac{22}{7}\right)$.

Solution

$$\Rightarrow r = \frac{360l}{2\pi\theta} = \frac{360(48)(7)}{2(22)(55)} = 50 \text{ cm}$$

Area of a Sector

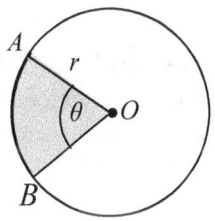

Figure 31:31

In the same way as the length of an arc of a circle, the area of a sector of a circle subtending an angle θ at the centre of the circle is the fraction $\dfrac{\theta}{360}$ of the area of the circle.
Area of sector OAB is given by

$$S = \frac{\theta}{360} \times \text{Area of the circle}$$

$$\Rightarrow S = \frac{\theta}{360} \times \pi r^2$$

$$\Rightarrow r = \sqrt{\frac{360S}{\pi\theta}} \text{ and } \theta = \frac{360S}{\pi r^2}.$$

Example 31:18
A sector of a circle of radius 21 cm subtends an angle of 120° at the centre of the

circle. Calculate the area of the sector. $\left(\text{Take } \pi = \frac{22}{7}\right)$.

Solution

$$S = \frac{\theta}{360} \times \pi r^2$$

$$\Rightarrow S = \frac{120°}{360} \times \frac{22}{7} \times 21^2 = 462 \text{ cm}^2$$

Example 31:19
The length of an arc of a circle of radius 14 cm is 21 cm. Calculate the area of the sector of the circle.

Solution

$$S = \frac{\theta}{360} \times \pi r^2 \ldots\ldots\ldots\ldots\ldots\ldots\ldots① $$

$$l = \frac{\theta}{360} \times 2\pi r \ldots\ldots\ldots\ldots\ldots\ldots② $$

$$① ÷ ② : \quad \frac{S}{l} = \frac{r}{2}$$

$$\Rightarrow S = \frac{rl}{2} = \frac{14(21)}{2} = 147 \text{ cm}^2$$

Example 31:20
The radius of a sector of area 176 cm² is 14 cm. Calculate the angle subtended by the sector at the centre of the circle $\left(\text{Take } \pi = \frac{22}{7}\right)$.

Solution

$$S = \frac{\theta}{360} \times \pi r^2$$

$$\Rightarrow \theta = \frac{360S}{\pi r^2} = \frac{360(176)(7)}{22(14)^2} = 102.9°$$

Examples 31:21
The area of a sector, which subtends an angle of 108° at the centre, is 198 cm². Find the radius of the sector. Take $\pi = \frac{22}{7}$.

Solution

$$S = \frac{\theta}{360} \times \pi r^2$$

$$\Rightarrow r = \sqrt{\frac{360S}{\pi\theta}} = \sqrt{\frac{360(198)(7)}{22(108)}} = 14.5 \text{ cm}$$

Area of a Segment

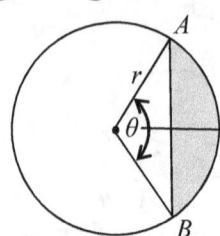

Figure 31:32
Area of Segment = Area of sector–Area of triangle

$$A_s = \frac{\theta}{360} \times \pi r^2 - \frac{1}{2}r^2 \sin\theta$$

Example 31:22
A chord subtends an angle of 135° at the centre of a circle of radius 21 cm. Calculate the area of the minor segment of the circle to the nearest square centimetre. $\left(\text{Take } \pi = \frac{22}{7}\right)$.

Solution

$$A_s = \frac{\theta}{360} \times \pi r^2 - \frac{1}{2}r^2 \sin\theta$$

$$= \frac{135}{360} \times \frac{22}{7}(21)^2 - \frac{1}{2}(21)^2 \sin 135$$

$$\Rightarrow A_s = 519.75 - 155.92 = 364 \text{ cm}^2$$

> ## EXERCISE 31:4

Where necessary in this exercise take $\pi = \frac{22}{7}$.

1. The arc of a circle subtends an angle of 57° at the centre of the circle. Find the length of the arc if the diameter of the circle is given as 7 cm.
2. Calculate the angle subtended by an arc of length 21 cm at the centre of a circle of radius 14 cm.
3. Find the perimeter of a sector, which subtends an angle of 108° at the centre of a circle of radius 7 cm.
4. Calculate the perimeter of a sector of a circle of radius 14 cm, which subtends an angle of 140° at the centre.
5. The arc length of a sector, which subtends an angle of 210° at the centre, is 88 cm. Find the radius of the sector.
6. In Figure 31:33, *AB* is a chord of a circle centre *O*. Given that |*AB*| = 24.2 cm and that the perimeter of △*AOB* is 52.2 cm.

Calculate angle *AOB* correct to the nearest degree.

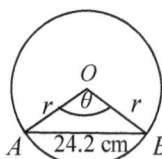

Figure 31:33

7. Calculate the length of a chord of a circle of radius 10 cm, given that the chord is 6 cm from the centre of the circle.

8. Calculate the area of the shaded segment of the sector shown in Figure 31:34.

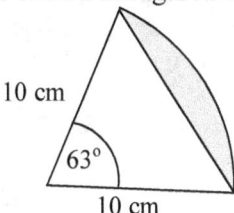

Figure 31:34

Composite Plane Figures

Problems on areas may involve a combination of two or more of the plane figures studied above. Such composite figures can be broken into recognizable figures.

Example 31:23

The diagonals of a kite are of length 60 cm and 36 cm. Calculate its area.

Solution

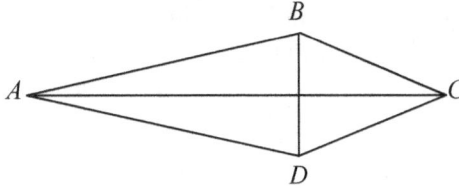

Figure 31:35

A kite *ABCD* can be considered as two congruent triangles *ABC* and *ADC*
Hence area of kite = 2 × Area of triangle *ABC*

$$= 2 \times \frac{1}{2}(AC)\frac{1}{2}(BD)$$

$$\Rightarrow A = \frac{1}{2}(AC)(BD) = \frac{1}{2}(60)(36) = 1080 \text{ cm}^2$$

EXERCISE 31:5

Where necessary in this exercise take $\pi = \frac{22}{7}$.

1. Figure 31:36, represents a miniature fan, which is in the shape of a third of a circle of radius 14 cm. The shaded part is painted and the remainder is a third of a circle of radius 10.5 cm. Calculate the area of the shaded part.

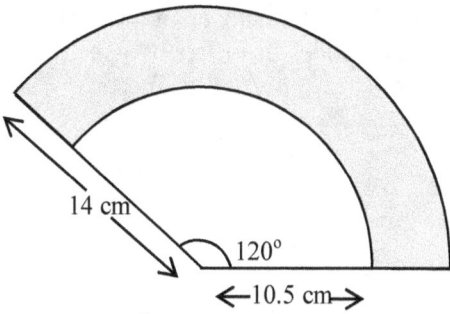

Figure 31:36

2. The area of a semi circle is 616 cm². Find to the nearest whole number the radius and the perimeter of the semi circle.

3. Figure 31:37 shows a rectangular farm 80 m long and 30 m wide, surrounded by a path of width 1 m. The farm is divided into four equal plots by paths of width 1m as shown. Calculate the total area of all the paths.

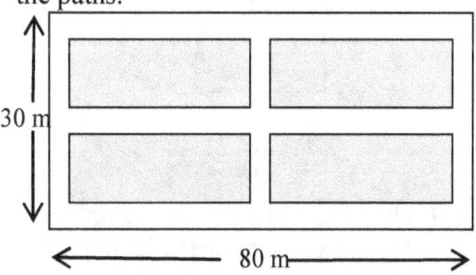

Figure 31:37

4. The circumference of a circle is 52.8 cm. Find to the nearest cm², the area of a semi circular arc of the circle.

5. Figure 31:38, shows the actual area of a sports stadium. Given that the ends are semi-circular, find the total area in m² to 3 significant figures of the stadium.

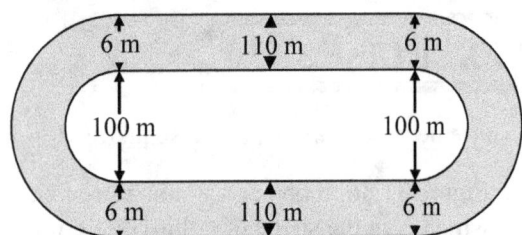

Figure 31:38

6. The area of a rectangular lawn is 50 m² and the perimeter of the lawn is 30 m. Find the length and width of the lawn.

7. Figure 31:39, is made up of a rectangle *ABCD* of sides $2x$ metres by y metres and a semi-circle on *AB* as diameter.

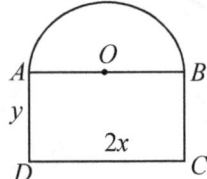

Figure 31:39

  (a) Write down, in terms of x, y and π an expression for the perimeter p of the figure.
  (b) Hence, find x in terms of p, y and π.

8. A circle of radius $3\frac{1}{2}$ cm is inscribed in a square *ABCD*, as in Figure 31:40.

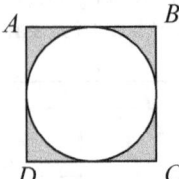

Figure 31:40

Find
  (a) The length of a side of the square.
  (b) The area of the shaded region.

9. Calculate the area of the shaded portion in Figure 31:41.

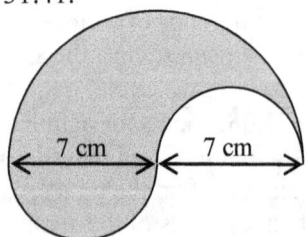

Figure 31:41

10. Calculate the area of the shaded portion in Figure 31:42.

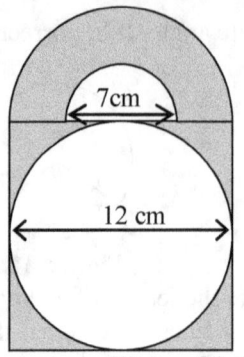

Figure 31:42

11. Figure 31:43, shows a circle of radius 6 cm inscribed in a square, which is again inscribed in a larger circle. Calculate the area of the un-shaded portion.

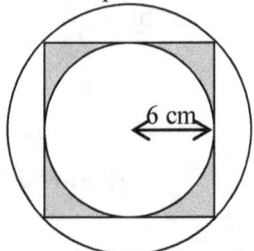

Figure 31:43

12. Calculate the area of the shaded portion in Figure 31:44.

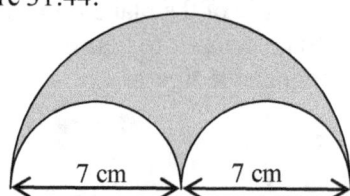

Figure 31:44

13. The design in Figure 31:45, is obtained by pinning sectors which subtend angles of 90° at the vertices *A, B, C* and *D* of a square of length 2 cm. Find
  (a) The perimeter of the shaded portion
  (b) The area of the shaded portion.

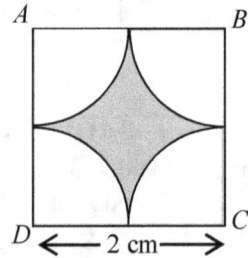

Figure 31:45

Finding Areas by Counting Squares

A method similar to that used in approximations can be employed in finding

326

the areas of plane figures by counting squares. Using this method, an area which is greater than half a unit square is counted as a full square while an area which is less than half is ignored. All the exact halves are counted and divided by 2.

Example 31:24
Find the area in square units, of the triangle in Figure 31:46 by counting squares.

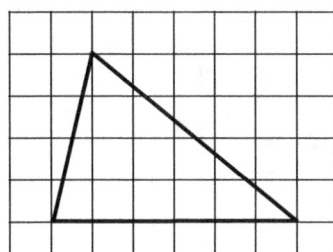

Figure 31:46

Solution
The area by counting squares as in Figure 31:47 is 12 square units.

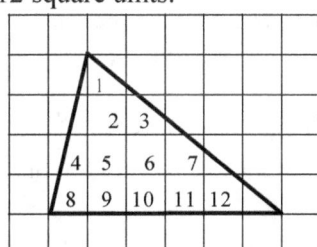

Figure 31:47
The answer can be verified by calculation as follows.

Area of triangle

$$= \frac{1}{2}bh = \frac{1}{2}(6)(4) == 12 \text{ un}^2$$

EXERCISE 31:6

1. Given that 1 square represents a square metre, find the areas in Figure 31:48 (a) to (c) by counting squares.

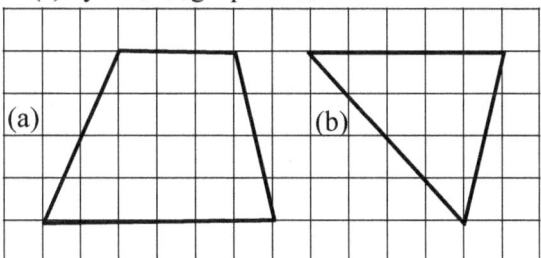

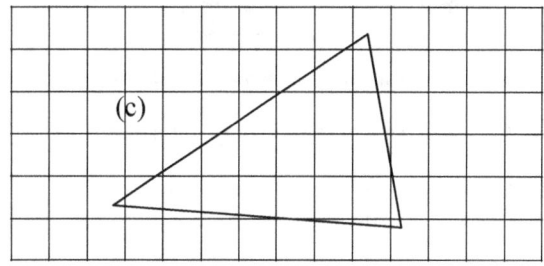

Figure 31:48

2. Figure 31:49 is the map of a certain village. Find its area, given that one square represents 2.57 square kilometres.

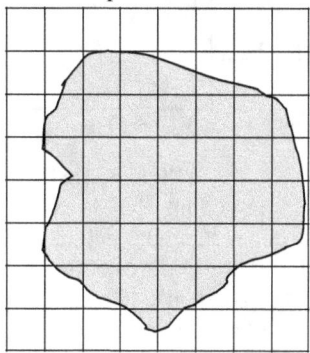

Figure 31:49

3. Figure 31:50 is the map of one of the regions of Cameroon.

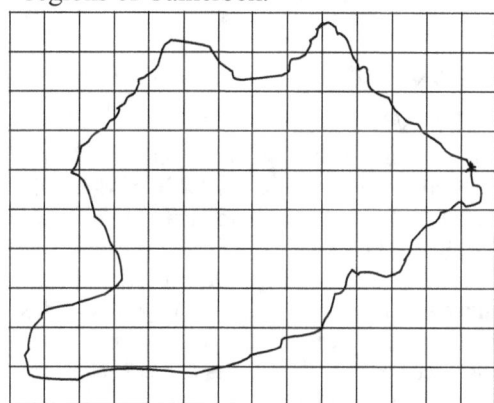

Figure 31:50
Identify the region and find its area to the nearest square kilometres, given that one square represents 266.15 square kilometre.

MULTIPLE CHOICE EXERCISE 31

In this exercise where necessary, take $\pi = \dfrac{22}{7}$.

1. If the perimeter of a square is 36 cm then the area of the square in square centimetres is:
   [A] 81 cm^2 [B] 36 cm^2
   [C] 9 cm^2 [D] 36^2 cm^2

2. The area of a square is x^2 cm^2. Its perimeter is:

3. Figure 31:51 is a rectangle. If the perimeter is 36 m, the area of the rectangle is:
   [A] x^4 cm   [B] $4x^2$ cm   [C] $4x$ cm   [D] $2x$ cm

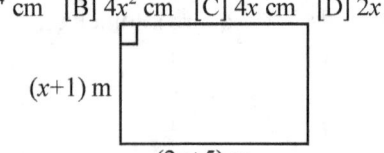

Figure 31:51

   [A] 64 m^2 [B] 65 m^2 [C] 84 m^2 [D] 124 m^2

4. One side of a rectangular field is 8 m and the diagonal is 10 m. The area of the rectangle is:
   [A] 80 m^2 [B] 36 m^2 [C] 40 m^2 [D] 48 m^2

5. The length of a rectangle is twice the width. If the length is 8 cm, the perimeter in centimetres is:
   [A] 24 cm   [B] 32 cm   [C] 48 cm   [D] 12 cm

6. The area of a rectangle with width 4 m and diagonal 8 m is:
   [A] $8\sqrt{3}$ m^2 [B] $12\sqrt{3}$ m^2
   [C] $16\sqrt{3}$ m^2 [D] 48 m^2

7. A rectangular photograph 15 cm by 9 cm is pasted on a rectangular card. If a margin of 2.5 cm is left round the photograph, the perimeter of the card is:
   [A] 58 cm [B] 68 cm [C] 98 cm [D] 228 cm

8. In a trapezium, the lengths of the parallel sides are 4 cm and 6 cm and the perpendicular distance between these sides is 3 cm. The area of the trapezium is:
   [A] 36 cm^2 [B] 18 cm^2
   [C] 30 cm^2 [D] 15 cm^2

9. The area in square units of the trapezium in Figure 31:52 is:
   [A] 24 un^2 [B] 40 un^2
   [C] 32 un^2 [D] 30 un^2

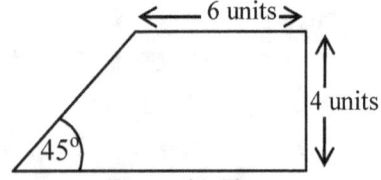

Figure 31:52

10. The perimeter of the trapezium PQRS in Figure 31:53 is:

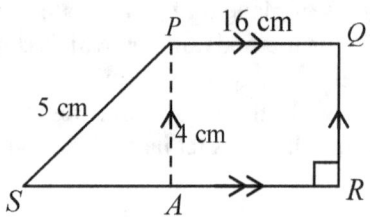

Figure 31:53

   [A] 24 cm [B] 44 cm [C] 36 cm [D] 70 cm

11. The area of the parallelogram in Figure 31:54 is:
   [A] 2736 cm^2 [B] 936 cm^2
   [C] 1368 cm^2 [D] 1872 cm^2

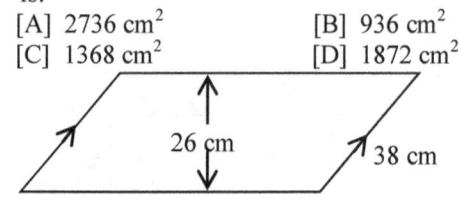

Figure 31:54

12. The lengths of the parallel sides of a trapezium are 5 cm and 7 cm. If its area is 120 cm^2, the perpendicular distance between the parallel sides is:
   [A] 5.0 cm   [B] 6.9 cm
   [C] 20.0 cm   [D] 10.0 cm

13. In Figure 31:55, PQRS is a trapezium in which $|PS| = 9$ cm , $|QR| = 15$ cm ,
   $|PR| = 2\sqrt{3}$ cm , $\angle PQR = 90°$ and
   $\angle QRS = 30°$. The area of the trapezium is:
   [A] $24\sqrt{3}$ cm^2 [B] $36\sqrt{3}$ cm^2
   [C] $42\sqrt{3}$ cm^2 [D] $72\sqrt{3}$ cm^2

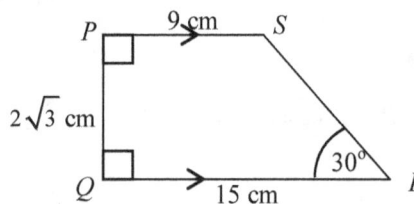

Figure 31:55

14. In Figure 31:56, ABCD is a parallelogram, in which DE is perpendicular to BC, DE = 5 cm, BC = 25 cm and AD = 45°. The size of ABC is:
   [A] 45°   [B] 90°   [C] 135°   [D] 145°

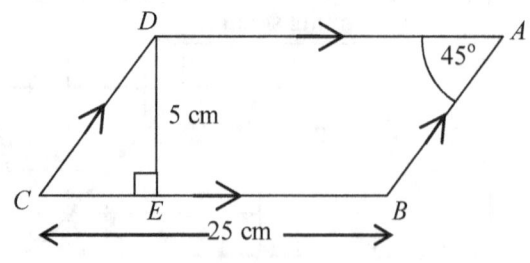

Figure 31:56

15. In Figure 31:56, *ABCD* is a parallelogram, in which *DE* is perpendicular to *BC*, *DE* = 5 cm, *BC* = 25 cm and *AD* = 45°. The length of *EC* is:
[A] 25 cm   [B] 20 cm   [C] 15 cm   [D] 5 cm

16. In Figure 31:56, *ABCD* is a parallelogram, in which *DE* is perpendicular to *BC*, *DE* = 5 cm, *BC* = 25 cm and *AD* = 45°. The area of *ABCD* is:
[A] 625 cm^2 [B] 250 cm^2

[C] 125 cm^2 [D] $62\frac{1}{2}$ cm^2

17. The area of triangle *PQR* in Figure 31:57 is:
[A] 24 cm^2 [B] 12 cm^2
[C] 10 cm^2 [D] 48 cm^2

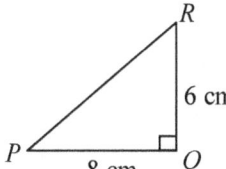

Figure 31:57

18. The area of a triangle, the sides of which are 5 cm, 4 cm and 3 cm long, is:
[A] 30 cm^2 [B] 15 cm^2
[C] 12 cm^2 [D] 6 cm^2

19. The area of an equilateral triangle of side 16 cm is:
[A] $64\sqrt{3}$ cm^2 [B] $32\sqrt{3}$ cm^2
[C] 96 cm^2 [D] 128 cm^2

20. Figure 31:58, is a triangle *XYZ* whose area is 23.5 cm^2. *XY* = 10 cm and *YZ* = 8 cm. The value of *θ*, correct to the nearest degree is:
[A] 34°   [B] 35°   [C] 36°   [D] 37°

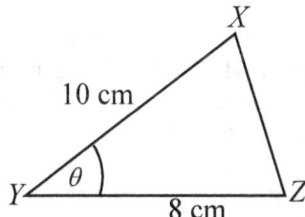

Figure 31:58

21. In Figure 31:59, *PS*∥*RQ*, |*RQ*| = 6.4 cm and perpendicular *PH* = 3.2 cm.

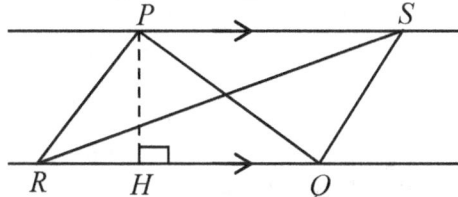

Figure 31:59

The area of triangle *SQR* is:
[A] 10.24 cm^2 [B] 9.60 cm^2

[C] 5.12 cm^2 [D] 20.48 cm^2

22. If in Figure 31:60, the area of the triangle *DCF* = 24 cm^2, the area of the quadrilateral *ABCD* is equal to:
[A] 24 cm^2 [B] 48 cm^2
[C] 80 cm^2 [D] 96 cm^2

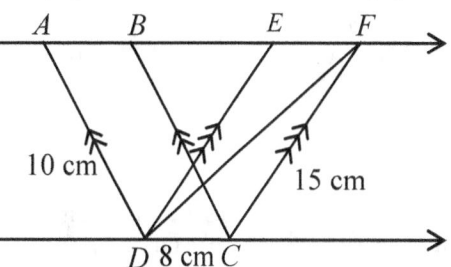

Figure 31:60

23. The diagonals *AC* and *BD* of a rhombus *ABCD* are 16 cm and 12 cm respectively. The area of the rhombus must be:
[A] 36 cm^2 [B] 48 cm^2
[C] 60 cm^2 [D] 96 cm^2

24. The area of a rhombus is 24 cm^2 and one of its diagonals is 8 cm. The side of the rhombus is:
[A] 4.3 cm [B] 5 cm [C] 6 cm [D 10 cm

25. If *O* is the centre of the circle in Figure 31:61, the area of the shaded part is:

[A] $\dfrac{3\pi r^2}{8}$ [B] $\dfrac{\pi r^2}{2}$

[C] $\dfrac{\pi r^2}{4}$ [D] $\dfrac{\pi r}{2}$

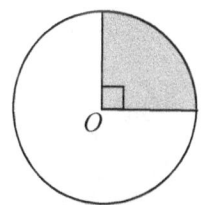

Figure 31:61

26. The area of a circular field is 154 m^2. The perimeter of the field is:
[A] 44 m [B] 49 m [C] 88 m [D] 176 m

27. The area of a circle is 38.5 cm^2, its diameter is?
[A] 22 m [B] 14 m [C] 7 m [D] 6 m

28. The diameter of a circular field whose area is 616 cm^2 is:
[A] 98.00 m   [B] 28.00 m
[C] 49.00 m   [D] 24.82 m

29. In Figure 31:62, *PXR* and *PYO* are two semicircles with diameters 14 cm and 7 cm respectively. The area of the enclosed region *PXROY* correct to the nearest whole number is:
[A] 96 cm^2 [B] 116 cm^2
[C] 154 cm^2 [D] 192 cm^2

30. In Figure 31:62, *PXR* and *PYO* are two semi-

circles with diameters 14 cm and 7 cm respectively. The perimeter of the region is:
[A] 2 cm [B] 33 cm [C] 40 cm [D] 66 cm

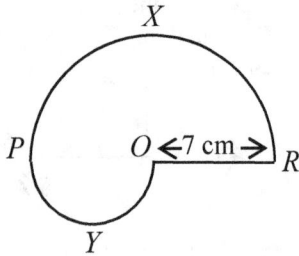

Figure 31:62

31. An arc of a circle of radius 7 cm is 14 cm long. The angle the arc subtends at the centre of the circle is:
[A] 44° [B] 51.43° [C] 98° [D] 114.55°

32. Correct to three significant figures the length of an arc, which subtends an angle of 70° at the centre of a circle of radius 4 cm is:
[A] 2.44 cm   [B] 4.89 cm
[C] 9.78 cm   [D] 25.1 cm

33. An arc of length 22 cm subtends an angle θ at the centre of a circle. If the radius of the circle is 15 cm, the value of θ will be:
[A] 84° [B] 70° [C] 96° [D] 156°

34. In Figure 31:63, O is the centre of the circle with radius 10 cm and $ABC = 30°$. The length of the arc AC, correct to one decimal place is:
[A] 5.2 cm   [B] 13.2 cm
[C] 10.5 cm   [D] 20.6 cm

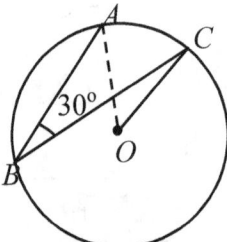

Figure 31:63

35. To 2 significant figures, the length of the arc of a circle of radius 3.5 cm that subtends an angle of 75° at the centre of the circle is:
[A] 2.3 cm [B] 4.6 cm [C] 8 cm [D] 16 cm

36. The length of the major arc PXQ in Figure 31:64 is:

[A] $18\frac{1}{3}$ cm   [B] 11 cm

[C] $9\frac{1}{6}$ cm   [D] $7\frac{1}{3}$ cm

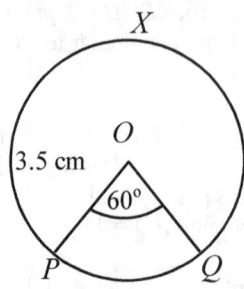

Figure 31:64

37. An arc of length 44 cm subtends an angle of 200° at the centre of a circle. The radius of the circle is:
[A] 3.9 cm   [B] 12.6 cm
[C] 25.2 cm   [D] 38.4 cm

38. In Figure 31:65, O is the centre of the circle PRQ. The radius is 3.5 cm and $\angle POQ = 50°$. The length of the arc PQ, correct to one decimal place is:
[A] 157.1 cm   [B] 37.7 cm
[C] 11.0 cm   [D] 3.1 cm

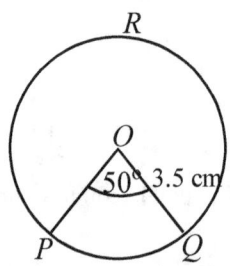

Figure 31:65

39. An arc of a circle 50 cm long subtends an angle of 75° at the centre of the circle. Correct to 3 significant figures the radius of the circle is:
[A] 8.74 cm   [B] 38.2 cm
[C] 61.2 cm   [D] 76.4 cm

40. In Figure 31:66, O is the centre of the circle. Reflex angle $XOY = 210°$ and the length of the minor arc is 5.5 cm. The length of the major arc correct to the nearest metre is:
[A] 8 cm [B] 9 cm [C] 10 cm [D] 13 cm

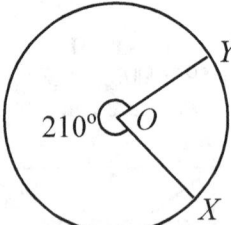

Figure 31:66

41. In Figure 31:67, AOB is a sector of the circle with centre O and radius 3.5 cm. If the reflex $\angle AOB = 290°$ the length of the minor arc AB is:
[A] 17.72 cm   [B] 14.97 cm
[C] 4.28 cm   [D] 2.14 cm

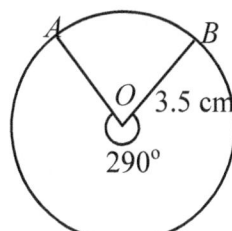

Figure 31:67

42. A rope of length 18 m is used to form a sector of circle of radius 3.5 m, on a school playing field. The size of the angle of the sector correct to the nearest degree is:
[A] 40° [B] 90° [C] 270° [D] 180°

43. The angle of a sector of a circle of radius 10.5 cm is 120°. The perimeter of the sector is:
[A] 22 cm   [B] 33.5 cm
[C] 43 cm   [D] 66 cm

44. The angle of a sector of a circle is 108°. If the radius of the circle is $3\frac{1}{2}$ cm , the perimeter of the sector is:

[A] $6\frac{4}{5}$ cm   [B] $7\frac{1}{10}$ cm

[C] $10\frac{2}{3}$ cm   [D] $13\frac{3}{5}$ cm

45. The angle of a sector of a circle of radius 35 cm is 28°. The perimeter of the sector is:

[A] $\frac{154}{9}$ cm   [B] $\frac{784}{9}$ cm

[C] $\frac{469}{9}$ cm   [D] $\frac{286}{9}$ cm

46. A sector of a circle of radius 14 cm subtends an angle of 135° at the centre of the circle. The perimeter of the sector is:
[A] 47 cm   [B] 61 cm
[C] 88 cm   [D] 231 cm

47. The angle of a sector of a circle of diameter 8 cm is 135°. The area of the sector is:

[A] $9\frac{3}{7}$ cm^2 [B] $12\frac{4}{7}$ cm^2

[C] $18\frac{6}{7}$ cm^2 [D] $25\frac{1}{7}$ cm^2

48. A sector of a circle of radius 7 cm has an area of 44 cm^2. The angle of the sector correct to the nearest degree is:
[A] 103°   [B] 26°   [C] 6°   [D] 206°

49. A sector of a circle of radius 9 cm subtends an angle of 120° at the centre of the circle. The area of the sector to the nearest cm^2 is:
[A] 75 cm^2 [B] 84 cm^2
[C] 85 cm^2 [D] 86 cm^2

50. A circle has radius x cm. The area of a sector of the circle with angle 135°, in terms of x and π is:

[A] $\frac{\pi x^2}{8}$ [B] $\frac{3\pi x^2}{8}$

[C] $\frac{5\pi x^2}{8}$ [D] $\frac{\pi x}{8}$

51. The area of the minor sector POQ in Figure 31:68 is:

[A] $148\frac{1}{2}$ cm^2 [B] $32\frac{1}{12}$ cm^2

[C] $6\frac{5}{12}$ cm^2 [D] $1\frac{5}{6}$ cm^2

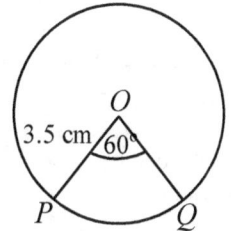

Figure 31:68

52. The length of an arc of a circle of radius 5 cm is 4 cm. The area of the sector is:
[A] 2 cm^2 [B] 8 cm^2
[C] 10 cm^2 [D] 20 cm^2

53. Correct to three significant figures, the area of the minor sector, OPQ in Figure 31:69 is approximately equal to:
[A] 3.41 cm^2 [B] 157 cm^2
[C] 10.9 cm^2 [D] 5.35 cm^2

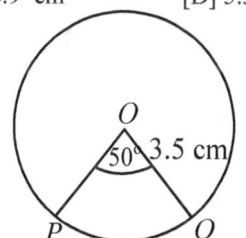

Figure 31:69

54. Figure 31:70, PQR is a circle centre O. $|OQ| = |OR| = 7$ cm .

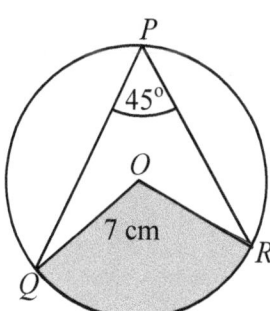

Figure 31:70

The area of the shaded portion is:
[A] 11 cm^2 [B] 22 cm^2
[C] 38.5 cm^2 [D] 77 cm^2

55. In Figure 31:71, O is the centre of the circle. Given that $OA = 5$ cm, $OD = 3$ cm and $\angle AOD = \angle BOD$, the length of the chord AB is:
[A] 8 cm   [B] 5 cm   [C] 3 cm   [D] 15 cm

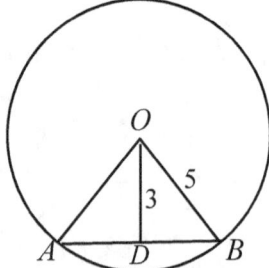

Figure 31:71

56. The angle subtended by a chord at the centre of a circle of radius 6 cm is 120°. The length of the chord should be:

[A] 6 cm   [B] 12 cm   [C] $6\sqrt{3}$ cm   [D] $2\sqrt{3}$ cm

57. Given that, a chord of a circle 8 cm long subtends an angle of 120° at the centre. The radius of the circle is:
[A] 6.93 cm   [B] 5.00 cm
[C] 4.62 cm   [D] 3.82 cm

58. Figure 31:72, shows the shaded segment of a circle of radius 7 cm. The area of the triangle OXY is cm^2. The area of the segment is:

[A] $\dfrac{5}{12}$ cm   [B] $\dfrac{7}{12}$ cm

[C] $1\dfrac{1}{6}$ cm   [D] $2\dfrac{1}{3}$ cm

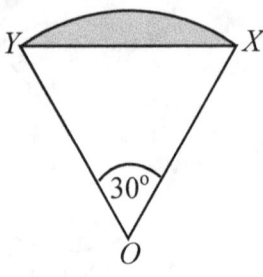

Figure 31:72

59. In Figure 31:73, $PQRO$ is one quarter of a circle with centre O. $\left|RQ\right| = \left|PR\right| = 7$ cm .
Correct to two decimal places, the area of the shaded portion is:
[A] 57.70 cm^2 [B] 38.50 cm^2
[C] 27.00 cm^2 [D] 19.25 cm^2

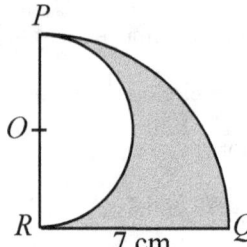

Figure 31:73

60. The area of a square is equal to that of a triangle of base 9 cm and altitude 32 cm. The length of the side of the square must be:
[A] 6 cm   [B] 6.2 cm   [C] 12 cm   [D] 2.2 cm

61. The length of the side of a square, which is equal in area to a rectangle measuring 45 cm by 5 cm, is:
[A] 25 cm   [B] 23 cm   [C] 16 cm   [D] 15 cm

62. A square has a diagonal of 10 cm. the length of a side of the square is:
[A] $\sqrt{10}$ cm   [B] $\sqrt{50}$ cm
[C] 10 cm   [D] 5 cm

TOPIC 32

MENSURATION OF SOLID FIGURES

OBJECTIVES

At the end of this topic, the learner should be able to:

1. Calculate the total surface area of prisms and cylinders.
2. Calculate the volumes of prisms and cylinders.
3. Calculate the total surface area of cones and pyramids.
4. Calculate the volumes of cones and pyramids.
5. Calculate the surface area and volume of a sphere.
6. Calculate the surface area and volume of any composite solid figures such as a frustum and a hemisphere, made up of identifiable solid figures.
7. Identify longitudes and latitudes and distinguish between great circles and small circles.
8. Calculate the shortest distance between two places on the earth surface which lie on the same latitude or the same great circle.
9. Determine the volume of flow of a liquid flowing through a pipe.

Gottfried Wilhelm Leibniz

The 17th-century thinker Gottfried Leibniz made contributions to a variety of subjects, including theology, history, and physics, although he is best remembered as a mathematician and philosopher. In mathematics his most important contribution was the discovery of the fundamental principles of infinitesimal calculus.

Corbis

For vocabulary associated with all the solid figures treated in this topic, consult TOPIC 30.

Prisms and Cylinders

Surface Area and Volume of Prisms

A right prism has a pair of parallel and congruent polygonal base.
Consider a prism of length l, cross-sectional area A and cross-sectional perimeter p.
The surface area S of the prism is given by

$$S = 2A + pl \quad ....①$$

The volume V of the prism is given by

$$V = Al \quad ....②$$

Example 32:1

A right-angled triangular prism has arms 4 cm and 3 cm and length 10 cm. Calculate
(a) its surface area.   (b) its volume.

Solution

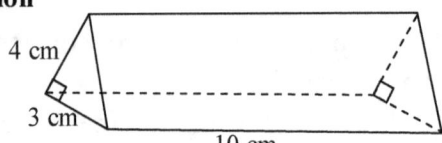

4 cm
3 cm
10 cm

Figure 32:1

$$S = 2A + pl$$

Hypotenuse of triangle $= \sqrt{3^2 + 4^2} = 5$ cm

$\Rightarrow p = 3$ cm $+ 4$ cm $+ 5$ cm $= 12$ cm

(a) $A = \dfrac{1}{2}bh = \dfrac{1}{2}(4)(3) = 6$ cm^2

$\therefore S = 2(6 \text{ cm}^2) + (12 \text{ cm})(10 \text{ cm}) = 132$ cm^2

(b) $V = Al$, $A = 6$ cm^2, $l = 10$ cm

$\Rightarrow V = 6$ cm$^2(10$ cm$) = 60$ cm^3

Example 32:2
The sides of a rectangular block (cuboids) are 5 cm, 11 cm and 40 cm. Calculate
(a) its surface area.   (b) its volume.

Solution

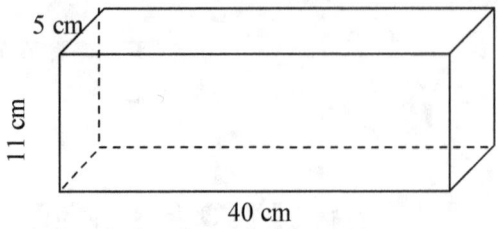

5 cm
11 cm
40 cm

Figure 32:2

(a) $S = 2A + pl$

$\Rightarrow S = 2(11)(5) + 2(5 + 11)40 = 1390$ cm^2

(b) $A = 5(11) = 55$ cm^2

$V = Al$
$= (55 \text{ cm}^2)(40 \text{ cm}) = 2200$ cm^2

Surface Area and Volume of Cubes

A cube is a special prism with the length, width and height equal.

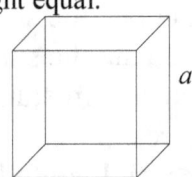

a

Figure 32:3

For a cube of side a, the perimeter and area of the cross-section are respectively

$p = 4a$ and $A = a^2$.

Applying equation ① and ② to the cube,

Surface area of cube, $S = 6a^2$

Volume of cube, $V = a^3$

Example 32:3
A cube has side 9 cm. Find
(a) its surface area.   (b) its volume.

Solution
(a) $S = 6a^2 = 6(9)^2 = 486$ cm^2
(b) $V = a^3 = 9^3 = 729$ cm^3

Surface Area and Volume of Cylinders
A cylinder is similar to a prism because they both have uniform cross-sections. The cross-section of a cylinder is circular. When a cylinder is closed at both ends, it is described as a **solid** cylinder.

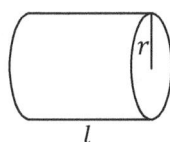

Figure 32:4

For a right-circular solid cylinder with radius r and length l, the perimeter (circumference) and area of the circular cross-section are respectively $p = 2\pi r$ and $A = \pi r^2$.

Applying equation ① and ② to a right-circular solid cylinder,

$$S = 2\pi r^2 + 2\pi r l \text{ or } S = 2\pi r(r + l)$$
$$V = \pi r^2 l$$

Example 32:4

A right-circular solid cylinder has a height of 30 cm and a radius of 7 cm. Calculate
(a) its surface area   (b)   its volume

Solution

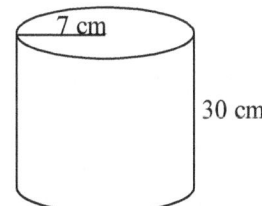

Figure 32:5

(a) $S = 2\pi r(r + l)$

$$\Rightarrow S = 2\left(\frac{22}{7}\right)(7)(7 + 30) = 1,628 \text{ cm}^2$$

(b) $V = \pi r^2 l$

$$\Rightarrow V = \left(\frac{22}{7}\right)(7)^2(30) = 4620 \text{ cm}^3$$

EXERCISE 32:1

In this exercise where necessary, take $\pi = \dfrac{22}{7}$.

1. A box is 35 cm long, 20 cm wide and 12 cm high. Calculate the volume of the box.
2. What volume of sand will fill a room of length 15 m, breadth 12 m and height 8 m?
3. A box has a volume of 140,000 cm^3. If its breadth is 50 cm, and its length is 70 cm, find its height.

4. The volume of a rectangular tank is 520 m^3. Find the length of the tank given that the width and the height are 8 m and 5 m respectively.
5. Find the level that 270 m^3 of water will rise in a rectangular tank which is 6 m by 4.5 m.
6. Find the volume of a rectangular block 0.5 m long and cross-sectional dimension of 3 cm by 20 cm.
7. Calculate the number of litres of water, which a rectangular tank with dimensions 8 m by 6 m by 3 m will hold.
8. How many litres of water will fill a rectangular tank of 500 cm by 100 cm by 200 cm?
9. Find the capacity in litres of a rectangular tank with dimensions 3 m by 4m by 5 m.
10. A cube measures 2 cm. Calculate
    (a) The volume of the cube
    (a) The total surface area of the cube
11. Calculate the number of blocks each of dimension 2.5 cm by 5 cm by 7.5 cm that can be stored in a box of dimensions 1 dm by 3 dm by 6 dm.
12. The end of a prism is a right-angled triangle, with the sides containing the right angle 8 cm by 12 cm. If the prism is 20 cm long, what is,
    (a) Its volume?   (b) Its surface area?
13. A prism is 10 cm long. Its cross-section is an equilateral triangle of side 8 cm calculated
    (a) The volume of the prism
    (b) The surface area of the prism
14. Water is full in a hollow cylinder of internal radius 10.5 cm and height 60 cm. A solid cone having the same height and radius is placed fully into the cylinder. Find the volume of water that will be left in the cylinder.
15. Table 32:1 shows the dimensions of a cylinder.   Complete the table

(a)	Base radius	5 cm	10 cm
(b)	Height	20 cm	30 cm
(c)	Surface area		
(d)	Volume		

Table 32:1

16. A cylindrical tank 3.5 m in diameter contains water to a depth of 4 m. Find the total area of the wetted surface of the tank.

17. A cuboids of sides 2 cm by 4 cm by 11 cm is full of water. If this water is poured into a cylindrical jar of diameter 8 cm, find the depth of the water.

18. Calculate the volume of a cylindrical container with diameter 14 cm and height 7 cm.

19. The volume of a cylinder is 396 cm^3. Calculate the radius of the cylinder given that the height is 14 cm.

20. Find the height of a cylinder with radius 7 cm if its volume is 770 cm^3.

21. A drum is 20 cm in diameter and 70 cm tall. Calculate the capacity of the drum.

22. How many litres of liquid can a cylindrical can 14 cm in diameter and 20 cm high contain?

23. 90 litres of water is poured into a cylindrical bucket, which is 30 cm in diameter. Find the depth of water in the bucket.

24. The height of a cylinder with radius 35 cm is 21 cm. Find the curved surface area of the cylinder.

25. Find the area of the curved surface of a cylinder whose radius is 7 cm and whose height is 5 cm.

Cones and Pyramids

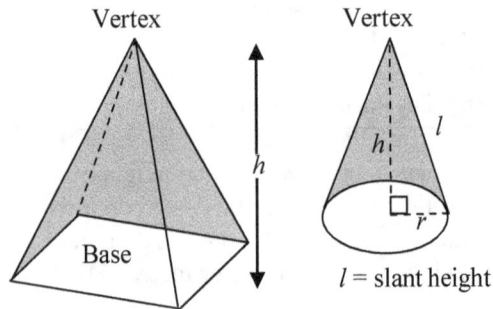

l = slant height

Figure 32:6

Volume of Cones and Pyramids

The volume V of any cone or pyramid with base area A and vertical height h is $\frac{1}{3}$ that of a cylinder or prism with the same base area and vertical height. Hence,

$$V = \frac{1}{3}\text{base area} \times \text{height}$$

$$V = \frac{1}{3}bh$$

Example 32:5

The base of a pyramid is a square of side 5 cm and its height 12 cm. Find its volume.

Solution

$$V = \frac{1}{3}Ah = \frac{1}{3}(5 \text{ cm})^2(12 \text{ cm}) = 100 \text{ cm}^3$$

Example 32:6

Calculate the volume of a cone whose height is 12 cm and whose base radius is 7 cm.

Solution

$$V = \frac{1}{3}Ah = \frac{1}{3}\pi r^2 h$$

$$\Rightarrow V = \frac{1}{3}\left(\frac{22}{7}\right)(7)^2(12)\text{ cm}^3 = 616 \text{ cm}^3$$

Surface Area of a Cone

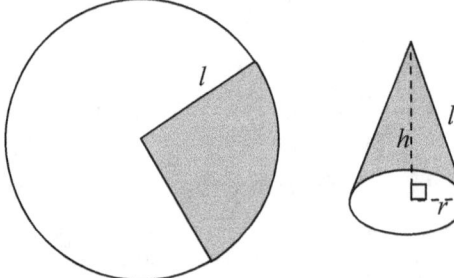

Figure 32:7

A cone can be made, by folding a sector, which subtends an angle θ at the center of a circle. In this way, the radius l of the circle will be the slant height of the cone with radius r. The lateral surface area of the cone will be the area S of the sector. The circumference $C = 2\pi r$ of the base of the cone will be the arc length of the sector.

$$\Rightarrow S = \frac{\theta}{360} \times \pi l^2 \ \text{.............................}①$$

$$2\pi r = \frac{\theta}{360} \times 2\pi l$$

$$\Rightarrow r = \frac{\theta}{360} \times l \dots\dots\dots\dots\dots\dots\dots\dots②$$

① ÷ ② : $\quad \frac{S}{r} = \pi l$

Therefore, the lateral surface area of a cone is given by

$$S = \pi r l$$

Since $l = \sqrt{h^2 + r^2}$, the lateral surface area of a cone is also given by

$$S = \pi r \sqrt{h^2 + r^2}$$

The total surface area of a solid cone will therefore be

$$S = \pi r^2 + \pi r l \quad \text{or} \quad S = \pi r (r + l)$$

In terms of h, the total surface area of a solid cone will be

$$S = \pi r^2 + \pi r \sqrt{h^2 + r^2} \quad \text{or} \quad S = \pi r \left(r + \sqrt{h^2 + r^2} \right)$$

From equation ②, it should be clear that the radius of the base of a cone, which is made out of a sector of radius l that subtends an angle θ at the center, is given by

$$r = \frac{\theta l}{360°}$$

Example 32:7

Calculate the surface area of a solid cone with base radius 3.5 cm and slant height 5.5 cm.

Solution

$$S = \pi r (r + l)$$

$$\Rightarrow S = \frac{22}{7}(3.5)(3.5 + 5.5) = 11(9) \text{ cm}^2 = 99 \text{ cm}^2$$

Surface Area of a Pyramid

To calculate the surface area S of a pyramid, calculate the area of all the faces independently and add to the area of the base.

Thus Surface area S of pyramid is given by

S = Area of base + Sum of area of all the triangular faces

Example 32:8

The faces of a pyramid are made of 4 isosceles triangles each with a base 8 cm and height 12 cm. Calculate the surface area of the pyramid.

Solution

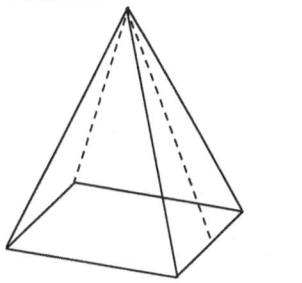 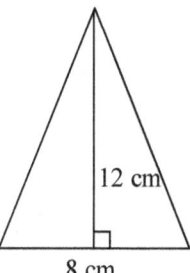

Figure 32:8

Area of each triangle $= \frac{1}{2}bh = \frac{1}{2}(8)(12) = 48 \text{ cm}^2$

∴ Area of 4 faces $= 4(48) = 192 \text{ cm}^2$

Area of base $= (8 \text{ cm})^2 = 64 \text{ cm}^2$

Surface area $= 64 \text{ cm}^2 + 192 \text{ cm}^2 = 256 \text{ cm}^2$

EXERCISE 32:2

1. A cone is 8 cm tall and has a base diameter of 12 cm. Find its slant height.
2. Find the surface area of a solid cone whose base radius is 7 cm and its slant height is 14 cm.
3. Find the surface area of a solid cone of radius 7 cm and slant height 9 cm.
4. The vertical angle of a cone is 70° and its slant height is 11 cm. Calculate the height of the cone correct to the nearest whole number.
5. The radius of a cone is 6 cm. Given that the height of the cone is 14 cm, calculate the volume of the cone.
6. Calculate the area of the surface of a cone with radius 14 cm and slant height 20 cm, when;
   (a) The base is open
   (b) the base is closed.
7. Find the volume of a cone with radius 7 cm and slant height 13 cm
8. A cone has a diameter of 7 cm and a height

of 12 cm. Calculate the volume of the cone.

9. Calculate the volume of a cone with base diameter 10 cm and slant height 13 cm.

10. The base of a pyramid is a square. Each face is made up of isosceles triangles with base 12 cm and height 16 cm. Calculate
    (a) its surface area   (b) its volume

11. Figure 32:9 shows a pyramid on a square base. If all the eight edges of the pyramid are equal, find, in degrees, the value of angle *OAC*.

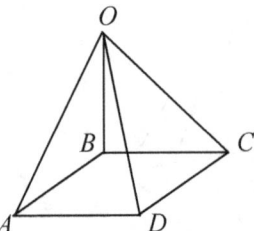

Figure 32:9

12. Figure 32:10 shows the shaded sector of a circle of radius 21 cm, which subtends an angle of 120° at the center. This sector is used to form a cone. Find the radius, in metres, of the base of the cone.

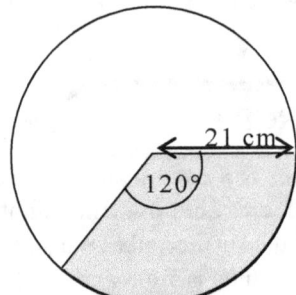

Figure 32:10

Surface Area and Volume of a Sphere

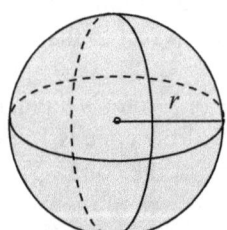

Figure 32:11

The surface area *S* of a sphere with radius *r* is four times the area of a circle with radius *r*.

Area of sphere, $S = 4 \times$ Area of circle $= 4\pi r^2$

The volume *V* of the sphere is given by, $V = \dfrac{4}{3}\pi r^3$

Example 32:9

Calculate
(a) the surface area of and
(b) The volume of a sphere whose radius is 14 cm.

Solution

(a) $S = 4\pi r^2 = 4\left(\dfrac{27}{7}\right)(10.5)^2 = 1386 \text{ cm}^2$

(b) $V = \dfrac{4}{3}\pi r^3 = \dfrac{4}{3}\left(\dfrac{22}{7}\right)(10.5)^3 = 4851 \text{ cm}^3$

EXERCISE 32:3

1. Find the surface area and volume of the spheres with the following radius.
   (a) 21 cm   (b) 10.5 cm

2. The surface area of a sphere is 616 cm², what is the volume of the sphere correct to two significant figures?

3. Find the ratio of volumes of two spheres whose radii are in the ratio 1:7.

4. Calculate to the nearest whole number the volume of air required to fill completely a ball which when fully expanded has a diameter of 15 cm.

5. A lead sphere of diameter 100 mm is melted down and made into spherical balls of diameter 1 mm. If no lead is lost in the process, calculate the number of small spheres that will be made.

6. A hemisphere has a diameter of 6 cm. Calculate
   (a) Its Volume
   (b) The area of its curved surface.
   (c) Its total surface area.

Composite Solid Figures

As with the case of composite plane figures, problems on solid figures may involve a combination of two or more of the solid figures studied above. Such composite figures must then be broken into recognisable figures. As examples of special composite solid figures, the frustum and the hemisphere shall be examined.

338

Volume and Surface Area of a Frustum

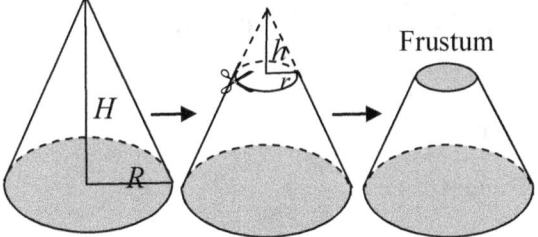

Figure 32:12

When the smaller end of a cone or pyramid is cut off through its cross-section by a plane parallel to the base as shown in Figure 32:12 the portion left is called a **frustum**.

Consider a conical frustum of height h, made by cutting off a small cone of radius r from a larger cone of radius R. Figure 32:13 is the diametrical cross-section of this cone.

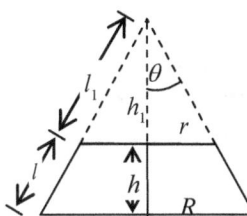

Figure 32:13

$$\tan \theta = \frac{R}{h + h_1} = \frac{r}{h_1}$$

$$\Rightarrow h_1 = \frac{rh}{R - r} \dots\dots\dots\dots\dots\text{①}$$

$$\sin \theta = \frac{R}{l + l_1} = \frac{r}{l_1}$$

$$\Rightarrow l_1 = \frac{rl}{R - r} \dots\dots\dots\dots\dots\text{②}$$

Since the volume of a cone with radius r and height h is given by $V = \frac{1}{3}\pi r^2 h$, the volume of this frustum will be

$$V = \frac{1}{3}\pi R^2 (h + h_1) - \frac{1}{3}\pi r^2 h_1 \dots\dots\text{③}$$

Substituting ① in ③ and rearranging gives

$$V = \frac{1}{3}\pi h \left(R^2 + rR + r^2 \right)$$

Since the lateral surface area of a cone with radius r and height h is given by $A = \pi r l$, the lateral surface area of the frustum will be

$$S_1 = \pi R (l + l_1) - \pi r l_1 \dots\dots\dots\text{④}$$

Substituting ② in ④ and rearranging gives

$$S_1 = \pi (R + r)l$$

But $l = \sqrt{h^2 + (R - r)^2}$

Therefore, $S_1 = \pi (R + r)\sqrt{h^2 + (R - r)^2}$

If the frustum is a solid, the area of the two circular ends will be added. Hence, surface area of a solid frustum is

$$S = \pi \left(R^2 + r^2 + (R + r)l \right)$$

OR

$$S = \pi \left(R^2 + r^2 + (R + r)\sqrt{h^2 + (R - r)^2} \right)$$

For a frustum open at the larger end such as a bucket the surface area will be

$$S = \pi \left(r^2 + (R + r)l \right)$$

OR

$$S = \pi \left(r^2 + (R + r)\sqrt{h^2 + (R - r)^2} \right)$$

Example 32:10

A cone with base radius 4 cm is cut off from a cone of radius 6 cm and height 4 cm. Find
(a) volume of the frustum left.
(b) the total surface area of the frustum left.

Solution

Let V, v, R, r and H, h be the volume, radius and height of the big and small cones respectively.

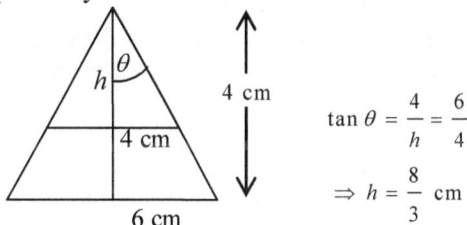

$$\tan \theta = \frac{4}{h} = \frac{6}{4}$$

$$\Rightarrow h = \frac{8}{3} \text{ cm}$$

Figure 32:14

(a) $V = \frac{1}{3}\pi h \left(R^2 + rR + r^2 \right)$

$$= \frac{1}{3}\left(\frac{22}{7} \right)\left(\frac{4}{3} \right)\left(6^2 + 6(4) + 4^2 \right)$$

$$= \frac{1}{3}\left(\frac{22}{7} \right)\left(\frac{4}{3} \right)(76) = 106.2 \text{ cm}^3$$

(b) $S = \pi\left(R^2 + r^2 + (R+r)\sqrt{h^2 + (R-r)^2}\right)$

$= \left(\frac{22}{7}\right)\left(6^2 + 4^2 + (6+4)\sqrt{\left(\frac{4}{3}\right)^2 + (6-4)^2}\right)$

$= 239 \text{ cm}^2$

Example 32:11

A bucket is in the form of a frustum. The diameters of the two ends are 14 cm and 21 cm. If the slant height of the bucket is 30 cm, calculate its surface area when it is

(a) open   (b) closed

Solution

(a) $S = \pi\left(r^2 + (R+r)l\right)$

$= \frac{22}{7}\left(7^2 + (10.5 + 7)30\right)$

$= 1804 \text{ cm}^2$

(b) $S = \pi\left(R^2 + r^2 + (R+r)l\right)$

$= \frac{22}{7}\left(10.5^2 + 7^2 + (10.5 + 7)30\right)$

$= 2150.5 \text{ cm}^2$

Volume and Surface Area of a Hemisphere

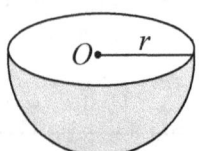

Figure 32:15

Since a hemisphere is half of a sphere, its volume V and curved surface area S are half that of the sphere. Hence,

$$V = \frac{2}{3}\pi r^3 \text{ and } S = 2\pi r^2$$

In addition, a hemisphere has a circular part so the total surface area of a solid hemisphere is

$$S = \pi r^2 + 2\pi r^2 = 3\pi r^2$$

Example 32:12

A hemispherical bowl with a flat lid has a

radius of 7 cm. Calculate the
(a) Volume of the bowl
(b) The surface area of the bowl when
  (i) Open.   (ii) Closed

Solution

Volume of bowl $V = \frac{2}{3}\pi r^3$

$\Rightarrow V = \frac{2}{3}\left(\frac{22}{7}\right)(7)^3 = 718.7 \text{ cm}^3$

(b) (i) $S_0 = 2\pi r^2 = 2\left(\frac{22}{7}\right)(7)^2 = 308 \text{ cm}^2$

  (ii) $S = 3\pi r^2 = 3\left(\frac{22}{7}\right)(7)^2 = 462 \text{ cm}^2$

EXERCISE 32:4

In this exercise, where necessary take $\pi = \frac{22}{7}$.

1. A bucket in the shape of a frustum has a base diameter of 4 cm and a brim diameter of 10 cm. Given that its height is 3 cm, calculate
   (a) The volume of the bucket.
   (b) The external surface area of the bucket.
2. A frustum is cut out from a cone. Given that, its diameters are 6 cm and 18 cm and that its height is 15 cm. Calculate
   (a) The volume of the bucket.
   (b) The external surface area of the bucket.
3. Figure 32:16, shows a frustum of a cone. The height of the frustum is 10 cm. If the diameters of the two ends are 8 cm and 6 cm respectively, calculate
   (a) The volume of the frustum,
   (b) The surface area of the frustum, leaving your answer in terms of π.

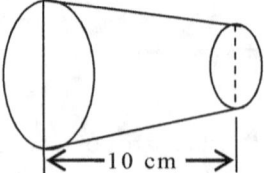

$\leftarrow$ 10 cm $\rightarrow$

Figure 32:16

4. Figure 32:17, shows a sharpened round pencil with a hemispherical eraser of radius

340

0.5 cm. Calculate
(a) The total volume of material contained in the pencil.
(b) The total surface area of the pencil.

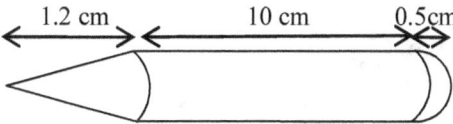

Figure 32:17

5. A cylindrical cake of radius 3 cm and height 4 cm is standing on a table. A slice of the cake is removed by cutting vertically through the radii *OA* and *OB*, where ∠*AOB* = 30° as shown in Figure 32:18. Calculate, leaving your answers in terms of π.
(a) The volume of the remaining cake.
(b) Its total surface area.

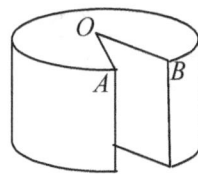

Figure 32:18

6. A cylindrical jar with internal diameter 15 cm contains a metal sphere of radius 6.3 cm. With the base of the cylinder horizontal, water is poured in until it covers the sphere as in Figure 32:19.
(i) The sphere is removed. Calculate the fall in the water level.
(ii) If a metal cube is now immersed in the cylinder and the water level rises to its original level, calculate the length of an edge of the metal cube.

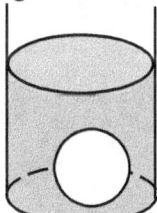 

Figure 32:19   **Figure 32:20**

7. A cone of radius 4 cm is attached to a hemisphere with equal radius (Figure 32:20). Given that the total height of the object is 10 cm, find
(a) The volume of the object.
(b) The total surface area of the object.

8. One corner of a solid cube of side 8 cm is cut off through the midpoints of three adjacent sides as in Figure 32:21. Calculate the volume of the remaining piece.

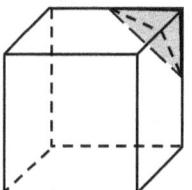 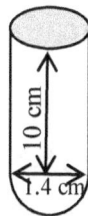

Figure 32:21   **Figure 32:22**

9. Figure 32:22, shows a diagram of a test tube. Calculate
(a) The volume of acid that will be required to fill it right to the brim.
(b) The total surface area, which is in contact with the acid when full to the brim. (Assume that the thickness of the tube is negligible)

10. The external radius of a ball is 14 cm and the internal radius 12.6 cm. Calculate to the nearest tenth of a centimeter,
(a) The volume of air that the ball can hold if fully inflated.
(b) The volume of material that the ball is made of
(c) The surface area of the ball.

THE EARTH AS A SPHERE

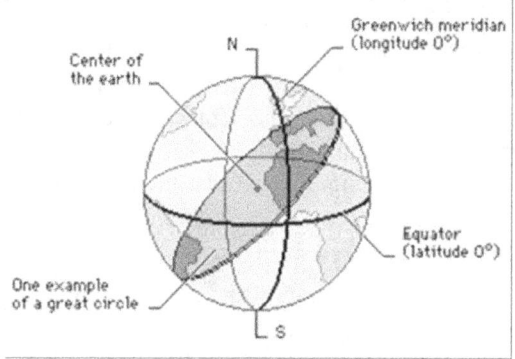

Figure 32:23

341

Longitudes and Latitudes

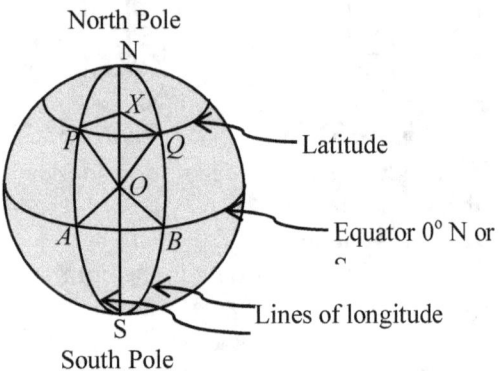

Figure 32:24

Points along the latitudes are measured from the Greenwich meridian, E or W of the Greenwich meridian.
Points along the longitudes are measured from the equator, N or S of the equator.
Longitudes and latitudes act like the coordinate plane.
Longitudes are measured from 0° to 90° N or S of the equator.
Latitudes are measured from 0° to 180° E or W of the Greenwich meridian.

Great Circles

A great circle is a circular section of a sphere whose radius is equal to that of the sphere. The equator and all the longitudes are called great circles of the earth.
Using the formula for calculating arc length, which was treated in Topic 29, the distance D along a great circle is given by,

$$D = \frac{\theta}{360} \times 2\pi R$$

Where $R \approx 6370$ km, is the radius of the earth, and θ is the angular distance subtended by the two points at the centre of the earth.

Small Circles

Any circle drawn round the earth's surface exempting the equator and the longitudes is called a small circle.

Example 32:13
Calculate in km the shortest distance, measured over the earth's surface between the points $P(40°N, 20°W)$ and $Q(20°S, 20°W)$.

Solution

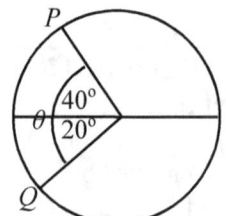

Figure 32:25

$$\text{Arc length} PQ = \frac{\theta}{360} \times 2\pi R$$
$$= \frac{60}{360} \times 2\left(\frac{22}{7}\right)(6370)$$
$$= 6673.3 \text{ km}$$

Example 32:14
Find in km the shortest distance, between the points $A(70°N, 50°W)$ and $B(20°N, 50°W)$.

Solution

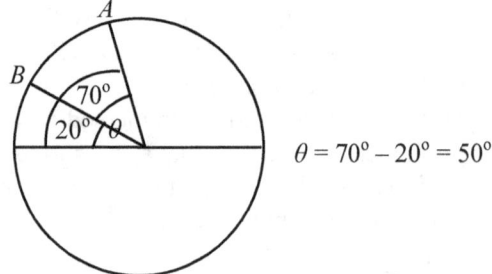

$\theta = 70° - 20° = 50°$

Figure 32:26

$$\text{Arc length } AB = \frac{\theta}{360} \times 2\pi R$$
$$= \frac{50}{360} \times 2\left(\frac{22}{7}\right)(6370)$$
$$= 5561.1 \text{ km}$$

EXERCISE 32:5

In this exercise take $\pi = \dfrac{22}{7}$ and R, the radius of the earth to be 6400 km.

1. Calculate in km the shortest distance, between the pair of points,
   (a) (50° N, 10° W) and (66° S, 10° W)
   (b) (10° N, 97° E) and (65° N, 97° E)
   (c) (32° S, 30° W) and (84° S, 30° W)

2. Find in km the shortest distance, between the following pair of points, which are along the equator.
   (a) 150° W and 15° E
   (b) 130° E and 92° E
   (c) 40° W and 130° W

3. Calculate the angular distance between two points on the same line of longitude if their distance apart is
   (a) 2816 km   (b) 10057 km

VOLUME OF FLOW

The time taken by a liquid to fill a tank depends on the volume of liquid flowing per unit time into the tank. The volume of liquid flowing per unit time is called the **rate of flow** and is usually stated in cm³/s or m³/min. Sometimes when liquid is flowing through a pipe, the rate of flow is stated in m/min or cm/s.

$$\text{Rate of flow} = \frac{\text{Volume of fluid}}{\text{Time taken}} \quad \text{or} \quad R = \frac{V}{t}$$

Suppose V m³ of liquid passes through a pipe of unit length and unit cross-sectional area in 1 second, then the total volume of liquid that will pass through a pipe of unit length and area A m² in 1 second, will be VA m³. It follows that the total volume of liquid that will pass through a pipe of length l and area A m² in 1 second, will be VAl m³

Example 32:15

A circular tank of diameter 2.8 m and height 4 m is to be filled by a pipe through which water is flowing at the rate of 8 m³/min. How long does it take to fill the tank?

Solution

$$R = \frac{V}{t} \Rightarrow t = \frac{V}{R}$$

But $V = \pi r^2 h, \quad \pi = \dfrac{22}{7}, r = \dfrac{2.8}{2}, h = 4$

$$= \frac{22}{7}\left(\frac{2.8}{2}\right)^2 (4)$$

$$= 24.64 \text{ m}^3$$

$$\therefore t = \frac{24.64}{8} = 3.08 \text{ mins.}$$

Example 32:16

A tank of volume 550 m³ filled with water is employed by a pipe of diameter 0.5 m in 10 minutes. Calculate the speed of the water in the pipe.

Solution

Let l be the length of pipe, which can contain 550 m³ of water. Then,

$$V = \pi r^2 l$$

$$550 = \frac{22}{7}\left(\frac{0.5}{2}\right)^2 l$$

$$\Rightarrow l = \frac{7}{22}\left(\frac{2}{0.5}\right)^2 (550) = 2800 \text{ m.}$$

Suppose that the water is emptied into a pipe of unit cross-sectional area. Then
Volume of water $= 2800$ m)(1 m²)
$$= 2800 \text{ m}^3$$

$$\text{Rate of flow} = \frac{\text{Volume of water}}{\text{time taken}}$$

$$= \frac{2800 \text{ m}^3}{10 \text{ min}} = 280 \text{ m}^3/\text{min}$$

EXERCISE 32:6

1. The speed of water through a pipe whose bore is 20 cm is 4 m/s. Calculate the rate of discharge of water from the pipe in,
   (a) Cubic metres per second
   (b) litres per minute.

2. Water flows through a circular pipe of radius 10 cm at 6 m/s. How many cubic metres of water does it discharge per minute?

3. An outlet issues out water at the rate of 6 m/s. Find the number of litres of water issuing through the outlet each minute.

4. (a) A cylindrical tank has diameter 2 m and height 7 m. Calculate its capacity when full.
   (b) Given that water flows into the tank at the rate of 70 cm³/s, determine in terms of π, how long it will take to fill the tank.

5. Water flows into a cylindrical reservoir 20 m in diameter at the rate of 7000 litres per minute. At what rate does the water rise in cm per minute?

6. A tank, which holds 1 m3 of water, is filled by a pipe of diameter 2 cm in 10 minutes. Find the speed of water in the pipe in m/s.

7. Water is flowing through a circular channel at 8 m/s. Find the number of m³ discharged per minute if the channel has a diameter of 8 cm and is always full.

8. A pipe of cross sectional area 10 cm² fills a rectangular tank with cross section 2 m by 1 m and height 1 m. If the pipe delivers 1 m³ per minute, find
   (a) in m/min, the rate of flow of water in the pipe.
   (b) the time in seconds taken to fill the tank.

◁ **MULTIPLE CHOICE EXERCISE 32** ▷

In this exercises, where necessary, take $\pi = \frac{22}{7}$.

1. The shape of each side of a cuboids is:
   [A] A triangle   [B] A trapezium
   [C] A circle   [D] A rectangle

2. The number of vertices in a cuboids is:
   [A] 4 [B] 6 [C] 8 [D] 12

3. The number of faces in a cuboids is:
   [A] 4 [B] 6 [C] 8 [D] 12

4. The number of edges in a cuboids is:
   [A] 12 [B] 8 [C] 6 [D] 4

5. The total surface area of a cube of edge 3 cm is:
   [A] 27 cm² [B] 27 cm³ [C] 54 cm² [D] 36 cm²

6. The sides of two cubes are in the ratio 2:5. The ratio of their volumes is:
   [A] 4:5 [B] 8:15 [C] 6:125 [D] 8:125

7. A rectangular tank 2.25 m long and 1.6 m wide contains 2800 litres of water. Correct to the nearest cm, the depth of water in the tank is:
   [A] 76 cm [B] 78 cm [C] 770 cm [D] 780 cm

8. A cylindrical container closed at both ends, has a radius of 7 cm and a height 5 cm. The total surface area of the container is:
   [A] 154 cm²   [B] 220 cm²
   [C] 528 cm²   [D] 770 cm²

9. A cylindrical container closed at both ends, has a radius of 7 cm and a height 5 cm. The volume of the container is:
   [A] 154 cm³   [B] 220 cm³
   [C] 528 cm³   [D] 770 cm³

10. The curved surface area of a cylindrical tin is 704 cm². The height when the radius is 8 cm is:
    [A] 3.5 cm [B] 7 cm [C] 14 cm [D] 28 cm

11. Correct to 1 decimal place the volume of a cylinder of height 8 cm and base radius 3 cm is:
    [A] 300.0 cm³   [B] 250.0 cm³
    [C] 226.2 cm³   [D] 150.9 cm³

12. Water flows into a cylindrical container at the rate of 5π cm³ per second. If the radius of the container is 3 cm, the level of the water in the container at the end of 9 seconds will be:
    [A] 2 cm [B] 5 cm [C] 8 cm [D] 15 cm

13. Figure 32:27 shows a rectangular sheet of thin metal from which a cylinder, 10 cm high, is to be made with no overlap. The radius of this cylinder is:

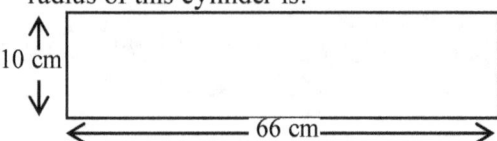

Figure 32:27

    [A] 3.3 cm   [B] 6.6 cm
    [C] 10.5 cm   [D] 21 cm

14. A solid cylinder of radius 7 cm is 10 cm long. Its total surface area is:
    [A] 70π cm²   [B] 18 π cm²
    [C] 210 π cm²   [D] 238 π cm²

15. The volume of a cylinder of radius 14 cm is 210 cm³. The curved surface area of the cylinder is:
    [A] 30 cm²   [B] 15 cm²
    [C] 616 cm²   [D] 1262 cm²

16. The internal and external radii of a cylindrical bronze pipe are 1.5 cm and 2

cm respectively. If the pipe is 10 cm long, the volume of the bronze used is:

[A] $5\dfrac{1}{2}$ cm³   [B] 55 cm³

[C] $196\dfrac{2}{5}$ cm³   [D] 550 cm³

17. A cone is made from a sector of a circle of radius 14 cm and angle 90°. The area of the curved surface of the cone is:
[A] 22 cm²   [B] 88 cm²
[C] 77 cm²   [D] 154 cm²

18. The volume of a cone of radius 3.5 cm and vertical height 12 cm is:
[A] 15.5 cm³   [B] 21.0 cm³
[C] 154.0 cm³   [D] 42.0 cm³

19. A sector is cut off from a circle of radius 8.2 cm to form a cone. If the radius of the resulting cone is 3.5 cm, then, the curved surface area of the cone is:
[A] 12.83 cm²   [B] 22.0 cm²
[C] 67.2 cm²   [D] 90.2 cm²

20. The angle of a sector of a circle of radius 8 cm is 240°. This sector is bent to form a cone. The radius of the base of the cone must be:

[A] $\dfrac{16}{3}$ cm   [B] $\dfrac{15}{3}$ cm

[C] $\dfrac{16}{5}$ cm   [D] $\dfrac{8}{3}$ cm

21. The volume of a cone of height 9 cm is 1848 cm³. Its radius is:
[A] 7 cm   [B] 14 cm
[C] 28 cm   [D] 98 cm

22. The total surface area of a cone whose height is 12 cm and whose base radius is 5 cm is:

[A] $240\dfrac{2}{7}$ cm²   [B] $235\dfrac{5}{7}$ cm²

[C] $282\dfrac{6}{7}$ cm²   [D] $251\dfrac{3}{7}$ cm²

23. A cone is 14 cm deep and the base radius is $4\dfrac{1}{2}$ cm.

The volume of water, which is exactly half the volume of the cone, is:
[A] 49.5 cm³   [B] 99 cm³
[C] 148.5 cm³   [D] 297 cm³

24. The total surface area of a solid circular cone with base radius 3 cm and slant height 4 cm is:

[A] 66 cm²   [B] $\dfrac{753}{7}$ cm²

[C] $\dfrac{782}{7}$ cm²   [D] 88 cm²

25. A 210° sector of a circle of radius 21 cm is bent to form a cone. The base radius of the cone is:

[A] $3\dfrac{1}{2}$ cm   [B] 7 cm

[C] $10\dfrac{1}{2}$ cm   [D] $12\dfrac{1}{4}$ cm

26. The total surface area of a solid cone of slant height 15 cm and base radius 8 cm in terms of π is:
[A] 120π cm²   [B] 184π cm²
[C] 200 π cm²   [D] 320 π cm²

27. The curved surface area of a cone of radius 3 cm and slant height 7 cm is:
[A] 22 cm²   [B] 44 cm²
[C] 66 cm²   [D] 132 cm²

28. The height of a right circular cone is 4 cm. The radius of the base is 3 cm. Its curved surface area is:
[A] 9π cm²   [B] 15 π cm²
[C] 16 π cm²   [D] 20 π cm²

29. The base diameter of a cone is 14 cm, and its volume is 462 cm³. Its height is:
[A] 3.5 cm [B] 5 cm [C] 7 cm [D] 9 cm

30. The total surface area of a solid right circular cone of base radius r cm and height r cm is:

[A] $2\pi r^2$ cm²   [B] $2\pi r^2$ cm²

[C] $\dfrac{7}{3}\pi r^2$ cm²   [D] $\dfrac{4}{3}\pi r^2$ cm²

31. The surface area of a sphere of radius 7 cm is:
[A] 86 cm²   [B] 154 cm²
[C] 616 cm²   [D] 143 cm²

32. Two solid spheres have volumes 250 cm³ and 128 cm³ respectively. The ratio of their radii is certainly:
[A] 5:4 [B] 25:16 [C] 2:1 [D] 4:3

33. A hollow sphere has a volume of k cm³ and a surface area of k cm². The diameter of the sphere is:
[A] 3 cm [B] 12 cm [C] 9 cm [D] 6 cm

34. A sphere has a surface area of 4312 cm². The radius of the sphere in cm correct to one decimal place is:
[A] 18.0 [B] 18.5 [C] 19.0 [D] 19.5

35. The cross-section of a prism is a right angled triangle 3 cm by 4 cm by 5 cm. The height of the prism is 8 cm. Its volume is:
[A] 48 cm³   [B] 60 cm³
[C] 96 cm³   [D] 120 cm³

36-37 Figure 32:28 shows a triangular prism of length 7 cm. The right angled triangle *PQR* is a cross section of the prism
$|PR| = 5$ cm. and $|RQ| = 3$ cm.

Use the information to answer questions 36 to 37.

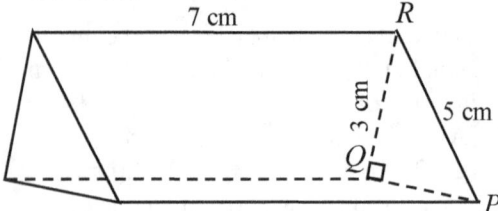

Figure 32:28

36. The area of the cross-section is:
[A] 4 cm² [B] 6 cm² [C] 15 cm² [D] 20 cm²

37. The volume of the prism is:
[A] 28 cm³   [B] 42 cm³
[C] 70 cm³   [D] 84 cm³

38. The height of a pyramid on a square base is 15 cm. Given that the volume is 80 cm³, the length of the side of the base in cm is:
[A] 3.3   [B] 5.3   [C] 4.0   [D] 8.0

39. The height of a pyramid on a square base is 15 cm. If the volume is 80 cm³, the area of the square base is:
[A] 16 cm²   [B] 9.6 cm²
[C] 8 cm²   [D] 25 cm²

40. A right pyramid is on a square base of side 4 cm. The slanting side of the pyramid is $2\sqrt{3}$ cm. The volume of the pyramid is:
[A] $\dfrac{10}{3}$ [B] $\dfrac{16}{3}$ [C] $\dfrac{32}{3}$ [D] $\dfrac{64}{3}$

41. A pyramid on a square base of side 10 cm has a height of 15 cm, its volume must be:
[A] 150 cm³   [B] 500 cm³
[C] 1500 cm³   [D] 5000 cm³

42. The base of a pyramid is a 12 cm by 12 cm. If its height is 20 cm, the volume of the pyramid in cm³ is:
[A] 960 [B] 80 [C] 1440 [D] 1600

43. The position of two countries *P* and *Q* are 15° N, 12° E and 65° N, 12° E respectively. Their difference in latitude is:
[A] 100° [B] 80° [C] 50° [D] 24°

44. Cotonou and Niamey are on the same line of longitude and Niamey is 7° north of Cotonou. If the radius of the earth is 6400 km, the distance of Niamey north of Cotonou along the line of longitude correct to the nearest kilometre is:
[A] 391 km   [B] 503 km
[C] 782 km   [D] 1006 km

45. *P* and *Q* are two places on the same circle of latitude 79° S. P is on longitude 68° E, while *Q* is on longitude 22° W. The angular distance between *P* and *Q* is:
[A] 12°   [B] 45°   [C] 48°   [D] 90°

46. Two ships on the equator are on longitude 45° W and 45° E respectively. Their distance apart along the equator, correct to 2 significant figures is:
[A] 3,200 km   [B] 10,000 km
[C] 6,400 km   [D] 5,000 km

47. Two points *P* and *Q* are on longitude 67° W. Their latitudes differ by 90°. Taking the radius of the earth as 6400 km, their distance apart in terms of π is:
[A] $6,400\pi$ km   [B] $\dfrac{6,400}{\pi}$ km
[C] $3,200\pi$ km   [D] $\dfrac{3,200}{\pi}$ km

48. Two places are 2816 km apart on the same line of longitude. The angular difference between their latitudes is: (Take $R = 6,400$ km)
[A] 25.2°   [B] 26.1°
[C] 51.3°   [D] 63.9°

49. Abijan is 4° west of Accra and on the same circle of latitude. If the radius of this circle of latitude is 6370 km, the distance of Abijan west of Accra, correct to the nearest km is:
[A] 222   [B] 445   [C] 890   [D] 5005

50. Figure 32:29 shows a cone with the dimensions of its frustum indicated. The height of the cone is:
[A] 12 cm [B] 15 cm [C] 18 cm [D] 24 cm

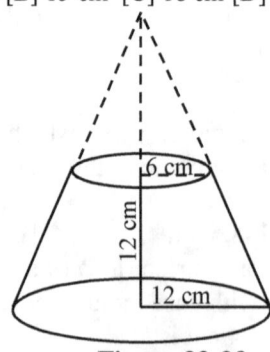

Figure 32:29

TOPIC 33

SIMILARITY AND CONGRUENCY

OBJECTIVES

At the end of this topic, the learner should be able to:

1. Identify congruent figures and corresponding sides.
2. State and use the properties of congruent triangles to calculate unknown lengths and angles.
3. Identify similar figures and corresponding sides.
4. State and use the properties of similar triangles to calculate unknown lengths and angles.
5. Find the sides and /or area of a plane figure from that of a similar figure.
6. Find the sides and /or volume of a solid figure from that of a similar figure.
7. Use scales to calculate real/actual lengths.

Gaspard Monge, Comte de Péluse

French mathematician Gaspard Monge began teaching physics at Collège de la Trinité at the age of 16. Between 1771 and 1789 Monge published a number of essays in which he pioneered a new branch of mathematics known as descriptive geometry.

Roger Viollet/Getty Images

Congruent Figures

Congruent figures are figures that have the same shape and the same size. Congruent plane figures are figures, which can fit on each other. The statement '*A* is congruent to *B*' is written $A \equiv B$.

Congruent Triangles

Conditions for Triangles to be Congruent

1. If two triangles are such that the three sides of one are equal to the three sides of the other as in Figure 33:1, then they are congruent by **side-side-side** abbreviated **SSS**.

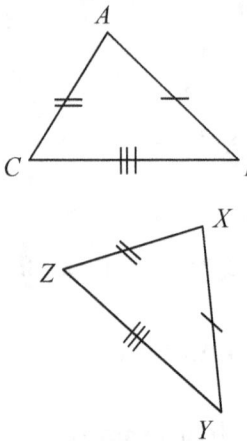

Figure 33:1

$$AB = XY$$
$$BC = YZ$$
$$AC = XZ$$

2. If two triangles are such that two sides of one are equal to two sides of the other and the included angles are equal as in Figure 33:2, then they are congruent by **side -included angle-side** abbreviated **SAS**.

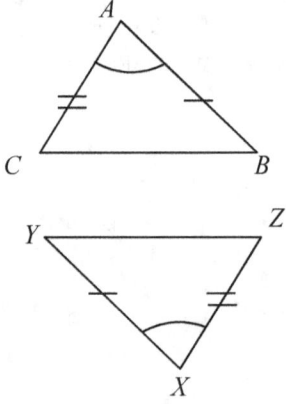

Figure 33:2

$$AB = XY$$
$$AC = XZ$$
$$\angle A = \angle X$$

3. If two triangles are such that two angles of one are equal to two angles of the other and one side of another is equal to one side of the other as in Figure 33:3 they are congruent by **angle-side-angle** abbreviated **ASA**.

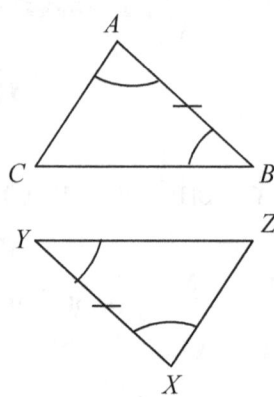

Figure 33:3

$$\angle A = \angle Y$$
$$\angle C = \angle Z$$
$$AC = YZ$$

4. If two right-angled triangles have equal hypotenuse and one arm of another is equal to one arm of the other as in Figure 33:4, then they are congruent by **right angle-hypotenuse- side** abbreviated **RHS**

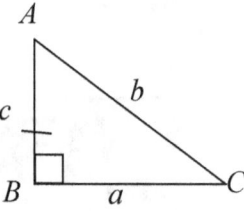

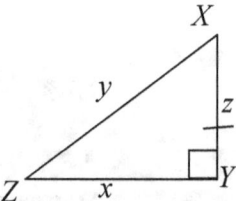

Figure 33:4

$$B = Y = 90°$$
$$AC = XZ$$
$$AB = XY$$

Example 33:1

In Figure 33:5, $PR = PS$ and Q and T are the mid-points of PR and PS respectively.

Determine the pairs of triangles, which are congruent giving arguments and reasons leading to your answer.

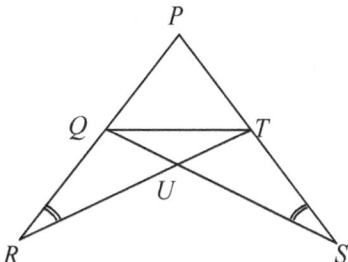

Figure 33:5

Solution

The solution is outlined in Table 33:1.

Argument	Reason
$PS = PR$	Given
$PQ = PT$	P and Q are mid-points of PR and PS
$\angle RPS$ is common to $\triangle PRT$ and $\triangle PSQ$	Shown on diagram
$\therefore \triangle PRT \equiv \triangle PSQ$	SAS
$RT = SQ$	$\triangle PRT = \triangle PSQ$ as proven above
$RQ = ST$	P and Q are mid-points of PR and PS
QT is common to $\triangle QRT$ and $\triangle TSQ$	Shown on diagram
$\therefore \triangle QRT \equiv \triangle TSQ$	SSS
$\angle QUR = \angle TUS$	Vert. opp. angles
$\angle QRT = \angle TSQ$	$\triangle QRT = \triangle TSQ$
$\angle RQU = \angle RTS$	Sum of $\angle$s of $\triangle$
$\therefore \triangle QRU \equiv \triangle TSU$	ASA

Table 33:1

EXERCISE 33:1

In problems 1 to 7, determine which pair of triangles, are congruent giving arguments leading to your answer.

1. In Figure 33:6 PQ is parallel to SR.

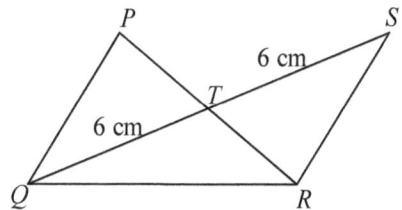

Figure 33:6

2. In Figure 33:7 triangle ABE is isosceles $BC = DE = 3$ cm .

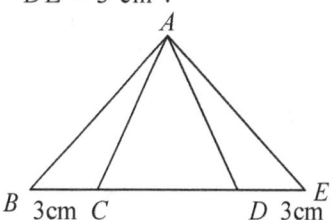

Figure 33:7

3. In Figure 33:8, O is such that $OY = OZ$ and $\angle YOZ = \angle ZOX = \angle XOY = 120°$.

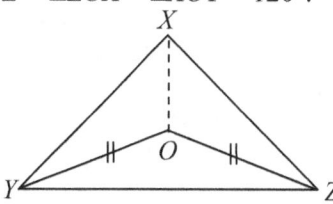

Figure 33:8

4. In Figure 33:9, $OM = NL$, $\angle OMP = \angle LNQ$ and $\angle POM = \angle QLN$.

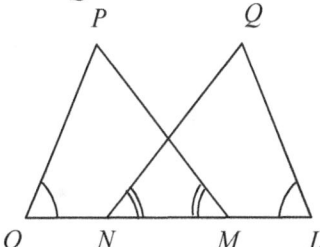

Figure 33:9

5. In Figure 33:10, $PQRS$ is a parallelogram.

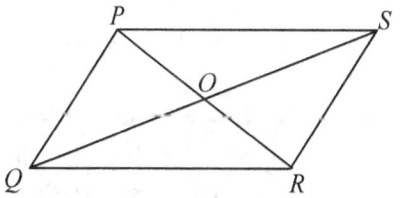

Figure 33:10

6. In Figure 33:11, $PT = TR$ and PQ is parallel to SR.

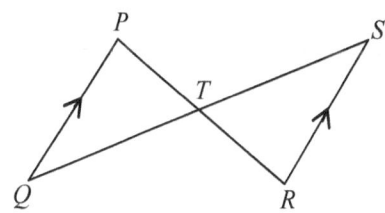

Figure 33:11

7. In Figure 33:12, $XO = WO$ and XW is parallel to YZ.

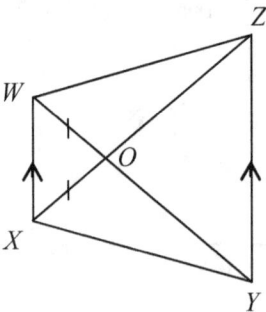

Figure 33:12

8. In Figure 33:13, triangle *ABC* is isosceles and *D*, *E* and *F* are mid-points of *AB*, *BC* and *AC* respectively. How many sets of congruent triangles are there in the figure? List each of these sets.

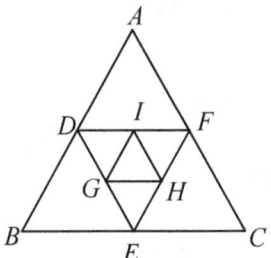

Figure 33:13

Similar Figures

Similar figures are figures that have the same shape but not necessarily the same size. The statement '*A* is similar to *B*' is written *A///B*.

Conditions for Triangles to be Similar

1. If two triangles are equiangular, then they are similar by **angle-angle-angle** abbreviated **AAA**.

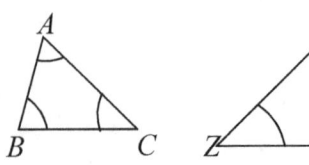

Figure 33:14

$$\angle A = \angle X$$

$$\angle B = \angle Y$$

$$\angle C = \angle Z$$

2. If the corresponding sides of two triangles are in a common ratio then they are

similar by **side- side-side** abbreviated **SSS**.

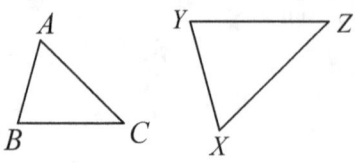

Figure 33:15

$$\frac{AB}{XY} = \frac{AC}{XZ} = \frac{BC}{YZ}$$

3. If an angle of one triangle is equal to an angle of another triangle and the sides containing this angle are in a common ratio then they are similar by **side- included angle-side** abbreviated **SAS**.

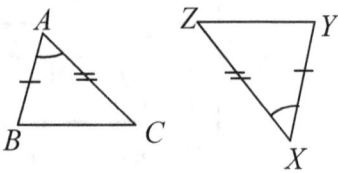

Figure 33:16

$$\angle A = \angle X$$

$$\frac{AB}{XY} = \frac{AC}{XZ}$$

Example 33:2
Given that in Figure 32:17, $\triangle ABC \sim \triangle XYZ$. Find the value of *x*.

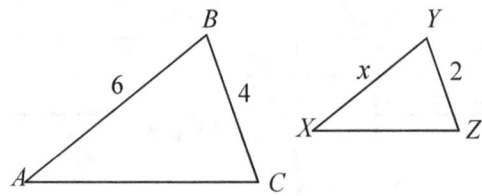

Figure 33:17

Solution

$$\frac{XY}{AB} = \frac{YZ}{BC}$$

$$\frac{x}{6} = \frac{2}{4}$$

$$x = 3$$

Example 33:3
In Figure 33:18, *PQ* ∥ *ST* calculate the value of *y*.

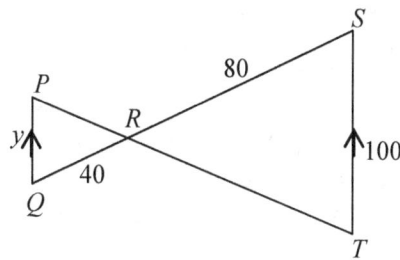

Figure 33:18

Solution

Since $PQ \parallel$ $T \sim \Delta RQP \Rightarrow \dfrac{PQ}{TS} = \dfrac{RQ}{RS}$

$$\dfrac{y}{100} = \dfrac{40}{80}$$

$$\Rightarrow y = 50$$

EXERCISE 33:2

1. In Figure 33:19, CAD is a right-angled triangle with BE drawn parallel to CD. Given that $BE = 10$ cm, $AE = 6$ cm and $ED = 12$ cm, find
   (i) AB (ii) CD.

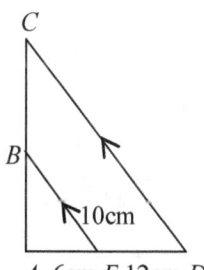

Figure 33:19

2. Determine which of the following sets consist entirely of elements, which are similar figures:
   (a) {triangles}
   (b) {squares}
   (c) {rectangles}
   (d) {parallelograms}
   (e) {rhombuses}
   (f) {trapeziums}
   (g) {hexagons}
   (h) {circles}
3. In Figure 33:20, $AB = 10$ cm, $PB = 3$ cm, $BC = 20$ cm and $AN = 9$ cm. AN is perpendicular to BC. If PQ is parallel to

BC, calculate
   (i)   the length of PQ
   (ii) the area of triangle APQ.

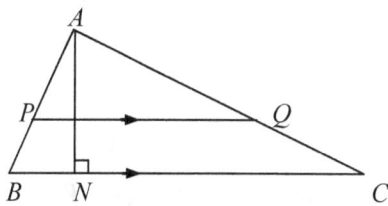

Figure 33:20

4. In Figure 33:21 the lengths of EC and AD are 8 cm and 6 cm respectively, $\angle DAE = 60^\circ$ and $\angle AED = \angle ACB = 90^\circ$. Calculate the length of BC in cm to one decimal place.

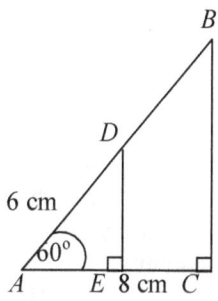

Figure 33:21

5. In Figure 33:22, triangle XYZ of sides 9, 15 and n cm is an enlargement of triangle PQR of sides' m, 5 and 4.5 cm. Evaluate m and n.

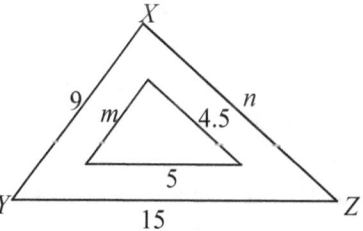

Figure 33:22

Ratio of Areas of Similar Figures

Consider the two rectangles in Figure 33:23, which are such that $PQRS$ is an enlargement of $ABCD$ with scale factor k.

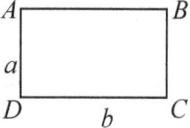

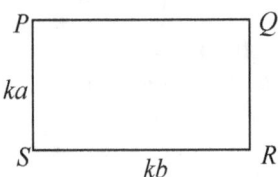

Figure 33:23

Area of $ABCD = ab$

Area of $PQRS = (ka)(kb) = k^2ab$

$$\frac{\text{Area of } PQRS}{\text{Area of } ABCD} = \frac{k^2ab}{ab} = k^2$$

$$\Rightarrow \frac{\text{Area of } PQRS}{\text{Area of } ABCD} = k^2$$

Therefore if the ratio of corresponding sides of two similar plane figures is $m:n$, then the ratio of their areas is $m^2 : n^2$

Example 33:4

The ratio of the corresponding sides of two similar triangles is 1:3. Calculate the area of the larger triangle given that the area of the smaller triangle is 8 cm^2.

Solution

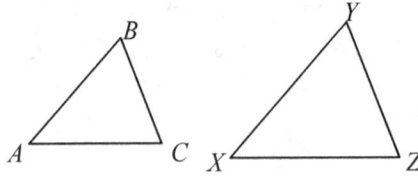

Figure 33:24

$$\frac{\text{Area of } \Delta XYZ}{\text{Area of } \Delta ABC} = \frac{3^2}{1^2}$$

Area of $\Delta XYZ = 9 \times$ Area $\Delta ABC = 9 \times 8$
$$= 72 \text{ cm}^2$$

EXERCISE 33:3

1. On a map in which 1 cm represents 2 km, a plot of land is represented by a square of length 2.5 cm. Calculate the actual area, in km^2, of the plot of land.
2. The sides of a triangle are 5 cm, 6 cm, and 7 cm. The longest side of a similar triangle is 28 cm. Find the lengths of the other sides and the ratio of the area of the smaller triangle to the larger one.

3. The sides of a triangle are 6 cm, 9 cm, and 15 cm. The shortest side of a similar triangle is 2 cm. Find the lengths of the other sides and the ratio of the area of the smaller triangle to the larger one.
4. Given that the two trapeziums in Figure 33:25 are similar, find the value of x.

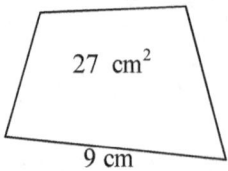

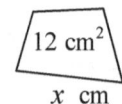

Figure 33:25

5. XY is parallel to BC and $\dfrac{AB}{AX} = \dfrac{3}{2}$. If the area of $\Delta AXY = 4$ cm^2, find the area of ΔABC.

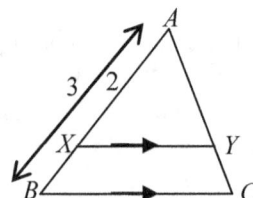

Figure 33:26

6. If the triangles in Figure 33:27 are similar, find the area of the smaller triangle.

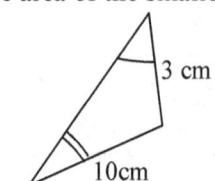

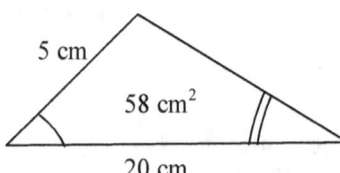

Figure 33:27

7. Two rectangles have areas 4 cm^2 and 9 cm^2 and their corresponding sides are x cm and 3 cm. Find the value of x.
8. In Figure 33:28, AB is parallel to DC, $AB = 32$ cm, $BE = 12$ cm and $DE = 3$ cm.
   (a) Show that the triangles ABE and CDE

352

are similar. Calculate:
(b) The length CD.
(c) The ratio of the areas of $\triangle ABE:\triangle CDE$.

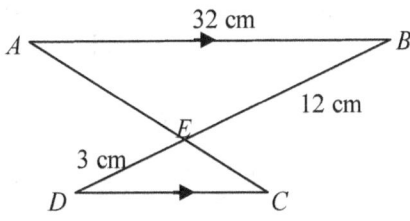

Figure 33:28

Ratio of Volumes of Similar Figures

Consider the two cylinders in Figure 32:29, which are such that C_2 is an enlargement of C_1, with scale factor k.

Figure 33:29

Volume of $C_1 = \pi r^2 h$

Volume of $C_2 = \pi(kr)^2(kh)$

$$\Rightarrow \frac{\text{Volume of } C_2}{\text{Volume of } C_1} = \frac{\pi(kr)^2(kh)}{\pi r^2 h}$$

$$\frac{\text{Volume of } C_2}{\text{Volume of } C_1} = k^3$$

Therefore if the ratio of corresponding sides of two similar solid figures is $m:n$, then the ratio of their volumes is $m^3:n^3$

Example 33:5
A cube whose volume is 20 cm^3, is enlarged such that its volume now is V cm^3. Given that, the scale factor of the enlargement is 4. Find the value of V.

Solution

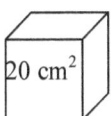 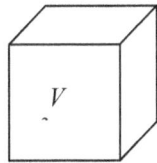

Figure 33:30

Let V_0 = volume of original cube

Then, $\dfrac{V}{V_0} = k^3 \Rightarrow V = k^3 V_0$

$k = 4$ and $V_0 = 20$ cm

$\Rightarrow V = 4^3 \times 20 = 1280$ cm^3

EXERCISE 33:4

1. Two similar containers have heights of 3 cm and 5 cm respectively. If the capacity of the smaller container is 54 cm^3, find the capacity of the larger container.
2. Two similar flasks are such that the ratio of the volume of the larger to the smaller is 7:2. Calculate the volume of the larger flask if that of the smaller is 56 cm^3. Given that the height of the smaller is 10 cm, find the height of the larger.
3. Figure 33:31, shows two similar bowls with their volumes and the diameter of the smaller one given. Calculate the diameter d cm of the larger bowl.

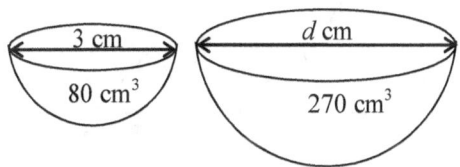

Figure 33:31

4. Two similar buckets have shapes in the form of a frustum. The volume of the smaller one is 24 cm^3 and its slant height is 6 cm. Given that the larger one has a slant height of 9 cm, calculate the volume V of the larger one.
5. In Figure 33:32, find the value of V, if the cones are similar.

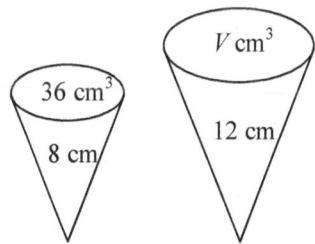

Figure 33:32

6. Given that the cylinders in Figure 33:33, are similar, find the value of y to the nearest tenth.

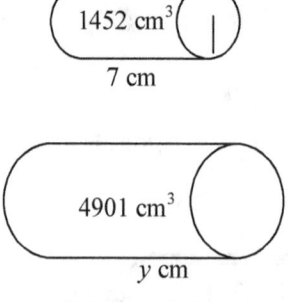

Figure 33:33

7. If the ratio of the diameters of two similar cups is 4:25, find the ratio of their volumes.

8. The surface area of a container of capacity 12.8 liters is 5000 cm². Find the surface area of a similar contain with a capacity of 5.5 liters.

MULTIPLE CHOICE EXERCISE 33

1. The pair of triangles in Figure 33:34 which is definitely congruent is:

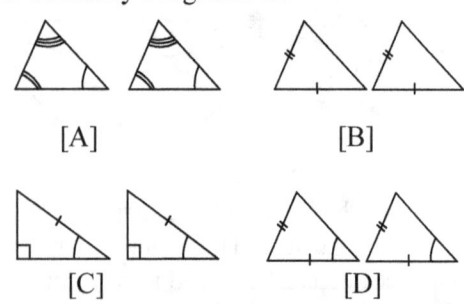

Figure 33:34

2. The pair of triangles in Figure 33:35 which is definitely congruent is:

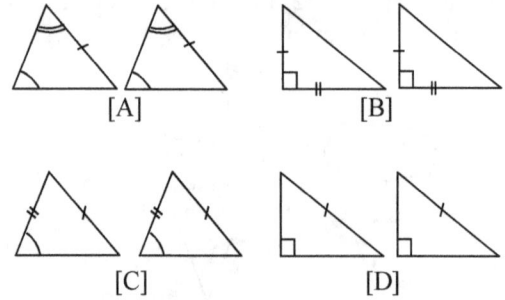

Figure 33:35

3. In three triangles *PQR*, *DEF* and *XYZ*: triangles *PQR* and *DEF* are equiangular but not congruent; triangles *DEF* and *XYZ* are congruent. It follows that triangles:

[A] *PQR* and *DEF* are equal in area
[B] *PQR* and *XYZ* are congruent
[C] *PQR* and *XYZ* are equal in area
[D] *PQR* and *XYZ* are similar

4. The pair of triangles in Figure 33:36 (not drawn to scale) which is similar is:

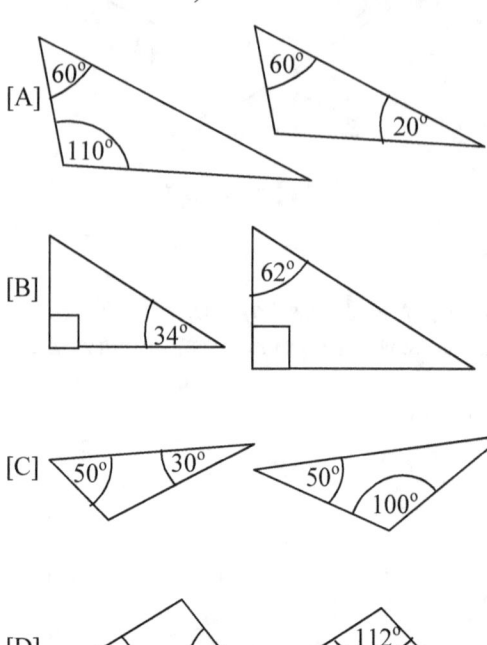

Figure 33:36

5. The pair of triangles in Figure 33:37 which is similar is:

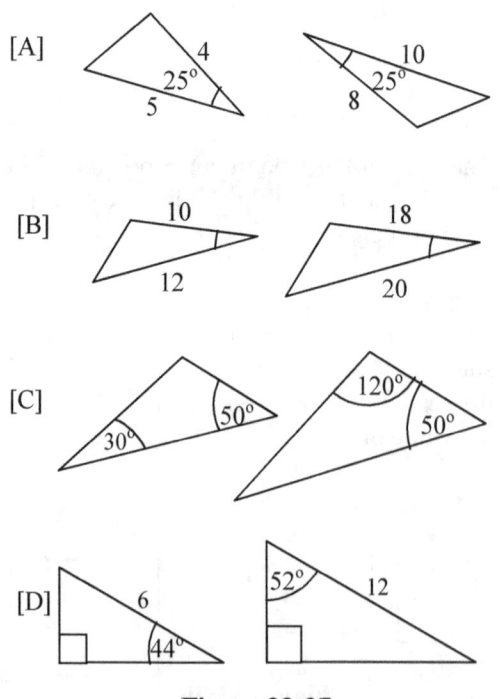

Figure 33:37

6. Given that the triangles in Figure 33:38 are similar. It follows that:

[A] $\dfrac{AC}{XY} = \dfrac{YZ}{XZ}$ [B] $\dfrac{AC}{XY} = \dfrac{BC}{YZ}$

[C] $\dfrac{BC}{AB} = \dfrac{YZ}{XZ}$ [D] $\dfrac{BC}{AB} = \dfrac{XZ}{YZ}$

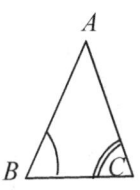

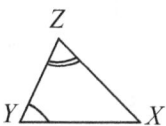

Figure 33:38

7. In Figure 33:39, if $\dfrac{AB}{XY} = \dfrac{AC}{XZ}$ and $\angle B = \angle Y$. Then:

[A] $\dfrac{AB}{XY} = \dfrac{BC}{YZ}$ [B] $\Delta A = \Delta X$ [C] $\Delta C = \Delta Z$

[D] None of the above is necessarily true.

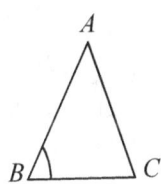

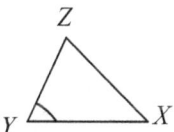

Figure 33:39

8. In Figure 33:40, $\angle A = \angle X$ and $\angle B = \angle Y$. Hence, XY is equal to:

[A] $6\dfrac{7}{8}$ cm   [B] $17\dfrac{3}{5}$ cm

[C] $19\dfrac{1}{5}$ cm   [D] $8\dfrac{1}{2}$ cm

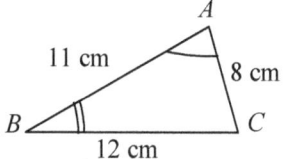

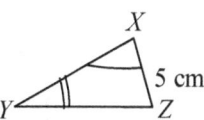

Figure 33:40

9. In Figure 33:41, $PS = 8$ cm and $QS = 2$ cm. Hence $\dfrac{ST}{QR}$ is equal to:

[A] $\dfrac{1}{4}$ [B] $\dfrac{4}{1}$ [C] $\dfrac{4}{5}$ [D] $\dfrac{5}{4}$

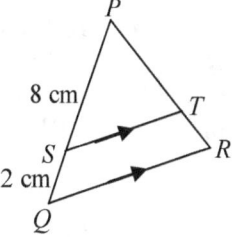

Figure 33:41

10. In Figure 33:42, ST and QR are parallel. $|PS| = 6$ cm, $|SQ| = 8$ cm, $|PR| = 18\dfrac{2}{3}$ cm. $|PT|$ is equal to:

[A] 7 cm   [B] 8 cm

[C] $8\dfrac{2}{3}$ cm   [D] 10 cm

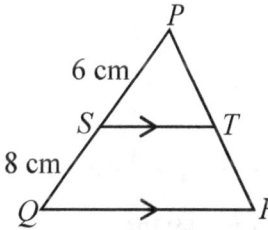

Figure 33:42

11. In Figure 33:43, $|AB| = 12$ cm, $|AE| = 8$ cm, $|DC| = 9$ cm and $AB\|DC$. The length $|EC|$ is:

[A] 10 cm   [B] 9 cm
[C] 8 cm   [D] 6 cm

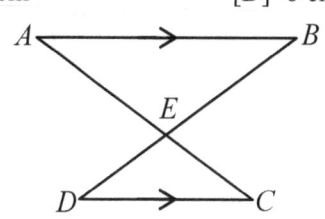

Figure 33:43

12. In Figure 33:44, $\angle PMN = \angle PRQ$ and $\angle PNM = \angle PQR$. If $|PM| = 3$ cm, $|MQ| = 7$ cm

and |PN|=5 cm, |NR|=

[A] 1 cm   [B] 3 cm

[C] $3\frac{1}{2}$ cm   [D] 5 cm

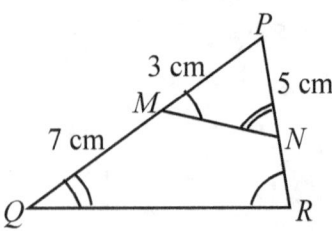

Figure 33:44

13. In Figure 33:45, $EF\|QR$, $PE = 2$ cm, $EQ = 4$ cm and $FR = 6$ cm. x should be:
[A] 2 cm   [B] 3 cm
[C] 4 cm   [D] 6 cm

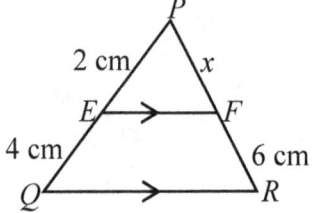

Figure 33:45

14. The value of x in Figure 33:46 is:
[A] 6.8 cm   [B] 6.6 cm
[C] 6.5 cm   [D] 5.6 cm

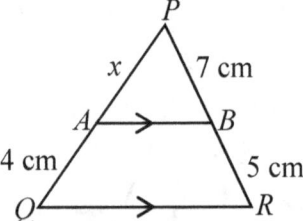

Figure 33:46

15. In Figure 33:47, XY is parallel to BC and AB is parallel to YZ.

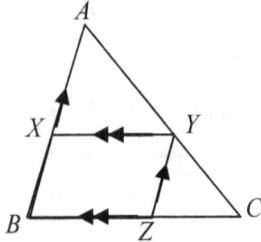

Figure 33:47

Therefore:

[A] $\angle ABC = \angle ZYC$

[B] $\frac{YZ}{ZC} = \frac{AC}{BC}$

[C] $\triangle ABC$ is similar to $\triangle ZYC$

[D] $\frac{ZC}{AC} = \frac{YZ}{AB}$

16. In Figure 33:48 AB is parallel to DC, $AB = 3$ cm and $DC = 5$ cm. Hence $\frac{XD}{XB}$ is equal to:

[A] $\frac{3}{5}$ [B] $\frac{5}{3}$ [C] $\frac{5}{8}$ [D] $\frac{8}{5}$

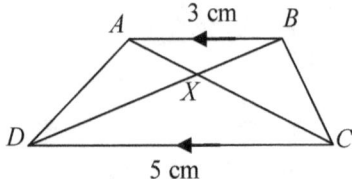

Figure 33:48

17. The triangles in Figure 33:49 are:
[A] congruent
[B] similar
[C] identical
[D] none of the above

Figure 33:49

18. The ratio of the areas of the two triangles shown in the Figure 33:50 is:

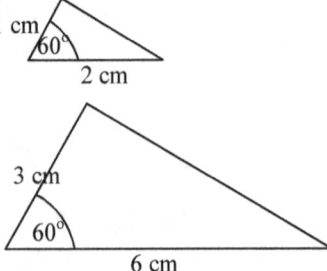

Figure 33:50

[A] 1:6   [B] 1:2   [C] 1:4 [D] 1:9

19. Figure 33:51, shows a right-angled triangle ACD.

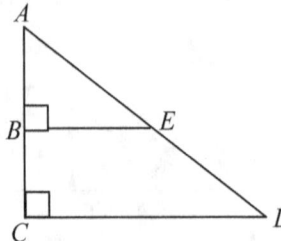

Figure 33:51

$AB = 6$ cm, $AC = 8$ cm, $CD = 6$ cm and

BE is parallel to *CD*. The value of the length *BE* is:

[A] 9 [B] 6 [C] 4.5 [D] 4

20. Figure 33:52, shows two right-angled triangles *OMN* and *OPQ*. Given $\frac{OM}{OP} = \frac{1}{4}$, *MN* is parallel to *PQ*, *OM* = 1 cm, *ON* = 2 cm. The area of *POQ* in cm² is:

[A] 20 [B] 16 [C] 12 [D] 8

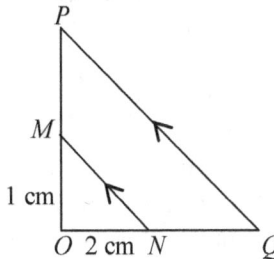

Figure 33:52

21. On a map drawn to a scale of 2 cm representing 1 km, the area represented by a square of side 4 cm is:

[A] 2 km² [B] 4 km [C] 4 km² [D] 1 km²

22. Two similar cylinders have heights of 3 cm and 6 cm respectively. The ratio of their volumes is:

[A] 1:4 [B] 1:8 [C] 2:5 [D] 1:2

23. In Figure 33:53 triangle *ABC* is similar to triangle *AED* and *AB*=16 cm, *AE*= 8 cm and *AC*=14 cm. The value of the length of the side marked *x* is:

[A] 7 cm

[B] $\frac{80}{7}$ cm

[C] $\frac{70}{8}$ cm

[D] 6 cm

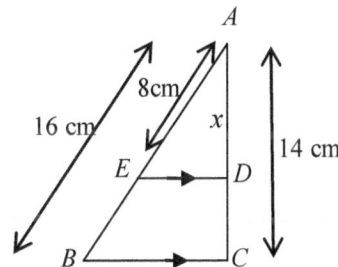

Figure 33:53

24. In Figure 33:54, if the area of *△XYZ* is 10 cm², then the area of *△ABC* is:

[A] 160 cm² [B] 40 cm² [C] 90 cm²
[D] insufficient information.

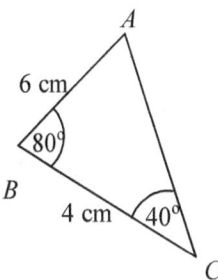

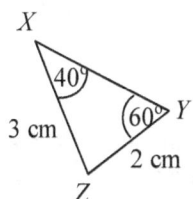

Figure 33:54

25. In Figure 33:55, *△ABC* is similar to *△DEF*. Given that the area of *△ABC* is 20 cm², then the area of *△DEF* is:

[A] 10 cm²
[B] 5 cm²
[C] 8 cm²
[D] None of the above

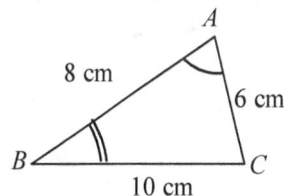

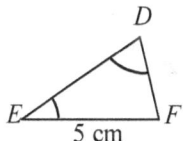

Figure 33:55

26. In Figure 33:56, *∠A*=*∠X* and *∠B*=*∠Y*. *△ABC* has an area of 36 cm² and *△XYZ* has an area of 4 cm². If *AB* = 4 cm , then *XY* is equal to:

[A] $\frac{3}{4}$ [B] $\frac{4}{3}$ [C] $\frac{4}{4}$ [D] $\frac{9}{4}$

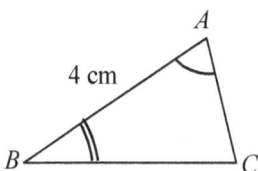

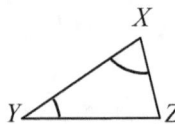

Figure 33:56

27. Two buckets *A* and *B*, identical in shape, are such that the dimensions of A are three times as large as the corresponding dimensions of *B*. The ratio of the volumes of *A:B* is:
    [A] 1:9   [B] 27:1
    [C] 9:1   [D] 1:27

28. The pairs of triangles *PQR*, *XYZ* are congruent is:
    [A] $XY = PQ, XZ = QR, \angle X = \angle Q$

    [B] $XZ = QR, YZ = PR, \angle Y = \angle P$

    [C] $\angle Y = \angle P, \angle Z = \angle Q, XZ = PQ$

    [D] $\angle Z = \angle P, \angle Y = \angle Q, XY = PR$

29. Similar triangles differ from congruent triangles in that:
    [A] The areas of congruent triangles are not necessarily equal but the areas of similar triangles are necessarily equal.
    [B] The areas of congruent triangles are necessarily equal but the areas of

similar triangles are not necessarily equal.
    [C] Similar triangles are necessarily congruent but congruent triangles are not necessarily similar.
    [D] The sides of similar triangles are necessarily equal but the sides of congruent triangles are not necessarily equal.

30. The set, which consist entirely of elements, which are similar figures is:
    [A]   Triangles
    [B]   Quadrilaterals
    [C]   Circles
    [D]   Hexagons

31. The false statement (s) is/are:
    [A] All Similar objects have the same shape but not necessarily the same size.
    [B] All similar objects are congruent.
    [C] All congruent objects are similar.
    [D] All similar objects are congruent and have the same shape but not necessarily the same size.

TOPIC 34

GRAPHS OF FUNCTIONS

OBJECTIVES

At the end of this topic, the learner should be able to:

1. Graph a straight line given a table of values.
2. Make a table of values and use it to plot a graph.
3. Draw and identify corresponding, inconsistent and intersecting lines.
4. Solve simultaneous linear equations using the graphical method.
5. Draw the graph of a quadratic function.
6. Plot smooth curves.
7. Identify maximum and minimum points and determine them.
8. Find the solutions of a quadratic equation from a graph.
9. Solve graphically simultaneous equations one linear-one quadratic.
10. Solve quadratic equations from a graph of any other quadratic function.
11. Draw tangents to a curve at a point and find their gradients.
12. Define gradient.
13. Appreciate gradient as a rate of Change.
14. Use elementary differentiation to find maximum and minimum values, velocity and acceleration of a moving object.
15. Draw speed time graphs and use them to find distance, speed and acceleration at a given time.

Augustin Louis Cauchy

French mathematician Augustin Louis Cauchy provided logical foundations to calculus by developing the theory of limits. Cauchy became a leader in 19th-century mathematics and was one of the first mathematicians to publish his work extensively.

Roger Viollet/Getty Images

Graphing Straight Lines from Tables

Given the table of values of x and y for a straight line, a graph of the line can be drawn.

Example 34:1

The following is a table of values for x and y. Draw a graph for these values.

x	-4	-2	0
y	3	0	-3

Solution

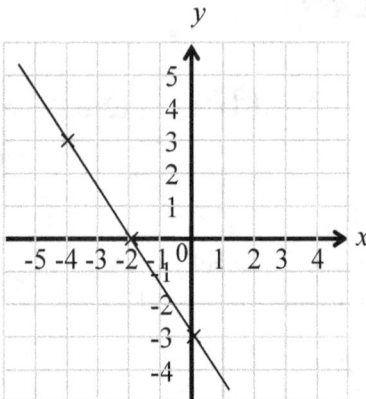

Figure 34:1

Example 34:2

Draw the straight line represented by the following table of values.

x	-4	-2	0
y	0	3.5	7

Solution

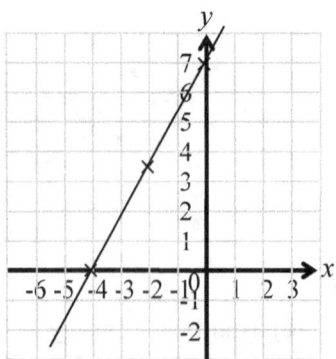

Figure 34:2

The table of values can equally be made from the equation of a straight line and the required straight line drawn. In this case write the equation in the form $y = mx + c$, and then substitute integral values of the independent variable x to obtain corresponding values of the dependent variable y.

For easy calculations, it is wiser to chose small integral values of x. Values within the range $-6 \le x \le 6, x \in \mathbb{R}$ are often very convenient. Unless where impossible avoid fractional values, for easy plotting. Two values are sufficient to draw a straight line, and a third point, acts as a check for the other two.

Example 34:3

Make a table of values of x against y and use this table to draw a graph of the line $y = 2x + 5$.

Solution

x	-2	0	1
y	1	5	7

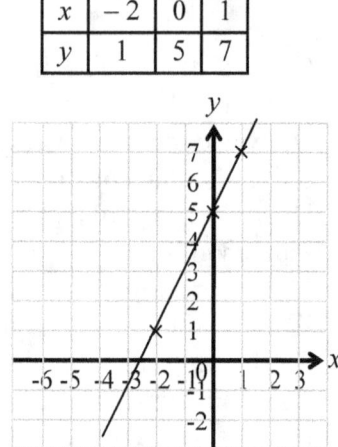

Figure 34:3

Example 34:4

The equation of a straight line is given as $y - 2 = -2(x - 3)$. Make a table of values and hence draw the graph of this equation.

Solution

$$y - 2 = -2(x - 3)$$
$$\Rightarrow y = -2x + 8$$

x	1	3	5
y	6	2	-2

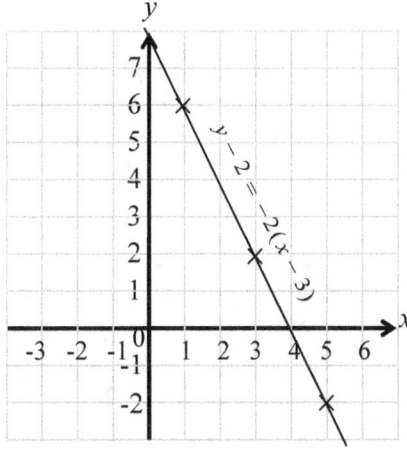

Figure 34:4

Corresponding, Inconsistent and Intersecting Lines

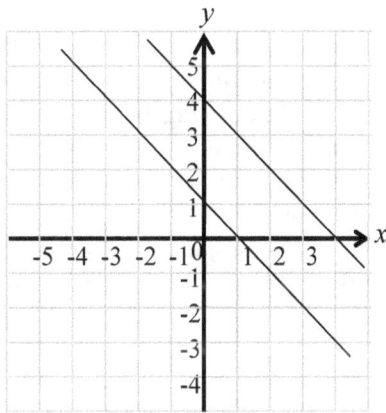

Figure 34:5

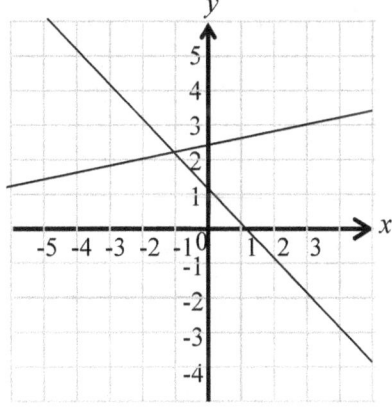

Figure 34:6

If two lines are **parallel** (Figure 34:5), they do not meet anywhere. Such lines are **inconsistent**. It is also possible for two lines to be identical. If the equations $y = x + 2$ and $3y = 3x + 6$ are graphed, they will be the same. This is because one equation is simply a multiple of the other. Such lines are **corresponding lines** and their equations are **dependent**. Some other lines meet at a point and are therefore called **intersecting lines**. This is illustrated in Figure 34:6.

Simultaneous Linear Equations (Graphical Method)

To solve simultaneous linear equations using the graphical method, plot the graphs of the two equations on the same Cartesian axes. The coordinates of the point of intersection of the lines gives the solution of the equation.

Example 34:5
Solve the simultaneous equations $y = 2x$ and $y + x = 3$, using the graphical method.

Solution

$y = 2x$

x	0	1	2
y	0	2	4

$y + x = 3$

x	0	2	4
y	3	1	-1

The graph is shown in Figure 34:7.

From Figure 34:7, the two lines meet at the point (1,2). Therefore, the solution of the simultaneous equations is $x = 1, y = 2$.

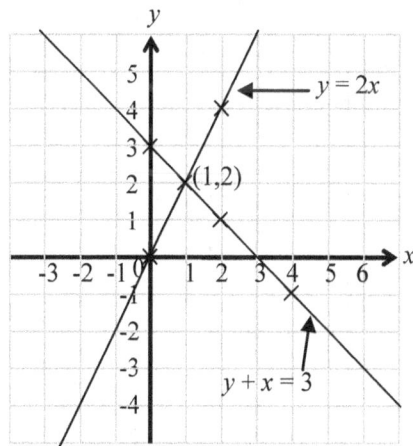

Figure 34:7

361

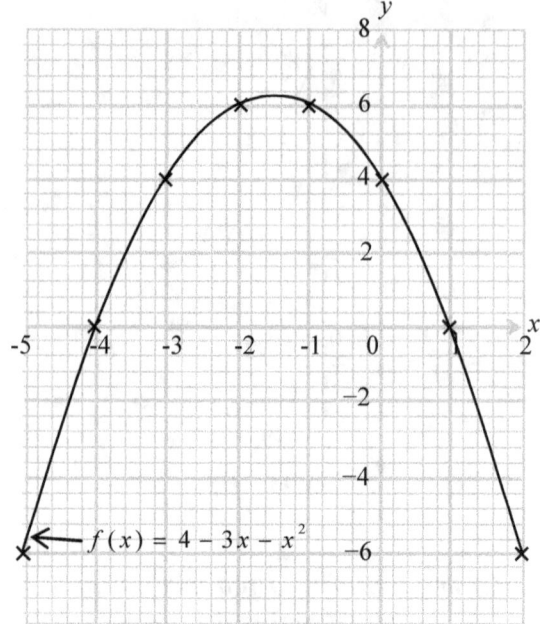

EXERCISE 34:1

Find graphically the solutions of the simultaneous equations.

1. $3x - 3y = 5, \quad 3x + 2y = 14$

2. $7x + 3y = 6, \quad 5y - 9x = 10$

3. $6x + 5y = 11, \quad 5x - 2y = 3$

4. $5x - y = 9, \quad y - 2x = 3$

5. $7x + 2y = -3, \quad 3x - 5y = 28$

6. $4x + 3y = 1, \quad 6x - 5y = -8$

7. $3x - 7 = y, \quad 4x - 5y = 2$

8. $y = 4x, \quad 3x + y = 21$

9. $5x - 2y = 9, \quad x - 5y = 7$

10. $3x + 4 = 0$

Hint: The solution will be the point where the line $y = 3x + 4$ cuts the x-axis.

Graphs of Quadratic Functions

Recall that a quadratic function is of the form $f(x) = ax^2 + bx + c$, where a, b and c are constants and $a \neq 0$.

To plot the graph, first make a table of values of y against x.

Example 34:6

Given that $f(x) = 4 - 3x - x^2$, make up a table of values of x against $y = f(x)$, for integral values of x in the range $-5 \leq x \leq 2, x \in \mathbb{R}$.
Taking 1 cm to represent 1 unit on both axes, draw the graph of $y = f(x)$.

Solution

x	$4 - 3x - x^2$	$y = f(x)$
-5	$4 + 15 - 25$	-6
-4	$4 + 12 - 16$	0
-3	$4 + 9 - 9$	4
-2	$4 + 6 - 4$	6
-1	$4 + 3 - 1$	6
0	$4 + 0 - 0$	4
1	$4 - 3 - 1$	0
2	$4 - 6 - 4$	-6

Figure 34:8

Hints for Drawing Smooth Curves

1. To ensure the smoothness of a curve, make sure that the drawing hand is always inside the curve when drawing the graph, even if it means turning the graph paper or book upside down.
2. Use a pencil to sketch the curve very lightly and retrace it when it is judged very smooth.

Example 34:7

Plot the graph of $y = x^2 - 5x + 4$ for integral values of x from 0 to $+5$.

Solution

x	$x^2 - 5x + 4$	y
0	$0 - 0 + 4$	4
1	$1 - 5 + 4$	0
2	$4 - 10 + 4$	-2
3	$9 - 15 + 4$	-2
4	$16 - 20 + 4$	0
5	$25 - 25 + 4$	4

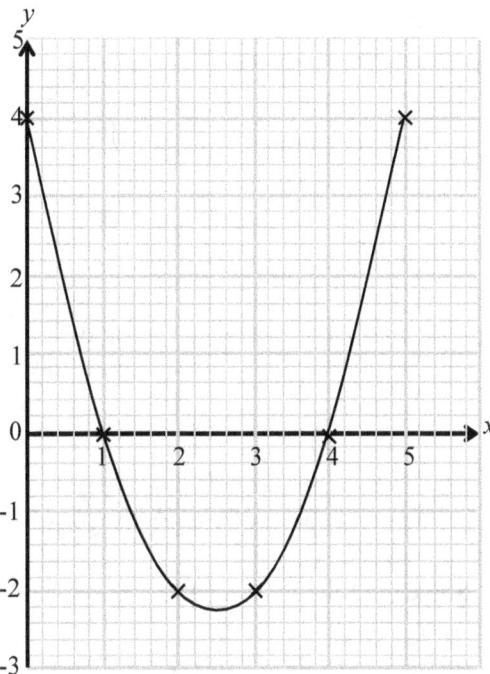

Figure 34:9

Turning Points

The **turning point** is the point where the curve changes direction. If any horizontal line is drawn to cut the curve at two points, the x-coordinate of the turning point must be between the points of intersection, since the quadratic curve is symmetrical about the perpendicular line through the turning point. This idea can be used to find the x-coordinate of the midpoint, which is equally the x-coordinate of the turning point. By substituting this value into the equation $y = f(x)$, the y-coordinate of the turning point can be found.

Example 34:8
Find the turning point on the quadratic curve $4 - 3x - x^2$ and the point where the curve cuts the y-axis.

Solution
The intercepts with the x-axis occur when,

$f(x) = 0$ i.e. $4 - 3x - x^2 = 0$

$$(4 + x)(1 - x) = 0$$

$\Rightarrow \quad x = -4$ or $x = 1$

But midpoint $= \dfrac{x_1 + x^2}{2} = \dfrac{-4 + 1}{2} = -\dfrac{3}{2}$

Therefore, the graph is symmetrical about the line $-\dfrac{3}{2}$.

Substitute in $f(x) = 4 - 3x - x^2$.

$$f\left(-\frac{3}{2}\right) = 4 - 3\left(-\frac{3}{2}\right) - \left(-\frac{3}{2}\right)^2 = \frac{25}{4}$$

Therefore, the turning point is $\left(-\dfrac{3}{2}, \dfrac{25}{4}\right)$.

The intercept with the y-axis occurs when $x = 0$.

$\Rightarrow f(0) = 4$ is the intercept with the y-axis.

From Figure 30:8 and Figure 30:9, it can be seen that;

1. A graph of quadratic function has either an ∩ shape or a ∪ shape. The graph has a maximum point if it has an ∩ shape and a minimum point if it has a ∪ shape. A maximum point exist if the coefficient of x^2 is less than zero (i.e. negative) whereas a minimum point exist if the coefficient of x^2 is greater than zero (i.e. positive).

2. A graph of quadratic function is symmetrical about an axis through the turning point.

3. The intercept with the y-axis is c.

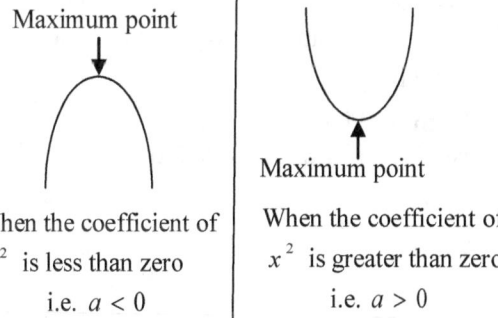

Maximum point

When the coefficient of x^2 is less than zero i.e. $a < 0$

Maximum point

When the coefficient of x^2 is greater than zero i.e. $a > 0$

Figure 34:10

There are six different types of graphs of quadratic functions depending on the discriminant ($\Delta = b^2 - 4ac$) and the value of the coefficient a of x^2.

For real and distinct roots, the graph cuts the x-axis at two points

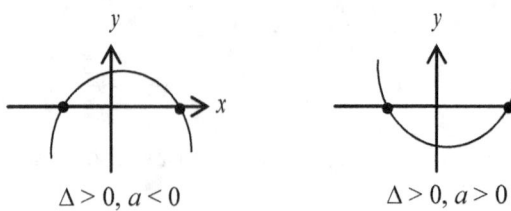

$\Delta > 0, a < 0$ $\Delta > 0, a > 0$

For real and equal roots, the graph touches the x-axis at one and only one point.

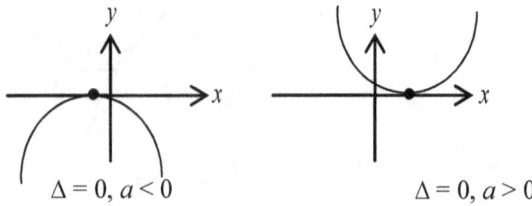

$\Delta = 0, a < 0$ $\Delta = 0, a > 0$

For imaginary roots, the graph will neither cut nor touch the x-axis

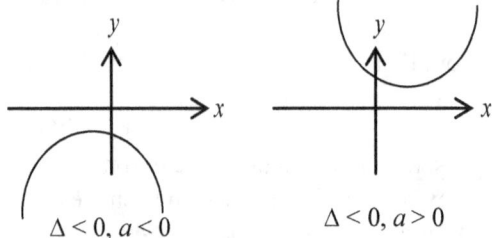

$\Delta < 0, a < 0$ $\Delta < 0, a > 0$

Figure 34:11

Graphical Solutions of Quadratic Equations

Draw the graph of $y = ax^2 + bx + c$, and find the roots of the equation $ax^2 + bx + c = 0$ from the graph. The roots are the values of x for which the line $y = 0$ (the x-axis) intersects the graph.

Example 34:9
Use Figure 34:8 to find the roots of the equation $0 = 4 - 3x - x^2$.

Solution
From Figure 34:8, the roots of the equation $y = 4 - 3x - x^2$ are $x = -4$ and $x = 1$ where the curve cuts the x-axis.

Example 34:10
Use Figure 34:9 to find the roots of the equation $x^2 - 5x + 4 = 0$.

Solution
From Figure 34:9 the roots of the equation $x^2 - 5x + 4 = 0$ are $x = 1$ or $x = 4$.

EXERCISE 34:2

1. Sketch the graph of $f(y) = 2y^2 + 2y - 15$, for integral values of y from -4 to $+4$.

2. If $f(x) = 5 + 4x - x^2$,
   (a) Construct a table of values of the function $y = f(x)$ for $-2 \leq x \leq 5$.
   (b) Using a scale of 2 cm to represent 1 unit on the x-axis and 2 cm to represent 2 units on the y-axis, draw the graph of $y = f(x)$ for $-2 \leq x \leq 5$.

   Use your graph to;
   (c) Find the minimum value of $f(x)$.
   (d) Find the range of values of x for which $5 + 4x - x^2 > 0$

3. Solve the following equations graphically
   (i) $x^2 - 1 = 0$ (ii) $4x^2 - 9 = 0$
   (iii) $x^2 - 4x = 0$ (iv) $2x^2 + 7x = 0$
   (v) $x^2 - 2x = 8$ (vi) $4x^2 = 12x - 9$

Simultaneous Equations - One Linear, One Quadratic - Graphical Method

Simultaneous equations-one linear, one quadratic can be solved graphically by determining the point of intersection of the curve and the straight line in the same way as simultaneous linear equations are solved graphically.

Example 34:11
Using the graph in Figure 34:9, solve the simultaneous equations $y = 2x - 6$ and $y = x^2 - 5x + 4$.

Solution
Draw the line $y = 2x - 6$ on the same axes as the graph of $y = x^2 - 5x + 4$. The point of intersection of the two graphs, gives the solution. The table of values of $y = 2x - 6$ is as shown below. Figure 34:12 shows the two graphs.

x	2	3	4
y	3	4	2

The graphs intersect at $(2, -2)$ and $(5,4)$. Hence the solutions of the simultaneous equations are $x = 2$, when $y = -2$ and $x = 5$, when $y = 4$.

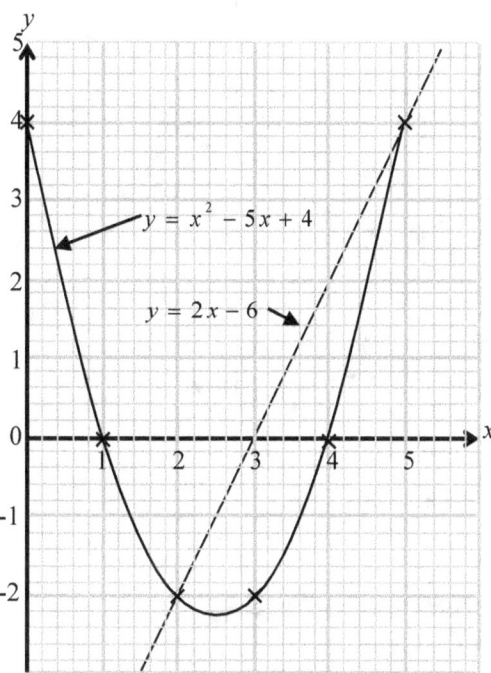

Figure 34:12

Solving Quadratic Equations from the Graph of any Other Quadratic Function

It is possible to solve quadratic equations using the graph any quadratic function. This is done by drawing a suitable straight line on the same Cartesian plane as the graph of quadratic function. The equation of the suitable straight line is obtained by combining the quadratic equation and the quadratic function in such a way that the squared terms are eliminated.

Example 34:12
Use the graph in Figure 34:8 to solve the quadratic equation $x^2 + x - 6 = 0$.

Solution

The equation of the quadratic graph is
Rearranging $x^2 + x - 6 = 0$

$0 = 6 - x - x^2 \ldots \ldots \textcircled{2}$

Subtracting $\textcircled{2}$ from $\textcircled{1}$: $f(x) = -2 - 2x$.

Therefore the required straight line is
$y = -2 - 2x$.

This straight line and the quadratic curve are shown in Figure 34:11.

From the graph, the solution of the quadratic $x^2 + x - 6 = 0$ is $x = -3$ or $x = 2$

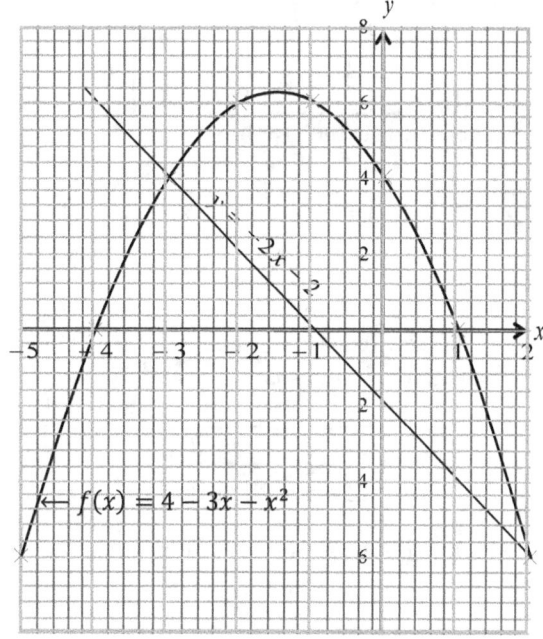

Figure 34:13

Example 34:13

The function f is defined on the set $\mathbb{R}$ of real numbers as $f(x) = x^2 + x - 12$.

(a) Using a scale of 1 cm to represent 2 units on the x-axis and 1 cm to represent 4 units on the y-axis, draw the graph of $y = f(x)$ for $x: -6 \leq x \leq 5$.

(b) By drawing a suitable straight line on your graph, solve the equation $x^2 - 4 = 0$.

(c) Also, solve from your graph the equation $2x^2 + 3x - 2 = 0$.

Solution
The table of values for the function is

x	$x^2 + x - 12$	$f(x)$
–6	$36 - 6 - 12$	18
–5	$25 - 5 - 12$	8
–4	$16 - 4 - 12$	0
–3	$9 - 3 - 12$	–6
–2	$4 - 2 - 12$	–10
–1	$1 - 1 - 12$	–12
0	$0 + 0 - 12$	–12
1	$1 + 1 - 12$	–10
2	$4 + 2 - 12$	–6
3	$9 + 3 - 12$	0
4	$16 + 4 - 12$	8
5	$25 + 5 - 12$	18

Figure 34:14, is the graph.

(b) $f(x) = x^2 + x - 12$..................①

               $0 = x^2 - 14$.........................②

   ① – ② : $f(x) = x + 2$............................③

The required straight line is $y = x + 2$ and its table of values is

x	0	–2	2
y	2	0	4

From the graph (Figure 34:14), the solution of $x^2 - 4 = 0$ is $x \approx -3.8$ and $x \approx 3.8$.

(c) The equation $2x^2 + 3x - 2 = 0$ can be solved in a similar way.

       $f(x) = x^2 + x - 12$.......................④

          $0 = 2x^2 + 3x - 2$...................⑤

④ $- \dfrac{1}{2} \times$ ⑤ : $f(x) = -\dfrac{1}{2}x - 11$

The required straight line is the line

$f(x) = -\dfrac{1}{2}x - 11$ whose table of values is

x	–6	2	6
y	–8	–12	–14

The line is shown in Figure 34:12 and from the graph, the solution of $2x^2 + 3x - 2 = 0$ is $x \approx -2$ and $x \approx 0.5$.

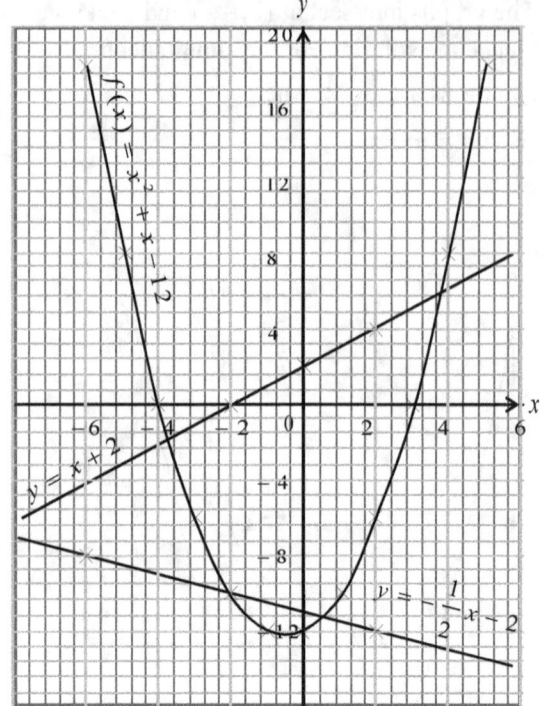

Figure 34:14

> **EXERCISE 34:3**

1. The function $f(x)$ is defined as follows:

   $f(x) = \dfrac{1}{2}(x^2 + 2x - 5)$

   (a) Copy and complete the table below.

x	–5	–4	–3	–2	–1	0	1	2	3
$f(x)$									5

   (b) Taking 2 cm to represent 1 unit on both axes draw the graph of $y = f(x)$.

   (c) Write down the coordinates of the turning point.

   (d) Use your graph to estimate the value of x when $f(x) = 1.8$.

   (e) On the same axes, draw the line $y = 3x - 4$.

   (f) Write down the equation whose solutions are the points of intersection of the curve and the line $y = 3x - 4$.

2. (a) Complete the table below for the function $f(x) = 2x + \dfrac{1}{x}$.

x	0.25	0.5	1.0	1.5	2	2.5
$f(x)$	4.5					5.4

   (b) Using 2 cm to represent 1 unit on the y-

366

axis and 4 cm to 1 unit on the *x*-axis draw the graph of $y = f(x)$.

Use your graph to answer the following questions.

(c) Write down the minimum value of $f(x)$ within the range of values 0.25 to 2.5.

(d) Find the gradient of the curve at the point where $x = 0.5$

(e) Solve for *x*, the equation $f(x) = 3.5$, giving your answer to 2 decimal places.

3. In an experiment involving two variables *x* and *y*, the following readings were recorded:

x	−3	−2	−1	0	1	2	3	4	5
y	−6.1			0.3	2	2.7	2.4	1.1	−1.2

Given that *x* and *y* are related by the formula $2 + kx - \frac{1}{2}x^2$.

(a) Find the value of *k*.

(b) Determine the two missing values in the table.

(c) Using 2 cm for 1 unit in the table, on both axes, draw the graph of this relationship.

(d) Solve using your graph the equation
$$\frac{1}{2} + kx - \frac{1}{2}x^2 = 0$$

(e) Using your graph find the gradient of the curve at the point where $x = 0$.

4. Given the function $f(x) = 3x^2 - 8x - 7$.

(a) Construct a table of values of *f*(*x*) for $-2 \le x \le 5$.

(b) Draw the graph of $y = f(x)$ for the range $-2 \le x \le 5$, taking 2 cm to represent 1 unit on the *x*-axis and 1 cm to represent 2 units on the *y*-axis. Use your graph to estimate the value of *x* for which:

(i) $3x^2 - 8x - 7 = 0$

(ii) $3x^2 - 8x - 7 = 19$

(iii) The gradient at $x = 3$.

5. Given that $f(x) = 2x^2 - 7x - 2$

(a) Copy and complete the following table

x	−2	−1	0	1	2	3	4	5
f(x)	20		−2					

(b) Taking 1 cm to represent 1 unit on the *x*-axis and 1 cm to represent 2 units on the *y*-axis, draw your graph to find

(i) The roots of the equation $f(x) = 0$.

(ii) The minimum value of $f(x)$ and the corresponding value of *x*.

(iii) The range of values of *x* for which $f(x) < 0$.

(c) By drawing a suitable straight line on your graph, estimate the roots of the equation $2x^2 - 8x + 1 = 0$.

Curves and Tangents

For a straight line to be a tangent to a curve at a point, the line must touch the curve at one and only one point. The implication of this is that equal roots will result when the equation of the curve and that of the straight line are solved simultaneously.

Example 34:14

Determine which of the following lines, is a tangent to the curve $y = x^2$.

(a) $y = x + 2$ (b) $y = 6x - 9$

Solution

(a) $x^2 = x + 2$

$x^2 - x - 2 = 0$

$(x - 2)(x + 1) = 0$

$\Rightarrow x = 2 \text{ or } x = -1$

Therefore, the line cuts the curve at two points.

Hence $y = x + 2$ is not a tangent to the curve.

(b) $x^2 = 6x - 9$

$x^2 - 6x + 9 = 0$

$(x - 3)^2 = 0 \Rightarrow x = 3$

Therefore, the line touches the curve at the point where $x = 3$. Hence, $y = 6x - 9$ is a tangent to the curve $y = x^2$.

The Gradient of a Curve at a Point

The gradient of a straight line is constant, but the gradient of a curve varies from point to point. The gradient of a curve at a particular point can be defined as the gradient of the tangent drawn to the curve at that point.

To find the gradient of a curve at a point,
(i) Draw the graph of the function,

(ii) Draw the tangent to the curve at that point.

(iii) Find the gradient of this tangent by choosing two points on the curve.

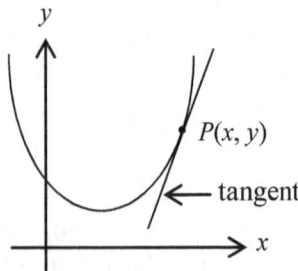

Figure 34:15

In finding the gradient of a function at a point using the graph of the function, the result depends on how well both the curve and the tangent have been drawn so utmost care must be taken.

Example 34:15

Find the gradient of the curve $y = x^2$ at the point $P(2,4)$.

Solution

The table of values and the graph of $y = x^2$ are as shown below.

x	−3	−2	−1	0	1	2	3
y	9	4	1	0	1	4	9

The graph is shown in Figure 34:16.

Gradient m at P $= \dfrac{y_2 - y_1}{x_2 - x_1} = \dfrac{8 - 0}{3 - 1} = 4$

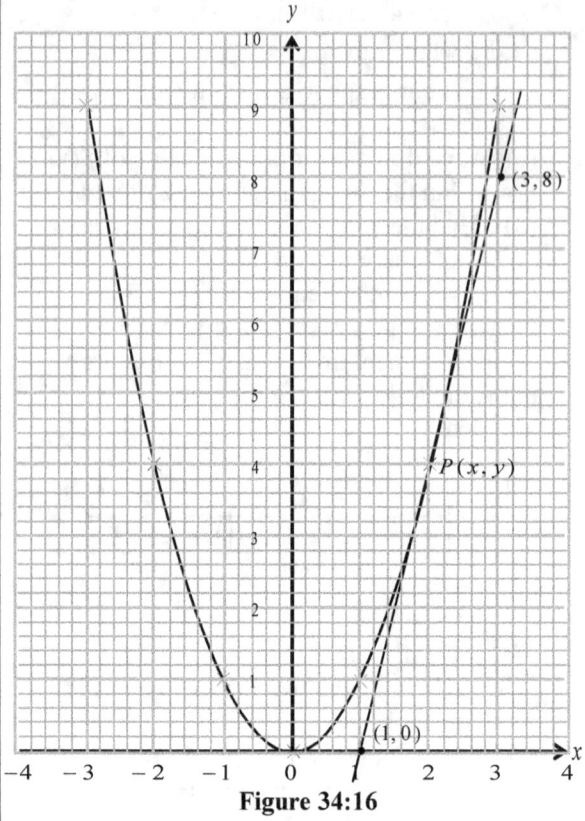

Figure 34:16

EXERCISE 34:4

1. By taking values of x in the range $-4 \le x \le 4$ and 1 cm to represent 1 unit on both axes, use a graphical method, to find the gradient of the curve $y = 2x^2 + x - 15$ at the point where $x = 1$.

2. If $y = 2x^2 + 3x - 2$,

   (a) Copy and complete the following table.

x	−4	−3	−2	−1	0	1	2	3
y		13						19

   (b) Using a scale of 1 cm to represent 1 unit on the x-axis and 1 cm to represent 2 units on the y-axis, draw graph $y = 2x^2 + 3x - 2$.

   (c) From your graph, find the gradient of the curve at the point where $x = 2$.

3. Given that $y = 2x^2 - 9x - 1$,

   (a) Make a table of values of y against x for the range $-1 \le x \le 6$.

   (b) Using a scale of 2 cm to represent 1 unit on the x-axis and 2 cm to represent

5 units on the *y*-axis, draw the graph of *y* against *x*.

(c) Use your graph, to find the gradient of the curve at $x = 3$.

4. If $y = 5 - 2x - x^2$,

(a) Using a scale of 2 cm to represent 1 unit on the *x*-axis and 2 cm to represent 2 units on the *y*-axis, draw the graph of $y = 5 - 2x - x^2$ for $-4 \leq x \leq 3$.

Gradient as a Rate of Change

The gradient of a function at any point is always a measure of the rate of change of some quantity with respect to another at that point. For instance, the gradient of a distance/time graph of a moving particle at any point gives the velocity of the particle at that point. Similarly, the gradient of a velocity/time graph of a moving particle at any point gives the acceleration of the particle at that point.

Example 34:16
When a stone is thrown vertically upwards, its distance *d* m after *t* seconds, is given by the equation $d = 6t - t^2$.

(a) Using a scale of 1 cm to represent 1 s on the *t*-axis and 1 cm to represent 1 m on the *d*-axis, draw the graph of $d = 6t - t^2$ for values of *t* from 0 to 6 seconds. From your graph, determine

(b) The maximum height attained by the stone.

(c) The time when the stone returns to its original position.

(d) The gradient of the curve when $t = 8$ seconds, stating it units and what the gradient stand for.

Solution

(a) The table of values and the graph are as below.

t (s)	0	1	2	3	4	5	6
d (m)	0	5	8	9	8	5	0

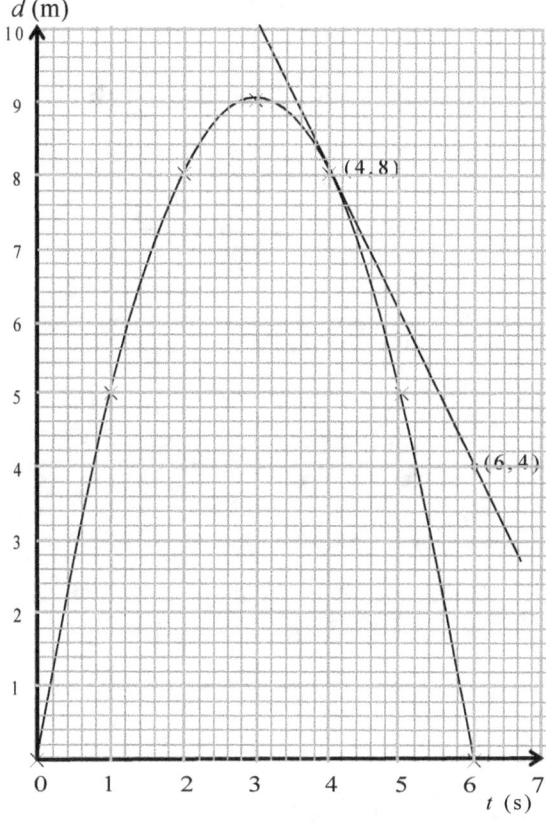

Figure 34:17

(b) Maximum height attained is 9 m.

(c) The stone returns to its original position at the sixth second.

(d) The gradient at
$$P(4,8) = \frac{d_2 - d_1}{t_2 - t_1} = \frac{8-4}{4-6} = -2.$$
The units of the gradient are m/s. This gives the velocity of the stone when $t = 8$ seconds, so the units of the gradient are correct.

Elementary Differentiation

Differentiation is very important in finding the gradient or rate of change of one quantity with respect to another and the maximum and minimum values of a given function. If *x* and *y* are variables, the rate of change of *y* with respect to (w.r.t) *x* is denoted by $\frac{dy}{dx}$. Other names for the rate of change of *y* w.r.t *x* are; the **gradient function**, **derivative**, **derived function** or **differential coefficient** of *y* w.r.t *x*. The process of finding $\frac{dy}{dx}$ is called **differentiation** of *y* w.r.t *x*.

Differentiation of some Simple Functions

If a, b, c, m and n are real constants:

1. The derivative of $y = c$ is $\dfrac{dy}{dx} = 0$.

2. The derivative of $y = x^n$ is $\dfrac{dy}{dx} = nx^{n-1}$.

3. The derivative of
$y = ax^n$ is $\dfrac{dy}{dx} = anx^{n-1}$.

Since $\dfrac{1}{x^n} = x^{-n}$,

4. The derivative of
$y = \dfrac{a}{x^n}$ is $\dfrac{dy}{dx} = -anx^{-n-1}$.

5. The derivative of a sum (or difference) is equal to the sum (or difference) of the derivatives of the terms.

Example 34:17

Find $\dfrac{dy}{dx}$ in each of the following.

(a) $y = x^3$ (b) $y = x$

(c) $y = 6$ (d) $y = x^5 + 3x^2$

(e) $y = x^2 - 4x + 2$

Solution

(a) $y = x^3 \Rightarrow \dfrac{dy}{dx} = 3x^2$

(b) $y = x \Rightarrow \dfrac{dy}{dx} = 1$

(c) $y = 6 \Rightarrow \dfrac{dy}{dx} = 0$

(d) $y = x^5 + 3x^2 \Rightarrow \dfrac{dy}{dx} = 5x^4 + 3(2)x = 5x^4 + 6x$

(e) $y = x^2 - 4x + 2 \Rightarrow \dfrac{dy}{dx} = 2x - 4$

Example 34:18

Find the gradient of the curve $y = 3x^2 - 11x + 9$ at the point where $x = 4$.

Solution

$\dfrac{dy}{dx} = 6x - 11 \Rightarrow$ when $x = 4$, $\dfrac{dy}{dx} = 6(4) - 11 = 13$

Therefore the gradient at $x = 4$ is 13.

Example 34:19

Determine the point at which the gradient of the curve $y = 2x^2 + 3x - 2$ is 7.

Solution

$\dfrac{dy}{dx} = 4x + 3 \Rightarrow 4x + 3 = 7 \Rightarrow x = 1$

When $x = 1$, $y = 2(1)^2 + 3(1) - 2 = 3$
Therefore, the point at which the gradient is 7 is (1,3).

EXERCISE 34:5

1. Find $\dfrac{dy}{dx}$ in each of the following.

   (a) $y = x^4$ (b) $y = x^6 + 4$

   (c) $y = 7x^4 - 2x^3$ (d) $y = x^4 + 5x^3$

   (e) $y = 4x^3 - 3x^2 - 2$ (f) $y = 3x^2$

   (g) $y = 5x - 3$ (h) $y = 2x^2 - 3x + 7$

   (i) $y = \dfrac{1}{4}x - 4x^2$

2. Find the gradient of the curve
   $y = 2x^2 - 3x + 7$ at the point where
   $x = 2$.

3. Find the gradient of the given curves at the points indicated.

   (a) $y = 4x^2 + 2x - 1$, at the point $(0,-1)$

   (b) $y = 6x - 3$, at the point $(2,4)$

   (c) $y = x^2 - 7$, at the point $(3,2)$

   (d) $y = 2 + x - 4x^2$, at the point $(3,-31)$

4. Determine the point at which the gradient of the curve $y = 3x^2 + 2x - 4$ is -4.

5. Calculate the gradient of the line joining the points
   (a) $A(1,3)$ and $B(-3,9)$
   (b) $P(7,4)$ and $Q(-5, -6)$
   (c) $R(3,5)$ and $S(6,9)$
   (d) $X(-3,8)$ and $Y(6, 4)$.

6. Given that $y = 3x^2 + 2x - 7$, find the
   value of $\dfrac{dy}{dx}$ and hence evaluate $\dfrac{dy}{dx}$ when
   $x = 4$.

7. If $y = 2x^2 - 4x + 3$, find an expression
   for $\dfrac{dy}{dx}$ and hence find $\dfrac{dy}{dx}$ when $x = -2$.

Maximum and Minimum Values

It was earlier mentioned that $\frac{dy}{dx}$ or the gradient is a measure of the rate of change of one quantity y w.r.t. another, x. Therefore, it is also possible to find the rate at which the gradient $\frac{dy}{dx}$ is changing w.r.t x or from one point on a curve to another. The derivative of the gradient w.r.t x is $\frac{dy}{dx}$ known as the second derivative of y w.r.t x, denoted by $\frac{d^2y}{dx^2}$ and is useful in finding the maximum and minimum values of a function. At both the maximum and minimum points, the gradient is zero since the tangent to the curve is horizontal i.e. $\frac{dy}{dx} = 0$.

At the maximum point, the gradient $\frac{dy}{dx}$ is changing from positive to negative. Therefore $\frac{d^2y}{dx^2} < 0$.

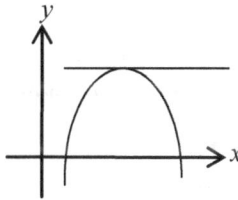

At the maximum point $\frac{dy}{dx} = 0$ and $\frac{d^2y}{dx^2} < 0$.

On the other hand, at the minimum point the gradient $\frac{dy}{dx}$ is changing from negative to positive i.e. $\frac{d^2y}{dx^2} > 0$.

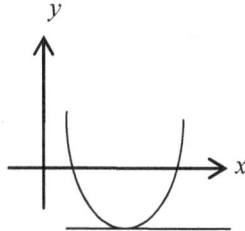

At a minimum value $\frac{dy}{dx} = 0$ and $\frac{d^2y}{dx^2} > 0$.

EXERCISE 34:6

1. Find the turning point on the curve
   $f(x) = 5 + 4x - x^2$ and say whether it is a maximum or minimum point.

2. The distance s in m of a particle from a fixed point O is given by
   $s = t^3 - \frac{13}{2}t^2 + 14x + 5$ where t is the time taken in seconds
   (a) What is the velocity of the particle when $t = 1$ second?
   (b) Find the acceleration of the particle after 2 seconds
   (c) Make a table of values of velocity against time for the first five seconds of its motion, hence
   (d) Plot a graph of velocity against time for the first five seconds for the particle.
   (e) From your graph, what is the minimum value of the velocity and at what time does it occur?
   (f) Use the method of differentiation to confirm this minimum value.
   (g) At what time does this occur?
3. A cardboard tube is such that the sum of its length and circumference is 8π cm. What is its maximum volume?

Velocity and Acceleration

The velocity v of a moving body is the rate of change of its displacement s. i.e. $v = \frac{ds}{dt}$ where, t is the time (in seconds).
The acceleration a of a body is the rate of change of its velocity v. i.e.

$$a = \frac{dv}{dt} \text{ or } a = \frac{d^2s}{dt^2} \text{ since } v = \frac{ds}{dt}.$$

Certain values of velocity and acceleration

have implications as follows,
1. When the velocity $v = 0$, it means that the particle is at rest.
2. When the velocity $v < 0$ (i.e. negative), it means the particle is moving in the opposite direction to that in which the displacement is measured.
3. When the velocity $v > 0$ (i.e. positive), it means the particle is moving in the same direction to that in which the displacement is measured.
4. When $a = 0$, the velocity of the particle is constant (i.e. not changing)
5. When $a > 0$ or a is positive, the particle is accelerating or speeding up.
6. When $a < 0$ or a is negative, the particle is retarding or slowing down.

Example 34:20
A particle moves in a straight line so that its distance from a fixed point O after t seconds is s metres, where $s = \frac{1}{3}t^3 - \frac{3}{2}t^2 + 2t$. Find the times when the particle is at rest. Find the acceleration of the particle at these times and interpret the result.

Solution
$$s = \frac{1}{3}t^3 - \frac{3}{2}t^2 + 2t$$

$$\frac{ds}{dt} = t^2 - 3t + 2$$

At rest, $\frac{ds}{dt} = 0 \Rightarrow t^2 - 3t + 2$

$$\Rightarrow (t - 1)(t - 2) = 0$$

$$\Rightarrow t = 1 \text{ or } t = 2$$

Therefore, the particle is at rest when $t = 1$ s and when $t = 2$ s.

From $\frac{ds}{dt} = t^2 - 3t + 2$,

$$\Rightarrow \frac{d^2s}{dt^2} = 2t - 3$$

When $t = 1$, $\frac{d^2s}{dt^2} = 2(1) - 3 = -1 \text{ ms}^{-2}$

Therefore, when $t = 1$, the particle is retarding or slowing down.

When $t = 2$, $\frac{d^2s}{dt^2} = 2(2) - 3 = 1 \text{ ms}^{-2}$

Therefore, when $t = 2$, the particle is accelerating or speeding up.

Speed time Graphs

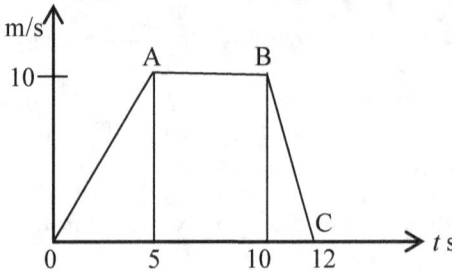

Figure 34:18

Figure 34:18, shows a speed time graph for a body that starts from rest (i.e. with zero velocity) at time $t = 0$ s and accelerates uniformly until it attains a velocity of 10 m/s at A. The body then moves with uniform velocity of 10 m/s for 5 s from A to B. At B, the velocity of the body decreases steadily until the body comes to rest again at $t = 12$ s . Such a graph gives two pieces of information.

(a) The total distance covered by the body is the area under the graph.
Thus distance = Area of trapezium

(b) The gradient $\frac{ds}{dt}$ of the graph in each section of its motion is a measure of the acceleration of the particle at that section.

Example 34:21
From Figure 34:16, calculate
(i) The total distance covered by the body.
(ii) The acceleration of the particle in each section.

Solution
(i) Total distance = Area of trapezium
$$= \frac{1}{2}(OC + AB)h$$

$$\Rightarrow \text{Total distance} = \frac{1}{2}(12 + 5)10 = 85 \text{ m}$$

(ii) From O to A, acceleration $= \frac{10}{5} = 2 \text{ m/s}^2$

From A to B, acceleration $= 0 \text{ m/s}^2$

From B to C, acceleration $= -\frac{10}{2} = -5 \text{ m/s}^2$

Distance Time Graphs

Distance-time graphs are very useful especially in determining the time or positions at which moving bodies meet. The distance is usually plotted on the *y*-axis against the time on the *x*-axis. The graph of each body does not necessarily start from (0, 0) as the bodies may start at different distances from the origin and at different times. On a distance time graph, the gradient of the line or curve at any point gives the speed of the moving object.

Example 34:22

A boy left a town at 8 a.m. trekking to his village 8 km away at a speed of 5 km/h. After trekking 2 km, he met a car coming from the village at a steady speed. Arriving in town at 8:30 a.m. the driver loaded passengers for 15 minutes then immediately returned to the village at the same speed. Draw a graph for these journeys and use it to determine

(a) The time at which the car left the village.
(b) The time when the car overtook the man.
(c) The distance covered by the man by the time the car overtakes him.

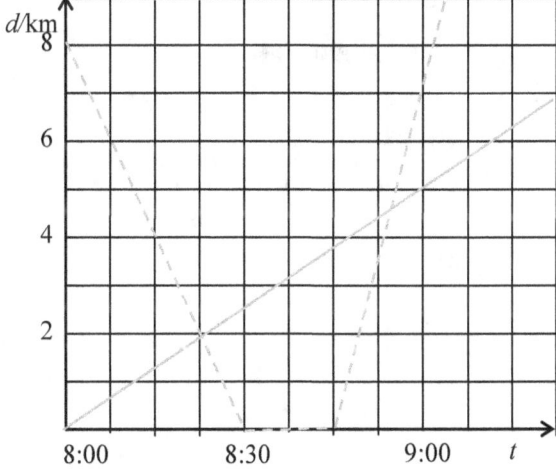

Figure 34:19

From the graph,
(a) The car left the village at 8:45 a.m.
(b) The car overtakes the man at 8:48 a.m.
(c) The distance covered by the man 4.6 km.

EXERCISE 34:7

1. A car starts from rest and is accelerated uniformly at the rate of 2 m/s for 6 s. It then maintains a constant speed for half a minute. The brakes are then applied and the vehicle uniformly retarded to rest in 5 s. Find the maximum speed reached in km/h and the total distance covered in metres.

2. A car is moving along a straight road. It is taken from rest to a velocity of 20 m/s by a constant acceleration of 5 m/s^2. It maintains a constant velocity of 20 m/s, for 5 s and then is brought to rest again by a constant acceleration of −2 m/s^2. Draw a velocity time graph and find the distance covered by the car.

3. A particle moves from rest in a straight line, with an acceleration of 4 m/s^2 for 3 seconds. It maintains a uniform velocity for 6 s and is then brought to rest again in 4 seconds with a uniform retardation. Draw a velocity time graph and find the final acceleration and the final displacement of the particle from its starting point.

4. (i) Joan left school at 10.00 a.m. for her home 8 km away walking at the rate of 6 km per hour. She spent half an hour at home. She was then given a ride back to school traveling at 16 km per hour.
   (a) Determine the time she arrived their house.
   (b) Determine the time when she took off from the house to go back to school.
   (c) Determine the time she got back to school.
   (d) Draw the travel graph of her journey.
   (e) Using your graph or otherwise, determine how far she was away from the school at 12.00 noon. Express your answer to 1 decimal place.

   (ii) Given the straight lines
   A : $y - 3x - 5 = 0$
   B : $x = 8 - 3y$
   C : $2y = 5 - 6x$

   Determine which of the lines (if any)
   (a) are perpendicular
   (b) are parallel
   (c) pass through the origin

MULTIPLE CHOICE EXERCISE 34

1. The root of the equation represented by the graph in Figure 34:19 is:
   [A]   4   [B]   7   [C]   –4   [D]   –7

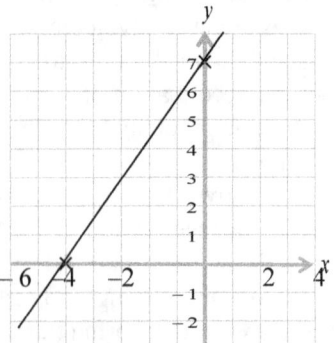

Figure 34:20

2. The graph of $f(x) = x^2 + x - 6$ is most likely:

[A]
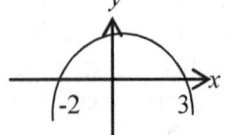

[B]
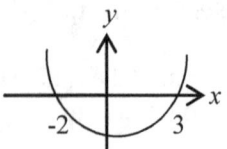

[C]
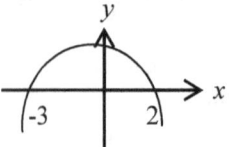

[D]
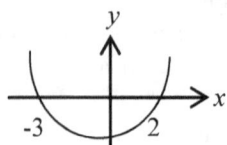

Figure 34:21

3. The graph of $y = 4 - 3x - x^2$ is most certainly:

[A]
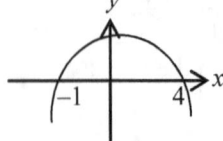

[B]
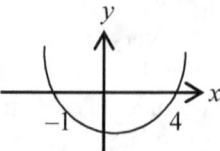

[C]
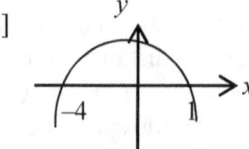

[D]
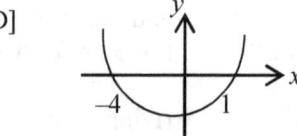

Figure 34:22

4. The equation, which represents the sketch graph in Figure 34:23, is:

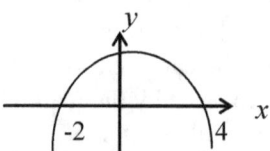

Figure 34:23

   [A] $y = 8 - 2x + x^2$ [B] $y = 8 + 2x + x^2$

   [C] $y = 8 - 2x - x^2$ [D] $y = 8 + 2x - x^2$

5. The equation, which represents the sketch graph in Figure 34:24, is:

   [A] $y = x^2 - 2x - 3$ [B] $y = x^2 + 2x - 3$

   [C] $y = x^2 + 2x + 3$ [D] $y = x^2 - 2x + 3$

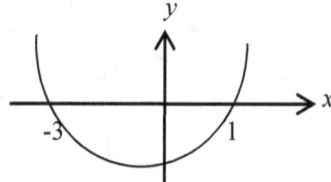

Figure 34:24

6. Figure 34:25 is the graph of the function:

   [A] $y = x^2 - x - 6$ [B] $y = x^2 + x - 6$

   [C] $y = x^2 + x + 6$ [D] $y = x^2 - x + 6$

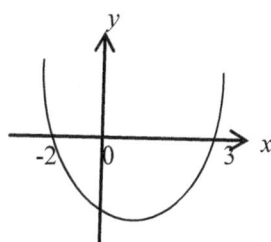

Figure 34:25

7. The graph representing inconsistent lines is:

[A]

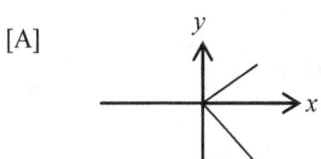

[B]

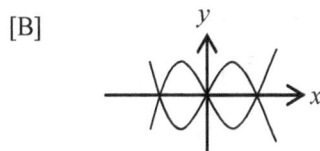

[C]

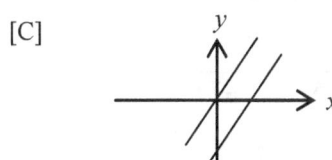

[D]

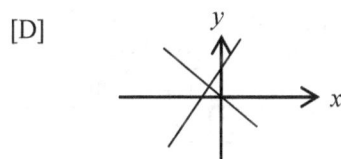

Figure 34:26

8. The line, which is a tangent to the curve, $y = x^2$ is:

   [A] $y = 2x - 3$ [B] $y = 2x - 2$

   [C] $y = 2x - 1$ [D] $y = 2x$

9. A stone thrown vertically upward moves s metres in t seconds, where $s = 80t - 5t^2$. The maximum height it reaches is:
   [A] 640 m   [B] 320 m
   [C] 75 m   [D] 40 m

10. The maximum value of $x^2 - 4x + 5$ is:
    [A] −5 [B] 0 [C] −1 [D] 1

11. The function whose curve has a maximum point is:

    [A] $y = 8 - 2x - x^2$ [B] $y = 8 + 2x + x^2$

    [C] $y = x^2 + 2x + 8$ [D] $y = x^2 - 2x + 8$

12. The function whose curve has a minimum point is:

    [A] $f(x) = 12 + 8x - x^2$ [B] $f(x) = 12 - 8x + x^2$

    [C] $f(x) = 12 - 8x - x^2$ [D] $f(x) = -12 + 8x - x^2$

13. The graph of the curve $y = 2x^2 - 5x - 1$ and a straight line PQ were drawn to solve the equation $y = 2x^2 - 5x - 1$. The equation of the straight line PQ is:
    [A] $y = 1$ [B] $y = 0$
    [C] $y = 3$ [D] $y = -3$

14. The equation of a curve is given by $y = 2x^2 - x - 1$.
    The intercept with the y-axis is:
    [A] $(-1, 0)$ [B] $(1, 0)$
    [C] $(0, -1)$ [D] $(0, 1)$

15. The equation of a curve is given by $y = 2x^2 - x - 1$.
    The intercepts with the x-axis is:

    [A] $(0, 1)$ and $\left(0, -\dfrac{1}{2}\right)$

    [B] $(-1, 0)$ and $\left(-\dfrac{1}{2}, 0\right)$

    [C] $(-1, 0)$ and $\left(\dfrac{1}{2}, 0\right)$

    [D] $(1, 0)$ and $\left(-\dfrac{1}{2}, 0\right)$

16. A particle P moves so that its distance, s metres, from a fixed point O, at time t seconds, $t \geq 0$ is given by $s = 3 + 8t - \dfrac{1}{12}t^2$.
    The velocity V of the particle when $t = 4$ is:

    [A] $\dfrac{77}{3}$ m/s   [B] $\dfrac{22}{3}$ m/s

    [C] $\dfrac{68}{3}$ m/s   [D] $\dfrac{72}{3}$ m/s

17. A particle P moves so that its distance, s metres, from a fixed point O, at time t seconds, $t \geq 0$ is given by $3 + 8t - \dfrac{1}{12}t^2$.
    The value of t when P finally comes to rest is:
    [A] $t = 24$ s   [B] $t = 32$ s
    [C] $t = 48$ s   [D] $t = 64$ s

18. A train of length 100 m is traveling at a speed of 40 km per hour. In minutes, the length of time taken for the train to pass completely over a bridge of length 0.7 km is:
    [A] 1.2 minutes   [B] 0.02 minutes
    [C] 12 minutes   [D] 0.2 minutes

19. A particle P moves so that its distance, s metres, from a fixed point O, at time t seconds is given by $s = 20 + 24t - t^2$. The distance traveled during the first 3 seconds of motion is:
    [A] 20 m   [B] 72 m
    [C] 83 m   [D] 92 m

20. A particle P moves so that its distance, s metres, from a fixed point O, at time t seconds is given by $s = 20 + 24t - t^2$. The initial velocity of the particle is:
    [A] 20 m/s   [B] 24 m/s
    [C] 22 m/s   [D] 45 m/s

21. A particle P moves so that its distance, s metres, from a fixed point O, at time t seconds is given by $s = 20 + 24t - t^2$. The velocity of the particle when $t = 3$ s is:
    [A] 20 m/s   [B] 45 m/s
    [C] 18 m/s   [D] 21 m/s

22. The distance s in metres of a particle from a fixed point O is given by
    $$s = t^2 = \frac{13}{2}t^2 + 14t + 5$$ where t is the time taken in seconds. The velocity of the particle when $t = 1$ second is:
    [A] 5 m/s   [B] −1 m/s
    [C] 14 m/s   [D] 4 m/s

23. Figure 34:27 not drawn to scale shows the speed time graph of a cyclist. The length of time in hours for which the cyclist rode at uniform speed is:
    [A] 3   [B] 4   [C] 6   [D] 18

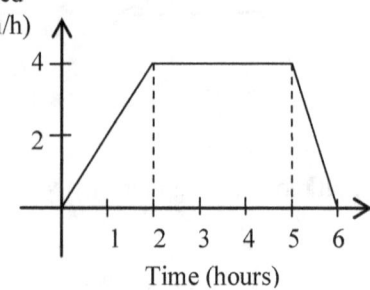

Figure 34:27

24. The table of values for $y = x - 6$ is:

[A]
x	−5	−8	−7
y	1	−14	−13

[B]
x	−5	−8	−7
y	−11	−2	−13

[C]
x	−5	−8	−7
y	−11	−14	−13

[D]
x	−5	−8	−7
y	1	−2	−1

25. The equation, which corresponds to the table of values, is:
    [A] $y = 4 + 5x$ [B] $y = 3 + 6x$
    [C] $y = 5 + 4x$ [D] $y = 6 + 3x$

Input (x)	1	2	3	4	5
Output (y)	9	12	15	18	21

26. The graph of a quadratic function is given as
    [A] $\Delta > 0, a > 0$ [B] $\Delta < 0, a < 0$
    [C] $\Delta > 0, a < 0$ [D] $\Delta < 0, a > 0$

TOPIC 35

VARIATIONS (PROPORTIONS)

OBJECTIVES

At the end of this topic, the learner should be able to:

1. State whether two variables are in direct or inverse variation.
2. Interpret, write and use the symbol $\propto$.
3. Solve problems on direct or inverse variation.
4. Interpreted and Solve simple problems on partial and joined Variation.

Leonhard Euler

Although hindered by loss of sight, Leonhard Euler was an important contributor to both pure and applied mathematics. Euler is best known for his analytical treatment of mathematics and his discussion of calculus concepts, but he is also credited for work in acoustics, mechanics, astronomy, and optics.

Culver Pictures

Real Life Examples of Variation

In real life, there are many quantities, which are related in one way or the other. Examples of such related quantities are:

1. The interest generated per year increases as the amount of money invested increases.
2. As the number of people increase, the time taken to complete a given piece of work decreases.
3. The electric bill increases as the amount of current used increases.
4. The area of the circle of a circle increases as the radius increases.
5. Temperature decreases with increasing altitude (i.e. the higher you go, the colder it becomes)
6. The amount of current passing in a wire increases with decreases in resistance.

The above are examples of variations and in each case; it is possible to find a mathematical relation or formula connecting the two quantities. There are two basic types of variation.

- Direct variation or direct proportion and
- Inverse variation or inverse proportion.

Direct variation

Suppose that one pen cost 100 FRS. Table 35:1 shows the prices of 1,2,3 ... 10 pens.

Number of pens, n	Cost of pens, FRS
1	100
2	200
3	300
4	400
5	500
6	600
7	700
8	800
9	900

Table 35:1

Notice that the number of pens increases with the price. This is an example of direct variation or direct proportion.
If two quantities x and y are related in this way, then y is said to be **directly proportional** to x or y varies directly as x.

Direct variation be expressed mathematically as

$$y \propto x \Rightarrow y = kx \Leftrightarrow \frac{y}{x} = k$$

Where k is a constant, called the **constant of proportionality**. This constant can be found if the value y for a given value of x is known.

Thus, when $y = 12$ and $x = 2, k = \frac{12}{2} = 6$

Example 35:1

Given that y is directly proportional to x and $y = 64$ when $x = 8$. Find the value of y when $x = 20$.

Solution

$$y \propto x \Rightarrow k = \frac{y}{x} \Rightarrow k = \frac{64}{8} = 8$$

When $x = 20, y = kx \Rightarrow y = 8(20) = 160$

Example 35:2

Two quantities x and y are such that y varies directly as x. Copy and complete Table 35:2, hence draw a graph to represent this variation. Use your graph to find the value of y when $x = 2.5$.

x	0	5	10	15	20	25
y				3		

Table 35:2

Solution

From the table, $x = 15$ when $y = 3$

$$\Rightarrow k = \frac{y}{x} = \frac{3}{15} = \frac{1}{5}.$$

So $y = kx = \frac{1}{5}x$ and the complete Table is

x	0	5	10	15	20	25
y	0	1	2	3	4	5

Table 35:3

The graph is shown in Figure 31:1. Notice that, the graph is a straight line passing through the origin. This is the usual nature of a graph of direct variation.

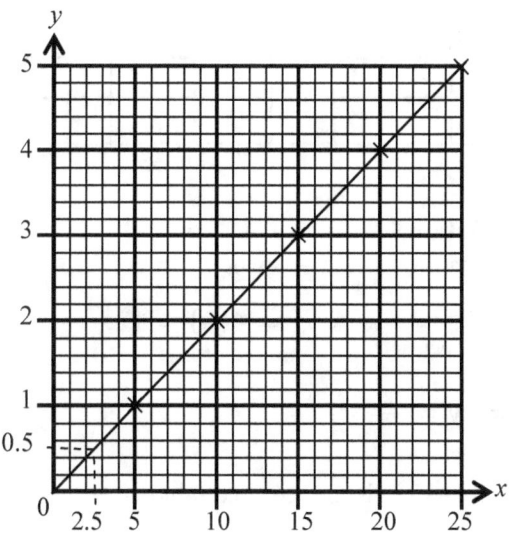

Figure 35:1

There are so many other cases of direct variation. For instance, one quantity may vary directly as the square or square root of the other. Thus, the area of a circle varies directly as the square of its radius.

Example 35:3

The variables x and y are known to be related by the formula $y = kx^2$. Given that $x = -2$ when $y = 1$, find the value of k and hence calculate the value x when $y = 16$. Also, make a table of values of x against y for integral values of x in the range $-4 \leq x \leq 4$. Use this table to draw the graph of y against x.

Solution

$$k = \frac{y}{x^2} = \frac{1}{(-2)^2} = \frac{1}{4} \Rightarrow y = \frac{1}{4}x^2$$

But $kx^2 = y \Rightarrow x = \pm\sqrt{\dfrac{y}{k}}$

When $y = 16, x = \pm\sqrt{16 \div \frac{1}{4}} = \pm 8$

x	-4	-3	-2	-1	0	1	2	3	4
y	4	2.25	1	0.25	0	0.25	1	2.25	4

Table 35:4

The graph in Figure 33:2 is the typical shape of the graph of $y = kx^2$. Notice that the function is a quadratic function with the

coefficient of x and the independent term both zero.

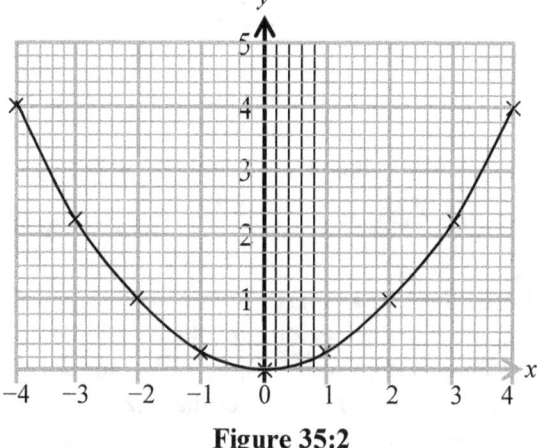

Figure 35:2

EXERCISE 35:1

1. Given that $c \propto n$ and $c = 28$ when $n = 4$. Find the formula connecting c and n.

2. Given that m is directly proportional to n and k is the constant of proportionality. Find a formula connecting m and n.

3. Given that x varies directly with y and $x = 7$ when $y = 35$. Find the relationship between x and y.

4. Given that $x \propto y$ and when $x = 4, y = 20$. Find x when $y = 5$.

5. Given that t varies directly as the square of m and $t = 10$ when $m = 25$. Calculate t when $m = 4$.

6. Given that x varies directly as the square of y and $x = 4$ when $y = 6$. Find the value of y when $x = 16$.

7. Given that s varies directly as T^2 and $T = 2$ cm when $s = 12$ cm^2. Find s when $T = 3$ cm.

8. Given that p varies directly as q and $p = 4.5$ when $q = 12$. Find the relationship between p and q and hence find p when $q = 16$.

9. Given that $y \propto x^2$ and that $y = 54$ when $x = 3$, find the constant of proportionality

hence the value of x when $y = 24$.

10. Given the mapping

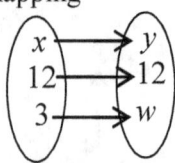

Figure 35:3

Find the value of w given that y varies directly as the square of x.

Inverse variation

Suppose that one person can take 60 minutes to fill a tank with water. Table 35:5, shows the amount of time (t) in minutes that 1,2,3,4,5 and 6 people all working at the same rate can use to fill a similar tank.

Number of People, P	1	2	3	4	5	6
Time taken, t (mins)	60	30	20	15	12	10

Table 35:5

Notice that the time decreases with the number of people. This is an example of inverse variation or inverse proportion described by 'y is inversely proportional to x' or 'y varies inversely with x' and symbolically expressed as

$$y \propto \frac{1}{x} \Leftrightarrow y = \frac{k}{x} \Leftrightarrow k = yx ,$$

Where, k is the constant of proportionality. Thus, when $x = 2$ and $y = 5$, $k = 2(5) = 10$.

Example 35:4

Given that y is inversely proportional to x and $y = 4$ when $x = 12$ find the value of y when $x = 3$.

Solution

$$y \propto \frac{1}{x} \Leftrightarrow k = yx$$

When $y = 4$ and $x = 12$, $k = 4(12) = 48$.

When $x = 4$, $y = \dfrac{k}{x} = \dfrac{48}{3} \Rightarrow y = 16$.

Example 35:5

Two quantities x and y are such that y varies inversely as x. Copy and complete Table 35:6 then, draw a graph to represent the variation.

Use your graph to find the value of y when $x = 1.5$.

x	−4	−3	−2	−1	0	1	2	3	4
y				−12					3

Table 35:6

Solution

From the table, $x = 4$ when $y = 3$.

$$y = \frac{k}{x} \Rightarrow k = yx = 4(3) = 12$$

$$\Rightarrow y = \frac{12}{x} .$$

Table 35:7, is the required table.

x	−4	−3	−2	−1	0	1	2	3	4
y	−3	−4	−6	−12	∞	12	6	4	3

Table 35:7

The graph follows as in Figure 33:4.

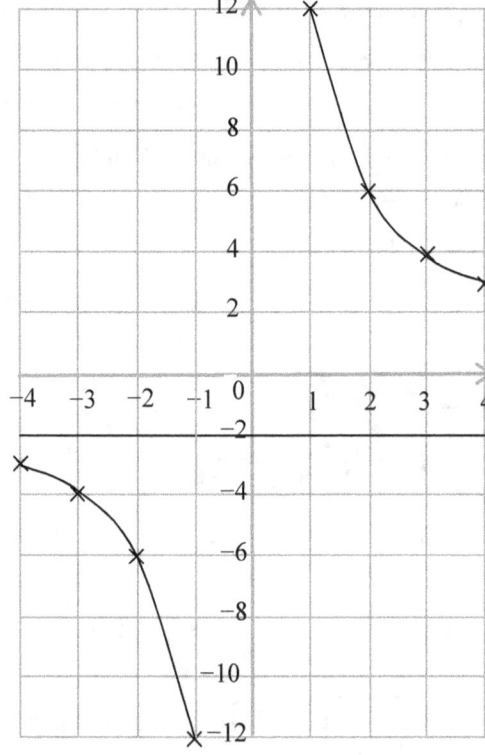

Figure 33:4

From the graph, when $x = 1.5, y = 8$.

This type of graph is called a rectangular hyperbola. The graph shows the typical nature of a graph of inverse variation.

Clearly, the curve approaches the x and the y-axes but never touching it. This is because neither x nor y can be exactly zero.

When $y = 0$, $x = \frac{k}{0} = \infty$ and when $x = 0$,

$y = \frac{k}{0} = \infty$.

When a curve approaches a line in this way but never touching it, such a line is called an **asymptote** of the curve. Therefore, the x and the y-axes are the asymptotes of the above curve.

Other Inverse Variations

Consider the statement "y is inversely proportional to the square of x"; this can be expressed symbolically as;

$$y \propto \frac{1}{x^2} \Rightarrow y = \frac{k}{x^2} \text{ or } k = yx^2$$

On the other hand "y is inversely proportional to the square root of x, can be expressed as,

$$y \propto \frac{1}{\sqrt{x}} \Rightarrow y = \frac{k}{\sqrt{x}} \text{ or } k = y\sqrt{x}$$

Example 35:6
The variables x and y are known to be related by the formula $x^2 y = k$. Given that $x = -2$ when $y = 36$, find the value of k and hence make a table of values of y against x for $-4 \le x \le 4$. Use this table to draw a graph of y against x.

Solution

$$k = x^2 y = (-2)^2 (36) = 144 \implies y = \frac{144}{x^2}$$

Table 35:8, is the required table.

x	–4	–3	–2	–1	0	1	2	3	4
y	9	16	36	144	∞	144	36	16	9

Table 35:8

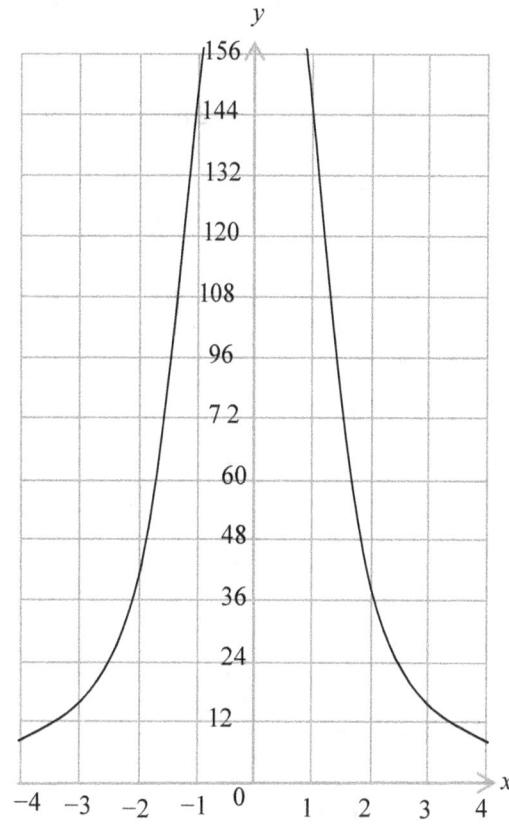

Figure 33:5

Again, the y-axis and the x-axis are asymptotes to the curve.

EXERCISE 35:2

1. Given that y is inversely proportional to the square of x and that when $x = \frac{1}{2}, y = 8$.
   (a) Write down an expression to show the relationship between x and y.
   (b) Calculate the value of y when $x = 2\sqrt{3}$.

2. Given that y varies inversely as the square of x, find the numerical values of a and b in Table 35:9.

x	a	2	8
y	16	144	b

Table 35:9

3. In Table 35:10, y varies inversely as x. Find the numerical values of a and b.

x	2	4	...	b
y	6	a	...	1

Table 35:10

4. Two quantities x and y are believed to be related by the relation $y \propto \dfrac{1}{x^2}$. In an experiment, x and y were measured and the results in Table 35:11 obtained.

x	2	3	4	6	8	12
y	36	16	9	4	3	1

Table 35:11

It is suspected that one of the values of y is wrong. Find the wrong one and give the correct value.

5. Given that x varies as $\dfrac{1}{\sqrt{y}}$ and $y = 4$ when $x = 4$, find

    (a) x when $y = 16$ (b) y when $x = \dfrac{1}{2}$.

6. Given that y varies inversely as the square of x, and that $x = 2$ when $y = 3$. Find the value of x when $y = 27$.

7. In Table 35:12, find the values of a and b if s is inversely proportional to t.

t	5	2	b
s	4	a	144

Table 35:12

8. x and y are known to be related by the formula $x^2 y = 16$. In Table 35:13, find which of the value(s) of y is/are incorrect and give the correct value.

x	-2	-1	1	2	3	4
y	4	16	16	4	2	1

Table 35:13

9. Given that x varies inversely as $\dfrac{1}{y}$, complete Table 33:14.

x	0	5	10	15	20	25
y				3		

Table 35:14

Joint or Combined Variation

A variation, in which a quantity varies as more than one quantity vary, is known as **join** or **combined variation**.

Examples 35:7
Write the following statement symbolically. The volume of a right circular cylinder is proportional to both the square of the radius of the cross-section and the height of the cylinder.

Solution
$V \propto r^2$ and $V \propto h$
Combining gives $V \propto r^2 h$
Introducing the constant of proportionality, k leads to $V = kr^2 h$.

It can be proven that this constant is π, hence $V = \pi r^2 h$.

Examples 35:8
The pressure of a fixed mass of gas varies directly as the temperature and inversely as the volume of the gas. Express this statement symbolically.

Solution
$$P \propto T \text{ and } P \propto \dfrac{1}{V}$$

Combining gives $P \propto \dfrac{T}{V}$.

Since the mass, m of the gas is fixed; it means the constant of proportionality is m, $P = \dfrac{mT}{V}$.

Examples 35:9
The height of a cone varies directly as its volume and inversely as the square of its radius. Using k as the constant of proportionality, deduce the formula for calculating the height of a cone.

Solution
$$h \propto V \text{ and } h \propto \dfrac{1}{r^2}$$
Combining and introducing the constant of proportionality leads to $h = \dfrac{kV}{r^2}$.

EXERCISE 35:3

1. Given that l varies jointly as m and n and that when $m = 2$, $n = 3$ and $l = 9$. Find the law connecting l, m, and n.

2. A is proportional to B and is inversely proportional to C. When $A = 8$ and $B = 4$, $C = 3$.

(a) Find the formula that connects A, B and C.

(b) Find A when $B = 5$ and $C = 6$.

3. Given that x varies directly as y and inversely as z and that when $y = 7$ and $z = 3$, $x = 42$. Find the relationship between x, y and z. Find x when $y = 5$ and $z = 9$.

4. Given that x is directly proportional to y and inversely proportional to z and that when $x = 9$, $y = 24$ and $z = 8$. What is the value of x when $y = 5$ and $z = 6$?

MULTIPLE CHOICE EXERCISE 35

1. Given that n varies directly as m and if $n = 8$ when $m = 20$. The value of m when $n = 7$ is:

[A] 13 [B] 15 [C] $17\frac{1}{2}$ [D] $18\frac{1}{2}$

2. Given that $(x+3)$ varies directly as y and $x = 3$ when $y = 12$, the value of x when $y = 8$ is:

[A] 1 [B] $\frac{1}{2}$ [C] $-\frac{1}{2}$ [D] -1

3. Given that y is directly proportional to x^2 and $y = 5$ when $x = 2$, then when $x = 6$, $y =$:

[A] 18 [B] 21 [C] 27 [D] 45

4. P varies inversely as the square of W. When $W = 4$, $P = 9$, then the value of P when $W = 9$ is:

[A] $\frac{4}{3}$ [B] 6 [C] 4 [D] $\frac{16}{9}$

5. Given that y varies inversely as x^2 then x varies:

[A] inversely as y^2 [B] inversely as $\sqrt{y}$

[C] directly as y^2 [D] directly as $\sqrt{y}$

6. Given that x varies inversely as y and $x = \frac{2}{3}$ when $y = 9$, the value of y when $x = \frac{3}{4}$ is:

[A] $\frac{1}{18}$ [B] $\frac{81}{8}$ [C] $\frac{9}{2}$ [D] 8

7. Given that $y \propto \dfrac{1}{\sqrt{x}}$ and $x = 16$ when $y = 2$ when $y = 24$, x will be:

[A] $\frac{1}{9}$ [B] $\frac{1}{6}$ [C] $\frac{1}{3}$ [D] $\frac{2}{3}$

8. Given that x is inversely proportional to m^2 and $x = 3$ when $m = 9$, the value of x when $m = 3$ is:

[A] 3 [B] 6 [C] 9 [D] 27

9. Given that $p \propto \dfrac{1}{\sqrt{r}}$ and $p = 3$ when $r = 16$ the value of r when $p = \dfrac{3}{2}$ is:

[A] 48 [B] 72 [C] 64 [D] 324

10. Given that R is inversely proportional to S and $R = 15$ when $S = 12$. The value of S when $R = 60$ is:

[A] $\frac{1}{4}$ [B] 3 [C] 4 [D] 5

11. m varies directly as n and inversely as the square of p; Given that $m = 3$, when $n = 2$ and $P = 1$. The value of m in terms of n and p is:

[A] $m = \dfrac{2n}{3p}$ [B] $m = \dfrac{3n}{2p}$

[C] $m = \dfrac{2n}{3p^2}$ [D] $m = \dfrac{3n}{2p^2}$

12. Given that p varies directly as q while q varies inversely as r. The statement, which is true:

[A] r varies directly as p.

[B] p varies inversely as r.

[C] p varies directly as r.

[D] q varies inversely as p.

13. Given that 20 men take 6 days to clear a field. The time it would take 12 men working at the same rate to clear a similar field is:

[A] 40 days [B] 2 days

[C] $3\frac{1}{2}$ days [D] 10 days

14. K varies directly as N and inversely as the square of L. Given that $L = 1$ when $N = 3$ and $K = 2$. The value of K in terms of N

and L is:

[A] $K = \dfrac{2N}{3L^2}$ [B] $K = \dfrac{3N}{2L^2}$

[C] $K = \dfrac{2L^2}{3N}$ [D] $K = \dfrac{3L^2}{2N}$

15. Given that $x \propto y$ and $y \propto \dfrac{1}{z^2}$. The way x varies with z is:

[A] $x \propto \dfrac{1}{z}$ [B] $x \propto \dfrac{1}{\sqrt{z}}$

[C] $x \propto \dfrac{1}{z^2}$ [D] $x \propto z^2$

16. $x \propto y$ and that $x = 28$ when $y = 4$. The formula connecting x and y is:

[A] $x = 2y$ [B] $x = 4y$

[C] $x = 7y$ [D] $x = 14y$

17. $x \propto y$ and when $x = 4$, $y = 20$. The value of x when $y = 5$ is:

[A] 4 [B] 3 [C] 2 [D] 1

18. $m \propto \dfrac{1}{n}$ and $m = 3$ when $n = 2$. The law connecting m and n is:

[A] $m = 6n$ [B] $m = 3n$ [C] $m = \dfrac{6}{n}$ [D] $\dfrac{3}{n}$

19. Given that $x \propto \dfrac{1}{y}$ and that $x = 9$ when $x = 4$. The formula, which connects, x and y is:

[A] $x = \dfrac{36}{y}$ [B] $x = \dfrac{13}{y}$

[C] $x = \dfrac{9}{y}$ [D] $x = \dfrac{5}{y}$

20. L varies jointly as M and N. When $M = 2$, $N = 3$ and $L = 9$. The law connecting L, M and N is:

[A] $M = \dfrac{2}{3}LN$ [B] $L = MN$

[C] $L = \dfrac{2}{3}MN$ [D] $M = \dfrac{3}{2}LN$

21. R varies directly as t and inversely as m. K is the constant of proportionality. The relationship between R, t and m is:

[A] $R = \dfrac{Km}{t}$ [B] $R = \dfrac{Kt}{m}$

[C] $R = K + \dfrac{m}{t}$ [D] $R = t + \dfrac{K}{m}$

22. The energy E of a moving body varies partly as the square of the height, H of the body above sea level and partly as the square root of its velocity, V. Given that a and b are constants, the equation representing the above expression is:

[A] $E = aH^2 + b\sqrt{V}$ [B] $E = a\sqrt{H} + bV^2$

[C] $E = \dfrac{a}{H^2} + \dfrac{b}{\sqrt{V}}$ [D] $E = \dfrac{a}{\sqrt{H}} + \dfrac{b}{V^2}$

23. The equation, which represents the relation in Table 35:15, is:

[A] $y = -3x + 8$ [B] $y = \dfrac{1}{3}x + 8$

[C] $y = -\dfrac{1}{3}x + 8$ [D] $y = 3x + 8$

x	−3	−2	−1	0	1
y	1	−2	−5	−8	−11

Table 35:15

24. The equation, which represents the relation in the Table 35:16, is:

[A] $y = 8x + 9$ [B] $y = \dfrac{1}{8}x + 9$

[C] $y = 8x + 9$ [D] $y = -\dfrac{1}{8}x + 9$

x	−2	−1	0	1	2
y	−7	1	9	17	25

Table 35:16

25. The equation, which best describes the relation between x and y in Table 35:17 is:

[A] $y = x^2 + 5$ [B] $y = -x^2 - 5$
[C] $y = x^2 - 5$ [D] $y = -x^2 + 5$

x	−1	0	1	2	3
y	4	5	4	1	−4

Table 35:17

TOPIC 36

GRAPHS OF INEQUALITIES

OBJECTIVES

At the end of this topic, the learner should be able to:

1. Identify and name the three sets of points into which a straight line divides a plane.
2. Identify and draw vertical and horizontal boundary Lines
3. Solve simultaneous linear inequations using the graphical method.

Blaise Pascal

Noted primarily as a mathematician, scientist, and author, Blaise Pascal focused on religion late in his short life. Pascal argued that faith in God is reasonable. He reasoned that, although no one can prove God's existence or nonexistence, the potentially infinite benefits of believing God exists far outweigh any finite benefits that might be gained by believing God does not exist.

Hulton Deutsch

Half planes

A straight line is considered to divide a plane into three sets of points, namely:
(1) The set of points on the line
(2) The set of points on either side of the line

The planes mentioned in (2) above are called **half planes**.

Example 36:1

Represent the inequation $2x + y \geq 3$ on a graph.

Solution

x	-1	0	1
y	5	3	1

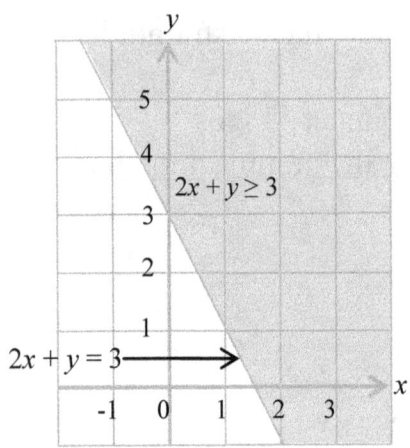

Figure 36:1

Since the points on the line $2x + y = 3$ are included, the line is drawn as a continuous line. By testing any point (a, b) on any side of the line, the region which satisfies the inequation, can be detected. The shaded region in the diagram satisfies the inequation. No point on the other side of the line whatsoever will satisfy the inequation. Testing using the point $(0,0)$ is very convenient since calculation with 0 is very easy to perform. There is no need testing using more than one point, for if an inequality is true for a point, it must be true for all other points in that region and false for all points on the other region and vice versa.

Testing using the point $(0,0)$,

L.H.S. $= 2(0) + 0 = 0$

Therefore, the point does not satisfy the inequality since $0 < 3$.

Testing using the point $(2, 2)$,

L.H.S. $= 2(2) + 2 = 6$

Therefore, the point $(2,2)$ satisfies the inequation since $6 > 3$.

Example 36:2

Represent the inequation $2x + y > 3$ on a graph.

Solution

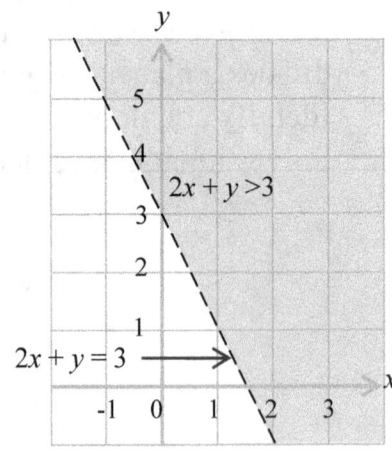

Figure 36:2

Notice that, the line $2x + y = 3$ is shown as a broken line because it is not included in the solution of the inequation.

Vertical and Horizontal Boundary Lines

Consider the inequations involving the lines $x = a$ and $y = b$, where a and b are constants. The line $x = a$ will be vertical and the line $y = b$ will be horizontal.

Example 36:3

Represent the following inequations on separate Cartesian planes.
(a) $x > 2$ (b) $x < 0$
(c) $y \leq 3$ (d) $y > 0$

Solution

(a)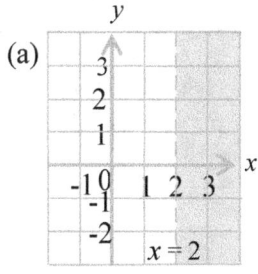

(b)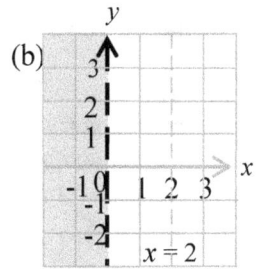

(c)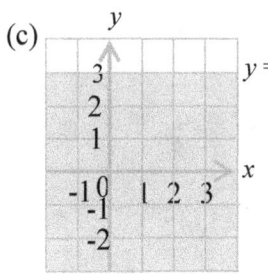

(d)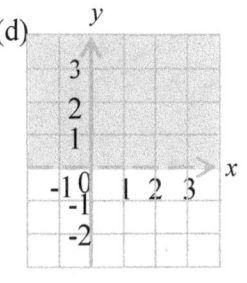

Figure 36:3

Simultaneous Linear Inequations

Recall that graphically, the point of intersection of the two lines representing the equations gives the solution of simultaneous linear equations. Equally, the region described by the intersection of two or more inequations gives the solution of the inequations.

Example 36:4

Shade the region in which the inequations $x + y < 7$, $x - y \geq 3$ and $y \geq -4$ are all satisfied.

Solution

$x + y < 7$			
x	0	2	7
y	7	5	0

$x - y \geq 3$			
x	0	2	3
y	-3	-1	0

The graph is shown in figure 35:4

$$R = \{(x,y) : x - y \geq 3\} \cap \{(x,y) : x + y < 7\} \cap \{(x,y) : y \geq -4\}$$

Since the line $x - y = 3$ satisfies the inequation $x - y \geq 3$, it is drawn as a continuous line. The line on the other hand does not satisfy the inequation $x + y < 7$; hence, it is drawn as a broken line.

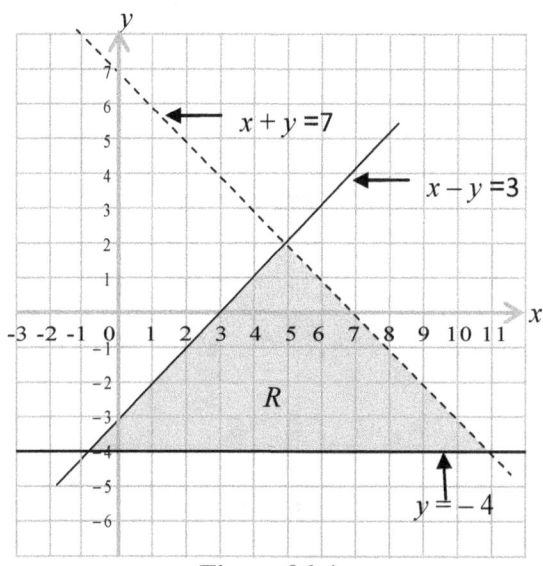

Figure 36:4

EXERCISE 36

1. By drawing suitable straight lines on a graph, show by shading the region R for which all the three $x + y \leq 8$, $y < x$, $y < 3$ inequalities are satisfied.

2. Shade the unwanted regions in each of the following.

   (a) $y \geq 0$, $x \geq 0$, $x + y \leq 9$

   (b) $x \geq 0$, $y \geq x$, $2x + y \leq 10$

   (c) $y < 6$, $x < 5$, $x + y \geq 5$

   (d) $y \geq \frac{1}{2}x$, $x + y \leq 10$, $y \leq 3x$

   (e) $x < 4$, $y < x + 3$, $2y + x > 4$

   (f) $y < 4x$, $x + y < 10$, $3y > x$

   (g) $x \geq 0$, $x + y < 11$, $2y > x + 5$

   (h) $x \geq 0$, $x + 2y \geq 12$, $x + y \geq 12$

   (i) $y \geq 0$, $x + y \leq 12$, $3x + y \geq 12$

3. Describe the region bounded by the shaded regions in each of the following.

   (a)
   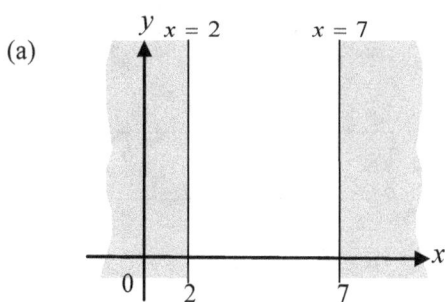

(b)

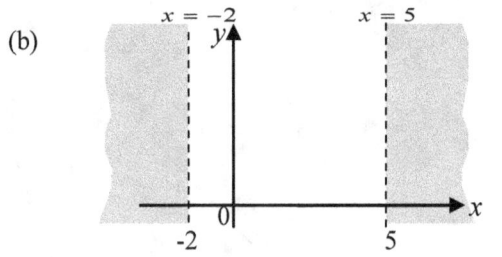

(c)

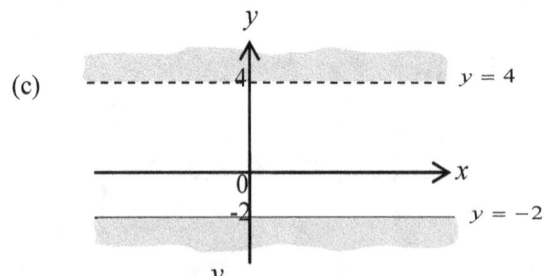

(d)

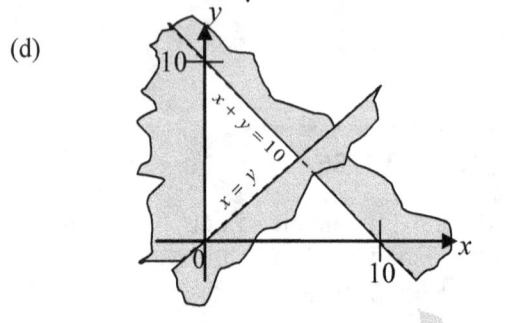

(e)

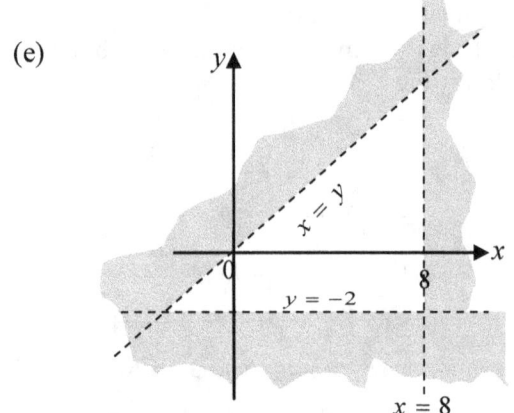

(f)

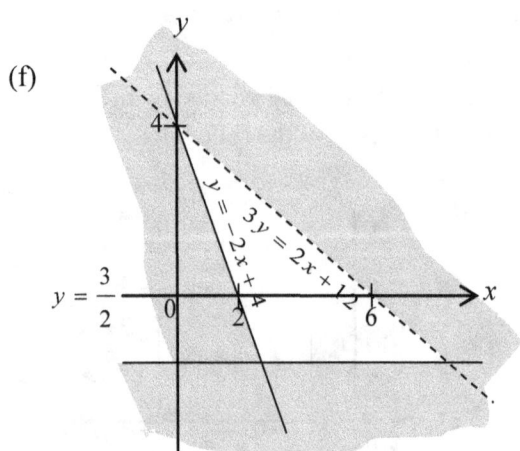

Figure 36:5

MULTIPLE CHOICE EXERCISE 36

1. In Figure 36:6, the shaded portion shows the boundary of the half plane defined by the inequality:

[A] $4x + 3y > 6$ [B] $4x + 3y < 6$

[C] $4x + 3y \geq 6$ [D] $4x + 3y \leq 6$

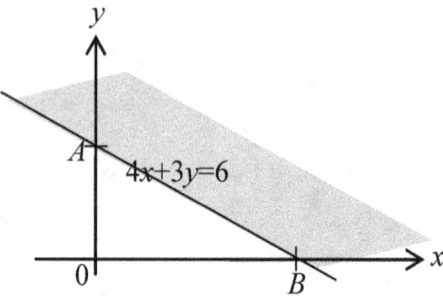

Figure 36:6

2. In Figure 36:6, the co-ordinates of point B is:

[A] $\left(0, 1\frac{1}{2}\right)$ [B] $(0, 2)$

[C] $(2, 0)$ [D] $\left(1\frac{1}{2}, 0\right)$

3. The inequality illustrated in the sketch graph in Figure 36:7 is:

[A] $y \geq -2x + 3$ [B] $y \geq -3x + 3$

[C] $y \geq 3x + 3$ [D] $y \geq 3x + 2$

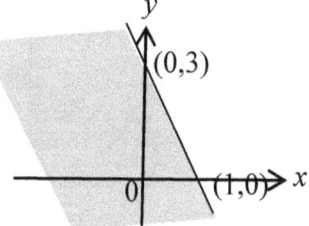

Figure 36:7

4. In Figure 36:8, the region P, Q, R or T which satisfies the inequalities $0 < y < 1$, $y < x + 2$ and $x < 0$ is:

[A] P [B] Q [C] R [D] S

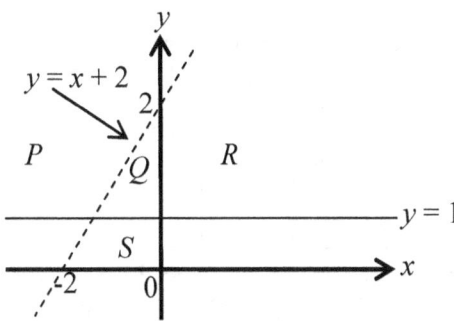

Figure 36:8

5. In Figure 36:9, the region defined by triangle *OPQ* can be represented by the inequalities:

[A] $y \geq x, y \leq 0, y + x \geq 7$

[B] $y \leq x, y \geq 0, y + x \leq 7$

[C] $y \geq x, y \geq 0, y + x \leq 7$

[D] $y \leq x, y \leq 0, y + x \leq 7$

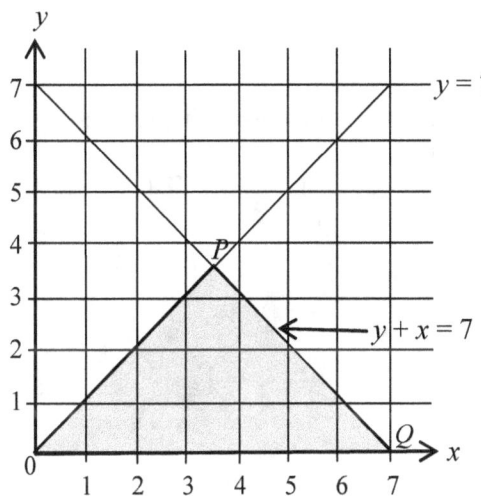

Figure 36:9

6. In Figure 36:10, the equations of the lines *AC*, *AB,* and *BC* are:

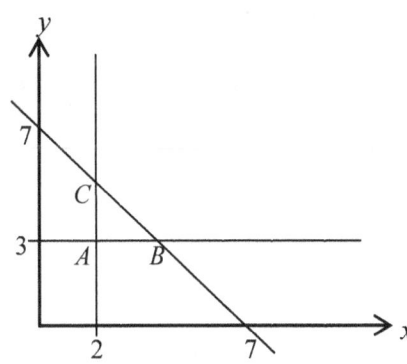

Figure 36:10

[A] $x = 2, y = 3, y + x = 7$

[B] $x = 3, y = 2, y + x = 7$

[C] $x + 2 = 0, y = 3, y + x = 7$

[D] $x = 2, y + 3 = 3, y + x = 7$

7. In Figure 36:10, the three inequalities, which define the triangle ABC, are:

[A] $x \geq 2, y \leq 3, y + x \leq 7$

[B] $x \geq 3, y \geq 2, y + x \leq 7$

[C] $x \geq 2, y \geq 3, y + x \leq 7$

[D] $x \geq 2, y + 3 \geq 0, y + x \geq 7$

8. The shaded region in Figure 36:11 is best described by:

[A] $x \leq 3, y \leq 4$ and $4x + 3y \leq 12$

[B] $x \leq 3, y < 4$ and $4x + 3y < 12$

[C] $x \geq 3, y \geq 4$ and $4x + 3y > 12$

[D] $x \leq 3, y < 4$ and $4x + 3y > 12$

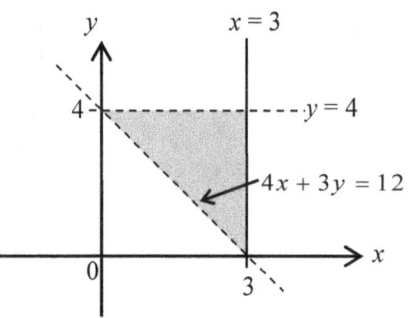

Figure 36:11

9. The graph, which represents the inequality, $x + 1 \leq 0$ is:

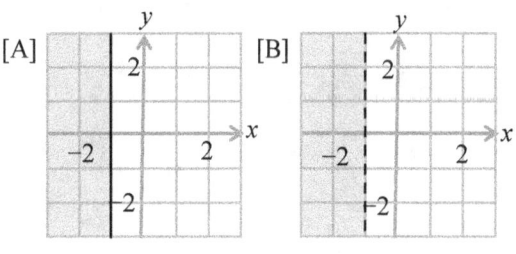

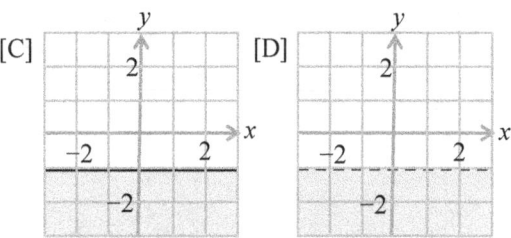

Figure 36:12

10. The inequalities $y > x + 2$ and

$y \geq 1 + 2x$ are represented by the graph:

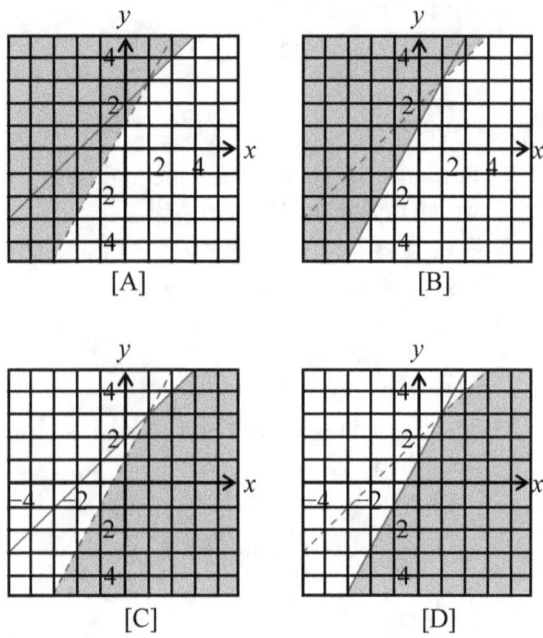

[A]   [B]

[C]   [D]

Figure 36:13

11. The graph in Figure 36:14, which represents the inequality $y \geq x-1$ is:

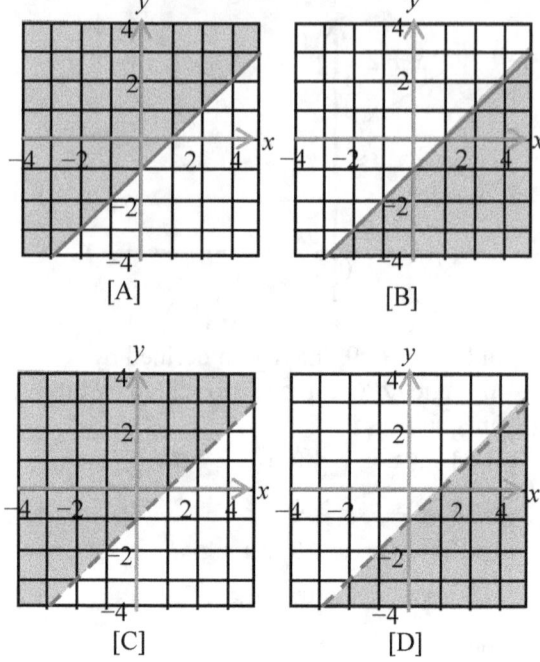

[A]   [B]

[C]   [D]

Figure 36:14

12. The linear inequality, which represents the graph, is:

[A] $y \leq -x + 1$ [B] $y \geq -x + 1$

[C] $y \leq x + 1$ [D] $y \geq x + 1$

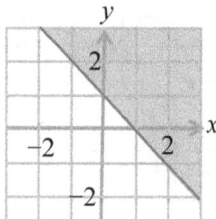

Figure 36:15

TOPIC 37

STATISTICS

OBJECTIVES

At the end of this topic, the learner should be able to:

1. Appreciate the importance of the topic for studies and in life.
2. Collect raw data, tally the data and tabulate it.
3. Represent data on statistical graphs and interpret statistical graphs.
4. Determine mode, median and mean of a small range of data considering, odd and even number of scores for median.
5. Determine which measure of central tendency is best for any given data or situation.
6. Analyse grouped data and distinguish between class Limits and class boundaries
7. Determine class Size and mid-interval value of a group.
8. Represent grouped data on histograms and use it to find the mode.
9. Determine the modal class of grouped data.
10. Draw frequency distribution curves for given data.
11. Draw cumulative frequency tables and use it to draw a cumulative frequency curve for given data.
12. Define and determine the median, quartiles, deciles and percentiles a from cumulative frequency curve.
13. Determine the range, the inter quartile range, semi-inter quartile range percentiles and the 10 - 90 percentile ranges for a given distribution.
14. Calculate the mean, mean deviation and standard deviation and variance of a given data.
15. Use the assumed or guess mean or the coding method, to simplify statistical calculations.

George Gallup, specialized in highly accurate public opinion surveys. His educational background in statistics helped him to revolutionize business and marketing. Gallup's methods allowed companies to discover people's interests before they marketed new products. His techniques also changed the nature of politics in the United States. Politicians use his polling systems to determine the views of a cross section of the general public on certain topics, which allow them to formulate campaign platforms that appeal to the desires of the public.

Herman Hollerith
American inventor Herman Hollerith developed the punch card system of recording data, a step that revolutionized the processing of census data.

THE BETTMANN ARCHIVE

The Concept of Statistics

Information such as shoe sizes, ages of people, number of students in each class of a school and so on, is known as **data**. The branch of mathematics, which deals with the collection, processing and analyzing of data, is called **statistics**, though the word statistics is sometimes used as a synonym for data.

Importance of Statistics

Statistics enables people to take proper decisions. For instance in order to make sure that the General Certificate Education questions are distributed to examination centres in such a way that excesses are minimized and shortages are avoided, the number of candidates who will go in for a particular subject in each centre, must be known. This is the main reason why during registration the subjects and the centre of any particular student must be well spelt out.

Raw Data

Raw data is unprocessed and unanalysed data. For instance, a shoe dealer bought a bag of shoes from Douala. On his arrival at Bamenda, he opened the bag and found out that the shoe-sizes were as follows:

```
42  39  38  45  44  44  40
37  40  43  41  39  43  44
43  42  39  43  45  41  46
44  39  40  43  37  42  41
43  41  44  39  42  45  40
42  38  43  40  44  40  39
```

The above, assorted display of data is an example of **raw data**. The data looks so disorganized and for it to make sense, it must be organized and processed.

UNGROUPED DATA

The following are the steps normally used to process raw data.
1. First, the data is tallied.
2. Next a frequency distribution table is drawn.

Tallying

In tallying strokes are made against each statistic (in this case shoe sizes) in such a way that each stroke represents one element of a specific class. The fifth stroke is horizontal across the first four. For instance, if a statistic occurs seven times, the tallying will be ||||| || . Each stroke is called a **tally**

and this method of processing raw data is known as the **tally method**.

Frequency-Distribution Tables

A tabular summary of statistical information is called a **frequency distribution table** or simply a **frequency distribution**. The process of construction of a frequency distribution is called **tabulation**. The number of times a particular statistical entity, x, occurs is called its **frequency**, usually denoted by f.

Example 37:1
(a) Tally the raw data above and
(b) Draw a frequency table for the data.

Solution

Shoe size	Tally
37	||
38	||
39	|||| |
40	|||| |
41	||||
42	||||
43	|||| ||
44	|||| |
45	|||
46	|

x	f
37	2
38	2
39	6
40	6
41	4
42	5
43	7
44	6
45	3
46	1

      (a)   (b)

Table 37:1

To crosscheck the information, add the frequency (number of shoes). The sum should correspond to the total (number of shoes). If not rectify it, by repeating the tallying or the survey.

EXERCISE 37:1

1. The following are the marks gained by 30 students in an examination.

```
57  60  58  62  56  59
62  60  62  63  64  56
60  62  63  58  59  63
66  66  56  58  51  58
53  57  59  57  53  54
49  54  60  64  60  64
```

392

Tally the marks and use them to draw a frequency distribution table of the data.

2. A boy measured to the nearest metre how far he could through a tennis ball. The results are as follows.

    66 69 70 68 71 68 69 70 67 68
    67 68 67 66 69 68 69 70 68 67

    Tally the raw data and hence draw a frequency-distribution table of the data.

3. In a certain restaurant, rice (r), beans (b), plantains (p) and yams (y) are sold. The food items ordered by the customers of the restaurant on a certain day are as followings.

    p y b r y p r y b b r y p p y
    b r y p p y p y b r y p r y b
    p y p r y b y b r y p b r y p

    Tally the data and draw a frequency-distribution table, which would be used to determine the number of customers that ordered each of the food items.

4. The weights in kilograms of 60 students are as follows.

    61 57 62 60 55 66 63 64 52 58
    64 59 64 57 62 61 61 55 62 60
    64 61 64 57 62 59 64 69 62 64
    63 64 66 56 62 66 58 63 58 58
    60 60 64 55 60 63 58 58 66 58
    60 63 61 59 66 63 63 61 58 64

    Tally the weights and draw a frequency-distribution table of the data.

Representation of Data-Statistical Graphs

The various ways of representing data include pictograms, bar charts, pie charts, histograms etc. The names 'chart', 'graph', or 'diagram' are synonymous in this context and shall be used in this topic interchangeably.

(i) Pictograms or Ideographs

Pictograms or ideographs are pictures used to represent data in such a way that one can see at a glance the relative frequency (the frequency of one statistic with respect to another), of any of the statistic. Pictograms are used for comparative data such as population of countries.

Example 37:2
A fruit dealer bought 200 pineapples, 400 Bananas, 500 watermelons and 900 Oranges. Represent this information using a pictogram.

Solution

Figure 37:1: 1 Fruit represents 100 fruits

(ii) Pie Charts or Circular Diagrams

In this type of graphs, a circle, divided into sectors is used to represent the frequencies. Each sector is proportional to the statistic representing it. It is therefore necessary that pie charts should always be drawn with a pair of compasses and the angles measured with a protractor. To draw a pie chart the angles must first be calculated.

Example 37:3
Represent the numbers in Example 37:2 on a pie chart.

Solution

Fruit	No.	Calculation	Angle
Pineapple(P)	200	$\dfrac{200}{2000} \times 360$	$36°$
Bananas (B)	400	$\dfrac{400}{2000} \times 360$	$72°$
Watermelons (W)	500	$\dfrac{500}{2000} \times 360$	$90°$
Oranges (O)	900	$\dfrac{900}{2000} \times 360$	$162°$
TOTAL	2000		$360°$

Table 37:2

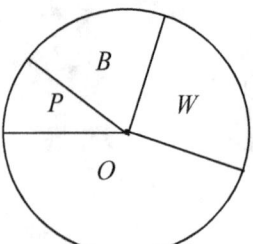

Figure 37:2

OR

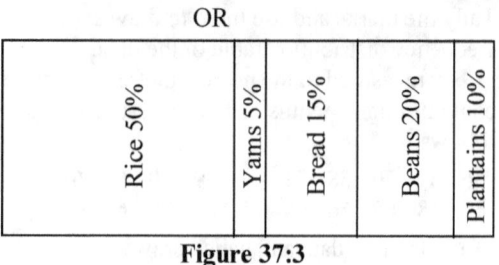

Figure 37:3

(iii) Bar Charts

In this case, bars of equal width represent information. Only the length or height of the bars has any significance. Therefore, the width of the bars can be any convenient size. The bars may be together or separated but must not overlap. Bar charts are of many types and are drawn horizontal or vertical.

(a) *The proportionate bar Chart*

In the proportionate bar chart, the bars may be placed one upon another and are of equal widths. In this case, the height of each bar is proportional to the frequency. Otherwise, the bars may be placed one besides the other and are of equal heights. In this case, the width of each bar is proportional to the frequency.

Example 37:4

A survey of the best dishes of a number of students showed that 50% like rice, 5% like yams, 15% like bread, 20% like beans, and 10% like plantains. Draw a proportionate bar chart to represent this information.

Solution

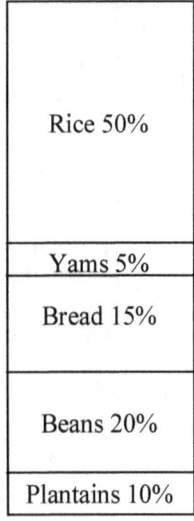

(b) *Chronological Bar Chart*

These types of bar chart usually represent the variation of some statistic over a period. It may be drawn vertically or horizontally.

Example 37:5

Table 37:3, shows the number of people who owned televisions in a certain town from 1985 to 1993. Draw a chronological bar chart to represent this data.

Year	Number of People
1985	600
1986	800
1987	900
1988	1100
1989	1300
1990	1400
1991	1200
1992	1500
1993	1600

Table 37:3

Solution

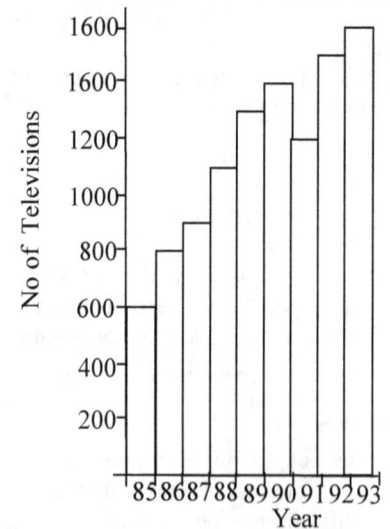

OR

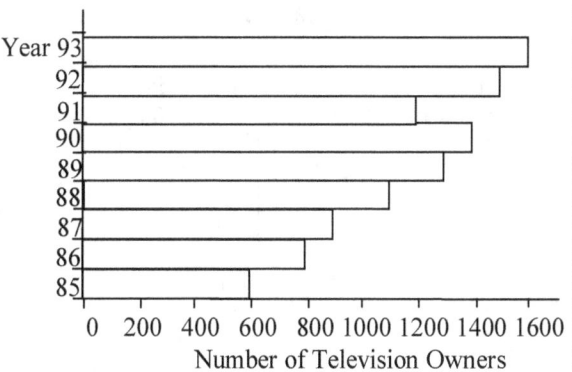

Figure 37:4

(c) *Simple Bar Charts*

The bars in this type of bar chart are separated from each other.

Example 37:6

Use Table 37:3 to draw a simple bar chart.

Solution

Number of TV Owners from 1985 to 1993

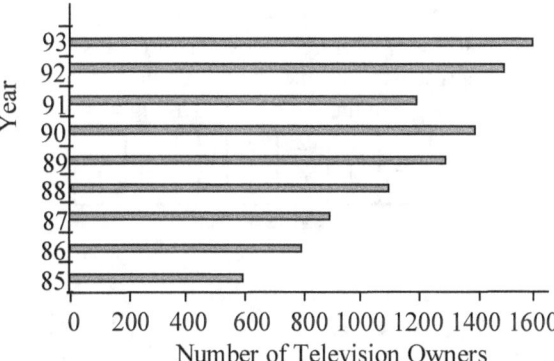

OR

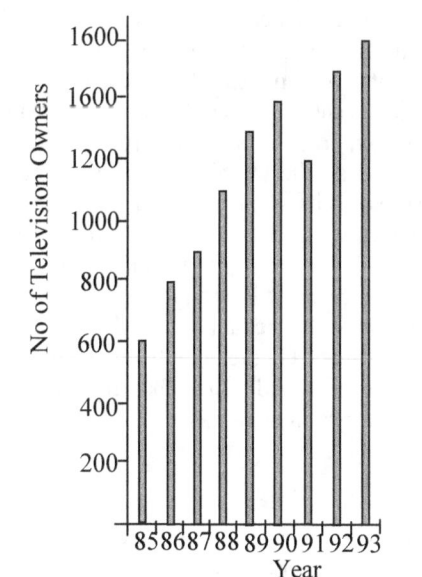

Figure 37:5

Histograms

Histograms are similar to chronological bar charts. The only difference is that while the frequency is proportional to the length of the bars in the case of the bar chart, the frequency is proportional to the areas of the rectangles in the case of the histogram. Therefore, for the histogram, both the length and the width of the rectangle are important. Though the width of the rectangle, may be of different sizes it is often more convenient to make them the same size. In this way, histograms are therefore very similar to chronological bar charts

Example 37:7

Table 37:4 shows the number of form 2 students in a certain school in the year 2002 who had the required textbooks for the subjects Mathematics (M), English (E), French (F), History (H), Geography (G), Chemistry (C), Physics (P), Biology (B) and Literature (L).

Draw a histogram to represent this information.

Textbook	M	E	F	H	G	C	P	B	L
No. of students	14	15	11	3	9	5	7	1	13

Table 37:4

Solution

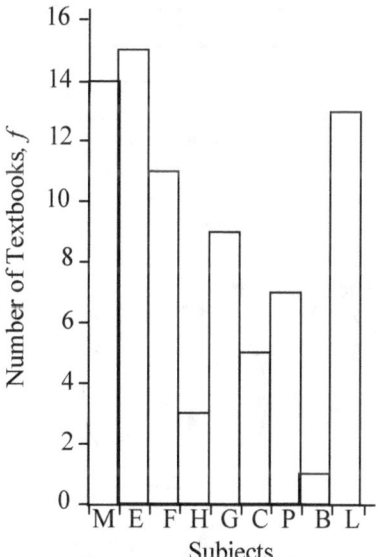

Figure 37:6

EXERCISE 37:2

1. The form three students of a certain school participated in team sports as shown in Table 37:5.

Team sport	Number of students
Handball	45
Basketball	60
Football	75

Table 37:5

Represent this information on a pie chart and state the angle for basketball.

2. Figure 37:7 is a pie chart (not drawn to scale) showing how a student spent his pocket money amounting to 27,000 FCFA. Given that he spent twice as much on books as he did on taxi, calculate:
   (a) How much he spent on books.
   (b) How much he spent on others.

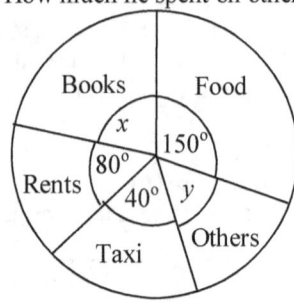

Figure 37:7

3. Table 33:6, shows a survey carried out on a group of students to find out what they ate for launch on a certain day.

Achu	15
Rice	9
Garri	4
Bread	2

Table 37:6

Draw a histogram to display this data

4. The pie chart in Figure 37:8 shows the number of votes for candidates A, B and C in an election. Calculate the percentage of the votes to the nearest tenth in favour of candidate B.

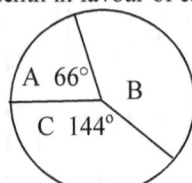

Figure 37:8

5. Draw a histogram for the distribution in Table 37:7.

Score, x	1	2	3	4	5	6
Frequency, f	4	6	7	3	3	1

Table 37:7

7. The livestock of a certain farm consist of 28 cows, 300 sheep, 74 pigs, 306 poultry, 9 dogs and 3 cats. If this information is recorded on a pie chart, calculate the angle in degrees, at the centre of the sector representing the cows.

8. Five boys A, B, C, D and E are of heights 160,

144, 120, 96 and 80 centimetres respectively. Represent this information on a bar chart.

9. Figure 37:9 shows a pie chart indicating the favourite colours of a group of 108 girls.

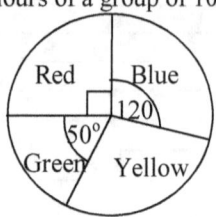

Figure 37:9

(a) Find the angle of the sector for girls who like yellow.
(b) Find the number of girls who like green.

10. A student used Table 37:8, to draw a pie chart. Find the values of w, x, y and z.

Item type	A	B	C	D
Frequency	48	104	x	y
Sector angle ($^\circ$)	72	w	108	z

Table 37:8

11. Table 37:9 shows a statistical table of a variable x with frequency f. This data is to be represented by a pie chart. Calculate in degrees, the angle of the sector representing $x = 3$.

x	1	2	3	4	5	6
f	4	6	7	3	3	1

Table 37:9

Measures of Central Tendencies

Mode, mean and **median**, which are very commonly used, are examples of averages, otherwise called measures of central tendencies. They are so called because their values are representative or typical of any given data, and tend to lie centrally when the data is ranked or arranged in order of magnitude (from highest to smallest or smallest to highest). Each of these measures has its advantages and disadvantages depending on the data and purpose for which it is intended. For instance, the mean has the disadvantage that it is strongly affected by extreme values, while the median is not affected by extreme values. On the other hand, the mode may or may not exist and turns to be very subjective. At times, there are many modes.

Mode

The mode of any given data is the variable or statistic that occurs most frequently.

Example 37:8

Find the mode of the following data.
1,2,4,6,2,7,7,2,2,7.

Solution

x	1	2	6	7
f	1	4	1	3

Table 37:10

The frequency of 2 is 4, which is the highest frequency, so 2 is the mode.

Example 37:9
Find the mode of the following data.
(i) 99, 100, 101, 102 and 101.
(ii) 99, 100, 101, 100, 102 and 101.

Solution
(i) Mode =101   (ii) Mode =100 and 101

Notice in Example 37:9(ii) that there are 2 modes, 100 and 101. In such a case, the distribution is said to be **bimodal**. If there are three modes, the distribution is said to be **trimodal** and generally if there are more than one mode the distribution is said to be **multimodal.**

Arithmetic Mean (Average or Mean)

The mean is usually denoted by $\bar{x}$, and is obtained by summing all the data and dividing by the frequency. Thus

$$\bar{x} = \frac{\text{sum of data}}{\text{total frequency}}$$

$$\bar{x} = \frac{\sum x}{\sum f}$$

Where $\sum x$ and $\sum f$, read 'summation x' and 'summation f', respectively meaning 'sum of data' and 'sum of the frequencies' respectively. $\sum$ is called the sigma notation and $\sum x$ and $\sum f$, can be read 'sigma x' and 'sigma f' respectively.

Example 37:10
Find the mean of 11, 9, 15, 12 and 13

Solution

$$\bar{x} = \frac{\sum x}{\sum f} = \frac{11 + 9 + 15 + 12 + 13}{5} = 12$$

If some of the data repeat themselves, advantage is taken of multiplication as repeated addition, to write the formula as

$$\bar{x} = \frac{\sum xf}{\sum f}$$

$\sum fx$, is read 'sigma fx', or 'summation fx', where fx means the product of each statistic and its frequency.

Example 37:11
The following shows the marks obtained by 30 students during a test. Calculate the average mark.

55	60	65	40	60	60
65	50	40	60	50	60
60	50	60	30	40	60
60	50	60	50	60	50
60	50	60	60	50	60

Solution
To ease the work a frequency distribution table can first be drawn.

Mark, x	Frequency, f	fx
30	1	30
40	3	120
50	8	400
55	1	55
60	15	900
65	2	130
	$\sum f = 30$	$\sum fx = 1635$

Table 37:11

$$\bar{x} = \frac{\sum fx}{\sum f} = \frac{1635}{30} = 54.5$$

Examples 35:12
Find the mean of the following data.
13, 13, 13, 13, 13, 13, 14, 14, 15, 15,
15, 16, 16, 16, 16, 16, 16, 16, 16, 16

Solution

x	f	fx
13	6	78
14	2	28
15	3	45
16	9	144
	$\sum f = 20$	$\sum fx = 295$

$$\bar{x} = \frac{\sum fx}{\sum f} = \frac{295}{20} = 14.75$$

Median

To obtain the median, first rank the data. In other words, arrange the data in order of magnitude. For an odd number of numbers, the median is the middle number and for an even number of numbers, the median is the average of the two middle numbers.

Examples 35:13
Find is the median of 12,2,7,13,6.

Solution
Ranking: 2, 6, 7, 12, 13

∴Median = 7

Examples 35:14
Find the median of 2, 7, 6, 13, 12, and 8

Solution
Ranking: 2, 6, 7, 8, 12, 13

$$\therefore \text{median} = \frac{7+8}{2} = 7.5$$

EXERCISE 37:3

1. Find the number that must be removed from the eight numbers 4, 11, 13, 8, 4, 5, 8 and 2, so that the mean of the remaining seven numbers is 6.
2. The mean of five numbers is 4. When a sixth number is added, the mean of the six numbers is $3\frac{1}{2}$. Find the sixth number.
3. Given that the mean of 3, 4 and m is 6, find the mean of 2, m and 14.
4. Table 37:13 represents the weights in kg of 11 students.

Weight, kg	45	53	54	49
No. of students	2	3	4	2

Table 37:13

(a) State the modal weight of the students.
(b) Find the mean weight of the students.
(c) Find the median of the distribution.

5. Use the frequency distribution in Table 37:14 to Calculate:
    (a) the mean   (b) the modal score
    (c) Find the median of the distribution.

Score (x)	1	2	3	4	5	6
Frequency (f)	4	6	7	3	3	1

Table 37:14

6. Table 37:15 shows the marks obtained by pupils in a mathematics test.

Marks (x)	0	3	5	6	8	9	10
No. of pupils (f)	2	4	6	2	4	1	1

Table 37:15

(a) State the mode of the distribution.
(b) Calculate, to 1 decimal place, the mean of the distribution.
(c) Find the median of the distribution.

7. Consider the frequency distribution in Table 37:16.

Score x	3	5	7	9	11
Frequency f	4	6	10	5	5

Table 37:16

(a) State the mode of the distribution.
(b) Calculate, to 1 decimal place, the mean of the distribution.
(c) Find the median of the distribution.

8. Table 37:17 shows the number of coins of six denominations in a bag. Find:
    (a) the average value of the coins in the bag.
    (b) the mode of the coins in the bag.
    (c) the median of the distribution.

Value of coin FRS	5	10	25	50	100	500
Number	4	10	6	8	15	7

Table 37:17

9. The weights of 8 teachers in a certain primary school were measured in kg as follows: 74,64,68,76,80,72,68 and 60 respectively. Find
    (a) the median.
    (b) the mode of the data.
    (c) their mean weight.
10. The frequency distribution in Table 37:18 shows the scores in a mathematics test in a certain class.

Score (x)	2	3	4	7	8	9
Frequency (f)	1	4	6	8	9	2

Table 37:18

(a) Find how many students wrote the test.
(b) Find the mode of this distribution.
(c) Find the mean mark for the test to 1 d.p.

11. The numbers of absences from a mathematics class registered within the first 20 lessons in the first term are 2, 3, 1, 0, 0, 4, 3, 2, 2, 2, 1, 4, 5, 5, 0, 0, 1, 1, 2, and 2. Find the
    (a) mode   (b) median
    (c) mean number of absences.
12. 10 packets of different sizes contain sweets as shown in Table 37:19.

Number of sweets	5	12	6	15
Number of packets	4	2	3	1

Table 37:19

(a) State the mode of the number of packets.
(b) Find the median of the number of packets.
(c) Calculate the mean number of sweets per packet.

Choicest Measure of Central Tendency

In choosing which measure of central tendency to use, two things have to be taken into account. These are:
1. The nature of the distribution,
2. The use for which the measure of central tendency is intended.

Mean
The mean has the advantage that:
(a) It can be expressed as a simple formula, which can easily be used.
(b) It takes account of all the values involved in

the distribution.

The disadvantage of the mean is that it is highly affected by extreme values. Therefore, it is not a good measure of central tendency if the distribution involves one or more extreme values in one direction.

For instance in the data 3, 5, 8, 36, the mean is 13. This will be a very misleading measure of central tendency.

Median

The median has the advantages that:

(a) It is very easy to calculate.

(b) It is the middle of any distribution and is not affected by extreme values.

The median has the disadvantage that it is greatly affected if there are too many values in one direction.

In the data 1, 3, 4, 6, 17, 18, and 19, the median is 6. This value is too low because most of the values are small. Therefore, the median is not a good representation of the distribution in this case.

Mode

The mode has the advantage that:

(a) It is far easier to determine.

(b) It is not affected by extreme values.

The disadvantage of the mode is that, it is meaningless when there are several values having the highest frequency of occurrence. For instance

2, 5, 3, 2, 2, 7, 5, 8, 5, 7, 3, 9, 7, 3

Ranking and tabulating the data

x	2	3	5	7	8	9
f	3	3	3	3	1	1

Table 37:20

The values 2, 3, 5, and 7 by definition are the modes, which has no significance.

Example 37:15

What is the most appropriate measure of central tendency for the following distribution?

3, 4, 4, 4, 4, 4, 6, 7, 9, 13

Solution

The median is 4, the mode is 4, and the mean is 5.8.

The mean is the best because it is almost central and therefore the best representation of the data.

Example 37:16

The hourly wages of five employees in an office are 252 FRS, 396 FRS, 328 FRS, 920 FRS and 325 FRS. Find

    (a) The median hourly wage.

    (b) The mean hourly wage.

    (c) Comment on your result.

Solution

(a) Arranging in ascending order, the wages are 252 FRS, 328 FRS, 375 FRS, 396 FRS and 920 FRS. This gives the median of 375 FRS.

(b) $\bar{x} = \frac{252+396+328+920+325}{5} = 444.2$

(c) The median is not affected by the extreme value 920 FRS while the mean is affected by it. In this case, the median is a better measure of the average hourly wage than the mean.

EXERCISE 37:4

1. Which measure of central tendency do you think the manager of New Life Supermarket will be most interested in? Give reasons for your answer.

2. In a certain week, a bus driver brought to his Patron the following balances: 20,000 FCFA, 20,000 FCFA, 22,000 FCFA, 24,000 FCFA, 24,000 FCFA, 36,000 FCFA, and 38,000 FCFA respectively on each day of the week. What measure of central tendency would you use to compare the balances, and why?

3. The data in Table 37:21 shows the number of students of LS1, who passed in the following subjects: Mathematics (M), Physics (P), Chemistry (C), Biology (B) and Further Mathematics (F).

Subject	M	P	C	B	F
No. of students	8	7	2	8	1

Table 37:21

Which measure of central tendency would be the most appropriate for the analysis of this data? Give reasons for your answer.

4. Table 37:22 is a survey of the number of pigs owned by a group of farmers. One of Nga's two pigs just had a litter of 7 and 8 piglets respectively, so he has 16 pigs.

Number of piglets	0	1	2	4	16
Number of students	10	10	6	3	1

Table 37:22

    (a) Explain why the average should not be used as a measure of central tendency in this case.

    (b) State the most appropriate measure of central tendency, which could have been used.

GROUPED DATA

To analyse very large masses of raw data, it is often necessary to distribute the data into **classes, class intervals** or **groups,** as shown in Table 37:23, which is frequency distribution of the scores of 50 students in a test. This is known as grouped data.

Mark, x	30-39	40-49	50-59
Frequency, f	10	14	26

Table 37:23

Though data grouped in this way is easier to analysed, this method has a disadvantage in that, most of the original details of the information is usually destroyed.

Class Limits and Class Boundaries

Consider the class 30-39. The smaller number, 30 is called the **lower class limit**, and the larger number 39, is called the **upper class limit.** Thus, 40 and 50 are the lower class limits, while 49 and 59, are the upper class limits for the classes 40-49 and 50-59 respectively.

If the data were rounded up to the nearest whole number, the true class limits (called **class boundaries**) will actually be 29.5, 39.5, 49.5 and 59.5. For the class 30-39, the lower and upper class boundaries will be 29.5 and 39.5 respectively. For the class 40-49, the lower and upper class boundaries will be 39.5 and 49.5 respectively.

Class Size

This is the difference between the upper class boundary and the lower class boundary. The class size is sometimes called the **class width** or **class length** or class interval denoted by c.

Mid-Interval Value

The mid-interval value is sometimes referred to as the class mark or mid-point and is the arithmetic mean of the upper and lower class limits. Thus for the class 40-49,

$$\text{Class mark} = \frac{40 + 49}{2} = 44.5$$

The class mark is the mark, which is representative of the given class.

Therefore, for the clarity and mathematical analysis, all observations within a given class are assumed to coincide with the class mark.

HISTOGRAMS FOR GROUPED DATA

When drawing histograms for grouped data, it is preferable to use the class boundaries and the class mark than the class limits.

(a) *Histograms with equal class widths*

These types of histograms are the most commonly used. In this case, the frequency is proportional to the height (length) of the rectangles.

Example 37:17

The marks obtained by 80 students in an examination are arranged in a frequency distribution in Table 37:24:

Marks, x	Frequency, f
1-10	3
11-20	5
21-30	5
31-40	9
41-50	11
51-60	15
61-70	14
71-80	8
81-90	6
91-100	4

Table 37:24

Draw a histogram to represent this data.

Solution

Marks Obtained by 80 Students in an Examination

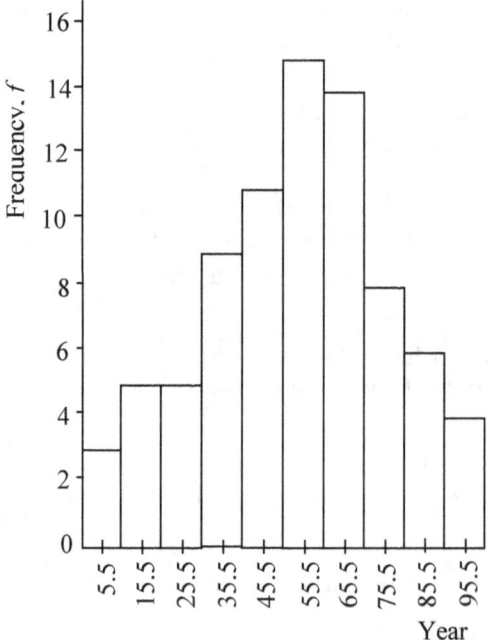

Figure 37:10

Finding the Mode from a Histogram

The Modal class is the class with the highest frequency represented by the tallest bar in the histogram. The mode can be obtained from a histogram as in Example 37:18.

Marks Obtained by 80 Students in an Examination

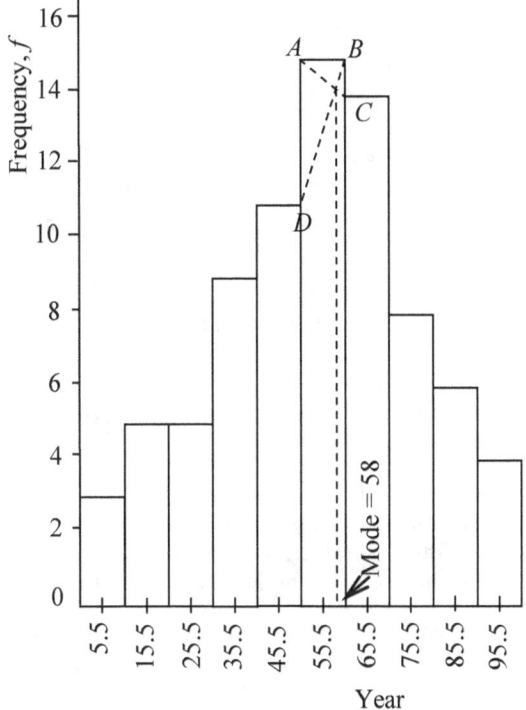

Figure 37:11

Example 37:18
Use the histogram in Figure 37:10 to obtain the mode.

Solution
The mode is obtained by reading the mark corresponding to the point of intersection of the dotted lines *AC* and *BD* as shown in Figure 37:11. From the figure, the mode is 58.5.

Finding the Mode by Calculation

If the class intervals are of equal sizes, the mode can also be obtained using the formula

$$\text{mode} = L_1 + \left(\frac{\Delta_1}{\Delta_1 + \Delta_2} \right) c \text{ , where}$$

L_1 = lower class boundary of the modal class

C = class width

Δ_1 = modal class frequency − next lower class

frequency

Δ_2 = next upper class frequency − modal class

frequency

Example 37:19
By calculation, find the mode of the data in Example 37:17.

Solution

$$\text{mode} = L_1 + \left(\frac{\Delta_1}{\Delta_1 + \Delta_2} \right) c = 50.5 + \left(\frac{4}{4+1} \right) 10 = 58.5$$

Frequency Distribution Curve (Frequency Polygon)

After drawing a histogram another type of graph called a **frequency distribution curve** or **frequency polygon** (Figure 37:12) can be drawn by joining the tips of the rectangles of the histogram.

Example 37:20
Use the histogram in Figure 37:10 to draw a frequency distribution curve.

Solution
Marks Obtained by 80 Students in an Examination

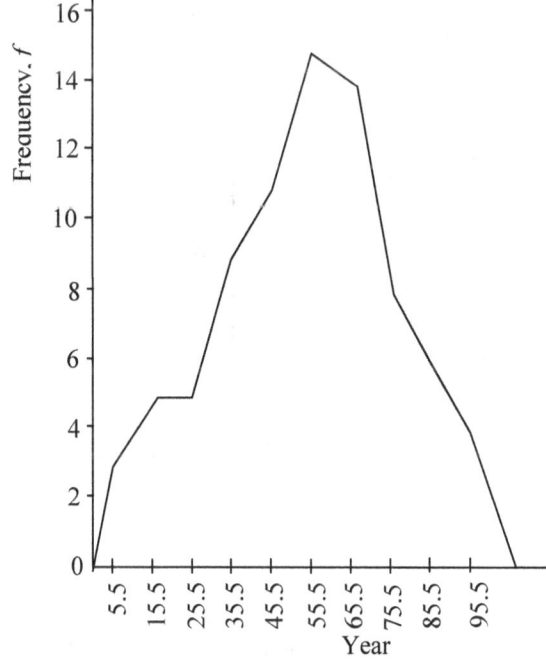

Figure 37:12

Example 37:21
The following scores out of 100 were obtained by

36 students in a test.

25	49	76	12	51	56
81	50	45	92	58	67
55	52	43	31	48	84
66	56	44	39	45	22
56	74	98	67	34	41
34	68	69	70	85	51

(a) Starting with 0-9, arrange the marks in a grouped frequency table with class intervals of size 10.
(b) State the modal class.
(c) Draw a histogram to represent this data hence, obtain the mode of the distribution.
(d) Draw a frequency polygon of the distribution.

Solution

(a)

Marks, x	Class mark	Frequency, f
0-9	4.5	0
10-19	14.5	1
20-29	24.5	2
30-39	34.5	4
40-49	44.5	7
50-59	54.5	9
60-69	64.5	5
70-79	74.5	3
80-89	84.5	3
90-99	94.5	2

Table 37:25

(b) The modal class is 50-59

Marks Obtained by 36 Students in a test

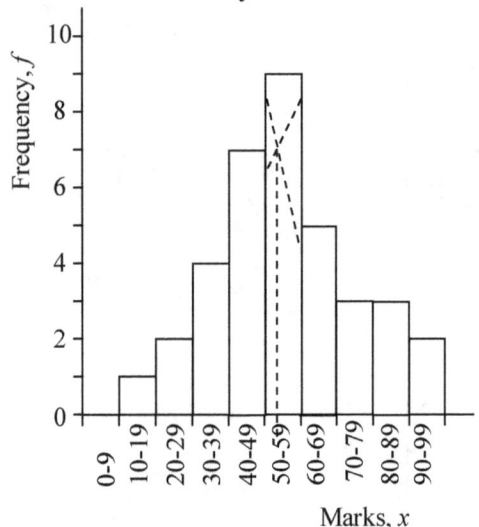

Figure 37:13

(c) From the histogram, the mode is 53.5.
(d) The frequency polygon is as shown in Figure 37:14.

Marks Obtained by 36 Students in a test

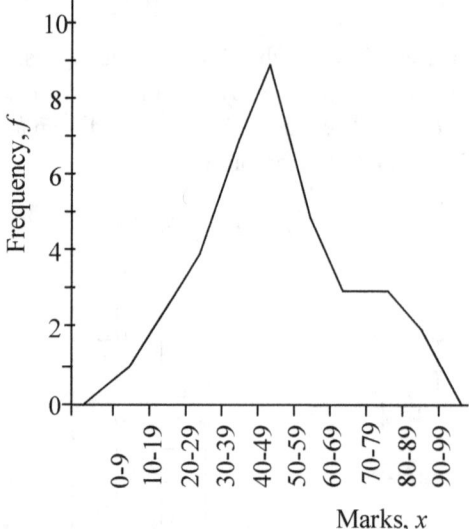

Figure 37:14

(b) *Histograms with unequal class widths*

Histograms with unequal class widths are less common than histograms with equal class widths. However, it is important that note should be taken of them. On such histograms,

Frequency = class width × standard frequency,

Or

$$\text{Standard frequency, S.F.} = \frac{\text{frequency of the class}}{\text{class width}}$$

Standard frequency is also referred to as relative frequency or frequency density and is represented by the height of each rectangle. Therefore, the frequency is proportional to the area of the rectangle for each class and not to the height of each bar.

Example 37:22

Table 37:26 shows the wage distribution amongst three groups of employees in a large company.
(a) Draw a histogram for the distribution.
(b) Draw the frequency polygon for the distribution.

Wage in thousand CFA	Frequency
0-9	8
10-19	16
20-39	10

Table 37:26

Solution

x	f	S.F.
0-9	8	0.8
10-19	16	1.6
20-39	10	0.5

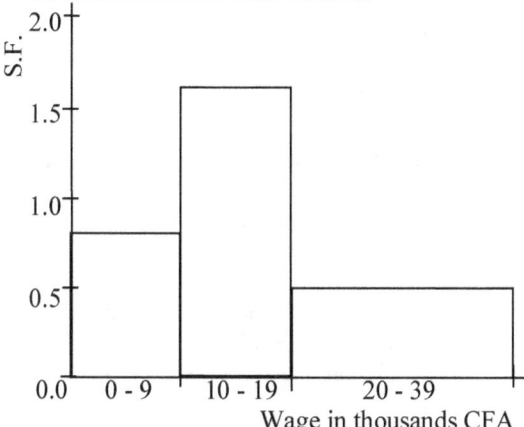

Figure 37:15

Cumulative Frequency

The cumulative frequency, C.F. is the total frequency of all the values less than the upper class boundary of a given class. For instance, in Example 37:20, the total frequency of all the classes up to and including the class 40-49 is $0 + 1 + 2 + 4 + 7 = 14$. This means that 14 students had marks less than 49.5.

A table showing cumulative frequencies is called a **cumulative frequency table**, a **cumulative frequency distribution**, or simply a **cumulative distribution.**

Cumulative Frequency Curves

Cumulative frequency curves, otherwise known as **ogives** are graphs of cumulative frequency against upper class boundary of each class.

Example 37:23
Draw a cumulative frequency table for the data in Example 37:21.

Solution

Marks, x	f	C.F.
< 9.5	0	0
< 19.5	1	1
< 29.5	2	3
< 39.5	4	7
< 49.5	7	14
< 59.5	9	23
< 69.5	5	28
< 79.5	3	31
< 89.5	3	34
< 99.5	2	36

Table 37:27

Quantiles

Recall that given ranked data, the median M, is the middle term for an odd number of terms and the arithmetic mean of the two middle terms for an even number of terms. Thus, the median divides the data into two equal parts. This idea can be extended to those values, which divide the data into four, ten or even more equal parts.

The three values that divide data into 4 equal parts when the data is ranked are called **quartiles,** denoted by Q_1, Q_2 and Q_3. The nine values, which divide data into 10 equal parts, when the data is ranked, are called **deciles**, denoted by D_1, D_2, D_3, D_4, $D_5...D_9$. In like manner, the ninety-nine values that divide data into 100 equal parts are called **percentiles** denoted by P_1, P_2, P_3, $P_4...P_{99}$. The 50th percentile P_{50}, the 5th decile, D_5 and the 2nd quartile Q_2, all correspond to the median. The 25th percentile P_{25} and the 75th percentile P_{75}, correspond to the first quartile Q_1 and the third quartile Q_3 respectively. The first and the third quartiles are also called the lower and upper quartiles respectively.

Quartiles, deciles, percentiles and other values obtained by equal divisions of ranked data are called **quantiles**. Quantiles are often easily obtained from cumulative frequency curves.

Example 37:24
The marks obtained by 80 students in an examination are arranged in a frequency distribution in Table 37:28;

Marks, x	Frequency, f
1-10	3
11-20	4
21-30	6
31-40	8
41-50	12
51-60	15
61-70	13
71-80	9
81-90	6
91-100	4

Table 37:28

(a) Make a cumulative frequency table for this distribution.
(b) Taking 1cm to represent 10 units on each axis, draw a graph of this distribution.
(c) Use your graph to obtain the
    (i) Median, M
    (ii) Lower quartile, Q_1
    (iii) Upper quartile, Q_3
    (iv) 90th percentile, P_{90}
    (v) 10th percentile, P_1

Solution

(a)

x	f	C.F.
≤ 10	3	3
≤ 20	4	7
≤ 30	6	13
≤ 40	8	21
≤ 50	12	33
≤ 60	15	48
≤ 70	13	61
≤ 80	9	70
≤ 90	6	76
≤ 100	4	80

Table 37:29

(b) See graph (Figure 37:16) on page 404.

(c) From the graph (Figure 37:16), the median, the lower and upper quartiles, the 90th percentile and any other quantile can be read from the x-axis by extrapolating from the y-axis. Thus;

(i) The cumulative frequency corresponding to median is $\frac{1}{2}$ of $80 = 40$, therefore by extrapolation, the median M = 55

(ii) The cumulative frequency corresponding to the lower quartile is 25% of $80 = 20$, therefore by extrapolation, the lower quartile, $Q_1 = 39$

(iii) The cumulative frequency corresponding to the upper quartile is 75% of $80 = 60$, therefore by extrapolation, the upper quartile, $Q_3 = 68$

(iv) The cumulative frequency corresponding to 90th percentile is 90% of $80 = 72$, therefore by extrapolation, the 90th percentile, $P_{90} = 82$.

(v) The cumulative frequency corresponding to 10th percentile is 10% of $80 = 8$, therefore by extrapolation, the 10th percentile, $P_{10} = 20$

(b) Marks out of 100 Obtained by 80 Students

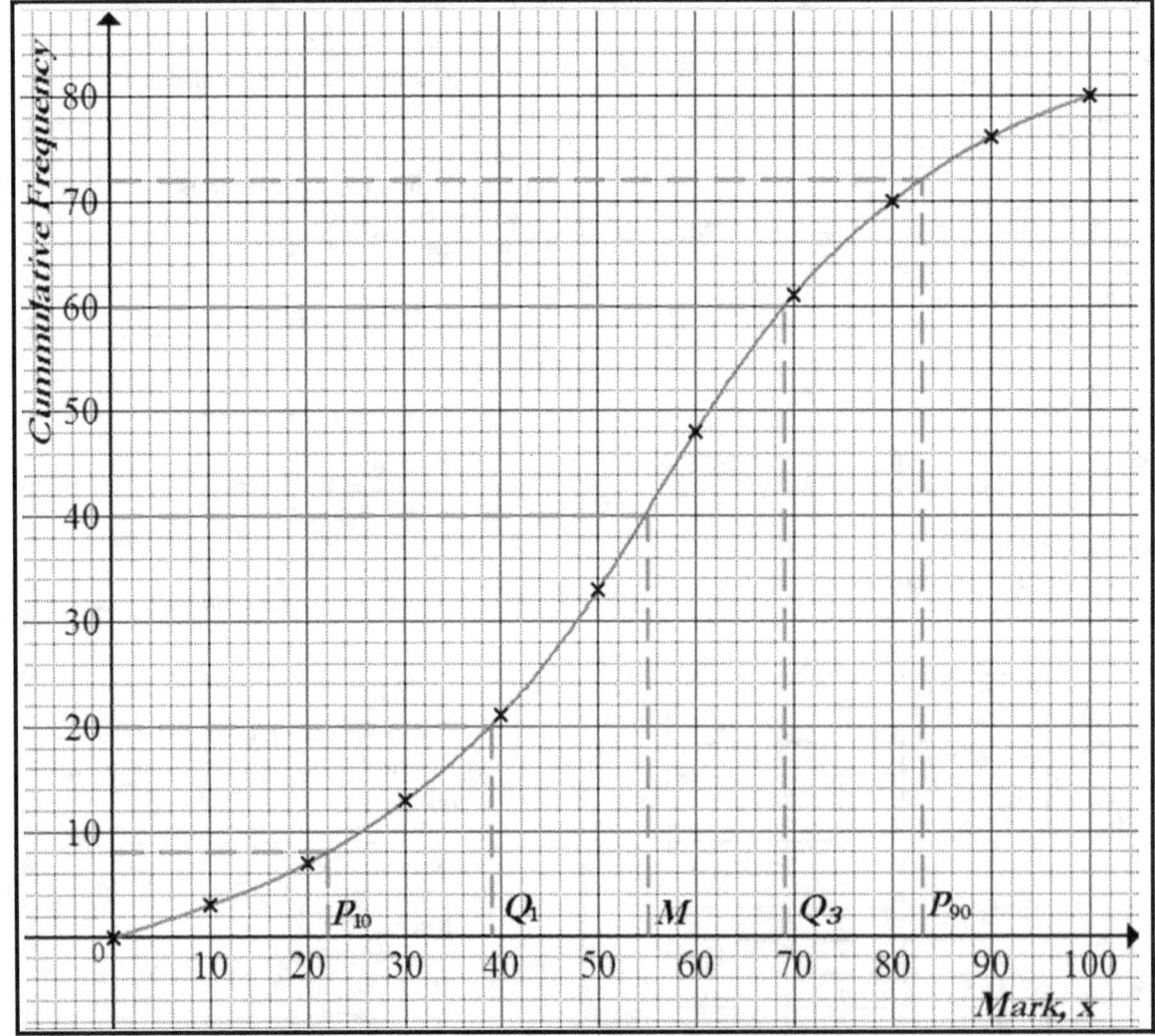

Figure 37:16

Median of Grouped Data by Calculation

The median of group data can also be found by calculation using the following formula, which is based on the principles of interpolation.

$$\text{Median} = L_1 + \left(\frac{\dfrac{\sum f}{2} - \left(\sum f \right)_1}{f_{median}} \right) c$$

Where,

L_1 = lower class boundary of the median class

$\sum f$ = total frequency

$\left(\sum f \right)_1$ = sum of frequency of all classes below the median class

f_{median} = median class frequency

c = median class size

Example 37:25
Calculate the median of the data in Example 37:24

Solution

$$\text{Median} = L_1 + \left(\frac{\dfrac{\sum f}{2} - \left(\sum f \right)_1}{f_{median}} \right) c$$

$$= 50.5 + \left(\frac{\dfrac{80}{2} - 33}{15} \right) 10 = 55.5$$

Measures of Dispersion (Variation, Spread or Scatter) of Data

The measures of dispersion are the degrees to which numerical data turns to spread about an average. The most common methods used to measure the spread of numerical data are the **range**, the **mean deviation**, the **inter quartile range** the **10- 90 percentile range**, and the **standard deviation**.

The Range

The range of a set of data is the difference between the smallest and the largest statistic in the set.

Example 37:26
Find the range of the following data. 2, 7, 3, 7, 8, 21, 17, 35, 4, 39.

Solution

Range = largest statistic − smallest statistic
  = 39−2 = 37

Inter Quartile Range

The inter quartile range is defined as

Inter quartile range = $Q_3 - Q_1$

Semi-inter Quartile Range

The semi-inter quartile range is defined as

$$\text{Semi-interquartile range} = \frac{Q_3 - Q_1}{2}$$

Example 37:27
Use Figure 37:16 to find the inter quartile range and hence calculate the semi-inter quartile range of the data.

Solution

Inter quartile range = $Q_3 - Q_1$ = 68−38 = 30

$$\text{Semi-interquartile range} = \frac{Q_3 - Q_1}{2} = \frac{30}{2} = 15$$

The 10 - 90 percentile ranges

The 10-90 percentile range of a set of data is defined as

10-90 percentile range = $P_{90} - P_{10}$

Example 37:28
Use Figure 37:16 to find the 10-90 percentile range of the data.

Solution

10-90 percentile range = $P_{90} - P_{10}$
  = 82−20 = 62

Mean of Grouped Data

It was mentioned earlier that the class mark is representative of any class and for any statistical analysis, all the observations within a given class can be assumed to coincide with the class mark. To find the mean of grouped data therefore, the class mark is taken to represent the mark for the whole class.

Example 37:29

The marks x, obtained by 80 students in an examination are arranged as in the frequency Table 37:30. Determine the mean of the distribution.

Marks, x	Frequency, f
1-10	3
11-20	5
21-30	5
31-40	9
41-50	11
51-60	15
61-70	14
71-80	8
81-90	6
91-100	4

Table 37:30

Solution

The class mark is chosen to represent the class.

Marks, x	Frequency, f	xf
5.5	3	16.5
15.5	5	77.5
25.5	5	127.5
35.5	9	319.5
45.5	11	500.5
55.5	15	832.5
65.5	14	917
75.5	8	604
85.5	6	513
95.5	4	382
	$\sum f = 80$	$\sum xf = 4290$

Table 37:31

$$\bar{x} = \frac{\sum xf}{\sum f} = \frac{4290}{80} = 53.625 = 53.63 \text{ to 2d.ps.}$$

Mean Deviation from the Mean

As the name suggests, the mean (average) deviation from the mean is the mean of the deviations of all the data x_1 from the mean.

The mean deviation from the mean denoted by $M.D.$ of a set of numbers can be defined as

$x_1, x_2, x_3, \ldots x_n$.

$$M.D. = \frac{\sum_{i=1}^{n} |x_i - \bar{x}|}{n}$$

Where $i = 1, 2, 3, \ldots n$ and $\bar{x}$ is the mean

$\sum_{i=1}^{n}$ Means the sum of the data from 1 to n

Example 37:30

Calculate the mean deviation from the mean of the set of numbers 2, 3, 6, 8, 11.

Solution

$$M.D. = \frac{\sum_{i=1}^{n} |x_i - \bar{x}|}{n}$$

$$\bar{x} = \frac{2 + 3 + 6 + 8 + 11}{5} = \frac{30}{5} = 6$$

$$\Rightarrow M.D. = \frac{|2-6| + |3-6| + |6-6| + |8-6| + |11-6|}{5}$$

$$= \frac{4 + 3 + 0 + 2 + 5}{5} = \frac{14}{5} = 2.8$$

If the frequency of any of the statistics in the data is greater than one, the formula above can be rewritten as

$$M.D. = \frac{\sum_{i=1}^{n} f_i |x_i - \bar{x}|}{\sum_{i=1}^{n} f_i}$$

Example 37:31

Find the mean deviation from the mean of the data in Example 37.27.

Solution

The mean was already calculated to be $\bar{x} = 53.625$.

Marks, x	f	$|x_i - \bar{x}|$	$f|x_i - \bar{x}|$
5.5	3	48.125	144.375
15.5	5	38.125	190.625
25.5	5	28.125	140.625
35.5	9	18.125	163.125
45.5	11	8.125	89.375
55.5	15	1.875	28.125
65.5	14	11.875	166.250
75.5	8	21.875	175.000
85.5	6	31.875	191.250
95.5	4	41.875	167.500
			1456.25

Table 37:32

$$M.D. = \frac{\sum_{i=1}^{5} f |x_i - \bar{x}|}{\sum_{i=1}^{5} f} = \frac{1456.25}{80} = 18.20 \text{ to 2 d.p.s}$$

Standard Deviation and Variance

The variance of a set of n numbers

$x_1, x_2, x_3, ..., x_n$ with mean $\bar{x}$ is denoted by

σ^2 or s^2 and is defined as the mean of the squares of the deviations of each of the data from the mean. Thus,

$$\sigma^2 = \frac{\sum_{i=1}^{n} (x_i - \bar{x})^2}{n}$$

The standard deviation of a set of n numbers

$x_1, x_2, x_3, ..., x_n$ with mean $\bar{x}$ is denoted by

σ or s and is defined as the square root of the variance (mean of the squares of the deviations of each of the data from the mean). Thus,

$$\sigma = \sqrt{\frac{\sum_{i=1}^{n} (x_i - \bar{x})^2}{n}}$$

Example 37:31
Find the variance and hence the standard deviation of the set of data 12, 6, 15, 7, 3, 10, 18, 5.

Solution

$$\sigma^2 = \frac{\sum_{i=1}^{n} (x_i - \bar{x})^2}{n}, \quad \bar{x} = \frac{\sum_{i=1}^{n} x_i}{n}.$$

x_i	$x_i - \bar{x}$	$(x_i - \bar{x})^2$
3	-6.5	42.25
5	-4.5	20.25
6	-3.5	12.25
7	-2.5	6.25
10	0.5	0.25
12	2.5	6.25
15	5.5	30.25
18	8.5	72.25
$\sum x_i = 76$		$\sum_{i=1}^{n} (x_i - \bar{x})^2 = 190$

Table 37:33

$$\therefore \bar{x} = \frac{76}{8} = 9.5$$

The mean $\bar{x}$ is then used in the table to

calculate $\bar{x} - x_i$ and $(\bar{x} - x_i)^2$.

$$\therefore \sigma^2 = \frac{190}{8} = 23.75 \text{ and } \sigma \approx 4.87$$

If the frequency of some of the statistics in the data is greater than one, the formula above can be rewritten as

$$\sigma = \sqrt{\frac{\sum_{i=1}^{n} f_i (x_i - \bar{x})^2}{\sum_{i=1}^{n} f_i}}$$

Example 37:33
Find the variance and hence the standard deviation of the set of data 9, 3, 8, 8, 9, 8, 9, 18.

Solution

$$\sigma^2 = \frac{\sum_{i=1}^{n} f_i (x_i - \bar{x})^2}{\sum_{i=1}^{n} f_i}, \quad \bar{x} = \frac{\sum_{i=1}^{n} f_i x_i}{\sum_{i=1}^{n} f_i}.$$

x_i	f_i	$x_i f_i$	$x_i - \bar{x}$	$f_i (x_i - \bar{x})^2$
3	1	3	-6	36
8	3	24	-1	3
9	3	27	0	0
18	1	18	9	81
$\sum$	8	72		120

Table 37:34

$$\therefore \bar{x} = \frac{72}{8} = 9$$

$$\therefore \sigma^2 = \frac{120}{8} = 15 \text{ and } \sigma \approx 3.87 \text{ to 2 d.ps.}$$

SIMPLIFYING STATISTICS

At times, the data is so cumbersome that certain skills are required to simplify its analyses. These skills include:

1. Choosing a number called the *guess* or *assumed mean* or sometimes, working *zero*.
2. Using a method called *coding*.

1. Assumed or Guess Mean (Working Zero)

As the name may suggest, the assumed or guess mean is an arbitrary number, careful chosen and assumed for the main time to be the mean. This number may be one of the statistics in the data or may not necessarily be.

The deviations d_i of the assumed mean A from any given statistic x_i in the data is defined as

$$d_i = x_i - A \Leftrightarrow x_i = A + d_i \Rightarrow \bar{x} = A + \bar{d}$$

By substituting $A + d_i$ for x_i all the statistical formulae so far met, the formulae can be transformed into the easier usable equivalent forms below.

$$\text{Mean, } \bar{x} = A + \frac{\sum f_i d_i}{\sum f_i} = A + \bar{d}$$

$$M.D. = \frac{\sum f_i \left| d_i - \bar{d} \right|}{\sum f_i}$$

$$\text{Variance, } \sigma^2 = \frac{\sum f_i d_i^2}{\sum f_i} - \left(\frac{\sum f_i d_i}{\sum f_i} \right)^2$$

Where $\bar{d} = \dfrac{\sum f_i d_i}{\sum f_i}$ is the mean of the deviations from the assumed mean.

Though these formulae at sight appear to be more complex than the previous ones, they are far easier to use if memorized.

Example 37:34

In a class of 70 students, the following scores were recorded in a test.

Score	No. of students
4	3
5	6
6	5
7	6
8	7
9	10
10	13
11	8
12	7
13	2
14	3

Table 37:35

Calculate
(a) the mean

(b) the mean deviation
(c) the standard deviation

Solution

$$\bar{x} = A + \frac{\sum f_i d_i}{\sum f_i} = A + \bar{d}, \quad M.D. = \frac{\sum f_i \left| d_i - \bar{d} \right|}{\sum f_i}$$

$$\sigma = \sqrt{ \frac{\sum f_i d_i^2}{\sum f_i} - \left(\frac{\sum f_i d_i}{\sum f_i} \right)^2 }$$

Let $A = 10$

The tabulation of the data is as in Table 37:36

(a) $\bar{x} = 10 + \dfrac{-70}{70} = 9$

(b) $\bar{d} = \dfrac{\sum fd}{\sum f} = \dfrac{-70}{70} = -1 \Rightarrow M.D. = \dfrac{146}{70} = 2.09$

(c) $\sigma = \sqrt{ \dfrac{532}{70} - (-1)^2 } = \sqrt{7.6 - 1} = \sqrt{6.6} = 2.57$

x_i	f_i	d_i	$f_i d_i$	$\left\|d_i - \bar{d}\right\|$	$f_i\left\|d_i - \bar{d}\right\|$	d_i^2	$f_i d_i^2$
4	3	-6	-18	+5	15	36	108
5	6	-5	-30	+4	24	25	150
6	5	-4	-20	+3	15	16	80
7	6	-3	-18	+2	12	9	54
8	7	-2	-14	+1	7	4	28
9	10	-1	-10	0	0	1	10
10	13	0	0	1	13	0	0
11	8	1	8	2	16	1	8
12	7	2	14	3	21	4	28
13	2	3	6	4	8	9	18
14	3	4	12	5	15	16	48
Σ	70		-70		146		532

Table 37:36

2. The Coding Method

To simplify the work more, the coding method is used when the data is grouped into a frequency distribution whose class intervals have equal size c. To use the coding method, the formulae are further transformed into the following equivalent forms.

$$\text{Mean, } \bar{x} = A + c\bar{u}, \quad \text{where } \bar{u} = \frac{\sum fu}{\sum f}$$

$$M.D. = c\frac{\sum f \left| u - \bar{u} \right|}{\sum f}$$

$$\text{Standard deviation, } \sigma = c\sqrt{ \frac{\sum fu^2}{\sum f} - \left(\frac{\sum fu}{\sum f} \right)^2 }$$

Example 37:35

The following marks out of 100 were obtained by 36 students in a test.

Mark	Frequency
0-9	0
10-19	1
20-29	2
30-39	4
40-49	7
50-59	9
60-69	5
70-79	3
80-89	3
90-99	2

Table 37:37

Calculate (a) the mean

(b) the mean deviation

(c) the standard deviation to 2 d.p.s

Solution

$$\bar{x} = A + c\bar{u}, \quad M.D. = c\frac{\sum f|u - \bar{u}|}{\sum f},$$

$$\sigma = c\sqrt{\frac{\sum fu^2}{\sum f} - (\bar{u})^2}, \quad \text{where } \bar{u} = \frac{\sum fu}{\sum f}$$

Table 37:38 is built to ensure that all the required sums in the formulae can be obtained from it.

Mark	x	d	u	u^2	f	fu	fu^2	$|u-\bar{u}|$	$f|u-\bar{u}|$
0-9	4.5	−50	−5	25	0	0	0	5.08	0
10-19	14.5	−40	−4	16	1	−4	16	4.08	4.08
20-29	24.5	−30	−3	9	2	−6	18	3.08	6.17
30-39	34.5	−20	−2	4	4	−8	16	2.08	8.33
40-49	44.5	−10	−1	1	7	−7	7	1.08	7.58
50-59	54.5	0	0	0	9	0	0	0.08	0.75
60-69	64.5	10	1	1	5	5	5	0.92	4.58
70-79	74.5	20	2	4	3	6	12	1.92	5.75
80-89	84.5	30	3	9	3	9	27	2.92	8.75
90-99	94.5	40	4	16	2	8	32	3.92	7.83
$\sum$					36	3	133		53.82

Table 37:38

(a) $\bar{x} = 54.5 + 10\left(\dfrac{3}{36}\right) = 55.33$ to 2d.p.s

(b) $\bar{u} = \dfrac{\sum fu}{\sum f} = 0.0833$

$\Rightarrow$ M. D. $= \dfrac{10(53.82)}{36} = 14.95$ to 2 d.p.s

(c) $\therefore \sigma = 10\sqrt{\dfrac{133}{36} - \left(\dfrac{3}{36}\right)^2} = 10\sqrt{3.6944 - 0.0069}$

$= 10\sqrt{3.686} = 10(1.920) = 19.20$ to 2 d.p.s

EXERCISE 37:5

1. A boy measured to the nearest metre, how far he could throw tennis on 20 successive trials, and obtained the following results:

   66 69 70 68 71 68 69 70 67 68

   68 68 67 66 69 68 69 70 68 67

   (a) Draw a frequency distribution table for the data.

   (b) State the mode of the distribution.

   (c) Find the median of the distribution

   (d) An approximate formula for determining the mean is given by

   Mean − Mode = 3(Mean − Median)

   Use this formula to calculate the mean of the distribution.

   (e) Calculate the exact mean of the distribution, to one decimal place.

   $\dfrac{1}{9}$ of the candidates obtained grade D and an equal number of the candidates were awarded grades E and F.

   (a) Copy and complete Table 37:40 that is used to draw up a pie chart.

GRADE	A	B	C	D	E	F
Angle of Sector					70	70

   Table 37:40

   (b) Draw the pie chart accurately on a circle of radius 5 cm, labeling the sectors.

   (c) Given that 6 candidates obtained grade A, find

   (i) The number of candidates that took the test.

   (ii) The number that obtained grade D.

   (d) Given that grades A, B, and C, are pass grades, state the ratio 'pass to fail' in its simplest form.

   (e) The distribution of the grades for boys is shown in Figure 37:17. On a similar set of axes, construct a bar chart to show the distribution of the grades for girls.

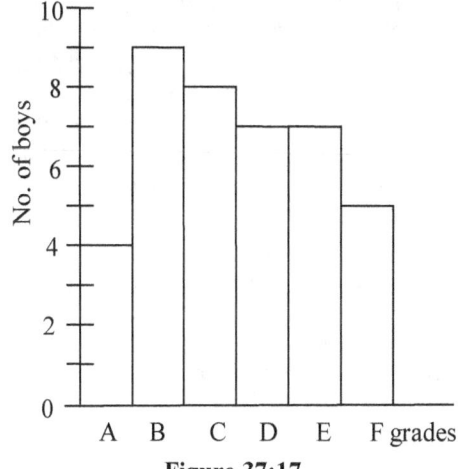

Figure 37:17

3. (i) Table 37:42 shows the group frequency distribution of examination marks for 120 students. Each mark is a whole number.
   (a) Construct a cumulative frequency table for this frequency distribution. Take the first class to be ≤10.

Marks	Number of candidates
1-10	0
11-20	2
21-30	6
31-40	7
41-50	14
51-60	20
61-70	35
71-80	29
81-90	6
91-100	1

Table 37:42

Candidates with a score of 50 or less will have to re-sit the examination and those with scores above 60 will be given credit certificates. Using your table or otherwise, determine, the number of candidates who
   (i) Will have to re-sit for the examination
   (ii) Earned credit certificates.

4. A mathematics teacher gave a test and promised to award prizes to the top 30 students of the class. The marks scored by the top 30 students were recorded as follows:
   10, 8, 4, 4, 5, 7, 4, 6, 5, 4, 9, 7, 6, 4, 8,
   8, 6, 7, 4, 6, 8, 5, 6, 4, 8, 7, 6, 4, 6, 4.
   (a) draw a frequency table to show the distribution of marks over the top 30 students in the class
   (b) state the mode of the distribution
   (c) Calculate the mean mark for the top 30 students.

5. The points obtained by 40 players in a world soccer were recorded as follows:
   82 42 61 57 55 39 67 78 65 66
   22 71 67 8 45 49 56 52 68 14
   60 57 18 64 58 38 50 46 83 76
   44 29 61 34 74 81 91 48 47 59
   (a) Using the intervals 0-9, 10-19, 20-29, etc, construct a frequency table to show this information.
   (b) State the modal class
   (c) Making use of the mid-interval values calculate to 1 decimal place the mean of the distribution.
   (d) Draw a cumulative frequency curve for this distribution.
   (e) From your graph, find the inter-quartile range

The following grades were assigned to the points above
≥ 70 (A grade), 50 – 69 (B grade),
30 – 49 (C grade), ≤ 30 (D grade),
   (f) Eto'o Fils had 79 points. What was his grade?

6. The following numbers were marks gained by 50 pupils in an examination:
   57 60 37 74 62 40 56 59 80 60
   62 94 78 73 56 68 67 79 83 87
   90 93 58 46 77 63 66 66 56 71
   51 77 53 69 70 69 70 70 47 54
   49 54 68 35 64 67 76 73 68 61
   (a) Reduce these marks to a frequency distribution with equal intervals, having as the first interval 35-44 inclusive.
   (b) Draw a cumulative frequency curve for this distribution.
   (c) What is the median mark?
   (d) Use your graph to estimate what percentage of candidates passed the examination if the pass mark is 55.

7. The mean of two numbers in a set A is 21.8. The mean of three numbers in a set B is 28.8. Find:
   (a) the sum of the two numbers in set A,
   (b) the sum of the three numbers in set B,
   (c) the mean of the five numbers put together.
   (d) If a sixth number x is introduced into the set in (c) above, and the mean of the six numbers is now 20, find x.

8. The time a sales girl used in serving 80 customers in a certain week were recorded in Table 37:43.

Time (min)	Number of customers
20 < t = 25	8
25 < t = 30	8
30 < t = 35	12
35 < t = 40	30
40 < t = 45	18
45 < t =5 0	4

Table 37:43

   (i) Calculate the mean time she spent for each customer.
   (ii) Draw a cumulative frequency table and use it to draw the cumulative frequency curve for the above data, using a scale of 1 cm to represent 5 minutes on the time axis and 1 cm to represent 10 customers.
   (iii) From your graph determine the maximum time that was used to serve 75% of the workers.

9. The Table 37:44 shows the time in minutes taken by 100 students to complete a quiz.

Time	11-20	21-30	31-40	41-50	51-60	61-70
Freq.	6	14	22	36	20	2

Table 37:44

Find to 3 significant figures, the mean and standard deviation of the distribution.

10. In a recruitment examination, the marks obtained out of 100 by 70 students are recorded in Table 37:45.
Calculate to two decimal places
(a) the mean of the distribution
(b) the standard deviation of the distribution.

Mark	Frequency
1-10	3
11-20	5
21-30	9
31-40	14
41-50	12
51-60	9
61-70	7
71-80	6
81-90	5
91-100	0

Table 37:45

11. The diameters in mm of 50 watermelons were measured and the distribution in Table 37:46 was obtained.

Diameter	Number of fruits
160-161	1
162-163	3
164-165	9
166-167	20
168-169	14
170-171	2
172-173	1
174-175	0

Table 37:46

Calculate the mean, median and standard deviation of the diameter.

12. The number of times the 200 students in a certain school were punished in an entire school year is shown in Table 37:48.
Arrange this data in class intervals 0-4,5-9, 10-14 etc. then calculate
(a) the mean
(b) variance
(c) standard deviation.

No. of times	No. of students
0	34
6	1
10	3
12	4
15	6
16	4
17	3
18	5
19	2
20	11
21	3
22	6
23	8
24	13
25	20
26	5
27	6
28	9
29	5
30	12
31	4
32	5
33	2
34	3
35	10
36	8
37	2
38	1
40	4
44	1

Table 37:48

13. Table 37:47, shows the time taken by 60 competitors to complete a given task. Estimate to one decimal place
(a) the mean   (b) variance
(c) standard deviation of the distribution.

Time	Frequency
10-19	3
20-29	7
30-39	12
40-49	18
50-59	10
60-69	6
70-79	3
80-89	1

Table 37:47

14. The number of civil servants in a certain community classified by age groups is as in Table 37:49.

Age	Frequency
15-19	66
20-24	65
25-29	56
30-34	50
35-39	42
40-44	37
45-49	35
50-54	30
55-59	24
60-64	22

Table 37:49

Calculate;
(a) the arithmetic mean, (b) the median,
(c) the mode   (d) the variance
   (e) the standard deviation of the ages of the civil servants.

15. The marks obtained by 500 candidates in an examination are shown in Table 37:50.

Mark	Frequency
20-29	66
30-39	65
40-49	56
50-59	50
60-69	42
70-79	37
80-89	35

Table 37:50

(a) Calculate the mean mark of the distribution.
(b) Make a table of cumulative frequencies for marks below 29.5, 39.5 etc.
(c) Draw a cumulative frequency graph and use it to find;
(i) the median mark
(ii) the approximate number of candidates who attained a mark of at least 55 %.

MULTIPLE CHOICE EXERCISE 37

1. A survey of people and/or property is called:
   [A] Census   [B] Data
   [C] Population   [D] Sample

2. The tally marks ‖‖ ‖‖ ‖‖ ‖‖ ‖‖ ||| represent the number:
   [A] 18 [B] 23 [C] 28 [D] 33

3. The correct tally representation of 17 students is:
   [A] ‖‖ ‖‖ ‖‖   [B] ‖‖ ‖‖ |||||||
   [C] ||||| |||| |||| || [D] ‖‖ ‖‖ ‖‖ ||

4. In recording data, the tally marks ‖‖ ‖‖ ‖‖ ||| will be recorded as:
   [A] 18 [B] 15 [C] 13   [D] 20

5. In a school examination, 480 out of 720 candidates were awarded a D Grade. On a pie chart showing all the grades the angle at the centre for the D grade is:
   [A] $270°$ [B] $240°$ [C] $210°$ [D] $180°$

6. The measure of central tendency a shoe company will be most interested in is:
   [A] mean   [B] mode
   [C] median   [D] mean and median

7. The average of 0, 1, 6, 7, 9 and 19 is:
   [A] 9 [B] 6   [C] 7 [D] 10

8. The average of 1, 2, 5, 7, and 15 is:
   [A] 6   [B] 30   [C] 7   [D] 15

9. A group of four people found that their heights were 1.38 m, 1.71 m, 1.23 m and 1.40 m. Their average height (in metres) is:
   [A] 1.145 [B] 1.18 [C] 1.39 [D] 1.43

10. The average wage bill in FCFA of 40 men who collectively earn 3,540,000 FCFA is:
    [A] 87,000   [B] 29,500
    [C] 88,500   [D] 31,700

11. The mean of 9,13,16,17,19,23,24 correct to two decimal places is:
    [A] 23.00 [B] 17.29 [C] 16.50 [D] 16.33

12. The average of the first four prime numbers greater than 10 is:
    [A] 20   [B] 19   [C] 17   [D] 15

13. The mean of 20 observations is 4. The observed largest value 23 is removed. The mean of the remaining observations is:
    [A] 4   [B] 3   [C] 2.85   [D] 2.60

14. The mean heights of the three groups of students consisting respectively of 20, 16 and 14 students are 1.67 m, 1.50 m and 1.40 m respectively. The mean height of all the students is:
    [A] 1.52 m   [B] 1.53 m
    [C] 1.54 m   [D] 1.55 m

15. The mean of 30 observations is 5. The

observed largest value of 34 is deleted. The mean of the remaining observations is:
[A] 4   [B] 3.8   [C] 3.4   [D] 5

16. Table 37:51, shows the scores of some students in a test. The average score is 3.5. The value of x is:

Scores	1	2	3	4	5	6
No. of students	1	4	5	6	x	2

Table 37:51

[A] 1   [B] 2   [C] 3   [D] 4

17. The mean of 9, x and 13 is 11. The value of x is:
[A] 7   [B] 8   [C] 11   [D] 13

18. The value of x which qualifies 4 as the mean of the data 4, $3x$, 0 and 3 is:
[A] 1   [B] 2   [C] 3   [D] 4

19. The mean of five observations is 15. Four of them are 11, 12, 19 and 20. The fifth is:
[A] 10   [B] 25   [C] 20   [D] 13

20. A pie chart is drawn to represent the percentages 20%, 50%, 25% and 5%. The angle which represents 5% is:
[A] 5°   [B] 18°   [C] 25°   [D] 126°

21. Given the scores −3, 4, 0, 4,−2,−5, 1, 7,10,5 the median of the scores is:
[A] 2.5   [B] 2   [C] 4   [D] 3.5

22. From Table 37:52, the mean number of male children per family is:
[A] 5   [B] 4   [C] 3   [D] 2

No. of male children	0	1	2	3	4
No. of families	4	8	6	2	7

Table 37:52

23. The mean of four numbers a, b, c and d is 6. The mean of 5 numbers a, b, c, d and e is 10. The value of e is:
[A] 24   [B] 25   [C] 26   [D] 27

24. The average age of five boys is 11 years. A sixth boy whose age is 17 years is added, the mean age in years will now be:
[A] 14   [B] 12   [C] 13   [D] 11

25. The median of 8, 10, 9, 6, 7, 10, 12, 8, 9, 8 is:
[A] 7.5   [B] 8   [C] 8.5   [D] 8.7

26. The median of the set of scores 65, 75, 55, 48, 78 is:

[A] 55   [B] 60   [C] 72   [D] 65

27. The median of the set of numbers 2.64, 2.50, 2.72, 2.91, 2.35 is:
[A] 2.72   [B] 2.64   [C] 2.50   [D] 2.35

28. Given the set of numbers 12,15,13,14,12 and 12. The median is:
[A] 12.5   [B] 12   [C] 13   [D] 13.5

29. Table 37:53, shows the age distribution of a group of children. Their median age is:
[A] 4 years   [B] 7 years
[C] 8 years   [D] 9 years

Age (in years)	4	5	6	7	8	9	10
Frequency	2	1	2	4	3	6	2

Table 37:53

30. Table 37:54, gives the marks scored by a group of students in a test. The median mark is:
[A] 4   [B] 3   [C] 2   [D] 1

Mark	0	1	2	3	4	5
Frequency	1	2	7	5	4	3

Table 37:54

31. The mode of the numbers 8, 10, 9, 9, 10, 8, 11, 8, 10, 9, 8 and 14 is:
[A] 8   [B] 9   [C] 10   [D] 11

32. A group of students measured a certain angle (to the nearest degree) and obtained the following results.
75° 76° 72° 73° 74° 79° 72°
72° 77° 72° 71° 70° 78° 73°
The mode of their measurements is:
[A] 78°   [B] 74°   [C] 73°   [D] 72°

33. The measure, which is not a measure of dispersion, is:
[A] Mode
[B] mean deviation
[C] Inter-quartile range
[D] standard deviation

34. It is true to say that the measure, which is not measure of dispersion, is:
[A] Range
[B] Variance
[C] Mode
[D] Percentile range

35. The Variance of a given distribution is 25. The standard deviation is:
[A] 625   [B] 75   [C] 25   [D] 5

36. The standard deviation of the marks 2, 3, 6, 2,

413

5, 0, 4, 2 is:
[A] 1.5 [B] 1.7 [C] 1.8 [D] 1.9

37. The standard deviation of the numbers 2, 5, 6, 4 and 8 is:
[A] 2 [B] 4 [C] 6 [D] 7

38. The mode of the distribution in Table 37:55 is:
[A] 2 [B] 3 [C] 4 [D] 5

Score	0	1	2	3	4	5
Frequency	2	3	4	2	7	2

Table 37:56

39. The mean score of the distribution in Table 37:55 is:
[A] 1.75 [B] 2 [C] 2.5 [D] 2.75

40. The median score of the distribution in Table 37:55 is:
[A] 0 [B] 2.5 [C] 3 [D] 5

41. For a class of 30 students, the scores in a test out of 10 marks were as in Table 37:56. The mode is:
[A] 3 [B] 5 [C] 6 [D] 7

4	5	7	2	3	6	5	5	8	9
5	4	2	3	7	9	8	7	7	7
3	4	5	5	2	3	6	7	7	2

Table 37:57

42. For a class of 30 students, the scores in a test out of 10 marks were as in Table 37:56. The median score is:
[A] 3 [B] 5 [C] 6 [D] 7

43. For a class of 30 students, the scores in a test out of 10 marks were as in Table 37:56. The range of the distribution is:
[A] 7 [B] 2 [C] 8 [D] 9

44. Table 37:57, shows the tithes in thousand FCFA, collected in a church. The mode is:
[A] 3 [B] 6 [C] 9 [D] 12

Amount (thousand FCFA)	3	6	9	12	15	18
No. of Christians	3	9	6	15	3	12

Table 37:58

45. Table 37:57, shows the tithes in thousand FCFA, collected in a church. The median of the distribution is:
[A] 3 [B] 9 [C] 12 [D] 15

46. Table 37:58 shows the frequency distribution of a number of chairs in each rooms of a hotel. The mean of the distribution is:
[A] 3.5 [B] 4.0 [C] 4.4 [D] 5.0

47. Table 37:58, shows the frequency distribution of a number of chairs in each rooms of a hotel. The mode of the distribution is:
[A] 2 [B] 5 [C] 7 [D] 9

48. Table 37:58, shows the frequency distribution of a number of chairs in each rooms of a hotel. The median of the distribution is:
[A] 4 [B] 4.5 [C] 5 [D] 5.5

49. Table 37:59, shows the frequency distribution of marks scored by a group of students in a test. The number of students who took the test is:
[A] 14 [B] 15 [C] 18 [D] 20

Marks	2	3	4	5	6
Frequency	2	4	5	3	1

Table 37:59

50. Table 37:59, shows the frequency distribution of marks scored by a group of students in a test. The modal score is:
[A] 2 [B] 3 [C] 4 [D] 5

51. Table 37:59, shows the frequency distribution of marks scored by a group of students in a test. The mean mark is:
[A] 1.3 [B] 2 [C] 3 [D] 3.8

52. Table 37:60, shows the scores of 15 students in a physics test. The number of students who scored at least 5 is:
[A] 6 [B] 8 [C] 9 [D] 7

Marks	1	2	3	4	5	6	7	8	9	10
No. of students	1	3	2	0	1	6	1	0	1	0

Table 37:60

53. Table 37:60, shows the scores of 15 students in a Physics test. The median score is:
[A] 5 [B] 6 [C] 7 [D] 8

54. Table 37:61, shows the scores of a group of 40 students in a Biology test. If the mode is m and the median is n then (m, n) as an ordered pair is:
[A] (5,5) [B] (5,6) [C] (6,5) [D] (9,4)

Score	1	2	3	4	5	6	7	8	9
Frequency	2	3	6	7	9	6	2	2	3

Table 37:61

55. Table 37:61 shows the scores of a group of 40

students in a physics test. The mean of the distribution is:

[A] 4.5 [B] 4.8 [C] 5.0 [D] 5.2

56. The number of goals scored by a football team in 20 matches is shown in Table 37:62. The mean number of goals scored is:

[A] 1.75 [B] 1.9 [C] 2 [D] 2.15

Number of goals	0	1	2	3	4	5
Number of matches	3	5	7	4	1	0

Table 37:62

57. The number of goals scored by a football team in 20 matches is shown in Table 37:62. The modal number of goals scored is:

[A] 1 [B] 2 [C] 5 [D] 7

58. The distribution by Region of 840 students in the faculty of science of the University of Buea in a certain session is as follows:

Adamawa Region   45
North West Region   410
Littoral Region   105
Western Region   126
South West Region   154

In a pie chart drawn to represent this distribution, the angle subtended by Western Region is:

[A] 42° [B] 45° [C] 48° [D] 54°

59. The pie chart in Figure 37:18, illustrates the amount of private time a student spends in a week studying various subjects. The value of k is:

[A] 15° [B] 30° [C] 60° [D] 90°

60. The pie chart in Figure 37:18, illustrates the amount of private time a student spends in a week studying various subjects. Given that, he spends 2 and a half hours on science, the total number of hours he studies in a week is:

[A] $3\frac{1}{2}$ [B] 5 [C] 8 [D] 12

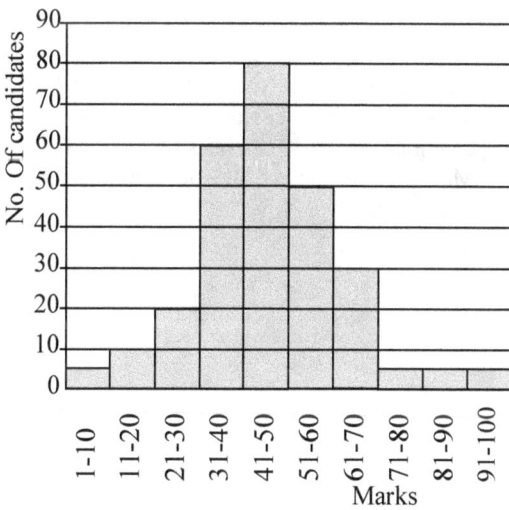

Figure 37:18

61. The pie chart in Figure 37:19 represent the

number of fruits on display in a grocery shop. Given that there are 60 oranges in display, the number of apples is:

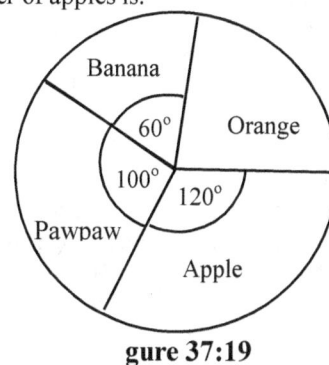

gure 37:19

[A] 40 [B] 80 [C] 90 [D] 120

62. The marks obtained by pupils of a certain class are grouped as shown below; 0-4, 5-9, 10-14, 15-19. It is true to say that:

I: The mid values of the grouped marks are 2,7,12, and 17.

II: The class interval is 4.

III: The class boundaries are 0.5, 4.5, 9.5, 14.5 and 19.5.

[A] I only [B] II only
[C] III only [D] I and II

63. The histogram in Figure 37:20 shows the number of candidates, in thousands who obtained given ranges of marks in an entrance examination.

The total number of candidates who sat for the examination is:

[A] 120,000 [B] 250,000
[C] 260,000 [D] 270,000

64. The histogram in Figure 37:20 shows the number of candidates, in thousands who obtained given ranges of marks in an entrance examination. The number of candidates who scored at most 30% is:
[A] 20,000   [B] 25,000
[C] 35,000   [D] 60,000

65. The histogram in Figure 37:21 shows the distribution of a group of students according to their ages. The range of their ages is:
[A] 14 years   [B] 20 years
[C] 30.5 years   [D] 31 years

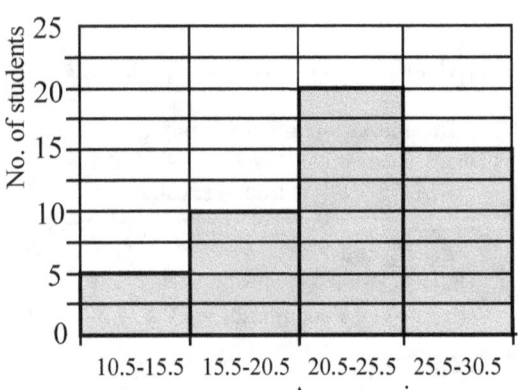

Figure 37:21

66. The histogram in Figure 37:21 shows the distribution of a group of students according to their ages. The mode of their ages is:
[A] 22.5 years   [B] 23.0 years
[C] 24.0 years   [D] 24.5 years

67. For six sequences, Ngange scored 76, 57, 97, 86, 86, 70 in Mathematics. If Ngange wants to convince his parents of his strength in Mathematics, the measure he should use should be:

[A] Mean [B] median [C] mode [D] range

68. Six employees earn 800 FCFA, 850 FCFA, 900 FCFA, 950 FCFA, 1000 FCFA, and 2350 per hour. The manager claims that the median of the hourly wages is 925 FCFA. The manager is:
[A] wrong because 925 FCFA is the mode.
[B] wrong because he seems to ignore the amount 2350 FCFA.
[C] wrong because 925 FCFA is the mean.
[D] right because 925 FCFA is the correct median.

69. The president of a certain Credit Union used the data in Table 37:63 to find the mean monthly salary of the Credit Union staff.

Monthly Salary	No. of workers
26,000 FCFA	7
30,000 FCFA	8
240,000 FCFA	1
260,000 FCFA	1
300,000 FCFA	3

Table 37:63

In a report the president said that, the typical salary at the Credit Union is about 92,000 FCFA. His statement is:
[A] misleading because salaries of five staff are far above those of the other fifteen.
[B] misleading because the mean of the data is not 92,000 FCFA.
[C] misleading because the president ignored the highest salary.
[D] right.

TOPIC 38

PROBABILITY

OBJECTIVES

At the end of this topic, the learner should be able to:

1. Define probability or say what probability is about.
2. Understand and the basic probability terminologies such as biased, unbiased, trial, event, outcome, sample space, event subset etc.
3. Calculate simple probabilities.
4. State and use the standard definition of probability.
5. Appreciate that probability is a number in the range $1 \leq P(E) \leq 0$ and that the probability of a sure event is 1, while that of an impossible event is 0.
6. Perform simple experiments using a die and/or playing cards.
7. Calculate the probabilities of an event, given the probability of its complementary.
8. Calculate the probability of intersecting events, mutually exclusive events, dependent events and independent events.
9. Calculate the conditional probability of repeated trials involving drawing with and without replacement.
10. Draw and use tree diagrams to calculate probabilities.
11. Calculate probabilities from data given as a frequency table.

Pierre de Fermat, French mathematician, with his friend the French scientist and philosopher Blaise Pascal, made a series of investigations into the properties of figurate numbers. From these studies Fermat later derived an important method of calculating probabilities. He was also greatly interested in the theory of numbers and made several discoveries in this field.

Corbis

The Concept of Probability

Consider the following statements:
(i) It will probably rain tonight.
(ii) It is likely that the principal will address the students tomorrow.
(iii) The chances of the indomitable lions of Cameroon winning the next world cup football tournament are very high.
How true is each of the above statements? What is the likelihood of it raining tonight? What is the chance of the indomitable lions wining the next world cup football tournament? Questions such as those above lead us into the subject of probability. Probability seeks to answer questions of chance, likelihood, possibility or the degree of truth in an event or something occurring. In other words, probability is a numerical measure of the degree of chance, possibility or likelihood of an event occurring or not occurring.

Probability is often used to take certain life decisions, which depend on chance. For instance, in order to predict the score of say a football march between Egypt and Cameroon, probability comes into play.

Some Basic Probability Terminology

Suppose a coin is tossed (or thrown), it will either turn up heads (H) or tails (T). If the coin is as likely to turn up heads as to turn up tails then the coin is said to be **fair** or **unbiased**. If the likelihood of the coin turning up heads is not equal to its likelihood of turning up tails the coin is said to be **unfair** or **biased**. The act of tossing the coin is called a **trial** or an **experiment**. The appearance of a head or tail is called an **event**. The event (H or T), one of which must occur in an experiment are called the **outcomes** and the set of all the possible outcomes in a particular experiment is called the **sample space**, usually denoted by S. The sample space S in probability is equivalent to the universal set in set theory.

Thus for the case of the coin, $S = (H, T)$ The set of all possible outcomes in an experiment under specified conditions is called the **event subset** or the **possibility space** and is a subset of the sample space.

Example 38:1

An unbiased die is tossed once find the sample space. State the event subset if the event is:
(i) A: obtaining an even number.
(ii) B: obtaining an odd number.
(iii) C: obtaining a number less than 3.

Solution

$S = \{1, 2, 3, 4, 5, 6\}$
(i) $A = \{2, 4, 6\}$
(ii) $B = \{1, 3, 5\}$
(iii) $C = \{1, 2\}$

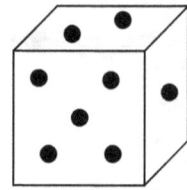

Figure 38:1

Example 38:2

A card is drawn at random from a well-shuffled pack of 15 cards numbered 1 to 15. Give the event subsets if the event is obtaining
(a) X: the number 7.
(b) Y: a number greater than or equal to 10
(c) Z: a multiple of 3

Solution

(a) $X = \{7\}$

(b) $Y = \{10, 11, 12, 13, 14, 15\}$

(c) $Z = \{3, 6, 9, 12, 15\}$

EXERCISE 38:1

1. A student has five pairs of socks of the following colours: blue, red, green, white, and black. On one dark morning, the student chooses a pair of socks at random to put on.
   (a) What is the sample space of this experiment?
   (b) What is the event subset for the event that he chooses a white pair of socks or a green pair of socks?
2. Two fair coins are tossed simultaneously. What is the set of possible outcome?
3. A die is tossed. What is the sample space?
4. A bag contains 10 tickets numbered 1-10. State the event subset for the event:
   (a) Drawing a prime number,
   (b) Drawing an even number,
   (c) Drawing an odd number,
   (d) Drawing a square number,

(e) Drawing a multiple of 3.
5. A letter is chosen at random from the letters of the word 'MATHEMATICS'. What is the possibility space for the event of choosing a vowel?
6. A bag contains 4 red marbles, 7 blue marbles and 1 yellow marble. How many elements, has the sample space?

Probability as a Number

The probabilities that any event E will occur (or will not occur) always lie between zero and one *inclusively*.
This means that the probability of an event occurring is always zero, one or any number between zero and 1. The probability of an event E occurring is denoted by $P(E)$.

Thus; $0 \le P(E) \le 1$

The probability of a **sure** or **certain** event A is 1.

i.e. $P(A) = 1 \Leftrightarrow$ Event A must occur

For instance, the probability that any living person will die one day is 1. The probability of an event B occurring is zero, when it is **impossible** for the event to occur.

i.e. $P(B) = 0$, Event B will not occur

For instance, the probability that someday a man will be pregnant is zero.
The probability of an event E occurring is any number between 0 and 1 exclusively (not including 0 and 1). This means that there are some chances of the event occurring and some chance of the event not occurring. For instance if it is stated that the probability of an event E occurring is $\frac{1}{4}$ it means that out of 4 trials, the event is expected to occur once and it is expected not to happen 3 times and out of 40 trials, the event is expected to happen 10 times and it is expected not to happen 30 times.

Equiprobable Outcomes

Equiprobable or **equally likely** events are events, which have equal chances of occurrence. If there are n such events, the probability $P(E)$ of one of the events occurring is given by $P(E) = \frac{1}{n}$.

Example 38:3
A fair coin is tossed once. State the probability of:
  (a) a head   (b) a tail

Solution
$S = \{H, T\}$

(a) $P(H) = \frac{1}{2}$ (b) $P(T) = \frac{1}{2}$

$\Rightarrow P(H) = P(T) = \frac{1}{2}$ or 0.5 or 50%

Example 38:4
State the probability of each of the faces showing 1, 2, 3, 4, 5, and 6 if a fair die is tossed.

Solution
$S = \{1, 2, 3, 4, 5, 6\}$

$P(1) = P(2) = P(3) = P(4) = P(5) = P(6) = \frac{1}{6}$

Standard Definition of Probability

Suppose a sample space S consists of a finite number of equiprobable out comes, then the probability of an event E occurring is defined as:

$$\text{Probability of } E = \frac{\text{No of outcomes in the event } E}{\text{Total number of outcomes } S}$$

i.e. $P(E) = \frac{n(E)}{n(S)}$

Suits of Playing Cards

An ordinary pack of playing cards contains 52 cards. There are four types of cards; hearts, clubs, diamonds and spades; each type having

13 members labelled A, 1, 2, 3, 4, 5, 6, 7, 8, 9, 10, Q, K, and J. Each type of card has 3 picture cards labeled **Q**, **K** and **J,** called queen, king or jack. Hearts and diamonds are red while clubs and spades are black.

Ace of clubs

Ace of spades

Queen of clubs

Jack of spades

Figure 38:2

Example 38:5
A card is picked at random from a well-shuffled pack of 52 playing cards. What is the probability that
(i) it is an Ace of heart   (ii) it is a king

Solution

$$n(S) = 52$$

(i) n(Ace of hearts) = 1

$$\therefore P\left(\text{Ace of hearts}\right) = \frac{n\left(\text{Ace of hearts}\right)}{n(S)} = \frac{1}{52}$$

(ii) n(Kings) = 4

$$\therefore P\left(\text{Kings}\right) = \frac{n\left(\text{Kings}\right)}{n(S)} = \frac{4}{52} = \frac{1}{13}$$

Example 38:6
Fourteen girls are sitting in a circle equally spaced. One is from form four, 2 are from form five, 6 are from lower sixth and 5 are from upper sixth. A girl is selected at random from amongst the girls. Find the probability that the girl is from.
(i) form four   (ii) form five
(iii) lower sixth   (iv) upper sixth

Solution

$n(S) = 14$, n(form four) = 1, n(form five) = 2

n(Lower 6th) = 6 and n(upper 6th) = 5

(i) $P(\text{form four}) = \dfrac{n(\text{form four})}{n(s)} = \dfrac{1}{14}$

(ii) $P(\text{form five}) = \dfrac{n(\text{form five})}{n(5)} = \dfrac{2}{14} = \dfrac{1}{7}$

(iii) $P(\text{lower } 6^{th}) = \dfrac{n(\text{lower } 6^{th})}{n(s)} = \dfrac{6}{14} = \dfrac{3}{7}$

(iv) $P(\text{upper } 6^{th}) = \dfrac{n(\text{upper } 6^{th})}{n(s)} = \dfrac{5}{14}$

EXERCISE 38:2

1. A die is tossed. Find the probability of:
   (a) obtaining a 2.
   (b) obtaining an even number.
   (c) obtaining an odd number.
   (d) obtaining a number less than 5.
   (e) obtaining a prime number.
2. Figure 38:3 shows a spinner. Find the probability of obtaining:
   (a) 3   (b) 1   (c) 2   (d) 5

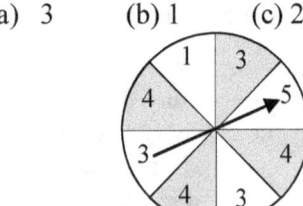

Figure 38:3

3. In a race of twenty horses, 7 of the horses are black, 8 are white and the rest are dotted. Find the probability that the winner will be dotted.
4. A conference is attended by 9 boys, 12 girls, 15 men and 14 women. Find the probability that a person elected as president will be a man.
5. A letter is chosen at random from the letters of the word 'PROBABILITY'. Find the probability of choosing the letter B.
6. A number is chosen at random from the integers 5 to 25 inclusive. Find the probability that it is a prime number.

Complementary Events

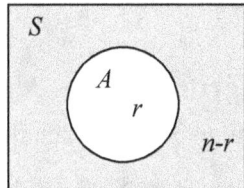

Figure 38:4

If A is an event in a sample space S, then the event that A does not occur is called "not A" or the "complement of A" denoted by $\bar{A}$ or A' and is defined as the union of all subsets of S whose elements do not belong to A. A is represented by the shaded portion in the Venn diagram above. Thus if the cardinality of S is n and the cardinality of A is r, then the cardinality of A' is given by

$$n(A') = n(s) - n(A)$$
$$n(A') = n - r$$

Therefore, if two events are complementary, the sum of their probabilities is 1.

$$P(A) + P(A') = 1 \Leftrightarrow P(A') = 1 - P(A)$$

Note!

A' is the complement of $A \Leftrightarrow A$ is the complement of A'.

Some examples of complementary events are:
1. The events "obtaining a head" and "obtaining a tail" when a coin is tossed.
2. The events "getting an even number" in one toss of die and "getting an odd number".

Example 38:7
The probability that a student will pass the GCE is $\dfrac{7}{11}$. What is the probability that he will fail?

Solution
Since the events, 'passing' and 'failing' are complementary,
$$P(\text{passing}) + P(\text{failing}) = 1$$
$$\Rightarrow P(\text{failing}) = 1 - P(\text{passing}) = 1 - \frac{7}{11} = \frac{4}{11}$$

Example 38:8
A fair die is tossed. Find the probability that a 2 will not be obtained.

Solution
Let the event of obtaining a two be T, and the event of not obtaining a two be T'. Then,
$$P(T') = 1 - P(T)$$
$$P(T) = \frac{1}{6}$$
$$\Rightarrow P(T') = 1 - \frac{1}{6} = \frac{5}{6}$$

EXERCISE 38:3

1. A bag contains 12 blue marbles, 14 red marbles and 9 green marbles. A marble is drawn at random from the bag. Find the probability that the marble is
   (a) Green (b) Not blue (c) Red
2. A bag of mangoes contains 30 mangoes 8 of which are bad. A mango is taken at random from the bag. What is the probability that the mango is good?
3. A grocer bought 100 oranges. Given that, 16 of them are bad. Find the probability that a mango chosen at random from the lot is good.
4. In a class of 50 students, 35 are girls. Find the probability that a boy will be voted as class prefect.
5. A poultry farm consists of 140 fowls 60 of which are hens. A man buys a fowl from the poultry; find the probability that it is a hen.
6. In a class of 80 students, 75 % of the students pass the examination. Find the probability that a student selected at random from the class failed.

Addition Laws of Probability (P (or))

The addition laws are used when the probability of one event **or** the other is required. Sometimes the word **either** is used for emphasis. There are two cases involved- intersecting events and mutually exclusive events.

Intersecting Events

Consider any two events X and Y of an experiment which are such that $P(X) \neq 0$ and $P(Y) \neq 0$. X and Y can be represented as in the following Venn diagram.

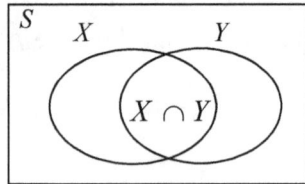

Figure 38:5

The probability of X or Y is given by

$$P(X \text{ or } Y) = P(X) + P(Y) - P(X \cap Y)$$

Since "X or Y" means, "only X occurs, or only Y occurs or both X and Y occur", X or Y in set notation is written as $X \cup Y$.

$$\therefore \ P(X \cup Y) = P(X) + P(Y) - P(X \cap Y)$$

Example 38:9

In a group of 20 adults, 4 out of the 7 women and 2 out of the 13 men wear eye glasses. What is the probability that a person chosen at random from the group is:
(i) a woman or someone who wears eye glasses?
(ii) a man or someone who wears eye glasses?

Solution

Let M, W and G be the events of choosing a man, a woman and someone who wears eyeglasses respectively. Then,

(i) $P(W \cup G) = P(W) + P(G) - P(W \cap G)$

$$= \frac{7}{20} + \frac{6}{20} - \frac{4}{20}$$

$$= \frac{9}{20}$$

The probability that the person chosen is a woman or someone who wears glasses is $\frac{9}{20}$.

(ii) $P(M \cup G = P(M) + P(G) - P(M \cap G)$

$$= \frac{13}{20} + \frac{6}{20} - \frac{2}{20}$$

$$= \frac{7}{20}$$

The probability that the person chosen is a man or someone who wears eyeglasses is $\frac{17}{20}$.

Example 38:10

If X and Y are two events such that

$$P(X \cap Y) = \frac{1}{9}, P(X) = \frac{1}{3} \text{ and}$$

$$P(X \cup Y) = \frac{4}{9}. \text{ Find } P(Y).$$

Solution

$$P(X \cup Y) = P(X) + P(Y) - P(X \cap Y)$$

$$\Rightarrow P(Y) = P(X \cup Y) + P(X \cap Y) - P(X)$$

$$= \frac{4}{29} + \frac{1}{9} - \frac{3}{9}$$

$$P(Y) = \frac{2}{9}$$

Mutually Exclusive Events

There are certain events for which the occurrence of one excludes the occurrence of the other. For instance, a tossed coin cannot land head up and tail up at the same time. Such events are said to be mutually exclusive. On a Venn diagram they are represented as disjoint sets, since cannot jointly occur at the same time. i.e. If two events X and Y are mutually exclusive,

$$X \cap Y = \varnothing \Leftrightarrow n(X \cap Y)$$

$$\therefore P(X \cap Y) = 0$$

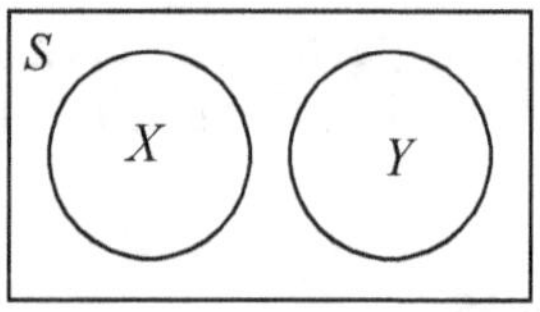

Figure 38:6

This means that the probability of one of them occurring or $P(X \text{ or } Y)$ is given by

$$P(X \cup Y) = P(X) + P(Y)$$

In general if there are n mutually exclusive events $E_1, E_2,\ldots,E_n$, then the occurrence of one excludes the occurrence of all the others. Hence,

$$P(E_1 \cup E_2 \cup \ldots \cup E_n) = P(E_1) + P(E_2) + \cdots + P(E_n)$$

Example 38:11

In an election for the president of a credit union, the probability that Shey (*S*) wins is 0.3, the probability that Chia (*C*) wins is 0.2 and the probability that Doh (*D*) wins is 0.4. Find the probability that:
(a) Shey or Chia wins.
(b) neither Shey nor Doh wins.

Solution
The events are mutually exclusive, so
(a) $P(S \text{ or } C) = P(S) + P(C)$
$$= 0.3 + 0.2$$
$$= 0.5$$
(a) $P(\text{neither } S \text{ nor } D) = 1 - \{P(S) + P(D)\}$
$$= 1 - (0.3 + 0.4)$$
$$= 0.3$$

Example 38:12

Find the probability that a card drawn at random from an ordinary deck of 52 cards is either a king or an Ace of spade.

Solution
Let *A* and *K* be the events drawing an Ace of spade and a king respectively. Then,

$$P(A) = \frac{n(A)}{n(S)} = \frac{1}{52}$$

$$P(K) = \frac{n(K)}{n(S)} = \frac{4}{52}$$

$$P(A \cup K) = P(A) + P(K) = \frac{1}{52} + \frac{4}{52} = \frac{5}{52}$$

Example 38:13

A coin is tossed twice. Find the probability that the result is either two tails or a tail and a head respectively.

Solution
$$S = \{HH, HT, TT, TH\}$$

$$P(TT) = \frac{1}{4}, P(TH) = \frac{1}{4}$$

$$\therefore P(TT \cup TH) = P(TT) + P(TH) = \frac{1}{4} + \frac{1}{4} = \frac{1}{2}$$

EXERCISE 38:4

1. There are 12 cards numbered 1-12. A card is selected at random. Find the probability that, the card is either even or a perfect square?

2. Two events *A* and *B* are such that
$$P(A) = \frac{1}{2}, \ P(B) = \frac{1}{2} \text{ and}$$
$$P(A \cap B) = \frac{1}{3}. \text{ Calculate the probability}$$
of *A* or *B* occurring.

3. Two fair dice are thrown at the same time. Find the probability that their sum will be 7 or 11.

4. A bag contains 5 red, 7 blue and 6 green marbles. What is the probability that a boy with closed eyes picks a green or red marble from it?

5. A number is chosen at random from the set {15, 16, 17,…, 32}. Find the probability that the chosen number is:
(i) a multiple of 3   (ii) a prime number
(iii) a multiple of 3 or a prime number

6. A bucket contains 6 mangoes, 11 bananas and 13 oranges. A fruit is chosen at random from the bucket. What is the probability that it is either a mango or a banana?

7. A bag contains just 2 red, 4 green and 3 white balls. A ball is selected at random from the bag. Find the probability that it is either red or green.

8. A woman is to select a day convenient for her to attend a meeting in the Town Hall. What is the probability that she will choose:
(a) a day that begins with the letter *T*.
(b) a day that begins with the letter *T* or *M*.

9. A letter is chosen at random from the 26 letters of the English alphabet. Find the probability that it is:
(a) Either letter *f* or *j*.
(b) One of the letters of the word 'TRIANGLE'

10. A number is chosen at random from the numbers 1 to 10. Find the probability that it is:
(a) a prime number   (b) a multiple of 3
(c) a prime number or a multiple of 3.

11. A dice is tossed once. Find the probability

of obtaining:
    (a)   5   (b)   2   (c)   5 or 2

12. A fair cubic die is tossed once. Find the probability of obtaining 5 or 4.

13. The probability that candidate A, wins an election is $\frac{4}{9}$, the probability that candidate B, wins is $\frac{2}{5}$ and the probability that candidate C, wins is $\frac{3}{7}$. Find the probability that:
    (a) A or B wins   (b) B or C wins
    (c) Neither A nor C wins.

14. There are 40 students in a class. All students offer Further Mathematics (F) or Human Biology (H). 27 offer Further Mathematics, and 24 offer Human Biology. Find the probability that a student chosen at random from the class offers:
    (a) Further Mathematics.
    (b) Human Biology.
    (c) Further Mathematics or Human Biology.

15. Given that $P(A) = \frac{3}{5}, P(B) = \frac{4}{7}$ and $P(A \cup B) = \frac{5}{8}$, find $P(A \cap B)$.

The Multiplication Law

The multiplication laws are used when the probability of one event **and** the other or **both** events is required. There are two cases involved-dependent events and independent events.

Dependent Events

If two events X and Y are such that $P(X) \neq 0$ and $P(Y) \neq 0$, then the probability of X given that Y has already occurred is denoted by $P(X/Y)$ and is given by

$$P(X/Y) = \frac{n(X \cap Y)}{n(Y)} = \frac{P(X \cap Y)}{P(Y)}$$

$$\Rightarrow P(X \cap Y) = P(Y) \cdot P(X/Y)$$

This is known as the multiplication law for dependent events

Example 38:14

Given that a heart is picked at random from a pack of 52 playing cards, find the probability that the next card chosen is a picture card.

Solution

$$P(P/H) = \frac{P(P \cap H)}{P(H)} = \frac{3}{52} \div \frac{13}{52} = \frac{3}{13}$$

Alternatively:

$$P(P/H) = \frac{\text{number of heart picture cards}}{\text{number of hearts}} = \frac{3}{13}$$

Independent Events

Consider the two events "a student eats in the morning" and "the teacher of the student is rich" Clearly, the event that a student eats in the morning does not in any way depend or affect the event that his teacher is rich. If two events X and Y are such that the occurrence or the non-occurrence of X does not in any way affect or depend on the occurrence or non-occurrence of Y, then the events X and Y are said to be **independent**.

If two events X and Y are independent then,

$$P(X/Y) = P(X) \text{ and } P(Y/X) = P(Y)$$

but $P(X/Y) = \dfrac{P(X \cap Y)}{P(Y)}$

but $P(X/Y) = \dfrac{P(X \cap Y)}{P(Y)}$

$$\Rightarrow P(X) = \frac{P(X \cap Y)}{P(Y)}$$

$$\Rightarrow P(X \cap Y) = P(X) \cdot P(Y)$$

This is called the **multiplication law for independent events.**

For n independent events

$$P(E_1 \cap E_2 \cap ... \cap E_n) = P(E_1) \times P(E_2) \times ... \times P(E_n)$$

Example 38:15

A die and a coin are thrown in succession. Find the probability that a head is shown on the die and a 2 is shown on the coin.

Solution

$$P(2 \cap H) = P(2) \cdot P(H) = \frac{1}{6} \cdot \frac{1}{2} = \frac{1}{12}$$

Example 38:16

Two coins are tossed simultaneously. One of the coins is fair and the other one is loaded so that a head is twice as likely to turn up as a tail. Find the probability that the two coins turn up heads.

Solution

Let H_f and H_u represent the events that a head is shown on the fair and unfair coins respectively then

$$P(H_f) = \frac{1}{2} \quad \text{and} \quad P(H_u) = \frac{2}{3}$$

Since these are independent events

$$P(H_f \cap H_u) = P(H_f) \cdot P(H_u) = \frac{1}{2} \cdot \frac{2}{3} = \frac{1}{3}$$

Example 38:17

An unbiased die numbered 1 to 6 is rolled twice. Find the probability of:
(i)   Rolling 2 sixes.
(ii) The second throw being a six given that the first throw is a six.
(iii) Getting a score of ten from the two throws.
(iv) Throwing at least one six.
(v)   Throwing exactly one six.

Solution

(i) $P(6) = \frac{1}{6} \Rightarrow P(2 \text{ sixes}) = P(6 \cap 6)$

$$= \left(\frac{1}{6}\right)\left(\frac{1}{6}\right) = \frac{1}{36}$$

(ii) $P(6/6) = \dfrac{P(6 \cap 6)}{P(6)} = \dfrac{1}{36} \div \dfrac{1}{6} = \dfrac{1}{36} \times \dfrac{6}{1} = \dfrac{1}{6}$

(iii) $P(\text{score of ten}) = P(4.6) + P(6,5) + P(5,5)$

$$= \frac{1}{36} + \frac{1}{36} + \frac{1}{36} = \frac{1}{12}$$

(iv) $P(\text{at least one six}) = \dfrac{11}{36}$

(v) $P(\text{exactly one six}) = \dfrac{10}{36} = \dfrac{5}{18}$

Conditional Probability

Drawing With and Without Replacement

Example 38:18

A bag contains 10 balls, 7 of which are red and 3 of which are white.
(a) What is the probability that a ball chosen at random from the bag is
     (i) Red?   (ii) White?

Solution

Let the events "choosing a red ball" and "choosing a white ball" be R and W respectively. Then,

(i) $P(R) = \dfrac{7}{10}$ (ii) $P(W) = \dfrac{3}{10}$

Example 38:19

Suppose that in Example 38:18 the ball chosen is red and is not replaced.
 (i)   How many red balls are left there altogether?
 (ii) How many balls are left there altogether?
 (iii) What is the probability that the second ball chosen at random from the bag is
         (a) Red?   (b) White?
 (iv) Suppose the second ball chosen was red again and was not replaced but 3 white balls were put inside the bag. What will be the probability that a third ball chosen at random from the bag will be
         (a) Red?   (b) White?

Solution

(i) $n(R) = 6$ (ii) $n(S) = 9$

(iii) (a) $P(R) = \dfrac{6}{9} = \dfrac{2}{3}$ (b) $P(W) = \dfrac{3}{9} = \dfrac{1}{3}$

(iv) $n(R) = 5$ (ii) $n(W) = 6$

(a) $P(R) = \dfrac{5}{11}$ (b) $P(W) = \dfrac{6}{11}$

Repeated Trials

At times, an experiment is repeated. If the conditions of the experiment are unaltered, the events can be considered independent.

Example 38:20

A biased coin is tossed five times. If $P(H) = 2P(T)$, find the probability of having a head, a head, a tail, a head, and a tail in that order.

Solution

$$P(H) = 2P(T) \Rightarrow P(T) = \frac{P(H)}{2}$$

But $P(H) + P(T) = 1 \Rightarrow 2P(T) + P(T) = 1$

$$\Rightarrow P(T) = \frac{1}{3}$$

$$P(HHTHT) = P(H).P(H).P(T).P(H).P(T)$$

$$= \left(2P(T)\right)^3 \left(P(T)\right)^2 = 8\left(\frac{1}{3}\right)^5 = \frac{8}{243}$$

Tree Diagrams

Tree diagrams are devices used to enumerate all the possible outcomes of a sequence of experiments, when each experiment occurs in a finite number of ways in solving probability problems especially when the problem consists of a mixture of mutually exclusive events and independent events.

Example 38:21

A fair coin is tossed twice. Draw a tree diagram to represent the possible outcomes. Hence, find the probability of obtaining:

(a) Two heads   (b) A head and a tail

Solution

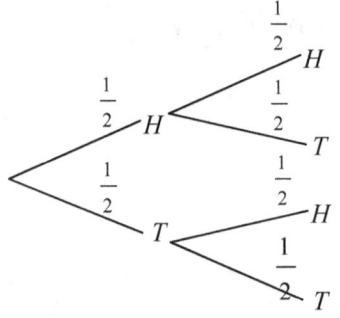

Figure 38:7

(a) $P(HH) = \frac{1}{2}\left(\frac{1}{2}\right) = \frac{1}{4}$

(b) $P\left(HT \text{ or } TH\right) = \left(\frac{1}{2}\right)\left(\frac{1}{2}\right) + \left(\frac{1}{2}\right)\left(\frac{1}{2}\right) = \frac{1}{2}$

Example 38:22

A game consists of tossing a coin. If a head shows, a fair die is thrown once and the score is noted. If a tail shows, a card is selected from a well-shuffled pack of 52 playing cards. Hearts score 1, diamonds score 2, clubs score 3 and spades score 4. Draw a tree diagram to show all the possible outcomes for a single throw of a coin and find the probability of scoring 3 or less.

Solution

For convenience

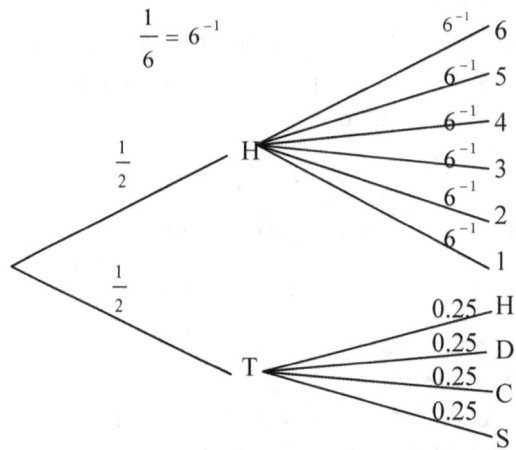

Figure 38:8

$$P(\leq 3) = 3\left(\frac{1}{2}\right)\left(\frac{1}{6}\right) + \frac{1}{2}\left(\frac{1}{4}\right)(3) = \frac{2}{8} + \frac{3}{8} = \frac{5}{8}$$

Example 38:23

A coin is tossed 4 times. Find the probability of obtaining 3 heads and one tail.

Solution

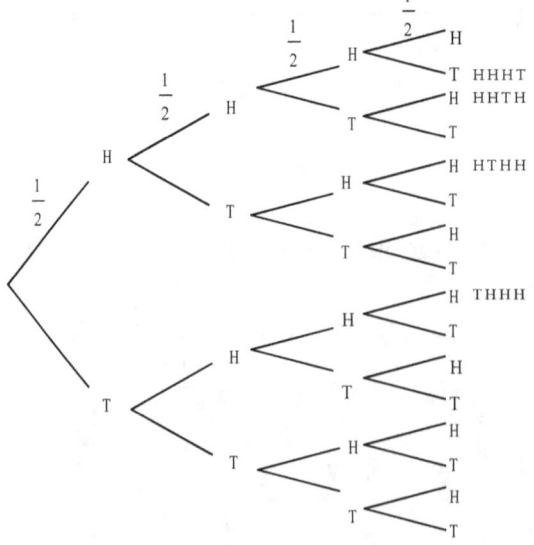

Figure 38:9

Let $P_{3,1}$ = Probability of 3 heads and one tail

$P_{3,1} = P(HHHT) + P(HHTH) + P(HTHH) + P(THHH)$

$P(HHHT) = \left(\dfrac{1}{2}\right)\left(\dfrac{1}{2}\right)\left(\dfrac{1}{2}\right)\left(\dfrac{1}{2}\right) = \dfrac{1}{16}$

$P(HHTH) = \left(\dfrac{1}{2}\right)\left(\dfrac{1}{2}\right)\left(\dfrac{1}{2}\right)\left(\dfrac{1}{2}\right) = \dfrac{1}{16}$

$P(HTHH) = \left(\dfrac{1}{2}\right)\left(\dfrac{1}{2}\right)\left(\dfrac{1}{2}\right)\left(\dfrac{1}{2}\right) = \dfrac{1}{16}$

$P(THHH) = \left(\dfrac{1}{2}\right)\left(\dfrac{1}{2}\right)\left(\dfrac{1}{2}\right)\left(\dfrac{1}{2}\right) = \dfrac{1}{16}$

$\therefore$ Probability of 3 heads and 1 tail $= 4\left(\dfrac{1}{16}\right) = \dfrac{1}{4}$.

Example 38:24

Three dice each numbered 1 to 6 are rolled. One die is fair and the others are biased so that for each of these a six is twice as likely as any other score. Find the probability of rolling exactly two sixes.

Solution

Let the event of obtaining a six be S, then the event of obtaining no six is $\bar{S}$.

On the fair die $P(S) = \dfrac{1}{6}$ and $P(\bar{S}) = \dfrac{5}{6}$

Let x be the probability of obtaining 1, 2, 3, 4 or 5 on the unfair die and $P(S)$ be probability of obtaining a 6.

Then, $5x + 2x = 1 \Rightarrow x = \dfrac{1}{7}$.

On the unfair die $P(S) = \dfrac{2}{7}$ and $P(\bar{S}) = \dfrac{5}{7}$.

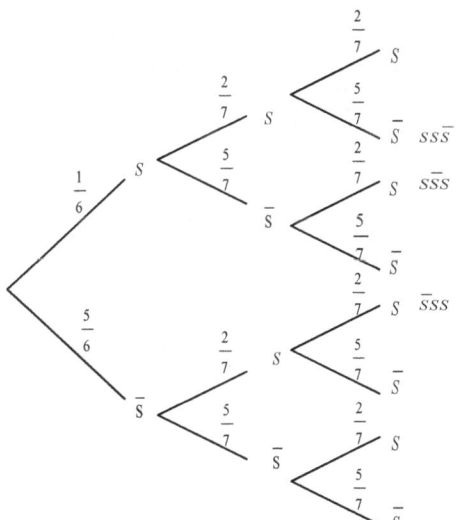

Figure 38:10

$P(\text{exactly 2 sixes})$

$= \left(\dfrac{1}{6}\right)\left(\dfrac{2}{7}\right)\left(\dfrac{5}{7}\right) + \left(\dfrac{1}{6}\right)\left(\dfrac{5}{7}\right)\left(\dfrac{2}{7}\right) + \left(\dfrac{5}{6}\right)\left(\dfrac{2}{7}\right)\left(\dfrac{2}{7}\right)$

$= \dfrac{20}{147}$

EXERCISE 38:5

1. A bag contains 3 black balls and 2 red balls. A ball is picked at random from the bag and then replaced before a second ball is drawn. What is the probability that
   (a) They are both black. (b) they are both red.
   (c) One is black and the other is red.

2. A pair of fair dice with each face numbered 1-6 is rolled once. Find the probability of scoring a 2 on one die and a 5 on the other.

3. A student has 3 khaki shirts and 2 white shirts; he has also 3 pairs of khaki trousers and 4 pairs of white trousers. One night in the dark, he chooses one shirt and one pair of trousers at random. Find the probability that the shirt and the trousers are of
   (a) The same colour. (b) Different colour.

4. A bag contains 5 white socks and 6 blue socks. Find the probability that if two socks are drawn at random in succession and without replacement, they will be
   (a) Both white. (b) Both blue.
   (c) Of the same colour (d) of different colour

5. There are four children in a certain family. Find the probability that the family has:
   (a) An equal number of boys and girls.
   (b) At least two boys.
   (c) More boys than girls
   (d) Children of the same sex.

6. A box contains 2 red balls and 4 blue balls. A ball is drawn at random from the box and replaced before a second ball is drawn. Find the probability of drawing 2 red balls.

7. A staff consists of 15 men and 10 women. 60 % of the men and 40% of the women can drive. A man and a woman are chosen at random from the staff. Find the probability that:
   (a) Both of them can drive.
   (b) Only the man can drive.
   (c) Only the woman can drive.
   (d) None of them can drive.
   (e) Only one of them can drive.

8. A bag contains 9 identical balls out of which 3 are blue, 2 are white and the rest are red. If two balls are drawn at random one after the other without replacement, what is the probability that:

427

(a) Both of them will be blue?
(b) They will be of the same colour?
(b) They will be of different colour?

9. A packet contains 4 red, 5 blue and 6 black identical balls. Two balls are picked at random from the bag without replacement. Find the probability of picking:
   (a) a red and a black ball.
   (b) Balls of the same colour.
   (c) Balls of different colour.

10. Two fair coins are tossed at once. Find the probability of getting at least one tail.

11. A pair of fair dice each numbered 1 to 6 are tossed. Find the probability of obtaining a sum of at least 9.

12. The probability that a civil servant owns a car is $\frac{3}{7}$.

   Find the probability that:
   (a) Two civil servants selected at random each own a car.
   (b) Of two civil servants selected at random only one owns a car.
   (c) Of three civil servants selected at random only one owns a car.

13. A number is chosen at random from the integers 5 to 25 inclusive. Find the probability that the number is a multiple of 5 or 3.

14. A bag contains 10 identical balls of which 4 are blue and 6 are red. Two balls are picked at random one after the other with replacement. What is the probability that:
   (a) Both are red?
   (b) Both are the same colour?

15. Two perfect dice are thrown once. Find the probability that their total score is 6.

16. A basket contains 24 mangoes, of which m are ripe. If 6 more ripe mangoes are added to the basket, find in terms of m:
   (a) The number of unripe mangoes in the basket.
   (b) The probability that a mango chosen at random from the basket is ripe.

17. A ball is to be drawn at random from a bag, which contains just red and green balls. There are 6 more green balls than red balls, and the probability of selecting a green ball is $\frac{2}{3}$. Find the number of balls in the bag.

18. A card is drawn at random from a pack of 52 well-shuffled cards. Find the probability that the card is:
   (a) a queen of clubs or hearts.

Probability from Frequency Tables

Example 38:25

The following scores out of 100 were obtained by 36 students in a test.

Marks	f
0-9	0
10-19	1
20-29	2
30-39	4
40-49	7
50-59	9
60-69	5
70-79	3
80-89	3
90-99	2

Table 38:1

(a) Make a cumulative frequency table for the data.
(b) Hence, calculate the probability that a student chosen at random from the class scored a mark of:
   (i) Less than 69.5%.
   (ii) Greater than or equal to 49%.
   (iii) From 30% to 79% inclusively.

Solution

(a)

Mark	Cumulative
< 9.5	0
< 19.5	1
< 29.5	3
< 39.5	7
< 49.5	14
< 59.5	23
< 69.5	28
< 79.5	31
< 89.5	34
< 99.5	36

Table 38:2

(b) (i) $P(X < 69.5) = \dfrac{28}{36} = \dfrac{7}{9}$

(ii) $P(X \geq 49.5) = 1 - P(X < 49.5)$

$\Rightarrow P(X \geq 49.5) = 1 - \dfrac{14}{36} = \dfrac{11}{18}$

(iii) $P(30 \leq X \leq 79) = \dfrac{31 - 3}{36} = \dfrac{7}{9}$

EXERCISE 38:6

1. A Mathematics Teacher drives to school every day. She records her journey time, to the nearest minutes, over a period of 60 days as follows:

Time (min)	8	9	10	11	12	13	14	15
No. of days	5	7	13	12	11	7	3	2

Table 38:3

(a) State the modal time.

(b) Calculate her mean journey time.

(c) Estimate, to 2 decimal places, the standard deviation. Find the probability that:

(d) She took 11 minutes to get to school on a certain day.

(e) She was not late on a certain day, given that she left her house 10 minutes before the start of her lesson.

2. The masses of 100 people were recorded to the nearest kg and displayed as follows:

Mass (kg)	Number of people
10-19	6
20-29	15
30-39	39
40-49	22
50-59	12
60-69	5
70-79	1

Table 38:4

(a) Determine the width of the class intervals.

(b) State the modal class.

(c) Calculate an estimate of the mean mass of the students to the nearest of 10th of a kg.

(d) Using the data draw a cumulative frequency curve.

(e) Find the inter-quartile range.

(f) Find the probability that a student chosen at random will weigh less than 55 kg.

3. In a research institute the approximate weights of 500 oranges were recorded as follows;

Weights (g)	Number of oranges
0-29	20
30-59	25
60-89	50
90-119	75
120-149	120
150-179	100
180-209	80
210-239	30

Table 38:5

(i) State the modal class of the distribution.

(ii) Estimate to 1 decimal place the mean of the distribution.

(iii) Construct a cumulative frequency table for the distribution. Taking 1cm to represent 25 oranges and 1 cm to represent 15 g draw a cumulative frequency graph to represent this data. Use your graph to estimate:

(a) The median of the distribution.

(b) The number of oranges which weigh less than 75 g each.

(c) The probability that an orange selected at random will have a weight greater than or equal to 175 g.

MULTIPLE CHOICE EXERCISE 38

1. There are m boys and 12 girls in class. The probability of selecting at random a girl from the class is:

   [A] $\dfrac{m}{12}$ [B] $\dfrac{12}{m}$ [C] $\dfrac{12}{m+12}$ [D] $\dfrac{m}{m+12}$

2. Table 38:6 gives the marks scored by a group of students in a test. The probability of selecting a student from the group that scored 2 or 3 is:

   [A] $\dfrac{1}{11}$ [B] $\dfrac{5}{25}$ [C] $\dfrac{7}{22}$ [D] $\dfrac{6}{11}$

Mark	0	1	2	3	4	5
Frequency	1	2	7	5	4	3

Table 38:6

3. The probability of having an odd number in a single toss of a fair die is:

[A] $\frac{2}{3}$ [B] $\frac{1}{6}$ [C] $\frac{1}{3}$ [D] $\frac{1}{2}$

4. The Table 38:7 gives the scores of a group of students in an English Language test. A student is chosen at random from the group. The probability that he scored at least 6 marks is:

[A] $\frac{3}{4}$ [B] $\frac{1}{5}$ [C] $\frac{1}{4}$ [D] $\frac{3}{10}$

Score	2	3	4	5	6	7
Number of students	2	4	7	2	3	2

Table 38:7

5. Two groups of students cast their votes on a particular proposal. The results are as in Table 38:8. A student in favour of the proposal is selected for a post, the probability that he is from group A is:

[A] $\frac{8}{9}$ [B] $\frac{16}{35}$ [C] $\frac{4}{5}$ [D] $\frac{4}{7}$

	In favour	Against
Group A	128	32
Group B	96	48

Table 38:8

6. Two groups of students cast their votes on a particular proposal. The results are as in Table 38:8. A student is chosen at random, the probability that he is against the proposal is:

[A] $\frac{3}{19}$ [B] $\frac{4}{19}$ [C] $\frac{5}{19}$ [D] $\frac{9}{19}$

7. The events X and Y are mutually exclusive and P $(X) = \frac{1}{3}$, P $(Y) = \frac{2}{5}$. P $(X \cap Y)$ is:

[A] 0 [B] $\frac{2}{15}$ [C] $\frac{4}{15}$ [D] $\frac{11}{15}$

8. The events X and Y are mutually exclusive and P $(X) = \frac{1}{3}$, P $(Y) = \frac{2}{5}$. P $(X \cup Y)$ is:

[A] 0 [B] $\frac{2}{15}$ [C] $\frac{4}{15}$ [D] $\frac{11}{15}$

9. A box contains 2 white and 3 blue identical marbles. Two marbles are picked at random one after the other without replacement. The probability of picking two marbles of different colours is:

[A] $\frac{2}{3}$ [B] $\frac{3}{5}$ [C] $\frac{2}{5}$ [D] $\frac{3}{10}$

10. Mrs. Ngala is expecting a baby. The probability of a boy is $\frac{1}{2}$ and the probability that the baby will have blue eyes is $\frac{1}{4}$. The probability that she will have a blue-eyed boy is:

[A] $\frac{1}{8}$ [B] $\frac{1}{4}$ [C] $\frac{3}{8}$ [D] $\frac{3}{4}$

11. A number is chosen at random from the set {1, 2, 3, ..., 9,10}. The probability that the number is greater than or equal to 7 is:

[A] $\frac{1}{10}$ [B] $\frac{3}{10}$ [C] $\frac{2}{5}$ [D] $\frac{3}{5}$

12. The probability of throwing a number greater than 2 with a single fair die is:

[A] $\frac{1}{6}$ [B] $\frac{1}{3}$ [C] $\frac{1}{2}$ [D] $\frac{2}{3}$

13. A fair die is rolled once. The probability of obtaining 4 or 6 is:

[A] $\frac{2}{3}$ [B] $\frac{1}{6}$ [C] $\frac{1}{3}$ [D] $\frac{1}{2}$

14. Three balls are drawn one after the other with replacement, from a bag containing 5 red, 9 white and 4 blue identical balls. The probability that they are one red, one white and one blue is:

[A] $\frac{5}{81}$ [B] $\frac{5}{27}$ [C] $\frac{5}{162}$ [D] $\frac{5}{243}$

15. The probability that an integer selected from the set of integers {20,21,...,30} is a prime number is:

[A] $\frac{2}{11}$ [B] $\frac{5}{11}$ [C] $\frac{6}{11}$ [D] $\frac{9}{11}$

16. A fair die is rolled once. The probability of obtaining a number less than 3 is:

[A] $\frac{1}{6}$ [B] $\frac{1}{3}$ [C] $\frac{1}{2}$ [D] $\frac{2}{3}$

17-18 The data below shows the number of workers employed in the various sections of a construction company in Yaounde. Use the information to answer questions 17 to 18. 24 Carpenters, 27 Labourers, 12 Plumbers, 15 Plasterers, 9 Painters, 3 Messengers and 18 Bricklayers.

17. One of the workers is absent on a day. The probability that he is a bricklayer is:

[A] $\frac{1}{9}$ [B] $\frac{2}{9}$ [C] $\frac{1}{6}$ [D] $\frac{1}{4}$

18. A worker is retrenched. The probability that he is a plumber or a plasterer is:

[A] $\dfrac{3}{4}$ [B] $\dfrac{1}{9}$ [C] $\dfrac{5}{36}$ [D] $\dfrac{1}{4}$

19. The probability that a total sum of seven would appear with two tosses of a fair die is:

[A] $\dfrac{5}{36}$ [B] $\dfrac{1}{6}$ [C] $\dfrac{7}{36}$ [D] $\dfrac{5}{6}$

20. A die is rolled 200 times. The outcomes obtained are shown in Table 38:9. The probability of obtaining a 2 is:
[A] 0.002 [B] 0.015
[C] 0.15 [D] 0.16

21. A die is rolled 200 times. The outcomes obtained are shown in Table 38:9. The probability of obtaining a number less than 3 is:
[A] 0.125 [B] 0.150
[C] 0.275 [D] 0.500

Number	1	2	3	4	5	6
Number of times	25	30	45	28	40	32

Table 38:9

22. Two cards are drawn one after the other with replacement from a well shuffled ordinary deck of 52 cards containing four aces. The probability that they are both aces is:

[A] $\dfrac{1}{13}$ [B] $\dfrac{1}{169}$ [C] $\dfrac{1}{52}$ [D] $\dfrac{1}{26}$

23. The probability that a number selected from the numbers 30 to 50 inclusive is a prime is:

[A] $\dfrac{1}{4}$ [B] $\dfrac{5}{21}$ [C] $\dfrac{3}{7}$ [D] $\dfrac{1}{3}$

24. Two fair dice are tossed together once. The probability that the sum of the outcome is at least ten is:

[A] $\dfrac{1}{12}$ [B] $\dfrac{5}{36}$ [C] $\dfrac{1}{6}$ [D] $\dfrac{5}{18}$

25. From a box containing 2 red, 6 white and 5 blackballs, a ball is randomly selected. The probability that the ball selected is black is:

[A] $\dfrac{2}{13}$ [B] $\dfrac{5}{13}$ [C] $\dfrac{5}{11}$ [D] $\dfrac{11}{13}$

26. A bag contains 3 red, 4 black and 5 green identical balls. Two balls are picked at random one after the other without replacement. The probability that one is red and the other is green is:

[A] $\dfrac{5}{48}$ [B] $\dfrac{5}{11}$ [C] $\dfrac{5}{22}$ [D] $\dfrac{5}{44}$

27. Table 38:10 gives the distribution of outcomes obtained when a die was roll 100 times. The experimental probability that it shows at most 4 when rolled again is:

[A] $\dfrac{8}{25}$ [B] $\dfrac{12}{25}$ [C] $\dfrac{13}{25}$ [D] $\dfrac{17}{25}$

Number of die	1	2	3	4	5	6
Frequency	18	14	20	16	15	17

Table 38:10

28. A bag contains red, black, and green identical balls. A ball is picked and replaced. Table 38:11, shows the result of 100 trials. The experimental probability of picking a green ball is:

Colour of ball	red	black	green
Number of occurrences	54	30	16

Table 38:11

[A] $\dfrac{4}{25}$ [B] $\dfrac{21}{25}$ [C] $\dfrac{1}{3}$ [D] $\dfrac{4}{21}$

29. A box contains 2 white and 3 blue identical balls. 2 balls are picked at random one after the other, without replacement. The probability of picking 2 balls of different colours is:

[A] $\dfrac{6}{25}$ [B] $\dfrac{7}{20}$ [C] $\dfrac{3}{5}$ [D] $\dfrac{2}{3}$

30. A group of eleven people can speak either English or French or both. Seven can speak English and six can speak French. The probability that a person chosen at random can speak both English and French is:

[A] $\dfrac{2}{11}$ [B] $\dfrac{4}{11}$ [C] $\dfrac{5}{11}$ [D] $\dfrac{11}{13}$

31. The probabilities that Awah and Suh pass an examination are $\dfrac{3}{4}$ and $\dfrac{3}{5}$ respectively. The probability of both boys failing the examination is:

[A] $\dfrac{2}{3}$ [B] $\dfrac{3}{10}$ [C] $\dfrac{9}{10}$ [D] $\dfrac{1}{10}$

32. A box contains 5 red, 3 green and 4 blue balls. A boy is allowed to take away two balls at random from the box. The probability that the two balls are red is:

[A] $\dfrac{5}{33}$ [B] $\dfrac{5}{36}$ [C] $\dfrac{103}{132}$ [D] $\dfrac{31}{36}$

33. A box contains 5 red, 3 green and 4 blue balls. A boy is allowed to take away two balls at random from the box. The probability that one is green and the other is blue is:

[A] $\dfrac{2}{11}$ [B] $\dfrac{5}{12}$ [C] $\dfrac{8}{12}$ [D] $\dfrac{7}{11}$

34. The probability that an event will occur is p and the probability that it will not occur is q. The true assertion is:

[A] $p - q = 1$ [B] $q - p = 0$

[C] $p + q = 1$ [D] $p + q = 0$

35. A number is selected at random from the set $Y = \{18, 19, 20,\ldots,28, 29\}$. The probability that the number is a prime number is:

[A] $\dfrac{1}{4}$ [B] $\dfrac{3}{11}$ [C] $\dfrac{1}{2}$ [D] $\dfrac{3}{4}$

36. The numbers of goals scored by a school team in 10 netball matches are: 3,5,7,7,8,8,8,11,11,12. The probability that in a match, the school team will score at most 8 goals is:

[A] $\dfrac{1}{5}$ [B] $\dfrac{2}{5}$ [C] $\dfrac{3}{5}$ [D] $\dfrac{7}{10}$

37. The probability that a number chosen at random from {2,3,4,...,9,10} is either a prime number or a multiple of 3 is:

[A] $\dfrac{5}{9}$ [B] $\dfrac{2}{3}$ [C] $\dfrac{6}{7}$ [D] $\dfrac{5}{7}$

38. The probabilities that Ade and his dog

will be alive in 10 years time are 0.8 and 0.3 respectively. The probability that they will both be alive in 10 years time is:
[A] 1.00 [B] 0.50 [C] 0.24 [D] 0.06

39. The probabilities of Fru and Nsang passing an examination are $\dfrac{3}{4}$ and $\dfrac{5}{8}$ respectively. The probability that the two boys fail the examination is:

[A] $\dfrac{3}{32}$ [B] $\dfrac{3}{8}$ [C] $\dfrac{15}{32}$ [D] $\dfrac{5}{8}$

40. Two beads are drawn at random, one after the other with replacement, from a box containing 5 red and 7 white identical beads. The probability that the beads are the same colour is:

[A] $\dfrac{119}{144}$ [B] $\dfrac{95}{144}$ [C] $\dfrac{37}{72}$ [D] $\dfrac{49}{144}$

41. The probability that John passes the GCE is $\dfrac{2}{3}$ and the probability that Paul fails the same exam is $\dfrac{1}{4}$. The probability that both John and Paul pass is:

[A] $\dfrac{1}{6}$ [B] $\dfrac{1}{2}$ [C] $\dfrac{3}{4}$ [D] $\dfrac{5}{12}$

42. Given that a coin is tossed twice, then the probability of obtaining exactly one head is:

[A] $\dfrac{1}{4}$ [B] $\dfrac{1}{2}$ [C] $\dfrac{3}{4}$ [D] 1

TOPIC 39

CONSTRUCTIONS AND LOCI

OBJECTIVES

At the end of this topic, the learner should be able to:

1. Construct a line segment of given length.
2. Construct a triangle with sides of given length.
3. Distinguish between the perpendicular bisector of a side of a given triangle, the altitude and the median that side of the triangle.
4. Construct the altitude and the median of a given triangle.
5. Construct the perpendicular bisector of a given line segment.
6. Construct the perpendicular to a given line segment from a given point.
7. Bisect a given angle.
8. Construct angles of 90°, 45°, 60° and 30° at a given point.
9. Construct the circumscribed circle of a given triangle.
10. Construct the inscribed circle of a given triangle.
11. Construct convex regular polygons with not more than six sides.
12. State and use the loci of points on a circle, a perpendicular bisector, angle bisector, parallel lines.

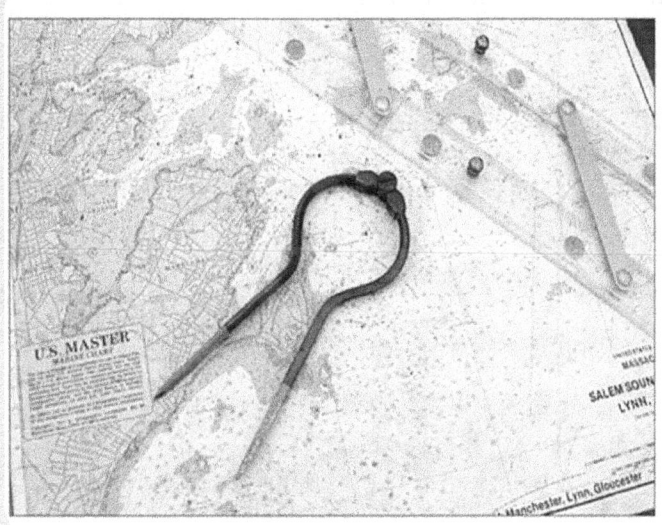

Marine navigators use **charts and plotting tools** to measure distances and to record a ship's progress as it travels through the water. This marine chart displays water depth and geographical features of a segment of the northeast coastline of the United States. Dividers centre, help the navigator measure distances on the chart. Parallel rulers, right, are used to transfer compass bearings from the compass rose, a diagram of a compass on the chart, to other parts of the chart. The navigator lays the parallel rulers over the compass rose, visible here through the transparent rulers, then walks the rulers, one leg at a time, across the chart and records the desired compass bearing on another section of the chart.

Richard Pasley/Stock Boston Inc.

CONSTRUCTIONS

Constructions are normally done with pencil, ruler and a pair of compasses, and unless otherwise stated no other device should be used. After a construction has been made the construction lines should never be erased.

Constructing a Line of Given Length

Example 39:1
Construct a line *AB* of length 6 cm.

Procedure
1. Draw a line of any length longer than 6 cm and mark a point *A* on it.
2. With open compass measure 6 cm from a ruler, (This is done by placing the pin end of the ruler at the zero mark and opening it up until the tip of the pencil is exactly on the 6 cm mark).
3. With the centre *A* trace an arc at *B*.

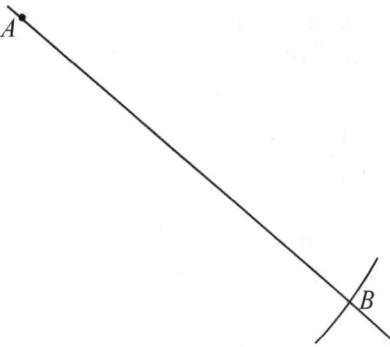

Figure 39:1

Constructing a Triangle *ABC* with Sides of Given Length

Example 39:2
Construct a triangle *ABC* with sides of length *AB* = 9 cm, *BC* = 6 cm and *AC* = 4 cm.

Procedure
(i) Following the steps in 1, construct the line *AB* of length 9 cm.
(ii) With compass, measure 6 cm and with centre *B*, draw an arc on one side of the line.
(iii) With compass, measure 4 cm and with centre *A* , draw another arc to cut the first. Mark their point of intersection *C*.
(iv) Using ruler and pencil join *AC* and *BC*

Note! Construction lines do not end at the vertices *A*, *B*, and *C*.

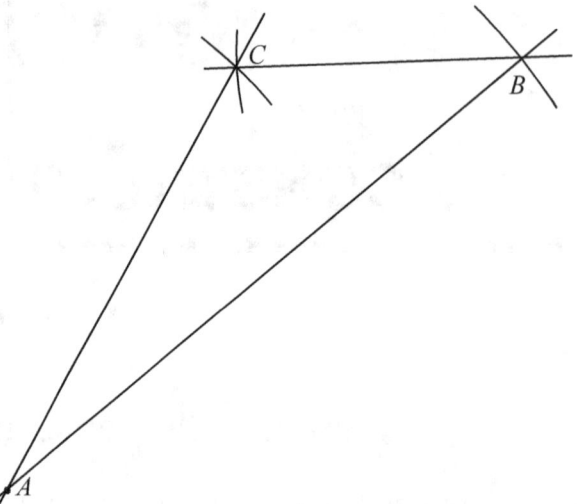

Figure 39:2

Height, Median and Perpendicular Bisector

The **altitude** (or **height**) of a triangle is the perpendicular distance from one side of the triangle to the opposite vertex.

The **perpendicular bisector** or **mediator** of a side of a triangle is a line that is perpendicular to the side and passes through its midpoint.

The **median** of a triangle is a line from the midpoint of a side of the triangle to the opposite vertex.

In Figure 39:3, *AM* is the altitude (or height), *AN* is the median and *XY* is the perpendicular bisector (or mediator) of the triangle *ABC* with respect to the side *BC*.

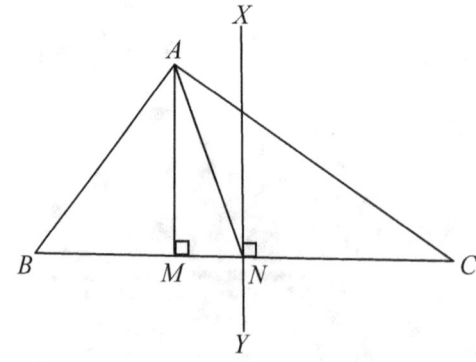

Figure 39:3

Constructing a Perpendicular Bisector (or Mediator) of a Given Line Segment

Example 39:3

Construct the perpendicular bisector of the line segment AB.

Figure 39:4

Procedure

(i) With centre A, draw two arcs of equal radii on the two sides of AB.

(ii) With centre B draw two arcs of equal radii as those in (a) to cut the first two at C and D.

(iii) Now join the points C and D of intersection of these arcs with ruler and pencil. The line CD is the perpendicular bisector of the line segment AB.

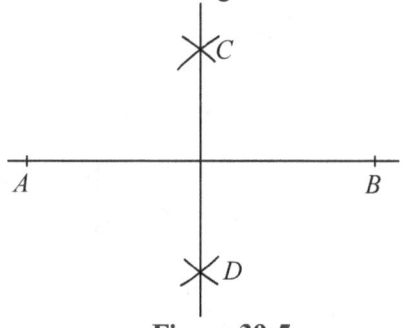

Figure 39:5

Constructing a Perpendicular to a Given Line Segment, from a Given Point

Example 39:4

Construct the perpendicular to the line segment AB passing through the point P.

Figure 39:6

Procedure

(i) With centre P draw two arcs of equal radii to cut AB at two points C and D.

(ii) With centres C and D draw two arcs of equal radii to intersect on the opposite side of P. Name this point of intersection Q.

(iii) Now join PQ. PQ is perpendicular to AB.

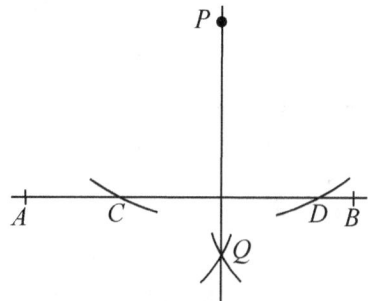

Figure 39:7

Note!!
The construction is the same event if P lies on AB.

Bisecting a Given Angle

Example 39:5

Construct the bisector of the angle PAR .

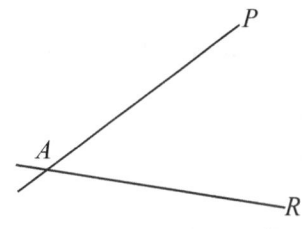

Figure 39:8

Procedure

(i) With centre A draw two arcs of equal radii to cut the adjacent sides to the given angle at two points C and B.

(ii) With centres C and B draw two arcs of equal radii to intersect at D.

(iii) Join A and D with ruler and pencil. AD is the bisector of the angle PAR.

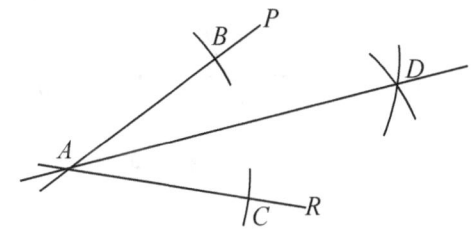

Figure 39:9

Constructing an Angle of 90° at a Given Point

Example 39:6

Construct an angle of $90°$ at the point A.

Figure 39:10

Procedure
(i) With centre A draw 2 arcs C and D of equal radii to cut the line passing through A.
(ii) With centres C and D draw two arcs of equal radii to intersect at B on one side of the line segment.
(iii) Now join AB. The angle AB makes with the line segment will be $90°$.

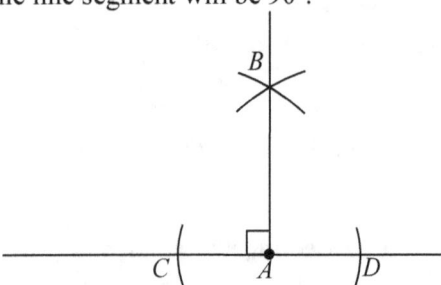

Figure 39:11

Constructing an Angle of 60°

Since all the three angles of an equilateral triangle are each $60°$, this idea can be used to construct an angle of $60°$.

Procedure
(i) With centre A draw two arcs of equal radii one to cut the line segment AB at B and the other on one side of AB.
(ii) With centre B and the same radius, draw another arc to intersect the first at C.
(iii) Join AC. The angle BAC is $60°$.

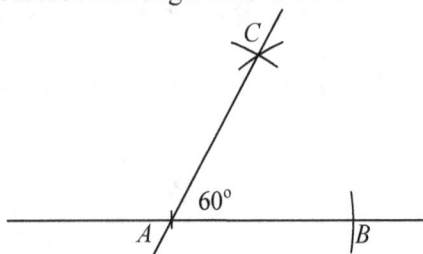

Figure 39:12

Note!
After constructing an angle of $60°$, a bonus is obtained by the $120°$ angle constructed.

Constructing an Angle of 45°

To construct an angle of $45°$ simply construct an angle of $90°$ and bisect it.

Constructing an Angle of 30°

To do this simply construct an angle of $60°$ and bisect it.

Trisecting a Right Angle

Suppose the right angle ABC has already been constructed. The following steps can be taken to trisect it.

Procedure
(i) With centre B draw a large arc to cut BC at C and AB at A.
(ii) With centre C draw an arc of the same radius to cut the large arc at D.
(iii) With centre A draw an arc of the same radius to cut the large arc at E.
(iv) Join BD and BE.

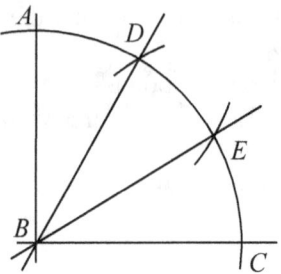

Figure 39:13

Constructing an Inscribed Circle of a Given Triangle

Example 39:7

Construct the inscribed circle of triangle ABC.

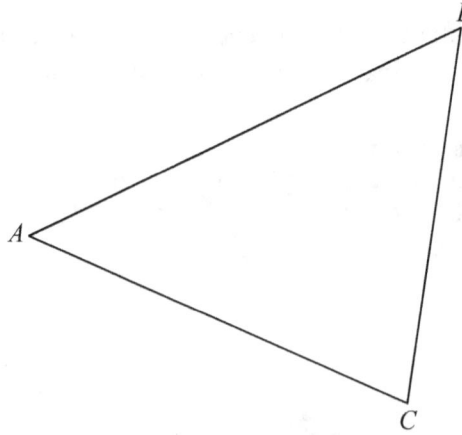

Figure 39:14

Procedure
(i) Construct the bisectors of each of the angles of the triangle ABC.
(ii) Where these bisectors intersect is the centre O of the circle.
(iii) Construct a perpendicular from O to any of the sides of the triangle. This helps to situate the radius of the circle.

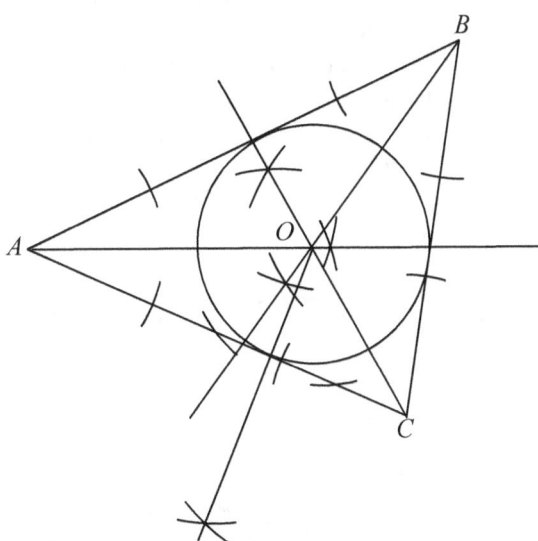

Figure 39:15

Constructing a Circumscribed Circle of a Given Triangle

Example 39:8

Construct the circumscribed circle of triangle *ABC*.

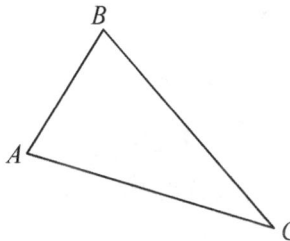

Figure 39:16

Procedure

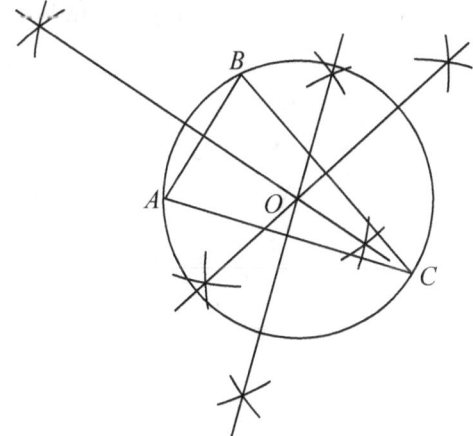

Figure 39:17

(i) Construct the perpendicular bisectors of each of the sides of the triangle *ABC*.

(ii) Where these bisectors intersect is the centre *O* of the circle.

Note!!

(a) Any point *P* on the arc *BAC* is such that $\angle PCA = \angle PBA$.

(b) Any point *Q* on the arc *CBA* is such that angle *QCB* = angle *QAB*.

(c) Any point *R* on the arc *ACB* is such that angle *RAC* = angle *RBC*.

Construction of Polygons

Example 39:9

Construct a regular hexagon with sides 3 cm.

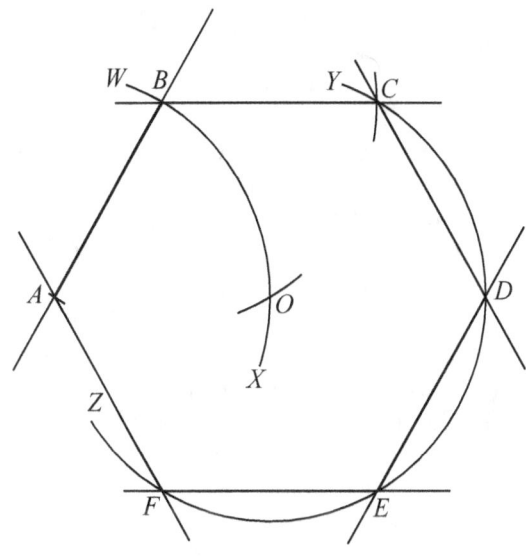

Figure 39:18

Procedure

1. Draw a straight line more than 3 cm.
2. Mark a point *A* on the line.
3. With a pair of compasses measure 3 cm from your ruler.
4. With centre *A*, draw a large arc *WX* of radius 3 cm to cut the line at *B*.
5. With centre *B*, draw another arc to cut *WX* at *O*.
7. With centre *B*, draw an arc to cut *YZ* at *C*.
8. With centre *C*, draw an arc to cut *YZ* at *D*.
9. With centre *D*, draw an arc to cut *YZ* at *E*.
10. With centre *E*, draw an arc to cut *YZ* at *F*.
11. Now connect the points by drawing the lines *BC*, *CD*, *DE*, *EF* and *FA*.

EXERCISE 39:1

In this exercise use only a ruler, a pair of compasses and a pencil, and show all construction lines.

1. (i) Draw a line *OA* of length 8 cm.
   (ii) Construct a line *OB* of length 6 cm perpendicular to *OA*.
   (iii) Measure *AB*.
   (iv) Bisect *AB* (by construction only) and mark the midpoint *M* of *AB*.
   (v) Measure *MO*.
   (vi) Construct the circle through *O*, *A* and *B*.

2. Draw a line segment, *PQ*, 7 cm long in the middle of a new page.
   (b) Bisect line *PQ* and label the mid-point, *X*.
   (c) Draw a circle such that *PQ* is the diameter
   (d) Locate a point *A* on the circumference of the circle such that *AP* = 4.5 cm.
   (e) Draw and measure line *AQ*.
   (f) Write down the value of angle *PAQ*.

3. (a) Draw a line *PQ* of length 8 cm.
   (b) Construct a line *OQ* of length 6 cm perpendicular to *PQ*.
   (c) Draw and measure the line *OP*.
   (d) Construct the perpendicular bisector of *OP* and label its foot *M*.
   (e) Construct the circle, which circumscribe triangle *OPQ*.

4. (a) Draw a line *AB* of length 8 cm.
   (b) Construct a line *AD* of length 6 cm such that angle *BAD* = 60°.
   (c) Construct the parallelogram *ABCD*.
   (d) Produce *DA* to *X* and bisect angle *BAX*.
   (e) Bisect the line segment *AB*.
      Given that the two bisecting lines meet at *O*,
   (f) Measure the distance of *O* from the line *AB* and from the point *A*.
   (g) Deduce the angle between the two bisecting lines.

5. (a) Draw a line *ST*, 5 cm long with *T* at the centre of a new page.
   (b) On the base *ST*, construct triangle *RST* such that ∠*S* = 60° and ∠*T* = 90° with the point R above the base *ST*.
   (c) Measure the length of *SR*.
   (d) The sides *RS* and *RT* are produced downwards to the points *M* and *N* such that *RSM* = 13 cm and *RTN* = 15 cm.
   (e) Construct the bisectors of angles *TSM*

and *STN* and where these lines intersect, mark the point *O*.
   (f) From *O* construct a perpendicular to the line *TN* to meet *TN* at *P*.
   (g) With centre O and radius OP, construct a circle.

6. Given that the triangle *ABC* in Figure 39:13.

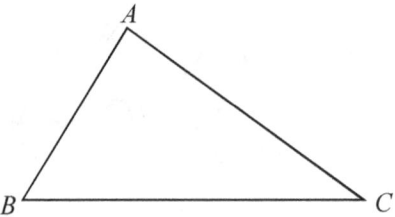

Figure 39:19

   (a) Using a ruler, pencil and compass only, copy triangle *ABC*.
   (b) (i) Construct the median passing through the point *A*.
      (ii) Construct the perpendicular bisector of the side *BC* of the triangle.
      (iii) Construct the perpendicular to the side *BC*, passing through the point *A*.

LOCI

A locus is the path of a moving point subject to some restrictions.
The following are some examples of loci.
(1) The locus of a point which is moving such that its distance from a fixed point is always constant (the same) is a circle.

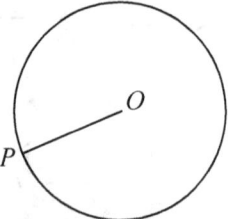

{*P*: *PO* = 1.8 cm}
Figure 39:20

(2) The locus of a point, which is moving such that its distance from two fixed points is always equal, is the mediator or perpendicular bisector of the line segment joining these two points.

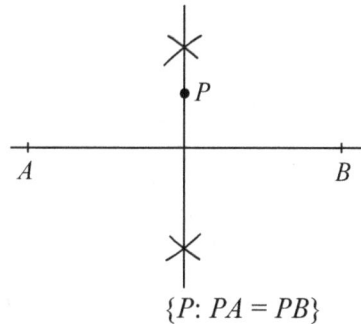

{P: PA = PB}
Figure 39:21

(3) The locus of a point, which is equidistant from two intersecting lines, is the bisector of the angle between the lines.

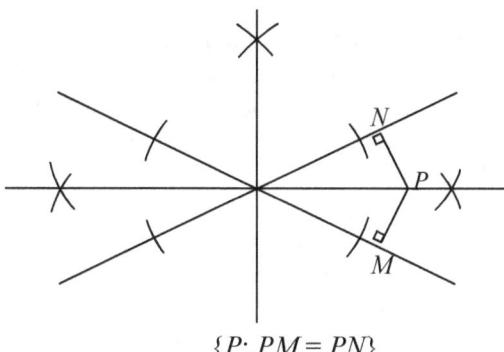

{P: PM = PN}
Figure 39:22

(4) The locus of a point, which is moving such that its distance d from a fixed line L_1 is always constant, is a line L_2 parallel to L_1.

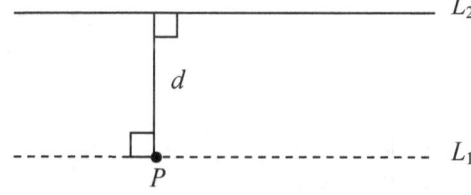

Figure 39:23

EXERCISE 39:2

1. State the locus of a point, which is equidistant from two fixed points.
2. $ABCD$ is a quadrilateral with unequal sides. Show how to find a point P such that $PA = PB$ and $PC = PD$.
3. YZ is a straight line 14 cm long. A point P moves so that $PY^2 - PZ^2 = 56$ cm^2. If X is the foot of the perpendicular from P to YZ,
   (a) Calculate the length of YX.
   (b) State the locus of P, and show it clearly on a diagram (not necessarily drawn accurately or to scale)

4. Draw a rectangle $ABCD$ with $AB = 10$ cm and $AD = 6$ cm. Draw the following loci in each case giving only that part of the locus which is <u>inside</u> the rectangle. The locus of the point, which is:
   (a) 2 cm from A.
   (b) equidistant from B and D.
   (c) 2 cm from BD.
   (d) equidistant from BD and DC.
5. Construct in a single diagram:
   (a) A triangle LMN with sides $LM = 7$ cm, $LN = 8$ cm and $MN = 6$ cm,
   (b) The locus of points, which are 5 cm from L.
   (c) The locus of points, which are equidistant from MN and LN. Given that,
   $$\mathscr{E} = \{P:P \text{ lies inside } \Delta LMN\}$$
   $$X = \{P : LP < 5 \text{ cm}\} \text{ and}$$
   $$Y = \{P:P \text{ is nearer to } MN \text{ than } LN\}$$
   (d) Indicate the region $X \cap Y$ by suitable shading of your diagram.

MULTIPLE CHOICE EXERCISE 39

1. AB bisects PQ at point N. The statement that is true of N is:
   [A] N is the midpoint of AB.
   [B] N is the midpoint of AB and the midpoint of PQ.
   [C] N is the midpoint of PQ.
   [D] N divides PQ in the ratio 2:1.
2. Point P is the midpoint of AB. Complete the statement: $PB = 7$ cm, AB is equal to:
   [A] 7 cm   [B] 14 cm   [C] 3.5 cm
   [D] none of the above
3. The diagram(s) in Figure 39:24 that demonstrate(s) the correct way of constructing an angle of 60° is:

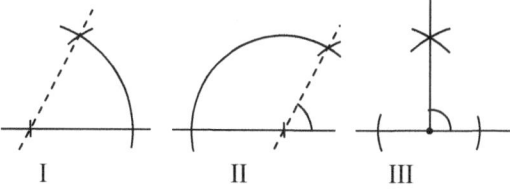

I   II   III

Figure 39:24

[A] I only   [B] II only

439

[C] III only   [D] I and II only

4. In the construction in Figure 39:25, the size of angle *BAC* is:
    [A] 60° [B] 90° [C] 120° [D] 150°

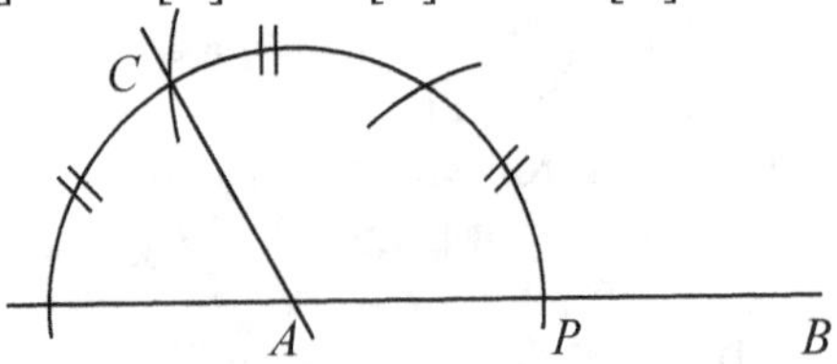

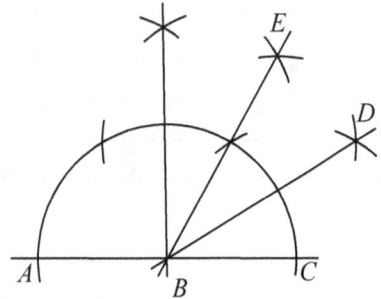

Figure 39:25

5. The angle, which can be constructed using a ruler and compass, only is:
    [A] 135° [B] 125° [C] 115° [D] 155°

6. Figure 39:26 shows the arcs used in constructing angles *DBC* and *EBC*. The size of angle *DBC* is:
    [A] 120° [B] 30° [C] 45° [D] 60°

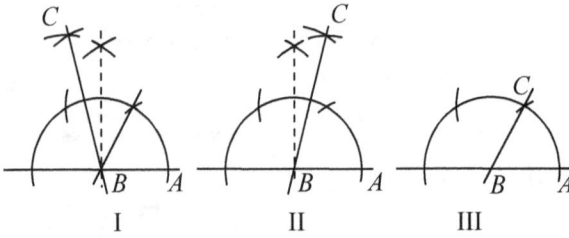

Figure 39:26

7. Figure 39:20 shows the arcs used in constructing angles *DBC* and *EBC*. The size of angle *EBC* is:
    [A] 120° [B] 30° [C] 45° [D] 60°

8. Figure 39:20 shows the arcs used in constructing angles *DBC* and *EBC*. The size of angle *ABE* is:
    [A] 120° [B] 30° [C] 45° [D] 60°

9. The diagram in Figure 39:27 which shows the construction of ∠ABC = 75° using a ruler and a pair of compass only is:
    [A] I only   [B] II only
    [C] III only   [D] I and II only

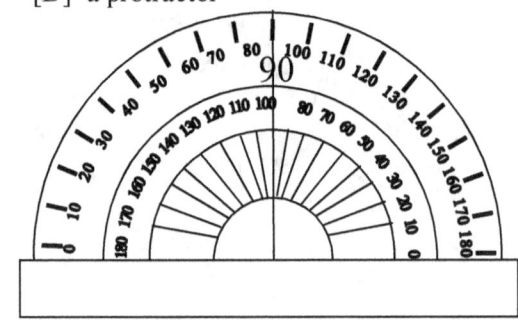

I   II   III

Figure 39:27

10. The name of the instrument in Figure 39:28 is:
    [A] a protractor

[B] a pair of dividers
[C] a pair of compass
[D] a set square

Figure 39:28

11. In order to construct an in-circle one needs at least:
    [A] a pair of compass, pencil and ruler
    [B] a protractor, pencil and ruler
    [C] a set square, pencil and ruler
    [D] a protractor, pencil and a set square

12. The name of the instrument in Figure 39:29 is:
    [A] a set square
    [B] a pair of compasses
    [C] a pair of dividers
    [D] a protractor

Figure 39:29

13. The locus of a point, which is moving such that its distance d from a fixed line l is always constant, is:
    [A] a mediator of *l*
    [B] a line parallel to *l*
    [C] a parabola
    [D] a circle

14. The locus of a point, which is moving such that its distance from a fixed point is always, constant (the same) is:
    [A] a mediator.
    [B] a line parallel to the point.
    [C] the mediator constant from the point.
    [D] a circle

15. The locus of a point, which is moving such that its distance from two fixed points is always equal, is:
    [A] a perpendicular bisector *l*.
    [B] a line parallel to *l*.

[C] the angle bisector from the two points.

[D] a circle.

16. The locus of a point, which is equidistant from two intersecting lines, is:

[A] the perpendicular bisector of the intersecting lines.

[B] a line parallel to the intersecting lines.

[C] a circle.

[D] the angle bisector between the intersecting lines

17. The diagram in Figure 39:30 that shows the construction of a perpendicular bisector is:

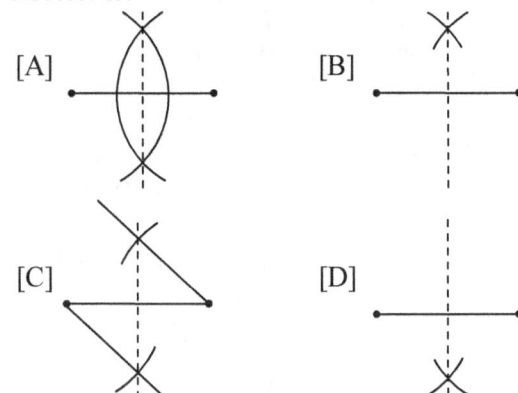

Figure 39:30

18. In Figure 39:31, the correct construction of the bisector of the angle *BAC* is:

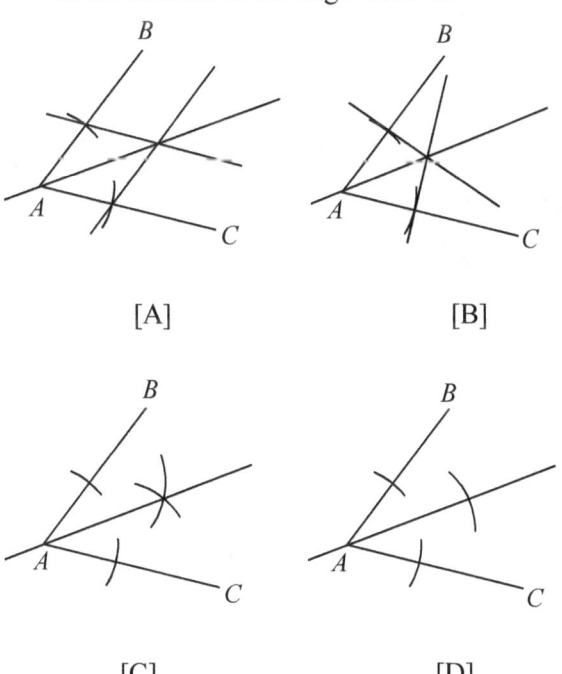

Figure 39:31

19. Another name for perpendicular bisector is:

[A] Altitude

[B] Midpoint

[C] Angle bisector

[D] Mediator

20. In Figure 39:32, *AM* is said to be:

[A] The median

[B] The mediator

[C] The altitude

[D] The angle bisector

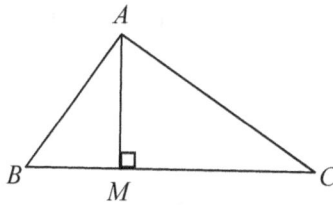

Figure 39:32

21. In Figure 39:33, given that M is the midpoint of *BC*, *AM* is called:

[A] The median

[B] The mediator

[C] The altitude

[D] The angle bisector

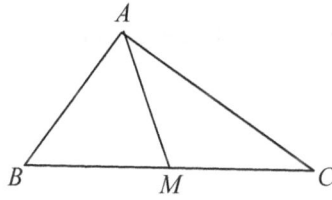

Figure 39:33

22. In Figure 39:34, given that, *M* is the midpoint of *BC*, *NM* is called:

[A] The median

[B] The mediator

[C] The altitude

[D] The angle bisector

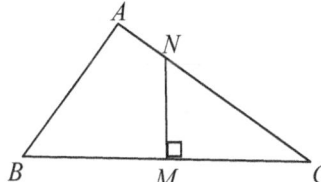

Figure 39:34

23. Figure 39:35 shows the construction of:

[A] A congruent segment

[B] A congruent angle

[C] A perpendicular bisector

[D] An angle bisector

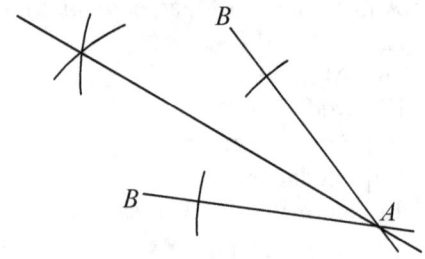

Figure 39:35

24. Figure 39:36 shows the construction of:
    [A] A mediator
    [B] A perpendicular to *AB*
    [C] An angle bisector
    [D] Intersecting lines

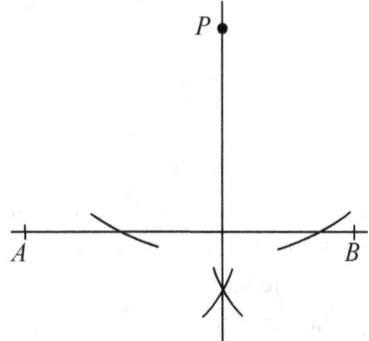

Figure 39:36

25. In Figure 39:37, the correct construction of a perpendicular bisector, which is equal in length to the line segment [XY], is:

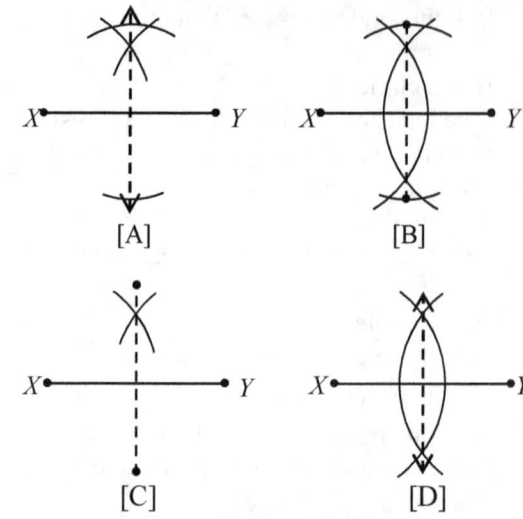

Figure 39:37

26. Given that *YN* is the bisector of ∠*XYZ* and $\lambda(\angle XYZ) = 88°$ then, $\lambda(\angle XYN)$ is equal to:
    [A] 176° [B] 88° [C] 44° [D] 22°

27. Given that *YN* is the trisector of ∠*XYZ* and $\lambda(\angle XYZ) = 72°$ then, $\lambda(\angle XYN)$ is equal to:
    [A] 24° [B] 48° [C] 72° [D] 144°

28. The angle bisector of ∠*ABC* is *BD*. If ∠*ABC* = 18°, ∠*ABD* is equal to:
    [A] 30° [B] 36° [C] 18° [D] 9°

TOPIC 40

CIRCLE GEOMETRY

OBJECTIVES

At the end of this topic, the learner should be able to:

1. State and use circle theorems involving the following to solve circle geometry problems.
    *   The symmetrical properties of circles.
    *   Angles in a semi-circle.
    *   Angles at the Centre and circumference.
    *   Angles in the same segment.
    *   Angles in opposite segment.
    *   Angles in a cyclic quadrilateral.
    *   Angle between Tangent and Radius.
    *   Alternate segment theorem.
    *   Intersecting chord theorem.
2. Perform the Cyclic Quadrilateral Test.

German mathematician and philosopher **David Hilbert** transformed the study of geometry. He also began an effort to establish a consistent basis for all of mathematics, but later mathematicians showed that this was impossible.

Science Source/Photo Researchers, Inc.

Circle Theorems

Symmetrical Properties of a Circle

(1) The perpendicular bisector of a chord passes through the centre (Figure 40:1).

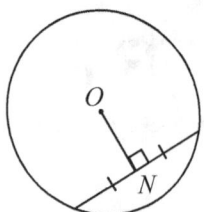

Figure 40:1

(2) Equal chords of a circle are equidistant from the centre (Figure 40:2).

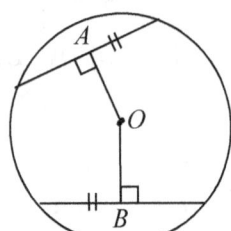

Figure 40:2

Angles in a Semi-circle

(3) The angle in a semi-circle is 90° (Figure 40:3)

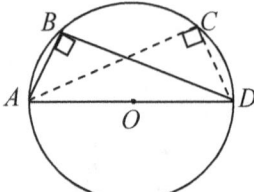

Figure 40:3

Example 40:1

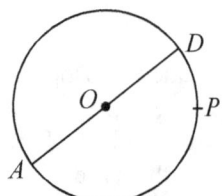

Figure 40:4

In Figure 40:4, *AOD* is a diameter of a circle centre *O* and radius 10 cm; *P* is a variable point on the circumference of the circle. Find the length of *AP* when $\triangle APD$ is isosceles.

Solution

$\triangle APD$ is isosceles $\Rightarrow AP = PD$

$\angle APD = 90°$ [Angle in a semi-circle]

$\angle DAP = ADP = 45°$ [*APD* is an isosceles $\triangle$]

$$AP = AD \sin 45° = 20\left(\frac{\sqrt{2}}{2}\right) = 10\sqrt{2} \text{ cm}$$

Alternatively,

Let the other two sides of the isosceles triangle be *x* cm.

Using the Pythagoras theorem,

$$x^2 + x^2 = 20^2 \Leftrightarrow x = 10\sqrt{2}$$

Angles at Centre and Circumference

(4) The angle subtended by an arc *PQ* at the centre of a circle is twice the angle subtended at the circumference by the same arc (Figure 40:5).

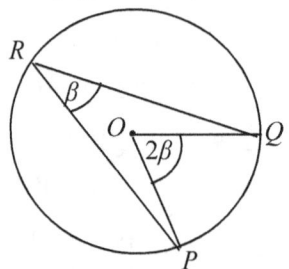

Figure 40:5

Example 40:2

In Figure 40:6 above, *A*, *B* and *C* are points on the circle whose centre is *O* and $O\hat{C}A = 25°$.

Calculate

    (i) Angle *AOC*   (ii) Angle *CBA*

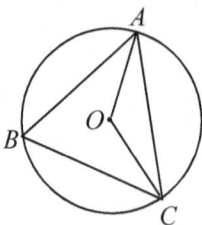

Figure 40:6

Solution

(i) *OA* = *OC* [radii of same circle]

   $\Rightarrow \triangle AOC$ is isosceles

   Hence, $\angle OAC = \angle OCA = 25°$

                 [base $\angle$s of isosceles $\triangle$]

$\angle OAC + \angle OCA + \angle AOC = 180°$

                       [angles in a $\triangle$]

$\angle AOC = 180° - 2(25°) = 130°$

(ii) $\angle CBA = \dfrac{1}{2} \angle AOC$

[∠at centre is twice ∠ at circumference]

$\Rightarrow \angle CBA = \dfrac{1}{2}(130°) = 65°$

Angles in the Same Segment

(5) The angles subtended in the same segment by the same arc are equal. In Figure 40:7 the ∠ *PSQ* and ∠ *PRQ*, are both angles subtended by the arc *PQ*, in the same segment. Hence, they are equal.

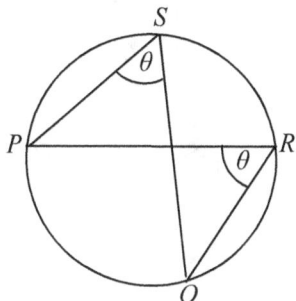

Figure 40:7

Angles in Opposite Segments

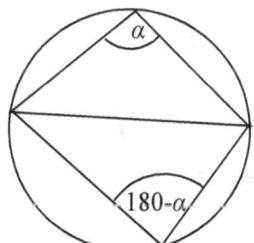

Figure 40:8

(6) The angles subtended by a chord in opposite segments are supplementary (Figure 40:8).

Angles in a Cyclic Quadrilateral

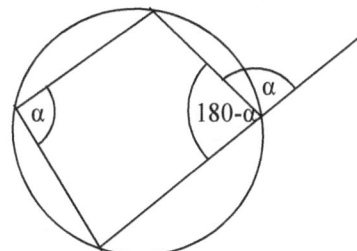

Figure 40:9

(7) Opposite angles of a cyclic quadrilateral are supplementary i.e. their sum is 180°.

(8) The exterior angle of a cyclic quadrilateral is equal to the opposite interior angle.

The Cyclic Quadrilateral Test

To test whether or not a quadrilateral is a cyclic quadrilateral, it suffices to show that its opposite angles are supplementary as in theorem (7).

EXERCISE 40:1

1. Two parallel chords of length 6 cm and 8 cm are drawn on opposite sides of the centre *O* of a circle of radius 5 cm. Calculate the distance between the chords. What would the distance have been if the two parallel chords had been on the same side of the centre?

2. *P* is the point of contact of a tangent *TP* to a circle, centre *R* and radius 8 cm. If *TR* = 17 cm, calculate the area of △*TPR* and the length of the perpendicular from *P* to *TR*.

3. *AB* is a chord of a circle and *C* is the midpoint of the minor arc *AB*. If *AC* = 5 cm and *AB* = 8 cm, calculate the length of the diameter of the circle.

4. In Figure 40:10, the chord *PQ* and *RS* intersect at *X* inside the circle, centre *T*. If ∠*PTR* = 64° and the minor arc *PR* is twice the minor arc *QS*, calculate ∠*QXS*.

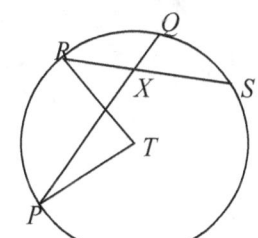

Figure 40:10

5.

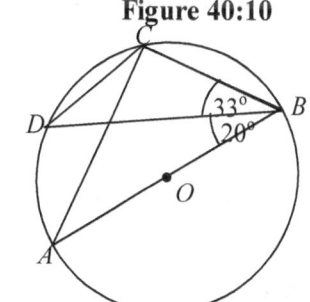

Figure 40:11

445

Figure 40:11 shows a circle with centre O angle $DBC = 33°$ and $\angle DBA = 20°$. Calculate angle CDB.

6. Figure 40:12 shows a cyclic quadrilateral $ABCD$ with $AD = AB$, $BC = DC$ and angle $BCE = 126°$. Determine, in degrees, the values of
    (a) angle ACD (b) angle BAD

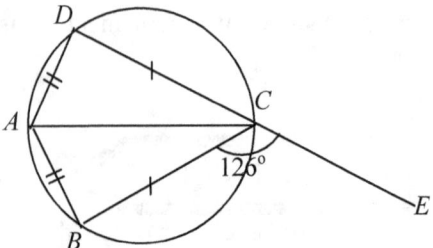

Figure 40:12

7. Find the values of x and y in Figure 40:13, showing your, working clearly.

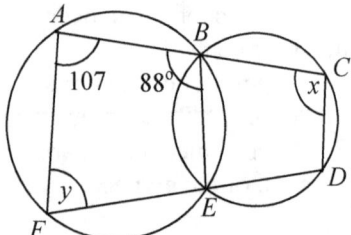

Figure 40:13

8. Find the angles marked x and y in Figure 40:14, where O is the centre of the circle.

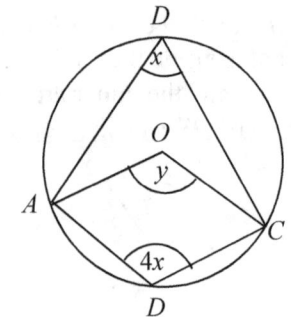

Figure 40:14

9.

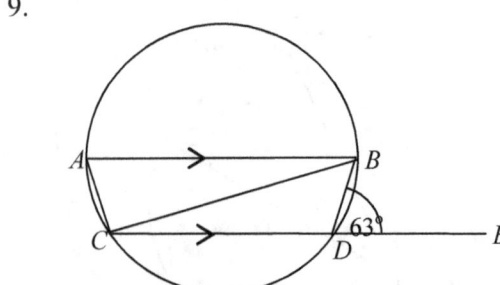

Figure 40:15

In Figure 40:15, $ABCD$ is a cyclic

quadrilateral in which AB is a diameter and DC is parallel to AB. The side CD is produced to E and $\angle BDE = 63°$. Calculate angle DBC.

10. Figure 40:16, (not drawn to scale) shows a circle with lines PS and QT passing through the centre O. Given that $\angle RQT = 30°$ and $\angle RPS = 15°$, find giving reasons for your arguments, the values of
    (a) $\angle PRS$ (b) $\angle RSP$ (c) $\angle RST$
    (d) $\angle TOR$

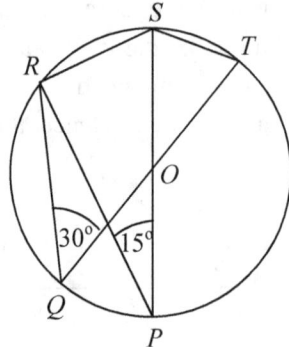

Figure 40:16

Angle between Tangent and Radius

(9) A tangent drawn to a circle is perpendicular to the radius of the circle at the point of contact (Figure 40:17).

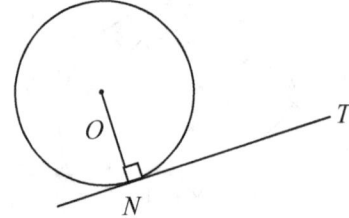

Figure 40:17

(10) Tangents to a circle from the same external point are equal in length (Figure 35:18).

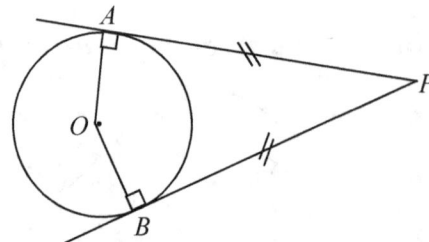

Figure 40:18

Alternate Segment Theorem

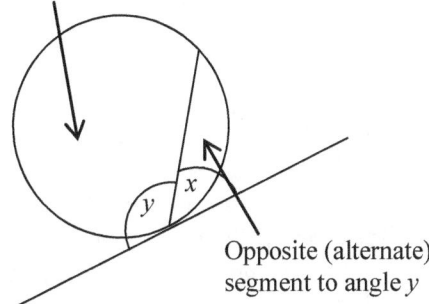

Figure 40:19

(11) The **alternate segment theorem** states that, the angle between a tangent and a chord at the point of contact is equal to the angle in the alternate (opposite) segment (Figure 40:20).

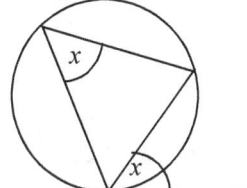

Figure 40:20

Examples 37:3

TA and *TB* are tangents drawn to a circle from a point *T*. *X* is a point on the major arc of the circle. If ∠*AXB* = 63° calculate ∠*ATB*.

Solution

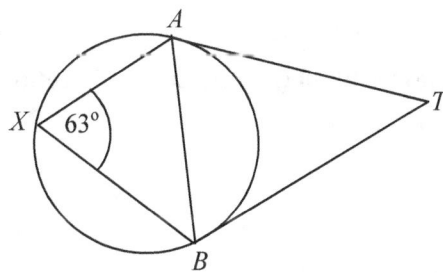

Figure 40:21

$AT = TB$ (tangents from same external point)

$\angle TAB = \angle TBA$ (ΔABT is isosceles)

Also $\angle TAB = \angle AXB = 63°$ ($\angle$ in alt segment)

$\angle TAB + \angle TBA + \angle ATB = 180°$ ($\angle$s in a Δ)

$\Rightarrow 63° + 63° + \angle ATB = 180°$

$\Rightarrow \angle ATB = 180° - 126°$

$= 54°$

1. In Figure 40:22, *ATM* is a tangent to the circle at *T*. The secant *ABC* cuts the circle at *B* and *C*. The diameter through *B* cuts the circle again at *D*. If ∠*DBC* = ∠*ATB* = 40°, calculate the measures of angles *BAT*, *CBT*, and *DTM* and hence say why:
   (i) The line *AC* is parallel to *TD*.
   (ii) The line *CT* is a diameter.

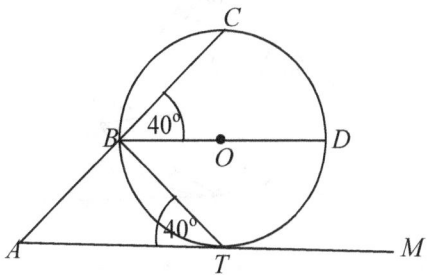

Figure 40:22

2.

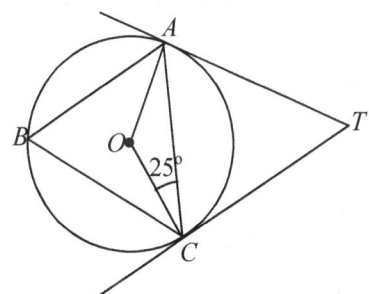

Figure 40:23

In Figure 40:23, *TA* and *TC* are tangents to a circle whose centre is *O*. *A*, *B* and *C* are points on the circle and angle OCA =25°. Given that the radius of the circle is 7 cm, calculate
   (i) ∠*OAC*   (ii) ∠*CBA*   (iii) ∠*ACT*
   (iv) the length of *AC* (v) the length of *OT*
   (vi) the area of *OATC*.

3. Find, giving reasons, the values of the angles marked *x*, *y* and *z* in Figure 40:24.

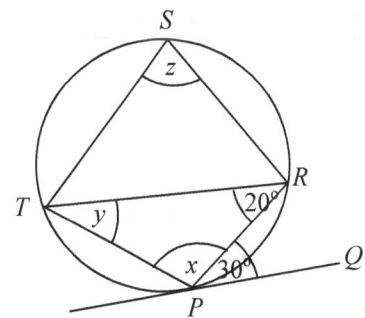

Figure 40:24

4. Figure 40:25, shows triangle *ABC* with
    ∠*CBA* = 50° and ∠*BCA* =70°; inscribed in
    a circle centre *O* of radius 5 cm. *DT* is a
    tangent. Find
    (a) ∠*CAB*   (b) Find ∠*CEA*
    (c) Show that ∠*CAE* = 30°, giving reasons.
    (d) Calculate to one decimal place, the
       lengths of *DE* and *DB*.

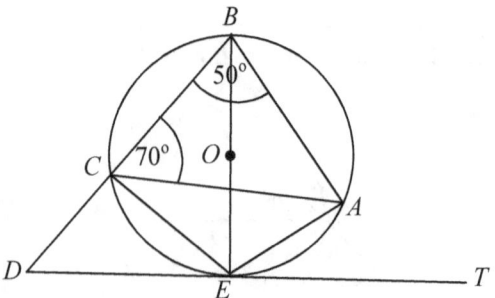

Figure 40:25

5. In Figure 40:26, *ABC* is a tangent to the
    circle. Given that ∠*DBA* = 50°,
    ∠*FBC* = 64° and *ED* = *EF.* Find the value
    of ∠*EDF*.

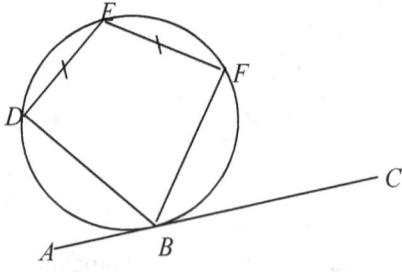

Figure 40:26

6.   In Figure 40:27, the lines *TA* and *TB* are
     tangents to the circle with centre *O*. Given
     that angle *ATB* = 30° find the angles
     denoted by the letters *a*, *b* and *c*.

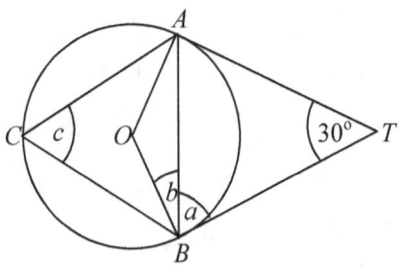

Figure 40:27

7. Figure 40:28, is a circle with centre *O*,
    tangent *PTQ*, and angle *ATQ* = 58°.
    Calculate the values of *x* and *y* where
    x = ∠*TBP* and *y* = ∠*TPB*.

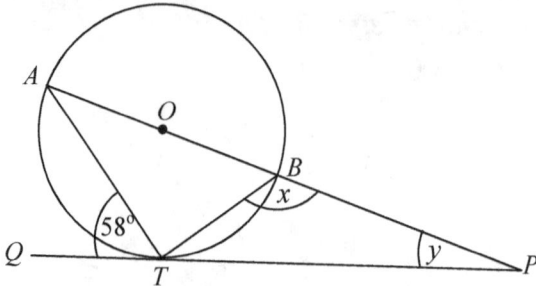

Figure 40:28

8. Given that *PQ* is parallel to *BA*, find the
    values of the angles *x*, *y* and *z* in Figure
    40:29.

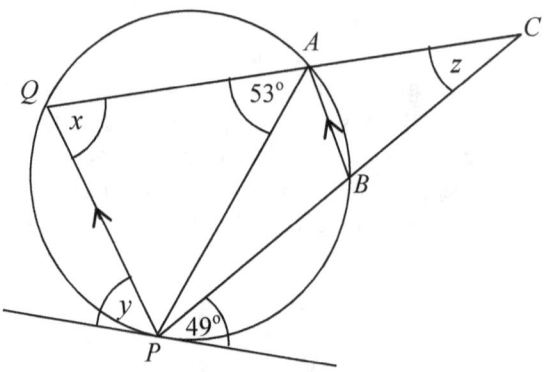

Figure 40:29

Intersecting chord theorem

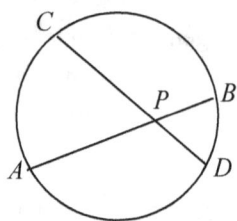

Figure 40:30 (i): Internal intersection

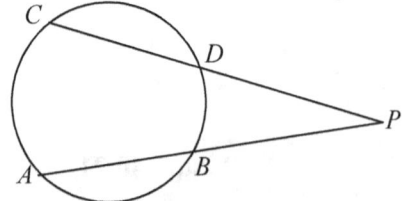

Figure 40:30 (ii): External intersection

In both cases, in Figure 40:30,
$$PA \cdot PB = PC \cdot PD$$

If in (ii) *PC* is a tangent then,
$$PA \cdot PB = PC^2$$

Note that in both cases all distances are
reckoned from *P*.

Example 40:4

Find the value of x in Figure 40:31.

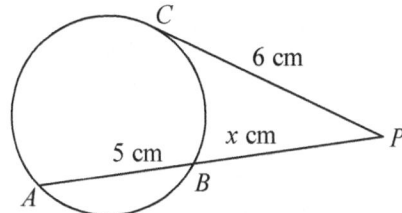

Figure 40:31

Solution

Using the intersecting chord theorem

$$x(x + 5) = 6^2$$

$$\Leftrightarrow x^2 + 5x - 36 = 0$$

$$(x + 9)(x - 4) = 0$$

$$x = -9 \text{ or } x = 4$$

Since $x \geq 0$, $x = 4$ cm.

Examples 37:5

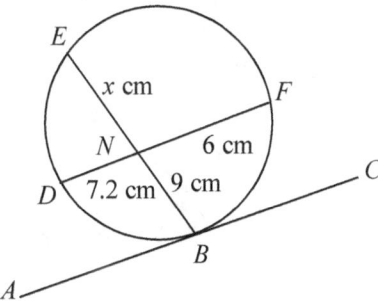

Figure 40:32

In Figure 40:32, ABC is a tangent to the circle.
Given that, EB is perpendicular to AC,
$DN = 7.2$ cm, $NF = 6$ cm, $BN = 9$ cm and
$NE = x$ cm. Find x.

Solution

Using the intersecting chord theorem

$$9x = 6(7.2)$$

$$x = \frac{6(7.2)}{9}$$

$$\Rightarrow x = 4.8 \text{ cm}$$

EXERCISE 40:3

1. Two chords BA and DC, of a circle when produced meet at O. If $AB = 5$ cm, $OA = 4$ cm and $OC = 3$ cm, calculate the length of the chord CD.

2. In Figure 40:33, find
    (a) The length of NC, (b) The angle ABC

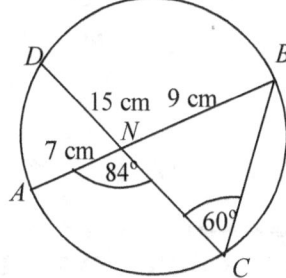

Figure 40:33

3. Find the value of x in Figure 40:34.

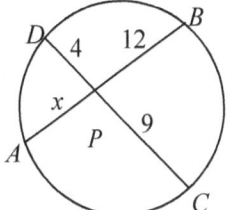

Figure 40:34

4. Given that O is the center of the circle, find the value of y in Figure 40:35.

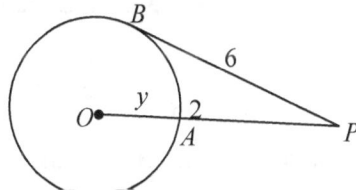

Figure 40:35

5. In Figure 40:36, the chords DC and FE are produced to meet at A, and AB is drawn to touch the circle at B. If $AC = 7$ cm, $CD = 21$ cm and $AE = 8$ cm, calculate the lengths of AB and EF.

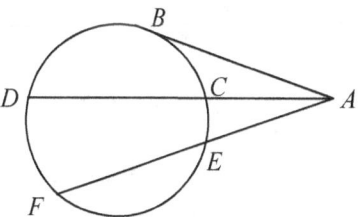

Figure 40:36

6. A chord AB of a circle is of length 48 cm, and 7 cm from the centre. AB is produced to P so that BP is of length 16 cm.

Calculate the radius of the circle and the length of the tangent from P to the circle.

7. In Figure 40:37, calculate the value of x.

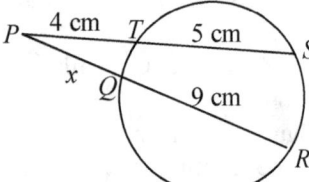

Figure 40:37

<< **MULTIPLE CHOICE EXERCISE 40** >>

1. In Figure 40:38, O is the centre of the circle through points L, M, and N. If $\angle MLN = 74°$ and $\angle MNL = 39°$. The value of $\angle LON$ is:
   [A] 100° [B] 113° [C] 126° [D] 134°

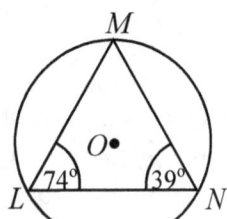

Figure 40:38

2. In Figure 40:39, O is the centre of the circle. If $\angle QRS = 62°$, the value of $\angle SQR$ is:
   [A] 14° [B] 28° [C] 31° [D] 90°

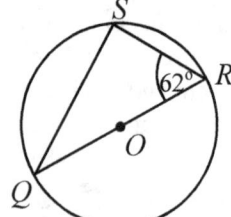

Figure 40:39

3. In Figure 40:40, $PQRS$ is a circle. $\angle SPR = p°$ and $\angle SQR = 2x°$. The value of x in terms of p is:
   [A] $x = 2p$ [B] $x = p - 2$

   [C] $x = p^2$ [D] $x = \dfrac{p}{2}$

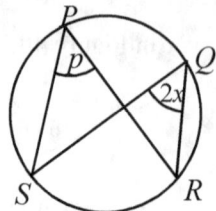

Figure 40:40

4. In Figure 40:41, O is the centre of the circle. If $\angle PAQ = 75°$, the value of $\angle PBQ$ is:
   [A] 51° [B] 75° [C] 105° [D] 150°

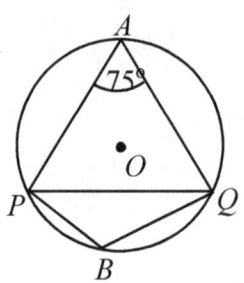

Figure 40:41

5. In Figure 40:42, O is the centre of the circle QRT and PT is the tangent to the circle at T. The angle x is:
   [A] 40°   [B] 35°   [C] 25°   [D] 20°

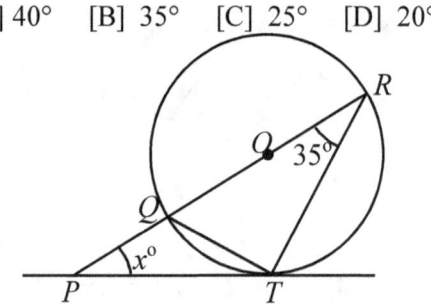

Figure 40:42

6. In Figure 40:43, O is the centre of the circle. Given that $\angle POR = 114°$ the value of $\angle PQR$ is:
   [A] 123°   [B] 118.5°   [C] 117°   [D] 114°

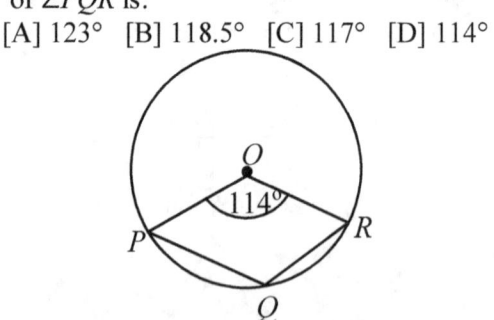

Figure 40:43

7. In Figure 40:44, O is the centre of the circle. If

$\angle POQ = 39°$ and $\angle PRQ = 5x°$ the value of x is:

[A] 4   [B] 8   [C] 16   [D] 20

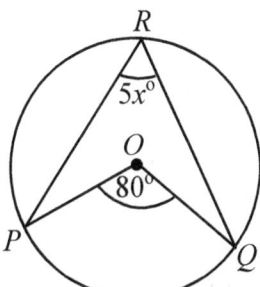

Figure 40:44

8. In Figure 40:45, SQ is the tangent to the circle at P. $XP\|YQ$, $\angle XPY = 56°$ and $\angle PXY = 80°$. The value of angle PQY is:
    [A] 34°   [B] 36°   [C] 44°   [D] 46°

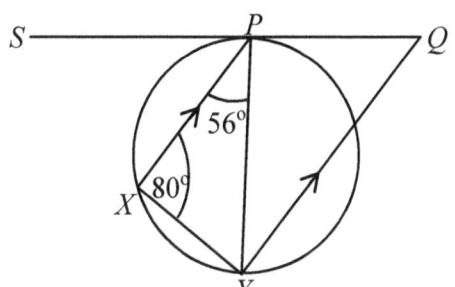

Figure 40:45

9. In Figure 40:46, O is the centre of the circle. It is true to say that:
    [A] $a = b$ [B] $b + c = 100$
    [C] $a + b = c$ [D] $a = b$ and $b + c = 100$

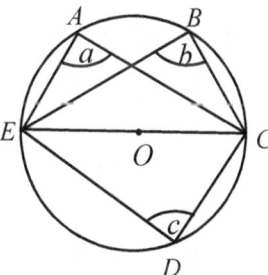

Figure 40:46

10. In Figure 40:47, O is the centre of the circle, $\angle SOR = 64°$ and $\angle PSO = 36°$. The value of $\angle PQR$ is:
    [A] 100°   [B] 96°   [C] 94°   [D] 86°

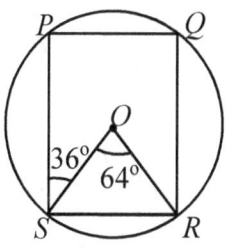

Figure 40:47

11. In Figure 40:48, $ABCD$ is a circle. The value of x is:

    [A] $\dfrac{20}{9}$ [B] $\dfrac{36}{5}$ [C] 3   [D] $\dfrac{45}{4}$

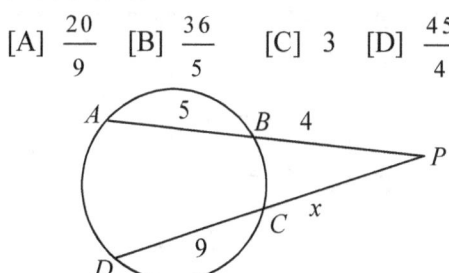

Figure 40:48

12. In Figure 40:49, $ABCD$ is a circle. The value of x is:

    [A] $\dfrac{48}{9}$ [B] 3   [C] 36   [D] $\dfrac{16}{3}$

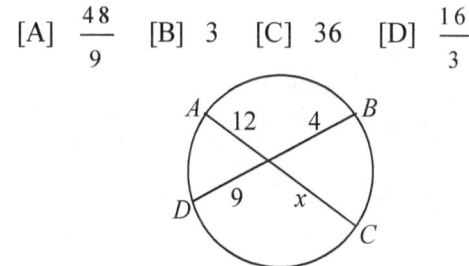

Figure 40:49

13. Given that in Figure 40:50, $PQ = 6$ cm, $TR = 5$ cm and $RQ = 7$ cm. The radius of the circle with centre O is:
    [A] 8 cm   [B] 6 cm   [C] 2 cm   [D] 4 cm

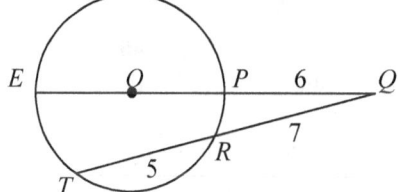

Figure 40:50

14. O is the centre of the circle in Figure 40:51. The size of the angle marked x is:
    [A] 120°   [B] 40°   [C] 50°   [D] 70°

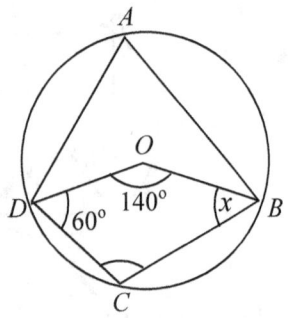

Figure 40:51

15. In Figure 40:52, *O* is the centre of the circle.

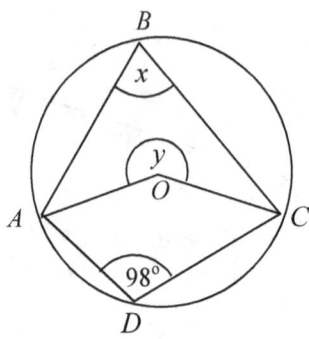

Figure 40:52

The value of *y – x* is:

[A] 164° [B] 114° [C] 66° [D] 16°

16. In Figure 40:53, ∠*RPS* = 32° and ∠*QSR* = 49° .The value of ∠*QOR* is:

[A] 64° [B] 82° [C] 98° [D] 116°

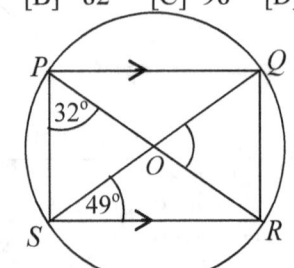

Figure 40:53

17. In Figure 40:54, *O* is the centre of the circle. ∠*BAO* = 30° and ∠*BCO* = 20°. The value of reflex angle *AOC* is:

[A] 330° [B] 300° [C] 270° [D] 260°

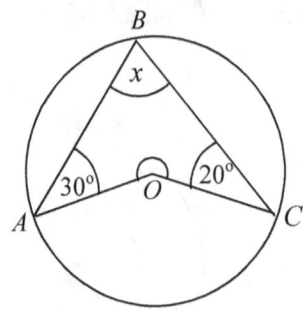

Figure 40:54

18. In Figure 40:55, *X*, *Y* and *Z* are points on a circle centre *O*. *WZ* is a tangent to the circle at the point *Z* and ∠*XYZ* = 22°. The value of *θ* is:

[A] 112° [B] 68° [C] 46° [D] 22°

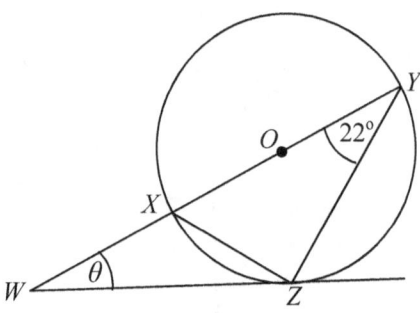

Figure 40:55

19. In Figure 40:56, *O* is the centre, ∠*DOE* = 100° and ∠*CBD* = 70°. The value of ∠*BEO* is:

[A] 20° [B] 30° [C] 40° [D] 60°

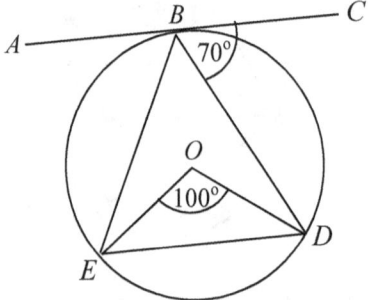

Figure 40:56

TOPIC 41

SOLID GEOMETRY AND TRIGONOMETRY

OBJECTIVES

At the end of this topic, the learner should be able to:

1. Identify and draw west-east lines, north-south lines, vertical lines, parallel lines, perpendicular lines and planes, intersecting planes, non perpendicular planes, line perpendicular to a plane, line of greatest slope.
2. Solve problems involving solid geometry and trigonometry.

Egyptian Pyramids

Construction of the pyramids at Giza, Egypt, required knowledge of mathematics. All Egyptian pyramids were aligned to the cardinal points, so that their sides faced due north, south, east, and west. In addition, each side had to slant inward at the same angle and had to narrow toward the top by the same amount so that the four sides met at a single point at the apex.

From Wikipedia, the free encyclopedia

René Descartes

The first modern philosopher, René Descartes believed science and mathematics could explain and predict events in the physical world. Descartes developed the Cartesian coordinate system for graphing equations and geometric shapes. Modern maps use a grid system that can be traced back to Cartesian graphing techniques.

Hulton Deutsch

Lines and Planes

In drawing three-dimensional diagrams, the following rules may be followed

I. *West-East lines*
West –east are drawn parallel to the top and bottom edges of the page.

$$W \text{————————} E$$

Figure 41:1

II. *North-South lines*
North south lines are drawn inclined to the right at an acute angle to the west east line.

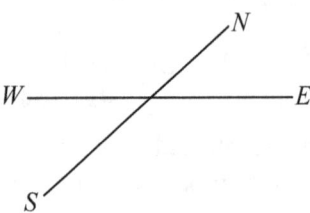

Figure 41:2

III. *Vertical lines*
Vertical lines are drawn parallel to the left and right edges of the page.

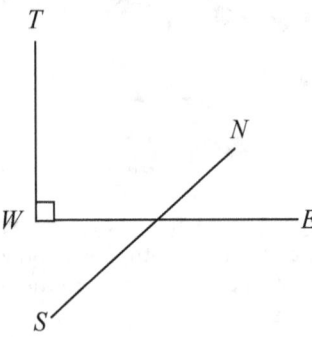

Figure 41:3

IV. *Parallel lines and Planes*

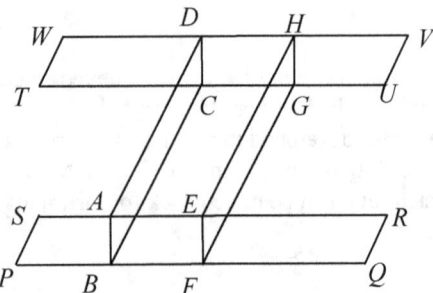

Figure 41:4

Parallel lines and planes are drawn in the required direction in such a way that they cannot meet.

In Figure 41:4, there are four sets of parallel lines as follows:
 (a) *PS, QR, TW* and *UV;*
 (b) *AB, EF, DC* and *HG;*
 (c) *AD, BC, EH* and *FG;*
 (d) *PQ, SR, TU* and *WV.*

There are two sets of parallel planes
(a) *PQRS* and *TUVW*
(b) *ABCD* and *EFGH*

V. *Perpendicular Lines and Planes*

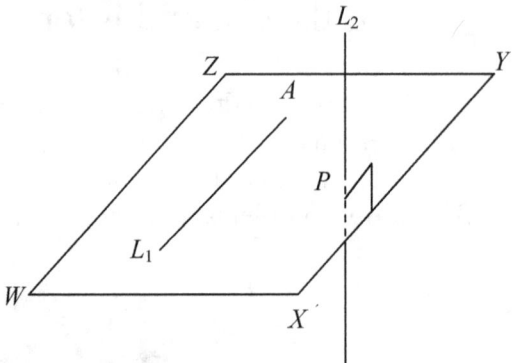

Figure 41:5

In Figure 41:5, the line L_2 is perpendicular to the plane *WXYZ* and passes through *P* in the plane. The line L_1 is in the plane and passes through the point *A*.

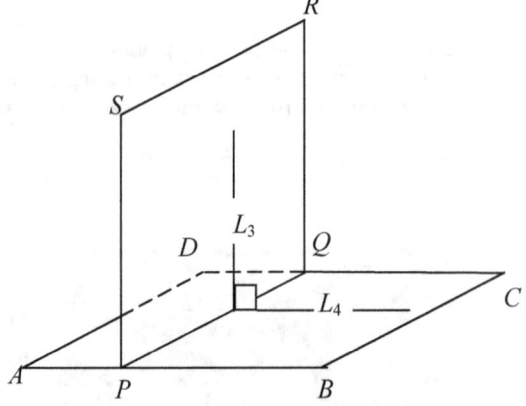

Figure 41:6

In Figure 41:6, the lines L_3 and L_4 lie on the planes *PQRS* and *ABCD* respectively. The planes *PQRS* and *ABCD* are perpendicular. Therefore, the lines L_3 and L_4 are perpendicular.

VI. Intersecting Planes

In Figure 41:7, the planes *I*, *J* and *K* are perpendicular to each other (they are mutually perpendicular). The vectors **i**, **j** and **k**, which lie on the planes *I*, *J* and *K*, are equally perpendicular to each other. The vectors **i** and **j** are perpendicular unit vectors in the plane *K*, while the vectors **j** and **k** are perpendicular unit vectors in the plane *J*, while the vectors **i** and **k** are perpendicular unit vectors in the plane *I*.

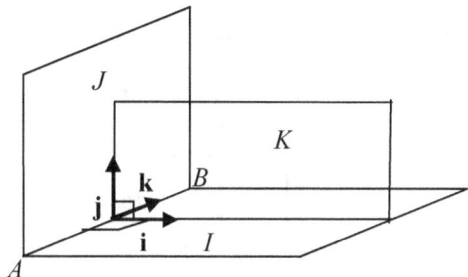

Figure 41:7

The three planes intersect at one and only one point.

VII. Non Perpendicular Planes

Not all planes intersect at right angles. Figure 41:8 shows two planes *ABCD* and *ABEF* which intersect on the line *AB*. Their angle of intersection is *θ*.

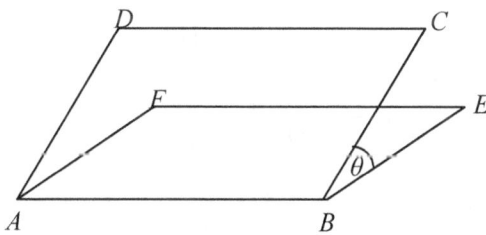

Figure 41:8

VII. Line Perpendicular to a Plane

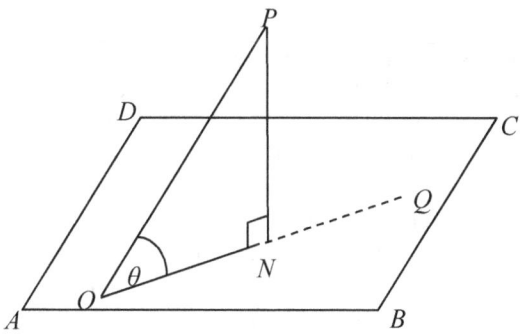

Figure 41:9

Figure 41:9, shows a line *OP* that intersects with the plane *ABCD* making an angle θ with the plane at *O*. The line *PN* is perpendicular to the plane as indicated. Any line such as *PN* that is perpendicular to a plane is necessarily perpendicular to any line on the plane.

IX. Skew lines

Skew lines are lines that are neither parallel nor intersecting. In Figure 41:9, examples of skew lines are *DC* and *OP*, *BC* and *OP*.

IX. Line of Greatest Slope

Take a piece of paper *CDEF* and draw a line *AB* across the center at right angle to the sides *CD* and *EF*. Mark a point *O* near the center of *AB*. Construct a line perpendicular to *AB* and passing through *O*. Label this line *PQ*. Fold the paper along *AB* so that it forms two planes as in Figure 41:10. The angle between the two planes is the angle *θ*. The lines *AD*, *OP*, *BC* or any other line parallel to these lines are called **lines of greatest slope**, because these are the only lines on the plane *ABCD* with maximum gradient.

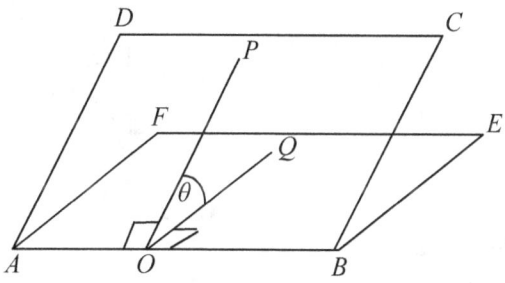

Figure 41:10

General Procedure for Solving Problems Involving Solid Geometry

The general procedure for solving problems involving solid geometry is to consider the triangles in different planes in turn. First draw a perspective figure with all the dimensions clearly marked. At each stage of a calculation, it is often helpful to draw the triangle, rectangle, or any other plane figure under consideration.

Figure 41:11, shows a cuboids *PQRSWXYZ* with the plane *PQYZ* highlighted. Study it very carefully.

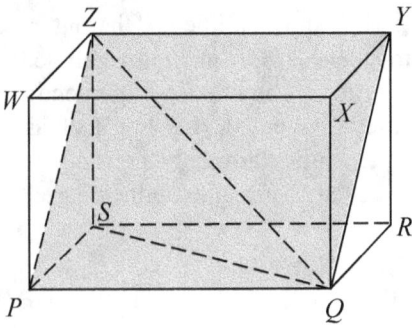

Figure 41:11

Examples 41:1
Figure 41:12 shows a cuboids. Given that
$AB = 8$ cm, $BC = 6$ cm and $CG = 4$ cm,
calculate:

(a) AH (b) BD

(c) $\angle DBH$ (d) $\angle CBG$

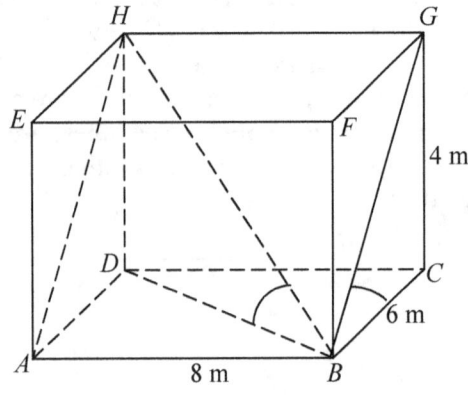

Figure 41:12

Solution
(a)

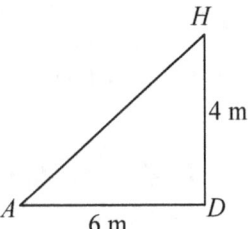

Figure 41:13

$$AH = \sqrt{AD^2 + DH^2}$$
$$= \sqrt{6^2 + 4^2}$$
$$= 7.21 \text{ cm}$$

(b)

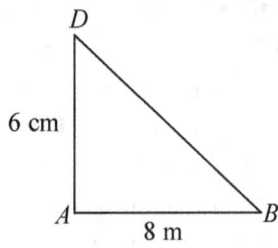

Figure 41:14

$$BD = \sqrt{AB^2 + AD^2}$$
$$= \sqrt{8^2 + 6^2}$$
$$= 10 \text{ cm}$$

(c)

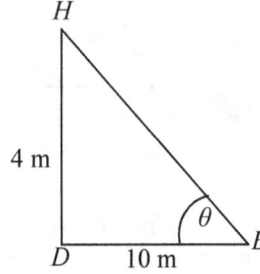

Figure 41:15

$$\theta = \tan^{-1}\left(\frac{4}{10}\right) = 21.8°$$

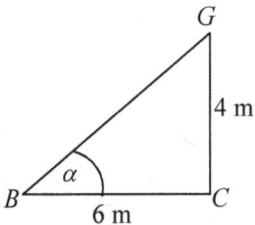

Figure 41:16

$$\alpha = \tan^{-1}\left(\frac{4}{6}\right)$$
$$= 33.7°$$

EXERCISE 41

1. In Figure 41:17, *ABCDEFGH* is a cube.
   Name the angle between:
   (a) The line *EB* and the plane *ABCD*.
   (b) The line *BF* and the plane *ABCD*.
   (c) The line *CH* and the plane *ADHE*.
   (d) The plane *ABGH* and the plane *ABCD*.
   (e) The plane *EHCB* and the plane *ABCD*.
   (f) The plane *ABGH* and the plane *GHDC*.

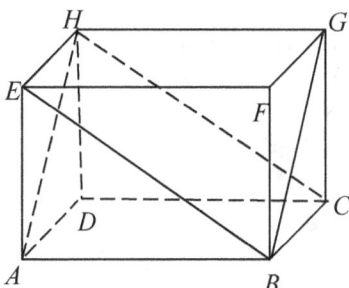

Figure 41:17

2. The flour of a room is a square of side 5 metres. The walls are vertical and of height 3 metres. A fly starting from X, one corner of the floor, walks along two adjacent walls and reaches the ceiling at a corner opposite to X. The fly walks on a path always inclined at $p°$ to the horizontal. Find
   (i) The value of p
   (ii) The total length of the fly's path.

3. Figure 41:18 is a rectangular box with top $WXYZ$ and base $ABCD$. $AB = 6$ cm, $BC = 8$ cm and $WA = 3$ cm.

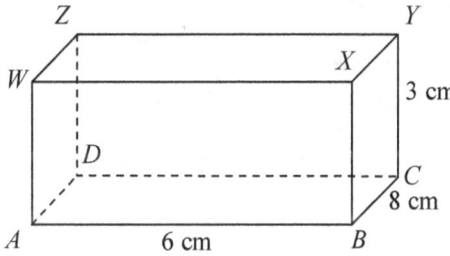

Figure 41:18

Calculate:
   (a) The length of AC
   (b) The angle between WC and AC

4. Figure 41:19 is a cuboids in which $AB = 8$ cm, $BC = 8$ cm and $YC = 5$ cm.

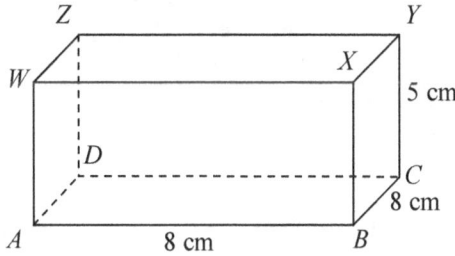

Figure 41:19

Calculate:
   (a) The length of AC
   (b) The length of AY.

   (c) The angle between AY and the plane $ABCD$.
   (d) The angle between the planes $WBCZ$ and $ABCD$.

5. In Figure 41:20, $VABCD$ is a pyramid with a rectangular base $ABCD$ in which $AB = 12$ cm and $BC = 5$ cm. V is vertically above the center of the rectangle and $VA = VB = VC = VD = 10$ cm. Find:
   (a) The angle between VA and the plane $ABCD$.
   (b) The angle between the planes VBC and the plane $ABCD$.

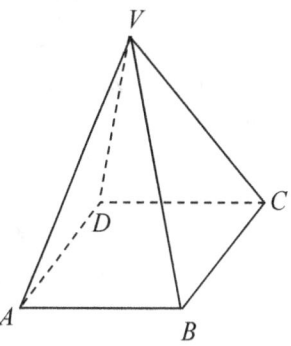

Figure 41:20

6. Figure 41:21 is a square base pyramid with V vertically above the middle of the base, $AB = 10$ cm and $VC = 20$ cm. Find:
   (a) AC (b) the height of the pyramid
   (c) The angle between VC and the base $ABCD$
   (d) The angle AVB (e) the angle AVC.

7. Figure 41:21 shows a triangular pyramid on a triangular base ABC. V is vertically above B where $VB = 10$ cm, $A\hat{B}C = 90°$ and $AB = BC = 15$ cm. Calculate the angle between the planes AVC and ABC.

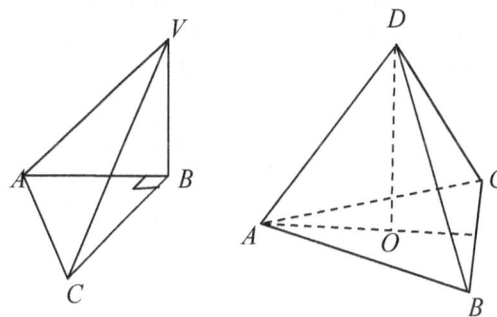

Figure 41:21   **Figure 41:22**

8. In Figure 41:22, $ABCD$ is a regular tetrahedron of side 10 cm. Calculate:
   (a) The height DO of the tetrahedron

(b) The angle between *DA* and the plane *ABC*

(c) The angle between the planes *DBC* and *ABC*.

9. In Figure 41:22, *ABCD* is a tetrahedron with *AB* = *BC* = *CA* = 8 cm and *DA* = *DB* = *DC* = 10 cm. Calculate:
   (a) The height *DO* of the tetrahedron
   (b) The angle between *DC* and the plane *ABC*
   (c) The angle between the planes *DBC* and *ABC*.

10. In Figure 41:22, *ABCD* is a tetrahedron with *AB* = *BC* = *CA* = 6 cm and *DA* = *DB* = *DC* = 9 cm. Calculate:
    (a) The height *DO* of the tetrahedron
    (b) The angle between *DA* and the plane *ABC*
    (c) The angle between the planes *DAB* and *ABC*.

11. In Figure 41:23, $B\hat{A}D = D\hat{C}A = 90°$, $C\hat{A}D = 32.4°$, $B\hat{D}A = 41°$ and $AD = 100$ cm .

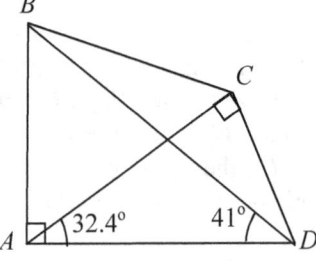

Figure 41:23

Calculate:
   (a) The length of *AB*
   (b) The length of *DC*
   (c) The length of *BD*

12. Fred the mathematical fly crawls from *X* to *Y* by the shortest route on the surface of the cuboids in Figure 41:24.

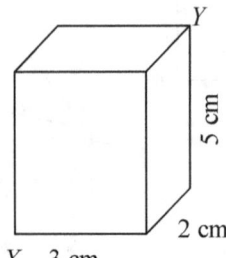

Figure 41:24

(a) Calculate the length of his journey.
(b) Sketch the cuboids marking on it his route.

13. A triangular pyramid *ABCD* (Figure 41:25) is cut off from the corner of a cube.

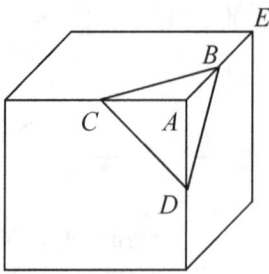

Figure 41:25

Given that $AB = AC = AD = 3$ cm , $AE = 5$ cm. Find the volume of the pyramid. Draw a net for the pyramid and one for the remaining solid.

⬡ MULTIPLE CHOICE EXERCISE 41 ⬡

1. In the cube in Figure 41:26, Δ*HDB* is:
   [A] Isosceles but not equilateral
   [B] Right-angled and scalene
   [C] Equilateral
   [D] Scalene but not right-angled

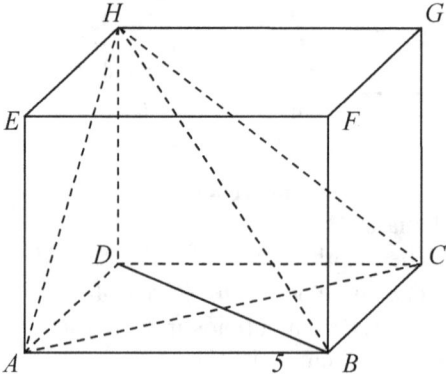

Figure 41:26

2. In the cube in Figure 41:26, Δ*HAC* is:
   [A] Isosceles but not equilateral
   [B] right-angled
   [C] Equilateral
   [D] scalene

3. In the cube in Figure 41:26, angle *HAC* is equal to:
   [A] 30°   [B] 45°   [C] 90°   [D] 60°

4. In the cube in Figure 41:26, The segment, which intersects with *DC*, is:
   [A] *EF*   [B] *GF*   [C] *BC*   [D] *AB*

5. In the cube in Figure 41:26, *EF* and *AE*:
   [A] have point *C* in common
   [B] are parallel
   [C] will intersect
   [D] will never meet
6. In Figure 41:26, the number of sets of parallel lines is:
   [A] 12   [B] 4   [C] 3   [D] 2
7. In Figure 41:27, the number of planes is:
   [A] 12   [B] 4   [C] 3   [D] 2

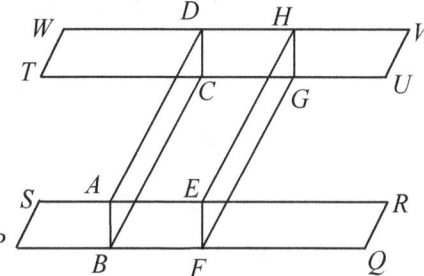

Figure 41:27

8. A fly crawls from *X* to *Y* by the shortest route on the surface of the cuboids in Figure 41:28. The length of this journey is:

   [A] $2\sqrt{34}$ cm   [B] $10\sqrt{2}$ cm

   [C] 20 cm   [D] $\sqrt{58}$ cm

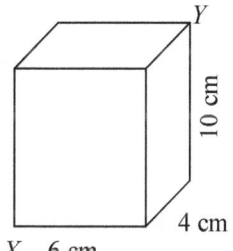

Figure 41:28

9. Figure 41:29 shows a pyramid on a square base. All the eight edges of the pyramid are equal. The value in degrees, of angle *OAC* is:
   [A] 30°   [B] 75°   [C] 60°   [D] 45°

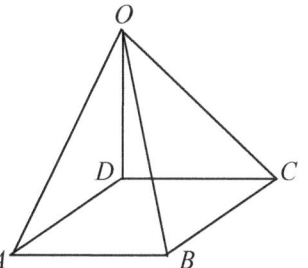

Figure 41:29

10. Figure 41:29, a line segment parallel to *DC* is:
    [A] *AB*   [B] *CO*   [C] *CB*   [D] *BO*
11. Figure 41:29, line segments that intersect *AD* are:
    [A] *AD, DO, CB, CO*
    [B] *DC, DO, AB, AO*
    [C] *BO, AO, DO, CD*
    [D] *BO, BC, DO, CD*
12. In Figure 41:29, a line segments skew to *DC* are:
    [A] *CO, DO*   [B] *AD, BC*
    [C] *BO, CO*   [D] *BO, AO*
13. In Figure 41:30, *PN* is a vertical pole standing on a horizontal plane *ABCD*. *XY* is a line on the plane. Therefore, the pole *PN* is:
    [A] Parallel to the line *XY*.
    [B] Perpendicular to the line *XY*.
    [C] Identical to the line *XY*.
    [D] Is in the same plane with the line *XY*.

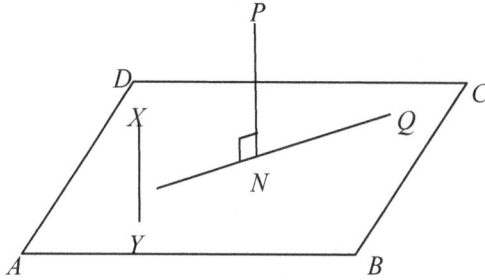

Figure 41:30

14. In Figure 41:31, *ABCD* is the square face of a desk, which slopes at 30° to the horizontal *ABEF*. The angle, which a diagonal of the square makes with the horizontal, is:
    [A] 30°   [B] 10.2°   [C] 20.7°   [D] 60°

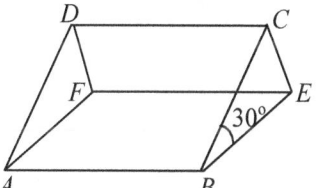

Figure 41:31

15. In Figure 41:31, *ABCD* is the square face of a desk, which slopes at 30° to the horizontal *ABEF*. Given that the side of the square is 70 cm and that DF and CE are upright , the height of *CE* is:

459

[A] 35 cm   [B] 17.5 cm
[C] 8.8 cm   [D] 52.5 cm

16. In Figure 41:32, *AB* = 20 cm, *BC* = 16 cm and *AE* = 12 cm. The length of the diagonal *AG* in cm is:

[A] $4\sqrt{41}$ [B]   20

[C] $20\sqrt{2}$ [D] $4\sqrt{34}$

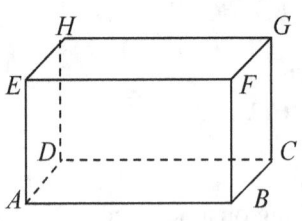

Figure 41:32

17. In Figure 41:32, *AB* = 20 cm, *BC* = 16 cm and *AE* = 12 cm. The angle, which AB makes with, *BE* is:
[A] 31°   [B] 53°   [C] 37°   [D] 59°

18. In Figure 41:32, *AB* = 20 cm, *BC* = 16 cm and *AE* = 12 cm. The angle, which the plane *ADGF* makes with the plane ADHE, is:
[A] 31°   [B] 53°   [C] 37°   [D] 59°

19. Figure 41:33 is a pyramid with a square base *ABCD* of side 5 cm.
OA = *OB* = *OC* = *OD* = 8 cm. The height of the pyramid to the nearest whole number is:
[A] 7 cm   [B] 13 cm   [C] 5 cm   [D] 8 cm

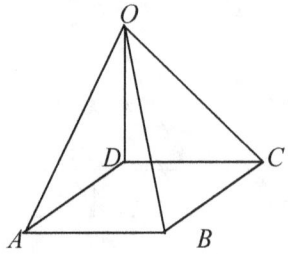

Figure 41:33

TOPIC 42

NETWORKS

OBJECTIVES

At the end of this topic, the learner should be able to:

1. Define the terms network, vertex or node, edge, arc or link and region.
2. Identify and count the number of regions, vertices and edges in a given network.
3. Use the relation $R + V - E = 2$ to find an unknown parameter, given 2 parameters.
4. List the set of vertices V and edges E given a diagrammatic network graph.
5. Distinguish between an ordered list and an unordered list.
6. Name and distinguish between the various types of network graphs.
7. Determine whether or not a given network is traversable.
8. Interpret simple networks in real life situations.

Leonhard Euler

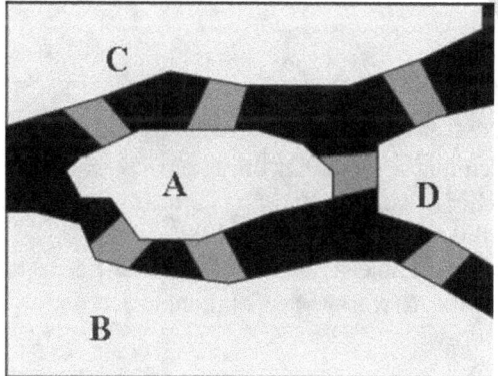

The Seven Bridges of Königsberg

Although hindered by loss of sight, Leonhard Euler was an important contributor to both pure and applied mathematics. Euler is best known for his analytical treatment of mathematics and his discussion of calculus concepts, but he is also credited for work in acoustics, mechanics, astronomy, and optics.

In the 18th century, his solution to the question concerning the seven bridges in the town of Königsberg, Germany marked the beginning of network theory. The question was "Is it possible to take a walk and cross each bridge only once?

Culver Pictures

461

Network Terminology

Consider the diagram in Figure 42:1. This diagram shows a set of objects or points 1, 2, 3, 4, 5, and 6, connected by lines. Such a diagram is called a network. Thus, a **network** is a collection of points, called **vertices** or **nodes** and lines, called **edges** or **arcs** or **links**, connecting these points. The area bounded by the vertices and edges is called a **region**.

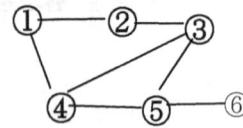

Figure 42:1

The network in Figure 42:1 has 6 vertices, 7 edges (arcs) and 3 regions. Notice that the area outside the network diagram is counted as one region.

Example 42:1

Find the number of vertices, regions, and edges in the network in Figure 42:2.

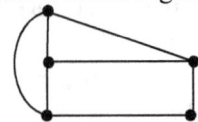

Figure 42:2

Solution

The network has 5 vertices, 4 regions, and 7 edges.

Euler's Formula

Investigative Exercise 42:1

Given that R = number of regions, V = number of vertices and E = number of edges; use the networks in Figure 42:3(a) to (f) to complete Table 42:1. (a) has already been filled for network (a). What conclusion do you draw?

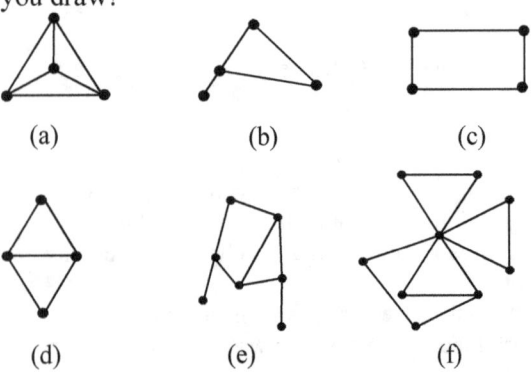

(a)   (b)   (c)

(d)   (e)   (f)

Figure 42:3

Network	R	V	E	R + V– E
(a)	4	4	6	2
(b)				
(c)				
(d)				
(e)				
(f)				

Table 42:1

In any network, the relation $R + V–E = 2$ is always true.

This relation is known as **Euler's formula** and is named after the Swiss mathematician Leonhard Euler who was the spark to network theory.

Ordered Lists and Unordered Lists

Consider a set X containing two elements a and b. This set is written using curly braces. Thus, $X = \{a, b\}$. The order in which the elements in any set are written is immaterial. Therefore, $X = \{a, b\} = \{b, a\}$. Since the two elements in a set containing two elements can be written in any order, a set of two elements is therefore an **unordered pair**.

It follows that, two sets $\{a, b\}$ and $\{c, d\}$ are equal if and only if either $a = c$ and $b = d$ or $a = d$ and $b = c$.

On the other hand, an **ordered pair** is a collection of two objects such that one can be distinguished as the *first element* and the other as the *second element*. Ordered pairs are written using parentheses instead of curly braces, which are used for sets. Thus, the ordered pair (a, b) consists of the first element a and the second element b.

The fundamental property of ordered pairs is that two ordered pairs are equal if and only if their *first elements* are equal and their *second elements* are equal. This means that, two ordered pairs (a, b) and (c, d) are equal if and only if $a = c$ and $b = d$.

With an intuitive extension of the above notation, ordered lists can also be written for any finite number of elements. Thus,

- (1, 2, 3) is an **ordered list** and is not the same as (3, 2, 1) whereas the set {1, 2, 3} is an **unordered list** and is the same as {3, 2, 1}.
- {(1, 2), (3, 4)} is a set of two elements, each of which is an ordered pair. One of the elements of the set is (1, 2) and the other is (3, 4).
- ({1, 2}, {3}, {2, 4}) is an ordered list of three sets. The first element of the list is the set {1, 2}; the second element is the set {3}; the third element is the set {2, 4}.

Odd and Even Vertices

If the number of arcs meeting at a vertex is *even*, the vertex is called an **even vertex** otherwise it is an **odd vertex**.

Example 42: 2

In Figure 42:4, list the set:
(a) *E* of even vertices. (b) *O* of odd vertices.

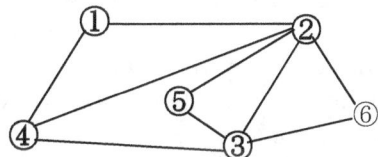

Figure 42:4

Solution
(a) *E* = {1, 3, 5, 6} (b) *O* = {2, 4}.

EXERCISE 42:1

In Figure 42:5 (a) to (g) list the set:
(a) *E* of even vertices. (b) *O* of odd vertices.

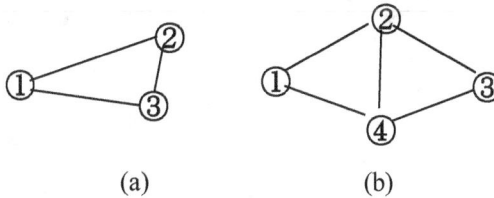

(a) (b)

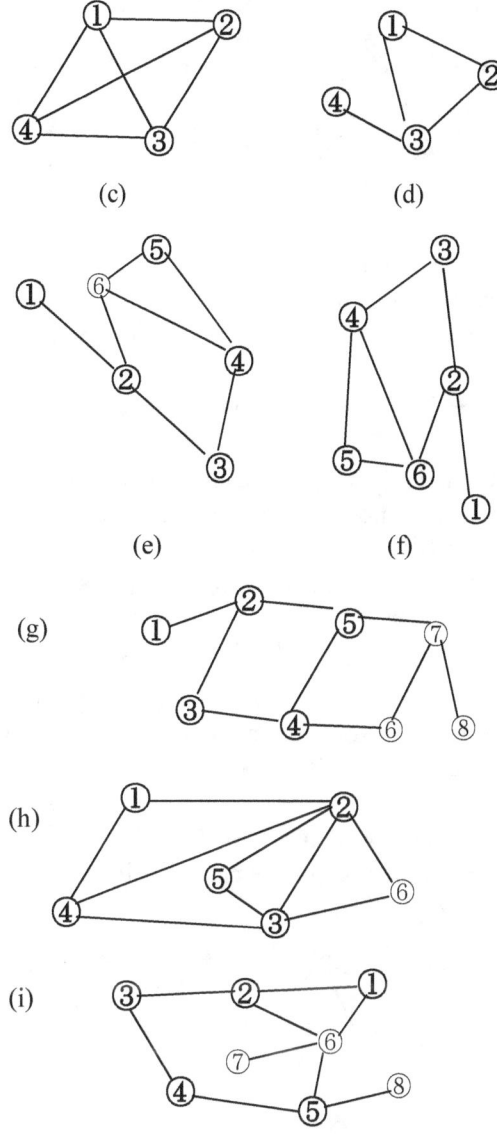

(c) (d)

(e) (f)

(g)

(h)

(i)

Figure 42:5

Traversable Networks

Figure 42:6 shows the path which participants in a certain race must follow. The rules of the race are that each participant has to run through all the paths shown in figure 6 without passing more than once through any path. What are the possibilities that this can be done?

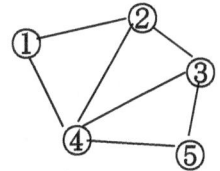

Figure 42:6

463

Figure 42:7 shows the possibilities.

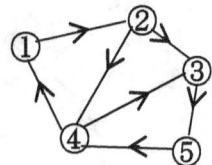 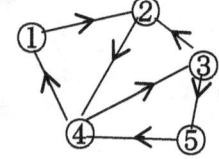

Figure 42:7

One can begin at 2 and end at 3. Two of the possible paths in this case are shown below.

$2 \to 4 \to 1 \to 2 \to 3 \to 5 \to 4 \to 3$ or (2, 4, 1, 2, 3, 5, 4, 3)

$2 \to 3 \to 5 \to 4 \to 1 \to 2 \to 4 \to 3$ or (2, 3, 5, 4, 1, 2, 4, 3)

Alternatively, one can begin at 3 and end at 2. Two of the possible paths in this case are shown below.

$3 \to 2 \to 4 \to 3 \to 5 \to 4 \to 1 \to 2$ or (3, 2, 4, 3, 5, 4, 1, 2)

$3 \to 5 \to 4 \to 3 \to 2 \to 4 \to 1 \to 2$ or (3, 5, 4, 3, 2, 4, 1, 2)

A network, which can be traced exactly once beginning at some point without retracing any arc, is said to be **traversable**.

Investigative Exercise 42:2

1. Examine each of the networks in Figure 42:8.

(a) 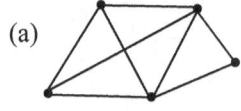   (b)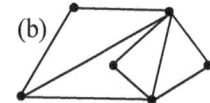

(c) 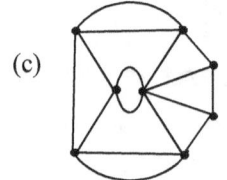   (d)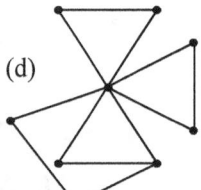

(e) 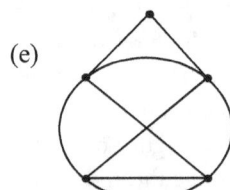   (f)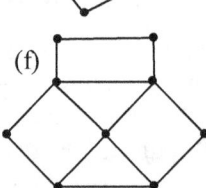

Figure 42:8

(i) How many odd vertices does each network have?

(ii) Is each network traversable?

(iii) By tracing each of these networks on a separate paper, deduce a conjecture about the beginning and ending points

of traversable networks with exactly the same number of odd vertices.

2. (i) Determine which of the networks in Figure 42:5 are traversable?

(ii) For each traversable network, write down two different sequences of nodes from the starting vertex to the ending vertex to show how the network can be traversed.

3. State the number of odd vertices in each of the traversable networks.

4. What is the relation between a traversable network and the number of odd vertices in the network?

5. Use the networks in Figure 42:5 to verify your conclusion.

Conditions for a Network to be Traversable

Whether a network is traversable depends on the number of odd vertices as follows.

1. A network, which has no odd vertices, is traversable. This means that a network whose vertices are all even is traversable. In this case, any vertex may be the beginning point, and the same vertex will be the ending point.

2. A network, which has exactly two odd vertices, is traversable. If one odd vertex is the beginning point, the other odd vertex is the ending point.

3. A network with more than two odd vertices is not traversable.

Example 42: 3

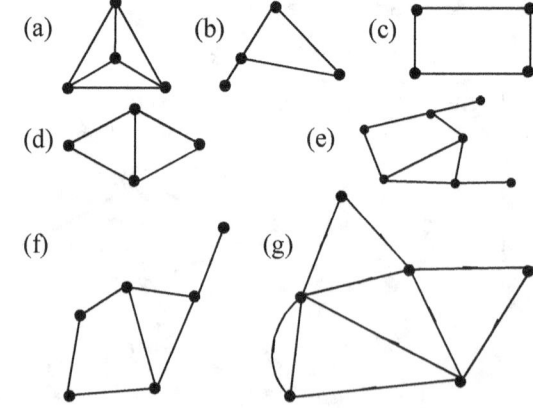

Figure 42:9

In Figure 42:9 (a) to (g), determine giving reasons for your answer which of the networks are traversable. Draw each of the traversable networks on a separate paper and mark the beginning point, the end-point and the direction in which the network can be traversed using arrows.

Solution

(a), (e) and (f) are not traversable because they each have more than 2 odd vertices.

(c) Traversable because all the vertices are even. In other words, it has no odd vertices.

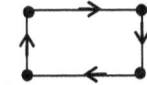

Figure 42:10

Any of the vertices may be the beginning or endpoint.

(b), (d) and (g) are traversable because each has exactly 2 odd vertices.

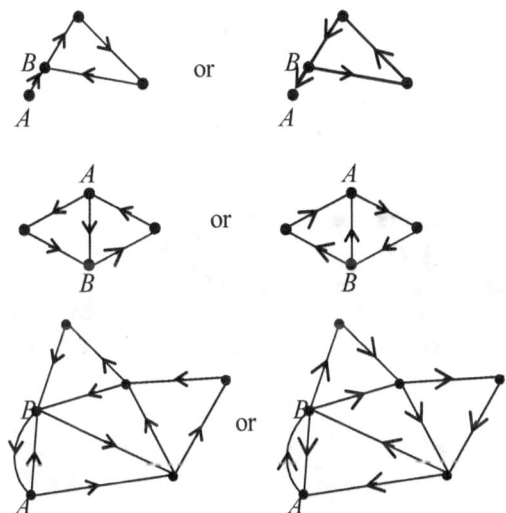

Figure 42:11

In all three cases, either of the odd vertices A or B can be the beginning point, and the other will be the endpoint.

EXERCISE 42:2

Determine which of the networks in Figure 42:12 (a) to (c) is/are traversable.

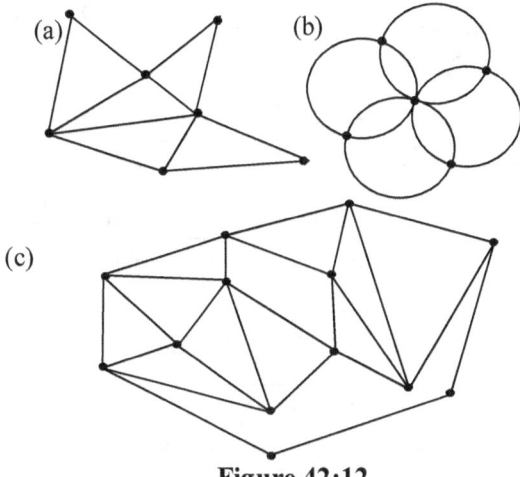

(a)

(b)

(c)

Figure 42:12

Networks in Real Life

The concept of network can be used to describe many different systems (computer networks, technical, physical, biological, sociological etc, etc.).

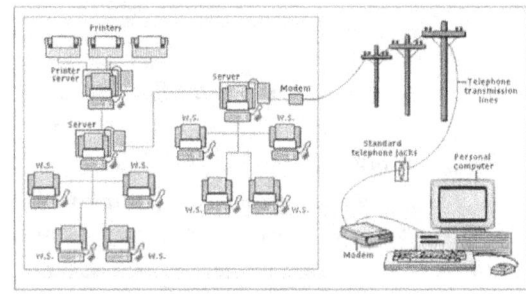

Figure 42:13

Computer Networking

Computer Networks are connections between groups of computers and associated devices that allow users to transfer information electronically. The local area network shown above is representative of the setup used in many offices and companies. Individual computers called workstations (W.S.), communicate to each other via cable or telephone lines linking to servers.

Table 42:1 shows the existence of networks in different disciplines and the terminology used to refer to them, their points and their connection lines.

Discipline	Name used	Point name	Line name
Mathematics	Graph	Vertex (node)	Edge or arc
Computer science	Network	Node	Link
Physics	system	Site	Bond
Chemistry	Bonding	Atom	Bond
Sociology	Social network	Actor (individual)	Tie (friendship)
Communication	World wide web	Webpage Website	Link (d)
	Internet	Network	Connection Bridge
	Road system (road network)	Crossing (junction)	Road

Table 42:2

Table 42:2 is not exhaustive. It illustrates the importance of networks and their applications in real life. Some applications of networks are in planning roads and museums. Museums are often planned in such a way that congestion can be minimized at doorways when so many visitors visit the museum. To realize this, visitors are directed to move through each door only once. This implies that such a museum must be traversable.

Example 42:4

Figure 42:14 (i) and (ii) shows the floor plans of two museums.

1. Draw network diagrams to represent each plan.
2. Use your network diagrams to determine giving reasons for your answer whether it is possible to visit all the rooms passing through each door exactly once.
3. If it is possible, confirm your results by sketching the plan and showing the route using arrows how this can be done. Your diagram should indicate the starting and the finishing points.

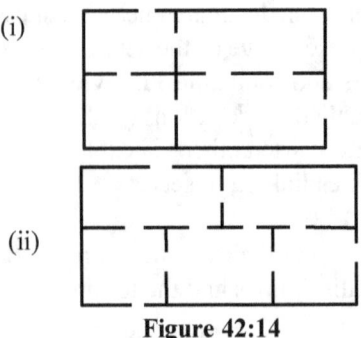

Figure 42:14

Solution

1. Taking each room and the outside region to be nodes and the doors to be edges, the network diagrams can be drawn as in Figure 42:15.

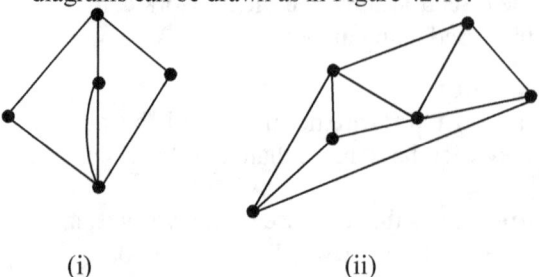

(i)   (ii)

Figure 42:15

2. In (i) it is possible because there are exactly two odd nodes, meaning that the network is traversable. In (ii) it is impossible because there are four odd nodes, implying that the network is not traversable.

3.

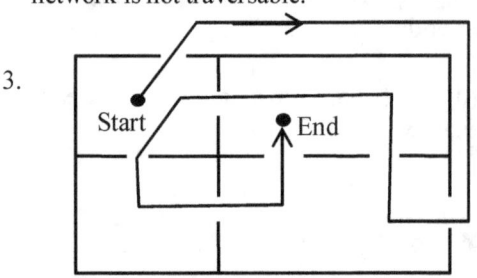

Figure 42:16

EXERCISE 42:3

1. (a) Mark the beginning point and the end point for traversing the network in Figure 42:17.

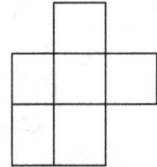

Figure 42:17

(b) Draw shapes showing the minimum number of squares that must be removed from the grids in Figure 42:18, in order for the remaining network to be traversable.

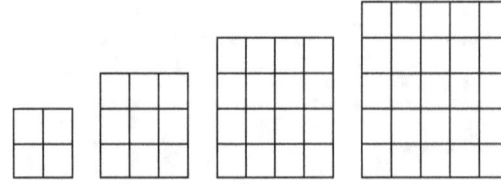

Figure 42:18

(c) Using your answer in (b), find a pattern and predict the minimum number of squares that must be removed from a 12 × 12 grid of squares in order for the remaining network to be traversable.

   (d) Write an algebraic expression for the minimum number of squares that must be removed from an *n* x *n* grid in order for the remaining network to be traversable.

2. Figure 42:19 shows the road network in a certain city. In order to reduce traffic, the city council decides to turn all roads into one way by building one more road linking two of the junctions in the city. Draw five different road network diagrams to show how this can be done.

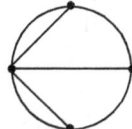

Figure 42:19

3. To organize a trade fair, a council builds a fence with gates as shown in the diagram in Figure 42:20. In order to visit all the stands, a visitor must pass through all the gates. However, to reduce congestion at the gates the council passes an ultimatum that no one passes through any gate more than once.

   (a) Given that a visitor is allowed to begin either from inside or outside the fence, is it possible for a visitor to visit all the stands?

   (b) If so, draw a diagram showing the starting and finishing points. Use arrows to show the direction of movement.

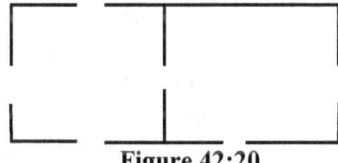

Figure 42:20

   (c) Using the apartments of the fence and the region outside the fence as nodes and the gates as arcs, draw a simple network diagram to represent the plan.

Network Graphs

A network graph is a set of objects connected by lines. The objects in a graph are usually called *nodes* or *vertices* and the lines connecting the objects are called *links* or *edges* or *arcs*.

More formally, a graph is an ordered pair $G = (V, E)$ comprising the set V of vertices (or nodes) together with the set E of edges (or

lines) such that each element of E is a doubleton (or paired) subset of V.

Example 42:5

Given the diagram in Figure 42:21 and the graph $G = (V, E)$. List the elements of V and E.

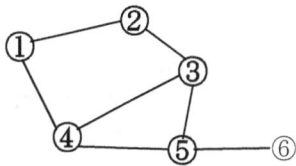

Figure 42:21

Solution

$V = \{1, 2, 3, 4, 5, 6\}$

$E = \{\{1,2\}, \{1,4\}, \{2,3\}, \{3,4\},\}, \{3,5\}, \{4,5\}, \{5,6\}\}$

Example 42:6

A network is defined by $G = (V, E)$ where

$V = \{a, b, c, d, e\}$ and

$E = \{\{a, b\}, \{a, c\}, \{b, c\}, \{b, d\}, \{c, d\}, \{a, e\}\}$.

   (a) Sketch the network diagrammatically.

   (b) State the number of regions, vertices and edges of the network.

Solution

(a)

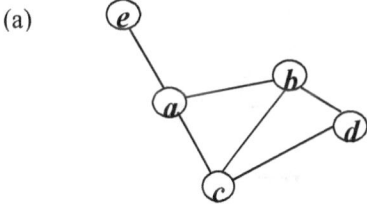

Figure 42:22

   (b) Number of regions = 3,
Number of vertices = 5
Number of edges = 6

Directed and Undirected Edges

An arc $a = (x, y)$ is conventionally considered to be directed from x to y and in its diagrammatic representation, the arrow points towards y.

Figure 42: 23

In this case, x is called the **tail** and y is called the **head**.

Since a path leads from x to y, y is said to be the successor of x and reachable from x and x is said to be the predecessor of y. Emphatically y is the **direct successor** of x and x is the **direct predecessor** of y. The arc (y, x) is called arc (x, y) **inverted.**

Figure 42: 24

An arc $a = \{x, y\} = \{y, x\}$ is conventionally considered to be directed in either ways. Both x and y are heads and tails. Diagrammatically, this is represented as

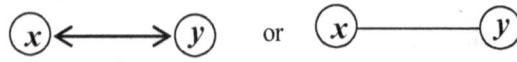

Figure 42: 25

Properties of Network Graphs

(i) **Adjacent or Coincident Edges:** If two edges of a graph share a common vertex, they are called **adjacent** or **coincident edges**. In Figure 42:26, the edges AX and BX are adjacent or coincident because they share the same vertex X.

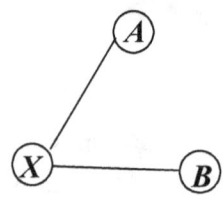

Figure 42:26

(ii) **Consecutive Arrows**

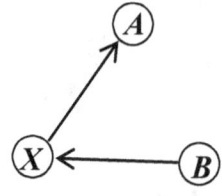

Figure 42:27

If the head of one arrow of a graph is at the nock (notch end) of another arrow of the graph the two arrows are called **consecutive arrows**. In Figure 42:27, the arrows BX and XA are consecutive the head of BX is at the tail of XA.

(iii) **Adjacent Vertices:** If two vertices of a graph share a common edge, they are called **adjacent vertices** and the common edge is said to join the two vertices.

Figure 42:28

(iv) **Consecutive Vertices:** If two adjacent vertices are such that the head of the edge is on one vertex and the tail of the edge is on the other, the vertices are called **consecutive vertices**.

Figure 42:29

(v) **Incident edge and Vertex:** An edge and a vertex on that edge are called **incident edge and vertex.**

Figure 42:30

> **EXERCISE 42:4**

1. In Figure 42:31, list
   (a) 8 pairs of coincident edges.
   (b) 3 pairs of consecutive arrows.
   (c) 7 pairs of adjacent vertices.
   (d) 5 pairs of consecutive vertices.
   (e) 8 pairs of incident edges and vertices.

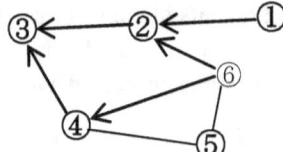

Figure 42:31

2. Write down the graph of the networks in Figure 42:31 symbolically.

3. Draw the network structure for the graph $D = (V, A)$ where $V = \{1, 2, 3, 4, 5, 6\}$ and $A = \{(1, 2), (2,1), (3,1), (3, 2), (3,4), (3, 5), (4, 1), (4,5), (5,6)\}$.

Types of Network Graphs

1. *Directed Graphs*

A **directed graph** sometimes referred to simply as a **digraph** is an ordered pair $D = (V, A)$ such that V is the set of vertices and A is the set of ordered edges. i.e. A is a set of ordered pairs (a, b).

Example 42:6

The graph of a network structure is $D = (V, A)$, where $V = \{1, 2, 3\}$ and $A = \{(1, 3), (2, 1), (3, 1), (3, 2)\}$. Represent this network diagrammatically.

Solution

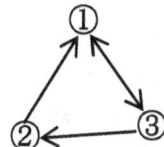

Figure 42:32

Figure 42:33 is an example of a digraph because each edge (line or arc) in a directed graph has at least an arrow at one end.

Example 42:7

Represent the networks in Figure 42:33 symbolically.

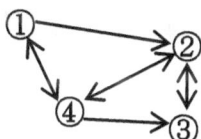

Figure 42:33

Solution

The symbolic representation of the graph is $D = (V, A)$ where $V = \{1, 2, 3, 4, 5\}$ and $A = \{(1, 2), (1,4), (2,3), (2,4), (3,2), (4, 1), (4, 2), (4, 3)\}$.

Example 42:8

Draw the network structure for the graph $D = (V, A)$ where $V = \{1, 2, 3, 4, 5, 6\}$ and $A = \{(1,2),(2,1),(3,1),(3,2),(3,4), (3,5),(4,1), (4,5),(5,6)\}$.

Solution

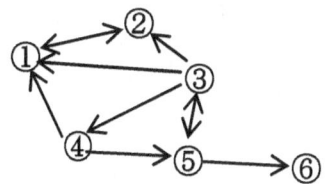

Figure 42:34

2. *Undirected Graphs*

An **undirected graph** is a graph for which the edges have no arrows. For such a graph, the edge connecting a and b is simply represented as $\{a, b\}$ or $\{b, a\}$. In other words an undirected graph is a set of unordered pairs.

Example 42:9

Draw the network structure for the graph $G = (V, E)$ where $V = \{1, 2, 3\}$ and $E = \{\{1, 2\}, \{1, 3\}, \{2, 3\}\}$.

Solution

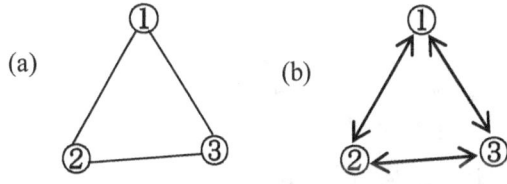

Figure 42:35

Figure 42:35 (a), is an example of an undirected graph.

Notice that though the graphs in Figure 42:35 (a) and (b) are equivalent, it is needless putting the arrows as in (b).

3. *Mixed Graph*

A **mixed graph** is a graph in which some edges are directed and some are undirected. This type of graph is written as an ordered triple $G = (V, E, A)$ where V is the set of vertices, E is the set of ordered edges and A is a set of ordered edges.

Example 42:10

Represent the networks in Figure 42:36 symbolically.

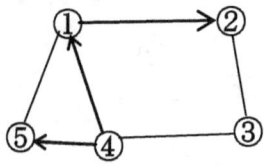

Figure 42:36

Solution

Symbolically, the graph of this network is
$G = (V, E, A)$, where $V = \{1, 2, 3, 4, 5\}$,
$E = \{\{1, 5\}, \{2, 3\}, \{3, 4\}\}$ and
$A = \{(1, 2), (4, 1), (4, 5)\}$.
Figure 42:36 is an example of a mixed graph.
In real life situations, the mixed graph is what
occurs predominantly. The directed and the
undirected graphs are actually very special
cases.

4. *Complete Graph*

A **complete graph** is one for which each pair
of vertices is connected by an edge. This
means that each vertex has an edge to every
other vertex. Figure 42:38 is an Example of a
complete graph.

Figure 42:37

5. *Weighted Graph*

A **weighted graph** is one for which a
number (weight) is assigned to each edge.
Such weights may represent cost, length,
capacity, importance etc. For instance, in
electricity networks, the hospital line, the
administrative lines and the commercial
lines usually carry more weights than
private lines. If there is low voltage and
electricity has to be ration, it is very
unlikely that these lines will be cut off.

Example 42:11

Given that the numbers in the network show
distances in km between towns in a
municipality, find the shortest route from A to
G, using the distances shown on the network
in Figure 42:38

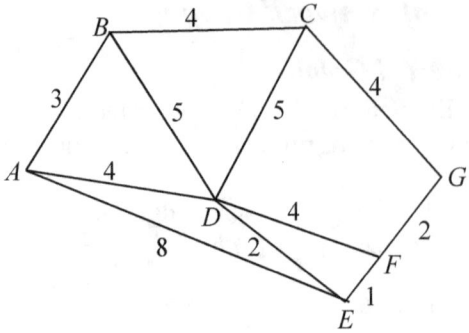

Figure 42:38

Solution

By adding the distances, the shortest distance
is 9 km and is the road $ADEFG$.

6. *Null Graph*

A **null graph** is a graph whose edge set is
empty. This means that all the vertices are
isolated from each other. A null graph on n
vertices is denoted by Nn. Figure 42:39 is
an example of a null graph on 5 vertices.

Figure 42:39

The graph in Figure 42:39 can be
represented symbolically by $G = (V, E)$
where $V = \{1, 2, 3, 4, 5\}$ and $E = \{ \ \}$.

Network Trees

A **network tree** can be defined in any of the
following ways:
(i) A tree is a connected graph with no cycle.
(ii) A tree is a graph in which any two
     vertices are connected by exactly one path.
(iii) A tree is a connected graph in which

$$n(E) = n(V) - 1$$

Figure 42:40 shows three different trees made
from the same set of nodes. Check that each
of these trees satisfies each of the above
definitions.

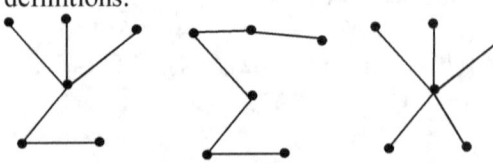

Figure 42:40

Examine the network in Figure 42:41. Why do you think it is not a tree?

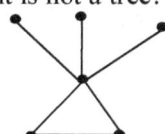

Figure 42:41

A **forest** is a collection of trees. This definition implies that a forest is a disjoint union of one or more trees with no cycles. Figure 42:42 is an example of a forest.

Figure 42:42

EXERCISE 42:5

1. The graph of a network structure is $D = (V, A)$, where $V = \{1, 2, 3, 4\}$ and $A = \{(1, 3), (2, 1), (3, 1), (3, 2), (1,4), (3,4)\}$.
   Represent this network diagrammatically.
2. Represent the networks in Figure 42:43 symbolically.

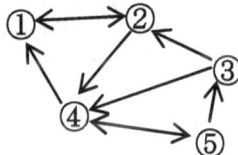

Figure 42:43

3. Draw the network structure for the graph $D = (V, A)$ where $V = \{1, 2, 3, 4\}$ and $A = \{(1,3),(2,1),(3,1),(3,2),(3,4), (4,1),(4,2)\}$.

4. Draw the network structure for the graph $G = (V, E)$ where $V = \{1, 2, 3, 4\}$ and $E = \{\{1,3\},\{2,1\},\{3,2\},\{3,4\}, \{4,1\},\{4,2\}\}$.
5. Represent the networks in Figure 42:44 symbolically.

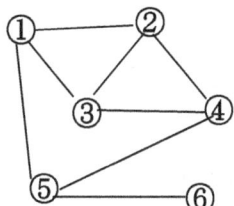

Figure 42:44

6. (a) Given that the numbers in the network in Figure 42:45, show the time in hours taken to travel from one town to another in a certain region, find the shortest time required to travel from town A to G.
   (b) Given that a litre of fuel cost 550 FCFA and that the amount of fuel consumed is proportional to the time taken, determine the least cost of travelling from A to G, if half a litre of fuel is consumed per hour.

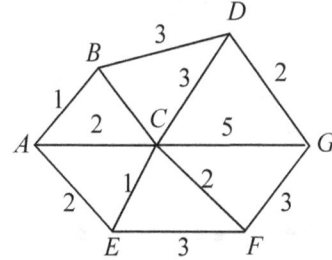

Figure 42:45

7. Draw five different trees with 6 vertices.
8. Determine which of the networks in Figure 42:46 (a) to (f) are trees, forests or neither giving reasons for answer.

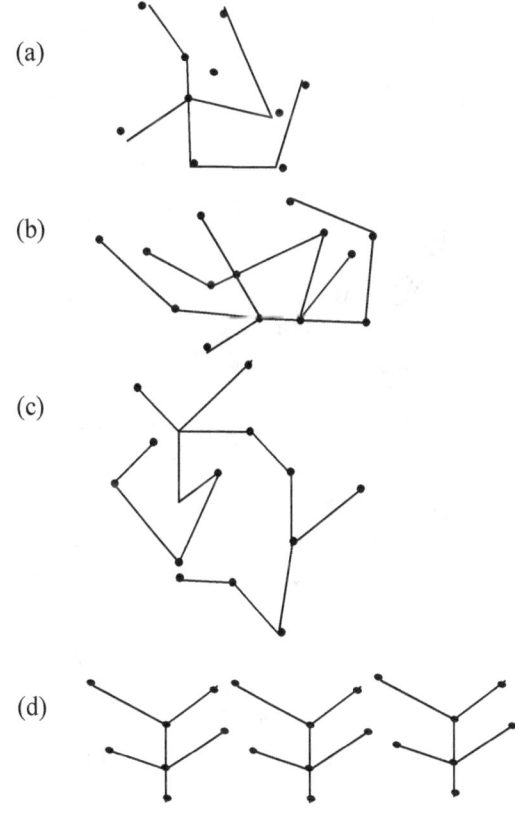

(a)

(b)

(c)

(d)

471

(e)

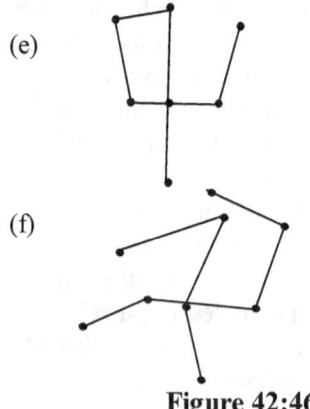

(f)

Figure 42:46

╱▔▔▔▔▔▔▔▔▔▔╲
MULTIPLE CHOICE EXERCISE 42
╲▁▁▁▁▁▁▁▁▁▁╱

1. The network in Figure 42:47 which is traversable is:

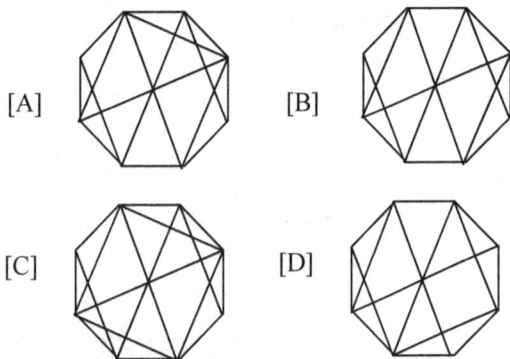

Figure 42:47

2. The network in Figure 42:48 which is not traversable is:

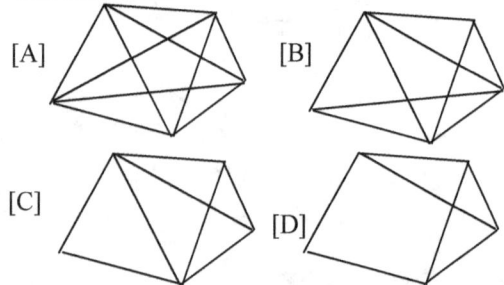

Figure 42:48

3. The network in Figure 42:49 which is traversable is:

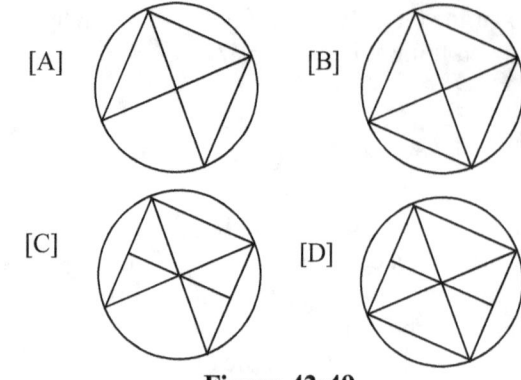

Figure 42:49

4. One of the possible routes to traverse the network in Figure 42:50 is:
   [A] (4, 2, 3, 5, 4, 3, 1, 4, 2)
   [B] (3, 2, 3, 5, 4, 3, 1, 4, 2)
   [C] (1, 2, 3, 5, 4, 3, 1, 4, 2)
   [D] (5, 2, 3, 1, 4, 3, 1, 4, 2)

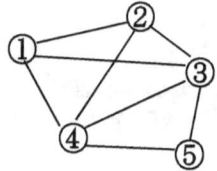

Figure 42:50

5. The number of vertices, regions and arcs the network in Figure 42:51 are respectively:
   [A] 6, 9, 5   [B] 6, 5, 9
   [C] 5, 9, 9   [D] 5, 6, 9

Figure 42:51

6. The smallest number of diagonals that can be drawn on the faces of a cube to make the network between its vertices and edges traversable is:
   [A] 1   [B] 2   [C] 3   [D] 4

7. The vertices and edges of polyhedra are three-dimensional networks. Among the four regular polyhedra in Figure 42:52 below, the one which is traversable is:

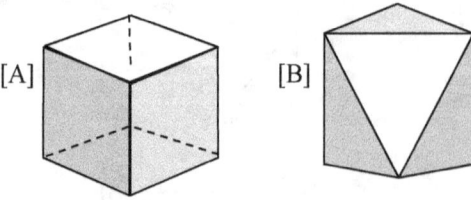

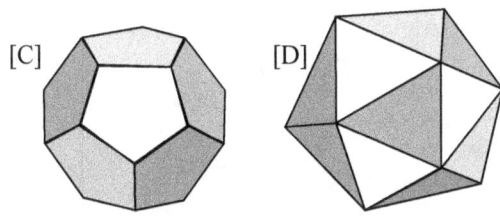

Figure 42:52

8. The floor plan of a house in Figure 42:53 forms a network which is:
   [A] traversable, with 16 nodes and 6 arcs.
   [B] traversable, with 6 nodes and 16 arcs.
   [C] not traversable, with 16 nodes and 6 arcs.
   [D] not traversable, with 6 nodes and 16 arcs

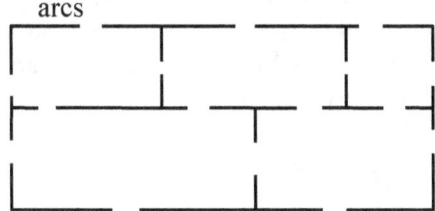

Figure 42:53

9. A house is said to be traversable if one can enter all the rooms without passing through any door more than once. This is possible if:
   [A] the network for the floor plan has no even vertices.
   [B] the network for the floor plan has no even edges.
   [C] the network for the floor plan has no odd vertices.
   [D] the network for the floor plan has no odd edges.

10. The statement which is certainly true of networks is:
    [A] A network with exactly two odd vertices is not traversable.
    [B] A network with no odd vertices is traversable.
    [C] A network with more than two odd vertices is traversable.
    [D] A network with more than two even vertices is traversable.

11. The statement which is certainly true of networks is that the starting point in a traversable network with:
    [A] all even vertices is also the ending point.
    [B] all even vertices cannot be the ending point.
    [C] two odd vertices is also the ending point.

[D] no odd vertices can be the ending points.

12. Given that $V = \{a, b, c, d, e\}$ and $E = \{\{a, b\}, \{a, c\}, \{b, c\}, \{b, d\}, \{c, d\}, \{a, e\}\}$. The number of regions in the network defined by $G = (V, E)$ is:
    [A] 3   [B] 4   [C] 5   [D] 6

13. Given that $V = \{1, 2, 3, 4, 5\}$ and $E = \{\{1, 2\}, \{1, 3\}, \{2, 3\}, \{2, 4\}, \{3, 4\}, \{1, 5\}\}$. The number of regions, nodes and arcs in the network defined by $G = (V, E)$ are respectively:
    [A] 3, 5 and 6   [B] 3, 6 and 5
    [C] 5, 3 and 6   [D] 6, 3 and 5

14. The graph in Figure 42:54 which represents the network $G = (V, E)$, $V = \{1, 2, 3\}$ and $E = \{\{1, 2\}, \{1, 3\}, \{2, 3\}\}$ is:

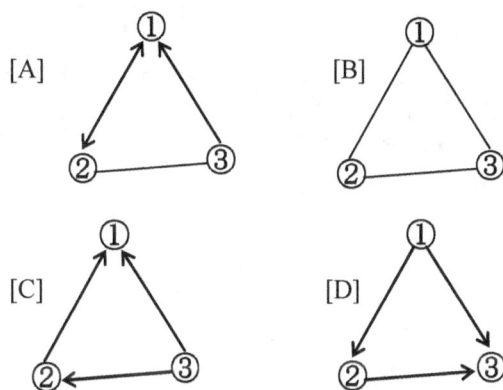

Figure 42:54

15. A complete graph among the graphs in Figure 42:55 is:

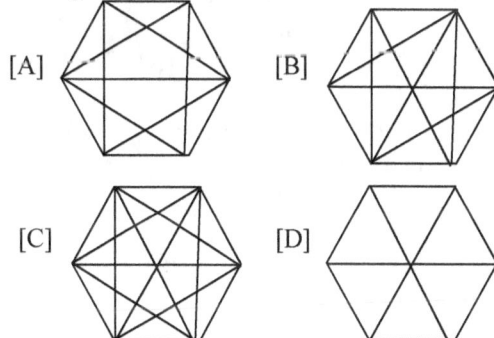

Figure 42:55

16. In the graph of network in Figure 42:56, the set of edges is:
    [A] {{1,2}, {3, 2}, {3,4}, {4, 1}}
    [B] { (1,2), (3, 2), (3,4), (4, 1)}.
    [C] { (2,1), (2, 3), (4,3), (1, 4)}.
    [D] {{2,1}, {2, 3}, {4,3}, {1, 4}}.

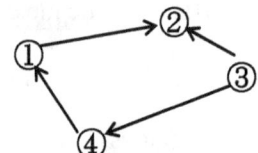

Figure 42:56

17. The network graph which is a tree in Figure 42:57 is:

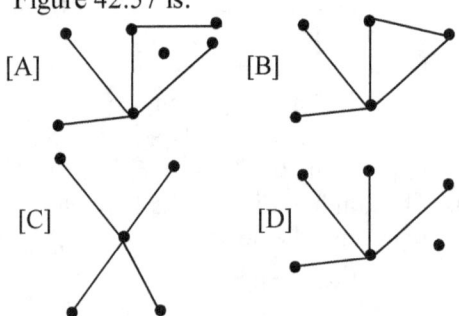

Figure 42:57

18. The number of regions in a network with 4 vertices and 6 edges is:
    [A] 3   [B] 4   [C] 5   [D] 6

19. In the net work in Figure 42:58, the set of odd vertices is:
    [A] {1, 2, 7, 8}   [B] {1, 3, 5, 7}
    [C] {3, 4, 5, 6}   [D] {1, 3, 6, 8}

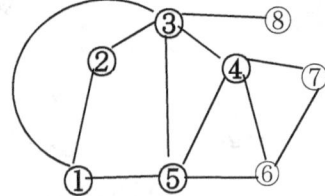

Figure 42:58

20. In the net work in Figure 42:58, the set of even vertices is:
    [A] {1, 2, 7, 8}   [B] {2, 4, 5, 7}
    [C] {3, 4, 5, 6}   [D] {2, 4, 6, 8}

21. In Figure 42:59, the edge {1,4} is incident to node:
    [A] 2   [B] 1   [C] 3   [D] 5

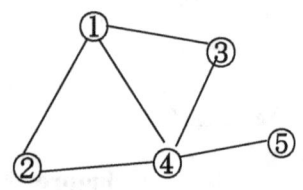

Figure 42:59

22. In Figure 42:60, node 3 and 4 are said to be:
    [A] consecutive   [B] coincident
    [C] adjacent   [D] incident

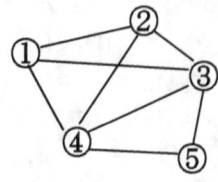

Figure 42:60

23. In Figure 42:60, edge {3,4} is said to be:
    [A] consecutive to node 3 and 4.
    [B] coincident to node 3 and 4.
    [C] adjacent to node 3 and 4.
    [D] incident to node 3 and 4.

24. In Figure 42:60, edge {3,4} and node 3 are said to be:
    [A] consecutive   [B] inverted
    [C] adjacent   [D] incident

25. In Figure 42:61, the edges {1,2} and {1,3} are said to be:
    [A] consecutive   [B] inverted
    [C] adjacent   [D] incident

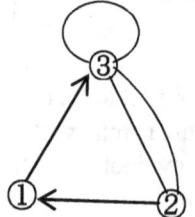

Figure 42:61

26. In Figure 42:61, the edges {1,2} and {2,3} are said to be:
    [A] consecutive   [B] inverted
    [C] adjacent   [D] incident

27. The network which is most likely to be a tree is the network which has:
    [A] 5 edges and 4 nodes
    [B] 4 edges and 5 nodes
    [C] 6 edges and 4 nodes
    [D] 4 edges and 6 nodes

28. The number of arcs in the network in Figure 42:62 is:
    [A] 2   [B] 3   [C] 4   [D] 5

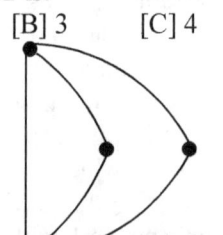

Figure 42:62

TOPIC 43

MATRICES

OBJECTIVES

At the end of this topic, the learner should be able to:

1. Represent information using matrices.
2. State the order or size of a matrix.
3. Identify and name different types of matrices.
4. Identify the diagonals of a square matrix and distinguish between a leading and minor diagonal.
5. State and use the conditions for equality of matrices.
6. Determine whether two or more matrices are conformable for matrix addition and subtraction, add and subtract matrices.
7. Multiply matrices by scalars.
8. Determine whether two matrices are conformable for matrix multiplication and multiply matrices.
9. Find the transpose of a matrix.
10. Find the inverse of a 2×2 matrix.
11. State the formula and use it to find the determinant of a 2×2 matrix.
12. Determine whether a 2×2 matrix is singular.
13. Find the adjoint of a 2×2 matrix.
14. Solve simultaneous linear equations in two unknowns using the matrix method.
15. Distinguish between route and incidence matrices.
16. Construct route and incidence matrices to show a given network.

A complex freeway interchange near Los Angeles, California, shows a typical cloverleaf pattern that facilitates easy and safe route changing. Interstate freeways have limited access and maintain low gradient, straight routings for maximum long-range visibility and safety.

Steven Frankel/Photo Researchers, Inc.

475

Introduction

A trader records his stock for three days as follows:

Monday   : 5 pens, 6 pencils, 2 rulers.
Tuesday   : 3 pens, 4 pencils, 1 ruler.
Wednesday: 1 pens, 0 pencils, 3 rulers.

This information can be displayed in the following ways.

	Pens	Pencils	Rulers
Mon	5	6	2
Tues	3	4	1
Wed	1	0	3

Or simply

$$\begin{pmatrix} 5 & 6 & 2 \\ 3 & 4 & 1 \\ 1 & 0 & 3 \end{pmatrix}$$

The second display of this information is an example of a matrix.

What is a Matrix?

A matrix is a rectangular array or arrangement of numbers in rows r and columns c. The numbers are called the **elements** or **entries** of the matrix. Matrices are denoted by capital letters A, B, C etc while the elements are denoted by lower case letters a, b, c,...if at all they are letters. In printed text, bold type **A, B, C** etc is used.

Order or Size of a Matrix

The size or order of a matrix with r rows and c columns is specified as $r \times c$, read 'r by c'.

Example 43:1

State the sizes of the following matrices.

(i) $\begin{pmatrix} 1 & 2 \\ 3 & 4 \end{pmatrix}$ (ii) $\begin{pmatrix} 2 & 3 \\ 4 & 6 \\ 7 & 5 \end{pmatrix}$ (iii) $\begin{pmatrix} 2 & 4 & 7 \\ 3 & 6 & 5 \end{pmatrix}$

Solution

(i) 2×2 (ii) 3×2 (iii) 2×3

The $\times$ in the notation for order of a matrix should not be confused with the multiplication sign 3×2 is read '3 by 2'.

Types of Matrices

1. A Square Matrix

A square matrix is a matrix with the number of rows and columns equal.

Examples of square matrices are $\begin{pmatrix} 0 & 1 & 2 \\ 8 & 3 & 7 \\ 5 & -4 & 6 \end{pmatrix}$,

$\begin{pmatrix} 2 & 1 \\ 4 & 3 \end{pmatrix}$.

In a square matrix such as $\begin{pmatrix} 0 & 1 & 2 \\ 8 & 3 & 7 \\ 5 & -4 & 6 \end{pmatrix}$, the

entries 0, 3 and 6 are said to be in the **leading diagonal**. The entries 2, 3 and 5 are said to be in the **minor diagonal**.

2. A Diagonal Matrix

A diagonal matrix is a square matrix with all the elements zeros except those in the leading diagonal.

Examples of diagonal matrices are $\begin{pmatrix} 1 & 0 & 0 \\ 0 & 3 & 0 \\ 0 & 0 & 6 \end{pmatrix}$,

$\begin{pmatrix} 2 & 0 \\ 0 & 3 \end{pmatrix}$.

Note that a diagonal matrix is necessarily a square matrix.

3. A Unit or Identity Matrix

The **unit** or **identity Matrix** usually denoted by **I** is a diagonal matrix with all the elements in the leading diagonal ones. The unit or identity Matrix is a very important matrix due to its applications.

Examples identity matrices are $\begin{pmatrix} 1 & 0 & 0 \\ 0 & 1 & 0 \\ 0 & 0 & 1 \end{pmatrix}$,

$\begin{pmatrix} 1 & 0 \\ 0 & 1 \end{pmatrix}$.

4. A rectangular Matrix

A rectangular matrix is a matrix with the number of rows different from the number of columns.

Examples of rectangular matrices are

$$\begin{pmatrix} -1 & 3 \\ 0 & -4 \\ 1 & 5 \end{pmatrix}, \begin{pmatrix} 0 & 1 & 2 \\ 3 & 0 & 4 \end{pmatrix}.$$

5. A Column Matrix

A column matrix is a matrix with all the elements in a single column.

Examples of column matrices are $\begin{pmatrix} 7 \\ 1 \\ -9 \end{pmatrix}, \begin{pmatrix} -2 \\ 4 \end{pmatrix}.$

6. A Row Matrix

A row matrix is a matrix with all the elements in a single row.

Examples of row matrices are $\begin{pmatrix} 7 & 1 & -9 \end{pmatrix},$

$\begin{pmatrix} -2 & 4 \end{pmatrix}.$

7. A Zero or Null Matrix

A zero or null matrix is one with all the elements zeros.

Examples of zero or null matrices are

$$\begin{pmatrix} 0 & 0 \\ 0 & 0 \end{pmatrix}, \begin{pmatrix} 0 & 0 & 0 \\ 0 & 0 & 0 \end{pmatrix}.$$

Some matrices can be classified in more than one group.

For instance $\begin{pmatrix} 0 & 0 & 0 \\ 0 & 0 & 0 \end{pmatrix}$ is a rectangular zero

matrix and $\begin{pmatrix} 0 & 0 \\ 0 & 0 \end{pmatrix}$ is a square zero matrix.

EXERCISE 43:1

1. State the sizes of the following matrices.

(a) $\begin{pmatrix} 1 & 0 \\ 0 & 0 \end{pmatrix}$ (b) $\begin{pmatrix} 1 & 5 & 2 \\ 3 & 1 & 6 \end{pmatrix}$ (c) $\begin{pmatrix} -2 \\ 1 \\ -4 \end{pmatrix}$

(d) $\begin{pmatrix} 3 & 0 & 0 \\ 0 & -5 & 0 \\ 0 & 0 & -10 \end{pmatrix}$ (e) $\begin{pmatrix} 0 & 0 \\ 0 & 0 \\ 0 & 0 \end{pmatrix}$ (f) (7)

(g) $\begin{pmatrix} 2 & 1 & 4 \\ 0 & -1 & 0 \\ 1 & 3 & -5 \end{pmatrix}$ (h) $\begin{pmatrix} 8 & 1 & 4 \end{pmatrix}$

(i) $\begin{pmatrix} 1 & 2 \\ 3 & 0 \\ 7 & -3 \end{pmatrix}$ (j) $\begin{pmatrix} 1 & 0 & 0 \\ 0 & 1 & 0 \\ 0 & 0 & 1 \end{pmatrix}$

2. Classify the matrices in question (1) above.

Equality of Matrices

Two matrices **A** and **B** are equal if and only if:
(i) They have the same size or order.
(ii) Their corresponding elements are equal.

Thus, $\begin{pmatrix} 1 & 3 \\ 2 & 5 \end{pmatrix} = \begin{pmatrix} a & b \\ c & d \end{pmatrix}$

$\Leftrightarrow a = 1, b = 3, c = 2$ and $d = 5$.

Note that $\begin{pmatrix} 1 & 3 \\ 2 & 5 \end{pmatrix} \neq \begin{pmatrix} 3 & 1 \\ 2 & 5 \end{pmatrix} \neq \begin{pmatrix} 1 & 2 \\ 3 & 5 \end{pmatrix},$

because though they have the same size and the same elements, corresponding elements are not equal.

Examples 43:2

Find x and y given that $\begin{pmatrix} 3x & y \\ 0 & 4x \end{pmatrix} = \begin{pmatrix} 6 & 2 \\ 0 & 8 \end{pmatrix}.$

Solution
Equating corresponding entries,
$3x = 6 \Rightarrow x = 2$ or $4x = 8 \Rightarrow x = 2$ and $y = 2$

EXERCISE 43:2

1. Find x and y given that

(a) $\begin{pmatrix} 0 & 2y \\ 5x & -y \end{pmatrix} = \begin{pmatrix} 0 & 8 \\ -15 & -4 \end{pmatrix}$

(b) $\begin{pmatrix} x+2 \\ y-1 \end{pmatrix} = \begin{pmatrix} 3 \\ -1 \end{pmatrix}$

(c) $\begin{pmatrix} 2x \\ 3y \end{pmatrix} = \begin{pmatrix} 10 \\ 12 \end{pmatrix}$

(d) $\begin{pmatrix} x+1 \\ x+y \end{pmatrix} = \begin{pmatrix} 4 \\ 1 \end{pmatrix}$

(e) $\begin{pmatrix} 0+3y \\ 4x+y \end{pmatrix} = \begin{pmatrix} 6 \\ 2 \end{pmatrix}$

2. Find x, y and z in each of the following cases.

(a) $\begin{pmatrix} 2x & y & 0 \\ x & -3y & 5z \end{pmatrix} = \begin{pmatrix} -6 & 4 & 0 \\ -3 & -12 & 2 \end{pmatrix}$

(b) $\begin{pmatrix} 4x & 3y \\ -2x & 0 \\ 0 & 5z \end{pmatrix} = \begin{pmatrix} -8 & 12 \\ 4 & 0 \\ 0 & -10 \end{pmatrix}$

3. Find x and y in the following:

(a) $\begin{pmatrix} x & 2y \\ 0 & -2 \end{pmatrix} = \begin{pmatrix} 1 & -8 \\ 0 & -2 \end{pmatrix}$

(b) $\begin{pmatrix} x+3 \\ 2-y \end{pmatrix} = \begin{pmatrix} 1 \\ -3 \end{pmatrix}$

(c) $\begin{pmatrix} 2x-y \\ x+y \end{pmatrix} = \begin{pmatrix} 3 \\ -9 \end{pmatrix}$

(d) $\begin{pmatrix} x+2y \\ 2y-x \end{pmatrix} = \begin{pmatrix} 0 \\ 2x-y \end{pmatrix}$

(e) $\begin{pmatrix} 7 & -6 \\ -15 & 3y \end{pmatrix} = \begin{pmatrix} 2x+y & 2y \\ -3x & -9 \end{pmatrix}$

(f) $\begin{pmatrix} 3 & x+1 \\ y & 5 \end{pmatrix} = \begin{pmatrix} x-1 & 5 \\ x-3y & 5 \end{pmatrix}$

Addition and Subtraction of Matrices

For addition and subtraction of matrices to be possible, the matrices must have the same size. If this condition is satisfied, the matrices are then said to be **compatible** for matrix addition or subtraction and corresponding elements are added or subtracted as the case may be.

Example 43:3

Let $\mathbf{A} = \begin{pmatrix} 1 & 3 \\ 1 & 2 \end{pmatrix}$ and $\mathbf{B} = \begin{pmatrix} 0 & -1 \\ 2 & 3 \end{pmatrix}$. Find

(i)　$\mathbf{A} + \mathbf{B}$　　　　(ii)　$\mathbf{A} - \mathbf{B}$

Solution

(i)　$\mathbf{A} + \mathbf{B} = \begin{pmatrix} 1 & 3 \\ 1 & 2 \end{pmatrix} + \begin{pmatrix} 0 & -1 \\ 2 & 3 \end{pmatrix} = \begin{pmatrix} 1 & 2 \\ 3 & 5 \end{pmatrix}$

(ii)　$\mathbf{A} - \mathbf{B} = \begin{pmatrix} 1 & 3 \\ 1 & 2 \end{pmatrix} - \begin{pmatrix} 0 & -1 \\ 2 & 3 \end{pmatrix} = \begin{pmatrix} 1 & 4 \\ -1 & -1 \end{pmatrix}$

EXERCISE 43:3

1. Evaluate the following.

(a) $\begin{pmatrix} -7 \\ 3 \end{pmatrix} + \begin{pmatrix} 9 \\ -5 \end{pmatrix}$　　(b) $\begin{pmatrix} 4 & 3 \\ 1 & 2 \end{pmatrix} + \begin{pmatrix} 5 & -2 \\ 2 & 6 \end{pmatrix}$

(c) $\begin{pmatrix} -7 \\ 3 \end{pmatrix} - \begin{pmatrix} 9 \\ -5 \end{pmatrix}$　　(d) $\begin{pmatrix} 4 & 3 \\ 1 & 2 \end{pmatrix} - \begin{pmatrix} 5 & -2 \\ 2 & 6 \end{pmatrix}$

2. Given that $\mathbf{E} = \begin{pmatrix} 1 & 2 & 4 \\ 3 & -1 & 8 \\ 2 & 5 & -7 \end{pmatrix}$ and

$\mathbf{F} = \begin{pmatrix} 0 & 3 & -5 \\ 2 & -3 & 2 \\ 1 & 4 & 7 \end{pmatrix}$, find:

(a)　$\mathbf{E} - \mathbf{F}$　　　　(b)　$\mathbf{F} - \mathbf{E}$

3. Given that $\mathbf{A} = \begin{pmatrix} 1 & 2 & 0 \\ 2 & 3 & 0 \end{pmatrix}$,

$\mathbf{B} = \begin{pmatrix} 0 & 1 & 2 \\ 0 & 2 & 3 \end{pmatrix}$ and $\mathbf{C} = \begin{pmatrix} 1 & 0 & 2 \\ 3 & 0 & 4 \end{pmatrix}$.

Find

(a)　$\mathbf{A} + \mathbf{B}$　　(b)　$\mathbf{B} + \mathbf{A}$　　(c)　$\mathbf{B} + \mathbf{C}$
(d)　$\mathbf{A} - \mathbf{B}$　　(e)　$\mathbf{B} - \mathbf{A}$　　(f)　$\mathbf{B} - \mathbf{C}$
(g)　$(\mathbf{A} + \mathbf{B}) + \mathbf{C}$　　(h)　$\mathbf{A} + (\mathbf{B} + \mathbf{C})$
(i)　$(\mathbf{A} - \mathbf{B}) - \mathbf{C}$　　(j)　$\mathbf{A} - (\mathbf{B} - \mathbf{C})$

Say whether or not:
(k)　Addition of matrices is commutative.
(l)　Subtraction of matrices is commutative.
(m)　Addition of matrices is associative.
(n)　Subtraction of matrices is associative.

4. Given that $\mathbf{A} = \begin{pmatrix} 3 & -1 \\ 2 & 5 \end{pmatrix}$, $\mathbf{B} = \begin{pmatrix} -3 & 1 \\ -2 & -5 \end{pmatrix}$,

find
(a) **A + B**   (b) **B + A**
(c) Comment on your result in (a) and (b).

5. Given that $\mathbf{A} = \begin{pmatrix} 3 & -1 \\ 2 & 5 \end{pmatrix}$ and $\varnothing = \begin{pmatrix} 0 & 0 \\ 0 & 0 \end{pmatrix}$

(a) **A** + $\varnothing$ (b) $\varnothing$ + **A**
(c) Suggest a name for $\varnothing$ with respect to its relationship to **A**.

6. Find *a*, *b*, *c*, *d* and *e*, given that

$$\begin{pmatrix} 3 & 2 & 5 \\ 1 & 4 & c \end{pmatrix} + \begin{pmatrix} a & b & c \\ 0 & a & b \end{pmatrix} = \begin{pmatrix} 4 & a & b \\ a & d & e \end{pmatrix}$$

7. Find the unknowns in each of the following:

(a) $\begin{pmatrix} 2 & 4 \\ 3 & z \end{pmatrix} + \begin{pmatrix} x & y \\ 3 & 4 \end{pmatrix} = \begin{pmatrix} 4 & 4 \\ w & 0 \end{pmatrix}$

(b) $\begin{pmatrix} x & y \\ z & w \end{pmatrix} - \begin{pmatrix} 2 & 1 \\ 5 & 3 \end{pmatrix} = \begin{pmatrix} 1 & 0 \\ 0 & 1 \end{pmatrix}$

Scalar Multiplication of Matrices

To multiply a matrix by a scalar, simply multiply each element of the matrix by the scalar.

Example 43:4

Evaluate the following:

(a) $2 \begin{pmatrix} 2 & 3 \\ 1 & 4 \end{pmatrix}$ (b) $-3 \begin{pmatrix} 4 & 1 & 0 \\ 5 & 2 & 7 \end{pmatrix}$

(c) $\frac{1}{3} \begin{pmatrix} 6 & 9 \\ -15 & 12 \end{pmatrix}$

Solution

(a) $2 \begin{pmatrix} 2 & 3 \\ 1 & 4 \end{pmatrix} = \begin{pmatrix} 2 \times 2 & 2 \times 3 \\ 2 \times 1 & 2 \times 4 \end{pmatrix} = \begin{pmatrix} 4 & 6 \\ 2 & 8 \end{pmatrix}$

(b) $-3 \begin{pmatrix} 4 & 1 & 0 \\ 5 & 2 & 7 \end{pmatrix} = \begin{pmatrix} -3 \times 4 & -3 \times 1 & -3 \times 0 \\ -3 \times 5 & -3 \times 2 & -3 \times 7 \end{pmatrix}$

$$= \begin{pmatrix} -12 & -3 & 0 \\ -15 & -6 & -21 \end{pmatrix}$$

(c) $\frac{1}{3} \begin{pmatrix} 6 & 9 \\ -15 & 12 \end{pmatrix} = \begin{pmatrix} \frac{1}{3} \times 6 & \frac{1}{3} \times 9 \\ \frac{1}{3} \times -15 & \frac{1}{3} \times 12 \end{pmatrix} = \begin{pmatrix} 2 & 3 \\ -5 & 4 \end{pmatrix}$

EXERCISE 43:4

(1) Evaluate $-\dfrac{1}{2} \begin{pmatrix} -4 & 0 & -1 \\ 3 & 1 & -2 \\ 0 & 8 & 5 \end{pmatrix}$.

(2) Given that $x \begin{pmatrix} 1 \\ 2 \end{pmatrix} + 3 \begin{pmatrix} 9 \\ -4 \end{pmatrix} = \begin{pmatrix} 30 \\ -6 \end{pmatrix}$, find *x*.

(3) Find *x* and *y* given that $x \begin{pmatrix} 1 \\ 0 \end{pmatrix} - y \begin{pmatrix} 0 \\ 4 \end{pmatrix} = \begin{pmatrix} 5 \\ 8 \end{pmatrix}$.

(4) Find the unknown if:

(i) $2 \begin{pmatrix} x \\ y \end{pmatrix} = \begin{pmatrix} -4 \\ -10 \end{pmatrix}$

(ii) $\begin{pmatrix} x \\ y \end{pmatrix} = -\dfrac{1}{3} \begin{pmatrix} -27 \\ 21 \end{pmatrix}$

(iii) $a \begin{pmatrix} 0 \\ 3 \end{pmatrix} + b \begin{pmatrix} 1 \\ 0 \end{pmatrix} = \begin{pmatrix} 11 \\ 6 \end{pmatrix}$

Multiplication of Matrices

If **A** and **B** are matrices, then the product **A** × **B** read "A cross B" is possible only if the number of columns in the matrix **A** are equal to the number of rows in the matrix **B**. This means that if the size of the matrix **A** is $m \times n$, the size of **B**, must be $n \times p$. If this condition is satisfied, **A** and **B** are said to be **compatible** or **conformable** for matrix multiplication and the product **A** × **B** exists. The size of **A** × **B** will then be $m \times p$

Therefore, *(m by n) × (n by m) = m by p*

A × **B** can simply be written as **AB**.

Example 43:5
Compute the following.

(1) $(a \quad b) \begin{pmatrix} x \\ y \end{pmatrix}$ (2) $(a \quad b) \begin{pmatrix} w & x \\ y & z \end{pmatrix}$

(3) $(4 \quad 5) \begin{pmatrix} 2 \\ 3 \end{pmatrix}$ (4) $(2 \quad 4) \begin{pmatrix} 1 & 4 \\ 2 & 0 \end{pmatrix}$

(5) $\begin{pmatrix} a & b \\ c & d \end{pmatrix} \begin{pmatrix} w & x \\ y & z \end{pmatrix}$ (6) $\begin{pmatrix} 1 & 0 \\ 2 & 3 \end{pmatrix} \begin{pmatrix} -3 & 2 \\ 1 & 7 \end{pmatrix}$

(7) $\begin{pmatrix} 2 & 7 \\ 1 & 3 \end{pmatrix}\begin{pmatrix} 1 & 4 & -3 \\ 5 & 2 & 0 \end{pmatrix}$

Solution

(1) $(a \quad b)\begin{pmatrix} x \\ y \end{pmatrix} = (ax + by)$

(2) $(a \quad b)\begin{pmatrix} w & x \\ y & z \end{pmatrix} = (aw + by \quad ax + bz)$

(3) $(4 \quad 5)\begin{pmatrix} 2 \\ 3 \end{pmatrix} = (4(2) + 5(3)) = (8 + 15) = (23)$

(4) $(2 \quad 4)\begin{pmatrix} 1 & 4 \\ 2 & 0 \end{pmatrix} = (2(1) + 4(2) \quad 2(4) + 4(0)) = (10 \quad 8)$

(5) $\begin{pmatrix} a & b \\ c & d \end{pmatrix}\begin{pmatrix} w & x \\ y & z \end{pmatrix} = \begin{pmatrix} aw + by & ax + bz \\ cw + dy & cx + dz \end{pmatrix}$

(6) $\begin{pmatrix} 1 & 0 \\ 2 & 3 \end{pmatrix}\begin{pmatrix} -3 & 2 \\ 1 & 7 \end{pmatrix} = \begin{pmatrix} 1(-3) + 0(1) & 1(2) + 0(7) \\ 2(-3) + 3(1) & 2(2) + 3(7) \end{pmatrix}$

$= \begin{pmatrix} -3 & 2 \\ -3 & 25 \end{pmatrix}$

(7) $\begin{pmatrix} 2 & 7 \\ 1 & 3 \end{pmatrix}\begin{pmatrix} 1 & 4 & -3 \\ 5 & 4 & 0 \end{pmatrix} = \begin{pmatrix} 2+35 & 8+14 & -6+0 \\ 1+15 & 4+6 & -3+0 \end{pmatrix}$

$= \begin{pmatrix} 37 & 22 & -6 \\ 16 & 10 & -3 \end{pmatrix}$

Example 43:6

Let $\mathbf{A} = \begin{pmatrix} 1 & 2 \\ 1 & 4 \end{pmatrix}$ and $\mathbf{B} = \begin{pmatrix} 4 & 1 \\ 3 & -2 \end{pmatrix}$. Find

(i) **AB**   (ii) **BA**

Solution

(i) $\mathbf{AB} = \begin{pmatrix} 1 & 2 \\ 1 & 4 \end{pmatrix}\begin{pmatrix} 4 & 1 \\ 3 & -2 \end{pmatrix} = \begin{pmatrix} 10 & -3 \\ 16 & -7 \end{pmatrix}$

(ii) $\mathbf{BA} = \begin{pmatrix} 4 & 1 \\ 3 & -2 \end{pmatrix}\begin{pmatrix} 1 & 2 \\ 1 & 4 \end{pmatrix} = \begin{pmatrix} 5 & 12 \\ 1 & -2 \end{pmatrix}$

In the product **AB**, **A** is said to be pre-multiplied by **B**, while **B** is said to be post-multiplied by **A**.
Notice that $\mathbf{AB} \neq \mathbf{BA}$. Therefore, matrix multiplication is not commutative.

Example 43:7

Given that $\mathbf{P} = \begin{pmatrix} 3 & 4 \\ 5 & 6 \end{pmatrix}$ and $\mathbf{Q} = \begin{pmatrix} 2 \\ 1 \end{pmatrix}$, find

PQ .

Solution

$\mathbf{PQ} = \begin{pmatrix} 3 & 4 \\ 5 & 6 \end{pmatrix}\begin{pmatrix} 2 \\ 1 \end{pmatrix} = \begin{pmatrix} 10 \\ 16 \end{pmatrix}$

Examples 43:8

Find $\mathbf{A} = \begin{pmatrix} 0 & -1 \\ 3 & -2 \end{pmatrix}$ and $\mathbf{B} = \begin{pmatrix} 2 & 4 & 3 \\ 1 & 5 & 0 \end{pmatrix}$

Solution

$\mathbf{AB} = \begin{pmatrix} 0 & -1 \\ 3 & -2 \end{pmatrix}\begin{pmatrix} 2 & 4 & 3 \\ 1 & 5 & 0 \end{pmatrix} = \begin{pmatrix} -1 & -5 & 0 \\ 4 & 2 & 9 \end{pmatrix}$

EXERCISE 43:5

1. Let $\mathbf{A} = \begin{pmatrix} 2 & -3 \\ 4 & 2 \end{pmatrix}$ and $\mathbf{B} = \begin{pmatrix} -1 & 0 \\ 2 & -3 \end{pmatrix}$. Find **AB**.

2. $\mathbf{C} = (5 \quad 7)$ and $\mathbf{D} = \begin{pmatrix} 3 & -7 & 2 \\ -4 & 0 & 1 \end{pmatrix}$. Evaluate **CD**.

3. Given that $\mathbf{A} = \begin{pmatrix} 3 & 7 \\ 5 & -2 \end{pmatrix}$ and $\mathbf{D} = \begin{pmatrix} 1 & 0 \\ 0 & 1 \end{pmatrix}$, find (i) $\mathbf{A} \times \mathbf{D}$ (ii) $\mathbf{D} \times \mathbf{A}$ (iii) $\mathbf{A}^2$

4. Evaluate the following and comment on your results.

(a) $\begin{pmatrix} 1 & 3 \\ 5 & 4 \end{pmatrix}\begin{pmatrix} 1 & 0 \\ 0 & 1 \end{pmatrix}$ (b) $\begin{pmatrix} 1 & 0 \\ 0 & 1 \end{pmatrix}\begin{pmatrix} 1 & 3 \\ 5 & 4 \end{pmatrix}$

(c) $\begin{pmatrix} 1 & 3 \\ 5 & 4 \end{pmatrix}\begin{pmatrix} 0 & 0 \\ 0 & 0 \end{pmatrix}$ (d) $\begin{pmatrix} 0 & 0 \\ 0 & 0 \end{pmatrix}\begin{pmatrix} 1 & 3 \\ 5 & 4 \end{pmatrix}$

(e) $\begin{pmatrix} 1 & 2 \\ 5 & 4 \\ 2 & 6 \end{pmatrix}\begin{pmatrix} 1 & 0 \\ 0 & 1 \end{pmatrix}$ (f) $\begin{pmatrix} 1 & 2 \\ 5 & 4 \\ 2 & 6 \end{pmatrix}\begin{pmatrix} 0 & 0 \\ 0 & 0 \end{pmatrix}$

5. Let $\mathbf{A} = \begin{pmatrix} 2 & -4 \\ 1 & 3 \end{pmatrix}$, $\mathbf{B} = \begin{pmatrix} 2 & 3 \\ 1 & 4 \end{pmatrix}$ and $\mathbf{C} = \begin{pmatrix} 0 & 2 \\ -1 & 3 \end{pmatrix}$

(i) Evaluate the following:
(a) $\mathbf{A}^2$ (b) $\mathbf{B}^2$ (c) $\mathbf{AB}$
(d) $\mathbf{BA}$ (e) $\mathbf{BC}$ (f) $(\mathbf{AB})\mathbf{C}$
(g) $\mathbf{A}(\mathbf{BC})$ (h) $\mathbf{A} + \mathbf{B}$ (i) $\mathbf{A} - \mathbf{B}$
(j) $(\mathbf{A} + \mathbf{B})^2$ (k) $(\mathbf{A} - \mathbf{B})^2$
(l) $\mathbf{A}^2 - \mathbf{B}^2$ (m) $(\mathbf{A} + \mathbf{B})(\mathbf{A} - \mathbf{B})$
(n) $\mathbf{A}^2 - 2\mathbf{AB} + \mathbf{B}^2$ (o) $\mathbf{A}^2 + 2\mathbf{AB} + \mathbf{B}^2$

(ii) State whether the following operations are true of matrix multiplication.

(a) $AB = BA$ (b) $(AB)C = A(BC)$
(c) $(A + B)^2 = A^2 + 2AB + B^2$
(d) $(A - B)^2 = A^2 - 2AB + B^2$
(e) $(A + B)(A - B) = A^2 - B^2$

6. Compute the following products:

(a) $\begin{pmatrix} 5 & -3 \\ -4 & 1 \end{pmatrix}\begin{pmatrix} x \\ y \end{pmatrix}$ (b) $\begin{pmatrix} 3 & 4 \\ 5 & 1 \end{pmatrix}\begin{pmatrix} a \\ b \end{pmatrix}$

(c) $\begin{pmatrix} 2 & 1 \\ -2 & 3 \end{pmatrix}\begin{pmatrix} r \\ s \end{pmatrix}$ (d) $\begin{pmatrix} 4 & 1 \\ 2 & 0 \end{pmatrix}\begin{pmatrix} p \\ q \end{pmatrix}$

(e) $\begin{pmatrix} 4 & 2 \\ 3 & 5 \end{pmatrix}\begin{pmatrix} x \\ y \end{pmatrix}$ (f) $\begin{pmatrix} -2 & -3 \\ -2 & 1 \end{pmatrix}\begin{pmatrix} x \\ y \end{pmatrix}$

The Transpose of a Matrix

The transpose A^T of a matrix A is obtained by interchanging the rows and columns of the matrix.

Example 43:9
Find the transpose of:

(a) $A = \begin{pmatrix} 1 & 4 \\ 2 & 5 \end{pmatrix}$ (b) $B = \begin{pmatrix} 1 & 3 \\ 2 & 4 \\ 7 & -9 \end{pmatrix}$

Solution

(a) $A^T = \begin{pmatrix} 1 & 2 \\ 4 & 5 \end{pmatrix}$ (b) $B^T = \begin{pmatrix} 1 & 2 & 7 \\ 3 & 4 & -9 \end{pmatrix}$

EXERCISE 43:6

Write down the transpose of each of the following matrices.

(a) $\begin{pmatrix} 1 & 3 \\ 0 & 4 \end{pmatrix}$ (b) $\begin{pmatrix} 1 & 2 & 7 \end{pmatrix}$

(c) $\begin{pmatrix} -1 \\ 0 \\ 1 \end{pmatrix}$ (d) $\begin{pmatrix} 2 & 0 & -4 \\ 1 & 8 & 6 \\ -3 & 5 & 1 \end{pmatrix}$

The Determinant of a 2×2 Matrix

The **determinant** of a 2×2 matrix $A = \begin{pmatrix} a & b \\ c & d \end{pmatrix}$, is denoted by det A, $\det\begin{pmatrix} a & b \\ c & d \end{pmatrix}$, $|A|$ or $\begin{vmatrix} a & b \\ c & d \end{vmatrix}$

and is obtained by finding the difference between the product of the leading diagonal elements and the minor diagonal elements. Hence,

$$\text{Det } A = ad - bc.$$

Note! $\begin{pmatrix} a & b \\ c & d \end{pmatrix} \neq \begin{vmatrix} a & b \\ c & d \end{vmatrix}$

Example 43:10
Evaluate (a) $\begin{vmatrix} 3 & 2 \\ 1 & 5 \end{vmatrix}$ (b) $\begin{vmatrix} 4 & 2 \\ -1 & 3 \end{vmatrix}$

Solution

(a) $\begin{vmatrix} 3 & 2 \\ 1 & 5 \end{vmatrix} = 3(5) = 2(1) = 13$

(b) $\begin{vmatrix} 4 & 2 \\ -1 & 3 \end{vmatrix} = 4(3) = 2(-1) = 14$

Singular Matrices

A Matrix whose determinant is equal to zero is called a **singular matrix**.

Example 43:11
Given that $A = \begin{pmatrix} 2 & -1 \\ 4 & -4 \end{pmatrix}$ and $B = \begin{pmatrix} 1 & 1 \\ 1 & 1 \end{pmatrix}$,

find
(i) det A   (ii) $|B|$.

Solution
(i) det $A = 2(-2) - 4(-1) = 0$
(ii) $|B| = 1(1) - 1(1) = 0$.

Therefore both A and B are singular matrices.

EXERCISE 43:7

1. Find the determinant of $A = \begin{pmatrix} 3 & 4 \\ 5 & 6 \end{pmatrix}$.

2. The matrices $P = \begin{pmatrix} m & 2 \\ -2 & 0 \end{pmatrix}$ and

$\mathbf{Q} = \begin{pmatrix} 2 & -m \\ -2 & 1 \end{pmatrix}$ are such that $\mathbf{P} + \mathbf{Q}$ is singular. Find the value of m.

3. Find the value of a for which the matrix $\begin{pmatrix} a & 3-a \\ 2 & 1 \end{pmatrix}$ is singular.

4. Given that the matrix $\begin{pmatrix} a & 1 \\ 4 & a \end{pmatrix}$ is singular, find the possible values of a.

5. Which of the following matrices is/are singular?

$\begin{pmatrix} -2 & 2 \\ 2 & 2 \end{pmatrix}$, $\begin{pmatrix} -2 & -2 \\ 2 & 2 \end{pmatrix}$, $\begin{pmatrix} -2 & -2 \\ 2 & -2 \end{pmatrix}$,

$\begin{pmatrix} -2 & -2 \\ -2 & 2 \end{pmatrix}$, $\begin{pmatrix} 2 & -2 \\ -2 & -2 \end{pmatrix}$.

6. Find $\begin{pmatrix} 2 & -3 \\ -2 & 4 \end{pmatrix}\begin{pmatrix} -2 & 1 \\ -5 & 3 \end{pmatrix}$ and calculate its determinant.

7. Find the values of m for which the matrix $\begin{pmatrix} 6-m & 2 \\ 25 & 1-m \end{pmatrix}$ is singular.

8. Find the values of x for which the matrix $\begin{pmatrix} 2x & 2 \\ 3 & x-2 \end{pmatrix}$ is singular.

9. Solve the equation $\det\begin{pmatrix} x & 10 \\ x & x \end{pmatrix} = -21$.

The Adjoint of a 2×2 Matrix

The adjoint of the 2×2 matrix $\mathbf{A} = \begin{pmatrix} a & b \\ c & d \end{pmatrix}$

is denoted by adj $\mathbf{A}$ or $\text{Adj}\begin{pmatrix} a & b \\ c & d \end{pmatrix}$ and is

obtained by interchanging the elements along the leading diagonal and changing the sign of the elements along the minor diagonal. Hence,

$$\text{Adj}\begin{pmatrix} a & b \\ c & d \end{pmatrix} = \text{Adj}\mathbf{A} = \begin{pmatrix} d & -b \\ -c & a \end{pmatrix}$$

Example 43:12
Find the adjoint of the following matrices.

(a) $\begin{pmatrix} 15 & 9 \\ 10 & 5 \end{pmatrix}$ (b) $\begin{pmatrix} 3 & 7 \\ -4 & 1 \end{pmatrix}$

Solution

(a) $\text{Adj}\begin{pmatrix} 15 & 9 \\ 10 & 5 \end{pmatrix} = \begin{pmatrix} 5 & -9 \\ -10 & 15 \end{pmatrix}$

(b) $\text{Adj}\begin{pmatrix} 1 & -7 \\ 4 & 3 \end{pmatrix} = \begin{pmatrix} 1 & -7 \\ 4 & 3 \end{pmatrix}$

EXERCISE 43:8

Write down the adjoint of the following matrices.

(a) $\begin{pmatrix} 5 & -3 \\ -4 & 1 \end{pmatrix}$ (b) $\begin{pmatrix} 3 & 4 \\ 5 & 1 \end{pmatrix}$ (c) $\begin{pmatrix} 2 & 1 \\ -2 & 3 \end{pmatrix}$

(d) $\begin{pmatrix} 4 & 1 \\ 2 & 0 \end{pmatrix}$ (e) $\begin{pmatrix} 4 & 2 \\ 3 & 5 \end{pmatrix}$ (f) $\begin{pmatrix} -2 & -3 \\ -2 & 1 \end{pmatrix}$

(g) $\begin{pmatrix} 1 & 0 \\ 0 & 1 \end{pmatrix}$ (h) $\begin{pmatrix} \frac{1}{2} & 1 \\ 2 & 4 \end{pmatrix}$ (i) $\begin{pmatrix} 2 & 5 \\ 3 & 7 \end{pmatrix}$

The Inverse of a 2×2 Matrix

The inverse or reciprocal of a square matrix $\mathbf{A}$ is denoted by $\mathbf{A}^{-1}$, read "$\mathbf{A}$ inverse". $\mathbf{A}^{-1}$ is the inverse of $\mathbf{A}$ if and only if

$$\mathbf{A}^{-1}\mathbf{A} = \mathbf{A}\mathbf{A}^{-1} = \mathbf{I}$$

Where, $\mathbf{I}$ is a compatible identity matrix.

Finding the Inverse of a 2×2 Matrix

Example 43:13

Find the inverse of the matrix $\begin{pmatrix} 4 & 1 \\ 6 & 2 \end{pmatrix}$.

Solution
The definition $\mathbf{A}^{-1}\mathbf{A} = \mathbf{I}$ may be used. Thus,

Let $\mathbf{A} = \begin{pmatrix} 4 & 1 \\ 6 & 2 \end{pmatrix}$ and $\mathbf{A}^{-1} = \begin{pmatrix} a & b \\ c & d \end{pmatrix}$

Then, $\begin{pmatrix} a & b \\ c & d \end{pmatrix}\begin{pmatrix} 4 & 1 \\ 6 & 2 \end{pmatrix} = \begin{pmatrix} 1 & 0 \\ 0 & 1 \end{pmatrix}$.

$\Rightarrow \begin{pmatrix} 4a+6b & a+2b \\ 4c+6d & c+2d \end{pmatrix} = \begin{pmatrix} 1 & 0 \\ 0 & 1 \end{pmatrix}$

$4a + 6b = 1$ ....①

$a + 2b = 0$ ....②

①$-3$② $\Rightarrow a = 1$

Substitute ② in $\Rightarrow b = -\dfrac{1}{2}$

Similarly,

$4c + 6d = 0$ ....③

$c + 2d = 1$ ....④

③$-3$④ $\Rightarrow c = -3$

Substitute ④ in $\Rightarrow -3 + 2d = 1 \Rightarrow d = 2$

Therefore the inverse of $\begin{pmatrix} 4 & 1 \\ 6 & 2 \end{pmatrix}$ is

$\begin{pmatrix} 1 & -\dfrac{1}{2} \\ -3 & 2 \end{pmatrix}$

Example 43:14

Find the inverse of $\mathbf{A} = \begin{pmatrix} a & b \\ c & d \end{pmatrix}$.

Solution

Let $\mathbf{A}^{-1} = \begin{pmatrix} w & x \\ y & z \end{pmatrix}$. Then by definition $\mathbf{A}^{-1}\mathbf{A}$

$= \mathbf{I}$.

$\Rightarrow \begin{pmatrix} w & x \\ y & z \end{pmatrix}\begin{pmatrix} a & b \\ c & d \end{pmatrix} = \begin{pmatrix} 1 & 0 \\ 0 & 1 \end{pmatrix}$

Multiplying and equating corresponding entries,

$aw + cx = 1$ ....①

$bw + dx = 0$ ....②

$ay + cz = 0$ ....③

$by + dz = 1$ ....④

Solving for w, x, y and z in these equations,

$w = \dfrac{d}{ad-bc}, \quad x = -\dfrac{b}{ad-bc},$

$y = -\dfrac{c}{ad-bc}, \quad z = \dfrac{a}{ad-bc}$

$\mathbf{A}^{-1} = \begin{pmatrix} \dfrac{d}{ad-bc} & -\dfrac{b}{ad-bc} \\ -\dfrac{c}{ad-bc} & \dfrac{a}{ad-bc} \end{pmatrix}$

$= \dfrac{1}{ad-bc}\begin{pmatrix} d & -b \\ -c & a \end{pmatrix}$

Since the adjoint and determinant of the 2×2 matrix $\mathbf{A} = \begin{pmatrix} a & b \\ c & d \end{pmatrix}$ are given by

$\text{Adj}\,\mathbf{A} = \begin{pmatrix} d & -c \\ -b & a \end{pmatrix}$ and $\det \mathbf{A} = ad - bc$

respectively,

$$\mathbf{A}^{-1} = \dfrac{1}{ad-bc}\begin{pmatrix} d & -b \\ -c & a \end{pmatrix} = \dfrac{1}{\text{Det}\,\mathbf{A}} \times \text{Adj}\,\mathbf{A}$$

This formula can be used to find the inverse of any 2×2 matrix, which has an inverse.

Example 43:15

Find the inverse of $\mathbf{A} = \begin{pmatrix} 4 & 1 \\ 6 & 2 \end{pmatrix}$.

Solution

$\mathbf{A}^{-1} = \dfrac{1}{ad-bc}\begin{pmatrix} d & -b \\ -c & a \end{pmatrix} = \dfrac{1}{\text{Det}\,\mathbf{A}} \times \text{Adj}\,\mathbf{A}$

$\mathbf{A}^{-1} = \dfrac{1}{4(2)-1(6)}\begin{pmatrix} 2 & -1 \\ -6 & 4 \end{pmatrix} = \dfrac{1}{2}\begin{pmatrix} 2 & -1 \\ -6 & 4 \end{pmatrix}$

$\Rightarrow \mathbf{A}^{-1} = \begin{pmatrix} 1 & -\dfrac{1}{2} \\ -3 & 2 \end{pmatrix}$

Example 43:16

Find the inverse of $\begin{pmatrix} 4 & 6 \\ 5 & 8 \end{pmatrix}$.

Solution

$\begin{pmatrix} 4 & 6 \\ 5 & 8 \end{pmatrix}^{-1} = \dfrac{1}{4(8)-6(5)}\begin{pmatrix} 8 & -6 \\ -5 & 4 \end{pmatrix} = \dfrac{1}{2}\begin{pmatrix} 8 & -6 \\ -5 & 4 \end{pmatrix}$

$$\Rightarrow \begin{pmatrix} 4 & 6 \\ 5 & 8 \end{pmatrix}^{-1} = \begin{pmatrix} 4 & -3 \\ -\dfrac{5}{2} & 2 \end{pmatrix}$$

EXERCISE 43:9

1. Find the inverses of the following matrices.

   (a) $\begin{pmatrix} 5 & -3 \\ -4 & 1 \end{pmatrix}$ (b) $\begin{pmatrix} 3 & 4 \\ 5 & 1 \end{pmatrix}$

   (c) $\begin{pmatrix} 2 & 1 \\ -2 & 3 \end{pmatrix}$ (d) $\begin{pmatrix} 4 & 1 \\ 2 & 0 \end{pmatrix}$

   (e) $\begin{pmatrix} 4 & 2 \\ 3 & 5 \end{pmatrix}$ (f) $\begin{pmatrix} -2 & -3 \\ -2 & 1 \end{pmatrix}$

   (g) $\begin{pmatrix} 1 & 0 \\ 0 & 1 \end{pmatrix}$ (h) $\begin{pmatrix} \dfrac{1}{2} & 1 \\ 2 & 4 \end{pmatrix}$

   (i) $\begin{pmatrix} 2 & 5 \\ 3 & 7 \end{pmatrix}$

2. Given that $\mathbf{A} = \begin{pmatrix} 3 & 1 \\ 2 & 1 \end{pmatrix}$, find $\mathbf{A}^{-1}$ and hence $\mathbf{A}^{-1}\mathbf{A}$.

3. If $\mathbf{B} = \begin{pmatrix} 1 & 3 \\ 1 & 5 \end{pmatrix}$, find $\mathbf{B}^{-1}$ and hence $\mathbf{B}^{-1}\mathbf{B}$.

4. Find the inverses of the following matrices.

   (i) $\mathbf{A} = \begin{pmatrix} 4 & -2 \\ 1 & 3 \end{pmatrix}$ (ii) $\mathbf{B} = \begin{pmatrix} 1 & 0 \\ 0 & 1 \end{pmatrix}$

   (iii) $\mathbf{C} = \begin{pmatrix} 3 & 0 \\ 0 & 2 \end{pmatrix}$ (iv) $\mathbf{D} = \begin{pmatrix} 2 & -2 \\ 3 & 1 \end{pmatrix}$

5. Show that $\begin{pmatrix} 1 & 0 \\ 0 & 1 \end{pmatrix}$ is the identity matrix under matrix multiplication.

6. Show that $\begin{pmatrix} -2 & 3 \\ 3 & -4 \end{pmatrix}$ is the inverse of $\begin{pmatrix} 4 & 3 \\ 3 & 2 \end{pmatrix}$ under matrix multiplication.

Simultaneous Equations (The Matrix Method)

Consider,

$$\begin{pmatrix} 4 & -1 \\ 3 & 1 \end{pmatrix}\begin{pmatrix} x \\ y \end{pmatrix} = \begin{pmatrix} 3 \\ 4 \end{pmatrix} \quad\text{......................①}$$

Computing the product in the LHS leads to

$$\begin{pmatrix} 4x - y \\ 3x + y \end{pmatrix} = \begin{pmatrix} 3 \\ 4 \end{pmatrix}$$

Equating corresponding entries the following simultaneous equations are obtained.

$$4x - y = 3 \quad\text{..........................②}$$
$$3x + y = 4 \quad\text{..........................③}$$

Therefore, the simultaneous equations in ② and ③ can be rearranged to obtain ①. Since multiplying a matrix by its inverse gives rise to the identity matrix and multiplying a matrix by the identity matrix leaves the matrix unchanged, this idea can be used to solve simultaneous equations.

Example 43:17
Using the matrix method, solve the simultaneous equations
$$4x - y = 3$$
$$3x + y = 4$$
Solution
$$4x - y = 3$$
$$3x + y = 4$$
$$\Rightarrow \begin{pmatrix} 4 & -1 \\ 3 & 1 \end{pmatrix}\begin{pmatrix} x \\ y \end{pmatrix} = \begin{pmatrix} 3 \\ 4 \end{pmatrix} \quad\text{....................①}$$
Let $\mathbf{A} = \begin{pmatrix} 4 & -1 \\ 3 & 1 \end{pmatrix}$,
Then,
$$\mathbf{A}^{-1} = \frac{1}{\det \mathbf{A}}(\text{Adj }\mathbf{A}) = \frac{1}{4(1) - 3(-1)}\begin{pmatrix} 1 & 1 \\ -3 & 4 \end{pmatrix}$$
$$\Rightarrow \mathbf{A}^{-1} = \frac{1}{7}\begin{pmatrix} 1 & 1 \\ -3 & 4 \end{pmatrix} = \begin{pmatrix} \dfrac{1}{7} & \dfrac{1}{7} \\ -\dfrac{3}{7} & \dfrac{4}{7} \end{pmatrix}$$

Pre-multiply both sides of equation ① $\mathbf{A}^{-1}$

$$\begin{pmatrix} x \\ y \end{pmatrix} = \begin{pmatrix} \dfrac{1}{7} & \dfrac{1}{7} \\ -\dfrac{3}{7} & \dfrac{4}{7} \end{pmatrix}\begin{pmatrix} 3 \\ 4 \end{pmatrix}$$

$$\Rightarrow \begin{pmatrix} x \\ y \end{pmatrix} = \begin{pmatrix} 1 \\ 1 \end{pmatrix} \text{ i.e. } x = 1 \text{ and } y = 1.$$

Example 43:18

Using the matrix method, solve the equations

$$4x + y = 3$$
$$6x + 2y = 5$$

Solution

$$\begin{pmatrix} 4 & 1 \\ 6 & 2 \end{pmatrix}\begin{pmatrix} x \\ y \end{pmatrix} = \begin{pmatrix} 3 \\ 5 \end{pmatrix} \dots\dots\dots\dots①$$

$$\begin{pmatrix} 4 & 1 \\ 6 & 2 \end{pmatrix}^{-1} = \begin{pmatrix} 1 & -\dfrac{1}{2} \\ -3 & 2 \end{pmatrix}$$

Pre-multiply both sides of equation

① by $\begin{pmatrix} 4 & 1 \\ 6 & 2 \end{pmatrix}^{-1}$.

$$\begin{pmatrix} x \\ y \end{pmatrix} = \begin{pmatrix} \dfrac{1}{2} \\ 1 \end{pmatrix} \text{ i.e } x = \dfrac{1}{2} \text{ and } y = 1$$

Example 43:19

Solve the following simultaneous equations using the matrix method.

(i) $2x + 3y = 2$ (ii) $3p + 4q = 10$

       $4x + 5y = 5$ $2p + 3p = 7$

Solution

(i) $2x + 3y = 2$

       $4x + 5y = 5$

$$\Rightarrow \begin{pmatrix} 2 & 3 \\ 4 & 5 \end{pmatrix}\begin{pmatrix} x \\ y \end{pmatrix} = \begin{pmatrix} 2 \\ 5 \end{pmatrix}$$

$$\begin{pmatrix} 2 & 3 \\ 4 & 5 \end{pmatrix}^{-1} = \begin{pmatrix} -\dfrac{5}{2} & \dfrac{3}{2} \\ 2 & -1 \end{pmatrix}$$

$$\therefore \begin{pmatrix} x \\ y \end{pmatrix} = \begin{pmatrix} -\dfrac{5}{2} & \dfrac{3}{2} \\ 2 & -1 \end{pmatrix}\begin{pmatrix} 2 \\ 5 \end{pmatrix}$$

$$\Rightarrow x = \dfrac{5}{2} \text{ and } y = -1$$

(ii) $3p + 4q = 10$

       $2p + 3p = 7$

$$\begin{pmatrix} 2 & 3 \\ 4 & 5 \end{pmatrix}^{-1} = \begin{pmatrix} -\dfrac{5}{2} & \dfrac{3}{2} \\ 2 & -1 \end{pmatrix}$$

$$\therefore \begin{pmatrix} x \\ y \end{pmatrix} = \begin{pmatrix} -\dfrac{5}{2} & \dfrac{3}{2} \\ 2 & -1 \end{pmatrix}\begin{pmatrix} 2 \\ 5 \end{pmatrix}$$

$$\Rightarrow x = \dfrac{5}{2} \text{ and } y = -1$$

$$\Rightarrow \begin{pmatrix} 3 & 4 \\ 2 & 3 \end{pmatrix}\begin{pmatrix} p \\ q \end{pmatrix} = \begin{pmatrix} 10 \\ 7 \end{pmatrix}$$

$$\begin{pmatrix} 3 & 4 \\ 2 & 3 \end{pmatrix}^{-1} = \begin{pmatrix} 3 & -4 \\ -2 & 3 \end{pmatrix}$$

$$\begin{pmatrix} p \\ q \end{pmatrix} = \begin{pmatrix} 3 & -4 \\ -2 & 3 \end{pmatrix}\begin{pmatrix} 10 \\ 7 \end{pmatrix} = \begin{pmatrix} 2 \\ 1 \end{pmatrix}$$

$$\therefore p = 2 \text{ and } q = 1$$

EXERCISE 43:10

Solve the following simultaneous equations using the matrix method.

1. $p - 3q = 10$ 2. $s = 2t - 1$

    $3p - 2q = 16$ $2s = 3t + 2$

3. $5x - 2y = 14$ 4. $4m + 4n = 3$

    $x + y = 7$ $m + 2n = 1$

5. $2x + y = 7$ 6. $3u - 7v = 1$

    $3x - 2y = 7$ $2u + v = 12$

Information Matrices

Network information is often represented on matrices called information matrices. The most popular information matrices related to networks are route matrices and incidence matrices.

Route and Incidence Matrices

Consider the following network made up of the regions w, x, y and z; the junctions A, B and C in the region and the routes 1,2,3,4 and 5 linking the junctions.

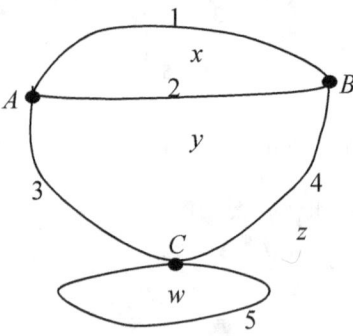

Figure 43:1

A **route matrix** (matrix i) is a matrix that represents the number of direct routes linking each node (junction, town or joint) in a network to another. For instance the following is a route matrix representing the number of direct routes, which link the junctions A, B and C in the network in Figure 43:1

An **incidence matrix** on the other hand is a matrix that shows the relationship between each node and each region (matrix ii) or each node and each route (matrix iii) or each region and each route (matrix iv) in a network.

$$
\begin{array}{c}
\begin{array}{ccc} A & B & C \end{array}\\
\begin{array}{c} A \\ B \\ C \end{array}
\begin{pmatrix}
0 & 2 & 1 \\
2 & 0 & 1 \\
1 & 1 & 1
\end{pmatrix}
\end{array}
$$

Matrix i

$$
\begin{array}{c}
\begin{array}{cccc} w & x & y & z \end{array}\\
\begin{array}{c} A \\ B \\ C \end{array}
\begin{pmatrix}
0 & 1 & 1 & 1 \\
0 & 1 & 1 & 1 \\
1 & 0 & 1 & 1
\end{pmatrix}
\end{array}
$$

Matrix ii

$$
\begin{array}{c}
\begin{array}{ccccc} 1 & 2 & 3 & 4 & 5 \end{array}\\
\begin{array}{c} A \\ B \\ C \end{array}
\begin{pmatrix}
1 & 1 & 1 & 0 & 0 \\
1 & 1 & 0 & 1 & 0 \\
0 & 0 & 1 & 1 & 1
\end{pmatrix}
\end{array}
$$

Matrix iii

$$
\begin{array}{c}
\begin{array}{cccc} w & x & y & z \end{array}\\
\begin{array}{c} 1 \\ 2 \\ 3 \\ 4 \\ 5 \end{array}
\begin{pmatrix}
0 & 1 & 0 & 0 \\
0 & 1 & 1 & 0 \\
0 & 0 & 1 & 0 \\
0 & 0 & 1 & 1 \\
1 & 0 & 0 & 1
\end{pmatrix}
\end{array}
$$

Matrix iv

EXERCISE 43:11

1. Figure 43:2 shows the roads s, t, u, v, w, x, y and z linking four cities A, B, C and D. Build up
   (a) A route matrix to show the network.
   (b) An incidence matrix to show the network

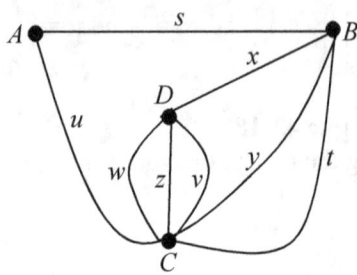

Figure 43:2

2. The network connecting towns P, Q and R shown in Figure 43:3 can be expressed as a route matrix, **M**, where the entries show the number of routes leading to the town. Complete the entries in the matrix, **M**.

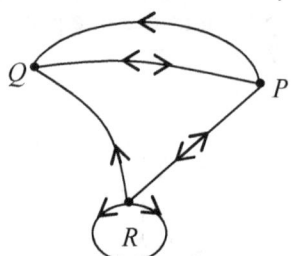

Figure 43:3

$$
\mathbf{M} = \text{from}
\begin{array}{c}
\quad\quad\quad \text{To}\\
\begin{array}{ccc} P & Q & R \end{array}\\
\begin{array}{c} P \\ Q \\ R \end{array}
\begin{pmatrix}
 & 2 & \\
 & 0 & \\
1 & & 2
\end{pmatrix}
\end{array}
$$

MULTIPLE CHOICE EXERCISE 43

1. The order of the matrices $\begin{pmatrix} 3 & 1 & 2 \\ 4 & 0 & 3 \end{pmatrix}$ and $\begin{pmatrix} 4 \\ 0 \\ 3 \end{pmatrix}$ are respectively:

   [A] 2×3 and 1×3   [B] 2×3 and 3×1
   [C] 3×2 and 1×3   [D] 3×2 and 3×1

2. **A** is a 2 by 3 matrix and **B** is a 4 by 2 matrix. The number of elements in **A** and **B** are respectively:
   [A] 4 and 6   [B] 6 and 4
   [C] 8 and 6   [D] 6 and 8

3. An example of a 3 by 2 matrix is:

[A] $\begin{pmatrix} 3 \\ 2 \end{pmatrix}$ [B] $\begin{pmatrix} 0 & 2 & 4 \\ 0 & 0 & 1 \end{pmatrix}$

[C] $\begin{pmatrix} 0 & 1 \\ 0 & -1 \\ 0 & 4 \end{pmatrix}$ [D] $\begin{pmatrix} 3 & 0 \\ 0 & 3 \end{pmatrix}$

4. Let $\mathbf{A} = \begin{pmatrix} 4 & 1 \\ -2 & 3 \end{pmatrix}$. The transpose of $\mathbf{A}$ is:

[A] $\begin{pmatrix} 4 & -2 \\ 1 & 3 \end{pmatrix}$ [B] $\begin{pmatrix} 3 & -2 \\ 1 & 4 \end{pmatrix}$

[C] $\begin{pmatrix} 3 & 1 \\ -2 & 4 \end{pmatrix}$ [D] $\begin{pmatrix} 4 & 1 \\ -2 & 3 \end{pmatrix}$

5. Given that $\mathbf{P} = \begin{pmatrix} 3 & 2 \\ 4 & 5 \end{pmatrix}$ then Adj $\mathbf{P}$ is equal to:

[A] $\begin{pmatrix} -3 & 4 \\ 2 & -5 \end{pmatrix}$ [B] $\begin{pmatrix} 5 & 4 \\ 2 & 3 \end{pmatrix}$

[C] $\begin{pmatrix} 5 & -2 \\ -4 & 3 \end{pmatrix}$ [D] $\begin{pmatrix} 5 & -4 \\ -2 & 3 \end{pmatrix}$

6. Given that $\mathbf{A}$ is a 3 by 2 matrix and $\mathbf{B}$ is a 2 by 4 matrix. The matrix product $\mathbf{AB}$ will have size:

[A] 2 by 3 [B] 3 by 4 [C] 3 by 2 [D] 2 by 4

7. The values of m for which the matrix

$\begin{pmatrix} 6-m & 2 \\ 25 & 1-m \end{pmatrix}$ is singular are:

[A] -11 and 4 [B] -11 and -4
[C] 11 and 4 [D] 11 and -4

8. Given that $\mathbf{A} = \begin{pmatrix} 2 & -3 \\ -2 & 4 \end{pmatrix}$ and $\mathbf{B} = \begin{pmatrix} -2 & 1 \\ -5 & 3 \end{pmatrix}$.

The matrix product $\mathbf{AB}$ is:

[A] $\begin{pmatrix} -11 & 7 \\ 16 & 10 \end{pmatrix}$ [B] $\begin{pmatrix} 11 & -7 \\ 16 & 10 \end{pmatrix}$

[C] $\begin{pmatrix} 11 & -7 \\ -16 & 10 \end{pmatrix}$ [D] $\begin{pmatrix} 11 & 7 \\ -16 & 10 \end{pmatrix}$

9. Given that $\mathbf{A} = \begin{pmatrix} 2 & -3 \\ -2 & 4 \end{pmatrix}$ and

$\mathbf{B} = \begin{pmatrix} -2 & 1 \\ -5 & 3 \end{pmatrix}$. The determinant of the product $\mathbf{AB}$ is:

[A] 2 [B] 222 [C] -222 [D] -2

10. Given that $\mathbf{M} = \begin{pmatrix} 4 & 2 \\ 6 & 1 \end{pmatrix}$, $\mathbf{M}^{-1}$ is equal to:

[A] $\dfrac{1}{8}\begin{pmatrix} 1 & -2 \\ -6 & 4 \end{pmatrix}$ [B] $-\dfrac{1}{8}\begin{pmatrix} -1 & 2 \\ 6 & -4 \end{pmatrix}$

[C] $-\dfrac{1}{8}\begin{pmatrix} 1 & 2 \\ 6 & -4 \end{pmatrix}$ [D] $\dfrac{1}{8}\begin{pmatrix} 1 & -2 \\ -6 & 4 \end{pmatrix}$

11. Given that $\mathbf{A} = \begin{pmatrix} 0 & 2 \\ 4 & -2 \\ -1 & 1 \end{pmatrix}$ and $\mathbf{B} = \begin{pmatrix} -2 & 4 & -1 & 2 \\ 0 & 3 & 0 & 1 \end{pmatrix}$.

The size of the matrix product $\mathbf{AB}$ is:
[A] 2×3 [B] 3×4
[C] 3×2 [D] 2×4

12. The singular matrix among the following matrices is:

[A] $\begin{pmatrix} -2 & 2 \\ 2 & 2 \end{pmatrix}$ [B] $\begin{pmatrix} -2 & -2 \\ 2 & 2 \end{pmatrix}$

[C] $\begin{pmatrix} -2 & -2 \\ 2 & -2 \end{pmatrix}$ [D] $\begin{pmatrix} -2 & -2 \\ -2 & 2 \end{pmatrix}$

13. Given that the matrix $\begin{pmatrix} a & 1 \\ 4 & a \end{pmatrix}$ is singular, the possible values of a are:
[A] 4 and -1 [B] 4 and 1
[C] -4 and 1 [D] -2 and 2

14. The inverse of the matrix $\begin{pmatrix} 5 & 2 \\ 7 & 3 \end{pmatrix}$ is:

[A] $\begin{pmatrix} -3 & -7 \\ -2 & -5 \end{pmatrix}$ [B] $\begin{pmatrix} -3 & 2 \\ 7 & -5 \end{pmatrix}$

[C] $\begin{pmatrix} 3 & -2 \\ -7 & 5 \end{pmatrix}$ [D] $\begin{pmatrix} 3 & -7 \\ -2 & 5 \end{pmatrix}$

15. The value of a for which the matrix

$\begin{pmatrix} a & 3-a \\ 2 & 1 \end{pmatrix}$ is singular is:

[A] 2 [B] -2 [C] 3 [D] -3

16. Given that $\mathbf{PQ = I}$, where $\mathbf{P} = \begin{pmatrix} 5 & -4 \\ -1 & 1 \end{pmatrix}$

and $\mathbf{I} = \begin{pmatrix} 1 & 0 \\ 0 & 1 \end{pmatrix}$. The matrix $\mathbf{Q}$ is:

[A] $\begin{pmatrix} 1 & 4 \\ -1 & 5 \end{pmatrix}$ [B] $\begin{pmatrix} 1 & -4 \\ -1 & 5 \end{pmatrix}$

[C] $\begin{pmatrix} 1 & -4 \\ 1 & 5 \end{pmatrix}$ [D] $\begin{pmatrix} 1 & 4 \\ 1 & 5 \end{pmatrix}$

17. The matrices $\mathbf{P} = \begin{pmatrix} m & 2 \\ 5 & 0 \end{pmatrix}$ and

$\mathbf{Q} = \begin{pmatrix} 2 & -m \\ -2 & 1 \end{pmatrix}$ are such that $\mathbf{P + Q}$ is singular. The value of m is:
[A] -1 [B] 1 [C] 4 [D] -4

18. The inverse of the 2×2 matrix $\begin{pmatrix} 6 & 10 \\ 2 & 4 \end{pmatrix}$ is:

[A] $\begin{pmatrix} 1 & -2\frac{1}{2} \\ -\frac{1}{2} & 1\frac{1}{2} \end{pmatrix}$ [B] $\begin{pmatrix} 4 & -10 \\ -2 & 6 \end{pmatrix}$

[C] $\begin{pmatrix} -2 & 6 \\ -10 & 4 \end{pmatrix}$ [D] $\begin{pmatrix} 1 & 2\frac{1}{2} \\ \frac{1}{2} & 1\frac{1}{2} \end{pmatrix}$

19. The transpose of the matrix $\begin{pmatrix} 7 & 3 \\ -1 & 5 \end{pmatrix}$ is:

[A] $\begin{pmatrix} 5 & -3 \\ 1 & 7 \end{pmatrix}$ [B] $\begin{pmatrix} 7 & -1 \\ 3 & 5 \end{pmatrix}$

[C] $\begin{pmatrix} -7 & -1 \\ 3 & -5 \end{pmatrix}$ [D] $\begin{pmatrix} -7 & 1 \\ -3 & -5 \end{pmatrix}$

20. Given that $\begin{pmatrix} 2 & 1 \\ 1 & -1 \end{pmatrix}\begin{pmatrix} 1 & -2 \\ 1 & 2 \end{pmatrix} = 3\mathbf{A}$. The matrix $\mathbf{A}$ is:

[A] $\begin{pmatrix} 3 & -2 \\ 0 & -4 \end{pmatrix}$ [B] $\begin{pmatrix} 1 & \frac{2}{3} \\ 0 & \frac{4}{3} \end{pmatrix}$

[C] $\begin{pmatrix} 3 & \frac{2}{3} \\ 0 & \frac{4}{3} \end{pmatrix}$ [D] $\begin{pmatrix} 3 & -\frac{2}{3} \\ 0 & -\frac{4}{3} \end{pmatrix}$

21. The determinant of the matrix $\mathbf{A} = \begin{pmatrix} x & 7 \\ 4 & 2 \end{pmatrix}$ is 6. The value of x is:

[A] -17 [B] -34 [C] 17 [D] 34

22. Let $\mathbf{M} = \begin{pmatrix} 3 & 1 \\ 1 & 2 \end{pmatrix}$ and $\mathbf{N} = \begin{pmatrix} 1 & 4 \\ -3 & 2 \end{pmatrix}$. As a single matrix, $\mathbf{M} + 3\mathbf{N}$ is equal to:

[A] $\begin{pmatrix} 10 & 7 \\ 1 & 4 \end{pmatrix}$ [B] $\begin{pmatrix} 10 & 7 \\ -1 & 4 \end{pmatrix}$

[C] $\begin{pmatrix} 6 & 13 \\ 8 & -8 \end{pmatrix}$ [D] $\begin{pmatrix} 6 & 13 \\ -8 & 8 \end{pmatrix}$

23. The inverse of the 2×2 matrix $\begin{pmatrix} 6 & 11 \\ 2 & 4 \end{pmatrix}$ is:

[A] $\begin{pmatrix} 2 & -5\frac{1}{2} \\ -1 & 3 \end{pmatrix}$ [B] $\begin{pmatrix} 4 & 2 \\ -11 & 6 \end{pmatrix}$

[C] $\begin{pmatrix} 3 & -5\frac{1}{2} \\ -1 & 2 \end{pmatrix}$ [D] $\begin{pmatrix} 0 & 0 \\ 0 & 0 \end{pmatrix}$

24. The adjoint of the matrix $\begin{pmatrix} 7 & 3 \\ -1 & 5 \end{pmatrix}$ is:

[A] $\begin{pmatrix} 5 & -3 \\ 1 & 7 \end{pmatrix}$ [B] $\begin{pmatrix} 7 & -1 \\ 3 & 5 \end{pmatrix}$

[C] $\begin{pmatrix} -7 & -1 \\ 3 & -5 \end{pmatrix}$ [D] $\begin{pmatrix} -7 & 1 \\ -3 & -5 \end{pmatrix}$

25. Given that $p = 2q - 1$ and $2p = 3q + 2$, then:

[A] $\begin{pmatrix} 1 & 2 \\ 2 & 3 \end{pmatrix}\begin{pmatrix} p \\ q \end{pmatrix} = \begin{pmatrix} -1 \\ 2 \end{pmatrix}$

[B] $\begin{pmatrix} 2 & -1 \\ 3 & 2 \end{pmatrix}\begin{pmatrix} p \\ q \end{pmatrix} = \begin{pmatrix} 1 \\ 2 \end{pmatrix}$

[C] $\begin{pmatrix} 1 & -1 \\ 2 & 2 \end{pmatrix}\begin{pmatrix} p \\ q \end{pmatrix} = \begin{pmatrix} 2 \\ 3 \end{pmatrix}$

[D] $\begin{pmatrix} 1 & -2 \\ 2 & -3 \end{pmatrix}\begin{pmatrix} p \\ q \end{pmatrix} = \begin{pmatrix} -1 \\ 2 \end{pmatrix}$

26. The incorrect statement among the following is:
   [A] A square matrix is necessarily a diagonal matrix.
   [B] A diagonal matrix is necessarily a square matrix.
   [C] A unit matrix is necessarily a square matrix.
   [D] A unit matrix is necessarily a diagonal matrix.

TOPIC 44

VECTORS

OBJECTIVES

At the end of this topic, the learner should be able to:

1. Distinguish between vector and scalar quantities.
2. Represent vectors by directed line segments.
3. State the direction and sense of a vector.
4. Identify collinear and non-collinear vectors.
5. Denote vectors using various notations.
6. Interpret vectors written in column vector form and in **i** and **j** vector form.
7. Find the magnitude of a vector
8. State and use the conditions for vectors to be equal.
9. Distinguish between fixed and free vectors and position vectors.
10. Add and subtract vectors written in column vector form and in **i** and **j** vector form.
11. Add and subtract vectors diagrammatically.
12. Identify a zero or null vector.
13. Use the parallelogram law of vectors to add and subtract vectors.
14. Multiply of vectors by scalars.
15. Apply distributive property of vectors.
16. State and apply the section theorem to solve problems on vectors.
17. Find the position vector of the midpoint of a vector.
18. Find the angle between two vectors.

Vector and Scalar Quantities

Figure 44:1, is a map of a village of four quarters, namely; north quarter N, west quarter W, south quarter S and east quarter E. A straight road links N and S and another links W and E in such a way that they meet at the junction J. Each quarter is 4 km away from J.

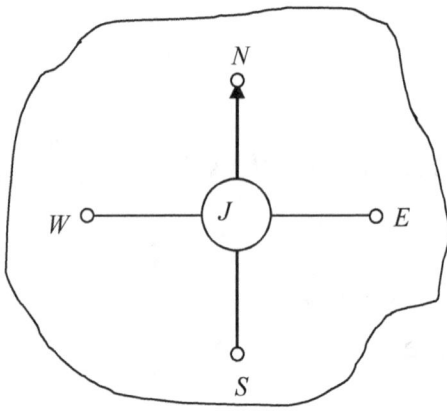

Figure 44:1

Three boys Nfor, Kanjo and Shey while standing at J, describe the journeys they will undertake. Nfor says "I will travel 4 km from here". Kanjo says "I will travel 4 km along the road linking W and E". Shey says "I will travel 4 km towards E".

Answer the following question giving reasons for your answers.

1. What distance will each of them cover?
2. To which quarter do you think each of them is going to?

From the statements they make it is clear that each of them will cover a **distance** of 4 km. 4 km is called the **magnitude, size, length, modulus or norm** of the journey.
One cannot be able to say exactly to which quarter Nfor will be going. From Kanjo's statement, he will be going to either W or E, though he has not stated precisely whether he will go to W or he will go to E. Kanjo has mentioned his **distance** and **direction**. If Kanjo travels from J to W, his direction is the same as if he travels from J to E.
It is certain that Shey will travel to E, since he is traveling towards E and is going to cover a

distance of 4 km. Shey's description towards E" gives the **sense** of his journey.
The distance covered in a specific direction is called **displacement.**
In the statement made by Shey above, three pieces of information have been given.
1. The **length**, **magnitude**, **size**, **norm** or **modulus** is 4 km.
2. The **direction** is along the road joining W and E.
3. The **sense** is towards E.

A quantity carrying all these three pieces of information is called a **vector quantity** or simply a **vector**. Examples of vector quantities are: displacement, velocity, momentum, force, acceleration etc.
A **scalar quantity** or **scalar** on the other hand has no direction and hence no sense. Examples of scalar quantities are mass, temperature, distance, speed, area, length, volume, time, age marks, price etc.

Vectors as Directed Line Segments

Figure 44:2

If L is a straight line and X and Y are any two points on the line, then XY is called a **line segment**. Therefore, *a line segment is simply part of a line*. A line segment with an arrow at one end is called a **directed line segment**.
Diagrammatically, vectors can be represented using directed line segments. The end carrying the arrow is called the **head** while the other end is called the **tail** of the vector.

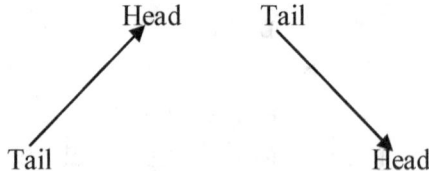

Figure 44:3

The length of a directed line segment, gives the magnitude of the vector, the orientation of the line gives the direction while the arrow gives the sense of the vector.

Direction and Sense of a Vector

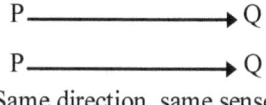

Same direction, same sense

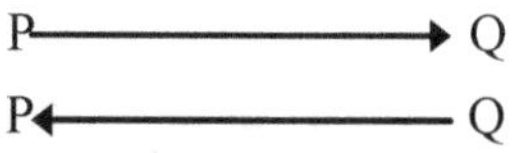

Same direction, opposite senses

Figure 44:4

Suppose P and Q are two points on a straight line, then the vector from P to Q is said to be in the same direction as the vector from Q to P, but in an opposite sense as illustrated below.

Thus for two vectors to be in the same direction, they must either line on the same straight line or they must at least be parallel to each other. If two vectors lie on the same straight line, they are said to be **collinear**. Otherwise they are **non collinear**.

Vector Notation

A vector represented by the directed line segment AB Figure 44:5, is denoted by $\underline{a}$, $\vec{a}$ or $\overrightarrow{AB}$. The arrow emphasizes that the sense is from A to B. In textbooks, bold print is used. Thus the in Figure 44:5, is denoted by **AB** or **a**. In using bold print arrows or bars are not necessary.

Figure 44:5

The magnitude of the vector **AB** or **a,** above is denoted by $|\underline{a}|$, $|\vec{a}|$ or $|\overrightarrow{AB}|$ or AB and is defined as the length of the line segment from A to B.

Column Vectors

A vector **AB** that is a units in the Ox direction and b units in the Oy direction is written as $\begin{pmatrix} a \\ b \end{pmatrix}$ in column vector form.

a is called the "x **component** of **AB**" while b is called the y **component** of **AB**".

Example 44:1

Write the vectors on the Cartesian plane in Figure 44:6 as column vectors.

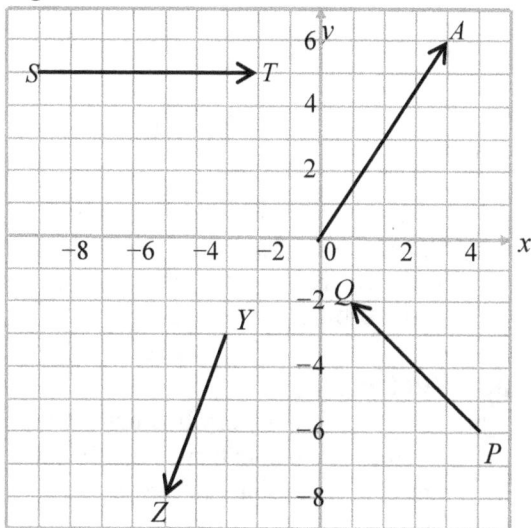

Figure 44:6

Solution

$$\overrightarrow{OA} = \begin{pmatrix} \\ 6 \end{pmatrix}, \overrightarrow{ST} = \begin{pmatrix} \\ 0 \end{pmatrix}, \overrightarrow{YZ} = \begin{pmatrix} \\ -5 \end{pmatrix}, \overrightarrow{PQ} = \begin{pmatrix} \\ 4 \end{pmatrix}$$

⮞ EXERCISE 44:1

1. Represent the following on a Cartesian plane.

   $\mathbf{OA} = \begin{pmatrix} 5 \\ 2 \end{pmatrix}$, $\mathbf{OB} = \begin{pmatrix} -5 \\ 3 \end{pmatrix}$, $\mathbf{OC} = \begin{pmatrix} -4 \\ 1 \end{pmatrix}$, $\mathbf{OD} = \begin{pmatrix} 6 \\ -2 \end{pmatrix}$

2. Write down the vectors in Figure 44:7 as column vectors.

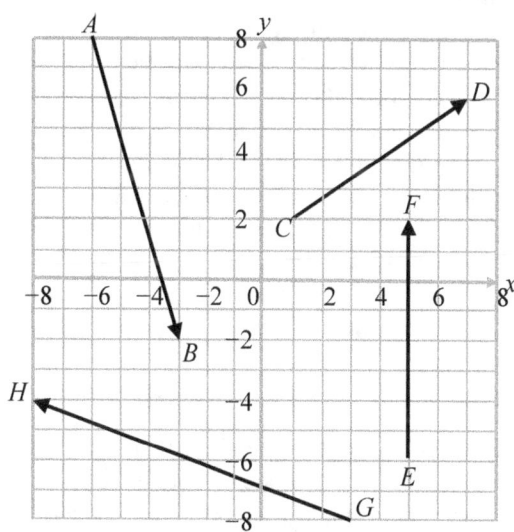

Figure 44:7

3. Given that A, B, C, D and E are the

491

points(2, 5), (−1, 4), (0, 3), (−5, −6), and (7, −2) respectively. Represent the following vectors on a Cartesian plane.

$$\mathbf{AV} = \begin{pmatrix} 2 \\ 4 \end{pmatrix}, \mathbf{BW} = \begin{pmatrix} -3 \\ 5 \end{pmatrix}, \mathbf{CX} = \begin{pmatrix} -2 \\ -3 \end{pmatrix},$$

$$\mathbf{DY} = \begin{pmatrix} 0 \\ 6 \end{pmatrix}, \mathbf{EZ} = \begin{pmatrix} -5 \\ 0 \end{pmatrix}$$

4. Given that A, B, C, D, E are the points (2, 5), (−1, 4), (0, 3), (−5, −6), and (7, −2) respectively. Represent the vectors **AB**, **CE** and **BD** on a Cartesian plane.

Base Vectors

Suppose that **a** and **b** are any two non-parallel vectors in a plane, then **a** and **b** can only intersect at one point I as shown in Figure 44:8.

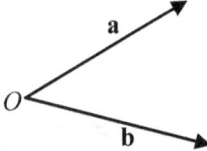

Figure 44:8

Any vector in the plane containing the vectors **a** and **b** and the point O can be expressed in terms of **a** and **b**. Vectors such as **a** and **b** which are used to express other vectors in the same plane, are called base vectors. If the magnitude of any vector is 1, the vector is called a **unit vector**. In the x-y plane a unit vector in the direction Ox is denoted by **i** while a unit vector in the direction Oy is denoted by **j**. Since the x and y axes meet at O the origin, and **i** and **j** are in the directions Ox and Oy respectively, **i** and **j** can be used as base vectors in the x-y plane. Base vectors such as **i** and **j** whose magnitude is 1 are called **unit base vectors**. Using **i** and **j** as base vectors in the x-y plane, a units in the direction OX is written a**i** while b units in the direction OY is written b**j**

Thus the column vectors $\begin{pmatrix} 2 \\ 3 \end{pmatrix}$ and $\begin{pmatrix} -4 \\ 3 \end{pmatrix}$ can be written as 2**i** + 3**j** and −4**i** + 3**j** respectively.

If two vectors are perpendicular they are said to be **orthogonal**. Since the x and y axes are perpendicular **i** and **j** are orthogonal vectors. Unit vectors such as **i** and **j**, which are perpendicular, are said to be **orthonormal**.

Thus orthonormal means the vectors are:
(i) Unit vectors
(ii) Perpendicular (orthogonal).

Therefore, the unit vectors **i** and **j** are **orthornomal base vectors**.

Example 44:2
If **a** and **b** are unit vectors in the direction **OA** and **OB** as shown in the Figure 44:9

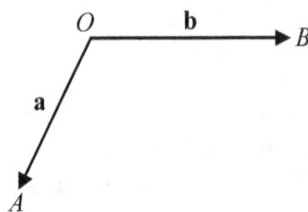

Figure 44:9

Write the following in terms of **a** and **b**.
(i) 7 units in the direction **OA** and 3 units in the direction **OB**.
(ii) −3 units in the direction **OA** and 4 units in the direction **OB**.
(iii) x units in the direction **OA** and y units in the direction **OB**.

Solution
(i) $7\mathbf{a} + 3\mathbf{b}$ (ii) $-3\mathbf{a} + 4\mathbf{b}$ (iii) $x\mathbf{a} + y\mathbf{b}$

> **EXERCISE 44:2**

1. Represent the following position vectors on a Cartesian plane.
   (i) $\mathbf{i} + 4\mathbf{j}$ (ii) $-2\mathbf{i} + 3\mathbf{j}$
   (iii) $5\mathbf{i} - \mathbf{j}$ (iv) $-3\mathbf{i} - 4\mathbf{j}$

2. Write the following column vectors in terms of **i** and **j**.
   (i) $\begin{pmatrix} 2 \\ 5 \end{pmatrix}$ (ii) $\begin{pmatrix} 5 \\ -3 \end{pmatrix}$ (iii) $\begin{pmatrix} -4 \\ -1 \end{pmatrix}$ (iv) $\begin{pmatrix} 6 \\ -2 \end{pmatrix}$

3. Using the unit base vectors **p** and **q** in Figure 44:10 above represent the following.
   (i) 3 units in the direction **OP** and 5 units in the direction **OQ**.
   (ii) −7 units in the direction **OP** and 2 units in the direction **OQ**.
   (iii) −4 units in the direction **OP** and −2 units in the direction **OQ**.

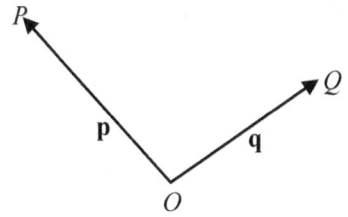

Figure 44:10

4. Write the following vectors as column vectors.
   (i) $-3\mathbf{i} + 2\mathbf{j}$ (ii) $2\mathbf{i} - 5\mathbf{j}$
   (iii) $9\mathbf{i} + 4\mathbf{j}$ (iv) $-7\mathbf{i} - 3\mathbf{j}$

The Magnitude of a Vector

Let $\mathbf{AB} = \begin{pmatrix} x \\ y \end{pmatrix} = x\mathbf{i} + y\mathbf{j}$

The Pythagoras theorem can be used to find the magnitude of **AB**, since **i** and **j** are perpendicular.

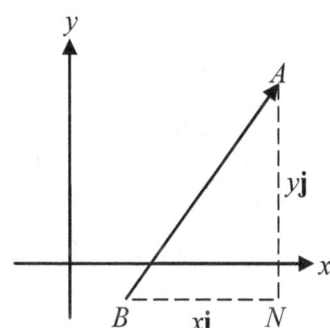

Figure 44:11

Thus, $\left| \mathbf{AB} \right| = \sqrt{x^2 + y^2}$.

Example 44:3
Find the modulus of the following vectors:

(i) $PQ = \begin{pmatrix} -4 \\ 3 \end{pmatrix}$ (ii) $\mathbf{XY} = 5\mathbf{i} - 12\mathbf{j}$

Solution

(i) $\mathbf{PQ} = \sqrt{(-4)^2 + 3^2} = \sqrt{25} = 5$ units

(ii) $\mathbf{XY} = \sqrt{5^2 + (-12)^2} = \sqrt{169} = 13$ units

Note!!
The magnitude of a vector is always positive.

If the end points of the vector **AB** are $A(x_1, y_1)$ and $B(x_2, y_2)$ the modulus will be the distance between the points A and B.

Thus, $AB = \sqrt{\left(x_2 - x_1 \right)^2 + \left(y_2 - y_1 \right)^2}$

Example 44:4
Calculate the magnitude of the vector **YZ** where Y and Z are the points $(-2, 2)$ and $(4, -6)$ respectively.

Solution
$\mathbf{YZ} = \sqrt{(-6 - 2)^2 + (4 - (-2))^2} = \sqrt{100} = 10$ units

EXERCISE 44:3

1. Find the magnitudes of the vectors:

   (i) $\begin{pmatrix} 5 \\ 12 \end{pmatrix}$ (ii) $2\mathbf{i} - 3\mathbf{j}$ (iii) $\begin{pmatrix} -7 \\ 3 \end{pmatrix}$

   (iv) $2\mathbf{i} - 3\mathbf{j}$ (v) $\begin{pmatrix} -8 \\ -6 \end{pmatrix}$ (vi) $-4\mathbf{i} + 3\mathbf{j}$

2. Calculate the modulus of the vectors joining the given points.
   (i) $(3, 5)$ and $(-2, 1)$ (ii) $(-4, 7)$ and $(0, 3)$
   (iii) $(-2, -1)$ and $(-5, -3)$

3. Which of the following are unit vectors?

   (i) $\mathbf{i} + \mathbf{j}$ (ii) $\mathbf{i}$ (iii) $\frac{1}{2}\mathbf{i} + \frac{1}{2}\mathbf{j}$ (iv) $\frac{3}{5}\mathbf{i} + \frac{4}{5}\mathbf{j}$

   (v) $\mathbf{j}$ (vi) $\frac{1}{\sqrt{2}}\mathbf{i} - \frac{1}{\sqrt{2}}\mathbf{j}$ (vii) $-\frac{\sqrt{3}}{2}\mathbf{i} + \frac{1}{2}\mathbf{j}$

4. Given that $\mathbf{u} = -3\mathbf{i} + 4\mathbf{j}$. Find the magnitude of **u**.

5. Find $\left| \mathbf{V} \right|$ if $\mathbf{V} = 3\mathbf{i} - 2\mathbf{j}$.

VECTOR ALGEBRA

Equality of Vectors
Two vectors **a** and **b** are equal if and only if they satisfy the following conditions.
 (i) They have the same magnitude
 (ii) They are in the same direction (parallel or collinear)
 (iii) They have the same sense.

Example 44:5

Which of the vectors in Figure 44:12 are equal?

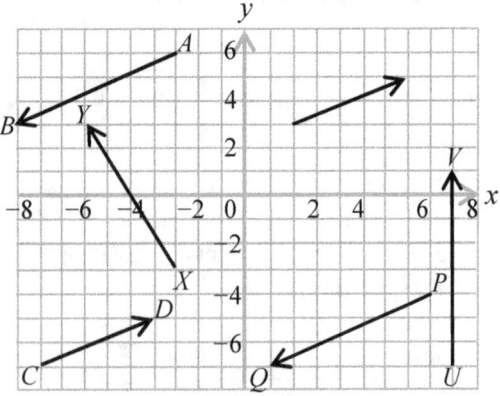

Figure 44:12

Solution

$$AB = PQ = \begin{pmatrix} -7 \\ -3 \end{pmatrix} \qquad CD = MN = \begin{pmatrix} 5 \\ 2 \end{pmatrix}$$

Fixed and Free Vectors

From Figure 44:12 it can be seen that though the vectors **AB** and **PQ** are equal, they can be represented at different positions on a Cartesian plane. This is equally true of **CD** and **MN**. A vector which can be situated anywhere in a plane is called **free vector**. Free vectors have no particular point of action and only their magnitude, direction and sense is important. The point of action of some other vectors is very important. These types of vectors therefore have a fixed position in a plane or space. They are thus called **fixed vectors**. A practical example of a fixed vector is the gravitation force acting on an object. The point of action of this force is the center of mass of the object and the direction and sense is towards the center of the earth.

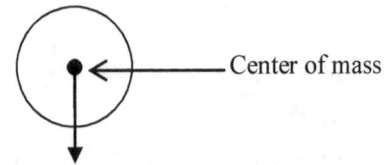

Figure 44:13

Another example is a vector **EF** whose end points are given such as E (2, 3) and F (–7, 4).

Position Vectors

If $P(x, y)$ is any point in the x-y plane, then the position vector of the point P is denoted by **r** and is given by

$$\mathbf{r} = \begin{pmatrix} x \\ y \end{pmatrix} \text{ or } \mathbf{r} = x\mathbf{i} + y\mathbf{j}$$

A position vector is another example of a fixed vector discussed earlier and is never a free vector. This is because a position vector is always tied to the origin. This means that tail must be at the origin.

Example 44:6

Draw the position vectors of the points A (3, 1) and B (–5, 4) on a Cartesian plane.

Solution

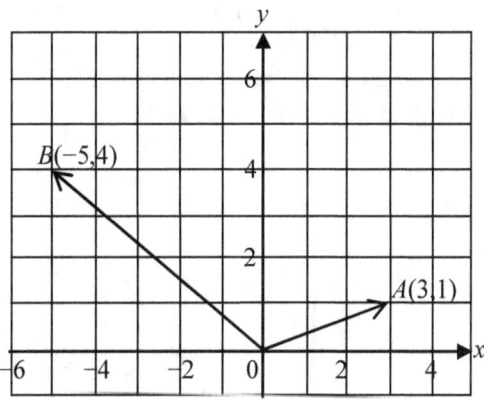

Figure 44:14

Example 44:7

Write the position vector of the point H (–7, 5) in
(a) Column vector form
(b) Component form.

Solution

(a) In column vector form, $\mathbf{r} = \begin{pmatrix} -7 \\ 5 \end{pmatrix}$

(b) In component form, $\mathbf{r} = -7\mathbf{i} + 5\mathbf{j}$.

▷ **EXERCISE 44:4**

1. Write the position vectors of the points labeled A to F in:
   (a) Component form
   (b) Column vector form.

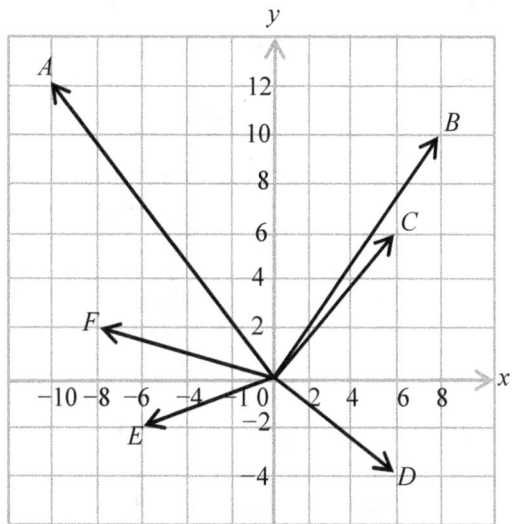

Figure 44:15

2. (i) Write the position vector of the points
   A(–1,5), *B* (4,–7), *C* (–9, 3) and *D* (–3, 6)
   (a) Component form
   (b) Column vector form.
   (ii) Represent these position vectors on the
   Cartesian plane.
3. Which of the vectors in Figure 44:16 are
   equal?

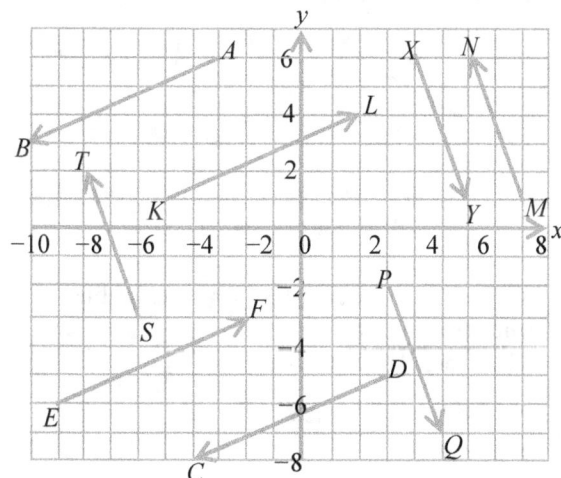

Figure 44:16

4. Which of the following vectors are equal?
   (a) $-5\mathbf{i} - 2\mathbf{j}$ (b) $2\mathbf{i} + 5\mathbf{j}$
   (c) $-2\mathbf{i} - 5\mathbf{j}$ (d) $5\mathbf{i} + 2\mathbf{j}$
   (e) $2(-5\mathbf{i} - 2\mathbf{j})$ (f) $4(2\mathbf{i} + 5\mathbf{j})$
   (g) $-1(-2\mathbf{i} - 5\mathbf{j})$ (h) $3(5\mathbf{i} + 2\mathbf{j})$
   (i) $\begin{pmatrix} -2 \\ -5 \end{pmatrix}$ (j) $\begin{pmatrix} 2 \\ 5 \end{pmatrix}$
   (k) $\begin{pmatrix} 5 \\ 2 \end{pmatrix}$ (l) $\begin{pmatrix} -5 \\ -2 \end{pmatrix}$ (m) $-1\begin{pmatrix} -2 \\ -5 \end{pmatrix}$

(n) $3\begin{pmatrix} 2 \\ 5 \end{pmatrix}$ (o) $4\begin{pmatrix} 5 \\ 2 \end{pmatrix}$ (p) $2\begin{pmatrix} -5 \\ -2 \end{pmatrix}$

5. Which of the vectors in Figure 44:17 are
   position vectors relative to the origin?

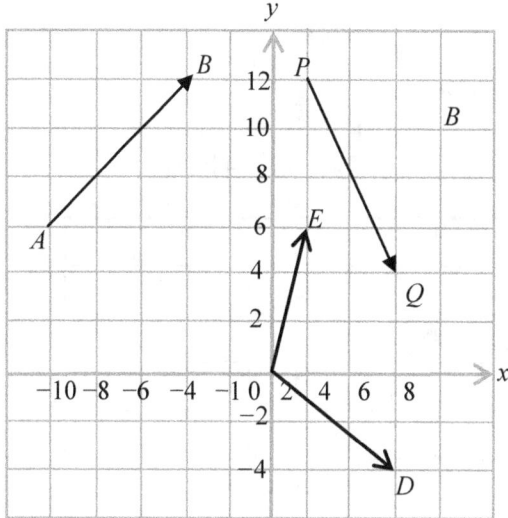

Figure 44:17

Addition of Vectors

Adding Column vectors
Corresponding entries are added just in the
same way as matrix addition is done.

Example 44:8

Find $\mathbf{x} + \mathbf{y}$ if $\mathbf{x} = \begin{pmatrix} 3 \\ 4 \end{pmatrix}$ and $\mathbf{y} = \begin{pmatrix} 1 \\ -2 \end{pmatrix}$.

Solution
$$\mathbf{x} + \mathbf{y} = \begin{pmatrix} 3 \\ 4 \end{pmatrix} + \begin{pmatrix} 1 \\ -2 \end{pmatrix} = \begin{pmatrix} 4 \\ 2 \end{pmatrix}$$

Adding Vectors Written in component form
To add vectors written in component form,
like components are added.

Example 44:9
If $\mathbf{a} = 3\mathbf{i} + 4\mathbf{j}$ and $\mathbf{b} = \mathbf{i} + 2\mathbf{j}$, find $\mathbf{a} + \mathbf{b}$.

Solution
$\mathbf{a} + \mathbf{b} = (3\mathbf{i} + 4\mathbf{j}) + (\mathbf{i} + 2\mathbf{j}) = 4\mathbf{i} + 6\mathbf{j}$

Adding Vectors Diagrammatically

Diagrammatically vectors are added by drawing to scale the vectors in such a way that the head of one vector is at the tail of the other. The sum is then the vector whose tail is at the tail of the first vector and whose head is at the head of the second vector.

Example 44:10
Find the sum of the vectors **a** and **b** shown diagrammatically (Figure 44:18).

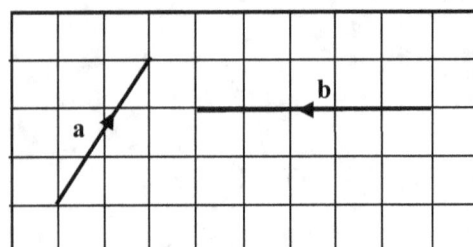

Figure 44:18

Solution

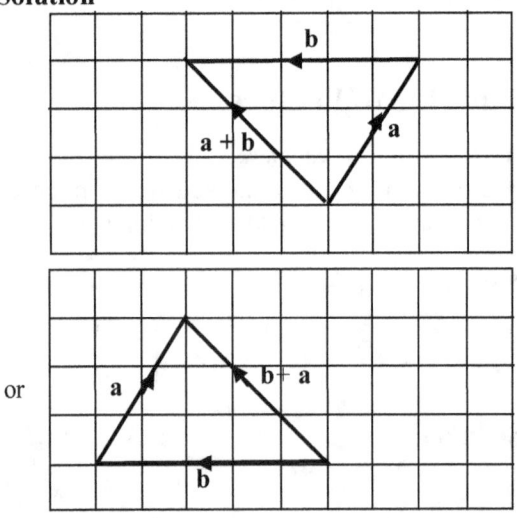

Figure 44:19

The sum is then obtained by completing and measuring the completed side of the triangle called the **triangle of vectors**.

By measuring **a** + **b** and **b** + **a** from the two triangles above it can be seen that **a + b = b + a**
.

This means that vector addition is commutative.

The polygon of vectors

The idea of the triangle of vectors can be extended to any other polygon, depending on the number of vectors to be added.

Example 44:11
Use Figure 44:20 to show that
AE = a + b + c + d.

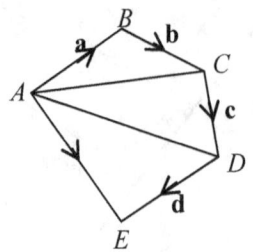

Figure 44:20

Solution

$$A\,C = A\,B + B\,C \Rightarrow A\,C = a + b$$
$$A\,D = A\,C + C\,D = a + b + c$$
$$A\,E = A\,D + D\,E = a + b + c + d$$

Hence, **A B + B C + C D + D E = A E** .

The Additive Inverse of a Vector

Consider the vector **AB** in Figure 44:21 (i).The sense of **AB** can be reversed to obtain **BA** as in Figure 44:21 (ii)

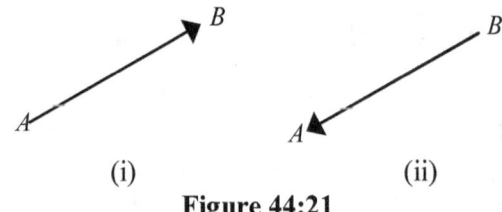

(i)   (ii)

Figure 44:21

BA is called the **additive inverse** (or the **opposite**) of **AB**.
Therefore, **BA = –BA**

The additive inverse of **a** is **–a**

The additive inverse of $\begin{pmatrix} x \\ y \end{pmatrix}$ is $\begin{pmatrix} -x \\ -y \end{pmatrix}$.

The additive inverse of $a\mathbf{i} + b\mathbf{j}$ is $-a\mathbf{i} - b\mathbf{j}$.

Subtraction of Vectors

Subtracting Column vectors
To subtract vectors written in component form, corresponding entries are subtracted.

Example 44:12

Find $\mathbf{a} - \mathbf{b}$ if $\mathbf{a} = \begin{pmatrix} 4 \\ 1 \end{pmatrix}$ and $\mathbf{b} = \begin{pmatrix} 7 \\ 3 \end{pmatrix}$.

Solution

$\mathbf{a} - \mathbf{b} = \begin{pmatrix} 4 \\ 1 \end{pmatrix} - \begin{pmatrix} 7 \\ 3 \end{pmatrix} = \begin{pmatrix} -3 \\ -2 \end{pmatrix}$

Subtracting component form Vectors

To subtract vectors written in component form, like components are subtracted.

Example 44:13

If $\mathbf{a} = 7\mathbf{i} + 3\mathbf{j}$ and $\mathbf{b} = 4\mathbf{i} - \mathbf{j}$. Find $\mathbf{a} - \mathbf{b}$.

Solution

$\mathbf{a} - \mathbf{b} = 7\mathbf{i} + 3\mathbf{j}) - (4\mathbf{i} - \mathbf{j}) = 3\mathbf{i} + 4\mathbf{j}$

Subtracting Vectors Diagrammatically

Subtraction of vectors is the same as the addition of inverses. This means that $\mathbf{a} - \mathbf{b} = \mathbf{a} + (-\mathbf{b})$.

This idea can be used to subtract vectors diagrammatically.

Example 44:14

Figure 44:22, shows the representation of the vectors $\mathbf{a}$ and $\mathbf{b}$. Draw a diagram to show the vector $\mathbf{a} - \mathbf{b}$.

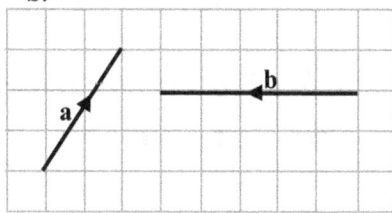

Figure 44:22

Solution

Figure 44:23, shows the vector $-\mathbf{b}$.

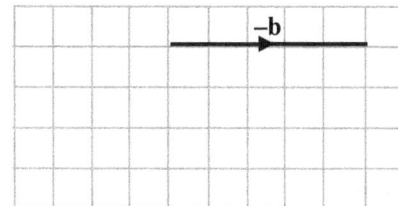

Figure 44:23

Figure 44:24 shows how $\mathbf{a}$ and $-\mathbf{b}$ are added diagrammatically to obtain $\mathbf{a} - \mathbf{b}$.

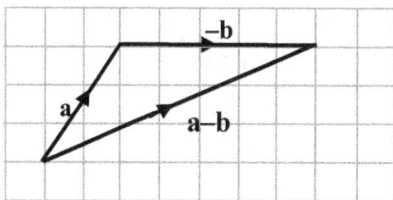

or

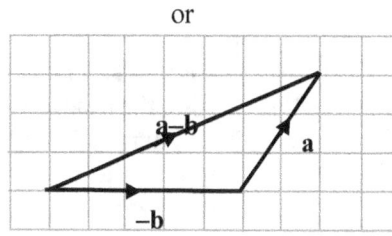

Figure 44:24

The Zero or Null Vector

Consider two vectors $\mathbf{a}$ and $\mathbf{b}$ such that $\mathbf{a} = \mathbf{b}$
Then $\mathbf{a} - \mathbf{b} = \mathbf{a} + (-\mathbf{b}) = \mathbf{a} + (-\mathbf{a})$
But $\mathbf{a} - \mathbf{b} = \mathbf{a} - \mathbf{a} = \mathbf{0}$
$\therefore \mathbf{a} - \mathbf{a} = \mathbf{0}$

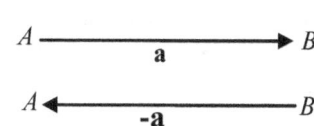

Figure 44:25

$\mathbf{a} + (-\mathbf{a})$ therefore, means a journey from A to B and back to A, or simply a zero displacement. Such a vector is called a **zero vector** or **null vector** and is denoted by $\mathbf{0}$.

Hence, $\mathbf{0} = \begin{pmatrix} 0 \\ 0 \end{pmatrix}$.

The Parallelogram Law of Vectors

If two vectors $\mathbf{OA} = \mathbf{a}$ and $\mathbf{OC} = \mathbf{c}$ have a common point of action O (i.e. if their tails are at the same point O) but are acting in different directions as shown in the diagram below, then their sum and difference can be obtained by completing the parallelogram $OABC$ and measuring its diagonals.

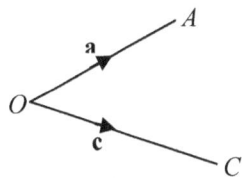

Figure 44:26

497

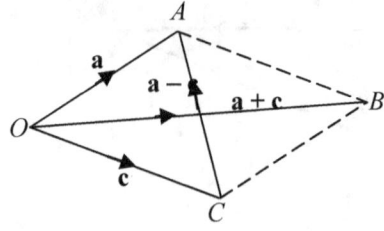

Figure 44:27

The diagonal AC joining the heads represents the difference while the diagonal OB, represents the sum of the vectors.

Thus $\mathbf{a} + \mathbf{c} = \mathbf{OB}$ and $\mathbf{a} - \mathbf{c} = \mathbf{CA}$

The sense of the difference $\mathbf{CA} = \mathbf{a} - \mathbf{c}$ is obtained by noting that $-\mathbf{c}$ means the additive inverse $+ (-\mathbf{c})$ of $\mathbf{c}$.

EXERCISE 44:5

1. Figure 44:28, shows two vectors $\mathbf{a}$ and $\mathbf{b}$. Illustrate on different Cartesian planes how you would diagrammatically find:
   (i) $\mathbf{a} + \mathbf{b}$ (ii) $\mathbf{a} - \mathbf{b}$ (iii) $\mathbf{b} - \mathbf{a}$

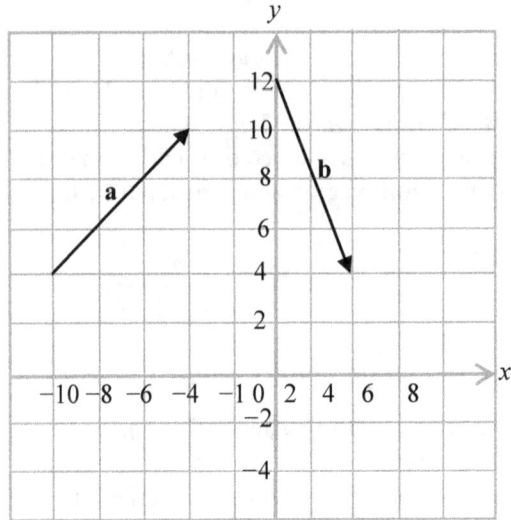

Figure 44:28

2. If $\mathbf{u} = 3\mathbf{i} - \mathbf{j}$ and $\mathbf{v} = -5 + 3\mathbf{j}$, find
   (i) $\mathbf{u} + \mathbf{v}$ (ii) $\mathbf{v} + \mathbf{u}$
   (iii) $\mathbf{u} - \mathbf{v}$ (iv) $\mathbf{v} - \mathbf{u}$

3. Give the additive inverse of each of the following vectors:
   (a) $-\mathbf{i} - 2\mathbf{j}$ (b) $2\mathbf{i} + 5\mathbf{j}$
   (c) $2\mathbf{i} - 7\mathbf{j}$ (d) $-\mathbf{i} + 2\mathbf{j}$

(e) $\begin{pmatrix} -2 \\ -4 \end{pmatrix}$ (f) $\begin{pmatrix} 7 \\ 5 \end{pmatrix}$ (g) $\begin{pmatrix} -3 \\ 2 \end{pmatrix}$ (h) $\begin{pmatrix} 8 \\ -3 \end{pmatrix}$

4. Given the vectors $\mathbf{a} = \begin{pmatrix} 2 \\ 5 \end{pmatrix}$ and $\mathbf{b} = \begin{pmatrix} -3 \\ 2 \end{pmatrix}$, evaluate:
   (i) $\mathbf{a} + \mathbf{b}$ (ii) $\mathbf{b} + \mathbf{a}$ (iii) $\mathbf{a} - \mathbf{b}$ (iv) $\mathbf{b} - \mathbf{a}$
   (v) What conclusions do you draw from your results in (i), (ii), (iii) and (iv)?

Multiplication of Vectors by Scalars

Multiplication of a vector by a scalar can be considered as repeated addition of the same vector. Thus if $\mathbf{d}$ is any vector then $5\mathbf{d} = \mathbf{d} + \mathbf{d} + \mathbf{d} + \mathbf{d} + \mathbf{d}$.

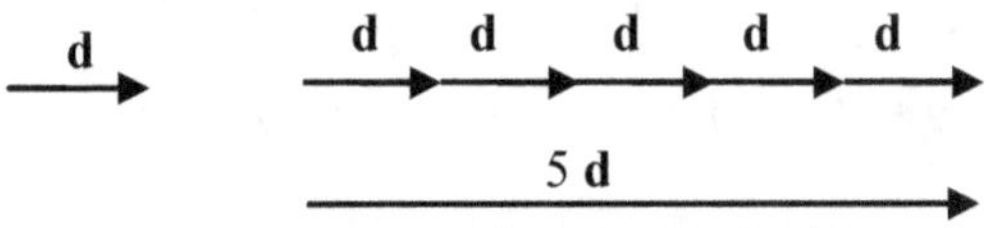

Figure 44:29

Generally for any scalar n,

$n\mathbf{d} = \mathbf{d} + \mathbf{d} + \mathbf{d} + \ldots +$ up to n times.

This generalization is true even if n is a fraction or n is negative. If n is negative, the sense of the vector will be opposite to the sense of the original vector.

Therefore, if $\mathbf{a}$ and $\mathbf{b}$ are parallel vectors then $\mathbf{a}$ can be expressed as a scalar multiple of $\mathbf{b}$.
i.e. $\mathbf{a} = n\mathbf{b}, n \in \mathbb{R}$.

Distributive Property of Vectors

If $\mathbf{a}$ and $\mathbf{b}$ are any two vectors the
   $k(\mathbf{a} + \mathbf{b}) = k\mathbf{a} + k\mathbf{b}$

Example 44:15

If $\mathbf{a} = \begin{pmatrix} -6 \\ 4 \end{pmatrix}$, evaluate the following and find their magnitudes.

   (i) $3\mathbf{a}$ (ii) $-2\mathbf{a}$ (iii) $\dfrac{1}{2}\mathbf{a}$ (iv) $-\dfrac{1}{2}\mathbf{a}$

Solution

(i) $3\mathbf{a} = 3\begin{pmatrix} -6 \\ 4 \end{pmatrix} = \begin{pmatrix} -18 \\ 12 \end{pmatrix}$

$\Rightarrow |3\mathbf{a}| = \sqrt{(-18)^2 + 12^2} = \sqrt{468} = 6\sqrt{13}$

(ii) $-2\mathbf{a} = -2\begin{pmatrix} -6 \\ 4 \end{pmatrix} = \begin{pmatrix} 12 \\ -8 \end{pmatrix}$

$\Rightarrow |-2\mathbf{a}| = \sqrt{12^2 + (-8)^2} = \sqrt{208} = 4\sqrt{13}$

(iii) $\frac{1}{2}\mathbf{a} = \frac{1}{2}\begin{pmatrix} -6 \\ 4 \end{pmatrix} = \begin{pmatrix} -3 \\ 2 \end{pmatrix}$

$\Rightarrow \left|\frac{1}{2}\mathbf{a}\right| = \sqrt{(-3)^2 + 2^2} = \sqrt{13}$

(iv) $-\frac{1}{2}\mathbf{a} = -\frac{1}{2}\begin{pmatrix} -6 \\ 4 \end{pmatrix} = \begin{pmatrix} 3 \\ -2 \end{pmatrix}$

$\Rightarrow \left|-\frac{1}{2}\mathbf{a}\right| = \sqrt{3^2 + (-2)^2} = \sqrt{13}$

Example 44:16
Given that $\mathbf{a} = 2\mathbf{i} + 5\mathbf{j}$ and $\mathbf{c} = \mathbf{i} + 3\mathbf{j}$. Find the vector $\mathbf{b}$ such that $\mathbf{b} = 2\mathbf{a} - 3\mathbf{c}$.

Solution
$\mathbf{b} = 2\mathbf{a} - 3\mathbf{c} = 2(2\mathbf{i} + 5\mathbf{j}) - 3(\mathbf{i} + 3\mathbf{j})$

$\Rightarrow \mathbf{b} = 4\mathbf{i} + 10\mathbf{j} - 3\mathbf{i} - 9\mathbf{j} = \mathbf{i} + \mathbf{j}$

Example 44:17
If $\mathbf{a} = \begin{pmatrix} 2 \\ 5 \end{pmatrix}$, $\mathbf{b} = \begin{pmatrix} 4 \\ 9 \end{pmatrix}$ and $\mathbf{c} = \begin{pmatrix} 1 \\ 3 \end{pmatrix}$, find a relationship between $\mathbf{a}$, $\mathbf{b}$ and $\mathbf{c}$ in the form $\mathbf{b} = u\mathbf{a} + v\mathbf{c}$ where u and v are integers.

Solution
$\begin{pmatrix} 4 \\ 9 \end{pmatrix} = u\begin{pmatrix} 2 \\ 5 \end{pmatrix} + v\begin{pmatrix} 1 \\ 3 \end{pmatrix}$

$\Rightarrow 2u + v = 4 \ldots\ldots\ldots\ldots\ldots\ldots①$

$\quad 5u + 3v = 9 \ldots\ldots\ldots\ldots\ldots\ldots②$

$①3 - 2 : u = 3$

Substitute in $①$: $2(3) + v = 4 \Rightarrow v = -2$

$\therefore \mathbf{b} = 3\mathbf{a} - 2\mathbf{c}$

EXERCISE 44:6

1. Given that $\mathbf{u} = \begin{pmatrix} 2 \\ 3 \end{pmatrix}$ and $\mathbf{v} = \begin{pmatrix} 0 \\ 1 \end{pmatrix}$, find numbers a and b such that $a\mathbf{u} + b\mathbf{v} = \begin{pmatrix} 4 \\ 5 \end{pmatrix}$

2. Given that $\mathbf{a} = \begin{pmatrix} 4 \\ -3 \end{pmatrix}$, $\mathbf{b} = \begin{pmatrix} 5 \\ 7 \end{pmatrix}$ and $\mathbf{c} = \begin{pmatrix} -6 \\ 4 \end{pmatrix}$, evaluate the following.

   (i) $\mathbf{a} + \mathbf{b}$ (ii) $\mathbf{a} - \mathbf{b}$ (iii) $\mathbf{b} - \mathbf{a}$
   (iv) $\mathbf{a} + (\mathbf{b} + \mathbf{c})$ (v) $(\mathbf{a} + \mathbf{b}) + \mathbf{c}$

   (vi) $3\mathbf{a}$ (vii) $\mathbf{b} - \frac{1}{2}\mathbf{c}$ (viii) $\mathbf{b} - 2\mathbf{a} + 3\mathbf{c}$

   (ix) $\frac{1}{3}(3\mathbf{a} + 2\mathbf{a} + \mathbf{c})$ (x) $2\mathbf{a} + \mathbf{b} - 3\mathbf{c}$

3. Given the vectors
   $\mathbf{x} = \begin{pmatrix} 2 \\ 1 \end{pmatrix}$, $\mathbf{y} = \begin{pmatrix} 1 \\ -4 \end{pmatrix}$, $\mathbf{z} = \begin{pmatrix} 1 \\ -1 \end{pmatrix}$, find the vector $\mathbf{x} + \mathbf{y}$ and give a reason why this vector is in the same direction as $\mathbf{z}$.

4. The position vectors of A, B, C and D are respectively $\mathbf{a}$, $\mathbf{b}$, $\mathbf{c}$ and $\mathbf{d}$, where $2\mathbf{c} = \mathbf{a}$ and $4\mathbf{d} = \mathbf{a} + \mathbf{b}$.
   (a) Show that $\mathbf{AB}$ is parallel to $\mathbf{CD}$.
   (b) Find the ratio $AB:CD$

5. Given that $\mathbf{a} = 3\mathbf{i} - 4\mathbf{j}$ and $\mathbf{b} = 4\mathbf{i} + 3\mathbf{j}$, find the magnitude of $\mathbf{a} + \mathbf{b}$, leaving your answer in surd form.

6. Given that $\mathbf{a} = \begin{pmatrix} 3 \\ 4 \end{pmatrix}$, $\mathbf{b} = \begin{pmatrix} 1 \\ 4 \end{pmatrix}$ and $\mathbf{c} = \begin{pmatrix} 5 \\ 12 \end{pmatrix}$ and that $u\mathbf{a} + v\mathbf{b} = \mathbf{c}$. Write simultaneous equations in u and v and solve them.

Vector Geometry
The Section Theorem

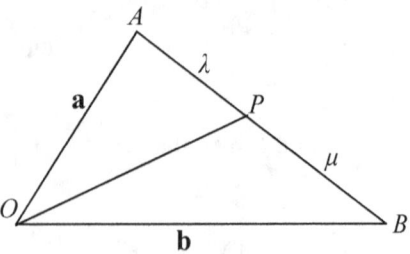

Figure 44:30

Let A and B be two points with position vectors **a** and **b** respectively and suppose P is a point dividing the line AB in the ratio $\lambda : \mu$. Then the position vector **OP** of P is given by

$$\mathbf{OP} = \mathbf{OA} + \mathbf{AP}$$

But $\dfrac{AP}{AB} = \dfrac{\lambda}{\lambda + \mu}$

$$\Rightarrow \mathbf{AP} = \frac{\lambda}{\lambda + \mu}\mathbf{AB}$$

$$\Rightarrow \mathbf{OP} = \mathbf{OA}\,\frac{\lambda}{\lambda + \mu}\mathbf{AB}$$

Also $\mathbf{AB} = \mathbf{AO} + \mathbf{OB}$
$$= -\mathbf{OA} + \mathbf{OB}$$
$$= -\mathbf{a} + \mathbf{b}$$

$$\Rightarrow \mathbf{OP} = \mathbf{a} + \frac{\lambda}{\lambda + \mu}(-\mathbf{a} + \mathbf{b})$$

i.e. $\boxed{\mathbf{OP} = \mathbf{a} + \dfrac{\lambda}{\lambda + \mu}(\mathbf{b} - \mathbf{a})}$

This is known as the **section theorem**.

The Position Vector of the Midpoint
If P is the midpoint M, of **AB** then $\lambda = \mu$ and the ratio becomes 1:1. Therefore the position vector of the midpoint M will be given by

$$\mathbf{OM} = \mathbf{a} + \frac{(\mathbf{b} - \mathbf{a})}{2}$$

$$\boxed{\mathbf{OM} = \frac{1}{2}(\mathbf{a} + \mathbf{b})}$$

Example 44:18

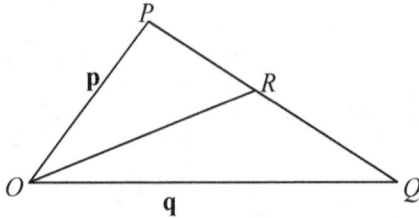

Figure 44:31

Given that $\mathbf{OP} = \mathbf{p}$, $\mathbf{OQ} = \mathbf{q}$, $\mathbf{OR} = \mathbf{r}$ and that R divides PQ in the ratio 2:3. Express **r** in terms of **p** and **q**.

Solution
By section theorem,

$$\mathbf{r} = \mathbf{p} + \frac{2}{2+3}(\mathbf{q} - \mathbf{p}) = \mathbf{p} + \frac{2}{5}\mathbf{q} - \frac{2}{5}\mathbf{p} \Rightarrow \mathbf{r} = \frac{3}{5}\mathbf{p} + \frac{2}{5}\mathbf{q}$$

Example 44:19

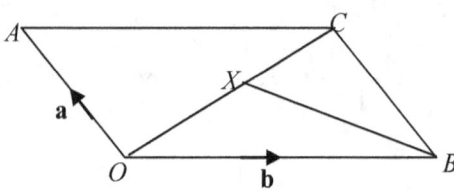

Figure 44:32

In the parallelogram $OACB$, the point X divides **OC** in the ratio 2:1. Given that $\mathbf{OA} = \mathbf{a}$ and $\mathbf{OB} = \mathbf{b}$, express the vector **XB** in terms of **a** and **b**.

Solution
By the section theorem,

$$\mathbf{BX} = \mathbf{BC} + \frac{1}{2+1}(-\mathbf{OB} - \mathbf{BC})$$

$\mathbf{OA} = \mathbf{BC} = \mathbf{a}$ [opposite sides of a parallelogram]

$$\mathbf{BX} = \mathbf{a} + \frac{1}{2+1}(-\mathbf{b} - \mathbf{a}) = \frac{2}{3}\mathbf{a} - \frac{1}{3}\mathbf{b}$$

$$\mathbf{XB} = -\mathbf{BX} \Rightarrow \mathbf{XB} = -\frac{2}{3}\mathbf{a} + \frac{1}{3}\mathbf{b}$$

Example 44:20

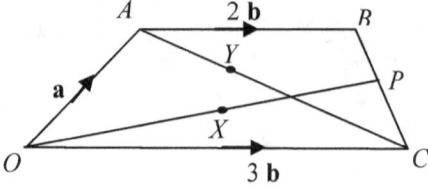

Figure 44:33

In the Figure 44:33 $OABC$ is a trapezium with $\mathbf{OA} = \mathbf{a}$, $\mathbf{AB} = 2\mathbf{b}$, $\mathbf{OC} = 3\mathbf{b}$. The point P is the midpoint of **BC**, the point X is the midpoint of **OP** and the point Y is the midpoint of **AC**. Express **OX** and **PY** in terms of **a** and **b**.

Solution
By the section theorem,

$$OY = a \frac{1}{2}(3b - a) = \frac{1}{2}a + \frac{3}{2}b$$

$$OX = \frac{1}{2}OP = \frac{1}{2}(OA + AB + BP)$$

$$\Rightarrow OX = \frac{1}{2}\left(a + 2b + \frac{1}{2}BC\right)$$

$$BC = -AB + (-OA) + OC$$

$$= -2b - a + 3b$$

$$= b - a$$

$$OX = \frac{1}{2}\left(a + 2b + \frac{1}{2}(b - a)\right)$$

$$= \frac{1}{2}\left(\frac{1}{2}a + \frac{5}{2}b\right)$$

$$= \frac{1}{4}(a + 5b)$$

$$PY = PC + CO + AY$$

$$= -\frac{1}{2}BC - 2b + \frac{1}{2}AC$$

$$= -\frac{1}{2}(b - a) - 2b + \frac{1}{2}(-a + 3b)$$

$$= -\frac{1}{2}b + \frac{1}{2}a - 2b - \frac{1}{2}a + \frac{3}{2}b$$

$$= -b$$

EXERCISE 44:7

1. In the quadrilateral *OABC*, *D* is the midpoint of *BC* and *G* is the point on *AD* such that *AG:GD* =2:1. Given that **OA** = **a**, **OB** = **b** and **OC** − **c**, express **OD** and **OG** in terms of **a**, **b** and **c**.
2. In Figure 44:34 , *OPQR* is a parallelogram, **TR** = 2**OT** and **RM** = **MQ**. If **OP** = **p** and **OR** = **r**, express the vectors **TM** and **PM** in terms of **p** and **r**.

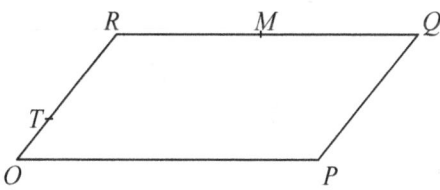

Figure 44:34

3. Using Figure 44:35,
   (a) express **AC** in terms of **a** and **d** only
   (b) find expressions for **OB** in terms of
      (i) **a** and **b**   (ii) **c** and **d**

Hence, express **c** in terms of **a**, and **d**, given that $b = \frac{3}{2}d$

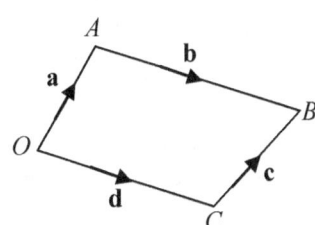

Figure 44:35

4. Figure 44:36 shows a triangle *OBA*. Given that **OA** = **a**, **OB** = **b** and AM:MB = 2:1. Find in terms of **a** and **b**, the vectors
   (a) **AB**   (b) **OM**

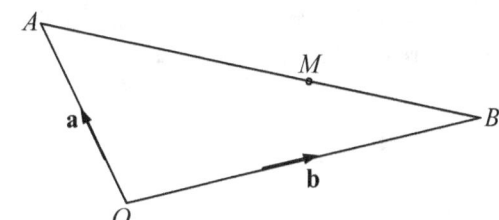

Figure 44:36

5. Figure 44:37 shows a parallelogram, divided into smaller congruent parallelograms. Given that **OA** = **a** and **OB** = **b**, express in terms of **a** and **b** the displacements **PQ**, **QR** and **RP**.

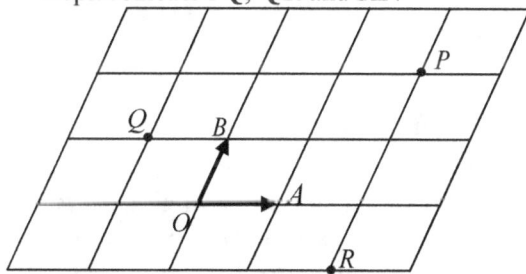

Figure 44:37

6. In the triangle *OAB* in Figure 39:38, the point C divides AB in the ratio 3:1. Given that **OA** = **a** and **OB** = **b**, express **OC** in terms of **a** and **b**.

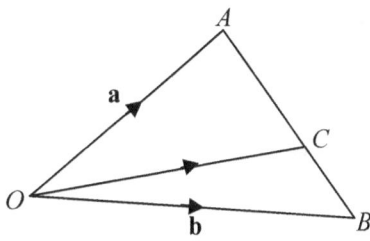

Figure 44:38

7. *A* **and** *B* are two points whose position vectors are $\mathbf{a} = -3\mathbf{i} + 6\mathbf{j}$ and $\mathbf{b} = 9\mathbf{i} + 3\mathbf{j}$ respectively. Points *L* and *M* divide **OA** and **OB** internally in the ratio *m : n* i.e.

$$OL : LA = OM : MB = m : n$$

(a) Find in terms of **i** and **j**
 (i) The vector **AB**.
 (ii) The position vectors of the points *L* and *M*.
 (iii) The vector **LM**.

(b) Hence or otherwise, show that **AB** is parallel to **LM**.

(c) Given that *m:n* = 2:1, find:
 (i) The ratio of the length of *LM* to the length of *AB*.
 (ii) The area of the quadrilateral *ALMB*.

8. In Figure 44:39, *OPQR* is a trapezium with *OP* and *RQ* parallel. Given that $\mathbf{RQ} = 2\mathbf{a}$,

$\mathbf{OR} = \mathbf{b}$, $\mathbf{RQ} = \dfrac{2}{3}$ **OP** and that *M* is the midpoint of *RQ*. Find the vector

(a) **OM** in terms of **a** and **b**
(b) **QP** in terms of **a** and **b**

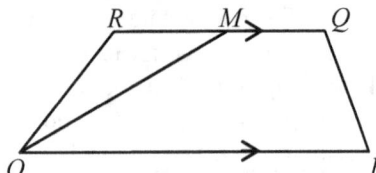

Figure 44:39

The Scalar or Dot Product

The scalar or dot product has many applications in geometry, dynamics and many other branches of science. It is especially used in finding the angle between two vectors, acting in the same plane.

The dot or scalar product of two vectors **a** and **b** acting on the same plane is denoted by **a·b**, (read 'a dot b') and is defined as

$\mathbf{a.b} = |\mathbf{a}||\mathbf{b}|\cos\theta$,

where $0° \le \theta \le 180°$ is the angle between the two vectors **a** and **b**.

From the definition, it follows that
(a) If **a** is perpendicular to **b**,
  $\theta = 90°$, $\cos 90° = 0 \Rightarrow \mathbf{a \cdot b = 0}$

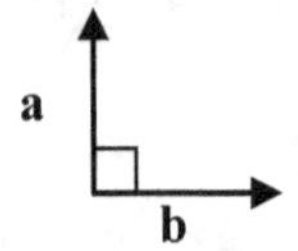

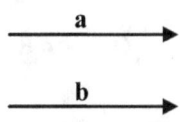

Figure 44:40

(b) If **a** is parallel to **b**, $\theta = 0°$,
  $\cos 0 = 1 \Rightarrow \mathbf{a \cdot b} = |\mathbf{a}||\mathbf{b}|$

$$\xrightarrow{\hspace{3cm}}$$
a

$$\xrightarrow{\hspace{3cm}}$$
b

Figure 44:41

From (a) and (b) above, it can be deduced that $\mathbf{i \cdot i} = 1$, $\mathbf{j \cdot j} = 1$ and $\mathbf{i \cdot j} = \mathbf{j \cdot i} = 0$.

Therefore, if $\mathbf{a} = x_1\mathbf{i} + y_1\mathbf{j}$ and $\mathbf{b} = x_2\mathbf{i} + y_2\mathbf{j}$,

Then, $\mathbf{a \cdot b} = (x_1\mathbf{i} + y_1\mathbf{j}) \cdot (x_2\mathbf{i} + y_2\mathbf{j})$

$$= x_1\mathbf{i} \cdot (x_2\mathbf{i} + y_2\mathbf{j}) + y_1\mathbf{j} \cdot (x_2\mathbf{i} + y_2\mathbf{j})$$

$$\Rightarrow \mathbf{a \cdot b} = x_1 x_2 + y_1 y_2$$

Example 44:21
The magnitudes of two vectors **a** and **b** are 5 and 8 and the angle between them is 60°. Find $\mathbf{a \cdot b}$.

Solution

$$\mathbf{a.b} = |\mathbf{a}||\mathbf{b}|\cos\theta = (5)(8)\cos 60° = (5)(8)\left(\frac{1}{2}\right) = 20$$

Example 44:22
Find the scalar product of the following pairs of vectors and state which of the pairs are parallel, perpendicular or not parallel or perpendicular.

1. $3\mathbf{i}+4\mathbf{j}$ and $2\mathbf{i}-\mathbf{j}$ 2. $2\mathbf{i}+3\mathbf{j}$ and $3\mathbf{i}-2\mathbf{j}$
3. $10\mathbf{i}+25\mathbf{j}$ and $6\mathbf{i}+15\mathbf{j}$

Solution

1. $(3\mathbf{i} + 4\mathbf{j}) \cdot (2\mathbf{i} - \mathbf{j}) = 3(2) + 4(-1) = 2$

$$|3\mathbf{i} + 4\mathbf{j}||2\mathbf{i} - \mathbf{j}| = \left(\sqrt{3^2 + 4^2}\right)\left(\sqrt{2^2 + (-1)^2}\right)$$

$$= \left(\sqrt{25}\right)\left(\sqrt{5}\right) = 5\left(\sqrt{5}\right)$$

Therefore $(3\mathbf{i}+4\mathbf{j})$ and $(2\mathbf{i}-\mathbf{j})$ are neither perpendicular nor parallel.

2. $(2\mathbf{i} + 3\mathbf{j}) \cdot (3\mathbf{i} - 2\mathbf{j}) = 2(3) + 3(-2) = 0$

$$\therefore (2\mathbf{i} + 3\mathbf{j}) \perp (3\mathbf{i} - 2\mathbf{j})$$

3. $(10\mathbf{i} + 25\mathbf{j}) \cdot (6\mathbf{i} + 15\mathbf{j}) = 10(6) + 25(15) = 435$

$$|10\mathbf{i} + 25\mathbf{j}||6\mathbf{i} + 15\mathbf{j}| = \left(\sqrt{10^2 + 25^2}\right)\left(\sqrt{6^2 + 15^2}\right)$$

$$= \left(\sqrt{725}\right)\left(\sqrt{261}\right) = \sqrt{189225}$$

$$\Rightarrow |10\mathbf{i} + 25\mathbf{j}||6\mathbf{i} + 15\mathbf{j}| = 435$$

Also notice that $10\mathbf{i} + 25\mathbf{j} = \dfrac{5}{3}(6\mathbf{i} + 15\mathbf{j})$

Therefore, $10\mathbf{i} + 25\mathbf{j} \parallel$.

Angle between two vectors

The angle θ between two vectors is the angle of rotation between the directions of the two vectors. In other words it is the angle θ between the two divergent rays or the two convergent rays representing the vectors. In Figure 44:42, the angle α is the wrong angle.

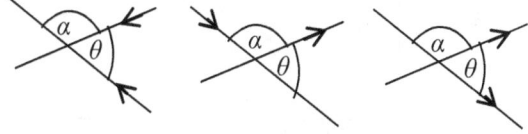

Figure 44:42

Finding the Angle between Two Vectors by the Scalar Product Method

Since $\mathbf{a} \cdot \mathbf{b} = |\mathbf{a}||\mathbf{b}|\cos\theta$, the scalar product can be used to find the angle between any two vectors on the same plane. Hence,

$$\cos\theta = \frac{\mathbf{a} \cdot \mathbf{b}}{|\mathbf{a}||\mathbf{b}|} \Leftrightarrow \theta = \cos^{-1}\left(\frac{\mathbf{a} \cdot \mathbf{b}}{|\mathbf{a}||\mathbf{b}|}\right)$$

This method is most recommended because it does not require any intuitive judgment.

Example 44:23
Use the scalar product method to find the angle between the vectors $6\mathbf{i} + 8\mathbf{j}$ and $4\mathbf{i} - 2\mathbf{j}$.
Solution

$$\theta = \cos^{-1}\left(\frac{\mathbf{a} \cdot \mathbf{b}}{|\mathbf{a}||\mathbf{b}|}\right) = \cos^{-1}\left(\frac{(6\mathbf{i} + 8\mathbf{j}).(4\mathbf{i} - 2\mathbf{j})}{|6\mathbf{i} + 8\mathbf{j}||4\mathbf{i} - 2\mathbf{j}|}\right)$$

$$\Rightarrow \theta = \cos^{-1}\left(\frac{(24 - 16)}{\left(\sqrt{6^2 + 8^2}\right)\left(\sqrt{4^2 + (-2)^2}\right)}\right)$$

$$\theta = \cos^{-1}(0.1789) = 79.7°$$

Finding the Angle between Two Vectors by the Graphical Method

Example 44:25

Using the graphical method find the angle θ between the vectors $6\mathbf{i} + 8\mathbf{j}$ and $4\mathbf{i} - 2\mathbf{j}$.

Solution
Plot the vectors as position vectors on graph or square paper as in Figure 44:43. Measure the angle θ using a protractor. Depending on the accuracy of the graph the angle should be found to be about 80°.

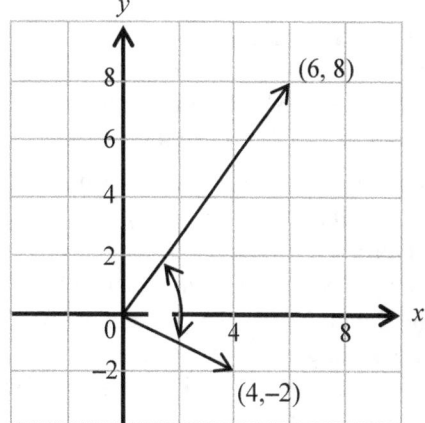

Figure 44:43

Formula Method

The angle θ between two vectors $\mathbf{a} = x_1\mathbf{i} + y_1\mathbf{j}$ and $\mathbf{b} = x_2\mathbf{i} + y_2\mathbf{j}$ is given by the formula

$$\theta = \left|\tan^{-1}\left(\frac{y_1}{x_1}\right) - \tan^{-1}\left(\frac{y_2}{x_2}\right)\right|$$

In using this method, a sketch of the vectors should first be drawn. Even after doing this serious care must still be taken to ensure that the angle is not the wrong angle. Unless the student has a very good sense of judgment, this method is least recommended.

Example 44:24
Calculate the angle between the vectors $6\mathbf{i} + 8\mathbf{j}$ and $4\mathbf{i} - 2\mathbf{j}$.

Solution

$$\theta = \left|\tan^{-1}\frac{y_1}{x_1} - \tan^{-1}\frac{y_2}{x_2}\right|$$

$$= \left|\tan^{-1}\left(\frac{8}{6}\right) - \tan^{-1}\left(\frac{-2}{4}\right)\right|$$

$$= 53.13° - (-26.57°)$$

$$= 79.7°$$

EXERCISE 44:8

1. Given that $\mathbf{u} = 3\mathbf{i}+4\mathbf{j}$ and $\mathbf{v} = 4\mathbf{i}-3\mathbf{j}$, find
   (a) The modulus of $\mathbf{u}$.
   (b) The angle between $\mathbf{u}$ and $\mathbf{v}$.
2. Given two vectors $\mathbf{u} = 2\mathbf{i}+3\mathbf{j}$ and $\mathbf{v} = 3\mathbf{i}-2\mathbf{j}$, find
   (a) The modulus of $\mathbf{u}$.
   (b) The angle between $\mathbf{u}$ and $\mathbf{v}$.
3. Given that $|\mathbf{x}| = 6$, $|\mathbf{y}| = 6$ and the angle between $\mathbf{x}$ and $\mathbf{y}$ is $45°$. Calculate $\mathbf{x} \cdot \mathbf{y}$.
4. Given that $|\mathbf{a}| = 9$, $|\mathbf{b}| = 3$, find $\mathbf{a} \cdot \mathbf{b}$ if:
   (a) $\mathbf{a}$ is perpendicular to $\mathbf{b}$.
   (b) $\mathbf{a}$ is parallel to $\mathbf{b}$.
5. The magnitudes of two vectors $\mathbf{a}$ and $\mathbf{b}$ are 6 and 2 and their scalar product is -6. Calculate the angle between these vectors.
6. Find the scalar product of $3\mathbf{i} +4\mathbf{j}$ and $4\mathbf{i} -3\mathbf{j}$ and state whether these vectors are parallel or perpendicular.
7. Find the angle between the vectors $\mathbf{u} = 3\mathbf{i}+4\mathbf{j}$ and $\mathbf{v} = -\mathbf{i} + 8\mathbf{j}$.
8. Find the angle between the vectors $\mathbf{u}$ and $\mathbf{v}$, where $\mathbf{u} = 3\mathbf{i} + 4\mathbf{j}$ and $\mathbf{v} = -6\mathbf{i} + 8\mathbf{j}$.

MULTIPLE CHOICE EXERCISE 44

1. The modulus of $6\mathbf{i} + 8\mathbf{j}$ is:
   [A] $2\sqrt{7}$ [B] 8   [C] 10   [D] 6
2. If $\mathbf{a} = 3\mathbf{i} - \mathbf{j}$ and $\mathbf{b} = \mathbf{i} +2\mathbf{j}$, then $\mathbf{a} \cdot \mathbf{b}$ is equal to:
   [A] 1   [B] 5   [C] -5 [D] -1
3. Given that A, B and C are collinear and $\mathbf{OA} = \mathbf{i} + \mathbf{j}$, $\mathbf{OB} = 2\mathbf{i} -\mathbf{j}$ and $\mathbf{OC} = 3\mathbf{i} + a\mathbf{j}$. The value of a is:
   [A] 1   [B] -1 [C] -3 [D] -2
4. The vector $\begin{pmatrix} -4 \\ 1 \end{pmatrix}$ in $\mathbf{i}$ and $\mathbf{j}$ component form is:
   [A] $4\mathbf{i} - \mathbf{j}$ [B] $\mathbf{i} - 4\mathbf{j}$
   [C] $-4\mathbf{i} + \mathbf{j}$ [D] $-\mathbf{i} - 4\mathbf{j}$
5. The vector $2\mathbf{i} -3\mathbf{j}$ in column vector form is:
   [A] $\begin{pmatrix} -3 \\ -2 \end{pmatrix}$ [B] $\begin{pmatrix} 3 \\ -2 \end{pmatrix}$ [C] $\begin{pmatrix} -2 \\ 3 \end{pmatrix}$ [D] $\begin{pmatrix} 2 \\ -3 \end{pmatrix}$
6. The additive inverse of $3\mathbf{i} -5\mathbf{j}$ is:
   [A] $3\mathbf{i}+5\mathbf{j}$ [B] $-3\mathbf{i}+5\mathbf{j}$ [C] $-3\mathbf{i}-5\mathbf{j}$ [D] $3\mathbf{i}-5\mathbf{j}$
7. A unit vector among the following is:

[A] $\dfrac{1}{\sqrt{2}}\mathbf{i} - \dfrac{1}{\sqrt{2}}\mathbf{j}$ [B] $\dfrac{1}{\sqrt{2}}\mathbf{i} + \dfrac{1}{\sqrt{2}}\mathbf{j}$
[C] $\mathbf{i} + \mathbf{j}$ [D] $\mathbf{i} - \mathbf{j}$

8. The additive inverse of $\begin{pmatrix} -3 \\ 2 \end{pmatrix}$ is:

[A] $\begin{pmatrix} -3 \\ -2 \end{pmatrix}$ [B] $\begin{pmatrix} 3 \\ -2 \end{pmatrix}$ [C] $\begin{pmatrix} -2 \\ 3 \end{pmatrix}$ [D] $\begin{pmatrix} 2 \\ -3 \end{pmatrix}$

9. As a column vector, the vector represented by the directed line segment in Figure 44:41 is:

[A] $\begin{pmatrix} -3 \\ 8 \end{pmatrix}$ [B] $\begin{pmatrix} 3 \\ -8 \end{pmatrix}$ [C] $\begin{pmatrix} -8 \\ 3 \end{pmatrix}$ [D] $\begin{pmatrix} 8 \\ -3 \end{pmatrix}$

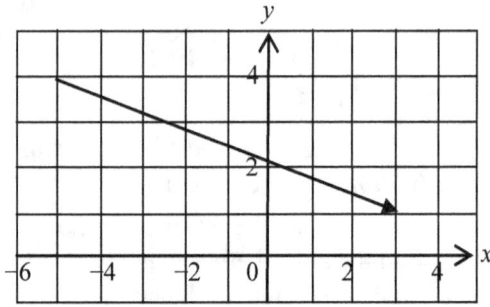

Figure 44:41

10. On the Cartesian plane, the free vector $-5\mathbf{i} +3\mathbf{j}$ could be represented as:

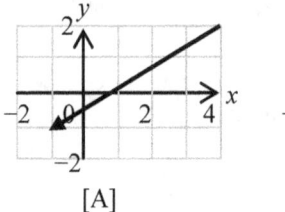
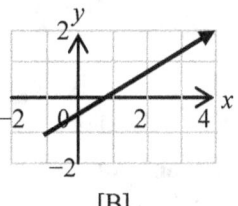

[A]   [B]

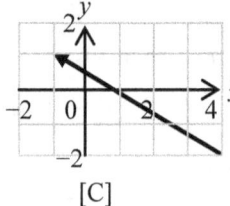
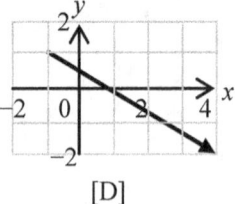

[C]   [D]

Figure 44:42

11. Given that $\mathbf{u} = \begin{pmatrix} 2 \\ 3 \end{pmatrix}$ and $\mathbf{v} = \begin{pmatrix} 0 \\ 1 \end{pmatrix}$, the numbers a and b such that

$$a\mathbf{u} + b\mathbf{v} = \begin{pmatrix} 4 \\ 5 \end{pmatrix}$$

[A] $a = -1, b = 2$ [B] $a = -2, b = -1$
[C] $a = 2, b = -1$ [D] $a = 1, b = 2$

12. In the quadrilateral $OABC$, D is the mid-

504

point of BC. Given that, $\mathbf{OA} = \mathbf{a}$, $\mathbf{OB} = \mathbf{b}$ and $\mathbf{OC} = \mathbf{c}$. In terms of $\mathbf{a}$, $\mathbf{b}$ and $\mathbf{c}$ $\mathbf{OD}$ is:

[A] $\frac{1}{2}(\mathbf{b} - \mathbf{c})$ [B] $\frac{1}{2}(\mathbf{b} + \mathbf{c})$

[C] $\frac{1}{2}(-\mathbf{b} + \mathbf{c})$ [D] $-\frac{1}{2}(\mathbf{b} - \mathbf{c})$

13. In the quadrilateral $OABC$, D is the mid-point of BC and G is the point on AD such that $AG : GD = 2 : 1$. Given that, $\mathbf{OA} = \mathbf{a}$, $\mathbf{OB} = \mathbf{b}$ and $\mathbf{OC} = \mathbf{c}$. $\mathbf{OG}$ in terms of $\mathbf{a}$, $\mathbf{b}$ and $\mathbf{c}$ is:

[A] $\frac{1}{3}(\mathbf{a} + \mathbf{b} - \mathbf{c})$ [B] $\frac{1}{3}(\mathbf{a} - \mathbf{b} + \mathbf{c})$

[C] $\frac{1}{3}(\mathbf{a} - \mathbf{b} - \mathbf{c})$ [D] $\frac{1}{3}(\mathbf{a} + \mathbf{b} + \mathbf{c})$

14-15 In Figure 41.43, $OPQR$ is a parallelogram $TR = 2OT$ and $RM = MQ$. Given that $\overline{OP} = \mathbf{p}$ and $\overline{OR} = \mathbf{r}$.

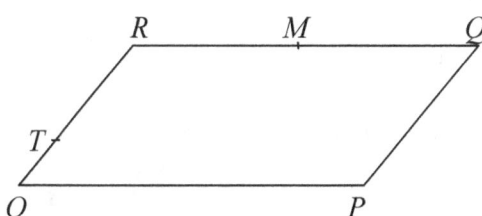

Figure 44:43

14. The vector $\mathbf{TM}$ in terms of $\mathbf{p}$ and $\mathbf{r}$ is:

[A] $\frac{2}{3}\mathbf{r} - \frac{1}{2}\mathbf{p}$ [B] $-\frac{2}{3}\mathbf{r} + \frac{1}{2}\mathbf{p}$

[C] $-\frac{2}{3}\mathbf{r} - \frac{1}{2}\mathbf{p}$ [D] $\frac{2}{3}\mathbf{r} + \frac{1}{2}\mathbf{p}$

15. The vector $\mathbf{PM}$ in terms of $\mathbf{p}$ and $\mathbf{r}$ is:

[A] $\frac{1}{2}\mathbf{p} + \mathbf{r}$ [B] $\mathbf{r} - \frac{1}{2}\mathbf{p}$

[C] $\frac{1}{2}\mathbf{r} - \mathbf{p}$ [D] $\mathbf{p} - \frac{1}{2}\mathbf{r}$

16. Let $\mathbf{a} = \begin{pmatrix} 2 \\ 5 \end{pmatrix}$, $\mathbf{b} = \begin{pmatrix} 4 \\ 4 \end{pmatrix}$ and $\mathbf{c} = \begin{pmatrix} 1 \\ 3 \end{pmatrix}$. The relationship between $\mathbf{a}$, $\mathbf{b}$, and $\mathbf{c}$, in the form $\mathbf{b} = u\mathbf{a} + v\mathbf{c}$, where u and v are integers is:

[A] $\mathbf{b} = 3\mathbf{a} - 2\mathbf{c}$ [B] $\mathbf{b} = 3\mathbf{a} + 2\mathbf{c}$

[C] $\mathbf{b} = -2\mathbf{a} + 3\mathbf{c}$ [D]

$\mathbf{b} = -3\mathbf{a} + 2\mathbf{c}$

17. The magnitude of $\mathbf{u} = 2\mathbf{i} + 3\mathbf{j}$ is:

[A] 5   [B] 13   [C] $\sqrt{13}$ [D] $\sqrt{5}$

18. The angle between the vectors $\mathbf{u} = 2\mathbf{i} + 3\mathbf{j}$

and $\mathbf{v} = 3\mathbf{i} - 2\mathbf{j}$ is:

[A] $30°$ [B] $45°$ [C] $60°$ [D] $90°$

19-20 In Figure 44:44, $OABC$ is a trapezium with $\mathbf{OA} = \mathbf{a}$ and $\mathbf{OB} = \mathbf{b}$. The point P is the midpoint of BC, the point X is the midpoint of OP and the point Y is the midpoint of AC.

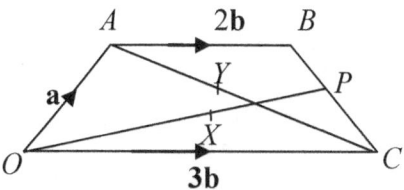

Figure 44:44

19. In terms of $\mathbf{a}$ and $\mathbf{b}$, $\mathbf{OX}$ is equal to:

[A] $\frac{1}{2}\mathbf{a} + \frac{5}{2}\mathbf{b}$ [B] $\frac{1}{4}(\mathbf{a} + 5\mathbf{b})$

[C] $\frac{1}{2}\mathbf{a} - \frac{5}{2}\mathbf{b}$ [D] $\frac{1}{4}(\mathbf{a} - 5\mathbf{b})$

20. In terms of $\mathbf{a}$ and $\mathbf{b}$, $\mathbf{OY}$ is equal to:

[A] $\frac{1}{4}(\mathbf{a} - 3\mathbf{b})$ [B] $\frac{1}{4}(\mathbf{a} + 3\mathbf{b})$

[C] $\frac{1}{2}\mathbf{a} - \frac{3}{2}\mathbf{b}$ [D] $\frac{1}{2}(\mathbf{a} + 3\mathbf{b})$

21. Given that $\mathbf{a} = 2\mathbf{i} + 5\mathbf{j}$, $\mathbf{b} = 4\mathbf{i} + 9\mathbf{j}$, the vector $\mathbf{c}$ such that $\mathbf{b} = 3\mathbf{a} - 2\mathbf{c}$ is:

[A] $\mathbf{i} - 3\mathbf{j}$ [B] $-\mathbf{i} + 3\mathbf{j}$

[C] $\mathbf{i} + 3\mathbf{j}$ [D] $2\mathbf{i} + 6\mathbf{j}$

22. In Figure 44:45, AC in terms of $\mathbf{a}$ and $\mathbf{d}$ only is:

[A] $-\mathbf{a} + \mathbf{d}$ [B] $-\mathbf{a} - \mathbf{d}$
[C] $\mathbf{a} - \mathbf{d}$ [D] $\mathbf{a} + \mathbf{d}$

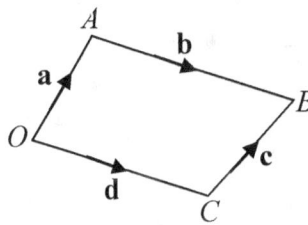

Figure 44:45

23. In Figure 44:45, the right expression of $\mathbf{OB}$ in terms of $\mathbf{a}$ and $\mathbf{b}$ is:

[A] $-\mathbf{a} + \mathbf{b}$ [B] $-\mathbf{a} - \mathbf{b}$
[C] $\mathbf{a} - \mathbf{b}$ [D] $\mathbf{a} + \mathbf{b}$

24. In Figure 44:45, the right expression of $\mathbf{OB}$ in terms of $\mathbf{c}$ and $\mathbf{d}$ is:

[A] $\mathbf{c} + \mathbf{d}$ [B] $-\mathbf{c} - \mathbf{d}$
[C] $-\mathbf{c} + \mathbf{d}$ [D] $\mathbf{c} - \mathbf{d}$

25. In Figure 44:45, given that $\mathbf{b} = \dfrac{3}{2}\mathbf{d}$. In terms of **a** and **d**, **c** is:

    [A] $-\mathbf{a} + \dfrac{1}{2}\mathbf{d}$ [B] $-\mathbf{a} - \dfrac{1}{2}\mathbf{d}$

    [C] $\mathbf{a} + \dfrac{1}{2}\mathbf{d}$ [D] $\mathbf{a} - \dfrac{1}{2}\mathbf{d}$

26. Given that the position vectors of A, B, C and D are respectively **a**, **b**, **c** and **d** where, $2\mathbf{c} = \mathbf{a}$ and $4\mathbf{d} = \mathbf{a} + \mathbf{b}$. **CD** can be written in terms of **AB** as:

    [A] $4\mathbf{CD} = \mathbf{AB}$ [B] $4\mathbf{CD} = -\mathbf{AB}$
    [C] $\mathbf{CD} = 4\mathbf{AB}$ [D] $\mathbf{CD} = -4\mathbf{AB}$

27-28 In Figure 44:46, $OPQR$ is a trapezium with OP and RQ parallel. Given that $\mathbf{RQ} =$ 2**a**, $\mathbf{OR} = \mathbf{b}$, $\mathbf{RQ} = \dfrac{2}{3}\mathbf{OP}$ and that M is the midpoint of RQ

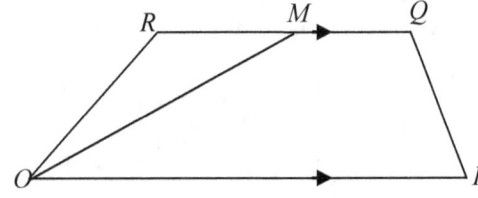

Figure 44:46

27. The vector **OM** in terms of **a** and **b** is:
    [A] $-\mathbf{a} + \mathbf{b}$ [B] $-\mathbf{a} - \mathbf{b}$
    [C] $\mathbf{a} - \mathbf{b}$ [D] $\mathbf{a} + \mathbf{b}$

28. The vector **QP** in terms of **a** and **b** is:
    [A] $-\mathbf{a} + \mathbf{b}$ [B] $-\mathbf{a} - \mathbf{b}$
    [C] $\mathbf{a} - \mathbf{b}$ [D] $\mathbf{a} + \mathbf{b}$

29. Figure 44:47 shows a triangle OBA.

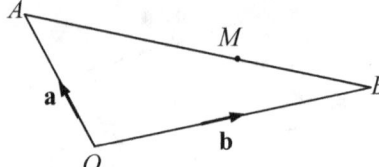

Figure 44:47
Given that $\mathbf{OA} = \mathbf{a}$, $\mathbf{OB} = \mathbf{b}$ and $AM : MB = 2 : 1$. In terms of **a** and **b**, **AB** is equal to:
    [A] $-\mathbf{a} + \mathbf{b}$ [B] $-\mathbf{a} - \mathbf{b}$
    [C] $\mathbf{a} - \mathbf{b}$ [D] $\mathbf{a} + \mathbf{b}$

30. Figure 44:47 shows a triangle OAB such that $\mathbf{OA} = \mathbf{a}$, $\mathbf{OB} = \mathbf{b}$ and $AM : MB = 2 : 1$. In terms of **a** and **b**, **OM** is equal to:

    [A] $-\dfrac{1}{3}(\mathbf{a} + 2\mathbf{b})$ [B] $\dfrac{1}{3}(\mathbf{a} + 2\mathbf{b})$

[C] $\dfrac{1}{3}(\mathbf{a} - 2\mathbf{b})$ [D] $\dfrac{1}{3}(-\mathbf{a} + 2\mathbf{b})$

31. In Figure 44:48, the pair of vectors, which are parallel, are:
    [A]   **AB** and **NM**   [B]   **XY** and **MN**
    [C]   **XY** and **PQ**   [D]   **BA** and **MN**

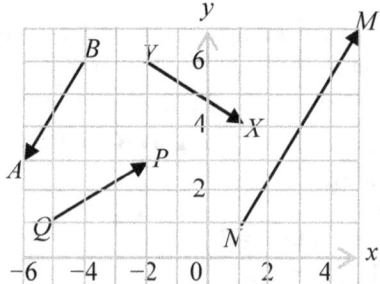

Figure 44:48

32. Given that, $\mathbf{a} = -5\mathbf{i} + 12\mathbf{j}$, $\mathbf{b} = 12\mathbf{i} - 5\mathbf{j}$, $\mathbf{c} = 5\mathbf{i} - 12\mathbf{j}$ and $\mathbf{d} = \begin{pmatrix} -5 \\ 12 \end{pmatrix}$. The vectors, which are equal, are:
    [A] **a** and **b**   [B] **a** and **c**

    [C] **a** and **d**   [D] **c** and **d**

33. Given that, $\mathbf{a} = \begin{pmatrix} -9 \\ 12 \end{pmatrix}$, $\mathbf{b} = \begin{pmatrix} -12 \\ 9 \end{pmatrix}$, $\mathbf{c} = -9\mathbf{i} + 12\mathbf{j}$ and $\mathbf{d} = 12\mathbf{i} - 9\mathbf{j}$. The vectors, which are equal, are:
    [A] **a** and **b**   [B] **a** and **c**
    [C] **a** and **d**   [D] **b** and **d**

34. The directed line segments $[AB]$, $[QP]$ and $[XY]$ are parallel and equal in magnitude. It is true to say that:
    [A]   **AB** = **XY**   [B]   **PQ** = **XY**
    [C]   **AB** = **PQ**   [D]   **AB** = **YX**

35. Two vectors are equal if and only if they have the same:
    [A] Magnitude and direction.
    [B] Magnitude and are parallel.
    [C] Magnitude and sense.

    [D] Sense and are parallel $\dfrac{1}{3}(\mathbf{a} + 2\mathbf{b})$

TOPIC 45

TRANSFORMATIONS AND SYMMETRY

OBJECTIVES

At the end of this topic, the learner should be able to:

1. Identify and draw the image of a figure under a transformation such as a translation, a reflection, a rotation or an enlargement.
2. Determine the scale factor of an enlargement.
3. Find the area of the image of an object under an enlargement.
4. Identify a shear and determine the shear factor of a shear.
5. Identify a stretch and determine the stretch factor of a stretch.
6. Identify, state and use the properties of line or reflective symmetry.
7. Identify, state and use the properties of Rotational or Radial symmetry.
8. Identify point and line symmetry in various polygons and locate the image of an object under point and line symmetry.
9. Identify radial or rotational symmetry in various figures and locate the image of an object under radial or rotational symmetry.

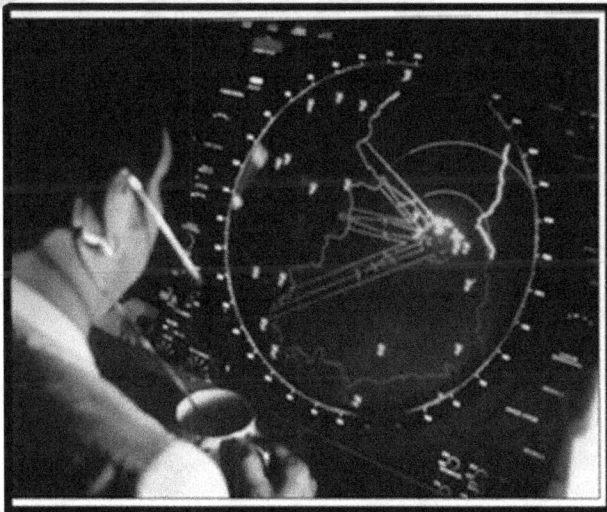

Radar Screen

Radar displays indicate the presence and movement of objects out of the range of vision, which is particularly useful for navigators. Electronic equipment records the behaviour of radio waves projected by the vessel; waves which do not encounter anything simply disperse, while waves bounced back reveal the shape and position of all objects in the region. The characteristic sweep of the radar display occurs because the area is continually reassessed for new information, and the screen is reprinted in response.

U.S. Air Force

507

TRANSFORMATIONS

A **transformation** is a change in position of an object.

Translation, Reflection, Rotation

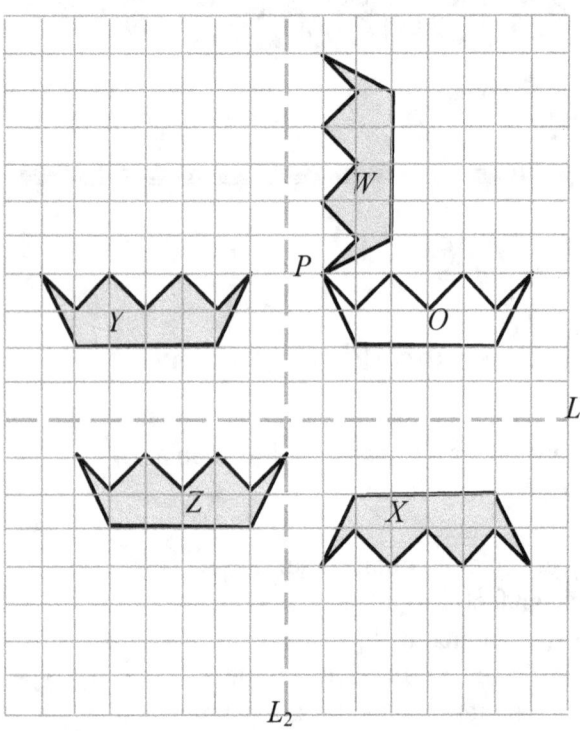

Figure 45:1: Isometry

1. In Figure 45:1, W is a rotation of the plane figure O through an angle of 90° in an anticlockwise sense about the point P as axis of rotation. Note that clockwise rotations are conventionally taken as negative.
2. X is a reflection of the plane figure O in the dotted line L_1 as mirror line. In a reflection, every point on the object is the same perpendicular distance from the mirror line as the corresponding point on the image.
3. Y is a reflection of the plane figure O in the dotted line L_2 as mirror line.
4. Z is a translation of the plane figure O, 7 units to the left and 5 units down the page. This translation can be described by the column vector $\begin{pmatrix} -7 \\ -5 \end{pmatrix}$ called a **translation vector**.

In a translation, every point moves the same distance in the same direction, without any rotation.

NOTE!

In each of the above transformations, the shape and size of the figure remains unchanged (shape and size are invariant). A transformation in which shape and size are invariant is called an **isometry**.

Enlargements

In Figure 45:2, $A'B'C'$ is an enlargement of ABC, with enlargement factor 3 and centre O. If corresponding sides are compared, it will be noticed that each side of $\triangle ABC$ is tripled in $\triangle A'B'C'$.

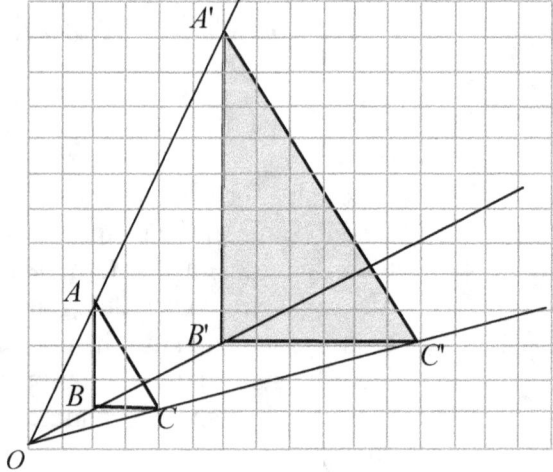

Figure 45:2

Remarks

1. If during an enlargement, each side of a figure increases k times, the enlargement is said to have a scale factor k.
2. By counting squares, it can be seen that the area of the image $\triangle A'B'C'$ is 9 (or 3^2) times the area of $\triangle ABC$. Thus, the area of an enlargement scale factor k is k^2 times that of the original object.

$$k = \frac{\text{image length}}{\text{object length}}$$

$$\Rightarrow \text{image length} = k \times \text{object length}$$

$$k^2 = \frac{\text{image area}}{\text{object area}}$$

$$\Rightarrow \text{image area} = k^2 \times \text{object area}$$

3. If the object is reduced k times, the scale

factor will be $\dfrac{1}{k}$ (i.e. a fraction). The then object is said to be enlarged by scale factor $\dfrac{1}{k}$ and this description is understood as a reduction.

Notice that in an enlargement object and image are similar.

Example 45:1

An irregular figure of area 252 cm^2 is enlarged by scale factor $\dfrac{5}{6}$; find the area of its image.

Solution

$$\text{Image area} = k^2 \times \text{object area}$$

$$= \left(\dfrac{5}{6}\right)^2 \times 252 = 175 \text{ cm}^2$$

Shear

In Figure 45:3, the unit square is mapped into the parallelogram shaded. This type of transformation is called a shear. In particular, this shear is parallel to the *x*-axis and points on this axis are invariant. On the other hand, points on *BC* move through the greatest distance of 2 units to the positive *x* direction. In such a case, the transformation is described as **a shear parallel to the *x*-axis with shear factor 2**.

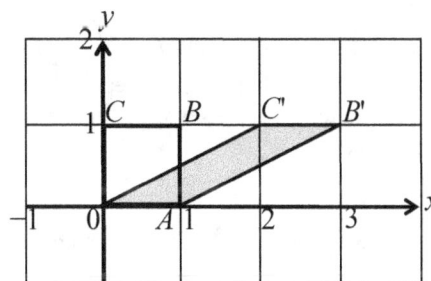

Figure 45:3

Figure 45:4, represents a shear parallel to the *x*- axis with shear factor –2.

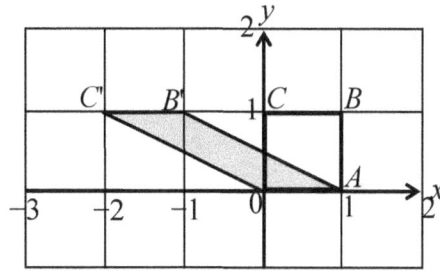

Figure 45:4

Example 45:2

In Figure 45:5 (a) and (b), the unit square *OABC* is transformed into the parallelograms *OAB′C′* and *OAB″C″*. Describe these transformations completely.

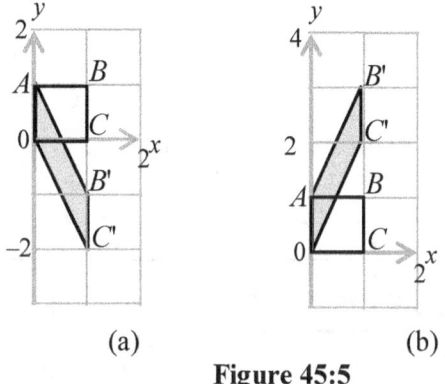

(a)   (b)

Figure 45:5

Solution

(a) A shear parallel to the *y*-axis with shear factor –2.

(b) A shear parallel to the *y*-axis with shear factor 2.

Stretch

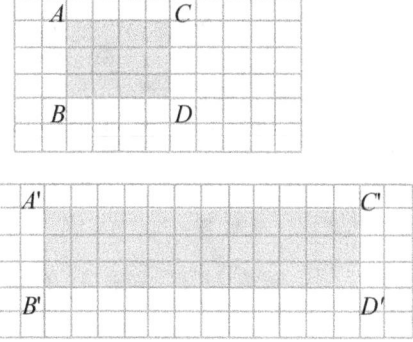

Figure 45:6

In Figure 45:6, $A'B'C'D'$ is a **stretch** of *ABCD* towards the right, **stretch factor** 3. The stretch factor is important because it gives us some valuable information. For a stretch with stretch factor k

1. $k = \dfrac{\text{length of stretched side}}{\text{length of corresponding object side}}$

2. $k = \dfrac{\text{image area}}{\text{object area}}$

3. When $k < -1$, the stretch is in the negative direction and when $k > 1$, the stretch is in the positive direction.

4. When $-1 < k < 1$, the object is compressed in one direction only.

EXERCISE 45:1

1. Write down the image of the point $(-2, 3)$ if it is reflected in the x-axis followed by a reflection in the y-axis.

2. The points in Table 45:1 undergo a reflection in the lines shown. Fill in the images of the resulting points.

	(2,3)	(3,0)	(0,-5)	(3,3)	(-4,-6)
x-axis					
y-axis					
$y = x$					
$y = -x$					

Table 45:1

3. Write down the image of the point $(4, 2)$ if it is reflected in the line $y = x$ followed by a reflection in the x-axis.

4. The triangle ABC with vertices $A(-3,4)$, $B(5,3)$ and $C(-4,-2)$ is reflected in the x-axis. Make a sketch of the triangle ABC and its image $A'B'C'$.

5. Each of the following points is reflected in the line $y = x \tan 60°$. Determine their images.
   (i) $(2, 3)$ (ii) $(3, 0)$ (iii) $(0,-5)$
   (iv) $(3, 3)$ (v) $(-4, -6)$

6. Find the image of the point $(3, -3)$ after a rotation of $180°$ anticlockwise about the point $(-2, 2)$.

7. Given that, a translation T takes the origin to the point $(2, 3)$; find the image of the following points under T.
   (a) $(1, 0)$ (b) $(0, 1)$ (c) $(-3,-1)$
   (d) $(-1, 2)$ (e) $(-2, -2)$ (f) $(-2, -3)$
   (g) (a, b) (h) $(1, 1)$ (i) $(-11,15)$
   (j) $(-4, 3)$

8. The triangle ABC with vertices $A(1,-2)$, $B(5,3)$ and $C(-3,4)$ is translated by T which takes the origin to the point $(2,3)$. Make a sketch of the

triangle ABC and its image $A'B'C'$. Given that T_1 is the translation that takes the origin to the point $(-2,-3)$, what will be the result of T followed by T_1? Name this type of transformation.

9. A translation takes the point $(2, 3)$ to $(3, -4)$. What is the image of the following points under the translation?
   (a) $(0, 0)$ (b) $(1,1)$ (c) $(1, 0)$
   (d) $(-1, 2)$ (e) $(4, -2)$ (f) $(-2, -3)$

10. An equilateral triangle ABC of side 1.5 cm is mapped onto $A_1 B_1 C_1$ by an anticlockwise rotation of $120°$ about A. $A_1 B_1 C_1$ is then mapped onto $A_2 B_2 C_2$ by an anticlockwise rotation of $120°$ about B. Finally, $A_2 B_2 C_2$ is mapped onto $A_3 B_3 C_3$ by an anticlockwise rotation of $120°$ about C. Using a plane paper accurately construct the complete figure and state a single geometrical transformation, which maps ABC onto $A_3 B_3 C_3$.

11. A translation takes the origin to the point $(2, 4)$. Find the image of the equation $y = x^2 + 4x$ under the transformation.

12. Given that $\mathbf{v}$ is the vector from the origin to the point $(3, 2)$, state the image of the point $(-4, 2)$ under the translation by the vector $\mathbf{v}$.

13. P and Q are transformations defined as follows:
    P: 'Reflect in the y-axis'
    Q: 'Translate +3 units parallel to the x-axis.
    If PQ means P followed by Q, find
    (a) $PQ(2,4)$ (b) $QP(2,4)$
    (c) (x, y) if $QP(x, y) = (4,3)$
    (d) (x, y) if $PQ(x, y) = (4,3)$

14. On graph paper, taking 1 cm to represent 1 unit, draw the x- and y-axes from -6 to $+12$ for each axis.
    (a) Draw the triangle T with vertices $P(2,1)$, $Q(2,3)$ and $R(5,1)$, constructing on the graph and writing down the coordinates of the vertices.
    (b) Translate by vector $\begin{pmatrix} -6 \\ 5 \end{pmatrix}$ the triangle T and label it T_1.
    (c) Reflect T in the y-axis and label it T_2.
    (d) Rotate T through $90°$ clockwise and label it T_3.
    (e) Enlarge T about the origin by scale factor of 2 and label it T_4.

SYMMETRY

Symmetry is the correspondence of parts on opposite sides of a point, line or plane. Symmetry is a very important phenomenon in disciplines such as architecture, mathematics, biology, physics, mineralogy etc. The bodies of many animals for instance exhibit bilateral symmetry on two opposite sides of a linear axis, or a median plane.

Wat Phra Kaeo

The Wat Phra Kaeo temple in Thailand built in 1782 is one of the greatest symmetrical architectural designs in the world.

Types of Symmetry

Objects exhibit two main types of symmetry. Namely, mirror symmetry and rotational symmetry.

(a)   Reflective or Mirror Symmetry

Mirror symmetry is that property of an object in which one of the halves of the object appears like the other part when placed against a mirror.
In plane figures, the mirror line is termed a **line of symmetry** illustrated by the dotted lines in the four-pointed star and regular octagon in Figure 45:7.

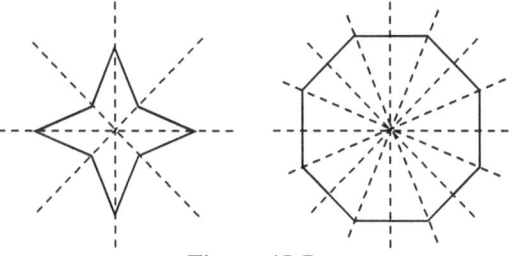

Figure 45:7

The star has four lines of symmetry. Can you count them? The regular octagon has eight lines of

symmetry. Also, count them.
Table 45:2, shows more plane figures and their lines of symmetry. The lines of symmetry are shown dotted.
   Notice that in mirror symmetry, we can superimpose one of the halves on the other by folding along the line of symmetry. Therefore, a line of symmetry of a plane figure is a line along which two halves of the figure can be superimposed by folding.

	Plane Figure	Number of lines of symmetry
Isosceles trapezium		1
Square		4
Rhombus		2
Rectangle		2
Kite		1
Equilateral triangle		3
Isosceles triangle		1
Parallelogram		none

Table 45:2

In solid figures, the analogue of the mirror line is termed a **plane of symmetry** illustrated by the shaded planes in the sphere and cuboids in Figure 45:8. The cuboids have three planes of symmetry. Another way of expressing this is to say cuboids exhibit mirror symmetry of order 3. Can you count them? How many planes of symmetry, has a sphere? Try to visualize them. A sphere has an infinite number of planes of symmetry.

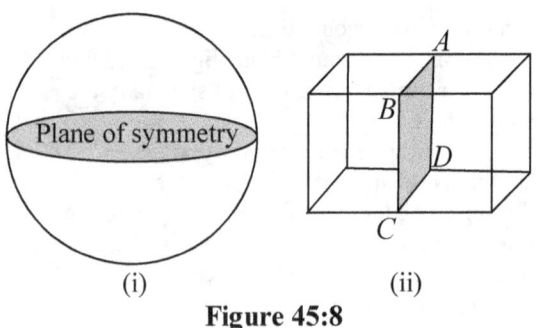

(i)   (ii)

Figure 45:8

(b) Rotational or Radial symmetry

The proportional arrangement of similar parts of a body around a central axis, as in the case of jellyfish or starfish, is known as radial symmetry. If a circle is rotated any amount about the centre, it remains unchanged. Therefore, a circle exhibits radial symmetry of infinite order about its centre.

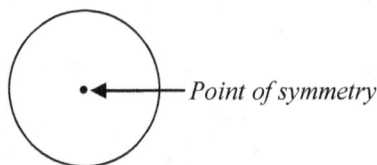

Figure 45:9

An equilateral triangle remains unchanged in three different positions when rotated about the point of symmetry. Hence, an equilateral triangle exhibits radial symmetry of order 3.

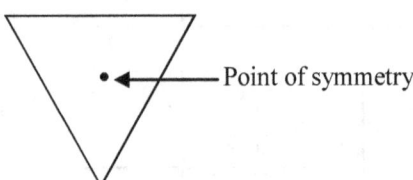

Figure 45:10

Any regular polygon with n sides has rotational symmetry of order n because when rotated, it shape is exactly repeated in n different positions as shown in Table 45:3.

REGULAR POLYGON	
Equilateral triangle	3
Square	4
Regular pentagon	5
Regular hexagon	6
n-gon	n

Table 45:3

In Table 45:3, it was seen that a parallelogram has no line of symmetry. Figure 45:11 shows that a parallelogram exhibits point symmetry about the point O. A_2, B_2, C_2, and D_2 can respectively fit on A_1, B_1, C_1 and D_1. In other words if the parallelogram is rotated through $180°$, it remains unchanged.

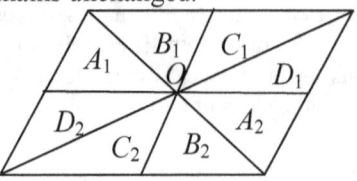

Figure 45:11

What is the order of rotational symmetry of swastika (Figure 45:12) the German army batch?

Figure 45:12

For solid figures, that exhibit radial or rotational symmetry, the centre of symmetry is called the **axis of symmetry**.

Figure 45:13(a) to (g) shows some solids with one axes of symmetry shown. Apart from the cone and pyramid, all the other solids have more than one axis of symmetry. Which of the figures have an infinite number of axes of symmetry? Copy the diagrams of the solids with a finite number of axes of symmetry and indicate these axes.

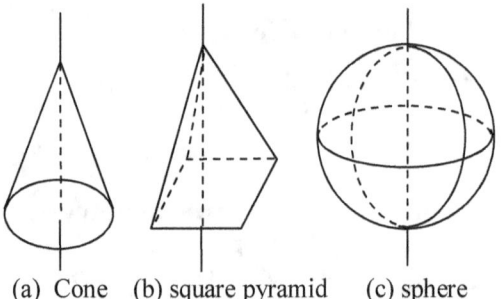

(a) Cone   (b) square pyramid   (c) sphere

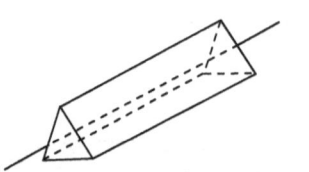
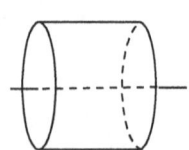

(d) Equilateral triangular prism   (e) cylinder

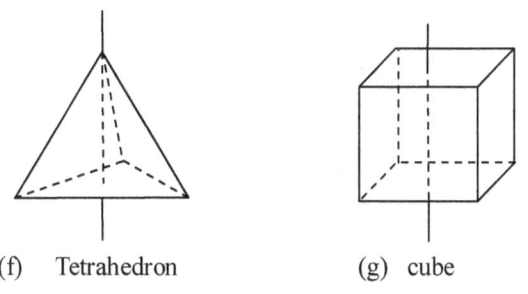

(f)   Tetrahedron   (g)   cube

Figure 45:13

> **EXERCISE 45:2**

1. How many lines of symmetry has
   (a) A regular hexagon
   (b) a regular pentagon
   (c) A regular octagon

2. By drawing and showing, using dotted lines state the number of axes (lines) of symmetry, if any which the following have:
   (a) A kite
   (b) An equilateral triangle
   (c) A square
   (d) A rectangle
   (e) A regular pentagon
   (f) A rhombus
   (g) A parallelogram
   (h) A regular hexagon

3. Plot the following points on square paper and connect them.
   (a) A (2, 5), B (4, 12), C (6, 10)
   (b) A (4, 5), B (6, 5), C (6, 10), D (4, 10)
   (c) A (1, 5), B (6, 5), C (6, 10), D (4, 12)
   (d) A (4,10), B (8,10), C (10,15), D (6,15)
   (e) A (4, 3), B (6, 10), C (8, 3), D (10, 10)
   (f) A (2, 2), B (6, 5), C (6, 15), D (2, 15)
   (g) A(0,10), B(5,5), C(10,5), D(15,10),
       E(15,15), F(10,20), G(5,20), H(0,15)
   (h) A (0, 5), B (5, 0), C (0,-5), D (-5, 0)
   (i) A(0,2), B(2,3), C(6,3), D(6,1), E(2,1)
   (i) List the shapes, which possess
      (i) Line symmetry
      (ii) Point symmetry
      (ii) Write down the order of rotational symmetry for each of the shapes or write N if the shape does not possess rotational symmetry.

4. Complete each of the diagrams in Figure 45:14, so that each will have rotational symmetry of order

(i)   Two   (ii)   three   (iii) four

5. Name all the capital letters of the English alphabet, which have one and only one axes of symmetry.

6. State the number of planes of symmetry which each of the following has.
   (a) A cube   (b) A cone
   (c) A sphere   (d) A square pyramid.

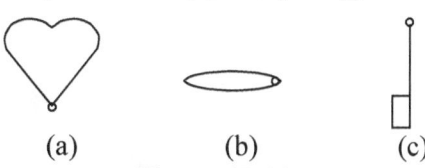

   (a)   (b)   (c)
Figure 45:14

7. List all the capital letters of the English alphabet, which have at least two axes of symmetry.

8. List all the capital letters of the English alphabet, which have no axes of symmetry.

9. Draw each of the following plane figures showing all the lines of symmetry.
   (a) a square
   (b) an equilateral triangle
   (c) a regular hexagon
   (d) an isosceles triangle
   (e) a parallelogram
   (f) an isosceles trapezium

10. $\mathscr{E}$ - capital letters of the English alphabet
    V = letters with a vertical line symmetry
    H = letters with a horizontal line symmetry
    R = letters which have rotational symmetry
    Find $H \cap V$. What conclusion do you draw?

MULTIPLE CHOICE EXERCISE 45

1. In Figure 45:15, Triangle $A'B'C'$ is an enlargement of triangle ABC. The scale factor of the enlargement is:

   [A] $\dfrac{3}{2}$ [B] 2   [C] $\dfrac{5}{2}$ [D] 3

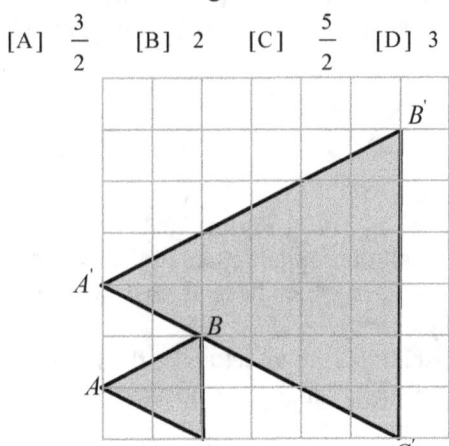

Figure 45:15

2. In the Figure 45:16 above, not drawn to scale, triangle $OA'B'$ is an enlargement of triangle OAB.

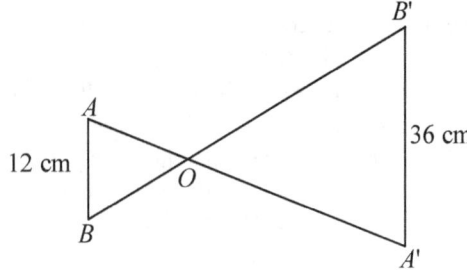

Figure 45:16

   The scale factor of the enlargement is:

   [A] $-\dfrac{1}{3}$ [B] -3 [C] $\dfrac{1}{3}$ [D] 3

3. The point $(4,3)$ is reflected in the x-axis followed by a reflection in the y-axis. The final image is:

   [A] $(4,-3)$ [B] $(-4,3)$
   [C] $(-4,-3)$ [D] $(-3,4)$

4. A certain transformation T is defined by $T: (x, y) \mapsto (2x + y, -x + y)$. The image of the point $(-2, 6)$ under T is:

   [A] $(-2,6)$ [B] $(2,8)$
   [C] $(2,-6)$ [D] $(2,-8)$

5. A transformation T is defined by $T: (x, y) \mapsto (2x, 2y)$.
   The image of $(-3,4)$ under T is:

   [A] $(-6,-8)$ [B] $(6,-8)$

   [C] $(6,8)$ [D] $(-6,8)$

6. A transformation T is defined by $T : (x, y) \mapsto$ (   ). The statement that correctly describes the transformation T is:

   [A] An enlargement scale factor 2 centre $(0, 0)$
   [B] An enlargement scale factor 2 centre $(0, 2)$
   [C] An enlargement scale factor 2 centre $(2, 2)$
   [D] A translation 2 units to the right

7. The square with vertices $W(0,0)$, $X(1,0)$, $Y(1,1)$, $Z(0,1)$ is transformed into a square with vertices $W(0,0)$, $X(1,0)$, $Y(3,1)$, $Z(2,1)$. The statements that correctly describes the transformation T is:

   [A] A shear, shear factor 2 where points on the y-axis are invariant.
   [B] A stretch, stretch factor 2 where points on the y-axis are invariant.
   [C] A shear, shear factor 2 where points on the x-axis are invariant.
   [D] A stretch, stretch factor 2 where points on the x-axis are invariant.

8. A triangle whose vertices are $A(1,2)$, $B(3,4)$ and $C(6,2)$ is transformed to a triangle whose vertices are $A'(1,-2)$, $B'(3,-4)$ and $C'(6,-2)$. The statement, which correctly describes the transformation, is:

   [A] T is a reflection in the line $y = -1$
   [B] T is a reflection in the line $y = x$
   [C] T is a reflection in the line $y = -x$
   [D] T is a reflection in the line $y = 0$

9. The order of rotational symmetry of a pyramid with a square base and axis through the vertex and centre of the square is:

   [A] 2   [B] 3   [C] 4   [D] 5

10. The graph in Figure 45:18 which shows the reflection of triangle ABC in Figure 45:17 in the line XY is:

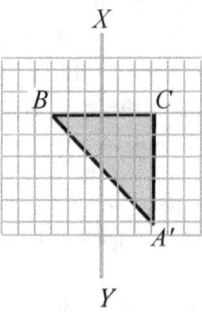

Figure 45:17

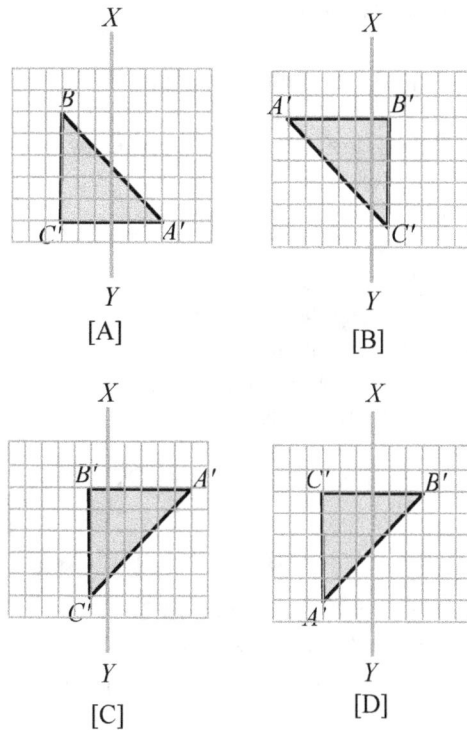

Figure 45:18

11. In Figure 45:19, the diagram that represents a shear, shear factor –3 parallel to the *y*-axis is:

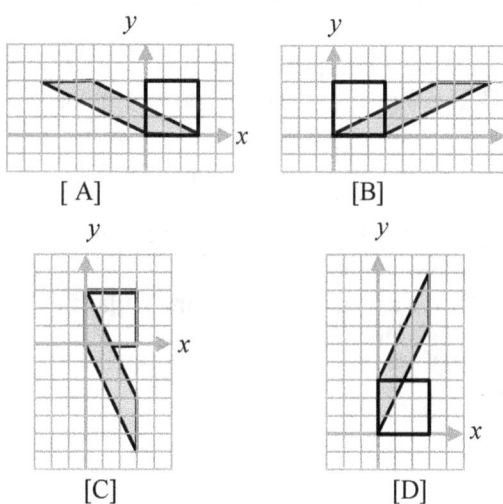

Figure 45:19

12. In Figure 45:20, the invariant line is:
   [A] *AD*   [B] *BC*   [C] *CD* [D] *AB*

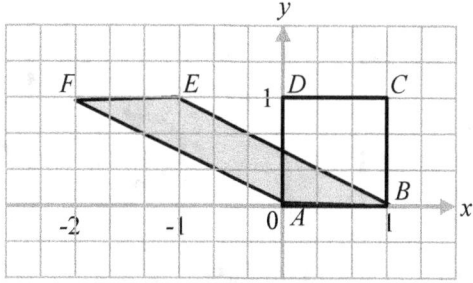

Figure 45:20

13. Figure 45:21, shows a cuboids with square faces *ABCD* and *EFGH*. The number of planes of symmetry in the cuboids is:
   [A] 3   [B] 4   [C] 5   [D] 6

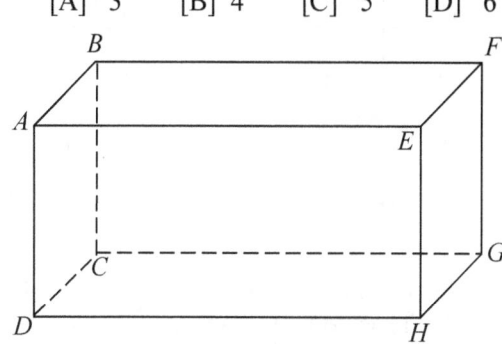

Figure 45:21

14. Using their symmetric properties, the odd plane figure among the following is:
   [A]   An isosceles triangle.
   [B]   A semi-circle.
   [C]   A rectangle.
   [D]   A pentagon with four sides equal.

15. The number of lines of symmetry in a rectangle is:
   [A] 1   [B] 2   [C] 4   [D] 8

16. Symmetrically, a square differs from a rectangle because:
   [A]   A square has 2 lines of symmetry while a rectangle has 4.
   [B]   A square has 4 lines of symmetry while a rectangle has 2.
   [C]   The 4 sides of a square are equal but a rectangle has a pair of opposite sides equal.
   [D]   The diagonals of a square intersect at right angles but those of a rectangle do not.

17. The shapes in Figure 45:22, which are mirror images of each other, are:

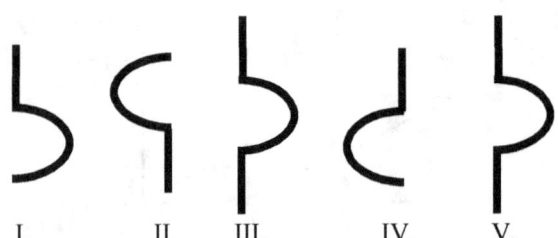

I.   II.   III.   IV.   V.

Figure 45:22

[A] I and II   [B] III and V
[C] I and IV   [D] II and IV

18. The net of a cube in Figure 45:23, which possess rotational symmetry of order greater than one is:

[A] 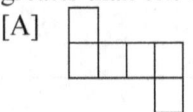   [B]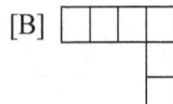

[C] 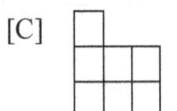   [D]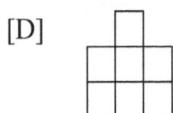

Figure 45:23

19. The net of a tetrahedron in Figure 45:24 which has no line symmetry is:

[A]    [B]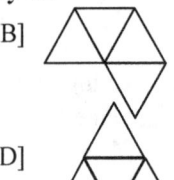

[C]   [D]

Figure 45:24

20. Among the following, the quadrilateral, which has exactly one line of symmetry, is:
[A] A kite   [B] A rectangle
[C] A Parallelogram   [D] A Rhombus

21. The number of lines of symmetry in the triangle in Figure 45:25 is:

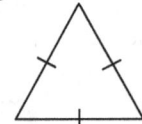

Figure 45:25
[A] 2   [B] 3   [C] 4   [D] 5

22. The figure, which has two lines of symmetry, is:

[A] An isosceles triangle
[B] A square
[C] An equilateral triangle
[D] A rhombus

23. The figure, which has 9 planes of symmetry, is:
[A] A regular nonagon
[B] A cube
[C] A regular octagon
[D] A regular hexagon

24. The diagrams in Figure 45:26 as drawn which is not symmetrical about a horizontal axis is:

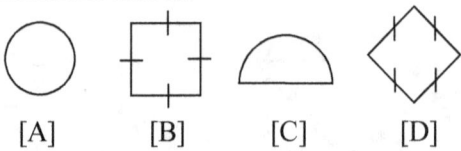

[A]   [B]   [C]   [D]

Figure 45:26

25. A plane figure may have:
[A] A line of symmetry and a plane of symmetry.
[B] A line of symmetry and a point of symmetry.
[C] A point of symmetry and a plane of symmetry.
[D] An axis of symmetry and a plane of symmetry.

26. A solid figure may have:
[A] A line of symmetry and a plane of symmetry.
[B] A line of symmetry and a point of symmetry.
[C] A point of symmetry and a plane of symmetry.
[D] An axis of symmetry and a plane of symmetry.

TOPIC 46

TRANSFORMATIONS WITH MATRICES

OBJECTIVES

At the end of this topic, the learner should be able to:

1. Use transformation matrices to transform points in two dimensions.
2. Describe transformations.
3. Find transformation matrices.
4. Find the inverse of a transformation.
5. Identify transformations by singular matrices.
6. Find the area of the image using the transformation matrix.
7. Determine the ratio of the image area to the object area.
8. Determine the result of a composite transformation by two or more matrices.

Portrait in London by Barraud & Jerrard

Arthur Cayley, a British Mathematician published a treatise on geometric transformations using matrices that were not rotated versions of the coefficients being investigated. Instead he defined operations such as addition, subtraction, multiplication, and division as transformations of those matrices and showed the associative and distributive properties held true. Cayley investigated and demonstrated the non-commutative property of matrix multiplication as well as the commutative property of matrix addition.

From Wikipedia, the free encyclopedia

Transformation Matrix (Matrix Operator)

Transformations can easily be performed using matrices. In the following pages, only transformations in the *x-y* plane will be considered. Consider the point $P(x, y)$. It was earlier stated that this point can be represented by the column matrix or vector $\mathbf{P} = \begin{pmatrix} x \\ y \end{pmatrix}$.

Pre-multiply $\mathbf{P} = \begin{pmatrix} x \\ y \end{pmatrix}$ by the matrix

$$\mathbf{A} = \begin{pmatrix} a & b \\ c & d \end{pmatrix}.$$

$$\mathbf{AP} = \begin{pmatrix} a & b \\ c & d \end{pmatrix}\begin{pmatrix} x \\ y \end{pmatrix} = \begin{pmatrix} ax + by \\ cx + dy \end{pmatrix} = \begin{pmatrix} X \\ Y \end{pmatrix}$$

Thus $\begin{pmatrix} X \\ Y \end{pmatrix}$ is the position vector of a new point.

Therefore, $P(x, y)$ has been transformed by the matrix $\mathbf{A}$ to the new point $P_1(X, Y)$. The matrix $\mathbf{A}$ is called a **transformation matrix** or a **matrix operator.** $P_1(X, Y)$ is said to be the **image** of $P(x, y)$. This transformation is denoted by $(x, y) \longmapsto (X, Y)$ and read (x, y) maps to (X, Y).

Example 46: 1
Find the image of the point $(2, -3)$ under the transformation defined by the matrix $\begin{pmatrix} 2 & 0 \\ 0 & 2 \end{pmatrix}$.

Solution

$$\mathbf{AP} = \begin{pmatrix} 2 & 0 \\ 0 & 2 \end{pmatrix}\begin{pmatrix} 2 \\ -3 \end{pmatrix} = \begin{pmatrix} 4 \\ -6 \end{pmatrix}$$

Example 46:2
Plot the position vectors of the points $A(2,0)$ and $B(-2,4)$. Find the images A' and B' of A and B under the matrix operato $\mathbf{M} = \begin{pmatrix} 1 & -1 \\ 1 & 1 \end{pmatrix}$ and plot them on the same graph as A and B.

Solution

$$\begin{pmatrix} 1 & -1 \\ 1 & 1 \end{pmatrix}\begin{pmatrix} 2 \\ 0 \end{pmatrix} = \begin{pmatrix} 2 \\ 2 \end{pmatrix}$$

$$\begin{pmatrix} 1 & -1 \\ 1 & 1 \end{pmatrix}\begin{pmatrix} -2 \\ 4 \end{pmatrix} = \begin{pmatrix} -6 \\ 2 \end{pmatrix}$$

$\therefore (2,0) \longmapsto (2,2)$ and $(-2,4) \longmapsto (-6,2)$

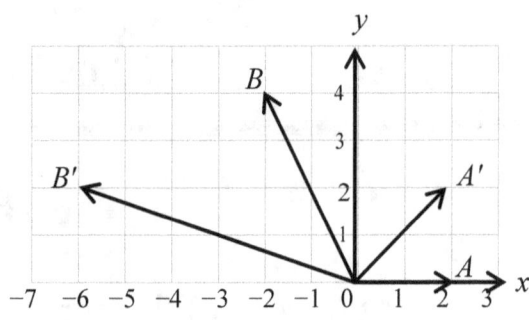

Figure 46:1

Example 46:3
Show that the origin $(0, 0)$ is invariant (i.e. remains unchanged) under any transformation defined by a 2×2 matrix.

Solution

Let the transformation matrix be $\begin{pmatrix} a & b \\ c & d \end{pmatrix}$.

Then, $\begin{pmatrix} a & b \\ c & d \end{pmatrix}\begin{pmatrix} 0 \\ 0 \end{pmatrix} = \begin{pmatrix} 0 \\ 0 \end{pmatrix}$

$\Rightarrow (0, 0) \longmapsto (0, 0)$

Therefore, origin $(0, 0)$ is invariant.

EXERCISE 46:1

1. Find the image of the point $(-5, 3)$ under the transformation defined by each of the following matrices.

   (a) $\begin{pmatrix} 1 & 0 \\ 0 & -1 \end{pmatrix}$ (b) $\begin{pmatrix} 2 & 0 \\ 0 & -2 \end{pmatrix}$

   (c) $\begin{pmatrix} \frac{1}{2} & 0 \\ 0 & \frac{1}{2} \end{pmatrix}$ (d) $\begin{pmatrix} 0 & -1 \\ 1 & 0 \end{pmatrix}$

2. Plot the object and image of the following set of points under the matrix indicated.
   (a) $A(-3,4)$, $B(-1,4)$, $C(-1,3)$ under

$$\begin{pmatrix} 0 & -1 \\ -1 & 0 \end{pmatrix}.$$

(b) $P(2,1)$, $Q(2,2)$, $R(0,2)$ under

$$\begin{pmatrix} 2 & 0 \\ 0 & -2 \end{pmatrix}.$$

Transformations Involving Many Points

If the points whose images are required are many, they can be written in a single matrix where each column represents the position vector of a single point. The product can then be computed at once.

Example 46:4
Find the image of the square with vertices $(0, 0)$, $(1, 0)$, $(1, 1)$ and $(0, 1)$ under the transformation represented by the matrix

$$\begin{pmatrix} 1 & 2 \\ 0 & 1 \end{pmatrix}.$$

Solution
As a single matrix the vertices of the square

are $\begin{pmatrix} 0 & 1 & 1 & 0 \\ 0 & 0 & 1 & 1 \end{pmatrix}.$

Pre-multiply by $\begin{pmatrix} 1 & 2 \\ 0 & 1 \end{pmatrix}.$

$$\begin{pmatrix} 1 & 2 \\ 0 & 1 \end{pmatrix}\begin{pmatrix} 0 & 1 & 1 & 0 \\ 0 & 0 & 1 & 1 \end{pmatrix} = \begin{pmatrix} 0 & 1 & 3 & 2 \\ 0 & 0 & 1 & 1 \end{pmatrix}$$

▷ EXERCISE 46:2

1. The transformation matrix $\begin{pmatrix} 1 & 0 \\ 1 & 1 \end{pmatrix}$ is applied to the quadrilateral whose vertices are $(0, 0)$, $(2, 0)$, $(2, 3)$, $(0, 3)$. Determine the vertices of the resulting quadrilateral.

2. Find and draw the image of the square with vertices $(0, 0)$, $(1, 1)$, $(0, 2)$, $(-1, 1)$ under the transformation represented by the

   matrix $\begin{pmatrix} 4 & 3 \\ -3 & -2 \end{pmatrix}.$

3. Plot the rectangle $P(0, 0)$, $Q(0, 2)$, $R(3, 2)$, $S(3, 0)$. Determine and draw the image of $PQRS$ under the following transformation matrices.

   (a) $\begin{pmatrix} 3 & 0 \\ 0 & 2 \end{pmatrix}$ (b) $\begin{pmatrix} 1 & 2 \\ 0 & 1 \end{pmatrix}$ (c) $\begin{pmatrix} 2 & 0 \\ 0 & 2 \end{pmatrix}$

Describing Transformation

Given any transformation matrix, it is possible to determine and describe completely the nature of the transformation. To do this, consider the first and second columns of the matrix operator to be the images of the base

vectors $\mathbf{i} = \begin{pmatrix} 1 \\ 0 \end{pmatrix}$ and $\mathbf{j} = \begin{pmatrix} 0 \\ 1 \end{pmatrix}$ respectively.

The vertices of a given figure can be pre-multiplied by the matrix operator to obtain the vertices of the image.

Example 46:5
Describe the transformation represented by

the matrix $\begin{pmatrix} -1 & 0 \\ 0 & 1 \end{pmatrix}.$

Solution

$$\begin{pmatrix} 1 \\ 0 \end{pmatrix} \mapsto \begin{pmatrix} -1 \\ 0 \end{pmatrix} \text{ and } \begin{pmatrix} 0 \\ 1 \end{pmatrix} \mapsto \begin{pmatrix} 0 \\ 1 \end{pmatrix}$$

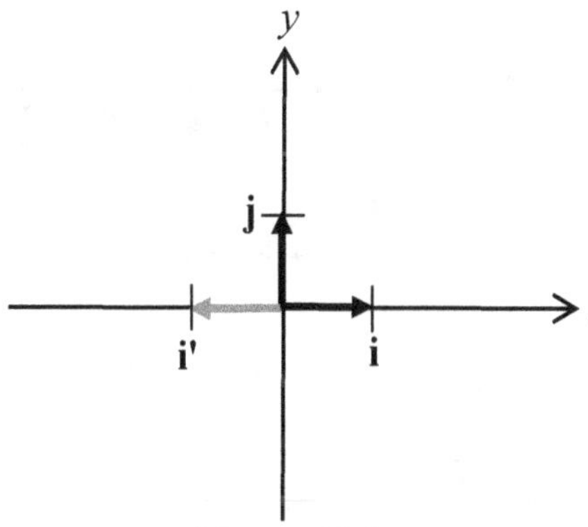

Figure 46:2

This shows that points on the positive x-axis are reflected on the negative x-axis and points on the y-axis are invariant. Therefore the transformation is a reflection in the y-axis (i.e. the line $x = 0$)

For more obscure cases, the vertices of a regular figure such as the rectangle with vertices $O(0,0)$, $A(2,0)$, $B(2,1)$ and $C(0,1)$ can be pre-multiplied by the transformation matrix to obtain the vertices of the image. The two figures are then sketched on the same Cartesian plane and compared.

Example 46:6
Using the rectangle with vertices $O(0,0)$, $A(3,0)$, $B(3,2)$ and $C(0,2)$ show that the matrix $\begin{pmatrix} 1 & 0 \\ 0 & -1 \end{pmatrix}$ is a matrix of reflection in the line $y = 0$ (i.e. the x-axis).

Solution
The single matrix, representing the vertices of the rectangle can be written as

$$R = \begin{pmatrix} 0 & 3 & 3 & 0 \\ 0 & 0 & 2 & 2 \end{pmatrix}.$$

Pre-multiplying R by $\begin{pmatrix} 1 & 0 \\ 0 & -1 \end{pmatrix}$ gives

$$\begin{pmatrix} 1 & 0 \\ 0 & -1 \end{pmatrix}\begin{pmatrix} 0 & 3 & 3 & 0 \\ 0 & 0 & 2 & 2 \end{pmatrix} = \begin{pmatrix} 0 & 3 & 3 & 0 \\ 0 & 0 & -2 & -2 \end{pmatrix}$$

Sketching R (not shaded) and its image (shaded) on the same Cartesian plane gives:

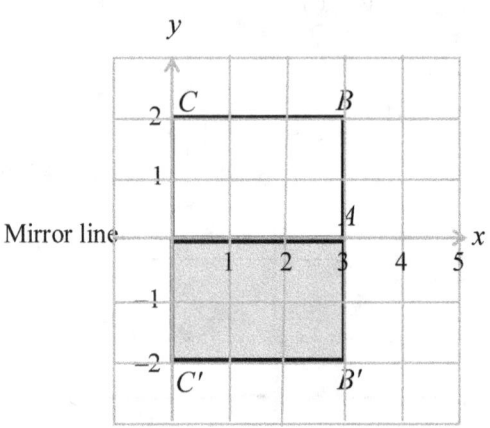

Figure 46:3

From the graph, $\begin{pmatrix} 1 & 0 \\ 0 & -1 \end{pmatrix}$ is clearly a matrix of reflection in the line $y = 0$ or the x-axis.

Example 46:7
The triangle with vertices $A(1, 1)$, $B(3, 1)$, $C(3, 2)$ is transformed by the matrix $\begin{pmatrix} 0 & 1 \\ -1 & 0 \end{pmatrix}$. Find and describe the image of the transformation.

Solution

The single matrix, representing the vertices of the triangle can be written as $T = \begin{pmatrix} 1 & 3 & 3 \\ 1 & 1 & 2 \end{pmatrix}$.

Pre-multiplying T by $\begin{pmatrix} 0 & 1 \\ -1 & 0 \end{pmatrix}$ gives

$$\begin{pmatrix} 0 & 1 \\ -1 & 0 \end{pmatrix}\begin{pmatrix} 1 & 3 & 3 \\ 1 & 1 & 2 \end{pmatrix} = \begin{pmatrix} 1 & 1 & 2 \\ -1 & -3 & -3 \end{pmatrix}$$

The object and image are plotted in Figure 46:4.

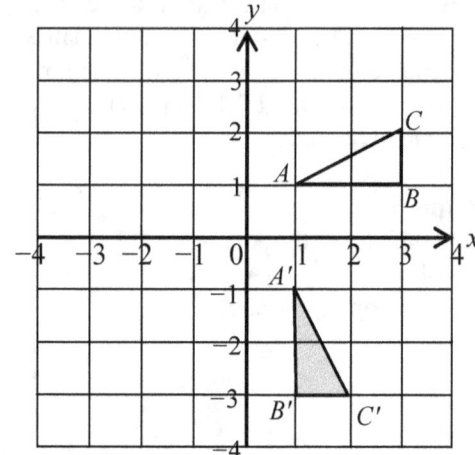

Figure 46:4

Therefore, from the graph, $\begin{pmatrix} 0 & 1 \\ -1 & 0 \end{pmatrix}$ is a matrix of rotation through 270° about $(0,0)$ as center.

Remarks!
1. Rotations in the anti-clockwise sense are regarded as positive, while rotations in the clockwise sense are regarded as negative.
2. A rotation through 270° is the same as a rotation through −90°.
3. A rotation through 180° is the same as a rotation through −180°
4. A rotation through 360° is an invariant transformation. Therefore, its transformation matrix is the identity matrix $\mathbf{I} = \begin{pmatrix} 1 & 0 \\ 0 & 1 \end{pmatrix}$.

Example 46:8
Find the image of the rectangle with vertices $O(0,0)$, $A(3,0)$, $B(3,2)$ and $C(0,2)$ under the

transformation with matrix $\begin{pmatrix} 4 & 0 \\ 0 & 4 \end{pmatrix}$. Hence, describe the transformation completely.

Solution

Pre-multiplying the vertices by $\begin{pmatrix} 4 & 0 \\ 0 & 4 \end{pmatrix}$ gives

$$\begin{pmatrix} 4 & 0 \\ 0 & 4 \end{pmatrix}\begin{pmatrix} 0 & 3 & 3 & 0 \\ 0 & 0 & 2 & 2 \end{pmatrix} = \begin{pmatrix} 0 & 12 & 12 & 0 \\ 0 & 0 & 8 & 8 \end{pmatrix}$$

The rectangle $OABC$ and its image $OA'B'C'$ are shown in Figure 46:5. Such a transformation is called an enlargement. Careful observation will reveal that each side of the image is 4 times longer. In this case, the enlargement is said to have scale factor 4.

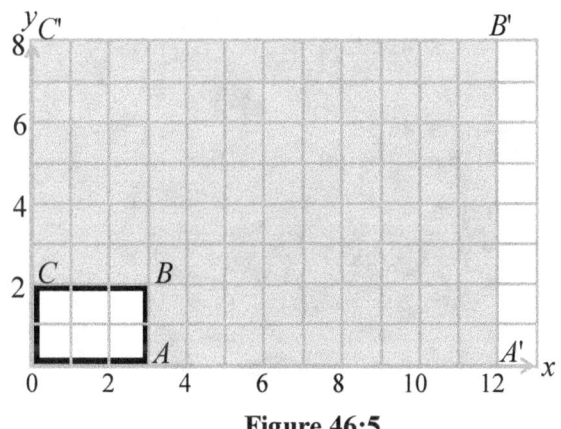

Figure 46:5

Remarks

1. If during an enlargement, each side increases k times, the enlargement is said to have a scale factor k.
2. Since in an enlargement object and image are similar,

$$k = \frac{\text{image length}}{\text{object length}}$$

$$\Rightarrow \text{image length} = k \times \text{object length}$$

3. The area of an enlargement with scale factor k is k^2 times the area of the object. Thus,

$$\text{image area} = k^2 \times \text{object area}$$

4. If the object is instead reduced k times, the scale factor will be $\frac{1}{k}$ and the object is said

to be enlarged by scale factor $\frac{1}{k}$ and this description is understood to be a reduction.

Example 46:9

Use the rectangle with vertices $O(0,0)$, $A(3,0)$, $B(3,2)$ and $C(0,2)$ to determine and describe each of the following transformations:

(i) $\begin{pmatrix} 1 & 2 \\ 0 & 1 \end{pmatrix}$ (ii) $\begin{pmatrix} 2 & 0 \\ 0 & 1 \end{pmatrix}$

Solution

(i) Pre-multiplying the vertices by $\begin{pmatrix} 1 & 2 \\ 0 & 1 \end{pmatrix}$ gives

$$\begin{pmatrix} 1 & 2 \\ 0 & 1 \end{pmatrix}\begin{pmatrix} 0 & 3 & 3 & 0 \\ 0 & 0 & 2 & 2 \end{pmatrix} = \begin{pmatrix} 0 & 3 & 7 & 4 \\ 0 & 0 & 2 & 2 \end{pmatrix}$$

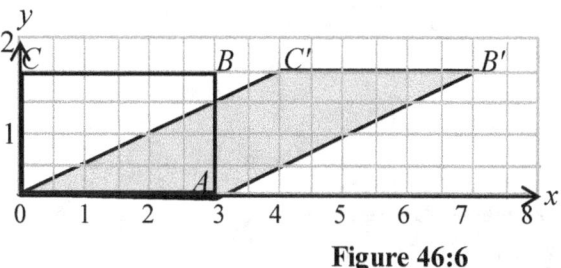

Figure 46:6

From the graph, it can be seen that, the transformation represents a shear with scale factor 2, parallel to the positive x- direction.

(ii) Pre-multiplying the vertices by $\begin{pmatrix} 2 & 0 \\ 0 & 1 \end{pmatrix}$ gives

$$\begin{pmatrix} 2 & 0 \\ 0 & 1 \end{pmatrix}\begin{pmatrix} 0 & 3 & 3 & 0 \\ 0 & 0 & 2 & 2 \end{pmatrix} = \begin{pmatrix} 0 & 6 & 6 & 0 \\ 0 & 0 & 2 & 2 \end{pmatrix}$$

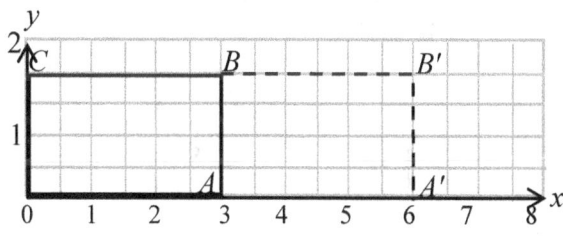

Figure 46:7

From the graph, it can be seen that, the transformation represents a stretch with scale factor 2, parallel to the positive x- direction.

Remarks

1. The transformation matrix representing a shear parallel to the x-axis is always of the form $\begin{pmatrix} 1 & k \\ 0 & 1 \end{pmatrix}$. When $k < 0$, the shear is in the negative x- direction and when $k > 0$, the shear is in the positive x- direction. It can be shown that the transformation matrix representing a shear parallel to the y-axis is of the form $\begin{pmatrix} 1 & 0 \\ k & 1 \end{pmatrix}$.

2. The transformation matrix representing a stretch parallel to the x-axis is always of the form $\begin{pmatrix} k & 0 \\ 0 & 1 \end{pmatrix}$. It can be shown that the transformation matrix representing a stretch parallel to the y-axis is of the form $\begin{pmatrix} 1 & 0 \\ 0 & k \end{pmatrix}$.

EXERCISE 46:3

1. Use base vectors to determine and describe the transformations whose matrices are as follows:

$$A = \begin{pmatrix} 1 & 0 \\ 0 & 1 \end{pmatrix}, \quad B = \begin{pmatrix} 0 & -1 \\ 1 & 0 \end{pmatrix}, \quad C = \begin{pmatrix} 0 & 1 \\ -1 & 0 \end{pmatrix}$$

$$D = \begin{pmatrix} -1 & 0 \\ 0 & -1 \end{pmatrix}, \quad E = \begin{pmatrix} 0 & 1 \\ 1 & 0 \end{pmatrix}, \quad F = \begin{pmatrix} -1 & 0 \\ 0 & 1 \end{pmatrix}$$

$$G = \begin{pmatrix} 0 & -1 \\ -1 & 0 \end{pmatrix}, \quad H = \begin{pmatrix} 1 & 0 \\ 0 & -1 \end{pmatrix}, \quad P = \begin{pmatrix} 2 & 0 \\ 0 & 2 \end{pmatrix},$$

$$Q = \begin{pmatrix} \frac{1}{2} & 0 \\ 0 & \frac{1}{2} \end{pmatrix}, \quad R = \begin{pmatrix} -\frac{1}{2} & 0 \\ 0 & -\frac{1}{2} \end{pmatrix}, \quad S = \begin{pmatrix} -2 & 0 \\ 0 & -2 \end{pmatrix}$$

$$T = \begin{pmatrix} 1 & 0 \\ 0 & 2 \end{pmatrix}, \quad U = \begin{pmatrix} 1 & 0 \\ 2 & 1 \end{pmatrix}, \quad V = \begin{pmatrix} 1 & 1 \\ 0 & 1 \end{pmatrix},$$

$$W = \begin{pmatrix} 3 & 0 \\ 0 & 1 \end{pmatrix}, \quad Y = \begin{pmatrix} 1 & 0 \\ 0 & 3 \end{pmatrix}.$$

2. Using the matrix in question (1) draw, the object and the image of the rectangle

represented by the matrix $\begin{pmatrix} 0 & 2 & 2 & 0 \\ 0 & 0 & 1 & 1 \end{pmatrix}$ and describe the transformation in each case.

Finding Transformation Matrices

There are two ways of finding transformation matrices.

1. Find the images of the base vectors $\begin{pmatrix} 1 \\ 0 \end{pmatrix}$ and $\begin{pmatrix} 0 \\ 1 \end{pmatrix}$ under the transformation. Thus if under the given transformation $\begin{pmatrix} 1 \\ 0 \end{pmatrix} \mapsto \begin{pmatrix} a \\ c \end{pmatrix}$ and $\begin{pmatrix} 0 \\ 1 \end{pmatrix} \mapsto \begin{pmatrix} b \\ d \end{pmatrix}$ then the transformation matrix is $\begin{pmatrix} a & b \\ c & d \end{pmatrix}$.

Example 46:10
Find the matrix for a reflection in the x-axis.

Solution
$$\begin{pmatrix} 1 \\ 0 \end{pmatrix} \mapsto \begin{pmatrix} 1 \\ 0 \end{pmatrix} \text{ and } \begin{pmatrix} 0 \\ 1 \end{pmatrix} \mapsto \begin{pmatrix} 0 \\ -1 \end{pmatrix}.$$

$\therefore$ the transformation matrix is $\begin{pmatrix} 1 & 0 \\ 0 & -1 \end{pmatrix}$.

Example 46:11
What is the transformation matrix for a stretch by a factor 5 parallel to Ox?

Solution
$$\begin{pmatrix} 1 \\ 0 \end{pmatrix} \mapsto \begin{pmatrix} 5 \\ 0 \end{pmatrix} \text{ and } \begin{pmatrix} 0 \\ 1 \end{pmatrix} \mapsto \begin{pmatrix} 0 \\ 1 \end{pmatrix}$$

$\therefore$ the transformation matrix is $\begin{pmatrix} 5 & 0 \\ 0 & 1 \end{pmatrix}$.

2. If two object points different from the origin and not on the invariant line are known and two corresponding image points are known pre-multiply the object points by $\begin{pmatrix} a & b \\ c & d \end{pmatrix}$ and equate to the image points.

Then solve for a, b, c and d.

Example 46:12

Under a certain transformation, $\begin{pmatrix} 3 \\ -2 \end{pmatrix} \mapsto \begin{pmatrix} 9 \\ 8 \end{pmatrix}$ and $\begin{pmatrix} 5 \\ 1 \end{pmatrix} \mapsto \begin{pmatrix} -3 \\ 4 \end{pmatrix}$. Find the transformation matrix.

Solution

Let the transformation matrix be $\begin{pmatrix} a & b \\ c & d \end{pmatrix}$.

Then, $\begin{pmatrix} a & b \\ c & d \end{pmatrix}\begin{pmatrix} 3 \\ -2 \end{pmatrix} = \begin{pmatrix} 9 \\ 8 \end{pmatrix}$.

$\Rightarrow 3a - 2b = 9$ ....①

$\quad\; 3c - 2d = 8$ ....②

$\begin{pmatrix} a & b \\ c & d \end{pmatrix}\begin{pmatrix} 5 \\ 1 \end{pmatrix} = \begin{pmatrix} -3 \\ 4 \end{pmatrix}$

$\Rightarrow 5a + b = -3$ ....③

$\quad\; 5c + d = 4$ ....④

$2 \times ③ + ① :\; 13a = 3 \Rightarrow a = \dfrac{3}{13}$

Substitute in ①:

$3\left(\dfrac{3}{13}\right) - 2b = 8 \Rightarrow b = -\dfrac{54}{13}$

$2 \times ④ + ② :\; 13c = 16 \Rightarrow c = \dfrac{16}{13}$

Substitute in ②:

$3\left(\dfrac{3}{13}\right) - 2d = 8 \Rightarrow d = -\dfrac{28}{13}$

Therefore, the transformation matrix is

$\begin{pmatrix} \dfrac{3}{13} & -\dfrac{54}{13} \\ \dfrac{16}{13} & -\dfrac{28}{13} \end{pmatrix}$.

```
╔══════════════════════╗
║ EXERCISE 46:4        ╲
╚══════════════════════╱
```

Determine the transformation matrix, which maps the unit square $\begin{pmatrix} 0 & 1 & 1 & 0 \\ 0 & 0 & 1 & 1 \end{pmatrix}$ into each of the following figures. Illustrate each mapping with a sketch and describe the transformation in each case.

1. $\begin{pmatrix} 0 & 3 & 5 & 2 \\ 0 & 1 & 2 & 1 \end{pmatrix}$ 2. $\begin{pmatrix} 0 & 3 & 4 & 1 \\ 0 & 0 & 1 & 1 \end{pmatrix}$

3. $\begin{pmatrix} 0 & 2 & 2 & 0 \\ 0 & 0 & 1 & 1 \end{pmatrix}$ 4. $\begin{pmatrix} 0 & 2 & 2 & 0 \\ 0 & 0 & 2 & 2 \end{pmatrix}$

5. $\begin{pmatrix} 0 & 4 & 5 & 1 \\ 0 & 0 & 2 & 1 \end{pmatrix}$ 6. $\begin{pmatrix} 0 & 1 & 1 & 0 \\ 0 & 0 & -1 & -1 \end{pmatrix}$

7. $\begin{pmatrix} 0 & 3 & 3 & 0 \\ 0 & 0 & -3 & -3 \end{pmatrix}$ 8. $\begin{pmatrix} 0 & 0 & -1 & -1 \\ 0 & 1 & 1 & 0 \end{pmatrix}$

9. $\begin{pmatrix} 0 & 3 & 3 & 0 \\ 0 & 0 & 3 & 3 \end{pmatrix}$ 10. $\begin{pmatrix} 0 & 1 & -1 & -2 \\ 0 & 0 & 1 & 1 \end{pmatrix}$

11. $\begin{pmatrix} 0 & 3 & 4 & 1 \\ 0 & 1 & 3 & 2 \end{pmatrix}$ 12. $\begin{pmatrix} 0 & 0 & 1 & 1 \\ 0 & \dfrac{1}{2} & 1\dfrac{1}{2} & 1 \end{pmatrix}$

Inverse Transformations

Suppose **A** is a transformation matrix. Provided the determinant of **A** is not zero i.e. $|A| \neq 0$, the inverse of **A** (i.e. A^{-1}) can be found. A^{-1} performs the reverse of what **A** does. For instance, if **A** is a rotation of $90°$ anticlockwise, then A^{-1} will be a rotation of $90°$ clockwise.

Therefore, if A maps $\begin{pmatrix} x \\ y \end{pmatrix}$ to $\begin{pmatrix} X \\ Y \end{pmatrix}$

Then A^{-1} maps $\begin{pmatrix} X \\ Y \end{pmatrix}$ to $\begin{pmatrix} x \\ y \end{pmatrix}$

Example 46:13

The matrix $\begin{pmatrix} 0 & -1 \\ 1 & 1 \end{pmatrix}$ transforms the point $(2,-3)$ onto the point $(3,2)$. Find the matrix, which transforms the point $(3,2)$ onto the point $(2,-3)$.

Solution

Let $\begin{pmatrix} 0 & -1 \\ 1 & 1 \end{pmatrix} = A$.

Then the required matrix is

$A^{-1} = \dfrac{1}{\det A}(\text{Adj}A) = \dfrac{1}{1}\begin{pmatrix} 0 & 1 \\ -1 & 1 \end{pmatrix} = \begin{pmatrix} 0 & 1 \\ -1 & 1 \end{pmatrix}$

Example 46:14

The matrix $\mathbf{M} = \begin{pmatrix} 1 & 0 \\ -2 & 1 \end{pmatrix}$ maps a point (x, y) to $(6, -9)$. Find the values of x and y.

Solution

$$\begin{pmatrix} 1 & 0 \\ -2 & 1 \end{pmatrix}\begin{pmatrix} x \\ y \end{pmatrix} = \begin{pmatrix} 6 \\ -9 \end{pmatrix}$$

$$\begin{pmatrix} 1 & 0 \\ -2 & 1 \end{pmatrix}^{-1} = \frac{1}{1}\begin{pmatrix} 1 & 0 \\ 2 & 1 \end{pmatrix} = \begin{pmatrix} 1 & 0 \\ 2 & 1 \end{pmatrix}$$

$$\Rightarrow \begin{pmatrix} x \\ y \end{pmatrix} = \begin{pmatrix} 1 & 0 \\ 2 & 1 \end{pmatrix}\begin{pmatrix} 6 \\ -9 \end{pmatrix}$$

$$\Rightarrow \begin{pmatrix} x \\ y \end{pmatrix} = \begin{pmatrix} 6 \\ 3 \end{pmatrix}$$

Therefore, the values of x and y are $x = 6$ and $y = 3$.

Transformations by Singular Matrices

Consider the transformation described by the matrix $\begin{pmatrix} 4 & 2 \\ 2 & 1 \end{pmatrix}$. The rectangle $\begin{pmatrix} 0 & 3 & 3 & 0 \\ 0 & 0 & 2 & 2 \end{pmatrix}$ will be transformed as follows:

$$\begin{pmatrix} 4 & 2 \\ 2 & 1 \end{pmatrix}\begin{pmatrix} 0 & 3 & 3 & 0 \\ 0 & 0 & 2 & 2 \end{pmatrix} = \begin{pmatrix} 0 & 12 & 16 & 4 \\ 0 & 6 & 8 & 2 \end{pmatrix}$$

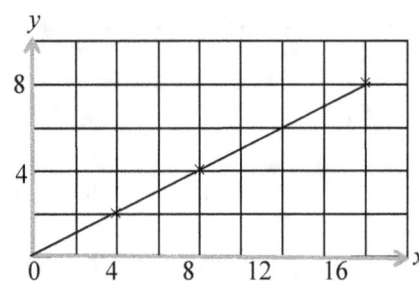

Figure 46:8

Verify that $\det\begin{pmatrix} 4 & 2 \\ 2 & 1 \end{pmatrix} = 0$.

Hence $\begin{pmatrix} 4 & 2 \\ 2 & 1 \end{pmatrix}$ is a non-zero singular matrix.

Therefore, non-zero singular matrices map all points to a straight line. Such a transformation, which maps geometrical shapes to a straight line, is called a **dilation**.

Transformations involving Change of Area

For any transformation, the ratio of the image area to the object area is equal to the absolute value of the determinant of the transformation matrix **A**.

$$\boxed{\text{image area} = |\det \mathbf{A}| \times \text{object area}}$$

Example 46:15

A rectangle with vertices $(4, 4)$, $(4, -1)$, $(10, -1)$, $(10, 4)$ is transformed by the matrix $\begin{pmatrix} 3 & 3 \\ 2 & 3 \end{pmatrix}$. Find the area of its image.

Solution

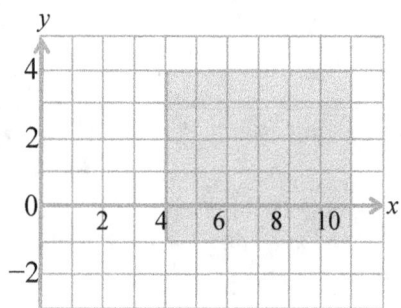

Figure 46:9

Image area = |det **A**|×object area
Object area = $5 \times 6 = 30$ un^2

$$\Rightarrow \text{Image area} = \left|\det\begin{pmatrix} 3 & 3 \\ 2 & 3 \end{pmatrix}\right| \times 30$$

$$= |9 - 6|30 = 90 \text{ un}^2$$

Example 46:16

A transformation is defined by the matrix $\begin{pmatrix} 4 & 2 \\ 1 & 2 \end{pmatrix}$. Find the ratio of the image area to the object area.

Solution

Let $\mathbf{A} = \begin{pmatrix} 4 & 2 \\ 1 & 2 \end{pmatrix}$

Image area = |det **A**|×object area
Image area: object area = |det **A**|:1
Det **A** = 8−2 = 6
∴Image area: object area = 6:1

Example 46:17

The base of a triangle is 6 cm and its height is 8 cm. Find the area of its image under a transformation defined by the matrix operator

$$\begin{pmatrix} 16 & 4 \\ 4 & 2 \end{pmatrix}.$$

Solution

Let $\mathbf{A} = \begin{pmatrix} 16 & 4 \\ 4 & 2 \end{pmatrix}$

Image area $= |\det \mathbf{A}| \times$ object area

$|\det A| = |16(2) - (4)(4)| = |32 - 16| = 16$

$\therefore$ Image area $= 16 \times 24 = 384 \text{ cm}^2$

Composite Transformations

Suppose $\mathbf{A}$ and $\mathbf{B}$ are two matrices representing two transformations for which the origin is invariant, then a transformation $\mathbf{A}$ followed by $\mathbf{B}$ can be performed by the matrix $\mathbf{BA}$.

Example 46:18

The triangle T_1 with vertices $A_1(1,1)$, $B_1(2,3)$ and $C_1(4,3)$, is mapped into triangle T_2 with vertices $A_2 B_2 C_2$ by the transformation P whose matrix is $\mathbf{P} = \begin{pmatrix} 0 & 1 \\ -1 & 0 \end{pmatrix}$.

The transformation Q whose matrix is

$\mathbf{Q} = \begin{pmatrix} 1 & 2 \\ 0 & 1 \end{pmatrix}$ further maps T_2 to T_3 whose

vertices are $A_3 B_3 C_3$. Find the image of T_3 without finding the image of T_2.

Solution

$$\mathbf{QP} = \begin{pmatrix} 1 & 2 \\ 0 & 1 \end{pmatrix}\begin{pmatrix} 0 & 1 \\ -1 & 0 \end{pmatrix} = \begin{pmatrix} -2 & 1 \\ -1 & 0 \end{pmatrix}$$

$$\therefore T_3 = \begin{pmatrix} -2 & 1 \\ -1 & 0 \end{pmatrix}\begin{pmatrix} 1 & 2 & 4 \\ 1 & 3 & 3 \end{pmatrix}$$

Translations

A translation is a transformation in which every object point moves the same distance in the same direction. Translations are the only transformations treated in this book, which cannot be represented by a 2×2 matrix. However, transformations are represented by translation vectors.

The translation vector $\mathbf{t} = \begin{pmatrix} p \\ q \end{pmatrix}$ transforms any

point $A(a, b)$ by moving it p units in the Ox direction and q units in the Oy direction. In this light, the image A' of the point $A(a,b)$ under a transformation represented by the

translation vector $\mathbf{t} = \begin{pmatrix} p \\ q \end{pmatrix}$ will be given by

$$\mathbf{OA'} = \mathbf{OA} + \mathbf{t} = \begin{pmatrix} a \\ b \end{pmatrix} + \begin{pmatrix} p \\ q \end{pmatrix} = \begin{pmatrix} a+p \\ b+q \end{pmatrix}$$

Example 46:19

Determine the image of the triangle ABC whose vertices are $A(1, 1)$, $B(-2, 1)$ and $C(3, -4)$ under the transformation whose

translation vector is $\begin{pmatrix} 4 \\ -7 \end{pmatrix}$.

Solution

$$\begin{pmatrix} 1 \\ 1 \end{pmatrix} \rightarrow \begin{pmatrix} 1 \\ 1 \end{pmatrix} + \begin{pmatrix} 4 \\ -7 \end{pmatrix} = \begin{pmatrix} 5 \\ -6 \end{pmatrix}$$

$$\begin{pmatrix} -2 \\ 1 \end{pmatrix} \rightarrow \begin{pmatrix} -2 \\ 1 \end{pmatrix} + \begin{pmatrix} 4 \\ -7 \end{pmatrix} = \begin{pmatrix} 2 \\ -6 \end{pmatrix}$$

$$\begin{pmatrix} 3 \\ -4 \end{pmatrix} \rightarrow \begin{pmatrix} 3 \\ -4 \end{pmatrix} + \begin{pmatrix} 4 \\ -7 \end{pmatrix} = \begin{pmatrix} 7 \\ -11 \end{pmatrix}$$

Therefore, the image of the triangle ABC is $A'(5, -6)$, $B'(2, -6)$ and $C'(7, -11)$

SUMMARY

The following transformation matrices are very common at this level and note should be taken of them.

	Operator	**Geometrical Effect**
1	$\begin{pmatrix} 1 & 0 \\ 0 & -1 \end{pmatrix}$	A reflection in the x-axis
2	$\begin{pmatrix} -1 & 0 \\ 0 & 1 \end{pmatrix}$	A reflection in the y-axis
3	$\begin{pmatrix} 0 & 1 \\ 1 & 0 \end{pmatrix}$	A reflection in the line $y = x$
4	$\begin{pmatrix} 0 & -1 \\ -1 & 0 \end{pmatrix}$	A reflection in line $y = -x$
5	$\begin{pmatrix} 0 & -1 \\ 1 & 0 \end{pmatrix}$	An anticlockwise rotation through $90°$ about the origin $(0,0)$
6	$\begin{pmatrix} -1 & 0 \\ 0 & -1 \end{pmatrix}$	An anticlockwise rotation through $180°$ about the origin $(0,0)$
7	$\begin{pmatrix} 0 & 1 \\ -1 & 0 \end{pmatrix}$	An anticlockwise rotation through $270°$ about the origin $(0,0)$
8	$\begin{pmatrix} 1 & 0 \\ 0 & 1 \end{pmatrix}$	The identity transformation leaves all points invariant
9	$\begin{pmatrix} k & 0 \\ 0 & 1 \end{pmatrix}$	A stretch in the direction Ox, stretch factor k
10	$\begin{pmatrix} 1 & 0 \\ 0 & k \end{pmatrix}$	A stretch in the direction Oy, stretch factor k
11	$\begin{pmatrix} k & 0 \\ 0 & k \end{pmatrix}$	An enlargement centre $(0,0)$, enlargement factor k
12	$\begin{pmatrix} 1 & k \\ 0 & 1 \end{pmatrix}$	A shear in the direction Ox, shear factor k
13	$\begin{pmatrix} 1 & 0 \\ k & 1 \end{pmatrix}$	A shear in the direction Oy, shear factor k

EXERCISE 46:5

1. (a) Draw the quadrilateral with vertices $A(0,0)$, $B(3,4)$, $C(4,0)$, $D(3,1)$. Find the image of $ABCD$ under the transformation represented by the matrix $\begin{pmatrix} -2 & 0 \\ 0 & -2 \end{pmatrix}$

   Find the ratio $\left(\dfrac{\text{area of image}}{\text{area of object}} \right)$.

   (b) Repeat (a) using the quadrilateral $P(-4,1)$, $Q(-2,4)$, $R(0,1)$, $S(-1,0)$.

2. (a) Draw the triangle $A(1,1)$, $B(4,1)$, $C(1,3)$ and find the image of ABC under the transformation with matrix $\begin{pmatrix} 2 & 0 \\ 0 & 2 \end{pmatrix}$.

   (b) Without calculating the actual areas of the triangles, determine the ratio

   $\left(\dfrac{\text{area of image}}{\text{area of object}} \right)$.

3. An irregular figure O whose area is 16 cm^2 is transformed to the image I by the matrix $\begin{pmatrix} 2 & 1 \\ 1 & 2 \end{pmatrix}$. Find the area of I.

4. The quadrilateral with vertices $A(0, 0)$, $B(3, 4)$, $C(7,4)$, $D(4, 0)$ undergoes a transformation represented by the matrix $\mathbf{Q} = \begin{pmatrix} 2 & 3 \\ 6 & 9 \end{pmatrix}$. Find the area of the image $A^1 B^1 C^1 D^1$ of $ABCD$.

5. (a) Taking 1 cm for 1 unit on each axis, ranging from -6 to $+6$, draw the figure $OABC$ with $A(3, 2)$, $B(5, 2)$ and $C(2, 0)$. Given the transformation matrix $\mathbf{T} = \begin{pmatrix} 0 & -1 \\ 1 & 0 \end{pmatrix}$.

   (b) Plot the image $OA'B'C'$ of $OABC$ under the transformation T.

   (c) Determine T^2 and plot the image $OA''B''C''$ of $OABC$ under T^2.

   (d) Describe the geometrical effects on $OABC$ of T and T^2.

6. A transformation is represented by the 2×2 matrix $\mathbf{M} = \begin{pmatrix} 0 & -1 \\ -1 & 0 \end{pmatrix}$.

   Describe this transformation completely. Use this matrix to find the image of a quadrilateral $ABCD$ where $A(3, 1)$, $B(0, 3)$, $C(-2, 1)$ and $D(1, -1)$. Give the coordinates of the vertices of the image of $ABCD$.

7. The triangle ABC with vertices A $(1, 0)$, B $(2, 1)$ and C $(4, 0)$ is mapped onto triangle $A'B'C'$ by the transformation matrix $\mathbf{T} = \begin{pmatrix} 0 & 1 \\ -1 & 0 \end{pmatrix}$.

(a) Find the coordinates of the images A', B' and C'.

(b) Using the scale of 2 cm to 1 unit on both axes plot the triangle ABC and its image $A'B'C'$ on graph paper.

(c) Describe the transformation T geometrically.

(d) Determine, the inverse of the transformation matrix. Hence, or otherwise, describe completely the transformation represented by this inverse matrix.

8. The matrix $\mathbf{M} = \begin{pmatrix} 0 & -1 \\ 1 & 0 \end{pmatrix}$ maps the point A on to the point B $(4, -6)$.

(a) Find the coordinates of A

(b) Describe completely the transformation whose matrix is $\mathbf{M}$.

9. Using graph paper and with a scale of 1 cm for 1 unit on both axes, plot the points A $(0, 2)$, B $(1, 2)$, C $(3, 1)$ and D $(2, 1)$. Name the type of quadrilateral $ABCD$.

Transform $ABCD$ using the matrix

$$\mathbf{M} = \begin{pmatrix} -2 & 0 \\ 0 & -2 \end{pmatrix}.$$

Plot the image $A'B'C'D'$. Describe completely the transformation $\mathbf{M}$.

By construction, find the image $A''B''C''D''$ of $A'B'C'D'$ under the transformation $\mathbf{M}$. Write on your graph the coordinates of the vertices.

Find the ratio of the area of $ABCD$ to that of $A''B''C''D''$.

◀ MULTIPLE CHOICE EXERCISE 46 ▶

1. A non-zero singular matrix:
   [A] has no inverse.
   [B] does not map any plane figure to a straight line.
   [C] does not have a determinant equal to zero.
   [D] can be a unit matrix.

2. A non-zero matrix whose determinant is zero:
   [A] has an inverse.
   [B] maps any plane figure to a straight line.
   [C] is not a singular matrix.
   [D] is an identity matrix.

3. The transformation matrix $\begin{pmatrix} 6 & 3 \\ 6 & 4 \end{pmatrix}$ maps the point $(x, 1)$ onto the point $(3, 4)$. The value of x is:
   [A] –3   [B] 4   [C] 0   [D] –6

4. The transformation matrix $\begin{pmatrix} a & 2a \\ b & 4b \end{pmatrix}$ maps the point $(1, 2)$ to $(5, 18)$. The values of a and b are:
   [A] $a = 2$ and $b = 1$ [B] $a = 1$ and $b = 2$
   [C] $a = 4$ and $b = 8$ [D] $a = 8$ and $b = 4$

5. The matrix, which represents a reflection in the line, $y = 0$ is:

   [A] $\begin{pmatrix} -1 & 0 \\ 0 & 1 \end{pmatrix}$ [B] $\begin{pmatrix} 0 & 1 \\ -1 & 0 \end{pmatrix}$

   [C] $\begin{pmatrix} 0 & -1 \\ 1 & 0 \end{pmatrix}$ [D] $\begin{pmatrix} 1 & 0 \\ 0 & -1 \end{pmatrix}$

6. The transformation with matrix $\begin{pmatrix} -1 & 0 \\ 0 & 1 \end{pmatrix}$ is:
   [A] A reflection in the line $x = 0$.
   [B] A reflection in the line $y = 0$.
   [C] A rotation through 180° center $(0, 0)$
   [D] A shear with shear factor -1

7. The transformation represented by the matrix $\begin{pmatrix} 1 & 0 \\ 2 & 1 \end{pmatrix}$ can completely be described as:
   [A] A stretch with stretch factor 2 and points on the x-axis invariant.
   [B] A stretch with stretch factor 2 and points on the y-axis invariant.
   [C] A shear with shear factor 2 and points on the x-axis invariant.
   [D] A shear with shear factor 2 and points on the y-axis invariant.

8. The transformation matrix $\begin{pmatrix} 2 & 4 \\ 0 & 2 \end{pmatrix}$ maps the point P onto P'. Given that P' is the point $(9, 4)$, then P must be the point:

[A] $\left(\dfrac{1}{2}, 0\right)$ 　　　　[B] $(0, 2)$

[C] $\left(2, \dfrac{1}{2}\right)$ 　　　　[D] $\left(\dfrac{1}{2}, 2\right)$

9. A rotation through $90°$ center $(0, 0)$ can be represented by the matrix:

[A] $\begin{pmatrix} -1 & 0 \\ 0 & 1 \end{pmatrix}$ 　　　　[B] $\begin{pmatrix} 0 & 1 \\ -1 & 0 \end{pmatrix}$

[C] $\begin{pmatrix} 0 & -1 \\ 1 & 0 \end{pmatrix}$ 　　　　[D] $\begin{pmatrix} 1 & 0 \\ 0 & -1 \end{pmatrix}$

10. The transformation matrix $\begin{pmatrix} 2 & 4 \\ 0 & 2 \end{pmatrix}$ maps the point A onto B. Given that, A is the point $(3, 2)$, then B is the point:
[A] $(14, 4)$ 　　　　[B] $(4, 14)$
[C] $(6, 8)$ 　　　　[D] $(0, 4)$

11. A linear transformation $\mathbf{M}: \mathbb{R}^2 \to \mathbb{R}^2$ is defined by $\mathbf{P} = \mathbf{Mq}$, where $\mathbf{M}$ is a 2 x 2 matrix and $\mathbf{p}$, $\mathbf{q}$ are 2×1 column vectors. Given that

$\mathbf{p} = \begin{pmatrix} 3 \\ 7 \end{pmatrix}$ when $\mathbf{q} = \begin{pmatrix} 1 \\ 0 \end{pmatrix}$ and $\mathbf{p} = \begin{pmatrix} 6 \\ -1 \end{pmatrix}$

when $\mathbf{q} = \begin{pmatrix} 2 \\ -3 \end{pmatrix}$, $\mathbf{M} = $:

[A] $\begin{pmatrix} 3 & 0 \\ 7 & 5 \end{pmatrix}$ 　　　　[B] $\begin{pmatrix} 5 & 7 \\ 0 & 3 \end{pmatrix}$

[C] $\begin{pmatrix} 0 & 5 \\ 3 & 7 \end{pmatrix}$ 　　　　[D] $\begin{pmatrix} 7 & 3 \\ 5 & 0 \end{pmatrix}$

12. The triangle with vertices $A(-1, -3)$, $B(2, 1)$ and $C(-2, 2)$ is transformed by the matrix $\begin{pmatrix} x & 0 \\ 0 & y \end{pmatrix}$ into the triangle with vertices $A^1(-2, -3)$, $B^1(4, 1)$, $C^1(-4, 2)$. The values of x and y is respectively:
[A] 2 and 0 　　　　[B] 0 and 1
[C] 1 and 2 　　　　[D] 2 and 1

13. A certain transformation T is defined by $T: (x, y) \mapsto (2x + y, -x + y)$. The 2×2 matrix representing T is:

[A] $\begin{pmatrix} 2 & 1 \\ -1 & 1 \end{pmatrix}$ 　　　　[B] $\begin{pmatrix} 1 & 1 \\ -1 & 2 \end{pmatrix}$

[C] $\begin{pmatrix} 1 & -1 \\ 1 & 2 \end{pmatrix}$ 　　　　[D] $\begin{pmatrix} 1 & 1 \\ 2 & -1 \end{pmatrix}$

14. The quadrilateral $A(2,0)$, $B(2,2)$, $C(-2,-2)$, $D(-2,0)$ is transformed by the matrix
$$\mathbf{P} = \begin{pmatrix} 2 & -1 \\ 1 & 0 \end{pmatrix}.$$

The image $A'B'C'D'$ of $ABCD$ under $\mathbf{P}$ is:
[A] $A'(2,2)$, $B'(6,4)$, $C'(10,-6)$, $D'(6,-4)$
[B] $A'(4,2)$, $B'(-6,-2)$, $C'(2,2)$, $D'(-4,-2)$
[C] $A'(4,2)$, $B'(2,2)$, $C'(-2,-2)$, $D'(-4,-2)$
[D] $A'(4,2)$, $B'(2,2)$, $C'(-4,-2)$, $D'(-6,-2)$

15. A transformation is defined by $T:(x, y) \mapsto (2x, 2y)$. The 2×2 matrix representing T is:

[A] $\begin{pmatrix} 2 & 2 \\ 1 & 1 \end{pmatrix}$ 　　　　[B] $\begin{pmatrix} 0 & 2 \\ 2 & 0 \end{pmatrix}$

[C] $\begin{pmatrix} 2 & 2 \\ 2 & 2 \end{pmatrix}$ 　　　　[D] $\begin{pmatrix} 2 & 0 \\ 0 & 2 \end{pmatrix}$

16. The points $A(2,0)$, $B(2,2)$, $C(-2,-2)$ and $D(-2,0)$ are transformed by the matrix $\mathbf{P} = \begin{pmatrix} 2 & -1 \\ 1 & 0 \end{pmatrix}$. The invariant line is:

[A] the line $x = 0$
[B] the line $y = 0$
[C] the line $y = x$
[D] the line $y = 1$

17. A transformation is defined by $T : (x, y) \mapsto (\qquad)$.

In words, the transformation T can be described as:
[A] A translation 2 units along Ox.
[B] A shear scale factor 2
[C] An enlargement scale factor 2
[D] A stretch scale factor 2.

TOPIC 47

FLOW DIAGRAMS

OBJECTIVES

At the end of this topic, the learner should be able to:

1. Appreciate the importance of the topic in computer technology and in real life.
2. Recognise and distinguish between various types of flow diagrams.
3. Distinguish between an instruction box and a decision box.
4. Interpret and draw flow charts or diagrams:
   (a) Representing or describing a process or the control processes of a system or program.
   (b) Showing the operations being carried out to solve simple problems.
   (c) Showing or representing the functioning of simple mechanical systems.
5. Understand the functioning of stores and use stores to store and retrieve data in flow charts.

Frank Bunker Gilbreth, Sr. (July 7, 1868 – June 14, 1924) was an early advocate of scientific management and a pioneer of motion study, and is perhaps best known as the father and central figure of *Cheaper by the Dozen*. He and his wife Lillian Moller Gilbreth were themselves industrial engineers and efficiency experts who contributed to the study of industrial engineering in fields such as motion study and human factors. They played a very significant role in the invention of flow diagrams.

From Wikipedia, the free encyclopedia

The Concept of a Flow Diagram

Computers process large amounts of data. To do this a list of instructions telling the computer the sequence of operations is first written in a language that computers can understand. The language used to write this list of instructions is called a **programming language**. Once the instructions have been well written, a computer can use these instructions to solve in seconds a problem that might otherwise have taken weeks to solve. Before a computer program is written, the instructions first have to be written in form of a **flow diagram**. *A flow diagram is a schematic or graphic representation of the steps leading to the solution of a given problem.*

TYPES OF FLOW DIAGRAMS

Flow diagrams are of different types and are used in almost all disciplines such as mathematics, computer programming, business, physics, geography, sociology, biology, literature, engineering, production etc. Below is a brief outline of a few common flow diagrams.

1. **A functional flow block diagram**
   This is a sketch showing the various parts of a system and their functions in relation to one another. Figure 47:1 is a functional flow block of a computer.

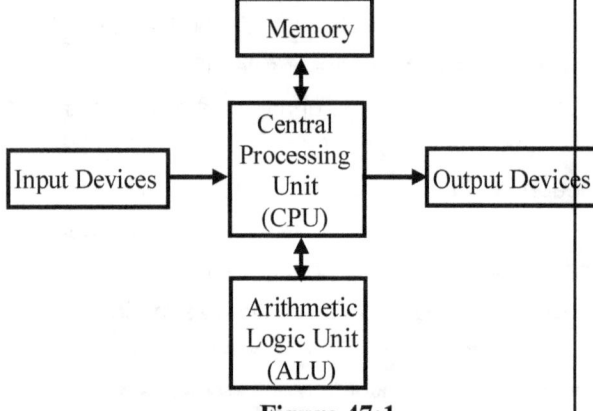

Figure 47:1

2. **A process flow diagram**
   This is usually used in operations as a graphical representation of a process.

Example 47:1
Draw a process flow diagram to show the production and distribution of drinks by a brewery industry.

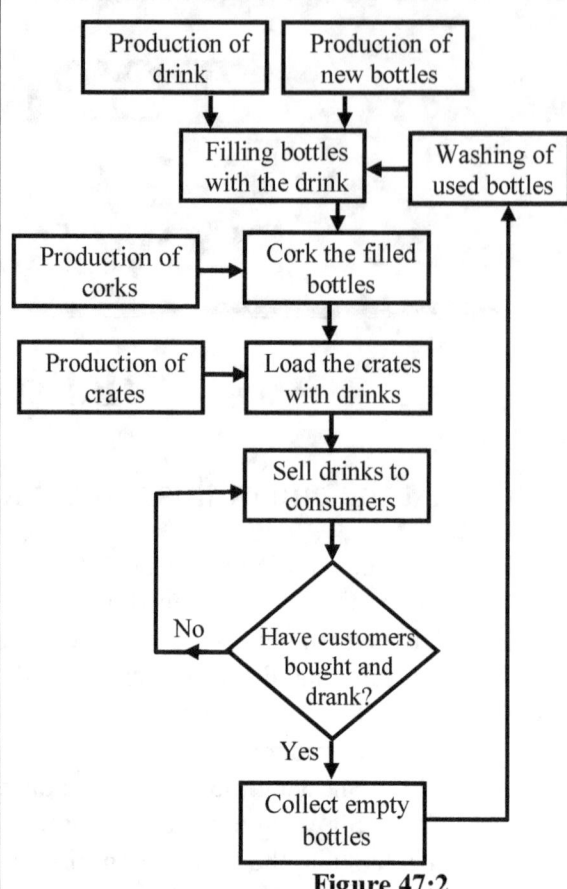

Figure 47:2

3. **An alluvial diagram**
   This is a graphical summary and highlight of the important structural changes in networks.
   Figure 47:3 and Figure 47:4 are examples of alluvial diagrams.

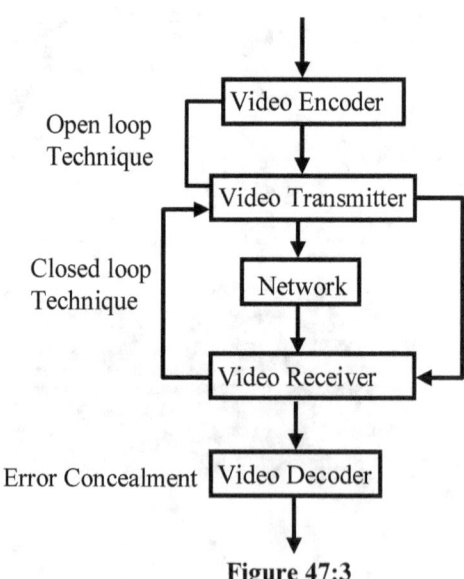

Figure 47:3

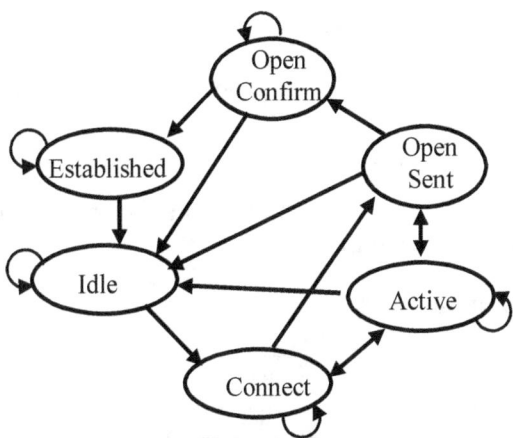

Figure 47:4

4. A control flow diagram

A control flow is a diagram describing the control flow of a business process, a program or any other process.

Figure 47:5 is a control flow diagram describing how an automated thermostat controls the temperature of a room.

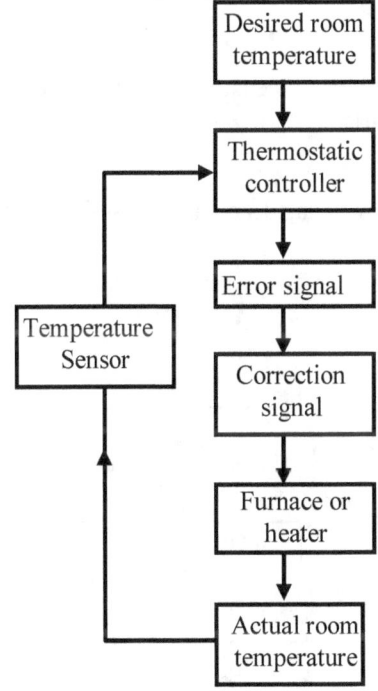

Figure 47:5

5. A data flow diagram

This is a graphical representation of the flow of data through an information system.

Figure 47:6, is a simplified data flow diagram.

Figure 47:6

6. A flow map

This is a mixture of maps and flow charts used in cartography, to show the movement of objects from one location to another.

Figure 47:7 is a flow map of the voyage of Vasco da Gama from 1497 to 1498.

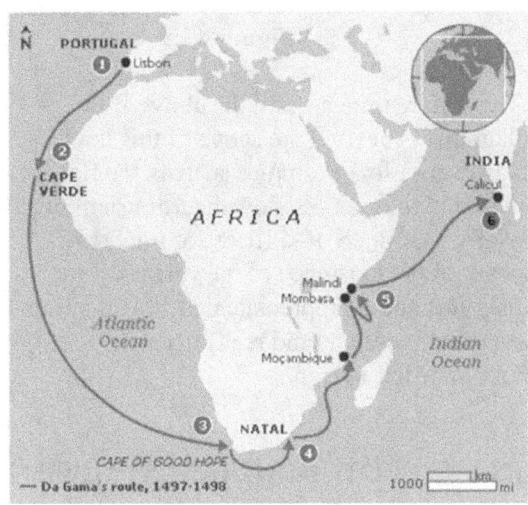

Figure 47:7

7. A flowchart is a graphical representation of the step-by-step solution to a given problem.

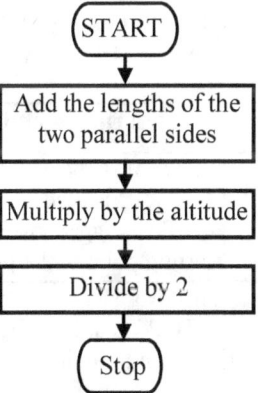

Figure 47:8

8. A signal flow graph

This is a graph used in mathematics to show the relations among the variables of a set of linear algebraic relations. A signal flow graph often looks something like Figure 47:9. Its treatment is beyond the scope of this book and for now, just be contented at being aware that such a graph exists.

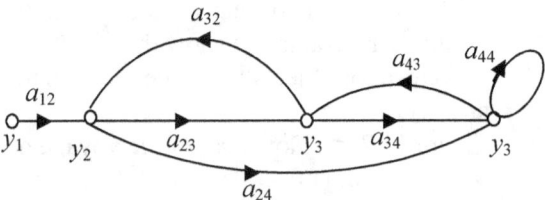

Figure 47:9

The full treatment of all the above flow diagrams is beyond the scope of this book. However, a few examples such as the flow chart and the process flow diagram shall be briefly examined. It suffices for now to be aware of the existence of the various flow diagrams and to appreciate that flow diagrams are usually drawn (and read) from top to bottom or left to right.

BASIC FLOW DIAGRAM SYMBOLS

On a flow diagram boxes each containing an instruction or question and connected by an arrow pointing to the next instruction or question are used. There are many symbols, but in this topic only the most common symbols are shown in Table 1 will be treated.

	Symbol	Description and Function
1	START	This oval shape with the instruction "START" is used at the beginning of a flow diagram to indicate the starting point of the program.
2	STOP	This oval shape with the instruction "STOP" is used at the end of a flow diagram to indicate the ending point of the program.
3	▭	This rectangular shape called the **instruction block** is used to issue intermediate instructions
4	◇	This diamond shape called the **decision block** is used to ask precise questions with "Yes" or "No" answers. A decision block usually has two exits, one for "Yes" and the other for "No".
5	↓	Arrows connecting the instruction and decision boxes are used to show the direction in which the instructions have to be followed.

Table 47:1

Instruction Boxes

Example 47:2 illustrates the use of instruction boxes.

Example 47:2

Draw a flow chart, which can be used to find the area of a trapezium.

Solution

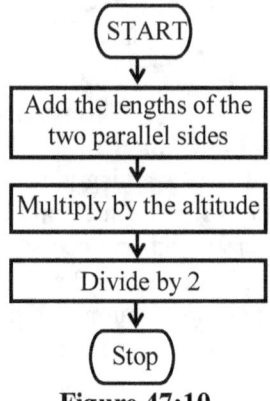

Figure 47:10

Calculations can easily be carried out using flow charts.

Example 45:3

Use a flow Chart to evaluate $9 - 6 + 4 \times 6 \div 3$.

Solution

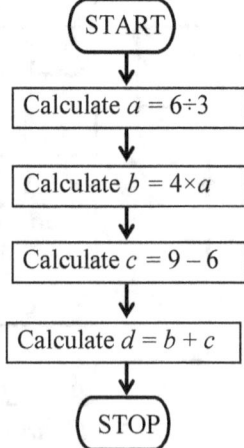

Figure 47:11

EXERCISE 47:1

1. Draw a flow chart for each of the following calculation.
   (a) $(4.7 \times 3 - 5.3) \div 2$
   (b) $4.7 \times 3 - 5.3 \div 2$

(c) $4.7 \times (3 - 5.3 \div 2)$

(d) $\sqrt{5.6 + 18.2 \div 7}$

(e) $\sqrt{5.6 + 18.2} \div 7$

2. Draw a flow chart to describe how to construct using ruler, pencil and compasses only:

   (a) the angle bisector between two rays OX and OY.

   (b) a line segment PQ, 6 cm long.

   (c) an equilateral triangle ABC of side 5 cm.

   (d) an angle BAC of 60°.

   (e) a line parallel to a given line and passing through a given point P.

3. Write down the expression, which is being evaluated by each of the following flow charts.

(a)

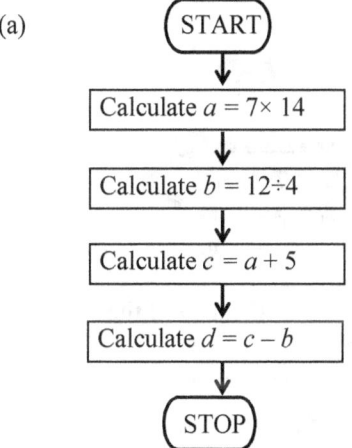

Figure 47:12

(b)

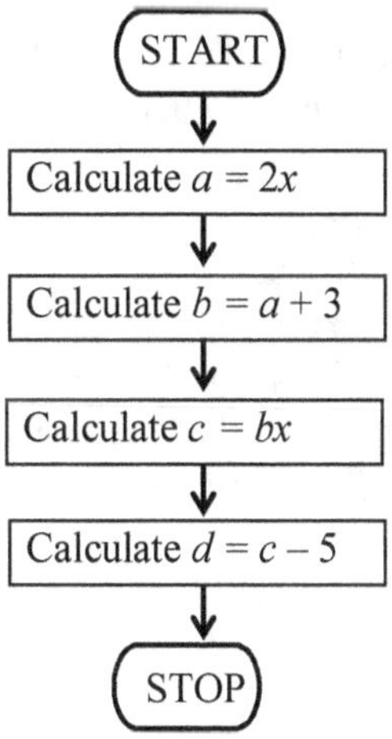

START

Calculate $a = 2x$

Calculate $b = a + 3$

Calculate $c = bx$

Calculate $d = c - 5$

STOP

Figure 47:13

(c)

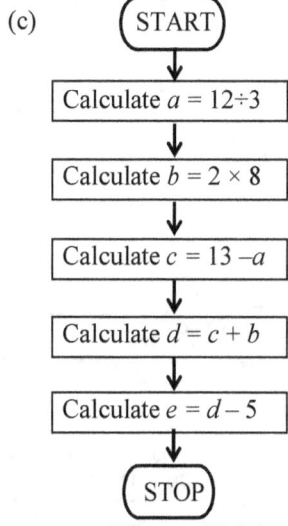

Figure 47:14

4. Draw a flow chart, which can be used to:

   (a) Find the L.C.M. of 12 and 42.

   (b) Find the HCF of 12 and 42

   (c) Add three equal vectors that have the same direction and sense.

Decision Boxes

Examples 45:4 and 45:5 illustrate the use of decision boxes. There are usually two outlets from a decision box, one for yes and the other for no. Sometimes, an outlet from a decision box leads to an earlier stage of the flow diagram as shown in Figure 47:6. Sometimes, an outlet meets the other at a later stage of the flow diagram as shown in Figure 47:7. Whichever way, such an outlet is known as a loop.

Example 47:4

Draw a flow diagram to show how you would close from school.

Solution

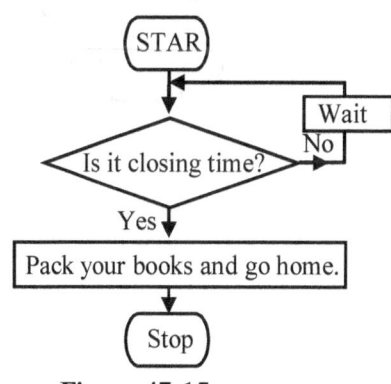

Figure 47:15

Example 47:5

Draw a flow diagram, which you would use to draw a diagram with a pencil.

Solution

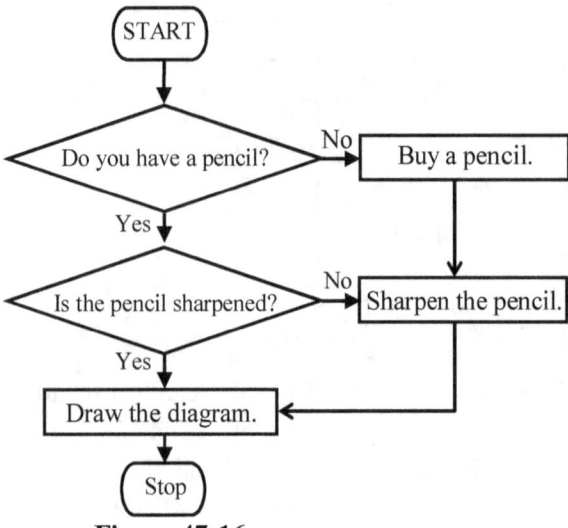

Figure 47:16

Example 47:6

Five times a certain positive integer plus seven equals twenty two. Find the number.

Solution

An inverse operation method may be used in solving the problem as follows.

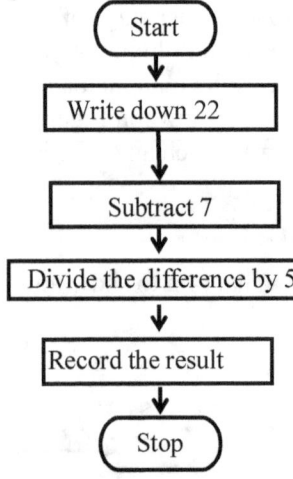

Figure 47:17

Alternatively, by a trial and error method, the positive integers 1, 2, 3,...can be tried in turns.

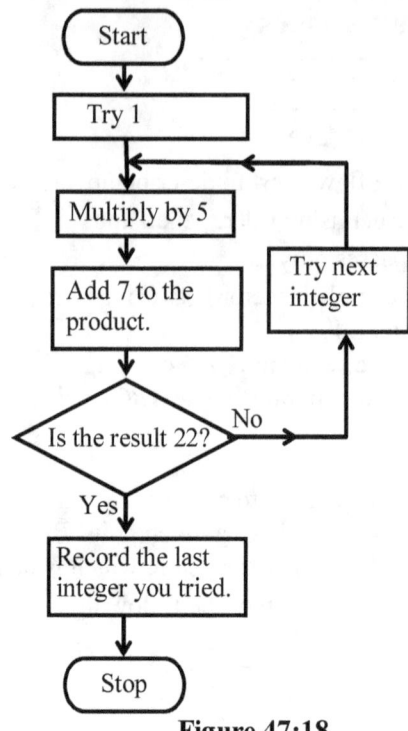

Figure 47:18

EXERCISE 47:2

Draw a flow diagram, which can be used to:

1. Find the arithmetic mean of the three numbers.
2. Find the number of elements in the union of two sets, A and B.
3. Factorise the algebraic expression $x^2 - 2x - 35$.
4. Change the subject in the formula $A = \pi r^2$.
5. Find the resultant of two vectors that are perpendicular to each other in direction and sense.
6. Find a number given that twice the number plus 3 is equal to 15.
7. Find a number given that, seven times the number, minus 9, is equal to 40.
8. Find a number given that the sum of the number, 2 and 5 is equal to 10.
9. Find a number given that, two and a half times the number plus 2.5, is equal to 7.5.
10. Find a number given that three times the number minus five is equal to that number plus five.
11. Find a number given that the number plus 2 is equal to twice the number minus 6.
12. Find the square root of 81 by listing factors.

Data and stores

In the arithmetic logic unit (ALU) of the computer, numbers are held in stores and each store has an address.

Each store is usually denoted by a single capital or small letter such as a, b, n, x, A, B etc. As in algebra, advantage is taken of the simplicity and clarity of symbolic language in mathematical and scientific works. The number in a particular store often changes as the calculation progresses. For this reason, in flow charts the symbolic language, $n: = 1$ is used to represent the number in store n and is interpreted as "n takes the value 1". If 1 is added to the number in store n, this is written as $n: = n + 1$, which means, "n takes the new value $n + 1$".

Example 47:7

Use symbolic language to do Example 47:6.

Solution

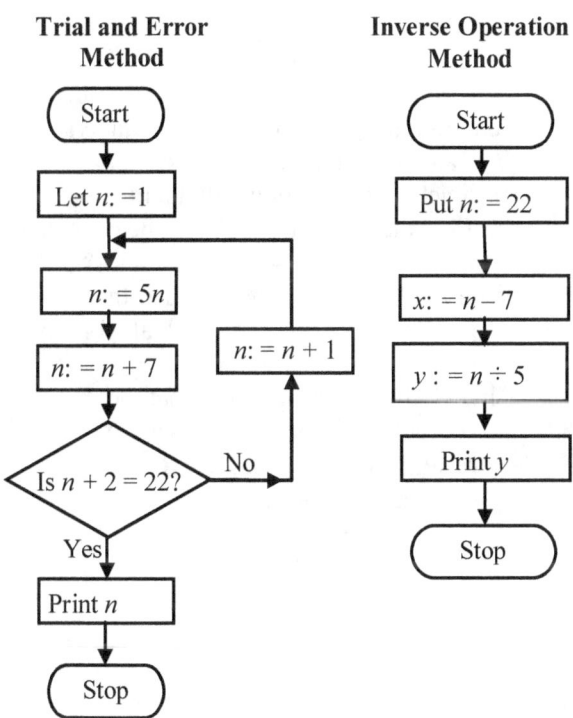

Figure 47:19

1. Draw a flow diagram, which can be used to convert 874_{ten} to a number in base six.
2. A flow chart showing how to write a

given number n ($n < 25$) as a product of prime factors contains:

  (i) The following instructions:

    Divide n by 2 and write 2 in the product.

    Divide the quotient by 2, and write 2 in the product.

    Divide the quotient by 3, and write 3 in the product.

    Write the last quotient in the product.

  (ii) The following questions:

    Is the quotient even?

    Is the quotient a multiple of 3?

  (iii) Two loops

    Draw the flow chart using all these facts and no more.

3. Draw a flow chart to describe the steps in the calculation of $\sqrt{b^2 - 4ac}$ for, given values of a, b, c.

4. The Fibonacci sequence consists of terms, each one of which is the sum of the two previous ones.

  If $u_1 = u_2 = 1$, draw a flow chart for calculating the later terms. Include the instruction record u_n and stop at u_{10}.

5. Repeat No. 4 taking $u_1 = 1$, $u_2 = 3$. Use your flow chart to calculate the first eight terms of this series.

  In Problem 6(a) to (c), follow through the instructions in the flow chart in Figure 47:20 and record the results as requested.

6. (a) What pattern of numbers is being generated in each of Figure 47:20 and Figure 47:22?

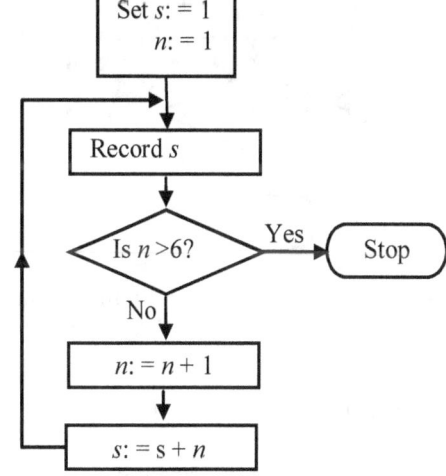

Figure 47:20

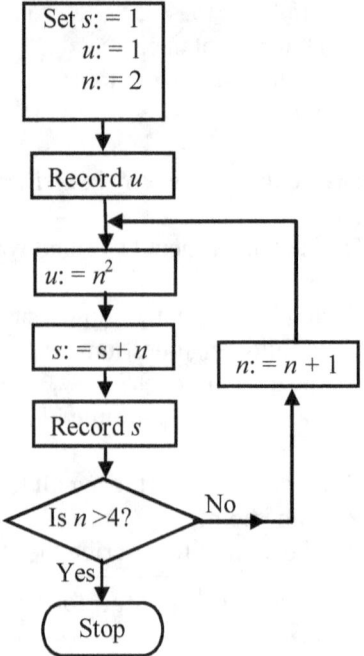

Figure 47:21

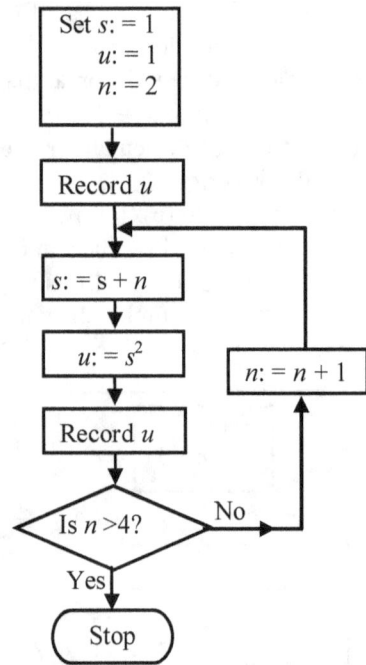

Figure 47:23

(b) Compare the pattern of numbers in Figure 47:20 and Figure 47:22.

7. What is the output from the flow chart in Figure 47:23 when X is read as
   (i) 12   (ii) 75   (iii) 49?
   What is the relationship between the number read and the number printed?

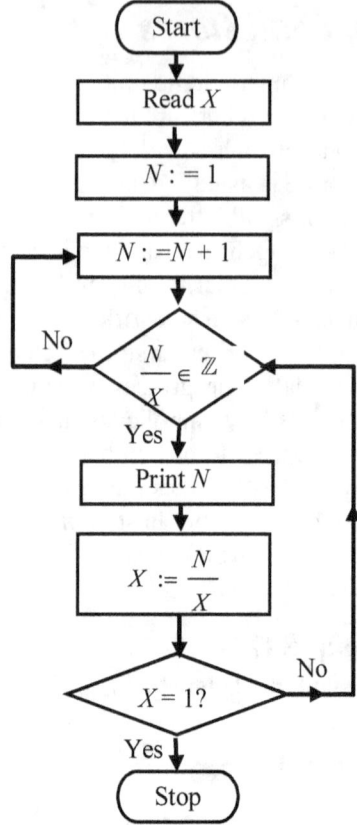

Figure 47:23

8. The flow chart in Figure 47:24 describes a computer program to read four numbers A, B, C and X, and to print the result of an arithmetical calculation with them.
   (i) Find the number which will be printed when $A = 3$, $B = -4$, $C = -1$ and $X = 2$.
   (ii) Find the general formula that the program is designed to evaluate for any values of A, B, C and X.
   (iii) Modify the program so that the computer will printout a series of results for $X = 1, 2, 3, \ldots, 10$ (A, B and C remaining constant)

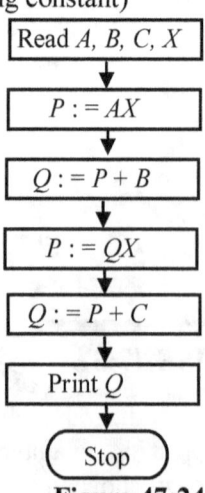

Figure 47:24

9. The program in Figure 47:25 is a design to print the highest common factor of two positive numbers x and y. Where $y > x$. Draw up a table of two columns showing the contents of stores A and B at each stage of the program when $x = 12$ and $y = 30$ and verify that the program does print the highest common factor.
Show that the program fails when $x = 8$ and $y = 24$. Explain why it fails and show clearly how you would amend the program so that it will work correctly.

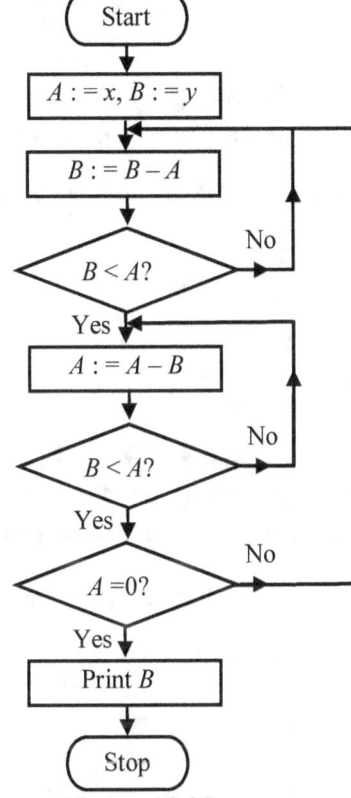

Figure 47:25

10. Work through the flow chart in Figure 47:26, showing the content of every store after each instruction, and displaying your answer in tabular form, beginning with:

A	B	C
1	0	0
...	...	...

Underline each number that will be printed. Do not carry out calculations to more than two decimal places.
If the original data store A had contained the number X instead of the number 1, find a formula involving x for the first n

umber to be printed.
The rows containing the printed values are given as guide:

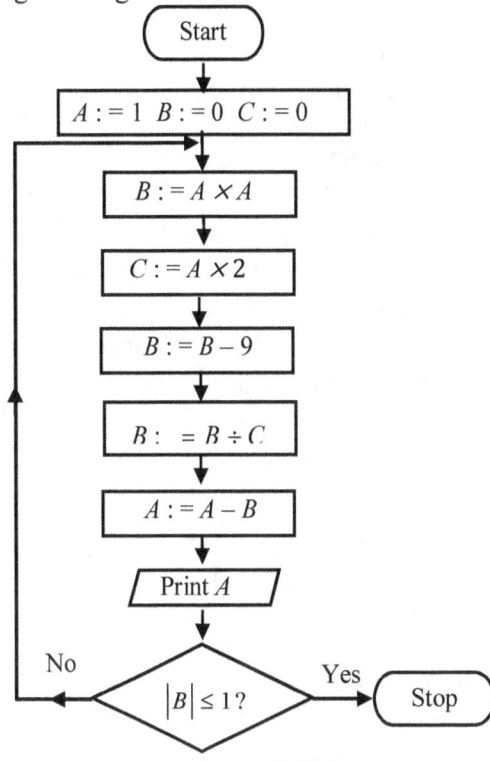

Figure 47:26

11. Figure 47:27 shows the flow chart for a simple computing device containing three stores A, B and C, each capable of storing one number only at a time.

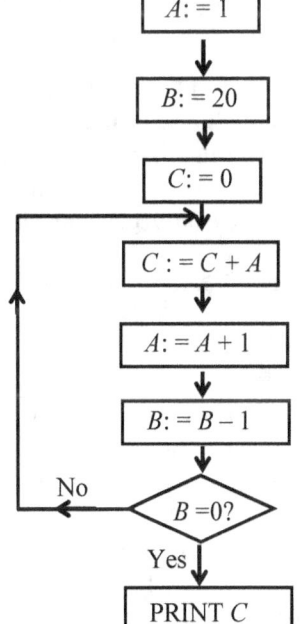

Figure 47:27

The instruction '$A := x$' is read 'A becomes x' and denotes that store A is to hold the number x in place of whatever number it held previously. The following flow diagram was intended to sum the series $1 + 3 + 5 +\ldots$ to 20 terms. It contains a mistake.

(a) Say what it will actually print.

(b) Which is the wrong instruction? Write a correct version

12. Draw up a flow chart, which can be used to add the terms of the series $12 + 15 + 18 + 21 + \ldots + 69$. Use three stores a, b and n and start with $a = b = 12$, $n = 1$.

13. Draw up a flow chart to show how a shuffled pack of cards may be sorted into four suits.

14. Temperature can be converted from the Celsius scale to the Fahrenheit and vice versa using the formula $F = \dfrac{9}{5}C + 32$ and an equivalent one with C as subject. Draw flow charts showing how the two formulae may be obtained from one another.

15. Make (a) r (b) v the subject of the formula $P = \dfrac{mv^2}{r}$. In each case, show the sequence of operations performed in flow chart form, and the sequence of inverse operations in a second flow chart.

16. If $r = \sqrt{(a^2 + b^2)}$, express b in terms of r and a.

   Assume that r, a and b are positive quantities. Show the sequence of operations performed in flow chart form. Show also in another flow chart, the reverse sequence of inverse operations.

17. If $x = \dfrac{\sqrt{b^2 - 4ac} - b}{2a}$, prove that $ax^2 + bx + c = 0$.

   Show in flow chart form the sequence of operations performed on this formula for x to obtain the final quadratic equation. Show also in another flow chart, the reverse sequence of inverse operations that leads from the general quadratic equation $ax^2 + bx + c = 0$ to the formula which gives the roots. Note that there should be two roots.

18. Before crossing the road, the Cameroon Highway Code recommends 'look left,

look right, look left again and cross if the road is clear.' Complete the flow diagram in Figure 47:5 for this operation:

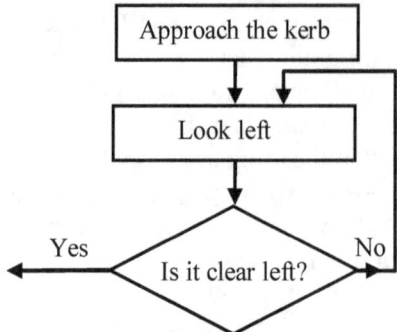

Figure 47:28

If the traffic flow in any road is such that at any instant the probability of there being traffic in one direction $\dfrac{1}{2}$ is and in the other is $\dfrac{2}{3}$, what is the probability of the road being completely clear?

═══════════════════════════════════
MULTIPLE CHOICE EXERCISE 47
═══════════════════════════════════

1. On a flow chart, the shape used for a decision block is:

   [A] (rounded rectangle)   [B] (rectangle)

   [C] (diamond)   [D] ⟶

2. The shape (rectangle) is used on a flow chart as:
   [A] a decision block
   [B] an instruction block
   [C] a question block
   [D] a description block

3. The type of flow diagram which highlights and summarizes the significant structural changes in networks is:
   [A] a data flow diagram
   [B] a process flow diagram
   [C] an alluvial diagram
   [D] a signal flow diagram

4. Given the line segment $XY = 8$ cm. The construction described by the flow chart in Figure 47:29 is:

[A] the inscribed circles C_1 and C_2
[B] the locus of C_1 and C_2
[C] the intersection of C_1 and C_2
[D] the mediator of XY

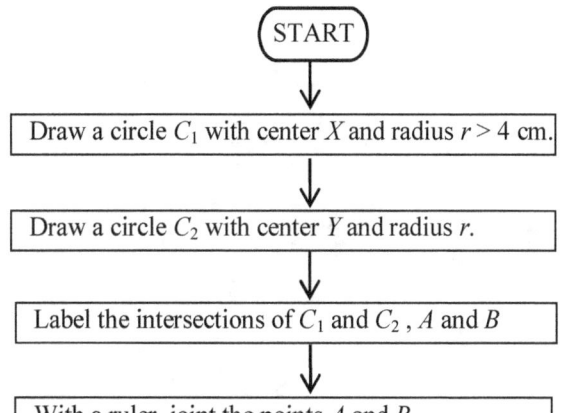

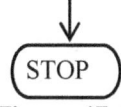

Figure 47:29

5. Given that $a, b, r \in \mathbb{N}$. A flow chart to show the sequence of instructions performed to make b the subject of the formula $r = \sqrt{a^2 + b^2}$ contains the following instructions.
I: Find the square root of both sides
II: Square both sides
III: Subtract a^2 from both sides.
The correct order of instructions is:
[A] I, II, III   [B] II, III, I
[C] III, I, I   [D] II, I, III

6. Given the instructions
I: Multiply both sides by r
II: Divide both sides by P
III: Divide both sides by r
IV: Multiply both sides by P
The sequence of instructions performed to make r the subject of the formula
$P = \dfrac{mv^2}{r}$ is:
[A] I, II   [B] I, III
[C] I, IV   [D] IV, III

7. In a flow chart to show the sequence of instructions performed to make C the subject of the formula $F = \dfrac{9}{5}C + 32$, the first two instructions are; multiply both sides by 5, divide both sides by 9 respectively. The third and final instruction

should be:

[A] Multiply both sides by $\dfrac{160}{9}$

[B] Divide both sides by $\dfrac{160}{9}$

[C] Add $\dfrac{160}{9}$ to both sides

[D] Subtract $\dfrac{160}{9}$ from both sides

8. In the flow diagram in Figure 47:32, the number of loops is:
[A] 9   [B] 4   [C] 2   [D] 3

9. In the flow diagram in Figure 47:30, the number of stores is:
[A] 9   [B] 4   [C] 2   [D] 3

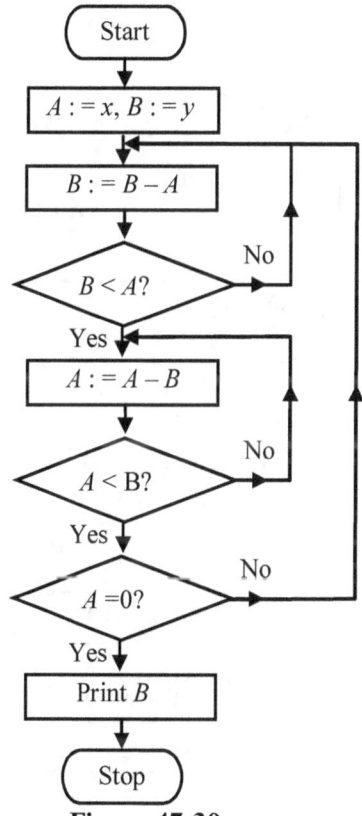

Figure 47:30

10. The instruction boxes for the flow chart which shows the sequence of operations to change the subject of the formula $A = \pi r^2$ chronologically contains contain the instructions:
I: Find the square root of both sides.
II: Divide both sides by π.
III: square both sides.
IV: Multiply both sides by π.
[A] I, II   [B] II, IV   [C] II, I   [D] I, IV

539

11. The flows chart in Figure 47:29 shows the probability $P(X)$ of obtaining a sum, which is a multiple of 6 when two dice are rolled. According to the flow chart if the sum is 9 then:

[A] the dice should be rolled again.

[B] subtracts 3 from 9 to continue.

[C] adds 3 to 9 to continue.

[D] the probability is $\dfrac{3}{4}$.

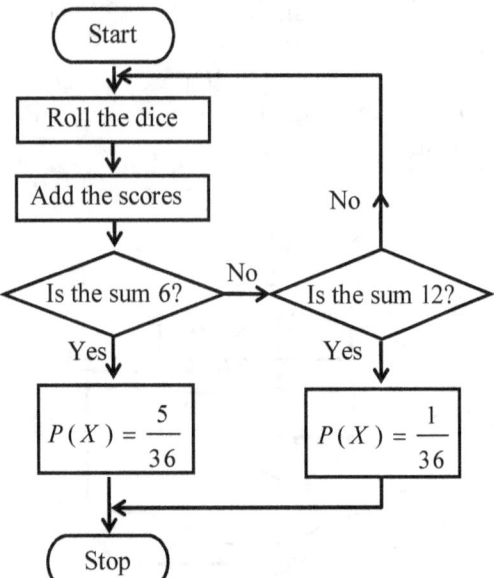

Figure 47:31

12. The expression which is being evaluated by the flow chart in Figure 47:31 is:

[A] $3x^2 + 2x - 5$ [B] $2x^2 + 3x - 5$

[C] $3a^2 + 2a - 5$ [D] $2a^2 + 3a - 5$

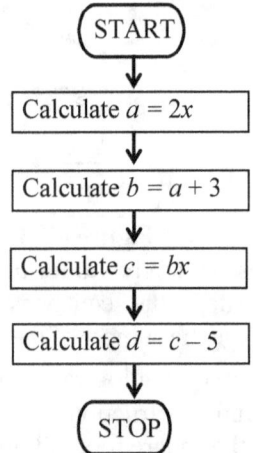

Figure 47:32

13. The expression which is being evaluated by the flow chart in Figure 47:33 is:

[A] $13 - 12 \div 3 + 2 \times 8$

[B] $(13 - 12) \div 3 + 2 \times 8$

[C] $13 - 12 \div (3 + 2) \times 8$

[D] $(13 - 12) \div (3 + 2) \times 8$

```
         START
           ↓
  Calculate a = 12÷3
           ↓
  Calculate b = 2 × 8
           ↓
  Calculate c = 13 −a
           ↓
  Calculate d = c + b
           ↓
         STOP
```

Figure 47:33

14. The flow chart in Figure 47:34 which represents the evaluation of

$$9 - 6 + 4 \times 6 \div 3 \text{ is:}$$

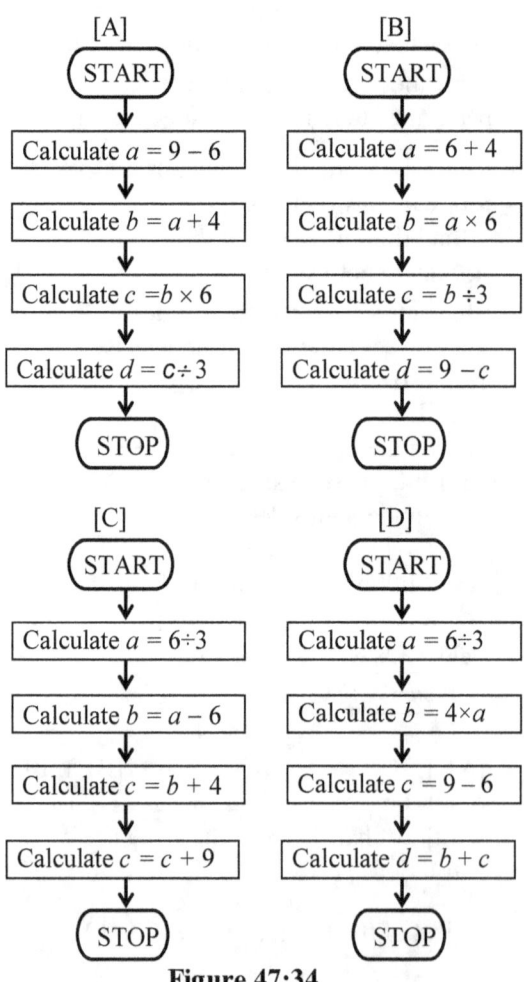

Figure 47:34

APPENDIX-THE SCIENTIFIC CALCULATOR

This is a practical lesson and the reader should have a scientific calculator and follow each step as described.

In this lesson, though to key a number such as 187.35 requires six keys; $\boxed{1}$,$\boxed{8}$,$\boxed{7}$, $\boxed{3}$ and $\boxed{5}$. This shall be abbreviated $\boxed{187.35}$, as if it were only one key. Appreciate that calculators perform operations following the order- function (log, trig etc), division, multiplication, addition and subtraction.

The following are the main keys discussed in this book.

$\boxed{+}$ Plus $\boxed{-}$ Minus $\boxed{\times}$ Times $\boxed{\div}$ Divided by
$\boxed{=}$ Equal to $\boxed{x^2}$ Square $\boxed{\sqrt{}}$ Square root

$\boxed{\sqrt[3]{}}$ Cube root $\boxed{\pi}$ Pie $\boxed{\dfrac{1}{x}}$ Reciprocal

$\boxed{10^x}$ Ten raised to the power x

$\boxed{y^x}$ The number y raised to the power x

$\boxed{EXP}$ A number times ten raised to the power x

$\boxed{MS}$ Alternatively, $\boxed{X-M}$ Store in the memory

$\boxed{M+}$ Add to the number in the memory

The Second Function or Inverse Key

Certain calculator functions do not appear on the button itself but above it. The label for this key is different on different calculator models. The different labels are:

$\boxed{SHIFT}$, $\boxed{2^{ND}F}$, or $\boxed{INV}$.

To use a function that appears above the button first press the second function key.
Since calculators are slightly different, the arrangement and the labeling of the keys must be well mastered for each made.
The starting point in any calculation is very essential and should be such that the user almost completely performs the calculations without use of a jotter. This maximizes the accuracy of the answer.

Example A:1

Use the calculator to evaluate the following to 4 significant figures.

(i) $72.94\left(\dfrac{26.59}{9.56-6.95}\right)$

(ii) $\left(\dfrac{73.139}{6.48+8.17}\right)\left(\dfrac{95.35-35.89}{5.32}\right)$

(iii) $\sqrt[3]{\left(\dfrac{2.63\times1.938}{0.917+1.62}\right)}$

541

(iv) $\dfrac{-9 + \sqrt{9^2 - 4(3)(2)}}{(3)(2)}$

(v) $\left(\dfrac{2}{4.23} - \dfrac{5}{3.76}\right)^4$

(vi) $\pi \times 7^2 \times \dfrac{23}{360}$

(vii) $\left(\dfrac{4 \times 10^5}{3 \times \times 10^5}\right)\left(\dfrac{2 \times 10^{-5}}{6 \times 10^{-3}}\right)$

Solutions

(i) Begin from the denominator in the bracket.

$\boxed{9.56}\ \boxed{-}\ \boxed{6.95}\ \boxed{=}\ \boxed{\tfrac{1}{x}}\ \boxed{\times}\ \boxed{26.59}\ \boxed{\times}\ \boxed{72.94}\ \boxed{=}$

Answer: 743.1 (to 4 sig. fig.)

Alternatively;

$\boxed{9.56}\ \boxed{-}\ \boxed{6.95}\ \boxed{=}\ \boxed{MS}\ \boxed{26.59}\ \boxed{\div}\ \boxed{MR}\ \boxed{\times}\ \boxed{72.94}\ \boxed{=}$

Answer: 743.1 (to 4 sig. fig.)

Solutions

(i) Begin from the denominator in the bracket.

$\boxed{9.56}\ \boxed{-}\ \boxed{6.95}\ \boxed{=}\ \boxed{\tfrac{1}{x}}\ \boxed{\times}\ \boxed{26.59}\ \boxed{\times}\ \boxed{72.94}\ \boxed{=}$

Answer: 743.1 (to 4 sig. fig.)

Alternatively;

$\boxed{9.56}\ \boxed{-}\ \boxed{6.95}\ \boxed{=}\ \boxed{MS}\ \boxed{26.59}\ \boxed{\div}\ \boxed{MR}\ \boxed{\times}\ \boxed{72.94}\ \boxed{=}$

Answer: 743.1 (to 4 sig. fig.)

(ii) $\boxed{6.48}\ \boxed{+}\ \boxed{8.17}\ \boxed{=}\ \boxed{\tfrac{1}{x}}\ \boxed{MS}\ \boxed{\times}\ \boxed{73.139}\ \boxed{=}\ \boxed{MS}$

$\boxed{95.35}\ \boxed{-}\ \boxed{35.89}\ \boxed{=}\ \boxed{\div}\ \boxed{5.32}\ \boxed{\times}\ \boxed{MR}\ \boxed{=}$

Answer: 55.80 (to 4 sig. fig.)

(iii) $\boxed{0.917}\ \boxed{+}\ \boxed{1.62}\ \boxed{=}\ \boxed{\tfrac{1}{x}}\ \boxed{\times}\ \boxed{2.63}\ \boxed{\times}\ \boxed{1.835}\ \boxed{=}\ \boxed{\sqrt[3]{x}}$

Answer: 2.305 (to 4 sig. fig.)

(iv) $\boxed{9^2}\ \boxed{-}\ \boxed{4}\ \boxed{\times}\ \boxed{3}\ \boxed{\times 2}\ \boxed{=}\ \boxed{\sqrt{x}}\ \boxed{-}\ \boxed{9}\ \boxed{=}\ \boxed{\div}\ \boxed{6}\ \boxed{=}$

Answer: 0.2416 (to 4 sig. fig.)

(v) $\boxed{5}\ \boxed{\div}\ \boxed{3.76}\ \boxed{=}\ \boxed{MS}\ \boxed{2}\ \boxed{\div}\ \boxed{4.23}\ \boxed{=}\ \boxed{-}\ \boxed{MR}\ \boxed{=}\ \boxed{y^n}\ \boxed{4}\ \boxed{=}$

Answer: 0.8570 (to 4 significant figures)

(vi) $\boxed{\pi}\ \boxed{\times}\ \boxed{7}\ \boxed{x^2}\ \boxed{\times}\ \boxed{23}\ \boxed{\div}\ \boxed{360}\ \boxed{=}$

Answer: 9.835 (to 4 significant figures)

(vii) $\boxed{6}\ \boxed{EXP}\ \boxed{3}\ \boxed{\pm}\ \boxed{\tfrac{1}{x}}\ \boxed{\times}\ \boxed{2}\ \boxed{EXP}\ \boxed{5}\ \boxed{\pm}\ \boxed{=}\ \boxed{MS}$

$\boxed{3}\ \boxed{EXP}\ \boxed{5}\ \boxed{\tfrac{1}{x}}\ \boxed{\times}\ \boxed{4}\ \boxed{EXP}\ \boxed{3}\ \boxed{=}\ \boxed{\div}\ \boxed{MR}\ \boxed{=}$

Answer: 4

Note! Always check the mode of the calculator before usage. The mode always appears on the screen.

For the use of the scientific calculator for number bases, trigonometry and statistics see TOPICS 18, 27 and 35 respectively.

EXERCISE A:1

Use your calculator to evaluate the following to 3 significant figures.

1. $\dfrac{2667 - 1143}{2667 \times 1143}$

2. $\dfrac{\sqrt{1.1} \times 14.23}{39.67}$

3. $\dfrac{1 + \sqrt{0.072}}{1 - \sqrt{0.072}}$

4. $2\pi\sqrt{\dfrac{98.1}{32.2}}$

5. $\dfrac{1 + \sqrt[3]{00075}}{1 - \sqrt[3]{00075}}$

6. $\sqrt{\dfrac{8.621 \times 27.34}{52.18 \times 0.0724}}$

7. $\dfrac{8.072 \times \sqrt{0.74}}{0.084}$

8. $2\pi(5.4^2 - 3.4^2)$

9. $\dfrac{187\{(1.3)^{12}\} - 1}{3426}$

10. $\dfrac{0.2463 \times (0.1721)^2}{0.7621}$

11. $\left(\dfrac{23}{641} - \dfrac{751}{864}\right)^3$

12. $\log\left(\dfrac{28.3}{13.4} - \dfrac{12.6}{\sqrt{27.1}}\right)$

Trigonometric Ratios from Calculators

There are two main ways of finding trigonometric ratios of an angle using scientific calculators depending on the type of calculator. For the purpose of reference, the calculators shall be referred to as Type 1 and Type 2. For Type 1, first key the value of the angle followed by the required trigonometric ratio function key. For Type 2, first key the trigonometric ratio function key followed by the value of the angle. In both cases, make sure that the calculator is set in the degree (DEG) mode.

Example A:2
Use a scientific calculator to find the following:
(a) sin 30°   (b)   cos 45°   (c)   tan 60°

Solution

	Trig ratio	Type 1	Type 2	Display
(a)	Sin 30°	30 sin	sin 30	0.5
(b)	cos 45°	45 cos	cos 30	0.707106...
(c)	tan 60°	60 tan	tan 30	1.73205...

If an angle is given in degrees and minutes, it is necessary to convert first the minute part to a decimal of a degree by dividing by 60.

Example A:3
Use a calculator to find the following
(a)   sin 47°26'   (b)   cos 53°47'   (c)   tan 72°13'

Solution
(a) sin 47°26'

Calculator	Procedure	Display
Type 1	26 ÷ 60 + 47 = sin	0.7364...
Type 2	sin (26 ÷ 60 + 47) =	0.7364...

(b) cos 53°47'

Calculator	Procedure	Display
Type 1	47 ÷ 60 + 53 = cos	0.5908...
Type 2	cos (47 ÷ 60 + 53) =	0.5908...

(c) tan 72°13'

Calculator	Procedure	Display
Type 1	13 ÷ 60 + 72 = tan	3.117...
Type 2	tan (13 ÷ 60 + 72) =	3.117...

To find the angle whose trigonometric ratio has been given, the procedures for the two types of calculators are as follows. For Type 1, first punch the value of the trig ratio, followed by the 2nd F (or INV or SHIFT) button followed by the trig function key in that order. For Type 2, first key the 2nd F (or INV or SHIFT) followed by the trig function key and then the value of the angle in that order. Remember that the calculator should be set in the degree (DEG) mode.

Example A:4
Use a calculator to find the following.
(a) $\sin^{-1} 0.7153$ (b) $\cos^{-1} 0.3431$ (c) $\tan^{-1} 1.34$

Solution
(a) $\sin^{-1} 0.7153$

Calculator	Procedure	Display
Type 1	$\sin^{-1}$ 0.7153 2nF sin	45.6677...
Type 2	$\sin^{-1}$ 2ndF sin 0.7153 =	45.6677...

(b) $\cos^{-1} 0.3431$

Calculator	Procedure	Display
Type 1	$\cos^{-1}$ 0.3431 2nF cos	69.934...
Type 2	$\cos^{-1}$ 2ndF cos 0.3431 =	69.934...

(c) $\tan^{-1} 1.34$

Calculator	Procedure	Display
Type 1	$\tan^{-1}$ 1.34 2nF tan	53.267...
Type 2	$\tan^{-1}$ 2ndF tan 1.34 =	53.267...

EXERCISE A:2

1. Use a calculator to find the following trigonometric ratios

(a) sin 34°   (b) sin 72°18'   (c) sin 81°20'

(d) cos 63°   (e) cos 54°36'   (f) cos 63°43'

(g) tan 46°   (h) tan 27°6'   (i) tan 82°16'

2. Use a calculator to find the angles whose sines are given below.
   (a) 0.1564 (b) 0.9135 (c) 0.9880 (d) 0.8020
   (e) 0.9814 (f) 0.7395 (g) 0.0500 (h) 0.2700

3. Use a calculator to find the angles whose cosines are given below.
   (a) 0.9135 (b) 0.3420 (c) 0.9673 (d) 0.4289
   (e) 0.9586 (f) 0.0084 (g) 0.2611 (h) 0.4700

4. Use calculators to find the angles whose tangents are given below.
   (a) 0.4452 (b) 3.2709 (c) 0.0769 (d) 0.3977
   (e) 0.3568 (f) 1.9251 (g) 0.0163 (h) 0.8263

Using Scientific Calculators for Statistics

The statistical buttons on a scientific calculator are as follows.

| STAT | Statistical mode key |

$\boxed{x}$ Data frequency key

| DATA | Data consolidation key

$\boxed{\bar{x}}$ Mean of data entered

$\boxed{\sum x}$ Sum of the variables

$\boxed{n}$ Number of samples entered

$\boxed{\sum x^2}$ Sum of the squares of the variables

$\boxed{\sigma}$ Standard deviation

$\boxed{s}$ Sample standard deviation

| CD | Cancel data

Example A:5
Calculate
(a) The number of samples
(b) The sum of the samples
(c) The mean of the samples
(d) The sum of the squares of the samples
(e) Statistical deviation of the samples
(f) The standard deviation of the following data.
     4, 5, 5, 6, 6, 8, 8, 8, 9, 10

Solution

STAT 4 DATA 5 DATA 5 DATA 6 DATA

6 DATA 8 DATA 8 DATA 9 DATA 10 DATA

(a) Press $\boxed{n}$, the calculator displays 10

(b) Press $\boxed{\sum x}$, the calculator displays 69

(c) Press $\boxed{\bar{x}}$, the calculator displays 6.9

(d) Press $\boxed{\sum x^2}$, the calculator displays 511

(e) Press $\boxed{s}$, the calculator displays 1.97

(f) Press $\boxed{\sigma}$, the calculator displays 1.87

Example A:6
The following are marks obtained by a group of students in a test.

Score (x)	2	3	4	5
Frequency (f)	6	7	3	4

Use the calculator to compute

(a) $\sum f$ (b) $\sum x$ (c) $\bar{x}$ (d) $\sum x^2$ (e) σ

Solution

STAT 2 × 6 DATA 3 × 7 DATA

4 × 3 DATA 5 × 4 DATA

(a) Press $\boxed{n}$, the calculator displays 20

(b) Press $\boxed{\sum x}$, the calculator displays 65

(c) Press $\boxed{\bar{x}}$, the calculator displays 3.25

(d) Press $\boxed{\sum x^2}$, the calculator displays 235

(e) Press $\boxed{\sigma}$, the calculator displays 1.09....

EXERCISE A:3

The following shows the marks obtained by some students during a test.

Mark, x	Frequency, f
30	1
40	3
50	8
55	1
60	15
65	2

Use a calculator to find:

(a) $\sum f$ (b) $\sum x$ (c) $\bar{x}$

(d) $\sum x^2$ (e) σ (f) s

ANSWERS TO STRUCTURAL EXERCISES

Exercise 1:1

1. (a) two threes and two
   (b) Two fives and three

2. (a) ‖‖ ‖‖ ‖‖ ‖‖ ‖‖ ‖‖ |||

   (b) ‖‖ ‖‖ ‖‖ ||

   (c) ‖‖ ‖‖ ‖‖ ‖‖ ‖‖ |||

   (d) ‖‖ ‖‖ ‖‖ ‖‖ ‖‖ ‖‖ ‖‖ ‖‖ |

   (e) ‖‖ ‖‖ ‖‖

3. (a) 27 (b) 18 (c) 39 (d) 21 (e) 32
4. (a) number (b) numeral (c) numeral

Exercise 1:2

1. (a) 9∧∧∧|||| ∧∧∧∧||||| (b) 𒌋𒌋99∧∧∧|||| 𒌋∧99∧∧∧||||

   (c) 𒌋𒌋𒌋99∧∧| 𒌋𒌋𒌋9∧∧∧ (d) 99∧| 99∧

   (e) 𒌋99∧∧|||| 𒌋99∧∧∧∧|| (f) 𒌋99∧|| 9∧∧||

   (g) ∧|||| |||| (h) 𒌋99999∧∧∧∧|||| 99999∧∧∧∧||||

   (i) 𒌋𒌋99∧∧∧∧|||| 𒌋𒌋99∧∧∧||||| (j) 9∧∧∧∧||| 9∧∧∧∧|||

2. (a) 42 (b) 3207 (c) 2142 (d) 2428 (e) 1161

   (f) 1334 (g) 734 (h) 6742 (i) 5523 (j) 16545

3. (a) 𒀹9999∧∧| 999∧∧∧|| (b) 𒀹𒌋9∧∧∧∧|| 𒌋9∧∧∧∧|||

4. 𒌋𒌋𒌋9999∧∧∧||||||| 𒌋𒌋𒌋9999∧∧∧∧|||

Exercise 1:3

1. (a) XXIX   (b) XLVIII
   (c) MMMMDCCCLXXIV
   (d) MMMCMXCIII
   (e) MCCCXXXVIII   (f) MCMXC
   (g) MMM   (h) MMCDXXXII
   (i) DXLIX   (j) MCDXXVIII
   (k) MMMMCDLXXXVIII   (l) CCLXXXVI

2. (a) One hundred and fifty four
   (b) Six hundred and nine
   (c) One thousand one hundred and fourteen
   (d) Seven hundred and sixty one
   (e) Five hundred and eighty nine
   (f) One thousand nine hundred ninety six
   (g) Fifty nine
   (h) One thousand five hundred and nine
   (i) Three thousand four hundred and forty nine
   (j) Eight hundred and one
   (k) Two thousand four hundred and fifty
   (l) One thousand four hundred and forty eight

3. I, II, III, IV, V, VI, VII, VIII, IX, X, XI, XII, XIII, XIV, XV, XVI, XVII, XVIII, XIX, XX, XXI, XXII, XXIII, XXIV, XXV, XXVI, XXVII, XXVIII, XXIX, XXX, XXXI, XXXII, XXXIII, XXXIV, XXXV, XXXVI, XXXVII, XXXVIII, XXXIX, XL, XLI, XLII, XLIII, XLIV, XLV, XLVI, XLVII, XLVIII, XLIX, L

4. (a) DXXII   (b) MMCCXXIII
   (c) MMMMDVII   (d) MCCCXXXIV
   (e) XLIV   (f) CCCXXV

5. (a) 𒌋99∧∧∧||| 999∧∧||||| (b) ∧∧∧∧|||| ∧∧∧∧|||

   (c) ∧∧∧||| ∧∧|| (d) 999∧∧∧||| 999∧∧∧||

   (e) 𒌋9∧∧∧| 𒌋9∧∧∧∧ (f) 𒌋9999∧∧∧∧∧||| 9999∧∧∧∧||||

Exercise 1:4

1. (a) 5 (b) 500 (c) 50 (d) 50,000
2. (a) 3405   (b) 9870
   (c) 71894   (d) 358027
3. (a) ٣٤٩٥٦   (b) ٣٠٤٧٨
   (c) ٣٢٦٠   (d) ٩٧٢٠٣
4. (a) 30,000 (b) 3000 (c) 300,000
5. 68 stands for six tens, eight units while 86 stands for eight tens, six units.
6. (a) Five thousand five hundred and seventy eight
   (b) Fifty thousand four hundred and forty eight

545

(c) Eight hundred and ninety three thousand two hundred and sixty one
(d) seventeen thousand two hundred and four.
7. (a) 538001   (b) 17004   (c) 9909   (d) 232
   (e) 111101   (f) 8080   (g) 10010
8. (a) Seven thousand five hundred and sixty four.
   (b) Six hundred and forty four thousand three hundred and twenty five.
   (c) Twenty nine million five hundred and seventy six thousand, five hundred and thirty two.
   (d) Six million, four hundred and thirty five thousands, five hundred and fifty three.
   (e) Fifty six million, four hundred and forty two thousand, four hundred and forty three.
9. (a) 5,246,831   (b) 43,207,019
10. Nine million   11. Forty thousand

Exercise 1:5
(a) Associative law of addition.
(b) Multiplicative property of zero.
(c) Distributive property of multiplication over addition.
(d) Commutative property of addition.
(e) Commutative property of multiplication.
(f) Associative law of multiplication.
(g) Additive (or identity) property of zero.
(h) Multiplicative (or identity) property of 1.
(i) Additive inverse property.
(j) Multiplicative inverse property.

Exercise 1:6
(1) 16   (2) 11 (3) 94 (4) 100 (5) 31 (6) 73

Exercise 2:1
1. (a) $7 \times 7 \times 7 \times 7 \times 7 \times 7 \times 7 \times 7 \times 7$
   (b) $2 \times 2 \times 2$ (c) 8×8 (d) $4 \times 4 \times 4 \times 4 \times 4$
2. (a) 6^5 (b) 12^3 3. (a) 7^7 (b) 13^5

Exercise 2:2
(1) 2^9 (2) 7^8 (3) 3^{17} (4) 10^{10} (5) 10^{16}

Exercise 2:3
(a) 10^5 (b) 10^5 (c) 10^4 (d) 10^2
(e) 3^3 (f) 5^4 (g) 6^2 (h) 13^2

Exercise 2:4
1. (a) {1,2,3,4,6,8,12,24}
   (b) {1,2,3,4,5,6,10,12,15,20,30}
   (c) {1,2,3,4,5,6,8,10,12,15,20,24,30,40,60,120}
   (d) {1,2,3,4,6,8,9,12,18,24,36,72}
   (e) {1,3,5,7,15,21,35,105}
   (f) {1,3,5,15,25,75}
2. (a) {2,4,6,8,10}   (b) {4,8,12,16,20}
   (c) {5,10,15,20,25}   (d) {7,14,21,28,35}
   (e) {8,16,24,32,40}   (f) {12,24,36,48,60}

3. {1, 2, 3, 4, 6,8}   4. {18, 24, 30,36}

Exercise 2:5
1. {5,29,47}.   2. {49,35,24}.
3. 15, {2,3,5,7,11,13,17,19,23,29,31,37,41,43,47}
4. {1,2,3,4,6,8,9,12,18,24,36,72}. 9 composite factors

Exercises 2:6
1. (a) 2^5 (b) 3^4 (c) $2^2 \times 3 \times 5$
   (d) $2^3 \times 3^2$ (e) $3^2 \times 5$ (f) $3^2 \times 7$
   (g) 3×17 (h) $2^4 \times 3$ (i) 3^5
2. (a) $2 \times 3^2 \times 5 \times 7$ (b) $2^3 \times 5^2 \times 11$
   (c) $2 \times 7 \times 11 \times 13$ (d) $2^6 \times 3^3$
   (e) $2^5 \times 3 \times 5 \times 11$

Exercise 2:7
(a) HCF = 6, LCM = 36   (b) HCF = 2, LCM = 144
(c) HCF = 24, LCM = 288 (d) HCF = 3, LCM = 168
(e) HCF = 4, LCM = 24   (f) HCF = 2, LCM = 24
(g) HCF = 6, LCM = 72   (h) HCF = 3, LCM = 105
(i) HCF = 5, LCM = 75   (j) HCF = 72, LCM = 4320

2. {1,2,3,6,9,18}.   3. {9,18,27,36}.

Exercise 2:8
(1)   (a) 169   (b) 961   (c) 289   (d) 400
(2)   (a) 324   (b) 784   (c) 361   (d) 1024
(3)   (a) 28   (b) 17   (c) 80   (d) 68
(4)   (a) 25   (b) 23   (c) 32   (d) 53
(5)   (a) 64   (b) 343   (c) 2197   (d) 8000
(6)   (a) 125   (b) 1331 (c) 512   (d) 1000
(7)   6

Exercise 2:9
1.

Number	Divisible by											
	2	3	4	5	6	8	9	10	25	50	100
12644750	x			x				x	x	x	
74319275				x					x		
1861425		x		x			x		x		
6671456300	x		x	x				x	x	x	x
925675435				x							

2. (a) 85564, 21342, 4378, 23490, 6936
     Last digit is even.
    (b) 64665, 21342, 97965, 23490, 6936
     Sum of digits is divisible by 3.
    (c) 85564, 6936
     Number formed last two digits is divisible by 4.

(d) 64665, 97965, 76445, 23490
   Last digit is 5 or 0.
(e) 23490, 6936
   Number is even and sum of digits
   is divisible by 3
(f) 3689, 97965
   The difference between the twice
   the last digit and the rest of the
   number is 0 or is divisible by 7.
(g) 6936
   Number formed by last three
   digits is divisible by 8.
(h) 64665, 97965, 23490
   Sum of digits is divisible by 9.
(f) 23490
   Last digit is 0.
(j) 4378
   The difference between the sum
   of the odd digits and the sum of
   the even digits is 0
(k) 6936
   Sum of digits is divisible by 3
   and number formed by last two
   digits is divisible by 4

Exercise 2:10

1. $P = \{0, 1, 4, 10, 20, 35, 56, ...\}$
2. (i) (a) $\{0,1,4\}$ (b) $\{0,1,6\}$ (c) $\{0,1,36\}$
   (ii) (a)

$$1 - 0 = 1$$
$$4 - 1 = 3$$
$$9 - 4 = 5$$
$$16 - 9 = 7$$
$$25 - 16 = 9$$
$$36 - 25 = 11$$

(b) Odd numbers

(c) (i)
$$1 + 0 = 1$$
$$3 + 1 = 4$$
$$6 + 3 = 9$$
$$10 + 6 = 16$$
$$15 + 10 = 25$$
$$21 + 15 = 36$$
   Square numbers

(ii)
$$1 - 0 = 1$$
$$3 - 1 = 2$$
$$6 - 3 = 3$$
$$10 - 6 = 4$$
$$15 - 10 = 5$$
$$21 - 15 = 6$$
   Natural numbers

(d) The square of y is equal to x ($y^2 = x$)

3.

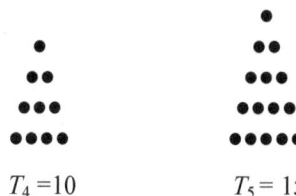

$T_4 = 10$ $T_5 = 15$

Exercise 3:1

1. (a) 60   (b) 36   (c) 25
   (d) 36   (e) 12   (f) 135
2. (a) $\frac{11}{40}$ (b) $\frac{3}{2}$ (c) $\frac{4}{5}$ (d) $\frac{4}{3}$ (e) $\frac{4}{3}$ (f) $\frac{20}{27}$

Exercise 3:2

1. (a) >   (b) <   (c) ≡   (d) <
2. (a) $\frac{2}{6}, \frac{3}{6}, \frac{4}{6}, \frac{5}{6}, \frac{7}{6}$ (b) $\frac{2}{8}, \frac{4}{8}, \frac{6}{8}, \frac{7}{8}$

   (c) $\frac{9}{11}, \frac{9}{8}, \frac{9}{7}, \frac{9}{5}, \frac{9}{2}$ (d) $\frac{10}{5}, \frac{12}{4}, \frac{11}{3}, \frac{16}{4}$

3. $\frac{1}{4}, \frac{1}{3}, \frac{2}{5}, \frac{1}{2}, \frac{2}{3}, \frac{3}{4}, \frac{4}{5}, \frac{6}{7}, \frac{8}{9}$

Exercise 3:3

1. (a) >   (b) >   (c) ≡   (d) <

2. (a) $3\frac{2}{3}$ (b) $7\frac{1}{2}$ (c) $2\frac{1}{4}$ (d) $2\frac{3}{5}$

3. (a) $\frac{8}{3}$ (b) $\frac{11}{2}$ (c) $\frac{7}{4}$ (d) $\frac{13}{5}$

4. $\frac{20}{3}, 5\frac{1}{3}, 4\frac{2}{5}, \frac{13}{3}, 3\frac{1}{2}, 3\frac{1}{4}, \frac{16}{5}, \frac{18}{6}, \frac{17}{8}, 1\frac{3}{5}$

Exercise 3:4

1. (a) $\frac{6}{7}$ (b) $\frac{10}{11}$ (c) 1 (d) $\frac{5}{9}$ (e) $\frac{2}{5}$

2. (a) $\frac{2}{9}$ (b) $\frac{2}{7}$ (c) $\frac{1}{4}$ (d) $\frac{5}{18}$ (e) $\frac{2}{5}$

3. (a) $1\frac{5}{12}$ (b) $1\frac{19}{63}$ (c) $1\frac{9}{40}$ (d) $\frac{29}{30}$ (e) $\frac{7}{10}$

4. (a) $\frac{4}{9}$ (b) $\frac{20}{63}$ (c) $\frac{37}{88}$ (d) $\frac{24}{91}$ (e) $\frac{7}{20}$

5. (a) $\frac{6}{7}$ (b) $1\frac{49}{60}$ (c) $1\frac{5}{63}$

   (d) $1\frac{7}{40}$ (e) $\frac{4}{13}$ (f) $\frac{127}{315}$

Exercise 3:5

1. $10\frac{1}{5}$ 2. $\frac{3}{7}$ 3. $1\frac{3}{5}$ 4. $1\frac{31}{35}$ 5. $10\frac{3}{7}$

6. 1   7. $\dfrac{23}{30}$ 8. $12\dfrac{29}{40}$ 9. $5\dfrac{1}{4}$

10. $5\dfrac{1}{8}$ 11. $6\dfrac{11}{12}$ 12. $1\dfrac{23}{24}$ 13. $4\dfrac{43}{70}$

Exercise 3:6

(1) 40   (2) 135   (3) $\dfrac{15}{32}$ (4) $\dfrac{3}{8}$

(5) 6   (6) 3   (7) 6   (8) 10

(9) $\dfrac{6}{11}$ (10) $\dfrac{1}{30}$ (11) $\dfrac{7}{12}$ (12) $\dfrac{3}{4}$

(13) $\dfrac{2}{15}$ (14) $\dfrac{1}{20}$ (15) $\dfrac{3}{10}$ (16) 10

(17) 14   (18) 78   (19) 30   (20) $3\dfrac{1}{8}$

(21) $5\dfrac{5}{6}$ (22) $12\dfrac{2}{15}$ (23) 56   (24) 56

(25) $21\dfrac{2}{3}$ (26) $7\dfrac{3}{20}$

Exercise 3:7

1. (a) $\dfrac{3}{7}$ (b) 9   (c) $\dfrac{4}{3}$ (d) $\dfrac{8}{5}$

    (e) 6   (f) $\dfrac{1}{8}$ (g) $\dfrac{1}{3}$

2. (a) 21   (b) $\dfrac{3}{20}$

3. (a) $\dfrac{1}{6}$ (b) $\dfrac{6}{7}$ (c) 1 (d) $1\dfrac{2}{3}$ (e) $\dfrac{7}{110}$

4. (a) $\dfrac{7}{9}$ (b) $\dfrac{3}{4}$ (c) $1\dfrac{1}{2}$

5. (a) $\dfrac{1}{9}$ (b) 36

Exercise 3:8

1. (a) $\dfrac{4}{5}$ (b) $\dfrac{7}{18}$ (c) $1\dfrac{13}{15}$ (d) $1\dfrac{5}{16}$

2. (a) 120 (b)(i) $\dfrac{7}{10}$ (ii) $\dfrac{3}{10}$ (iii) $\dfrac{1}{25}$ (c) $\dfrac{3}{7}$

Exercise 3:9

1. (a) 0.675   (b) 5.86   (c) 0.389896
   (d) 0.76587   (e) 0.4535
2. (a) seven millionth   (b) twenty thousandth
   (c) two hundredth   (d) three thousandth

Exercise 3:10

1. (a) 0.75 (b) 0.5   (c) 0.4   (d) 0.625
   (e) 1.5 (f) 1.25   (g) 2.25   (h) 5.48

2. (a) $\dfrac{4}{5}$ (b) $\dfrac{13}{20}$ (c) $1\dfrac{23}{100}$ (d) $3\dfrac{3}{4}$

Exercise 3:11

(1) 144.3   (2) 2.973   (3) 24.21 (4) 14.35
(5) 37.4958 (6) 176.7522 (7) 16.86 (8) 150.79

(9) 6.52   (10) 3.8585 (11) 6.1225 (12) 15.4

Exercise 3:12
(1) 0.48   (2) 0.036   (3) 6300
(4) 934   (5) 2750   (6) 65

Exercise 3:13
(1) 1.35   (2) 70.5042   (3) 24.65991
(4) 199.4905   (5) 274.428   (6) 0.197508
(7) 0.2795   (8) 230.7102   (9) 33.07392
(10) 1.20995

Exercise 3:14
(1) 0.716   (2) 0.019   (3) 1.3   (4) 2.337
(5) 0.502 (6) 0.16   (7) 13   (8) 11.28
(9) 390   (10) 0.1036   (11) 1.52
(12) 0.2395   (13) 3.4   (14) 0.35

Exercise 3:15
1. (a), (c), (d), (f), (h), (j), (l)
2. (a) 0.25   (b) $0.77\overline{7}$ (c) 0.375   (d) 1.5

   (e) 0.2727   (f) 0.44   (g) 0.166   (h) 0.325
   (i) $0.\overline{571428}$ (j) 0.15   (k) $4.16\overline{6}$ (l) 0.7

Exercise 3:16

1. (a) $\dfrac{1}{4}$ (b) $\dfrac{18}{25}$ (c) $\dfrac{41}{50}$ (d) $\dfrac{19}{20}$

   (e) $\dfrac{27}{200}$ (f) $\dfrac{1}{2000}$ (g) $\dfrac{29}{400}$ (h) $\dfrac{139}{400}$

   (i) 4   (j) $2\dfrac{1}{2}$ (k) 6   (l) $7\dfrac{1}{2}$

2. (a) 75%   (b) 80%   (c) 300%
   (d) 34%   (e) 45%   (f) 52%

Exercise 3:17
1. (a) 40% (b) 75%   (c) 250% (d) 4035%
2. (a) 0.2 (b) 0.35   (c) 1.15 (d) 2.5

Exercise 3:18
(1) (a) 15 (b) 5527.5 (c) 5.6 (d) 81000
(2) 18   (3) 16%   (4) 2.5%   (5) 8%
(6) $33\dfrac{1}{3}\%$

Exercise 3:19
1. (a) 10   (b) 3   (c) 8   (d) 6
   (e) 8   (f) 1.5   (g) 8   (h) 1.5
   (i) 7
2. (a) 3:5   (b) 2:3   3.(a) 3:4 (b) 3:7 (c) 4:7
4. (a) 2:3 (b) 1:2 (c) 2:3   5. 800 Frs., 1200 Frs.
6. Bih =20 kg, Manka = 50 kg
7. 10,000 FCFA, 8000 FCFA, 6000 FCFA.
8. Ndi = 12, Shey = 18, Nfor = 24.

9. (a) 3375000 CFA
   (b) 1125000 CFA, 1500000 CFA
10. 6 hours 24 minutes 11. 33
12. (a) 2000 Frs. and 1500 Frs. (b) 36 bags at
    54000 Frs. is better because its unit price is
    smaller.

Exercise 4:1
1. (a) 1080 (b) 500 (c) 9500 (d) 5140
   (e) 22100 (f) 360 (g) 470 (h) 47400
2. (a) 3 (b) 12 (c) 4 (d) 6
3. (a) 1.1 (b) 14.6 (c) 0.6 (d) 3.9

Exercise 4:2
1. (a) 1,000 (b) 38,000 (c) 30
   (d) 0.8 (e) 54 (f) 24
2. (a) 3 (b) 10 (c) 2.1
   (d) 0.1 (e) 20 (f) 0.08

Exercise 4.3

1. Number		Number of significant figures		
		1	2	3	4
a	0.0068398	0.007	0.0068	0.00684	0.006840
b	2.0068398	2	2.0	2.01	2.007
c	4.69768	5	4.7	4.70	4.698
d	1.006127	1	1.0	1.01	1.006

2. Number		Number of Decimal places		
		1	2	3	4
a	0.0068398	0.0	0.01	0.007	0.0068
b	2.0068398	2.0	2.01	2.007	2.0068
c	4.69768	4.7	4.70	4.698	4.6977
d	1.006127	1.0	1.01	1.006	1.0061

3. (a) 0.004 (b) 0.457 (c) 0.505
4. (a) 14.90 (b) 23.11 (c) 6.04
5. (a) 0.0249 (b) 4.03
6. (a) 550 (b) 60 (c) 5400
7. (a) 0.009 (b) 5 (c) 0.2

Exercise 4:4
(a) 5×10^3 (b) 4.8×10^2 (c) 1.02×10^4
(d) 7×10^5 (e) 3.2×10^{-3} (f) 7.3×10^{-5}
(g) 9.25×10^{-1} (h) 1×10^{-3} (i) 5.6×10^{-1}
(j) 3×10^{-5} (k) 1.96×10^{-3} (l) 3.4×10^{-10}

Exercise 4:5
1. (a) 9.192×10^7 (b) 3.5×10^1 (c) 5.4×10^{-3}
   (d) 2×10^1 (e) 3.5×10^2 (f) 6×10^{-2}
   (g) 4×10^4 (h) 9.687×10^2 (i) 3.66×10^{-2}
   (j) 1.33×10^0 (k) 6×10^{-3} (l) 4.37×10^0
2. 3×10^4 3. 1.7×10^{-24} g
4. (a) 9.73×10^{-1} (b) 1.0 (c) 0.97 or 9.7×10^{-1}
5. (a) 1.42×10^{-2} (b) 1.4×10^{-2} or 0.014 (c) 0.014

Exercise 5:1
1. (a) 6380 m (b) 82.3 m (c) 14352 m
   (d) 0.02435 m (e) 28000 m (f) 52.97 m
   (g) 0.007 m (h) 0.249 m (i) 3.128 m
   (j) 43.51 m (k) 0.72 m (l) 13.79 m
2. (a) 382 mm (b) 7342.3 mm (c) 221 mm
   (d) 820000 mm (e) 4370 mm
3. (a) 1.7943 km (b) 0.078 km (c) 0.14873 km
   (d) 0.023 km (e) 0.00371 km
4. (a) 831.2 cm (b) 512.8 cm (c) 4.5 cm
   (d) 324000 cm (e) 99.1 cm

Exercices 5 :2
1. 5264 mg 2. 18000 kg 3. 7.148 g
4. 560 kg 5. 7.342 tons
Exercise 5:3
1. (a) 20,000 cm² (b) 0.03 cm²
   (c) 10,000 cm² (d) 4 cm²
2. (a) 0.0009 km² (b) 0.000004 km²
   (c) 0.0005 km²
3. (a) 600 mm² (b) 12,000,000 mm²
   (c) 400 mm² (d) 1,000,000 mm²
4. (a) 3 m² (b) 0.0002 m² (c) 30,000 m²
   (d) 500 m² (e) 4,000,000 m² (f) 0.04 m²
   (g) 0.000053 m² (h) 7 m²
5. (a) 700,000,000 cm² (b) 400,000,000 cm²
6. (a) 7 a (b) 40,000 a
7. 0.0096 a 8. 2,250,000 cm² 9. 1.892 m²

Exercise 5:4
1. (a) 0.7 m³ (b) 0.5 m³ (c) 0.008 m³ (d) 504 m³
2. (a) 400,000,000 cm³ (b) 3,140 cm³
   (c) 3,000,000 cm³ (d) 64,000,000 cm³
3. (a) 0.2 liters (b) 3.5 liters (c) 7 liters
   (d) 4.8 liters
4. (a) 17.6 ml (b) 850 ml (c) 350 ml (d) 0.174 ml
5. 12000 cm³ 6. 10 liters
7. 11000 cm³ 8. 72000 cm³
9. (a) 7000 cm³ (b) 40,000 cm³
   (c) 30 cm³ (d) 17.4 cm³

Exercise 5:5
1. (a) 9 Hours (b) 540 minutes (c) 32400 secs
2. 144 days 3. 730 seconds 4. 65 minutes
5. 6.50 a.m.
6. (a) 28 hours (b) 1680 minutes (c) 100800 secs
7. 443 days 8. 10080 minutes

Exercise 5:6
1. (a) 5:25 p.m.   (b) 11:36 p.m.
   (c) 2:20 p.m.   (d) 11:10 a.m.
2. (a) 20:34 (b) 5:56 (c) 12:45 (d) 00:45
3. (a) 9 hours 30 minutes (b)14 hours 30 minutes
   (c) 19 hour 2 minutes (d) 10 hours 42 minutes
4.(a) ante meridiem (between midnight and
     midday)
   (b) post meridiem (between midday and
       midnight)

Exercise 5:7
1. (a) 140°F   (b) 95°F   (c) 68°F
   (d) 104°F   (e) 167°F
2. (a) 50°C   (b) 34°C   (c) 17°C
   (d) 93 °C   (e) 45°C
3. (a) hot   (b) warm   (c) cold
   (d) warm   (e) cold   (f) cold
   (g) hot   (h) cold   (i) warm

Exercise 5:8
1. 779280 FCFA 2. 2454861.6 FCFA 3. £2251

Exercise 6:1
1. (a) 180,000 FCFA (b) 9600 FCFA
   (c) 60,000 FCFA
2. 1,000,000 FCFA   3. 10%   4. 6 months
5. (a) 12% (b) 300,000 FCFA   (c) 2 years

Exercise 6:2
1. 3,340,810 FCFA 2. 61,800 FCFA
3. 5,701,440 FCFA 4. 331013 FCFA 5. 4 years
6. 562,432 FCFA   7. 374,562 FCFA

Exercise 6:3
1. 2,048,000 FCFA   2.   81,920 FRS

Exercise 6:4
1. 24%   2. $16\frac{2}{3}$ %   3. 29.2% 4. 10%

5. 12.5% 6. 22.8 % 7. 10 % 8. $3\frac{1}{3}$ %

9. 300 students   10. 10%   11. 16 %
12. 5640000 FCFA   13.   5980000 FCFA
14. 2000 FCFA   15. 1300 FCFA

Exercise 6:5
1. 84000 FCFA   2. 20 %   3. 25 %

Exercise 6:6
1. (a) 240,000 FCFA (b) 414,000 FCFA
2. 253120 FCFA

Exercise 7.1
1. (a) 20   (b) 13   (c) 5.8   (d) π
2. (a) <   (b) <   (c) >   (d) >
   (e) =   (f) >   (g) =   (h) =

(i) >   (j) =   (k) >   (l) <
3. (a) True   (b) False (c) True   (d) True
   (e) True   (f) False (g) False (h) True
   (i) True   (j) False (k) True   (l) True
4. (a) +13   (b) −14   (c) −20   (d) −10
   (e) +28   (f) 0   (g) +18   (h) −3.9

(i) $-2\frac{3}{4}$ (j) -0.05 (k) $-3\frac{1}{4}$ (l) +2.7

5. (a) +20 (b) −6   (c) +22   (d) +14 (e) +19
   (f) −85 (g) −29 (h) +7.98 (i) −3   (j) +1

Exercise 7.2
1. (a) -25 (b) $+53$ (c) $+42$ (d) 2π
   (e) $+\frac{5}{13}$ (f) $-\frac{4}{11}$ (g) $\frac{26}{19}$ (h) $-\frac{\sqrt{3}}{2}$
2. (a) +27   (b) −20   (c) +10
   (d) +20   (e) −12   (f) −16
   (g) +3   (h) −1.1   (i) $9\frac{3}{4}$
   (j) −0.14   (k) $-12\frac{1}{4}$ (l) −13.7
3. (a) +10   (b) +4   (c) +24
   (d) +4   (e) +5   (f) −1
   (g) $+35\frac{1}{2}$ (h) −10.94 (i) +25 (j) +45

Exercise 7.3
1. (a) −6   (b) −12   (c) +15
   (d) −30   (e) +16   (f) −64
   (g) $\frac{63}{8}$ (h) 3.5   (i) −0.75
   (j) −0.0414   (k) $\frac{21}{8}$ (l) −4.8
2. (a) +15   (b) −6   (c) −6 (d) −40
   (e) −8   (f) +40   (g) $-\frac{3}{32}$
   (h) +6.72   (i) +24   (j) +36
3. (a) +6   (b) −6   (c) +6
   (d) −6   (e) +0.06   (f) −0.06
   (g) +0.6   (h) −60
4. (a) +16   (b) −16   (c) −9
   (d) +9   (e) 0   (f) 0
   (g) +90   (h) +15

Exercise 8:1
1. $2x$ 2. $x+3$ 3. $x-5$ 4. $x+7$

5. $x+8$ 6. $x-9$ 7. $2x$ 8. $\frac{1}{2}x$

9. $x-10$ 10. $2x+7$ 11. $\frac{x}{2}$ 12. $4x$

13. $\frac{x}{4}$ 14. x^2 15. x^3 16. $3x-4$

17. $\sqrt{2x}$ 18. x^2+2 19. $15-x$

20. $2x-8$ 21. $\frac{1}{2}x-7$ 22. $\frac{1}{2}x+7$

Exercise 8:2

1. 7 2. 1 3. 10 cm^2 4. -15
5. 36 6. 88 cm^3 7. 154 cm^2

Exercise 8:3

1) (a) $6x, -2xy, 3y$ (b) $px, \dfrac{p}{y}$

(c) $w, 5pz, r$ (d) $3pt, -\dfrac{8x}{y}, \dfrac{1}{w}$

2) (a) 7 (b) -2 (c) 5

3) (a) p, t, pt, (b) $x, \dfrac{1}{y}, \dfrac{x}{y}$ (c) p, w, pw

Exercise 8:4

(1) $21x - 17y$ (2) $s - t$ (3) $5p - 22q$
(4) $-x - 7$ (5) $u - 3v$ (6) $20a + 10b$
(7) $2x - 2y + 2$ (8) b (9) $99x$
(10) $-pq + 2qr$

Exercise 8:5

1. (a) $8b$ (b) $3x^2$ (c) $20pq$ (d) $18ab^2$
(e) $24x^2 y$ (f) $28x^2 y$ (g) $50ab$ (h) $30a^2$
(i) $20x^2 y$ (j) $27u^2 v^2$

2. (a) $7xy$ (b) $6p$ (c) $7x$ (d) $6x$
(e) $11b$ (f) $6p$ (g) $\dfrac{7}{8}xy$ (h) $4u$
(i) $5y$ (j) $3m$ (k) $4\dfrac{p}{q} + 4\dfrac{q}{p}$ (l) x

Exercise 8:6

1. HCF $= 3a$, LCM $= 12a^2$
2. HCF $= 2a$, LCM $= 40a^2 b$
3. HCF $= a$, LCM $= abxy$
4. HCF $= a$, LCM $= 420a^2$
5. HCF $= 2a^2$, LCM $= 24a^2 b^2$
6. HCF $= ab$, LCM $= 15a^2 b^2$
7. HCF $= a$, LCM $= 14a^2$
8. HCF $= 3pq$, LCM $= 45p^2 q^2$
9. HCF $= 4x^2 y$, LCM $= 24x^3 y^2$
10. HCF $= 2ab^2$, LCM $= 120a^2 b^2 c^2$

Exercise 9:1

(1) $\dfrac{(x+3)}{2} = 4$ (2) $7d = 42000$ (3) $x + 3 = 14$
(4) $4w = 16$ or $\dfrac{4}{3}l = 16$ (5) $3(x+6) = 33$
(6) $x + 20 = 3x$ (7) $x + 5 = 2x - 4$
(8) $12h = 120 - 36$ (9) $250m = 450$
(10) $16960 + t = 17810$

Exercise 9:2

1. (a) 5 (b) 8 2. (a) 10 (b) 17
3. (a) Divide by 3 (b) Divide by 7
(c) Multiply by 5 (d) Multiply by 14
(e) Divide by 0.5 (f) Multiply by 0.02
(g) Subtract 5 (h) Multiply by 13 and add 2

Exercise 9:3

(1) $x = 6$ (2) $t = 7$ (3) $x = 10$ (4) $a = 8$
(5) $x = 7$ (6) $y = 10$ (7) $x = 9$ (8) $x = 4$
(9) $y = 4$ (10) $x = \dfrac{16}{5}$ (11) $n = 4$ (12) $x = 40$

Exercise 9:4

1. $x = 8$ 2. $t = 3$ 3. $p = -6$ 4. $u = 42$
5. $z = 36$ 6. $w = -33$ 7. $r = -27$ 8. $q = -6$

Exercise 9:5

1. $x = 2$ 2. $t = \dfrac{5}{2}$ 3. $q = 4$ 4. $p = 1$
5. $a = 7$ 6. $x = 8$ 7. $d = 8$ 8. $y = \dfrac{9}{2}$
9. $u = 16$ 10. $m = 40$ 11. $n = 4$ 12. $x = 230$
13. $k = 40$ 14. $z = 60$ 15. $n = 4$ 16. $u = 9$
17. $x = 5$ 18. $x = 10$ 19. $t = \frac{11}{2}$ 20. $a = 3$
21. $u = 15$ 22. $y = \frac{3}{2}$ 23. $y = 35$ 24. $t = 10$

Exercise 9:6

1. $p = \frac{5}{2}$ 2. $m = 2$ 3. $y = 10$
4. $x = 3$ 5. $x = 2$ 6. $x = 2$
7. $m = -42$ 8. $x = \frac{10}{3}$ 9. $t = -20$
10. $x = 1$ 11. $p = -1$ 12. $x = 4$

Exercise 9:7

(1) 11 cm (2) 8 (3) $y = 100$ (4) $x = 2$ (5) 1
(6) $x = 5$ (7) 6000 (8) 7 years (9) 4
(10) 5 (11) 10 (12) 9 (13) 7 (14) $t = 850$
(15) $3000n + 6000 = 27000$, $n = 7$ (16) 9 (17) 7

Exercise 10:1

(a) 1. $3x + 15$ 2. $4y + 8$ 3. $4x + xy$

4. $3v + uv$ 5. $2x - 2$ 6. $5y - 15$

7. $6p - 3pq$ 8. $4t - st$ 9. $10x - 2x^2$

10. $x^2 + 4x + 3$ 11. $x^2 + x - 2$

12. $pq - p + 3q - 3$ 13. $x^2 + 6x + 9$

14. $x^2 - 4x + 4$ 15. $16 + 8a + a^2$

16. $9 - 6y + y^2$

(b) 1. 2401   2. 10201   3. 39204   4. 195364

Exercise 10:2

1. (a) $p^2 - 20p + 100$ (b) $y^2 - 14y + 49$

(c) $x^2 + 5x + \dfrac{25}{4}$ (d) $u^2 + \dfrac{4}{3}u + \dfrac{4}{9}$

(e) $x^2 + 18x + 81$ (f) $x^2 - 4x + 4$

2. (a) $(t + 6)^2$ (b) $(x - 9)^2$

(c) $\left(y + \dfrac{7}{2}\right)^2$ (d) $\left(u - \dfrac{11}{2}\right)^2$

(e) $(x + 10)^2$ (f) $(m - 15)^2$

Exercise 10:3

(1) $3(x + y)$ (2) $2(p + 3q)$ (3) $5(x - 2y)$

(4) $4y(x + 2)$ (5) $3u(2v - f)$ (6) $\dfrac{1}{4}x(y + p)$

(7) $\dfrac{1}{2}x(a - b)$ (8) $\dfrac{1}{3}(2y - x)$

Exercise 10:4

(a) $(y + 1)(x + 1)$ (b) $(a + 3)(x + 1)$

(c) $(2p + 1)(3x + 2)$ (d) $(3x + 2)(1 - 2y)$

(e) $(3x - 2)(2y - 3)$ (f) $(u - v)(5 - t)$

(g) $(a - 3)(m - n)$ (h) $(r + 3s)(p - 2q)$

Exercise 10:5

1. (a) $(x + y)(x - y)$ (b) $(1 + 3y)(1 - 3y)$

(c) $3(3)(2 + x)(2 - x)$ (d) $(2a + 1)(2a - 1)$

(e) $(2y - 3x)(2y + 3x)$ (f) $(2u - 5y)(2u + 5y)$

(g) $(6a - 7b)(6a + 7b)$ (h) $(x - yz)(x + yz)$

(2) (a) 81   (b) 976   (c) 115000

(d) 0.75   (e) 135000   (f) 0.000005

Exercise 10:6

1. $(x + 3)(x - 2)$ 2. $(p + 1)(p + 11)$

3. $(y - 4)(y - 3)$ 4. $(x + 8)(x - 2)$

5. $(x - 5)(x + 3)$ 6. $(5x - 3)(2x + 1)$

7. $(4a + 5)(a - 2)$ 8. $(5 - 6y)(1 - 2y)$

9. $(3 + 2x)(1 - x)$ 10. $(5 - 3x)(x + 5y)$

11. $(x + 4y)(x - 3y)$ 12. $(2x - 3y)(x + 5y)$

13. $(6k - x)(k + 3x)$ 14. $(2p - q)(6p - 5q)$

Exercise 11:1

1. $x = 1, y = 2$ 2. $x = 1, y = 3$

3. $a = 2, b = \dfrac{3}{2}$ 4. $x = 3, y = 1$

5. $x = 4, y = 1$ 6. $x = \dfrac{47}{37}, y = \dfrac{25}{37}$

7. $x = 0, y = 2$ 8. $x = 3, y = 12$

Exercise 11:2

See Exercise 11:1 above

Exercise 11:3

1. (a) $p = 4, q = -2$ (b) $s = 7, t = 4$

(c) $x = 4, y = 3$ (d) $m = \dfrac{1}{2}, n = \dfrac{1}{4}$

(e) $x = 3, y = 1$ (f) $u = 5, v = 2$

2. See 1. Above

3. (a) $a = 17, b = -10$ (b) $x = 2, y = -1$

(c) $x = -2, y = \dfrac{3}{2}$ (d) $x = 1, y = 2$

(e) $x = 6, y = 1$ (f) $x = -4, y = -2$

(g) $x = 1, y = -3$ (h) $x = 4, y = 3$

(i) $x = 1, = 3$ (j) $a = 3, b = -1$

(k) $x = 5, y = 2$ (l) $x = 5, y = 2$

(m) $a = 2, b = -3$ (n) $x = 8, y = 3$

(o) $a = -1, b = -1$ (p) $r = 2, s = 3$

Exercise 11:4

1. $x = 9, y = 1$ 2. $x = -\dfrac{1}{2}, y = 1$

3. $a = 500, b = 1000$ 4. $x = 1400, y = 450$

5. $m = 700, n = 200$ 6. $x = 800, y = 500$

5. $m = 700, n = 200$ 6. $x = 800, y = 500$

7. $p = 700, q = 600$ 8. $u = 4000, v = 7000$

Exercise 11:5

1. Book = 100 Frs.,   pencil = 20 Frs.

2. $m = 5, n = -6$ 3. $27, -1$ 4. $16, 10$

5. bottle = 17g, cork = 1g   6. (a) 16g   (b) 6g

7. Coconut = 130 Frs., orange = 20 Frs.

8. Man = 36 years, son = 4 years.

9. (a) Number of first-class rooms = 25
      Number of second-class rooms = 7

(b) (i) First-class rooms = 112500 FRS
(ii) second -class rooms = 17500 FRS
10. 16 shirts and 16 trousers   11.   4 km

Exercise 12:1
1. $x = 4$ or $x = -2$ 2. $x = -2$ or $x = 1$

3. $x = 2$ or $x = 3$ 4. $x = -\dfrac{1}{2}$ or $x = \dfrac{3}{2}$

5. $x = 7$ or $x = -3$ 6. $x = -\dfrac{2}{3}$ or $x = \dfrac{3}{2}$

7. $x = \dfrac{5}{6}$ or $x = -\dfrac{3}{2}$ 8. $x = -\dfrac{1}{2}$ or $x = 5$

Exercise 12:2
1. $x = 3$ or $x = -15$ 2. $u = -19$ or $u = -1$
3. $p = -8$ or $p = 1$ 4. $y = 13$ or $y = 5$
5. $a = 4$ or $a = 7$ 6. $x = 33$ or $x = -3$
7. $x = -16$ or $x = 2$ 8. $x = -19$ or $x = 1$

Exercise 12:3
1. $\dfrac{5 \pm \sqrt{57}}{4}$ 2. $\dfrac{-7 \pm \sqrt{29}}{2}$ 3. $\dfrac{-5 \pm \sqrt{17}}{4}$

4. $-3 \pm \sqrt{19}$ 5. $\dfrac{2 \pm \sqrt{10}}{3}$ 6. $\dfrac{5 \pm \sqrt{7}}{5}$

7. $\dfrac{5 \pm \sqrt{5}}{5}$ 8. $\dfrac{-1 \pm \sqrt{73}}{18}$

Exercise 12:4
1. $u = -2$ or $p = -4$ 2. $a = 8$ or $a = -3$

3. $x = \dfrac{1}{3}$ or $x = -7$ 4. $m = -\dfrac{5}{2}$ or $m = -1$

5. $p = 2$ or $p = -10$ 6. $x = \dfrac{7}{2}$ or $x = 1$

7. $x = \dfrac{3}{2}$ or $x = -4$ 8. $a = -\dfrac{1}{2}$ or $a = 10$

Exercise 12:5
1. $x = \pm 8$ 2. $y = 0$ or $y = 7$ 3. $a = \pm\dfrac{1}{2}$

4. $p = 0$ or $p = \dfrac{1}{9}$ 5. $x = 0$ or $x = 4$

6. $y = \pm 3$ 7. $b = 0$ or $b = 6$ 8. $p = \pm 5\sqrt{2}$

Exercise 12:6
1. Imaginary   2. Real and distinct
3. Imaginary   4. Imaginary
5. Real and distinct   6. Real and equal
7. Imaginary   8. Real and distinct
9. Real and equal   10. Imaginary

11. Real and distinct   12. Real and equal

Exercise 12:7
1. -3 and -2 or 3 and 2
2. -3 and $-3^2 = 9$ or 4 and $4^2 = 16$
3. 15, 135 FCFA   4. 15 m by 16.5 m
5. $n = -13$ or $n = 12$ 6. 3 and 9
7. 16   8. 12 m by 2 m
9. $h = 8$ m, $b = 12$ m   10. 8
11. 11 and 3   12. 12 rows

Exercise 12:8
1. $x = 4, y = -\dfrac{3}{2}$ or $x = 3, y = -2$

2. $x = \dfrac{4}{3}, y = -\dfrac{4}{3}$ or $x = 4, y = 28$

3. $x = 3, y = 4$ or $x = 4, y = 3$

4. $x = \dfrac{1}{2}, y = \pm 1$

5. $x = 1, y = 3$ or $x = -\dfrac{95}{41}, y = \dfrac{21}{41}$

6. $x = \dfrac{27}{4}, y = \dfrac{49}{6}$ or $x = -4, y = 1$

7. $x = -2, y = 1$ or $x = -\dfrac{3}{11}, y = -\dfrac{62}{33}$

8. $x = 1, y = 2$ or $x = -\dfrac{3}{2}, y = -\dfrac{1}{2}$

9. $x = 2, y = \pm 3$

10. $x = 1, y = 2$ or $x = \dfrac{1}{9}, y = \dfrac{22}{9}$

Exercise 13:1
1. $\dfrac{y}{4z}$ 2. $\dfrac{2}{x}$ 3. $\dfrac{2a - b}{3g}$

4. $\dfrac{2q}{p + 4}$ 5. $\dfrac{1}{3}$ 6. 2

7. $\dfrac{1}{2}$ 8. $\dfrac{x - 3}{7}$ 9. $\dfrac{p - 5}{4}$

10. $\dfrac{x + y}{x - y}$ 11. $\dfrac{x}{q}$ 12. $\dfrac{b}{x}$

13. $-\dfrac{1}{5}$ 14. $-\dfrac{x}{7}$ 15. $-\dfrac{p + 2}{x}$

16. $-\dfrac{y}{3x + 2}$ 17. $-\dfrac{1}{a}$ 18. $-\dfrac{3}{2(u + v)}$

19. $-\dfrac{m(x + y)}{y}$ 20. $\dfrac{b - a}{b + a}$ 21. $\dfrac{x}{x + 7}$

22. $\dfrac{y-1}{y-7}$ 23. $\dfrac{2a+b}{2a-b}$ 24. $\dfrac{3}{2-1}$

25. $\dfrac{2(u-1)}{(u-5)}$ 26. $\dfrac{y-4}{2(3y+1)}$ 27. $\dfrac{x-5}{3x+2}$

28. $\dfrac{r+1}{3r-1}$ 29. $\dfrac{2x+3}{x+2}$

30. $\dfrac{y+1}{2y-3}$ 31. $\dfrac{y-1}{2y+3}$

Exercise 13:2

1. $\dfrac{3}{7}$ 2. $\dfrac{y}{3x}$ 3. $\dfrac{7}{p}$ 4. $\dfrac{6}{4y-3}$

5. $\dfrac{a}{3}$ 6. $\dfrac{5(a-b)}{2}$ 7. $\dfrac{10(10+a)}{p}$

8. $\dfrac{x-7}{x^2}$ 9. $\dfrac{15}{u}$ 10. $\dfrac{xy(x-3)}{4(y-3)}$

Exercise 13:3

1. $\dfrac{7}{4}$ 2. $\dfrac{5}{7}$ 3. $\dfrac{3u}{2v}$ 4. $\dfrac{3}{a}$

5. $\dfrac{ax}{2b}$ 6. $\dfrac{3}{5}$ 7. $\dfrac{3}{5}$ 8. $\dfrac{9}{4}$

9. $\dfrac{8}{3}$ 10. $\dfrac{7}{6y}$ 11. $\dfrac{1}{3}$ 12. $\dfrac{p-q}{21}$

13. $\dfrac{2}{x-2y}$ 14. $\dfrac{a+c}{b-c}$ 15. $\dfrac{(x+5)}{2}$

16. $2u$ 17. $\dfrac{1}{3}$ 18. $\dfrac{1}{6}$

Exercise 13:4

1. $\dfrac{3x}{2}$ 2. $\dfrac{4a}{5}$ 3. $\dfrac{13}{x}$ 4. $\dfrac{1}{p}$ 5. $\dfrac{2x}{y^2}$

6. $\dfrac{6u-3}{v}$ 7. $2x$ 8. $\dfrac{3}{5}$ 9. -3 10. 5

11. -2 12. $\dfrac{2}{a+2}$ 13. $\dfrac{5x+1}{(x-1)(x+1)}$

14. $\dfrac{2y^2}{(2+y)(2-y)}$ 15. $\dfrac{3a+8b-4c}{24}$

16. $\dfrac{3}{t}$ 17. $\dfrac{10-4x+5y}{x^2y^2}$ 18. $\dfrac{1}{15(x+y)}$

19. $\dfrac{24}{(x-6)(x+6)}$ 20. $\dfrac{4ab}{(3b-4a)(3b+4a)}$

19. $\dfrac{24}{(x-6)(x+6)}$ 20. $\dfrac{4ab}{(3b-4a)(3b+4a)}$

21. $\dfrac{2x}{7-4x}$ 22. $\dfrac{x^2+y^2}{x^2-y^2}$

Exercise 13:5

(a) 1. $x=0$ 2. $x=3$ 3. $x=-3$

4. $x=0$ 5. $x=5$ 6. $x=0, x=-5$

7. $x=-1, x=3$ 8. $x=-4, x=5$

9. $x=-5, x=-7$ 10. $x=2a$

(b) 1. $\{3\}$ 2. $\{-1,-5\}$ 3. $\{5\}$

4. $\{4,-2\}$ 5. $\left\{1,-\dfrac{1}{3}\right\}$ 6. $\left\{-\dfrac{2}{3}\right\}$

(c) 1. $x=0$ or $y=0$ 2. $m=0$ or $n=0$

3. $x=y$ 4. $x=2y$ 5. $x=2$ 6. $x=9$

Exercise 14:1

1. $w=\dfrac{bTu}{a-bT}$ 2. $-40°, C=\dfrac{5F-160}{9}$

3. $W=\dfrac{10R}{T-5S}, 10$ 4.(a) 1.5×10^{-5} (b) $h=\dfrac{cV}{\pi r^2}$

5. $\dfrac{x}{y}=\dfrac{1}{5}$ 6. $p=4m-2n=2(2m-n)$

7. (i) $g=\dfrac{hkL}{kL-h}$ (ii) $k=2$

Exercise 14:2

1. $A=\left(\dfrac{\pi n}{V}\right)^2+3$ 2. $t=\dfrac{p^2r}{(a-p)^2}$

3. $p=\dfrac{4n^2\pi^2(k+1)}{m^2}, p=64$ 4. $q=\dfrac{p^2(an+b)}{n}$

5. (a) $v=\left(\dfrac{k}{y}\right)^n$ (b) $v=\pm2$

6. $l=4\pi^2T^2g$ 7. $a=\dfrac{v^2s+t}{r}$

8. $m=\dfrac{n}{p^2-1}$ 9. $h=\dfrac{3V-\pi r^3}{\pi r^2}$

Exercise 14:3

1. (a) $r=\dfrac{-\pi l\pm\sqrt{\pi^2l^2-4A\pi}}{2\pi}$

(b) $t=\dfrac{-u\pm\sqrt{u^2\,2as}}{a}$

(c) $x=\dfrac{-b\pm\sqrt{b^2-4ac}}{2a}$

(d) $t=\dfrac{-\pi r\pm\sqrt{A\pi+\pi r^2}}{\pi}$

2. $\dfrac{r}{s}=1$ or $\dfrac{r}{s}=3$ 3. $G=-\dfrac{1}{4}T$ or $G=\dfrac{T}{3}$

4. $r = \dfrac{6p}{q}$ or $r = -\dfrac{p}{3q}$ 5. $n = 2m$ or $n = \dfrac{6}{5}m$

Exercise 15

1. (a) $(x - 2), (x - 1)(x + 1)$

    (b) $(2x + 1), (x - 1)(x + 1)$

2. (a) $x = \dfrac{1}{2}$ (b) $x = 1, -1$ or $\dfrac{1}{6}$ 3. -8

4. $k = -7, f(x) = (x + 3)(x - 2)(x - 1)$

5. $a = 3, b = -5, c = 8$ 6. $x^2 - 7x - 6$

7. $a = 3, b = 12$

8. $k = 2$, roots are $x = 2$ and $x = 3$

9. $k = -7,\ (x + 3)(x - 1)(x - 2)$

10. $k = 2, f(x) = (x + 1)(x + 2)(x - 1)$

11. Remainder = 5, required number = -5

Exercise 16.1

(c) 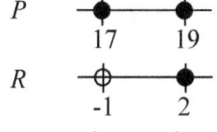   (d)

1. (a)   (b)

(e)

2. P is closed. Both 17 and 19 are included.
    Q is closed. Both 1 and 3 are included
    R is half opened, half closed, -1 is excluded
       but 2 is included.
    S is half opened, half closed, -2 is included but
       2 is excluded.
    T is closed. Both 2 and 5 are included.
    U is opened. Both -2 and 0 are excluded.
    V is opened. Both 5 and 8 are excluded.
    W is opened. Both -7 and -2 are excluded.

3. P   17 19   Q   1   3
    R   -1 2   S   -2 2
    T   2   5   U   -2 0
    V   5   8   W   -7 -2

4. (a) $(-2,2],\ ]-2,2]$
    (b) $[98,101],\ 98 \le x \le 101$
    (c) $(2,5),\ 2 < x < 5,\ ]2,5[$
    (d) $[1,3),\ [1,3[$
    (e) $(-6, -1),\ -6 < x < -1$
    (f) $]14, 19],\ 14 < x \le 19$
    (g) $[0, 7[,\ 0 \le x < 7$
    (h) $[-3, 1),\ -3 \le x < 1$
    (i) $(11, 13],\ 11 < x \le 13$

5. (a) closed   (b) Half open, half closed

(c) Half open, half closed   (d) Opened
(e) Closed   (f) Open   (g) Closed

Exercise 16.2

1. $x + 8 \le 14$ 2. $\dfrac{40}{n} > 4$ 3. $3(x + 7) \le 27$

4. $4m \ge 20$ 5. $\dfrac{3}{5}y - 10 < 12$

6. $3x - \dfrac{1}{2}x < x + 6$ 7. $2(m + 7) \le m + 12$

8. $3(p + 5) < 4p + 2$ 9. $30 - 5n \ge 4$

10. $4x - 60,000 \le 600,000$

Exercise 16:3

1. $x < 2$ 2. $x > 9$ 3. $x \ge 4$

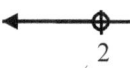

4. $y \ge 4$ 5. $x \le \dfrac{16}{5}$ 6. $n > 4$

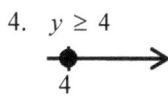
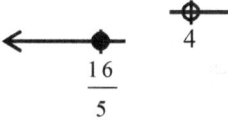

7. $x < 40$ 8. $x > 5$ 9. $y \ge 35$

10. $t < 10$ 11. $u < 9$ 12. $x \le 10$

13. $x > \dfrac{11}{2}$ 14. $x \ge 3$ 15. $y \le \dfrac{3}{2}$

16. $x < 5$ 17. $x \le 15$ 18. $p < \dfrac{5}{2}$

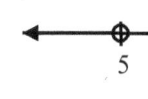
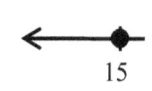

19. $m < 2$

20. $x \le 7$

21. $x \ge 5$

22. $x \le 1$

23. $-\dfrac{13}{4} \le x \le 0$

24. $-1 \le w \le 8$

25. $-1 < x < 2$

26. $-1 < x < 2$

27. $-1 \le x \le 2$

Exercise 16:4

1. $-4 < x < -2$

2. $a < -3$ or $a > 9$

3. $x \le -7$ or $x > \dfrac{1}{3}$

4. $m \le -\dfrac{5}{2}$ or $m \le 1$

5. $p < -10$ or $p > 2$

6. $x < 1$ or $x > \dfrac{7}{2}$

7. $-4 < x < \dfrac{3}{2}$

8. $-\dfrac{1}{3} \le a \le 10$

Exercise 17:1

1. 216   2. 64   3. 72   4. 675   5. y^{10}

6. $8x^{10}$ 7. $60x^9$ 8. a 9. $2a^3$ 10. $\dfrac{2x}{3y}$

11. $\dfrac{2}{3b}$ 12. $\dfrac{1}{a^2}$ 13. $5x$ 14. $4ab^2$ 15. $\dfrac{b}{a}$

16. $-\dfrac{x}{3}$ 17. $2a^2$ 18. $5x$

Exercise 17:2

(i)   1. 4   2. $\dfrac{5}{4}$ 3. 5   4. $\dfrac{1}{36}$

5. $\dfrac{3}{2}$ 6. 4   7. 2250   8. $\dfrac{8}{5}$

9. $-\dfrac{31}{28}$ 10. $-\dfrac{17}{12}$ 11. $\dfrac{1}{12}$ 12. $\dfrac{1}{20}$

13. 100   14. $25a^2$ 15. 3   16. $\dfrac{25}{3}$

(ii) (1) $x = \dfrac{7}{4}$ (2) $x = -\dfrac{5}{4}$ (3) $x = -\dfrac{3}{5}$

(4) $x = \dfrac{1}{2}$ (5) $x = -6$ (6) $x = \dfrac{5}{3}$

(7) $x = 3$ (8) $x = 5$ (9) $x = -\dfrac{4}{5}$

(10) $x = 6$ (11) $y = -\dfrac{3}{2}$ (12) $x = \dfrac{7}{3}$

(13) $x = \dfrac{9}{5}, y = \dfrac{27}{5}$ (14) $x = 2, y = 1$

(15) $x = 1, y = -2$

Exercise 17.3

(1) n (2) 2   (3) -1

(4) 3   (5) $\dfrac{1}{3}$ (6) 4

(7) 5   (8) $\dfrac{3}{2}$ (9) $\dfrac{3}{2}$

(10) $\dfrac{1}{3}$ (11) $\dfrac{1}{2}$ (12) $\dfrac{3}{2}$

(13) $y = 3^x$ (14) $10^x = 3$ (15) $n = 1$

(16) $n = 10$ (17) $y = 1$ (18) $x = 16$

(19) $x = \dfrac{1}{10}$ (20) $x = 3$

Exercise 17:4

1. (a) 3   (b) 7   (c) 3   (d) 1   (e) 2

(f) 3   (g) 2   (h) 2   (i) -4 (j) $\dfrac{3}{2}$

(k) 2   (l) -1 (m) 2   (n) 6

2. (a) 1.8060   (b) 0.3010   (c) 0.1505

3. (i) 0.3891　(ii) -0.1743　4. 2.130

Exercise 17:5

1. (a) 5　(b) $\dfrac{13}{3}$　(c) $\dfrac{3\sqrt{2}}{2}$　(d) 6

2. (a) $2\sqrt{6}$　(b) $\dfrac{\sqrt{30}}{10}$　(c) $6\sqrt{2}$

(d) $1 + \dfrac{3\sqrt{2}}{2}$　(e) $\sqrt{3}$　(f) $\sqrt{c^3}$

(g) $\sqrt{7}$　(h) $2\sqrt{2}$

3. (a) $\dfrac{4}{7}$　(b) $\dfrac{41}{4}$　(c) $x = -5$ or $x = -6$

(d) $x = 1$ or $x = -7$

4. (a) 4　(b) 12　(c) 1

(d) 5　(e) 444　(f) $\dfrac{3}{5}$

5. (a) $\dfrac{5}{11}$　(b) 6　(c) $\dfrac{\sqrt[3]{3}}{3}$　(d) 16

6. (a) $\dfrac{2\sqrt{3} + 1}{11}$　(b) $\sqrt{2} + \sqrt{3}$

Exercise 18:1

1. (a) Yes, all digits are less than 6
   (b) No, 7 does not exist in base seven
   (c) Yes, all digits are less than 2
   (d) No, 9 does not exist in base nine
   (e) Yes, all digits are less than 10
   (f) Yes, all digits are less than 9
   (g) Yes, all digits are less than 5
   (h) No, 8 does not exist in base eight
   (i) Yes, all digits are less than 5
   (j) Yes, all digits are less than 3
   (k) No, 6 does not exist in base six
   (l) Yes, all digits are less than 10

Exercise 18:2

1. $8 \times 10 + 7$
2. $1 \times 10^2 + 2 \times 10 + 4$
3. $8 \times 10^2 + 4 \times 10 + 5$
4. $1 \times 10^3 + 3 \times 10^2 + 7 \times 10 + 4$
5. $1 \times 10^4 + 4 \times 10^3 + 8 \times 10^2 + 9 \times 10 + 1$
6. $7 \times 10^2 + 3 \times 10 + 5$
7. $4 \times 10^2 + 3 \times 10$
8. $6 \times 10^3 + 7$
9. $3 \times 10^3 + 2 \times 10^2 + 4$
10. $9 \times 10^4 + 6 \times 10^3 + 8 \times 10^2$

Exercise 18:3

1. $3 \times 5^3 + 2 \times 5^2 + 1 \times 5 + 4$
2. $1 \times 2^6 + 1 \times 2^4 + 1 \times 2^3 + 1 \times 2^2 + 1$
3. $5 \times 7^4 + 2 \times 7^3 + 3 \times 7^2 + 1$

4. $4 \times 6^4 + 3 \times 6^3 + 2 \times 6 + 1$
5. $5 \times 8^4 + 1 \times 8^3 + 6 \times 8^2 + 3 \times 8 + 2$
6. $2 \times 7^4 + 4 \times 7^3 + 1 \times 7^2 + 6 \times 7 + 2$
7. $2 \times 3^4 + 1 \times 3^2 + 2 \times 3 + 1$
8. $3 \times 8^4 + 4 \times 8^3 + 7 \times 8^2 + 2 \times 8 + 1$
9. $7 \times 9^3 + 4 \times 9^2 + 8 \times 9 + 1$
10. $1 \times 2^3 + 1 \times 2^2 + 1$
11. $4 \times 9^3 + 8 \times 9^2 + 3 \times 9 + 6$
12. $2 \times 4^3 + 3 \times 4^2 + 1 \times 4 + 1$
13. $4 \times 6^2 + 2 \times 6 + 1$
14. $7 \times 8^3 + 1 \times 8^2 + 4$
15. $1 \times 4^4 + 3 \times 4^3 + 2 \times 4^2 + 2 \times 4 + 1$
16. $2 \times 3^3 + 1 \times 3^2 + 2$
17. $1 \times 3^3 + 1 \times 3^2 + 2$
18. $4 \times 6^4 + 3 \times 6^3 + 2 \times 6^2 + 5 \times 6 + 1$

Exercise 18:4

(1) 47　(2) 59　(3) 167　(4) 961
(5) 412　(6) 93　(7) 1703　(8) 11635
(9) 27　(10) 2590　(11) 127　(12) 148
(13) 5041　(14) 1399　(15) 1846　(16) 353029

Exercise 18:5

1. 1100010101_{two}　　2. 11443_{eight}
3. 1244_{six}　　4. 233301_{four}
5. 1211222_{three}　　6. 656_{nine}
7. 1011100_{two}　　8. 102523_{six}
9. 1430_{eight}　　10. 10663_{seven}

Exercise 18:6

(i) 1. 65_{seven}　2. 33_{eight}　　3. 333020_{five}
4. 2131_{four}　5. 21112201_{three}　6. 145_{nine}
7. 5252_{seven}　8. 4051_{six}　　9. 1123111_{four}
10. 122_{seven}　11. 662_{eight}　　12. 501_{six}
13. 1201323_{four}　14. 113_{nine}　15. 12000021_{three}
(ii) (a) 742000 m　(b) 142221000_{five} m

Exercise 18:7

1. 1013_{four}　2. 715_{eight}　3. 2000_{three}　4. 435_{six}
5. 1242_{five}　6. 34_{five}　7. 1635_{eight}　8. 1100_{three}
9. 13_{four}　10. 12222_{three}　11. 32042_{five}　12. 5551_{eight}
13. 54606_{seven}　14. 63_{eight}　15. 115_{six}　16. 253_{seven}

Exercise 18:8

1. (a) 100010_{two}　(b) 111001_{two}　(c) 101110_{two}
   (d) 111111_{two}　(e) 1001000_{two}
2. (a) 53　(b) 27　(c) 109　(d) 85　(e) 29
3. (a) 1000_{two}　(b) 1000001_{two}　(c) 101100_{two}
   (d) 101001_{two}　(e) 11111_{two}　(f) 11101_{two}
   (g) 11110_{two}　(h) 100100110_{two}　(i) 10011100_{two}
   (j) 10_{two}　(k) 10_{two}　(l) 100_{two}

Exercise 19:1

1. (a) Tataw　(b) Njah　(c) Lum　(d) Abe
   (e) Feh　(f) Tata
2. No. (3,4) stand for column 3, row 4 while (4,3) stands for column 4, row 3.

Exercise 19:2

1.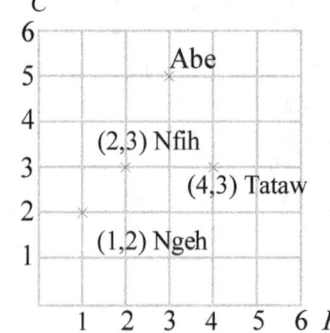

2.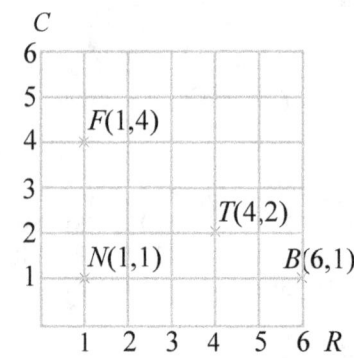

3. Bamenda (13,22), Limbe (8,8) Mbengwi (13,20), Mutengene (8,10), Kumba (8,12), Fundong (12,24), Nkambe (18,26), Buea (6,8), Nguti (8,14), Eyumojock (5,16)

4. (12,26) Wum, (15,22) Ndop, (10,16) Fontem, (6,8) Buea, (17,22) Kumbo.

Exercise 19:3

1.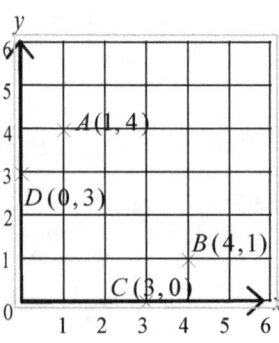

2.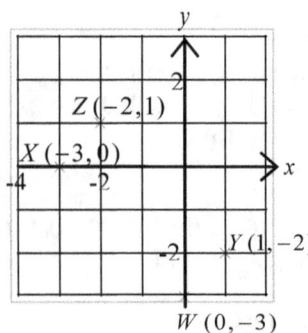

3.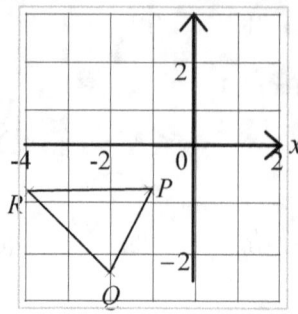

4. $A(-4,5)$, $B(3,3)$, $C(-3,0)$, $D(-5,-4)$, $E(6,-5)$

Exercise 19 :4

(a) 13 cm   (b) 10 cm   (c) 9 cm
(d) 8 cm   (e) 17 cm   (f) 45 cm
(g) 4 cm   (h) 4 cm   (i) 8 cm

Exercise 19 :5

1. (a) $\sqrt{2}$ (b) 13 (c) 5 (d) $\sqrt{205}$ (e) 2

2. $AB = 13$, $AC = 17$, $BC = 5\sqrt{2}$

3. $-3 \pm 3\sqrt{3}$ 4. $2x + y = 14$ 5. $a(t^2 + 1)$

6. $x^2 + y^2 - 6x - 8y + 9 = 0$

7. $k = -17$ 8. $-\frac{11}{2}$

Exercise 19:6

1. (a) $(-2, -2)$ (b) $\left(-\frac{15}{2}, -4\right)$

2. (a) $\left(x, \frac{y_2 + y_1}{2}\right)$ (b) $\left(\frac{x_2 + x_1}{2}, y\right)$

3. (a) $(8,1)$ (b) $\left(\frac{1}{2}, \frac{1}{2}\right)$ (c) $\left(-\frac{7}{2}, -\frac{7}{2}\right)$

(d) $\left(\frac{3}{4}, -\frac{1}{2}\right)$ (e) $\left(\frac{3\sqrt{2}}{2}, 2\sqrt{3}\right)$ 4. 13 units

Exercise 19 :7

1. $(1,5)$ 2. $\left(6, \frac{19}{2}\right)$ 3. $(5,1)$ 4. $(-16, 20)$

Exercise 19 :8

1. (a) $\frac{3}{7}$ (b) ∞ (c) 1 (d) $\frac{3}{4}$ (e) $\frac{8}{15}$ (f) $\frac{3}{4}$

2. $k = 11$ 3. $k = \pm 4$

4. (a) The points lie on a straight line

(b) At $(0,1)$ (c) $\left(-\frac{1}{2}, 0\right)$

(d) (i) 2   (ii) 2   (iii) 2

(e) The gradients are the same
This suggests that the gradient of a line can be

calculated using any two points on the line.

Exercise 19 :9

1. (a) $y = -\dfrac{2}{3}x + \dfrac{11}{3}$ (b) $13y = 15x - 1$

   (c) $4y = 26x + 29$ (d) $5y - x = 33$

   (e) $y = x + \sqrt{2}$

2. $y = -\dfrac{1}{3}x + 1$ 3. $5y = -2x + 11$

4. (a) $3y = x - 5$ (b) $y = -3x + 5$

5. (a) $7x = -3x + 1$ (b) $3y = 7x + 17$

6. $5y = -12x + 29$

7. (a) $(2,3)$ (b) $8 : 5$ (c) $3y = 5x - 33$

8. $y = -4x + 11$ 9. $b = \pm 4$

10. $n^2 = 9m^2$ 11. $an = mb$

Exercise 20:1

1. (a) pen (b) ink (c) London
    (d) i (e) white paper

2. (a) $x \in A$ (b) $y \notin B$ (c) $n(G) = 3$

3. (a) Correct because b is an element of F
    (b) Not correct because $\{b\}$ is a set and F is not a set of sets.

4. A is a set. A = {September, April, June, November}
   B is not a set. Beauty is relative.
   X is a set. Its members can be listed.
   G is not a set. Goodness is relative.
   P is a set. Its members can be listed.
   I is not a set. Not well defined.
   C is not a set. Members are of different families.

5. No. X is a set , but y is an element.

6. $M = \{2,4,6,8,10,12,14,16,18\}$
   $F = M = \{$spades, hearts, diamonds, clubs$\}$
   $V = \{a, e, i, o, u\}$
   $A = \{1,2,3,4,6,9,12,18,36\}$
   $N = \{2,3,4,5,6,7,8,9\}$

7. A = Odd numbers less than 14.
   B = Even numbers between 1 and 21.
   C = Fruits
   D = Adjectives
   E = Games

8. $X = \{x : x$ is a prime number less than 20$\}$
   $M = \{x : x$ is a factor of 42$\}$
   $C = \{x : x$ is a division in the North West Region of Cameroon$\}$
   $Y = \{x : x$ is a renowned river in Cameroon$\}$
   $T = \{x : x$ is a multiple of 3 less than 31$\}$

Exercise 20:2

1. (a) Trebleton (b) singleton (c) singleton
   (d) Doubleton (e) doubleton (f) Trebleton

(g) Doubleton (h) none of the above
(i) None of the above (j) empty set
(k) Empty set

2. (a) 7 (b) 2 (c) 0 (d) 1 (e) 1
   (f) count and state the number of students in your class.
   (g) 8 (h) 5 (i) 14 (j) 20

3. (a) infinite (b) finite (c) finite (d) Infinite
   (e) finite (f) infinite (g) Finite
   (h) finite (i) finite (j) Infinite

4. (a) $\notin$ (b) $\in$ (c) $\in$ (d) $\notin$ (e) $\in$ (f) $\notin$

Exercise 20:3

1. Yes

2. $M = \{2, 4, 6, 8, 10, 12, 14, 16, 18\}$
   $F = \{$Club, diamond, heart, spade$\}$
   $V = \{a, e, i, o, u\}$
   $A = \{1, 2, 3, 4, 6, 9, 12, 18, 36\}$
   $N = \{2, 3, 4, 5, 6, 7, 8, 9\}$

3. A = The set of Odd numbers less than 14
   B = The set of Even numbers between 1 and 21
   C = The set of Fruits
   D = The set of adjectives
   E = The set of sporting activities

4. $X = \{x : x$ is a prime number less than 20$\}$
   $M = \{x : x$ is a factor of 42$\}$
   $C = \{d : d$ is a division in the NWR of Cameroon$\}$
   $Y = \{r : r$ is a river in Cameroon$\}$
   $T = \{x : x$ is a multiple of 3 less than 31$\}$

Exercise 20:4

1. (a) $\subset$ (b) $\in$ (c) $\notin$ (d) = (e) $\neq$ (f) $\neq$

2. (a) False (b) False (c) False (d) True
   (e) False (f) True (g) False (h) True
   (i) True (j) False (k) False (l) False
   (m) False

3. $\{ \varnothing , \{1\}, \{2\}, \{3\}, \{1,2\}, \{1,3\}, \{2,3\}, P\}$.
   8 subsets

4. All are equal

5. $\{1, 2, 3, 4, 5\} \sim \{a, b, c, d, e\}$,
   $\{1, 2, 3, 4, 5\} = \{4, 2, 5, 2, 3\}$
   $\{2, 4, 6\} \sim \{$Biology, Chemistry, Mathematics$\}$

6. None

7. (a) False (b) True (c) False (d) False (e) False
   (f) False (g) True (h) True

8. (a) $a \in C$ (e) $e \notin A$ (f) $f \notin C$

9. $A = D,$ $B = C,$ $E = F$
   $A \sim B,$ $A \sim C,$ $A \sim E, A \sim F,$
   $B \sim A,$ $B \sim D,$ $B \sim E,$ $B \sim F,$
   $D \sim B,$ $D \sim C,$ $D \sim E,$ $D \sim F$
   $E \sim A,$ $E \sim B,$ $E \sim C,$ $E \sim D,$
   $F \sim A,$ $F \sim B,$ $F \sim C,$ $F \sim D$

Exercise 20:5

1. (i) $A \cup B$ (ii) $A \cap B'$ (iii) $(A \cup B)'$
   (iv) $A \cup B'$ (v) $(A \cap B) \cup (A \cup B)'$

(iii) 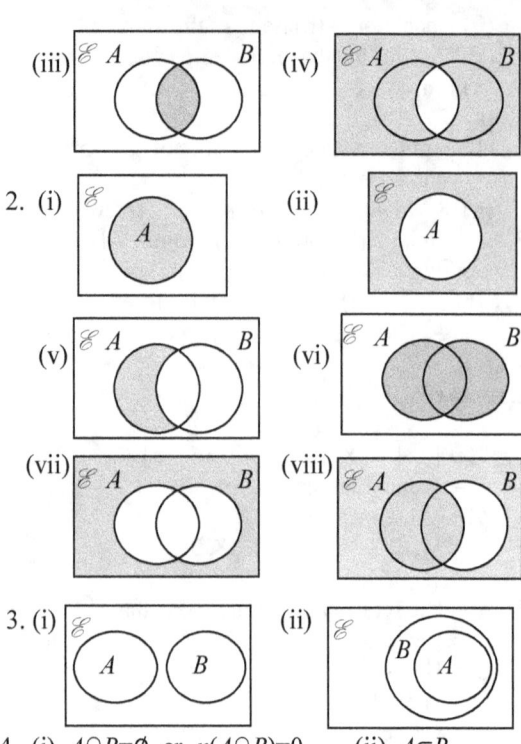 (iv)

2. (i) (ii)

(v) (vi)

(vii) (viii)

3. (i) 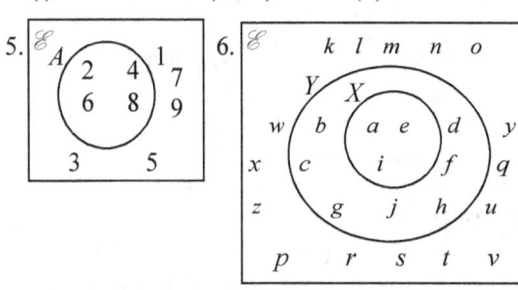 (ii)

4. (i) $A \cap B = \emptyset$ or $n(A \cap B) = 0$ (ii) $A \subset B$

5. 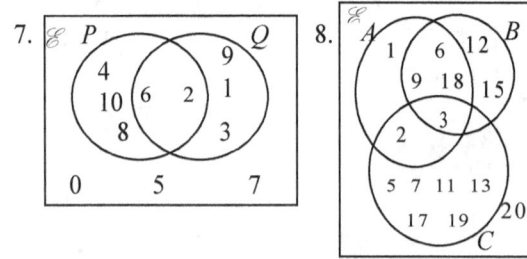 6.

7. 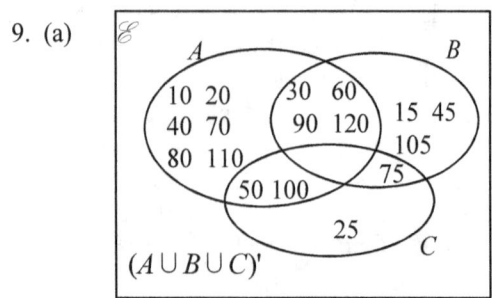 8.

(a) {3}   (b) {3,6,9,18}   (c) {3}   (d) {2,3}

9. (a)

$(A \cup B \cup C)'$

(b) $n(A \cup B) = 16$, $n(A \cap B \cap C) = 0$

Exercise 20:6

1.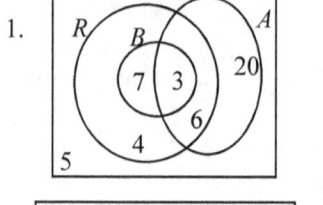
(a) 29
(b) 10
(c) 38
(d) 35

2.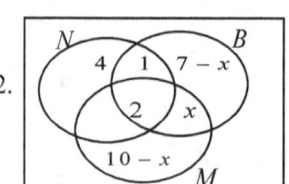
$n(N \cap B \cap M)' = 4$

3.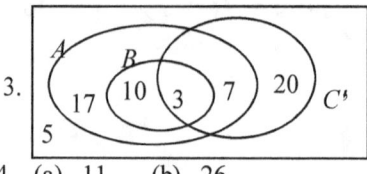
(a) 37
(b) 7
(c) 62

4. (a) 11   (b) 26

Exercise 21:1

1. (a) −23   (b) 3,−1
2. (a) and (b) are both commutative and associative.
   (c) and (d) are neither commutative nor associative.
3. (a) 2 (b) 3   (c) 1,   A is not closed since $1 \notin A$.
4. (a) 7 (b) 11   (c) 12   (d) 8   (e) 13   (f) 1
5. (a) 11 (b) 7   (c) 28   (d) 786 (e) 36

   R is closed under ∗. ∗ is neither commutative nor associative.
6. {(−1,4),(2,4),(5,4),(−1,1),(2,1),(5,1)}
7. (a) closure   (b) commutative property
   (c) associative property
   (d) identity element property
8. (a) closed since $\forall a, b \in S, a_* b \in S$. (b)(i) c (ii) a
9. (a) (i) $\dfrac{4}{x}$ (b)(ii) 0

10. (a)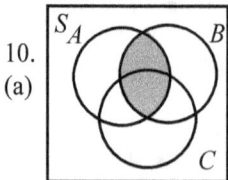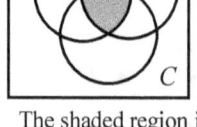

The shaded region is
$A * B = A \cap B$

The shaded region is
$(A*B)*C = (A \cap B) \cap C$

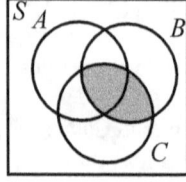   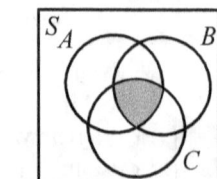

The shaded region is
$B * C = B \cap C$

The shaded region is
$A*(B*C) = A \cap (B \cap C)$

$\therefore (A \cap B) \cap C = A \cap (B \cap C)$

(a)
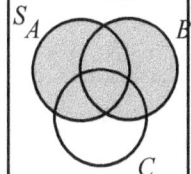
Shaded region is
$A*B = A\cup B$

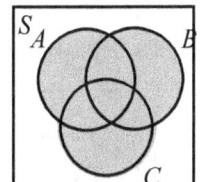
Whole Shaded region is
$(A*B)*C = (A\cup B)\cup C$

(b)
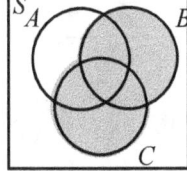
Shaded region is
$B*C = B\cup C$

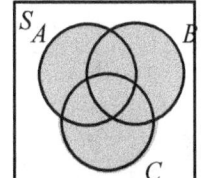
Whole Shaded region is
$A*(B*C) = A\cup(B\cup C)$

$$\therefore (A\cup B)\cup C = A\cup(B\cup C)$$

12. (a) 0   (b) $-\dfrac{4}{7}$

13. (a) Not closed   (b) Not commutative

14. (a)(i) 1   (ii) 2   (iii) 0   (b) No   (c) 0

15. $x = -12$

16. (a)(i) 2   (ii) 12   (iii) 0   (c) $x = -2$

Exercise 21:2

1.

*	1	2	3	4	5
1	1	2	3	4	5
2	2	4	0	2	4
3	3	0	3	0	3
4	4	2	0	4	2
5	5	4	3	2	1

$(S,*)$ is not a group.

2. (a)

*	0	1	2	3	4	5
0	0	1	2	3	4	5
1	1	2	3	4	5	0
2	2	3	4	5	0	1
3	3	4	5	0	1	2
4	4	5	0	1	2	3
5	5	0	1	2	3	4

(b) Yes, $(S,*)$ is an Abelian group.

3. (a)

$\otimes$	1	3	5	7
1	1	3	5	7
3	3	1	7	5
5	5	7	1	3
7	7	5	3	1

$\oplus$	1	3	5	7
1	2	4	6	0
3	4	6	0	2
5	6	0	2	4
7	0	2	4	6

(b) The set S is closed under $\otimes$, because no new element is introduced.

The set S is not closed under $\oplus$, because new elements $\{0,2,4,6\}$ are introduced.

(c) $e = 1$ (d) Every element is self inverse.

(e) The operation is commutative but not associative.

(f) The operation does not form a group because it is not associative.

Exercise 22:1

1. (a) True   (b) True   (c) True   (d) True
   (e) True   (f) True   (g) False   (h) False
   (i) False (j) True   (k) False

2. (a) $\{7\}$ (b) $\{8,9\}$ (c) $\{2,4,6,8\}$ (d) $\{5\}$ (e) $\varnothing$
   (f) $\{8\}$ (g) $\varnothing$ (h) $\varnothing$ (i) $\{5\}$
   (j) $\{1,2,3,4,6,7,8,9\}$(k) $\{7,8,9\}$ (l) $\{5,6,7,8,9\}$

3. (b) false, (c) false, (d) false, (g) false (h) true

4. (a) closed   (b) closed   (c) open   (d) closed
   (e) closed   (f) closed   (g) closed   (h) open
   (i) closed   (j) open   (k) open

5. (a) $\{6,7,8,9\}$ (b) $\{1,2,5\}$ (c) $\{1,3,5,7,9\}$
   (d) $\{2,4,6,8,10\}$ (e) $\{2,3,5,7\}$ (f) $\{3,6,7\}$

6. (a) T (b) T (c) F (d) F (e) T (f) F (g) T

Exercise 22:2

1.(a) It is not true that Mr. Fonche died two years ago.

(c) They are not lazy.

(d) Not all Bamenda people eat Achu.

(e) Loh can drive.

(f) Nigeria is not an African Country

(g) History is not a science subject.

(h) He was not the president of Cameroon.

(i) It is not true that everyone loves Mr. Paul Biya.

(j) She does not come from Nkambe.

(k) It is not true that Science has done more harm than good.

2. (a) $4y \not> 12$ (b) $3+5 \neq 9$ (c) $3p-1 \not> 17$
   (d) $4 \times 3 \neq 12$ (e) $6x+1 \not< 19$ (f) $2x-1 \neq 0$
   (g) $A \subset B$ (h) $A \cap B = \varnothing$

Exercise 22:3

1. (i) $\sim p$ (ii) $q \wedge p$ (iii) $\sim p \wedge \sim q$
   (iv) $p \vee q$ (v) $\sim (p \wedge q)$ (vi) $\sim (p \vee q)$

2. (a) Mrs. Ngwa visited me.
       Mrs. Tayong visited me.
   (b) Nfor likes rice.
       Nfor likes beans.
   (c) Mr. Nkwain is a Cameroonian.
       Mr. Nkwain is an ambassador.
   (d) Bamenda is a big city.
       Bafoussam is a big city.

3. (i) Nfor is not hungry.

(ii) Nfor is hungry and thirsty.

561

(iii) Nfor is hungry or thirsty.

(iv) If Nfor is hungry then he is thirsty.

(v) Nfor is hungry if and only if he is thirsty.

(vi) If Nfor is hungry then he is not thirsty.

(vii) Nfor is neither hungry nor thirsty.

(viii) Nfor is hungry if and only if he is not thirsty.

4. (a) If Fombe is rich then he is happy.
   (b) If he was drunk then he drank alcohol.
   (c) If it is night in Cameroon then places are dark.
   (d) If she performed well then she had a prize.

6. (a) Fombe is rich if and only if he is happy.
   (b) He was drunk is a necessary and sufficient condition that he drank alcohol.
   (c) It is night in Cameroon if and only if places are dark.
   (d) She performed well if and only if she had a prize.

6.

p	q	$p \to q$	$q \to p$	$(p \to q) \wedge (q \to p)$
T	T	T	T	T
T	F	F	F	F
F	T	T	F	F
F	F	F	T	F

p	q	$p \to q$	$q \to p$	$p \leftrightarrow q$
T	T	T	T	T
T	F	F	F	F
F	T	T	F	F
F	F	F	T	T

Therefore, $(p \to q) \wedge (q \to p) \equiv p \leftrightarrow q$ is false.

7.

p	q	$p \to q$	p	q	$\sim p$	$\sim p \vee q$
T	T	T	T	T	F	T
T	F	F	T	F	F	T
F	T	T	F	T	T	T
F	F	F	F	F	T	F

p	q	$\sim p$	$\sim p \vee q$
T	T	F	T
T	F	F	T
F	T	T	T
F	F	F	F

Therefore, $p \to q \not\equiv \sim p \vee q$.

8.

(a)

p	q	r	$p \wedge q$	$q \wedge r$	$(p \wedge q) \wedge r$	$p \wedge (q \wedge r)$
T	T	T	T	T	T	T
T	T	F	T	F	F	T
T	F	T	F	F	F	F
T	F	F	F	F	F	F
F	T	T	F	T	F	F
F	T	F	F	F	F	F
F	F	T	F	F	F	F
F	F	F	F	F	F	F

Therfore, $(p \wedge q) \wedge r \equiv p \wedge (q \wedge r)$

(b)

p	q	r	$q \wedge r$	$p \vee q$	$p \vee r$	$p \vee (q \wedge r)$	$(p \vee q) \wedge (p \vee r)$
T	T	T	T	T	T	T	T
T	T	F	F	T	T	T	T
T	F	T	F	T	T	T	T
T	F	F	F	T	T	T	T
F	T	T	T	T	T	T	T
F	T	F	F	T	F	F	F
F	F	T	F	F	T	F	F
F	F	F	F	F	F	F	F

Therfore, $p \vee (q \wedge r) \equiv (p \vee q) \wedge (p \vee r)$

9. (a)

p	q	$p \vee$	$p \wedge$	$(p \vee q) \to (p \wedge r)$
T	T	T	T	T
T	F	T	F	F
F	T	T	F	F
F	F	F	F	T

(b)

p	q	$p \vee q$	$\sim p$	$(p \vee q) \wedge \sim p$
T	T	T	F	F
T	F	T	F	F
F	T	T	T	T
F	F	F	T	F

(c)

p	q	$\sim q$	$p \wedge \sim q$	$p \to (p \wedge \sim q)$
T	T	F	F	F
T	F	T	T	T
F	T	F	F	T
F	F	T	F	T

(d)

p	q	$p \to q$	$(p \to q) \wedge q$
T	T	T	T
T	F	F	F
F	T	T	T
F	F	T	F

(e)

p	q	$p \rightarrow q$	$\sim p$	$(p \rightarrow q) \rightarrow \sim p$
T	T	T	F	F
T	F	F	F	T
F	T	T	T	T
F	F	T	T	T

(f)

p	q	$p \rightarrow q$	$p \wedge (p \rightarrow q)$
T	T	T	T
T	F	F	F
F	T	T	F
F	F	T	F

10. (b)

11.

A	B	$A \vee B$
T	T	T
T	F	T
F	T	T
F	F	F

12.

S	P	$S \wedge P$
T	T	T
T	F	F
F	T	F
F	F	F

Exercise 22:4

1. $\exists! 2 \in E, p(x)$
2. $\forall x \in D, p(x)$
3. $\exists v \in F, p(x)$
4. $\not\exists x \in S, p(x)$
5. $\exists! 0 \in \mathbb{R}$
6. $\forall s \in F, p(x)$
7. $\exists p \in R, s(x)$
8. $\not\exists x \in \mathbb{R}$

Exercise 22:5

1. r: Ngoh is a liar.
2. r: Some polygon are rectangles.
3. r: x is not in the range $4 \le x \le 7$ or $x \notin 4 \le x \le 7$
4. r: Nfor is never happy.
5. r: Bamenda is in Europe.
6. r: 18 is a multiple of 12.
7. r: Each angle in the quadrilateral $ABCD$ is a right angle.
8. r: $ABCD$ does not have equal sides.
9. r: Bih is not beautiful.
10. r: $A = B$.

Exercise 23:1

1. (a) $D = \{0, 1, 2, 3\}$, $R = \{0, 3, 6, 9\}$

   (b) $D = \{0, 1, 4\}$, $R = \{-2, -1, 0, 1, 2\}$

   (c) $D = \{1, 2, 3\}$, $R = \{15, 20, 25\}$

2. (a) $\{(1, 2), (2, 3), (3, 4), (4, 5), (5, 6)\}$

   (b) $\{(1, 3), (2, 2), (3, 1), (4, 0), (5, -1)\}$

   (c) $\{(1, -1), (2, -2), (3, 0), (4, 2), (5, 4)\}$

   (d) $\{(1, 2), (2, 5), (3, 10), (4, 17), (5, 26)\}$

3. (a) $\{(-4, 4), (-2, 2), (0, 0), (2, -2), (4, -4)\}$

   (b) $\{(-4, -8), (-2, 4), (0, 0), (2, 4), (4, 8)\}$

   (c) $\{(-4, 2), (-2, 0), (0, -2), (2, 0), (4, 2)\}$

   (d) $\{(-4, -3), (-2, 1), (0, 1), (2, 3), (4, 5)\}$

4.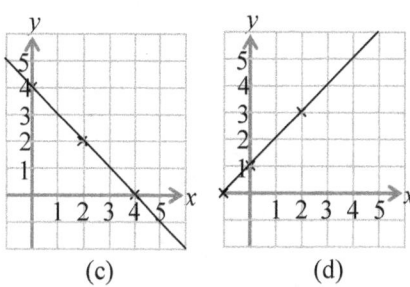

5.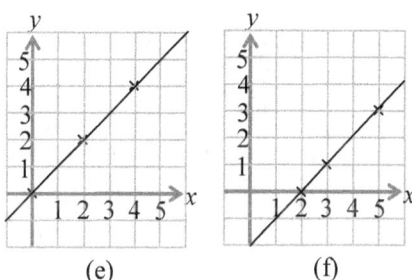

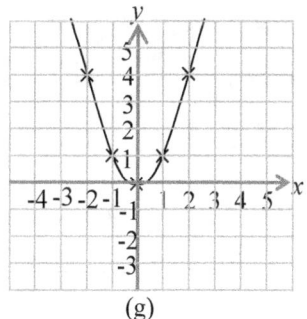

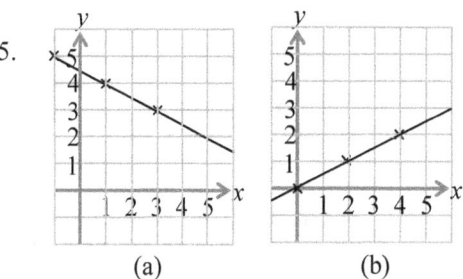

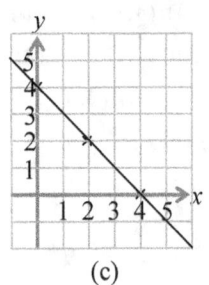
(c)

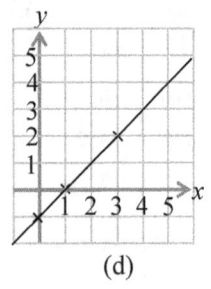
(d)

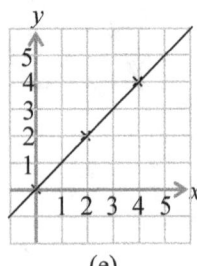
(e)

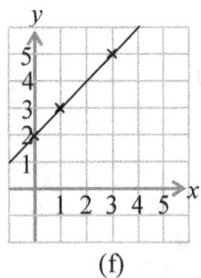
(f)

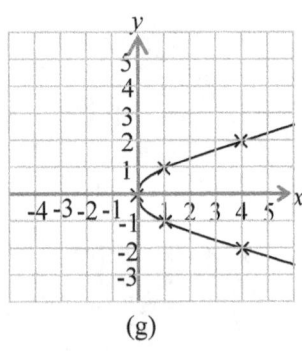
(g)

6. {(1,2),(1,4),(1,6),(3,2),(3,4),(3,6),(5,2),(5,4),(5,6)}

7. (a) {(−2,10),(−2,20),(1,10),(1,20),(4,10),(4,20)}

   (b) {(10,−2),(10,1),(10,4),(20,−2),(20,1),(20,4)}

   (c) {(−2,−2),(−2,1),(−2,4),(1,−2),(1,1),

       (1,4,),(4,−2),(4,1),(4,4)}

   (d) {(10,10),(10,20),(20,10),(20,20)}

8.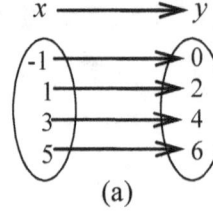
   (a)

   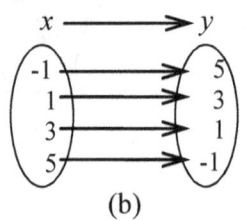
   (b)

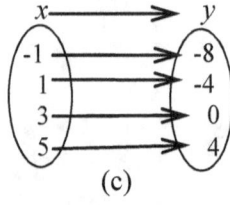
(c)

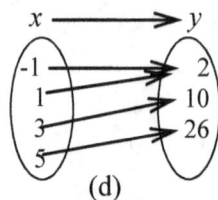
(d)

9. (a) 'is a husband of'   (b) 'is a factor of'
   (c) 'is a student of'   (d) 'is one fifth of'
   (e) 'is two less than thrice'

---

10. (a) Any number is a factor of itself.
    (b) $u = 4, v = 36, w = 12, x = 16$

Exercise 23:2

1. (i)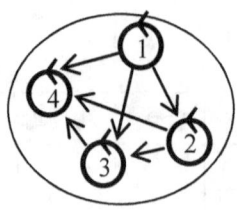

   (ii) $\forall x, y \in A, xRy \not\Leftrightarrow yRx$ i.e. the relation is not symmetric.

2. (i)

Reflexive	(a), (b), (c), (d), (f), (g)
Symmetric	(e), (f), (g), (h)
Transitive	(a), (b), (c), (d), (f), (g)
Anti-symmetric	(a), (b), (c), (d)

   (ii) (f) and (g)

Exercise 23:3

1. (a), (b), (f), (g), (h), (i)

2. (a)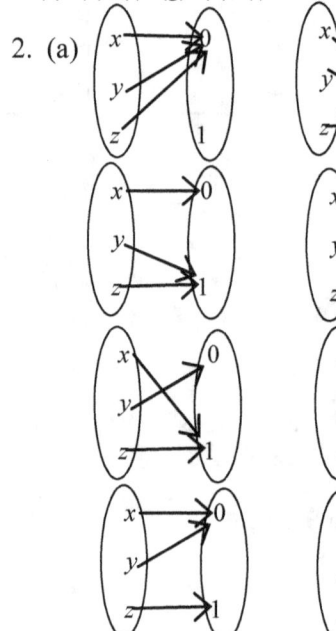

   (b)   8

3.(i)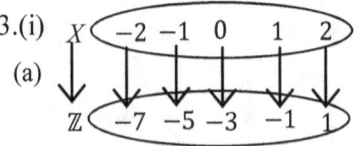
   (a)

(b)
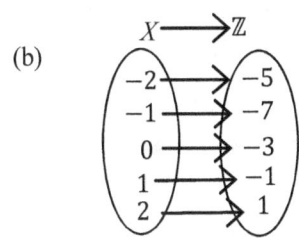

(c)
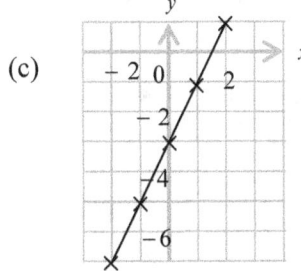

(ii) Domain = X, Range = $\{-7,-5,-3,-1,1\}$

Exercise 23:4

1. (a) (ii)   (b) (i), (iii)   (c) none   (d) all
2. (i) many-one   (ii) one-one
   (iii) many-one   (iv) one-one
3. (i) none   (ii) all
4. (a) (i) range = $\{2,3,5\}$
   (ii) f is an 'into' but not an 'onto'
5. injections: None
   Surjections: (a), (b), (f), (g), (h)
   Bijections: (i)
6. (a) Not a function   (b) Function, surjective
   (c) Function, surjective (d) Function, surjective
   (e) Not a function   (f) Function, surjective

Exercise 23:5

1. (a) a (b) $R = \{2,3,5\}$
2. (i) $f^{-1}:x \longmapsto \dfrac{x-1}{2}$ (ii) $\dfrac{4}{3}$ 3. 0 or $\dfrac{2}{3}$
4. (a) $gf:x \longmapsto \dfrac{6x-1}{3}$ (b) $g^{-1}:x \longmapsto \dfrac{3x-1}{2}$
5. (i) 1   (ii) ± 3
6. (i) 2   (ii) $x = -1$ or -2
7. (a) $\{10,0,-2\}$ (b) $D = \{3,0,1\}$
8. (a) -5 (b) ± 2
9. (a) 1   (b) $f \circ g(x) = x^2 + 4x + 1$
   (c) $g^{-1}:x \longmapsto x - 2$ (d) $p = 3, q = -1$
10. (a) $f^{-1}:x \longmapsto x - 2$
    (b) $g(\sqrt{2}) = 1,\ hf(0) = 28$
    (c) $x = 0$ or $x = \dfrac{2}{7}$
11. (i) $gf(x) = 3x^2 - 18$
    (ii) $9x^2 - 36x + 50 = 0$
    (iii) $x = 0$ or $x = 4$
12. (i) Domain = $\mathbb{R} - \{2\}$, Range = $\mathbb{R} - \{-1\}$

(ii) $g^{-1}:x \longmapsto \dfrac{x+1}{2}$ (iii) $h:x \longmapsto \dfrac{x+2}{x-1}$

13. (a) $pr(x) = 4x^2 + 1$
    (b) $rp(x) = 2x^2 + 2$
    (c) $x = 1$ or $x = 2$ (d) $x \neq 1$
    (e) $p^{-1}:x \longmapsto \pm\sqrt{x-1}$
        $q^{-1}:x \longmapsto 3(x - 1)$
14. (a) -1 (b) $f^{-1}:x \longmapsto 1 - x$
    (c) $\dfrac{1 \pm \sqrt{5}}{2}$
15. (a) $\mathbb{R} - \{-3,3\}$
    (b) $\{x: x \leq -2$ or $x \geq 2, x \in \mathbb{R}\}$ (c) $\mathbb{R}$
    (d) $\mathbb{R} - \{0\}$ (e) $\mathbb{R} - \{-3,2\}$ (f) $\mathbb{R}$

Exercise 24 :1

(1) Even numbers; 10,12,14
(2) Odd multiples of 3; 27,33,39
(3) Consecutive square numbers; 25,36,49
(4) Adding 3 to preceding term; 20,2326
(5) Dividing preceding term by 2; 10,5, $\dfrac{5}{2}$
(6) Adding 1,2,3,4...; 11,16,22
(7) Dividing preceding term by 3; 1, $\dfrac{1}{3}$, $\dfrac{1}{9}$
(8) Adding 3 to preceding term; 1, $\dfrac{1}{3}$, $\dfrac{1}{9}$
(9) Adding 10 to preceding term; 42,52, 62
(10) Subtracting 4 from preceding term; 25,21,17
(11) Adding 10 and subtracting 3 alternately; 17,27,24
(12) Subtracting 4 from preceding term; $-16,-20,-24$
(13) Consecutive even numbers to the preceding term; 33,45,59
(14) Adding 4 to preceding term; 19, 23, 27

Exercise 24:2

1. (a) $3(2^{n-1})$ (b) $3n + 2$ (c) n^3
   (d) $10(4^{n-1})$ (e) 2^{1-n} (f) $\dfrac{n}{n+1}$
2. (a) $5,12,21$ (b) $2,6,12$
   (c) $\dfrac{11}{10}, \dfrac{7}{10}, \dfrac{19}{30}$ (d) $-4,-6,-6$ (e) $6,12,24$
3. $2,6,12,20;\ U_n = n(n+1)$
4. $4,6,8;\ U_n = 2(n+1)$
5. $S_1 = 1,\ S_2 = 5, S_3 = 14,$
   $S_4 = 30, S_5 = 55$
   $U_1 = 1,\ U_2 = 4, U_3 = 9,$
   $U_4 = 16, U_5 = 25$
6. $1,3,6;\ U_n = n$

7. (a) 31, 41; adding consecutive even numbers

(b) $\dfrac{1}{720}, \dfrac{1}{4840}$ Multiply the preceding term by

$\dfrac{1}{2}, \dfrac{1}{3}, \dfrac{1}{4}, \dfrac{1}{5}, \dfrac{1}{6}$, respectively.

8. 8, 3, 0, -1, 0

9. 16

10. (a) 5, 11, 17 (b) 456

11. (a) -10, -6, -2, 2, 6 (b) 110

Exercise 24:3

1. 145   2. $a = -27, U_{28} = 27$ 3. 24

4. $5, \dfrac{7}{3}, -\dfrac{1}{3}, -3 \cdots$ 167 (b) $\dfrac{14}{83}$ (c) 11

6. 13, 17, 21, 25   7. 210   8. 10

9. 16, 13, 10, 7, 4 or 4, 7, 10, 13, 16   10. 12

11. $\dfrac{3a+b}{4}, \dfrac{a+b}{2}, \dfrac{a+3b}{4}$

Exercise 24:4

1. (a) $-\dfrac{12}{7}, 3$ (b) $r = \dfrac{3}{2}, U_4 = \dfrac{81}{4}$

2. $r = -\dfrac{3}{2}, U_n = 4\left(-\dfrac{3}{2}\right)^{n-1}$ 3. $\dfrac{3}{2}$

4. $U_{12} = \dfrac{29296875}{2048}, S_{12} = \dfrac{244136529}{102040}$

5. (a) ± 12 (b) ± 27 (c) $\pm x^3$

6. (a) 4, 8, 16 (b) $4, 1, \dfrac{1}{4}$ 7. (a) 1023 (b) -340

(c) $\dfrac{2186}{243}$ (d) $\dfrac{(3a)^{10} - 1}{3a - 1}$

8. (a) $-\dfrac{7}{6}$ (b) $\dfrac{1}{2}$

(c) A.P.: $U_2 = \dfrac{17}{6}, U_5 = -\dfrac{2}{3}$,

G.P.: $U_2 = 2, U_5 = \dfrac{1}{4}$ (d) 7

Exercise 25:1

4. (a) The length of a line AB.
(b) A ray beginning from B and passing through A
(c) A ray beginning from A and passing through B.
(d) A line passing through the points A and B.
(e) A line segment with end points A and B.
(f) A line segment with end points A and B.

5. (a) 4 unit (b) 12 units (c) 8 units
(d) 16 units (e) 12 units

6. (a) [AB); A ray from A, passing through B.
(b) [PQ]; A line segment with end points P and Q.
(c) (LM); A line passing through the points L and M
(d) (XY]; A ray from Y, passing through X.

7. (a) 10 (b) 6 (c) 3 (d) 12 (e) 5 (f) 8 (g) 6 (h) 8

Exercise 25:2

(1)(a) 36° (b) 30° (c) 15° (d) 30° (e) 29° (f) 85°
(2) 30°   (3) 22.5
(4) $p + q + r + t = 360°, p = 172.8°, q = 86.4°$,
    $r = 57.6°, s = 43.2°$
(5) (a) $90° < x < 180°$ (b) $180° < x < 360°$

Exercise 25:3

(1) (a) $a = 100°, b = 80°$ (b) $x = 60°, y = 43°$
(2) $a + b + c = 180°$

Exercise 25:4

(1) $x = 60°, y = 120°$ (2) $a = 82°$ (3) $r = 220°$
(4) $b = 60°$ (5) $x = 60°, y = 35°$

Exercise 26:1

1.

Property	equilateral	equiangular	regular	irregular	convex	Re-entrant
All the sides are equal	x	x	x			
All the angles are equal		x	x			
All the angles are less than 180°	x		x	x	x	
At least one angle is greater than 180°						x

(Column header title: Name of Polygon)

2. (a) quadrilateral (b) Rectangle (c) Hexagon
(d) Triangle   (e) Pentagon

Exercises 26:2

1. (i) $S \subset R_e, S \subset R_h, S \subset P, S \subset K, R_e \subset P$,
    $R_h \subset P, R_h \subset k$
(ii) See Figure 26:15
(iii) (a) $\{R_h, S\}$ (b) {all irregular rectangles, T}
2. $x = y = 110°$
3. (a){square}   (b) {rectangle, trapezium, kite}
4. (a) trapezium (b) rectangle (c) parallelogram
(d) rhombus   (e) square

5.

	Property	Square	Rectangle	Rhombus	Parallelogram	Trapezium
a	All the sides are equal	Y	N	Y	N	N
b	All the diagonals are equal	Y	Y	N	N	N
c	Diagonals bisect each other	Y	Y	Y	Y	N
d	Diagonals are perpendicular	Y	N	Y	N	N
e	Diagonals bisect opposite angles	Y	Y	Y	N	N
f	Adjacent sides are equal	Y	N	Y	N	N
g	Opposite sides are equal and parallel	Y	Y	Y	Y	N
h	Only two sides are parallel	N	N	N	N	Y
i	Adjacent sides are perpendicular	Y	Y	N	N	N

6. (i) {square, rhombus}   (ii) {square, rectangle}

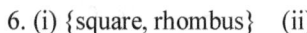

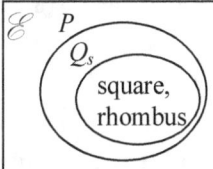
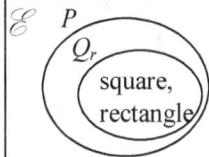

(iii) {square}

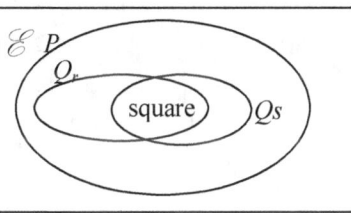

Exercise 26:3
1. (ii) A and I (iii) Q, R and S
2. (a) Right-angled triangle
   (b) Obtuse-angled triangle
   (c) Acute-angled triangle
3. (a) Acute-angled isosceles triangle
   (b) Right-angled triangle
   (c) Acute-angled scalene triangle.
4. (a) Scalene triangle   (b) isosceles triangle
   (c) Equilateral triangle (d) Equilateral triangle
   (e) Scalene triangle   (f) isosceles triangle

Exercise 26:4
1. (i) (a) chord (b) radius (c) diameter (d) minor arc
       (e) secant (f) tangent (g) major arc
   (ii) (a) semi-circle (b) minor sector
        (c) minor segment
2. (a) Minor arc (b) radius   (c) isosceles
   (d) chord   (e) Minor sector
3. (a) Minor segment (b) Minor arc (c) radius
   (d) diameter   (e) Chord   (f) Minor sector

Exercise 27:1
1. (a) $x = 70°$ (b) $x = 100°, y = 40°$
   (c) $x = 20°, y = 44°$ (d) $x = 80°, y = 40°$
   (e) $x = 72°, y = 54°$ (f) $x = 75°, y = 30°$
   (g) $p = 115°$

2(c)

No of sides of polygon	No of triangles insides polygon	Sum of interior angles
3	1	180°
4	2	360°
5	3	540°
6	4	720°
7	5	900°
8	6	1080°

(d)(i) 7 (ii) 1260° (iii) $n-2$ (iv) $(n-2)180°$
   (v) 360°

3(c)

No of sides of polygon	No of triangles insides polygon	Sum of interior angles
3	3	180°
4	4	360°
5	5	540°
6	6	720°
7	7	900°
8	8	1080°

(d)(i) 9 (ii) 1620° (iii) 360° (iv) 1260°
   (v) n (vi) $180n$ (vii) $(n-2)180°$ (viii) 360°

Exercise 27:2
(1) 89°   (2) 40°   (3) 6   (4) 80
(5) 12   (6) 12   (7) 144°   (8) 12
(9) (a) Yes, 18   (b) No   (c) Yes, 24.
(10) 10   (11) 108°   (12) $\angle AFC = 45°$
(13) 12, $\angle ACD = 135°$
(14) (a) $\angle ABC = 140°$, $\angle CAE = 40°$, $\angle ACG = 60°$
     (b) 100°
(15) $\angle PQR = 135°$, $\angle RSP = 45°$ (16) 60°

Exercise 28:1
1. 10 cm   2. 4 cm   3. 41 cm   4. 18 cm
5. 12 cm and 16 cm   6. (a), (c), (e), (h)

Exercise 28:2
1. (a) $\dfrac{12}{5}$ (b) $\dfrac{12}{13}$ (c) $\dfrac{5}{13}$ (d) $\dfrac{5}{12}$ (e) $\dfrac{5}{13}$ (f) $\dfrac{12}{13}$

(ii) (a) $\dfrac{40}{9}$ (b) $\dfrac{40}{41}$ (c) $\dfrac{9}{41}$ (d) $\dfrac{9}{40}$ (e) $\dfrac{9}{41}$ (f) $\dfrac{40}{41}$

2. (a) $\frac{3}{4}$ (b) $\frac{4}{5}$ (c) $\frac{3}{5}$ (d) $\frac{15}{8}$ (e) $\frac{8}{17}$ (f) $\frac{15}{17}$

3. (a) $\frac{7}{25}$ (b) $\frac{24}{25}$ 4. (i) $\frac{\sqrt{3}}{2}$ (ii) $\frac{1}{2}$ (iii) $\sqrt{3}$

5. (i) $\frac{\sqrt{2}}{2}$ (ii) $\frac{\sqrt{2}}{2}$ (iii) 1

6. $\cos x = \frac{\sqrt{n^2 - m^2}}{n}$, $\tan x = \frac{m}{\sqrt{n^2 - m^2}} = \frac{m\sqrt{n^2 - m^2}}{n^2 - m^2}$

7. (a) $\frac{3}{5}$ (b) $\frac{3}{4}$

8. (a) $\frac{4}{5}$ (b) $\frac{3}{5}$ (c) $\frac{3}{4}$ (d) $\frac{3}{5}$ (e) $\frac{4}{5}$ (f) $\frac{4}{3}$

(g) $\frac{5}{13}$ (h) $\frac{12}{13}$ (i) $\frac{12}{5}$ (j) $\frac{12}{13}$ (k) $\frac{5}{13}$ (l) $\frac{5}{12}$

Exercise 28:3
1. (a) 0.5592 (b) 0.9527 (c) 0.9886
   (d) 0.4540 (e) 0.5793 (f) 0.4428
   (g) 1.0355 (h) 1.0392 (i) 7.3639
2. (a) $9°$ (b) $66°$ (c) $81°7^1$ (d) $53°19^1$
   (e) $78°55^1$ (f) $47°43^1$ (g) $2°52^1$ (h) $15°40^1$
3. (a) $24°$ (b) $70°$ (c) $14°42'$
   (d) $64°36'$ (e) $16°32'$ (f) $89°29'$
   (g) $74°52'$ (h) $61°56'$
4. (a) $24°$ (b) $73°$ (c) $4°24'$ (d) $21°41'$
   (e) $19°37'$ (f) $62°52'$ (g) $0°56'$ (h) $39°34'$

Exercise 28:4
1. $\sin \alpha = 0.36$ and $\tan \alpha = 0.5624$
2. $\sin x = \frac{12}{13}$, $\tan x = \frac{5}{12}$ 3. $\sin A = \frac{4}{5}$, $\cos A = \frac{3}{5}$

Exercise 28:5
1. (i) $18°$ (ii) $65°$ (iii) $46.5°$
   (iv) $90°$ (v) $67.5°$ (vi) $90°$ 2. 0.5736
3. (a) $38°$ (b) $67°$ (c) $50°$ (d) $20°$
4. (a) $30°$ (b) $56°$ (c) $45°$ (d) $0°$ (e) $90°$
5. 0.7431 6. 0.4540
7. Any two positive numbers x and y such that
   $x + y = 90°$

Exercise 28:6
1. (i) (a) $\frac{5}{12}$ (b) $\frac{12}{13}$ (c) $\frac{5}{13}$ (d) $\frac{12}{5}$ (e) $\frac{5}{13}$ (f) $\frac{12}{13}$

(ii) (a) $\frac{9}{40}$ (b) $\frac{40}{41}$ (c) $\frac{9}{41}$ (d) $\frac{40}{9}$ (e) $\frac{9}{41}$ (f) $\frac{40}{41}$

2. (a) $\frac{4}{3}$ (b) $\frac{4}{5}$ (c) $\frac{3}{5}$ (d) $\frac{8}{15}$ (e) $\frac{8}{17}$ (f) $\frac{15}{17}$

(ii) (a) $\frac{3}{4}$ (b) $\frac{4}{5}$ (c) $\frac{3}{5}$ (d) $\frac{15}{8}$ (e) $\frac{8}{17}$ (f) $\frac{15}{17}$

3. (a) $\frac{7}{25}$ (b) $\frac{24}{25}$ 4. (i) $\frac{\sqrt{3}}{2}$ (ii) $\frac{1}{2}$ (iii) $\frac{\sqrt{3}}{3}$

5. (i) $\frac{\sqrt{2}}{2}$ (ii) $\frac{\sqrt{2}}{2}$ (iii) 1

6. $\sec x = \frac{\sqrt{n^2 - m^2}}{n}$, $\cot x = \frac{\sqrt{n^2 - m^2}}{m}$

7. (a) $\frac{3}{5}$ (b) $\frac{4}{3}$

8. (a) $\frac{4}{5}$ (b) $\frac{3}{5}$ (c) $\frac{4}{3}$ (d) $\frac{3}{5}$ (e) $\frac{4}{5}$ (f) $\frac{3}{4}$

(g) $\frac{5}{13}$ (h) $\frac{12}{13}$ (i) $\frac{5}{12}$ (j) $\frac{12}{13}$ (k) $\frac{5}{13}$ (l) $\frac{12}{5}$

Exercise 28:7
1. (a) 1.7883 (b) 1.0497 (c) 1.0125
   (d) 1.6243 (e) 1.7263 (f) 2.2583
   (g) 0.9004 (h) 1.9542 (i) 0.1358
2. (a) $38°$ (b) $16°12'$ (c) $67°46'$
   (d) $9°40'$ (e) $81°44'$ (f) $22°23'$
   (g) $42°42'$ (h) $13°50'$
3. (a) $34°$ (b) $19°12'$ (c) $76°46'$
   (d) $4°40'$ (e) $61°44'$ (f) $27°23'$
   (g) $31°42'$ (h) $84°50'$
4. (a) $37°$ (b) $22°12'$ (c) $62°46'$
   (d) $8°40'$ (e) $63°44'$ (f) $24°23'$
   (g) $33°42'$ (h) $82°50'$

Exercise 28:8
1. (a) $\frac{5}{4}$ (b) $-\frac{1}{4}$ (c) 1
   (d) $\frac{3}{2}$ (e) $1 - \sqrt{2}$ (f) 1 (g) $\frac{4}{3}$
   (h) 1 (i) 1 (j) 7
2. (a) 1 (b) $\sqrt{3}$

Exercise 28:9
1. (a) $\frac{1}{2}$ (b) $\frac{\sqrt{3}}{3}$ (c) $\sqrt{3}$
   (d) -1 (e) $\frac{\sqrt{2}}{2}$ (f) $\sqrt{3}$
2. (a) -0.1736 (b) 1.4281 (c) -1.4281
   (d) -0.9848 (e) 0.342 (f) -0.9848
3. (a) $-\frac{2\sqrt{3}}{3}$ (b) $\sqrt{3}$ (c) $-\frac{\sqrt{3}}{3}$
   (d) ∞ (e) $-\sqrt{2}$ (f) 2
4. (a) 1.0154 (b) 0.7002 (c) -0.7002
   (d) -5.7588 (e) -1.0642 (f) 5.7588

Exercise 28:10
1. (a) $\frac{1}{2}$ (b) $-\frac{\sqrt{3}}{3}$ (c) $\sqrt{3}$

(d) 1   (e) $-\dfrac{\sqrt{2}}{2}$ (f) $-\dfrac{\sqrt{3}}{2}$

2. (a) -0.1736 (b) 1   (c) 1.4281
   (d) 0.9848   (e) -0.3420 (f) -0.9848
3. (a) -0.1736 (b) 1   (c) 1.4281
   (d) 0.9848   (e) -0.3420 (f) -0.9848
4. (a) $\dfrac{2\sqrt{3}}{3}$ (b) $-\sqrt{3}$ (c) $\dfrac{\sqrt{3}}{3}$

   (d) ∞ (e) $-\sqrt{2}$ (f) -2
5. (a) -1.0154 (b) -0.7002 (c) 0.7002
   (d) -5.7588 (e) -1.0642 (f) -5.7588
6. (a) -1.0154 (b) -0.7002 (c) 0.7002
   (d) -5.7588 (e) -1.0642 (f) -5.7588

Exercise 29:1
1. (a) $AB = 15.59$ cm, $BC = 9$ cm
   (b) $AB = 15.05$ cm, $BC = 17.04$ cm
2. (a) $90°, 22.06°, 67.4°$ (b) $90°, 28.1°, 61.9°$
   (c) $90°, 16.3°, 73.7°$ (d) $90°, 79.6°, 10.4°$
3. 8 m   4. 3 m   5. 11 m
6. (a) 6 m   (b) $120°$ (c)
7. (a) 1124 m   (b) 2782 m

Exercise 29:2
1. (a) $\sqrt{7}$, $\angle X = 41.4°$, $\angle Y = 48.6°$
   (b) $\sqrt{1649}$, $\angle X = 52°$, $\angle Y = 38°$
   (c) 9, $\angle X = 53.1°$, $\angle Y = 36.9°$
   (d) $\sqrt{111}$, $\angle X = 31.8°$, $\angle Y = 58.2°$
   (e) 3, $\angle X = 36.9°$, $\angle Y = 53.1°$
   (f) $3\sqrt{231}$, $\angle X = 62.9°$, $\angle Y = 27.1°$
2. $40°$ 3. $3\sqrt{2}$ cm , $\angle X = \angle Z = 45°$

Exercise 29:3
1. 6.57 cm
2. $4\sqrt{2}$ cm , $\angle ABC = 38°$, $\angle BAC = BCA = 70.5°$
3. (a) $\sqrt{91}$ cm, $\angle A = \angle C = 72.5°$, $\angle B = 34.9°$
   (b) $\sqrt{7}$ cm, $\angle A = \angle C = 41.4°$, $\angle B = 97.2°$
   (c) $\dfrac{\sqrt{171}}{2}$ cm, $\angle A = \angle C = 69.1°$, $\angle B = 41.8°$
4. (a) 21 cm   (b) 79 cm
5. $\angle BAC = 36.4°$, $\angle ABC = ACB = 71.8°$

Exercise 29:4
1. (a) 18 m (b) 47 m   2. (a) $13\sqrt{3}$ m (b) 26 m
3. 8 m   4. (a) 69 m   (b) 57 m   (c) 224 m
5. (a) 23 m (b) 39 m   (c) 4 m   6. 82 m
7. (a) 32 m   (b) 35 m   8. 61.4 m

Exercise 29:5
1. (a) 5 km   (b) $76°$ 2. (a) $\sqrt{13}$ km (b) $73.3°$
3. (a) 71 km   (b) 92 km
4. $BC = 7.8$ cm, $AC = 4.8$ cm   2. 6.7 km, $061°$
5. $\sqrt{19}$ km, $36.6°$ 6. (a) 185.2 km (b) $071°$

Exercise 29:6
1. $AC = 4.8$ km, $BC = 7.8$ km
2. Distance = 6.7 km, bearing = $61.1°$
3. Distance = 4.4 km, bearing = $36.7°$
4. Distance = 185.2 km, bearing = $71.1°$
5. (a) $120°$ (b) (i) 15.6 km   (ii) $76.3°$

Exercise 30
1. (i) 1   (ii) 6   (iii) 2   (iv) 3
   (v) 5   (vi) 6   (vii) 4
2. (i) BC (ii) C, F and G
   (iii) 9 edges, 5 vertices
   (iv) Equilateral triangular hexahedron

3.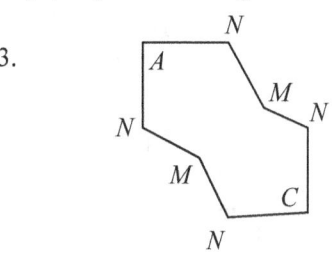

4. (a) 2 triangles, 3 rectangles

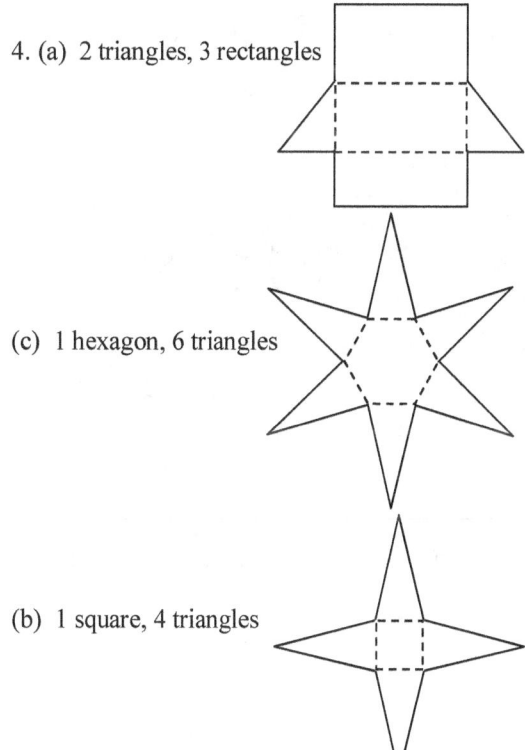

(c) 1 hexagon, 6 triangles

(b) 1 square, 4 triangles

5. A triangular prism has two triangular bases,
while a triangular pyramid has only one. The
lateral faces of a triangular prism are rectangles,

while the lateral faces of a triangular pyramid are triangles. A triangular pyramid has a vertex but a triangular prism does not.

6. (a) cone   (b) rectangular pyramid
7. (a) Tetrahedron   (b) octahedron
    (c) Dodecahedron   (d) Icosahedron

Exercise 31:1
1. (i) (a) 1750 m^2 (b) 170 m
    (ii) (a) 5400 m^2 (b) 300 m
    (iii) (a) 512 m^2 (b) 96 m
    (iv) (a) 3000 m^2 (b) 70 m
2. 4   3. 26.3   4. $45\,000 \text{ m}^2$ 5. 5   6. 15 m, 54 m
7. 11 cm   8. 26 m   9. $(21 - x)$ cm
10. (a) $x = 2.5$ (b) 73 cm^2 (c) 40 cm
11. (i) (a) 2601 cm^2 (b) 204 cm
    (ii) (a) 144 cm^2 (b) 48 cm
    (iii) (a) 225 cm^2 (b) 60 cm
    (iv) (a) 900 cm^2 (b) 120 cm
12. 56 cm   13. (a) 289 cm^2 (b) 68 cm
14. (a) $\dfrac{5\sqrt{2}}{2}$ cm   (b) $30\sqrt{2}$ cm

Exercise 31:2
1. 40 cm^2 2. 6   3. 84 cm^2
4. (a) 3 cm   (b) $\dfrac{\sqrt{55}}{2}$ cm   (c) $\dfrac{3\sqrt{55}}{2} \text{ cm}^2$
5. 30 cm^2 6. 6 cm   7. $16\sqrt{3} \text{ cm}^2$
8. (a) 160 cm^2 (b) $2\sqrt{41}$ cm
9. (a) $8\sqrt{22}$ cm   (b) $36\sqrt{22} \text{ cm}^2$ 10. 21 cm
11. (a) 10 cm   (b) 42 cm^2
12. (a) $\sqrt{29}$ cm   (b) 40 cm^2 13. 126 cm^2
14. $40 \text{ cm}^2, \sqrt{24}$ cm   15. 54 cm
16. (a) 96 cm^2 (b) 192 cm^2 (c) 96 cm^2
    (d) 96 cm^2 (e) 192 cm^2 (f) 288 cm^2

Exercise 31:3
1. 38.5 cm^2 2. 154 cm^2 3. 3.5 cm   4. 28 cm
5. 22 cm   6. 88 cm   7. 7 cm   8. 1886.5 cm^2
9. 132 m   10. 7 cm   11. 110 m^2 12. 115.5 cm^2

Exercise 31:4
1. 3.5 cm   2. 85.9 cm   3. 27.2   4. 62.2 cm
5. 24 cm   6. 30° 7. 16 cm   8. 10.4 cm^2

Exercise 31:5
1. 89.8 cm^2 2. 20 cm, 102 cm   3. 321 m^2
4. 111 cm^2 5. 22200 m^2 6. 10 m, 5 m
7. (a) $P = \pi x + 2x + 2y$ (b) $x = \dfrac{P - 2y}{\pi + 2}$
8. (a) 7 cm   (b) 10.5 cm^2 9. 68.2 cm^2
10. 70.3 cm^2 11. 195.3 cm^2 12. 115.5 cm^2
13. (a) 12.57 cm,   (b) 0.86 cm^2

Exercise 31:6
1. (a) 16 m^2 (b) 10 m^2 (c) 18 m^2
2. 84.81 km   3. 17.3 km^2

Exercise 32:1
1. 8400 cm^3 2. 1440 m^3 3. 40 cm   4. 13 m
5. 10 m   6. 3000 cm^3 7. $144000\ l$ 8. $10,000\ l$
9. $60,000\ l$ 10. (a) 8 cm^3 (b) 24 cm^2 11. 192
12. (a) 960 cm^3 (b) 784 cm^3
13. (a) 277.1 cm^3 (b) 295.4 cm^2 14. 6927.2 cm^3

15.

	Base radius	5 cm	10 cm
	Height	20 cm	30 cm
	Surface area	785.7 cm^2	2514.3 cm^2
	Volume	1571.4 cm^3	9428.6 cm^3

16. 53.6 m^2 17. 1.8 cm   18. 1078 cm^3 19. 3 cm
20. 5 cm   21. 22000 cm^3 22. 3080 cm^3
23. 127.3 cm   24. 4620 cm^3 25. 220 cm^3

Exercise 32:2
1. 10 cm   2. 462 cm   3. 352 cm   4. 4 cm
5. 528 cm^3 6(a) 880 cm^2 (b) 1496 cm^3
7. 562 cm^3 8. 462 cm^3 9. 314.3 cm^3
10. (a) 528 cm^2 (b) 712 cm^3 11. 45° 12. 0.007 m

Exercise 32:3
1. (a) $5544 \text{ cm}^2, 38808 \text{ cm}^3$
    (b) $1386 \text{ cm}^2, 4850 \text{ cm}^3$
2. 1400 cm^2 3. 1:343   4. 1767 cm^3 5. 10,000
6. (a) 56.6 cm^2 (b) 56.5 cm^2 (c) 84.9 cm^2

Exercise 32:4
1. (a) 122.6 cm^3 (b) 106 cm^2
2. (a) 1685.1 cm^3 (b) 594 cm^2 (c) 7.5 cm
3. (a) $\dfrac{370}{3}\pi \text{ cm}^3$ (b) $12\pi\sqrt{101} \text{ cm}^2$
4. (a) 8.2 cm^3 (b) 34.9 cm^2
5. (a) $3\pi \text{ cm}^3$ (b) $\left(\dfrac{7}{2}\pi + 24\right) \text{ cm}^2$
6. (i) 5.9 cm   (ii) 10.2 cm
7. (a) 234.7 cm^3 (b) 191.2 cm^2
8. 5011.3 cm^3 9. (a) $16..1 \text{ cm}^3$ (b) 47.08 cm^2
10. (a) 8382.5 cm^3 (b) 3116.1 cm^3 (c) 2464 cm^2

Exercise 32:5
1. (a) 12962.5 km   (b) 6146 km   (c) 5810.8 km
2. (a) 18438.1 km   (b) 4246.3 km   (c) 10057.1 km
3. (a) 25.2° (b) 90°

Exercise 32:6
1. (a) $0.127 \text{ m}^3/\text{s}$ (b) 754.9 liters/min.
2. $11.3 \text{ m}^3/\text{min.}$ 3. 7,200,000 liters/min.
4. (a) $7\pi \text{ m}^3$ (b) $100,000\pi$ s
5. 3.5 cm/min.   6. 5.31 m/s
7. 2.41 m^3 8. (a) 1000 m/min.   (b) 120 s

Exercise 33:1

1. $\Delta PTQ \equiv \Delta RTS$, $\Delta PQR \equiv \Delta SRQ$, ASA

2. $\Delta ABC \equiv \Delta AED$, $\Delta ABD \equiv \Delta AEC$, SAS

3. $\Delta XOY \equiv \Delta XOZ$, SAS

4. $\Delta POM \equiv \Delta QLN$, ASA

5. $\Delta POS \equiv \Delta ROQ$, SAS or ASA or SSS

   $\Delta POQ \equiv \Delta ROS$, ASA

   $\Delta PQS \equiv \Delta RSQ$, SAS

   $\Delta PRS \equiv \Delta RPQ$, SAS

6. $\Delta PTQ \equiv \Delta RTS$, ASA

7. $\Delta WOZ \equiv \Delta XOY$, ASA

   $\Delta WYZ \equiv \Delta XZY$, ASA

8. Two sets by SAS or ASA or SSS

   $\Delta ADF \equiv \Delta DBE \equiv \Delta FEC \equiv \Delta DFE$

   $\Delta DIG \equiv \Delta IFH \equiv \Delta IGH \equiv \Delta GHE$

Exercise 33:2

1. (i) 8 cm (ii) 30 cm   2. (b), (h)
3. (i) 14 cm (ii) 44.1 cm^2 4. 19.1 cm
5. $m = 3$, $n = 13.5$

Exercise 33:3

1. 25 km^2 2. 20 cm and 24 cm; 1:16
3. 3 cm and 5 cm; 1:9   4. 4   5. 9 cm^2
6. 20.88 cm^2 7. 2   8. (b) 8 cm   (c) 16:1

Exercise 33:4

1. 250 cm^3 2. 268 cm^3, 35 cm 3. 4.5 4. 81 cm^3

5. 121.5 cm^3 6. 10.5 7. 64:15625 8. 2847 cm^3

Exercise 34:1

1. $x = 3.5, y = 1.8$ 2. $x = 0, y = 2$

3. $x = 1, y = 1$ 4. $x = 4, y = 11$

5. $x = 1, y = -5$ 6. $x = -0.5, y = 1$

7. $x = 3, y = 2$ 8. $x = 3, y = 12$

9. $x = 1.3, y = -1.1$ 10. $x = -1.3$

Exercise 34:2

2. (c) 9   (d) $-1 < x < 5$ 3. (i) ± 1 (ii) $x = \pm 1.5$

(iii) $x = 0$ or $x = 4$ (iv) $x = 0$ or $x = -3.5$

(v) $x = -2$ or $x = 4$ (vi) $x = 1.5$

Exercise 34:3

1.(a)

x	−5	−4	−3	−2	−1	0	1	2	3
y	5	1.5	−1	−2.5	−3	−2.5	−1	1.5	5

(c) $(-1,-3)$ (d) -4.1 or 2.1

(f) $f(x) = x^2 - 5 - \dfrac{1}{x}$

2.(a)

x	0.25	0.5	1.0	1.5	2	2.5
y	4.5	3	3	3.7	4.5	5.4

(c) 2.8   (d) −2   (e) 0.35 or 1.39

3. (a) 1.2   (b) −2.4, −16.5   (d) −0.4, 2.7   (e) 1.2

4.(a)

x	−2	−1	0	1	2	3	4	5
y	21	4	−7	−12	−11	−4	9	28

(c) −0.7, 3.4   (d) −1.9 or 4.6   (e) 10

5.(a)

x	−2	−1	0	1	2	3	4	5
$f(x)$	20	7	−2	−7	−8	−5	2	13

(b) (i) −0.3 or 3.8   (ii) $f(x)_{min} = -8.1$, $x = 1.8$
     (iii) $-0.3 < x < 3.8$ (c) −2.9 or 0.1

Exercise 34:4

1. 1
2.(a)

x	−4	−3	−2	−1	0	1	2	3
y		13		−1	−2	1	8	19

(c) 11

3.(a)

x	−1	0	1	2	3	4	5	6
y	10	−1	−8	−11	−10	−5	4	17

(c) 3   4. (b) −4

Exercise 34:5

1. (a) $4x^2$ (b) $6x^2$ (c) $28x^2 - 6x^2$

   (d) $4x^2 + 15x^2$ (e) $12x^2 - 6x$ (f) $6x$

   (g) 5   (h) $4x - 3$ (i) $\dfrac{1}{4} - 8x$

2. 17   3. (a) 2   (b) 6   (c) 6   (d) −23

4. $(-1,-3)$ 5. (a) $-\dfrac{3}{2}$ (b) $\dfrac{5}{6}$ (c) $\dfrac{4}{3}$ (d) $-\dfrac{4}{9}$

6. $6x + 2$, 26   7. $6x^2 - 4$, 20

Exercise 34:6

1. (2,9), minimum
2. (a) 4 m/s   (b) −1 m/s^2 (e) 0 m/s at $t = 2$ s

3. $\dfrac{512}{27}\pi^2$ m^2

Exercise 34:7

1. 7.2 km/h, 77 m

2.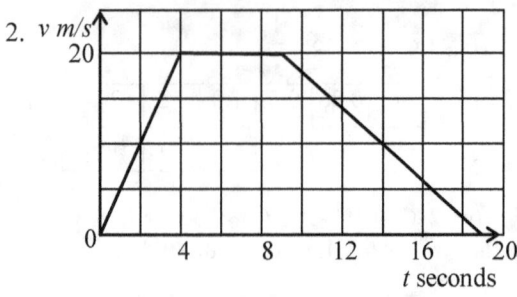

Distance covered = 240 m

3.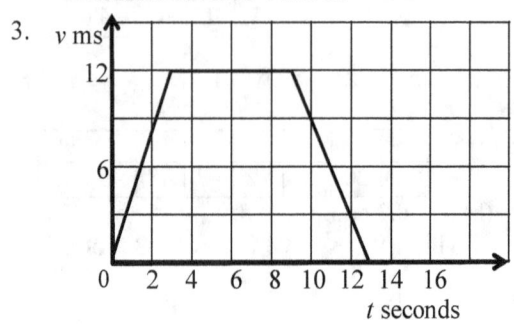

Final acceleration = 3 m/s²,
Final displacement = 114 m

4. (i) (a) 11:20 a.m.   (b) 11:50 a.m.   (c) 12:20 p.m.
(e) 5.3 km
(ii) (a) *A* and *B*   (b) *A* and *C*   (c) None

Exercise 35:1

1. $c = 7n$ 2. $m = nk$ 3. $x = \frac{1}{5}y$ 4. $x = 1$

5. $\frac{32}{125}$ 6. $y = \pm 12$ 7. $s = 27$

8. $p = \frac{3}{8}q, p = 6$ 9. (a) $k = 6$ (b) $x = \pm 2$ 10. $\frac{3}{4}$

Exercise 35:2

1. (a) $y = \frac{2}{x^2}$ (b) $\frac{1}{6}$ 2. $a = \pm 6, b = 9$

3. $a = 3, b = 12$ 4. $y \neq 3, y = 2.25$

5. (a) 2   (b) $y = 256$ 6. $x = \pm\frac{2}{3}$

7. $a = 275, b = \frac{5}{6}$ 8. $y \neq 2, y = \frac{16}{9}$

9.

x	0	5	10	15	20	25
y	0	1	2	3	4	5

Exercise 35:3

1. $l = \frac{3}{2}mn$ 2. (a) $A = 6\frac{B}{C}$ (b) $A = 5$

3. (a) $y = 21xz$ (b) $x = \frac{5}{189}$ 4. $x = \frac{5}{2}$

Exercise 36

1.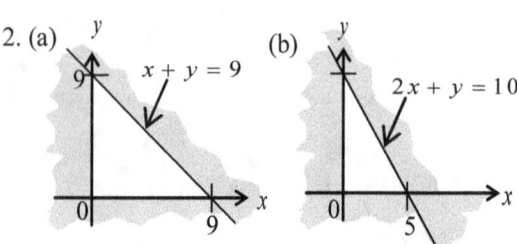

2. (a) 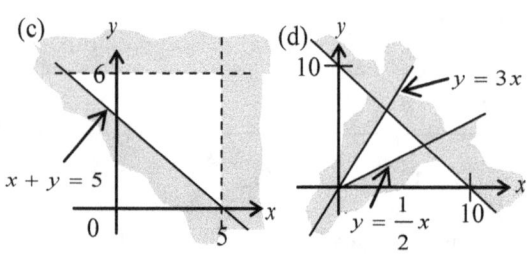   (b)

(c)   (d)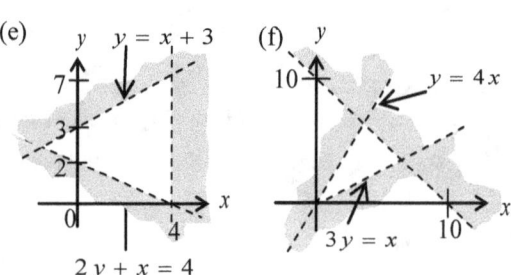

(e)   (f)

(g)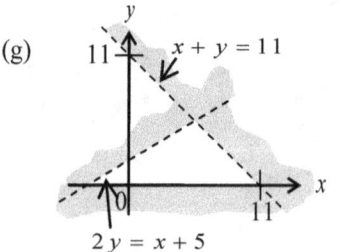

(h)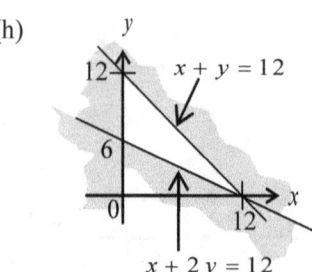

(i)

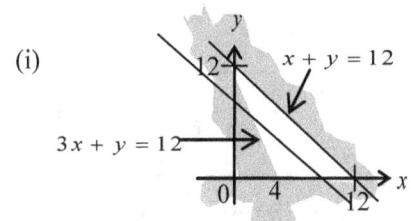

3. (a) $2 \le x \le 7$ (b) $-2 < x < 5$
    (c) $-2 \le y < 4$ (d) $x \le 0, x + y < 10, y > x$
    (e) $y \le -2, x < 8, y < x$
    (f) $y \ge 0, 3y < 2x + 12, y \ge -2x + 4$

Exercise 37:1

1.

Marks, x	Tally
49	|
51	|
53	||
54	||
56	|||
57	|||
58	|||
59	|||
60	卌
62	||||
63	|||
64	|||
66	||

Marks, x	f
49	1
51	1
53	2
54	2
56	3
57	3
58	3
59	3
60	5
62	4
63	3
64	3
66	2

2.

Distance, x	66	67	68	69	70	71
Tally	|||	||||	卌|	||||	|||	|

Distance, x	66	67	68	69	70	71
Frequency, f	2	4	6	4	3	1

3.

Food item	Tally
r	卌 ||||
b	卌 ||||
p	卌 卌 ||
y	卌 卌 卌

Food item	f
r	9
b	9
p	12
y	15

4.

weight	Tally
52	|
55	|||
56	|
57	|||
58	卌 |||
59	|||
60	卌 |
61	卌 |
62	卌 |
63	卌 ||
64	卌 卌
66	卌
69	|

weight	f
52	1
55	3
56	1
57	3
58	8
59	3
60	6
61	6
62	6
63	7
64	10
66	5
69	1

Exercise 37:2

1.

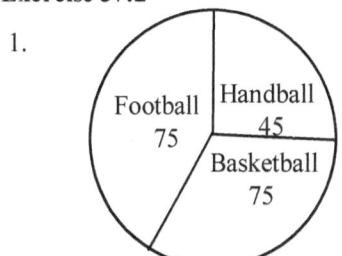

2. (a) 600 FCFA   (b) 75 FCFA.

3.

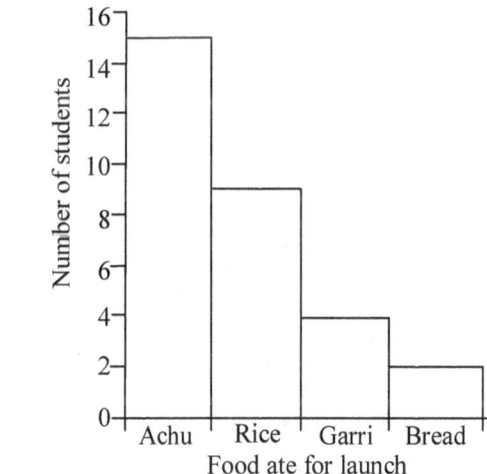

4. 41.7 %

573

5.

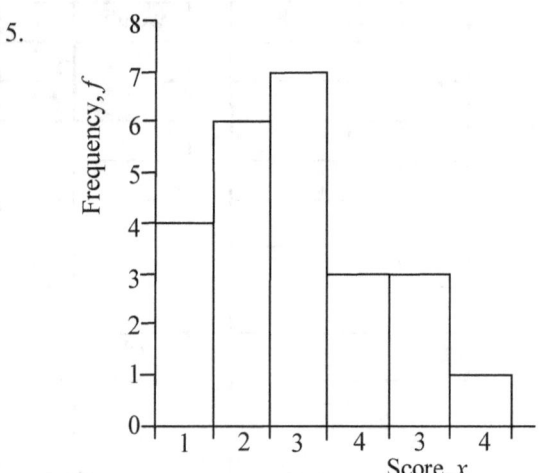

6. 14°   7. 56

8.

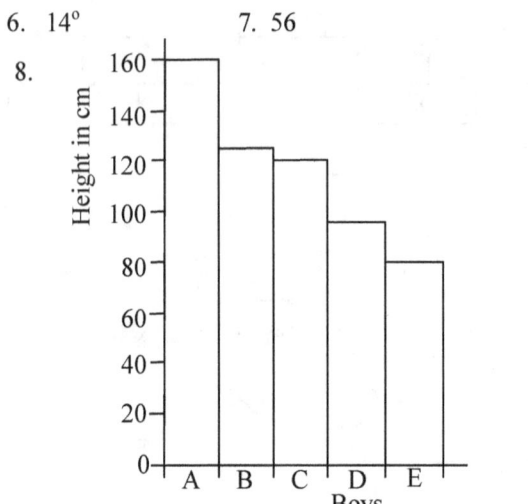

9.(a) 100°   (b) 15
10. $w = 156°, x = 72, y = 16, z = 24°$ 11. 105°

Exercise 37:3
1. 13   2. 1   3. 9
4. (a) 54 kg (b) 51.2 kg (c) 54 kg
5. (a) 2.92   (b) 3   (c) 3
6. (a) 5.3   (b) 5   (c) 5
7. (a) 7   (b) 7.1 (c) 7
8. (a) 113.4 FRS (b) 100 FRS   (c) 50 FRS
9. (a) 70 kg   (b) 68 kg   (c)   70.25 kg
10. (a) 30   (b) 8   (c) 6.1
11. (a) 2   (b) 2   (c) 2
12. (a) 5   (b) 12   (c) 7.7

Exercise 37:4
1. Mode. Stock goods are most needed.
2. Mean. It takes into account all the values.
3. Mean. It takes into account all the values.
4. The average is a misleading statistic because so
   many farmers have no pigs and one farmer has
   so many.

Exercise 37:5
1. (a)

x	66	67	68	69	70	71
f	2	3	7	4	3	1

   (b) 68   (c) 68   (d) 68   (e) 68.3

2. (a)

GRADE	A	B	C	D	E	F
Angle of Sector	30	60	90	40	70	70

   (b)

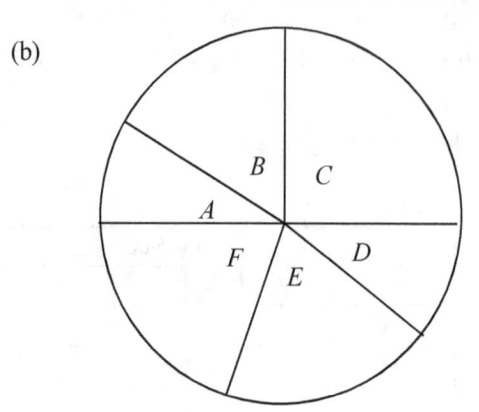

   (c) (i) 72   (ii) 8   (d) 1:1

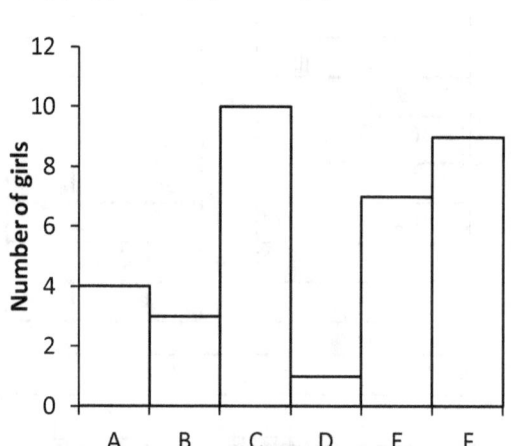

3. (i) (a)

Marks	Cum freq
≤ 10	0
≤ 20	2
≤ 30	8
≤ 40	15
≤ 50	29
≤ 60	49
≤ 70	84
≤ 80	113
≤ 90	119
≤ 100	120

4. (a)

x	4	5	6	7	8	9	10
f	9	3	7	4	5	1	1

   (b) 4   (c) 6

5. (a)

x	f
0- 9	1
10-19	2
20-29	2
30-39	3
40-49	7
50-59	8
60-69	9
70-79	4
80-89	3
90-99	1

(b)

x	f	Cum freq
≤ 10	1	1
≤ 20	2	3
≤ 30	2	5
≤ 40	3	8
≤ 50	7	15
≤ 60	8	23
≤ 70	9	32
≤ 80	4	36
≤ 90	3	39
≤ 100	1	40

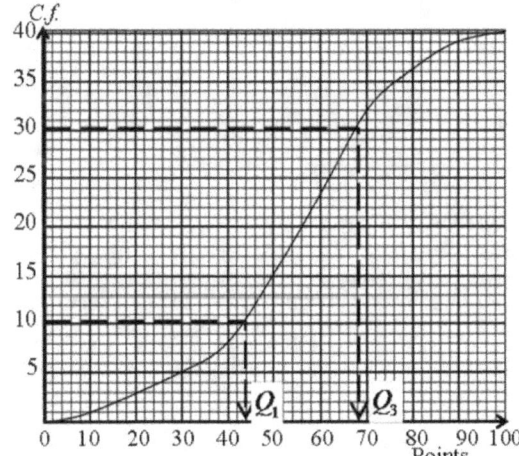

(c) 60-69   (d) 54.2   (e) 24   (f) A grade

6. (a)

x	f
35-44	3
45-54	7
55-64	13
65-74	16
75-84	7
85-94	4

(b)

x	f	
< 45	3	3
< 55	7	10
< 65	13	23
< 75	16	39
< 85	7	46
< 95	4	50

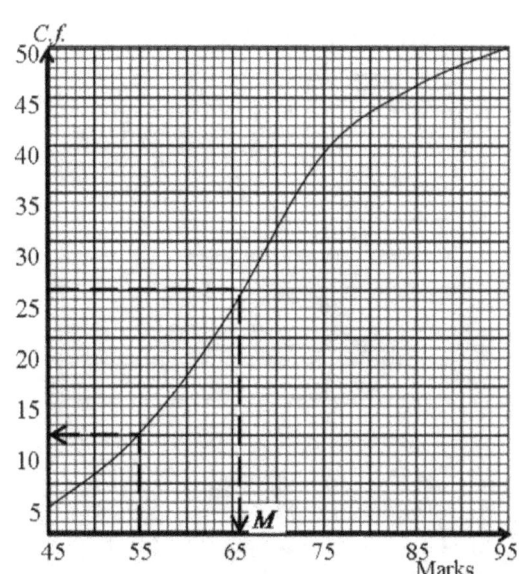

(c) 66   (d) 80%

7. (a) 43.6,   (b) 86.4   (c) 26   (d) −10

8. (i) 34.5 minutes

(ii)

Time (min)	Cum freq
t ≤ 20	0
t ≤ 25	8
t ≤ 30	16
t ≤ 35	28
t ≤ 40	58
t ≤ 45	76
t ≤ 50	80

(ii)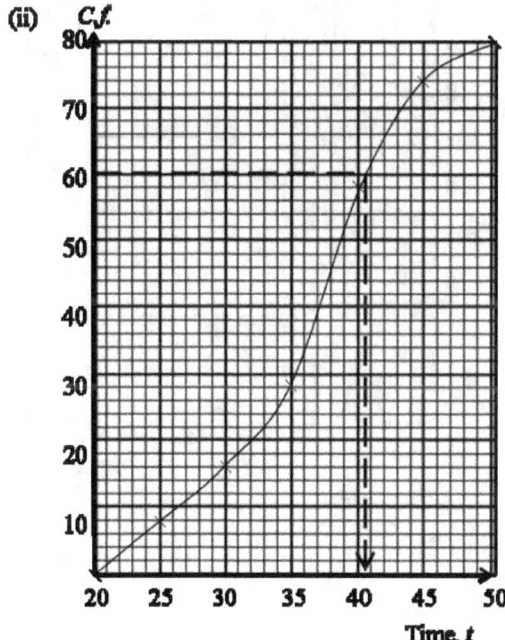

(iii) 40.5 minutes

9. 41.1, 11.9   10. (a) 45.79   (b) 21.18

11. 166.62, 166.55, 2.21,

12.

No. of times	No. of students
0 -5	34
6-10	4
11-15	10
16-20	25
21-25	50
26-30	37
31-35	24
36-40	15
41-45	1

   (a) 21.5   (b) 113.7   (c) 10.7

13. (a) 44.5   (b) 240   (c) 15.5

14. (a) 34.4 years   (b) 32 years   (d) 17 years
   (d) 186.5   (e) 13.7 years

15.(a) 44.58   (b)

No. of times	Cum frequency
< 29.5	40
< 39.5	160
< 49.5	360
< 59.5	460
< 69.5	480
< 79.5	496
< 89.5	500

(c)

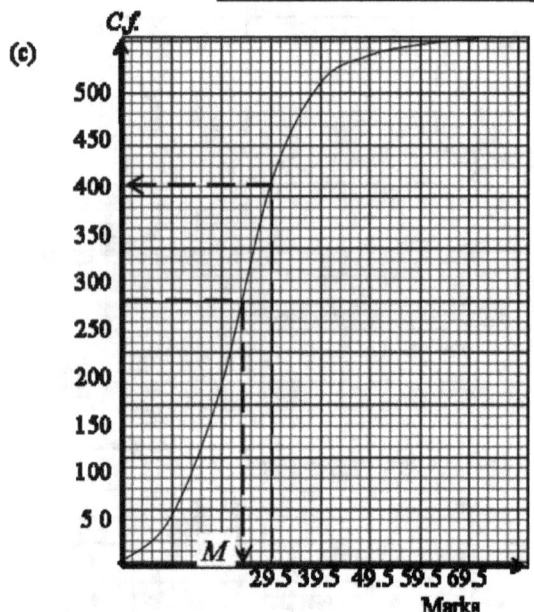

(i) 43.6   (ii) 360

Exercise 38:1

1. (a) {blue, red, green, white, black}
   (b) {white, green}
2. {(H,H),(H,T),(T,H),(T,T)}
3. {1,2,3,4,5,6}
4. (a) {2,3,5,7}   (b) {2,4,6,8,10}   (c) {1,3,5,7,9}
   (d) {1,4,9}   (e) {3,6,9}

5. {a, e, a, i}   6. 12

Exercise 38:2

1. (a) $\dfrac{1}{6}$ (b) $\dfrac{1}{2}$ (c) $\dfrac{1}{2}$ (d) $\dfrac{2}{3}$ (e) $\dfrac{1}{2}$

2. (a) $\dfrac{3}{8}$ (b) $\dfrac{1}{8}$ (c) 0 (d) $\dfrac{1}{8}$

3. $\dfrac{1}{4}$ 4. $\dfrac{3}{10}$ 5. $\dfrac{2}{11}$ 6. $\dfrac{1}{3}$

Exercise 38:3

1. (a) $\dfrac{9}{35}$ (b) $\dfrac{23}{35}$ (c) $\dfrac{2}{5}$

2. $\dfrac{11}{15}$ 3. $\dfrac{21}{25}$ 4. $\dfrac{3}{10}$ 5. $\dfrac{3}{7}$ 6. $\dfrac{1}{4}$

Exercise 38:4

1. $\dfrac{2}{3}$ 2. $\dfrac{2}{3}$ 3. $\dfrac{2}{9}$ 4. $\dfrac{11}{18}$

5.(i) $\dfrac{1}{3}$ (ii) $\dfrac{5}{18}$ (iii) $\dfrac{11}{18}$ 6. $\dfrac{17}{30}$ 7. $\dfrac{2}{3}$

8.(a) $\dfrac{2}{7}$ (b) $\dfrac{3}{7}$ 9.(a) $\dfrac{1}{13}$ (b) $\dfrac{4}{13}$

10.(a) $\dfrac{2}{5}$ (b) $\dfrac{3}{10}$ (c) $\dfrac{3}{5}$ 11.(a) $\dfrac{1}{6}$ (b) $\dfrac{1}{6}$ (c) $\dfrac{1}{3}$

12. $\dfrac{1}{3}$ 13. 1 14. 1 15. $\dfrac{13}{15}$

Exercise 38:5

1.(a) $\dfrac{9}{25}$ (b) $\dfrac{4}{25}$ (c) $\dfrac{12}{25}$ 2. $\dfrac{1}{36}$

3.(a) $\dfrac{17}{35}$ (b) $\dfrac{18}{35}$

4.(a) $\dfrac{2}{11}$ (b) $\dfrac{3}{11}$ (c) $\dfrac{5}{11}$ (d) $\dfrac{6}{11}$

5.(a) $\dfrac{3}{8}$ (b) $\dfrac{11}{16}$ (c) $\dfrac{5}{16}$ (d) $\dfrac{1}{8}$

6.(a) $\dfrac{1}{9}$ (b) $\dfrac{4}{9}$ (c) $\dfrac{4}{9}$

7.(a) $\dfrac{6}{25}$ (b) $\dfrac{9}{25}$ (c) $\dfrac{4}{25}$ (d) $\dfrac{6}{25}$ (e) $\dfrac{13}{25}$

8.(a) $\dfrac{1}{12}$ (b) $\dfrac{7}{9}$ (c) $\dfrac{5}{18}$

9.(a) $\dfrac{8}{35}$ (b) $\dfrac{32}{105}$ (c) $\dfrac{79}{105}$ 10. $\dfrac{3}{4}$ 11. $\dfrac{5}{18}$

12.(a) $\dfrac{9}{49}$ (b) $\dfrac{24}{49}$ (c) $\dfrac{144}{343}$ 13. $\dfrac{11}{21}$

14.(a) $\dfrac{9}{25}$ (b) $\dfrac{13}{25}$ 15. $\dfrac{5}{36}$

16.(a) $24 - m$ (b) $\dfrac{m + 6}{30}$ 17. 18

18.(a) $\dfrac{1}{26}$ (b) $\dfrac{6}{13}$ 19. 1   20. 1

Exercise 38:6
1. (a) 10   (b) 30-39   (c) 47.7 kg

(d)

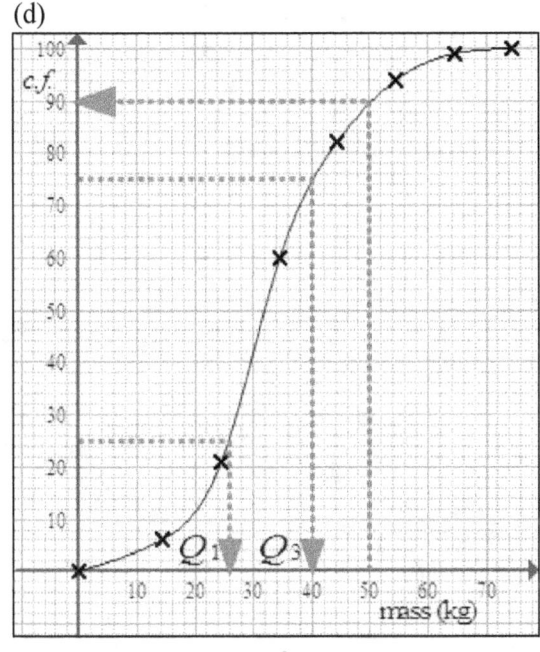

(e)   14 kg   (f) $\dfrac{9}{10}$

2. (a) 10 minutes   (b) 11 minutes   (c) 1.75
(d) $\dfrac{1}{5}$ (e) $\dfrac{5}{12}$

3. (i)   120-149   (ii) 135.7 kg

(iii)

Weights(g)	C.F.
<30	20
<60	45
<90	95
<120	170
<150	290
<180	390
<210	470
<240	500

(iii) See graph on the next page

(a) 140 g   (b) 70   (c) $\dfrac{6}{25}$

Exercise 39:1
1. (iii) 10 cm (v) 6.2 cm   2. (e) 5.4 cm   (f) 90°
3. (c) 10 cm   4. (f) 14.1 cm (g) 23°
5. (c) 10 cm

Exercise 39:2
1. mediator or perpendicular bisector.
3. (a) $YX = 9$ cm
   (b) The locus of P is a straight line
       perpendicular to YZ and distant 9 cm from
       Y.

Exercise 38:6 question 3. (iii)

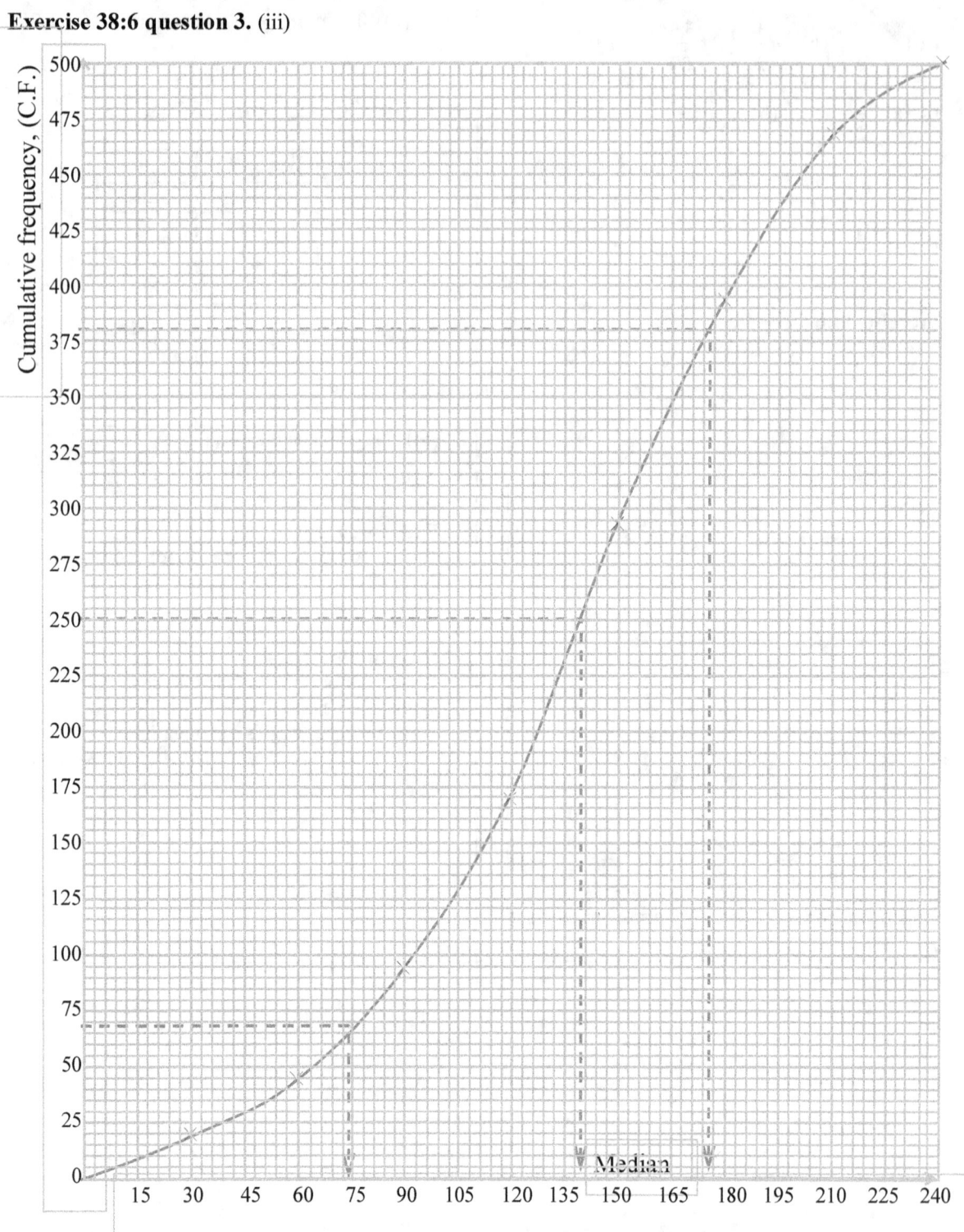

Exercise 40:1
1. 7 cm, 1 cm   2. 60 cm², 7.06 cm   3. 8.3 cm
4. 48° 5. 37°   6. (a) 27°   (b) 126°
7. $x = 73°, y = 92°$ 8. $x = 36°, y = 72°$ 9. 36°
10. (a) 90°   (b) 75°   (c) 150°   (d) 60°

Exercise 40:2
1. $\angle BTA = 50°$, $\angle CBT = 90°$, $\angle DTM = 50°$
   (a) $\angle ABT = \angle BTD = 90°$ (alt. $\angle$s btw ∥

(b) $\angle CBT = 90°$
2. (i) 25°   (ii) 65°   (iii) 65°   (iv) 12.7 cm
   (v) 16.6 cm   (vi) 105.1 cm
3. $x = 130°$, $y = 30°$, $z = 50°$
4.(a) 60° (b) 130° (d) $DE = 5.8$ cm, $DB = 11.5$ cm
5. 57°   6. $a = 75°$, $b = 15°$, $c = 75°$
7. $x = 122°, y = 26°$ 8. $x = 78°, y = 53°, z = 49°$

Exercise 40:3
1. 9 cm   2.(a) 4.2 cm (b) 24°   3. 3 cm
4. 8 cm   5. AB = 14 cm, EF = 16.5 cm
6. r = 25 cm, length of tangent = 32 cm   7. 3 cm
8. Radius = 3.5 cm, Area = 38.5 cm^2

Exercise 41
1. (a) $\angle ABC$ or $\angle DCH$
   (b) $\angle ABF$ or $\angle CBF$ or $\angle DBF$
   (c) $\angle DHC$ (d) $\angle CBG$ or $\angle DAH$
   (e) $\angle ABE$ or $\angle DCH$ (f) $\angle BGC$ or $\angle AHD$

2.(i) 16.7° (ii) $\sqrt{109}$ cm   3.(a) 10 cm (b) 16.7°

4. (a) $8\sqrt{2}$ cm   (b) $3\sqrt{17}$ cm (c) 23.8° (d) 32°

5. (a) 49.5°   (b) 51.7°

6. (a) $10\sqrt{2}$ cm   (b) $15\sqrt{2}$ cm   (c) 69.3°
   (d) 29° (e) 41.4°

7. 54.7°   8. (a) 8.16 cm (b) 54.7° (c) 70.5°

9. (a) 8.87 cm   (b) 62.5°   (c) 75.4°

10. (a) 8.31 cm   (b) 67.4°   (c) 78.2°

11. (a) 86.9 cm   (b) 53.6 cm (c) 132.5 cm

12. (a) $\sqrt{50}$ cm   13.   4.5 cm

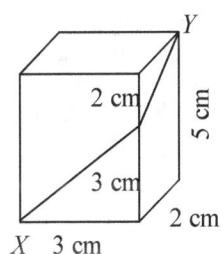

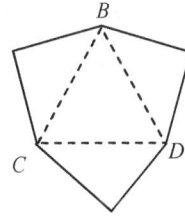

Net of pyramid

Net of remaining solid

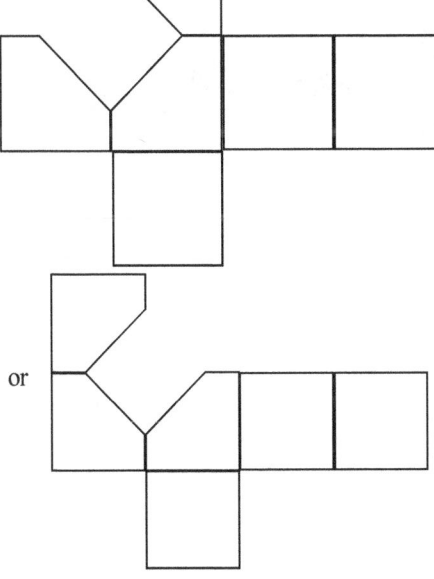

or

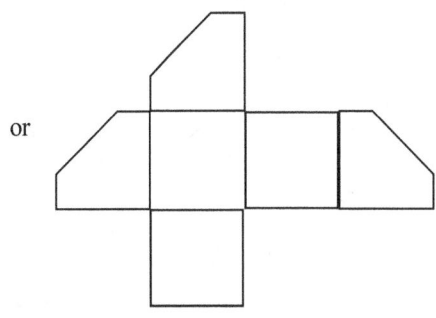

or

Exercise 42:1
(a) $E = \{1, 2, 3\}$, $O = \emptyset$ or $\{\ \}$
(b) $E = \{1, 3\}$, $O = \{2, 4\}$
(c) $E = O = \emptyset$ or $\{\ \}$, $O = \{1, 2, 3, 4\}$
(d) $E = \{1, 2\}$, $O = \{3, 4\}$
(e) $E = \{3, 5\}$, $O = \{1, 2, 4, 6\}$
(f) $E = \{3, 5\}$, $O = \{1, 2, 4, 6\}$
(g) $E = \{3, 6\}$, $O = \{1, 2, 4, 5, 7, 8\}$
(h) $E = \{1, 3, 5, 6\}$, $O = \{2, 4\}$
(i) $E = \{1, 3, 4, 6\}$, $O = \{2, 5, 8\}$

Exercise 42:2
(a) traversable because there are exactly 2 odd vertices.
(b) traversable because all vertices are even.
(c) not traversable because there are 4 odd vertices and 4 is greater than 2.

Exercise 42:3
1. (a)

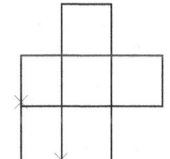

The beginning point can be either of the two odd vertices; the endpoint will be the other odd vertex.

(b)

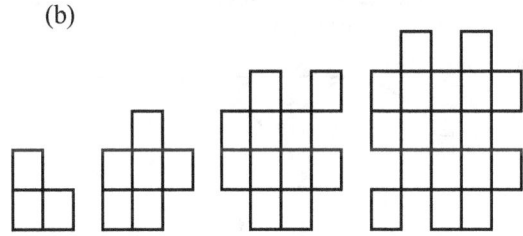

Remove 1 square from the 2 × 2 grid; remove 3 squares from the 3 × 3 grid; remove 5 squares from the 4 × 4 grid; remove 7 squares from the 5 × 5 grid.

(c) The number of squares to remove from a 12 × 12 grid is the 11th odd number, 21.

(d) The number of squares to remove from an n × n grid is the $(n-1)^{\text{th}}$ odd number, $2n - 3$.

2.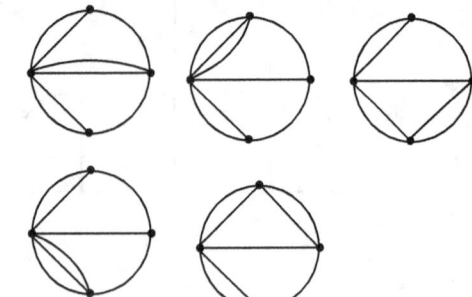

3. (a) Yes

(b)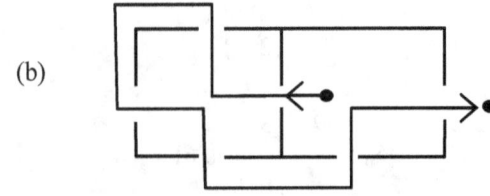

or

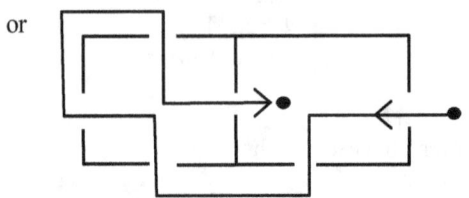

Exercise 42:4

1. (a) {1,2} and {2,6}, {2,3} and {3,4}, {4,6} and {4,5}, {4,5} and {5,6}, {4,6} and {5,6}, {4,6} and {2,6}, {1,2} and {1,3}, {2,3} and {2,6}.

(b) (1,2) and (2,3), (6,2) and (2,3), (6,4) and (4,3).

(c) (1,2), (2,3), (4,3), (6,4), (6,2).

(d) 1 and 2, 2 and 3, 3 and 4, 4 and 5, 4 and 6, 5 and 6, 2 and 6.

(e) 1 and {1,2}, 6 and {2,6}, 2 and {2,3}, 3 and {3,4}, 4 and {4,6}, 4 and {4,5}, 5 and {5,6}, 2 and {1,2} etc.

2. $G = (V, E)$ where $V = \{1, 2, 3, 4, 5, 6\}$ and $E = \{(1,2), (2,3), (4,3), (6,4), (6,2), \{4,5\}, \{5,6\}\}$.

3.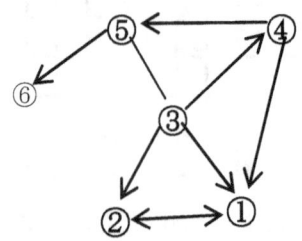

Exercise 42:5

1.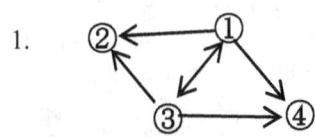

2. $D = (V, A)$, where $V = \{1, 2, 3, 4, 5\}$ and $A = \{(1, 2), (2, 1), (2, 4), (3, 2), (3,4), (4,1),$

(4,5),(5,3), (5,4)}.

3.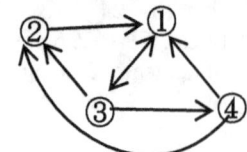

4.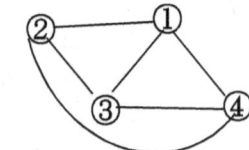

5. $G = (V, E)$ where $V = \{1, 2, 3, 4, 5, 6\}$ and $E = \{\{1,2\}, \{1,3\}, \{1,5\}, \{2,3\}, \{2,4\}, \{3,4\}, \{4,5\}, \{5,6\}\}$.

6. (a) 6 hours   (b) 1650 FCFA

7. Answers vary.

8. Trees :   (b), (c), (f). $n(E) = n(V) - 1$
   Forest : (a), (d). Combination of trees.
   Neither : (e) $n(E) \neq n(V) - 1$

Exercise 43:1

1. (a) 2×2 (b) 2×3 (c) 3×1 (d) 3×3 (e) 3×2
   (f) 1×1 (g) 3×3 (h) 1×3 (i) 3×2 (j) 3×3

2. (a) square matrix   (b) rectangular matrix
   (c) column matrix   (d) diagonal matrix
   (e) rectangular zero matrix (f) trivial
   (g) square matrix   (h) row matrix
   (i) rectangular matrix   (j) unit matrix

Exercise 43:2

1. (a) $x = -3, y = 4$ (b) $x = 1, y = 0$

   (c) $x = 5, y = 4$ (d) $x = 3, y = -2$

   (e) $x = 0, y = 2$

2. (a) $x = -3, y = 4, z = \dfrac{2}{5}$ (b) $x = -2, y = 4, z = -2$

3. (a) $x = -3, y = -4$ (b) $x = -2, y = 5$

   (c) $x = -2, y = -7$ (d) $x = 0, y = 0$

   (e) $x = 5, y = -3$ (f) $x = 4, y = 1$

Exercise 43:3

1. (a) $\begin{pmatrix} 2 \\ -2 \end{pmatrix}$ (b) $\begin{pmatrix} 9 & 1 \\ 3 & 8 \end{pmatrix}$

   (c) $\begin{pmatrix} -16 \\ 8 \end{pmatrix}$ (d) $\begin{pmatrix} -1 & 5 \\ -1 & -4 \end{pmatrix}$

2. (a) $\begin{pmatrix} 1 & -1 & 9 \\ 1 & 2 & 6 \\ 1 & 1 & -14 \end{pmatrix}$ (b) $\begin{pmatrix} -1 & 1 & -9 \\ -1 & -2 & -6 \\ -1 & -1 & 14 \end{pmatrix}$

3 (a). $\begin{pmatrix} 1 & 3 & 2 \\ 2 & 5 & 3 \end{pmatrix}$ (b) $\begin{pmatrix} 1 & 3 & 2 \\ 2 & 5 & 3 \end{pmatrix}$

(c) $\begin{pmatrix} 1 & 1 & 4 \\ 3 & 2 & 7 \end{pmatrix}$ (d) $\begin{pmatrix} 1 & 1 & -2 \\ 2 & 1 & -3 \end{pmatrix}$

(e) $\begin{pmatrix} -1 & -1 & 2 \\ -2 & -1 & 3 \end{pmatrix}$ (f) $\begin{pmatrix} -1 & 1 & 0 \\ -3 & 2 & -1 \end{pmatrix}$

(g) $\begin{pmatrix} 2 & 3 & 4 \\ 5 & 5 & 7 \end{pmatrix}$ (h) $\begin{pmatrix} 2 & 3 & 4 \\ 5 & 5 & 7 \end{pmatrix}$

(i) $\begin{pmatrix} 0 & 1 & -4 \\ -1 & 1 & -7 \end{pmatrix}$ (j) $\begin{pmatrix} 2 & 1 & 0 \\ 5 & 1 & 1 \end{pmatrix}$

(k) Addition of matrices is commutative.
(l) Subtraction of matrices is not commutative.
(m) Addition of matrices is associative.
(n) Subtraction of matrices is not associative.

4. (a) $\begin{pmatrix} 0 & 0 \\ 0 & 0 \end{pmatrix}$ (b) $\begin{pmatrix} 0 & 0 \\ 0 & 0 \end{pmatrix}$

(c) **A** is the additive inverse of **B**.

5. (a) $\begin{pmatrix} 3 & -1 \\ 2 & 5 \end{pmatrix}$ (b) $\begin{pmatrix} 3 & -1 \\ 2 & 5 \end{pmatrix}$

(c) $\varnothing$ is the additive identity.

6. $a=1, b=-1, c=-6, d=5, e=-7$

7. (a) $w=6, x=2, y=0, z=-4$

(b) $w=4, x=3, y=1, z=5$

Exercise 43:4

1. $\begin{pmatrix} 2 & 0 & \frac{1}{2} \\ -\frac{3}{2} & -\frac{1}{2} & 1 \\ 0 & -4 & -\frac{5}{2} \end{pmatrix}$ 2. $x=3$

3. $x=3, y=-2$

4. (i) $x=-2, y=55$ (ii) $x=9, y=-7$

(iii) $a=2, b=11$

Exercise 43:5

1. $\begin{pmatrix} -8 & 9 \\ 0 & -6 \end{pmatrix}$ 2. $(-13 \; -35 \; 17)$

3. (i) $\begin{pmatrix} 3 & 7 \\ 5 & -2 \end{pmatrix}$ (ii) $\begin{pmatrix} 3 & 7 \\ 5 & -2 \end{pmatrix}$ (iii) $\begin{pmatrix} 44 & 7 \\ 5 & 39 \end{pmatrix}$

4. (a) $\begin{pmatrix} 1 & 3 \\ 5 & 4 \end{pmatrix}$ (b) $\begin{pmatrix} 1 & 3 \\ 5 & 4 \end{pmatrix}$ (c) $\begin{pmatrix} 0 & 0 \\ 0 & 0 \end{pmatrix}$

(d) $\begin{pmatrix} 0 & 0 \\ 0 & 0 \end{pmatrix}$ (e) $\begin{pmatrix} 1 & 2 \\ 5 & 4 \\ 2 & 6 \end{pmatrix}$ (f) $\begin{pmatrix} 0 & 0 \\ 0 & 0 \\ 0 & 0 \end{pmatrix}$

For any compatible matrices, multiplication by

(i) $\begin{pmatrix} 1 & 0 \\ 0 & 1 \end{pmatrix}$ leaves the matrix unchanged.

(ii) $\begin{pmatrix} 0 & 0 \\ 0 & 0 \end{pmatrix}$ gives the result $\begin{pmatrix} 0 & 0 \\ 0 & 0 \end{pmatrix}$.

5. (i) (a) $\begin{pmatrix} 0 & -20 \\ 5 & 5 \end{pmatrix}$ (b) $\begin{pmatrix} 7 & 18 \\ 6 & 19 \end{pmatrix}$ (c) $\begin{pmatrix} 0 & -10 \\ 5 & 15 \end{pmatrix}$

(d) $\begin{pmatrix} 7 & 1 \\ 6 & 8 \end{pmatrix}$ (e) $\begin{pmatrix} -3 & 13 \\ -4 & 14 \end{pmatrix}$ (f) $\begin{pmatrix} 10 & -30 \\ -15 & 55 \end{pmatrix}$

(g) $\begin{pmatrix} 10 & -30 \\ -15 & 55 \end{pmatrix}$ (h) $\begin{pmatrix} 4 & -1 \\ 2 & 7 \end{pmatrix}$ (i) $\begin{pmatrix} 0 & -7 \\ 0 & -1 \end{pmatrix}$

(j) $\begin{pmatrix} 14 & -11 \\ 22 & 47 \end{pmatrix}$ (k) $\begin{pmatrix} 0 & 7 \\ 0 & 1 \end{pmatrix}$ (l) $\begin{pmatrix} -7 & -38 \\ -1 & -4 \end{pmatrix}$

(m) $\begin{pmatrix} -14 & -49 \\ -2 & -7 \end{pmatrix}$ (n) $\begin{pmatrix} 7 & -22 \\ 21 & 54 \end{pmatrix}$ (o) $\begin{pmatrix} 7 & 18 \\ 1 & -6 \end{pmatrix}$

(ii) (a) $\mathbf{AB} \neq \mathbf{BA}$ (b) $(\mathbf{AB})\mathbf{C} \neq \mathbf{A}(\mathbf{BC})$

(c) $(\mathbf{A}+\mathbf{B})^2 \neq \mathbf{A}^2 + 2\mathbf{AB} + \mathbf{B}^2$

(d) $(\mathbf{A}-\mathbf{B})^2 \neq \mathbf{A}^2 - 2\mathbf{AB} + \mathbf{B}^2$

(e) $(\mathbf{A}+\mathbf{B})(\mathbf{A}-\mathbf{B}) \neq \mathbf{A}^2 - \mathbf{B}^2$

6. (a) $\begin{pmatrix} 5x-3y \\ -4x+y \end{pmatrix}$ (b) $\begin{pmatrix} 3a+4b \\ 5a+b \end{pmatrix}$ (c) $\begin{pmatrix} 2r+s \\ -2r+3s \end{pmatrix}$

(d) $\begin{pmatrix} 4p+q \\ 2p \end{pmatrix}$ (e) $\begin{pmatrix} 4x+2y \\ 3x+5y \end{pmatrix}$ (f) $\begin{pmatrix} -2x-3y \\ -2x+y \end{pmatrix}$

Exercise 43:6

(a) $\begin{pmatrix} 1 & 0 \\ 3 & 4 \end{pmatrix}$ (b) $\begin{pmatrix} 1 \\ 2 \\ 7 \end{pmatrix}$

(c) $(-1 \; 0 \; 1)$ (d) $\begin{pmatrix} 2 & 1 & -3 \\ 0 & 8 & 5 \\ -4 & 6 & 5 \end{pmatrix}$

Exercise 43:7

1. -2 2. $\frac{10}{3}$ 3. 2 4. ± 2

5. $\begin{pmatrix} -2 & -2 \\ 2 & 2 \end{pmatrix}$ 6. $\begin{pmatrix} 11 & -7 \\ -16 & 10 \end{pmatrix}, -2$

7. −4 or 11   8. −1 or 3   9. 3 or 7

Exercise 43:8

(a) $\begin{pmatrix} 1 & 3 \\ 4 & 5 \end{pmatrix}$ (b) $\begin{pmatrix} 1 & -4 \\ -5 & 3 \end{pmatrix}$ (c)

$\begin{pmatrix} 3 & -1 \\ 2 & 2 \end{pmatrix}$

(d) $\begin{pmatrix} 0 & -1 \\ -2 & 4 \end{pmatrix}$ (e) $\begin{pmatrix} 5 & -2 \\ -3 & 4 \end{pmatrix}$ (f) $\begin{pmatrix} 1 & 3 \\ 2 & -2 \end{pmatrix}$

(g) $\begin{pmatrix} 1 & 0 \\ 0 & 1 \end{pmatrix}$ (h) $\begin{pmatrix} 4 & -1 \\ -2 & \frac{1}{2} \end{pmatrix}$ (i) $\begin{pmatrix} 7 & -5 \\ -3 & 2 \end{pmatrix}$

Exercise 43:9

1. (a) $\begin{pmatrix} -\dfrac{1}{7} & -\dfrac{3}{7} \\ -\dfrac{4}{7} & \dfrac{5}{7} \end{pmatrix}$ (b) $\begin{pmatrix} -\dfrac{1}{17} & -\dfrac{3}{17} \\ -\dfrac{4}{17} & \dfrac{5}{17} \end{pmatrix}$

(c) $\begin{pmatrix} \dfrac{3}{8} & -\dfrac{1}{8} \\ \dfrac{1}{4} & \dfrac{1}{4} \end{pmatrix}$ (d) $\begin{pmatrix} 0 & \dfrac{1}{2} \\ 1 & -2 \end{pmatrix}$

(e) $\begin{pmatrix} \dfrac{5}{14} & -\dfrac{1}{7} \\ -\dfrac{3}{14} & \dfrac{2}{7} \end{pmatrix}$ (f) $\begin{pmatrix} -\dfrac{1}{8} & -\dfrac{3}{8} \\ -\dfrac{1}{4} & \dfrac{1}{4} \end{pmatrix}$

(g) $\begin{pmatrix} 1 & 0 \\ 0 & 1 \end{pmatrix}$ (h) No inverse   (i) $\begin{pmatrix} -7 & 5 \\ 3 & -2 \end{pmatrix}$

2. $\begin{pmatrix} 1 & -1 \\ -2 & 3 \end{pmatrix}, \begin{pmatrix} 1 & 0 \\ 0 & 1 \end{pmatrix}$ 3. $\begin{pmatrix} \dfrac{5}{2} & -\dfrac{3}{2} \\ -\dfrac{1}{2} & \dfrac{1}{2} \end{pmatrix}, \begin{pmatrix} 1 & 0 \\ 0 & 1 \end{pmatrix}$

4. (i) $\begin{pmatrix} \dfrac{3}{14} & \dfrac{1}{7} \\ -\dfrac{1}{14} & \dfrac{2}{7} \end{pmatrix}$ (ii) $\begin{pmatrix} 1 & 0 \\ 0 & 1 \end{pmatrix}$

(iv) $\begin{pmatrix} \dfrac{1}{3} & 0 \\ 0 & \dfrac{1}{2} \end{pmatrix}$ (v) $\begin{pmatrix} \dfrac{1}{8} & \dfrac{1}{4} \\ -\dfrac{3}{8} & \dfrac{1}{4} \end{pmatrix}$

Exercise 43:10

1. $p = 4, q = -2$ 2. $s = 7, t = 4$

3. $x = 4, y = 3$ 4. $m = \dfrac{1}{2}, y = \dfrac{1}{4}$

5. $x = 3, y = 1$ 6. $u = 5, v = 2$

Exercise 43:11

1. (a)
$\begin{array}{c} \\ A \\ B \\ C \\ D \end{array} \begin{array}{cccc} A & B & C & D \\ \begin{pmatrix} 0 & 1 & 1 & 0 \\ 1 & 0 & 2 & 1 \\ 1 & 2 & 0 & 3 \\ 0 & 1 & 3 & 0 \end{pmatrix} \end{array}$

(b)
$\begin{array}{c} \\ A \\ B \\ C \\ D \end{array} \begin{array}{cccccccc} s & t & u & v & w & x & y & z \\ \begin{pmatrix} 1 & 0 & 1 & 0 & 0 & 0 & 0 & 0 \\ 1 & 1 & 0 & 0 & 0 & 1 & 1 & 0 \\ 0 & 1 & 1 & 1 & 1 & 0 & 1 & 1 \\ 0 & 0 & 0 & 1 & 1 & 1 & 0 & 1 \end{pmatrix} \end{array}$

2.
$\begin{array}{c} \\ P \\ Q \\ R \end{array} \begin{array}{ccc} P & Q & R \\ \begin{pmatrix} 0 & 2 & 1 \\ 1 & 0 & 0 \\ 1 & 1 & 2 \end{pmatrix} \end{array}$

Exercise 44:1

1.

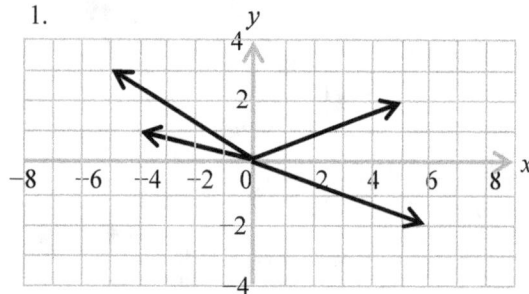

2. $\mathbf{AB} = \begin{pmatrix} 3 \\ -11 \end{pmatrix}$, $\mathbf{CD} = \begin{pmatrix} 7 \\ 6 \end{pmatrix}$, $\mathbf{EF} = \begin{pmatrix} 0 \\ 8 \end{pmatrix}$, $\mathbf{GH} = \begin{pmatrix} 0 \\ 8 \end{pmatrix}$

3.

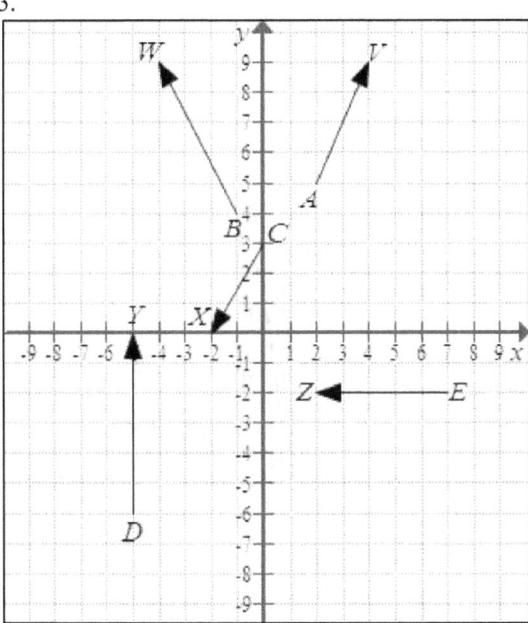

4.

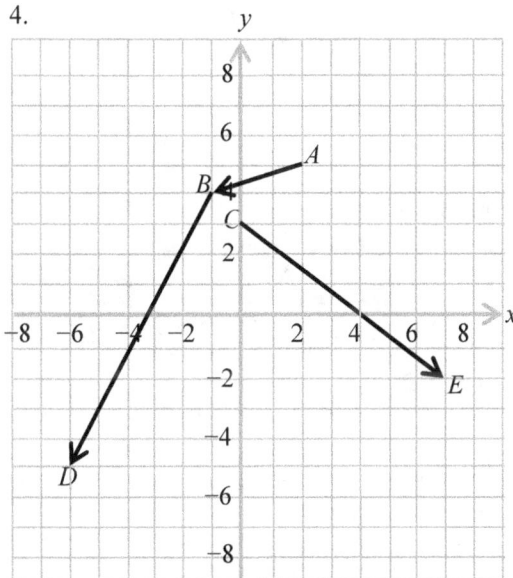

Exercise 44:2

1.

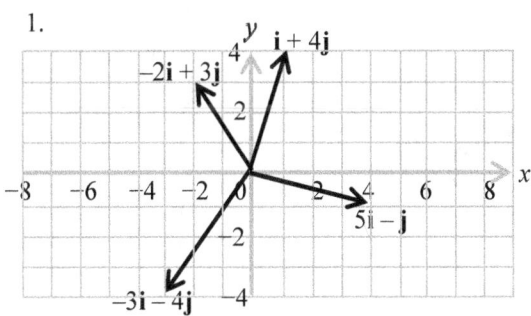

2. (i) $2\mathbf{i} + 5\mathbf{j}$ (ii) $-5\mathbf{i} + 3\mathbf{j}$

  (iii) $-4\mathbf{i} - \mathbf{j}$ (iv) $6\mathbf{i} - 2\mathbf{j}$

3. (i) $3\mathbf{p} + 5\mathbf{q}$ (ii) $-7\mathbf{p} + 2\mathbf{q}$ (iii) $-4\mathbf{p} - 2\mathbf{q}$

4. (i) $\begin{pmatrix} -3 \\ 2 \end{pmatrix}$ (ii) $\begin{pmatrix} 2 \\ -5 \end{pmatrix}$ (iii) $\begin{pmatrix} 9 \\ 4 \end{pmatrix}$ (iv) $\begin{pmatrix} -7 \\ -3 \end{pmatrix}$

Exercise 44:3

1. (i) 13 (ii) $\sqrt{13}$ (iii) $\sqrt{58}$

  (iv) 13 (v) 10 (vi) 5

2. (i) $\sqrt{41}$ (ii) $4\sqrt{2}$ (iii) $\sqrt{13}$

3. $\mathbf{i}$, $\frac{3}{5}\mathbf{i} + \frac{4}{5}\mathbf{j}$, $\mathbf{j}$, $\frac{1}{\sqrt{2}}\mathbf{i} - \frac{1}{\sqrt{2}}\mathbf{j}$, $-\frac{\sqrt{3}}{2}\mathbf{i} + \frac{1}{2}\mathbf{j}$

4. 5 5. $\sqrt{13}$

Exercise 44:4

1. (a) $\mathbf{OA} = \begin{pmatrix} -10 \\ 12 \end{pmatrix}$, $\mathbf{OB} = \begin{pmatrix} 8 \\ 10 \end{pmatrix}$, $\mathbf{OC} = \begin{pmatrix} 6 \\ 6 \end{pmatrix}$,

$\mathbf{OD} = \begin{pmatrix} 6 \\ -4 \end{pmatrix}$, $\mathbf{OE} = \begin{pmatrix} -6 \\ -2 \end{pmatrix}$, $\mathbf{OF} = \begin{pmatrix} -8 \\ 2 \end{pmatrix}$,

  (b) $\mathbf{OA} = -10\mathbf{i}+12\mathbf{j}$, $\mathbf{OB} = 8\mathbf{i}+10\mathbf{j}$, $\mathbf{OC} = 6\mathbf{i}+6\mathbf{j}$, $\mathbf{OD} = 6\mathbf{i}-4\mathbf{j}$, $\mathbf{OE} = -6\mathbf{i}-2\mathbf{j}$, $\mathbf{OF} = -8\mathbf{i}+2\mathbf{j}$

2. (i) (a) $\mathbf{OA} = \begin{pmatrix} -1 \\ 5 \end{pmatrix}$, $\mathbf{OB} = \begin{pmatrix} 4 \\ -7 \end{pmatrix}$, $\mathbf{OC} = \begin{pmatrix} -9 \\ 3 \end{pmatrix}$,

$\mathbf{OD} = \begin{pmatrix} -3 \\ 6 \end{pmatrix}$

  (b) $\mathbf{OA} = -\mathbf{i} + 5\mathbf{j}$, $\mathbf{OB} = 4\mathbf{i} - 7\mathbf{j}$, $\mathbf{OC} = -9\mathbf{i}+3\mathbf{j}$, $\mathbf{OD} = -3\mathbf{i} + 6\mathbf{j}$

(ii)

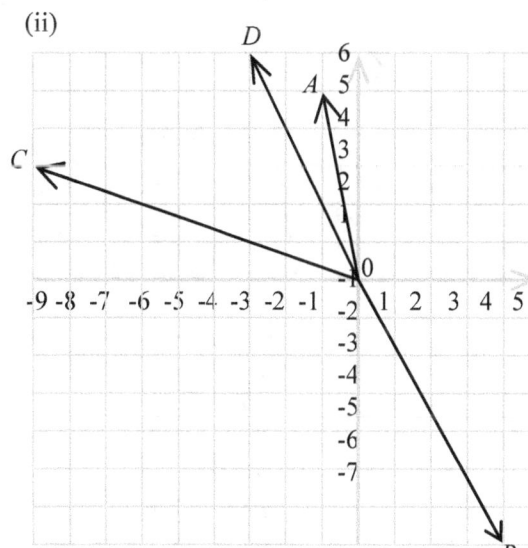

3. $\mathbf{AB} = \mathbf{DC}$, $\mathbf{EF} = \mathbf{KL}$, $\mathbf{ST} = \mathbf{MN}$, $\mathbf{PQ} = \mathbf{XY}$.
4. a and l, b, g, j and m, c and i, d and k, e and p.
5. $\mathbf{OD}$ and $\mathbf{OE}$.

Exercise 44:**5**

1. (i)

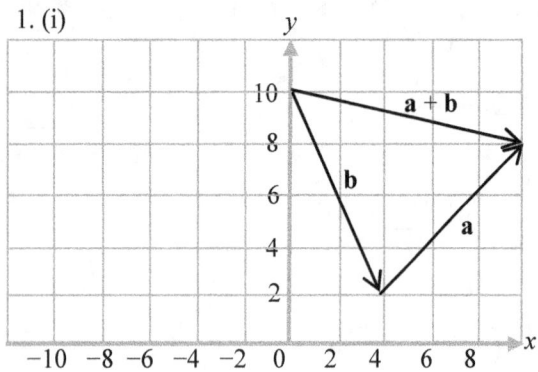

Or

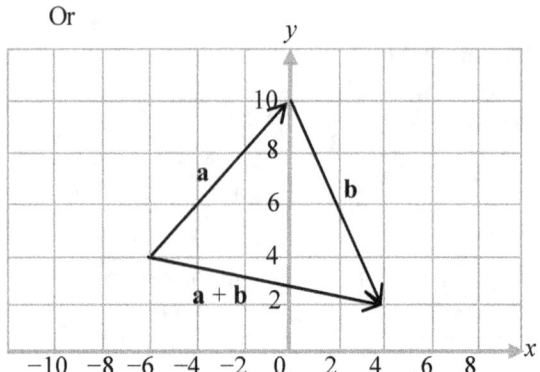

(ii)

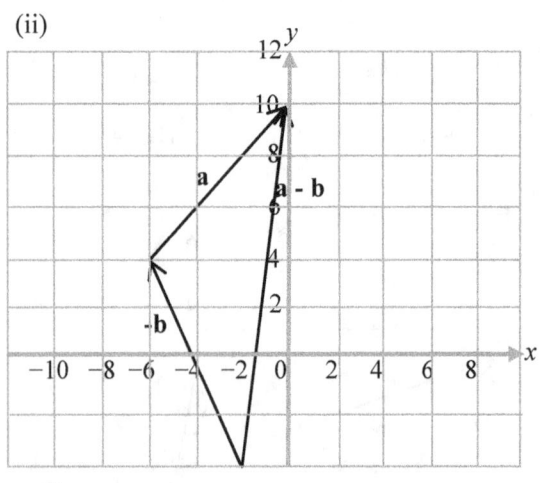

Or

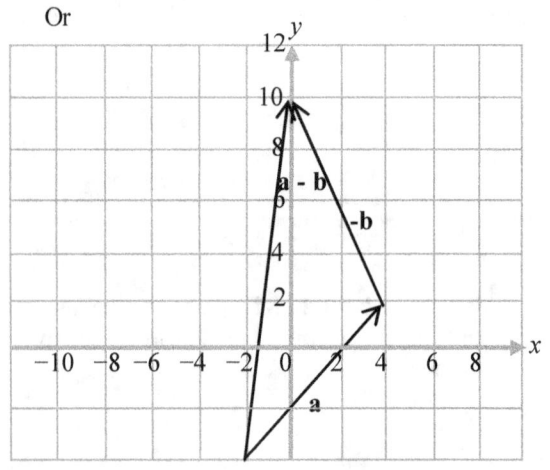

(iii)

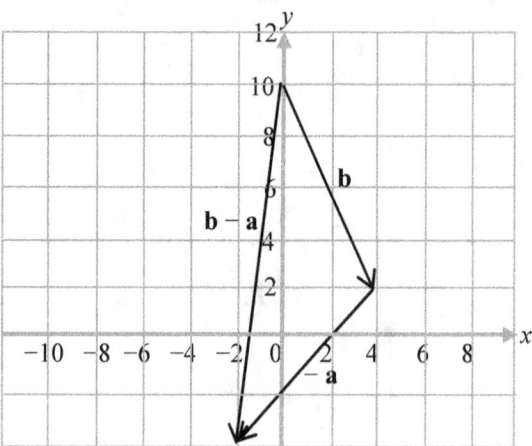

Or

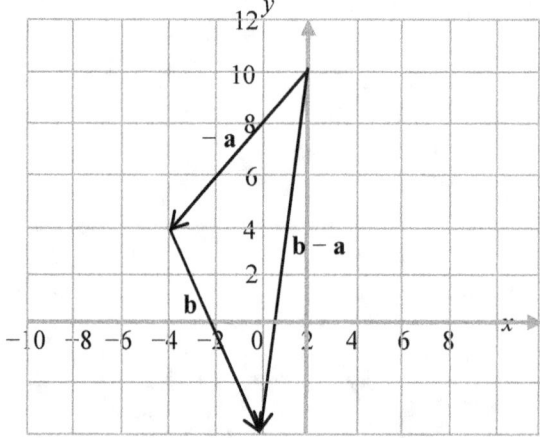

2. (i) $-2\mathbf{i} + 2\mathbf{j}$ (ii) $-2\mathbf{i} + 2\mathbf{j}$ (iii) $8\mathbf{i} - 4\mathbf{j}$ (iv) $-8\mathbf{i} + 4\mathbf{j}$

3. (a) $\mathbf{i} + 2\mathbf{j}$ (b) $-2\mathbf{i} - 5\mathbf{j}$ (c) $-2\mathbf{i} + 7\mathbf{j}$

   (d) $\mathbf{i} - 2\mathbf{j}$ (e) $\begin{pmatrix} 2 \\ 4 \end{pmatrix}$ (f) $\begin{pmatrix} -7 \\ -5 \end{pmatrix}$

   (g) $\begin{pmatrix} 3 \\ -2 \end{pmatrix}$ (h) $\begin{pmatrix} -8 \\ 3 \end{pmatrix}$

4. (i) $\begin{pmatrix} -1 \\ 7 \end{pmatrix}$ (ii) $\begin{pmatrix} -1 \\ 7 \end{pmatrix}$ (iii) $\begin{pmatrix} 5 \\ 3 \end{pmatrix}$ (iv) $\begin{pmatrix} -5 \\ -3 \end{pmatrix}$

Conclusions:

$\mathbf{a} + \mathbf{b} = \mathbf{b} + \mathbf{a}$ and $\mathbf{a} - \mathbf{b} = -(\mathbf{b} - \mathbf{a})$

Exercise 44:**6**

1. $a = 2$ and $b = -1$

2. (i) $\begin{pmatrix} 9 \\ 4 \end{pmatrix}$ (ii) $\begin{pmatrix} -1 \\ -10 \end{pmatrix}$ (iii) $\begin{pmatrix} 1 \\ 10 \end{pmatrix}$ (iv) $\begin{pmatrix} 3 \\ 8 \end{pmatrix}$

   (v) $\begin{pmatrix} 3 \\ 8 \end{pmatrix}$ (vi) $\begin{pmatrix} 12 \\ -9 \end{pmatrix}$ (vii) $\begin{pmatrix} 8 \\ 5 \end{pmatrix}$ (viii) $\begin{pmatrix} -21 \\ 25 \end{pmatrix}$

   (ix) $\begin{pmatrix} \frac{14}{3} \\ -\frac{11}{3} \end{pmatrix}$ (x) $\begin{pmatrix} 31 \\ -11 \end{pmatrix}$

3. $\begin{pmatrix} 3 \\ -3 \end{pmatrix}$, $x + y = 3z$ i.e. $x + y$ is a scalar multiple of **z**.

4. (a) Since $4\mathbf{CD} = \mathbf{AB}$, **AB** is parallel to **CD**. (b) $4 : 1$

5. $5\sqrt{2}$ 6. $3u + v = 5$, $u + v = 3$; $u = 1, v = 2$

Exercise 44:7

1. $\mathbf{OD} = \dfrac{1}{2}(\mathbf{b} + \mathbf{c})$, $\mathbf{OG} = \dfrac{1}{3}(\mathbf{a} + 2\mathbf{c})$

2. $\mathbf{TM} = \dfrac{2}{3}\mathbf{r} + \dfrac{1}{2}\mathbf{p}$, $\mathbf{PM} = \mathbf{r} - \dfrac{1}{2}\mathbf{p}$

3. (a) $\mathbf{d} - \mathbf{a}$ (b) (i) $\mathbf{a} + \mathbf{b}$ (ii) $\mathbf{d} + \mathbf{c}, \mathbf{c} = \dfrac{1}{2}\mathbf{d}$

4. (a) $\mathbf{a} + \mathbf{b}$ (b) (i) $\mathbf{a} - \mathbf{b}$

5. $\mathbf{PQ} = -3\mathbf{a} - \mathbf{b}, \mathbf{QR} = 3\mathbf{a} - 2\mathbf{b}, \mathbf{RP} = 3\mathbf{b}$

6. $\dfrac{1}{4}(\mathbf{a} + 3\mathbf{b})$

7. (a)(i) $12\mathbf{i} - 3\mathbf{j}$

(ii) $\mathbf{OL} = \dfrac{m}{m + n}(-3\mathbf{i} + 6\mathbf{j})$, $\mathbf{OM} = \dfrac{m}{m + n}(9\mathbf{i} + 3\mathbf{j})$

   (iii) $\mathbf{LM} = \dfrac{m}{m + n}(12\mathbf{i} - 3\mathbf{j})$

  (b) $\dfrac{m}{m + n}\mathbf{AB}$

   (c) (i) $2:3$ (ii) 17.5 square units

8. (a) $\mathbf{a} + \mathbf{b}$ (b) $\mathbf{a} - \mathbf{b}$

Exercise 44:8

1. (a) 5 (b) $90°$ 2. (a) $\sqrt{13}$ (b) $90°$

3. $18\sqrt{2}$ 4. (a) 0 (b) 27   5. $120°$

6. 0, perpendicular   7. $44°$ 8. $73.7°$

Exercise 45:1

1. $(2, -3)$

2.
	(2,3)	(3,0)	(0,-5)	(3,3)	(-4,-6)
x-axis	(2,-3)	(3,0)	(0,5)	(3,-3)	(-4,6)
y-axis	(-2,3)	(-3,0)	(0,-5)	(-3,3)	(4,-6)
$y = x$	(3,2)	(0,3)	(-5,0)	(3,3)	(-6,-4)
$y = -x$	(-3,-2)	(0,-3)	(5,0)	(-3,-3)	(6,4)

3. $(2, -4)$

4.

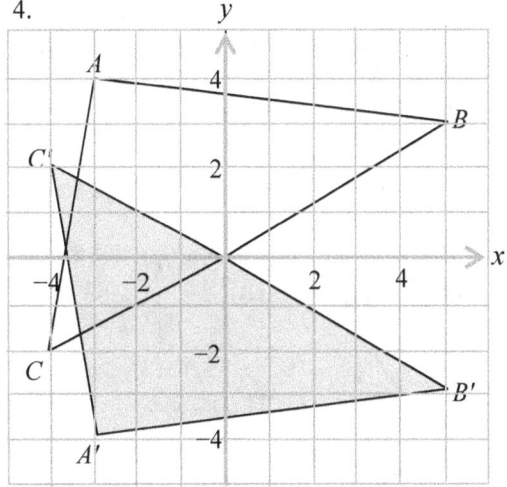

5. (i) $\left(\dfrac{3\sqrt{3}}{2} - 1, \sqrt{3} + \dfrac{3}{2} \right)$ (ii) $\left(\dfrac{3}{2}, \dfrac{3\sqrt{3}}{2} \right)$

(iii) $\left(-\dfrac{5\sqrt{3}}{2}, \dfrac{5}{2} \right)$ (iv) $\left(\dfrac{3\sqrt{3}}{2} - \dfrac{3}{2}, \dfrac{3\sqrt{3}}{2} + \dfrac{3}{2} \right)$

(v) $\left(2 - 3\sqrt{3}, 3 - 2\sqrt{3} \right)$

6. $(-7, 7)$

7. (a) $(3,3)$ (b) $(2,4)$ (c) $(-1,2)$ (d) $(1, 5)$
   (e) $(0, 1)$ (f) $(0,0)$ (g) $(a+2, b+3)$
   (h) $(3,4)$ (i) $(-9, 18)$ (j) $(-2, 6)$

8.

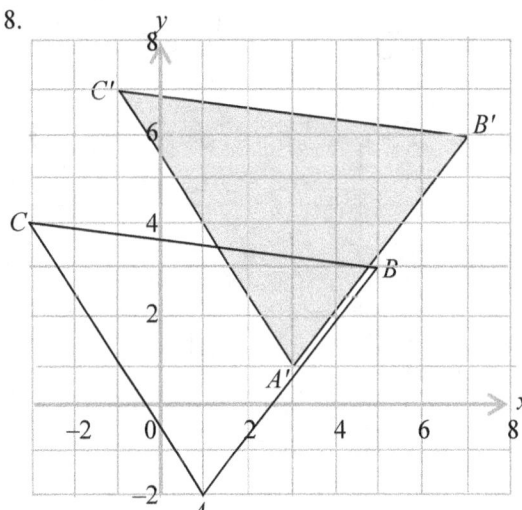

T_1 is the inverse of T. TT_1 is the identity transformation, because it leaves all points unchanged.

9. (a) $(1, -7)$ (b) $(2, -6)$ (c) $(2, -7)$
   (d) $(0, -5)$ (e) $(5, -9)$ (f) $(-1, -10)$

10.

The single transformation is a translation of 4.5 cm along AC.

11. $y = x^2 + 8x + 8$ 12. $(-1, 4)$

13. (a) $(1,4)$ (b) $(-5,4)$ (c) $(-7,3)$ (d) $(-1,3)$

Exercise 45:2

1. (a) 6   (b) 5   (c) 8

2. See next column

3. (i) (a) b, g, h, I   (b) b, d, e, g, h

   (ii) (a) N (b) 2 (c) N (d) 1 (e) 1

       (f) N (g) 8 (h) 4 (i) N

4. (i)

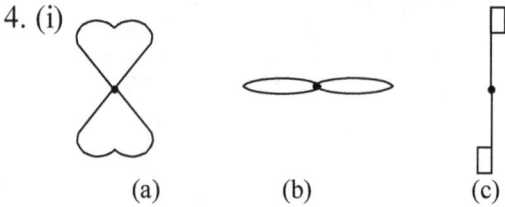

       (a)   (b)   (c)

   (ii)

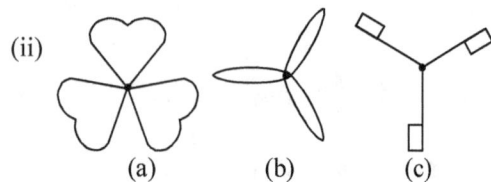

       (a)   (b)   (c)

   (iii)

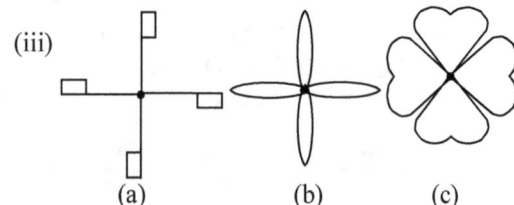

       (a)   (b)   (c)

2.

Name of figure	Diagram	No. of lines of symmetry
(a) Kite		1
(b) Equilateral triangle		3
(c) Square		4
(d) Rectangle		2
(e) Regular pentagon		5
(f) Rhombus		2
(g) Parallelogram		0
(h) Regular hexagon		6

5.   A, B, C, D, E, I, K, M, T, U, V, W, Y.

6. (a) 9   (b) 8   (c) 8   (d) 4

7. H, O, X

8. F, G, J, L, N, P, Q, R, S, Z.

9.   (a)   (b)   (c)

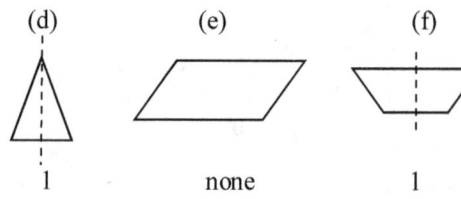

        4   3   6

        (d)   (e)   (f)

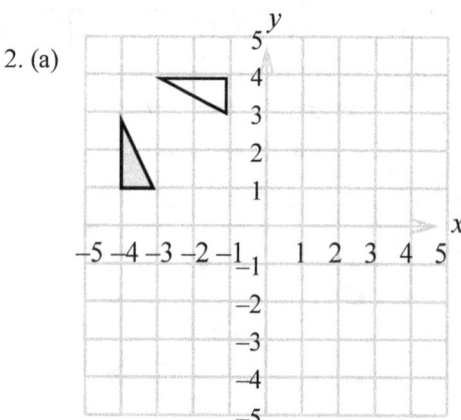

        1   none   1

10. $H \cap V = \{ H, I, O, X \}$

    $H \cap V = R$ and all Members of $H \cap V$ have at least two lines of symmetry.

Exercise 46:1

1. (a) $(-5, -3)$ (b) $(-10, -6)$

   (c) $\left(-\dfrac{5}{2}, \dfrac{3}{2}\right)$ (d) $(-3, -5)$

2. (a)

(b)

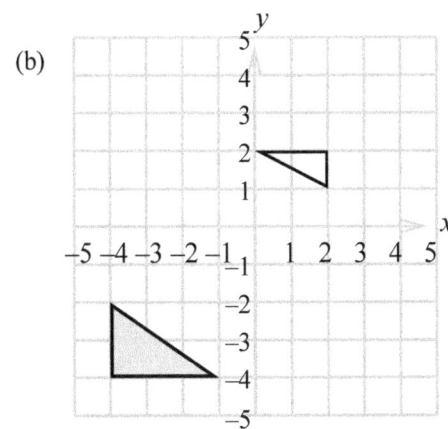

Exercise 46:2

1. $(0,0), (2,2), (2,5), (0,3)$

2. $(0,0), (7,-5), (6,-4), (-1,1)$

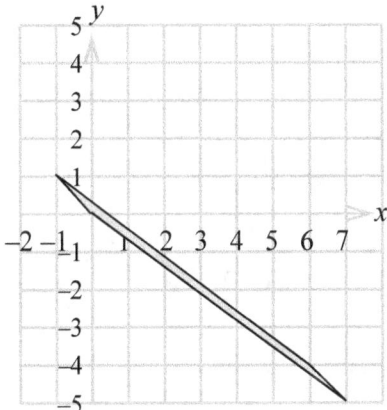

3. (a)

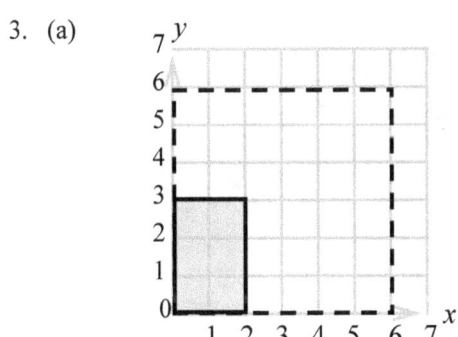

(b)

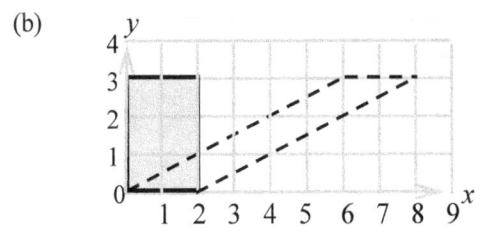

(c)

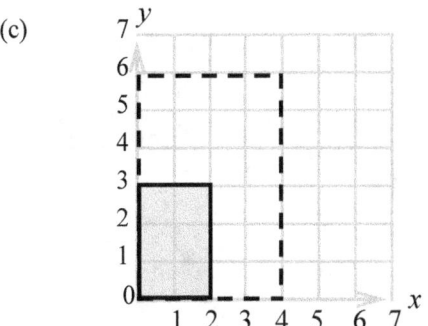

Exercise 46:3

1. **A** is an identity transformation; object and image are identical.
   B is a rotation about the origin through 90° in an anticlockwise sense.
   C is a rotation about the origin through 90° in an clockwise sense.
   D is a rotation about the origin through 180° in the clockwise or anticlockwise sense.
   E is a reflection in the line $y = x$.
   F is a reflection in the y-axis.
   G is a reflection in the line $y = -x$.
   H is a reflection in the x-axis.
   P is an enlargement scale factor 2, center $(0,0)$.
   Q is an enlargement scale factor $\dfrac{1}{2}$, center $(0,0)$.
   R is a rotation about the origin through 180° in the clockwise or anticlockwise sense followed by an enlargement scale factor $\dfrac{1}{2}$.
   S is a rotation about the origin through 180° in the clockwise or anticlockwise sense followed by an enlargement scale factor 2.
   T is a stretch, stretch factor 2 in the Oy direction
   U is a shear, shear factor 1 in the Ox direction with points on the x-axis invariant.
   V is a shear, shear factor 2 in the Oy direction with points on the y-axis invariant.
   W is a stretch, stretch factor 3 in the Ox direction.
   X is a stretch, stretch factor 2 in the negative Ox direction.
   Y is a stretch, stretch factor 3 in the Oy direction.

2.

587

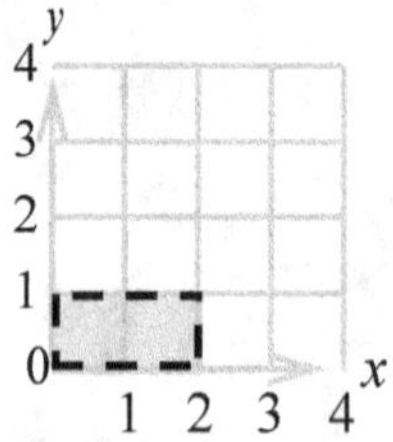

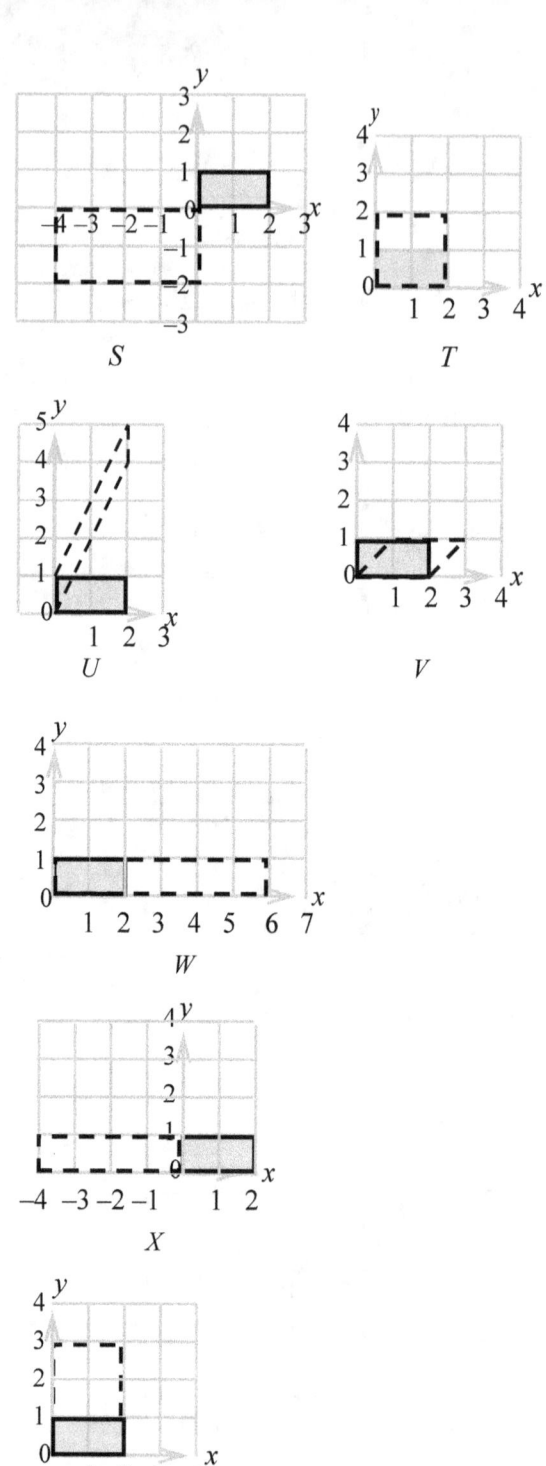

A

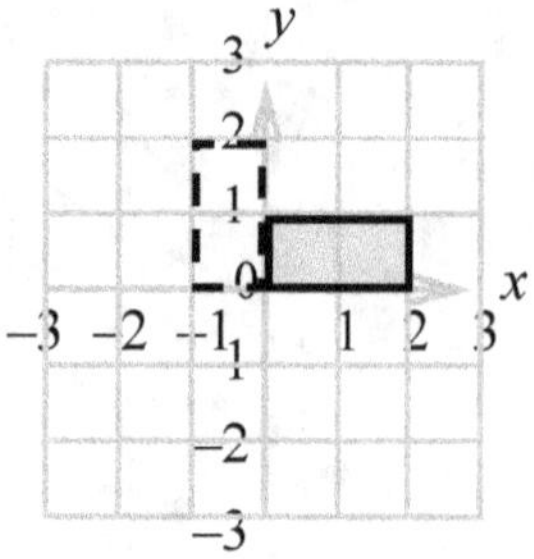

B

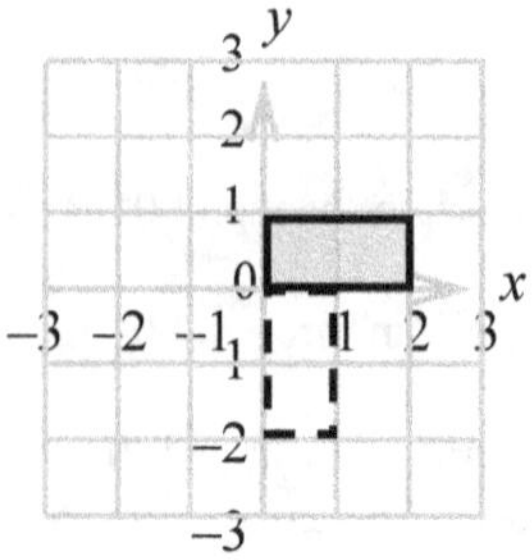

C

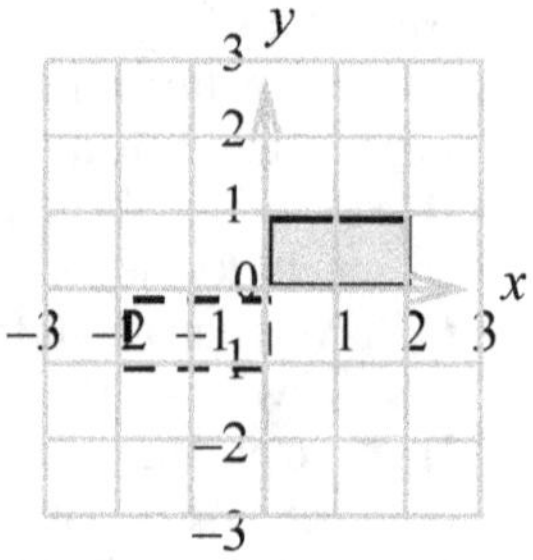

D

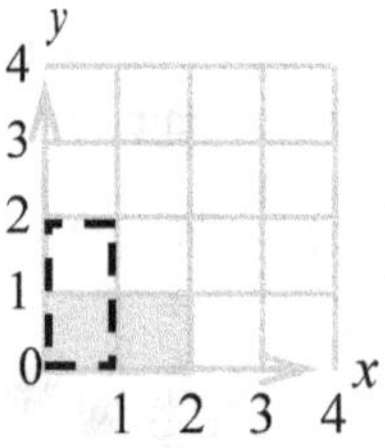

E

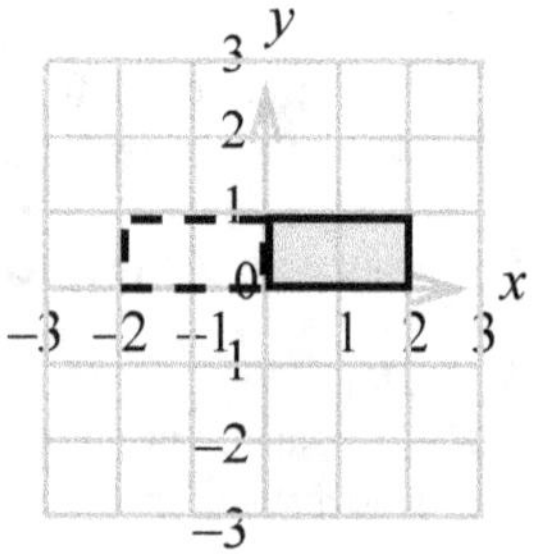

F

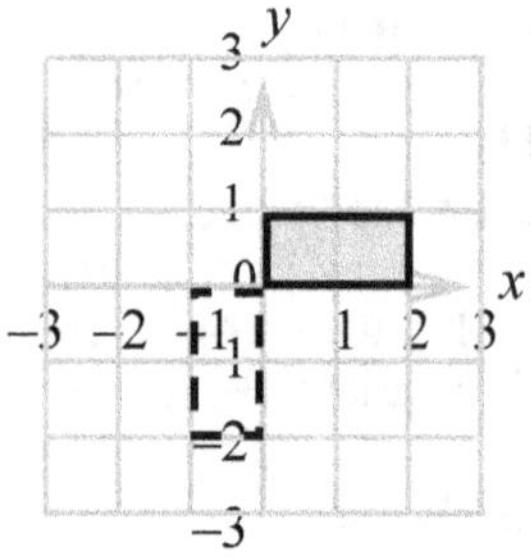

G

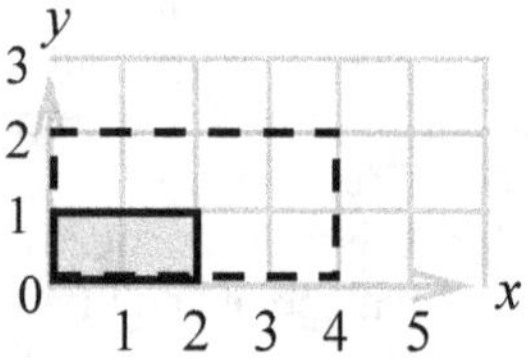

P

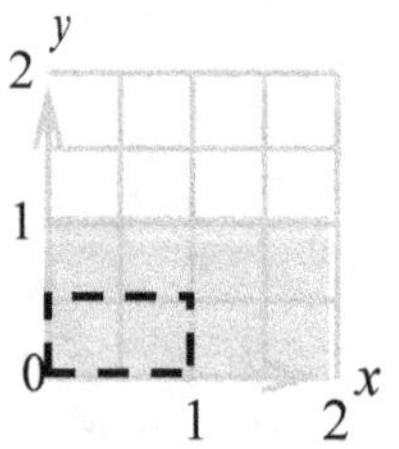

Q

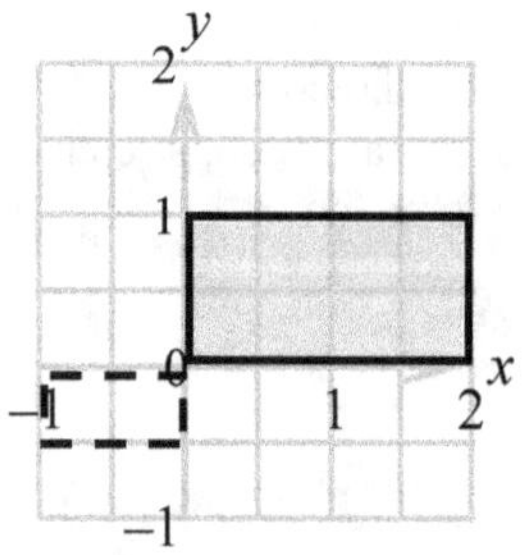

R

S

T

U

V

W

X

Y

Exercise 46:4

1. $\begin{pmatrix} 3 & 2 \\ 1 & 1 \end{pmatrix}$

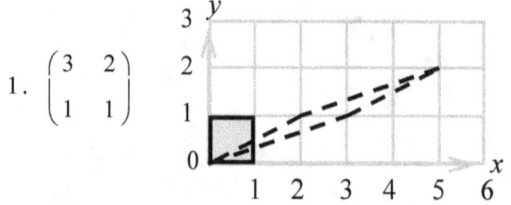

A shear along the line $y = \dfrac{5}{2}x$ with shear factor 5.

2. $\begin{pmatrix} 3 & 1 \\ 0 & 1 \end{pmatrix}$

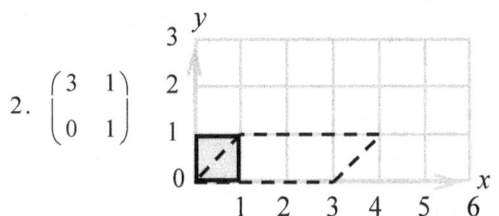

A shear in the direction Ox, shear factor 3

3. $\begin{pmatrix} 2 & 1 \\ 0 & 0 \end{pmatrix}$

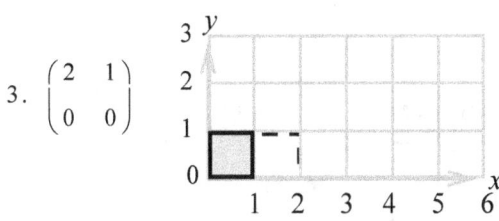

4. $\begin{pmatrix} 2 & 0 \\ 0 & 2 \end{pmatrix}$

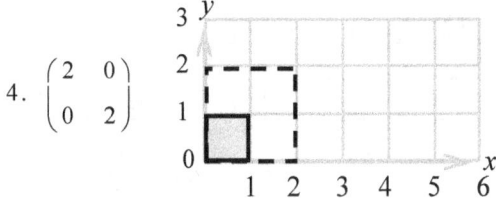

5. $\begin{pmatrix} 4 & 1 \\ 0 & 1 \end{pmatrix}$

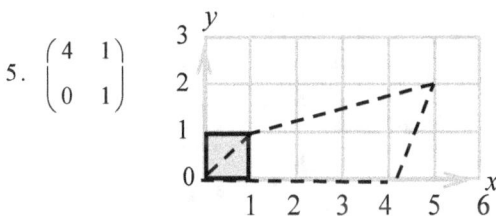

6. $\begin{pmatrix} 1 & 0 \\ 0 & -1 \end{pmatrix}$

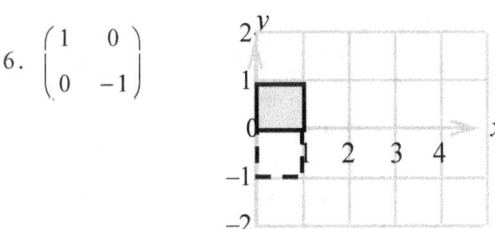

7. $\begin{pmatrix} 3 & 0 \\ 0 & -3 \end{pmatrix}$

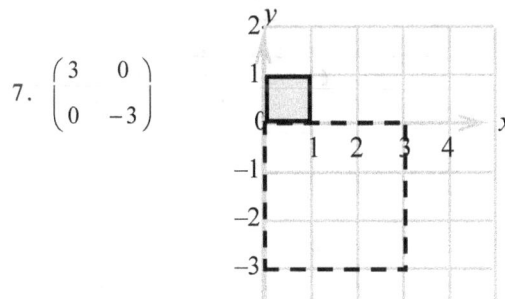

8. $\begin{pmatrix} 0 & -1 \\ 1 & 1 \end{pmatrix}$

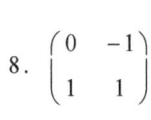 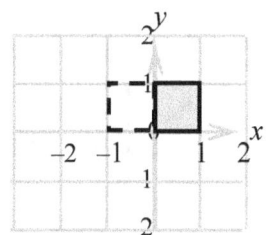

9. $\begin{pmatrix} 3 & 0 \\ 0 & 3 \end{pmatrix}$

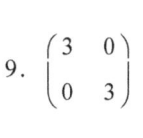

10. $\begin{pmatrix} 1 & -2 \\ 0 & 1 \end{pmatrix}$

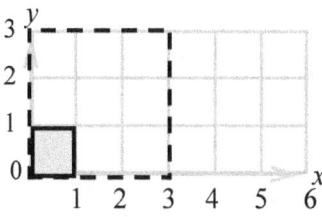

11. $\begin{pmatrix} 3 & 1 \\ 1 & 2 \end{pmatrix}$

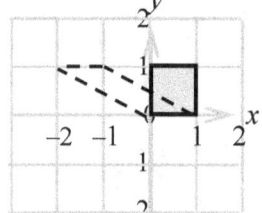

12. $\begin{pmatrix} 0 & 1 \\ \frac{1}{2} & 1 \end{pmatrix}$

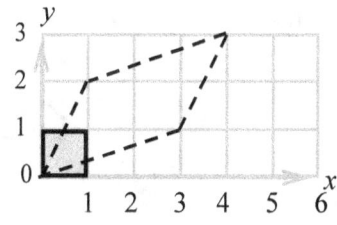 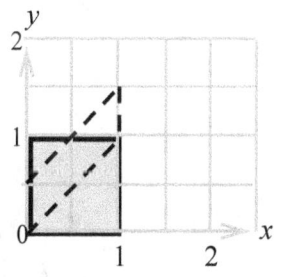

Exercise 46:5

1. (a) $\begin{pmatrix} 0 & -6 & -8 & -6 \\ 0 & -8 & 0 & -2 \end{pmatrix}$, 8:1

   (b) $\begin{pmatrix} 8 & 4 & 0 & 2 \\ -2 & -8 & -2 & 0 \end{pmatrix}$, 8:1

2. (a) $\begin{pmatrix} 2 & 8 & 2 \\ 2 & 2 & 6 \end{pmatrix}$ (b) 8:1   3. 48 cm^2

4. 0 un^2

589

5.

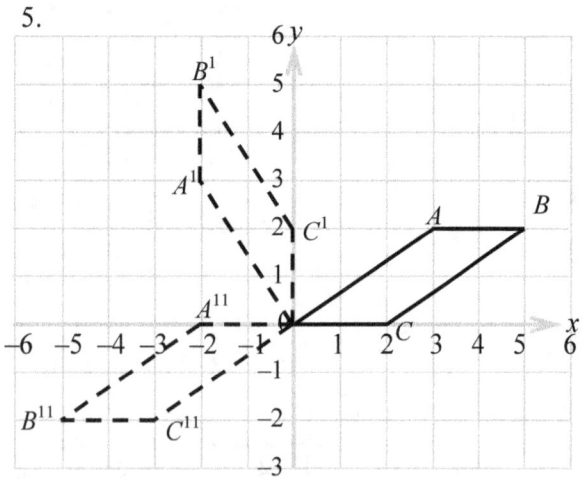

(d) *T* is a rotation anti-clockwise through 90° about (0,0).
T^2 is a rotation anti-clockwise through 180° about (0,0)

6. *M* represents a reflection in the *x*-axis.
$A^1(-1,-3)$, $B^1(-3,0)$, $C^1(-1,2)$ and $D^1(1,-1)$.

7. (a) $A^1(0,-1)$, $B^1(1,-2)$, $C^1(0,-4)$.

(b)

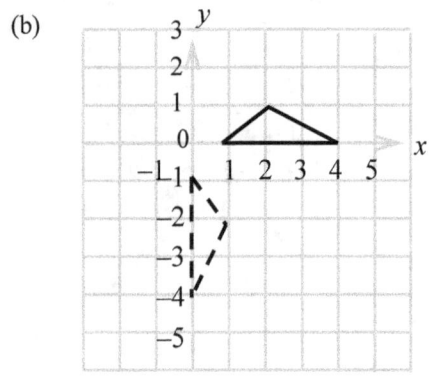

(c) *T* is a rotation clockwise through 90° about (0,0).

(d) $\mathbf{T}^{-1} = \begin{pmatrix} 3 & 1 \\ 1 & 2 \end{pmatrix}$. T^{-1} is a rotation anti-clockwise through 90° about (0,0).

8. (a) $A^1(-6,-4)$.
(b) *M* is a rotation anti-clockwise through 90° about (0, 0).

9. *ABCD* is a parallelogram.
M represents a rotation anti-clockwise through 180° about (0,0), followed by an enlargement scale factor 2.
$A''(0,8)$, $B''(4,8)$, $C''(12,4)$, $D''(8,4)$

Area of $ABCD : A''B''C''D'' = 16:1$

Exercise 47:1

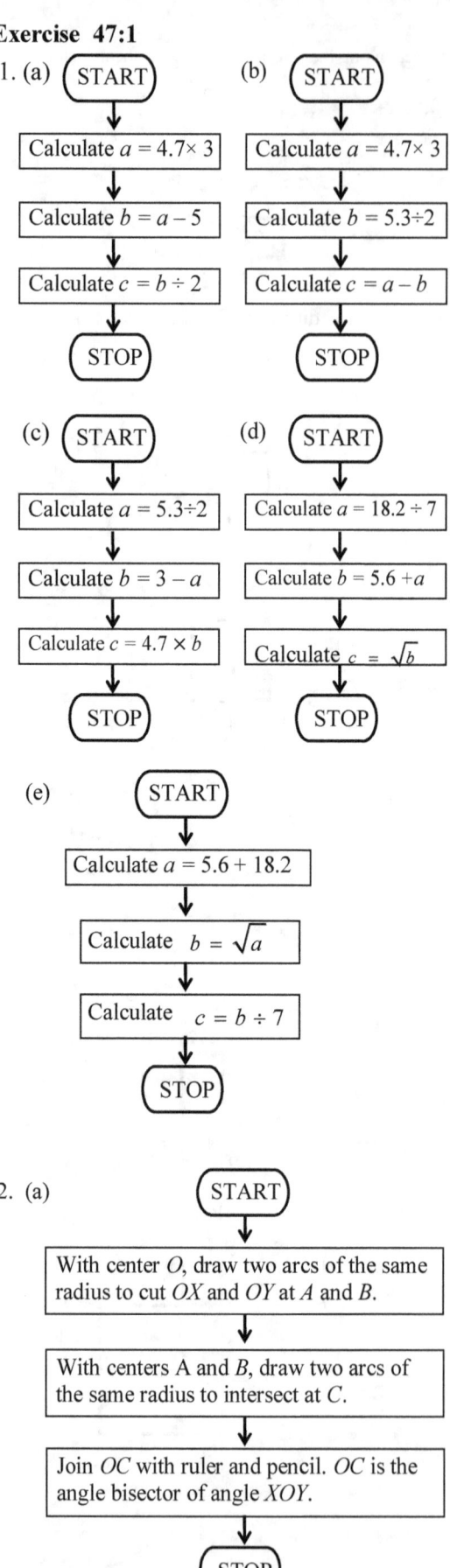

1. (a) START → Calculate $a = 4.7 \times 3$ → Calculate $b = a - 5$ → Calculate $c = b \div 2$ → STOP

(b) START → Calculate $a = 4.7 \times 3$ → Calculate $b = 5.3 \div 2$ → Calculate $c = a - b$ → STOP

(c) START → Calculate $a = 5.3 \div 2$ → Calculate $b = 3 - a$ → Calculate $c = 4.7 \times b$ → STOP

(d) START → Calculate $a = 18.2 \div 7$ → Calculate $b = 5.6 + a$ → Calculate $c = \sqrt{b}$ → STOP

(e) START → Calculate $a = 5.6 + 18.2$ → Calculate $b = \sqrt{a}$ → Calculate $c = b \div 7$ → STOP

2. (a) START → With center *O*, draw two arcs of the same radius to cut *OX* and *OY* at *A* and *B*. → With centers A and B, draw two arcs of the same radius to intersect at *C*. → Join *OC* with ruler and pencil. *OC* is the angle bisector of angle *XOY*. → STOP

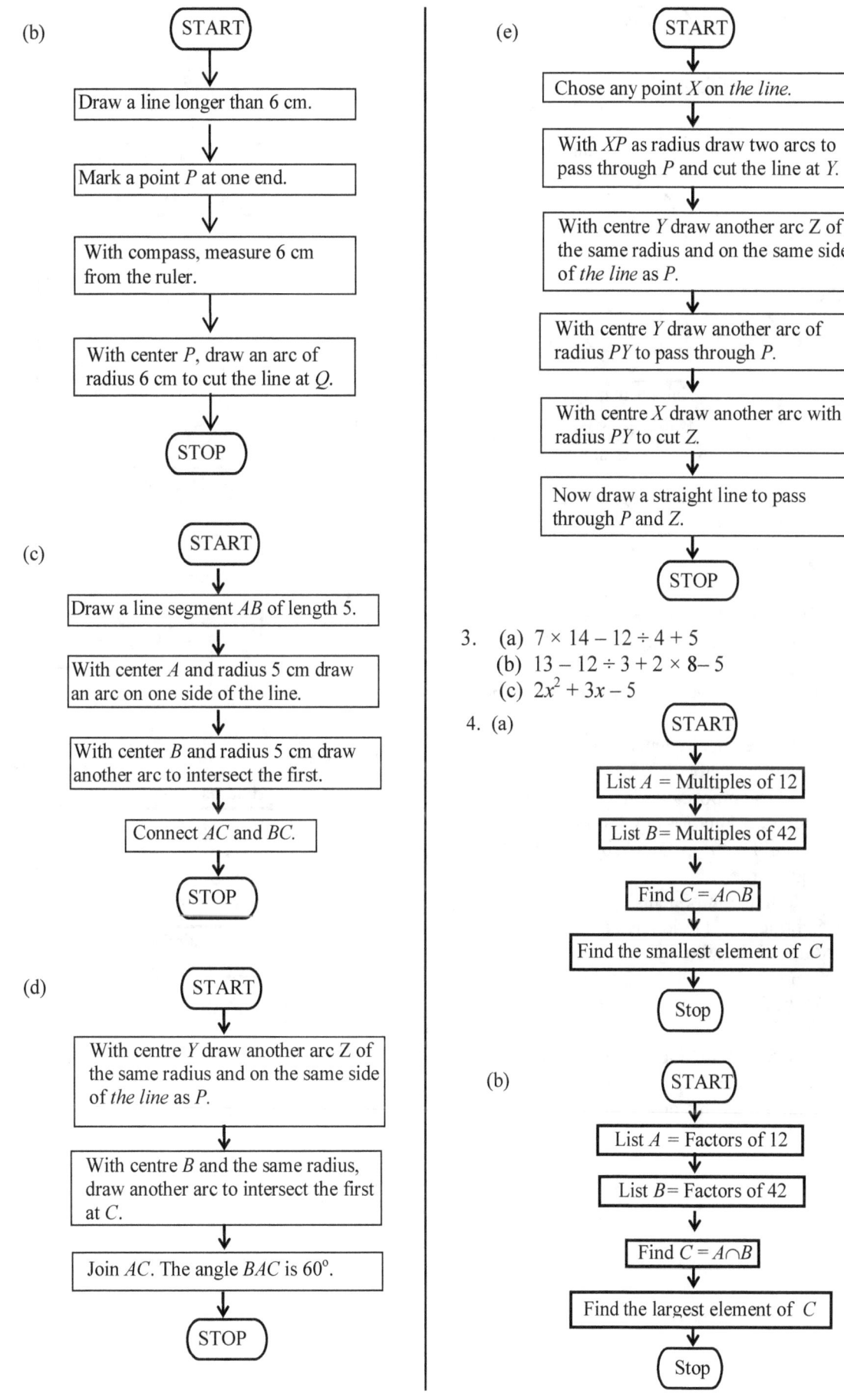

(b)

START

Draw a line longer than 6 cm.

Mark a point P at one end.

With compass, measure 6 cm from the ruler.

With center P, draw an arc of radius 6 cm to cut the line at Q.

STOP

(c)

START

Draw a line segment AB of length 5.

With center A and radius 5 cm draw an arc on one side of the line.

With center B and radius 5 cm draw another arc to intersect the first.

Connect AC and BC.

STOP

(d)

START

With centre Y draw another arc Z of the same radius and on the same side of *the line* as P.

With centre B and the same radius, draw another arc to intersect the first at C.

Join AC. The angle BAC is $60°$.

STOP

(e)

START

Chose any point X on *the line*.

With XP as radius draw two arcs to pass through P and cut the line at Y.

With centre Y draw another arc Z of the same radius and on the same side of *the line* as P.

With centre Y draw another arc of radius PY to pass through P.

With centre X draw another arc with radius PY to cut Z.

Now draw a straight line to pass through P and Z.

STOP

3. (a) $7 \times 14 - 12 \div 4 + 5$
   (b) $13 - 12 \div 3 + 2 \times 8 - 5$
   (c) $2x^2 + 3x - 5$

4. (a)

START

List A = Multiples of 12

List B = Multiples of 42

Find $C = A \cap B$

Find the smallest element of C

Stop

(b)

START

List A = Factors of 12

List B = Factors of 42

Find $C = A \cap B$

Find the largest element of C

Stop

591

(c)

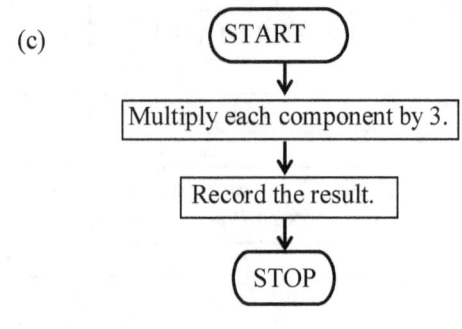

Exercise 47:2

1.

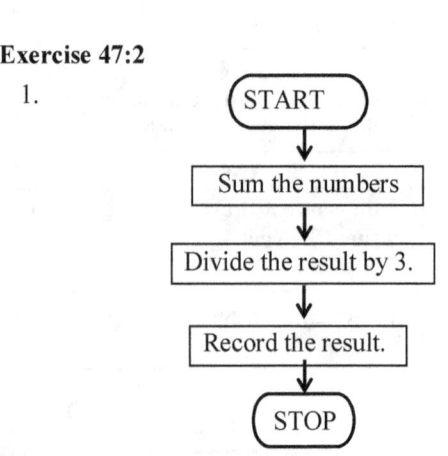

The instructions in the boxes between start and stop for question 2 to 11 are as stated.

2. 'List all the elements in both sets A and B', 'Separate the elements by commas', 'Enclose the elements in braces'.

3. 'Find two factors of -35 whose sum is -2', 'Decompose $-2x$ in terms of these factors', 'Factorise by grouping'.

4. 'Write $A = \pi r^2$', 'Divide by π', 'Find the square root'.

5. 'Add like components'.

6. 'Subtract 3', 'Divide by 2'.

7. 'Add 9', 'Divide by 7'.

8. 'Subtract 5', 'Subtract 2'.

9. 'Multiply by 2', 'Subtract 5', 'Divide by 5'.

10. 'Add 5 to both sides', 'Subtract x from both sides', 'Divide both sides by 2'.

11. 'Add 6 to both sides', 'Subtract x from both sides',

12.

Exercise 47:3

1.

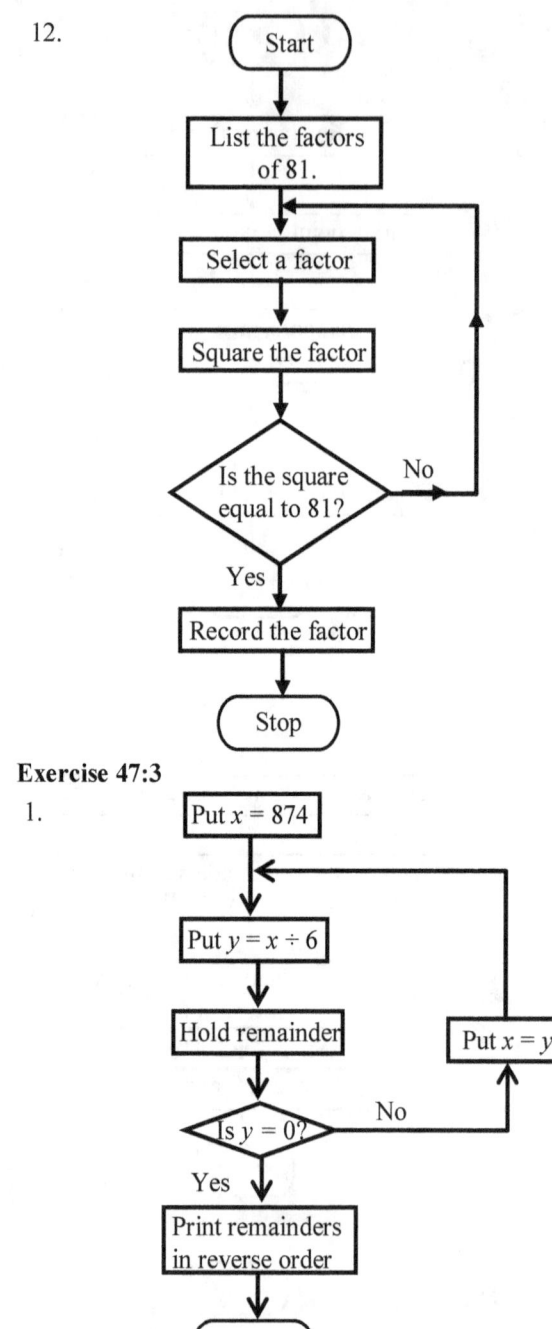

2.

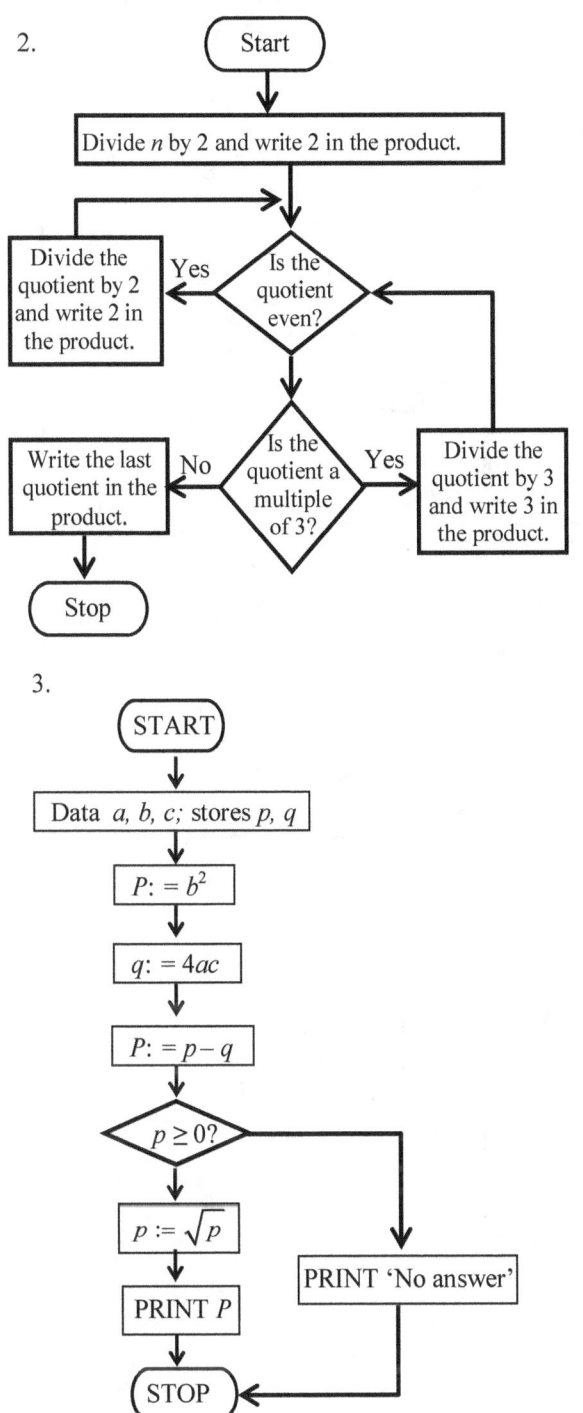

3.

4.

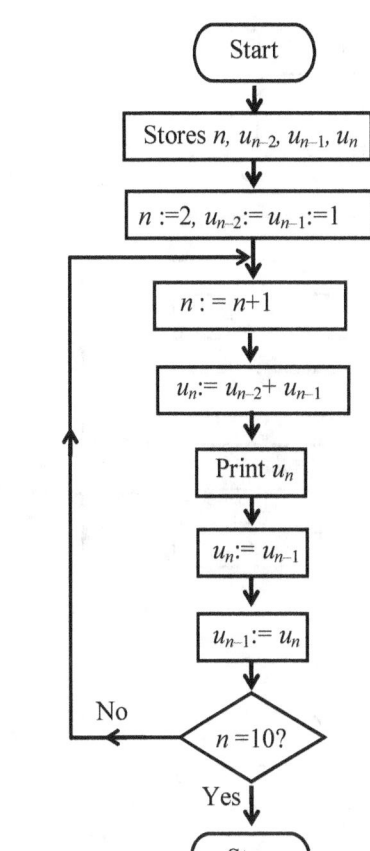

5. 1, 3, 4, 7, 11, 18, 29, 47,…
6. (a) (i) 1, 3, 6, 10, 15, 21, 28, …
        (ii) 1, 9, 36, 100,…
        (iii) The cubes 1, 8, 27, 64, 125,...
    (b) The sequences in (a) (i) and (a) (ii) are
         the triangular numbers and the squares
         of the triangular numbers respectively.
7. (i)   2, 2, 3   (ii) 3, 5, 5 (iii) 7, 7
         The prime factors are printed.
8. (i) 3   (ii) $AX^2 + BX + C$

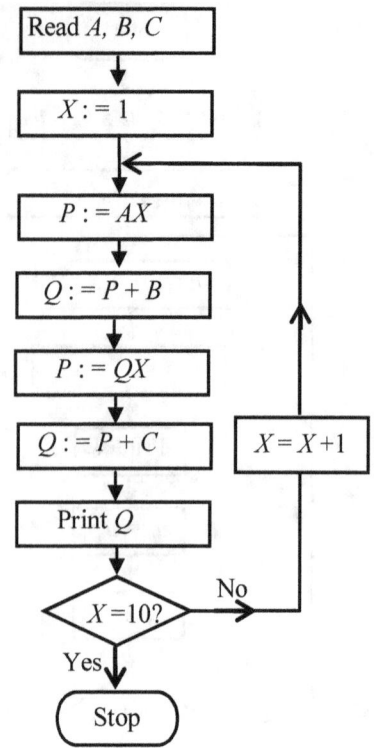

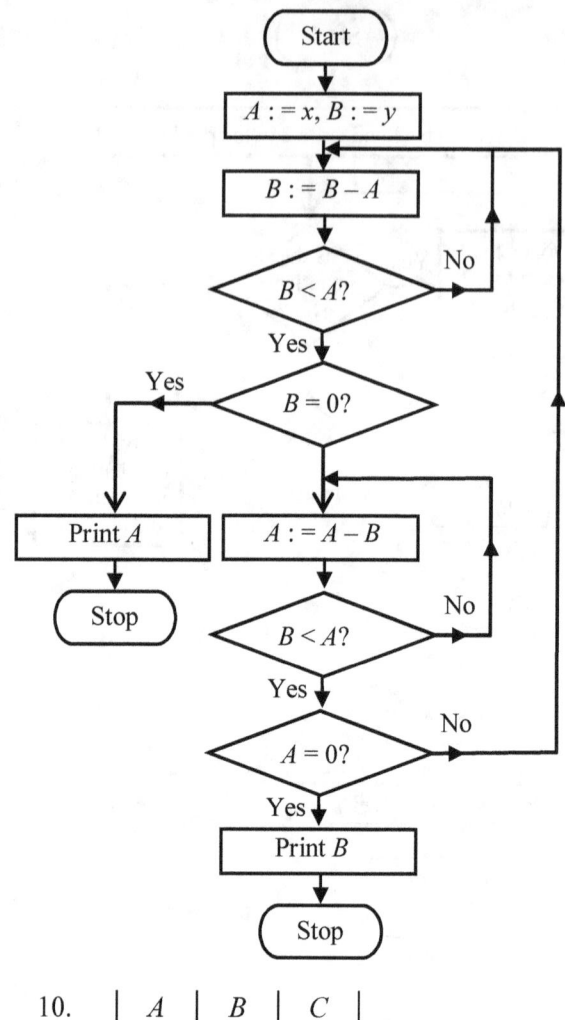

9.

A	B
12	30
12	18
12	6
6	6
0	6

The program fails because it this not take care of the case when $B = 0$. The program can be amended as in the figure below.

10.

A	B	C
5	–4	2
3.4	1.6	10
3.02	0.38	6.8

$$\frac{x^2 + 9}{2x}$$

11. (a) $1 + 2 + 3 + 4 + \ldots + 20$
    (b) $A := A + 2$ (for $A + 1$)

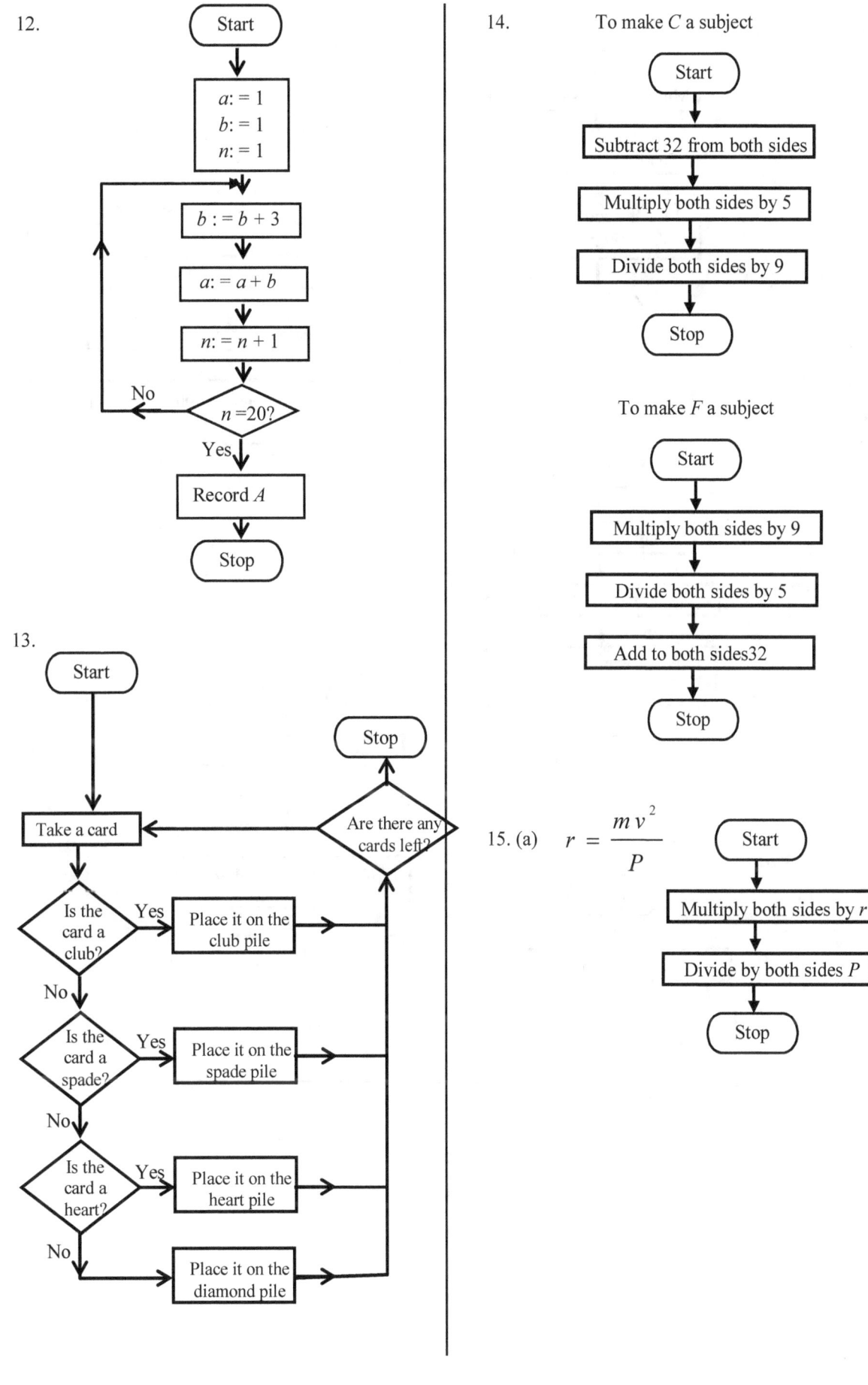

12.

Start

a: = 1
b: = 1
n: = 1

b : = b + 3

a: = a + b

n: = n + 1

n = 20?

No

Yes

Record A

Stop

13.

Start

Take a card

Is the card a club?   Yes → Place it on the club pile

No

Is the card a spade?   Yes → Place it on the spade pile

No

Is the card a heart?   Yes → Place it on the heart pile

No

Place it on the diamond pile

Are there any cards left?

Stop

14.   To make C a subject

Start

Subtract 32 from both sides

Multiply both sides by 5

Divide both sides by 9

Stop

To make F a subject

Start

Multiply both sides by 9

Divide both sides by 5

Add to both sides 32

Stop

15. (a) $r = \dfrac{m v^2}{P}$

Start

Multiply both sides by r

Divide by both sides P

Stop

(b) $v = \sqrt{\dfrac{rP}{m}}$

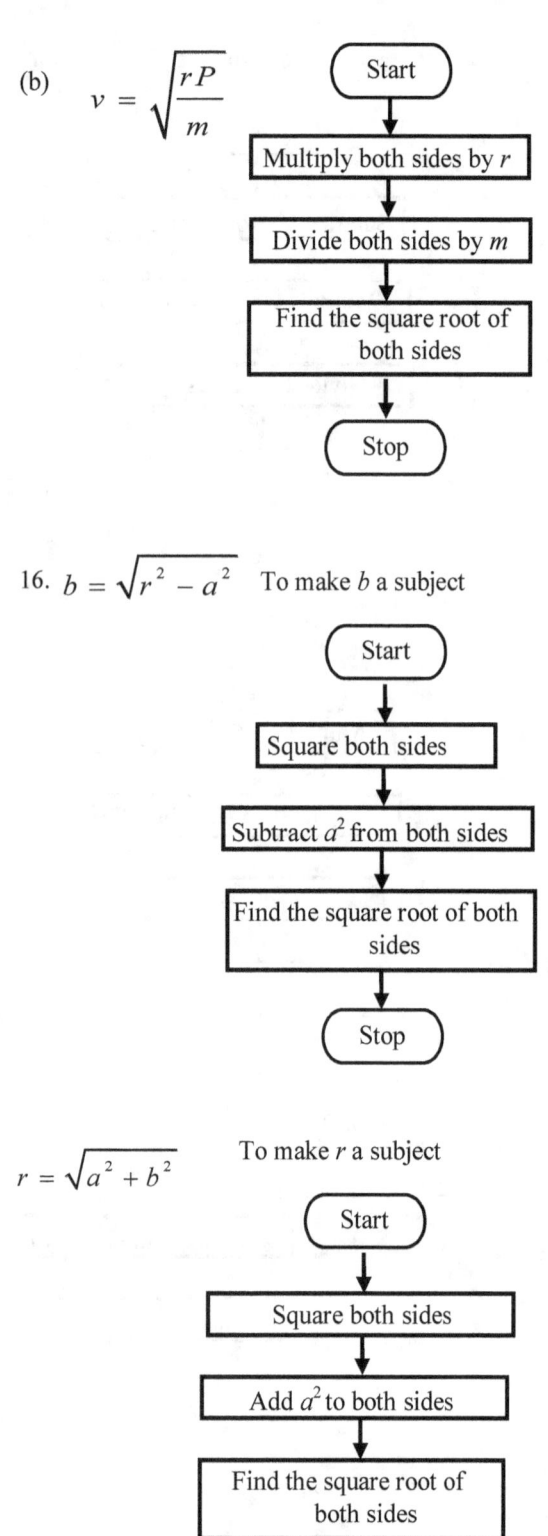

16. $b = \sqrt{r^2 - a^2}$ To make b a subject

$r = \sqrt{a^2 + b^2}$ To make r a subject

17. To prove that $ax^2 + bx + c = 0$

To prove that $x = \dfrac{\sqrt{b^2 - 4ac} - b}{2a}$

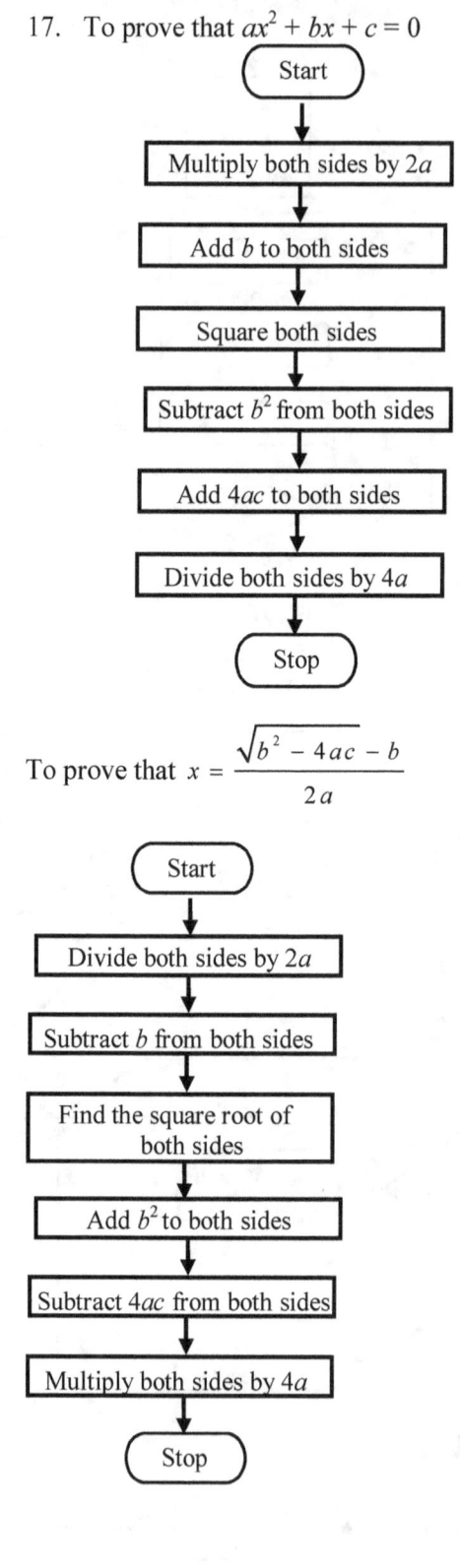

18.

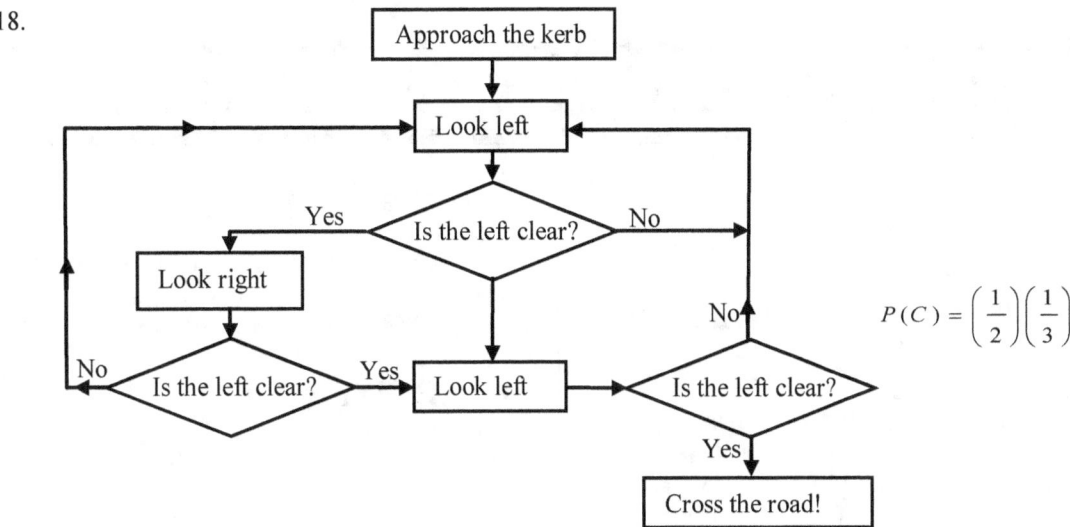

$$P(C) = \left(\frac{1}{2}\right)\left(\frac{1}{3}\right)$$

ANSWERS TO APPENDIX EXERCISES

EXERCISE A:1

1. 0.050 2. 1.19 3. 1.73 4. 11.0
5. 1.49 6. 79.0 7. 82.7 8. 111
9. 1.27 10. 0.00836 11. 21.8 12. 0.129

EXERCISE A:2

1. (a) 0.5592 (b) 0.9518 (c) 0.9886
   (d) 0.4540 (e) 0.9793 (f) 0.4438
   (g) 1.0355 (h) 0.5117 (i) 7.3639
2. (a) 8.998° (b) 65.99° (c) 81.11° (d) 53.32°

(e) 78.93 (f) 47.69° (g) 2.87° (h) 15.66°
3. (a) 24.01° (b) 70° (c) 14.69° (d) 64.6°
   (e) 16.54° (f) 89.52° (g) 74.86 (h) 61.97°
4. (a) 24° (b) 73° (c) 4.4° (d) 21.69°
   (e) 19.64° (f) 62.55° (g) 0.93° (h) 39.57°

EXERCISE A:1

(a) 30 (b) 1635 (c) 45.5
(d) 91175 (e) 8.302 (f) 8.44

ANSWERS TO MULTIPLE CHOICE EXERCISES

MULTIPLE CHOICE EXERCISE 1

1. C 2. B 3. D 4. A 5. D
6. B 7. A 8. D 9. A 10. C
11. D 12. C 13. D 14. B 15. C
16. A 17. D 18. C 19. B 20. B
21. D 22. C 23. A 24. C 25. B
26. C 27. B 28. D 29. C 30. D
31. D

MULTIPLE CHOICE EXERCISE 2

1. B 2. C 3. C 4. B 5. D
6. A 7. C 8. B 9. D 10. B
11. C 12. C 13. B 14. A 15. B
16. D 17. B 18. C 19. B 20. A
21. C 22. D 23. B 24. A 25. C
26. A 27. D 28. C 29. A 30. D
31. C 32. B 33. A

MULTIPLE CHOICE EXERCISE 3

1. C 2. A 3. D 4. C 5. B
6. D 7. C 8. D 9. B 10. B
11. D 12. C 13. B 14. B 15. D
16. A 17. C 18. A 19. B 20. C
21. D 22. D 23. A 24. C 25. D
26. D 27. A 28. A 29. C 30. B
31. A 32. C 33. D 34. C 35. D
36. D 37. B 38. C 39. A 40. D
41. D 42. B 43. C 44. A 45. C
46. A 47. D 48. B 49. A 50. D
51. D 52. A 53. B 54. A 55. B
56. D 57. C 58. D 59. D 60. B
61. C 62. B 63. C 64. B 65. C
66. C 67. D 68. D 69. B 70. D
71. A 72. D 73. D 74. A 75. A
76. A 77. B 78. C 79. D 80. A
81. C 82. D 83. A 84. D 85. A
86. B 87. C 88. B 89. D 90. C

91. C 92. D 93. C 94. D 95. C
96. C 97. D 98. B 99. A 100. B

MULTIPLE CHOICE EXERCISE 4

1. A 2. B 3. C 4. D 5. B
6. B 7. B 8. D 9. C 10. A
11. D 12. C 13. A 14. B 15. A
16. C 17. D 18. C 19. D 20. C
21. C 22. A 23. D 24. A 25. B
26. D 27. D 28. D 29. C 30. C
31. A 32. D 33. B 34. B 35. C
36. B 37. C 38. A 39. D 40. B
41. C 42. B 43. D 44. B 45. A
46. C 47. B 48. D 49. A 50. C
51. A 52. A 53. B 54. B 55. D
56. A 57. B 58. B

MULTIPLE CHOICE EXERCISE 5

1. B 2. B 3. A 4. C 5. D
6. D 7. B 8. B 9. B 10. D
11. C 12. B 13. D 14. C 15. C
16. A 17. A

MULTIPLE CHOICE EXERCISE 6

1. B 2. A 3. A 4. C 5. D
6. B 7. C 8. A 9. B 10. D
11. A 12. B 13. C 14. C 15. A
16. D 17. B 18. B 19. C 20. A
21. A 22. D 23. C 24. B 25. B
26. D 27. B 28. C 29. B 30. C
31. B 32. D

MULTIPLE CHOICE EXERCISE 7

1. A 2. C 3. B 4. A 5. A
6. C 7. B 8. D 9. B 10. C
11. B 12. A 13. C 14. D 15. A
16. C 17. B 18. B 19. A 20. C
21. D 22. C 23. B 24. D 25. A

26. C 27. B 28. A

MULTIPLE CHOICE EXERCISE 8

1. C 2. B 3. C 4. B 5. B
6. A 7. A 8. B 9. B 10. A
11. C 12. B 13. B 14. A 15. A
16. C 17. D 18. A 19. C 20. B
21. D 22. A 23. B 24. A 25. B
26. C 27. D 28. C 29. D 30. B

MULTIPLE CHOICE EXERCISE 9

1. A 2. B 3. C 4. B 5. D
6. B 7. A 8. B 9. D 10. D
11. A 12. B 13. C 14. C 15. D
16. D 17. C 18. D 19. A 20. D
21. B 22. B 23. D 24. B 25. B
26. A 27. C 28. D 29. A 30. B
31. C 32. C 33. D 34. D 35. B
36. D 37. A 38. D 39. B 40. B
41. C 42. B 43. A 44. C 45. A
46. A 47. B

MULTIPLE CHOICE EXERCISE 10

1. D 2. D 3. A 4. D 5. D
6. C 7. A 8. A 9. B 10. D
11. D 12. B 13. C 14. D 15. B
16. B 17. A 18. D 19. D 20. B
21. D 22. C 23. B 24. A 25. D
26. B 27. C 28. B 29. A 30. C
31. A 32. B 33. C 34. D 35. A
36. D 37. B 38. D 39. C 40. B
41. A 42. A 43. C 44. D 45. B
46. D 47. C 48. A

MULTIPLE CHOICE EXERCISE 11

1. C 2. D 3. A 4. D 5. B
6. D 7. B 8. A 9. C 10. B
11. A 12. D 13. A 14. C 15. A
16. D 17. B 18. B 19. C

MULTIPLE CHOICE EXERCISE 12

1. D 2. C 3. B 4. D 5. A
6. A 7. D 8. C 9. A 10. C
11. B 12. A 13. B 14. A 15. C
16. D 17. B 18. A 19. C 20. C
21. D 22. B 23. D 24. A 25. D
26. A 27. D 28. C 29. B 30. C
31. A 32. B 33. D 34. A 35. D
36. B 37. C 38. B 39. D 40. B
41. B 42. C

MULTIPLE CHOICE EXERCISE 13

1. B 2. B 3. D 4. C 5. C
6. C 7. D 8. A 9. D 10. A
11. A 12. C 13. B 14. C 15. C
16. C 17. D 18. A 19. C 20. D
21. B 22. B 23. D 24. A 25. D
26. C 27. D 28. B 29. A 30. D
31. A 32. B

MULTIPLE CHOICE EXERCISE 14

1. C 2. D 3. B 4. A 5. C
6. A 7. C 8. C 9. D 10. B
11. A 12. D 13. B 14. B

MULTIPLE CHOICE EXERCISE 15

1. C 2. C 3. D 4. B 5. C
6. B 7. C 8. D 9. B 10. A
11. B 12. C 13. B 14. A 15. D
16. A 17. C 18. C 19. D 20. C
21. A

MULTIPLE CHOICE EXERCISE 16

1. A 2. B 3. C 4. B 5. D
6. C 7. A 8. D 9. B 10. C
11. D 12. D 13. B 14. A 15. D
16. C 17. B 18. B 19. A 20. D
21. A 22. B 23. C 24. B

MULTIPLE CHOICE EXERCISE 17

1. D	2. A	3. C	4. D	5. C
6. D	7. B	8. A	9. D	10. C
11. A	12. D	13. C	14. C	15. D
16. D	17. A	18. C	19. A	20. C
21. C	22. A	23. A	24. C	25. D
26. D	27. A	28. B	29. D	30. C
31. A	32. C	33. B	34. D	35. B
36. D	37. B	38. B	39. D	40. C
41. C	42. D	43. A	44. C	45. A
46. C	47. B	48. A	49. B	50. D
51. A	52. B	53. C	54. B	55. C
56. D	57. A	58. B	59. A	60. B
61. A	62. A	63. D	64. B	65. A
66. C	67. B	68. C	69. D	70. B
71. C	72. C	73. D	74. B	75. C
76. A	77. B			

MULTIPLE CHOICE EXERCISE 18

1. A	2. C	3. D	4. A	5. B
6. C	7. D	8. C	9. B	10. B
11. C	12. D	13. C	14. D	15. B
16. B	17. C	18. D	19. B	20. C
21. D	22. D	23. A	24. C	25. B
26. B	27. C	28. C	29. A	30. A
31. C	32. D	33. C	34. B	

MULTIPLE CHOICE EXERCISE 19

1. D	2. C	3. A	4. D	5. A
6. C	7. D	8. D	9. A	10. D
11. C	12. A	13. C	14. B	15. C
16. C	17. D	18. A	19. B	20. C
21. C	22. A	23. D	24. C	25. D
26. B	27. A	28. B	29. A	30. C
31. D	32. B	33. C	34. D	35. B
36. D	37. D	38. A	39. C	40. B
41. A	42. B	43. D	44. D	45. D
46. C	47. A	48. D	49. C	50. B
51. D	52. D	53. C	54. D	55. B
56. C	57. C	58. D	59. B	60. A

MULTIPLE CHOICE EXERCISE 20

1. C	2. A	3. D	4. B	5. B
6. A	7. D	8. B	9. C	10. D
11. A	12. D	13. B	14. D	15. D
16. A	17. C	18. A	19. B	20. C
21. B	22. C	23. A	24. A	25. B
26. D	27. B	28. D	29. A	30 B

MULTIPLE CHOICE EXERCISE 21

1. B	2. B	3. C	4. A	5. D
6. C	7. A	8. B	9. D	10. A
11. A	12. B	13. C	14. C	15. D
16. C	17. B	18. A	19. B	20. B
21. C	22. D	23. D		

MULTIPLE CHOICE EXERCISE 22

1. C	2. A	3. B	4. D	5. D
6. C	7. D	8. C	9. D	10. A
11. B	12. C	13. A	14. C	15. A
16. D	17. A	18. C	19. B	20. C
21. C	22. B	23. A	24. D	25. B

MULTIPLE CHOICE EXERCISE 23

1. A	2. C	3. C	4. C	5. B
6. B	7. C	8. C	9. A	10. A
11. B	12. B	13. D	14. D	15. B
16. D	17. C	18. D	19. C	20. D
21. B	22. A	23. D	24. B	25. A
26. C	27. C	28. D	29. B	30. D
31. A	32. A	33. B	34. C	35. C
36. A	37. D			

MULTIPLE CHOICE EXERCISE 24

1. C	2. B	3. A	4. D	5. D
6. B	7. C	8. B	9. C	10. B
11. C	12. A	13. B	14. D	15. C
16. D	17. A	18. B	19. B	20. C
21. C	22. D	23. C	24. B	25. C
26. D	27. A	28. C	29. B	30. D
31. D	32. C	33. B	34. A	35. B

36. A   37. C

MULTIPLE CHOICE EXERCISE 25

1. B	2. D	3. D	4. C	5. C
6. B	7. A	8. C	9. D	10. A
11. A	12. C	13. A	14. B	15. A
16. B	17. D	18. D	19. D	20. A
21. D	22. A	23. D	24. B	25. C
26. B	27. B	28. C		

MULTIPLE CHOICE EXERCISE 26

1. B	2. C	3. D	4. A	5. D
6. A	7. D	8. B	9. A	10. C
11. D	12. A	13. D	14. C	15. D
16. C	17. A	18. B	19. A	20. A
21. D	22. A	23. B	24. D	25. C
26. D	27. C	28. B	29. D	30. A
31. D	32. A	33. C	34. D	

MULTIPLE CHOICE EXERCISE 27

1. B	2. C	3. D	4. B	5. A
6. B	7. C	8. B	9. D	10. A
11. B	12. C	13. B	14. B	15. A
16. C	17. C	18. D	19. A	20. A

MULTIPLE CHOICE EXERCISE 28

1. C	2. D	3. A	4. D	5. B
6. C	7. B	8. C	9. A	10. A
11. D	12. C	13. A	14. B	15. D
16. C	17. C	18. A	19. C	20. A
21. B	22. C	23. D	24. D	25. A
26. D	27. C	28. B	29. A	30. B
31. A	32. B	33. D	34. C	35. D
36. A	37. C	38. C	39. A	40. B
41. C	42. D	43. B	44. C	45. A
46. D				

MULTIPLE CHOICE EXERCISE 29

1. C	2. D	3. A	4. C	5. B
6. A	7. A	8. C	9. B	10. A
11. B	12. D	13. B	14. A	15. D
16. C	17. D	18. B	19. C	20. B

21. D	22. A	23. B	24. C	25. B
26. A	27. B	28. C	29. C	30. A
31. D	32. D	33. C	34. D	35. D
36. C	37. C	38. D	39. A	40. C
41. D				

MULTIPLE CHOICE EXERCISE 30

1. A	2. B	3. D	4. C	5. D
6. B	7. A	8. B		

MULTIPLE CHOICE EXERCISE 31

1. A	2. C	3. B	4. D	5. A
6. C	7. A	8. D	9. C	10. B
11. D	12. C	13. A	14. C	15. D
16. C	17. A	18. D	19. B	20. C
21. A	22. B	23. D	24. B	25. C
26. A	27. C	28. B	29. A	30. C
31. D	32. B	33. A	34. B	35. C
36. A	37. B	38. D	39. B	40. A
41. C	42. D	43. C	44. D	45. B
46. B	47. C	48. A	49. C	50. B
51. C	52. B	53. D	54. C	55. A
56. C	57. D	58. B	59. D	60. C
61. D	62. B			

MULTIPLE CHOICE EXERCISE 32

1. D	2. C	3. B	4. A	5. C
6. D	7. B	8. C	9. D	10. D
11. C	12. B	13. C	14. D	15. A
16. B	17. D	18. C	19. D	20. A
21. B	22. C	23. D	24. A	25. D
26. C	27. C	28. B	29. D	30. B
31. C	32. A	33. D	34. B	35. A
36. B	37. B	38. C	39. A	40. D
41. B	42. A	43. C	44. C	45. D
46. B	47. C	48. A	49. B	50. D

MULTIPLE CHOICE EXERCISE 33

1. C	2. B	3. D	4. A	5. C
6. C	7. D	8. A	9. C	10. B
11. D	12. A	13. B	14. D	15. C

16. B 17. B 18. D 19. C 20. B
21. C 22. B 23. A 24. C 25. B
26. B 27. D 28. A 29. B 30. C
31. D

MULTIPLE CHOICE EXERCISE 34

1. C 2. D 3. C 4. D 5. B
6. A 7. C 8. D 9. B 10. D
11. A 12. B 13. D 14. C 15. D
16. B 17. C 18. A 19. C 20. B
21. C 22. D 23. A 24. C 25. D
26 D

MULTIPLE CHOICE EXERCISE 35

1. C 2. A 3. D 4. C 5. B
6. D 7. A 8. D 9. C 10. B
11. D 12. C 13. D 14. A 15. D
16. C 17. D 18. C 19. A 20. B
21. B 22. A 23. C 24. A 25. D

MULTIPLE CHOICE EXERCISE 36

1. C 2. D 3. B 4. D 5. B
6. A 7. C 8. D 9. C 10. B
11. C 12. B

MULTIPLE CHOICE EXERCISE 37

1. C 2. D 3. C 4. A 5. B
6. B 7. C 8. A 9. D 10. C
11. B 12. D 13. B 14. C 15. A
16. B 17. C 18. B 19. D 20. B
21. A 22. A 23. C 24. B 25. C
26. D 27. B 28. A 29. C 30. B
31. A 32. D 33. A 34. C 35. D
36. C 37. A 38. C 39. D 40. C
41. D 42. B 43. A 44. D 45. C
46. C 47. B 48. C 49. B 50. C
51. D 52. C 53. B 54. A 55. B
56. A 57. B 58. D 59. B 60. D
61. C 62. D 63. D 64. C 65. B
66. B 67. B 68. D 69. A

MULTIPLE CHOICE EXERCISE 38

1. C 2. D 3. D 4. C 5. A
6. C 7. A 8. D 9. B 10. A
11. C 12. D 13. C 14. C 15. A
16. B 17. C 18. D 19. B 20. C
21. C 22. B 23. B 24. C 25. B
26. D 27. D 28. A 29. C 30. A
31. D 32. A 33. B 34. C 35. A
36. D 37. B 38. C 39. A 40. C
41. B 42. B

MULTIPLE CHOICE EXERCISE 39

1. C 2. B 3. D 4. C 5. A
6. B 7. D 8. A 9. B 10. C
11. A 12. D 13. B 14. D 15. C
16. D 17. A 18. C 19. D 20. C
21. A 22. B 23. D 24. B 25. B
26. C 27. A 28. D

MULTIPLE CHOICE EXERCISE 40

1. D 2. B 3. D 4. C 5. D
6. A 7. B 8. C 9. A 10. D
11. C 12. B 13. D 14. C 15. B
16. C 17. D 18. C 19. B

MULTIPLE CHOICE EXERCISE 41

1. B 2. C 3. D 4. C 5. C
6. C 7. B 8. B 9. D 10. A
11. B 12. D 13. B 14. A 15. A
16. C 17. A 18. D 19. A

MULTIPLE CHOICE EXERCISE 42

1. B 2. D 3. A 4. C 5. B
6. C 7. B 8. B 9. C 10. B
11. A 12. A 13. A 14. B 15. C
16. B 17. C 18. B 19. D 20. B
21. B 22. C 23. D 24. D 25. C
26. A 27 B 28 D

MULTIPLE CHOICE EXERCISE 43

1. B 2. B 3. C 4. A 5. C
6. B 7. D 8. C 9. D 10. A

11. B	12. B	13. D	14. B	15. A
16. D	17. B	18. A	19. B	20. D
21. C	22. D	23. A	24. A	25. D
26. A				

11. C	12. D	13. A	14. C	15. B
16. B	17. C	18. A	19. A	20. A
21. B	22. D	23. B	24. C	25. B
26. D				

MULTIPLE CHOICE EXERCISE 44

1. D	2. A	3. C	4. C	5. D
6. B	7. A	8. B	9. D	10. D
11. C	12. B	13. D	14. D	15. B
16. A	17. C	18. D	19. B	20. D
21. C	22. A	23. D	24. A	25. C
26. A	27. D	28. C	29. A	30. B
31. D	32. C	33. B	34. A	35. C

MULTIPLE CHOICE EXERCISE 45

1. D	2. D	3. C	4. B	5. D
6. A	7. C	8. D	9. C	10. D

MULTIPLE CHOICE EXERCISE 46

1. A	2. B	3. C	4. B	5. D
6. A	7. D	8. D	9. C	10. A
11. A	12. D	13. A	14. A	15. C
16. C	17. C			

MULTIPLE CHOICE EXERCISE 47

1. C	2. B	3. C	4. D	5. B
6. A	7. D	8. D	9. C	10. C
11. A	12. B	13. A	14 D	

GLOSSARY

Absolute value of a number	The positive value of any real number.
Acute angle	An angle whose value is less than 90°.
Acute angle triangle	A triangle whose angles are all less than 90°.
Additive inverse	A number, vector or some other quantity that gives zero when added to another. The additive inverse of a is $(-a)$ and has the property that $a + (-a) = 0$. The sum of a number or quantity and its additive inverse is called the additive identity.
Adjacent edges	Two edges of a network graph which share a common vertex. Adjacent edges are also called coincident edges. The edges AX and BX are adjacent or coincident because they share the same vertex X.

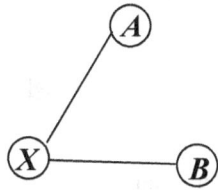

Adjacent vertices	Two vertices x and y of a network graph which share a common edge.

Alluvial diagram	A graphical summary and highlight of the important structural changes in networks.
Alternate (interior) angles	The internal angles on opposite sides of the transversal, which intersects two parallel lines.
Altitude of a triangle	The perpendicular distance between a vertex of a triangle and the opposite side to the vertex.
Amount	The sum of the interest and the principal invested in a business.
Angle sum of a polygon	The sum of the interior angles of a polygon with n sides is $180° \times (n-2)$. The sum of the exterior angles of a polygon is 360° no matter the number of sides of the polygon.
Angle	The amount of turn between two divergent lines. It is measured in degrees, minutes and seconds or in radians.
Arc (x, y) inverted	Given two arcs arc (x, y) represented by $\textcircled{X} \rightarrow \textcircled{Y}$ and arc (y, x) represented by $\textcircled{X} \leftarrow \textcircled{Y}$, the arc (y, x) is called arc (x, y) inverted.
Arc	A part of the circumference of a circle.
Arcs	see edges
Area	The number of square units covered by a surface.
Associative property	The characteristic of mathematical operations that are independent of grouping applied to the numbers involved. The Associative property of addition is $(a + b) + c = a + (b + c)$. The Associative property of multiplication is $(a \times b) \times c = a \times (b \times c)$.
Average	A number regarded as typical of a group of numbers. The common averages in data procession are the arithmetic mean, the mode and the median.
Axes of a graph	The two lines drawn at right angles to each other on a graph. The vertical axis is called the y-axis while the horizontal axis is called the x-axis.
Bar chart	A statistical diagram in which data is represented by means of bars with equal widths. The height (or length) of each bar represents the magnitude of the data.
Base	The number of different single-digit symbols used in a particular number system. In the usual counting system of numbers, the decimal system (with symbols 0, 1, 2, 3, 4, 5, 6, 7, 8, 9), the base is 10.
Bicimal	A number containing a binary point (or binary marker).
Biconditional Statement	A proposition in logic involving two statements such that one is true only when

	the other is true.
Binary system	A number system in which counting is done in groups of two. The digits used in this system are only 0 and 1.
Binomial expression	An algebraic expression, which contains two terms.
Bisect	Divide into two equal parts. The line, which is used to bisect, is called a bisector.
Capacity	The quantity of liquid, which a container will hold. Capacity is measured in liters.
Cartesian coordinates	The perpendicular distances of a point from the x and y axes respectively. Cartesian coordinates are sometimes called rectangular coordinates.
Chord	A straight line, which joints two points on the circumference of a circle.
Circle	The locus of points, which are equidistant from a fixed point (called the center of the circle).
Circumference	The perimeter of a circle. The formula for calculating the circumference is $C = 2\pi r^2$, where r is the radius of the circle.
Closed statement	A statement concerning a definite object.
Coefficient of x	The number which multiplies x in a term.
Coincident edges	(see adjacent edges).
Column matrix	A matrix, which has only one column.
Complementary angles	Two angles whose sum is 90°.
Complete graph	A network graph in which each vertex is connected to every other vertex by an edge.
Composite Statements	(see Compound Statements)
Compound Statements	A statement, which is made by combining two or more statements. A Compound Statements is also called a Composite Statements.
Conclusion	The final proposition leading from two or more propositions called hypotheses or premises
Conditional Statement	A composite statement obtained by joining two statements p and q in such a way that if p is true q must be true.
Congruent triangles	Triangles, which are identical in size and shape.
Conjunction	A composite statement made by combining two statements with the use of the preposition "and"
Connector	Any of the logical symbols ~, ∧, ∨, → or ↔.
Consecutive arrows	Two network edges, which are such that the head of one is at the tail of the other.

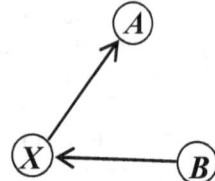

| Consecutive vertices | Two adjacent vertices, which are such that the head of the edge is on one vertex and the tail of the edge is on the other. |

Contradictions	A proposition that is always false.
Control flow diagram	A diagram describing the control flow of a business process, a program or any other process.
Corresponding angles	A pair of equal angles lying on the same side of a transversal and making an interior and exterior angle with the intersected lines.
Cosine of an angle	The trigonometric ratio of the length of the opposite side (to an angle) to the length of the hypotenuse of a right angled triangle.
Counting numbers	The positive whole numbers.
Cube of a number	The product of a number multiplied by itself three times.
Data flow diagram	A graphical representation of the flow of data through an information system.
De Morgan's Laws	The logically equivalent statements $\sim p \wedge \sim q = \sim (p \vee q)$ and $\sim p \vee \sim q = \sim (p \wedge q)$ or their corresponding set algebra equivalence.
Decimal places	The number of digits following the decimal point in a number.
Decimal system	A number system in which counting is done in groups of ten. The digits used in the decimal system are 0, 1, 2, 3, 4, 5, 6, 7, 8 and 9.

Decision block or box	A diamond shape used flow diagrams to ask precise questions with "Yes" or "No" answers.
Denominator	The bottom part of a vulgar fraction. The denominator tells the number of parts that the whole quantity is divided into.
Diagonal	A line segment joining two non consecutive sides of a polygon.
Diagonal matrix	A square matrix in which all the elements except those in the leading diagonal are zeros.
Diameter	The longest chord of a circle. The diameter passes through the center of the circle and its length is twice the radius of the circle.
Difference	The result obtained from the subtraction of two quantities.
Difference of two squares	Any expression of the form $a^2 - b^2$. The factors of $a^2 - b^2$ are (a + b)(a – b).
Digits	The numbers 0, 1, 2, 3, 4, 5, 6, 7, 8 and 9.
Directed edges	An arc with an arrow at one end showing its direction.
Directed graph	An ordered pair $D = (V, A)$ such that V is the set of vertices and A is the set of ordered edges. A directed graph is sometimes referred to simply as a digraph.
Directed numbers	Numbers with a + or – sign attached to them.
Directly proportional	A mathematical relationship in which an increase in one variable by a given factor brings about an increase by the same factor in another.
Discount	The amount, which a business person takes of his selling price when payment is made in cash.
Disjunction	A composite statement made by combining two statements with the use of the preposition "or"
Distributive property	The property which allows the expression $a(b+c)$ to be written as $ab +ac$.
Dividend	The number to be divided.
Divisible	If a number is divided by a second number and the remainder is zero, the number is said to be divisible by the second.
Divisor	The number by which the dividend is divided.
Domain	The set of all the possible values which the variable can take is called the domain
Edges	The lines joining the vertices of a network, a plane figure or plane figure.
Element	The different objects or numbers that make up a set or a matrix.
Equilateral triangle	A triangle, which has all it sides equal, and all its angles equal.
Equivalent fractions	Two or more fractions, which represent the same quantity.
Euler's formula	The relation $R + V–E = 2$ between the number of regions vertices and edges of a network
Euler's law	(see Euler's formula
Even number	A number, which is divisible by two.
Even vertex	A vertex at which an even number of edges meet.
Exchange rate	The equivalence (or the rate of conversion) of one currency to another.
Exclusive disjunction	A disjunction composed of two statements p and q, which cannot both be true.
Existential Quantifier	The quantifier denoted by ∃, read "there exists" or "for at least one" or "for some".
Exponent	The number of times that a quantity is to be multiplied by itself. The expression x^n is called the power, n is called the exponent and x is called the base.
Extremes	The first and fourth terms in a proportion.
Factor	A number or algebraic expression, which will divide exactly into another number or expression.
Finite set	A set in which all the elements can be listed.
Flow chart	A graphical representation of the step-by-step solution to a given problem.
Flow diagrams	A schematic or graphic representation of the steps leading to the solution of a given problem.
Flow map	A mixture of maps and flow charts used in cartography, to show the movement of objects from one location to another.
Forest	A collection of network trees or a disjoint union of one or more trees with no cycles.
Frequency	The number of times an event or a particular data occurs.
Frequency-distribution table	A tabular summary of statistical information.
Function in algebra	A variable quantity whose value depends upon the varying values of one or more

	other quantities.
Function in set theory	A numerical relationship between two sets, in which each member of one set corresponds uniquely to a member of the other set. A function is denoted by f.
Functional flow block diagram	A sketch showing the various parts of a system and their functions in relation to one another.
Gradient	The slope or steepness of a straight or the slope of a curve at any given point usually obtained by measuring the slope of the tangent at that point.
Graph of an equation	A diagrammatic representation of the solutions of an equation.
Group frequency	One of the categories formed when raw data is classified into different classes.
Head	(1) The end of a line segment or edge with an arrow. (2) The side of a coin with a logo.
Hexagon	A polygon with six sides.
Highest common factor(H.C.F.)	The largest common factor of two or more numbers or algebraic expressions.
Histogram	A statistical representation of frequency distribution. The area of each rectangle in the diagram represents the class frequency.
Hypotenuse	The longest side of a right angle triangle. The hypotenuse is always opposite the right angle.
Hypothesis	A preliminary proposition in an argument, which may be true or false pending justification.
Improper fraction	A fraction with the numerator greater than the denominator.
Incident edge and vertex	An edge and a vertex on that edge.

Inclusive disjunction	A disjunction composed of two statements p and q, which can both be true.
Index	The power to which a number is raised.
Inequality	A statement that one thing is greater or less than another.
Infinite	Endless. Continues without end.
Infinite set	A set in which it is impossible to list all the elements.
Inscribed circle	A circle inside which is inside a triangle and touches all the three sides of the triangle. The center of an in circle can be found by bisecting the angles of the triangle.
Instruction block or box	A rectangular shape used to issue intermediate instructions in a flow diagram
Integers	All positive and negative whole numbers including zero.
Interest	Money paid when a sum called the principal is invested.
Intersection	The set of all the elements common to two or more sets.
Inverse function	A function that exactly reverses the transformation produced by a function. The inverse of the function f is usually denoted by f^{-1}. An inverse function f^{-1} has the property that $f \circ f^{-1}(x) = x$.
Inverse of a matrix	A matrix that exactly reverses the transformation produced by another matrix. The inverse of the matrix $\mathbf{A}$ is usually denoted by $\mathbf{A}^{-1}$. An inverse matrix $\mathbf{A}^{-1}$ has the property that $\mathbf{A}\,\mathbf{A}^{-1} = \mathbf{I}$, where $\mathbf{I}$ is the identity matrix.
Inversely proportional	A Mathematical relationship in which an increase in one variable by a given factor brings about a decrease by the same factor in another.
Irrational number	A number, which cannot be expressed as a vulgar fraction.
Isometry	A transformation in which shape and size remain unchanged.
Isosceles triangle	A triangle with two equal sides and the angles adjacent to the equal sides equal.
Kilogram	The standard unit for measuring the mass of an object.
Least common multiple (L.C.M.)	The smallest and common multiples of a number or an algebraic expression.
Linear equation	Another name for a simple equation. An equation in which the unknown is of first degree.
Links	(see edges)
Liter	The standard unit for capacity measurements.
Logarithm	The exponent or power to which a stated number, called the *base*, is raised to yield a specific number.
Logic	A science, which deals with the principles of valid reasoning and argument.
Logically equivalent	Two statements p and q whose truth tables are the same.

statements

Lowest term	A fraction in which numerator and denominator have no common factors.
Mapping	A relationship between two sets, in which each member of one set corresponds uniquely to a member of the other set. A The members of the set may or may not be numbers.
Matrix	A rectangular arrangement of numbers in rows and columns.
Mean	A value obtained by dividing the sum of quantities by the number of quantities.
Means	The second and third terms in a proportion.
Median	A statistical average obtained when numbers are arranged in ascending or descending order and the middle or the mean of the two middle numbers is found.
Median of a triangle	A straight line joining a vertex to the midpoint of the opposite side of a triangle.
Mediator	The perpendicular bisector of a straight line.
Meter	The standard unit of length.
Mixed graph	A graph in which some edges are directed and some are undirected. A mixed graph is usually written as an ordered triple $G = (V, E, A)$ where V is the set of vertices, E is the set of ordered edges and A is a set of ordered edges.
Mixed number	A number consisting of a whole number part and a fractional part.
Mode	The most occurring value in a set of quantities.
Multiple	A number, which can divide into another without a remainder, is said to be a multiple of the other.
Multiplicative inverse	A number or quantity that gives 1 when multiplied with another. The multiplicative inverse of a is a^{-1} and has the property that $a(a^{-1}) = 1$. The product of a number or quantity and its multiplicative inverse is called the multiplicative identity.
Negation	The negation of a statement p denoted by $\sim p$ is the statement formed from p by inserting the word "not" into p or placing the phrases "it is false that"…or "it is not true that"…before the statement p. If p is, true $\sim p$ is false; if p is false, $\sim p$ is true.
Net	A plane figure, which can be folded to make a solid figure.
Network	A collection of points called vertices or nodes by lines called edges or arcs or links.
Network graphs	A graph of an ordered pair $G = (V, E)$ comprising the set V of vertices together with the set E of edges such that each element of e is a doubleton subset of V. Informally, a network graph is a diagram representing a network.
Network tree	A connected graph with no cycle or network graph in which any two vertices are connected by exactly one path or a connected graph in which $n(E) = n(V) - 1$
Nodes	(see vertices 1)
Null graph	A network graph whose edge set is empty or all the vertices are isolated from each other.
Null set	This is a set with no elements. Another name of a null set is an empty set.
Number line	An ordered representation of real numbers on a straight line.
Number	The idea of how many things has been counted.
Numeral	A symbol used to represent a number.
Numerator	The top part of a vulgar fraction.
Obtuse angle	An angle whose value is greater than 90° but less than 180°.
Octagon	A polygon with eight sides.
Octal system	A number system in which counting is done in groups of eight.
Odd vertex	A vertex at which an odd number of edges meet.
Open statement	A statement that does not concern a definite object.
Ordered list	A list in which the order of the elements is very important. The elements of an unordered list are usually enclosed in parentheses ( ).
Ordered pair	The Cartesian coordinate of a point on a graph. The order of the points is of prime importance.
Origin	The starting point on a number line or the intersection of the x- and the y- axes on a graph.
Parallel lines	Two or more straight lines in a plane that do not intersect.
Parallelogram	A quadrilateral with opposite sides equal and parallel.
Pentagon	A polygon with five sides.

Percent	The ratio of two numbers expressed as a fraction with denominator 100.
Perfect square	A number or algebraic expression that can be written as a product of two equal factors.
Perimeter	The distance round a plane figure.
Perpendicular lines	Two lines that intersect at an angle of $90°$.
Pie chart	A circular statistical diagram divided into sectors which represents statistical quantities.
Place value	The value of a digit due to its location in numeral or the value of the place that a digit occupies in a numeral.
Plane	A flat surface.
Polygon	A closed plane figure bounded by straight lines
Powers	The power of a number is written as a superscript and represents the number of times the number has to be multiplied by itself.
Predecessor (direct predecessor)	A vertex x is said to be the predecessor of y if the edge leads from x to y.
Premises	(see hypotheses)
Prime factorization	A factorization in which all the factors are prime numbers.
Prime number	A whole number other than zero and one which has only two factors, 1 and the number itself.
Principal	The sum of money invested for the sake of generating interest.
Probability	Branch of mathematics that deals with measuring or determining quantitatively the likelihood that an event or experiment will have a particular outcome.
Process flow diagram	A graphical representation of a process.
Product	The result obtained when two or more quantities are multiplied together.
Programming language	A language used to write a list of instructions to be followed by a computer.
Proportion	A statement equating two ratios.
Propositions	(see Statements)
Pyramid	A solid figure with a polygonal base and triangular faces.
Pythagoras theorem	The Pythagoras theorem states that in a right angle triangle, the sum of the squares on the two arms equal the square on the hypotenuse.
Pythagorean triples	Three numbers which are such that the sum of the squares of the two smaller ones is equal to the square of the largest.
Quadrants	The four sections into which the perpendicular axes of a coordinate system divide a two-dimensional surface.
Quadratic equation	An equation with the highest power of the unknown 2.
Quadrilateral	Any four sided figure.
Quantifier	A Logical operator which is used to stand for a word such as; for all, for every, for some, for each, for any, there exist, for at least etc.
Quotient	The result when quantities are divided.
Radian	A unit of measurement of angles. A radian is equal to the angle subtended at the center of a circle by an arc whose length is equal to the radius of the circle.
Radius	A line segment from a point on the circle to the center of the circle.
Range	The difference between the largest number and the smallest in a set of data.
Rate per cent	The rate of interest paid per hundred per annum.
Ratio	A numerical scale or fraction, which compares two quantities.
Rational number	A number, which can be expressed as a vulgar fraction or the ratio of two integers.
Raw data	Collected statistical information that is still unorganized.
Real numbers	All rational and irrational numbers put together.
Reciprocal of a number	The multiplicative inverse of a number. The reciprocal of a number is obtained by dividing 1 by the number.
Rectangle	A quadrilateral with all its four angles equal to $90°$.
Reflection	A translation in which an object is as far behind a line as it is in front of the line.
Reflex angle	An angle greater than $180°$ but less than $360°$.
Region	The area bounded by the vertices and edges of a network.
Regular polygon	A polygon with all sides equal and all angles equal.
Repeating decimal	A decimal with one or more digits repeating endlessly.
Replacement set	(see domain)
Rhombus	A parallelogram with all sides equal.

Right angle	An angle whose value is a quarter turn or 90°.
Right angle triangle	A triangle with one angle equal to 90°.
Rotational transformation	An isometry with an invariant point, which is the center of the transformation.
Sample space	The set of all the possible outcomes of an experiment.
Scale of a map or graph	The number of units to a unit of the real length on a graph or map.
Scalene triangle	A triangle with all the three sides different in length.
Sector of a circle	A portion of a circle bounded by two radii and an arc of the circle.
Sequence	A list of numbers which such that any one of them is defined by a common rule.
Series	An expression in which the elements of a sequence are added.
Set	A collection of numbers or objects with similar attributes or properties.
Signal flow graph	A graph used in mathematics to show the relations among the variables of a set of linear algebraic relations.
Significant figures	The figures that are required to state a numerical quantity to a given degree of accuracy.
Similar triangles	Triangles, which have the same shape but not necessarily the same size. Similar triangles are equiangular.
Simple equation	Another name for a linear equation. An equation in which the unknown is of first degree.
Simplest form	A fraction in which numerator and denominator have no common factors.
Sine of an angle	The trigonometric ratio of the length of the adjacent side (to an angle) to the length of the hypotenuse of a right angled triangle.
Square of a number	The product of a number by itself.
Standard form	A number written in the form $A \times 10^n$, where A is a number greater than or equal to 1 but less than 10 and n is an integer.
Statements	A sentence that is either true or false but not both. A statement is also called a propositions
Straight angle	An angle whose value is equal to 180°.
Subset	A set A whose elements are contained in another set B is said to be the subset of B.
Substitution	The process of finding the numerical value of an algebraic expression for given values of the variables it contains.
Successor (direct successor)	A vertex y is said to be the successor of x if the edge leads from x to y.

$$x \longrightarrow y$$

Sum	The result obtained when two or more quantities are added.
Supplementary angles	Two angles whose sum is equal to 180° are said to be supplementary.
Surface area	The total area of the surface of a solid figure.
Syllogisms	An argument made up of statements which are of the forms "All A's are B's", "No A's are B's", "Some A's are B's" or "Some A's are not B's".
Symmetry	The correspondence of parts on opposite sides of a point, line or plane.
Tail	(1) The end of a line segment or edge without an arrow. (2) The side of a coin on which the value of the coin is written.
Tangent of an angle	The trigonometric ratio of the opposite side (to an angle) to the adjacent side (to the angle) of a right angled triangle.
Tautology	A proposition that is always true.
Terminating decimal	A decimal which can be expressed as a vulgar fraction and whose digits do not repeat.
Theorem	An established statement, which have been and can be proven to be absolutely true
Transformation	The mapping, which maps an object to its image.
Translation	An isometry in which every point on an object is moved the same distance in the same direction.
Transpose of a matrix	A matrix obtained by interchanging the rows and columns of another matrix.
Transversal	A straight line drawn to intersect parallel lines.
Trapezium	A quadrilateral with two of its sides parallel.
Traversable network	A network, which can be, traced exactly once beginning at some point without retracing any edge.
Trisect	Divide into three equal parts. The line, which is used to trisect, is called a trisector.
Truth Value	The truthfulness or falsity of a statement

Undirected edges	An arc with no arrows or arrows at both ends.
Undirected graphs	A graph for which the edges have no arrows. A set of unordered pairs $\{\{a, b\}, \{c, d\}...\}$.
Union	The set of all the elements contained in two or more sets is called the union of the sets.
Unitary Existential Quantifier	The quantifier denoted by $\exists!$ read "there exists one and only one".
Universal Quantifier	The quantifier denoted by $\forall$, read "for all" or "for every".
Universal set	A set containing all the subsets and elements in a particular set problem.
Unordered list	A list in which the order of the elements is immaterial. The elements of an unordered list are usually enclosed in braces $\{\ \}$.
Variable	A letter or symbol, which can take any value in a given range.
Variable	The open part in an open statement which must be closed to make it true or false
Venn diagram	A diagrammatic representation of the relationship between sets using plane figures, usually circles for the subsets and a rectangle for the universal set.
Vertex	A point common to two rays of an angle or two sides of a polygon.
Vertically opposite angles	Two equal angles formed opposite to each other when two straight lines are drawn to intersect.
Vertices	The points connected by the edges in a network diagram, a plane figure or plane figure or vertices the coordinates of points on a Cartesian plane
Volume	The number of cubic units contained in a solid.
Weighted graph	A network graph in which a number (weight) is assigned to each edge.
Whole number	Any of the numbers 0, 1, 2, 3, 4,…, which is not a fraction.

AAA .... See angle-angle-angle
Abelian Group 214
Abscissa 183
Absolute Inequalities 156
Absolute Value 78
    Inequalities 154
Acceleration 371
Acute angle..... 260, 263, 273, 285, 298
Acute Angle Trigonometric
    Ratios 285
Adding vectors 495, 496
Adding mixed numbers 30
Addition 6, 9, 175
    Law of logarithms 163
    Laws of Probability 421
    of Decimals 36
    of Directed Numbers 79
    of fractions 29
    of Inequalities 152
    of Matrices 478
    of Vectors 495
Additive Inverse 81, 95
Additive Inverse of a Vector ... 496
Adjacent angles 259, 261, 263, 273
Adjacent Edges 468
Adjacent Vertices................... 468
Adjoint of a 2×2 Matrix 482
Algebra tiles 78-80
Algebraic equations 94
Algebraic expression 86, 87
Algebraic Fractions 130-132
Algebraic Laws of Sets 205
Algebraic Rules 88
Algebraic Sentences 86
Alluvial diagram 530, 605
Alternate angles 262, 264
Alternate Segment Theorem ... 447
Altitude 434, 441
Amount 68, 70
An acute angle 282, 285, 289
Angle between Tangent and
    Radius 446
Angle between two vectors 503
Angle-angle-angle 350
Angles 259
    at a point 259
    at Centre and Circumference
    444
    between Intersecting Lines 262
    in a Cyclic Quadrilateral 445

in a Semi-circle 444
in Opposite Segments 445
in the Same Segment 445
of elevation and depression 301
Angle-side-angle 348
Anti-symmetric 234
Approximation 48
Approximations 49
Arc Length 323
Arcs 462, 463, 467, 472-474
Area .316-319, 322-324, 327, 334, 336-340
Area measure 59
Area of a Circle 322
Area of a plane figure 59
Area of a Sector 323
Area of a Segment 324
Areas by Counting Squares 326
Arithmetic
    in Non-Denary Bases 175
    Mean 251, 397
    progression 250
Arms 282, 283
ASA .... See angle-side-angle
Associated angle 292, 294
Associative
    Laws 205
    property of addition 7
    property of multiplication 8
Associativity 211
Assumed Mean 408
Asymptote 381
Average 397
Axes
    Extension of- 184
Axis of symmetry 512
Balancing Method 96
Banking 78
Bar Charts 394
Base five system 4
Base ten system 3, 172
Bearings in two dimensions 302
BEODMAS 9
Biased 418, 426, 427
Biconditional Statement . 222, 605
Bijections 239
Binary marker 177
Binary Operations 211
Binary point 177
Binary System 177, 178
    Importance of the - 178
Binomial 88

Binomial expression 104
Bisecting a Given Angle 435
Brackets 9
Buying and selling 72
Calculation of Taxes 74
Calculations in Standard Form .. 51
Capacity 61
Cardinal Points 302
Cardinality Logic 205
Cardinality of a Power Set 204
Cardinality of a Set 199
Cartesian plane 183
Cartesian product of Two Sets 230
Celsius Thermometer 65, 78
Centigrade Thermometer 78
Change of Base Formula 165
Changing Decimals
    to Fractions 35
    to Percentages 39
Changing Fractions
    to Decimals 35
    to Percentages 39
Changing Percentages to
    Decimals 39
Chasles' Theorem 276
Chronological Bar Chart 394
Circle Theorems 444
Circular Diagrams 393
Class Boundaries 400
Class intervals 399, 401, 408, 411, 429
Class length 400
Class Limits 400
Class Size 400
Class width 400-402
Classes 399
Clocks and Watches 63
Closed Intervals 150
Closed Statements 218
Closed-open Interval 150
Closure 210
Coding Method 408
Codomain 230, 235, 238, 239, 243
Coefficients 88
Coincident Edges 468
Co-interior angles 262
Collinear 491, 493, 504
Column Matrix 477
Column Vectors 491
Combination Tables 210
Common Base Parallelograms 319

Common difference 250-252, 255, 256
Common ratio 251- 255
Commutative Group 214
Commutative Laws 205
Commutative property
   of addition 7
   of multiplication 8
Commutativity 211
Comparing fractions 27
Compass Bearing 302
Compatible 478, 479, 482
Complement Laws 205
Complement of a Set 203
Complementary 260- 264, 273
Complementary Angles 288
Complementary Events 421
Complete Graph 470
Completing the square 105
Composite Functions 241
Composite number 14
Composite Plane Figures 325
Composite Solid Figures 338
Composite Statements 220, 606
Composite Transformations 525
Compound Inequalities 154
Compound Interest 69, 70
Compound Statements ... 220, 606
Concave polygon 268
Concept of a Set 198
Concept of Logic 218
Conclusion 226
Condensed form 173
Conditional Inequalities 152
Conditional Probability 425
Conditional Statements 222
Conditions for a Network to be
   Traversable 464
Cones 311, 336
Conformable 479
Congruent Figures 348
Congruent modulo 212
Congruent Triangles 348
Conjugate of a Surd 166
Conjunction 220, 606
Connectors 224, 606
Consecutive Arrows 468
Consecutive Vertices 468
Constant of proportionality 378
Constant term 88
Constants 88
Constructing
   a Circumscribed Circle 437
   a Line of Given Length 434
   a Perpendicular 435
   a Perpendicular Bisector 435
   a Triangle 434

   an Angle 435, 436
   an Inscribed Circle 436, 437
Construction of Polygons 437
Constructions 434
Contradictions 225, 606
Control flow diagram 531, 606
Conversion of Currency 65
Conversion of Number Bases . 173
Converting
   Denary to Non-Denary 174
   Non-Denary to Non-Denary
   175
   None Denary to Denary 173
Convex polygon 268
Coordinate plane 183
Coordinates 183, 184, 187-196
Coordinates 182
Corresponding angles 262, 263, 264
Corresponding lines 361
Corresponding, Inconsistent and
   Intersecting Lines 361
Cost price 72, 73, 76
Counting Systems 2
Critical values 155
Cube roots 18
Cube Roots by Prime
   Factorization 18
Cubes 18
Cubic Measure 61
Cubic numbers 18
Cumulative Frequency 403
Cumulative Frequency Curves 403
Cumulative frequency distributio
   403
Cumulative frequency table 403
Curves 362, 367
Cylinders 310, 334
Data and stores 535
Data flow diagram .. 531, 606, 608
De Morgan's laws 224
Deciles 403
Decimal Currency 65
Decimal form 50
Decimal fraction 34, 35, 53
Decimal marker see decimal point
Decimal Places 49
Decimal point 34
Decimal system .... see base ten system
Decimals
   operations with- 36
Decimals 34
Decision block 532, 607
Decision Boxes 533
Defined Operations 210
Defining Relations 230

Definition of Sets 198
De-Morgan's Laws 205
Denary system .... See base ten system
Denominator 26-39
Depreciation 70
Derivative 369
Derived function 369
Derived Set 204
Describing Transformation 519
Determinant of a 2×2 Matrix .. 481
Diagonal Matrix 476
Difference 7
Difference of two squares 106, 109
Difference of Two Squares 106
Differential coefficient 369
Differentiation 369
Digraph 469
Dilation 524
Direct variation 378
Directed Edges 467
Directed Graphs 469
Directed Numbers
   meaning of - 78
Directed numbers
   Operation with - 78
direction 490
Direction of a Vector 491
Directly proportional 378, 379, 383
Discount 73
Disjoint Sets 202
Disjunction 221, 607
Distance 490
Distance between Two Points . 186
Distances 78
Distance-time graphs 373
Distributive law 104
Distributive Laws 205
Distributive property 89
Distributive Property 8
Distributive Property of Vectors
   498
Dividend 7, 26, 37
Dividing by Algebraic Fractions
   132
Dividing Decimals 37
Divisibility 19
Divisibility Rules 19
Divisible 13
Division 7, 9, 174, 177, 187
Division by Fractions 33
Division Law of Indices ... 13, 160
Division of Directed Numbers .. 82
Division of Inequalities 153
Division of Polynomials 144
Divisor 7, 26, 37, 40, 144, 145

Domain ..218, 219, 230, 233, 235, 238, 239, 241-244, 607
Dot Product 502
Dots patterns.............20
Double Intercept Form 191
Doubleton or Pair set.............199
Drawing With and Without
Replacement.................425
EARTH AS A SPHERE........341
Edges 462, 466-469, 472-474
Egyptian Numerals 3
Elements. 462, 463, 467, 476-478, 481, 482, 486
Empty set 199
Enlargements.................508
Entries.... 476, 477, 483-486, 495, 496
Equality of Matrices.............477
Equality of Ratios 40
Equality of Sets.................201
Equality of Vectors 493
Equally likely 419
Equation
what is an- 94
Equation of a Straight Line.....190
Equations and Expressions
Difference between-........95
Equiangular polygon 268
Equilateral polygon................268
Equiprobable Outcomes 419
Equivalence relation.............234
Equivalent fractions 27, 40
Equivalent Fractions 26
Equivalent numbers 212
Equivalent Sets 201
Estimating
products and quotients.........48
sums and differences...........48
Estimations................48
Euler's Formula 462
Even Numbers 21
Even Vertices 463
Event....... 418-421, 424, 427, 432
Event subset 418
Exclusive disjunction 221, 607
Existential Quantifier 225, 607, 612
Expanded Form
Method............................174
in Base Ten.....................173
in other Bases 173
Expansions 104
Experiment..... 418, 422, 425, 426
Exponent 160, 161, 169
Exponential
Equations.........................162
Law of Logarithms...........164

notation 12
Exponents.....................9
Exterior angles of polygons 276
External division...................188
Extremes 41
Factor theorem...............145-147
Factor Tree 15
Factorisation 106
Factorisation method..............120
Factorising by Grouping.........106
Factorising Quadratic Expressions
...................................107
Factors...........................13, 15
Factors of a Polynomial..........144
Fahrenheit Thermometers.........65
Fair.................................418-431
Finding Transformation Matrices
.................................522
Finite sequence 248
Finite sets 200
Fixed Vectors 494
Flow chart method 96
Flow Charts 240
Flow Diagram.......................530
Flow diagram symbols 532
Flow map 531, 607
Flowchart 531, 607
Formulae
Containing Quadratics.......140
Containing Square Roots...139
without Square Roots........138
Fraction
Expressing one quantity as
a- 34
Fractional index laws 161
Fractions
Practical examples of- 26
Types of- 26
-with zero denominator ...26
Fractions and Decimals
Simple Linear equations
involving - 99
Fractions
with equal denominators 29
with unequal denominators..29
Free Vectors 494
Frequency........392-414, 428, 429
Frequency distribution .. 392, 393, 397-403, 408-410, 414
Frequency Distribution
Curve 401
table 392, 393, 397, 409
Frequency Polygon 401
Frustums.............................311
Full turn.............................259
Function Notation 235

Functional flow block diagram
.... 530, 608
Functions
Representation of 236
Functions.....................235
Gain 71, 72
GCD.........................15
General Angle.....................292
General form 192
Geometric Mean 254
Geometric progression ... 250, 252
Gradient 189-192
Gradient and one point form...190
Gradient as a Rate of Change..369
Gradient function 369
Gradient –Intercept form 190
Gradient of a Straight line 189
Graphical Solutions of Quadratic
Equations.........................364
Graphing Straight Lines 360
Graphs of Quadratic Functions362
Great Circles.......................342
Group theory 214
Grouped data 399
Groups.....399, 402, 411, 412, 430
Guess Mean...........................408
Guess method 96
Half Closed Intervals 150
Half Open Intervals...............150
Half planes 386
Half Turn.............................259
HCF 15, 90, 92
Head.............. 467, 468, 490, 496
Height 434, 457-460
Hemisphere 311, 340
Highest common factor 15
Hindu-Arabic Numerals 5
Histograms 395, 400, 402
Histograms
for grouped data................400
with equal class widths......400
with unequal class widths..402
Historical Time.......................62
Hypotenuse.............282, 283, 291
Hypotheses 226, 610
Idea of a Function 235
Idea of a Mapping.................235
Idempotent Laws 205
Identities.............................108
Identity Element 211
Identity function 241
Identity Laws.........................205
Identity Matrix.......................476
Identity property of one............ 8
Identity property of zero........... 7
Ideographs.............................393
Image 518

Implication 222
Improper Subsets 201
Incidence Matrices................485
Incidence matrix 486
Incident edge and Vertex........468
Inclusive disjunction 221, 228
Incomplete Quadratic Equations
....123
Inconsistent 361, 374
Independent Events................424
Index 160-162, 169
Index Equations.....................162
Index form..............12, 13, 15, 23
Index notation.........................12
Indices
    Division law of- 13
    Multiplication law of- 12
    Theory of-.....................160
Inequality 150
Inequations....................150-155
Infinite sequence...................248
Infinite sets............................200
Information Matrices..............485
Injections...............................239
Instruction block 532, 608
Instruction Boxes...................532
Integers 22
Inter Quartile Range..............405
Intercepts...............................190
Interest 68, 69, 70
Internal division.....................188
Intersecting chord theorem 448
Intersecting lines....................361
Intersecting Planes 455
Intersection of Sets................202
Into Mappings........................238
Inverse Element 212
Inverse Function 240
Inverse of a 2×2 Matrix..........482
Inverse of a function 240
Inverse Operations 95
Inverse Relation.....................232
Inverse Transformations.........523
Inverse Trigonometric Ratios .286
Inverse variation 378, 380
Inverse Variations..................381
Irrational numbers....................22
Irregular polygon 268
Isometry 508
Isosceles Triangle 300
Joint or Combined Variation...382
Kite..269
Lateral surface .310, 311, 336-339
Latitudes................................342
Law of Indices 160
Laws of Inequalities...............152
Laws of Logarithms 163

Laws of Surds........................166
LCM 16, 17, 23, 90, 92
Leading diagonal 476
Least common multiple...........16
Left-unbounded Open Intervals
....151
Legs 282, 291
Length 490
Like signs..............................83
like terms...........................88, 91
Line.......................................258
Line of Greatest Slope...........455
Line of symmetry....511, 512, 516
Line Perpendicular to a Plane .455
Line segment .. 185, 187,258, 272,
273, 490
Linear Inequalities 153
Linear relation 231
Lines and Planes 454
Links......................462, 467, 490
Literal coefficients 88, 89
Loci.......................................438
Logarithm....... 160, 162, 165, 169
    of 1...................................165
    Definition of - 162
Logarithmic Equations...........163
Logical Equivalence...............223
Longitudes.............................342
Lower class limit....................400
Magnitude 490
    of a Vector.......................493
Many-one mapping................238
Mappings...............................235
Mass Measure.........................58
Mathematical Relation 230
Matrix operator 518
Maximum Values...................371
Mean....... 397, 398, 405-409, 416
Mean deviation 405-409, 413
    from the Mean 406
Mean of Grouped Data...........405
Means......... 26, 27, 40, 41, 62, 63
Measures of Central Tendencies
....396
Measures of Dispersion or
    Variation (Spread or Scatter of
    Data) 405
Median....396-399, 403-405, 409-
416, 429, 434, 438, 441
Median of Grouped Data by
    Calculation 405
Mediator..........434, 438, 440-442
Membership Notation.............199
Method of Completing the Square
....121
Method of Elimination 112-116
Method of Substitution....113-116

Metric Units 59
Mid-Interval Value................400
Midpoint of a Line 187
Minimum Values 371
Minor diagonal 476
Minuend 7
Mirror Symmetry 511
Mixed Graph 469
Mixed Numbers 28
Mode..............396-401, 409, 413
Modulus 490
Money Measure.......................65
Monomial...............................88
Monomial and a Binomial......104
Multiples 13, 15
Multiples of Pythagorean Triples
....283
Multiplicand 7
Multiplication 7, 9, 176
    Properties of.......................7
    and Division of Terms.........89
    Law of Indices 12, 160
    of Directed Numbers...........82
    of Inequalities 153
    of Matrices 479
    of Vectors by Scalars 498
Multiplicative inverse.........32, 33
Multiplicative property of zero .. 8
Multiplicative property of zero .89
Multiplier 7
Multiplying Algebraic Fractions
....131
Multiplying Decimals
    by powers of 10.............. 36
    by whole numbers 36
    by Decimals....................37
Multiplying fractions...............31
    by whole numbers 31
    by mixed Numbers 32
Multi-Step Simple Linear
    Equations...........................98
Mutually Exclusive Events.....422
Naming Polygons...................268
Natural numbers 12
Nature of roots of a Quadratic
    Equation...........................124
Negation........................ 219, 609
Negative Angles 294
Negative index law 161
Nets of Solid Figures.............312
Network Graphs.....................467
Network Terminology...........462
Network Trees 470
Networks.................... 463, 465
Networks in Real Life...........465
Neutral Element.....................211

Nodes.....462, 464, 466, 467, 470, 473, 474
Non collinear 491
Non Perpendicular Planes.......455
Non-Adjacent angles..............261
Non-numerical coefficients 88
Non-Standard Quadratic Equations.........................123
Nontaxable Income 74
Non-terminating Decimals 38
Non-terminating non-recurring decimals 38
Norm.....................................490
Normal form............................50
North-South lines...................454
n^{th} term of a G.P.....................252
n^{th} Term of a Sequence..........248
n^{th} term of an A.P..................250
Null Graph.............................470
Null Matrix............................477
Null set.................................199
Null Vector............................497
Number bases 172
Number line............22, 78, 79, 80
Number Sequence...................248
Number System Vocabulary...172
Numbers and Numerals............. 2
Numbers as dots patterns..........20
Numerals................................. 2
Numerator 26, 27, 31, 33, 39
Numerical coefficient.........88, 89
Obtuse angle... 260, 261, 263, 273
Odd Numbers 21
Odd Vertices..........................463
Of 9
One Step Simple Linear Equations....................97
One-one mapping...................238
Onto mappings......................238
Open Statements.......218
Open-Closed Interval 150
Operation...............................210
    Rules..................... 6
    Tables....................210
Operations
    Properties of- 7
Operations with directed numbers
   78
Order of a Matrix 476
Order Relation 235
Order Symbols...................27, 81
Ordered Lists.........................462
Ordered pair.... 462, 463, 467, 469
Ordering...............................150
Ordinate 183
Origin....................................183
Orthogonal 183, 492

Orthonormal 492
Orthornomal base vectors.......492
Parallel Lines................ 192, 262
    and Planes 454
    and Transversals 262
Parallelogram. 269, 270, 274, 317, 321
Parallelogram Law of Vectors 497
Peeling method 14, 16, 17
Percentage
    Expressing one quantity as a- 40
Percentage Change..................71
Percentage Difference 71
Percentage Profit And Loss 72
Percentages...........................38
Percentages to Fractions
    Changing-........................38
Percentiles 403
Perfect number.......................21
Perfect square 17, 18, 23
Perfect squares.....................105
Perimeter..............................316
Perpendicular......... 183, 192-196
Perpendicular bisector... 434, 435, 438, 440-442, 444
Perpendicular Lines 192, 262
    and Planes 454
Pictograms...........................393
Pie Charts 393
Place Value System.................. 5
    -and decimals.................34
Plane 258
Plane of symmetry 511, 516
Playing Cards 419
Plus 6
Point....................................258
Polygon 268
    of vectors......................496
    Theorems.......................278
Polyhedrons..........................311
polynomial 88
Position Vector
    of the Midpoint................500
Position Vectors....................494
Possibility space 418, 419
Power 160, 162
Power or Derived Set of a Set.204
Powers of 10
    Multiplying decimals by- 36
Powers of Ten................. 12, 172
Prefixes in the SI Units.............56
Premises....................... 226, 610
Prime factorization..................14
Prime number 14
Principal 68, 70
Prisms 310, 334

Probability....... 418-421, 425-428
    as a Number....................419
    from Frequency Tables......428
    Terminology....................418
Process flow diagram 530, 610
Product............................... 7
    index law..........................161
    of Numbers with the same Power..............................161
    of prime Factors.................16
    of Two Binomials 104
Profit 71-73, 76, 84
Profit And Loss.......................72
Programming language .. 530, 610
Pro-numeral......................86, 88
Proper fractions 26
Proper Subsets 201
Properties of addition 89
Properties of multiplication 89
Properties of Network Graphs 468
Properties of Numbers............. 8
Properties of Relations 234
Proportional parts....................41
Proportionate bar Chart 394
Proportions 40
Propositions.................. 218, 611
Pyramids 310, 311, 336
Pythagoras Theorem 185, 282
Pythagorean Triples 283
Quadratic Equations 120, 123, 124
Quadratic Expressions............108
Quadratic formula
Quadratic Inequations 155
Quadrilaterals 269
Quantifiers.................. 225, 610
Quantiles 403
Quarter Turn.........................260
Quartiles....................... 403, 404
Quotient 7, 144, 145, 152
Radial symmetry....................512
Radical equations...................166
Range..... 230, 233, 238-405, 410, 413-416, 429
Rate.....................................68
Rate of Change 369
Ratio of
    Areas of Similar Figures....351
    Volumes of Similar Figures
   353
Rational numbers....................22
Rationalizing the Denominator
   166
Ratios.....................................40
Ratios and proportions 40
Raw Data.............................392
Real Life Examples of Variation
   378

Real Number 22, 78
Reciprocal Trigonometric Ratios
....289
Reciprocals.............................32
Rectangle.................... 310, 316
Rectangular Matrix 477
Rectangular numbers.........20, 21
Recurring decimals 38
Re-entrant Polygons..............268
Reflection............................508
Reflective Symmetry..............511
Reflexive...............................234
Region....462, 466, 467, 471, 485, 486
Regular polygon 268
Relation
   The idea of a 230
Remainder 144-147, 157
Remainder theorem................145
Removal of brackets..............104
Repeated Division Method 174
Repeated Trials.....................425
Repeating decimals.................38
Replacement set............. 218, 610
Representation of Data...........393
Representation of inequalities.150
Representation of Sets............202
Residue Classes 212
Restricted Domain 241
Restricted Function 241
Revolution............................259
Rhombus .269, 270, 273, 274, 318
RHS . See right angle-hypotenuse-side
Right angle 260, 282, 296, 306, 307
Right angle-hypotenuse- side..348
Right circular................ 310, 345
Right cone 311
Right prism................... 310, 334
Right pyramid.......................310
Right-Angled Triangle .. 282, 283, 299
Right-angled triangular prism.310
Right-unbounded
   Closed Interval.................151
   Open Intervals 151
Roman Numerals 4
Rotation...............................508
Rotational symmetry..............512
Rounding Down.......................48
Rounding Up 48
Route Matrices.............. 485, 486
Row Matrix 477
Rule Definition Method..........198
S.I. Unit................................63
Sample 418, 419, 421

space 418
SAS. See side-included angle-side
Saving and Borrowing..............78
Savings and loans analogy.. 79, 80
Scalar 479, 490, 498, 502-504
   Multiplication of Matrices.479
   Product............................502
   quantity 490
Scientific notation 50
Section Theorem....................500
Self inverse...........................233
Selling price 72-76
Semi-inter Quartile Range......405
Sense...................................490
Sense of a Vector..................491
Sequence 249
   of Operations 9
   recognition 248
Series 248, 249, 254, 258
Set Builder Notation..............198
Set Notation..........................198
Shear...................................509
SI Units 56
Side -included angle-side 348, 350
Side- side-side 348, 350
Sign of Products and Quotients.83
Signal flow graph........... 531, 611
Significant Figures..................49
Similar Figures 350, 351, 353
Simple Bar Charts..................395
Simple interest........................68
Simple Linear Equations ...94-100
Simplifying
   Algebraic Fractions...........130
   fractions to their lowest terms
....27
Simultaneous Equations 112
   Involving fractions and
      decimals.....................115
   One Linear one quadratic
     125
   with non-uniform
      coefficients...............113
   with uniform coefficients
     112
Simultaneous Equations - One
   Linear, One Quadratic -
   Graphical Method 364
Simultaneous Equations by the
   Matrix Method.................484
Simultaneous Linear
   Equations.................. 116, 361
   Inequations.......................387
Sine and cosine formulae........304
Singleton or unit set 199
Singular Matrices...................481
Size 490

Size of a Matrix 476
Skew lines 455
Small Circles 342
Solid cylinder . 310, 334, 335, 344
Solid Geometry......................455
Solution set................ 96, 98, 219
Solving Inequations 153
Solving Quadratic Equations from
   the Graph of any Other
   Quadratic Function 365
Solving Simple Linear Equations
....96
Special Angles......................290
Speed time Graphs 372
Sphere 311, 338
Square 310, 314, 316
Square Matrix........................476
Square numbers 20, 21
Square of a Binomial.............104
Square Roots 17
Squares...................................17
SSS .... See side-side-side
Standard Deviation 407
Standard form 50-53
   Quadratic Equations..........120
Standard Notation for Triangles
....282
Standard Units 56
Statements 218, 220, 222, 230, 606, 610, 611
Statistical Graphs...................393
Statistics 392
Straight line 258, 259, 360, 264
Stretch..................................509
   factor.................509, 514, 526
Subject of a Formula.............138
Subsets.......................... 200, 201
Subtracting
   Column vectors.................496
   component form Vectors ...497
   mixed numbers 30
   Vectors Diagrammatically.497
Subtraction 7, 9, 175
   Law of logarithms.............164
   of Decimals 36
   of Directed numbers....... 81, 79
   of fractions 29
   of Inequalities...................152
   of Matrices 478
   of Vectors.........................496
Subtrahend 7
Sum..................................... 6
   and Difference of Algebraic
      Fractions.....................132
   of n terms of a G.P. 252
   of ratio...............................41
   of the First n Terms of a

of the first n terms of an A.P. 250
to infinity 253
Supplementary 260-264, 269
Surd Equations 166
Surface Area and Volume
of a Sphere 338
of Cubes 334
of Cylinders 334
of Prisms 334
Surface Area of a Pyramid 337
Surjections 239
Syllogisms 226, 611
Symbolic Expressions 86
Symmetric 234, 244
Symmetrical Properties of a
Circle 444
Symmetry 511
Tabulation 392, 408
Tail 467, 468, 490, 494, 496
Tallies 2
Tallying 2, 392
Tangents 367
Tautologies 224
Taxable Income 74
Taxes 74
Temperature 64
Term 248
Terminating Decimals 38
Terminating decimals 38
Terms 7
Terms 68, 88, 89
Theory of indices 160
Theory of logarithms 162
Three Digit Bearings 302
Time 68
as a Non-Metric S.I. Unit 63
Measure 62
Units 62
Times 7
Transformation matrix 518
TRANSFORMATIONS 508
by Singular Matrices 524
involving Change of Area ..524
Involving Many Points 519
Transitive 234
Translation 508

Translation vector 508
Translations 525
Transpose of a Matrix 481
Transversal 262, 263
Trapezium 269, 270, 273, 274, 319, 321
Trapezoidal prism 310
Traversable Networks 463
Trebleton 199
Tree Diagrams 426
Trial 418
Trial and Error Method 146
Triangle 310, 318, 321
Triangle of vectors 496
Triangular numbers 20-22
Trig ratios 284, 288, 299
Trigonometric Ratios 284- 286, 289-291
Trigonometric Ratios from Tables 286
Trigonometry 282
Trinomial 88
Trinomial Perfect Square Test 105
Trisecting a Right Angle 436
Truth set 219
Truth Value of a Statement 218, 611
Turning Points 363
Two point form 191
Two-Step Simple Linear
Equations 98
Unbiased 418, 425
Unbounded Intervals 151
Undefined Expressions 134
Undirected Edges 467
Undirected Graphs 469
Unfair 418, 425, 427
Ungrouped data 392
Union of Sets 202
Unit base vectors 492
Unit Matrix 476
Unit vector 492
Unitary Existential Quantifier 225, 612
Units of Area 60
Units of Length 56
Units of Mass 58

Universal Quantifier 225, 612
Universal Set 200
Unknowns 88
Unlike signs 83
Unlike terms 88, 91
Unordered Lists 462
Upper class limit 400
Values 71, 86-89
Variable 218
ubstitution 87
Variables 86
Variance 407, 413
Vector 490
algebra 493
Geometry 500
Notation 491
quantity 490
Vectors as Directed Line
Segments 490
Velocity 371
Venn Diagrams 202
Vertical and Horizontal Boundary
Lines 386
Vertical lines 454
Vertices 462- 474
Volume and Surface Area
of a Frustum 339
of a Hemisphere 340
Volume
Measure 61
of Cones and Pyramids 336
of Cubes and Cuboids 61
of flow 343
Vulgar fractions 26
Weighted Graph 470
West-East lines 454
Word Equation 94
Working Zero 408
x-axis 183
x-coordinate 183
x-y plane 183
y-coordinate 183
Zero index law 161
Zero Matrix 477
Zero Values 144
Zero Vector 497

www.ingramcontent.com/pod-product-compliance
Lightning Source LLC
Chambersburg PA
CBHW080755180526
45168CB00006B/2213